Principles of Plasma Discharges and Materials Processing

Principles of Plasma Discharges and Materials Processing

Third Edition

Michael A. Lieberman
University of California
Berkeley, CA, USA

Allan J. Lichtenberg (Deceased)

Published by John Wiley & Sons, Inc., Hoboken, New Jersey.
Published simultaneously in Canada.

For general information on our other products and services or for technical support, please contact our Customer Care Department within the United States at (800) 762-2974, outside the United States at (317) 572-3993 or fax (317) 572-4002.

Wiley also publishes its books in a variety of electronic formats. Some content that appears in print may not be available in electronic formats. For more information about Wiley products, visit our web site at www.wiley.com.

Library of Congress Cataloging-in-Publication Data applied for:
Hardback ISBN: 9781394245376

Cover Design: Wiley
Cover Image: © Cover art provided by Michael A. Lieberman

Set in 9.5/12.5pt STIXTwoText by Straive, Chennai, India

SKY10087391_100924

To my colleague, coauthor and friend,
Al Lichtenberg

Contents

List of Figures *xxi*
List of Tables *xlv*
Preface to Third Edition *xlvii*
Preface to Second Edition *xlix*
Preface to the First Edition *li*
Symbols, Abbreviations, and Acronyms *lv*

1 Introduction *1*
1.1 Materials Processing *1*
1.2 Plasmas and Sheaths *5*
1.2.1 Plasmas *5*
1.2.2 Sheaths *10*
1.3 Discharges *12*
1.3.1 RF Diodes *12*
1.3.2 High-Density Sources *17*
1.4 Symbols and Units *20*

2 Basic Plasma Equations and Equilibrium *21*
2.1 Introduction *21*
2.2 Field Equations, Current, and Voltage *22*
2.2.1 Maxwell's Equations *22*
2.3 The Conservation Equations *25*
2.3.1 Boltzmann's Equation *25*
2.3.2 Macroscopic Quantities *26*
2.3.3 Particle Conservation *27*
2.3.4 Momentum Conservation *27*
2.3.5 Energy Conservation *29*
2.3.6 Summary *30*
2.4 Equilibrium Properties *30*
2.4.1 Boltzmann's Relation *32*
2.4.2 Debye Length *32*
2.4.3 Quasineutrality *34*
Problems *34*

3 Atomic Collisions *37*
3.1 Basic Concepts *37*
3.1.1 Elastic and Inelastic Collisions *37*
3.1.2 Collision Parameters *37*
3.1.3 Differential Scattering Cross Section *39*
3.2 Collision Dynamics *42*
3.2.1 Center-of-Mass Coordinates *42*
3.2.2 Energy Transfer *44*
3.2.3 Small Angle Scattering *44*
3.3 Elastic Scattering *46*
3.3.1 Coulomb Collisions *46*
3.3.2 Polarization Scattering *48*
3.4 Inelastic Collisions *53*
3.4.1 Atomic Energy Levels *53*
3.4.2 Electric Dipole Radiation and Metastable Atoms *55*
3.4.3 Electron Ionization Cross Section *58*
3.4.4 Electron Excitation Cross Section *59*
3.4.5 Ion–Atom Charge Transfer *60*
3.4.6 Ion–Atom Ionization *64*
3.5 Averaging Over Distributions and Surface Effects *64*
3.5.1 Averaging Over a Maxwellian Distribution *64*
3.5.2 Energy Loss per Electron–Ion Pair Created *67*
3.5.3 Surface Effects *68*
Problems *68*

4 Plasma Dynamics *73*
4.1 Basic Motions *73*
4.1.1 Motion in Constant Fields *73*
4.1.2 $E \times B$ Drifts *75*
4.1.3 Energy Conservation *76*
4.2 Nonmagnetized Plasma Dynamics *77*
4.2.1 Plasma Oscillations *77*
4.2.2 Dielectric Constant and Conductivity *79*
4.2.3 Ohmic Heating *81*
4.2.4 Electromagnetic Waves *82*
4.2.5 Electrostatic Waves *83*
4.3 Guiding Center Motion *84*
4.3.1 Parallel Force *85*
4.3.2 Adiabatic Constancy of the Magnetic Moment *86*
4.3.3 Drift Due to Motion Along Field Lines (Curvature Drift) *87*
4.3.4 Drift Due to Gyration (Gradient Drift) *88*
4.3.5 Polarization Drift *88*
4.4 Dynamics of Magnetized Plasmas *90*
4.4.1 Dielectric Tensor *90*
4.4.2 The Wave Dispersion *92*
4.5 Waves in Magnetized Plasmas *93*
4.5.1 Principal Electron Waves *94*

4.5.1.1 k || B_0 94
4.5.1.2 k ⊥ B_0 95
4.5.2 Principal Waves Including Ion Dynamics 96
4.5.2.1 k || B_0 96
4.5.2.2 k ⊥ B_0 97
4.5.3 The CMA Diagram 98
4.6 Microwave and RF Field Diagnostics 100
4.6.1 Interferometer 100
4.6.2 Hairpin Resonator Probe 103
4.6.3 Magnetic (B-dot) Probes 104
4.6.4 Cavity Perturbation 105
4.6.5 Wave Propagation 106
Problems 107

5 Diffusion and Transport 111
5.1 Basic Relations 111
5.1.1 Diffusion and Mobility 111
5.1.2 Free Diffusion 112
5.1.3 Ambipolar Diffusion 112
5.2 Diffusion Solutions 113
5.2.1 Boundary Conditions 113
5.2.2 Time-Dependent Solution 114
5.2.3 Steady-State Plane-Parallel Solutions 115
5.2.4 Steady-State Cylindrical Solutions 117
5.3 Low-Pressure Solutions 119
5.3.1 Variable Mobility Model 119
5.3.2 Langmuir Solution 120
5.3.3 Heuristic Solutions 122
5.4 Diffusion Across a Magnetic Field 123
5.4.1 Nonambipolar Diffusion 126
5.4.2 PIC Simulations 127
5.5 Magnetic Multipole Confinement 129
5.5.1 Magnetic Fields 129
5.5.2 Plasma Confinement 131
5.5.3 Leak Width w 132
Problems 133

6 DC Sheaths 137
6.1 Basic Concepts and Equations 137
6.1.1 The Collisionless Sheath 138
6.2 The Bohm Sheath Criterion 139
6.2.1 Plasma Requirements 140
6.2.2 The Presheath 141
6.2.3 Sheath Potential at a Floating Wall 142
6.2.4 Collisional Sheaths 143
6.2.5 Simulation Results 144

6.3 The High-Voltage Sheath 145
6.3.1 Matrix Sheath 145
6.3.2 Child Law Sheath 145
6.4 Generalized Criteria for Sheath Formation 147
6.4.1 Electronegative Gases 148
6.4.2 Multiple Positive Ion Species 149
6.5 High-Voltage Collisional Sheaths 152
6.6 Electrostatic Probe Diagnostics 153
6.6.1 Planar Probe with Collisionless Sheath 155
6.6.2 Non-Maxwellian Electrons 156
6.6.3 Cylindrical Probe with a Collisionless Sheath 158
6.6.4 Double Probes and Emissive Probes 161
6.6.5 Effect of Collisions and DC Magnetic Fields 163
6.6.6 Probe Construction and Circuits 164
6.6.7 Probes in Time-Varying Fields 166
Problems 167

7 Chemical Reactions and Equilibrium 171
7.1 Introduction 171
7.2 Energy and Enthalpy 172
7.3 Entropy and Gibbs Free Energy 179
7.3.1 Gibbs Free Energy 182
7.4 Chemical Equilibrium 184
7.4.1 Pressure and Temperature Variations 186
7.5 Heterogeneous Equilibrium 187
7.5.1 Equilibrium Between Phases 187
7.5.2 Equilibrium at a Surface 190
Problems 191

8 Molecular Collisions 195
8.1 Introduction 195
8.2 Molecular Structure 195
8.2.1 Vibrational and Rotational Motion 197
8.2.2 Optical Emission 198
8.2.3 Negative Ions 200
8.3 Electron Collisions with Molecules 202
8.3.1 Dissociation 202
8.3.2 Dissociative Ionization 204
8.3.3 Dissociative Recombination 205
8.3.4 Example of Hydrogen 206
8.3.5 Dissociative Electron Attachment 206
8.3.6 Polar Dissociation 208
8.3.7 Metastable Negative Ions 209
8.3.8 Electron Impact Detachment 209
8.3.9 Vibrational and Rotational Excitations 210
8.3.10 Elastic Scattering 211
8.4 Heavy-Particle Collisions 211

8.4.1 Resonant and Nonresonant Charge Transfer *212*
8.4.2 Positive–Negative Ion Recombination *213*
8.4.3 Associative Detachment *215*
8.4.4 Transfer of Excitation *216*
8.4.5 Penning, Associative, and Pooling Ionization *217*
8.4.6 Rearrangement of Chemical Bonds *219*
8.4.7 Ion–Neutral Elastic Scattering *220*
8.4.8 Three-Body Processes *220*
8.5 Reaction Rates and Detailed Balancing *221*
8.5.1 Temperature Dependence *222*
8.5.2 The Principle of Detailed Balancing *222*
8.5.3 A Data Set for Oxygen *225*
8.6 Optical Emission and Actinometry *229*
8.6.1 Optical Emission *231*
8.6.2 Optical Actinometry *232*
8.6.3 O Atom Actinometry *233*
8.6.4 Phase-Resolved Optical Emission Spectroscopy (PROES) *234*
Problems *237*

9 Chemical Kinetics and Surface Processes *243*
9.1 Elementary Reactions *243*
9.1.1 Relation to Equilibrium Constant *245*
9.2 Gas-Phase Kinetics *246*
9.2.1 First-Order Consecutive Reactions *246*
9.2.2 Opposing Reactions *249*
9.2.3 Bimolecular Association with Photon Emission *249*
9.2.4 Three-Body Association *250*
9.2.5 Three-Body Positive–Negative Ion Recombination *252*
9.2.6 Three-Body Electron–Ion Recombination *253*
9.3 Surface Processes *253*
9.3.1 Positive Ion Neutralization and Secondary Electron Emission *254*
9.3.2 Adsorption and Desorption *257*
9.3.3 Fragmentation *261*
9.3.4 Sputtering *261*
9.4 Surface Kinetics *263*
9.4.1 Diffusion of Neutral Species *264*
9.4.2 Loss Rate for Diffusion *264*
9.4.3 Adsorption and Desorption *266*
9.4.4 Dissociative Adsorption and Associative Desorption *267*
9.4.5 Physical Adsorption *267*
9.4.6 Reaction with a Surface *268*
9.4.7 Reactions on a Surface *268*
9.4.8 Surface Kinetics and Loss Probability *269*
9.5 Showerhead Gas Flow *270*
9.5.1 Approximate Solution *271*
Problems *273*

10 Particle and Energy Balance in Discharges *279*
10.1 Introduction *279*
10.2 Electropositive Plasma Equilibrium *281*
10.2.1 Basic Properties *281*
10.2.1.1 Low Pressure: $\lambda_i \gtrsim (R,\ l)$ *282*
10.2.1.2 Intermediate Pressures: $(R,\ l) \gtrsim \lambda_i \gtrsim (T_i/T_e)(R,\ l)$ *282*
10.2.1.3 High Pressures: $\lambda_i \lesssim (T_i/T_e)(R,\ l)$ *282*
10.2.2 Uniform Density Discharge Model *283*
10.2.2.1 Low-Density Discharges *286*
10.2.2.2 High-Density Discharges *286*
10.2.3 Nonuniform Discharge Model *286*
10.2.4 Neutral Radical Generation and Loss *288*
10.3 Electronegative Plasma Equilibrium *289*
10.3.1 Differential Equations *290*
10.3.2 Boltzmann Equilibrium for Negative Ions *293*
10.3.3 Conservation Equations *295*
10.3.4 Validity of Reduced Equations *296*
10.4 Approximate Electronegative Equilibria *297*
10.4.1 Global Models *297*
10.4.1.1 Particle and Energy Balance *297*
10.4.1.2 Ion Loss Flux Γ_{+s} and Recombination Volume $\mathcal{V}_{rec}$ *299*
10.4.2 Parabolic Approximation for Low Pressures *301*
10.5 Electronegative Discharge Experiments and Simulations *304*
10.5.1 Oxygen Discharges *304*
10.5.1.1 Measurements and Global Models *305*
10.5.1.2 PIC Simulations *308*
10.5.2 Chlorine Discharges *310*
10.5.2.1 Measurements *310*
10.5.2.2 Global Models and Simulations *311*
10.6 Pulsed Discharges *313*
10.6.1 Pulsed Electropositive Discharges *314*
10.6.2 Pulsed Electronegative Discharges *319*
10.6.3 Neutral Radical Dynamics *323*
Problems *324*

11 Low-Pressure Capacitive Discharges *329*
11.1 Homogeneous Model *330*
11.1.1 Plasma Admittance *332*
11.1.2 Sheath Admittance *332*
11.1.2.1 Displacement Current *333*
11.1.2.2 Conduction Current *334*
11.1.3 Particle and Energy Balance *336*
11.1.3.1 Ohmic Heating *337*
11.1.3.2 Stochastic Heating *337*
11.1.4 Discharge Parameters *338*
11.2 Inhomogeneous Model *340*
11.2.1 Collisionless Sheath Dynamics *340*

11.2.2 Child Law *342*
11.2.3 Sheath Capacitance *342*
11.2.4 Bulk Ohmic Heating *343*
11.2.5 Stochastic Heating *344*
11.2.6 Self-Consistent Model Equations *345*
11.2.7 Scaling *347*
11.2.8 Collisional Sheaths *348*
11.2.9 Low and Moderate Voltages *350*
11.2.10 Ohmic Heating in the Sheath *350*
11.2.11 Self-Consistent Collisionless Heating Models *351*
11.2.12 Electronegative Plasmas *352*
11.3 Experiments and Simulations *353*
11.3.1 Experimental Results *354*
11.3.2 Particle-in-Cell Simulations *357*
11.3.3 Secondaries, Gas Heating, and Excited Neutral States *361*
11.3.4 Boltzmann Term Analysis for PIC Simulations *362*
11.4 Asymmetric Discharges *365*
11.4.1 Capacitive Voltage Divider *365*
11.4.2 Spherical Shell Model *367*
11.5 Voltage-Driven Sheaths and Series Resonance *369*
11.5.1 Bias Voltage *370*
11.5.2 Wall Ion Flux Probe *371*
11.5.3 Nonlinear Excitation of Series Resonance *371*
11.6 Multi-frequency Capacitive Discharges *372*
11.6.1 Dual-Frequency Sheaths *374*
11.6.2 Dual-Frequency Discharges *376*
11.6.2.1 Homogeneous Model *376*
11.6.2.2 Child Law Models *378*
11.6.3 Electrical Asymmetry Effect (EAE) *380*
11.6.4 Tailored Voltage Waveforms *382*
11.7 Standing Wave and Skin Effects *383*
11.7.1 Experiments and Simulations *388*
11.8 Low-Frequency Sheaths *391*
11.9 Ion-Bombarding Energy at Electrodes *394*
11.10 Magnetically Enhanced Discharges *401*
11.10.1 Parallel Plate Homogeneous Model *402*
11.10.2 Measurements and Simulations *404*
11.11 Matching Networks and Power Measurements *406*
11.11.1 Power Measurements *408*
Problems *410*

12 Inductive Discharges *415*
12.1 High-Density, Low-Pressure Discharges *415*
12.1.1 Inductive Source Configurations *416*
12.1.2 Power Absorption and Operating Regimes *416*
12.1.3 Discharge Operation and Coupling *419*
12.1.4 Matching Network and Power Measurements *421*

12.1.4.1 Power Measurements 422
12.2 Other Operating Regimes 422
12.2.1 Low-Density Operation 422
12.2.2 Capacitive Coupling 423
12.2.3 Hysteresis and Instabilities 425
12.2.4 Power Transfer Efficiency 427
12.2.5 Exact Solutions 428
12.2.6 Helical Resonator Discharges 429
12.3 Planar Coil Configuration 430
12.3.1 Neutral Gas Depletion and Heating 435
12.4 High-Efficiency Planar Discharges 436
12.4.1 Low Frequencies 436
12.4.2 Close Coupling 437
12.4.3 Ferrite Enhancement 438
12.4.4 Experimental Results 440
Problems 441

13 Wave-Heated Discharges 445
13.1 Electron Cyclotron Resonance Discharges 445
13.1.1 Characteristics and Configurations 445
13.1.2 Electron Heating 450
13.1.2.1 Collisionless Heating Calculation 451
13.1.2.2 Collisional Heating Calculation 453
13.1.3 Resonant Wave Absorption 453
13.1.4 Model and Simulations 458
13.1.5 Plasma Expansion 458
13.1.6 Measurements 462
13.2 Helicon Discharges 464
13.2.1 Helicon Modes 464
13.2.2 Antenna Coupling 467
13.2.3 Power Absorption 469
13.2.4 Neutral Gas Depletion 473
13.3 Surface Wave Discharges 473
13.3.1 Planar Surface Waves 473
13.3.2 Cylindrical Surface Waves 474
13.3.3 Power Balance 475
Problems 477

14 DC Discharges 479
14.1 Qualitative Characteristics of Glow Discharges 479
14.1.1 Positive Column 479
14.1.2 Cathode Sheath 480
14.1.3 Negative Glow and Faraday Dark Space 481
14.1.4 Anode Fall 481
14.1.5 Other Effects 481
14.1.6 Sputtering and Other Configurations 482
14.2 Analysis of the Positive Column 482

14.2.1 Calculation of T_e 483
14.2.2 Calculation of E and n_0 484
14.2.3 Kinetic Effects 485
14.3 Analysis of the Cathode Region 485
14.3.1 Breakdown of a Gas-Filled Gap 486
14.3.2 Cathode Sheath 489
14.3.3 The Negative Glow and Faraday Dark Space 491
14.4 Hollow Cathode Discharges 492
14.4.1 Simple Discharge Model 493
14.4.2 Finite Sheath Effects 496
14.4.3 RF-Driven Hollow Cathodes 497
14.4.3.1 Experiments 497
14.4.3.2 Simulations 498
14.5 Planar Magnetron Discharges 498
14.5.1 Limitations of Glow Discharge Sputtering Source 499
14.5.2 Magnetron Configuration 499
14.5.3 Discharge Model 501
14.5.3.1 Magnetron Voltage V_{dc} 501
14.5.3.2 Ring Width w 502
14.5.3.3 Ion Current Density $\bar{J}_i$ and Sheath Thickness s 502
14.5.3.4 Ring Plasma Density n_i 503
14.5.3.5 Sputtering Rate R_{sput} and Absorbed Power P_{abs} 503
14.5.3.6 Ohmic Heating in the Plasma 503
14.5.4 Time-Varying Power Sources 504
14.5.4.1 Gas and Sputtered Atom Depletion and Recycling 505
14.5.4.2 PIC Simulations 506
14.6 Ionized Physical Vapor Deposition 507
Problems 510

15 High-Pressure Capacitive Discharges 513
15.1 Introduction 513
15.2 Intermediate Pressure RF Discharges 514
15.2.1 Energy Relaxation Length $\lambda_\mathcal{E}$ 515
15.2.2 Passive Bulk Plasma Model 516
15.2.3 Simulation Results 519
15.2.4 Metastables and Secondary Electrons 520
15.3 Alpha-to-Gamma (α–γ) Transition 524
15.3.1 Qualitative Description of α and γ Modes 524
15.3.2 Gamma Mode Model 526
15.3.3 Experimental Results 529
15.3.4 Particle-in-Cell Simulations 532
15.3.5 Low Pressures and α–γ Transition Curve 533
15.4 Atmospheric Pressure RF Discharges 534
15.4.1 The Atmospheric Pressure RF Regime 534
15.4.2 Homogeneous Model 536
15.4.2.1 Energy Balance 537
15.4.2.2 Particle Balance 539

15.4.3 Simulations and the α–γ Transition *541*
15.4.3.1 α-Mode PIC Simulations *542*
15.4.3.2 γ-Mode PIC Simulations *544*
15.4.4 Experimental Results *545*
15.5 Atmospheric Pressure Low-Frequency Discharges *548*
15.5.1 Discharge Regimes *548*
15.5.2 Streamer Breakdown *548*
15.5.3 Glow Discharge Regime *551*
15.5.3.1 Experimental Results *551*
15.5.3.2 Circuit Model *552*
15.5.4 Filamentary Regime *554*
15.5.4.1 Discharge Properties *555*
15.5.4.2 DBD Circuit Model *555*
Problems *556*

16 Etching *561*
16.1 Etch Requirements and Processes *561*
16.1.1 Plasma Etch Requirements *561*
16.1.2 Etch Processes *565*
16.2 Etching Kinetics *568*
16.2.1 Surface Kinetics *568*
16.2.2 Discharge Kinetics and Loading Effect *571*
16.2.3 Chemical Framework *572*
16.2.4 Pattern Transfer and Aspect-Ratio-Dependent Etching *573*
16.2.4.1 Aspect-Ratio-Dependent Etching (ARDE) *574*
16.3 Halogen Atom Etching of Silicon *575*
16.3.1 Pure Chemical F-Atom Etching *575*
16.3.2 Ion Energy-Driven F Atom Etching *579*
16.3.3 CF_4 Discharges *580*
16.3.4 O_2 and H_2 Feedstock Additions *584*
16.3.5 Cl-Atom Etching *586*
16.3.5.1 Pure Chemical Etching *586*
16.3.5.2 Ion-Assisted Etching *587*
16.3.5.3 Photon-Assisted Etching *588*
16.4 Other Etch Systems *588*
16.4.1 F and CF_x Etching of SiO_2 *588*
16.4.2 Si_3N_4 Etching *590*
16.4.3 Fluorocarbon Plasma Etch Selectivities *590*
16.4.4 Aluminum Etching *591*
16.4.5 Copper Etching *592*
16.4.6 Resist Etching *592*
16.4.7 Other Materials *594*
16.5 Atomic Layer Etching (ALE) *595*
16.5.1 Introduction and History *595*
16.5.2 Experimental Results *597*
16.5.3 Molecular Dynamics Simulations *597*
16.5.4 Model of Atomic Layer Etching *600*
16.5.4.1 Saturated Chlorination Step *600*

16.5.4.2 Ion-Enhanced Etch Step *601*
16.5.4.3 Saturated Etch *602*
16.5.4.4 Unsaturated Etch *602*
16.5.4.5 Other Anisotropic ALE Processes *604*
16.5.5 Quasi-Atomic Layer Etching *605*
16.5.6 Thermal Atomic Layer Etching *606*
16.5.6.1 Experimental Results *607*
16.6 Substrate Charging *608*
16.6.1 Gate Oxide Damage *608*
16.6.2 Grounded Substrate *609*
16.6.3 Nonuniform Plasmas *610*
16.6.4 Transient Damage During Etching *612*
16.6.5 Electron Shading Effect *612*
16.6.6 RF Biasing *614*
16.6.7 Etch Profile Distortions *615*
Problems *616*

17 Deposition and Implantation *619*
17.1 Introduction *619*
17.2 Plasma-Enhanced Chemical Vapor Deposition *621*
17.2.1 Amorphous Silicon *621*
17.2.2 Silicon Dioxide and Conformality *624*
17.2.2.1 Conformality *625*
17.2.3 Silicon Nitride *627*
17.2.4 Large-Area PECVD *628*
17.3 Atomic Layer Deposition *628*
17.3.1 Introduction *628*
17.3.2 Thermal ALD of Al_2O_3 *630*
17.3.3 Plasma-Enhanced ALD of Al_2O_3 *631*
17.3.4 Conformality of ALD *632*
17.3.4.1 Diffusion-Limited Regime *632*
17.3.4.2 Reaction-Limited Regime *633*
17.3.4.3 Recombination-Limited Regime *635*
17.4 Sputter Deposition *636*
17.4.1 Physical Sputtering *636*
17.4.2 Reactive Sputtering *638*
17.5 Plasma-Immersion Ion Implantation *640*
17.5.1 Collisionless Sheath Model *642*
17.5.1.1 Sheath Motion *643*
17.5.1.2 Matrix Sheath Implantation *644*
17.5.1.3 Child Law Sheath Implantation *644*
17.5.2 Collisional Sheath Model *646*
17.5.3 Applications of PIII to Materials Processing *649*
17.5.3.1 Semiconductor Processes *649*
17.5.3.2 Metallurgical Processes *650*
Problems *651*

18 Dusty Plasmas *655*
18.1 Qualitative Description of Phenomena *655*
18.2 Particle Charging and Discharge Equilibrium *656*
18.2.1 Equilibrium Potential and Charge *656*
18.2.2 Discharge Equilibrium *660*
18.3 Particulate Equilibrium *662*
18.4 Formation and Growth of Dust Grains *665*
18.5 Physical Phenomena and Diagnostics *670*
18.5.1 Strongly Coupled Plasmas and Dust Crystals *670*
18.5.2 Dust Acoustic Waves *671*
18.5.3 Driven Particulate Motion *671*
18.5.4 Laser Light Scattering *672*
18.5.4.1 LLS Visualizations *674*
18.6 Removal or Production of Particulates *675*
Problems *677*

19 Kinetic Theory of Discharges *681*
19.1 Basic Concepts *681*
19.1.1 Two-Term Approximation *682*
19.1.2 The Krook Collision Operator *682*
19.1.3 Two-Term Collisional Kinetic Equations *683*
19.1.4 Diffusion and Mobility *685*
19.1.5 Druyvesteyn Distribution *685*
19.1.6 Electron Distribution in an RF Field *686*
19.1.7 Effective Electrical Conductivity *687*
19.1.8 LXCat Database and Bolsig+ Solver *689*
19.2 Local Kinetics *690*
19.3 Nonlocal Kinetics *693*
19.4 Quasilinear Diffusion and Stochastic Heating *697*
19.4.1 Quasilinear Diffusion Coefficient *699*
19.4.2 Stochastic Heating *701*
19.4.3 Relation to Velocity Kick Models *701*
19.4.4 Two-Term Kinetic Equations *702*
19.4.5 Energy Relaxation Length *702*
19.5 Energy Diffusion in a Skin Depth Layer *703*
19.5.1 Stochastic Heating *703*
19.5.2 Effective Collision Frequency *705*
19.5.3 Energy Distribution *706*
19.6 Kinetic Modeling of Discharges *707*
19.6.1 Non-Maxwellian Global Models *707*
19.6.2 Inductive Discharges *708*
19.6.3 Capacitive Discharges *710*
Problems *714*

Appendix A Collision Dynamics *717*
A.1 Coulomb Cross Section *718*

Appendix B The Collision Integral *721*
B.1 Boltzmann Collision Integral *721*
B.2 Maxwellian Distribution *722*

Appendix C Diffusion Solutions for Variable Mobility Model *723*

References *727*
Index *749*

List of Figures

Figure 1.1 High aspect ratio anisotropic etches, showing the extraordinary capabilities of plasma processing; such etched features are used for device isolation, charge storage capacitors, channel holes, and many other purposes in integrated circuits: (*a*) trench etch (0.2 μm wide by 4 μm deep) in single-crystal silicon, circa 2004; (*b*) set of channel holes (each approximately 0.1 μm in diameter by 7.5 μm deep) etched into a stacked set of dielectric layers, circa 2023. Source: (*a*) M. A. Lieberman (Book Author). (*b*) Courtesy of Lam Research Corporation *2*

Figure 1.2 Deposition and pattern transfer in manufacturing an integrated circuit: (*a*) metal deposition; (*b*) photoresist deposition; (*c*) optical exposure through a pattern; (*d*) photoresist development; (*e*) anisotropic plasma etch; (*f*) remaining photoresist removal *2*

Figure 1.3 Plasma etching in integrated circuit manufacture: (*a*) example of isotropic etch; (*b*) sidewall etching of the resist mask leads to a loss of anisotropy in film etch; (*c*) illustrating the role of bombarding ions in anisotropic etch; (*d*) illustrating the role of sidewall passivating films in anisotropic etch *3*

Figure 1.4 Experimental demonstration of ion-enhanced plasma etching. Source: Adapted from Coburn and Winters (1979) *5*

Figure 1.5 Illustrating ion implantation of an irregular object: (*a*) in a conventional ion beam implanter, the beam is electrically scanned and the target object is mechanically rotated and tilted to achieve uniform implantation; (*b*) in plasma-immersion ion implantation (PIII), the target is immersed in a plasma, and ions from the plasma are implanted with a relatively uniform spatial distribution *6*

Figure 1.6 Schematic view of (*a*) a plasma and (*b*) a discharge *6*

Figure 1.7 Energy coupling between electrons and heavy particles in a low-pressure plasma *7*

Figure 1.8 Space and laboratory plasmas on a $\log_{10} n$ versus $\log_{10} \mathrm{T_e}$ diagram (Source: Book (1987)/Naval Research Laboratory); the electron Debye length λ_{De} is defined in Section 2.4 *8*

Figure 1.9 Densities and energies for various species in a low-pressure capacitive rf discharge *9*

Figure 1.10 Electron energy distribution function $g_e(\mathcal{E})$ in a weakly ionized discharge *9*

Figure 1.11 The formation of plasma sheaths: (*a*) initial ion and electron densities and potential; (*b*) densities, electric field, and potential after the formation of the sheath *10*

Figure 1.12 PIC simulation of positive ion sheath formation: (*a*) v_x–x electron phase space, with horizontal scale in meters; (*b*) electron density n_e; (*c*) electric field E_x; (*d*) potential Φ; (*e*) electron number $\mathcal{N}$ versus time t in seconds; (*f*) right hand potential v_x versus time t *11*

Figure 1.13 Typical multi-wafer capacitive rf discharge in plane-parallel geometry, used for anisotropic etching. Source: Lieberman and Gottscho (1994)/with permission of Elsevier *13*

Figure 1.14 The physical model of an rf diode. Source: Lieberman and Gottscho (1994)/ with permission of Elsevier *14*

Figure 1.15 Some modern capacitive discharges are used for etching and deposition; (*a*) single frequency, (*b*) dual frequency, and (*c*) magnetically enhanced *16*

Figure 1.16 Some non-capacitive, high-density discharges used for etching and deposition; (*a*) planar inductive; (*b*) electron cyclotron resonance; (*c*) helicon *17*

Figure 1.17 A dc planar magnetron discharge, used for thin film deposition *19*

Figure 1.18 The central problem of discharge analysis *19*

Figure 2.1 Kirchhoff's circuit laws: the total current J_T flowing across a nonuniform one-dimensional discharge is independent of x; the sum of the currents entering a node is zero ($I_{rf} = I_T + I_1$); the sum of voltages around a loop is zero ($V_{rf} = V_1 + V_2 + V_3$) *23*

Figure 2.2 PIC simulation of ion loss in a plasma containing ions only: (*a*) v_x–x ion phase space, showing the ion acceleration trajectories; (*b*) number $\mathcal{N}$ of ion sheets versus t, with the steps indicating the loss of a single sheet; (*c*) the potential Φ versus x during the first 10^{-10} s of ion loss *24*

Figure 2.3 One-dimensional v_x–x phase space, illustrating the derivation of the Boltzmann equation and the change in f due to collisions *25*

Figure 2.4 The force density due to the pressure gradient *28*

Figure 2.5 Calculation of the electron Debye length λ_{De}. A negatively charged sheet is introduced into a plasma containing electrons in thermal equilibrium *33*

Figure 3.1 A flux of incident particles collides with a population of target particles in the half-space $x > 0$ *38*

Figure 3.2 Hard-sphere scattering *39*

Figure 3.3 Definition of the differential scattering cross section *40*

Figure 3.4 The relation between the scattering angles in (*a*) the laboratory system and (*b*) the center of mass (CM) system *42*

Figure 3.5 Calculation of the differential scattering cross section for small-angle scattering. The center of mass trajectory is practically a straight line *45*

Figure 3.6 The processes that lead to large-angle Coulomb scattering: (*a*) single large-angle event; (*b*) cumulative effect of many small-angle events *47*

Figure 3.7 Polarization of an atom by a point charge q *49*

Figure 3.8 Scattering in the polarization potential, showing (*a*) hyperbolic and (*b*) captured orbits *50*

Figure 3.9 Probability of collision P_c for electrons in H_2 and He; the cross section is $\sigma \approx 2.87 \times 10^{-17} P_c$ cm^2 (Brown, 1959/MIT Press) *52*

Figure 3.10 Probability of collision P_c for electrons in Ne, Ar, Kr, and Xe, showing the Ramsauer minima for Ar, Kr, and Xe; the cross section is $\sigma \approx 2.87 \times 10^{-17} P_c$ cm^2 (after Brown, 1959) *52*

Figure 3.11 Atomic energy levels for the central field model of an atom, showing the dependence of the energy levels on the quantum numbers n and l; the energy levels are shown for sodium, without the fine structure (Thorne, 1988/Springer Nature) *54*

Figure 3.12 The energy levels of the argon atom, showing (*a*) the ($3p^5nl$) configurations and (*b*) details of the $3p^54s$ and $3p^54p$ configurations, with the two metastable levels shown as heavy solid lines (Edgell, 1961/Interscience Publishers) *57*

Figure 3.13 Ionization, excitation, and elastic scattering cross sections for electrons in argon gas (Vahedi, 1993) *60*

Figure 3.14 Illustrating the calculation of ion–atom charge transfer *61*

Figure 3.15 Experimental values for elastic scattering (*s*), charge transfer (*T*), and the sum of the two mechanisms (*t*) for (*a*) helium, (*b*) neon, and (*c*) argon ions in their parent gases (McDaniel et al., 1993/John Wiley & Sons) *63*

Figure 3.16 Electron collision rate constants K_{iz}, K_{ex}, and K_m versus T_e in argon gas (Vahedi, 1993) *65*

Figure 3.17 Collisional energy loss $\mathcal{E}_c$ per electron–ion pair created versus T_e in argon and oxygen. Source: Adapted from Gudmundsson, 2002b *67*

Figure 4.1 Charged particle gyration in a uniform magnetic field; **B** is directed out of the page *74*

Figure 4.2 Motion of electrons and ions in uniform crossed **E** and **B** fields *76*

Figure 4.3 Plasma oscillations in a slab geometry: (*a*) displacement of electron cloud with respect to ion cloud; (*b*) calculation of the resulting electric field *77*

Figure 4.4 Rf current and electric field amplitudes and phases in the sheath and plasma regions of an rf discharge *81*

Figure 4.5 Dispersion ω versus k for electromagnetic and electrostatic electron plasma waves in an unmagnetized plasma *83*

Figure 4.6 Calculation of the parallel force due to a magnetic field gradient $\partial B_z/\partial z$ *86*

Figure 4.7 Calculation of the curvature drift due to a magnetic field gradient $\partial B_x/\partial z$ *87*

Figure 4.8 Calculation of the perpendicular gradient drift due to a magnetic field gradient $\partial B_z/\partial x$: (*a*) the magnetic field lines; (*b*) the motion viewed in the x–y plane *88*

Figure 4.9 Dispersion ω versus k for the principal waves in a magnetized plasma with immobile ions for $\omega_{ce} > \omega_{pe}$ *96*

Figure 4.10 Dispersion ω versus k for the principal waves in a magnetized plasma with mobile ions *98*

Figure 4.11 The CMA diagram for waves in a magnetized plasma. The cutoffs and resonances are indicated by the lines labeled $u = \infty$ and $u = 0$, respectively, where u denotes the phase velocity and the subscripts label the principal waves. Source: Allis et al., 1963/MIT Press *99*

Figure 4.12 A microwave interferometer for plasma density measurement *101*

Figure 4.13 Mean electron density versus incident power at the midplane of an rf inductive discharge as measured by a microwave interferometer, compared with ion density as measured by a Langmuir probe. Source: Hopwood et al., 1993b/American Vacuum Society *102*

Figure 4.14 Design of (*a*) dc or grounded and (*b*) floating hairpin resonator probes. Source: Piejak et al. (2005). © IOP Publishing. Reproduced with permission. All rights reserved. https://doi.org/10.1088/0963-0252/14/4/012 *103*

Figure 4.15 Electron density versus absorbed power in a 10 mTorr argon discharge. Data from 443 MHz cavity resonance (circles), 506 MHz cavity resonance (squares), and Langmuir probe (triangles). Source: Moroney et al., 1989/with permission of AIP Publishing *105*

Figure 5.1 High-pressure diffusion solution for density n versus position x *117*

Figure 5.2 Low-pressure diffusion solutions for variable mobility model: (*a*) normalized ionization rate $\alpha = (\nu_{iz}l/2u_B)(\pi l/4\lambda_i)^{1/2}$ versus $2\lambda_i/l$; (*b*) normalized density $n/n(0)$ versus normalized position $(2x/l)(\alpha/\alpha_0)^{2/3}$. Source: Godyak, 1986/Delphic Associates *120*

Figure 5.3 Free-fall solution: variation of the normalized density $n/n(0)$ versus normalized position $1.75x\nu_{iz}/u_B$ *121*

Figure 5.4 Edge-to-center density ratio h_l versus l/λ_i, illustrating the three regimes of collisionless flow, variable mobility diffusion, and constant diffusion coefficient models *123*

Figure 5.5 PIC simulation results showing the h_l factor versus $l/\lambda_i \propto$ pressure p for various inductively and capacitively coupled discharges, with the secondary electron emission coefficient γ indicated. Source: Lafleur and Chabert (2015a). ©IOP Publishing. Reproduced with permission. All rights reserved. https://doi.org/10.1088/0963-0252/24/2/025017 *123*

Figure 5.6 A plasma-filled conducting box in a dc magnetic field, illustrating the calculation of ambipolar diffusion in a magnetized plasma *126*

Figure 5.7 Two-dimensional simulation results for electron and ion fluxes to the rectangular conducting walls with $d = 5$ cm and $l = 10$ cm at 1 mTorr argon; (*a*) perpendicular flux $\Gamma_\perp$ versus x at $B_0 = 0$ G; (*b*) parallel flux $\Gamma_\parallel$ versus y at $B_0 = 0$ G; (*c*) perpendicular flux $\Gamma_\perp$ versus x at $B_0 = 50$ G; (*d*) parallel flux $\Gamma_\parallel$ versus y at $B_0 = 50$ G). Source: Reproduced from Lafleur and Boswell (2012a), with the permission of AIP Publishing. https://doi.org/10.1063/1.4719701 *128*

Figure 5.8 Magnetic multipole confinement in cylindrical geometry, illustrating the magnetic field lines and the $|\mathbf{B}|$ surfaces near the circumferential walls *129*

Figure 5.9 Schematic for determining multipole fields in rectangular geometry *130*

Figure 6.1 Qualitative behavior of sheath and presheath in contact with a wall *139*

Figure 6.2 $\Phi/\mathrm{T_e}$ versus position within the presheath, showing (*a*) the geometric presheath; (*b*) a planar collisional presheath; and (*c*) a planar ionization presheath. The sheath–presheath edge is at the right. Source: Riemann, 1991/IOP Publishing *141*

Figure 6.3 Particle-in-cell simulation showing sheath formation from warm, initially uniform electron–proton plasma between short-circuited parallel plates: (a) density profiles at time $t = 4 \times 10^{-8}$ s; (b) electric field profile; (c) potential profile; (d) midpotential versus time *144*

Figure 6.4 Negative ion sheath solutions; (*a*) α_s/α_b and (*b*) $\Phi_p/\mathrm{T_e}$ versus α_b, with γ as a parameter. Source: Boyd and Thompson, 1959/The Royal Society *150*

Figure 6.5 Definition of voltage and current for a Langmuir probe *153*

Figure 6.6 Typical I–V_B characteristic for a Langmuir probe *153*

Figure 6.7 Ion orbital motion within the sheath of a cylindrical Langmuir probe *158*

Figure 6.8 Schematic of double probe measurement: (*a*) definition of voltage and currents; (*b*) typical current–voltage characteristic. Source: Chen, 1965/with permission of Elsevier *161*

Figure 6.9 Typical collecting and emitting current–voltage characteristics for an emissive wire probe in a plasma; the electron and wire temperatures are $\mathrm{T_e} = 3$ V and $\mathrm{T_w} = 0.3$ V. Source: Hershkowitz, 1989/with permission of Elsevier *163*

Figure 6.10 Construction of a cylindrical probe for rf discharge measurements. Source: Godyak et al., 1992/with permission of IOP Publishing *165*

Figure 6.11 Simple Langmuir probe biasing circuit *165*

Figure 6.12 Probe characteristics I versus V_B in a plasma with an oscillating space potential $\Phi_p(t)$, showing (heavy solid line) a time-averaged probe characteristic having an apparent electron temperature much higher than the actual $\mathrm{T_e}$ *166*

Figure 6.13 Probe circuit elements and blocking inductor used to measure the current–voltage characteristics in an rf discharge *167*

Figure 7.1 Typical materials processing reactor *172*

Figure 7.2 State space for a chemically reactive system *172*

Figure 7.3 (*a*) Specific heat C_p at constant pressure and (*b*) entropy S versus temperature T *181*

Figure 7.4 Gibbs free energy G versus composition *184*

Figure 7.5 Phase diagram p versus T for a pure substance *189*

Figure 7.6 The Langmuir isotherm *191*

Figure 8.1 Potential energy curves for the electronic states of a diatomic molecule *196*

Figure 8.2 Vibrational and rotational levels of two electronic states A and B of a molecule; the three double arrows indicate examples of transitions in the pure rotation spectrum, the rotation–vibration spectrum, and the electronic spectrum (Herzberg, G. (1971)/with permission of Dover Publications) *197*

Figure 8.3 Potential energy curves for H_2^-, H_2, and H_2^+. Source: Steinfeld (1985)/ with permission of Dover Publications *200*

Figure 8.4 Potential energy curves for O_2^-, O_2, and O_2^+. Source: Steinfeld (1985)/ with permission of Dover Publications *201*

Figure 8.5 Simplified potential energy curves for $He_2^+(X^2\Sigma_u^+)$ and the four lowest-energy excimers of He_2; the energies (in volts) above the He ground state are given at the right. The repulsive $He_2(X^1\Sigma_g^+)$ ground state is not shown. Source: Golubovskii et al. (2002). ©IOP Publishing. Reproduced with permission. All rights reserved. https://doi.org/10.1088/0022-3727/36/1/306 *202*

Figure 8.6 Illustrating the variety of dissociation processes for electron collisions with molecules *203*

Figure 8.7 Illustrating dissociative ionization and dissociative recombination for electron collisions with molecules *205*

Figure 8.8 Illustrating a variety of electron attachment processes for electron collisions with molecules; (*a*) capture into a repulsive state; (*b*) capture into an attractive state; (*c*) capture of slow electrons into a repulsive state; (*d*) polar dissociation *207*

Figure 8.9 Cross section for production of negative ions by electron impact in O_2. Source: Rapp and Briglia, 1965/with permission of AIP Publishing *209*

Figure 8.10 Illustrating nonresonant charge transfer processes for heavy-particle collisions *212*

Figure 8.11 Illustrating positive–negative ion recombination for heavy-particle collisions *214*

Figure 8.12 Illustrating associative detachment processes for heavy particle collisions; (*a*) the AB^- ground state lies above the AB ground state; (*b*) the AB^- ground state lies below the AB ground state *215*

Figure 8.13 Illustrating transfer of excitation for heavy-particle collisions *217*

Figure 8.14 Illustrating Penning and associative ionization for heavy-particle collisions *218*

Figure 8.15 Schematic representation of Penning ionization; (*a*) direct and (*b*) exchange processes *218*

Figure 8.16 A plot of the dipole locking constant C; 1 debye $\approx 3.34 \times 10^{-30}$ C-m. Source: Su and Bowers, 1973/with permission of Elsevier *220*

Figure 8.17 Cross sections for electron excitation of O_2 (Lawton and Phelps, 1978; Phelps, 1985; compiled by Vahedi, 1993) *226*

Figure 8.18 Energy level diagram for emission of radiation from an excited state *231*

Figure 8.19 Overlap of excitation cross sections and electron velocity distribution *233*

Figure 8.20 Comparison of actinometric measurements with a two-photon absorption laser-induced fluorescence (TALIF) measurement of oxygen atom density in an O_2/CF_4 discharge. Source: Walkup et al., 1986/with permission of AIP Publishing *233*

Figure 8.21 Schematic of a typical PROES measurement system. Source: Reproduced from Gans et al. (2006a), with the permission of AIP Publishing. https://doi.org/10.1063/1.2406035 *236*

Figure 8.22 PROES measurement of the time- and space-resolved excitation to the Paschen $2p_1$ level of neon, in a 13.56 MHz, 2 Pa neon discharge with a 5 cm gap; the dotted lines indicate the sheath motions, and the arrows indicate the trajectories of energetic electron groups generated by the expanding sheaths. Source: Schulze et al. (2010). ©IOP Publishing. Reproduced with permission. All rights reserved. https://doi.org/10.1088/0022-3727/43/12/124016 *236*

Figure 9.1 Transient kinetics for gas-phase reaction A → B → C; (*a*) $K_A = 1, K_B = 5$; (*b*) $K_A = 5, K_B = 1$ *247*

Figure 9.2 Illustrating ion neutralization and secondary emission at a metal surface; (*a*) the work function $\mathcal{E}_\phi$ and the Fermi energy $\mathcal{E}_F$; (*b*) Auger emission due to electron tunneling *255*

Figure 9.3 Illustrating the method of images for a metal surface to determine (*a*) the work function and (*b*) the van der Waals force *255*

Figure 9.4 Schematic diagrams of the potential energy near a surface for adsorption: (*a*) dissociative chemisorption; (*b*) physisorption; and (*c*) molecular chemisorption *259*

Figure 9.5 Relative sputtering yields for photoresist and aluminum versus angle of incidence θ. Source: Flamm and Herb, 1989/with permission Elsevier *262*

Figure 9.6 Illustrating the processes that can occur for reaction of an etchant with a surface *263*

Figure 9.7 Idealized, two-dimensional reactor configuration for showerhead gas flow calculation, with no variation along the y-coordinate *270*

Figure 9.8 Showerhead streamlines for the idealized, two-dimensional reactor configuration; (*a*) pure H_2, no substrate deposition; (*b*) 5%/95% SiH_4/H_2, with a loss probability $\gamma = 0.107$ on the electrodes. Source: Howling et al. (2012). ©IOP Publishing. Reproduced with permission. All rights reserved. https://doi.org/10.1088/0963-0252/21/1/015005 *272*

Figure 10.1 T_e versus $n_g d_{eff}$ for Maxwellian electrons in argon *284*

Figure 10.2 Positive ion, negative ion, and electron densities versus position for a plane-parallel electronegative discharge, showing the electronegative, electropositive, and sheath regions *290*

Figure 10.3 Density profiles versus position in oxygen; negative ions (dots) measured at 10 W input power at (*a*) 10 mTorr, with the electron density shown by the dashed line; and (*b*) at 40 mTorr. Source: Vender et al., 1995/with permission of American Physical Society *305*

Figure 10.4 Density profiles versus position in oxygen; measured negative (squares, open circles) and positive (triangles) ions at (*a*) 45 mTorr, and (*b*) 150 mTorr; solid and dashed lines are calculated positive and negative ion densities. Source: Berezhnoj et al., 2000/with permission of AIP Publishing *306*

Figure 10.5 (*a*) Electronegativity α and (*b*) electron density n_e versus discharge pressure p in oxygen at various discharge powers, in a cylindrical stainless steel chamber with $l = 7.6$ cm and $R = 15.2$ cm; 50 sccm flow rate, $\gamma_q = 0.007$, $\gamma_O = 0.5$, $T_g = 600$ K. Source: Gudmundsson (2004). ©IOP Publishing. Reproduced with permission. All rights reserved. https://doi.org/10.1088/0022-3727/37/15/005 *307*

Figure 10.6 Central electronegativity α_0 versus $O_2(a)$ wall quenching coefficient γ_q at pressures of 10 (pluses), 25 (crosses), and 50 (diamonds) mTorr, for a 4.5 cm gap, 222 V, 13.56 MHz plane-parallel electronegative discharge in oxygen. Source: Proto and

Gudmundsson (2018). ©IOP Publishing. Reproduced with permission. All rights reserved. https://doi.org/10.1088/1361-6595/aaca06 *309*

Figure 10.7 Simulation results for recombination-dominated 0.5/0.5 Ar/O_2 (top panel) and detachment-dominated 0.4/0.4/0.2 $Ar/O_2/O_2(a^1\Delta_g)$ (bottom panel) for 1, 10, and 100 mTorr, for a 5 cm gap, low power, 13.56 MHz plane-parallel inductively driven discharge. Source: Monahan and Turner (2008). ©IOP Publishing. Reproduced with permission. All rights reserved. https://doi.org/10.1088/0963-0252/17/4/045003 *309*

Figure 10.8 Positive ion and electron density versus discharge incident power in chlorine at 20 mTorr, in a cylindrical stainless steel chamber with $l = 20$ cm and $R = 18.5$ cm. Source: Malyshev and Donnelly, 2001/with permission of AIP Publishing *311*

Figure 10.9 Global model (lines) and measured (symbols) atomic chlorine density at 1 and 10 mTorr, and electron density at 10 mTorr, versus discharge power. Source: Thorsteinsson and Gudmundsson (2010a). ©IOP Publishing. Reproduced with permission. All rights reserved. https://doi.org/10.1088/0963-0252/19/1/015001 *312*

Figure 10.10 Time evolution of the plasma density n_e, the electron temperature T_e, and the excited atom (4s and 4p) densities for different periods τ, for a time-averaged power of 500 W and a duty ratio of 0.25: (*a*) $\tau = 10$ μs; (*b*) $\tau = 100$ μs; (*c*) $\tau = 1$ ms. Source: Ashida et al., 1995/with permission of AIP Publishing *318*

Figure 10.11 Time variation of (*a*) electron density n_e and (*b*) electron temperature T_e for 100 μs period and 0.50 duty ratio in chlorine (8 mTorr, 400 W) and in argon (6 mTorr, 200 W); the open and closed circles indicate the data for Ar and Cl_2, respectively; the crosses in (*a*) indicate the data obtained after photodetachment. Source: Ahn et al., 1995/with permission of IOP Publishing *321*

Figure 10.12 Global simulation results for $n_e(t)$, $n_-(t)$, and $T_e(t)$ (lines) for the chlorine measurements (symbols) of Ahn et al. (1995) shown in Figure 10.11. Source: Reproduced with permission from Kemaneci et al. (2014)/IOP Publishing *322*

Figure 11.1 The basic rf discharge model: (*a*) sheath and plasma thicknesses; (*b*) electron and ion densities *331*

Figure 11.2 Sheath voltages V_{ap}, V_{pb}, and their sum V_{ab} versus time; the time-average value $\overline{V}$ of V_{pb} is also shown *335*

Figure 11.3 Spatial variation of the total potential Φ (solid curves) for the homogeneous model of Section 11.2, at four different times during the rf cycle. The dashed curve shows the spatial variation of the time-averaged potential $\overline{\Phi}$ *336*

Figure 11.4 Nonlinear circuit model of the homogeneous rf plasma discharge. The dashed lines indicate that the series connection of the nonlinear elements C_a and C_b, and R_a and R_b, yield the corresponding linear elements C_s and R_s, respectively *339*

Figure 11.5 Schematic plot of the densities in a high-voltage, capacitive rf sheath *341*

Figure 11.6 Sketch of the electron sheath thickness s versus ωt, showing the definition of the phase $\phi(x)$ *342*

Figure 11.7 Effective collision frequency ν_{eff} versus pressure p, for a mercury discharge driven at 40.8 MHz. The solid line shows the collision frequency due to ohmic dissipation alone. Source: Popov and Godyak, 1985/with permission of AIP Publishing *354*

Figure 11.8 Discharge power absorbed P_{abs}, and rf voltage V_{rf} versus discharge current I_{rf} at (*a*) $p = 10$ mTorr and (*b*) $p = 100$ mTorr in argon. Source: Reproduced from Godyak et al. (1991)/IEEE *355*

Figure 11.9 Electron energy probability function g_p versus $\mathcal{E}_e$ for various discharge currents for argon gas with $f = 13.56$ MHz and $l = 6.7$ cm: (*a*) $p = 10$ mTorr and (*b*) $p = 100$ mTorr. Source: Godyak, 1990b/with permission of American Physical Society *356*

Figure 11.10 (*a*) Central plasma density versus pressure and (*b*) average electron temperature versus pressure, for an rms discharge current of 0.3 A; open circles are measurements from Godyak and Piejak (1990a) for a 2 cm gap; closed triangles are corresponding PIC simulations from Lafleur et al. (2014a); open squares are measurements from Lafleur et al. (2014a) for a 2.5 cm gap. Source: Reproduced from Lafleur et al. (2014a)/with permision of IOP Publishing *357*

Figure 11.11 One-dimensional electron velocity distribution function $f_e(x, v_x, t)$ for a 10 cm electrode spacing; each plot covers a time window of $\frac{1}{32}$ of an rf cycle. Each line on a plot represents a spatial window of 2 mm. Source: Wood, 1991/with permission of Wood, B. P *358*

Figure 11.12 Spatiotemporal distribution of ionizing collisions collected over 20 rf cycles, for a 10 MHz, 1 kV, 20 mTorr hydrogen discharge. Source: Reproduced from Vender and Boswell (1990)/with permission of IEEE *359*

Figure 11.13 Central plasma potential V_{pb} (dashed), driving voltage V_{rf} (dotted), and electron (positive) and ion (negative) currents to the electrode. The ion current is plotted 10 times enlarged to show modulation within the rf cycle. Source: Reproduced from Vender and Boswell (1990)/with permission of IEEE *360*

Figure 11.14 Symmetric discharge experiments compared to 1D PIC simulations, for central electron density n_e versus argon pressure p; (*a*) Langmuir probe measurements (closed squares with error bars), PIC with "effective" electron reflection coefficient γ_e, ion-induced secondary emission coefficient $\gamma_{se} = 0.7$, and gas heating (Schulenberg et al., 2021); (*b*) electron-induced γ_e (Vaughan model), ion-induced $\gamma_{se} = 0.7$, excited neutral-induced γ_{exc}, photon-induced γ_{ph} (Wen et al., 2023), and measured gas heating; 12 cm diameter stainless steel electrodes, 4 cm gap, 13.56 MHz, 250 V rf amplitude. Source: (*a*) Schulenberg et al. (2021)/IOP Publishing/CC BY 4.0. https://doi.org/10.1088/1361-6595/ac2222. (*b*) Wen et al. (2023)/IOP Publishing/CC BY 4.0 *363*

Figure 11.15 Electron heating fractions S/S_{tot} as a function of argon pressure; (*a*) apparent S_{ohmic}/S_{tot} and S_{stoc}/S_{tot}, from the data in Godyak and Piejak (1990b) (closed symbols), and from the PIC simulations (dashed curves), both analyzed using the homogeneous model of Section 11.2; and (*b*) S_{in}/S_{tot}, S_{press}/S_{tot} and S_{ohmic}/S_{tot} (symbols) from the PIC simulations using Boltzmann term analysis expression (11.3.6); the open and closed symbols correspond to two different, but equivalent, models of electron scattering. Source: Lafleur et al. (2014a)/with permission of IOP Publishing *364*

Figure 11.16 Capacitive voltage divider model of bias voltage formation in an asymmetric discharge *365*

Figure 11.17 Spherical shell model of an asymmetric rf discharge. Source: Lieberman, 1989b/with permission of AIP Publishing *367*

Figure 11.18 Voltage-driven rf sheath, showing (*a*) normalized voltage V_{norm}, (*b*) normalized charge Q_{norm}, and (*c*) normalized current I_{norm}, over two rf cycles; the solid lines show $V_B = V_0$, and the dot-dashed lines show $V_B = 1.05\,V_0$ *370*

Figure 11.19 Normalized discharge current over two rf cycles; the dot-dashed line shows the current for the nonlinear sheath alone, in the absence of the bulk plasma; the solid line shows the series resonance oscillations that are excited in an asymmetric discharge; $V_{rf} = 200$ V, $f = 13.56$ MHz, $l = 5.7$ cm, $n = 10^{15}\ m^{-3}$, $p = 3$ mTorr argon. Source: Reproduced from Lieberman et al. (2008), with the permission of AIP Publishing. https://doi.org/10.1063/1.2928847 *372*

Figure 11.20 Measured sensor current (proportional to the discharge current), showing the nonlinearly excited series resonance oscillations over two rf cycles, for a 13.56 MHz asymmetric discharge with a gap length of 6.7 cm; (upper curve) argon pressure 2 Pa, dc bias voltage 520 V, rf power 75 W; (lower curve) argon pressure 10 Pa, dc bias voltage 570 V, rf power 125 W. Source: Klick et al. (1997). ©IOP Publishing. Reproduced with permission. All rights reserved. https://doi.org/10.1143/JJAP.36.4625 *373*

Figure 11.21 Diode (*a*) and triode (*b*) configurations of a dual-frequency capacitive discharge *373*

Figure 11.22 Typical multi-frequency-driven discharge waveforms over three low-frequency periods; (*a*) widely separated incommensurate frequencies, with $V_l = 800$ V, $V_h = 200$ V, $f_l = 2$ MHz, $f_h = 27.12$ MHz, $\chi = 0$; (*b*) fundamental and second harmonic, with $V_l = V_h = 500$ V, $\chi = 0$; (*c*) fundamental and second harmonic, with $V_l = V_h = 500$ V, $\chi = \pi/2$; (*d*) tailored voltage, with $V_{rf} = 400\cos\omega_l t + 300\cos 2\omega_l t + 200\cos 3\omega_l t + 100\cos 4\omega_l t$; (*e*) rectangular waveform with 30% duty ratio *374*

Figure 11.23 Homogeneous model solution for a 2 MHz/27.12 MHz, 100 mTorr argon, dual-frequency discharge with 4.5 cm gap; (*a*) ion acceleration voltage $\overline{V}_a$ versus high-frequency voltage amplitude V_h, for various low-frequency voltages V_l; and (*b*) density n versus low-frequency voltage V_l, for various high-frequency voltages V_h *378*

Figure 11.24 Illustrating the high- and low-frequency sheath motions in a dual-frequency-driven Child law sheath *379*

Figure 11.25 Contour plot of space- and time-resolved optical emission for a dual-frequency discharge, illustrating the electron dynamics within the low-frequency rf cycle; only the 2 MHz motion is resolved; 490 mTorr He/O_2, 2 MHz at 800 W, 27 MHz at 200 W. Source: Reproduced from Gans et al. (2006b), with the permission of AIP Publishing. https://doi.org/10.1063/1.2425044 *380*

Figure 11.26 Electrical asymmetry effect for $V_h = V_l = 500$ V and $\omega_h = 2\omega_l$, showing $V_{rf\,max}$ and $-V_{rf\,min}$ versus the second harmonic phase χ; note that $V_{rf\,min}(\chi) = -V_{rf\,max}(\chi + \pi)$ *381*

Figure 11.27 Electrical asymmetry effect for $V_h = V_l = 500$ V and $\omega_h = 2\omega_l$, showing (*a*) bias voltage V_B versus the second harmonic phase χ; (*b*) ion acceleration voltage $\overline{V}_a$ versus χ *381*

Figure 11.28 (*a*) Measured mean energy $\langle \epsilon_i \rangle$ with the gap size indicated, and (*b*) relative ion flux at the grounded electrode versus phase angle χ in degrees; 10 cm diameter argon discharge driven at 13.56 and 27.12 MHz with equal amplitude voltages. Source: Schulze et al. (2009b). ©IOP Publishing. Reproduced with permission. All rights reserved. https://doi.org/10.1088/0022-3727/42/9/092005 *382*

Figure 11.29 Schematic of high-frequency excitation of a capacitive discharge *383*

Figure 11.30 Simple model for surface waves in a symmetrically driven capacitive discharge *384*

Figure 11.31 Symmetry properties of the symmetric (*a*) and antisymmetric (*b*) surface wave modes *385*

Figure 11.32 Low-density ($n = 10^{15}$ m^{-3}) standing wave model results for capacitive (E_z) and inductive (E_r) normalized power deposition versus radius r; (*a*) $f = 13.56$ MHz; (*b*) $f = 40.12$ MHz; the edge effects result from evanescent modes excited at the outer radius $R = 50$ cm; $s = 0.4$ cm and bulk width $d = 2d' = 4$ cm. Source: Adapted from Lieberman et al. (2002) ©IOP Publishing *387*

Figure 11.33 High-density ($n = 10^{16}$ m^{-3}) standing wave model results for capacitive (E_z) and inductive (E_r) normalized power deposition versus radius r; $f = 13.56$ MHz; the edge effects result from evanescent modes excited at the outer radius $R = 50$ cm; $s = 0.4$ cm and bulk width $d = 2d' = 4$ cm. Source: Adapted from Lieberman et al. (2002) ©IOP Publishing *388*

Figure 11.34 Experimental results in a 40 cm × 40 cm, 4.5 cm gap, 150 mTorr, symmetrically driven argon discharge, showing the two-dimensional normalized ion fluxes to the upper electrode; (left panel) 13.56, 60, and 81.36 MHz at 50 W rf power; (right panel) 50, 170, and 265 W rf power at 60 MHz. Source: Reproduced from Perret et al. (2003), with the permission of AIP Publishing. https://doi.org/10.1063/1.1592617 *389*

Figure 11.35 Two-dimensional fluid simulation results for the spatial distribution of the electron density in a symmetric cylindrical discharge ($R = 20$ cm and $l = 2l' = 4.8$ cm) at 80 MHz and 150 mTorr argon, for rf powers of (*a*) 40 W, (*b*) 110 W, and (*c*) 190 W. Source: Adapted from Lee et al. (2008) ©IOP Publishing *389*

Figure 11.36 Electron density contours for various rf powers in an asymmetric, 100 mTorr argon capacitive discharge driven at 180 MHz; 3.8 cm gap, 15 cm lower electrode radius, and 17.8 cm upper electrode radius. Source: Rauf et al. (2008). ©IOP Publishing. Reproduced with permission. All rights reserved. https://doi.org/10.1088/0963-0252/17/3/035003 *390*

Figure 11.37 Asymmetric low-frequency capacitive discharge *392*

Figure 11.38 Model of low-frequency asymmetric capacitive discharge *392*

Figure 11.39 Time-varying sheath voltages and currents *393*

Figure 11.40 Symmetric low-frequency capacitive discharge showing total current I_{rf} and displacement current I_{displ} versus time. Source: Kawamura et al., 1999/with permission of IOP Publishing *394*

Figure 11.41 Illustrating the formation of the ion energy distribution $g_i(\mathcal{E})$ on an electrode *395*

Figure 11.42 Ion energy distribution $g_i(\mathcal{E})$ for a symmetrically driven capacitive discharge with a low-frequency and/or high-density sheath; $V_{rf} = 200$ V *396*

Figure 11.43 Simulation results showing ion energy distributions $g_i(\mathcal{E})$ for a single sheath in a current-driven helium discharge at frequencies from 1 to 100 MHz; the maximum sheath voltage drop was about 200 V in every case *398*

Figure 11.44 Measured ion energy distributions $g_i(\mathcal{E})$ for H_3^+, H_2O^+, and Eu^+ ions at the grounded electrode of a 75 mTorr argon rf discharge driven at 13.56 MHz. Source: Coburn and Kay, 1972/with permission of AIP Publishing *399*

Figure 11.45 Measured ion energy distributions $g_i(\mathcal{E})$ at the powered electrode of a CF_4 discharge driven at 13.56 MHz. Source: Kuypers and Hopman, 1990/with permission of AIP Publishing *399*

Figure 11.46 Comparison of experimental and theoretical ion energy distributions $g_i(\mathcal{E})$ in an argon discharge driven at 13.56 MHz at various pressures (1 μbar = 0.76 mTorr). Source: Wild and Koidl, 1991/with permission of AIP Publishing *400*

Figure 11.47 Ion-bombarding energy distribution versus ion energy; comparison of Fourier transform relaxation model and PIC simulations for a 3 cm gap, 30 mTorr argon discharge driven by 800 V at 2 MHz and 400 V at 64 MHz. Source: Reproduced from Wu et al. (2007), with the permission of AIP Publishing. https://doi.org/10.1063/1.2435975 *401*

Figure 11.48 Two-dimensional hybrid-fluid simulation of a magnetically enhanced capacitive discharge, showing (*a*) Ar^+ production rate versus height (distance from powered electrode) and (*b*) Ar^+ ion density versus height, for magnetic field strengths varied from 0 to 220 G; rf frequency 10 MHz, pressure 40 mTorr argon, powered electrode radius 10 cm, grounded electrode radius ≈ 20 cm, gap height 4 cm. Source: Reproduced from Kushner (2003), with the permission of AIP Publishing. https://doi.org/10.1063/1.1587887 *405*

Figure 11.49 Equivalent circuit for matching the rf power source to the discharge using an L-network *407*

Figure 12.1 Schematic of inductively driven sources in (*a*) cylindrical and (*b*) planar geometries *417*

Figure 12.2 Equivalent transformer-coupled circuit model of an inductive discharge *420*

Figure 12.3 Equivalent circuit for matching an inductive discharge to a power source *422*

Figure 12.4 Absorbed power versus density from the inductive source characteristics (curves) for two different values of the driving current I_{rf}, and power lost versus density (straight line); the dotted curve includes the additional capacitive power at low density for $I_{rf} < I_{min}$ *423*

Figure 12.5 Positive ion, negative ion, and electron densities as a function of time for 1:1 Ar/SF_6 mixture; the total pressure is 5 mTorr, the average power absorbed is 550 W *426*

Figure 12.6 Absorbed electron power P_{abs} versus electron density n_e and two different curves of electron power lost versus n_e (P_{loss1} at a low negative ion density n_- and P_{loss3} at a high n_-) *427*

Figure 12.7 Schematic of a helical resonator plasma source *429*

Figure 12.8 Schematic of the rf magnetic field lines near a planar inductive coil: (*a*) without nearby plasma and (*b*) with nearby plasma. Source: Wendt, 1993/with permission of IOP Publishing *431*

Figure 12.9 Rf magnetic induction amplitude $|\tilde{B}_r|$ versus z in a 5 mTorr oxygen discharge. The solid lines are a least-squares fit to the data. Source: Hopwood et al., 1993a/with permission of AIP Publishing *432*

Figure 12.10 Rf magnetic induction amplitude $|\tilde{B}_r|$ versus diagonal radius r at three different distances below the window as measured in a 5 mTorr, 500 W argon discharge. Source: Hopwood et al., 1993a/with permission of AIP Publishing *432*

Figure 12.11 Ion density versus rf power at various argon pressures. Source: Hopwood et al., 1993b/with permission of AIP Publishing *433*

Figure 12.12 Ion density, electron temperature, and plasma potential versus argon pressure in a 500 W discharge with magnetic multipole confinement. Source: Hopwood et al., 1993b/with permission of AIP Publishing *433*

Figure 12.13 Normalized ion saturation current measured across the diagonal of the plasma chamber with and without magnetic multipole confinement. Source: Hopwood et al., 1993b/with permission of AIP Publishing *434*

Figure 12.14 Measured ratios (symbols) of Lorentz-to-electric-field force, F_L/F_E at $r = 4$ cm and $z = 1$ cm, versus frequency, at various argon pressures; 19.8 cm diameter, 10.5 cm length planar ICP. Source: Godyak, 2003/with permission of IOP Publishing *436*

Figure 12.15 Illustrating the use of ferrite materials to increase the coupling efficiency of a planar inductive discharge *439*

Figure 12.16 Measured coil voltage and current versus absorbed discharge power at 1, 10, 100, and 1000 mTorr argon in a 2 MHz, 20 cm diameter, close-coupled, ferrite-enhanced planar inductive discharge. Source: Godyak (2011). ©IOP Publishing. Reproduced with permission. All rights reserved. https://doi.org/10.1088/0963-0252/20/2/025004 *440*

Figure 12.17 Measured central plasma densities (*a*) and discharge power efficiencies (*b*) versus absorbed discharge power at 1, 10, 100, and 1000 mTorr argon in a 2 MHz, 20 cm diameter, close-coupled, ferrite-enhanced planar inductive discharge. Source: Godyak (2011). ©IOP Publishing. Reproduced with permission. All rights reserved. https://doi.org/10.1088/0963-0252/20/2/025004 *441*

Figure 13.1 A typical high-profile ECR system: (*a*) geometric configuration; (*b*) axial magnetic field variation, showing one or more resonance zones. Source: Lieberman and Gottscho (1994)/with permission of Elsevier *446*

Figure 13.2 Typical ECR microwave system. Source: Lieberman and Gottscho (1994)/with permission of Elsevier *447*

Figure 13.3 Microwave field patterns for ECR excitation; (*a*) TE_{10} rectangular to TE_{11} circular mode; (*b*) TE_{10} rectangular to TM_{01} circular mode. Source: Lieberman and Gottscho (1994)/with permission of Elsevier *448*

Figure 13.4 Common ECR configurations: (*a*) high aspect ratio; (*b*) low aspect ratio; (*c*) low aspect ratio with multipoles; (*d*) close-coupled; (*e*) distributed (DECR); (*f*) microwave cavity excited. Source: Lieberman and Gottscho (1994)/with permission of Elsevier *449*

Figure 13.5 Basic principle of ECR heating: (*a*) continuous energy gain for right-hand polarization; (*b*) oscillating energy for left-hand polarization. Source: Lieberman and Gottscho (1994)/with permission of Elsevier *450*

Figure 13.6 Energy change in one pass through a resonance zone. Source: Lieberman and Gottscho (1994)/with permission of Elsevier *451*

Figure 13.7 k/k_0 versus ω_{ce}/ω for (*a*) low density $\omega_{pe}/\omega \ll 1$ and (*b*) high density $\omega_{pe}/\omega \gg 1$. The heavy dashed curves denote imaginary values for k *456*

Figure 13.8 Parameters for good ECR source operation: pressure p versus incident power S_{inc}. Source: Lieberman and Gottscho (1994)/with permission of Elsevier *456*

Figure 13.9 Schematic of ECR configuration used to compare model with hybrid simulation *459*

Figure 13.10 Comparison between spatially averaged model and hybrid simulation predictions of (*a*) electron temperature, (*b*) ion impact energy, and (*c*) plasma density, versus neutral gas pressure, for $P_{abs} = 850$ W. Source: Adapted from Porteous et al. (1994) *460*

Figure 13.11 Model used to calculate the distributed potential V_d and the sheath potential V_s in a diverging field ECR system *461*

Figure 13.12 Potential drops V_T, V_s, and V_d versus A_s/A_0 for a diverging field ECR system *461*

Figure 13.13 Change in the bombarding ion energy distribution as the wafer-level coil current i_m is varied. Source: Matsuoka and Ono (1988)/with permission of AIP Publishing *462*

Figure 13.14 Commercial ECR reactor for phase-resolved optical emission spectroscopy. Source: Milosavljević et al. (2013)/with permission of AIP Publishing *463*

Figure 13.15 Transverse electric fields of helicon modes at five different axial positions: (*a*) $m = 0$; (*b*) $m = 1$. Source: Chen (1991)/with permission of IOP Publishing *465*

Figure 13.16 $k_\perp R$ versus k_z/k for helicon modes. Source: Lieberman and Gottscho (1994)/with permission of Elsevier *466*

Figure 13.17 The antenna for $m = 1$ helicon mode excitation. Source: Lieberman and Gottscho (1994)/with permission of Elsevier *468*

Figure 13.18 The quasistatic antenna coupling field $\tilde{E}_y$: (*a*) ideal and actual field; (*b*) spatial power spectrum of a typical field. Source: Lieberman and Gottscho (1994)/with permission of Elsevier *468*

Figure 13.19 Saddle-antenna excited helicon discharges; (*a*) measured density (dots) as a function of 13.56 MHz input power for $B_0 = 80$ G in a 15-cm diameter chamber at 5 mTorr argon, showing transitions from E-mode to two different helicon wave (W) modes; (*b*) measured density (dots) as a function of magnetic field at 180 W of 8.8 MHz input power in a 10 cm diameter chamber at 1.5 mTorr argon, showing three different helicon wave mode transitions; the dashed line represents the resonance condition imposed by the antenna. Source: Perry et al. (1991)/with permission of AIP Publishing *469*

Figure 13.20 Langmuir probe measurements (×) of the central downstream density versus power, in an 18-cm diameter helicon discharge excited at 13.56 MHz by a double half-turn antenna (shown upper left); $B_0 = 50$ G at 3 mTorr argon pressure; the transition from capacitive (E) to inductive (H) to helicon wave (W) modes is clearly seen. Source: Adapted from Degeling et al. (1996). *470*

Figure 13.21 Index surface N_z versus $N_\perp$ for the helicon-TG mode system with $f = 13.56$ MHz, $B_0 = 100$ G, and $n_e = 10^{12}$ cm^{-3}; the group velocity vector directions at a given N_z are indicated; the dashed line shows the TG mode limit for $N \gg 1$ *472*

Figure 13.22 Surface wave dispersion k_z versus ω. Source: Lieberman and Gottscho (1994)/with permission of Elsevier *475*

Figure 13.23 Determination of the equilibrium density in a surface wave discharge. The high-density intersection of P'_{abs} and P'_{loss} gives the equilibrium density. Source: Moisan and Zakrzewski (1991)/with permission of IOP Publishing *476*

Figure 13.24 Comparison of theory (dashed) and experiment (solid) of density n_0 and wave power P_w versus z for a typical surface wave source. Source: Moisan and Zakrzewski (1991)/with permission of IOP Publishing *477*

Figure 14.1 Qualitative characteristics of a dc glow discharge. Source: Brown (1959)/with permission of Massachusetts Institute of Technology *480*

Figure 14.2 Typical voltage–current characteristic of a dc glow discharge *482*

Figure 14.3 Field-intensified ionization cross section α/n_g versus reduced field E/n_g (1 Td ≡ 10^{-21} V-m^2). Source: (Data provided by Petrović and Marić, 2004). *487*

Figure 14.4 Breakdown voltage for plane-parallel electrodes at 20 °C: (*a*) noble gases; (*b*) molecular gases. Source: (Data supplied by Petrović and Marić, 2004). *488*

Figure 14.5 Cathode voltage drop versus discharge current, illustrating the normal and abnormal glow; $C = 2A/B \ln[1 + (1/\gamma_{se})]$. Source: Cobine (1958)/with permission of Dover Publications *491*

Figure 14.6 Cylindrical configuration of a hollow cathode discharge *493*

Figure 14.7 Hollow cathode-enhanced rf capacitive discharge, showing the measured electron density n_e versus dc cathode sheath voltage, in a 129 mTorr argon discharge driven at 13.56 MHz, for various hole radii R; the upper electrode contains a hexagonal pattern of uniformly spaced holes of length $h = 10$ mm and hole spacing $4R$. Source: Lee et al. (2011a)/with permission of Elsevier *498*

Figure 14.8 Dc discharges used for sputtering: (*a*) low aspect ratio dc glow discharge; (*b*) planar magnetron discharge *500*

Figure 14.9 Calculation of planar magnetron ring width *502*

Figure 14.10 The relative contributions to the total ionization due to ohmic heating and sheath energization. The curves show (14.5.17) using the measured G_a and G_b from the four combinations of pressure and discharge current in the dc magnetron sputtering discharge studied by Depla et al. (2009); the lines are solid only in the range of γ_{se} where they are supported by the measurements; the circle in the upper left-hand corner marks the HiPIMS study by Huo et al. (2013). Source: Brenning et al. (2016)/with permission of IOP Publishing *504*

Figure 14.11 3D spatial profiles of (*a*) the electron density in a range greater than 3×10^{16} m^{-3} and (*b*) the electron temperature in a range greater than 1 V. Source: Jo et al. (2022)/with permission of Springer Nature *506*

Figure 14.12 Schematic of an I-PVD system, showing a planar magnetron source and a supplementary inductively coupled rf source that is embedded between the magnetron cathode target and the substrate. Source: Gudmundsson (2020)/IOP Publishing/CC BY 4.0 *507*

Figure 14.13 Electron impact ionization is the primary path for metal ion production in a high electron density plasma; Penning ionization dominates under conditions of low electron density. Source: Hopwood (2000)/with permission of Elsevier *509*

Figure 15.1 One-dimensional PIC simulation of (*a*) time-averaged normalized ionization rate and (*b*) normalized electron density versus x for a capacitive discharge at various argon gas pressures; the driving current density is $J = 2.56$ A/m^2 at 13.56 MHz, and the secondary electron emission coefficient is $\gamma = 0.1$. Source: Lafleur and Chabert (2015a)/with permission of IOP Publishing *515*

Figure 15.2 Comparison of one-dimensional PIC simulation results (stars) with a passive bulk model (triangles), versus pressure p, for a 2.5 cm-gap argon discharge, showing (*a*) plasma–sheath edge density n_s; (*b*) fundamental component of rf sheath voltage V_1; (*c*) normalized sheath edge ion speed $g_l = u_s/u_B$; and (*d*) edge-to-center density ratio $h_l = n_s/n_0$; the driving current density is $J = 50$ A/m^2 at 13.56 MHz. Source: Adapted from Kawamura et al. (2020) *521*

Figure 15.3 Percentage of each ionization reaction versus gas pressure p for a 2.5-cm gap argon discharge driven at 13.56 MHz by a 50 A/m^2 current source; the shaded region labeled I shows a transition from direct ionization to Ar_m–Ar_m pooling ionization with increasing pressure; at the highest pressure, Ar_m stepwise ionization is also significant. Source: Wen et al. (2021)/with permission of IOP Publishing *522*

Figure 15.4 Ionization rates with no secondary emission versus gap position x for direct electron–neutral ionization, stepwise ionization, and metastable pooling ionization between excited state atoms, Ar_m, Ar_r and Ar(4p), for a pressure of 1.6 Torr and current density amplitude of $J = 50$ A/m^2. Source: Wen et al. (2021). ©IOP Publishing. Reproduced with permission. All rights reserved. https://doi.org/10.1088/1361-6595/ac1b22 *522*

Figure 15.5 Ionization rates including excited state kinetics, energy-dependent secondary electron emission due to ion and excited atom bombardment of the electrodes, and electron reflection, for a 1.6 Torr argon discharge with a gap separation of 2.54 cm driven by a 50 A/m^2 rf current source at 13.56 MHz. Source: Gudmundsson et al. (2022)/with permission of IOP Publishing *523*

Figure 15.6 Sheath breakdown voltage (Paschen voltage) V_b versus ps_m, with p the gas pressure at 20 °C and s_m the maximum sheath thickness, for argon (solid line) and oxygen (dashed line), showing the α–γ transition; at low ps_m, the transition is smooth; at high ps_m, there is a discontinuous jump to the Paschen minimum voltage *524*

Figure 15.7 Evolution of (*a*) discharge rf current I (mA), (*b*) central plasma density n_0 (cm^{-3}), (*c*) central bulk electron temperature T_e (V), and (*d*) central electric field E_0 (V/cm), versus discharge voltage V_{rf} (V) during the α–γ transition, for a 3 Torr helium discharge of radius $R = 3$ cm and gap length $l = 7.8$ cm driven at 3.2 MHz; solid lines with circles give the measurements; dashed lines show the γ-mode theory. Source: Godyak and Khanneh (1986)/with permission of IEEE *530*

Figure 15.8 Measured dependences of the α-mode minimum maintenance voltage V_{main} (bottom curve) and the α–γ transition voltage $V_{\alpha\gamma}$ (top curve), versus pressure p (Torr), for a helium discharge of radius $R = 3$ cm and gap length $l = 7.8$ cm driven at 3.2 MHz. Source: Godyak and Khanneh (1986)/with permission of IEEE *530*

Figure 15.9 Measured root-mean-square (rms) rf current–voltage (I–V) characteristics for discharges driven at 13.6 MHz, with 10 cm electrode diameter; (1) helium, 30 Torr, 0.9 cm gap; (2) air, 30 Torr, 0.9 cm gap; (3) air, 30 Torr, 3 cm gap; (4) CO_2, 30 Torr, 0.9 cm gap; (5) CO_2, 15 Torr, 3 cm gap; (6) air, 7.5 Torr, 1 cm gap, glass-coated electrodes; (7) air, 7.5 Torr, 1 cm gap, teflon-coated electrodes. Source: Raizer et al. (1995)/with permission of Taylor & Francis *531*

Figure 15.10 α–γ transition for a 2.5-cm gap, 635 K, 1.6 Torr nitrogen discharge driven at 13.56 MHz from the passive bulk α-mode model without secondary emission (dashed line) and the Paschen curve γ-mode model (solid line), along with the PIC simulations with γ_{se}=0.15 (stars), showing (*a*) the single-sheath rf voltage amplitude V_{sh} versus the maximum sheath width s_m, and (*b*) the corresponding curve V_{sh} versus the discharge current density J *532*

Figure 15.11 Paschen breakdown curve (dashed) from (15.3.6) and α–γ transition curve (solid) from (15.3.22), with p the argon gas pressure at 20°C; the dotted line shows $V_{rf,max} = \mathcal{E}_c/\gamma_{se}$ *534*

Figure 15.12 Discharge extinguishing voltage (1) and α–γ transition voltage (2), versus pressure p, for a 10-cm diameter, 2.2 cm-gap argon discharge. Source: Lisovskii (1998)/with permission of Springer Nature *534*

Figure 15.13 Comparison between a low-pressure argon rf discharge (first column) and an atmospheric pressure helium discharge (second column) driven at 13.56 MHz; (*a*, *b*) electron–neutral collision frequencies ν_{elas}, ν_{ex}, and ν_{iz} and driving frequency ω; (*c*, *d*) electron mean free path λ_m, energy relaxation length $\lambda_\mathcal{E}$ and gap length l; (*e*, *f*) electron energy relaxation time $\tau_\mathcal{E}$ and rf period $\tau_{rf} = 2\pi/\omega$. Source: Iza et al. (2008)/with permission of John Wiley & Sons *535*

Figure 15.14 Atmospheric pressure He/0.1%N_2, α-mode PIC results for a 1-mm gap driven at 27.12 MHz with an rf current density of 400 A/m^2, showing (*a*) time-average densities versus position x, (*b*) time-averaged temperatures versus x, (*c*) the space–time variation of the helium metastable excitation rate, with space and time variation shown on the vertical and horizontal axes, respectively, and (*d*) the space–time variation of the Penning ionization rate; in (*a*) and (*b*), the maximum sheath width s_m is indicated by the vertical dotted lines; in (*c*) and (*d*), the dark solid curves show the sheath edge positions at the opposing walls. Source: Kawamura et al. (2014)/with permission of IOP Publishing *543*

Figure 15.15 Atmospheric pressure He/0.1%N_2 γ-mode PIC results for a 1-mm gap driven at 27.12 MHz with an rf current density of 2000 A/m^2, showing (*a*) time-average densities versus position x, (*b*) time-averaged temperatures versus x, (*c*) the space–time variation of the helium metastable excitation rate, with space and time variation shown on the vertical and horizontal axes, respectively, and (*d*) the space–time variation of the Penning ionization rate; in (*a*) and (*b*), the maximum sheath width s_m is indicated by the vertical dotted lines; in (*c*) and (*d*), the dark solid curves show the sheath edge positions at the opposing walls. Source: Kawamura et al. (2014)/with permission of IOP Publishing *544*

Figure 15.16 Root-mean-square (rms) voltage–current experimental characteristics for an atmospheric pressure, 6 cm diameter helium discharge driven at 13.56 MHz, with gaps of (*a*) 1 mm, (*b*) 2 mm, (*c*) 3 mm, and (*d*) 4 mm; at 4 mm, normal (constant voltage contracted discharge) α and γ regimes appear. Source: Moon et al. (2006)/with permission of AIP Publishing *546*

Figure 15.17 Voltage–current characteristics in an rf atmospheric pressure helium discharge with 2 mm gap; the lines with solid symbols give the hybrid simulation results, with the peak corresponding to the α–γ transition; the open symbols give the experimental α-mode results. Source: Ding et al. (2014)/with permission of IOP Publishing *546*

Figure 15.18 O ($^3P \rightarrow {}^3S°$) line emission pattern in a He/0.5%O_2 α-mode atmospheric pressure Penning discharge, within one 13.56 MHz rf cycle and within the 1 mm electrode gap, on a linear gray scale starting from zero; (*a*) fluid simulation and (*b*) phase-resolved optical emission spectroscopy (PROES) measurement. Source: Waskoenig et al. (2010)/with permission of IOP Publishing *547*

Figure 15.19 Electric fields in a cathode (C) – anode (A) gap containing a cathode-emitted electron avalanche; (*a*) external field $\mathbf{E}_0$ and space charge field of the avalanche $\mathbf{E}'$, shown separately; (*b*) the resulting field $\mathbf{E} = \mathbf{E}_0 + \mathbf{E}'$. Source: Raizer (1991)/with permission of Springer Nature *549*

Figure 15.20 Streamer/Paschen breakdown voltage ratio versus *pl*; the streamer breakdown is calculated assuming a multiplication of 10^8. Source: Becker et al. (2005)/with permission of Taylor & Francis *551*

Figure 15.21 Time variation over one 10 kHz cycle of the measured values of applied voltage V_T, gas (gap) voltage V_g, memory (dielectric) voltage V_d, and discharge current I for a helium atmospheric pressure glow discharge. Source: Massines et al. (1998)/with permission of AIP Publishing *552*

Figure 15.22 Simple circuit model for an atmospheric pressure glow discharge *552*

Figure 15.23 Circuit model results for the time variation over one 10 kHz cycle for a helium atmospheric pressure glow discharge; (*a*) source voltage V_T, gap voltage V_g, and dielectric voltage V_d; (*b*) gap current I *553*

Figure 15.24 End-on photograph of dielectric barrier microdischarges in atmospheric pressure air; original size 6 cm × 6 cm; exposure time 20 ms. Source: Kogelschatz (2003)/with permission of Springer Nature *555*

Figure 16.1 Calculation of plasma etch requirements: (*a*) a typical set of films; (*b*) anisotropy requirement for polysilicon etch; (*c*) uniformity requirement, including the effect of photoresist erosion *562*

Figure 16.2 Acceptable trade-offs for plasma etching: (*a*) anisotropy versus photoresist selectivity and (*b*) uniformity versus oxide selectivity *564*

Figure 16.3 Four basic plasma etching processes: (*a*) sputtering; (*b*) pure chemical etching; (*c*) ion energy-driven etching; (*d*) ion-enhanced inhibitor etching (Flamm and Herb, 1989) *565*

Figure 16.4 The development of facets due to sputtering of photoresist: (*a*) before sputtering and (*b*) after sputtering *566*

Figure 16.5 Surface etch model assuming Langmuir kinetics and rate-limiting desorption *569*

Figure 16.6 Normalized vertical (E_v) and horizontal (E_h) etch rates versus normalized gas-phase density n_{OS}, for $K_i Y_i n_{is}/K_d = 5$ *570*

Figure 16.7 Illustrations of pattern fidelity during trench etches; (*a*) ideal etch, (*b*) microtrenching, (*c*) mask undercutting, (*d*) aspect-ratio-dependent etch (ARDE), (*e*) tapering, and (*f*) etch stop notching *574*

Figure 16.8 Probabilities $\gamma_{Si/F}$ of etching a silicon atom with a fluorine atom, versus the total fluorine atom flux Γ_F(tot), from various published sources; Source: Donnelly (2017)/AIP Publishing/CC BY 4.0 *576*

Figure 16.9 Probabilities $\gamma_{Si/Hal}$ of etching a silicon atom with a halogen atom, versus $1000/T$(K), from various published sources. Source: Reproduced from Donnelly and Kornblit (2013)/with permission of AIP Publishing *578*

Figure 16.10 The influence of fluorine to carbon (F/C) ratio and electrode bias voltage on etching and polymerization processes in a fluorocarbon discharge (Coburn and Winters, 1979) *584*

Figure 16.11 Locus of silicon etch rate E_{Si} and F-atom concentration n_F as the $\%O_2$ is varied in a CF_4/O_2 feedstock mix (Mogab et al., 1979) *584*

Figure 16.12 Etch rate E_{Si} versus doping level n_D and crystallographic orientation for Cl atom etching of n-type silicon at 400 K; p_{Cl} is the partial pressure of Cl atoms (after Ogryzlo et al., 1990) *586*

Figure 16.13 Cl_2-saturated, ion-assisted etching yields of silicon by Cl_2^+ and Ar^+; the pure sputtering yield by Ar^+ is also shown. Source: Levinson et al. (1997), with the permission of AIP Publishing *587*

Figure 16.14 (*a*) SiO_2 etch rate and (*b*) plasma-induced emission for CF_2 and F/Ar actinometric emission ratio, versus $\%H_2$ and $\%O_2$ addition to a CF_4 parallel plate discharge. Source: Flamm (1989)/with permission of Springer Nature *589*

Figure 16.15 A typical process flow for a silylated surface imaged resist dry development scheme *593*

Figure 16.16 Ideal and non-ideal atomic layer etching of silicon by ion bombardment; in the ideal case (*a*), the surface is chlorinated with Cl_2; and (*b*) a single silicon monolayer is removed by a saturated Ar^+ bombardment etch step with ion energies $\mathcal{E}_i \lesssim 10$–15 V during each cycle, leaving a pristine silicon surface; however, for $\mathcal{E}_i > 15$ V, a damaged surface is left, and the etch is non-ideal; in subsequent etch cycles (*c*) with $\mathcal{E}_i > 10$ V, a chlorinated, damaged layer many monolayers thick is formed by ion bombardment during the previous cycle; a part of this layer is removed during each saturated etch step, leaving a chlorinated damaged surface. Source: Vella and Graves (2023)/with permission of AIP Publishing *596*

Figure 16.17 Measurements of $Si/Cl/Ar^+$ ALE, showing (*a*) the etch depth per cycle (Å) versus the etch step time t ("bias phase duration") for $\mathcal{E}_i = 28.7$ and 46.2 V ion energies, and (*b*) etch depth per cycle versus $\mathcal{E}_i$ for a chlorination time ("passivation time") of 10 s and an etch time per cycle ("activation time") of 40 s. Source: Dorf et al. (2017)/with permission of IOP Publishing *598*

Figure 16.18 Visualization from a molecular dynamics simulation of $Si/Cl_2/Ar^+$ ALE, showing the damaged layer (*a*) just after the fourth chlorination step, and (*b*) just after the fourth Ar^+ etch step (*b*); $\mathcal{E}_i = 70$ V; the chlorine atoms are shown in dark grey, and the silicon atoms are shown semi-transparently. Source: Vella et al. (2022)/with permission of IOP Publishing *599*

Figure 16.19 Model results for atomic layer etching of chlorinated silicon for various Ar^+ bombarding energies $\mathcal{E}_i$; for $\mathcal{E}_i = 16$, 25, and 48 V, a saturated chlorination step followed by a saturated ion-enhanced etch step is shown in the cyclical steady state; at 53 V, an unsaturated ion-enhanced etch step is shown; (*a*) gives the increase in Cl surface coverage θ versus the normalized Cl_2 fluence $\varphi = n_0'^{-1} \int_0^t \Gamma_{Cl_2}(t')\,dt'$ during the saturated chlorination step; (*b*) gives the decreasing Cl atom surface coverage θ (after an initial redistribution of the saturated surface chlorine into the chlorinated layer) versus the normalized ion fluence $\varphi_i = n_0'^{-1} \int_0^t \Gamma_i(t')\,dt'$ during the etch step; and (*c*) gives the silicon monolayers N_{Si} etched versus the normalized ion fluence φ_i during the etch step; the parameters used are $s_0 = 0.4$, $K_{Si} = K_{SiCl_x} = 0.02$, $\mathcal{E}_{Si} = 49$ V, $\mathcal{E}_{SiCl_x} = 13$ V, $x = 2$, $K_\Delta = 0.36\ \text{nm/V}^{1/2}$, and $\mathcal{E}_\Delta = 9.6$ V; the solid dot in (*c*) gives the transition to silicon sputtering for $\mathcal{E}_i > \mathcal{E}_{Si}$ *603*

Figure 16.20 Measured film thickness change (Å) versus time for the $SiO_2/CF_x/Ar^+$ quasi-ALE etch process; eight etch cycles are shown. Source: Metzler et al. (2014)/with permission of AIP Publishing *606*

Figure 16.21 Film thickness etched (Å) versus number of thermal ALE etch cycles at 200 °C for various materials, using the two-step, $Sn(acac)_2$/HF cycle. Source: Lee et al. (2016)/with permission of American Chemical Society *607*

Figure 16.22 An antenna structure for an MOS transistor on a grounded silicon substrate *609*

Figure 16.23 Plasma current I_p and oxide current I_{FN} versus antenna voltage V_g for various gate oxide thicknesses. Source: Shin and Hu (1996)/with permission of IOP Publishing *610*

Figure 16.24 Gate oxide damage mechanisms in a nonuniform plasma; (*a*) thick oxide with insulated substrate, showing formation of an open circuit voltage V_{oc}; (*b*) thin oxide with insulated substrate, showing formation of a short circuit current I_{sc}. Source: Adapted from Cheung and Chang (1994) *611*

Figure 16.25 Transient damage of gate oxide during polysilicon etching in a nonuniform plasma; (*a*) during most of the etch time, the film is continuous and currents do not flow in the gate oxide; (*b*) near the end of the etch time, there are isolated gates with large antenna ratios, and large currents flow in the gate oxide *612*

Figure 16.26 Gate oxide damage in a uniform plasma due to the electron shading effect. Adapted from Hashimoto, 1994 *613*

Figure 16.27 Peak-to-peak 1 MHz rf charging voltage versus antenna ratio A_R for various plasma densities, with electron temperature $T_e = 5$ V and a field/gate oxide thickness ratio $T_R = 50$ (after Cheung and Chang, 1994) *615*

Figure 16.28 Location of the notch and the mechanisms proposed to contribute to the notching effect; (*a*) ion trajectory bending due to open area charging and direct bombardment of the polysilicon; (*b*) forward ion deflection due to SiO_2 charging under the etched area; (*c*) near grazing ion–SiO_2 surface collision, followed by forward scattering and bombardment of the notch apex Source: Hwang and Giapis (1997)/with permission of AIP Publishing *615*

Figure 17.1 Surface coverage model for amorphous silicon deposition; θ_a and θ_p are the fractions of the surface that are active and passive, respectively *623*

Figure 17.2 Chemical structure of TEOS *624*

Figure 17.3 Nonconformal deposition within a trench, illustrating formation of a void as deposition proceeds: (*a*) before deposition, with the dashed lines giving the deposition flux incident on the sidewall and bottom; (*b*) midway during deposition; (*c*) just after the keyhole-shaped void has formed *626*

Figure 17.4 A compilation (different symbols) from the published literature of growth per cycle (GPC) in Al atoms per (nm)^2 versus substrate temperature for ALD of Al_2O_3 : (*a*) thermal ALD and (*b*) plasma-enhanced ALD. Source: Vandalon and Kessels (2017)/with permission of AIP Publishing *631*

Figure 17.5 Top panel: Gordon's model simulation for Al_2O_3 deposition into rectangular test structures with aspect ratios $LP/(4A)$ of 33 (solid), 50 (dashed), and 100 (dotted); following four panels: Langmuir adsorption model simulations with $s_0 = 1, 0.1$, 0.01, and 0.001, respectively; L, P, and A are the depth, perimeter, and cross-sectional area of the hole. Source: Dendooven et al. (2009)/with permission of IOP Publishing. *634*

Figure 17.6 (*a*) Measured Al_2O_3 deposition depth profile measurements for 5 mm×20 mm rectangular macroscopic hole structures with aspect ratios $LP/(4A)$ of 33 (solid), 50 (dashed), and 100 (dotted), and (*b*) corresponding Langmuir model simulations using a zero-coverage sticking coefficient $s_0 = 0.1$; L, P, and A are the depth, perimeter, and cross-sectional area of the hole, respectively. Source: Dendooven et al. (2009)/with permission of IOP Publishing *634*

Figure 17.7 Monte Carlo simulation results for (*a*) the normalized saturation dose versus trench aspect ratio L/w for a recombination probability $\gamma_{rec} = 0$ and various zero-coverage reaction probabilities s_0, and (*b*) for values of γ_{rec} from 0.01 to 0.9 and various values of s_0; in (*b*), the trends with the same γ_{rec} value but different s_0 values overlap. Source: Knoops et al. (2010)/with permission of IOP Publishing *635*

Figure 17.8 Structure zone diagram for the morphology of sputtered films. Source: Thornton (1986)/with permission of AIP Publishing *638*

Figure 17.9 Reactive sputter deposition of TiN films, showing the optical emission signal for titanium versus the reactive gas flow rate (*a*) with and (*b*) without feedback control. Source: Berg et al. (1989)/ with permission of AIP Publishing *641*

Figure 17.10 Planar PIII geometry (*a*) just after formation of the matrix sheath and (*b*) after evolution of the quasistatic Child law sheath *642*

Figure 17.11 Normalized implantation current density $\bar{J} = J/(en_0u_0)$ versus normalized time $\bar{t} = \omega_{pi}t$. The dashed lines show the analytical solution for $\bar{t} < 2.7$ and $\bar{t} > 3.0$, and the solid line is the numerical solution of the fluid equations *645*

Figure 17.12 Ion velocity distribution at the target for a collisional sheath; the maximum velocity for collisionless acceleration to the target is roughly 5×10^4 m/s *648*

Figure 17.13 Schematic showing diode and triode configurations of PIII for semiconductor implantation *649*

Figure 18.1 Normalized floating potential Φ_d/T_e versus ion-to-electron temperature ratio T_i/T_e, for different values of the ion mass; the right axis gives the corresponding value of the number of electrons on the dust particle, normalized to its radius a (in nm) times the electron temperature T_e (in volts); the solid lines correspond to numerical solutions; the dashed lines correspond to the approximate analytical solution. Source: Adapted from Matsoukas and Russell (1995) *658*

Figure 18.2 Electron density (a) and electron temperature (b) versus time in an 30 sccm argon + 1.2 sccm silane discharge; the plasma reactor is 13 cm diameter and 3 cm in height, driven at 13.56 MHz with a peak-to-peak voltage of 600 V, with a total pressure of 150 mTorr. Source: Boufendi et al. (1996)/with permission of AIP Publishing *661*

Figure 18.3 Time development of the particle radius a (open circles) and the number density n_d (solid circles) for early discharge times, obtained from Rayleigh scattering; solid lines show the best fit of the Brownian free molecular motion coagulation model. Source: Courteille et al. (1996)/with permission of AIP Publishing *668*

Figure 18.4 Particle diameter versus discharge on-time. Source: Adapted from Böhme et al. (1994) *669*

Figure 18.5 Time evolution of particulate size $d = 2a$ at 34–42 mm above the grounded electrode of a capacitive discharge after power turn-on; 43 mm electrode spacing, 40 W rf power, 4 s on-time in He/SiH_4 (5%) at 30 sccm and 600 mTorr. Source: Shiratani et al. (1994)/with permission of IEEE *670*

Figure 18.6 Scattered intensities versus plasma on-time at 6° and 90° from the incident direction, in the plane perpendicular to the direction of polarization of the electric field; the corresponding measured particle size is indicated at the Rayleigh–Mie transition. Source: Boufendi et al. (1999)/with permission of John Wiley & Sons *673*

Figure 18.7 Visualization of particulates, showing (a) schematic of the electrode, wafers, and discharge features observed during forward direction LLS; the two foreground wafers are 5.7 cm diameter, the background wafer is 8.3 cm diameter; the dashed lines show the limits of the scanned laser beam; (b) photograph of scattered light from particulates in a 200 mTorr argon plasma; trapped particle structures including a dome and rings are seen over each of the wafers. Source: Selwyn et al. (1991)/AIP Publishing *674*

Figure 18.8 Examples of particulates on plasma-etched features. Source: Lee et al. (2014)/ Reproduced with permission of IOP Publishing *675*

Figure 19.1 Variations of ν_{eff}/ν_{dc} (solid lines) and ω_{eff}/ω (dashed lines) as a function of pressure for different electron temperatures T_e; here, $\nu_{dc} = \nu_{eff}(\omega = 0)$. Source: Lister et al. (1996)/with permission of AIP Publishing *688*

Figure 19.2 Ratios of ν_{eff}/ν_m (open circles) and exact-to-approximate ohmic heating power (open triangles) versus argon gas pressure, determined by Boltzmann term analysis of PIC simulation results, for a 2-cm gap capacitive discharge driven by 25.6 A/m^2 at 13.56 MHz. Source: Lafleur et al. (2014a)/with permission of IOP Publishing *689*

Figure 19.3 Schematic of electron energy probability function $g_p(\mathcal{E}, r)$ versus $\mathcal{E}_T^2 = (\mathcal{E} - \Phi(r))^2$ at a fixed heating field E and gas pressure p in the positive column of a glow discharge; (dashed line) local kinetics and (solid line) nonlocal kinetics *692*

Figure 19.4 Schematic showing the definition of the accessible volume $\mathcal{V}_{ac} = \pi r_{ac}^2$ in an infinitely long cylindrical discharge *695*

Figure 19.5 Illustrating the transformation of the EEPF g_{T0}, a function of the total energy $\mathcal{E}_T$, to the EEPF g_p, a function of the kinetic energy $\mathcal{E}$ *697*

Figure 19.6 $\mathcal{I}$ and v_{stoc} (normalized to $\overline{v}_e/4\delta$) versus α, where v_{stoc} is defined in the following subsection *705*

Figure 19.7 Comparison of global model results including Druyvesteyn ($x = 2$) and Maxwellian ($x = 1$) distributions; (*a*) T_{eff} versus energy exponent x at various pressures; (*b*) n_e versus p for various x's. Source: Gudmundsson (2001)/with permission of IOP Publishing *708*

Figure 19.8 Experimental EEPF as a function of total energy $\mathcal{E}_T$ at different radial positions in a 100 W, 50 mTorr argon planar inductive discharge; curve labels correspond to radius (in cm) from the center at a fixed axial distance (4.4 cm) from the dielectric window. Source: Kolobov et al. (1994)/with permission of AIP Publishing *709*

Figure 19.9 Comparison between measured and calculated EEPF in a planar inductive discharge; the measurements are performed in argon in the center of the discharge. Source: Adapted from Kortshagen et al. (1995) *710*

Figure 19.10 Comparison of EEPFs obtained experimentally and theoretically in an argon capacitive discharge at $p = 10$ mTorr and $V_{rf} = 425$ V. Source: Adapted from Wang et al. (1999) *713*

Figure A.1 Illustrating the exact classical calculation of the differential scattering cross section *718*

Figure A.2 The potential functions used for the calculation of elastic scattering in (*a*) an attractive inverse first power potential and (*b*) an attractive inverse third power potential *719*

List of Tables

Table 1.1 Typical Range of Parameters for RF Diode and High-Density Discharges *13*

Table 3.1 Scaling of Cross Section σ, Interaction Frequency ν, and Rate Constant K, with Relative Velocity v_R, for Various Scattering Potentials U *46*

Table 3.2 Relative Polarizabilities $\alpha_R = \alpha_p / a_0^3$ of Some Atoms and Molecules, Where a_0 is the Bohr Radius *50*

Table 3.3 Selected Reaction Rate Constants for Argon Discharges *66*

Table 4.1 Summary of Guiding Center Drifts ($\mathbf{R}_c / R_c^2 = -\nabla B / B$) *90*

Table 4.2 Summary of Cutoffs and Resonances for the Principal Waves *97*

Table 7.1 Thermodynamic Properties *175*

Table 7.2 Enthalpies of Formation *177*

Table 7.3 Bond Dissociation Enthalpies *178*

Table 7.4 Enthalpies of Formation of Gaseous Atoms *179*

Table 7.5 Vapor Pressures *189*

Table 8.1 Basic Constants for Oxygen Discharges *226*

Table 8.2 Selected Second-Order Reaction Rate Constants for Oxygen Discharges *228*

Table 8.3 Selected Third-Order Reaction Rate Constants for Oxygen Discharges *230*

Table 9.1 Work Functions and Secondary Emission Coefficients *257*

Table 9.2 Measured Sputtering Yields for Ar^+ at 600 V *263*

Table 9.3 Gas Kinetic Cross Sections in Units of 10^{-15} cm^2 *264*

Table 11.1 Scaling Exponent q for the Equation $\overline{V}_a / \overline{V}_b = (A_b / A_a)^q$ *369*

Table 14.1 Constants of the Equation $\alpha / p = A \exp(-Bp/E)$ *487*

Table 14.2 Normal Cathode Fall in Volts *490*

Table 14.3 Normal Cathode Fall Thickness pd in Torr-cm *490*

Table 15.1 Selected Reaction Rate Constants for a He/N_2 Atmospheric Pressure Plasma *541*

Table 16.1 Etch Chemistries Based on Product Volatility *567*

Table 16.2 Selected Second-Order Reaction Rate Constants for Electron Impact Collisions in CF_4 Discharges *581*

Table 16.3 Selected Values of Rate Constants K_2 and K_3' for Association Reactions in CF_4 Discharges *582*

Table 16.4 Coefficients of the Modified Arrhenius Form for Cl Atom Etching of *n*-Type Silicon *586*

Table 17.1 Selected Reaction Rate Constants for SiH_4 Discharges *622*

Preface to Third Edition

This third edition is long overdue. During the last 20 years, smartphones powered by advanced integrated circuit technology have appeared on the market. Cathode-ray tube (CRT) TV sets disappeared, replaced by thin film transistor (TFT) and large-area flat-panel displays. A new large-area solar panel industry based on TFT technology was created. Minimum integrated circuit feature sizes, such as metal-to-metal interconnect separations, decreased by a factor of ten, from 200 nm in 2004 to 20 nm in 2024. Substrate areas as large as 10 m^2 (mainly glass) are routinely processed.

Plasma processing technology has advanced dramatically to support these changes. The evolution of the field has been so rapid and encompassing that complete coverage of the new developments is impossible. And yet, the fundamentals still abide. The first nine chapters, on the basic principles of plasma physics and gas- and surface-phase chemistry, are relatively unchanged. Major changes from the second edition include:

- a revised Section 5.4 on diffusion across magnetic fields
- a new Section 9.5 on showerhead gas flows
- a revised Section 10.4 on approximate electronegative equilibria
- a new Section 11.5 on voltage-driven sheaths and series resonance
- a new Section 11.6 on multi-frequency capacitive discharges
- a new Section 11.7 on standing wave and skin effects
- a new Section 12.4 on high-efficiency planar inductive discharges
- a revised Section 14.4 on hollow cathode discharges, including rf-driven
- a revised Section 14.5 on planar magnetron discharges, including pulsed discharges
- a new Chapter 15 on high-pressure capacitive discharges:
  - Section 15.1 covering intermediate pressures
  - Section 15.2 on α–γ transitions
  - Section 15.3 on atmospheric pressure rf discharges
  - Section 15.4 on atmospheric pressure low-frequency discharges
- a new Section 16.5 on atomic layer etching
- a new Section 17.3 on atomic layer deposition

In addition, coverage is added on topics such as hairpin resonator and magnetic (B-dot) probes, phase-resolved optical emission spectroscopy, Boltzmann term analysis, neutral gas depletion and heating, and pattern transfer. Many other additions and changes are made. Coverage of some topics, such as models of electronegative equilibria and helical resonator discharges, is shortened. All second edition errors are corrected. Seventy new figures are added. Citations to recently published literature are emphasized; over 350 new citations appear. Readers should note that Chapters 15–18 in the second edition are Chapters 16–19 in this third edition. The particle-in-cell (PIC/MCC) software used for many of the simulations in this third edition can be downloaded from J. P. Verboncour's Plasma Theory and Simulation Group web site https://ptsg.egr.msu.edu at Michigan State University.

In preparing this revision, I have benefited from discussions with many colleagues. I am deeply indebted to Emi Kawamura for many insightful discussions on intermediate and atmospheric pressure discharge equilibria, and on particle-in-cell simulations of these and related phenomena. J. T. Gudmundsson carefully read Chapters 9–17 and provided many suggestions for corrections and improvements. P. Chabert read Chapters 5 and 10–15 during his sabbatical visit to Berkeley, and his comments have greatly improved the manuscript. V. M. Donnelly carefully read Chapter 16 and provided many suggestions for improvements and clarifications. D. B. Graves and J. R. Vella critically read the section on atomic layer etching. V. A. Godyak provided much useful information on Langmuir probes and magnetized plasma diffusion. I am indebted to Yi-Kang Pu and Jie Qiu for their list of second edition errors.

With deep regrets, I have to record the death of my coauthor Allan J. Lichtenberg in February 2023, just as I was to give him my first draft of the new chapter on high-pressure capacitive discharges. Therefore, all errors of commission and omission in this third edition are mine.

January 2025 *Michael A. Lieberman*
Berkeley, California

Preface to Second Edition

While the state of the art has advanced dramatically in the 10 years since the publication of our first edition, the fundamentals still abide. The first nine chapters on fundamentals of low-pressure partially ionized plasmas (Chapters 2–6) and gas-phase and surface physics and chemistry (Chapters 7–9) have been revised mainly to clarify the presentation of the material, based on the authors' continuing teaching experience and increased understanding. For plasmas, this includes significant changes and additions to Sections 5.2 and 5.3 on diffusion and diffusion solutions, 6.2 on the Bohm criterion, 6.4 on sheaths with multiple positive ions, and 6.6 on Langmuir probes in time-varying fields. For gas phase and surface physics and chemistry, it includes revised presentations in Sections 9.2 and 9.3 of sputtering physics, loss rates for neutral diffusion, and loss probabilities. The argon and oxygen rate coefficient data sets in Chapters 3 and 8 have been brought up to date.

Chapters 10–14 on discharges have been both revised and expanded. During the last decade, the processing community has achieved a more thorough understanding of electronegative discharge equilibrium, which lies at the core of the fluorine-, chlorine-, and oxygen-containing plasmas used for processing. Electronegative discharges are described in the new or revised Sections 10.3–10.5. An important new processing opportunity is the use of pulsed power discharges, which are described in a new Section 10.6. Chapter 11 on capacitive discharges has been expanded to incorporate new material on collisionless sheaths, dual frequency, high frequency, and electronegative discharges. New Sections 11.8 and 11.9 have been added on high-density rf sheaths and ion energy distributions, which are important for rf-biased, high-density processing discharges. Chapter 12 on inductive discharges now incorporates the electron inertia inductance in the discharge model and includes a new subsection on hysteresis and instabilities, whose effects can limit the performance of these discharges for processing. Section 13.2 on helicon discharges has been expanded to incorporate new understanding of helicon mode absorption and neutral gas depletion, both important for helicon discharge modeling. Two Sections 14.4 and 14.6 have been added on hollow cathode discharges and on ionized physical vapor deposition. Hollow cathode discharges have important applications both in processing and for gas lasers and serve as an example of low-pressure dc discharge analysis. Ionized physical vapor deposition has some important applications for thin film deposition and illustrates the combined use of dc and rf discharges for processing.

Chapters 16–17 on etching, deposition, and implantation have been brought up to date. In Section 16.4, a brief subsection on copper etching has been included. A new Section 16.6 on charging effects has been added, since differential substrate charging is now fairly well understood and is known to damage thin film oxides.

During the last decade, particulates in discharges have been studied both with a view to controlling their formation, to avoid generating defects during processing, and for producing powders and nanocrystalline materials. In a new Chapter 18 on dusty plasmas, the physics and technology of this important area are described, including particulate charging and discharge equilibrium, particulate equilibrium, particulate formation and growth, diagnostics, and removal and production techniques.

Also during the last decade, discharge analysis based on kinetic theory has advanced considerably, and kinetic techniques have found increasing use. In a new Chapter 19, we give an introduction to the kinetic theory of discharges, including the basic concepts, local and nonlocal kinetics, quasilinear diffusion and stochastic heating, and examples of discharge kinetic modeling.

Errors in the first and second printings of the first edition have been corrected. All topics treated have been brought up to date and incorporate the latest references to the literature. The list of references has been expanded from about six to fourteen pages.

Because we emphasize the development of a strong foundation in the fundamental physical and chemical principles, our one-semester course teaching this material to a mixed group of mainly graduate students in electrical, chemical, and nuclear engineering, materials science, and physics has not changed much over the years. The outline in the first preface for a 30, $1\frac{1}{2}$-hour lecture course is still relevant, with, perhaps, some additional emphasis on electronegative plasma equilibria and on pulsed plasmas. (Some sections have been renumbered.)

Our colleagues C.K. Birdsall and J.P. Verboncoeur and the plasma theory and simulation group (PTSG) at Berkeley continue to maintain a set of user-friendly programs for PCs and workstations for computer-aided instruction and demonstrations. The software and manuals can be downloaded from their web site http://ptsg.eecs.berkeley.edu.

In preparing this revision, we have received encouragement and benefited from discussions with many friends and colleagues. We thank I.D. Kaganovich for carefully reviewing Chapter 19 on kinetic theory. We are indebted to J.T. Gudmundsson for assistance in updating the argon and oxygen rate coefficient data sets (for more complete data, see his web site http://www.raunvis.hi.is/~tumi/) and to Z. Petrović and D. Marić, who provided assistance in updating the field-intensified ionization coefficient and the breakdown voltages given in Chapter 14. We thank B. Cluggish, R.N. Franklin, V.A. Godyak, and M. Kilgore for their comments clarifying various calculations. We have benefited greatly from the insight and suggestions of our colleagues C.K. Birdsall, J.P. Booth, R.W. Boswell, P. Chabert, C. Charles, S. Cho, T.H. Chung, J.W. Coburn, R.H. Cohen, D.J. Economou, D. Fraser, D.A. Graves, D.A. Hammer, Y.T. Lee, L.D. Tsendin, M. Tuszewski, J.P. Verboncoeur, A.E. Wendt, and H.F. Winters. Our recent postdoctoral scholars S. Ashida, J. Kim, T. Kimura, K. Takechi, and H.B. Smith and recent graduate students J.T. Gudmundsson, E. Kawamura, S.J. Kim, I.G. Kouznetsov, A.M. Marakhtanov, K. Patel, Z. Wang, A. Wu, and Y. Wu have taught us much, and some of their work has been incorporated into our revised text. The authors gratefully acknowledge the hospitality of R.W. Boswell at the Australian National University, Canberra, and M.G. Haines at Imperial College, London, where considerable portions of the revision were written.

September 2004
Berkeley, California

Michael A. Lieberman
Allan J. Lichtenberg

Preface to the First Edition

This book discusses the fundamental principles of partially ionized, chemically reactive plasma discharges and their use in thin-film processing. Plasma processing is a high-technology discipline born out of the need to access a parameter space in materials processing unattainable by strictly chemical methods. The field is interdisciplinary, combining the areas of plasma physics, surface science, gas-phase chemistry, and atomic and molecular physics. The common theme is the creation and use of plasmas to activate a chain of chemical reactions at a substrate surface. Our treatment is mainly restricted to discharges at low pressures, <1 Torr, which deliver activation energy, but not heat, to the surface. Plasma-based surface processes are indispensable for manufacturing the integrated circuits used by the electronics industry, and we use thin-film processes drawn from this field as examples. Plasma processing is also an important technology in the aerospace, automotive, steel, biomedical, and toxic waste management industries.

In our treatment of the material, we emphasize the development of a strong foundation in the fundamental physical and chemical principles that govern both discharges and gas- and surface-phase processes. We place little emphasis on describing state-of-the-art discharges and thin-film processes; while these change with time, the fundamentals abide. Our treatment is quantitative and emphasizes the physical insight and skills needed both to do back-of-the-envelope calculations and to do first-cut analyses or designs of discharges and thin-film processes. Practical graphs and tables are included to assist in the analysis. We give many examples throughout the book.

The book is both a graduate text, including exercises for students, and a research monograph for practicing engineers and scientists. We assume that the reader has the usual undergraduate background in mathematics (2 years), physics ($1\frac{1}{2}$ years), and chemistry ($\frac{1}{2}$ or 1 year). Some familiarity with partial differential equations as commonly taught in courses on electromagnetics or fluid dynamics at the junior or senior undergraduate level is also assumed.

After an introductory chapter, the book is divided into four parts: low-pressure partially ionized plasmas (Chapters 2–6); gas and surface physics and chemical dynamics (Chapters 7–9); plasma discharges (Chapters 10–14); and plasma processing (Chapters 16 and 17). Atomic and molecular collision processes have been divided into two relatively self-contained chapters (Chapters 3 and 8, respectively) inserted before the corresponding chapters on kinetics in each case. This material may be read lightly or thoroughly as desired. Plasma diagnostics appear in concluding sections (Sections 4.6, 6.6, 8.6, and 11.9) of various chapters and often also serve as applications of the ideas developed in the chapters.

For the last five years, the authors have taught a one-semester course based on this material to a mixed group of mainly graduate students in electrical, chemical, and nuclear engineering, materials science, and physics. A typical syllabus follows for 30 lectures, each $1\frac{1}{2}$ hours in length:

Chapter	Lectures
1	1
2	2
3	2 (light coverage)
4	1 (Sections 4.1 and 4.2 excluding waves, only)
5	2 (Sections 5.1–5.3 only)
6	3 (omit Section 6.4)
7	2
8	2 (light coverage, omit Section 8.6)
9	3
10	1 (omit Section 10.3)
11	2 (Sections 11.1 and 11.2 only)
12	1 (Section 12.1 only)
13	1 (Section 13.1 only)
14	2
15	3
16	2 (omit Section 17.5)

The core ideas of the book are developed in the sections of Chapters 2, 4–7, 9, and 10 listed in the syllabus. Atomic and molecular collisions (Chapters 3 and 8) can be emphasized more or less, but some coverage is desirable. The remaining chapters (Chapters 11–17), as well as some sections within each chapter, are relatively self-contained, and topics can be chosen according to the interests of the instructor. More specialized material on guiding center motion (Section 4.3), dynamics (Section 4.4), waves (Section 4.5), and diffusion in magnetized plasmas (Sections 5.4 and 5.5) can generally be deferred until familiarity with the core material has been developed.

Our colleagues C.K. Birdsall and V. Vahedi and the plasma simulation group at Berkeley have developed user-friendly programs for PCs and workstations for computer-aided instruction and demonstrations. A number of concepts in discharge dynamics have been illustrated using various output results from these programs (see Figures 1.12, 2.2, and 6.3). We typically do four or five 20-minute simulation demonstrations in the course during the semester using this software. The software and manuals can be obtained by contacting the Software Distribution Office, Industrial Liaison Program, Department of Electrical Engineering and Computer Sciences, University of California, Berkeley, CA 94720; the electronic mail address, telephone, and fax numbers are software@eecs.berkeley.edu, (510) 643-6687, and (510) 643-6694, respectively.

This book has been three years in writing. We have received encouragement and benefited from discussions with many friends and colleagues. We acknowledge here those who contributed significantly to our enterprise. We are indebted to D.L. Flamm who was a MacKay Visiting Lecturer at Berkeley in 1988–1989 and co-taught (with A.J.L.) an offering of our course in which he emphasized the chemical principles of plasma processing. One of the authors (M.A.L.) has taught abbreviated versions of the material in this book to process engineers in various short courses, along with his colleagues C.K. Birdsall, D.B. Graves, and V. Vahedi. We have benefited greatly from their insight and suggestions. Our colleagues N. Cheung, D. Graves, D. Hess, and S. Savas, our postdoctoral scholars C. Pico and R. Stewart, and our graduate students D. Carl, K. Kalpakjian, C. Lee, R. Lynch,

G. Misium, R. Moroney, K. Niazi, A. Sato, P. Wainman, A. Wendt, M. Williamson, and B. Wood have taught us much, and some of their work has been incorporated into our text. Some of the material in Chapters 10, 12, and 13 is based on a review article by R.A. Gottscho and one of the authors (M.A.L.) in *Physics of Thin Films*, Vol. 18, edited by M. Francombe and J.L. Vossen, Academic Press, New York, 1994. We thank V.A. Godyak, M.B. Lieberman, and S. Brown for reviewing several chapters and suggesting clarifications of the text. W.D. Getty has used a preprint of our manuscript to teach a course similar to ours, and the final text has benefited from his comments and suggestions. Many of the ideas expressed in the book were developed by the authors while working on grants and contracts supported by the National Science Foundation, the Department of Energy, the Lawrence Livermore National Laboratory, the State of California MICRO Program, the California Competitive Technology Program, SEMATECH and the Semiconductor Research Corporation, IBM, Applied Materials, and Motorola. The authors gratefully acknowledge the hospitality of M.G. Haines at Imperial College, London (M.A.L.) and of R. Boswell at the Australian National University, Canberra (A.J.L.), where much of the manuscript was developed. We gratefully thank E. Lichtenberg and P. Park for typing portions of the manuscript.

June, 1993
Berkeley, California

Michael A. Lieberman
Allan J. Lichtenberg

Symbols, Abbreviations, and Acronyms

Symbols

a	radius (m); atomic radius; a_0, Bohr radius; a_j, chemical activity of species j; a_v, etching anisotropy
$\mathbf{a}$	acceleration (m/s^2)
A	area (m^2); a constant; A_R, reduced mass (amu)
b	impact parameter (m); radius (m)
B	magnetic induction (T); a constant; susceptance (Ω^{-1}); B_{rot}, rotational constant of molecule
c	velocity of light in vacuum
C	a constant; capacitance (F); C_V, specific heat at constant volume (J/mol-K); C_p, specific heat at constant pressure
$\mathcal{C}$	a contour or closed loop
d	denotes an exact differential
đ	denotes a nonexact differential (Chapter 7)
d	distance (m); plasma size (m); d', plasma half-width (m)
D	diffusion coefficient (m^2/s); displacement vector (C/m^2); D_a, ambipolar diffusion coefficient; D_{a+}, ambipolar diffusion coefficient in the presence of negative ions; D_v, velocity space diffusion coefficient (m^3/s^3); $D_\mathcal{E}$, energy diffusion coefficient (V^2/s); D_{SiO_2}, deposition rate of silicon dioxide (m/s)
e	unsigned charge on an electron (1.602×10^{-19} C)
e	the natural base (2.718)
E	electric field (V/m); etch (or deposition) rate (Å/min)
$\mathcal{E}$	the voltage equivalent of the energy (V); i.e., energy (J) = $e\mathcal{E}$ (V)
f	frequency (Hz); distribution function (m^{-6}-s^3); f_m, Maxwellian distribution; f_{pe}, electron plasma frequency; f_{pi}, ion plasma frequency
$\mathbf{f}_c$	collisional force per unit volume (N/m^3)
F	force (N)
g	degeneracy; $\bar{g}$, statistical weight; energy distribution function; gravitational constant
g	denotes a gas
G	Gibbs free energy (J); volume ionization rate (m^{-3}-s^{-1}); conductance (Ω^{-1}); G_f, Gibbs free energy of formation; G_r, Gibbs free energy of reaction
h	edge-to-center density ratio; h_l, axial ratio; h_R, radial ratio
H	enthalpy (J); magnetic field (A/m); height (m); H_f, enthalpy of formation; H_r, enthalpy of reaction

H Boltzmann H function

i integer

I electrical current (A); differential scattering cross section (m^2/sr); I_{AB}, I_{mol}, moment of inertia of molecule (kg-m^2)

I modified Bessel function of the first kind

j $\sqrt{-1}$; integer

J electrical current density (A/m^2); rotational quantum number

J Bessel function of the first kind

$\mathcal{J}$ $\mathcal{J}_j$ denotes chemical species j

k Boltzmann's constant (1.381×10^{-23} J/K); wave number or wave vector (m^{-1})

K first-order (s^{-1}), second-order (m^3/s), or third-order (m^6/s) rate constant

K modified Bessel function of the second kind

$\mathcal{K}$ equilibrium constant

l discharge length (m); antenna length (m); quantum number; integer; l', discharge half-width (m)

l denotes a liquid

ℓ denotes length for a line integral

L length (m); volume loss rate (m^{-3}-s^{-1}); inductance (H); particle density sink (m^{-3}-s^{-1})

m electron mass (9.11×10^{-31} kg); mass (kg); azimuthal mode number; m_l, m_s, and m_J, quantum numbers for axial component of orbital, spin, and total angular momentum

M ion mass (kg)

$\mathcal{M}$ number of chemical species; secondary electron multiplication factor

n particle density (m^{-3}); principal quantum number (an integer); n_i, ion density; n_e, electron density; n_g, neutral gas density

n' area density (m^{-2}); n'_0, area density of surface sites

N quantity of a substance (mol); index of refraction of a wave

$\mathcal{N}$ number of turns

p pressure (N/m^2); particle momentum (kg-m/s); $p^{\ominus}$, standard pressure (1 bar or 1 atm); p_d, electric dipole moment (C-m); p_{ohm}, ohmic power density (W/m^3)

P power (W); probability; perimeter (m)

q electric charge (C)

$\mathbf{q}$ heat flow vector (W/m^2)

Q heat (J); electric charge (C)

$\mathcal{Q}$ resonant circuit or cavity quality factor

r radial position (m); r_c, gyroradius; r_{ce}, electron gyroradius

R gas constant (8.314 J/K-mol); cylinder radius (m); center of mass coordinate (m); nuclear separation (m); reaction rate (m^{-3}-s^{-1}); resistance (Ω)

s sheath thickness (m); sticking coefficient; $\bar{s}$, thermal sticking coefficient; s_v or s_h, etching selectivity

s denotes a solid

S energy flux (W/m^2-s); entropy (J/K); closed surface area (m^2); S_p, pumping speed (m^3/s)

$\mathcal{S}$ denotes a closed surface

t time (s)

T temperature (K); T_0, standard temperature (298 K)

T temperature in units of volts (V)

u average velocity (m/s); u_B, Bohm velocity; u_E, $\mathbf{E} \times \mathbf{B}$ velocity; u_D, diamagnetic drift velocity

U	energy (J); internal energy (J); potential energy (J)
v	velocity (m/s); vibrational quantum number; $\bar{v}$, average speed; v_{th}, thermal velocity; v_R, relative velocity; v_{ph}, phase velocity
V	voltage or electric potential (V); $\tilde{V}$, rf voltage; $\overline{V}$, dc or time-average voltage
$\mathcal{V}$	volume (m^3)
w	energy per unit volume (J/m^3); width (m)
W	kinetic energy (J); work (J)
x	rectangular coordinate (m); x_j, mole fraction of species j; x_{iz}, fractional ionization
X	reactance (Ω)
y	rectangular coordinate (m)
Y	etch yield; admittance (Ω^{-1})
z	rectangular or axial cylindrical coordinate (m)
Z	relative charge on an ion, in units of e; impedance (Ω)
α	spatial rate of variation (m^{-1}); spatial attenuation or decay constant (m^{-1}); first Townsend coefficient (m^{-1}); ratio of negative ion to electron density; α_j, stoichiometric coefficient of species j; α_p, atomic or molecular polarizability (m^3)
β	spatial rate of variation (m^{-1}); a constant
γ	secondary electron emission coefficient; wall loss probability; ratio of electron-to-ion temperature; ratio of specific heats; complex propagation constant; γ_{se}, secondary electron emission coefficient; γ_{sput}, sputtering coefficient
Γ	particle flux (m^{-2}-s^{-1})
$\mathit{\Gamma}$	the Gamma function
δ	Dirac delta function; layer thickness (m); δ_p, collisionless skin depth (m); δ_c, collisional skin depth (m); δ_e, anomalous skin depth (m)
Δ	denotes the change of a quantity
ϵ	dielectric constant (F/m); ϵ_0, vacuum permittivity (8.854×10^{-12} F/m); ϵ_p, plasma dielectric constant
ζ	a small displacement (m); ζ_L, fractional energy loss for elastic collision
θ	angle (rad); spherical polar angle; scattering angle in laboratory system; fractional surface coverage
Θ	scattering angle in center of mass system (rad)
η	efficiency factor; asymmetry factor
κ	relative dielectric constant; κ_p, relative plasma dielectric constant; κ_T, thermal conductivity; κ_C, Clausing factor transmission probability
λ	mean free path (m); λ_c, collisional mean free path; λ_e, electron mean free path; λ_i, ion mean free path; λ_{De}, electron Debye length (m); $\lambda_{\mathcal{E}}$, energy relaxation length
Λ	diffusion length (m); ratio of Debye length to minimum impact parameter
μ	mobility (m^2/V-s); chemical potential (J/mol); μ_0, vacuum permeability ($4\pi \times 10^{-7}$ H/m); μ_{mag}, magnetic moment
ν	collision or interaction frequency (s^{-1} or Hz); ν_c, collision frequency
ξ	a constant
π	3.1416
Π	stress tensor (N/m^2)
ρ	volume charge density (C/m^3); ρ_S, surface charge density (C/m^2)
σ	cross section (m^2); σ_{dc}, dc electrical conductivity (Ω^{-1}-m^{-1}); σ_{rf}, rf electrical conductivity
τ	mean free time (s); time constant (s); τ_c, collision time
ϕ	angle (rad); spherical azimuthal angle

φ	magnetic flux (T-m^2); normalized fluence (number of monolayers)
Φ	electric potential (V); Φ_p, plasma potential; Φ_w, wall potential
$\mathit{\Phi}$	fluence (m^{-2})
χ	angle (rad); χ_{01}, first zero of zero-order Bessel function
ψ	spherical polar angle in velocity space
Ψ	helix pitch (rad)
ω	radian frequency (rad/s); ω_{pe}, electron plasma frequency; ω_c, gyration frequency; ω_{ce}, electron gyration frequency
Ω	solid angle (sr)
∇, $\nabla_{\mathbf{r}}$	vector spatial derivative; $\nabla_{\mathbf{v}}$, vector velocity derivative; ∇_T, vector derivative in total energy coordinates
A	scalar
$\mathbf{A}$	vector
$\hat{A}$	unit vector (has unit magnitude)
$\tilde{A}$	oscillating or rf part
$\bar{A}$	average or dc part; equilibrium value
$\dot{A}$	$\mathrm{d}A/\mathrm{d}t$
$\ddot{A}$	$\mathrm{d}^2A/\mathrm{d}t^2$
$\langle A \rangle$	average
A'	areal density (m^{-2}); variable of integration
$\lvert A \rvert$	absolute magnitude

Unit Abbreviations

A	ampere
Å	angstrom
amu	atomic mass unit
atm	atmosphere
C	coulomb
°C	celsius
cal	calorie
eV	electronvolt
F	farad
g	gram
G	gauss
H	henry
Hz	hertz
J	joule
K	kelvin
l	liter
m	meter
min	minute
mol	mole
N	newton
Pa	pascal
rad	radian

s second
sr steradian
T tesla
Td townsend
V volt
W watt
Ω ohm

Subscript Abbreviations

a activation; adsorption
abs absorbed
adet associative detachment
ads adsorbed
aff affinity
appl applied
at atomic, atom
att attachment
c denotes collision or collisional, except ω_c and r_c denote gyration frequency and gyration radius, respectively
chemi chemisorption
cond conduction
cx charge transfer (charge exchange)
d desorption; denotes dust particles
dc constant in time (direct current)
desor desorption
det detachment
dex de-excitation
diss dissociation, dissipation
diz dissociative ionization
D diffusion
e denotes electron
ecr electron cyclotron resonance
edet electron detachment
eff effective or effective value
el elastic
esc escape
ex excitation
ext external
f formation
fin final
g denotes gas atom
h denotes hot or tail electrons; denotes horizontal; denotes high
i denotes positive ion
in in
inel inelastic

init	initial
inc	incident
ind	induced
iz	ionization
l	left-hand circularly polarized; denotes low
L	Langevin (capture)
loss	loss
m	momentum transfer; metal; H_m, S_m, and G_m denote per mole
mag	magnetic; magnetization
max	maximum of a quantity
min	minimum of a quantity
mol	molecule
ohm	ohmic
out	out
ox	oxide
p	usually denotes plasma; pumping
ph	phase
physi	physisorption
pol	polarization
poly	polysilicon
pr	photoresist
q	quenching
QL	quasilinear
r	right-hand circularly polarized; reaction
R	denotes reduced or relative value
rad	radiation
rec	recombination
refl	reflected
res	resonance
rf	radio frequency
rot	rotational
s	denotes sheath edge
S	denotes surface
sc	scattering
se	denotes secondary electron
sh	sheath
sput	sputtering
stoc	stochastic
subl	sublimation
T	denotes total
th	thermal
thr	threshold
trans	transmitted
v	denotes vertical
vap	vaporization
vib	vibrational
w	denotes wall

α	denotes presence of negative ions
0	denotes initial value, uniform value, or central value; zero-order quantity
1	first order quantity
$\parallel$	parallel
$\perp$	perpendicular
$\times$	cross term (off-diagonal term in matrix)
*	denotes excited states
+	denotes positive ion quantities
−	denotes negative ion quantities

Acronyms

ALD	atomic layer deposition
ALE	atomic layer etching
ARDE	aspect-ratio-dependent etching
BET	Brunauer–Emmett–Teller
CCD	charge-coupled device
CCP	capacitively-coupled plasma
CM	center of mass
CMA	Clemmow–Mullaly–Allis
CVD	chemical vapor deposition
DBD	dielectric barrier discharge
dc, DC	direct current; non-time-varying
DECR	distributed electron cyclotron resonance
EAE	electrical asymmetry effect
ECR	electron cyclotron resonance
EEDF	electron energy distribution function
EEPF	electron energy probability function
EN	electronegative
GPC	growth per cycle
HCE	hollow cathode effect
HiPIMS	high-power impulse magnetron sputtering
IC	integrated circuit
ICCD	image-intensified charge-coupled device
ICP	inductively-coupled plasma
I-PVD	ionized physical vapor deposition
LHP	left-hand polarized
LHS	left-hand side
LIF	laser-induced fluorescence
LLS	laser light scattering
MD	molecular dynamics
MEMS	micro-electromechanical system
MERIE	magnetically-enhanced reactive ion etching
ML	monolayer
MOS	metal–oxide–semiconductor
MRAM	magnetoresistive random access memory

OML	orbital-motion limited
PEALD	plasma-enhanced atomic layer deposition
PECVD	plasma-enhanced chemical vapor deposition
PIC	particle-in-cell
PIC/MCC	particle-in-cell, Monte-Carlo collisions
PIII	plasma immersion ion implantation
PROES	phase-resolved optical emission spectroscopy
PSII	plasma source ion implantation
PVD	physical vapor deposition
rf, RF	radio frequency
RFI	radio-frequency inductive
RHP	right-hand polarized
RHS	right-hand side
STP	standard temperature and pressure
TALIF	two-photon absorption, laser-induced fluorescence
TCP	transformer-coupled plasma
TG	Trivelpiece–Gould
UV	ultraviolet
VUV	vacuum ultraviolet
WKB	Wentzel–Kramers–Brillouin

Physical Constants and Conversion Factors

Quantity	Symbol	Value
Boltzmann constant	k	1.3807×10^{-23} J/K
Elementary charge	e	1.6022×10^{-19} C
Electron mass	m	9.1095×10^{-31} kg
Proton mass	M	1.6726×10^{-27} kg
Proton/electron mass ratio	M/m	1836.2
Planck constant	h	6.6262×10^{-34} J-s
	$\hbar = h/2\pi$	1.0546×10^{-34} J-s
Speed of light in vacuum	c	2.9979×10^{8} m/s
Permittivity of free space	ϵ_0	8.8542×10^{-12} F/m
Permeability of free space	μ_0	$4\pi \times 10^{-7}$ H/m
Bohr radius	$a_0 = 4\pi\epsilon_0\hbar^2/e^2m$	5.2918×10^{-11} m
Atomic cross section	πa_0^2	8.7974×10^{-21} m^2
Temperature T associated with T = 1 V		11,605 K

Quantity	Symbol	Value
Energy associated with $\mathcal{E} = 1$ V		1.6022×10^{-19} J
Avogadro number		
(molecules/mol)	N_A	6.0220×10^{23}
Gas constant	$R = kN_A$	8.3144 J/K-mol
Atomic mass unit (amu)		1.6606×10^{-27} kg
Standard temperature (25 °C)	T_0	298.15 K
Standard pressure		
(760 Torr = 1 atm)	$p^{\ominus}$	1.0133×10^{5} Pa
Gas density at STP	$n^{\ominus}$	2.4615×10^{25} m^{-3}
Loschmidt's number		
(gas density at 273.15 K, 1 atm)	n_L	2.6868×10^{25} m^{-3}
Pressure of 1 Torr		133.32 Pa
Pressure of 1 Pa		7.50 mTorr
Energy per mole at T_0	RT_0	2.4789 kJ/mol
1 calorie (cal)		4.1868 J
1 townsend (Td)	E/n_g	1×10^{-21} V-m^2

Practical Formulae

In the following practical formulae, n_e is in units of cm^{-3}, T_e is in volts, and B is in gauss (1 tesla = 10^4 gauss).

Electron plasma frequency	$\omega_{pe} = (e^2 n_e/\epsilon_0 m)^{1/2}$	$f_{pe} = 9000\sqrt{n_e}$ Hz
Electron gyration frequency	$\omega_{ce} = eB/m$	$f_{ce} = 2.8B$ MHz
Electron Debye length	$\lambda_{De} = (\epsilon_0 T_e/en_e)^{1/2}$	$\lambda_{De} = 740\sqrt{T_e/n_e}$ cm
Mean electron speed	$\bar{v}_e = (8eT_e/\pi m)^{1/2}$	$\bar{v}_e = 6.7 \times 10^7\sqrt{T_e}$ cm/s
Bohm velocity	$u_B = (eT_e/M)^{1/2}$	$u_B = 9.8 \times 10^5\sqrt{T_e/A_R}$ cm/s

1

Introduction

1.1 Materials Processing

Chemically reactive plasma discharges are widely used to modify the surface properties of materials. Plasma processing technology is vitally important to several of the largest manufacturing industries in the world. Plasma-based surface processes are indispensable for manufacturing the very large-scale integrated circuits (ICs) used by the electronics industry. Such processes are also critical for the flat panel display, solar panel, aerospace, automotive, steel, biomedical, and toxic waste management industries. Materials and surface structures can be fabricated that are not attainable by any other commercial method, and the surface properties of materials can be modified in unique ways. For example, 0.2-μm-wide, 4-μm-deep trenches can be etched into silicon films or substrates (Figure 1.1*a*). A human hair is 50–100 μm in diameter, so hundreds of these trenches would fit endwise within a human hair. A more recent example in Figure 1.1*b* shows a set of approximately 0.1 μm in diameter by 7.5 μm deep channel holes etched into a stacked set of dielectric layers. Unique materials such as diamond films and amorphous silicon for solar cells have also been produced, and plasma-based hardening of surgically implanted hip joints and machine tools have extended their working lifetimes manyfold.

It is instructive to look closer at IC fabrication, which is the key application driving the field that we describe in this book. As a very incomplete list of plasma processes, argon or oxygen discharges are used to sputter-deposit aluminum, tungsten, or high-temperature superconducting films; oxygen discharges can be used to grow SiO_2 films on silicon; SiH_2Cl_2/NH_3 and $Si(OC_2H_5)_4/O_2$ discharges are used for the plasma-enhanced chemical vapor deposition (PECVD) of Si_3N_4 and SiO_2 films, respectively; BF_3 discharges can be used to implant dopant (B) atoms into silicon; $CF_4/Cl_2/O_2$ discharges are used to selectively remove silicon films; and oxygen discharges are used to remove photoresist or polymer films. These types of steps (deposit or grow, dope or modify, etch or remove) are repeated again and again in the manufacture of a modern IC. They are the equivalent, on a submicrometer-size scale, of centimeter-size manufacture using metal and components, bolts and solder, and drill press and lathe. For microfabrication of an IC, one-third of the tens to hundreds of fabrication steps are typically plasma-based.

Figure 1.2 shows a typical set of steps to create a metal film patterned with submicrometer features on a large-area (300 mm diameter) wafer substrate. In (*a*), the film is deposited; in (*b*), a photoresist layer is deposited over the film; in (*c*), the resist is selectively exposed to light through a pattern; and in (*d*), the resist is developed, removing the exposed resist regions and leaving behind

Principles of Plasma Discharges and Materials Processing, Third Edition. Michael A. Lieberman and Allan J. Lichtenberg.

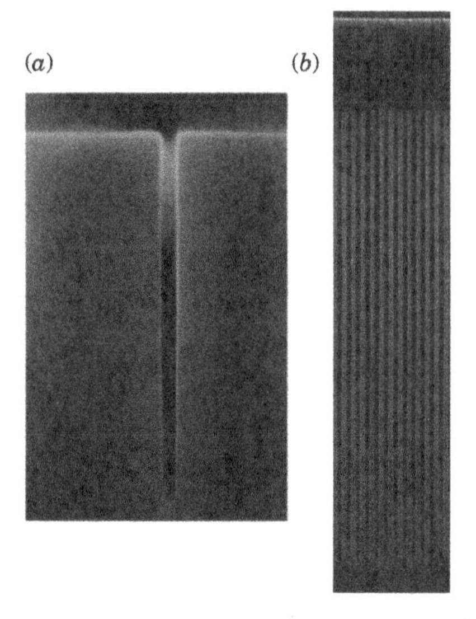

Figure 1.1 High aspect ratio anisotropic etches, showing the extraordinary capabilities of plasma processing; such etched features are used for device isolation, charge storage capacitors, channel holes, and many other purposes in integrated circuits: (*a*) trench etch (0.2 μm wide by 4 μm deep) in single-crystal silicon, circa 2004; (*b*) set of channel holes (each approximately 0.1 μm in diameter by 7.5 μm deep) etched into a stacked set of dielectric layers, circa 2023. Source: (*a*) M. A. Lieberman (Book Author). (*b*) Courtesy of Lam Research Corporation.

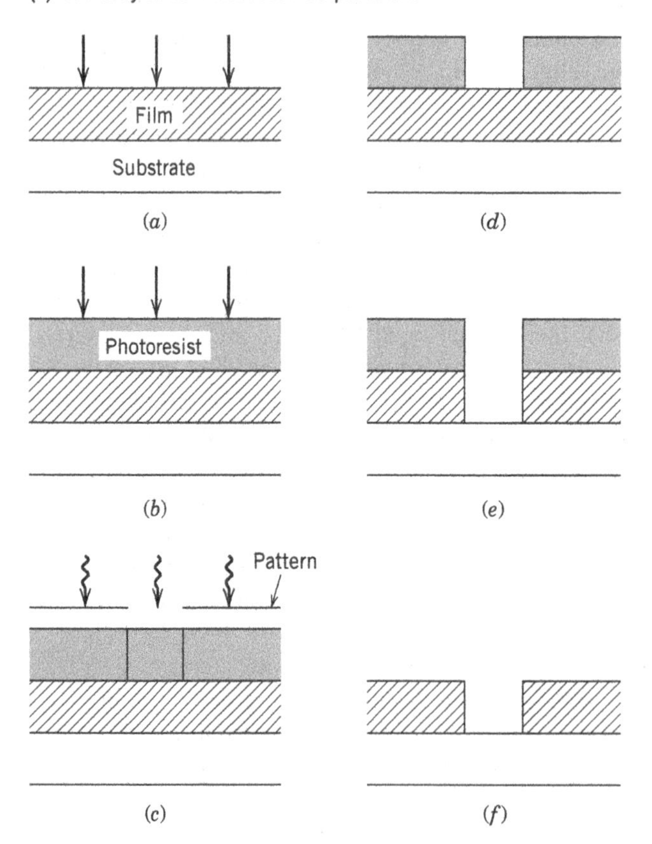

Figure 1.2 Deposition and pattern transfer in manufacturing an integrated circuit: (*a*) metal deposition; (*b*) photoresist deposition; (*c*) optical exposure through a pattern; (*d*) photoresist development; (*e*) anisotropic plasma etch; (*f*) remaining photoresist removal.

a patterned resist mask. In (*e*), this pattern is transferred into the film by an etch process; the mask protects the underlying film from being etched. In (*f*), the remaining resist mask is removed. Of these six steps, plasma processing is generally used for film deposition (*a*) and etch (*e*) and may also be used for resist development (*d*) and removal (*f*).

The etch process in (*e*) is illustrated as leading to vertical sidewalls aligned with the resist mask; i.e., the mask pattern has been faithfully transferred into the metal film. This can be accomplished by an etch process that removes material in the vertical direction only. The horizontal etch rate is zero. Such *anisotropic* etches are easily produced by plasma processing. On the other hand, one might imagine that exposing the masked film (*d*) to a liquid (or vapor phase) etchant will lead to the undercut *isotropic* profile shown in Figure 1.3*a* (compare to Figure 1.2*e*), which is produced by equal vertical and horizontal etch rates. Many years ago, feature spacings (e.g., between trenches) were tens of micrometers, much exceeding the required film thicknesses. Undercutting was then acceptable. This is no longer true with submicrometer feature spacings. The reduction in feature sizes and spacings makes anisotropic etch processes essential. In fact, strictly vertical etches are sometimes not desired; one wants controlled sidewall angles. Plasma processing is the only commercial technology capable of such control. Anisotropy is a critical process parameter in IC manufacture and has been a major force in driving the development of plasma processing technology.

The etch process applied to remove the film in Figure 1.2*d* is shown in Figure 1.2*e* as not removing either the photoresist or the underlying substrate. This selectivity is another critical process parameter for IC manufacture. Whereas wet etches have been developed having essentially infinite selectivity, highly selective plasma etch processes are not easily designed. Selectivity and anisotropy often compete in the design of a plasma etch process, with results as shown in Figure 1.3*b*. Compare this to the idealized result shown in Figure 1.2*e*. Assuming that film-to-substrate selectivity

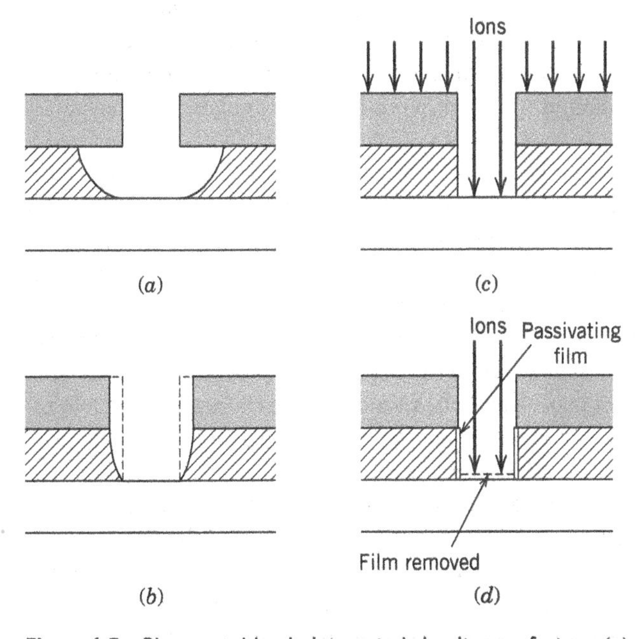

Figure 1.3 Plasma etching in integrated circuit manufacture: (*a*) example of isotropic etch; (*b*) sidewall etching of the resist mask leads to a loss of anisotropy in film etch; (*c*) illustrating the role of bombarding ions in anisotropic etch; (*d*) illustrating the role of sidewall passivating films in anisotropic etch.

is a critical issue, one might imagine simply turning off the plasma after the film has been etched through. This requires a good endpoint detection system. Even then, variations in film thickness and etch rate across the area of the wafer imply that the etch cannot be stopped at the right moment everywhere. Hence, depending on the process uniformity, there is a need for some selectivity. These issues are considered further in Chapter 16.

Here is a simple recipe for etching silicon using a plasma discharge. Start with an inert molecular gas, such as CF_4. Excite the discharge to sustain a plasma by electron–neutral dissociative ionization,

$$e + CF_4 \rightarrow 2e + CF_3^+ + F$$

and to create reactive species by electron–neutral dissociation,

$$\begin{aligned} e + CF_4 &\rightarrow e + F + CF_3 \\ &\rightarrow e + 2F + CF_2 \end{aligned}$$

The etchant F atoms react with the silicon substrate, yielding the volatile etch product SiF_4:

$$Si(s) + 4F(g) \rightarrow SiF_4(g)$$

Here, s and g indicate solid and gaseous forms, respectively. Finally, the product is pumped away. It is important that CF_4 does not react with silicon and that the etch product SiF_4 is volatile, so that it can be removed. This process etches silicon isotropically. For an anisotropic etch, there must be high-energy ion (CF_3^+) bombardment of the substrate. As illustrated in Figure 1.3*c* and *d*, energetic ions leaving the discharge during the etch bombard the bottom of the trench but do not bombard the sidewalls, leading to anisotropic etching by one of the two mechanisms. Either the ion bombardment increases the reaction rate at the surface (Figure 1.3*c*) or it exposes the surface to the etchant by removing passivating films that cover the surface (Figure 1.3*d*).

Similarly, Cl and Br atoms created by dissociation in a discharge are good etchants for silicon, F atoms and CF_2 molecules for SiO_2, O atoms for photoresist, and Cl atoms for aluminum. In all cases, a volatile etch product is formed. However, F atoms do not etch aluminum, and there is no commercially viable etchant for copper, because the etch products are not volatile at reasonable substrate temperatures.

We see the importance of the basic physics and chemistry topics treated in this book: (1) plasma physics (Chapters 2, 4–6, and 19) to determine the electron and ion densities, temperatures, and ion bombardment energies and fluxes for a given discharge configuration; and (2) gas-phase chemistry and (3) surface physics and chemistry (Chapters 7 and 9) to determine the etchant densities and fluxes and the etch rates with and without ion bombardment. The database for these fields of science is provided by (4) atomic and molecular physics, which we discuss in Chapters 3 and 8. We also discuss applications of equilibrium thermodynamics (Chapter 7) to plasma processing. The measurement and experimental control of plasma and chemical properties in reactive discharges is itself a vast subject. We provide brief introductions to some simple plasma diagnostic techniques throughout the text.

We have motivated the study of the fundamentals of plasma processing by examining isotropic and anisotropic etches for IC manufacture. These are discussed in Chapter 16. Other characteristics motivate its use for deposition and surface modification. For example, a central feature of the *low-pressure* processing discharges that we consider in this book is that the plasma itself, as well as the plasma–substrate system, is not in thermal equilibrium. This enables substrate temperatures to be relatively low, compared to those required in conventional thermal processes, while maintaining adequate deposition or etch rates. Putting it another way, plasma processing rates are greatly

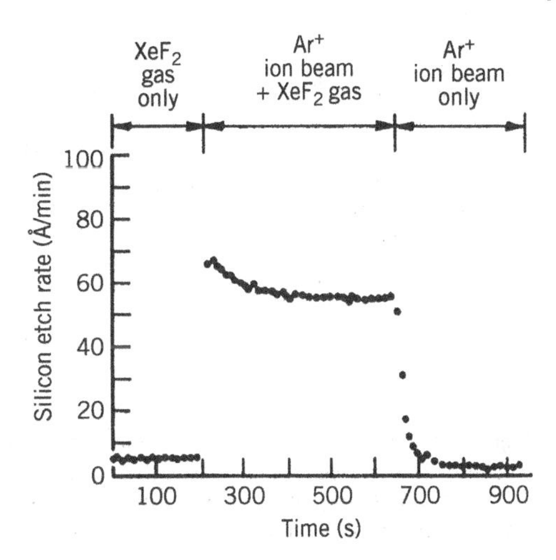

Figure 1.4 Experimental demonstration of ion-enhanced plasma etching. Source: Adapted from Coburn and Winters (1979).

enhanced over thermal processing rates at the same substrate temperature. For example, Si_3N_4 films can be deposited over aluminum films by PECVD, whereas adequate deposition rates cannot be achieved by conventional chemical vapor deposition (CVD) without melting the aluminum film. Chapter 17 gives further details.

Particulates or "dust" can be a significant component in processing discharges and a source of substrate-level contamination in etch and deposition processes. One can also control dust formation in useful ways, for example, to produce powders of various sizes or to incorporate nanoparticles during deposition to modify film properties. Dusty plasmas are described in Chapter 18.

The nonequilibrium nature of plasma processing has been known for many years, as illustrated by the laboratory data in Figure 1.4. In time sequence, this shows first, the equilibrium chemical etch rate of silicon in the XeF_2 etchant gas; next, the tenfold increase in etch rate with the addition of argon ion bombardment of the substrate, simulating plasma-assisted etching; and finally, the very low "etch rate" due to the physical sputtering of silicon by the ion bombardment alone.

Another application is the use of plasma-immersion ion implantation (PIII) to implant ions into materials at dose rates that are tens to hundreds of times larger than those achievable with conventional (beam-based) ion implantation systems. In PIII, a series of negative high-voltage pulses are applied to a substrate that is immersed directly into a discharge, thus accelerating plasma ions into the substrate. The development of PIII has opened a new implantation regime characterized by very high dose rates, even at very low energies, and by the capability to implant both large-area and irregularly shaped substrates, such as flat panel displays or machine tools and dies. This is illustrated in Figure 1.5. Further details are given in Chapter 17.

1.2 Plasmas and Sheaths

1.2.1 Plasmas

A plasma is a collection of free charged particles moving in random directions, that is, on an average, electrically neutral (see Figure 1.6*a*). This book deals with weakly ionized plasma discharges, which are plasmas having the following features: (1) they are driven electrically; (2) charged particle collisions with neutral gas molecules are important; (3) there are boundaries at which surface

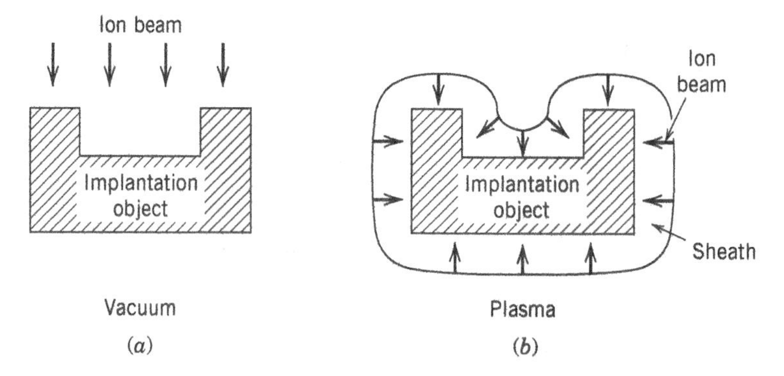

Figure 1.5 Illustrating ion implantation of an irregular object: (*a*) in a conventional ion beam implanter, the beam is electrically scanned and the target object is mechanically rotated and tilted to achieve uniform implantation; (*b*) in plasma-immersion ion implantation (PIII), the target is immersed in a plasma, and ions from the plasma are implanted with a relatively uniform spatial distribution.

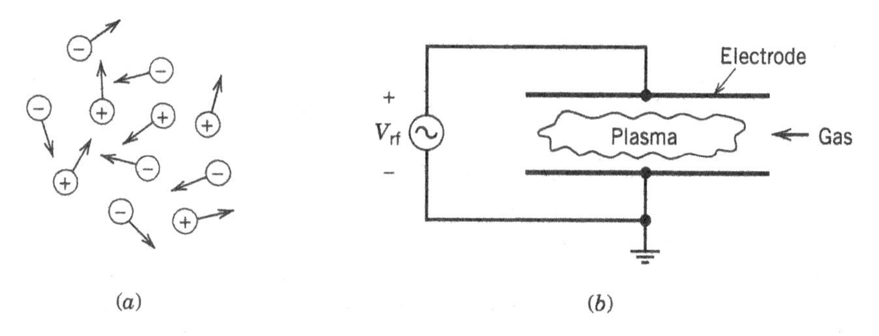

Figure 1.6 Schematic view of (*a*) a plasma and (*b*) a discharge.

losses are important; (4) ionization of neutrals sustains the plasma in the steady state; and (5) the electrons are not in thermal equilibrium with the ions.

A simple discharge is shown schematically in Figure 1.6*b*. It consists of a voltage source that drives current through a low-pressure gas between two parallel conducting plates or electrodes. The gas "breaks down" to form a plasma, usually weakly ionized, i.e., the plasma density is only a small fraction of the neutral gas density. We describe some qualitative features of plasmas in this section; discharges are described in the following section.

Plasmas are often called a fourth state of matter. As we know, a solid substance in thermal equilibrium generally passes into a liquid state as the temperature is increased at a fixed pressure. The liquid passes into a gas as the temperature is further increased. At a sufficiently high temperature, the molecules in the gas decompose to form a gas of atoms that move freely in random directions, except for infrequent collisions between atoms. If the temperature is further increased, then the atoms decompose into freely moving charged particles (electrons and positive ions), and the substance enters the plasma state. This state is characterized by a common charged particle density $n_e \approx n_i \approx n$ particles/m^3 and, in equilibrium, a common temperature for all species. The temperatures required to form plasmas from pure substances in thermal equilibrium range from roughly 4000 K for easy-to-ionize elements like cesium to 20,000 K for hard-to-ionize elements like helium. The fractional ionization of a plasma is

$$x_{iz} = \frac{n_i}{n_g + n_i}$$

where n_g is the neutral gas density, x_{iz} is near unity for fully ionized plasmas, and $x_{iz} \ll 1$ for weakly ionized plasmas.

Much of the matter in the universe is in the plasma state. This is true because stars, as well as most interstellar matter, are plasmas. Although stars are plasmas in thermal equilibrium, the light and heavy charged particles in low-pressure processing discharges are *almost never* in thermal equilibrium, either between themselves or with their surroundings. Because these discharges are electrically driven and are weakly ionized, the applied power preferentially heats the mobile electrons, while the heavy ions efficiently exchange energy by collisions with the cold background gas. Hence, $\mathrm{T_e} \gg \mathrm{T_i}$ for these plasmas.

Figure 1.8 identifies different kinds of plasmas on a $\log_{10} n$ versus $\log_{10} \mathrm{T_e}$ diagram. There is an enormous range of densities and temperatures for both laboratory and space plasmas. Two important types of processing discharges are indicated in the figure. Low-pressure glow discharges are characterized by $\mathrm{T_e} \approx 1$–10 V, $\mathrm{T_i} \ll \mathrm{T_e}$, and $n \approx 10^8$–10^{13} cm^{-3}. These discharges are used as miniature chemical factories in which feedstock gases are broken into positive ions and chemically reactive etchants, deposition precursors, etc., which then flow to and physically or chemically react at the substrate surface. While energy is delivered to the substrate also, e.g., in the form of bombarding ions, the energy flux is there to promote the chemistry at the substrate and not to heat the substrate. The gas pressures for these discharges are low: $p \approx 1$ mTorr–1 Torr. *These discharges and their use for processing are the principal subject of this book.* We give the quantitative framework for their analysis in Chapter 10.

As shown in Figure 1.7, at low pressures, the electrons are *not* in thermal equilibrium with the ions. Because of the very light mass of electrons compared to ions and neutrals, $m/M \sim 10^{-4}$, the energy transfer during electron collisions with heavy particles is very weak. Since mainly electrons are heated by the applied fields, $\mathrm{T_e} \gg \mathrm{T_i}$ in the plasma. At the same time, ions can acquire high energies $\mathcal{E}_i$ by acceleration across sheaths near the substrate surface, $\mathcal{E}_i \gg \mathrm{T_e}$. This enables "high temperature processing at low temperatures." The substrate can be kept near room temperature, with the hot electrons producing both chemically reactive species and energetic ions bombarding the substrate.

High-pressure arc discharges, shown in Figure 1.8, are also used for processing. These discharges have $\mathrm{T_e} \approx 0.1$–2 V and $n \approx 10^{14}$–10^{19} cm^{-3}, and the light and heavy particles are more nearly in thermal equilibrium, with $\mathrm{T_i} \lesssim \mathrm{T_e}$. These discharges are used mainly to deliver heat to the substrate, e.g., to increase surface reaction rates, to melt, sinter, or evaporate materials, or to weld or cut refractory materials. Operating pressures are typically near atmospheric pressure (760 Torr). High-pressure discharges of this type are beyond the scope of this book.

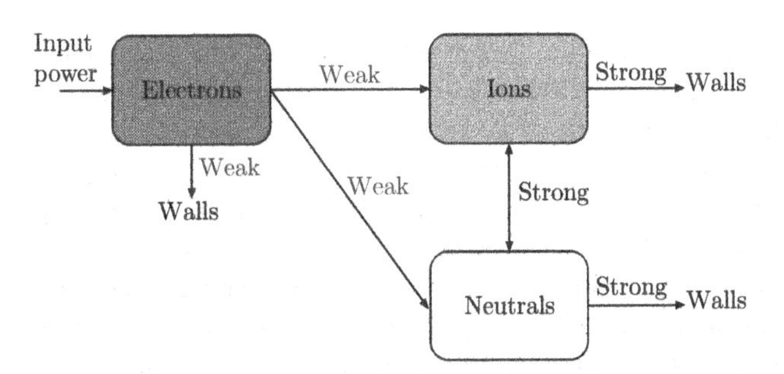

Figure 1.7 Energy coupling between electrons and heavy particles in a low-pressure plasma.

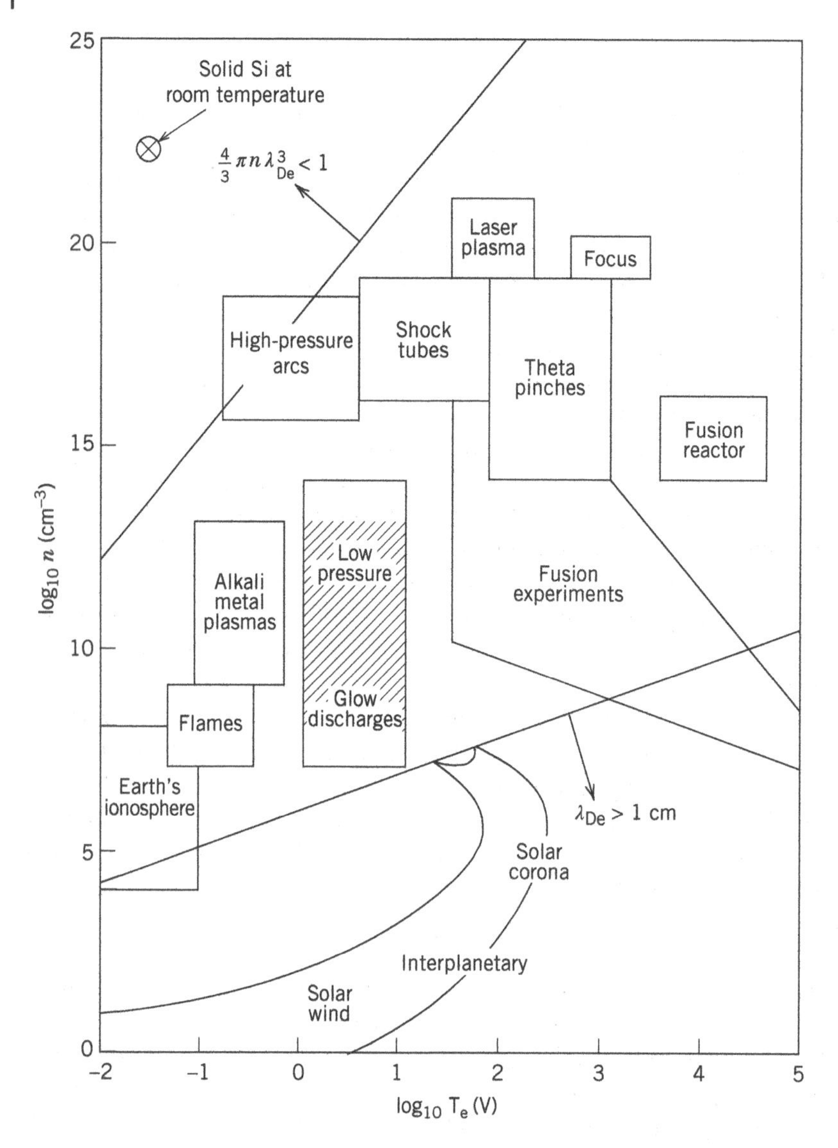

Figure 1.8 Space and laboratory plasmas on a $\log_{10} n$ versus $\log_{10} T_e$ diagram (Source: Book (1987)/Naval Research Laboratory); the electron Debye length λ_{De} is defined in Section 2.4.

Figure 1.9 shows the densities and temperatures (or average energies) for various species in a typical rf-driven, capacitively coupled low-pressure discharge, e.g., for silicon etching using CF_4, as described in Section 1.1. We see that the feedstock gas, etchant atoms, etch product gas, and plasma ions have roughly the same temperature, which does not exceed a few times room temperature (0.026 V). The etchant F and product SiF_4 densities are significant fractions of the CF_4 density, but the fractional ionization is very low: $n_i \sim 10^{-5} n_g$. The electron temperature T_e is two orders of magnitude larger than the ion temperature T_i. However, we note that the energy of ions bombarding the substrate can be 100–1000 V, much exceeding T_e. The acceleration of low-temperature ions across a thin *sheath* region where the plasma and substrate meet is central to all processing discharges. We describe this qualitatively next and quantitatively in later chapters.

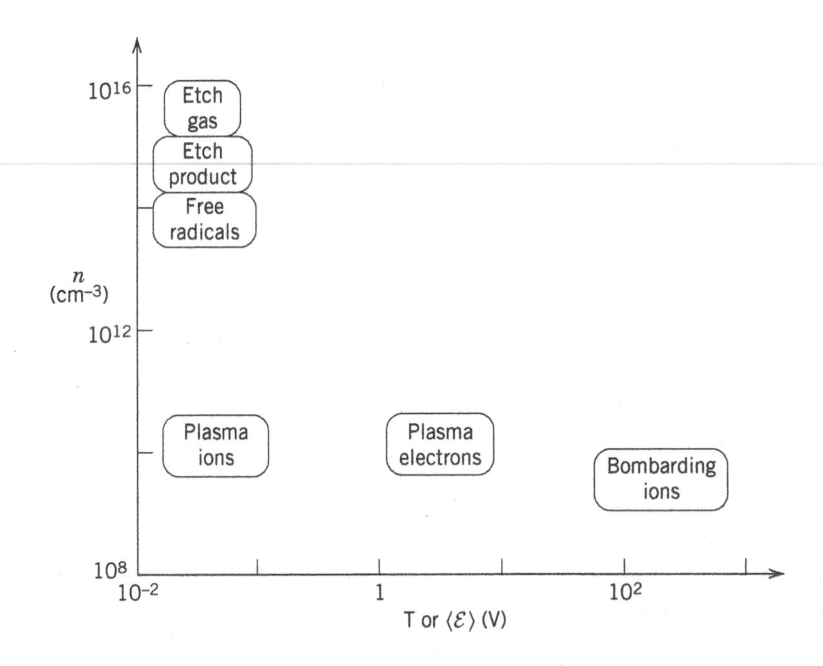

Figure 1.9 Densities and energies for various species in a low-pressure capacitive rf discharge.

Although n_i and n_e may be five orders of magnitude lower than n_g, the charged particles play central roles in sustaining the discharge and in processing. Because $T_e \gg T_i$, it is the electrons that dissociate the feedstock gas to create the free radicals, etchant atoms, and deposition precursors, required for the chemistry at the substrate. Electrons also ionize the gas to create the positive ions that subsequently bombard the substrate. As we have seen, energetic ion bombardment can increase chemical reaction rates at the surface, clear inhibitor films from the surface, and physically sputter materials from or implant ions into the surface.

T_e is generally less than the threshold energies $\mathcal{E}_{diss}$ or $\mathcal{E}_{iz}$ for dissociation and ionization of the feedstock gas molecules. Nevertheless, dissociation and ionization occur because electrons have a distribution of energies. Let $g_e(\mathcal{E})\,d\mathcal{E}$ be the number of electrons per unit volume with energies lying between $\mathcal{E}$ and $\mathcal{E} + d\mathcal{E}$, then the distribution function $g_e(\mathcal{E})$ is sketched in Figure 1.10. Electrons having energies below $\mathcal{E}_{diss}$ or $\mathcal{E}_{iz}$ cannot dissociate or ionize the gas. We see that dissociation and ionization are produced by the high-energy tail of the distribution. Although the distribution is sketched in the figure as if it were Maxwellian at the bulk electron temperature T_e, this may

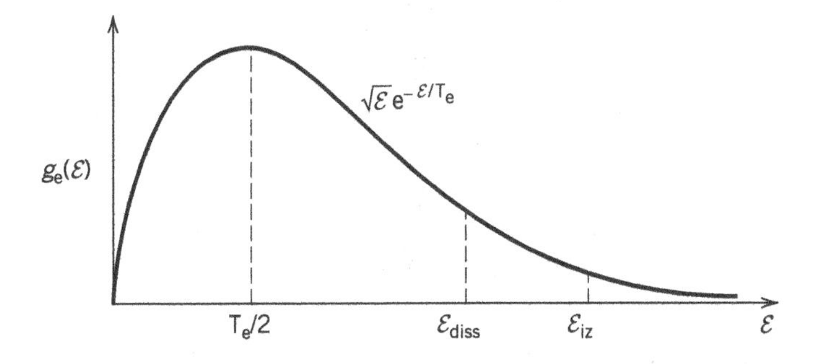

Figure 1.10 Electron energy distribution function $g_e(\mathcal{E})$ in a weakly ionized discharge.

not be the case. The tail distribution might be depressed below or enhanced above a Maxwellian by electron heating and electron–neutral collision processes. Two temperature distributions are sometimes observed, with T_e for the bulk electrons lower than T_h for the energetic electron tail. Non-Maxwellian distributions can only be described using the kinetic theory of discharges, which we introduce in Chapter 19.

1.2.2 Sheaths

Plasmas, which are quasineutral ($n_i \approx n_e$), are joined to wall surfaces across thin positively charged layers called *sheaths*. To see why, first note that the electron thermal velocity $(eT_e/m)^{1/2}$ is at least 100 times the ion thermal velocity $(eT_i/M)^{1/2}$ because $m/M \ll 1$ and $T_e \gtrsim T_i$. (Here, T_e and T_i are given in units of volts.) Consider a plasma of width l with $n_e = n_i$ initially confined between two grounded ($\Phi = 0$) absorbing walls (Figure 1.11*a*). Because the net charge density $\rho = e(n_i - n_e)$ is zero, the electric potential Φ and the electric field E_x are zero everywhere. Hence, the fast-moving electrons are not confined and will rapidly be lost to the walls. On a very short timescale, however, some electrons near the walls are lost, leading to the situation shown in Figure 1.11*b*. Thin ($s \ll l$) positive ion sheaths form near each wall in which $n_i \gg n_e$. The net positive ρ within the sheaths leads to a potential profile $\Phi(x)$ that is positive within the plasma and falls sharply to zero near both walls. This acts as a confining potential "valley" for electrons and a "hill" for ions because the electric fields within the sheaths point from the plasma to the wall. Thus, the force $-eE_x$ acting on electrons is directed into the plasma; this reflects electrons traveling toward the walls back into the plasma. Conversely, ions from the plasma that enter the sheaths are accelerated into the walls. If the plasma potential (with respect to the walls) is V_p, then we expect that $V_p \sim$ a few T_e in order to confine most of the electrons. The energy of ions bombarding the walls is then $\mathcal{E}_i \sim$ a few T_e. Charge uncovering is treated quantitatively in Chapter 2 and sheaths in Chapter 6.

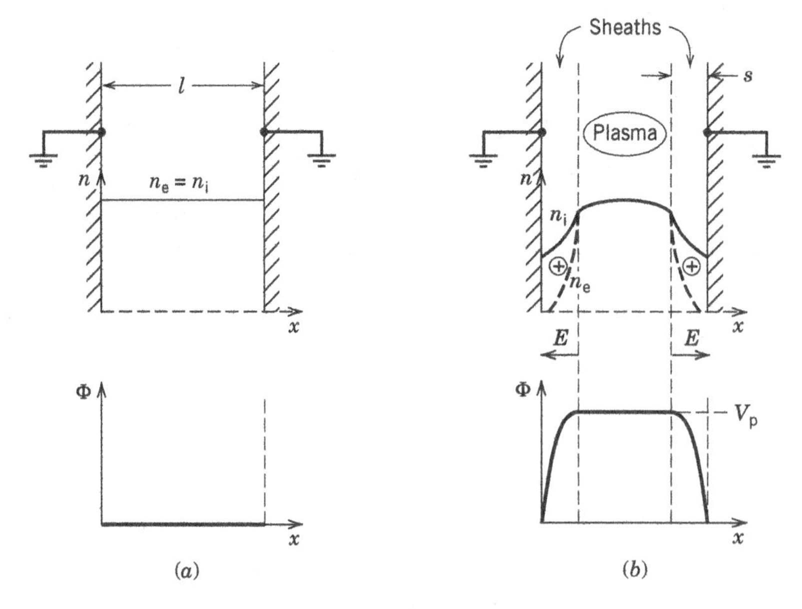

Figure 1.11 The formation of plasma sheaths: (*a*) initial ion and electron densities and potential; (*b*) densities, electric field, and potential after the formation of the sheath.

Figure 1.12 shows sheath formation as obtained from a particle-in-cell (PIC) plasma simulation. We use PIC results throughout this book to illustrate various discharge phenomena. In this simulation, the left wall is grounded, the right wall is floating (zero net current), and the positive ion density is uniform and constant in time. The electrons are modeled as $\mathcal{N}$ sheets having charge-to-mass ratio $-e/m$ that move in one dimension (along x) under the action of the time-varying fields produced by all the other sheets, the fixed ion charge density, and the charges on the walls. Electrons do not collide with other electrons, ions, or neutrals in this simulation. Four thousand sheets were used with $T_e = 1$ V and $n_i = n_e = 10^{13}$ m^{-3} at time $t = 0$. In (*a*), (*b*), (*c*), and (*d*), we see the v_x–x electron phase space, electron density, electric field, and potential after the sheath has formed, at $t = 0.77$ μs. The time history of $\mathcal{N}$ is shown in (*e*); 40 sheets have been lost to form the sheaths. Figure 1.12*a*–*d* shows the absence of electrons near each wall over a sheath width $s \approx 6$ mm. Except for fluctuations due to the finite $\mathcal{N}$, the field in the bulk plasma is near zero, and the fields in the sheaths are large and point from the plasma to the walls. (E_x is negative at the left wall and positive at the right wall to repel plasma electrons.) The potential in the center of the discharge is $V_p \approx 2.5$ V and falls to zero at the left wall (this wall is grounded by definition). The potential at the right wall is also low, but we see in (*f*) that it oscillates in time. We will see in Chapter 4 that these are *plasma oscillations*. We would not see them if the initial sheet positions and velocities were chosen exactly symmetrically about the midplane, or if many more sheets were used in the simulation.

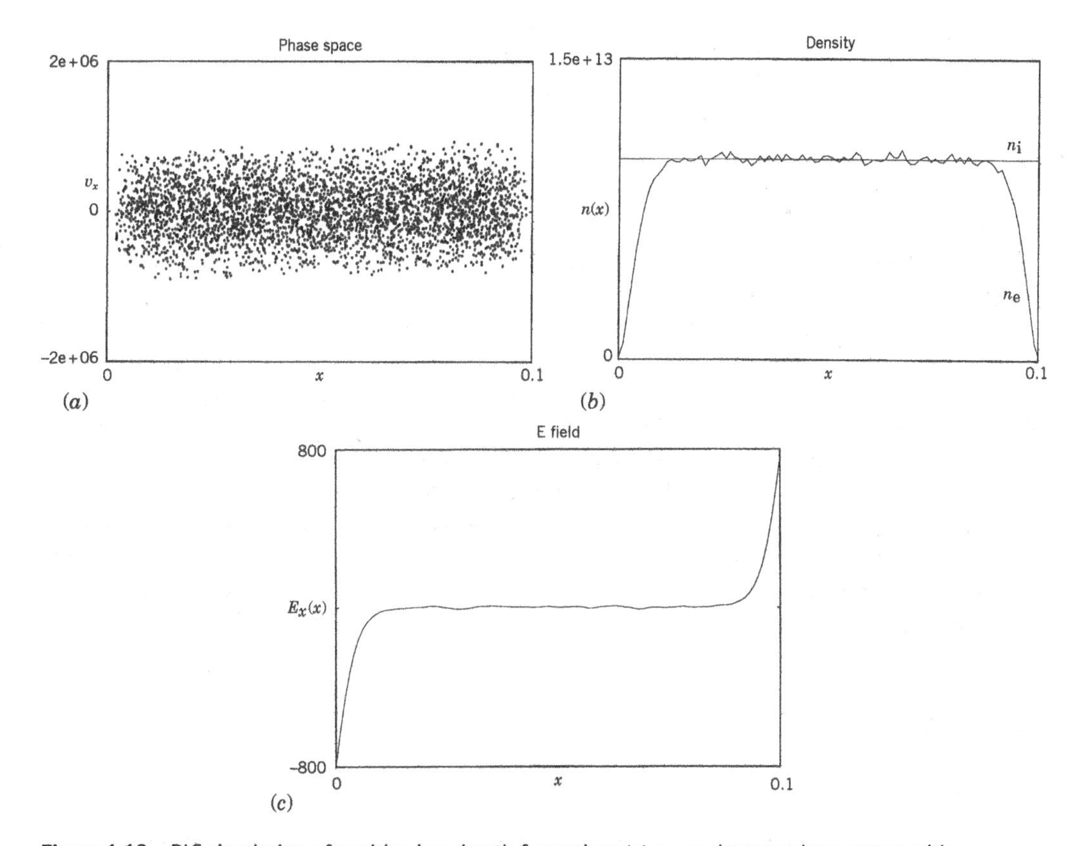

Figure 1.12 PIC simulation of positive ion sheath formation: (*a*) v_x–x electron phase space, with horizontal scale in meters; (*b*) electron density n_e; (*c*) electric field E_x; (*d*) potential Φ; (*e*) electron number $\mathcal{N}$ versus time t in seconds; (*f*) right hand potential v_x versus time t.

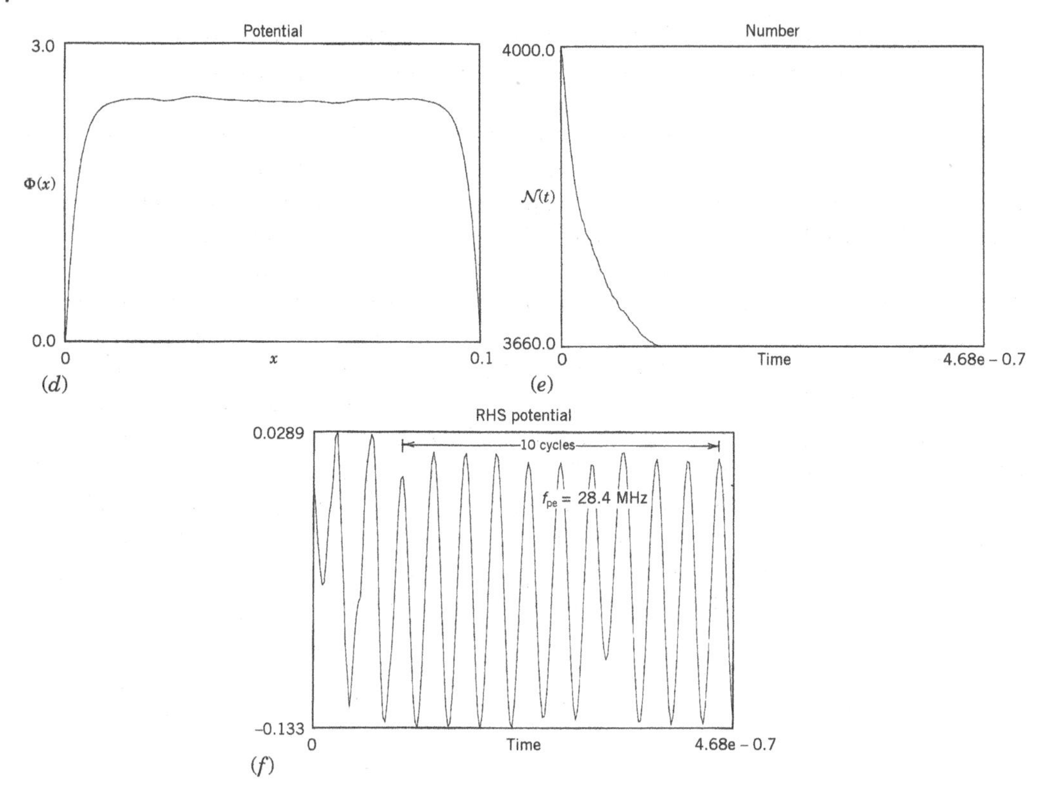

Figure 1.12 *(Continued)*

If the ions were also modeled as moving sheets, then on a longer timescale, we would see ion acceleration within the sheaths, and a consequent drop in ion density near the walls, as sketched in Figure 1.11*b*. We return to this in Chapter 6.

The separation of discharges into bulk plasma and sheath regions is an important paradigm that applies to all discharges. The bulk region is quasineutral, and both instantaneous and time-averaged fields are low. The bulk plasma dynamics are described by diffusive ion loss at high pressures and by free-fall ion loss at low pressures. In the positive space charge sheaths, high fields exist, leading to dynamics that are described by various ion space charge sheath laws, including low-voltage sheaths and various high-voltage sheath models, such as collisionless and collisional Child laws and their modifications. The plasma and sheath dynamics must be joined at their interface. As will be seen in Chapter 6, the usual joining condition is to require that the mean ion velocity at the plasma–sheath edge be equal to the ion-sound (Bohm) velocity: $u_B = (e\mathrm{T}_e/M)^{1/2}$, where e and M are the charge and mass of the ion and T_e is the electron temperature in volts.

1.3 Discharges

1.3.1 RF Diodes

Planar capacitively driven radio frequency (rf) discharges—so-called *rf diodes*—are a mainstay for materials processing. Early planar systems, as shown in Figure 1.13, were typically multi-wafer and used for anisotropic etches. This idealized discharge consists of a vacuum chamber containing two

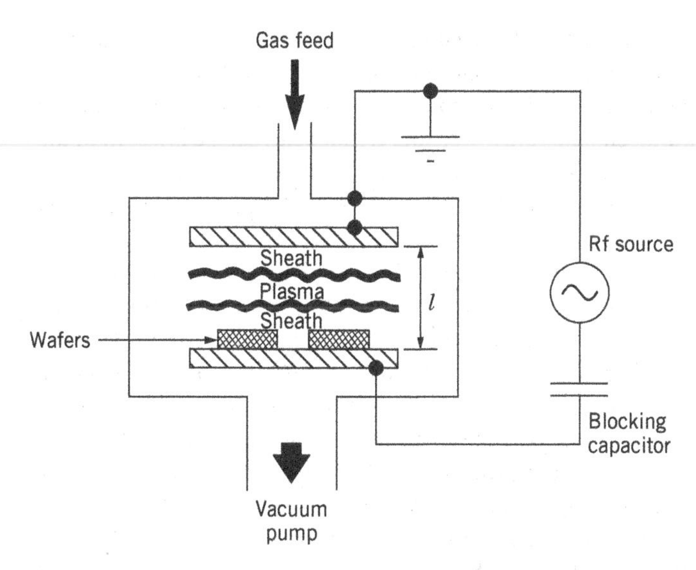

Figure 1.13 Typical multi-wafer capacitive rf discharge in plane-parallel geometry, used for anisotropic etching. Source: Lieberman and Gottscho (1994)/with permission of Elsevier.

Table 1.1 Typical Range of Parameters for RF Diode and High-Density Discharges.

Parameter	RF Diode	High-Density Source
Pressure p (mTorr)	3–3000	0.5–50
Power P (kW)	0.5–20	0.5–5
Frequency f (MHz)	0.05–60	2–2450
Cross-sectional area A (m^2)	0.1–10	0.1–1
Magnetic field B (kG)	0	0–1
Plasma density n (cm^{-3})	10^9–10^{11}	10^{10}–10^{12}
Electron temperature T_e (V)	1–5	2–7
Ion acceleration energy $\mathcal{E}_i$ (V)	200–1000	20–1000
Fractional ionization x_{iz}	10^{-6}–10^{-3}	10^{-4}–10^{-1}

Source: M. A. Lieberman, 2Ed modified

planar electrodes separated by a gap width l and driven by an rf power source. The substrates are placed on one electrode, feedstock gases are admitted to flow through the discharge, and effluent gases are removed by the vacuum pump. Typical parameters are shown in Table 1.1. The typical rf driving voltage is $V_{rf} = 100$–1000 V, and the plate separation is $l = 2$–10 cm. When operated at low pressure, with the wafer mounted on the powered electrode, and used to remove substrate material, such reactors are sometimes called reactive ion etchers (RIEs)—a misnomer, since the etching is a chemical process enhanced by energetic ion bombardment of the substrate, rather than a removal process due to reactive ions alone.

For anisotropic etching, typically pressures are in the range of 3–100 mTorr, powers per unit area are 3–30 kW/m^2, and driving frequencies are 13.56–60 MHz. Typical plasma densities are relatively low, 10^9–10^{11} cm^{-3}, and the electron temperature is of order 3 V. Ion acceleration energies (sheath voltages) are high, greater than 200 V, and fractional ionization is low. The degree of dissociation of the molecules into reactive species can range widely from less than 0.1% to nearly 100% depending on gas composition and plasma conditions. For deposition and isotropic etch applications, pressures tend to be higher, ion bombarding energies are lower, and frequencies can be lower than the commonly used standard of 13.56 MHz. The very large cross-sectional areas and powers shown for rf diodes in Table 1.1 are for processing large-area glass substrates for solar panels and flat panel displays.

The operation of capacitively driven discharges is reasonably well understood. As shown in Figure 1.14 for a symmetrically driven discharge, the mobile plasma electrons, responding to the instantaneous electric fields produced by the rf driving voltage, oscillate back and forth within the positive space charge cloud of the ions. The massive ions respond only to the time-averaged electric fields. Oscillation of the electron cloud creates sheath regions near each electrode that contains net positive charge when averaged over an oscillation period; i.e., the positive charge exceeds the negative charge in the system, with the excess appearing within the sheaths. This excess produces a strong time-averaged electric field within each sheath directed from the plasma to the electrode. Ions flowing out of the bulk plasma near the center of the discharge can be accelerated by the sheath fields to high energies as they flow to the substrate, leading to energetic-ion-enhanced processes. Typical ion-bombarding energies $\mathcal{E}_i$ can be as high as $V_{rf}/2$ for symmetric systems

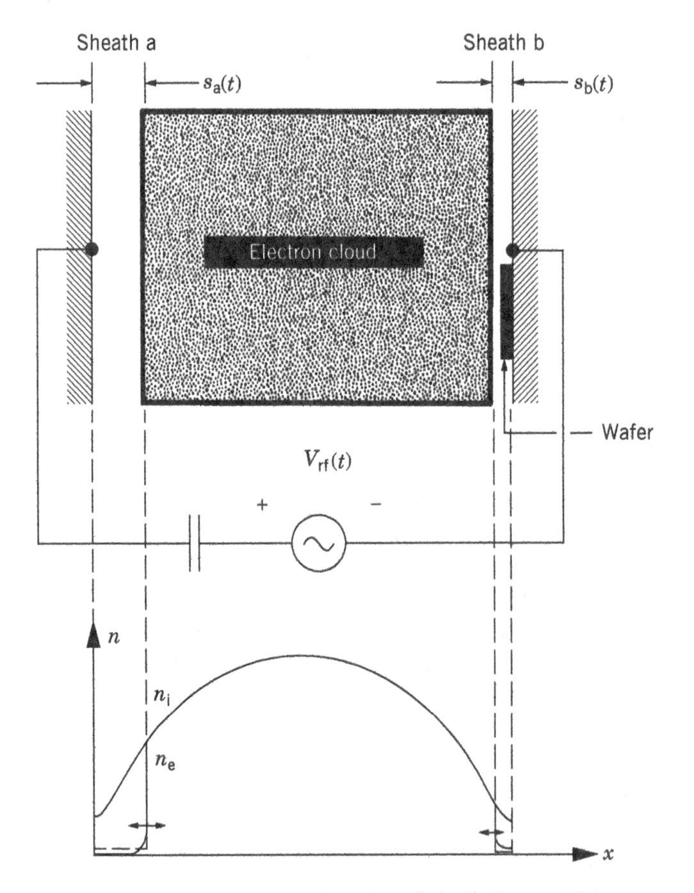

Figure 1.14 The physical model of an rf diode. Source: Lieberman and Gottscho (1994)/with permission of Elsevier.

(Figure 1.14) and as high as V_{rf} at the powered electrode for asymmetric systems, which have a powered electrode area greater than the area of the grounded electrode. A quantitative description of capacitive discharges is given in Chapter 11.

We note that the positive ions continuously bombard the electrode over an rf cycle. In contrast, electrons are lost to the electrode only when the oscillating cloud closely approaches the electrode. During that time, the instantaneous sheath potential collapses to near zero, allowing sufficient electrons to escape to balance the ion charge delivered to the electrode. Except for such brief moments, the instantaneous potential of the discharge must always be positive with respect to any large electrode and wall surface; otherwise, the mobile electrons would quickly leak out. Electron confinement is ensured by the presence of positive space charge sheaths near all surfaces.

We will see that a crucial limiting feature of rf diodes is that the ion-bombarding flux $\Gamma_{\text{i}} = nu_{\text{B}}$ and bombarding energy $\mathcal{E}_{\text{i}}$ cannot be varied independently. The situation is analogous to the lack of independent voltage and current control in diode vacuum tubes or semiconductor pn junctions. For a reasonable (but relatively low) ion flux, as well as a reasonable dissociation of the feedstock gas, sheath voltages at the driven electrode are high. For wafers placed on the driven electrode, this can result in undesirable damage, or loss of linewidth control. Furthermore, the combination of low ion flux and high ion energy leads to a relatively narrow process window for many applications. The low process rates resulting from the limited ion flux in rf diodes often mandate multi-wafer or batch processing, with consequent loss of wafer-to-wafer reproducibility. Higher ion and neutral fluxes are sometimes required for single-wafer processing in a clustered tool environment, in which a single wafer is moved by a robot through a series of process chambers. Clustered tools are used to control interface quality and offer significant cost savings in fabricating ICs. Finally, low fractional ionization poses a significant problem for processes where the feedstock costs and disposal of effluents are issues.

As shown in Figure 1.15*a*, modern capacitive discharges used for processing are almost always single-wafer systems. Furthermore, to meet the linewidth, selectivity, and damage control demands for critical etching and deposition steps, the mean ion bombarding energy, and perhaps its energy distribution, must be controllable independently of the ion and neutral fluxes. Some control over ion bombarding energy can be achieved by putting the wafer on the undriven electrode and independently biasing this electrode with respect to the chamber walls with a second rf source. However, processing rates are still low at low pressures and sputtering contamination is an issue. A more successful approach, shown in Figure 1.15*b*, is *dual frequency* operation, in which a high- and a low-frequency rf source are used to drive one or both plates of an rf diode. For well-separated frequencies, the high frequency mainly controls the ion flux and the low frequency controls the ion bombarding energy. Using a frequency higher than the conventional frequency of 13.56 MHz for the high-frequency drive results in an increased ion flux to the substrate for a fixed power input and allows the low-frequency drive to better control the ion energy. High frequencies of 27.12, 60, or 160 MHz, and low frequencies of 2 or 13.56 MHz, are used commercially.

Various magnetically enhanced rf diodes and triodes have also been developed to improve the performance of capacitive rf reactors. These include, for example, magnetically enhanced reactive ion etchers (MERIEs), shown in Figure 1.15*c*, in which a controllable dc magnetic field of 50–300 G is applied parallel to the powered electrode, on which the wafer sits. The magnetic field increases the efficiency of power transfer from the source to the plasma and also enhances plasma confinement. This results in a reduced sheath voltage and an increased plasma density when the magnetic field is applied. However, the plasma generated can be significantly nonuniform both radially and azimuthally. To increase process uniformity (at least azimuthally), the magnetic field is slowly rotated in the plane of the wafer, e.g., at a frequency of 0.5 Hz.

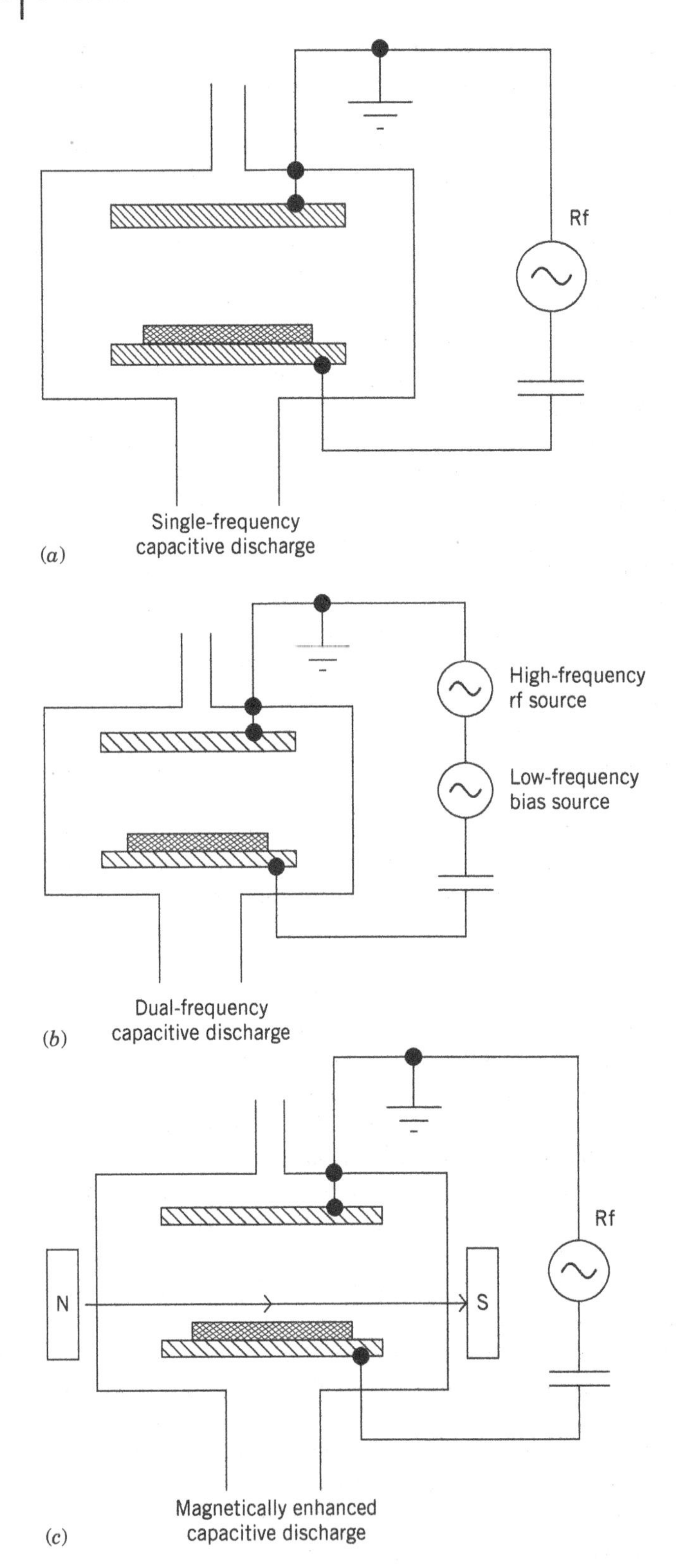

Figure 1.15 Some modern capacitive discharges are used for etching and deposition; (*a*) single frequency, (*b*) dual frequency, and (*c*) magnetically enhanced.

Low-pressure capacitive discharges are described quantitatively in Chapter 11. At higher pressures, typically ~ 0.2–10 Torr, the basic single-frequency capacitive configuration in Figure 1.15*a* is commonly used for PECVD. Important physical and chemical phenomena appear in this pressure range, including spatially varying electron temperatures and secondary electron emission from surfaces. At the higher pressures, the secondary emission can cause a transition to a new mode of operation in which the secondary electrons alone sustain the discharge, the so-called α–γ transition. Capacitive discharges can also be operated at atmospheric pressures with very narrow gap lengths, $l \sim 0.2$–3 mm, in a diffuse mode with relatively uniform transverse density profiles. We examine this high-pressure capacitive discharge regime quantitatively in Chapter 15.

1.3.2 High-Density Sources

The limitations of rf diodes and their magnetically enhanced variants have led to the development of new kinds of low-pressure, high-density plasma sources. In addition to dual-frequency capacitive discharges, three examples of non-capacitive discharges are shown schematically in Figure 1.16, and typical source and plasma parameters are given in Table 1.1. In addition to high density and low pressure, a common feature is that the rf or microwave power is coupled to the plasma across a dielectric window, rather than by direct connection to an electrode in the plasma, as for an rf diode. This noncapacitive power transfer is the key to achieving low voltages across all plasma sheaths at electrode and wall surfaces. Dc voltages, and, hence, ion acceleration energies, are then

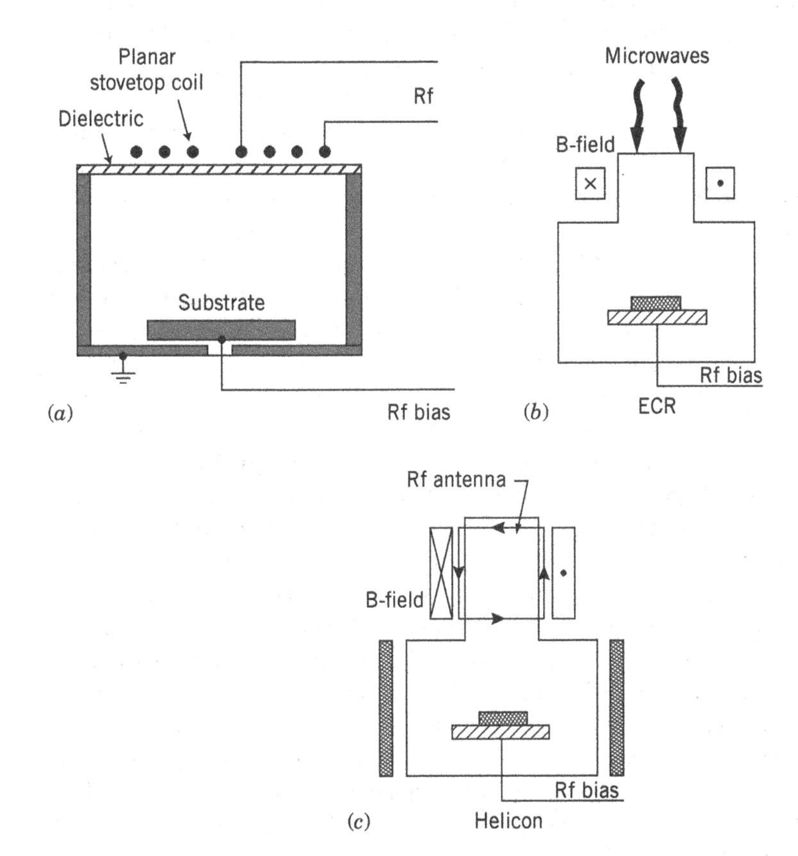

Figure 1.16 Some non-capacitive, high-density discharges used for etching and deposition; (*a*) planar inductive; (*b*) electron cyclotron resonance; (*c*) helicon.

typically 20–30 V at all surfaces. To control the ion energy, the electrode on which the wafer is placed is independently driven by a capacitively coupled rf source. Hence, independent control of the ion/radical fluxes (through the source power) and the ion-bombarding energy (through the wafer electrode power) is possible.

The common features of power transfer across dielectric windows and separate bias supply at the wafer electrode are illustrated in Figure 1.16. However, sources differ significantly in the means by which power is coupled to the plasma. A planar inductive (or transformer) coupled source is shown in Figure 1.16*a*. Here, the plasma acts as a single-turn, lossy conductor that is coupled to a multi-turn nonresonant rf coil across the dielectric discharge chamber; rf power is inductively coupled to the plasma by transformer action. We describe inductive discharges quantitatively in Chapter 12.

For the electron cyclotron resonance (ECR) source shown in Figure 1.16*b*, one or more electromagnet coils surrounding the cylindrical source chamber generate an axially varying dc magnetic field. Microwave power is injected axially through a dielectric window into the source plasma, where it excites a right-hand circularly polarized wave that propagates to a resonance zone, for cold electrons at $\omega = \omega_{ce}$, where the wave is absorbed. Here, $\omega = 2\pi f$ is the applied radian frequency and $\omega_{ce} = eB/m$ is the electron gyration frequency at resonance. For the typical microwave frequency used, $f = 2450$ MHz, the resonant magnetic field is $B \approx 875$ G. The plasma streams into the process chamber in which the wafer is located.

A helicon source is shown in Figure 1.16*c*. A weak (50–200 G) dc axial magnetic field together with an rf-driven antenna placed around the dielectric cylinder that forms the source chamber allows excitation of a helicon wave within the source plasma. Resonant wave–particle interaction is believed to transfer the wave energy to the plasma. ECR and helicon sources are described quantitatively in Chapter 13.

Figure 1.16 also illustrates the use of high-density sources to feed plasma into a relatively distinct, separate process chamber in which the wafer is located. As shown in Figure 1.16*c*, the process chamber can be surrounded by dc multipole magnetic fields to enhance plasma confinement near the process chamber surfaces, while providing a magnetic near-field-free plasma environment at the wafer. Such configurations are often called "remote" sources, a misnomer since at low pressures considerable plasma and free radical production occurs within the process chamber near the wafer. Sometimes, the source and process chambers are more integral, e.g., the wafer is placed very near to the source exit, to obtain increased ion and radical fluxes, reduced spread in ion energy, and improved process uniformity. But the wafer is then exposed to higher levels of damaging radiation.

Although the need for low pressures, high fluxes, and controllable ion energies has motivated high-density source development, there are many issues that need to be resolved. A critical issue is achieving the required process uniformity over 300-mm wafer diameters. In contrast to the nearly one-dimensional geometry of typical rf diodes (two closely spaced parallel electrodes), high-density cylindrical sources can have length-to-diameter ratios of order or exceeding unity. Plasma formation and transport in such geometries are inherently radially nonuniform. Another critical issue is efficient power transfer (coupling) across dielectric windows over a wide operating range of plasma parameters. Degradation of and deposition on the window can also lead to irreproducible source behavior and the need for frequent, costly cleaning cycles. Low-pressure operation leads to severe pumping requirements for high deposition or etching rates and, hence, to the need for large, expensive vacuum pumps. Furthermore, plasma and radical concentrations become strongly sensitive to reactor surface conditions, leading to problems of reactor aging and process irreproducibility. Finally, dc magnetic fields are required for some source concepts. These can lead to magnetic field-induced process nonuniformities and damage.

In addition to rf and microwave discharges, magnetically enhanced dc discharges are widely used for sputtering metal and dielectric films from a cathode "target" onto a substrate placed on the anode electrode. Sputtering of target atoms onto the substrate is induced by ions that flow across the 500–1000 V cathode sheath. Figure 1.17 illustrates a dc planar magnetron discharge, which is widely used for such depositions and also has important applications for the coating of large rectangular sheets of architectural glass. Such discharges can also be driven by rf and pulsed voltages. Various other types of dc discharges are also used. We examine dc discharges, quantitatively, in Chapter 14.

Figure 1.18 illustrates schematically the central problem of discharge analysis, using the example of an rf diode. Given the *control* parameters for the power source (frequency ω, driving voltage V_{rf}, or absorbed power P_{abs}), the feedstock gas (pressure p, flow rate, and chemical composition), and the geometry (simplified here to the discharge length l), then find the *plasma* parameters, including the plasma density n_i, the etchant density n_F, the ion and etchant fluxes Γ_i and Γ_F hitting the substrate, the electron and ion temperatures T_e and T_i, the ion bombarding energy $\mathcal{E}_i$, and the sheath thickness s. The control parameters are the "knobs" that can be "turned" in order to "tune" the properties of the discharge.

The tuning range for a given discharge is generally limited. Sometimes one type of discharge will not do the job no matter how it is tuned, so another type must be selected. As suggested in Figures 1.15–1.17, a bewildering variety of discharges are used for processing. Some are driven by rf, some by dc, and some by microwave power sources. Some use magnetic fields to increase the plasma confinement or the efficiency of power absorption. One purpose of this book is to guide the reader toward making wise choices when designing discharges used for processing.

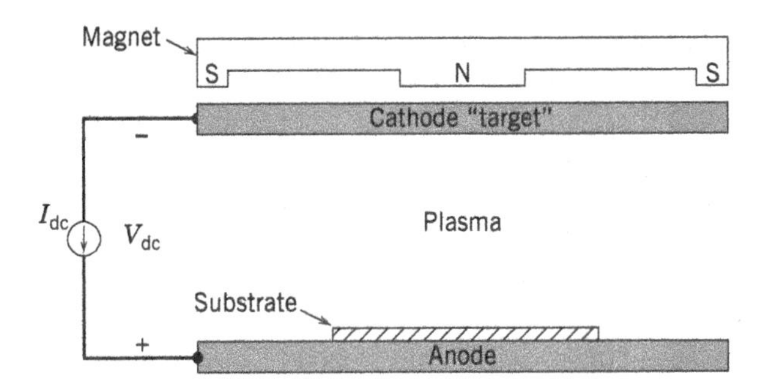

Figure 1.17 A dc planar magnetron discharge, used for thin film deposition.

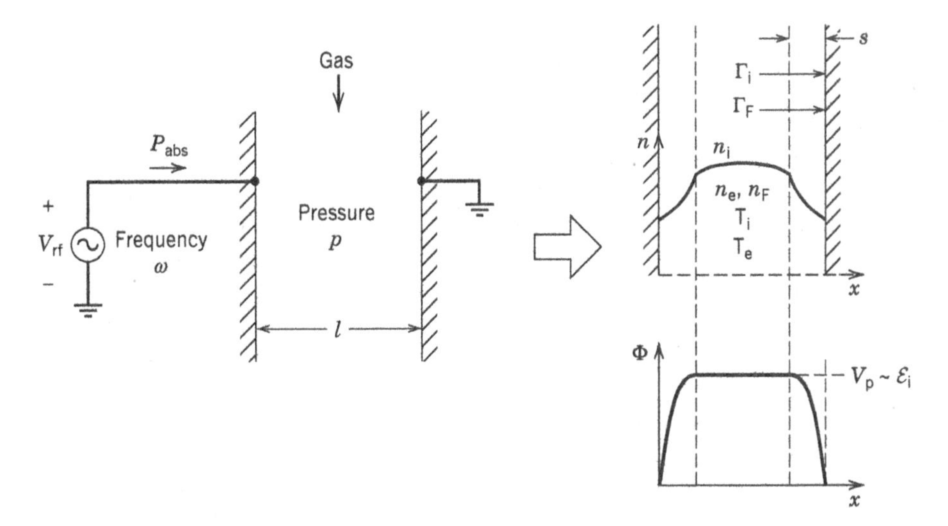

Figure 1.18 The central problem of discharge analysis.

1.4 Symbols and Units

The choice of symbols is always vexing. While various fields each have their consistent set of symbols to represent physical quantities, these overlap between different fields, e.g., plasma physics and gas-phase chemistry. For example, H is not only standard for enthalpy in chemistry but also standard for magnetic field in plasma physics. This also occurs within a given field, e.g., k is standard not only for Boltzmann's constant but also for wave number. Then, there is always the occasional symbol that must stand for many things in different contexts. We sometimes distinguish these by using different lettering (Roman, italic, script, and boldface), e.g., I is a current and I is a modified Bessel function; M is an ion mass and $\mathcal{M}$ is the number of chemical species. We can often distinguish commonly used symbols by the use of subscripts, e.g., σ denotes a cross section, but σ_{rf} and σ_{dc} denote electrical conductivities; we have done this whenever the notation is not too cumbersome. The meaning should be clear from the context, in most cases. To help avoid confusion, we have provided a table of symbols and abbreviations in the front matter of this book. These give the normal usage of symbols and their units.

As far as possible, we use the SI (MKS) system of units: meters (m), kilograms (kg), seconds (s), and coulombs (C) for charge. In these units, the charge on an electron is $-e \approx -1.602 \times 10^{-19}$ C. The unit of energy is the joule (J), but we often use the symbol $\mathcal{E}$ for the voltage that is the equivalent of the energy, i.e.,

$$U(\text{joules}) = e\mathcal{E}$$

where $\mathcal{E}$ is in volts. We also occasionally use the calorie (cal): 1 cal $\approx$ 4.187 J. The SI unit of pressure is the pascal (Pa), but we more commonly give gas pressures in Torr:

$$1\ \text{Torr} \approx 133.3\ \text{Pa}$$

We occasionally use 1 atm $\approx 1.013 \times 10^5$ Pa $\approx$ 760 Torr and 1 bar $= 10^5$ Pa to refer to gas pressures. The SI unit for the magnetic induction B is tesla (T), but we often give B in gauss (G): 1 T $= 10^4$ G. We use the italic typeface symbol T to refer to the temperature in kelvins (K). The energy equivalent temperature in joules is kT, where $k \approx 1.381 \times 10^{-23}$ J/K is Boltzmann's constant. We often use the roman typeface symbol T for the voltage equivalent of the temperature, where

$$e\mathrm{T}(\text{volts}) = kT(\text{kelvins})$$

Hence, room temperature $T = 297$ K is equivalent to $\mathrm{T} \approx 0.026$ V. Even within the standard unit system, quantities are often designated by subunits. For example, cross sections are often given in cm^2 rather than m^2 in tables, and wavelengths at microwave frequencies are commonly given in cm rather than in meters.

To assist our readers in making calculations, we give the commonly used constants in the SI system of units and the most common conversions between units in the front matter of the book. It is sometimes tempting to make a calculation in nonstandard units. For example, the collision frequency $\nu = n\sigma v$, which has units $(\text{m}^{-3} \cdot \text{m}^2 \cdot \text{m}\,\text{s}^{-1})$, could equally well be calculated in the commonly used units $(\text{cm}^{-3} \cdot \text{cm}^2 \cdot \text{cm}\,\text{s}^{-1})$, since the length units cancel. However, we urge the student not to take such shortcuts, but to systematically convert to standard units, before making a calculation.

2

Basic Plasma Equations and Equilibrium

2.1 Introduction

The plasma medium is complicated in that the charged particles are affected by both external electric and magnetic fields and contribute to them. The resulting self-consistent system is nonlinear and very difficult to analyze. Furthermore, the interparticle collisions, although also electromagnetic in character, occur on space and timescales that are usually much shorter than those of the applied fields or the fields due to the average motion of the particles.

To make progress with such a complicated system, various simplifying approximations are needed. The interparticle collisions are considered independently of the larger-scale fields to determine an *equilibrium distribution* of the charged-particle velocities. The velocity distribution is averaged over velocities to obtain the *macroscopic motion*. The macroscopic motion takes place in externally applied fields and the macroscopic fields generated by the average particle motion. These self-consistent fields are nonlinear, but may be linearized in some situations, particularly when dealing with waves in plasmas. The effect of spatial variation of the distribution function leads to pressure forces in the macroscopic equations. The collisions manifest themselves in particle generation and loss processes, as an average friction force between different particle species, and in energy exchanges among species. In this chapter, we consider the basic equations that govern the plasma medium, concentrating attention on the macroscopic system. The complete derivation of these equations, from fundamental principles, is beyond the scope of the text. We shall make the equations plausible and, in the easier instances, supply some derivations in appendices. For the reader interested in more rigorous treatment, references to the literature will be given.

In Section 2.2, we introduce the macroscopic field equations and the current and voltage. In Section 2.3, we introduce the fundamental equation of plasma physics, for the evolution of the particle distribution function, in a form most applicable for weakly ionized plasmas. We then define the macroscopic quantities and indicate how the macroscopic equations are obtained by taking moments of the fundamental equation. References given in the text can be consulted for more details of the averaging procedure. Although the macroscopic equations depend on the equilibrium distribution, their form is independent of the equilibrium. To solve the equations for particular problems, the equilibrium must be known. In Section 2.4, we introduce the equilibrium distribution and obtain some consequences arising from it and from the field equations. The form of the equilibrium distribution will be shown to be a consequence of the interparticle collisions in Appendix B.

Principles of Plasma Discharges and Materials Processing, Third Edition. Michael A. Lieberman and Allan J. Lichtenberg.

2.2 Field Equations, Current, and Voltage

2.2.1 Maxwell's Equations

The usual macroscopic form of Maxwell's equations is

$$\nabla \times \mathbf{E} = -\mu_0 \frac{\partial \mathbf{H}}{\partial t} \tag{2.2.1}$$

$$\nabla \times \mathbf{H} = \mathbf{J} + \epsilon_0 \frac{\partial \mathbf{E}}{\partial t} \tag{2.2.2}$$

$$\epsilon_0 \nabla \cdot \mathbf{E} = \rho \tag{2.2.3}$$

and

$$\mu_0 \nabla \cdot \mathbf{H} = 0 \tag{2.2.4}$$

where $\mathbf{E}(\mathbf{r}, t)$ and $\mathbf{H}(\mathbf{r}, t)$ are the electric and magnetic field vectors and where $\mu_0 = 4\pi \times 10^{-7}$ H/m and $\epsilon_0 \approx 8.854 \times 10^{-12}$ F/m are the permeability and permittivity of free space. The sources of the fields, the charge density $\rho(\mathbf{r}, t)$ and the current density $\mathbf{J}(\mathbf{r}, t)$, are related by the charge continuity equation (Problem 2.1):

$$\frac{\partial \rho}{\partial t} + \nabla \cdot \mathbf{J} = 0 \tag{2.2.5}$$

In general,

$$\mathbf{J} = \mathbf{J}_{\text{cond}} + \mathbf{J}_{\text{pol}}$$

where the conduction current density $\mathbf{J}_{\text{cond}}$ is due to the motion of the free charges, and the polarization current density $\mathbf{J}_{\text{pol}}$ is due to the motion of bound charges in a dielectric material. In a plasma in vacuum, $\mathbf{J}_{\text{pol}}$ is zero and $\mathbf{J} = \mathbf{J}_{\text{cond}}$.

If (2.2.3) is integrated over a volume $\mathcal{V}$, enclosed by a surface $\mathcal{S}$, then we obtain its integral form, Gauss's law:

$$\epsilon_0 \oint_{\mathcal{S}} \mathbf{E} \cdot \mathrm{d}\mathbf{A} = q \tag{2.2.6}$$

where q is the total charge inside the volume. Similarly, by integrating (2.2.5), we obtain

$$\frac{\mathrm{d}q}{\mathrm{d}t} + \oint_{\mathcal{S}} \mathbf{J} \cdot \mathrm{d}\mathbf{A} = 0$$

which states that the rate of increase of charge inside $\mathcal{V}$ is supplied by the total current flowing across $\mathcal{S}$ into $\mathcal{V}$, i.e., that charge is conserved.

In (2.2.2), the first term on the RHS is the displacement current density flowing in the vacuum, and the second term is the conduction current density due to the free charges. We can introduce the total current density

$$\mathbf{J}_{\text{T}} = \mathbf{J} + \epsilon_0 \frac{\partial \mathbf{E}}{\partial t} \tag{2.2.7}$$

and taking the divergence of (2.2.2), we see that

$$\nabla \cdot \mathbf{J}_{\text{T}} = 0 \tag{2.2.8}$$

In one dimension, this reduces to $\mathrm{d}J_{\text{T}x}/\mathrm{d}x = 0$, such that $J_{\text{T}x} = J_{\text{T}x}(t)$, independent of x. Hence, for example, the total current flowing across a spatially nonuniform one-dimensional discharge

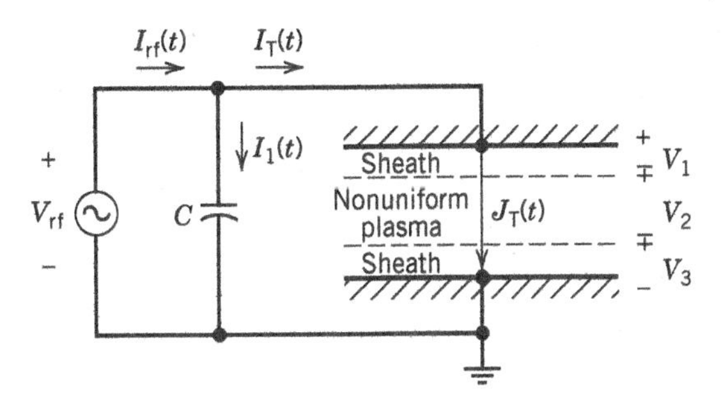

Figure 2.1 Kirchhoff's circuit laws: the total current J_T flowing across a nonuniform one-dimensional discharge is independent of x; the sum of the currents entering a node is zero ($I_{rf} = I_T + I_1$); the sum of voltages around a loop is zero ($V_{rf} = V_1 + V_2 + V_3$).

is independent of x, as illustrated in Figure 2.1. A generalization of this result is *Kirchhoff's current law*, which states that the sum of the currents entering a *node*, where many current-carrying conductors meet, is zero. This is also shown in Figure 2.1, where $I_{rf} = I_T + I_1$.

If the time variation of the magnetic field is negligible, as is often the case in plasmas, then from Maxwell's equations $\nabla \times \mathbf{E} \approx 0$. Since the curl of a gradient is zero, this implies that the electric field can be derived from the gradient of a scalar potential,

$$\mathbf{E} = -\nabla\Phi \tag{2.2.9}$$

Integrating (2.2.9) around any closed loop C gives

$$\oint_C \mathbf{E} \cdot d\ell = -\oint_C \nabla\Phi \cdot d\ell = -\oint_C d\Phi = 0 \tag{2.2.10}$$

Hence, we obtain *Kirchhoff's voltage law*, which states that the sum of the voltages around any loop is zero. This is illustrated in Figure 2.1, for which we obtain

$$V_{rf} = V_1 + V_2 + V_3$$

i.e., the source voltage V_{rf} is equal to the sum of the voltages V_1 and V_3 across the two sheaths and the voltage V_2 across the bulk plasma. Note that currents and voltages can have positive or negative values; the directions for which their values are designated as positive must be specified, as shown in the figure.

If (2.2.9) is substituted into (2.2.3), we obtain

$$\nabla^2\Phi = -\frac{\rho}{\epsilon_0} \tag{2.2.11}$$

Equation (2.2.11), *Poisson's equation*, is one of the fundamental equations that we shall use. As an example of its application, consider the potential in the center ($x = 0$) of two grounded ($\Phi = 0$) plates separated by a distance $l = 10$ cm and containing a uniform ion density $n_i = 10^{10}$ cm^{-3}, without the presence of neutralizing electrons. Integrating Poisson's equation

$$\frac{d^2\Phi}{dx^2} = -\frac{en_i}{\epsilon_0}$$

using the boundary conditions that $\Phi = 0$ at $x = \pm l/2$ and that $d\Phi/dx = 0$ at $x = 0$ (by symmetry), we obtain

$$\Phi = \frac{1}{2}\frac{en_i}{\epsilon_0}\left[\left(\frac{l}{2}\right)^2 - x^2\right]$$

The maximum potential in the center is 2.3×10^5 V, which is impossibly large for a real discharge. Hence, the ions must be mostly neutralized by electrons, leading to a quasineutral plasma.

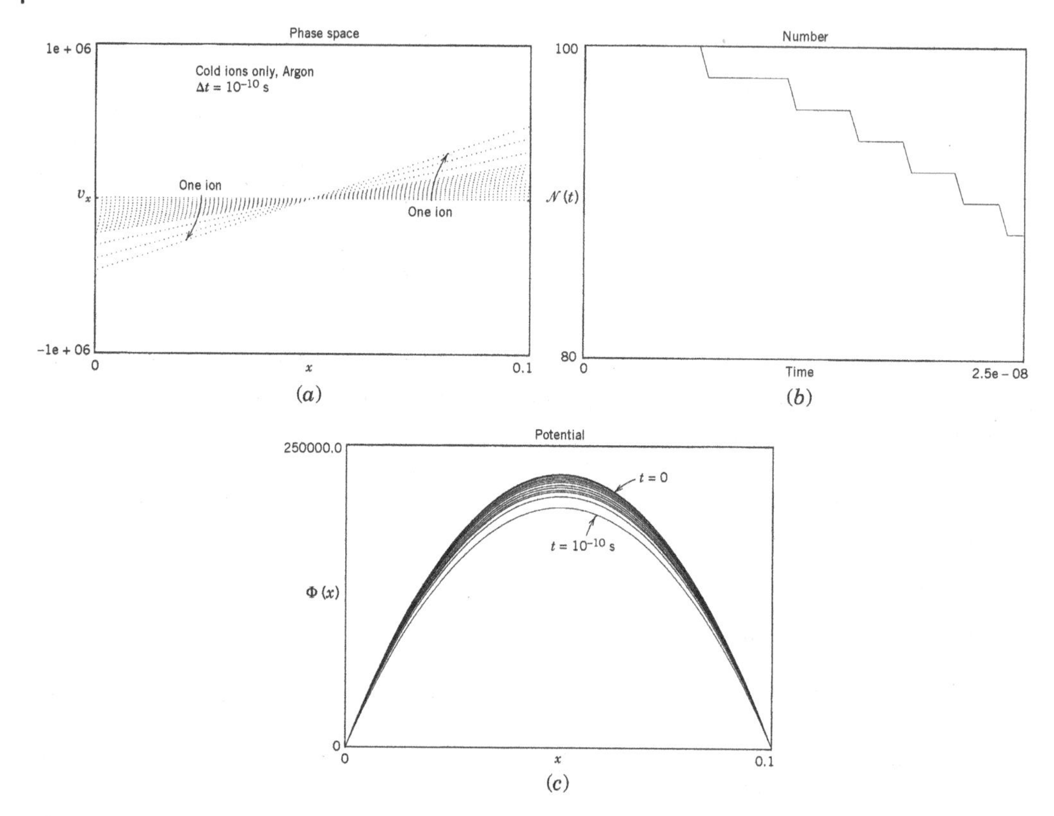

Figure 2.2 PIC simulation of ion loss in a plasma containing ions only: (*a*) v_x–x ion phase space, showing the ion acceleration trajectories; (*b*) number $\mathcal{N}$ of ion sheets versus t, with the steps indicating the loss of a single sheet; (*c*) the potential Φ versus x during the first 10^{-10} s of ion loss.

Figure 2.2 shows a PIC simulation time history over 10^{-10} s of (*a*) the v_x–x phase space, (*b*) the number $\mathcal{N}$ of sheets versus time, and (*c*) the potential Φ versus x for 100 unneutralized ion sheets (with e/M for argon ions). We see the ion acceleration in (*a*), the loss of ions in (*b*), and the parabolic potential profile in (*c*); the maximum potential decreases as ions are lost from the system. We consider quasineutrality further in Section 2.4.

Electric and magnetic fields exert forces on charged particles given by the *Lorentz force law*:

$$\mathbf{F} = q(\mathbf{E} + \mathbf{v} \times \mathbf{B}) \tag{2.2.12}$$

where $\mathbf{v}$ is the particle velocity and $\mathbf{B} = \mu_0 \mathbf{H}$ is the *magnetic induction vector*. The charged particles move under the action of the Lorentz force. The moving charges in turn contribute to both ρ and $\mathbf{J}$ in the plasma. If ρ and $\mathbf{J}$ are linearly related to $\mathbf{E}$ and $\mathbf{B}$, then the field equations are linear. As we shall see, this is not generally the case for a plasma. Nevertheless, linearization may be possible in some cases for which the plasma may be considered to have an *effective dielectric constant*, that is, the "free charges" play the same role as "bound charges" in a dielectric. We consider this further in Chapter 4.

2.3 The Conservation Equations

2.3.1 Boltzmann's Equation

For a given species, we introduce a *distribution function* $f(\mathbf{r}, \mathbf{v}, t)$ in the six-dimensional *phase space* $(\mathbf{r}, \mathbf{v})$ of particle positions and velocities, with the interpretation that

$$f(\mathbf{r}, \mathbf{v}, t)\,\mathrm{d}^3r\,\mathrm{d}^3v = \text{number of particles inside a six-dimensional phase space volume } \mathrm{d}^3r\,\mathrm{d}^3v \text{ at } (\mathbf{r}, \mathbf{v}) \text{ at time } t$$

The six coordinates $(\mathbf{r}, \mathbf{v})$ are considered to be independent variables. We illustrate the definition of f and its phase space in one dimension in Figure 2.3. As particles drift in phase space or move under the action of macroscopic forces, they flow into or out of the fixed volume $\mathrm{d}x\,\mathrm{d}v_x$. Hence, the distribution function f should obey a continuity equation which can be derived as follows. In a time $\mathrm{d}t$,

$f(x, v_x, t)\,\mathrm{d}x\,a_x(x, v_x, t)\,\mathrm{d}t$ particles flow into $\mathrm{d}x\,\mathrm{d}v_x$ across face 1

$f(x, v_x + \mathrm{d}v_x, t)\,\mathrm{d}x\,a_x(x, v_x + \mathrm{d}v_x, t)\,\mathrm{d}t$ particles flow out of $\mathrm{d}x\,\mathrm{d}v_x$ across face 2

$f(x, v_x, t)\,\mathrm{d}v_x\,v_x\,\mathrm{d}t$ particles flow into $\mathrm{d}x\,\mathrm{d}v_x$ across face 3

$f(x + \mathrm{d}x, v_x, t)\,\mathrm{d}v_x\,v_x\,\mathrm{d}t$ particles flow out of $\mathrm{d}x\,\mathrm{d}v_x$ across face 4

where $a_x \equiv \mathrm{d}v_x/\mathrm{d}t$ and $v_x \equiv \mathrm{d}x/\mathrm{d}t$ are the flow velocities in the v_x and x directions, respectively. Hence,

$$\begin{aligned}
&f(x, v_x, t + \mathrm{d}t)\,\mathrm{d}x\,\mathrm{d}v_x - f(x, v_x, t)\,\mathrm{d}x\,\mathrm{d}v_x \\
&\quad = [f(x, v_x, t)a_x(x, v_x, t) - f(x, v_x + \mathrm{d}v_x, t)a_x(x, v_x + \mathrm{d}v_x, t)]\,\mathrm{d}x\,\mathrm{d}t \\
&\quad\quad + [f(x, v_x, t)v_x - f(x + \mathrm{d}x, v_x, t)v_x]\,\mathrm{d}v_x\,\mathrm{d}t
\end{aligned}$$

Dividing by $\mathrm{d}x\,\mathrm{d}v_x\,\mathrm{d}t$, we obtain

$$\frac{\partial f}{\partial t} = -\frac{\partial}{\partial x}(f v_x) - \frac{\partial}{\partial v_x}(f a_x) \tag{2.3.1}$$

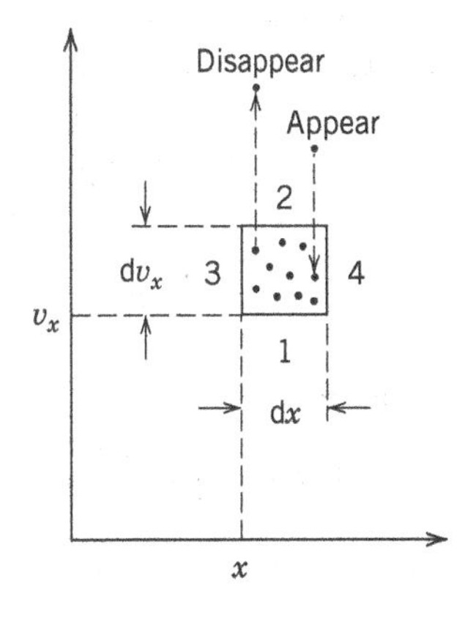

Figure 2.3 One-dimensional v_x–x phase space, illustrating the derivation of the Boltzmann equation and the change in f due to collisions.

Noting that v_x is independent of x and assuming that the acceleration $a_x = F_x/m$ of the particles does not depend on v_x, then (2.3.1) can be rewritten:

$$\frac{\partial f}{\partial t} + v_x \frac{\partial f}{\partial x} + a_x \frac{\partial f}{\partial v_x} = 0$$

The three-dimensional generalization,

$$\frac{\partial f}{\partial t} + \mathbf{v} \cdot \nabla_{\mathbf{r}} f + \mathbf{a} \cdot \nabla_{\mathbf{v}} f = 0 \tag{2.3.2}$$

with $\nabla_{\mathbf{r}} = (\hat{x}\partial/\partial x + \hat{y}\partial/\partial y + \hat{z}\partial/\partial z)$ and $\nabla_{\mathbf{v}} = (\hat{x}\partial/\partial v_x + \hat{y}\partial/\partial v_y + \hat{z}\partial/\partial v_z)$ is called the *collisionless Boltzmann equation* or *Vlasov equation*.

In addition to flows into or out of the volume across the faces, particles can "suddenly" appear in or disappear from the volume due to very short timescale interparticle collisions, which are assumed to occur on a timescale shorter than the evolution time of f in (2.3.2). Such collisions can practically instantaneously change the velocity (but not the position) of a particle. Examples of particles suddenly appearing or disappearing are shown in Figure 2.3. We account for this effect, which changes f, by adding a "collision term" to the right-hand side of (2.3.2), thus, obtaining the *Boltzmann equation*:

$$\frac{\partial f}{\partial t} + \mathbf{v} \cdot \nabla_{\mathbf{r}} f + \frac{\mathbf{F}}{m} \cdot \nabla_{\mathbf{v}} f = \left.\frac{\partial f}{\partial t}\right|_{\mathrm{c}} \tag{2.3.3}$$

The collision term in integral form will be derived in Appendix B. The preceding heuristic derivation of the Boltzmann equation can be made rigorous from various points of view, and the interested reader is referred to texts on plasma theory, such as Holt and Haskell (1965). A kinetic theory of discharges, accounting for non-Maxwellian particle distributions, must be based on solutions of the Boltzmann equation. We give an introduction to this analysis in Chapter 19.

2.3.2 Macroscopic Quantities

The complexity of the dynamical equations is greatly reduced by averaging over the velocity coordinates of the distribution function to obtain equations depending on the spatial coordinates and the time only. The averaged quantities, such as species density, mean velocity, and energy density, are called macroscopic quantities, and the equations describing them are the macroscopic conservation equations. To obtain these averaged quantities, we take *velocity moments* of the distribution function, and the equations are obtained from the moments of the Boltzmann equation.

The average quantities that we are concerned with are the particle density,

$$n(\mathbf{r}, t) = \int f \, \mathrm{d}^3 v \tag{2.3.4}$$

the particle flux

$$\Gamma(\mathbf{r}, t) = n\mathbf{u} = \int \mathbf{v} f \, \mathrm{d}^3 v \tag{2.3.5}$$

where $\mathbf{u}(\mathbf{r}, t)$ is the mean velocity, and the particle kinetic energy per unit volume

$$w = \frac{3}{2}p + \frac{1}{2}mu^2 n = \frac{1}{2}m \int v^2 f \, \mathrm{d}^3 v \tag{2.3.6}$$

where $p(\mathbf{r}, t)$ is the isotropic pressure, which we define next. In this form, w is sum of the *internal* energy density $\frac{3}{2}p$ and the *flow* energy density $\frac{1}{2}mu^2 n$.

2.3.3 Particle Conservation

The lowest moment of the Boltzmann equation is obtained by integrating all terms of (2.3.3) over velocity space. The integration yields the macroscopic *continuity equation*:

$$\frac{\partial n}{\partial t} + \nabla \cdot (n\mathbf{u}) = G - L \tag{2.3.7}$$

The collision term in (2.3.3), which yields the right-hand side of (2.3.7), is equal to zero when integrated over velocities, except for collisions that create or destroy particles, designated as G and L, respectively (e.g., ionization, recombination). In fact, (2.3.7) is transparent since it physically describes the conservation of particles. If (2.3.7) is integrated over a volume $\mathcal{V}$ bounded by a closed surface S, then (2.3.7) states that the net number of particles per second generated within $\mathcal{V}$ either flows across the surface S or increases the number of particles within $\mathcal{V}$. For common low-pressure discharges in the steady state, G is usually due to ionization by electron–neutral collisions:

$$G = \nu_{\text{iz}} n_{\text{e}}$$

where ν_{iz} is the ionization frequency. The volume loss rate L, usually due to recombination, is often negligible. Hence,

$$\nabla \cdot (n\mathbf{u}) = \nu_{\text{iz}} n_{\text{e}} \tag{2.3.8}$$

in a typical discharge. However, note that the continuity equation is clearly not sufficient to give the evolution of the density n, since it involves another quantity, the mean particle velocity $\mathbf{u}$.

2.3.4 Momentum Conservation

To obtain an equation for $\mathbf{u}$, a first moment is formed by multiplying the Boltzmann equation by $\mathbf{v}$ and integrating over velocity. The details are complicated and involve evaluation of tensor elements. The calculation can be found in most plasma theory texts, e.g., Krall and Trivelpiece (1973). The result is

$$mn\left[\frac{\partial \mathbf{u}}{\partial t} + (\mathbf{u} \cdot \nabla)\mathbf{u}\right] = qn(\mathbf{E} + \mathbf{u} \times \mathbf{B}) - \nabla \cdot \mathbf{\Pi} + \mathbf{f}\Big|_{\text{c}} \tag{2.3.9}$$

The left-hand side is the species mass density times the convective derivative of the mean velocity, representing the mass density times the acceleration. The convective derivative has two terms: the first term $\partial\mathbf{u}/\partial t$ represents an acceleration due to an explicitly time-varying $\mathbf{u}$; the second "inertial" term $(\mathbf{u} \cdot \nabla)\mathbf{u}$ represents an acceleration even for a steady fluid flow ($\partial/\partial t \equiv 0$) having a spatially varying $\mathbf{u}$. For example, if $\mathbf{u} = \hat{x}\, u_x(x)$ increases along x, then the fluid is accelerating along x (Problem 2.4). This second term is nonlinear in $\mathbf{u}$ and can often be neglected in discharge analysis.

The mass times acceleration is acted upon, on the right-hand side, by the body forces, with the first term being the electric and magnetic force densities. The second term is the force density due to the divergence of the pressure tensor, which arises due to the integration over velocities

$$\Pi_{ij} = mn\big\langle (v_i - u)\,(v_j - u)\big\rangle_{\text{v}} \tag{2.3.10}$$

where the subscripts i, j give the component directions and $\langle\cdot\rangle_{\text{v}}$ denotes the velocity average of the bracketed quantity over f.[1] For weakly ionized plasmas, it is almost never used in this form, but rather an isotropic version is employed:

$$\mathbf{\Pi} = \begin{pmatrix} p & 0 & 0 \\ 0 & p & 0 \\ 0 & 0 & p \end{pmatrix} \tag{2.3.11}$$

1 We assume f is normalized so that $\langle f\rangle_{\text{v}} = 1$.

such that

$$\nabla \cdot \mathbf{\Pi} = \nabla p \tag{2.3.12}$$

a pressure gradient, with

$$p = \frac{1}{3} mn \langle (v - u)^2 \rangle_{\mathrm{v}} \tag{2.3.13}$$

being the scalar pressure. Physically, the pressure gradient force density arises as illustrated in Figure 2.4, which shows a small volume acted upon by a pressure that is an increasing function of x. The net force on this volume is $p(x)\,\mathrm{d}A - p(x + \mathrm{d}x)\,\mathrm{d}A$ and the volume is $\mathrm{d}A\,\mathrm{d}x$. Hence, the force per unit volume is $-\partial p/\partial x$.

The third term on the right in (2.3.9) represents the time rate of momentum transfer per unit volume due to collisions with other species. For electrons or positive ions, the most important transfer is often due to collisions with neutrals. The transfer is usually approximated by a Krook collision operator

$$\mathbf{f}|_{\mathrm{c}} = -\sum_{\beta} mn\nu_{\mathrm{m}\beta}(\mathbf{u} - \mathbf{u}_\beta) - m(\mathbf{u} - \mathbf{u}_{\mathrm{G}})G + m(\mathbf{u} - \mathbf{u}_{\mathrm{L}})L \tag{2.3.14}$$

where the summation is over all other species, $\mathbf{u}_\beta$ is the mean velocity of species β, $\nu_{\mathrm{m}\beta}$ is the momentum transfer frequency for collisions with species β, and $\mathbf{u}_{\mathrm{G}}$ and $\mathbf{u}_{\mathrm{L}}$ are the mean velocities of newly created and lost particles. Generally $|\mathbf{u}_{\mathrm{G}}| \ll |\mathbf{u}|$ for pair creation by ionization, and $\mathbf{u}_{\mathrm{L}} \approx \mathbf{u}$ for recombination or charge transfer loss processes. We discuss the Krook form of the collision operator further in Chapter 19. The last two terms in (2.3.14) are generally small and give the momentum transfer due to the creation or destruction of particles. For example, if ions are created at rest, then they exert a drag force on the moving ion fluid because they act to lower the average fluid velocity.

A common form of the average force (momentum conservation) equation is obtained from (2.3.9) neglecting the magnetic forces and taking $\mathbf{u}_\beta = 0$ in the Krook collision term for collisions with one neutral species. The result is

$$mn\left[\frac{\partial \mathbf{u}}{\partial t} + \mathbf{u} \cdot \nabla \mathbf{u}\right] = qn\mathbf{E} - \nabla p - mn\nu_{\mathrm{m}}\mathbf{u} \tag{2.3.15}$$

where only the acceleration ($\partial \mathbf{u}/\partial t$), inertial ($\mathbf{u} \cdot \nabla \mathbf{u}$), electric field, pressure gradient, and collision terms appear. For slow time variation, the acceleration term can be neglected. For high pressures, the inertial term is small compared to the collision term and can also be dropped.

Equations (2.3.7) and (2.3.9) together still do not form a closed set, since the pressure tensor $\mathbf{\Pi}$ (or scalar pressure p) is not determined. The usual procedure to close the equations is to use

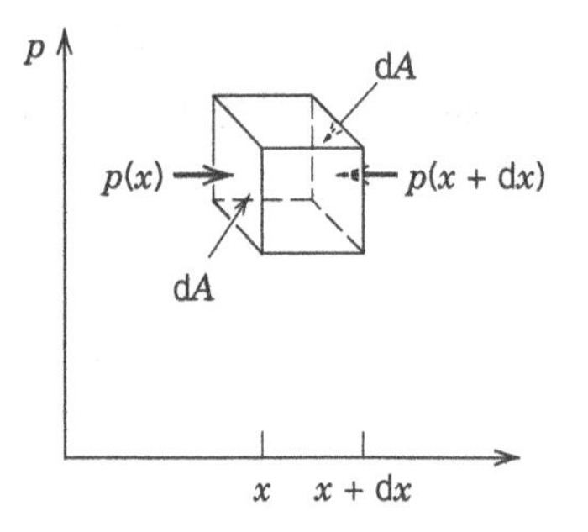

Figure 2.4 The force density due to the pressure gradient.

a thermodynamic *equation of state* to relate p to n. The *isothermal* relation for an equilibrium Maxwellian distribution is

$$p = nkT \tag{2.3.16}$$

so that

$$\nabla p = kT\nabla n \tag{2.3.17}$$

where T is the temperature in kelvins and k is Boltzmann's constant ($k = 1.381 \times 10^{-23}$ J/K). This holds for slow time variations, where temperatures are allowed to equilibrate. In this case, the fluid can exchange energy with its surroundings, and we also require an energy conservation equation (see next) to determine p and T. For a room temperature (297 K), neutral gas having density n_g and pressure p (2.3.16) yields

$$n_g(\text{cm}^{-3}) \approx 3.250 \times 10^{16}\ p(\text{Torr}) \tag{2.3.18}$$

Alternatively, the *adiabatic* equation of state is

$$p = Cn^\gamma \tag{2.3.19}$$

such that

$$\frac{\nabla p}{p} = \gamma\frac{\nabla n}{n} \tag{2.3.20}$$

where γ is the ratio of specific heat at constant pressure to that at constant volume. The specific heats are defined in Section 7.2; $\gamma = 5/3$ for a perfect gas; for one-dimensional adiabatic motion, $\gamma = 3$. The adiabatic relation holds for fast time variations, such as in waves, when the fluid does not exchange energy with its surroundings; hence, an energy conservation equation is not required. For almost all applications to discharge analysis, we use the isothermal equation of state.

2.3.5 Energy Conservation

The energy conservation equation is obtained by multiplying the Boltzmann equation by $\frac{1}{2}mv^2$ and integrating over velocity. The integration and some other manipulations yield

$$\frac{\partial}{\partial t}\left(\frac{3}{2}p\right) + \nabla\cdot\frac{3}{2}(p\mathbf{u}) + p\nabla\cdot\mathbf{u} + \nabla\cdot\mathbf{q} = \frac{\partial}{\partial t}\left(\frac{3}{2}p\right)\Big|_c \tag{2.3.21}$$

Here, $\frac{3}{2}p$ is the thermal energy density (J/m^3), $\frac{3}{2}p\mathbf{u}$ is the macroscopic thermal energy flux (W/m^2), representing the flow of the thermal energy density at the fluid velocity $\mathbf{u}$, $p\nabla\cdot\mathbf{u}$ (W/m^3) gives the heating or cooling of the fluid due to compression or expansion of its volume (Problem 2.5), $\mathbf{q}$ is the heat flow vector (W/m^2), which gives the microscopic thermal energy flux, and the collisional term includes all collisional processes that change the thermal energy density. These include ionization, excitation, elastic scattering, and frictional (ohmic) heating. The equation is usually closed by setting $\mathbf{q} = 0$ or by letting $\mathbf{q} = -\kappa_T\nabla T$, where κ_T is the thermal conductivity. For most steady-state discharges, the macroscopic thermal energy flux is balanced against the collisional processes, giving the simpler equation

$$\nabla\cdot\left(\frac{3}{2}p\mathbf{u}\right) = \frac{\partial}{\partial t}\left(\frac{3}{2}p\right)\Big|_c \tag{2.3.22}$$

Equation (2.3.22), together with the continuity equation (2.3.8), will often prove sufficient for our analysis.

2.3.6 Summary

Summarizing our results for the macroscopic equations describing the electron and ion fluids, we have in their most usually used forms the continuity equation

$$\nabla \cdot (n\mathbf{u}) = \nu_{\mathrm{iz}} n_{\mathrm{e}} \tag{2.3.8}$$

the force equation,

$$mn\left[\frac{\partial \mathbf{u}}{\partial t} + \mathbf{u}\cdot\nabla\mathbf{u}\right] = qn\mathbf{E} - \nabla p - mn\nu_{\mathrm{m}}\mathbf{u} \tag{2.3.15}$$

the isothermal equation of state

$$p = nkT \tag{2.3.16}$$

and the energy-conservation equation

$$\nabla\cdot\left(\frac{3}{2}p\mathbf{u}\right) = \frac{\partial}{\partial t}\left(\frac{3}{2}p\right)\Big|_{\mathrm{c}} \tag{2.3.22}$$

These equations hold for each charged species, with the total charges and currents summed in Maxwell's equations. For example, with electrons and one positive ion species with charge Ze, we have

$$\rho = e\left(Zn_{\mathrm{i}} - n_{\mathrm{e}}\right) \tag{2.3.23}$$

$$\mathbf{J} = e\left(Zn_{\mathrm{iui}} - n_{\mathrm{e}}\mathbf{u}_{\mathrm{e}}\right) \tag{2.3.24}$$

These equations are still very difficult to solve without simplifications. They consist of 18 unknown quantities $n_{\mathrm{i}}, n_{\mathrm{e}}, p_{\mathrm{i}}, p_{\mathrm{e}}, T_{\mathrm{i}}, T_{\mathrm{e}}, \mathbf{u}_{\mathrm{i}}, \mathbf{u}_{\mathrm{e}}, \mathbf{E}$, and $\mathbf{B}$, with the vectors each counting for three. Various simplifications used to make the solutions to the equations tractable will be employed as the individual problems allow.

2.4 Equilibrium Properties

Electrons are generally in near-thermal equilibrium at temperature $\mathrm{T_e}$ in discharges, whereas positive ions are *almost never* in thermal equilibrium. Neutral gas molecules may or may not be in thermal equilibrium, depending on the generation and loss processes. For a single species in thermal equilibrium with itself (e.g., electrons), in the absence of time variation, spatial gradients, and accelerations, the Boltzmann equation (2.3.3) reduces to

$$\frac{\partial f}{\partial t}\Big|_{\mathrm{c}} = 0 \tag{2.4.1}$$

where the subscript c here represents the collisions of a particle species with itself. We show in Appendix B that the solution of (2.4.1) has a Gaussian speed distribution of the form

$$f(v) = C\mathrm{e}^{-\xi^2 mv^2} \tag{2.4.2}$$

The two constants C and ξ can be obtained by using the thermodynamic relation

$$w = \frac{1}{2}mn\langle v^2\rangle_{\mathrm{v}} = \frac{3}{2}nkT \tag{2.4.3}$$

i.e., that the average energy of a particle is $\frac{1}{2}kT$ per translational degree of freedom, and by using a suitable normalization of the distribution. Normalizing $f(v)$ to n, we obtain

$$C\int_0^{2\pi} d\phi \int_0^{\pi} \sin\theta \, d\theta \int_0^{\infty} \exp\left(-\xi^2 m v^2\right) v^2 \, dv = n \tag{2.4.4}$$

and using (2.4.3), we obtain

$$\frac{1}{2}mC\int_0^{2\pi} d\phi \int_0^{\pi} \sin\theta \, d\theta \int_0^{\infty} \exp\left(-\xi^2 m v^2\right) v^4 dv = \frac{3}{2}nkT \tag{2.4.5}$$

where we have written the integrals over velocity space in spherical coordinates. The angle integrals yield the factor 4π. The v integrals are evaluated using the relation[2]

$$\int_0^{\infty} e^{-u^2} u^{2i} du = \frac{(2i-1)!!}{2^{i+1}}\sqrt{\pi},$$

$$\text{where } i \text{ is an integer} \geq 1 \tag{2.4.6}$$

Solving for C and ξ, we have

$$f(v) = n\left(\frac{m}{2\pi kT}\right)^{3/2} \exp\left(-\frac{mv^2}{2kT}\right) \tag{2.4.7}$$

which is the *Maxwellian distribution.*

Similarly, other averages can be performed. The average speed $\bar{v}$ is given by

$$\bar{v} = (m/2\pi kT)^{3/2} \int_0^{\infty} v \left[\exp\left(-\frac{v^2}{2v_{\text{th}}^2}\right)\right] 4\pi v^2 \, dv \tag{2.4.8}$$

where $v_{\text{th}} = (kT/m)^{1/2}$ is the thermal velocity. We obtain

$$\bar{v} = \left(\frac{8kT}{\pi m}\right)^{1/2} \tag{2.4.9}$$

The directed flux Γ_z in (say) the $+z$ direction is given by $n\langle v_z \rangle_v$, where the average is taken over $v_z > 0$ only. Writing $v_z = v\cos\theta$, we have in spherical coordinates

$$\Gamma_z = n\left(\frac{m}{2\pi kT}\right)^{3/2} \int_0^{2\pi} d\phi \int_0^{\pi/2} \sin\theta \, d\theta \int_0^{\infty} v\cos\theta \exp\left(-\frac{v^2}{2v_{\text{th}}^2}\right) v^2 \, dv$$

Evaluating the integrals, we find

$$\Gamma_z = \frac{1}{4}n\bar{v} \tag{2.4.10}$$

Γ_z is the number of particles per square meter per second crossing the $z = 0$ surface in the positive direction. Similarly, the average energy flux $S_z = n\langle \frac{1}{2}mv^2 v_z \rangle_v$ in the $+z$ direction can be found: $S_z = 2kT\Gamma_z$. We see that the average kinetic energy $\overline{W}$ per particle crossing $z = 0$ in the positive direction is

$$\overline{W} = 2kT \tag{2.4.11}$$

It is sometimes convenient to define the distribution in terms of other variables. For example, we can define a distribution of energies $W = \frac{1}{2}mv^2$ by

$$g(W)\,dW = 4\pi f(v)\, v^2 dv$$

2 !! denotes the double factorial function, e.g., $7!! = 7 \times 5 \times 3 \times 1$.

Evaluating $\mathrm{d}v/\mathrm{d}W$, we see that g and f are related by

$$g(W) = 4\pi \frac{v(W) f[v(W)]}{m} \tag{2.4.12}$$

where $v(W) = (2W/m)^{1/2}$.

2.4.1 Boltzmann's Relation

A very important relation can be obtained for the density of electrons in thermal equilibrium at varying positions in a plasma under the action of a spatially varying potential. In the absence of electron drifts ($\mathbf{u}_\mathrm{e} \equiv 0$), the inertial, magnetic, and frictional forces are zero, and the electron force balance is, from (2.3.15) with $\partial/\partial t \equiv 0$,

$$en_\mathrm{e}\mathbf{E} + \nabla p_\mathrm{e} = 0 \tag{2.4.13}$$

Setting $\mathbf{E} = -\nabla\Phi$ and assuming $p_\mathrm{e} = n_\mathrm{e}kT_\mathrm{e}$, (2.4.13) becomes

$$-en_\mathrm{e}\nabla\Phi + kT_\mathrm{e}\nabla n_\mathrm{e} = 0$$

or, rearranging,

$$\nabla(e\Phi - kT_\mathrm{e}\ln n_\mathrm{e}) = 0 \tag{2.4.14}$$

Integrating, we have

$$e\Phi - kT_\mathrm{e}\ln n_\mathrm{e} = \mathrm{const}$$

or

$$n_\mathrm{e}(\mathbf{r}) = n_0 \mathrm{e}^{e\Phi(\mathbf{r})/kT_\mathrm{e}} \tag{2.4.15}$$

which is *Boltzmann's relation* for electrons. We see that electrons are "attracted" to regions of positive potential. We shall generally write Boltzmann's relation in more convenient units

$$n_\mathrm{e} = n_0 \mathrm{e}^{\Phi/\mathrm{T}_\mathrm{e}} \tag{2.4.16}$$

where T_e is now expressed in volts, as is Φ.

For positive ions in thermal equilibrium at temperature T_i, a similar analysis shows that

$$n_\mathrm{i} = n_0 \mathrm{e}^{-\Phi/\mathrm{T}_\mathrm{i}} \tag{2.4.17}$$

Hence, positive ions in thermal equilibrium are "repelled" from regions of positive potential. However, positive ions are almost never in thermal equilibrium in low-pressure discharges because the ion drift velocity $\mathbf{u}_\mathrm{i}$ is large, leading to inertial or frictional forces in (2.3.15) that are comparable to the electric field or pressure gradient forces.

2.4.2 Debye Length

The characteristic length scale in a plasma is the electron Debye length λ_De. As we will show, the Debye length is the distance scale over which significant charge densities can spontaneously exist. For example, low-voltage (undriven) sheaths are typically a few Debye lengths wide. To determine the Debye length, let us introduce a sheet of negative charge having surface charge density $\rho_\mathrm{S} < 0\ \mathrm{C/m^2}$ into an infinitely extended plasma having equilibrium densities $n_\mathrm{e} = n_\mathrm{i} = n_0$. For simplicity, we assume immobile ions, such that $n_\mathrm{i} = n_0$ after the sheet is introduced. However, the negative sheet "repels" nearby electrons, leading to a reduced electron density near the sheet. The

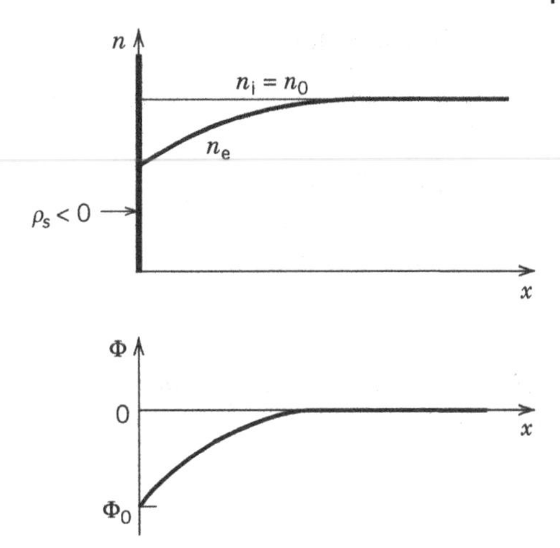

Figure 2.5 Calculation of the electron Debye length λ_{De}. A negatively charged sheet is introduced into a plasma containing electrons in thermal equilibrium.

situation after introduction of the sheet is shown in Figure 2.5. To determine the potential and density variation, we use Poisson's equation, which in one dimension can be written

$$\frac{d^2\Phi}{dx^2} = -\frac{e}{\epsilon_0}\left(n_i - n_e\right) \tag{2.4.18}$$

Setting $n_e = n_0 \exp\left(\Phi/T_e\right)$, from the Boltzmann relation (2.4.16), and taking $n_i = n_0$, Poisson's equation becomes

$$\frac{d^2\Phi}{dx^2} = \frac{en_0}{\epsilon_0}\left(e^{\Phi/T_e} - 1\right) \tag{2.4.19}$$

Expanding $\exp\left(\Phi/T_e\right)$ in a Taylor series for $\Phi \ll T_e$, (2.4.19) becomes, to lowest order in Φ/T_e,

$$\frac{d^2\Phi}{dx^2} = \frac{en_0}{\epsilon_0}\frac{\Phi}{T_e} \tag{2.4.20}$$

The symmetric solution of (2.4.20) that vanishes at $x = \pm\infty$ is

$$\Phi = \Phi_0 e^{-|x|/\lambda_{De}} \tag{2.4.21}$$

where

$$\lambda_{De} = \left(\frac{\epsilon_0 T_e}{en_0}\right)^{1/2} \tag{2.4.22}$$

In practical units, we find

$$\lambda_{De}\ (\text{cm}) \approx 743\sqrt{T_e/n_e} \tag{2.4.23}$$

with T_e in volts and n_e in cm^{-3}. We find for $T_e = 4$ V and $n_e = 10^{10}\ \text{cm}^{-3}$ that $\lambda_{De} = 0.14$ mm. It is on space scales larger than a Debye length that the plasma will tend to remain neutral.

The Debye length is useful in many contexts. In the next chapter, we shall see that it serves as a characteristic scale length to shield the Coulomb potentials of individual charged particles when they collide. Although we have calculated the above effect for electron shielding, it is also possible on slower timescales for the ions to contribute. We leave the calculation for a problem. Ion shielding plays a key role in dusty plasmas, which we treat in Chapter 18.

2.4.3 Quasineutrality

The potential variation across a plasma of length $l \gg \lambda_{De}$ can be estimated from Poisson's equation (2.2.11):

$$\nabla^2\Phi \sim \frac{\Phi}{l^2} \sim \left|\frac{e}{\epsilon_0}(Zn_i - n_e)\right| \tag{2.4.24}$$

We generally expect that

$$\Phi \lesssim T_e = \frac{e}{\epsilon_0} n_e \lambda_{De}^2 \tag{2.4.25}$$

where the equality on the right follows from the definition of λ_{De}. Combining (2.4.24) and (2.4.25), we have

$$\frac{|Zn_i - n_e|}{n_e} \lesssim \frac{\lambda_{De}^2}{l^2} \tag{2.4.26}$$

For $\lambda_{De}^2/l^2 \ll 1$, (2.4.26) implies that

$$|Zn_i - n_e| \ll n_e \tag{2.4.27}$$

such that we can set

$$Zn_i = n_e \tag{2.4.28}$$

except when used in Poisson's equation. Relation (2.4.27) is the basic statement of *quasineutrality* of a plasma and is often called the *plasma approximation.* We shall see in Chapter 6 that the plasma approximation is violated within a plasma–sheath, in proximity to a material wall, either because the sheath thickness $s \approx \lambda_{De}$ or because $\Phi \gg T_e$.

Problems

2.1 Charge Conservation Derive the conservation of charge law (2.2.5) from Maxwell's equations.

2.2 Homogeneous Discharge Model A plasma is confined between two grounded ($\Phi = 0$) parallel plates located at $x = 0$ and $x = l$. The ion density is $n_i(x) = n_0$ for $0 < x < l$. The electron density is $n_e(x) = n_0$ for $s < x < l - s$ and is $n_e(x) = 0$ in the "sheath" regions $0 < x < s$ and $l - s < x < l$.

(a) Solve Poisson's equation to determine the potential $\Phi(x)$ everywhere within the discharge $0 < x < l$. Find $\Phi_0 = \Phi(l/2)$ in the center of the discharge. Plot $\Phi(x)$ versus x for $0 < x < l$ for $s = l/8$.

(b) Plot the electric field E_x versus x and show that it acts to confine electrons within the bulk plasma at both sheaths.

(c) Choosing $\Phi_0 = 4T_e$, find an expression for s and show that s is of the order of an electron Debye length.

2.3 Potential in Asymmetric Discharge A plasma is confined between two grounded ($\Phi = 0$) parallel conducting plates located at $x = 0$ and $x = l$. The ion density is $n_i(x) = n_0$ for $0 < x < l$. The electron density is $n_e(x) = n_0$ for $l/4 < x < l$ and is $n_e(x) = 0$ in the "sheath" region $0 < x < l/4$ near the left-hand plate.

(a) Plot the volume charge density $\rho(x)$ within the plates.
(b) Solve Poisson's equation to determine the potential $\Phi(x)$ everywhere within the discharge $0 < x < l$. Plot $\Phi(x)$ versus x for $0 < x < l$. (Make sure that Φ and $d\Phi/dx$ are continuous functions at $x = l/4$ and that $\Phi = 0$ at the two plates $x = 0, l$, consistent with Maxwell's equations.)
(c) Plot the electric field E_x versus x within the plates.
(d) Find the surface charge density ρ_S on each of the plates. (Since both plates are grounded, there is no electric field outside the plates.)

2.4 Bernoulli's Law Starting from the force equation (2.3.9), derive Bernoulli's law for an incompressible fluid in steady one-dimensional flow:

$$\frac{1}{2}mnu^2(x) + p(x) = \text{const}$$

How would you use this effect to measure the change in the velocity of a fluid as it flows through a constriction in a pipe?

2.5 Compressional Heating of a Fluid Show using a one-dimensional calculation that the relative rate of change with time of a small volume $\Delta\mathcal{V}$ moving with the fluid velocity $\mathbf{u}$ can be written as

$$\frac{1}{\Delta\mathcal{V}}\frac{d(\Delta\mathcal{V})}{dt} = \nabla \cdot \mathbf{u}$$

Hence, show from (2.3.21) that if the fluid expands, its internal energy decreases.

2.6 Adiabatic Equation of State Derive the adiabatic equation of state (2.3.19) using particle conservation (2.3.7) and energy conservation (2.3.21), by assuming that the heat flow vector $\mathbf{q}$ and all collision terms in these equations are zero.

2.7 Averages Over a Maxwellian Distribution
(a) Show by integrating (2.4.8) that the average speed of electrons in a Maxwellian distribution is $\bar{v}_e = (8e\mathrm{T}_e/\pi m)^{1/2}$.
(b) Show by integrating the equation above (2.4.10) that the average one-way particle flux is $\Gamma_e = n_e\bar{v}_e/4$.
(c) Find the average one-way energy flux S_e by integrating the energy flux over a Maxwellian distribution. Comparing S_e to Γ_e, show that (2.4.11) holds, i.e., the average kinetic energy per particle crossing a surface is $W_e = 2kT_e$.

2.8 Debye Length Including Ions In the derivation of the Debye length in Section 2.4, it was assumed that the ions were immobile. Assuming mobile electrons *and* ions with densities given by the Boltzmann factors (2.4.16) and (2.4.17), derive an expression for the Debye length λ_D. For $\mathrm{T}_e \gg \mathrm{T}_i$, show that the Debye length depends on the *ions* alone. (However, note that in a typical discharge, the ions are *not* in thermal equilibrium, and (2.4.17) is *not* valid. The effective Debye length is then usually determined by the electrons alone: $\lambda_D \approx \lambda_{De}$.)

2.9 Sphere Immersed in a Plasma A conducting sphere of radius a is immersed in an infinite uniform plasma having density n_0, electrons in thermal equilibrium at temperature T_e, and infinite mass (immobile) ions. A small dc voltage $V_0 \ll \mathrm{T}_e$ is applied to the sphere with respect to the plasma.

(a) Starting from Poisson's equation in spherical coordinates and using Boltzmann's relation for the electrons at temperature T_e, derive an expression for the potential $\Phi(r)$ everywhere in the plasma.

(b) Find an expression for the Debye length from your expression for $\Phi(r)$.

(c) The capacitance of the sphere (with respect to the plasma) is $C = q/V_0$, where q is the total charge on the sphere and V_0 is the voltage of the sphere with respect to the plasma. Show that $C = 4\pi\epsilon_0 a(1 + a/\lambda_{De})$.

(d) Repeat part (c) for both electrons in thermal equilibrium and mobile ions in thermal equilibrium with temperature T_i.

Hint: Note that for spherical symmetry, $\nabla^2\Phi = (1/r)\mathrm{d}^2(r\Phi)/\mathrm{d}r^2$.

3

Atomic Collisions

3.1 Basic Concepts

When two particles collide, various phenomena may occur. As examples, one or both particles may change their momentum or their energy, neutral particles can become ionized, and ionized particles can become neutral. We introduce the fundamentals of collisions between electrons, positive ions, and gas atoms in this chapter, concentrating on simple classical estimates of the important processes in noble gas discharges such as argon. For electrons colliding with atoms, the main processes are elastic scattering, in which primarily the electron momentum is changed, and inelastic processes such as excitation and ionization. For ions colliding with atoms, the main processes are elastic scattering in which momentum and energy are exchanged, and resonant charge transfers. Other important processes occur in molecular gases. These include dissociation, dissociative recombination, processes involving negative ions, such as attachment, detachment, and positive–negative ion charge transfer, and processes involving excitation of molecular vibrations and rotations. We defer consideration of collisions in molecular gases to Chapter 8.

3.1.1 Elastic and Inelastic Collisions

Collisions conserve momentum and energy: the total momentum and energy of the colliding particles after collision are equal to that before collision. Electrons and fully stripped ions possess only kinetic energy. Atoms and partially stripped ions have internal energy level structures and can be excited, de-excited, or ionized, corresponding to changes in potential energy. It is the total energy, which is the sum of the kinetic and potential energy, that is conserved in a collision.

If the internal energies of the collision partners do not change, then the sum of kinetic energies is conserved and the collision is said to be *elastic*. Although the total kinetic energy is conserved, kinetic energy is generally exchanged between particles. If the sum of kinetic energies is not conserved, then the collision is *inelastic*. Most inelastic collisions involve excitation or ionization, such that the sum of kinetic energies after collision is less than that before collision. However, *superelastic* collisions can occur in which an excited atom can be de-excited by a collision, increasing the sum of kinetic energies.

3.1.2 Collision Parameters

The fundamental quantity that characterizes a collision is its *cross section* $\sigma(v_R)$, where v_R is the relative velocity between the particles before collision. To define this, we consider first the simplest

Principles of Plasma Discharges and Materials Processing, Third Edition. Michael A. Lieberman and Allan J. Lichtenberg.

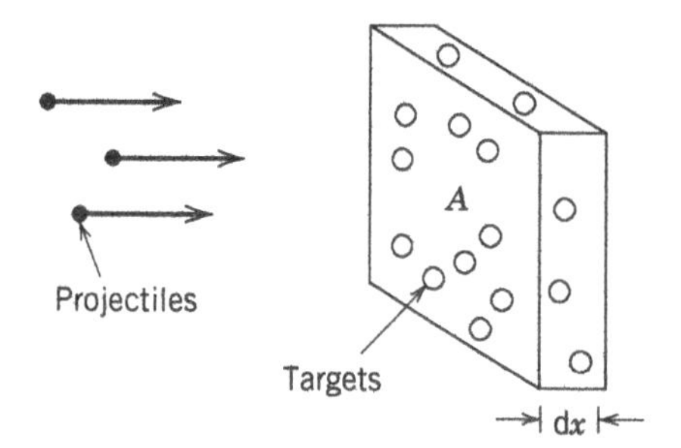

Figure 3.1 A flux of incident particles collides with a population of target particles in the half-space $x > 0$.

situation shown in Figure 3.1, in which a flux $\Gamma = nv$ of particles having mass m, density n, and fixed velocity v is incident on a half-space $x > 0$ of stationary, infinitely massive "target" particles having density n_g. In this case, $v_R = v$. Let dn be the number of incident particles per unit volume at x that undergo an "interaction" with the target particles within a differential distance dx, removing them from the incident beam. Clearly, dn is proportional to n, n_g, and dx for infrequent collisions within dx. Hence, we can write

$$dn = -\sigma n n_g \, dx \tag{3.1.1}$$

where the constant of proportionality σ that has been introduced has units of area and is called the cross section for the interaction. The minus sign denotes *removal* from the beam. To define a cross section, the "interaction" must be specified, e.g., ionization of the target particle, excitation of the incident particle to a given energy state, or scattering of the incident particle by an angle exceeding $\pi/2$. Multiplying (3.1.1) by v, we find a similar equation for the flux:

$$d\Gamma = -\sigma \Gamma n_g \, dx \tag{3.1.2}$$

For a simple interpretation of σ, let the incident and target particles be hard elastic spheres of radii a_1 and a_2, and let the "interaction" be a collision between the spheres. In a distance dx, there are $n_g \, dx$ targets within a unit area perpendicular to x. Draw a circle of radius $a_{12} = a_1 + a_2$ in the $x =$ const plane about each target. A collision occurs if the centers of the incident and target particles fall within this radius. Hence, the fraction of the unit area for which a collision occurs is $n_g \, dx \, \pi a_{12}^2$. The fraction of incident particles that collide within dx is then

$$\frac{d\Gamma}{\Gamma} = \frac{dn}{n} = -n_g \sigma \, dx \tag{3.1.3}$$

where

$$\sigma = \pi a_{12}^2 \tag{3.1.4}$$

is the hard sphere cross section. In this particular case, σ is independent of v.

Equation (3.1.2) is readily integrated to give the collided flux

$$\Gamma(x) = \Gamma_0(1 - e^{-x/\lambda}) \tag{3.1.5}$$

with the uncollided flux $\Gamma_0 e^{-x/\lambda}$. The quantity

$$\lambda = \frac{1}{n_g \sigma} \tag{3.1.6}$$

is the *mean free path* for the decay of the beam, that is, the distance over which the uncollided flux decreases by 1/e from its initial value Γ_0 at $x = 0$. If the velocity of the beam is v, then the mean time between interactions is

$$\tau = \frac{\lambda}{v} \tag{3.1.7}$$

Its inverse is the *interaction* or *collision frequency*

$$\nu \equiv \tau^{-1} = n_g \sigma v \tag{3.1.8}$$

and is the number of interactions per second that an incident particle has with the target particle population. We can also define the collision frequency per unit density, which is called the *rate constant*

$$K = \sigma v \tag{3.1.9}$$

and, trivially, from (3.1.8) and (3.1.9)

$$\nu = K n_g \tag{3.1.10}$$

3.1.3 Differential Scattering Cross Section

Let us consider only those interactions that scatter the particles by $\theta = 90°$ or more. For hard spheres, taking the angle of incidence equal to the angle of reflection, the 90° collision occurs on the $\chi = 45°$ diagonal (see Figure 3.2), therefore, having a cross section

$$\sigma_{90} = \frac{\pi a_{12}^2}{2} \tag{3.1.11}$$

which is a factor of two smaller than (3.1.4). Of course, multiple collisions at smaller angles (radii larger than $a_{12}/\sqrt{2}$) also eventually scatter incident particles through 90°. This indeterminacy indicates that a more precise way of determining the scattering cross section is required. For this purpose, we introduce a *differential scattering cross section* $I(v, \theta)$. Consider a beam of particles incident on a scattering center (again assumed fixed), as shown in Figure 3.3. We assume that the scattering force is symmetric about the line joining the centers of the two particles. A particle incident at a distance b off-center from the target particle is scattered through an angle θ, as shown in Figure 3.3. The quantity b is the *impact parameter* and θ is the *scattering angle* (see also Figure 3.2). Now, flux conservation requires that for incoming flux Γ,

$$\Gamma\, 2\pi b\, \mathrm{d}b = -\Gamma I(v, \theta) 2\pi \sin\theta\, \mathrm{d}\theta \tag{3.1.12}$$

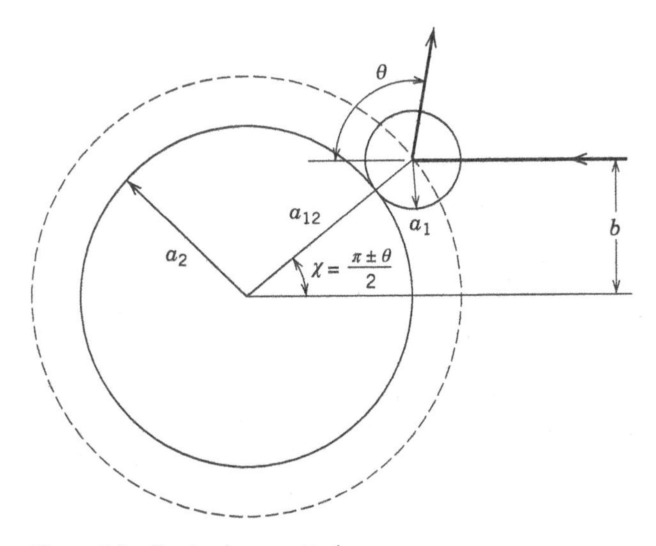

Figure 3.2 Hard-sphere scattering.

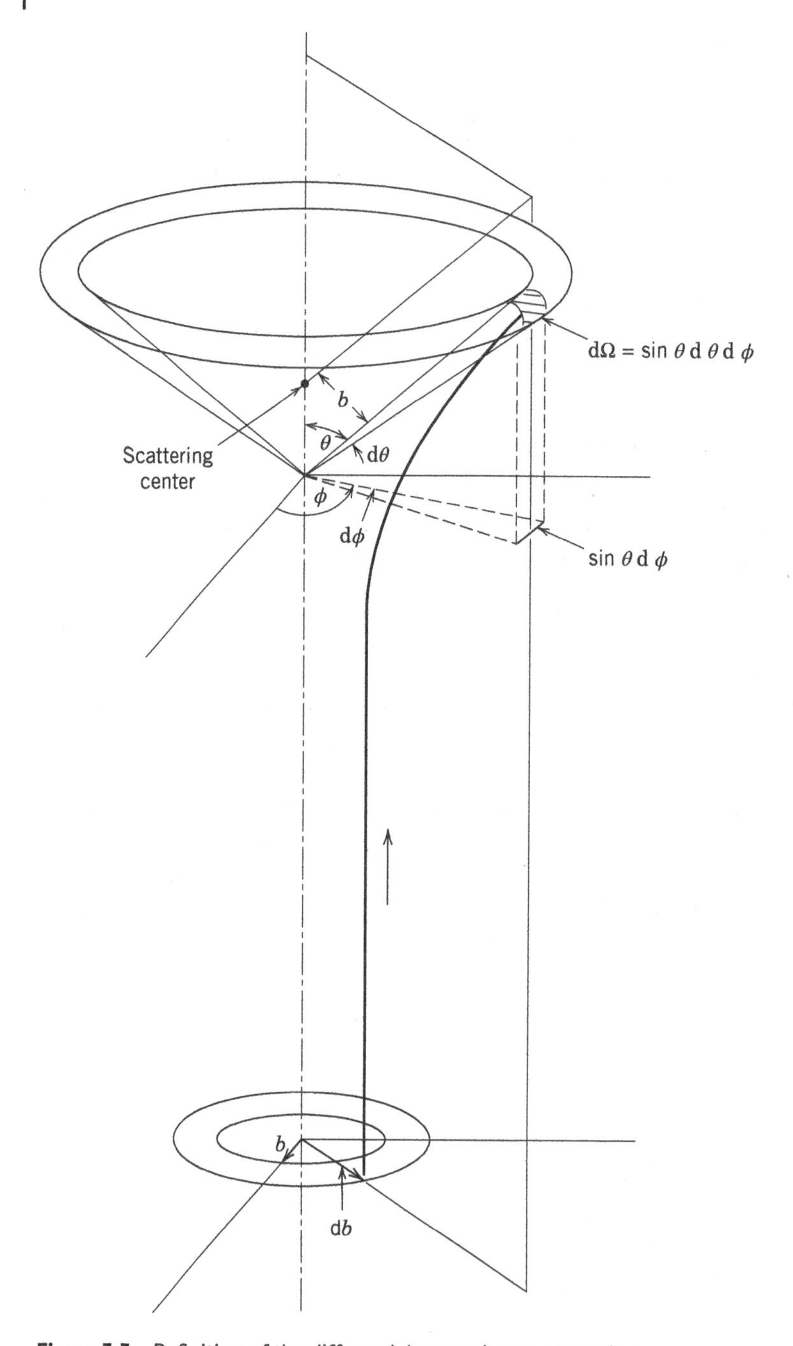

Figure 3.3 Definition of the differential scattering cross section.

i.e., that all particles entering through the differential annulus $2\pi b\, db$ leave through a differential solid angle $d\Omega = 2\pi \sin\theta\, d\theta$. The minus sign is because an increase in b leads to a decrease in θ. The proportionality constant is just $I(v, \theta)$, which has the dimensions of area per steradian. From (3.1.12) we obtain

$$I(v, \theta) = \frac{b}{\sin\theta} \left| \frac{db}{d\theta} \right| \tag{3.1.13}$$

The quantity $db/d\theta$ is determined from the scattering force, and the absolute value is used since $db/d\theta$ is negative. We will calculate $I(v, \theta)$ for various potentials in Section 3.2.

We can calculate the *total scattering cross section* σ_{sc} by integrating I over the solid angle

$$\sigma_{sc} = 2\pi \int_0^{\pi} I(v, \theta) \sin\theta \, d\theta \tag{3.1.14}$$

It is clear that $\sigma_{sc} = \sigma$ for scattering through any angle, as defined in (3.1.2). It is often useful to define a different cross section

$$\sigma_m = 2\pi \int_0^{\pi} (1 - \cos\theta) \, I(v, \theta) \sin\theta \, d\theta \tag{3.1.15}$$

The factor $(1 - \cos\theta)$ is the fraction of the initial momentum mv lost by the incident particle, and thus (3.1.15) is the *momentum transfer cross section.* It is σ_m that is appropriate for calculating the frictional drag in the force equation (2.3.9). For a single velocity, we would just have $\nu_m = \sigma_m v$, where σ_m is generally a function of velocity. In the macroscopic force equation (2.3.15), ν_m must be obtained by averaging over the particle velocity distributions, which we do in Section 3.5.

We illustrate the use of the differential scattering cross section to calculate the total scattering and momentum transfer cross sections for the hard-sphere model shown in Figure 3.2. The impact parameter is $b = a_{12} \sin\chi$, and differentiating, $db = a_{12} \cos\chi \, d\chi$, so that

$$b \, db = a_{12}^2 \sin\chi \cos\chi \, d\chi = \frac{1}{2} a_{12}^2 \sin 2\chi \, d\chi \tag{3.1.16}$$

From Figure 3.2, the scattering angle $\theta = \pi - 2\chi$, such that (3.1.16) can be written

$$b \, db = -\frac{1}{4} a_{12}^2 \sin\theta \, d\theta \tag{3.1.17}$$

Substituting (3.1.17) into (3.1.13), we have

$$I(v, \theta) = \frac{1}{4} a_{12}^2 \tag{3.1.18}$$

Using the definitions of σ_{sc} and σ_m in (3.1.14) and (3.1.15), respectively, we find

$$\sigma_{sc} = \sigma_m = \pi a_{12}^2 \tag{3.1.19}$$

for hard-sphere collisions. In general, $\sigma_{sc} \neq \sigma_m$ for other scattering forces. For electron collisions with atoms, the electron radius is negligible compared to the atomic radius so that $a_{12} \approx a$, the atomic radius. Although the value of $a \approx 10^{-8}$ cm gives $\sigma_{sc} = \sigma_m \approx 3 \times 10^{-16}$ cm^2, which is reasonable, it does not capture the scaling of the cross section with speed.

In the following sections of this chapter, we consider collisional processes in more detail. Except for Coulomb collisions, we confine our attention to electron–atom and ion–atom processes. After a discussion of collision dynamics in Section 3.2, we describe elastic collisions in Section 3.3 and inelastic collisions in Section 3.4. We reserve a discussion of some aspects of inelastic collisions until Chapter 8, in which a more complete range of atomic and molecular processes is considered. In Section 3.5, we describe the averaging over particle velocity distributions that must be done to obtain the collisional rate constants. Experimental values for argon are also given in Section 3.5; these are needed for discussing energy transfer and diffusive processes in the succeeding chapters. A more detailed account of collisional processes, together with many results of experimental measurements, can be found in McDaniel (1989), McDaniel et al. (1993), Massey et al. (1969–74), Smirnov (1981), and Raizer (1991).

3.2 Collision Dynamics

3.2.1 Center-of-Mass Coordinates

In a collision between projectile and target particles, there is recoil of the target as well as deflection of the projectile. In fact, both may be moving, and, in the case of like-particle collisions, not distinguishable. To describe this more complicated state, a center of mass (CM) coordinate system can be introduced in which projectiles and targets are treated equally. Without loss of generality, we can transform to a coordinate system in which one of the particles is stationary before the collision. Hence, we consider a general collision in the laboratory frame between two particles having mass m_1 and m_2, position $\mathbf{r}_1$ and $\mathbf{r}_2$, velocity $\mathbf{v}_1$ and $\mathbf{v}_2 \equiv 0$, and scattering angle θ_1 and θ_2, as shown in Figure 3.4*a*. We assume that the force $\mathbf{F}$ acts along the line joining the centers of the particles, with $\mathbf{F}_{12} = -\mathbf{F}_{21}$.

The center-of-mass coordinates may be defined by the linear transformation

$$\mathbf{R} = \frac{m_1\mathbf{r}_1 + m_2\mathbf{r}_2}{m_1 + m_2} \tag{3.2.1}$$

and

$$\mathbf{r} = \mathbf{r}_1 - \mathbf{r}_2 \tag{3.2.2}$$

with the accompanying CM velocity

$$\mathbf{V} = \frac{m_1\mathbf{v}_1 + m_2\mathbf{v}_2}{m_1 + m_2} \tag{3.2.3}$$

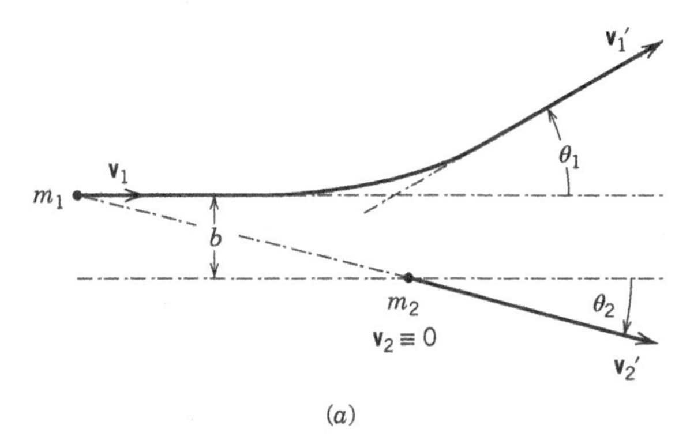

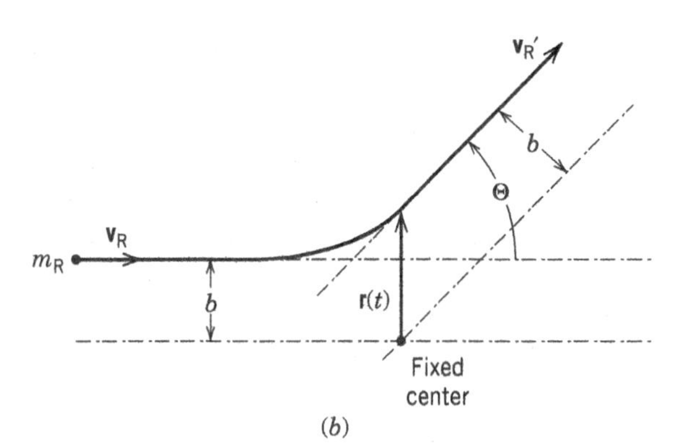

Figure 3.4 The relation between the scattering angles in (*a*) the laboratory system and (*b*) the center of mass (CM) system.

and the relative velocity

$$\mathbf{v}_R = \mathbf{v}_1 - \mathbf{v}_2 \tag{3.2.4}$$

The force equations for the two particles are

$$m_1\dot{\mathbf{v}}_1 = \mathbf{F}_{12}(r), \qquad m_2\dot{\mathbf{v}}_2 = \mathbf{F}_{21}(r) = -\mathbf{F}_{12}(r) \tag{3.2.5}$$

Adding these equations, we get the result for the CM motion that $\dot{\mathbf{V}} = 0$, such that the CM moves with constant velocity throughout the collision. Now dividing the first of (3.2.5) by m_1 and the second by m_2, and using the definition in (3.2.4), we have

$$m_R\dot{\mathbf{v}}_R = \mathbf{F}_{12}(r) \tag{3.2.6}$$

which is the equation of motion of a "fictitious" particle with a reduced mass

$$m_R = \frac{m_1 m_2}{m_1 + m_2} \tag{3.2.7}$$

in a fixed central force $\mathbf{F}_{12}(r)$. The fictitious particle has mass m_R, position $\mathbf{r}(t)$, velocity $\mathbf{v}_R(t)$, and scattering angle Θ, as shown in Figure 3.4*b*. This result holds for any central force, including the hard-sphere, Coulomb, and polarization forces that we subsequently consider. If (3.2.6) can be solved to obtain the motion, including Θ, then we can transform back to the laboratory frame to get the actual scattering angles θ_1 and θ_2. It is easy to show from momentum conservation (Problem 3.2) that

$$\tan\theta_1 = \frac{\sin\Theta}{(m_1/m_2)(v_R/v_R') + \cos\Theta} \tag{3.2.8a}$$

and

$$\tan\theta_2 = \frac{\sin\Theta}{v_R/v_R' - \cos\Theta} \tag{3.2.8b}$$

where v_R and v_R' are the speeds in the CM system before and after the collision, respectively.

For an elastic collision, the scattering force can be written as the gradient of a potential that vanishes as $r = |\mathbf{r}| \to \infty$:

$$\mathbf{F}_{12} = -\nabla U(r) \tag{3.2.9}$$

It follows that the kinetic energy of the particle is conserved for the collision in the CM system. Hence $v_R' = v_R$, and we obtain from (3.2.8) that

$$\tan\theta_1 = \frac{\sin\Theta}{m_1/m_2 + \cos\Theta} \tag{3.2.10}$$

and, using the double-angle formula for the tangent,

$$\theta_2 = \frac{1}{2}(\pi - \Theta) \tag{3.2.11}$$

For electron collisions with ions or neutrals, $m_1/m_2 \ll 1$ and we obtain $m_R \approx m_1$ and $\theta_1 \approx \Theta$. For collision of a particle with an equal mass target, $m_1 = m_2$, we obtain $m_R = m_1/2$ and $\theta_1 = \Theta/2$. Hence, for hard-sphere elastic collisions against an initially stationary equal mass target, the maximum scattering angle is 90°.

Since the same particles are scattered into the differential solid angle $2\pi \sin\Theta\, d\Theta$ in the CM system as are scattered into the corresponding solid angle $2\pi \sin\theta_1\, d\theta_1$ in the laboratory system, the differential scattering cross sections are related by

$$I(v_R, \Theta)\, 2\pi \sin\Theta\, d\Theta = I(v_R, \theta_1)\, 2\pi \sin\theta_1\, d\theta_1 \tag{3.2.12}$$

where $d\Theta/d\theta_1$ can be found by differentiating (3.2.10).

3.2.2 Energy Transfer

Elastic collisions can be an important energy transfer process in gas discharges, and can also be important for understanding inelastic collision processes such as ionization, as we will see in Section 3.4. For the elastic collision of a projectile of mass m_1 and velocity $\mathbf{v}_1$ with a stationary target of mass m_2, the conservation of momentum along and perpendicular to $\mathbf{v}_1$ and the conservation of energy can be written in the laboratory system as

$$m_1 v_1 = m_1 v_1' \cos\theta_1 + m_2 v_2' \cos\theta_2 \tag{3.2.13}$$

$$0 = m_1 v_1' \sin\theta_1 - m_2 v_2' \sin\theta_2 \tag{3.2.14}$$

$$\frac{1}{2} m_1 v_1^2 = \frac{1}{2} m_1 v_1'^2 + \frac{1}{2} m_2 v_2'^2 \tag{3.2.15}$$

where the primes denote the values after the collision. We can eliminate v_1' and θ_1 and solve (3.2.13)–(3.2.15) to obtain

$$\frac{1}{2} m_2 v_2'^2 = \frac{1}{2} m_1 v_1^2 \frac{4 m_1 m_2}{(m_1 + m_2)^2} \cos^2\theta_2 \tag{3.2.16}$$

Since the initial energy of the projectile is $\frac{1}{2} m_1 v_1^2$ and the energy gained by the target is $\frac{1}{2} m_2 v_2'^2$, the fraction of energy lost by the projectile *in the laboratory system* is

$$\zeta_L = \frac{4 m_1 m_2}{(m_1 + m_2)^2} \cos^2\theta_2 \tag{3.2.17}$$

Using (3.2.11) in (3.2.17), we obtain

$$\zeta_L = \frac{2 m_1 m_2}{(m_1 + m_2)^2} (1 - \cos\Theta) \tag{3.2.18}$$

where Θ is the scattering angle *in the CM system.* We average over the differential scattering cross section to obtain the average loss:

$$\begin{aligned} \langle \zeta_L \rangle_\Theta &= \frac{2 m_1 m_2}{(m_1 + m_2)^2} \frac{\int (1 - \cos\Theta) I(v_R, \Theta)\, 2\pi \sin\Theta \, d\Theta}{\int I(v_R, \Theta)\, 2\pi \sin\Theta \, d\Theta} \\ &= \frac{2 m_1 m_2}{(m_1 + m_2)^2} \frac{\sigma_m}{\sigma_{sc}} \end{aligned} \tag{3.2.19}$$

where σ_{sc} and σ_m are defined in (3.1.14) and (3.1.15).

For hard-sphere scattering of electrons against atoms, we have $m_1 = m$ (electron mass) and $m_2 = M$ (atom mass), and $\sigma_{sc} = \sigma_m$ by (3.1.19), such that $\langle \zeta_L \rangle_\Theta = 2m/M \sim 10^{-4}$. Hence, electrons transfer little energy due to elastic collisions with heavy particles, allowing $T_e \gg T_i$ in a typical discharge. On the other hand, for $m_1 = m_2$, we obtain $\langle \zeta_L \rangle_\Theta = \frac{1}{2}$, leading to strong elastic energy exchange among heavy particles and, hence, to a common temperature.

3.2.3 Small Angle Scattering

In the general case, (3.2.6) must be solved to determine the CM trajectory and the scattering angle Θ. We outline this approach and give some results in Appendix A. Here, we restrict attention to small-angle scattering ($\Theta \ll 1$) for which the fictitious particle moves with uniform velocity v_R along a trajectory that is practically unaltered from a straight line. In this case, we can calculate the transverse momentum impulse $\Delta p_\perp$ delivered to the particle as it passes the center of force at

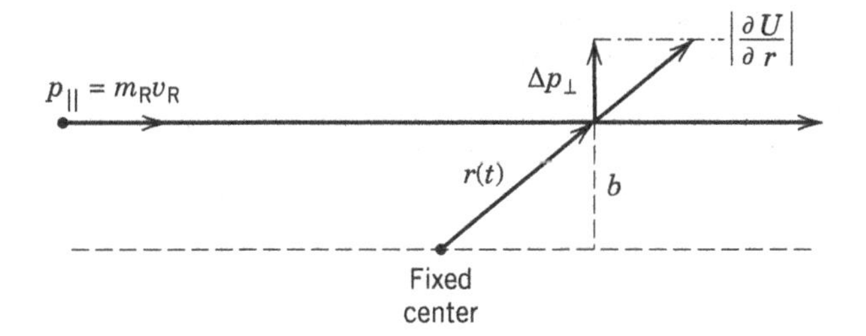

Figure 3.5 Calculation of the differential scattering cross section for small-angle scattering. The center of mass trajectory is practically a straight line.

$r = 0$ and use this to determine Θ. For a straight-line trajectory, as shown in Figure 3.5, the particle distance from the center of force is

$$r = (b^2 + v_R^2 t^2)^{1/2} \tag{3.2.20}$$

where b is the impact parameter and t is the time. We assume a central force of the form (3.2.9) with

$$U(r) = \frac{C}{r^i} \tag{3.2.21}$$

where i is an integer. The component of the force acting on the particle perpendicular to the trajectory is $(b/r)\,|\mathrm{d}U/\mathrm{d}r|$. Hence, the momentum impulse is

$$\Delta p_\perp = \int_{-\infty}^{\infty} \frac{b}{r}\left|\frac{\mathrm{d}U}{\mathrm{d}r}\right| \mathrm{d}t \tag{3.2.22}$$

Differentiating (3.2.20) to obtain

$$\mathrm{d}t = \frac{r}{v_R} \frac{\mathrm{d}r}{(r^2 - b^2)^{1/2}}$$

substituting into (3.2.22) and dividing by the incident momentum $p_{||} = m_R v_R$, we obtain

$$\Theta = \frac{\Delta p_\perp}{p_{||}} = \frac{2b}{m_R v_R^2} \int_b^{\infty} \left|\frac{\mathrm{d}U}{\mathrm{d}r}\right| \frac{\mathrm{d}r}{(r^2 - b^2)^{1/2}} \tag{3.2.23}$$

The integral in (3.2.23) can be evaluated in closed form (Smirnov, 1981, p. 384) to obtain

$$\Theta = \frac{A}{W_R b^i} \tag{3.2.24}$$

where $W_R = \frac{1}{2} m_R v_R^2$ is the CM energy and

$$A = \frac{Ci\sqrt{\pi}\Gamma\left[(i+1)/2\right]}{2\Gamma\left[(i+2)/2\right]} \tag{3.2.25}$$

with Γ, the Gamma function.[1] Inverting (3.2.24), we obtain

$$b = \left(\frac{A}{W_R \Theta}\right)^{1/i} \tag{3.2.26}$$

and differentiating, we obtain

$$\mathrm{d}b = -\frac{1}{i}\left(\frac{A}{W_R}\right)^{1/i} \frac{\mathrm{d}\Theta}{\Theta^{1+1/i}} \tag{3.2.27}$$

1 $\Gamma(l) = (l-1)! = (l-1)\Gamma(l-1)$, with $\Gamma(1/2) = \sqrt{\pi}$.

Table 3.1 Scaling of Cross Section σ, Interaction Frequency ν, and Rate Constant K, with Relative Velocity υ_R, for Various Scattering Potentials U

Process	$U(r)$	σ	ν or K
Coulomb	$1/r$	$1/\upsilon_R^4$	$1/\upsilon_R^3$
Permanent dipole	$1/r^2$	$1/\upsilon_R^2$	$1/\upsilon_R$
Induced dipole	$1/r^4$	$1/\upsilon_R$	Const
Hard sphere	$1/r^i, i \to \infty$	Const	υ_R

Substituting (3.2.26) and (3.2.27) into (3.1.13), with $\sin \Theta \approx \Theta$, we obtain the differential scattering cross section for small angles:

$$I(\upsilon_R, \Theta) = \frac{1}{i}\left(\frac{A}{W_R}\right)^{2/i} \frac{1}{\Theta^{2+2/i}} \tag{3.2.28}$$

The variation of σ, ν, and K with υ_R is determined from (3.2.28) and the basic definitions in Section 3.1. If (3.2.28) is substituted into (3.1.14) or (3.1.15), then we see that a scattering potential $U \propto r^{-i}$ leads to $\sigma \propto \upsilon_R^{-4/i}$ and $\nu \propto K \propto \upsilon_R^{-(4/i)+1}$. These scalings are summarized in Table 3.1 for the important scattering processes, which we describe in the next section.

3.3 Elastic Scattering

3.3.1 Coulomb Collisions

The most straightforward elastic scattering process is a Coulomb collision between two charged particles q_1 and q_2, representing an electron–electron, electron–ion, or ion–ion collision. The Coulomb potential is $U(r) = q_1 q_2 / 4\pi\epsilon_0 r$ such that $i = 1$ and we obtain

$$A = C = \frac{q_1 q_2}{4\pi\epsilon_0}$$

from (3.2.25). Using this in (3.2.28), we find

$$I = \left(\frac{b_0}{\Theta^2}\right)^2 \tag{3.3.1}$$

where

$$b_0 = \frac{q_1 q_2}{4\pi\epsilon_0 W_R} \tag{3.3.2}$$

is called the *classical distance of closest approach*. The differential scattering cross section can also be calculated exactly, which we do in Appendix A, obtaining the result

$$I = \left[\frac{b_0}{4\sin^2(\Theta/2)}\right]^2 \tag{3.3.3}$$

However, due to the long range of the Coulomb forces, the integration of I over small Θ (large b) leads to an infinite scattering cross section and to an infinite momentum transfer cross section, such that an upper bound to b, b_{max}, must be assigned. This is done by setting $b_{max} = \lambda_{De}$, the Debye shielding distance for a charge immersed in a plasma, which we calculated in Section 2.4. For momentum transfer, the dependence of σ_m on λ_{De} is logarithmic (Problem 3.5), and the exact choice of b_{max} (or Θ_{min}) makes little difference. For scattering, $\sigma_{sc} \sim \pi\lambda_{De}^2$, which is a very large cross section that depends sensitively on the choice of b_{max}. However, we are generally not interested in scattering through very small angles, which do not appreciably affect the discharge properties. The cross section for scattering through a large angle, say $\Theta \geq \pi/2$, is of more interest.

There are two processes that lead to a large scattering angle Θ for a Coulomb collision: (1) a single collision scatters the particle by a large angle; (2) the cumulative effect of many small-angle collisions scatters the particle by a large angle. The two processes are illustrated in Figure 3.6; the latter process is diffusive and, as we will see, dominates the former.

To estimate the cross section $\sigma_{90}(\text{sgl})$ for a single large-angle collision, we integrate (3.3.3) over solid angles from $\pi/2$ to π to obtain (Problem 3.6)

$$\sigma_{90}(\text{sgl}) = \frac{1}{4}\pi b_0^2 \tag{3.3.4}$$

To estimate $\sigma_{90}(\text{cum})$ for the cumulative effect of many collisions to produce a $\pi/2$ deflection, we first determine the mean square scattering angle $\langle\Theta^2\rangle_1$ for a single collision by averaging Θ^2 over all permitted impact parameters. Since the collisions are predominantly small angle for Coulomb collisions, we can use (3.2.24), which is $\Theta = b_0/b$. Hence,

$$\langle\Theta^2\rangle_1 = \frac{1}{\pi b_{max}^2}\int_{b_{min}}^{b_{max}} \left(\frac{q_1 q_2}{4\pi\epsilon_0 W_R}\right)^2 \frac{2\pi b\,db}{b^2} \tag{3.3.5}$$

The integration has a logarithmic singularity at both $b = 0$ and $b = \infty$ which is cut off by the finite limits. The singularity at the lower limit is due to the small-angle approximation. Setting $b_{min} = b_0/2$ is found to approximate a more accurate calculation. The upper limit, as already mentioned, is $b_{max} = \lambda_{De}$. Using these values and integrating, we obtain

$$\langle\Theta^2\rangle_1 = \frac{2\pi b_0^2}{\pi b_{max}^2} \ln\Lambda \tag{3.3.6}$$

where $\Lambda = 2\lambda_{De}/b_0 \gg 1$.

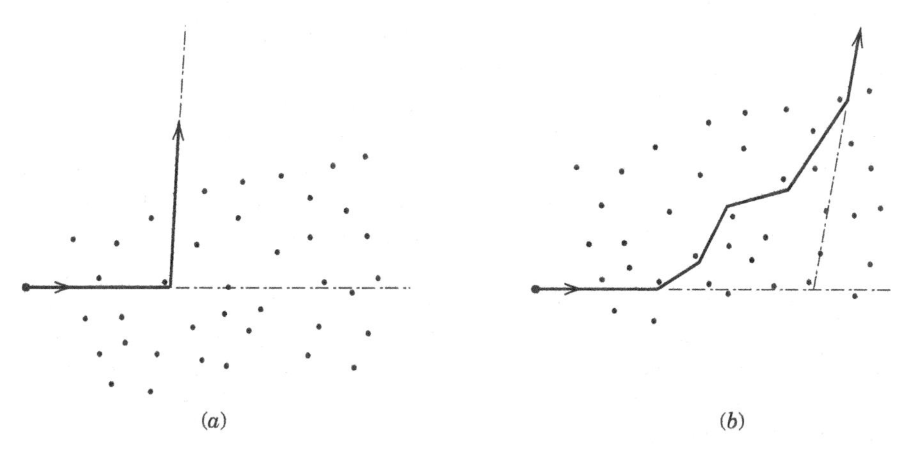

Figure 3.6 The processes that lead to large-angle Coulomb scattering: (*a*) single large-angle event; (*b*) cumulative effect of many small-angle events.

The number of collisions per second, each having a cross section of $\pi b_{\max}^2$ or smaller, is $n_g \pi b_{\max}^2 v_R$, where n_g is the target particle density. Since the spreading of the angle is diffusive, we can then write

$$\langle \Theta^2 \rangle (t) = \langle \Theta^2 \rangle_1 n_g \pi b_{\max}^2 v_R t$$

Setting $t = \tau_{90}$ at $\langle \Theta^2 \rangle = (\pi/2)^2$ and using (3.3.6), we obtain (see also Spitzer, 1956, chapter 5)

$$\nu_{90} = \tau_{90}^{-1} = n_g v_R \frac{8}{\pi} b_0^2 \ln \Lambda$$

Writing $\nu_{90} = n_g \sigma_{90} v_R$, we see that

$$\sigma_{90} = \frac{8}{\pi} b_0^2 \ln \Lambda \tag{3.3.7}$$

Although Λ is a large number, typically $\ln \Lambda \approx 10$ for the types of plasmas, we are considering.

Comparing $\sigma_{90}(\text{sgl})$ to σ_{90}, we see that due to the large range of the Coulomb fields, the effective cross section for many small-angle collisions to produce an rms deflection of $\pi/2$ is larger by a factor $(32/\pi^2) \ln \Lambda$. Because of this enhancement, it is possible for electron–ion or ion–ion particle collisions to play a role in weakly ionized plasmas (say one percent ionized). Another important characteristic of Coulomb collisions is the strong velocity dependence. From (3.3.2), we see that $b_0 \propto 1/v_R^2$. Thus, from (3.3.4) or (3.3.7)

$$\sigma_{90} \propto \frac{1}{v_R^4} \tag{3.3.8}$$

such that low-velocity particles are preferentially scattered. The temperature of the species is therefore important in determining the relative importance of the various species in the collisional processes, as we shall see in subsequent sections.

3.3.2 Polarization Scattering

The main collisional processes in a weakly ionized plasma are between charged and neutral particles. For electrons at low energy and for ions scattering against neutrals, the dominant process is relatively short-range polarization scattering. At higher energies for electrons, the collision time is shorter and the atoms do not have time to polarize. In this case, the scattering becomes more Coulomb-like, but with $b_{\max}$ at an atomic radius, and inelastic processes such as ionization become important also. The condition for polarization scattering is $v_R \lesssim v_{at}$, where v_{at} is the characteristic electron velocity in the atom, which we obtain in the next section. Because of the short range of the polarization potential, we need not be concerned with an upper limit for the integration over b, but the potential is more complicated. We determine the potential from a simple model of the atom as a point charge of value $+q_0$, surrounded by a uniform negative charge sphere (valence electrons) of total charge $-q_0$, such that the charge density is $\rho = -q_0/\frac{4}{3}\pi a^3$, where a is the atomic radius. An incoming electron (or ion) can polarize the atom by repelling (or attracting) the charge cloud quasistatically. The balance of forces on the central point charge due to the displaced charge cloud and the incoming charged particle, taken to have charge q, is shown in Figure 3.7, where the center of the charge cloud and the point charge are displaced by a distance d. Applying Gauss' law to a sphere of radius d around the center of the cloud,

$$4\pi\epsilon_0 d^2 E_{\text{ind}} = -q_0 \frac{d^3}{a^3}$$

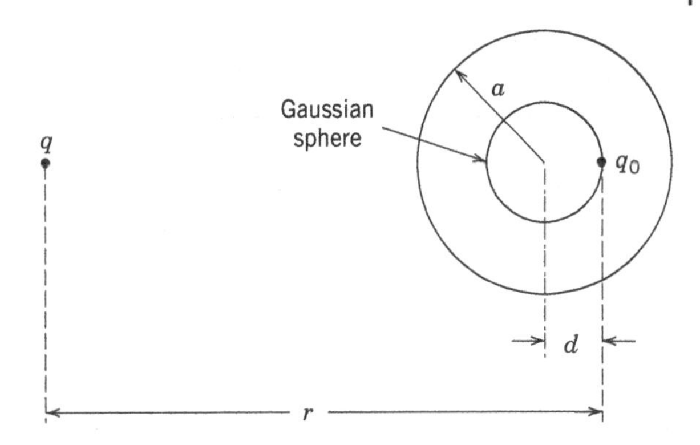

Figure 3.7 Polarization of an atom by a point charge q.

we obtain the induced electric field acting on the point charge due to the displaced cloud

$$E_{\mathrm{ind}} = -\frac{q_0 d}{4\pi\epsilon_0 a^3}$$

The electric field acting on the point charge due to the incoming charge is

$$E_{\mathrm{appl}} = \frac{q}{4\pi\epsilon_0 r^2}$$

For force balance on the point charge, the sum of the fields must vanish, yielding an induced dipole moment for the atom:

$$p_{\mathrm{d}} = q_0 d = \frac{qa^3}{r^2} \tag{3.3.9}$$

The induced dipole, in turn, exerts a force on the incoming charged particle:

$$\mathbf{F} = \frac{2p_{\mathrm{d}}q}{4\pi\epsilon_0 r^3}\hat{\mathbf{r}} = \frac{2q^2a^3}{4\pi\epsilon_0 r^5}\hat{\mathbf{r}} \tag{3.3.10}$$

Integrating **F** with respect to r, we obtain the attractive potential energy:

$$U(r) = -\frac{q^2a^3}{8\pi\epsilon_0 r^4} \tag{3.3.11}$$

The *polarizability* for this simple atomic model is defined as $\alpha_{\mathrm{p}} = a^3$. The relative polarizabilities $\alpha_{\mathrm{R}} = \alpha_{\mathrm{p}}/a_0^3$, where a_0 is the Bohr radius, for some simple atoms and molecules are given in Table 3.2.

The orbits for scattering in the polarization potential are complicated (McDaniel, 1989). As shown in Figure 3.8, there are two types of orbits. For impact parameter $b > b_{\mathrm{L}}$, the orbit has a hyperbolic character, and, for $b \gg b_{\mathrm{L}}$, the straight-line trajectory analysis in Section 3.2 can be applied (Problem 3.7). For $b < b_{\mathrm{L}}$, the incoming particle is "captured" and the orbit spirals into the core, leading to a large scattering angle. Either the incoming particle is "reflected" by the core and spirals out again or the two particles strongly interact, leading to inelastic changes of state.

The critical impact parameter b_{L} can be determined from the conservation of energy and angular momentum for the incoming particle having mass m and speed v_0, with the mass of the scatterer taken to be infinite for ease of analysis. In cylindrical coordinates (see Figure 3.8*a*), we obtain

$$\frac{1}{2}mv_0^2 = \frac{1}{2}m(\dot{r}^2 + r^2\dot{\phi}^2) + U(r) \tag{3.3.12a}$$

$$mv_0 b = mr^2\dot{\phi} \tag{3.3.12b}$$

Table 3.2 Relative Polarizabilities $\alpha_R = \alpha_p/a_0^3$ of Some Atoms and Molecules, Where a_0 is the Bohr Radius

Atom or molecule	α_R
H	4.5
C	12.
N	7.5
O	5.4
Ar	11.08
CCl_4	69.
CF_4	19.
CO	13.2
CO_2	17.5
Cl_2	31.
H_2O	9.8
NH_3	14.8
O_2	10.6
SF_6	30.

Source: Smirnov (1981).

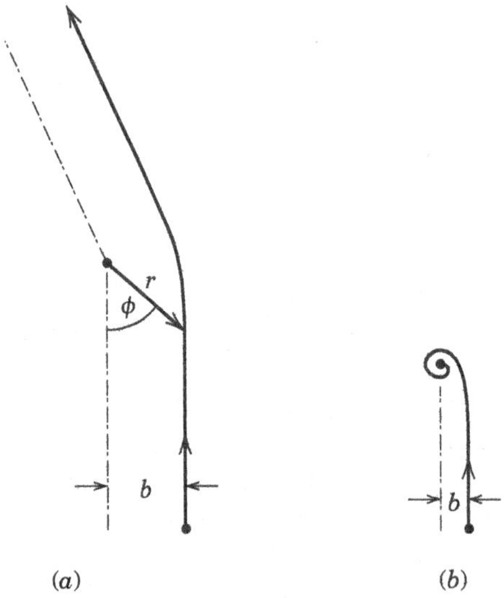

(a) (b)

Figure 3.8 Scattering in the polarization potential, showing (*a*) hyperbolic and (*b*) captured orbits.

At closest approach, $\dot{r} = 0$ and $r = r_{\mathrm{min}}$. Substituting these into (3.3.12) and eliminating $\dot{\phi}$, we obtain a quadratic equation for r_{min}^2:

$$v_0^2 r_{\mathrm{min}}^4 - v_0^2 b^2 r_{\mathrm{min}}^2 + \frac{\alpha_p q^2}{4\pi\epsilon_0 m} = 0$$

Using the quadratic formula to obtain the solution for $r^2_{\rm min}$, we see that there is no real solution for $r^2_{\rm min}$ when

$$(v_0^2 b^2)^2 - 4v_0^2 \frac{\alpha_{\rm p} q^2}{4\pi\epsilon_0 m} \leq 0$$

Choosing the equality at $b = b_{\rm L}$, we solve for $b_{\rm L}$ to obtain

$$\sigma_{\rm L} = \pi b_{\rm L}^2 = \left(\frac{\pi \alpha_{\rm p} q^2}{\epsilon_0 m}\right)^{1/2} \frac{1}{v_0} \tag{3.3.13}$$

which is known as the *Langevin* or *capture* cross section. If the target particle has a finite mass m_2 and velocity $\mathbf{v}_2$ and the incoming particle has a mass m_1 and velocity $\mathbf{v}_1$, then (3.3.13) holds provided m is replaced by the reduced mass $m_{\rm R} = m_1 m_2/(m_1 + m_2)$ and v_0 is replaced by the relative velocity $v_{\rm R} = |\mathbf{v}_1 - \mathbf{v}_2|$. We note that the cross section $\sigma_{\rm L} \propto 1/v_{\rm R}$. Hence, the collision frequency for capture is

$$\nu_{\rm L} = n_{\rm g} \sigma_{\rm L}\, v_{\rm R} = n_{\rm g} K_{\rm L} \tag{3.3.14}$$

where

$$K_{\rm L} = \left(\frac{\pi \alpha_{\rm p} q^2}{\epsilon_0 m_{\rm R}}\right)^{1/2} \tag{3.3.15}$$

is the rate constant for capture and $n_{\rm g}$ is the target particle density. Both $\nu_{\rm L}$ and $K_{\rm L}$ are independent of velocity. In practical units, the rate constants for electrons and ions are (with $q = \pm e$)

$$K_{\rm Le} = 3.85 \times 10^{-8} {\alpha_{\rm R}}^{1/2}\, {\rm cm^3/s} \tag{3.3.16}$$

$$K_{\rm Li} = 8.99 \times 10^{-10} \left(\frac{\alpha_{\rm R}}{A_{\rm R}}\right)^{1/2} {\rm cm^3/s} \tag{3.3.17}$$

where $A_{\rm R}$ is the reduced mass in atomic mass units (hydrogen $\approx$ 1 amu) and $\alpha_{\rm R}$ is the relative polarizability. Because $\sigma_{\rm L} \propto 1/\sqrt{\mathcal{E}}$, where $\mathcal{E}$ is the collision energy in the center of mass system, the Langevin cross section dominates the elastic and inelastic collisional behavior at thermal energies ($\mathcal{E} \sim 0.026$ V), especially for ion–neutral collisions. Some molecules (but not atoms) have permanent dipole moments, leading to a scattering potential $U \propto 1/r^2$ and an enhanced Langevin cross section. We describe this briefly in Chapter 8.

What is the actual velocity dependence of elastic electron–atom collisions? At low energies, we might expect quantum effects to be significant, which is indeed the case, such that some gases show low-energy resonances in their cross sections.

An example of a simple velocity dependence is shown for hydrogen and helium in Figure 3.9. Here, a normalized cross section unit is used called the *probability of collision* $P_{\rm c}$, defined as the average number of collisions in 1 cm of path through a gas at 1 Torr at 273 K. The elastic collision frequency in these units is

$$\nu_{\rm el} = v p_0 P_{\rm c}$$

where $p_0 = 273p/T$. We see from the figure that at low energy, the cross section is hard-sphere-like, being independent of velocity. At higher energies, $\sigma \propto v^{-1}$ and thus the polarization potential governs the behavior.

The low-energy cross sections can, in fact, be quite complicated, depending on quantum mechanical effects. For example, in many gases, the quantum mechanical wave diffraction of the electron

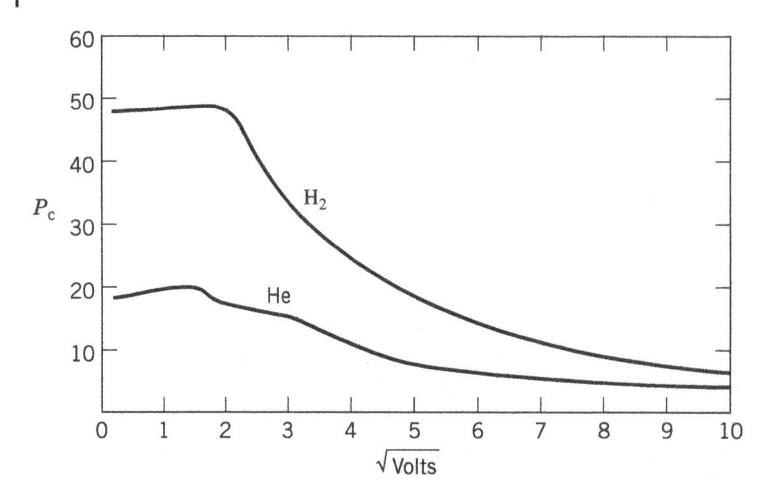

Figure 3.9 Probability of collision P_c for electrons in H_2 and He; the cross section is $\sigma \approx 2.87 \times 10^{-17}\, P_c$ cm^2 (Brown, 1959/MIT Press).

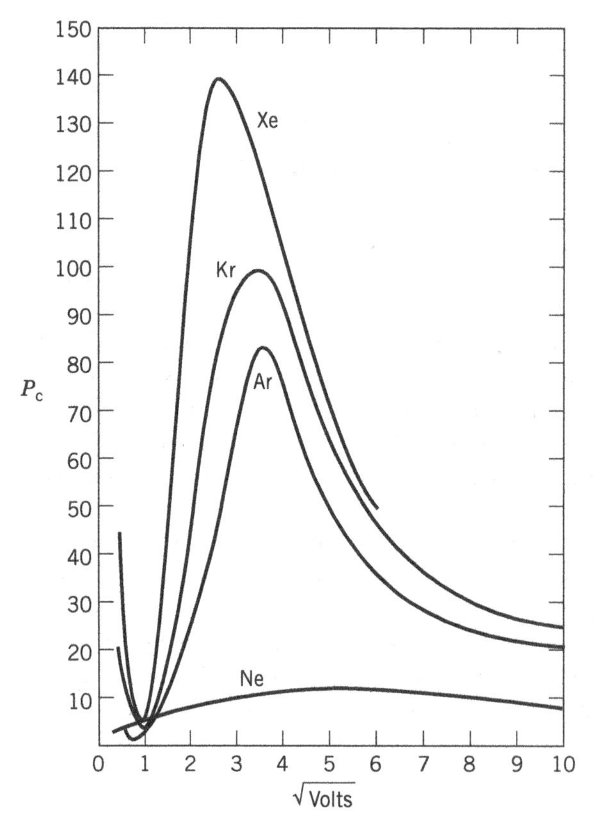

Figure 3.10 Probability of collision P_c for electrons in Ne, Ar, Kr, and Xe, showing the Ramsauer minima for Ar, Kr, and Xe; the cross section is $\sigma \approx 2.87 \times 10^{-17}\, P_c$ cm^2 (after Brown, 1959).

around the atom at low energy leads to a "hole" in the elastic collision frequency at some low energy. This is true for some noble gases, as seen in Figure 3.10, as well as some processing gases, such as CF_4. At higher (but still moderate) energy, the approximate proportionality for polarization scattering $\sigma \propto v^{-1}$ is still found.

3.4 Inelastic Collisions

3.4.1 Atomic Energy Levels

The physics and spectroscopy of atoms is a vast area, and we give only a brief summary here. The reader should consult textbooks such as Bransden and Joachain (1983) and Thorne (1988) for a more thorough treatment. Atoms consist of one or more electrons bound to a heavy positive nucleus.

In a classical description, electrons move in circular orbits whose radii a are set by the balance between the inward electrostatic and the outward centrifugal forces. For the hydrogen atom, the inward force is the Coulomb force of the proton, leading to the force balance:

$$\frac{e^2}{4\pi\epsilon_0 a^2} = \frac{mv^2}{a} \tag{3.4.1}$$

From (3.4.1), all radii (and corresponding velocities) are possible. A quantum description limits the orbits to those for which the angular momentum is an integral multiple of $\hbar$,

$$mva = n\hbar \tag{3.4.2}$$

where here $n \geq 1$ is an integer called the *principal quantum number*, and $\hbar = h/2\pi$, with Planck's constant $h \approx 6.626 \times 10^{-34}$ J-s. Solving (3.4.1) and (3.4.2) yields the quantized radii

$$a_n = n^2 a_0 \tag{3.4.3}$$

where for the lowest level ($n = 1$),

$$a_0 = \frac{4\pi\epsilon_0 \hbar^2}{e^2 m} \approx 5.29 \times 10^{-11}\ \text{m} \tag{3.4.4}$$

is the *Bohr radius*. The velocity is

$$v_n = \frac{v_{\text{at}}}{n}$$

where

$$v_{\text{at}} = \frac{e^2}{4\pi\epsilon_0 \hbar} \approx 2.19 \times 10^6\ \text{m/s} \tag{3.4.5}$$

is the electron velocity in the first Bohr orbit. The characteristic atomic timescale is then

$$t_{\text{at}} = \frac{a_0}{v_{\text{at}}} \approx 2.42 \times 10^{-17}\ \text{s} \tag{3.4.6}$$

The electron energy W_n is the sum of the kinetic and potential energy,

$$W_n = \frac{1}{2} m v_n^2 - \frac{e^2}{4\pi\epsilon_0 a_n} \tag{3.4.7}$$

Defining $W_n(\text{J}) = e\mathcal{E}_n(\text{V})$, we obtain

$$\mathcal{E}_n = -\frac{\mathcal{E}_{\text{at}}}{n^2} \tag{3.4.8a}$$

where

$$\mathcal{E}_{\text{at}} = \frac{1}{2}\frac{m}{e}\left(\frac{e^2}{4\pi\epsilon_0 \hbar}\right)^2 \approx 13.61\ \text{V} \tag{3.4.8b}$$

is the ionization potential of the hydrogen atom in its lowest energy state ($n = 1$).

For a many-electron atom, a valence electron sees some effective positive charge $Z_{\text{eff}}e$. This leads to a radius for the first Bohr orbit $a_{\text{eff}} = a_0/Z_{\text{eff}}$ and to an ionization potential $\mathcal{E}_{\text{iz}} = Z_{\text{eff}}^2\mathcal{E}_{\text{at}}$. When we combine these expressions, the radius of an atom is found to scale as

$$a_{\text{eff}} \approx a_0 \left(\frac{\mathcal{E}_{\text{at}}}{\mathcal{E}_{\text{iz}}}\right)^{1/2} \tag{3.4.9}$$

where $\mathcal{E}_{\text{at}}$ is given by (3.4.8b).

This picture, while qualitatively correct, is incomplete. Quantum mechanics specifies the state of each electron in an atom in terms of four quantum numbers, n, l, m_l, and m_s (n, l, and m_l are integers), with the restrictions $l + 1 \leq n$, $|m_l| \leq l$, and with $m_s = \pm\frac{1}{2}$. The quantum numbers l and m_l specify the total orbital angular momentum and its component in a particular direction; the quantum number m_s specifies the direction of the electron spin.

For the preceding model, the energy of each level depends only on n. By the restrictions on l, m_l, and m_s, there are $2n^2$ electron states having the same energy $\mathcal{E}_n$. The energy level $\mathcal{E}_n$ is said to have *degeneracy* $2n^2$. For an atom with more than one electron, the force balance includes not only the attractive force of the nucleus but also the repulsive forces of the other electrons. In the *central field model*, each electron moves under the influence of a spherically symmetric potential that includes the average effects of all the other electrons. This breaks the degeneracy such that the energy is a function of both n and l. Figure 3.11 shows a typical energy level diagram with the different l values displaced to the right. For historical reasons, electrons having $l = 0$, 1, 2, and 3 are known as s, p, d, and f electrons, respectively.

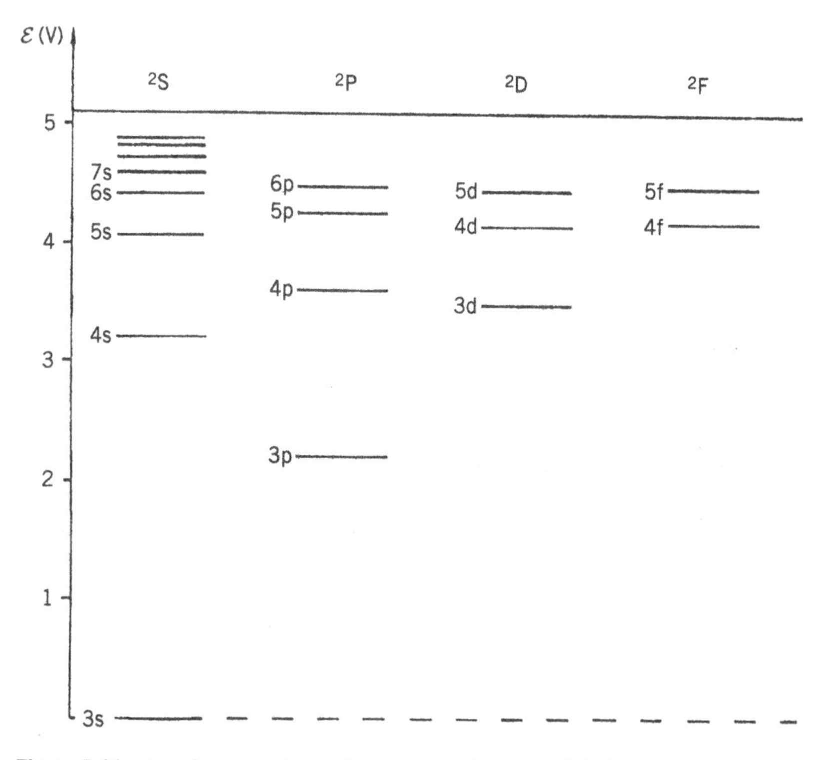

Figure 3.11 Atomic energy levels for the central field model of an atom, showing the dependence of the energy levels on the quantum numbers n and l; the energy levels are shown for sodium, without the fine structure (Thorne, 1988/Springer Nature).

The Pauli exclusion principle states that no two electrons can have the same state. Hence, stable atoms are built by placing electrons into the available states in order of increasing energy. For example, the *electronic configurations* of the lowest-energy states (ground states) of hydrogen, oxygen, and argon are 1s, $1s^2 2s^2 2p^4$, and $1s^2 2s^2 2p^6 3s^2 3p^6$. In this notation, the values of n and l specify a given electron *subshell*, and the superscript indicates the number of electrons in each subshell, which holds a maximum of $2(2l + 1)$ electrons. The *valence electrons*, which are those in the last (usually incomplete) subshell, determine the collisional and other behavior of atoms. For example, an electron collision with an argon atom can excite the atom to a higher energy level,

$$\mathrm{e} + \mathrm{Ar} \rightarrow \mathrm{Ar}^* + \mathrm{e}$$

corresponding to a change of state

$$3\mathrm{p}^6 \rightarrow 3\mathrm{p}^5 4\mathrm{s}^1$$

for the valence electrons.

For the light elements (roughly $Z \lesssim 40$), the energy levels are usually labeled by the values of the permitted orbital and spin angular momentum L and S for the sum of all the valence electrons. Levels with different L values are known as S, P, D, and F levels for $L = 0, 1, 2$, and 3, by analogy with single-electron terminology. The integer or half-integer value of S is indicated by a superscript $2S + 1$, the *multiplicity*, placed to the left of the L value. The degeneracy (number of states) for a level with a given L and S is $(2L + 1)(2S + 1)$. Part of the degeneracy is usually removed by weak magnetic interactions between the spin and orbital motions, giving rise to additional small splittings of the degenerate energy levels, the so-called *fine structure*. This is specified by a quantum number J for the sum of the total orbital and spin angular momentum, which can have integer or half-integer value, and which is written as a subscript to the right of the L value. The remaining degeneracy for each level with a given L, S, and J is $2J + 1$. The ground-state energy levels of hydrogen, oxygen, and argon in this notation are ${}^2\mathrm{S}_{1/2}$, ${}^3\mathrm{P}_2$, and ${}^1\mathrm{S}_0$, respectively. For heavy elements, roughly $Z \gtrsim 40$, the L and S values are no longer meaningful quantum numbers, and the n and J values alone, along with the j values of the individual electrons, can be used to specify a level.

3.4.2 Electric Dipole Radiation and Metastable Atoms

Atoms in their ground states can be excited by collisions or radiation to higher energy-bound states. In most cases, only a single-valence electron is excited. Most bound states can emit a photon by electric dipole radiation and return to some lower energy state or to the ground state:

$$\mathrm{e} + \mathrm{Ar} \rightarrow \mathrm{Ar}^* + \mathrm{e} \rightarrow \mathrm{Ar} + \mathrm{e} + \hbar\omega$$

Here, $\hbar\omega$ is the photon energy and ω is its radian frequency. The radiation is usually in the visible or ultraviolet. Electric dipole radiation is permitted between two states only if the selection rules

$$\Delta l = \pm 1$$

$$\Delta J = 0, \pm 1 \qquad (\text{but } J = 0 \rightarrow J = 0 \text{ forbidden}) \tag{3.4.10a}$$

are satisfied. For the light elements, with L and S also good quantum numbers, the additional rules

$$\Delta L = 0, \ \pm 1 \qquad (\text{but } L = 0 \rightarrow L = 0 \text{ forbidden})$$
$$\Delta S = 0 \tag{3.4.10b}$$

must also be satisfied.[2]

We can estimate the timescale for electric dipole radiation from the time-average energy per unit time radiated by a classical oscillating dipole $p_{\rm d}(t) = p_{\rm d0}\cos\omega t$ (Jackson, 1975, chapter 14):

$$P_{\rm rad} = \frac{\omega^4 p_{\rm d0}^2}{12\pi\epsilon_0 c^3} \tag{3.4.11}$$

Dividing the energy radiated $\hbar\omega$ by $P_{\rm rad}$, we obtain the radiation time

$$t_{\rm rad} = \frac{12\pi\epsilon_0 \hbar c^3}{\omega^3 p_{\rm d0}^2} \tag{3.4.12}$$

Taking the simple estimates $p_{\rm d0} = ea_0$ and $\omega = e\mathcal{E}_{\rm at}/\hbar$, with $\mathcal{E}_{\rm at}$ given by (3.4.8b), and using (3.4.6), we obtain

$$\begin{aligned} t_{\rm rad} &= 24\left(\frac{4\pi\epsilon_0\hbar c}{e^2}\right)^3 t_{\rm at} \approx 6.2\times 10^7\, t_{\rm at} \\ &\approx 1.5\times 10^{-9}\,{\rm s} \end{aligned} \tag{3.4.13}$$

We see that $t_{\rm rad}$ is long compared to the characteristic atomic timescale $t_{\rm at}$. However, the characteristic time between collisions is

$$\tau \sim (n_{\rm g}\pi a_0^2 \bar{v})^{-1}$$

For electrons with $T_{\rm e} \sim 3$ V and $n_{\rm g} \sim 3.3\times 10^{14}$ cm^{-3} (corresponding to a gas pressure of 10 mTorr), we obtain

$$\tau_{\rm e} \sim 3\times 10^{-7}\,{\rm s} \tag{3.4.14}$$

For heavy particle collisions, we estimate for $T_{\rm i} \sim T_{\rm e}$ and $M/m \sim 10^4$ that $\tau_{\rm i} \sim 100\tau_{\rm e}$. Hence, we have $t_{\rm rad} \ll \tau_{\rm e}, \tau_{\rm i}$ in low-pressure discharges. This implies that excited states will generally be de-excited by electric dipole radiation rather than by collisions.

Certain excited states, however, cannot satisfy the selection rules (3.4.10) for electric dipole radiation; for these states, $p_{\rm d0} \equiv 0$. While other radiative transitions may occur, such as electric quadrupole or magnetic dipole radiation, or radiationless transitions may occur, to states of nearly equal energy that subsequently do radiate, these mechanisms are generally weak, leading to transition times that can be long compared to the collision times $\tau_{\rm e}$ and $\tau_{\rm i}$. The energy levels from which electric dipole radiation is forbidden are called *metastable*, and the excited atoms are called *metastable atoms.* Metastable atoms are often present at considerable densities in weakly ionized discharges, where they can be further excited, ionized, or de-excited by collisions.

Figure 3.12*a* shows the energy levels for argon. The energies are given with respect to the energy required to create a $^2P_{3/2}$ Ar^+ ion from the $^1S_0$ neutral ground state. A ground-state 3p electron ($l = 1$) can be excited into one of the states in the 4s configuration ($l = 0$), the 4p configuration ($l = 1$), and so on. Details of the 4s and 4p states are shown in Figure 3.12*b*, with the energy spacings not drawn to scale, but with the energies given in parentheses. There are ten levels in the 4p

2 If L and S are good quantum numbers, then the atomic state has both a unique orbital and spin angular momentum.

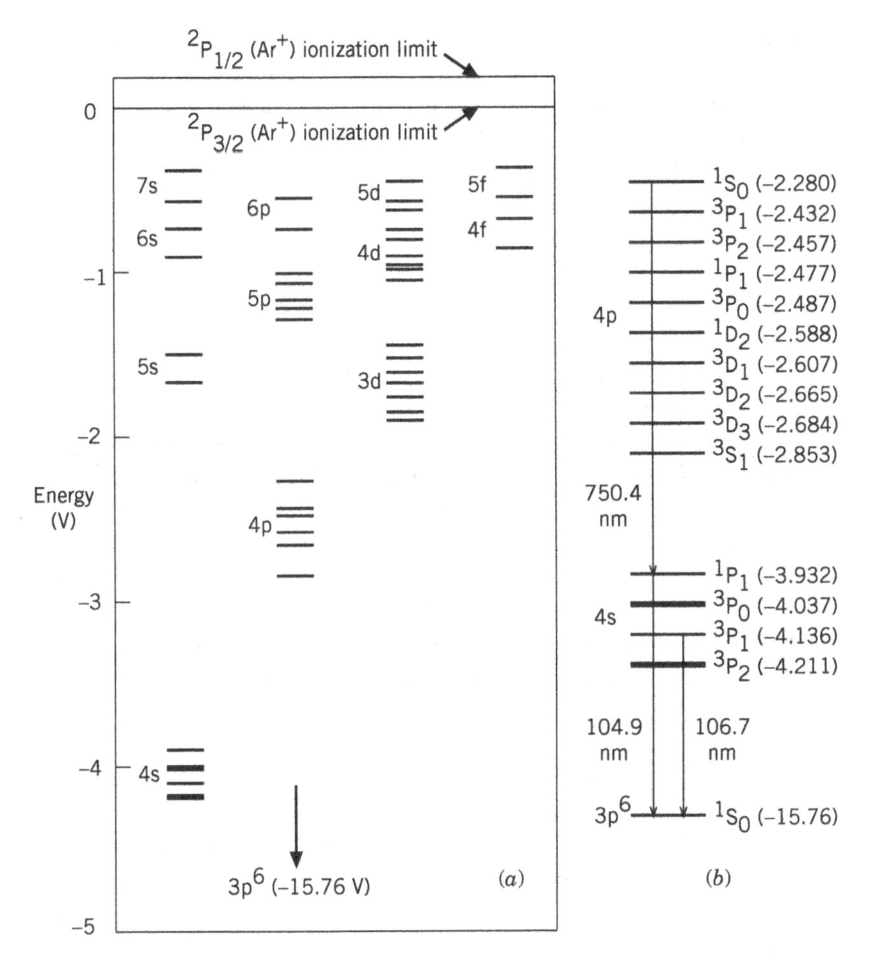

Figure 3.12 The energy levels of the argon atom, showing (*a*) the ($3p^5nl$) configurations and (*b*) details of the $3p^54s$ and $3p^54p$ configurations, with the two metastable levels shown as heavy solid lines (Edgell, 1961/Interscience Publishers).

configuration and four levels in the 4s configuration. The $^3P_0$ and $^3P_2$ levels of the 4s configuration, shown as heavy lines, are metastable, because (a) they do not satisfy the selection rule on J given in (3.4.10a) for electric dipole radiation to the ground state, and (b) electric dipole radiation from these levels to a lower energy 4s level does not satisfy the selection rule on l given in (3.4.10a). Accounting for the degeneracies $2J + 1$, there is one metastable state in the $^3P_0$ level and five in the $^3P_2$ level. The remaining two levels ($^1P_1$ and $^3P_1$) in the 4s configuration, each containing three states, can radiate to the $3p^6$ ground state. The resulting radiation is in the ultraviolet and is called *resonance radiation.* The $3p^54s$ ($^1P_1$) → $3p^6$($^1S_0$) radiative transition at 104.9 nm is very strong, with a lifetime of 2.5 ns. The $3p^54s$ ($^3P_1$) → $3p^6$ ($^1S_0$) transition at 106.7 nm is also strong, with a lifetime of 10.4 ns, even though this radiation is "prohibited" by the selection rule on ΔS given in (3.4.10b). This is because the additional selection rules in (3.4.10b) apply most strongly only to the light elements. Argon (mass = 40 amu) satisfies them only marginally (see Bransden and Joachain, 1983, for further details). The excited levels are sometimes denoted using "Paschen labels." In order of decreasing energy, the 4s configuration levels $^1P_1 \ldots {}^3P_2$ are labeled "$1s_2 \ldots 1s_5$," and the 4p configuration levels $^1S_0 \ldots {}^3S_1$ are labeled "$2p_1 \ldots 2p_{10}$."

Another example of metastable levels is for the two-valence electron helium system. Since electric dipole transitions between $S = 0$ and $S = 1$ states are forbidden by (3.4.10b), the energy level diagram decomposes into two nearly independent energy level systems: the *singlets* ($2S + 1 = 1$) and the *triplets* ($2S + 1 = 3$). Because $\Delta l = 0$ and $L = 0 \rightarrow L = 0$ are forbidden, the 2s ($^1S_0$) and 2s ($^3S_1$) levels are metastable. These states find application in He–Ne gas lasers, where they are excited by e–He collisions and are collisionally de-excited by He*–Ne collisions to create excited Ne* atoms that subsequently radiate, leading to laser action.

3.4.3 Electron Ionization Cross Section

Quantum mechanics is needed to properly treat electron–atom ionization. We give here a simple classical description (Thomson, 1912) that provides a qualitative treatment. The basic idea is to determine the condition for the incident electron (having velocity v) to transfer to a valence electron (assumed to be at rest) an energy equal to the ionization energy. Using (3.3.1) with $q_1 = q_2 = -e$ and $m_1 = m_2 = m$, the electron charge, and mass, we have for a small angle collision that

$$I(v, \Theta) = \left(\frac{e^2}{4\pi\epsilon_0}\right)^2 \frac{1}{W_R^2} \frac{1}{\Theta^4} \tag{3.4.15}$$

where $W_R = \frac{1}{2} m_R v^2$ is the CM energy and $m_R = m/2$ is the reduced mass. Substituting $\theta = \Theta/2$ into (3.4.15), we transform to the scattering angle in the laboratory frame, and using (3.2.12), we obtain

$$d\sigma = I(v, \theta)\, 2\pi \sin\theta\, d\theta = 2\pi \left(\frac{e^2}{4\pi\epsilon_0}\right)^2 \frac{1}{W^2} \frac{d\theta}{\theta^3} \tag{3.4.16}$$

where $W = \frac{1}{2} m v^2$ is the energy in the laboratory system. The energy transfer to a stationary target from a moving one is

$$W_L = \zeta_L(\Theta) W \tag{3.4.17}$$

where ζ_L is given by (3.2.18). Again, making the small-angle assumption, $\cos\Theta \approx 1 - \Theta^2/2$, with equal mass electrons, we obtain

$$W_L = \frac{1}{4}\Theta^2 W = \theta^2 W \tag{3.4.18}$$

and

$$dW_L = 2\theta\, d\theta\, W \tag{3.4.19}$$

Substituting (3.4.18) and (3.4.19) into (3.4.16), we have

$$d\sigma = \pi \left(\frac{e^2}{4\pi\epsilon_0}\right)^2 \frac{1}{W} \frac{dW_L}{W_L^2} \tag{3.4.20}$$

For ionization, we integrate W_L from the ionization energy U_{iz} (for $W > U_{iz}$) to W, obtaining

$$\sigma_{iz} = \pi \left(\frac{e^2}{4\pi\epsilon_0}\right)^2 \frac{1}{W} \left(\frac{1}{U_{iz}} - \frac{1}{W}\right) \tag{3.4.21a}$$

or, using voltage units $W = e\mathcal{E}$, $U_{iz} = e\mathcal{E}_{iz}$,

$$\sigma_{iz} = \pi \left(\frac{e}{4\pi\epsilon_0}\right)^2 \frac{1}{\mathcal{E}} \left(\frac{1}{\mathcal{E}_{iz}} - \frac{1}{\mathcal{E}}\right) \qquad \mathcal{E} > \mathcal{E}_{iz} \tag{3.4.21b}$$

which is the *Thomson cross section*. For $\mathcal{E} < \mathcal{E}_{iz}$, $\sigma_{iz} = 0$. The ionization cross section reaches its maximum value for $\mathcal{E} = 2\mathcal{E}_{iz}$

$$\sigma_{iz}(\max) = \frac{\pi}{4}\left(\frac{e}{4\pi\epsilon_0}\right)^2 \frac{1}{\mathcal{E}_{iz}^2}$$

and falls proportional to $\mathcal{E}^{-1}$ for $\mathcal{E} \gg \mathcal{E}_{iz}$. The cross section in (3.4.21) should be multiplied by the number of valence electrons if there is more than one.

Another classical estimate for σ_{iz} is found if the orbital electron motion and its radial distribution are taken into account. Smirnov (1981, p. 253) gives the result

$$\sigma_{iz} = \frac{\pi}{4}\left(\frac{e}{4\pi\epsilon_0}\right)^2 \frac{1}{\mathcal{E}}\left(\frac{5}{3\mathcal{E}_{iz}} - \frac{1}{\mathcal{E}} - \frac{2\mathcal{E}_{iz}}{3\mathcal{E}^2}\right), \qquad \mathcal{E} > \mathcal{E}_{iz} \tag{3.4.22}$$

which has twice the maximum value of the Thomson cross section at $\mathcal{E} \approx 1.85\,\mathcal{E}_{iz}$. Practical formulae for cross sections can be found in Barnett (1989). A quantum mechanical calculation shows that $\sigma_{iz} \propto \ln \mathcal{E}/\mathcal{E}$ at high energies.

The ionization rate, at a given energy, is obtained from the cross section as

$$\nu_{iz} = n_g \sigma_{iz} v$$

which falls as v^{-1} for $\mathcal{E} \gg \mathcal{E}_{iz}$. As with the collision frequency, the ionizations are usually caused by a distribution of electron energies, and particularly for a low-temperature Maxwellian (say $T_e = 4$ V), ν_{iz} is very sensitive to the exponential tail of the distribution. This also implies great sensitivity to the form of the distribution function. We shall encounter this effect, and the problems of analysis arising from it, in calculating the particle balance in discharges. In the next section, we consider the effective collision parameters when integrated over the particle distributions.

3.4.4 Electron Excitation Cross Section

A simple classical estimate for excitation to a given energy level $\mathcal{E}_n$ can be obtained by following the Thomson procedure but integrating $d\sigma$ over the energy W_L transferred from $e\mathcal{E}_n$ (for $W > e\mathcal{E}_n$) to min $(W, e\mathcal{E}_{n+1})$. For the total excitation cross section σ_{ex}, $d\sigma$ can be integrated from $e\mathcal{E}_2$ (for $W > e\mathcal{E}_2$) to min (W, U_{iz}). We leave this as an exercise for the reader. Quantum mechanics shows that the cross sections to levels that are optically forbidden (electric dipole radiation to the ground state is forbidden) are smaller and fall off faster with energy above the peak than for electron impact excitation to optically allowed levels.

For real gases, the atomic cross sections are only approximated by the analytic expressions found here. More accurate determinations are made experimentally using crossed beam techniques. As an example, for argon, which is a commonly used gas in discharges, the electron elastic, ionization, and excitation cross sections are shown in Figure 3.13. The ionization cross section reasonably follows the analytic estimates with $\mathcal{E}_{iz} = 15.76$ V. The analytic form (3.4.21b) with six valence electrons has $\sigma_{iz}(\max) \approx 3.9 \times 10^{-16}$ cm^2 at $\mathcal{E} \approx 31.6$ V, while the experimental values, from Figure 3.13, are $\sigma_{iz}(\max) \approx 3.9 \times 10^{-16}$ cm^2 at $\mathcal{E} \approx 60$ V. The total excitation cross section roughly follows the ionization cross section, except that it extends to lower energies, because the average excitation energy is roughly $\mathcal{E}_{ex} \approx \frac{3}{4}\mathcal{E}_{iz}$; for argon $\mathcal{E}_{ex} \approx 12.14$ V. The elastic scattering cross section, on the other hand, has a low energy dependence due to a quantum mechanical resonance, the Ramsauer minimum, and therefore follows neither the hard-sphere nor the polarization models. At the higher energies, the electrons can penetrate into the atomic cloud and a cross section $\sigma_{el} \propto v^{-2}$ is found, which implies an admixture of polarization and Coulomb scattering.

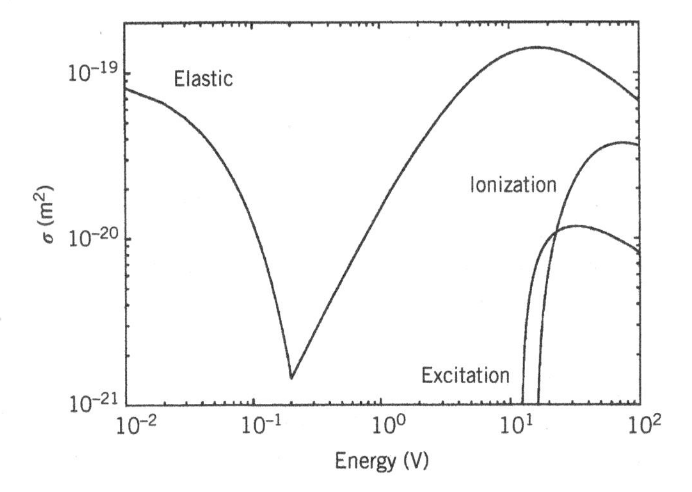

Figure 3.13 Ionization, excitation, and elastic scattering cross sections for electrons in argon gas (Vahedi, 1993).

3.4.5 Ion–Atom Charge Transfer

A positive ion can collide with an atom so as to capture a valence electron, resulting in a transfer of the electron from the atom to the ion. In general, the energy of the level from which the electron is released is not equal to the energy of the level into which the electron is captured, leading to an *energy defect* ΔW, which may be positive or negative. For $\Delta W \neq 0$, the kinetic energy of the colliding particles is not conserved in the collision. If, however, the atom and ion are parent and child, then the transfer can occur with zero defect, e.g.,

$$\mathrm{Ar}^+ \text{ (fast)} + \mathrm{Ar} \text{ (slow)} \to \mathrm{Ar} \text{ (fast)} + \mathrm{Ar}^+ \text{ (slow)} \tag{3.4.23}$$

and the process is said to be *resonant*. Although the ion and atom change their internal states, their kinetic energy is conserved. The cross section for resonant charge transfer is large at low collision energies, making this an important process in weakly ionized plasmas. Here, we give a simple classical estimate of charge transfer that provides a qualitative picture of the process. A more complete understanding depends on molecular phenomena that will be considered further in Chapter 8. For a more thorough treatment of the phenomena, the reader should consult the monograph by Bransden and McDowell (1992).

For the reaction

$$\mathrm{A}^+ + \mathrm{B} \text{ (at rest)} \to \mathrm{A} + \mathrm{B}^+ \tag{3.4.24}$$

the transfer from level n of B requires two steps: release from B and capture by A^+. For a center-to-center separation a_{12} of A^+ and B, the potential energy of the electron in level n of B is

$$W = -\frac{U_{\mathrm{izB}}}{n^2} - \frac{e^2}{4\pi\epsilon_0 a_{12}} \tag{3.4.25}$$

where the first term, from (3.4.8), is the energy when A^+ is not present, and the second term is the additional electrostatic energy due to the nearby positive charge A^+. The potential energy $U(z)$ of an electron in the Coulomb fields of the A^+ and B^+ ions is

$$U(z) = -\frac{e^2}{4\pi\epsilon_0 z} - \frac{e^2}{4\pi\epsilon_0 |a_{12} - z|} \tag{3.4.26}$$

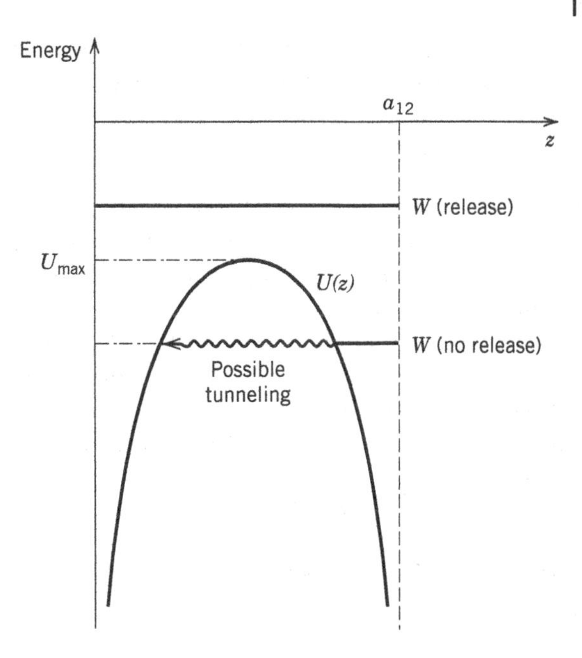

Figure 3.14 Illustrating the calculation of ion–atom charge transfer.

where z is the distance from the center of A^+ toward B. As sketched in Figure 3.14, $U(z) \to -\infty$ at the centers of A^+ and B^+ and has its maximum value

$$U_{max} = -\frac{e^2}{\pi\epsilon_0 a_{12}} \tag{3.4.27}$$

at $z = a_{12}/2$. The condition for release from B is found by equating W to U_{max} (see Figure 3.14), giving

$$a_{12} = \frac{3e^2 n^2}{4\pi\epsilon_0 U_{izB}} \tag{3.4.28}$$

For capture into level n' of A, the energy defect is

$$\Delta W_{AB} \approx \frac{U_{izB}}{n^2} - \frac{U_{izA}}{n'^2} \tag{3.4.29}$$

The capture is energetically possible only if

$$\frac{1}{2} m_R v_{A+}^2 \geq \Delta W_{AB} \tag{3.4.30}$$

with m_R the reduced mass. At the low incident velocities of interest in weakly ionized discharges, we have $v_{A+} \ll v_{at}$, where v_{at} given by (3.4.5) is the characteristic electron velocity in the atom. In this case, capture of the released electron occurs with high probability because the collision time $t \sim a_{12}/v_{A+}$ is long compared to the atomic timescale t_{at} given in (3.4.6). Hence, we estimate

$$\begin{aligned} \sigma_{cx} &\approx \pi a_{12}^2 \quad \text{for } \tfrac{1}{2} m_R v_{A+}^2 \gtrsim \Delta W_{AB} \\ &\approx 0 \qquad \text{otherwise} \end{aligned} \tag{3.4.31}$$

with a_{12} given by (3.4.28). For ground-state resonant transfer ($A \equiv B$), (3.4.28) gives a cross section that is independent of energy:

$$\sigma_{cx} \approx 36\pi \left(\frac{e^2}{8\pi\epsilon_0 U_{iz}} \right)^2 \tag{3.4.32}$$

where the quantity in parentheses is approximately the atomic radius of the ground-state atom.

The cross section (3.4.32) does not show a velocity dependence. However, more detailed theoretical calculations and experiments show that σ_{cx} varies as (Rapp and Francis, 1962)

$$\sigma_{cx} \sim \frac{1}{\mathcal{E}_{iz}}(C_1 - C_2 \ln v_{A+})^2 \tag{3.4.33}$$

in the range of v_{A+} from 10^5 to 10^8 cm/s, with $C_1 \approx 1.58 \times 10^{-7}$, $C_2 \approx 7.24 \times 10^{-8}$, $\mathcal{E}_{iz}$ the ionization potential of A in volts, and σ_{cx} in cm^2. The explanation is indicated in Figure 3.14. Even though electron release from B is not permitted classically, the electron can tunnel through the potential barrier quantum mechanically.

We can understand the form of (3.4.33) as follows (Smirnov, 1981): the ground-state valence electron in B oscillates in the Coulomb field of the nucleus with a period $\tau \approx h/e\mathcal{E}_{iz}$. The probability P that the electron tunnels across a potential barrier of height $\mathcal{E}_{iz}$ from $x = 0$ to $x = b_0$ in one oscillation is found by solving the Schrödinger equation for the electron wave function $\Psi(x)$,

$$-\frac{\hbar^2}{2m}\frac{d^2\Psi}{dx^2} = -e\mathcal{E}_{iz}\Psi \tag{3.4.34}$$

within this interval of x. We obtain $P = |\Psi(b_0)/\Psi(0)|^2 = e^{-2\alpha b_0}$, where

$$\alpha = \left(\frac{2me\mathcal{E}_{iz}}{\hbar^2}\right)^{1/2} \tag{3.4.35}$$

The time for the electron to tunnel from B to A$^+$ is then τP. Equating this time to the collision time b_0/v_{A+} and solving for b_0, we obtain

$$b_0 \approx \frac{1}{2\alpha}\ln\left(\frac{hv_{A+}}{e\mathcal{E}_{iz}b_0}\right) \tag{3.4.36}$$

Estimating the cross section as $\sigma_{cx} \approx \pi b_0^2$ and rearranging, we obtain the form (3.4.33).

The cross section (3.4.33) is based on the assumption of straight-line trajectories for the collision. At low collision energies, the trajectories are strongly perturbed by the polarization force and the collision partners can be "captured," as described in Section 3.3. The cross section σ_L for capture is given by (3.3.13). For such a capture, the probability of resonant charge transfer is $\frac{1}{2}$ (equal probability that the electron is found on either particle). Hence, we can estimate

$$\sigma_{cx} \approx \frac{1}{2}\sigma_L \tag{3.4.37}$$

for low collision energies. The condition that the trajectories be strongly perturbed can be estimated from the dynamics in the polarization potential for typical polarizabilities to be $v_{A+} \lesssim 10^5/A_R^{1/2}$ cm/s, where A_R is the reduced mass in amu.

Experimental values for resonant charge transfer and elastic (polarization) scattering of noble gas ions in their parent gases are shown in Figure 3.15. Because kinetic energy is conserved, resonant charge transfer acts as an elastic collision. At low energies, the cross sections are large. Because the resonant charge transfer cross section is large, the particles are practically undeflected in the CM system, leading after the charge transfer to an effective scattering angle for the ion, in the CM system, of 180° and a momentum transfer of $2m_R v_R$ for every collision. Hence, the momentum transfer cross section for resonant charge transfer is

$$\sigma_{mi} = 2\sigma_{cx} \tag{3.4.38}$$

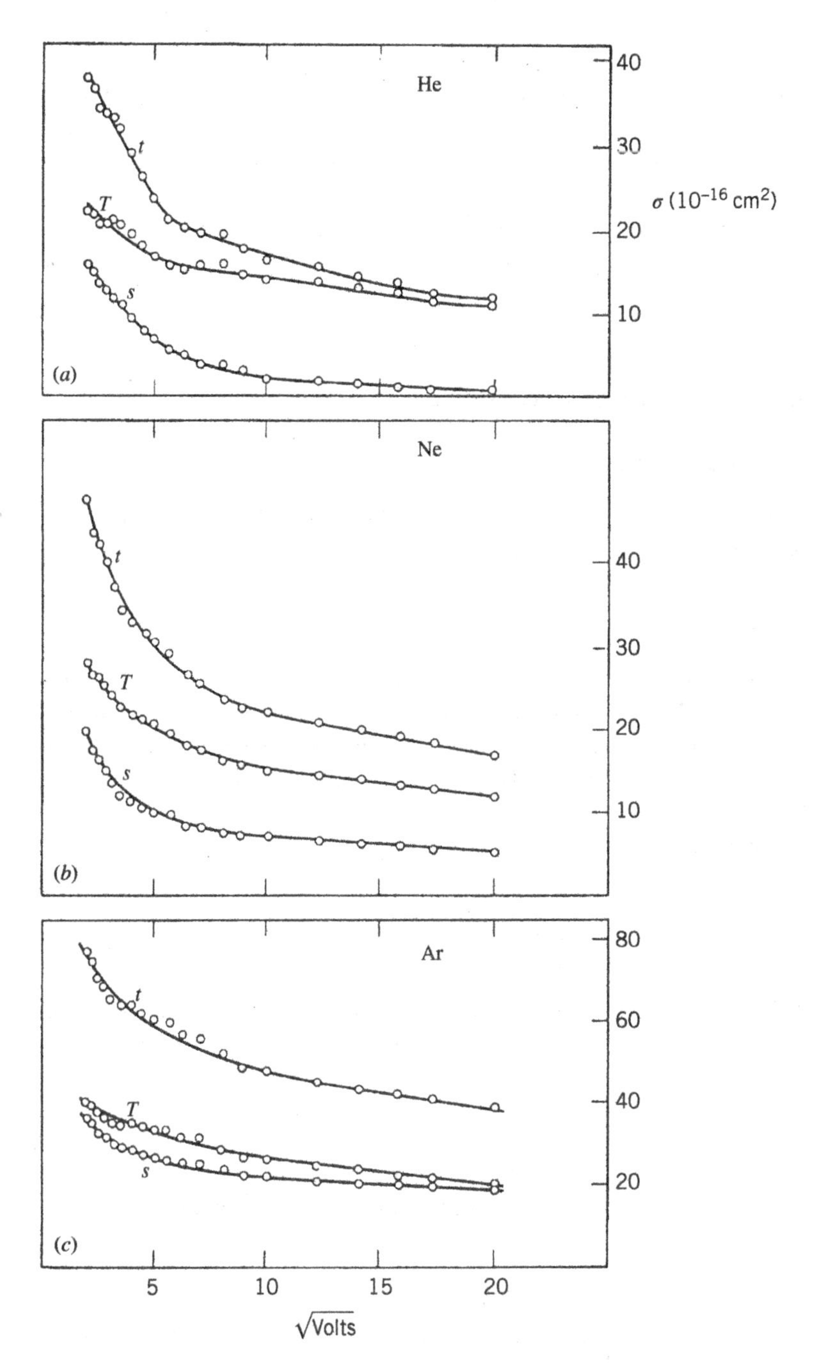

Figure 3.15 Experimental values for elastic scattering (s), charge transfer (T), and the sum of the two mechanisms (t) for (a) helium, (b) neon, and (c) argon ions in their parent gases (McDaniel et al., 1993/John Wiley & Sons).

3.4.6 Ion–Atom Ionization

An ion colliding with an atom would be expected to transfer only a small fraction $\sim 2m/M$ of its kinetic energy $\mathcal{E}$ to a valence electron. Hence, one might expect significant ionization only for $\mathcal{E} \gtrsim (M/2m)\mathcal{E}_{\text{iz}} \sim 10^4$–$10^5$ V. Experimentally, however, significant ionization is seen for $\mathcal{E} \gtrsim 100$ V, e.g., for argon ions in argon gas, $\sigma_{\text{iz,i}} \approx 10^{-16}$ cm^2 at $\mathcal{E} \sim 200$ V (Haugsjaa and Amme, 1970). This phenomenon may be due to the formation of an unstable Ar_2^+ molecular complex. We consider processes such as this in Chapter 8. Such a process may be important in the high-voltage sheaths of capacitive rf discharges.

3.5 Averaging Over Distributions and Surface Effects

3.5.1 Averaging Over a Maxwellian Distribution

To obtain the collision quantities in a plasma, we integrate over the velocity distribution functions of the particles. The collision frequency and rate constant are then

$$\begin{aligned} \nu &= n_g K = n_g \langle \sigma(v_R) v_R \rangle_{\mathbf{v}_1, \mathbf{v}_2} \\ &= n_g \int d^3 v_1\, d^3 v_2 f_1(\mathbf{v}_1) f_2(\mathbf{v}_2) \sigma(v_R) v_R \end{aligned} \tag{3.5.1}$$

where the distributions f_1 and f_2 have been normalized to unity and $v_R = |\mathbf{v}_1 - \mathbf{v}_2|$. If the characteristic velocities of the target particles are much less than those of the incident particles, which is often the case, then $v_R \approx |\mathbf{v}_1|$, and the $\mathbf{v}_2$ integration is trivially done. We usually take the incident distribution to be an isotropic Maxwellian, since this is the natural outcome of collisional processes, as derived in Appendix B.

The rate constant is then (writing v for v_1)

$$\begin{aligned} K(T) &= \langle \sigma(v) v \rangle_{\mathbf{v}} \\ &= \left(\frac{m}{2\pi kT} \right)^{3/2} \int_0^\infty \sigma(v) v \exp\left(-\frac{mv^2}{2kT} \right) 4\pi v^2\, dv \end{aligned} \tag{3.5.2}$$

where m and T are the incident particle mass and temperature (kelvins), respectively. Transforming (3.5.2) from speed to energy variables using $\mathcal{E} = \frac{1}{2} mv^2/e$, we obtain

$$K(\mathrm{T}) = \frac{\bar{v}_e}{\mathrm{T}^2} \int_0^\infty \sigma(\mathcal{E}) \mathcal{E}\, e^{-\mathcal{E}/\mathrm{T}} d\mathcal{E} \tag{3.5.3}$$

with $\mathrm{T} = kT/e$ (equivalent voltage units) and $\bar{v} = (8e\mathrm{T}/\pi m)^{1/2}$, the mean speed from (2.4.9).

For a hard-sphere collision, for which $\sigma = \pi a_{12}^2$ independent of v, the integration in (3.5.2) is easily performed, yielding

$$K(T) = \pi a_{12}^2 \bar{v} \tag{3.5.4}$$

where $\bar{v} \propto T^{1/2}$. For polarization scattering with $\sigma \propto 1/v$, we find $K(T) = \text{const}$, independent of T. For Coulomb scattering that has a velocity dependence $\sigma_{\text{el}} \propto 1/v^4$, from (3.3.7) (if we consider $\ln \Lambda$ as a constant for purposes of integration), calculating K_{el} as in (3.5.2) leads to a logarithmic infinity at $v = 0$. This is apparent, rather than real, as the momentum transfer rate constant K_{m} obtained from (3.1.15), which we use in the force equation, remains finite (see, for example, Holt and Haskell, 1965, Chapter 10). For electron–atom ionization and excitation, with $\mathrm{T_e} \sim 4\,\text{V} \ll \mathcal{E}_{\text{iz}}, \mathcal{E}_{\text{ex}}$, the threshold energies, only the tail of the Maxwellian and the behavior of $\sigma(v)$ near threshold, contribute to the rate constant, as shown in Figure 1.10.

For ionization, we can expand the Thomson cross section (3.4.21b) near $\mathcal{E} = \mathcal{E}_{iz}$ to obtain

$$\sigma_{iz}(\mathcal{E}) = \begin{cases} \sigma_0 \frac{\mathcal{E}-\mathcal{E}_{iz}}{\mathcal{E}_{iz}} & \mathcal{E} > \mathcal{E}_{iz} \\ 0 & \mathcal{E} \leq \mathcal{E}_{iz} \end{cases}$$

where $\sigma_0 = \pi(e/4\pi\epsilon_0\mathcal{E}_{iz})^2$. Inserting this into (3.5.3) and integrating, we obtain

$$K_{iz}(T_e) = \sigma_0\bar{v}_e\left(1 + \frac{2T_e}{\mathcal{E}_{iz}}\right) e^{-\mathcal{E}_{iz}/T_e} \tag{3.5.5}$$

where $\bar{v}_e = (8eT_e/\pi m)^{1/2}$. We leave the details to a problem.

In general, for electron collisions with atoms, the experimentally determined cross sections can be weighted by the electron distribution function and numerically integrated. Using the measured ionization, excitation, and elastic scattering cross sections for argon, given in Figure 3.13, we obtain the rate constants $K(T_e)$ shown in Figure 3.16. The rate constants are smoothed by the integration. Below the ionization and excitation threshold energies, there is an exponential decrease in the ionization and excitation rate constants with T_e, describing the exponentially decreasing number of electrons that are able to ionize or excite the atom.

A common simple approximation used for cross sections having a threshold energy $\mathcal{E}_{thr}$ is the Arrhenius form (see Section 8.5 for justification and further details)

$$\sigma = \sigma_0\left(1 - \frac{\mathcal{E}_{thr}}{\mathcal{E}}\right), \qquad \mathcal{E} \geq \mathcal{E}_{thr} \tag{3.5.6}$$

which increases from zero at $\mathcal{E}_{thr}$ and saturates at the value σ_0 as $\mathcal{E} \to \infty$. Integrating (3.5.3) for this cross section gives (Problem 3.11) the *Arrhenius rate constant*

$$K = \sigma_0\bar{v}_R\, e^{-\mathcal{E}_{thr}/T} \tag{3.5.7}$$

where $\bar{v}_R$ is the mean relative particle speed and T is the temperature.

As crude analytical approximations to K_{iz} and K_{ex}, over a limited range of T_e, we can fit these rate coefficient data to the Arrhenius form, obtaining, e.g.,

$$K_{iz} \approx K_{iz0}\, e^{-\mathcal{E}_{iz}/T_e} \tag{3.5.8}$$

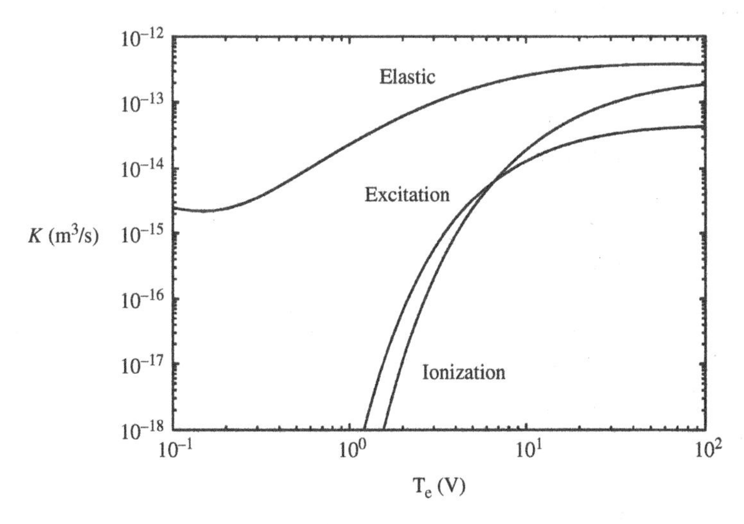

Figure 3.16 Electron collision rate constants K_{iz}, K_{ex}, and K_m versus T_e in argon gas (Vahedi, 1993).

where $\mathcal{E}_{iz}$ is the ionization energy and where the preexponential factor for argon is $K_{iz0} \approx 5 \times 10^{-14}$ m^3/s. For elastic scattering, we can do a similar fit, but we most often approximate

$$K_{el} \approx K_{el0} \approx 10^{-13}\,\text{m}^3/\text{s} \tag{3.5.9}$$

For ion–atom collisions, we most often require the total ion–atom scattering cross section for low-energy ions ($T_i \sim 0.05$ V), which we estimate from the data in Figure 3.15 to be

$$\sigma_i \approx 10^{-14}\,\text{cm}^2$$

Using (3.1.6), we obtain

$$\lambda_i = \frac{1}{n_g\sigma_i} \approx \frac{1}{330p}\,\text{cm}, \qquad (p\,\text{in Torr}) \tag{3.5.10}$$

A more complete and accurate set of rate constants for argon is given in Table 3.3. The first three collision processes describe elastic scattering, ionization, and average energy loss-weighted excitation, with the corresponding rate constants K_{el}, K_{iz}, and K_{ex}. These are fits to the numerically determined rate constants in the range $1 \leq T_e \leq 7$ V, based on the measured cross sections. The remaining processes describe excitations and de-excitations among the ground state, 4s metastable and resonance levels, and 4p levels (see Figure 3.12).

Table 3.3 Selected Reaction Rate Constants for Argon Discharges

Number	Reaction	Rate constant (m^3/s)	Source
1	e + Ar elastic scattering	$2.336\text{E}{-14}\,T_e^{1.609} \times e^{0.0618(\ln T_e)^2 - 0.1171(\ln T_e)^3}$	a)
2	$e + Ar \rightarrow Ar^+ + 2e$	$2.34\text{E}{-14}\,T_e^{0.59}\,e^{-17.44/T_e}$	a)
3	$e + Ar \rightarrow Ar^* + e$	$2.48\text{E}{-14}\,T_e^{0.33}\,e^{-12.78/T_e}$	a,b)
4	$e + Ar \rightarrow Ar(4s) + e$	$5.0\text{E}{-15}\,T_e^{0.74}\,e^{-11.56/T_e}$	c)
5	$e + Ar(4s) \rightarrow Ar + e$	$4.3\text{E}{-16}\,T_e^{0.74}$	d)
6	$e + Ar \rightarrow Ar(4p) + e$	$1.4\text{E}{-14}\,T_e^{0.71}\,e^{-13.2/T_e}$	c)
7	$e + Ar(4p) \rightarrow Ar + e$	$3.9\text{E}{-16}\,T_e^{0.71}$	d)
8	$Ar(4s) + e \rightarrow Ar(4p) + e$	$8.9\text{E}{-13}\,T_e^{0.51}\,e^{-1.59/T_e}$	c)
9	$Ar(4p) + e \rightarrow Ar(4s) + e$	$3.0\text{E}{-13}\,T_e^{0.51}$	d)
10	$e + Ar(4s) \rightarrow Ar^+ + 2e$	$6.8\text{E}{-15}\,T_e^{0.67}\,e^{-4.20/T_e}$	c)
11	$e + Ar(4p) \rightarrow Ar^+ + 2e$	$1.8\text{E}{-13}\,T_e^{0.61}\,e^{-2.61/T_e}$	c)
12	$e + Ar_m \rightarrow Ar_r + e$	2E−13	c)
13	$Ar_r \rightarrow Ar + h\nu$	3.0E7 s^{-1}	d,e)
14	$Ar(4p) \rightarrow Ar + h\nu$	3.2E7 s^{-1}	d,e)

Notes. T_e in volts. The notation E−8 means 10^{-8}. Subscripts m and r denote metastable and resonance 4s levels.

a) Fit by Gudmundsson (2002a) in the range $1 \leq T_e \leq 7$ V.
b) Average energy loss-weighted excitation rate constant for $\mathcal{E}_{ex} = 12.14$ V.
c) Kannari et al. (1985).
d) Ashida et al. (1995).
e) Average first-order rate constant in units of s^{-1}.

Rate constants for argon and many other gases can be calculated from data available on the open-access LXCat website https://www.lxcat.net/home, using the online program Bolsig+; this is briefly described at the end of Section 19.2. The student should consult Carbone et al. (2021) for a review and tutorial of the LXCat project, and Hagelaar and Pitchford (2005) for a description of Bolsig+.

3.5.2 Energy Loss per Electron–Ion Pair Created

A very important quantity that we use in subsequent chapters is the collisional energy loss per electron–ion pair created, $\mathcal{E}_c(\mathrm{T_e})$, which is defined as

$$K_{iz}\mathcal{E}_c = K_{iz}\mathcal{E}_{iz} + K_{ex}\mathcal{E}_{ex} + K_{el}\frac{3m}{M}\mathrm{T_e} \tag{3.5.11}$$

The terms on the RHS of (3.5.11) account for the loss of electron energy due to ionization, excitation, and elastic (polarization) scattering against neutral atoms. These are usually the dominant energy losses in weakly ionized electropositive discharges. The quantity $(3m/M)\mathrm{T_e}$ is the mean energy lost per electron for a polarization scattering, as determined using (3.2.19). The resultant values of $\mathcal{E}_c$ for argon and oxygen shown in Figure 3.17 are obtained using data such as that given in Figure 3.16 for argon. Because $\mathcal{E}_c$ depends on ratios of rate constants that have sensitive dependence on $\mathrm{T_e}$, accurate values must be used. A reasonable set for argon in the range $1 \lesssim \mathrm{T_e} \lesssim 7$ V is the first three rate constants in Table 3.3, with $\mathcal{E}_{iz} = 15.76$ V and $\mathcal{E}_{ex} = 12.14$ V. At high temperatures, $\mathcal{E}_c$ asymptotes to about 18 V. At temperatures below $\mathcal{E}_{iz}$, the energy loss per ionizing collision rises as the excitation energy loss exceeds that due to ionizations, and, at temperatures below about two volts, the elastic energy transfer becomes important. For a typical discharge with a temperature $\mathrm{T_e} = 3$ V, approximately 61 V of energy is lost per ionizing collision in argon.

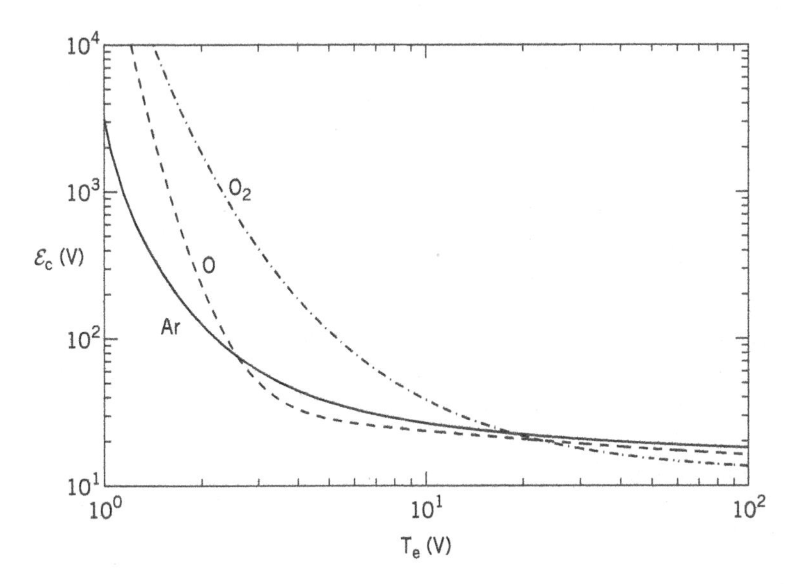

Figure 3.17 Collisional energy loss $\mathcal{E}_c$ per electron–ion pair created versus $\mathrm{T_e}$ in argon and oxygen. Source: Adapted from Gudmundsson, 2002b.

For molecular gases, additional collisional energy losses include excitation of vibrational and rotational energy levels, molecular dissociation, and, for electronegative gases, negative ion formation. We discuss these processes in Chapter 8. As shown in Figure 3.17, $\mathcal{E}_c$ is generally a factor of 2–10 times higher in a molecular gas than in an atomic gas for electron temperatures below 7 V.

3.5.3 Surface Effects

A few facts must be described about collisions of particles with surfaces. Averaged over short time scales, electrons, and positive ions arrive at surfaces in equal numbers, and almost all electron–ion pairs recombine on surfaces, leading to the reinjection of neutral atoms back into the discharge. Hence, we will treat surfaces as "black holes" for charged particles. High-energy ions can also sputter neutral atoms from surfaces or can cause secondary electrons to be emitted from surfaces. If Γ_i is the incident ion flux, then, with secondary emission coefficients γ,

$$\Gamma_{sput} = \gamma_{sput}\Gamma_i$$

$$\Gamma_{se} = \gamma_{se}\Gamma_i$$

For incident ion energies of order 1 kV, we find $\gamma_{sput} \sim 1$, $\gamma_{se} \sim 0.1$–0.2 for metals, and $\gamma_{se} \sim 1$ for some insulators. Sputtering is an important process by which films are deposited on substrates, and secondary emission is a critical process for maintaining dc glow discharges. We describe surface processes in detail in Chapter 9. Applications of secondary electron emission and sputtering are described in Chapters 14, 15, and 17.

Problems

3.1 Mean Free Path An electron beam having density n_e and velocity v_e along x is incident on a slab of thickness L along x consisting of a mixture of gases A, B, and C having densities n_A, n_B, and n_C. The collision cross sections for electrons with each type of gas molecule are σ_A, σ_B, and σ_C.
(a) Assuming that v_e is much greater than the thermal velocities of gas molecules, find the mean free path λ of electrons in the gas mixture.
(b) What is the probability that a beam electron entering the slab suffers at least one collision inside the slab?

3.2 Scattering Angle Transformations Show using momentum conservation that for collision of a projectile with an initially stationary target (3.2.8) holds for the transformation of scattering angles between the laboratory and CM systems.

3.3 Hard Sphere Scattering Using (3.2.10) and (3.2.12), find the differential scattering cross section in the laboratory system for a hard-sphere elastic collision of a projectile of mass m_1 with an initially stationary target of mass m_2.

3.4 Differential Scattering Cross Section Using (A.13), (3.2.10), and (3.2.12), find the differential scattering cross section in the laboratory system for Coulomb scattering of an electron with an initially stationary electron.

3.5 Momentum Transfer for Coulomb Collisions Calculate the momentum transfer cross section $\sigma_m(v_R)$ for Coulomb collisions.

(a) Use the small angle scattering result (3.3.1) in (3.1.15) and integrate from Θ_{min} to Θ_{max} to estimate $\sigma_m(v_R)$, where Θ_{min} and Θ_{max} are determined by setting $b = \lambda_{De}$ and $b = b_0$ in (3.2.24), respectively.

(b) Using the exact (Rutherford) cross section (3.3.3), show that $\sigma_m(v_R) = \pi b_0^2 \ln(2/\Theta_{min})$, if a lower limit for the scattering angle of $\Theta = \Theta_{min}$ and an upper limit of $\Theta = \pi$ is assumed.

3.6 Large-Angle Coulomb Scattering Integrate (3.3.3) over the appropriate solid angles to obtain (3.3.4).

3.7 Small-Angle Polarization Scattering For small-angle polarization scattering, determine the differential scattering cross section (3.2.28) in the CM system using the potential (3.3.11).

3.8 Cross Sections A point mass m having incoming speed v is scattered by a fixed (infinite mass) elastic hard sphere of radius a.

(a) Show that the differential elastic scattering cross section is $I(v, \theta) = a^2/4$.

(b) Find the elastic scattering cross section σ_{el} and the momentum transfer cross section σ_m and compare.

(c) Modeling electron–neutral elastic scattering in 20 mTorr argon gas at 25 °C as hard-sphere scattering with $a = \alpha_p^{1/3}$, where $\alpha_p = 11.08\, a_0^3$ is the polarizability of argon atoms ($a_0 \approx 0.53 \times 10^{-8}$ cm is the Bohr radius), and with v corresponding to a 5 V electron, find the mean free path λ_{el} and the collision frequency ν_{el} for scattering.

3.9 Elastic Scattering Power Losses Consider the average power p_{el} per unit volume lost by a Maxwellian distribution of electrons at temperature T_e due to elastic scattering of the electrons against a population of cold neutral gas atoms having a density n_g.

(a) Calculate p_{el} if the elastic scattering is due to polarization scattering with a polarization rate constant $K_{el} = K_L$, given by (3.3.15). Note that in this case, K_L is a constant, independent of electron speed. (To find p_{el}, you must integrate the electron energy loss over the Maxwellian distribution of electron speeds.) Show that your answer agrees with the last term in (3.5.11).

(b) Repeat part (a) if the elastic scattering is due to hard sphere scattering with a constant cross section σ_0.

3.10 Excitation Cross Section Estimate the total cross section σ_{ex} for electron impact excitation of an atom having one valence electron in the $n = 1$ ground state to the $n > 1$ bound states. As a simple model (the Bohr atom), if $\mathcal{E}_{iz}$ is the ionization potential of the ground state, then the $n > 1$ states have energies lying between $3\mathcal{E}_{iz}/4$ and $\mathcal{E}_{iz}$.

(a) To do this, integrate the differential cross section $I(v, \theta)$ for small-angle Coulomb scattering of the incoming electron (energy $\mathcal{E}$ in volts) by the (initially stationary) valence electron over all scattering angles θ for which the energy transfer $\mathcal{E}_L$ to the valence electron lies in the energy range from $3\mathcal{E}_{iz}/4$ to $\mathcal{E}$ for $\mathcal{E} < \mathcal{E}_{iz}$, and from $3\mathcal{E}_{iz}/4$ to $\mathcal{E}_{iz}$ for $\mathcal{E} > \mathcal{E}_{iz}$. Note that $\sigma_{ex} = 0$ for $\mathcal{E} < 3\mathcal{E}_{iz}/4$. The required procedure is similar to that used to obtain the Thomson ionization cross section σ_{iz}.

(b) Plot (linear scales) $\sigma_{ex}(\mathcal{E})$ and the Thomson cross section $\sigma_{iz}(\mathcal{E})$ versus $\mathcal{E}/\mathcal{E}_{iz}$ on the same graph and compare.

3.11 Arrhenius Rate Constant Using the Arrhenius cross section (3.5.6), derive the rate constant (3.5.7).

3.12 Ionization Rate Constant For most gas discharges, the electron temperature $T_e \ll \mathcal{E}_{iz}$, the ionization energy of the gas atoms. Thus, electrons in the tail of the Maxwellian distribution are responsible for the ionization of the gas.

(a) Using the Thomson formula for the ionization cross section near the threshold energy $\mathcal{E} = \mathcal{E}_{iz}$, obtain the ionization rate constant K_{iz} given in (3.5.5).

(b) Plot K_{iz} (log scale) versus T_e (linear scale, in V) for $\mathcal{E}_{iz} = 15.8$ V (argon gas) and T_e in the range 1 to 6 V.

(c) Show that for $T_e \ll \mathcal{E}_{iz}$, the maximum contribution to K_{iz} comes from electrons having energies $\mathcal{E} \approx \mathcal{E}_{iz} + 2T_e$.

3.13 Ionization from Metastable State Suppose that an $n = 2$ metastable level has an energy $\mathcal{E}^* = 3\mathcal{E}_{iz}/4$ above the ground state, such that the metastable ionization energy is $\mathcal{E}^*_{iz} = \mathcal{E}_{iz}/4$.

(a) Following the Thomson procedure, estimate the ionization cross section per valence electron from the metastable level, and find the ratio of the maximum metastable-to-ground-state Thomson ionization cross sections.

(b) Using your results in (a) and the expression (3.5.5) for the ionization rate constant, find the ratio of the metastable to the ground-state ionization rate constants for argon with $\mathcal{E}_{iz} = 15.8$ V and $T_e = 3$ V.

3.14 Excitation of Metastable State Ground-state argon atoms have six valence electrons (3p electrons with orbital angular momentum quantum number $l = 1$) in a ${}^1S_0$ energy level configuration. The first (lowest energy) excited levels of argon are a group of four closely spaced energy levels (${}^1P_1$, ${}^3P_{0,1,2}$) at $\mathcal{E}^* \approx 11.6$ V from the ground state (see Figure 3.12), with the excited electron having $l = 0$ (a 4s electron). Recall that the number of quantum states per level is $2J + 1$, where J is the total angular momentum quantum number. The next higher group is the 4p levels at $\mathcal{E}^{**} \approx 13.2$ V.

(a) Which of the first excited levels are metastable? What fraction of the total number of quantum states in this group of levels is metastable?

(b) Estimate the total cross section $\sigma^*(\mathcal{E})$ for electron impact excitation of ground-state argon to a metastable state. To do this, integrate the differential cross section $I(v, \theta)$ for small-angle Coulomb scattering of the incoming electron (energy $\mathcal{E}$ in volts) by a (initially stationary) valence electron over all scattering angles θ for which the energy transfer $\mathcal{E}_L$ to the valence electron lies in the energy range from $\mathcal{E}^*$ to $\mathcal{E}$ for $\mathcal{E} < \mathcal{E}^{**}$, and from $\mathcal{E}^*$ to $\mathcal{E}^{**}$ for $\mathcal{E} > \mathcal{E}^{**}$. Note that $\sigma^* = 0$ for $\mathcal{E} < \mathcal{E}^*$.

(c) Plot (linear scales) $\sigma^*(\mathcal{E})$ versus $\mathcal{E}$ for $0 < \mathcal{E} < 20$ V. Make sure your answer is reasonable. The maximum cross section should be of order 10^{-16} cm^2 (see Figure 3.13).

3.15 Charge Transfer to a Multiply Ionized Ion Following the approach used in Section 3.4, determine the maximum charge transfer cross section from the ground state of an atom to an ion having a positive charge of $+Ze$, where $Z > 1$.

3.16 Energy Transfer Consider the inelastic collision of two bodies A and B to form a single body AB*, where AB* is an excited state of AB having excitation energy $\mathcal{E}_{ex}$. Let A and B have masses m_A and m_B and initial speed v_A and $v_B \equiv 0$. Using momentum and energy conservation, find the speed v_{AB*} and the excitation energy $\mathcal{E}_{ex}$ after the collision. Hence, show that $\mathcal{E}_{ex}$ can never be zero; i.e., two bodies cannot collide elastically to form one body.

3.17 **Collisional Energy Losses** Using the rate constants for the first three collision processes in Table 3.3, along with $\mathcal{E}_{iz} = 15.76$ V and $\mathcal{E}_{ex} = 12.14$ V,

(a) Calculate $\mathcal{E}_c$ versus T_e using (3.5.11) and compare with Figure 3.17.

(b) Show that elastic scattering energy losses are small compared to excitation energy losses for $T_e \gtrsim 2$ V.

4

Plasma Dynamics

4.1 Basic Motions

The equations of motion for a particle acted on by electric and magnetic fields are

$$m\frac{\mathrm{d}\mathbf{v}}{\mathrm{d}t} = q\,[\mathbf{E}(\mathbf{r},t) + \mathbf{v}\times\mathbf{B}(\mathbf{r},t)] \tag{4.1.1a}$$

$$\frac{\mathrm{d}\mathbf{r}}{\mathrm{d}t} = \mathbf{v}(t) \tag{4.1.1b}$$

where the RHS of (4.1.1a) is the Lorentz force (2.2.12) and $\mathbf{v}(t)$ is the Lagrangian velocity. These equations cannot be solved for the general case where the force is a nonlinear function of $\mathbf{r}$, but solutions for various special cases can be found.

4.1.1 Motion in Constant Fields

For a constant electric field $\mathbf{E} = \mathbf{E}_0$ with $\mathbf{B} \equiv 0$, the particle moves with a constant acceleration along $\mathbf{E}_0$:

$$\mathbf{r}(t) = \mathbf{r}_0 + \mathbf{v}_0 t + \frac{1}{2}\mathbf{a}_0 t^2 \tag{4.1.2}$$

where $\mathbf{r}_0$ and $\mathbf{v}_0$ are the particle position and velocity at $t = 0$ and $\mathbf{a}_0 = q\mathbf{E}_0/m$, respectively. For a constant magnetic field $\mathbf{B} = \hat{z}B_0$ which we take to lie along z, with $\mathbf{E} \equiv 0$, the components of (4.1.1a) are

$$m\frac{\mathrm{d}v_x}{\mathrm{d}t} = qv_yB_0 \tag{4.1.3a}$$

$$m\frac{\mathrm{d}v_y}{\mathrm{d}t} = -qv_xB_0 \tag{4.1.3b}$$

$$m\frac{\mathrm{d}v_z}{\mathrm{d}t} = 0 \tag{4.1.3c}$$

The trivial z motion is decoupled from the x and y motions. Differentiating (4.1.3a) and eliminating v_y using (4.1.3b), we obtain

$$\frac{\mathrm{d}^2v_x}{\mathrm{d}t^2} = -\omega_c^2 v_x \tag{4.1.4}$$

Principles of Plasma Discharges and Materials Processing, Third Edition. Michael A. Lieberman and Allan J. Lichtenberg.

where

$$\omega_c = \frac{qB_0}{m} \tag{4.1.5}$$

is the *gyration* or *cyclotron frequency*. Solving (4.1.4) and using (4.1.3a) to obtain v_y, we find

$$v_x = v_{\perp 0}\cos(\omega_c t + \phi_0) \tag{4.1.6a}$$

$$v_y = -v_{\perp 0}\sin(\omega_c t + \phi_0) \tag{4.1.6b}$$

$$v_z = v_{z0} \tag{4.1.6c}$$

where $v_{\perp 0}$ is the speed perpendicular to $\mathbf{B}_0$, and ϕ_0 is an arbitrary phase. Integrating (4.1.1b) yields the particle position

$$x = r_c \sin(\omega_c t + \phi_0) + (x_0 - r_c \sin\phi_0) \tag{4.1.7a}$$

$$y = r_c \cos(\omega_c t + \phi_0) + (y_0 - r_c \cos\phi_0) \tag{4.1.7b}$$

$$z = z_0 + v_{z0}t \tag{4.1.7c}$$

where

$$r_c = \frac{v_{\perp 0}}{|\omega_c|} \tag{4.1.8}$$

is the *gyration radius.* Equations (4.1.6) and (4.1.7) show that the particle moves in a circular orbit perpendicular to $\mathbf{B}$ having frequency ω_c and radius r_c about a *guiding center*, $x = x_0, y = y_0, z = z_0 + v_{z0}t$, that moves uniformly along z. Positive charges gyrate around the magnetic field according to the *left-hand rule*, and negative charges gyrate according to the *right-hand rule*. We can understand the motion by equating the inward Lorentz force to the outward centrifugal force:

$$|qv_{\perp 0}B_0| = \frac{mv_{\perp 0}^2}{r_c}$$

as shown in Figure 4.1, which yields circular motion with a radius given by (4.1.8).

The gyrofrequency and radius are important frequency and length scales for magnetized plasmas. In practical units, for electrons,

$$f_{ce} = \frac{\omega_{ce}}{2\pi} \approx 2.80 \times 10^6 B_0 \text{Hz} \qquad (B_0 \text{ in gauss}) \tag{4.1.9}$$

$$r_{ce} \approx \frac{3.37\sqrt{\mathcal{E}}}{B_0}\text{ cm} \qquad (\mathcal{E} \text{ in volts}) \tag{4.1.10}$$

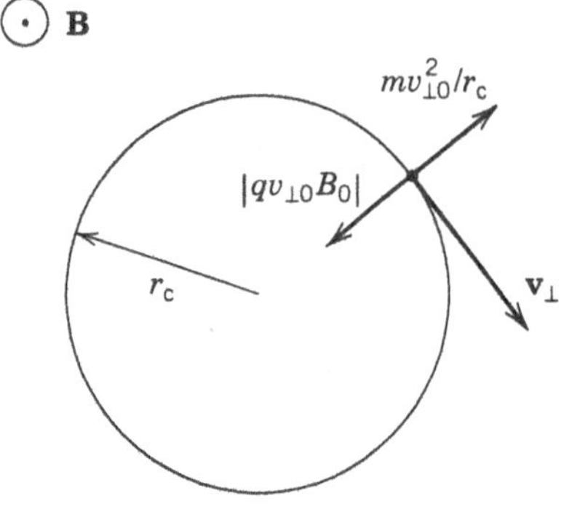

Figure 4.1 Charged particle gyration in a uniform magnetic field; **B** is directed out of the page.

and for singly charged ions,

$$f_{ci} = \frac{\omega_{ci}}{2\pi} \approx \frac{1.52 \times 10^3 B_0}{A_R} \text{ Hz} \qquad (B_0 \text{ in gauss}) \tag{4.1.11}$$

$$r_{ci} \approx \frac{1.44 \times 10^2 \sqrt{\mathcal{E} A_R}}{B_0} \text{ cm} \qquad (\mathcal{E} \text{ in volts}) \tag{4.1.12}$$

where A_R is the ion mass in atomic mass units (amu). At $B_0 = 100$ G (0.01 T) and for a 15 V (ionizing) electron, we find $f_{ce} \approx 280$ MHz and $r_{ce} \approx 1.3$ mm, showing that electrons are well confined perpendicular to **B**.

An argon ion ($A_R = 40$) in thermal equilibrium with neutrals ($\mathcal{E} = 0.026$ V) has $f_{ci} \approx 3.8$ kHz and $r_{ci} \approx 1.4$ cm and is more weakly confined. With ambipolar acceleration (see Chapter 5), the ion can take on the electron temperature, which at 5 V would give $r_{ci} \approx 20$ cm, which is larger than a typical discharge. Hence, ions are not well confined by the magnetic field. We will often model electrons as confined and ions as not confined in weakly magnetized discharges.

4.1.2 $E \times B$ Drifts

A simple solution is obtained for a particle moving in uniform **E** and **B** fields. Without loss of generality, we take $\mathbf{B} = \hat{z}B_0$ and $\mathbf{E} = \mathbf{E}_\perp + \hat{z}E_{z0} = \hat{x}E_{\perp 0} + \hat{z}E_{z0}$. Let $\mathbf{v} = \hat{z}v_z(t) + \mathbf{v}_\perp(t)$ in the Lorentz force equation (4.1.1a), we obtain a uniform acceleration along z, as in (4.1.2), and the equation for the transverse motion:

$$m\frac{d\mathbf{v}_\perp}{dt} = q(\hat{x}E_{\perp 0} + \mathbf{v}_\perp \times \hat{z}B_0) \tag{4.1.13}$$

We let

$$\mathbf{v}_\perp(t) = \mathbf{v}_E + \mathbf{v}_c(t) \tag{4.1.14}$$

where $\mathbf{v}_E$ is a constant velocity. Using this in (4.1.13), we find

$$m\frac{d\mathbf{v}_c}{dt} = q(\hat{x}E_{\perp 0} + \mathbf{v}_E \times \hat{z}B_0 + \mathbf{v}_c \times \hat{z}B_0)$$

Choosing the first two terms on the RHS to cancel, we obtain

$$\mathbf{v}_E = \frac{\mathbf{E} \times \mathbf{B}}{B_0^2} \tag{4.1.15}$$

and

$$m\frac{d\mathbf{v}_c}{dt} = q\mathbf{v}_c \times \hat{z}B_0 \tag{4.1.16}$$

We can write **E** rather than $\mathbf{E}_\perp$ in (4.1.15) because $\hat{z}E_{z0} \times \mathbf{B} \equiv 0$. We have seen that the solution to (4.1.16) is gyration at frequency ω_c with gyration radius r_c. Hence, the transverse motion is the sum of a guiding center drift $\mathbf{v}_E$ and a gyration:

$$\mathbf{v}_\perp(t) = \mathbf{v}_E + \text{Re}\left(\mathbf{v}_{c0} e^{j\omega_c t}\right) \tag{4.1.17}$$

We note from (4.1.15) that $\mathbf{v}_E$ is perpendicular to both **E** and **B** and is independent of the mass and charge of the particles; hence, electrons and ions drift with the same speed in the same direction. If $n_i = n_e$, there is no net current. Integrating (4.1.1b) using (4.1.17), we obtain

$$\mathbf{r}_\perp(t) = \mathbf{r}_{\perp 0} + \mathbf{v}_E t + \text{Re}\left(\frac{1}{j\omega_c}\mathbf{v}_{c0} e^{j\omega_c t}\right) \tag{4.1.18}$$

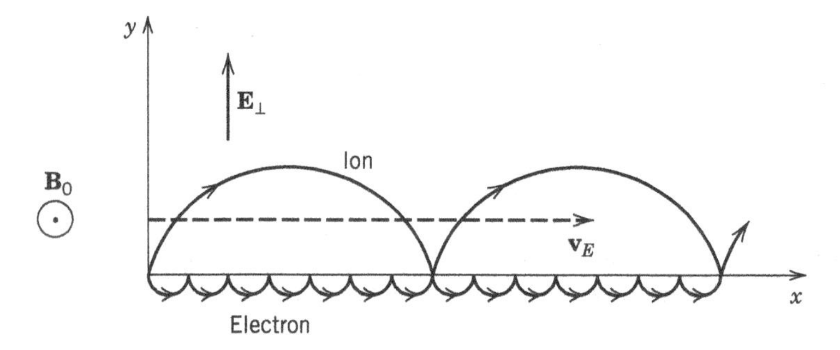

Figure 4.2 Motion of electrons and ions in uniform crossed **E** and **B** fields.

for the particle position. The orbits for electrons and ions are shown in Figure 4.2 for the case where the particles are initially at rest. In this case, $|\mathbf{v}_{c0}| = |\mathbf{v}_E|$ and the kinetic energies of the drift and gyration motions are equal. The orbits are cycloids with maximum displacement $2|\mathbf{v}_E/\omega_c|$ along y as shown. Physically, $\mathbf{E}_\perp$ initially accelerates the particles along y; as they gain speed, the $\mathbf{v} \times \mathbf{B}$ force turns them back toward their initial y positions.

It is clear from the procedure used to solve (4.1.13) that *any* constant transverse force $\mathbf{F}_\perp$ acting on a gyrating particle in a constant magnetic field will give rise to a drift perpendicular to both $\mathbf{F}_\perp$ and $\mathbf{B}$:

$$\mathbf{v}_F = \frac{(\mathbf{F}_\perp/q) \times \mathbf{B}}{B_0^2} \tag{4.1.19}$$

Nonuniform magnetic fields can give rise to additional forces both along (F_z) and perpendicular ($\mathbf{F}_\perp$) to $\mathbf{B}$. We consider these forces and the resulting particle motion in Section 4.3.

4.1.3 Energy Conservation

Dot multiplying (4.1.1a) by $\mathbf{v}$, we obtain

$$\frac{\mathrm{d}}{\mathrm{d}t}\left(\frac{1}{2}mv^2\right) = q\mathbf{v} \cdot \mathbf{E}[\mathbf{r}(t), t] \tag{4.1.20}$$

which shows that the magnetic field does no work on the particle. The rate of change of kinetic energy is equal to the power $q\mathbf{v} \cdot \mathbf{E}$ transferred from the electric field to the particle. For a static field, $\mathbf{E}(\mathbf{r}) = -\nabla\Phi$, (4.1.20) can be written as

$$\frac{\mathrm{d}}{\mathrm{d}t}\left(\frac{1}{2}mv^2\right) = -q\frac{\mathrm{d}\mathbf{r}}{\mathrm{d}t} \cdot \nabla\Phi[\mathbf{r}(t)] = -q\frac{\mathrm{d}}{\mathrm{d}t}\Phi[\mathbf{r}(t)]$$

which can be integrated to yield

$$\frac{1}{2}mv^2(t) + q\Phi[\mathbf{r}(t)] = \text{const} \tag{4.1.21}$$

This expresses the energy conservation for a particle in a static electric field.

For a collection of particles (a fluid consisting of one species), the force equation

$$mn\frac{\mathrm{d}\mathbf{u}}{\mathrm{d}t} = qn(\mathbf{E} + \mathbf{u} \times \mathbf{B}) - \nabla p - mn\nu_{\mathrm{m}}\mathbf{u} \tag{4.1.22}$$

repeated here from (2.3.15) is more complicated, with additional terms due to pressure gradients and collisions with particles of other species. Recall that $\mathrm{d}/\mathrm{d}t \equiv \partial/\partial t + \mathbf{u} \cdot \nabla$ is the convective derivative and that $\mathbf{u}(\mathbf{r}, t)$ is the Eulerian fluid velocity, which is related to the Lagrangian particle velocity by $\mathbf{v}(t) = \mathbf{u}[\mathbf{r}(t), t]$. Equation (4.1.22) cannot generally be solved, even when the fields are known. Furthermore, in most cases the fields themselves are functions of the particle motions, which act as charge and current sources in the Maxwell or Poisson equations. These must be determined self-consistently with the particle motions. This coupling of particles and fields lies at the root of all plasma phenomena.

In this chapter, we describe various solutions to (4.1.1) or (4.1.22), coupling the particle motions to the fields as needed. In Section 4.2, we consider a uniform unmagnetized plasma and introduce the coupling to describe such collective phenomena as plasma oscillations, the plasma dielectric constant, and, equivalently, the plasma conductivity. The conductivity determines the ohmic power dissipation, which is an important mechanism for electron heating in discharges. We also introduce wave phenomena, which can be important for plasma heating. The remainder of the chapter is devoted to magnetized plasmas, which are finding increasing application in materials processing. Guiding center motion in nonuniform magnetic fields is described in Section 4.3. Guiding center concepts play an important role in hot electron confinement in several low-pressure, high-density source concepts, which we describe in Chapters 11, 13, and 14. The dielectric tensor for magnetized plasmas is introduced in Section 4.4 and used in Section 4.5 to describe waves in uniform magnetized plasmas. These waves play a critical role in energy deposition in several high-density sources, such as ECRs and helicons, which we discuss in Chapter 13, and are also important for plasma diagnostics, which we introduce in Section 4.6. Wave phenomena in nonuniform or bounded plasmas will be dealt with in the application chapters that follow, when the need for the material naturally arises. The subject of waves in plasmas is vast, and the reader should consult more specialized monographs (Allis et al., 1963; Stix, 1992; Ginzburg, 1964) for more thorough treatments. We defer the study of steady-state solutions in nonuniform plasmas, which are important for particle diffusion and transport, to Chapter 5.

4.2 Nonmagnetized Plasma Dynamics

4.2.1 Plasma Oscillations

As the simplest example of the coupling of particles and fields, we consider the *undriven* motion of a plasma slab of finite width l containing a density $n_e = n_i = n_0$ of cold ($T_e = 0$) electrons and infinite mass (stationary) ions. Since $n_e = n_i$, the electric field $\mathbf{E} = 0$ in the slab. Now let the slab of electrons be displaced to the right with respect to the ions by a small distance $\zeta_e(t) \ll l$ at time t, as shown in Figure 4.3a. This leads to a surface charge density $\rho_S = en_0\zeta_e$ at the left edge due to

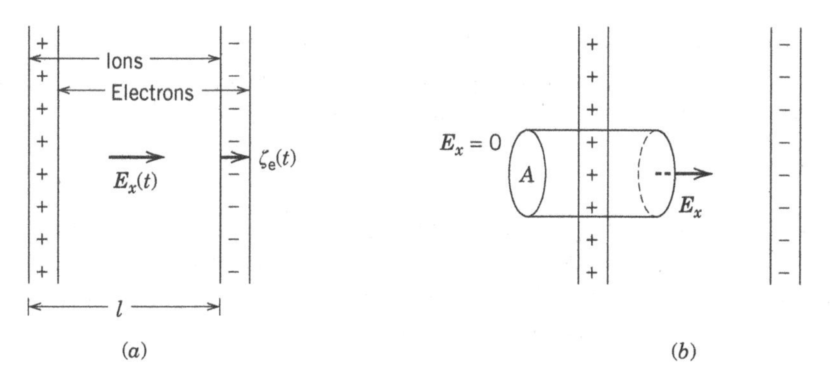

Figure 4.3 Plasma oscillations in a slab geometry: (a) displacement of electron cloud with respect to ion cloud; (b) calculation of the resulting electric field.

the uncovering of the stationary ion cloud. We similarly obtain $\rho_S = -en_0\zeta_e$ at the right edge. Using Gauss's law (2.2.6) applied to the pillbox shown in Figure 4.3*b*, these equal and opposite surface charges lead to an electric field within the slab:

$$E_x = \frac{en_0\zeta_e}{\epsilon_0} \tag{4.2.1}$$

The force equation for the electrons is[1]

$$m\frac{d^2\zeta_e}{dt^2} = -eE_x \tag{4.2.2}$$

Substituting (4.2.1) into (4.2.2) yields

$$\frac{d^2\zeta_e}{dt^2} = -\omega_{pe}^2\zeta_e \tag{4.2.3}$$

where

$$\omega_{pe} = \left(\frac{e^2n_0}{\epsilon_0 m}\right)^{1/2} \tag{4.2.4}$$

the *electron plasma frequency*, is the fundamental characteristic frequency of a plasma. The solution of (4.2.3) is

$$\zeta_e(t) = \zeta_{e0}\cos(\omega_{pe}t + \phi_0) \tag{4.2.5}$$

which represents a sinusoidal oscillation of the electron cloud with respect to the ion cloud at the natural frequency ω_{pe}. In practical units,

$$f_{pe} = \frac{\omega_{pe}}{2\pi} \approx 8980\sqrt{n_0}\,\text{Hz}, \qquad (n_0 \text{ in cm}^{-3}) \tag{4.2.6}$$

Plasma frequencies for discharges are typically in the microwave region (1–10 GHz).

If the assumption of infinite mass ions is not made, then the ions also move slightly and we obtain (Problem 4.1) the natural frequency

$$\omega_p = (\omega_{pe}^2 + \omega_{pi}^2)^{1/2} \tag{4.2.7}$$

where

$$\omega_{pi} = \left(\frac{e^2n_0}{\epsilon_0 M}\right)^{1/2} \tag{4.2.8}$$

is the ion plasma frequency. For $M \gg m$, $\omega_p \approx \omega_{pe}$.

The existence of plasma oscillations does not depend on the assumption of a slab geometry. It can be shown that any perturbed charge density oscillates at the plasma frequency (Problem 4.2). Note that the characteristic plasma scale length, velocity, and frequency are related by

$$\lambda_{De} = \frac{v_{th}}{\omega_{pe}} \tag{4.2.9}$$

Plasma oscillations are damped in time by collisions (Problem 4.3) and can also be damped collisionlessly by a mechanism known as *Landau damping*, which we describe below, when considering electrostatic waves. Collisional damping usually dominates Landau damping in discharges, and the oscillations generally fall to noise levels if there are no external drives.

1 Since ζ_e is small, the $\mathbf{u}\cdot\nabla\mathbf{u}$ term in (4.1.22) is small and there is no difference between Eulerian and Lagrangian velocities.

4.2.2 Dielectric Constant and Conductivity

We now consider a uniform plasma in the presence of a background gas that is *driven* by a small amplitude time-varying electric field:

$$E_x(t) = \tilde{E}_x \cos \omega t = \mathrm{Re}\, \tilde{E}_x \mathrm{e}^{j\omega t} \tag{4.2.10}$$

where $\tilde{E}_x$ is the electric field amplitude. We again let the ion mass be infinite for ease of calculation, and we assume that all quantities vary sinusoidally in time at frequency ω. The electron force equation is

$$m\frac{\mathrm{d}u_x}{\mathrm{d}t} = -eE_x - m\nu_\mathrm{m} u_x \tag{4.2.11}$$

where ν_m is the electron–neutral collision frequency. Let

$$u_x(t) = \mathrm{Re}\, \tilde{u}_x \mathrm{e}^{j\omega t} \tag{4.2.12}$$

and using this and (4.2.10) in (4.2.11), we obtain the complex velocity amplitude

$$\tilde{u}_x = -\frac{e}{m}\frac{1}{j\omega + \nu_\mathrm{m}}\tilde{E}_x \tag{4.2.13}$$

From (2.2.7), the total current is

$$J_{\mathrm{T}x} = \epsilon_0 \frac{\partial E_x}{\partial t} + J_x \tag{4.2.14}$$

where the conduction current J_x is due to the electron motion only, which, in the cold plasma approximation, is

$$\tilde{J}_x = -en_0\tilde{u}_x \tag{4.2.15}$$

We also have that

$$\frac{\partial E_x}{\partial t} = \mathrm{Re}\, j\omega \tilde{E}_x \mathrm{e}^{j\omega t}$$

such that the total current amplitude is

$$\tilde{J}_{\mathrm{T}x} = j\omega\epsilon_0\tilde{E}_x - en_0\tilde{u}_x \tag{4.2.16}$$

Using (4.2.13) in (4.2.16), we obtain

$$\tilde{J}_{\mathrm{T}x} = j\omega\epsilon_0 \left[1 - \frac{\omega_\mathrm{pe}^2}{\omega(\omega - j\nu_\mathrm{m})}\right]\tilde{E}_x \tag{4.2.17}$$

which relates the total current to the electric field in the sinusoidal steady state. Hence, we can introduce an effective *plasma dielectric constant*

$$\epsilon_\mathrm{p} = \epsilon_0\kappa_\mathrm{p} = \epsilon_0 \left[1 - \frac{\omega_\mathrm{pe}^2}{\omega(\omega - j\nu_\mathrm{m})}\right] \tag{4.2.18}$$

where κ_p is the relative dielectric constant. Maxwell's equation (2.2.2) can then be written

$$\nabla \times \tilde{\mathbf{H}} = j\omega\epsilon_\mathrm{p}\tilde{\mathbf{E}} \tag{4.2.19}$$

where we can introduce the displacement vector $\tilde{\mathbf{D}} = \epsilon_\mathrm{p}\tilde{\mathbf{E}}$, showing the correspondence of a plasma to a dielectric material.

We can also introduce a *plasma conductivity* by writing (4.2.17) in the form $\tilde{J}_{Tx} = (\sigma_p + j\omega\epsilon_0)\tilde{E}_x$, with

$$\sigma_p = \frac{\epsilon_0 \omega_{pe}^2}{j\omega + \nu_m} \tag{4.2.20}$$

such that (2.2.2) becomes

$$\nabla \times \tilde{\mathbf{H}} = (\sigma_p + j\omega\epsilon_0)\tilde{\mathbf{E}} \tag{4.2.21}$$

Equations (4.2.19) and (4.2.21) are equivalent. Hence, we can consider a plasma to be either a dielectric ϵ_p or a conductor σ_p, as we find useful. For low frequencies $\omega \ll \nu_m,\ \omega_{pe}$, we find that $\sigma_p \to \sigma_{dc}$, where

$$\sigma_{dc} = \frac{\epsilon_0 \omega_{pe}^2}{\nu_m} = \frac{e^2 n_0}{m\nu_m} \tag{4.2.22}$$

which is the *dc plasma conductivity* in the cold plasma approximation. For electron–ion rather than electron–neutral collisions (4.2.22), is replaced by the parallel Spitzer conductivity

$$\sigma_{ei} \approx \frac{0.019\,\mathrm{T_e}^{3/2}}{\ln\Lambda} \quad \Omega^{-1}-\mathrm{m}^{-1} \qquad (\mathrm{T_e}\ \text{in volts}) \tag{4.2.23}$$

where $\ln\Lambda$ is defined in (3.3.6). For high frequencies, it is more useful to consider ϵ_p rather than σ_p. For $\omega \gg \nu_m$, (4.2.18) reduces to the collisionless plasma dielectric constant

$$\epsilon_p = \epsilon_0\kappa_p = \epsilon_0\left(1 - \frac{\omega_{pe}^2}{\omega^2}\right) \tag{4.2.24}$$

At very high driving frequencies (in the high microwave regime) where $\omega > \omega_{pe}$, ϵ_p is positive but less than ϵ_0; hence, the plasma acts as a dielectric with a relative dielectric constant less than unity. At lower frequencies, $\omega < \omega_{pe}$, which is true for most discharges driven at rf frequencies, we see that $\epsilon_p < 0$. A slab of such a plasma of width l and cross-sectional area A then has a capacitance $C = \epsilon_p A/l$ that is negative, corresponding to an impedance $Z = 1/(j\omega C)$ that is inductive (positive imaginary). Hence, the plasma behaves like an inductor in this frequency regime.

Figure 4.4 illustrates the rf current and electric field amplitudes and phases in the sheath and plasma regions in the regime $\nu_m \ll \omega \ll \omega_{pe}$, which is typical for low-pressure rf discharges. From (2.2.8), $\tilde{J}_{Tx}$ is the same in the sheath and plasma regions. In the sheath regions, there is only displacement current and

$$\tilde{E}_x(\text{sheath}) = \frac{\tilde{J}_{Tx}}{j\omega\epsilon_0} \tag{4.2.25a}$$

In the bulk plasma region,

$$\tilde{E}_x(\text{plasma}) = \frac{\tilde{J}_{Tx}}{j\omega\epsilon_p} \tag{4.2.25b}$$

Since $\epsilon_p < 0$ and $|\epsilon_p| \gg \epsilon_0$, the field in the bulk plasma is much smaller than, and 180° out of phase with, the fields in the sheaths, as shown in the figure. Hence, almost all of the rf voltage is dropped across the sheath regions, and comparatively little voltage appears across the bulk plasma.

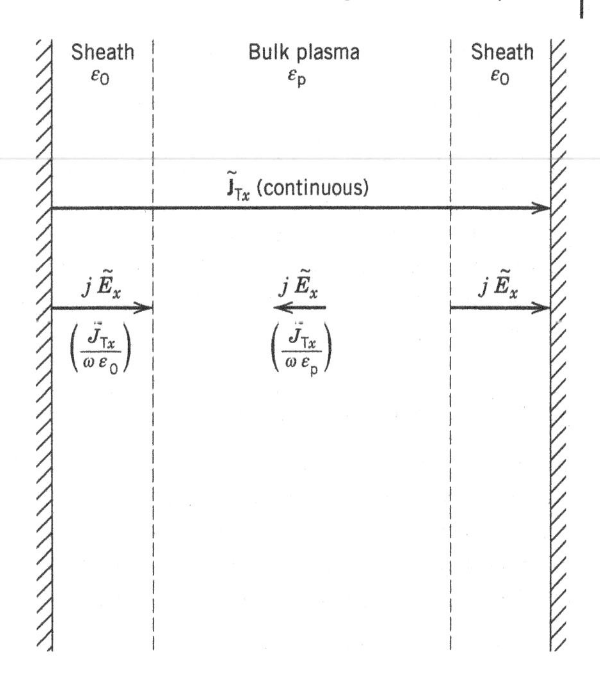

Figure 4.4 Rf current and electric field amplitudes and phases in the sheath and plasma regions of an rf discharge.

4.2.3 Ohmic Heating

Although the electric field within the bulk plasma is small, it gives rise to a significant electron heating due to electron–neutral collisions. The time-averaged power per unit volume absorbed by the plasma, p_{abs}, is given by

$$p_{\text{abs}} = \frac{1}{T}\int_0^T \mathbf{J}_{\text{T}}(t)\cdot\mathbf{E}(t)\,\mathrm{d}t = \frac{1}{2}\text{Re}\left(\tilde{\mathbf{J}}_{\text{T}}\cdot\tilde{\mathbf{E}}^*\right) = \frac{1}{2}\text{Re}\left(\tilde{\mathbf{J}}_{\text{T}}^*\cdot\tilde{\mathbf{E}}\right) \tag{4.2.26}$$

where $T = 2\pi/\omega$ is the period, the asterisk denotes complex conjugation, and the latter forms follow from (4.2.10) and the equivalent expression for $\mathbf{J}_{\text{T}}(t)$ (Problem 4.5). If we substitute $\tilde{\mathbf{J}}_{\text{T}} = (\sigma_{\text{p}} + j\omega\epsilon_0)\tilde{\mathbf{E}}$ into (4.2.26), then we obtain the collisional (ohmic) power absorbed by the electrons in terms of the electric field amplitude $\tilde{\mathbf{E}}$:

$$p_{\text{ohm}} = \frac{1}{2}|\tilde{\mathbf{E}}|^2\sigma_{\text{dc}}\frac{\nu_{\text{m}}^2}{\omega^2+\nu_{\text{m}}^2} \tag{4.2.27}$$

In many cases, the current density is known rather than the electric field. Let $\tilde{\mathbf{E}} = \tilde{\mathbf{J}}_{\text{T}}/(\sigma_{\text{p}} + j\omega\epsilon_0)$ in (4.2.26), we obtain

$$p_{\text{ohm}} = \frac{1}{2}|\tilde{\mathbf{J}}_{\text{T}}|^2\,\text{Re}\left(\frac{1}{\sigma_{\text{p}}+j\omega\epsilon_0}\right) \tag{4.2.28}$$

Taking the real part of $1/(\sigma_{\text{p}} + j\omega\epsilon_0)$, we obtain

$$\text{Re}\left(\frac{1}{\sigma_{\text{p}}+j\omega\epsilon_0}\right) = \frac{1}{\sigma_{\text{dc}}}\left(\frac{\omega_{\text{pe}}^4}{(\omega_{\text{pe}}^2-\omega^2)^2+\omega^2\nu_{\text{m}}^2}\right) \tag{4.2.29}$$

For $\omega \ll \omega_{pe}$, the term in parentheses is unity and we obtain the simple result

$$p_{ohm} = \frac{1}{2}|\tilde{\mathbf{J}}_T|^2 \frac{1}{\sigma_{dc}} \tag{4.2.30}$$

We shall apply (4.2.27) or (4.2.30) to find ohmic power absorption from waves as well as from oscillating fields. We will return to the calculation in Chapter 11 on rf discharges, where we determine $\tilde{\mathbf{J}}_T$, given the external driving source. However, we shall also find that, for low-pressure discharges, the ohmic power may not be the main source of power absorption by the plasma electrons. Rather, a mechanism of electron collisions with the oscillating sheaths can provide the principal electron heating.

4.2.4 Electromagnetic Waves

Waves can be important to carry energy from the surface of a plasma, where the wave is excited, into the bulk plasma, where the wave energy can be absorbed. Plasmas support both electromagnetic and electrostatic waves. Electromagnetic waves in plasmas are similar to those in dielectric materials and propagate due to the exchange of energy between electric and magnetic forms. Let the electric and magnetic fields of the wave vary as

$$\mathbf{E},\ \mathbf{H} \sim \exp j(\omega t - \mathbf{k}\cdot\mathbf{r}) \tag{4.2.31}$$

where $\mathbf{k}$ is the propagation vector, then for a uniform, isotropic (no applied dc magnetic field) plasma, the waves are transverse, with $\mathbf{E}$, $\mathbf{H}$, and $\mathbf{k}$ mutually perpendicular. To obtain the dispersion relation, we use (4.2.31) in (2.2.1) and (4.2.19) to obtain

$$\mathbf{k}\times\tilde{\mathbf{E}} = \omega\mu_0\tilde{\mathbf{H}} \tag{4.2.32}$$

and

$$\mathbf{k}\times\tilde{\mathbf{H}} = -\omega\epsilon_p\tilde{\mathbf{E}} \tag{4.2.33}$$

Cross multiplying (4.2.32) by $\mathbf{k}$ and using (4.2.33), we obtain

$$\mathbf{k}\times(\mathbf{k}\times\tilde{\mathbf{E}}) = -\omega^2\epsilon_p\mu_0\tilde{\mathbf{E}} \tag{4.2.34}$$

Expanding the triple cross-product[2] and noting that $\mathbf{k}\cdot\tilde{\mathbf{E}} = 0$ for transverse waves, we obtain

$$k^2\tilde{\mathbf{E}} = \kappa_p\frac{\omega^2}{c^2}\tilde{\mathbf{E}}$$

where k is the wave-vector magnitude, where we have written $\epsilon_p = \epsilon_0\kappa_p$ from (4.2.18), and where we have used $c = 1/\sqrt{\mu_0\epsilon_0}$ for the speed of light in vacuum. A nonzero $\tilde{\mathbf{E}}$ exists only if

$$k = \pm\frac{\sqrt{\kappa_p}\,\omega}{c} \tag{4.2.35}$$

which is the dispersion relation for transverse waves. Using κ_p for a cold collisionless plasma with infinite mass ions from (4.2.24), we see that the waves propagate (k is real) for $\kappa_p > 0$; i.e., for $\omega > \omega_{pe}$ and are cut off for $\omega < \omega_{pe}$. We plot ω versus k in Figure 4.5. Because ω is generally less than ω_{pe} in a discharge, electromagnetic waves excited at the plasma surface are not able to propagate into the plasma. In this case, the fields decay exponentially into the plasma. In general, if (4.2.18) is inserted in (4.2.35), we find that k separates into real and imaginary parts, $k = \beta - j\alpha$, with β the

2 $\mathbf{k}\times(\mathbf{k}\times\mathbf{E}) \equiv (\mathbf{k}\cdot\mathbf{E})\mathbf{k} - k^2\mathbf{E}$.

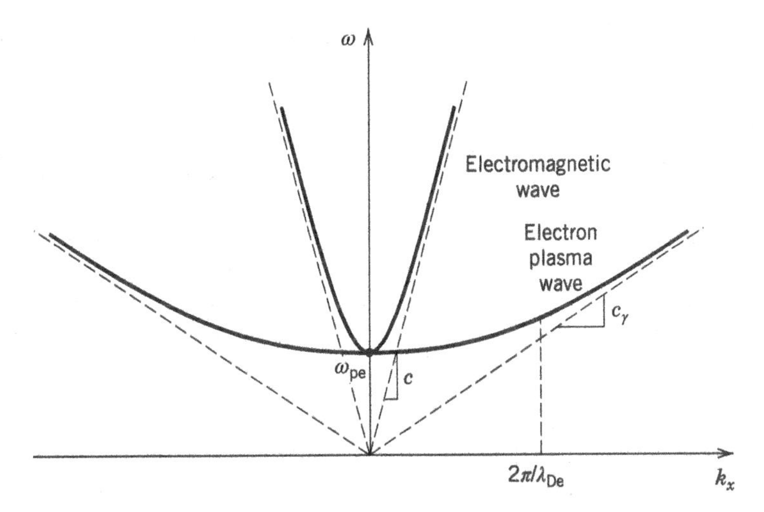

Figure 4.5 Dispersion ω versus k for electromagnetic and electrostatic electron plasma waves in an unmagnetized plasma.

real propagating part and α the real decay constant. An explicit calculation of α is in Problem 12.1, related to the determination of the power transfer in inductive discharges. On the other hand, we show in Section 4.5 that electromagnetic waves *can* propagate into a *magnetized* plasma.

The two independent polarizations have the same propagation constant k. Let $\mathbf{k} = \hat{x}k_x$, the most general transverse wave propagates along x with a polarization that is the superposition,

$$\tilde{\mathbf{E}} = \hat{y}\tilde{E}_y + \hat{z}\tilde{E}_z$$

which specifies a general elliptical polarization. As will be seen in Section 4.5, this is *not* true for waves in a magnetized plasma.

4.2.5 Electrostatic Waves

In a warm plasma, waves can propagate having $\mathbf{k} \parallel \mathbf{E}$. Such waves, which are not possible in a vacuum (or dielectric), are similar to sound waves in a gas. The waves propagate due to an exchange of energy between thermal and electric forms. Thermal electron motion, not considered in deriving the dielectric constant (4.2.24), leads to an additional term in the force equation due to ∇p_e, the gradient of the electron pressure. As a result of this, the electron plasma oscillations described by (4.2.3), for which $\mathbf{k} \parallel \mathbf{E}$, are converted into electron plasma waves.

To derive the dispersion relation, we use the property of an adiabatic equation of state (2.3.20), with $\nabla p_e/p_e = \gamma \nabla n_e/n_e$, to describe the variation of p_e, together with the usual Maxwellian relation $p_e = n_e k T_e$, with T_e constant. Substituting these quantities into (4.1.22), in the absence of a magnetic field and assuming that collisions are unimportant, we have

$$mn_e\left[\frac{\partial \mathbf{u}_e}{\partial t} + (\mathbf{u}_e \cdot \nabla)\mathbf{u}_e\right] = -en_e\mathbf{E} - \gamma k T_e \nabla n_e \tag{4.2.36}$$

where, here, k is Boltzmann's constant. We now make the usual assumptions of small signal quantities n_1, E_1, and u_1,

$$n_e = n_0 + n_1, \qquad \mathbf{E} = \hat{x}E_1, \qquad \mathbf{u}_e = \hat{x}u_1 \tag{4.2.37}$$

with no steady fields or drifts. We also assume sinusoidal wave motion, with all quantities varying as

$$n_1,\ E_1,\ u_1\ \sim\ \exp j(\omega t - k_x x) \tag{4.2.38}$$

where k_x is the propagation constant. Unlike electromagnetic waves, the electric field is parallel to $\mathbf{k}$ so that of the field equations only the divergence equation (2.2.3) is required. We further consider that the ions are essentially stationary on the time scale of the wave frequency. Assuming that all quantities vary as in (4.2.37) and (4.2.38), substituting into the continuity equation (2.3.7) (but without sources or sinks), the force equation (4.2.36), and the divergence equation (2.2.3), we obtain the first-order equations:

$$\omega n_1 - k_x n_0 u_1 = 0 \tag{4.2.39}$$

$$j\omega m n_0 u_1 = -en_0 E_1 + jk_x \gamma k T_e n_1 \tag{4.2.40}$$

$$jk_x \epsilon_0 E_1 = en_1 \tag{4.2.41}$$

Combining (4.2.39) through (4.2.41), we can factor out the first-order quantities to find the dispersion equation:

$$\omega^2 = \omega_{pe}^2 + k_x^2 c_\gamma^2 \tag{4.2.42}$$

where

$$c_\gamma = \left(\frac{\gamma k T_e}{m}\right)^{1/2} \tag{4.2.43}$$

is the adiabatic electron sound speed. For the one-dimensional motion considered here, $\gamma = 3$. The dispersion (4.2.42) is plotted in Figure 4.5, with the value of $k_x = 2\pi/\lambda_{De}$ indicated in the figure. As one might expect, for $k_x \gtrsim 2\pi/\lambda_{De}$, thermal disruption of the collective process would be expected to be very important, and the waves are strongly damped. This collisionless damping, called Landau damping, is discussed in most books on fully ionized plasmas (e.g., Chen, 1984, chapter 7). For long wavelengths, $k_x \ll 2\pi/\lambda_{De}$, the waves are not strongly damped, but they may be only weakly excited.

If the ions are also considered to be mobile, under certain circumstances, new waves can appear. For cold plasmas, the electron motion dominates the behavior of the waves, such that the plasma frequency in (4.2.42) is only slightly modified, as given by (4.2.7). For equal-temperature electrons and ions, this small modification still holds. However, for $T_i \ll T_e$, as usually exists in weakly ionized discharges, the electron random motion prevents the electrons from neutralizing independent ion motion, and short wavelength ion sound waves can exist. These are usually heavily damped and therefore not of great significance, but can become important if ions are streaming through electrons or other plasma species. Then, if the ion streaming velocity exceeds the local ion acoustic velocity, instabilities or nonlinear potential structures (shocks) can appear in the plasma. We leave details of an ion wave calculation to Problem 4.8. A discussion of ion waves and shocks can be found in many texts on fully ionized plasmas, e.g., Chen (1984, chapter 4 and section 8.3).

4.3 Guiding Center Motion

If the electric or magnetic field varies in space, the charged particle motion becomes much more complicated, and generally analytic solutions cannot be found. One very important configuration is that of a spatially varying magnetic field in which the gyration radius is much smaller than the scale length of the field variation. In that situation, an expansion in the gyroradius can be performed that allows separation into the fast gyromotion and slow drifts of the guiding center across field lines. We have already seen this separation in Section 4.1 for the trivial case of uniform **B**, where the guiding center moves uniformly along **B**. The separation of the motion is particularly useful for calculating particle confinement in fully ionized plasmas (see, e.g., Chen (1984, chapter 2) or Schmidt (1979, chapter 2)), but can also be applied to a number of high-density source concepts for materials processing. Here, we introduce the subject and point out a few implications for weakly ionized plasmas. A more complete derivation can be found in Schmidt (1979, chapter 2).

The basic procedure is to expand the instantaneous position into a guiding center and a gyroradius about that center,

$$\mathbf{r} = \mathbf{r}_{\mathrm{g}}(t) + \mathbf{r}_{\mathrm{c}}(t) \tag{4.3.1}$$

with an accompanying velocity,

$$\mathbf{v} = \mathbf{v}_{\mathrm{g}} + \mathbf{v}_{\mathrm{c}} \tag{4.3.2}$$

where $\mathbf{v}_{\mathrm{g}} = \mathrm{d}\mathbf{r}_{\mathrm{g}}/\mathrm{d}t$ and $\mathbf{v}_{\mathrm{c}} = \mathrm{d}\mathbf{r}_{\mathrm{c}}/\mathrm{d}t$. The magnetic field in the neighborhood of the guiding center is expanded as

$$\mathbf{B}(\mathbf{r}) = \mathbf{B}_0(\mathbf{r}) + (\mathbf{r}_{\mathrm{c}} \cdot \nabla)\mathbf{B}(\mathbf{r}) \tag{4.3.3}$$

with

$$|\mathbf{r}_{\mathrm{c}}\nabla\mathbf{B}/\mathbf{B}_0| \ll 1 \tag{4.3.4}$$

With this approximation, $|\mathbf{r}_{\mathrm{c}}(t)|$ can be taken as a constant over a gyroperiod. Then, averaging over a gyroperiod, the rapidly rotating terms average to zero in lowest order, resulting in an equation for the drift motion:

$$m\frac{\mathrm{d}\mathbf{v}_{\mathrm{g}}}{\mathrm{d}t} = \mathbf{F}_{\mathrm{ext}} + q\mathbf{v}_{\mathrm{g}} \times \mathbf{B} + q\langle \mathbf{v}_{\mathrm{c}} \times (\mathbf{r}_{\mathrm{c}} \cdot \nabla)\mathbf{B}\rangle \tag{4.3.5}$$

where $\langle\,\rangle$ denotes an average over a gyroperiod. The third term on the right-hand side has a product of rapidly oscillating quantities and therefore a first-order average value, giving, after some algebra,

$$m\frac{\mathrm{d}\mathbf{v}_{\mathrm{g}}}{\mathrm{d}t} = \mathbf{F}_{\mathrm{ext}} + q\mathbf{v}_{\mathrm{g}} \times \mathbf{B}_0 - \frac{\frac{1}{2}mv_\perp^2}{B_0}\nabla B \tag{4.3.6}$$

Here, $\mathbf{F}_{\mathrm{ext}}$ includes all external forces, $B_0 = |\mathbf{B}_0|$, and $v_\perp = |\mathbf{v}_{\mathrm{c}}|$, the velocity perpendicular to the field line. All quantities are calculated on the guiding center of the orbit. We indicate the effect of the various terms in (4.3.6) with some simple examples.

4.3.1 Parallel Force

We justify (4.3.6) for a particle gyrating in a magnetic field $\hat{z}B_z(z)$ that is increasing along z. The magnetic field lines converge as shown in Figure 4.6, and the Lorentz force $q\mathbf{v}_\perp \times \mathbf{B}$ has a component along z given by

$$F_z = -qv_\phi B_r \tag{4.3.7}$$

where $v_\phi = -v_{\perp 0}$, and B_r is obtained from (2.2.4), which is, in cylindrical coordinates,

$$\frac{1}{r}\frac{\partial}{\partial r}(rB_r) + \frac{\partial B_z}{\partial z} = 0$$

This yields B_r upon integrating with respect to r:

$$B_r \approx -\frac{r_c}{2}\frac{\partial B_z}{\partial z} \tag{4.3.8}$$

Substituting (4.3.8) into (4.3.7) and taking all quantities as constant over a gyro-orbit, in keeping with our expansion, we obtain the average force acting on the guiding center to be

$$F_z = -\frac{\frac{1}{2}mv_\perp^2}{B_z}\frac{\partial B_z}{\partial z} \tag{4.3.9}$$

We see that (4.3.9) corresponds to the z component of the third term on the RHS of (4.3.6). The force F_z pushes the particle into regions of smaller B and is independent of charge. From the averaging procedure, it is seen to be valid only for

$$r_c \ll \left(\frac{1}{B_z}\frac{dB_z}{dz}\right)^{-1} \tag{4.3.10}$$

which is equivalent to (4.3.4).

4.3.2 Adiabatic Constancy of the Magnetic Moment

For the field of Figure 4.6, we introduce the quantity

$$\mu_{\text{mag}} = \frac{\frac{1}{2}mv_\perp^2}{B_z} = \frac{W_\perp}{B_z} \tag{4.3.11}$$

which can be shown to be the *magnetic moment* of the particle (see Problem 4.10). As the particle moves, both B_z and $W_\perp$ can change; however, the total kinetic energy of the particle is conserved because the magnetic field does no work. For the above example,

$$W_\perp(z) + W_z(z) = \text{const} \tag{4.3.12}$$

where $W_z = \frac{1}{2}mv_z^2$. If the particle moves a distance dz, then

$$dW_z = F_z\,dz = -\frac{W_\perp}{B_z}\,dB_z \tag{4.3.13}$$

Differentiating (4.3.12) yields $dW_z = -dW_\perp$; hence, (4.3.13) becomes

$$\frac{dW_\perp}{W_\perp} = \frac{dB_z}{B_z} \tag{4.3.14}$$

which can be integrated to obtain

$$\frac{W_\perp}{B_z} \equiv \mu_{\text{mag}} = \text{const} \tag{4.3.15}$$

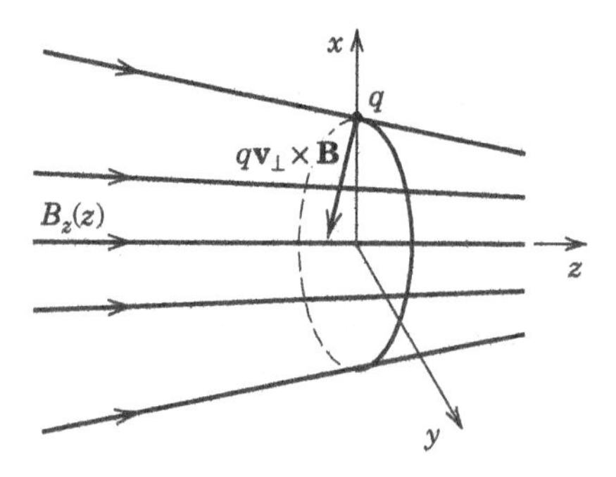

Figure 4.6 Calculation of the parallel force due to a magnetic field gradient $\partial B_z/\partial z$.

The magnetic moment is one example of an *adiabatic invariant,* a quantity that is approximately conserved in the motion if the scale length condition (4.3.4) is satisfied.

The constancy of μ_{mag} has an important consequence in the magnetic mirroring of charged particles in an increasing magnetic field. As B_z increases, $W_\perp$ increases to keep μ_{mag} constant, reflecting the particle when $W_\perp = W$ ($W_z = 0$). Although this property is of primary concern in nearly collisionless plasmas where plasma confinement is of greatest interest, it can also play a significant role in confining the higher-energy electrons in cyclotron resonance or magnetron discharges, which we consider in Chapters 13 and 14.

4.3.3 Drift Due to Motion Along Field Lines (Curvature Drift)

Consider a curved field line in the x–z plane. As shown in Figure 4.7, although $B_x = 0$ at the origin, $\partial B_x/\partial z$ is nonzero. The radius of curvature R of the field line is found from (see figure)

$$\frac{\mathrm{d}z}{R} = -\frac{\mathrm{d}B_x}{B_z}$$

which yields

$$\frac{1}{R} = -\frac{1}{B_z}\frac{\partial B_x}{\partial z} \tag{4.3.16}$$

The centrifugal force acting on the particle is

$$\mathbf{F}_R = \frac{mv_z^2}{R}\hat{x} = \frac{2W_z}{R}\hat{x} = -\frac{2W_z}{B_z}\frac{\partial B_x}{\partial z}\hat{x} \tag{4.3.17}$$

Since the force in (4.3.17) is an average force, we can substitute it into (4.1.19) to obtain the drift of the guiding center due to the field line curvature:

$$\mathbf{v}_R = \frac{2W_z}{qB_z^2}\frac{\partial B_x}{\partial z}\hat{y} \tag{4.3.18}$$

We see that electrons and ions drift in opposite directions perpendicular to both $\mathbf{B}$ and the curvature force, giving rise to a net current. The drift given in (4.3.18) is not immediately seen in the averaged equation (4.3.6). To obtain the drifts, (4.3.6) is cross-multiplied by $\mathbf{B}/qB^2$, such that the second term on the right is $\mathbf{v}_{g\perp}$. The drift (4.3.18) is then obtained from the vector decomposition of $(m\,\mathrm{d}\mathbf{v}_g/\mathrm{d}t) \times \mathbf{B}/qB^2$. We leave the calculation to Problem 4.11.

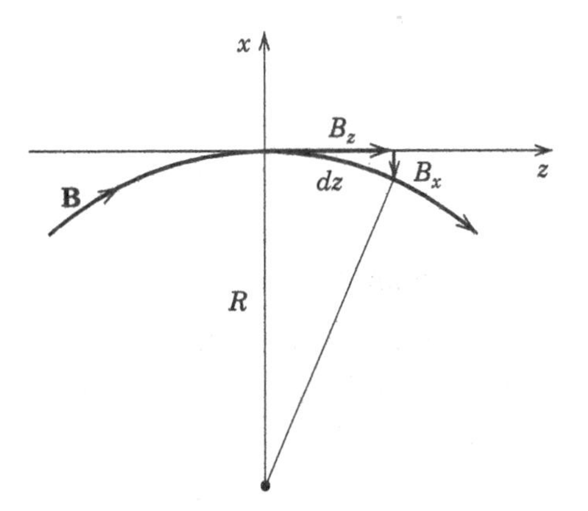

Figure 4.7 Calculation of the curvature drift due to a magnetic field gradient $\partial B_x/\partial z$.

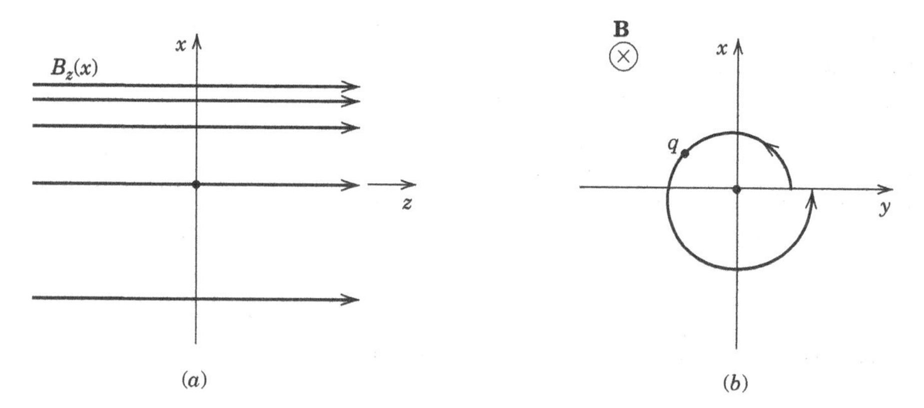

Figure 4.8 Calculation of the perpendicular gradient drift due to a magnetic field gradient $\partial B_z/\partial x$: (*a*) the magnetic field lines; (*b*) the motion viewed in the x–y plane.

4.3.4 Drift Due to Gyration (Gradient Drift)

Consider a magnetic field $B_z(x)$ with a gradient perpendicular to the lines of **B**, as shown in Figure 4.8*a*. Viewing the motion of a gyrating particle in the x–y plane (Figure 4.8*b*), we see that there is a stronger Lorentz force at the upper half of the orbit than at the lower half, producing a smaller gyration radius at the upper half than at the lower, and leading to a net drift along y. This drift can be obtained directly from the third term on the right in (4.3.6) with $\nabla B = \nabla B_z = \partial B_z/\partial x$, which, as an average force, can be substituted into (4.1.19) to give (see also Problem 4.12)

$$\mathbf{v}_{\nabla B} = -\frac{W_\perp}{qB_z^2}\nabla_\perp B_z \times \hat{z} \tag{4.3.19}$$

Electrons and ions drift in opposite directions, giving rise to a net current, as with the curvature drift.

Note that although the two drifts found in (4.3.18) and (4.3.19) are commonly called curvature drift and gradient drift, they are really distinguished by a velocity parallel to field lines and perpendicular to field lines, respectively. Both drifts arise due to field gradients. If the zero-order magnetic fields are produced by currents external to the plasma, then from (2.2.2),

$$\nabla \times \mathbf{B} \approx 0 \tag{4.3.20}$$

inside the plasma, where we have neglected the first-order (weak) currents produced by the moving charges in the plasma. In this case,

$$\frac{\partial B_x}{\partial z} = \frac{\partial B_z}{\partial x} \tag{4.3.21}$$

and the curvature and gradient drifts can be expressed in terms of a single gradient.

4.3.5 Polarization Drift

Consider a uniform magnetic field $\hat{z}B_0$ and a transverse electric field $\hat{x}E(t)$ that varies slowly with time. Then, the $\mathbf{E} \times \mathbf{B}$ drift velocity also varies slowly with time:

$$\mathbf{v}_E(t) = -\frac{E(t)}{B_0}\hat{y} \tag{4.3.22}$$

Hence, the guiding center accelerates along $\hat{y}$. The acceleration in the lab frame is

$$\mathbf{a}(t) = -\frac{1}{B_0}\frac{\partial E}{\partial t}\hat{y}$$

In the frame of the particle, there is therefore an average inertial force transverse to $\mathbf{B}$:

$$\mathbf{F}_\mathrm{p} = -\mathbf{F} = -m\mathbf{a} = \frac{m}{B_0}\frac{\partial E}{\partial t}\hat{y} \tag{4.3.23}$$

Using (4.1.19), this gives rise to a guiding center drift

$$\mathbf{v}_\mathrm{p} = \frac{m}{qB_0^2}\frac{\partial \mathbf{E}}{\partial t} \tag{4.3.24}$$

that lies along $\mathbf{E}$ itself. Again ions and electrons drift in opposite directions, giving an additive conduction current, which for $n_\mathrm{i} = n_\mathrm{e} = n_0$ is

$$\mathbf{J}_\mathrm{p} = \frac{(M+m)n_0}{B_0^2}\frac{\partial \mathbf{E}}{\partial t} \tag{4.3.25}$$

We see that the electron drift component of the current is negligible due to the mass dependence in (4.3.25). Introducing a low-frequency *perpendicular dielectric constant* $\epsilon_\perp$ through the relation $\mathbf{J}_\mathrm{p} = \epsilon_\perp \partial\mathbf{E}/\partial t$, and dropping the electron mass term, we obtain

$$\epsilon_\perp = \epsilon_0\left(1 + \frac{Mn_0}{\epsilon_0 B_0^2}\right) = \epsilon_0\left(1 + \frac{\omega_\mathrm{pi}^2}{\omega_\mathrm{ci}^2}\right) \tag{4.3.26}$$

For $n_0 \sim 10^{10}$ cm^{-3} and $B_0 \sim 100$ G, we obtain $\epsilon_\perp \sim 10^6\epsilon_0$. At low frequencies, $\omega \ll \omega_\mathrm{ci}$, this very large positive dielectric constant perpendicular to $\mathbf{B}$ shields a magnetized plasma from external electric fields perpendicular to $\mathbf{B}$. For electric fields along $\mathbf{B}$, we can introduce $\epsilon_\| = \epsilon_\mathrm{p}$ as given in (4.2.24), which at low frequencies is large and negative, also shielding the plasma from electric fields lying along $\mathbf{B}$.

The gyration motion itself also produces currents in a nonuniform plasma. To see this, we form

$$\mathbf{M} = -\hat{z}n(\mathbf{r})\mu_\mathrm{mag} \tag{4.3.27}$$

the magnetization of the plasma, such that the magnetic induction $\mathbf{B}$ and the magnetic field $\mathbf{H}$ are related by

$$\mathbf{B} = \mu_0(\mathbf{H} + \mathbf{M}) \tag{4.3.28}$$

Then, substituting (4.3.28) into (2.2.2) with $\mathbf{J} = \mathbf{J}_\mathrm{p}$ yields

$$\nabla \times \left(\frac{\mathbf{B}}{\mu_0}\right) = \mathbf{J}_\mathrm{p} + \mathbf{J}_\mathrm{mag} + \epsilon_0\frac{\partial \mathbf{E}}{\partial t} \tag{4.3.29}$$

where

$$\mathbf{J}_\mathrm{mag} = \nabla \times \mathbf{M} \tag{4.3.30}$$

is the magnetization current. Equation (4.3.29) shows explicitly the three sources of $\mathbf{B}/\mu_0$ in a magnetized plasma: the conduction, magnetization, and displacement currents. Since the currents of the gyrating charges act to weaken the applied field, the plasma is *diamagnetic*. As we can see from (4.3.27), the diamagnetism depends on both the plasma density and particle energies and becomes important only in dense energetic plasmas, primarily those encountered in fusion research. In all but the highest density discharges, the weakening of an applied magnetic field due to plasma diamagnetism is small.

Table 4.1 Summary of Guiding Center Drifts ($\mathbf{R}_c/R_c^2 = -\nabla B/B$)

General force drift	$\mathbf{v}_F = \dfrac{(\mathbf{F}/q)\times\mathbf{B}}{B^2}$
Electric field drift	$\mathbf{v}_E = \dfrac{\mathbf{E}\times\mathbf{B}}{B^2}$
Curvature drift	$\mathbf{v}_R = \dfrac{2W_\parallel}{q}\dfrac{\mathbf{R}_c\times\mathbf{B}}{R_c^2B^2}$
Grad-B drift	$\mathbf{v}_{\nabla B} = \dfrac{W_\perp}{q}\dfrac{\mathbf{B}\times\nabla B}{B^3}$
Polarization drift	$\mathbf{v}_p = \dfrac{m}{qB^2}\dfrac{\partial\mathbf{E}}{\partial t}$

The guiding center motion is derived by a formal expansion of (4.1.1) in most books on fully ionized plasmas, e.g., Schmidt (1979), rather than from the more physical approach given here. A summary of the drifts is given in Table 4.1.

4.4 Dynamics of Magnetized Plasmas

The response of a plasma immersed in a steady uniform magnetic field $\mathbf{B}_0$ and subject to time-varying electric and magnetic fields is very complicated. The fact that the gyromotion converts velocities being acted on by one field component to another velocity component leads to a *gyrotropic* dielectric tensor, having complex conjugate off-diagonal elements in the absence of dissipation. Furthermore, the inhibition of the electron motion perpendicular to $\mathbf{B}_0$ gives rise to an important ion response, particularly at low frequencies. Collisional dissipation further complicates the picture. Fortunately, for consideration of electromagnetic waves, the wave velocities are generally much higher than the thermal velocities, and thus the effects of the electron and ion thermal velocities can be ignored.

On the other hand, we have seen in Section 4.2, in the absence of $\mathbf{B}_0$, that electrostatic waves can resonate with thermal velocities, leading to strong temperature effects. Similarly, in magnetized plasmas, there are electrostatic waves that propagate across the magnetic field, whose nature depends on thermal effects. These waves are generally of little interest for weakly ionized plasmas and will not be considered here. The interested reader is directed to the literature (e.g., Stix, 1992). Our approach in this section will first be to derive the dielectric tensor in the simplest case where only electrons participate, and the electron fluid is considered to be cold and collisionless. It is then straightforward to include the effect of collisions and the addition of a mobile ion species. Using this dielectric tensor, in any of the above approximations, we can derive the dispersion relation for waves propagating at an arbitrary angle to $\mathbf{B}_0$. Because of the complexity of the wave problem, we leave a detailed consideration of the waves to Section 4.5.

4.4.1 Dielectric Tensor

We begin with the force equation in rectangular coordinates as in (4.1.1), with $\mathbf{B}_0 = \hat{z}B_{z0}$. Assuming sinusoidal variation (4.2.31) of the electric field, the linearized equations for the electron motion are then

$$j\omega\tilde{v}_x = -\frac{e}{m}\tilde{E}_x - \omega_{ce}\tilde{v}_y \tag{4.4.1a}$$

$$j\omega\tilde{v}_y = -\frac{e}{m}\tilde{E}_y + \omega_{ce}\tilde{v}_x \tag{4.4.1b}$$

$$j\omega\tilde{v}_z = -\frac{e}{m}\tilde{E}_z \tag{4.4.1c}$$

where we have chosen $\omega_{ce} = eB_{z0}/m$ to be explicitly positive. Solving (4.1.1a) and (4.4.1b) simultaneously, for $\tilde{v}_x$ and $\tilde{v}_y$, we have

$$\tilde{v}_x = \frac{e}{m}\frac{j\omega\tilde{E}_x - \omega_{ce}\tilde{E}_y}{\omega^2 - \omega_{ce}^2} \tag{4.4.2a}$$

$$\tilde{v}_y = \frac{e}{m}\frac{j\omega\tilde{E}_y + \omega_{ce}\tilde{E}_x}{\omega^2 - \omega_{ce}^2} \tag{4.4.2b}$$

Using our previous assumption that $\mathbf{J} = -en_0\mathbf{v}$ and defining the dielectric properties from Maxwell's equation,

$$\nabla \times \tilde{\mathbf{H}} = j\omega\epsilon_0\tilde{\mathbf{E}} + \tilde{\mathbf{J}} \equiv j\omega\overline{\overline{\epsilon}}_p \cdot \tilde{\mathbf{E}} \tag{4.4.3}$$

we obtain

$$\overline{\overline{\epsilon}}_p = \epsilon_0\overline{\overline{\kappa}}_p = \epsilon_0\begin{pmatrix} \kappa_\perp & -j\kappa_\times & 0 \\ j\kappa_\times & \kappa_\perp & 0 \\ 0 & 0 & \kappa_\parallel \end{pmatrix} \tag{4.4.4}$$

where

$$\kappa_\perp = 1 - \frac{\omega_{pe}^2}{\omega^2 - \omega_{ce}^2} \tag{4.4.5a}$$

$$\kappa_\times = \frac{\omega_{ce}}{\omega}\frac{\omega_{pe}^2}{\omega^2 - \omega_{ce}^2} \tag{4.4.5b}$$

$$\kappa_\parallel = 1 - \frac{\omega_{pe}^2}{\omega^2} \tag{4.4.5c}$$

The z or $\parallel$ component is the same as the dielectric constant (4.2.24) in the absence of $\mathbf{B}_0$. The other components are characteristic of a lossless gyrotropic medium, with $\epsilon_{ij} = \epsilon_{ji}^*$.

Given the collisionless electron dielectric tensor components (4.4.5), it is rather simple to include the effect of collisions, or the contribution of mobile ions. To include collisions, we recognize that each ω originating from the force equation is transformed as $\omega \to \omega - j\nu_m$. The ωs arising from Maxwell's equations, however, remain unchanged. Performing this operation, we obtain

$$\kappa_\perp = 1 - \frac{\omega - j\nu_m}{\omega}\frac{\omega_{pe}^2}{(\omega - j\nu_m)^2 - \omega_{ce}^2} \tag{4.4.6a}$$

$$\kappa_\times = \frac{\omega_{ce}}{\omega}\frac{\omega_{pe}^2}{(\omega - j\nu_m)^2 - \omega_{ce}^2} \tag{4.4.6b}$$

$$\kappa_\parallel = 1 - \frac{\omega_{pe}^2}{\omega(\omega - j\nu_m)} \tag{4.4.6c}$$

The dielectric tensor, including ion dynamics, is also easily obtained by generalizing (4.4.3). To do this, we recognize that the electron and ion currents add. Then, each term in the dielectric tensor consists of a sum of electrons and ion components of the same form, but with the parameters appropriate to that species. Thus, again ignoring collisions,

$$\kappa_\perp = 1 - \frac{\omega_{pe}^2}{\omega^2 - \omega_{ce}^2} - \frac{\omega_{pi}^2}{\omega^2 - \omega_{ci}^2} \tag{4.4.7a}$$

$$\kappa_\times = \frac{\omega_{ce}}{\omega}\frac{\omega_{pe}^2}{\omega^2 - \omega_{ce}^2} - \frac{\omega_{ci}}{\omega}\frac{\omega_{pi}^2}{\omega^2 - \omega_{ci}^2} \tag{4.4.7b}$$

$$\kappa_\parallel = 1 - \frac{\omega_p^2}{\omega^2} \tag{4.4.7c}$$

where $\omega_{ci} = eB_{z0}/M$ is defined to be explicitly positive, and we have combined the electron and ion plasma frequencies in (4.4.7c) using (4.2.7). Examining the size of the terms in (4.4.7a), we are often considering situations in which $\omega_{pe} \sim \omega_{ce}$. In that case, we see that $\omega_{pi} \sim (M/m)^{1/2}\omega_{ci}$ such that, depending on the range of frequencies being considered, the ion motion can dominate the transverse dielectric components. We have already seen an example of this for low frequencies, $\omega \ll \omega_{ci}$, where (4.4.7a) reduces to (4.3.26). We shall return to this point in considering the wave spectrum.

4.4.2 The Wave Dispersion

Returning to consideration of waves of the form $\exp j(\omega t - \mathbf{k}\cdot\mathbf{r})$, Maxwell's curl equations become

$$\mathbf{k} \times \tilde{\mathbf{E}} = \omega\mu_0\tilde{\mathbf{H}} \tag{4.4.8}$$

and

$$\mathbf{k} \times \tilde{\mathbf{H}} = -\omega\epsilon_0\overline{\overline{\kappa}}_p \cdot \tilde{\mathbf{E}} \tag{4.4.9}$$

where $\overline{\overline{\kappa}}_p$ is given by one of the forms in the previous subsection. Taking the cross product of $\mathbf{k}$ with (4.4.8) and substituting for $\mathbf{k} \times \tilde{\mathbf{H}}$ from (4.4.9), we obtain the equation describing electromagnetic waves in a magnetized plasma:

$$\mathbf{k} \times (\mathbf{k} \times \tilde{\mathbf{E}}) + k_0^2\overline{\overline{\kappa}}_p \cdot \tilde{\mathbf{E}} = 0 \tag{4.4.10}$$

where $k_0 = \omega/c$ is the propagation constant of a plane wave of frequency ω in free space, with c the velocity of light.

The vector equation (4.4.10) is very complicated because all of the components of $\tilde{\mathbf{E}}$ couple together. In deriving the dielectric tensor we used rectangular coordinates with $\mathbf{B}_0$ taken along the z direction for concreteness. We have one more direction to define, that of the wave vector, which we can take to lie in the x–z plane, without loss of generality. Doing this, (4.4.10) can be written as

$$\begin{bmatrix} k_z^2 & 0 & -k_xk_z \\ 0 & k_x^2 + k_z^2 & 0 \\ -k_xk_z & 0 & k_x^2 \end{bmatrix}\begin{bmatrix} \tilde{E}_x \\ \tilde{E}_y \\ \tilde{E}_z \end{bmatrix} = k_0^2\begin{bmatrix} \kappa_\perp & -j\kappa_\times & 0 \\ j\kappa_\times & \kappa_\perp & 0 \\ 0 & 0 & \kappa_\parallel \end{bmatrix}\begin{bmatrix} \tilde{E}_x \\ \tilde{E}_y \\ \tilde{E}_z \end{bmatrix} \tag{4.4.11}$$

If the angle between $\mathbf{k}$ and $\mathbf{B}$ is defined as θ, then $k_z = k\cos\theta$ and $k_x = k\sin\theta$, where here $k = |\mathbf{k}|$. Furthermore, it is usual to normalize the magnitude of k as $N = k/k_0$, where N here is the *index of refraction* of the wave. Using this notation, and requiring that the determinant of the coefficients of the equation for $\tilde{\mathbf{E}}$ vanishes for a nontrivial solution, we obtain

$$\det\begin{bmatrix} N^2\cos^2\theta - \kappa_\perp & j\kappa_\times & -N^2\cos\theta\sin\theta \\ -j\kappa_\times & N^2 - \kappa_\perp & 0 \\ -N^2\cos\theta\sin\theta & 0 & N^2\sin^2\theta - \kappa_\parallel \end{bmatrix} = 0 \tag{4.4.12}$$

Equation (4.4.12) is the dispersion equation, which relates $k \equiv k_0N$, ω, and θ.

4.5 Waves in Magnetized Plasmas

In this section, we first describe some general properties of waves in magnetized plasmas and then consider in some detail the *principal* waves, i.e., those traveling parallel to and perpendicular to $\mathbf{B}_0$. We then give a qualitative description of propagation at an arbitrary angle in the various regimes of frequency, density, and magnetic field.

Evaluating the determinant in (4.4.12), we find that the cubic terms in N^2 cancel, reducing the equation to a biquadratic form:

$$aN^4 - bN^2 + c = 0 \tag{4.5.1}$$

where

$$a = \kappa_\perp \sin^2\theta + \kappa_\| \cos^2\theta \tag{4.5.2a}$$

$$b = (\kappa_\perp^2 - \kappa_\times^2)\sin^2\theta + \kappa_\| \kappa_\perp (1 + \cos^2\theta) \tag{4.5.2b}$$

$$c = (\kappa_\perp^2 - \kappa_\times^2)\kappa_\| \tag{4.5.2c}$$

Hence, there are in general two different solutions for N^2 for each angle θ. These solutions correspond to the two allowed polarizations for the electric field of the wave. Because the discriminant $b^2 - 4ac$ of (4.5.1) is always positive, N^2 is real, and N is either real and the wave propagates or imaginary and the wave is cutoff. In the latter case, which may occur for one or both solutions, depending on the parameters, the wave of that polarization does not propagate but decays exponentially. The two wave polarizations are determined by the relative magnitudes of the components of the electric field. These are given by the ratios of the cofactors of any row in the matrix (4.4.12). Taking the first row, we obtain

$$\begin{aligned} \tilde{E}_x : \tilde{E}_y : \tilde{E}_z \; :: \; &(\kappa_\perp - N^2)(\kappa_\| - N^2\sin^2\theta) : j\kappa_\times(N^2\sin^2\theta - \kappa_\|) \\ &: (N^2 - \kappa_\perp)N^2 \sin\theta\cos\theta \end{aligned} \tag{4.5.3}$$

which gives two different ratios of the field components for the two values of N^2. Since the two waves generally have different propagation constants, their electric fields do not have the same spatial variation and their polarizations cannot be summed to determine a resultant polarization that remains fixed as the waves propagate.

Although (4.5.1) can be solved for N^2 as a function of θ, the results are not particularly illuminating. It is more useful to solve for θ as a function of N^2. Before doing this, it is convenient to introduce two combinations of the dielectric components,

$$\kappa_\mathrm{r} = \kappa_\perp - \kappa_\times \tag{4.5.4a}$$

and

$$\kappa_\mathrm{l} = \kappa_\perp + \kappa_\times \tag{4.5.4b}$$

such that $\kappa_\perp^2 - \kappa_\times^2 = \kappa_\mathrm{r}\kappa_\mathrm{l}$ in (4.5.2b) and (4.5.2c). For the simplest case of no collisions and infinite mass ions, we use (4.4.5a) and (4.4.5b) to obtain

$$\kappa_\mathrm{r} = 1 - \frac{\omega_{\mathrm{pe}}^2}{\omega(\omega - \omega_{\mathrm{ce}})} \tag{4.5.5a}$$

and

$$\kappa_\mathrm{l} = 1 - \frac{\omega_{\mathrm{pe}}^2}{\omega(\omega + \omega_{\mathrm{ce}})} \tag{4.5.5b}$$

with the obvious extensions $\omega \pm \omega_{ce} \to \omega \pm \omega_{ce} - j\nu_m$ in (4.5.5) for adding collisions and

$$\kappa_r = 1 - \frac{\omega_{pe}^2}{\omega(\omega - \omega_{ce})} - \frac{\omega_{pi}^2}{\omega(\omega + \omega_{ci})} \tag{4.5.6a}$$

and

$$\kappa_l = 1 - \frac{\omega_{pe}^2}{\omega(\omega + \omega_{ce})} - \frac{\omega_{pi}^2}{\omega(\omega - \omega_{ci})} \tag{4.5.6b}$$

for a collisionless plasma with mobile ions. Substituting $\sin^2\theta + \cos^2\theta$ for 1 into (4.5.2b) and (4.5.2c), substituting a, b, and c into (4.5.1), and dividing (4.5.1) by $\cos^2\theta$, we can solve to obtain

$$\tan^2\theta = -\frac{\kappa_\parallel (N^2 - \kappa_r)(N^2 - \kappa_l)}{(N^2 - \kappa_\parallel)(\kappa_\perp N^2 - \kappa_r \kappa_l)} \tag{4.5.7}$$

4.5.1 Principal Electron Waves

4.5.1.1 k || B₀

For this case ($\theta = 0$), the numerator of (4.5.7) vanishes, yielding

$$\kappa_\parallel (N^2 - \kappa_r)(N^2 - \kappa_l) = 0 \tag{4.5.8}$$

The first solution $\kappa_\parallel = 0$ gives the plasma oscillations for $\mathbf{E} \parallel \mathbf{B}_0$ discussed in Section 4.2. The second and third solutions give the principal waves. Using (4.5.5), these are

$$N_r^2 = 1 - \frac{\omega_{pe}^2}{\omega(\omega - \omega_{ce})} \tag{4.5.9a}$$

and

$$N_l^2 = 1 - \frac{\omega_{pe}^2}{\omega(\omega + \omega_{ce})} \tag{4.5.9b}$$

where ω_{ce} is explicitly positive. The first wave has a resonant denominator for $\omega = \omega_{ce}$, which gives the dispersion for the right-hand polarized (RHP) wave. At $\omega = \omega_{ce}$, the wave rotates in synchronism with the gyrating electrons, which then see a constant field leading to resonant energy absorption, as we will see in Chapter 13. The second wave is the left-hand polarized (LHP) wave, which is nonresonant.

To see that (4.5.9a) represents a right circularly polarized wave, we let $N^2 = \kappa_r$ in (4.5.3) to obtain

$$E_x : E_y :: \kappa_\perp - \kappa_r : -j\kappa_\times$$

and using (4.5.4a),

$$E_x : E_y :: \kappa_\times : -j\kappa_\times$$

Hence, the field is given by

$$\mathbf{E} = \mathrm{Re}\left[\tilde{E}_r(\hat{x} - j\hat{y})\exp j(\omega t - \mathbf{k}_r \cdot \mathbf{r})\right] \tag{4.5.10a}$$

which at fixed $\mathbf{r}$ has a constant amplitude and rotates in the right-hand sense around $\mathbf{B}_0$ at frequency ω. Similarly, the LHP wave has

$$\mathbf{E} = \mathrm{Re}\left[\tilde{E}_l(\hat{x} + j\hat{y})\exp j(\omega t - \mathbf{k}_l \cdot \mathbf{r})\right] \tag{4.5.10b}$$

and rotates in the left-hand sense around $\mathbf{B}_0$. The most general solution propagating along z is a sum of the RHP and LHP waves given above.

The wave dispersion is easily described by first computing the resonances, $N \to \infty$, and cutoffs, $N \to 0$. Besides the resonance of the RHP wave at $\omega = \omega_{ce}$, there is a cutoff at

$$1 - \frac{\omega_{pe}^2}{\omega(\omega - \omega_{ce})} = 0$$

or, solving for ω,

$$\omega_R = \frac{\omega_{ce} + \sqrt{\omega_{ce}^2 + 4\omega_{pe}^2}}{2} \tag{4.5.11}$$

Only the + solution corresponds to positive ω, leading to an upper cutoff frequency above both ω_{pe} and ω_{ce}. For the LHP wave, a similar calculation gives a cutoff at

$$\omega_L = \frac{-\omega_{ce} + \sqrt{\omega_{ce}^2 + 4\omega_{pe}^2}}{2} \tag{4.5.12}$$

Again the + solution has been taken, which leads to a lower cutoff frequency below ω_{pe}. We should, however, have some doubts about this part of the solution, because it can occur at low frequencies where ion dynamics may be important. It is now possible to sketch an ω–k or dispersion diagram for the waves. We first obtain the other principal waves, so that we can sketch the results on a single diagram.

4.5.1.2 $\mathbf{k} \perp \mathbf{B}_0$

For this case ($\theta = \pi/2$), the denominator of (4.5.7) vanishes, yielding

$$(N^2 - \kappa_\parallel)(\kappa_\perp N^2 - \kappa_r\kappa_l) = 0 \tag{4.5.13}$$

The first solution is just the wave (4.2.35) for propagation in an unmagnetized plasma. It corresponds to a linearly polarized wave electric field lying along the dc magnetic field direction $\hat{z}$, so that the motion is unaffected by $\mathbf{B}_0$, and is called the ordinary (o) wave. The second solution gives a wave having electric fields that are perpendicular to $\mathbf{B}_0$, but with components both perpendicular ($\hat{y}$) and parallel ($\hat{x}$) to $\mathbf{k}$. Solving for N, we have the extraordinary (x) wave dispersion:

$$N_x^2 = \frac{\left[1 - \omega_{pe}^2\omega(\omega - \omega_{ce})\right]\left[1 - \omega_{pe}^2\omega(\omega + \omega_{ce})\right]}{1 - \dfrac{\omega_{pe}^2}{\omega^2 - \omega_{ce}^2}} \tag{4.5.14}$$

We see that the numerator has the same two cutoff solutions that we found for the RHP and LHP waves. The resonance at $\omega = \omega_{ce}$ disappears, because of cancellation of the factor $\omega - \omega_{ce}$. However, a new resonance appears at the upper hybrid frequency ω_{UH} given by

$$\omega_{UH}^2 = \omega_{pe}^2 + \omega_{ce}^2 \tag{4.5.15}$$

when the numerator of $\kappa_\perp$ is zero.

The dispersion (ω–k) diagrams for the principal waves in an electron plasma are sketched in Figure 4.9. All the results above the lower cutoff frequencies are reasonably representative of the dispersion when ions are also present. However, at lower frequencies, particularly near ω_{pi} and below, we expect the ion dynamics to be important. We discuss these additional wave solutions below.

First, however, we point out some important characteristics of the less cluttered dispersion of Figure 4.9. Considering the RHP wave with $\omega_{ce} > \omega$ the wave is propagating. Now let B_0 decrease slowly in the direction of propagation until $\omega_{ce}(z) = \omega$. At this value, there is a resonance at which

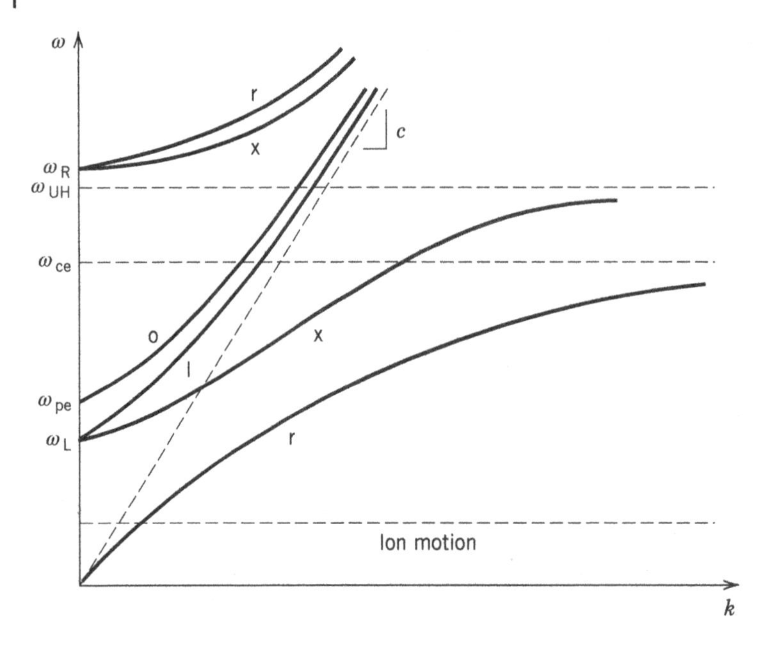

Figure 4.9 Dispersion ω versus k for the principal waves in a magnetized plasma with immobile ions for $\omega_{ce} > \omega_{pe}$.

$k_r = \infty$, and both the phase and group velocity go to zero. A careful analysis reveals that the wave energy is strongly absorbed in this field strength provided certain conditions on the scale length of the field variation and the density are satisfied. This phenomenon of absorption on a "magnetic beach" is an important mechanism for plasma heating and is a major subject of Chapter 13. A similar phenomenon occurs at the upper hybrid resonance for the x wave. However, this wave may not be accessible from outside the plasma, particularly at high density for which $\omega_{pe} > \omega_{ce}$ (not shown in Figure 4.9), if the decreasing magnetic field requires the wave to pass through the upper hybrid cutoff frequency ω_R. Similarly, the left-hand wave can also be cutoff at high densities if $\omega < \omega_L$.

4.5.2 Principal Waves Including Ion Dynamics

4.5.2.1 $\mathbf{k} \parallel \mathbf{B}_0$

Adding the ion dynamics into the dispersion equation using (4.5.6), we obtain, for the two polarizations, corresponding to (4.5.9),

$$N_r^2 = 1 - \frac{\omega_{pe}^2}{\omega(\omega - \omega_{ce})} - \frac{\omega_{pi}^2}{\omega(\omega + \omega_{ci})} \tag{4.5.16a}$$

and

$$N_l^2 = 1 - \frac{\omega_{pe}^2}{\omega(\omega + \omega_{ce})} - \frac{\omega_{pi}^2}{\omega(\omega - \omega_{ci})} \tag{4.5.16b}$$

where again ω_{ce} and ω_{ci} are positive. Considering first the RHP wave, we put the plasma terms under a common denominator and take $n_i = n_e$ to get

$$N_r^2 = 1 - \frac{\omega_p^2}{(\omega - \omega_{ce})(\omega + \omega_{ci})} \tag{4.5.17}$$

Similarly for the LHP wave, we have

$$N_l^2 = 1 - \frac{\omega_p^2}{(\omega + \omega_{ce})(\omega - \omega_{ci})} \tag{4.5.18}$$

4.5.2.2 $\mathbf{k} \perp \mathbf{B}_0$

In a similar manner, using the dispersion for the extraordinary (x) wave from (4.5.13), with (4.4.7a) and (4.5.6), we have

$$N_x^2 = \frac{\left[1 - \omega_{pe}^2\omega(\omega - \omega_{ce}) - \omega_{pi}^2\omega(\omega + \omega_{ci})\right]\left[1 - \omega_{pe}^2\omega(\omega + \omega_{ce}) - \omega_{pi}^2\omega(\omega - \omega_{ci})\right]}{1 - \dfrac{\omega_{pe}^2}{\omega^2 - \omega_{ce}^2} - \dfrac{\omega_{pi}^2}{\omega^2 - \omega_{ci}^2}} \tag{4.5.19}$$

The important properties of the waves are distinguished by their cutoffs and resonances. Comparing the numerator factors of (4.5.19) with (4.5.16), it is easy to see that the two cutoffs of the x wave correspond to the cutoffs of the RHP and LHP waves. In addition to the upper hybrid resonance ω_{UH}, a second resonance at the lower hybrid frequency ω_{LH} appears. For $\omega_{pi}^2 \gg \omega_{ci}^2$, (usual for materials processing discharges), we find

$$\frac{1}{\omega_{LH}^2} \approx \frac{1}{\omega_{pi}^2} + \frac{1}{\omega_{ce}\omega_{ci}} \tag{4.5.20}$$

Low-frequency wave energy can be strongly absorbed by the plasma at this resonance.

We list all of the cutoffs and resonances of these waves in Table 4.2. With these values, and noting where the propagation constant changes from real to imaginary, the dispersion diagram for the principal waves can be qualitatively sketched, as in Figure 4.10. The high-frequency range is, of course, similar to Figure 4.9. Near ω_{pi} and below, the waves are strongly modified by the ion

Table 4.2 Summary of Cutoffs and Resonances for the Principal Waves

Wave	Cutoffs ($k = 0$)	Resonances ($k = \infty$)
r wave	$(\omega - \omega_{ce})(\omega + \omega_{ci}) = \omega_p^2$ or $\omega \approx \dfrac{\omega_{ce} + \sqrt{\omega_{ce}^2 + 4\omega_p^2}}{2}$	$\omega = \omega_{ce}$
l wave	$(\omega + \omega_{ce})(\omega - \omega_{ci}) = \omega_p^2$ or $\omega \approx \dfrac{-\omega_{ce} + \sqrt{\omega_{ce}^2 + 4\omega_p^2}}{2}$	$\omega = \omega_{ci}$
x wave	Both as above	$\omega_{UH}^2 \approx \omega_p^2 + \omega_{ce}^2$ and $\dfrac{1}{\omega_{LH}^2} \approx \dfrac{1}{\omega_{pi}^2} + \dfrac{1}{\omega_{ce}\omega_{ci}}$ for $\omega_{pi} \gg \omega_{ci}$
o wave	$\omega = \omega_p$	None

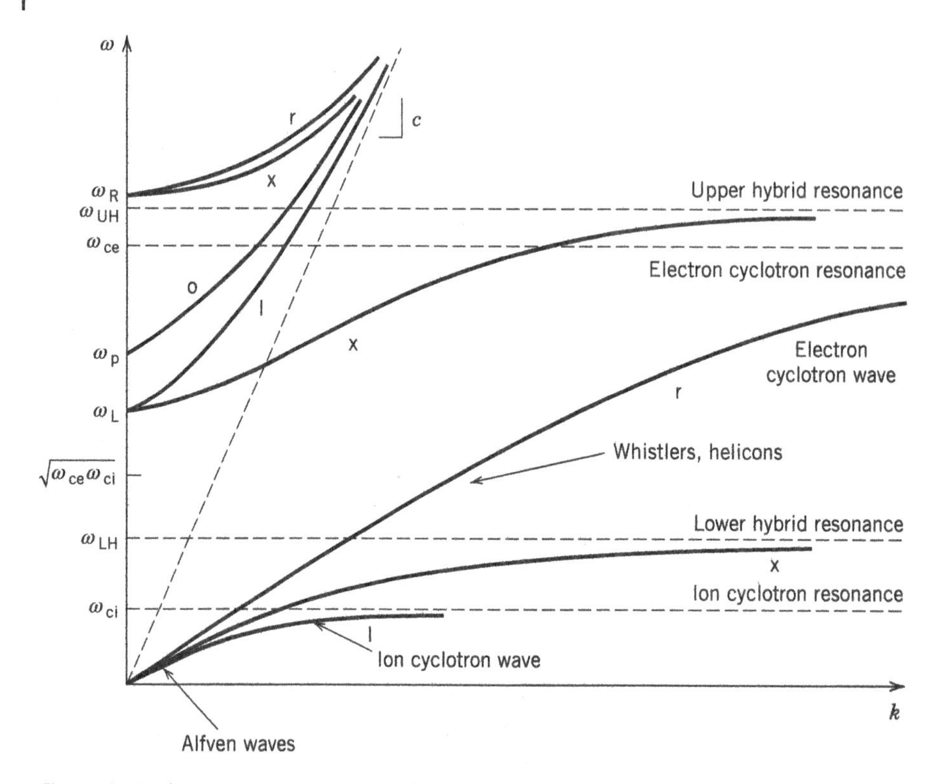

Figure 4.10 Dispersion ω versus k for the principal waves in a magnetized plasma with mobile ions.

dynamics. Of particular note is that for very low frequencies $\omega \ll \omega_{\text{ci}}$, the wave dispersions for RHP, LHP, and x waves all reduce to

$$k^2 = k_0^2 \left(1 + \frac{\omega_{\text{pi}}^2}{\omega_{\text{ci}}^2}\right) \tag{4.5.21}$$

which propagate down to zero frequency. The term in parentheses is just the low-frequency perpendicular dielectric constant defined in (4.3.26). For reasonably high density with $\omega_{\text{pi}} \gg \omega_{\text{ci}}$, the 1 can be discarded. The phase velocity of this wave is then

$$v_{\text{ph}} = \frac{\omega}{k} = \frac{\omega_{\text{ci}}}{\omega_{\text{pi}}} c \equiv v_{\text{A}} \tag{4.5.22}$$

where v_{A} is known as the Alfven velocity. Alfven waves were first described in connection with wave propagation in the earth's magnetosphere and play important roles in low-frequency phenomena in magnetized plasmas.

4.5.3 The CMA Diagram

The preceding gives a far from complete picture of the waves that can propagate at an arbitrary angle to the magnetic field. The complete dispersion equation (4.5.1) can be solved numerically to obtain the propagation constant for each of the waves at an arbitrary angle to the magnetic field. A convenient presentation of the results can be described in the Clemmow–Mullaly–Allis (CMA) diagram given in Figure 4.11. The relative phase velocities $v_{\text{ph}}/c = \omega/kc$ of the two waves are plotted in polar coordinates versus θ for various magnetic fields ($\omega_{\text{ce}}\omega_{\text{ci}}/\omega^2$) and densities ($\omega_{\text{p}}^2/\omega^2$) on the

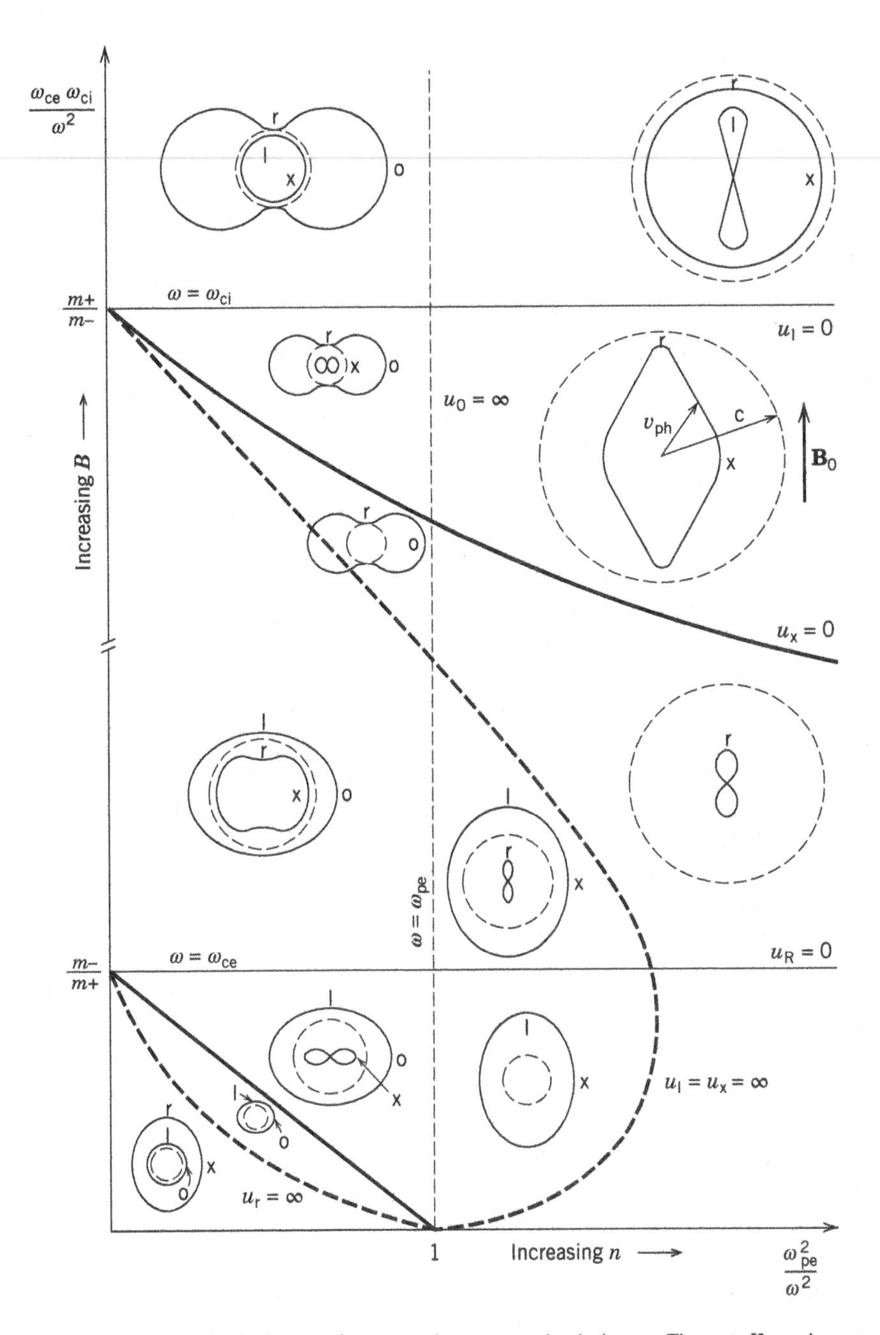

Figure 4.11 The CMA diagram for waves in a magnetized plasma. The cutoffs and resonances are indicated by the lines labeled $u = \infty$ and $u = 0$, respectively, where u denotes the phase velocity and the subscripts label the principal waves. Source: Allis et al., 1963/MIT Press.

ordinate and abscissa, with the B-field direction for the polar $(v_{ph}(\theta)/c)$ phase velocity surfaces being vertical. The principal propagating waves are indicated using the notation r, l, o, and x for the RHP, LHP, ordinary, and extraordinary waves, respectively. The velocity of light circle is shown dashed to give the radial scale of the surfaces. The cutoffs (labeled $u = \infty$) and resonances (labeled $u = 0$) of the principal waves divide the diagram into various regions, each having its own topology for the two-phase velocity surfaces. The topologies are either ellipsoids, dumbbells, or toroids, with the latter two indicating resonance ($k \to \infty$ or $v_{ph} \to 0$) at a nonzero propagation angle.

In this presentation, the high-frequency region where the propagation is like that of free space is in the lower left-hand corner, while the three Alfven waves are in the upper right. Most of the information can be understood by continuation of the principal wave solutions to arbitrary angles, as the reader is invited to confirm. Since the distance from the origin to the encircling surface represents the phase velocity in that direction with respect to the velocity of light, the CMA diagram has been described as a "plasma pond" in which the shape of each surface corresponds to the outward ripple for a disturbance at its center.

Although the CMA diagram gives a reasonable picture of the electromagnetic waves in an unbounded plasma, it neither gives a complete catalog of the waves that can propagate nor accounts for boundary conditions on the wave fields or spatial variations of the plasma and magnetic field. We have already discussed electrostatic electron and ion plasma waves that can propagate in the absence of or along a magnetic field. There is also a large class of electrostatic cyclotron waves that can propagate across the B field. These latter waves are not of great interest in the context of our applications. Analysis can be found in advanced books on plasma wave theory such as Stix (1992, chapter 9).

Variations in the plasma and B field play essential roles in plasma heated by electron cyclotron resonance interaction, as considered in Sections 11.7 and 13.3. The boundary conditions on the electromagnetic fields can also play an important role, as discussed in that chapter. Plasma boundaries can also support additional waves. These bounded plasma waves can be of importance in various contexts, as will be described in Chapter 13. The interested reader can find a description of some of them in Krall and Trivelpiece (1973, chapter 4), and more briefly in Chen (1984, chapter 4).

4.6 Microwave and RF Field Diagnostics

Because the propagation constant of a wave is dependent on the plasma frequency $\omega_{pe}^2 = e^2 n_e/\epsilon_0 m$, propagation measurements have been used to measure plasma density. In principle, the wave attenuation can also be used to measure the collision frequency, but this method has not been generally employed. Because the plasma frequency is often in the microwave (or submicrowave) range of frequencies, the waves used tend to have frequencies in that range, and the diagnostics are often referred to as *microwave diagnostics*. The methods of using the waves for electron density measurements vary with the plasma configuration. A few such methods are described below. A particular advantage of wave methods is that they are, in principle, noninvasive, and therefore can be used in situations where probe diagnostics (described in Section 6.6) would not be appropriate. Comprehensive accounts of plasma diagnostics, including wave diagnostics, can be found in Huddlestone and Leonard (1965) and Stenzel and Urrutia (2021).

4.6.1 Interferometer

A commonly used wave diagnostic is the microwave interferometer. The principle is that the change in phase shift across a region with and without a plasma can be measured. This in turn can be related to the change in propagation constant and, hence, to the plasma frequency. Starting from a wave propagating in a uniform plasma without an applied dc magnetic field, or with a linear polarization such that the electric field is directed along the dc magnetic field, the propagation constant is given by (4.2.35) as

$$k = \left(1 - \frac{\omega_{\mathrm{p}}^2}{\omega^2}\right)^{1/2} k_0 \tag{4.6.1}$$

where $k_0 = \omega/c$ is the free-space propagation constant. We ignore collisions in this approximation. Now consider that the wave propagates across a region of length l in which the density may be changing slowly compared to a wavelength. The WKB solution (see Section 13.1) is that k also changes slowly such that the phase shift can be written in the form

$$\phi = \int_0^l k(x)\,\mathrm{d}x \tag{4.6.2}$$

Substituting (4.6.1) into (4.6.2), and subtracting the free space phase shift $k_0 l$, the change in phase shift is

$$\Delta\phi = k_0 \left\{ \int_0^l \left[1 - \frac{\omega_{\mathrm{p}}^2(x)}{\omega^2}\right]^{1/2} \mathrm{d}x - l \right\} \tag{4.6.3}$$

It is often possible to choose the diagnostic frequency sufficiently high compared to the plasma frequency that the square root can be expanded. The free-space part of the phase shift then conveniently cancels from (4.6.3) leaving

$$\Delta\phi \approx -k_0 \int_0^l \frac{\omega_{\mathrm{p}}^2(x)}{2\omega^2}\,\mathrm{d}x = -\frac{k_0 e^2}{2\epsilon_0 m\omega^2} \int_0^l n(x)\,\mathrm{d}x \tag{4.6.4}$$

In this approximation, we see that the line integral of the density can be directly measured in terms of a phase shift. In many configurations, the density can be measured quite accurately by this method, serving as a check on local probe methods, described next and in Section 6.6. If the approximation in (4.6.4) cannot be made, it is still possible to determine the same information from (4.6.3), but the calculation is not straightforward.

The actual measurement technique uses an interferometer that compares signals going through the plasma region and around it. A schematic of such an interferometer is shown in Figure 4.12. In the absence of the plasma, the reference leg is adjusted to have a 180° phase shift at the same

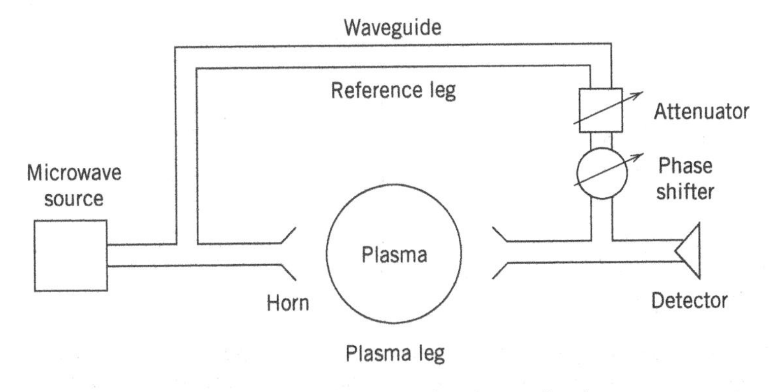

Figure 4.12 A microwave interferometer for plasma density measurement.

amplitude as the plasma leg, giving a null output. With the plasma present, the phase shift across the plasma leg changes and a signal is observed. The most convenient way of using the interferometer is to have $l \gg \lambda$, such that $\Delta\phi$ can change through more than 360° (a fringe shift) for $\omega_p^2/\omega^2 \ll 1$ (see (4.6.4)). For $\Delta\phi = 180°$, the signals through the two legs are in phase and the signal is a maximum, returning to a near null signal at $\Delta\phi = 360°$. Very accurate measurements can be made in this regime in which the plasma is turned on sufficiently slowly that the number of fringe shifts and fractions thereof can be measured. Often, however, the plasma size and available detection frequencies make $l \lesssim \lambda$, and fractional fringe shifts must be measured. This can be relatively straightforward if (4.6.4) holds such that $\Delta\phi \propto n$. However, the signal amplitude must be known, and this is complicated by reflection and refraction of the wave at the plasma–dielectric interfaces.

The finite size of the plasma, compared to the wavelength of the interferometer, has other consequences that can be more serious than the limited phase shift. If the transverse dimension of the plasma is also comparable to a wavelength, then diffraction around the plasma becomes a serious problem. This is often significant when diagnosing plasma cylinders. Small transverse plasma dimensions have tended to push the interferometer frequency up, such that $\omega_p^2/\omega^2 \ll 1$. In this case, the phase shift, which is proportional to this ratio, becomes small. This has led to more complicated methods of detection. For dense plasmas, laser interferometers have been used to obtain small but measurable phase shifts. The microwave interferometer has been a mainstay of fusion plasma diagnostics from their inception, since noninvasive measurement techniques are required on such plasmas. An early monograph (Heald and Wharton, 1965) recounted these techniques in detail.

An example of a 35 GHz microwave interferometer measurement of density and its comparison to density measurements using Langmuir probes (see Section 6.6) is shown in Figure 4.13 for a planar coil, rf-driven inductive discharge. The transmitting and receiving horn antennae

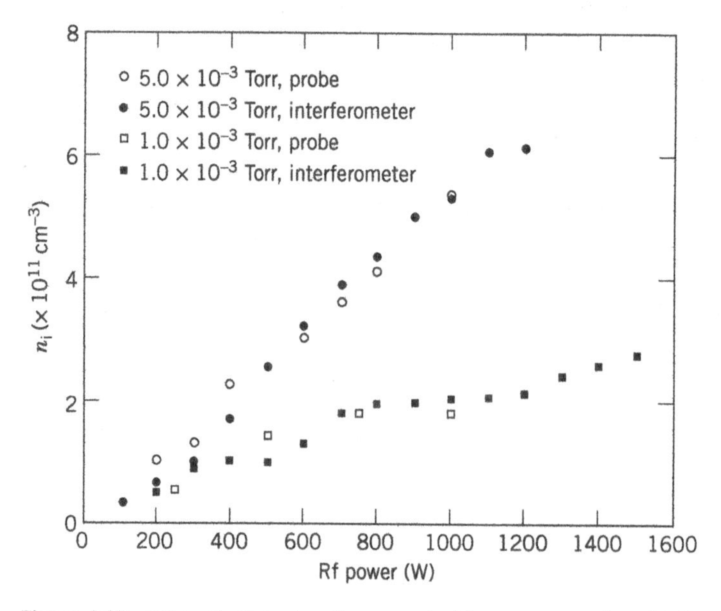

Figure 4.13 Mean electron density versus incident power at the midplane of an rf inductive discharge as measured by a microwave interferometer, compared with ion density as measured by a Langmuir probe. Source: Hopwood et al., 1993b/American Vacuum Society.

were placed externally to the chamber, with the microwaves transmitted through the rectangular chamber parallel to the surface of the planar coil (see Section 12.3 for further description of the discharge configuration).

The ordinary wave is not suitable for an interferometer if $\omega_p > \omega$, because the wave will not propagate. In time-varying plasmas, the cutoff itself can be used as a benchmark of qualitative plasma behavior. In a magnetic field, it is still possible to have a propagating wave along the field, provided $\omega_{ce} > \omega$, as given by (4.5.17). Although this wave is very important for plasma heating, as described in Section 13.1, it has only occasionally been used for plasma diagnostics.

4.6.2 Hairpin Resonator Probe

Hairpin resonator plasma probes were first used by Stenzel (1976). A short length of wire bent into the shape of a hairpin can act as a two-wire transmission line of length L with one shorted and one open-circuited termination. This structure resonates when $L = \lambda/4$, with $\lambda = 2\pi/k$ the transmission line wavelength and k the wavenumber. This puts a voltage zero (current maximum) at the shorted end and a voltage maximum (current zero) at the open-circuited end (Ramo et al., 1994, chapter 5). In vacuum, $k = \omega/c$, and the resonance frequency is then $\omega_0 = \pi c/(2L)$. If the hairpin is embedded in a non-magnetized plasma, using k given by (4.6.1), we obtain the resonance condition (see Problem 4.15)

$$\omega^2 = \omega_0^2 + \omega_p^2 \tag{4.6.5}$$

with ω the resonance frequency with the plasma present. The hairpin is typically driven with a variable frequency microwave source or network analyzer. The resonance frequency is determined from the minimum in reflected power versus microwave frequency, closely indicating a matched load. Alternatively, an increase in transmitted power can be used.

Some designs of hairpin resonator probes are shown in Figure 4.14. Typically L = 15–25 mm, with the 0.1 mm diameter tungsten wires separated by H = 2–4 mm. The hairpin can either be attached (soldered) directly to the outer coaxial probe conductor, as shown in (*a*), or can be held to the probe by an insulator, as shown in (*b*). In the former case, the dc hairpin voltage is that of the outer coax, which can be dc-biased or grounded. In the latter case, the hairpin voltage "floats" 3–6 T_e below the plasma potential, which can be very useful for rf-driven plasmas having large time-varying plasma potentials. (See Section 6.6 for the issues involved here.) As shown in the figure, for a reflection measurement the hairpin is usually driven by a nearby small loop (~ 1.5 mm diameter) terminating the coax. For a transmission measurement two nearby coaxial probes can be used, the second picking up a portion of the transmitted power.

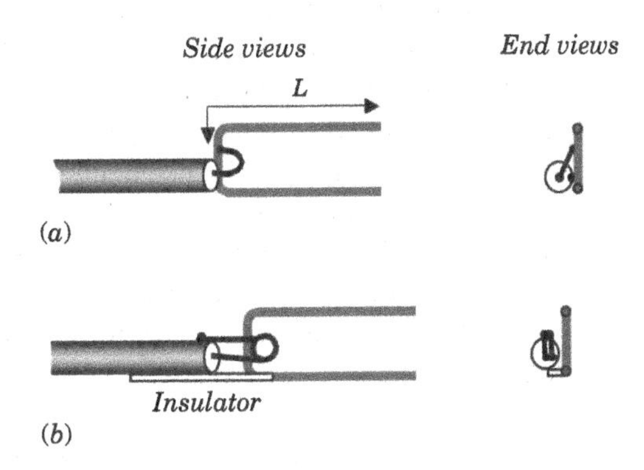

Figure 4.14 Design of (*a*) dc or grounded and (*b*) floating hairpin resonator probes. Source: Piejak et al. (2005). © IOP Publishing. Reproduced with permission. All rights reserved. https://doi.org/10.1088/0963-0252/14/4/012

An improved design with analysis incorporating an estimated sheath width surrounding the hairpin wires is given by Piejak et al. (2004). For a vacuum sheath width s_m between the wire and the plasma, they find

$$\omega^2 = \omega_0^2 + \omega_p^2 \left(1 - \beta \frac{\omega_0^2}{\omega^2} \right) \tag{4.6.6}$$

where $\beta = \ln(1 + s_m/a)/\ln(H/a)$ is a correction factor for the finite sheath width. This accounts for the series combination of the sheath and plasma capacitances surrounding the wire.

The effects of time-varying (rf) plasma potentials were examined by Piejak et al. (2005); these can be significantly alleviated using a floating hairpin design. Measurement inaccuracies due to density depletions around the probe and its holder, and due to inaccurate estimates of the probe sheath width, are discussed in Godyak (2017, 2021) and Godyak and Sternberg (2024).

The hairpin resonator can give good estimates of local density in etching or deposition plasmas where Langmuir probes (see Section 6.6) might suffer serious surface contamination. They are also useable in weakly magnetized plasmas, $\omega_{ce} \ll \omega_p$, where the basic relation (4.6.1) is still approximately valid.

4.6.3 Magnetic (B-dot) Probes

Magnetic (B-dot) probes have been extensively used in plasma diagnostics since the early days of controlled fusion research to measure time-varying magnetic fields in plasmas (Huddlestone and Leonard, 1965, chapter 3). The operating principle is based on Faraday's law (2.2.1), which in integral form is

$$V(t) \equiv \oint_C \mathbf{E} \cdot d\ell = -\mu_0 \frac{\partial}{\partial t} \int_A \mathbf{H} \cdot d\mathbf{A} \tag{4.6.7}$$

This implies that the voltage $V(t)$ induced at the terminals of a small loop of wire of area A immersed in the plasma is proportional to the time rate of change of the magnetic flux $\Phi = \int_A \mathbf{B} \cdot d\mathbf{A}$ linking the loop. Here, $\mathbf{B} = \mu_0 \mathbf{H}$ is the magnetic induction vector.

A common early measurement configuration was a movable coaxial cable terminated in a one- or multi-turn, small diameter planar loop of area A, which was inserted into a quartz tube immersed in the plasma. Such probes have also been used to measure the rf field components in rf-driven plasma processing discharges. An example is shown in Figures 12.9 and 12.10 for a planar inductive discharge. However, as discussed by Piejak et al. (1997), while convenient, the presence of the quartz tube can obstruct the flow of rf plasma currents, leading to measurement inaccuracies. More accurate measurements are obtained by directly immersing the probe into the plasma. A typical magnetic probe of this type uses a metal-shielded, multi-bore, 0.8 mm diameter ceramic tube terminated in a 1-turn, 4 mm diameter loop of thin wire (Piejak et al., 1997).

In an axisymmetric (in z) rf discharge driven at radian frequency ω, with the loop area lying in the r–θ plane, (4.6.7) gives $\tilde{V}_{rf} = -j\omega\mu_0\tilde{H}_z A$, with $\tilde{V}_{rf}(r, z)$ and $\tilde{H}_z(r, z)$ the complex induced voltage and z-component of the magnetic field vector, respectively. For a loop lying in the θ–z plane, the $\tilde{H}_r(r, z)$ field can be similarly measured. From these two measurements, Maxwell's equations determine the complex electric field components. These in turn give important information about discharge properties such as the current density distribution and the rf power absorption. The use of magnetic probes to determine rf magnetic fields in discharges is reviewed by Godyak (2021).

4.6.4 Cavity Perturbation

Another relatively straightforward technique for diagnosing a plasma is by the shift in frequency of a microwave cavity when a plasma fills part of the cavity. Slater's perturbation formula (Harrington, 1961, chapter 7) can be applied to an unmagnetized plasma in the frequency range where the plasma frequency $\omega_p \ll \omega_0$, the resonant frequency, (and $\nu_m \ll \omega_0$) giving the relative shift in resonance frequency:

$$\frac{\Delta\omega}{\omega_0} = \frac{1}{2\omega_0^2} \frac{\int \omega_p^2 |\mathbf{E}|^2 d\mathcal{V}}{\int |\mathbf{E}|^2 d\mathcal{V}} \tag{4.6.8}$$

where $\mathbf{E}$ is the unperturbed resonance electric field, and the integrals are over the total cavity volume. The formula can also be modified to include higher-density plasmas, provided the plasma dimensions are small compared to λ. For evaluating the integrals, most measurements have used cylindrical cavity modes such as the TM_{010} mode (see Ramo et al., 1994, chapter 10), for which $\mathbf{E} = \hat{z}\tilde{E}_z$, where

$$\tilde{E}_z = E_0 J_0 \left(\frac{\chi_{01} r}{R} \right) \tag{4.6.9}$$

where $\chi_{01} \approx 2.405$ and R is the cavity radius. Processing chambers usually have more complicated geometry. In this case, one can experimentally determine the electric field profiles of several modes. The spatial density profile can also be measured to evaluate the integral in the numerator of (4.6.8), although different estimates of the profile only slightly modify the results. Reasonable consistency in density measurements can be obtained by using more than one mode. An example of results and their comparison to Langmuir probe measurements is shown in Figure 4.15 for a particular process chamber. Interferometer and cavity perturbation methods are often used in conjunction with probe measurements to improve the reliability of the results.

Another easily measured quantity in a cavity is the Q defined by

$$Q \equiv \omega \frac{\text{Energy stored}}{\text{Power dissipated}} = \frac{\omega_0}{\Delta\omega} \tag{4.6.10}$$

where $\Delta\omega$ is the frequency shift between the half power points on each side of the resonance. The second equality, given in all circuit texts, follows directly from the definition. The cavity Q with

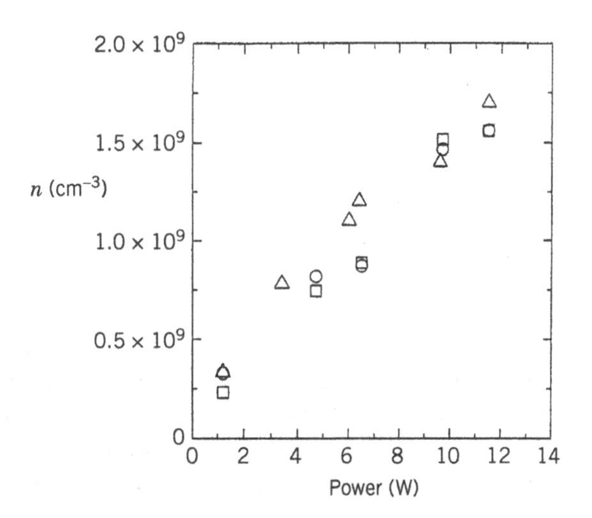

Figure 4.15 Electron density versus absorbed power in a 10 mTorr argon discharge. Data from 443 MHz cavity resonance (circles), 506 MHz cavity resonance (squares), and Langmuir probe (triangles). Source: Moroney et al., 1989/with permission of AIP Publishing.

plasma is lower than that without plasma due to dissipation within the plasma. Provided $\nu_m \ll \omega_0$, the microwave cavity measurement of density is not significantly modified. However, the change in Q can be used to directly determine the collision frequency of the plasma electrons if ohmic heating is the main source of energy absorption. Experiments of this nature have been successfully performed, but have not come into general use as a plasma diagnostic.

4.6.5 Wave Propagation

An interesting type of diagnostic is one that uses intrinsic properties of wave propagation in bounded plasmas. For example, one method of plasma heating, described in Section 13.3, is by surface waves. The propagation properties of these waves can be measured and related to the average plasma density over which the fields are important. For waves whose fields are confined close to the plasma–dielectric interface, the propagation can give information about the edge density, in contrast to the average density obtained from the methods described above.

Although more difficult to measure, it is also possible to obtain information on the electron collisionality from the wave decay. This is also considered in Section 13.3. We note, however, that the decay constant involves collisionless (Landau) damping as well as collisional damping, so that the results must be interpreted with care.

A particularly simple situation for obtaining the plasma frequency is that for which the plasma is transversely resonant. A simple calculation then yields the plasma frequency. For example, for a parallel plane geometry, let d be the length of the plasma and $2s_m$ be the total length of both sheaths. The discharge can be modeled as two capacitors in series, where the capacitances per unit area are

$$C_s \approx \frac{\epsilon_0}{2s_m} \tag{4.6.11}$$

and

$$C_p \approx \frac{\epsilon_p}{d} \approx \frac{\epsilon_0(1 - \omega_p^2/\omega^2)}{d} \tag{4.6.12}$$

Note that C_p is inductive ($C_p < 0$) for $\omega < \omega_p$. The total capacitance is then

$$C_T = \left(\frac{1}{C_p} + \frac{1}{C_s}\right)^{-1} \tag{4.6.13}$$

Substituting the expressions for C_s and C_p into (4.6.13), we obtain

$$C_T = \frac{\epsilon_0\left(\omega^2 - \omega_p^2\right)}{2s_m\left(\omega^2 - \omega_p^2\right) + d\omega^2} \tag{4.6.14}$$

This expression will have a resonance when the denominator vanishes, or

$$\omega = \omega_p\left(\frac{2s_m}{2s_m + d}\right)^{1/2} \tag{4.6.15}$$

The resonance has been observed in both capacitive and inductive discharges. The densities obtained from (4.6.15) agree reasonably well with other density measurements made on the same discharge. The method can also be applied to cylindrical plasmas, and configurations in which there are dielectrics, giving somewhat more complicated expressions replacing (4.6.15).

Indeed, the first application of the method was to a plasma cylinder, surrounded by a dielectric tube with split cylinder exciting electrodes. (See Parker et al. (1964) for details, including thermal effects.) The lowest-order "dipole" resonance can be approximated by the simple form

$$\omega = \frac{\omega_p}{\left(1 + \kappa_{eff}\right)^{1/2}} \tag{4.6.16}$$

where κ_{eff} is the effective relative dielectric constant of the region between the plasma and the electrode.

Problems

4.1 **Plasma Oscillations with Mobile Ions** Show in a slab geometry that the plasma oscillation frequency is given by (4.2.7) if the ions are permitted to be mobile.

4.2 **Plasma Oscillations for a Perturbed Charge Density** For a plasma with immobile uniform density ions, show that an arbitrary displacement $\zeta_e(\mathbf{r}, t)$ of the electron fluid with respect to the ions leads to a perturbed charge density $\rho = en_0\nabla \cdot \zeta_e$. Using the divergence equation for the electric field and the equation of motion for the electron fluid, show that the charge density oscillates sinusoidally at the electron plasma frequency ω_{pe}.

4.3 **Damped Electron Plasma Oscillations** Consider electron plasma oscillations in a slab geometry with a uniform electron density n_0 as shown in Figure 4.3 of the text, with infinite mass ions but in the presence of a background density of neutral gas. The gas atoms exert a frictional force on the moving electrons, such that the equation of motion (4.2.2) is modified to

$$m\frac{d^2\zeta_e}{dt^2} = -eE_x - m\nu_m\frac{d\zeta_e}{dt}$$

where ν_m is a constant electron–neutral momentum transfer frequency. Assume that the slab of electrons is displaced to the right with respect to the ions by a small distance ζ_0 and that the slab velocity is zero at time $t = 0$. Show that for $\nu_m < 2\omega_{pe}$, the motion consists of a damped plasma oscillation. Find the damping rate and the oscillation frequency for this case. Find the motion of the slab $\zeta_e(t)$ for these initial conditions.

4.4 **A Particle-in-Cell Simulation with One Electron Sheet** A plasma having uniform density n_0 is confined between two parallel perfectly conducting planes separated by a distance l. For computer simulation, the plasma is modeled as follows: the ions are assumed to be fixed and have a uniform density n_0. The electrons are all gathered into a single sheet of charge of surface charge density $\rho_S = -en_0 l$ C/m^2, which is allowed to move in response to the electric fields seen by the sheet.

(a) Show that the equilibrium position of the electron sheet is in the center of the plasma.
(b) If the sheet is given a small displacement about its equilibrium position and then released, what happens? Find the subsequent motion of the sheet.
(c) Suppose the two parallel planes are connected together (grounded). Repeat part (b) to determine the motion.

4.5 Time-Averaged Power in the Sinusoidal Steady State Show that (4.2.26) holds; i.e., if $\mathbf{J}_T(t)$ and $\mathbf{E}(t)$ are sinusoids having complex vector amplitudes $\tilde{\mathbf{J}}_T$ and $\tilde{\mathbf{E}}$, then the time-averaged absorbed power per unit volume can be written as

$$p_{abs} = \frac{1}{2}\mathrm{Re}\left(\tilde{\mathbf{J}}_T \cdot \tilde{\mathbf{E}}^*\right) = \frac{1}{2}\mathrm{Re}\left(\tilde{\mathbf{J}}_T^* \cdot \tilde{\mathbf{E}}\right)$$

4.6 Ohmic Heating Power in a Nonuniform RF Discharge An rf discharge with a nonuniform density $n(x)$ is ignited between two plane-parallel electrodes located at $x = \pm l/2$. The total rf current density (conduction + displacement) is $\mathbf{J}_T(x, t) = \hat{x}J_0 \cos \omega t$. The rf electric field in the discharge is similarly given by $\mathbf{E}(x, t) = \hat{x}E_0(x)\cos(\omega t + \phi_0)$.

(a) Prove from Maxwell's equations that J_0 is a constant, independent of x.

(b) Writing $\mathbf{J}_T = \hat{x}\mathrm{Re}(\tilde{J}e^{j\omega t})$ and $\mathbf{E} = \hat{x}\mathrm{Re}[\tilde{E}(x)e^{j\omega t}]$, find expressions for the complex amplitudes $\tilde{J}$ and $\tilde{E}$.

(c) For a high-pressure (collisional) discharge such that $\omega \ll \nu_m \ll \omega_{pe}$, with a plasma density $n(x) = n_0 \cos(\pi x/l)$, find an expression for $E_0(x)$ and $\tilde{E}(x)$ in terms of J_0, n_0, ν_m, l, and other constants. Use the expression (4.2.22) for the dc plasma conductivity $\sigma_{dc}(x)$ with $n_0 \to n(x)$.

(d) In the limit of (c), integrate p_{ohm} over x to find the ohmic power per unit area within a discharge volume $|x| \lesssim d/2$, where $d < l$.

(e) Note that your result in (d) tends to infinity as $d \to l$. Comment on the correctness of this result.

4.7 Series Resonance Discharge A one-dimensional slab model of an rf discharge between two parallel perfectly conducting electrodes of area A consists of a uniform plasma slab ($n_e = n_i = n_0$) of thickness d with two sheaths, each of thickness s, one near each electrode. An rf voltage source is connected across the electrodes, such that an rf current $I(t) = I_0 \cos \omega t$ flows across the plates. Neglect ion motions and assume that $\omega_p \gg \omega$, ν_m, where ω_p is the plasma frequency and ν_m is the electron–neutral momentum transfer frequency. Writing the voltage across the plates in the form $V(t) = \mathrm{Re}\,(V_0\, e^{j\omega t})$, then one can introduce

$$V_0 = I_0 Z = I_0(R + jX)$$

where Z, R, and X are the impedance, resistance, and reactance of the discharge, respectively. (V_0 and Z are complex numbers; I_0, R, and X are real numbers.)

(a) Find R and X for this discharge model. Sketch R and X versus ω for $0 \le \omega \le \omega_p$.

(b) Find the real power $P_{abs} = \frac{1}{2}\mathrm{Re}\,(V_0^* I_0)$ flowing into the discharge, and find the frequency ω_{res} for which $X = 0$ (the series resonance frequency).

4.8 Electrostatic Ion Plasma Waves Derive the dispersion relation for electrostatic ion plasma waves in a uniform collisionless plasma containing mobile ions with $T_e \gg T_i$, and show that for long wavelengths (low frequencies), the waves propagate at the ion sound speed $(kT_e/M)^{1/2}$. Use the Boltzmann relation to relate n_e to Φ.

4.9 Guiding Center Motion Consider a cylindrically symmetric, time-varying magnetic field that varies parabolically with axial distance z as $\mathbf{B} = \hat{z}B(t)(1 + z^2/l^2)$. Assume that $B(t)$ increases slowly from the value B_0 at time $t = 0$ to B_1 at $t = t_1$. A charged particle of mass m located at $z = 0$ has perpendicular energy $W_{\perp 0}$ and parallel energy W_{z0} at $t = 0$. Assume that the guiding center equations of motion are valid and that μ_{mag} = const.

(a) Give the final perpendicular energy $W_{\perp 1}$ at $z = 0$ (after a time t_1).

(b) Write the equation for the motion along z, assuming that the motion is fast compared to the time variation of $B(t)$. Show that the motion is a sinusoidal oscillation along z and calculate the oscillation frequency ω_b. This shows that the particle is confined axially in the magnetic field.

(c) Assume now that $W_{z0} = 0$ and $R(t) \ll l$, where $R(t)$ is the radial distance of the guiding center of the particle from the z-axis. By using Faraday's law (2.2.1) to find the induced electric field $E_\phi(t)$ and calculating the resulting $\mathbf{E} \times \mathbf{B}$ drift, show that $B(t)R^2(t) = \text{const}$ during the slow change from B_0 to B_1.

4.10 Magnetic Moment The magnetic moment of a charged particle gyrating in a magnetic field is defined as the product of the current generated by the rotating particle times the area enclosed by the rotation. Show that this is equal to μ_{mag} defined in (4.3.11).

4.11 Magnetic Drifts

(a) Cross multiplying (4.3.6) by $\mathbf{B}/qB^2$, obtain the three general guiding center drifts (force, curvature, and grad-B drifts in Table 4.1). This calculation is not straightforward, see Chen (1984).

(b) Show that an expansion of the left-hand side of (4.3.6) gives rise to (4.3.18).

4.12 Calculating the Gradient Drift For the geometry in Figure 4.8b, derive (4.3.19), starting from first principles, with the magnetic field $B_z = B_0 + r_c(\partial B/\partial x)\sin \omega_c t$ and $\mathbf{v}$ as given in (4.1.6) ($\phi_0 = 0$). To do this, first find the time-averaged Lorentz force and then use (4.1.19).

4.13 Waves in Magnetized Plasmas Sketch the wave dispersion ω versus k for the principal waves in an electron plasma (immobile ions) for high densities $\omega_{\text{pe}} > \omega_{\text{ce}}$ and compare to Figure 4.9.

4.14 Whistler Waves The RHP wave is known as the whistler wave in the frequency range for which $\omega_{\text{ci}} \ll \omega \ll \omega_{\text{ce}}$. Using these approximations in the dispersion relation, find the dependence of the phase velocity of the wave on the frequency.

4.15 Hairpin Resonator Probe For the hairpin resonator probe of Figure 4.14 with $L = 25$ mm, wire separation $H = 4$ mm, and wire radius $a = 0.1$ mm

(a) Show that the resonance condition $L = \lambda/4$ yields (4.6.5).

(b) Show that the unperturbed (no plasma) resonance frequency is $f_0 = 3$ GHz.

(c) The resonant frequency f when immersed in a plasma was measured to be 3.3 GHz. Ignoring finite sheath width effects, use (4.6.5) to find the electron density.

(d) For this density, find the Debye length λ_{De} given by (2.4.22). Using this, estimate a sheath length $s_m \approx 3\,\lambda_{\text{De}}$. Then, use (4.6.6) to find a corrected value for the electron density.

4.16 Microwave Diagnostic Consider a 3-cm-diameter uniform plasma column. It is desired to measure the plasma density either by measuring the perturbation of the resonant frequency of a 6-cm-diameter TM_{010} mode cavity or by measuring the phase shift of the ordinary wave using a $\lambda = 1.5$ cm interferometer.

(a) What is the approximate unperturbed resonant frequency of the cavity?

(b) Using the perturbation formula, calculate the frequency shift due to the plasma column for $n = 10^{10}\text{cm}^{-3}$ and $n = 10^{11}\text{cm}^{-3}$.
(c) Sketch the cross section of the electric field magnitude for each case.
(d) Find the phase shift for the 1.5-cm interferometer for each case.
(e) Explain which method you would use to find the density if it was expected to lie in the range of each of the two cases.

5

Diffusion and Transport

5.1 Basic Relations

5.1.1 Diffusion and Mobility

We have already seen in Section 4.2 that adding a friction term to the force equation, in a cold uniform plasma with an applied electric field, gives rise to a conductivity. The friction term, arising from collisions with a background species, also leads to diffusion in a nonuniform warm plasma. To see this, we start with the steady-state macroscopic force equation (2.3.15), neglecting the acceleration and inertial force terms

$$qn\mathbf{E} - \nabla p - mn\nu_{\mathrm{m}}\mathbf{u} = 0 \tag{5.1.1}$$

where we assume that the background species is at rest and that the momentum transfer frequency ν_{m} is a constant, independent of the drift velocity $\mathbf{u}$. Taking an isothermal plasma, such that $\nabla p = kT\nabla n$, and solving (5.1.1) for $\mathbf{u}$, we obtain

$$\mathbf{u} = \frac{q\mathbf{E}}{m\nu_{\mathrm{m}}} - \frac{kT}{m\nu_{\mathrm{m}}}\frac{\nabla n}{n} \tag{5.1.2}$$

Equation (5.1.2) can be written

$$\Gamma = \pm\mu n\mathbf{E} - D\nabla n \tag{5.1.3}$$

where $\Gamma = n\mathbf{u}$ is the particle flux, and

$$\mu = \frac{|q|}{m\nu_{\mathrm{m}}} \qquad \mathrm{m^2/V\text{-}s} \tag{5.1.4}$$

and

$$D = \frac{kT}{m\nu_{\mathrm{m}}} \qquad \mathrm{m^2/s} \tag{5.1.5}$$

are the macroscopic *mobility* and *diffusion* constants. These are determined separately for each species. In (5.1.3), the positive sign is for q positive and the negative sign is for q negative. Ion mobilities depend on the local ion energy distribution, which is determined by the local electric field. The measured mobilities μ_{i} of various ions in various gases are tabulated as a function of field in Ellis et al. (1976, 1978, 1984).

Principles of Plasma Discharges and Materials Processing, Third Edition. Michael A. Lieberman and Allan J. Lichtenberg.

Using the definition of the mean speed $\bar{v} = (8kT/\pi m)^{1/2}$ and a mean free path (for hard-sphere scattering) $\lambda = \bar{v}/\nu_m$, we can write D as

$$D = \frac{\pi}{8}\lambda\bar{v} = \frac{\pi}{8}\lambda^2\nu_m \tag{5.1.6}$$

For monoenergetic particles, $\bar{v} \to v$ and elementary kinetic theory gives a factor of $1/3$ rather than $\pi/8$ (Atkins, 1986, section 26.3c). Notice that D in (5.1.6) is in the form $(\Delta x)^2/\tau$, where Δx is the step length and τ is the time between steps of a *random walk*. This is the basic structure of a diffusion process.

5.1.2 Free Diffusion

From (5.1.3), in the absence of an electric field, we can directly obtain the diffusion law, relating the flux $\Gamma = n\mathbf{u}$ to the density gradient,

$$\Gamma = -D\nabla n \tag{5.1.7}$$

which is called *Fick's law*. Substituting (5.1.7) into the continuity equation (2.3.7),

$$\frac{\partial n}{\partial t} + \nabla \cdot \Gamma = G - L$$

with G and L the volume source and sink and with D independent of position, we obtain the diffusion equation for a single species:

$$\frac{\partial n}{\partial t} - D\nabla^2 n = G - L \tag{5.1.8}$$

Finally, we note that the *transport coefficients* μ and D are related by the *Einstein relation*:

$$D = \mu\frac{kT}{|q|} = \mu\mathrm{T} \tag{5.1.9}$$

5.1.3 Ambipolar Diffusion

Returning to the more general relation (5.1.3), we consider this to hold separately for electrons and ions. Furthermore, in the steady state, we make the *congruence assumption* that the flux of electrons and ions out of any region must be equal, $\Gamma_e = \Gamma_i$, such that charge does not build up (see Problem 5.1). This is still true in the presence of ionizing collisions, which create equal numbers of both species. Since the electrons are lighter, and would tend to flow out faster (in an unmagnetized plasma), an electric field must spring up to maintain the local flux balance. That is, a few more electrons than ions initially leave the plasma region to set up a charge imbalance and consequently an electric field. Using (5.1.3) for both species, with $\Gamma_e = \Gamma_i = \Gamma$ and $n_e \approx n_i = n$, we have

$$\mu_i n\mathbf{E} - D_i\nabla n = -\mu_e n\mathbf{E} - D_e\nabla n$$

from which we can solve for $\mathbf{E}$ in terms of ∇n:

$$\mathbf{E} = \frac{D_i - D_e}{\mu_i + \mu_e}\frac{\nabla n}{n} \tag{5.1.10}$$

Substituting this value of $\mathbf{E}$ into the common flux relation, we have (in the ion equation)

$$\begin{aligned}\Gamma &= \mu_i\frac{D_i - D_e}{\mu_i + \mu_e}\nabla n - D_i\nabla n \\ &= -\frac{\mu_i D_e + \mu_e D_i}{\mu_i + \mu_e}\nabla n\end{aligned} \tag{5.1.11}$$

which is symmetric in the coefficients and (of course) holds for both ions and electrons. Introducing the ambipolar diffusion coefficient

$$D_a = \frac{\mu_i D_e + \mu_e D_i}{\mu_i + \mu_e} \tag{5.1.12}$$

we see that (5.1.11) again has the form of Fick's law $\Gamma = -D_a \nabla n$. Substituting (5.1.11) into the continuity equation, and assuming that all coefficients are independent of position, we obtain

$$\frac{\partial n}{\partial t} - D_a \nabla^2 n = G - L \tag{5.1.13}$$

the ambipolar diffusion equation.

The ambipolar diffusion coefficient can usually be simplified by noting that $\mu_e \gg \mu_i$ in a weakly ionized discharge. Dropping μ_i in the denominator of (5.1.12), we have

$$D_a \approx D_i + \frac{\mu_i}{\mu_e} D_e$$

and using the Einstein relation (5.1.9), we obtain

$$D_a \approx D_i \left(1 + \frac{T_e}{T_i}\right) \tag{5.1.14}$$

For $T_e \gg T_i$, we find that $D_a \approx \mu_i T_e$. From (5.1.14), we see that the ambipolar diffusion is tied to the slower species, in this case, the ions, but that it is increased by a term proportional to the ratio of temperatures. Thus, in the usual case in weakly ionized plasmas, in which $T_e \gg T_i$, the ions and electrons both diffuse at a rate that greatly exceeds the ion free diffusion rate.

Let us note that in the regime where $\mu_e \gg \mu_i$ and $T_e \gg T_i$, the pressure gradient term in (5.1.3) is small compared to the flux and field terms for ions, such that

$$\Gamma_i = \Gamma \approx \mu_i n \mathbf{E} \tag{5.1.15}$$

On the other hand, for electrons the flux term is small compared to the field and pressure gradient terms, such that

$$\Gamma_e = \Gamma = -\mu_e n \mathbf{E} - D_e \nabla n \approx 0 \tag{5.1.16}$$

Hence ion motion is mobility-dominated and electron motion is determined by a Boltzmann equilibrium. Substituting (5.1.16) into (5.1.15) to eliminate $\mathbf{E}$ and using (5.1.9), we obtain $\Gamma = -D_a \nabla n$ with $D_a = \mu_i T_e$ (Problem 5.2).

In the above calculations, we have considered only unmagnetized plasmas. In a magnetic field, the motion of electrons is strongly confined perpendicular to the field, as we have already seen in Chapter 4, which can lead to quite different diffusion rates parallel to and perpendicular to the applied magnetic field. We shall discuss this situation in Section 5.4.

5.2 Diffusion Solutions

5.2.1 Boundary Conditions

With the appropriate boundary conditions, Equations (5.1.8) or (5.1.13) for free or ambipolar diffusion can be solved to determine the transport of various species, including positive ions and neutral atoms. In the following, let n be the appropriate diffusing species density and D be the (constant) diffusion coefficient. A common choice for the boundary condition at a perfectly absorbing wall is

$$n \approx 0 \tag{5.2.1}$$

However, this condition is not self-consistent because a finite particle flux $\Gamma = nu$ flowing to the wall would imply an infinite flow velocity u at the wall. The velocity into the wall is generally limited to some finite value u_{w0}. The boundary condition is then

$$-D(\nabla n)_w = n_w u_{w0} \tag{5.2.2}$$

where $(\nabla n)_w$ is the normal component of ∇n at the wall. For positive ion diffusion, (5.2.2) is still not correct because the diffusion equation is generally not valid in the sheath regions of low-pressure discharges, due to the neglect of the inertial term $Mn\mathbf{u} \cdot \nabla \mathbf{u}$ in the ion force equation (2.3.9). As will be shown in Section 6.2, a boundary condition of the form (5.2.2) can be applied at the plasma–sheath edge,

$$-D(\nabla n)_s = n_s u_B \tag{5.2.3}$$

where $u_B = (e\mathrm{T}_e/M)^{1/2}$ is called the Bohm velocity, with M the ion mass. We give an example in the following subsection. Let us note that $D \propto p^{-1}$; hence, at high pressures, the left-hand sides of (5.2.2) and (5.2.3) are small, and the simpler boundary condition (5.2.1) can often be used.

In some cases, the boundary is not wholly absorbing or can even be a source of diffusing particles. As we discuss in Section 9.4, this is commonly the case for diffusion of neutrals. The boundary condition (5.2.2) is then modified to (Chantry, 1987)

$$-D(\nabla n)_w = \frac{\gamma}{2(2-\gamma)} n_w \bar{v} \tag{5.2.4}$$

or

$$-D(\nabla n)_w = \Gamma_{w0} \tag{5.2.5}$$

In (5.2.5), Γ_{w0} is a specified flux. In (5.2.4), $\bar{v}$ is the mean speed given by (2.4.9), and γ is the probability that a molecule incident on the wall is lost to the wall. The coefficient γ is called a *sticking, recombination,* or *reaction coefficient* depending on the loss mechanism at the wall. The factor $2 - \gamma$ in the denominator of (5.2.4) accounts approximately for the change in the density at the wall for a given random thermal flux, as γ varies from zero, corresponding to a full Maxwellian distribution at the wall, to unity, corresponding to a half-Maxwellian distribution at the wall. For $\gamma \ll 1$, (5.2.4) reduces to a zero gradient condition and, for $\Gamma_{w0} < 0$, the wall is a source of diffusing particles. We will see examples of this in Section 9.4.

5.2.2 Time-Dependent Solution

Solutions to the diffusion equation (5.1.8) or (5.1.13) with no sources or sinks are easily obtained for spatial variation in one dimension. Because there are no sources, the solution must decay in time. For simplicity, taking a plane-parallel geometry of width l, we introduce a separation of variables,

$$n(x,t) = X(x)T(t)$$

which when substituted in (5.1.8) gives

$$X\frac{\mathrm{d}T}{\mathrm{d}t} = DT\frac{\mathrm{d}^2X}{\mathrm{d}x^2} \tag{5.2.6}$$

Dividing by XT, we obtain on the left-hand side a function of time alone and on the right-hand side a function of space alone. Consequently, both must equal a constant which we call $-1/\tau$. The function of T then is determined from

$$\frac{\mathrm{d}T}{\mathrm{d}t} = -\frac{T}{\tau} \tag{5.2.7}$$

which integrates to

$$T = T_0 e^{-t/\tau} \tag{5.2.8}$$

Similarly, the spatial part is determined by

$$\frac{d^2X}{dx^2} = -\frac{X}{D\tau} \tag{5.2.9}$$

which has a solution of the form

$$X = A\cos\frac{x}{\Lambda} + B\sin\frac{x}{\Lambda} \tag{5.2.10}$$

where $\Lambda = (D\tau)^{1/2}$ is the diffusion length, and A and B are constants. Taking boundary conditions of $X = 0$ at $x = \pm l/2$, then the lowest-order solution is symmetric ($B = 0$) and

$$\Lambda_0 = (D\tau_0)^{1/2} = \frac{l}{\pi}$$

Solving for $\tau = \tau_0$, we have the decay constant

$$\tau_0 = \left(\frac{l}{\pi}\right)^2 \frac{1}{D} \tag{5.2.11}$$

Combining the solutions for T and X, the complete solution is

$$n = n_0 e^{-t/\tau_0} \cos\frac{\pi x}{l} \tag{5.2.12}$$

with τ_0 given from (5.2.11). This gives the decay of the lowest-order mode. For an arbitrary initial value of the density within $-l/2 < x < l/2$, the initial density can be written as a Fourier series, which, with $n(x) = 0$ at $x = \pm l/2$, is

$$n = n_0 \left[\sum_{i=0}^{\infty} A_i \cos\frac{(2i+1)\pi x}{l} + \sum_{i=1}^{\infty} B_i \sin\frac{2i\pi x}{l}\right] \tag{5.2.13}$$

Then, assuming that each mode decays at its own characteristic rate, the symmetric ith mode has a product solution:

$$n_i = n_0 A_i e^{-t/\tau_i} \cos\frac{(2i+1)\pi x}{l} \tag{5.2.14}$$

where from the diffusion equation, as above, we find

$$\tau_i = \left[\frac{l}{(2i+1)\pi}\right]^2 \frac{1}{D} \tag{5.2.15}$$

From (5.2.15), we see that the higher modes, $i > 0$, decay more rapidly than the lowest mode, which becomes the dominant decay mode after sufficient time.

5.2.3 Steady-State Plane-Parallel Solutions

The diffusion solutions used for analyzing steady discharges are ones without time dependence. In these cases, it is necessary to either have flow into the region or a source within the region, to balance the diffusion out of the region. A simple case relevant to diffusion of neutrals or to a discharge containing negative ions (see Chapter 10) is with a specified flux entering on the one side and leaving on the other and with no volume source or sink. Taking a plane-parallel geometry, we have

$$-D\frac{d^2n}{dx^2} = 0 \tag{5.2.16}$$

The solution is just a linear decay across the region of interest:

$$n = Ax + B \tag{5.2.17}$$

If we specify that the flux is $\Gamma = \Gamma_0$ at $x = 0$ and the density is $n = 0$ at $x = l/2$, then

$$n = \frac{\Gamma_0}{D}\left(\frac{l}{2} - x\right) \tag{5.2.18}$$

The flux $\Gamma = -D\nabla n$ is independent of x.

Another interesting case is with a uniform specified source of diffusing particles. The steady-state diffusion equation has the form of *Poisson's equation*

$$-D\nabla^2 n = G_0 \tag{5.2.19}$$

The solution in a plane-parallel system is a parabola, and taking a symmetric solution with $n = 0$ at $x = \pm l/2$, we have

$$n = \frac{G_0 l^2}{8D}\left[1 - \left(\frac{2x}{l}\right)^2\right] \tag{5.2.20}$$

with the center density $G_0 l^2/8D$.

The most common case is for a plasma consisting of positive ions and an equal number of electrons which are the source of ionization. Then, with $n_e = n_i \equiv n$, the diffusion equation has the form of the *Helmholtz equation*

$$\nabla^2 n + \frac{\nu_{iz}}{D} n = 0 \tag{5.2.21}$$

with $D = D_a$ and with ν_{iz} the ionization frequency. Equation (5.2.21) has a homogeneous source (proportional to n). With the appropriate boundary conditions, the solution of (5.2.21) that is everywhere positive is the lowest-order eigenfunction, with the corresponding eigenvalue $\beta^2 = \nu_{iz}/D$.

For a plane-parallel geometry over the region $-l/2 \leq x \leq l/2$, (5.2.21) becomes

$$\frac{d^2 n}{dx^2} + \frac{\nu_{iz}}{D} n = 0$$

Taking the lowest-order symmetric eigenfunction, we obtain

$$n = n_0 \cos \beta x \tag{5.2.22}$$

where

$$\beta = \left(\frac{\nu_{iz}}{D}\right)^{1/2} \tag{5.2.23}$$

The flux is

$$\Gamma = -D\frac{dn}{dx} = Dn_0\beta \sin \beta x \tag{5.2.24}$$

and the diffusion velocity is

$$u = \frac{\Gamma}{n} = D\beta \tan \beta x \tag{5.2.25}$$

The ambipolar electric field, given by (5.1.10), points toward the walls, thus confining the more mobile electrons. With boundary conditions (5.2.1) that $n(\pm l/2) = 0$, (5.2.22) gives the spatial dependence

$$n = n_0 \cos \frac{\pi x}{l} \tag{5.2.26}$$

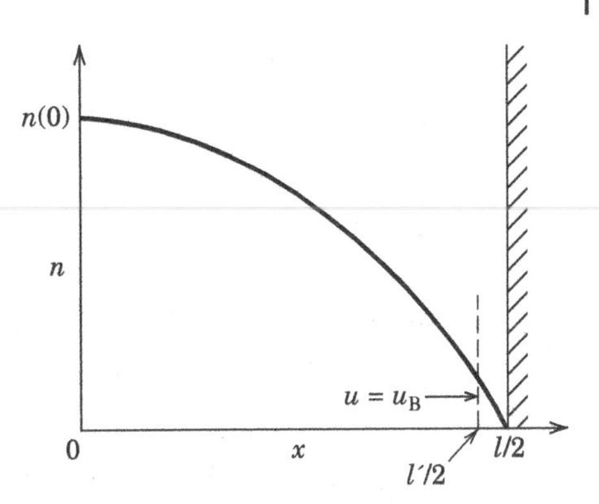

Figure 5.1 High-pressure diffusion solution for density n versus position x.

with the eigenvalue

$$\beta = \left(\frac{\nu_{iz}}{D}\right)^{1/2} = \frac{\pi}{l} \tag{5.2.27}$$

as shown in Figure 5.1. The reader may well ask how it is possible to have a relation of the type (5.2.27) when ν_{iz} and D are both given functions of the medium. The answer is that they are both temperature dependent, with ν_{iz} an exponentially sensitive function of T_e, as we have seen in Chapter 3. Thus, (5.2.27) is an equation for the electron temperature. We shall make this quite explicit in our discharge models in Chapters 10–15.

As we noted in the discussion following (5.2.1), the boundary conditions $n(\pm l/2) = 0$ that we have specified for the diffusion solutions are not self-consistent because the finite flux combined with the zero edge density leads to an infinite macroscopic edge velocity. Applying instead the boundary condition (5.2.3) at the plasma–sheath edge $x = l'/2$, where $l'/2 = l/2 - s$, with s the sheath thickness,

$$D\left.\frac{dn}{dx}\right|_{l'/2} = -n(l'/2)u_B \tag{5.2.28}$$

we obtain

$$u_B = D\beta \tan\frac{\beta l'}{2} \tag{5.2.29}$$

where from (5.2.23), $\beta = (\nu_{iz}/D)^{1/2}$. For a thin sheath, $s \ll l$, we have $l' \approx l$. Since u_B, ν_{iz}, and D are all functions of the electron temperature alone (with the neutral density specified), (5.2.29) again is an equation for T_e.

5.2.4 Steady-State Cylindrical Solutions

The preceding analysis is easily performed in cylindrical or spherical geometries. The cylindrical geometry is typical for the analysis of the positive column of a dc glow discharge, which we analyze in Section 14.2. The spherical geometry is a useful approximation to a small driving electrode in an rf-excited plasma, which we discuss in Section 11.4. Both cylindrical and spherical geometries are useful in analyzing electrostatic probes, which are considered in Section 6.6.

For an infinite cylinder with a specified uniform source term, the diffusion equation (5.2.19) in azimuthally symmetric coordinates is

$$\frac{d^2n}{dr^2} + \frac{1}{r}\frac{dn}{dr} + \frac{G_0}{D} = 0 \tag{5.2.30}$$

The homogeneous solution has the form

$$n_{\mathrm{h}} = c_1 \ln r + c_2 \tag{5.2.31}$$

and a particular solution is

$$n_{\mathrm{p}} = -\frac{G_0}{4D} r^2 \tag{5.2.32}$$

The complete solution $n = n_{\mathrm{h}} + n_{\mathrm{p}}$ is then easily obtained. For the boundary conditions that n is finite on-axis and $n = 0$ at $r = R$, we find that $c_1 = 0$ and $c_2 = G_0 R^2/4D$. The complete solution is parabolic

$$n = \frac{G_0 R^2}{4D}\left(1 - \frac{r^2}{R^2}\right) \tag{5.2.33}$$

as in the plane-parallel case.

In the more usual electropositive plasma, the electron and positive ion densities are equal and the ionization source is $\nu_{\mathrm{iz}} n$. The diffusion equation (5.2.21) in cylindrical coordinates with azimuthal symmetry is then

$$\frac{\mathrm{d}^2 n}{\mathrm{d}r^2} + \frac{1}{r}\frac{\mathrm{d}n}{\mathrm{d}r} + \frac{\mathrm{d}^2 n}{\mathrm{d}z^2} + \frac{\nu_{\mathrm{iz}}}{D} n = 0 \tag{5.2.34}$$

With no axial variation ($\mathrm{d}^2 n/\mathrm{d}z^2 = 0$), (5.2.34) is Bessel's equation, with solution

$$n = n_0 \mathrm{J}_0(\beta r) \tag{5.2.35}$$

where J_0 is the zero-order Bessel function. For the boundary condition $n(R) = 0$, we find

$$\beta = \left(\frac{\nu_{\mathrm{iz}}}{D}\right)^{1/2} = \frac{\chi_{01}}{R} \tag{5.2.36}$$

where $\chi_{01} \approx 2.405$ is the first zero of the J_0 Bessel function. Equation (5.2.36) determines $\mathrm{T_e}$. If there is also variation in z, then the variables can be separated in the usual way by assuming a product solution, which, with zero density on all boundaries, gives

$$n(r, z) = n_0 \mathrm{J}_0(\chi_{01} r/R) \cos(\pi z/l) \tag{5.2.37}$$

with

$$\beta^2 \equiv \frac{1}{\Lambda_0^2} = \frac{\nu_{\mathrm{iz}}}{D} = \frac{\chi_{01}^2}{R^2} + \frac{\pi^2}{l^2} \tag{5.2.38}$$

Λ_0 is the characteristic scale length for diffusion. The ion flux is

$$\Gamma_{\mathrm{iz}}(r) = -D\frac{\partial n}{\partial z} = \frac{\pi D}{l} n_0 \mathrm{J}_0\left(\chi_{01} r/R\right) \tag{5.2.39}$$

at the endwall $z = l/2$ and is

$$\Gamma_{\mathrm{i}r}(z) = -D\frac{\partial n}{\partial r} = \frac{\chi_{01} D}{R} n_0 \mathrm{J}_1\left(\chi_{01}\right) \cos\left(\pi z/l\right) \tag{5.2.40}$$

at the radial wall $r = R$.

One should note a fundamental difference between (5.2.19) for ambipolar diffusion with a specified source G_0, and (5.2.21) for diffusion with a source $\nu_{\mathrm{iz}} n$ proportional to n. In the former case, the density profile and the peak density n_0 are determined, *but* $\mathrm{T_e}$ *is not determined.* In the latter case, the density profile and $\mathrm{T_e}$ are determined, *but* n_0 *is not determined.* We will make this more explicit in our models in Chapter 10, where we will see the role of energy balance in specifying the remaining undetermined quantity.

If $\lambda_i \ll l$ and ν_m is independent of the ion flow velocity $\mathbf{u}$, the diffusion equations (5.2.19) or (5.2.21) are usually adequate. The condition that ν_m be independent of $\mathbf{u}$, however, usually limits the applicability of these diffusion models to quite high pressures, such that $\lambda_i/l \lesssim T_i/T_e$. If, on the other hand, we have $\lambda_i/l \gtrsim T_i/T_e$, then the assumptions of the constant D macroscopic diffusion theory begin to break down, and other approximations must be employed. The resulting equations are generally nonlinear and difficult to solve. In addition, the nonlinearity prevents a product solution in more than one spatial dimension. We treat some of the more important of these situations in Section 5.3.

5.3 Low-Pressure Solutions

5.3.1 Variable Mobility Model

Many discharges are run at low pressure where the assumptions used to obtain the solutions in Section 5.2 break down. In particular, at low pressure, the effective ion velocity for collision of ions with neutrals is the ion drift velocity $|\mathbf{u}|$ rather than the ion thermal velocity v_{thi}, i.e., for the pressure regime of interest $|\mathbf{u}| \gg v_{thi}$ over most of the discharge region. In this case, the ion-neutral collision rate can be written as $\nu_m \approx |\mathbf{u}_i|/\lambda_i$, where λ_i is the ion mean free path. Hence, we can replace the mobility from (5.1.4) by the relation (Smirnov, 1981, Problem 4.5)

$$\mu_i = \frac{2e\lambda_i}{\pi M|\mathbf{u}_i|} \tag{5.3.1}$$

Experimentally, over usual velocity ranges, λ_i is found to be reasonably approximated by a constant, and we assume this to be the case for the following analysis (see Figure 3.15 for some typical data). For the regime of interest here, $\mu_e \gg \mu_i$ and $T_e \gg T_i$, the basic equations can be simplified in a manner similar to that used to obtain (5.1.15) and (5.1.16). We make the assumption that the ion drift velocity due to the electric field dominates over the velocity due to the pressure gradient, such that

$$\mathbf{u}_i = \mu_i \mathbf{E} \tag{5.3.2}$$

For the electrons, we make the opposite assumption, namely that the drift velocity is negligible, to obtain

$$\mathbf{E} = -T_e \frac{\nabla n}{n} \tag{5.3.3}$$

This is equivalent to assuming that the electrons are governed by a Boltzmann distribution, as we have already described in Section 2.4.

With the above assumptions and the steady-state ion continuity equation,

$$\nabla \cdot (n\mathbf{u}_i) = \nu_{iz} n \tag{5.3.4}$$

we can derive a differential equation for the density profile. Taking a parallel plane geometry, as in Section 5.2 and solving for u_i in terms of $\nabla n/n$ from (5.3.2) and (5.3.3), we have, for $u_i > 0$,

$$u_i^2 = -u_B^2 \frac{2}{\pi} \frac{\lambda_i}{n} \frac{dn}{dx} \tag{5.3.5}$$

where $u_B = (eT_e/M)^{1/2}$ is the *Bohm velocity*. Taking the square root of (5.3.5) and substituting into (5.3.4), we obtain

$$u_B \left(\frac{2\lambda_i}{\pi}\right)^{1/2} \frac{d}{dx}\left(-n\frac{dn}{dx}\right)^{1/2} = \nu_{iz} n \tag{5.3.6}$$

Equation (5.3.6), which is nonlinear, has been solved by Godyak and Maximov (see Godyak, 1986) for the boundary conditions that $u_i = 0$ at the plasma center and $u_i = u_B$ at the sheath edge. The solution is given in Appendix C. The density profile found from inserting (C.11) into (C.8) is implicitly given by

$$\alpha^{2/3}\xi = \frac{1}{2}\ln\left[(1-y^3)^{1/3}+y\right] + \frac{1}{\sqrt{3}}\tan^{-1}\left[\frac{2(y^{-3}-1)^{1/3}-1}{\sqrt{3}}\right] + \frac{\pi}{6\sqrt{3}} \tag{5.3.7}$$

where $\xi = 2x/l$, $y = n/n(0)$, and

$$\alpha = \frac{\nu_{iz} l}{2u_B}\left(\frac{\pi l}{4\lambda_i}\right)^{1/2} \approx 1.25 \tag{5.3.8}$$

To a very good approximation, (5.3.7) with $\alpha = 1.25$ is the equation of a circle

$$\left(\frac{n}{n(0)}\right)^2 + \left(\frac{2x}{l}\right)^2 \approx 1$$

Actually, α varies slightly with ν_{iz} as shown in Figure 5.2*a*. We see from (5.3.7) that $n/n(0)$ is a function of a single parameter $2\alpha^{2/3}x/l$ where ν_{iz} and therefore α is determined from (5.3.8), which expresses the balance of ionization and loss at the boundary. The result for y is shown in Figure 5.2*b*. The density profile is roughly similar to the cosine profile (5.2.22) of the constant diffusion coefficient process discussed in Section 5.2, but is flatter in the middle and steeper at the edge.

The solution (5.3.7) has been generalized by Kouznetsov et al. (1996) to include a specified input flux at $x = 0$ in addition to the volume ionization. An important application is diffusion in a discharge containing negative ions. This low-pressure solution is also presented in Appendix C.

5.3.2 Langmuir Solution

At very low pressures, ($\lambda_i > l$), there is a limiting regime in which ions created at some location x' within the discharge half-space $0 < x' < l/2$ flow collisionlessly to the wall. Consequently, all ions

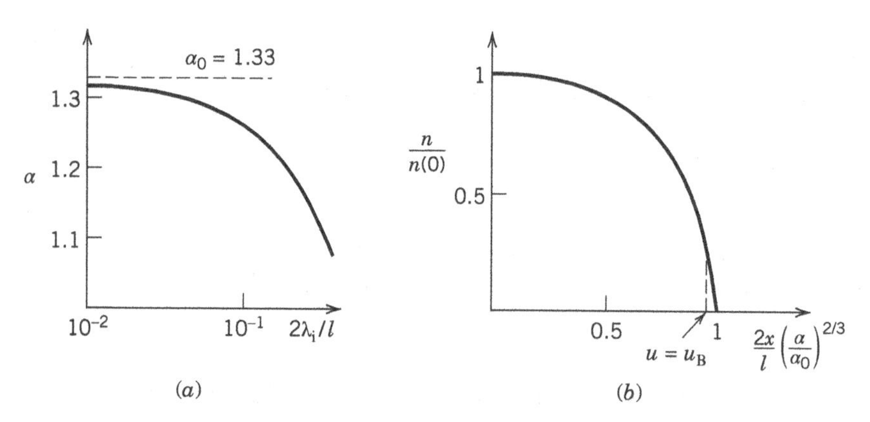

Figure 5.2 Low-pressure diffusion solutions for variable mobility model: (*a*) normalized ionization rate $\alpha = (\nu_{iz}l/2u_B)(\pi l/4\lambda_i)^{1/2}$ versus $2\lambda_i/l$; (*b*) normalized density $n/n(0)$ versus normalized position $(2x/l)(\alpha/\alpha_0)^{2/3}$. Source: Godyak, 1986/Delphic Associates.

born within a region $0 < x' < x$ contribute to the density at position x. In this situation, we replace the ion drift equation (5.3.2) by an ion velocity governed by energy conservation:

$$\frac{1}{2}Mu_i^2(x', x) = e[\Phi(x') - \Phi(x)] \tag{5.3.9}$$

This is equivalent to keeping the inertial and field terms and neglecting the acceleration and collision terms in the force equation (2.3.15). In (5.3.9), we have dropped the negligible thermal energy of created ions. The potential Φ is taken to be zero in the plasma center, with $\Phi < 0$ for $x > 0$. For the electrons, we keep the Boltzmann relation (5.3.3)

$$n(x) = n_0 e^{\Phi(x)/T_e}$$

To determine the plasma profile, we let $d\Gamma_i = \nu_{iz} n(x')dx'$ be the flux of ions created within a layer of thickness dx' at x'. This flux flows collisionlessly to position x, where it appears with a velocity u_i determined from the particle balance relation

$$\nu_{iz} n(x')dx' = dn\, u_i(x', x) \tag{5.3.10}$$

with dn the density produced at x by the flux created at x'. Inserting u_i from (5.3.9) into (5.3.10), solving for dn, and integrating dn over all positions $0 < x' < x$ contributing to the density at x, we obtain

$$n(x) = \left(\frac{M}{2e}\right)^{1/2} \int_0^x \frac{\nu_{iz} n(x')dx'}{[\Phi(x') - \Phi(x)]^{1/2}}$$

Eliminating n in the preceding equation using the Boltzmann relation for electrons yields a nonlinear integral equation for Φ

$$\exp\left(\frac{\Phi(\xi)}{T_e}\right) = \left(\frac{T_e}{2}\right)^{1/2} \int_0^\xi \frac{\exp\left(\Phi(\xi')/T_e\right)}{[\Phi(\xi') - \Phi(\xi)]^{1/2}} d\xi' \tag{5.3.11}$$

where $\xi = x\nu_{iz}/u_B$. This equation was first obtained for various geometries in a seminal paper by Tonks and Langmuir (1929), which included matching to the sheath region. The equation has a closed-form solution in terms of Dawson functions, but the solution was originally obtained by Tonks and Langmuir in the form of a power series and is shown in Figure 5.3. We note that the variable ξ is a function of the ionization, but ξ/ξ_s is not, where ξ_s is the value of ξ at the plasma edge $x = l'/2$, and so the solution is valid as $\nu_{iz} \to 0$. The endpoint, where there is a singularity (infinite derivative) of n, occurs at $\xi_s = 0.572$, $n_s/n_0 = 0.425$, $\Phi/T_e = 0.854$. The solution yields the velocity $u_s \approx 1.3\, u_B$ at the sheath edge, see also Section 6.2.

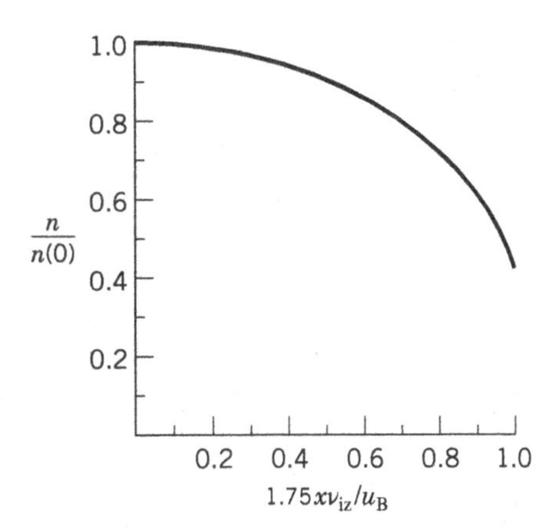

Figure 5.3 Free-fall solution: variation of the normalized density $n/n(0)$ versus normalized position $1.75x\nu_{iz}/u_B$.

5.3.3 Heuristic Solutions

The solution in (5.3.7) for $\lambda_i/l \lesssim 1$ does not join smoothly with the collisionless solution shown in Figure 5.3. It is possible to construct a heuristic solution that closely approximates the low-pressure constant λ_i solution for $\lambda_i/l \lesssim 1$ but has a transition to the approximate collisionless solution as $\lambda_i/l \to \infty$. Godyak (1986) has done this, obtaining an approximate result useful for calculations:

$$\nu_{iz} \approx \frac{2u_B}{l}\left(3 + \frac{l}{2\lambda_i}\right)^{-1/2} \tag{5.3.12}$$

and

$$h_l = \frac{n(l/2)}{n(0)} \approx 0.86\left(3 + \frac{l}{2\lambda_i}\right)^{-1/2} \tag{5.3.13}$$

A similar result for diffusion in a infinitely long cylinder of radius R was obtained (see Godyak, 1986):

$$\nu_{iz} \approx 2.2\frac{u_B}{R}\left(4 + \frac{R}{\lambda_i}\right)^{-1/2} \tag{5.3.14}$$

and

$$h_R = \frac{n(R)}{n(0)} \approx 0.8\left(4 + \frac{R}{\lambda_i}\right)^{-1/2} \tag{5.3.15}$$

For intermediate pressures, $T_i/T_e \lesssim \lambda_i/l \lesssim 1$, (5.3.13) reduces to the result $h_l \approx 0.86\,(2\lambda_i/l)^{1/2}$, which scales with pressure as $h_l \propto p^{-1/2}$. We give a derivation in Appendix C.

The preceding heuristic solutions joining the variable mobility diffusion model to the collisionless flow Langmuir result are not valid in the high-pressure regime $\lambda_i/l \lesssim T_i/T_e$, where a constant diffusion coefficient model is more appropriate. The three regimes can also be joined heuristically, giving the result (Lee and Lieberman, 1995)

$$h_l \approx \frac{0.86}{[3 + l/2\lambda_i + (0.86 l u_B/\pi D_a)^2]^{1/2}} \tag{5.3.16}$$

for parallel-plane geometry and

$$h_R \approx \frac{0.8}{[4 + R/\lambda_i + (0.8 R u_B/\chi_{01} J_1(\chi_{01}) D_a)^2]^{1/2}} \tag{5.3.17}$$

for cylindrical geometry. In the high-pressure regime, $h_l \approx \pi D_a/l u_B$ (Problem 5.7), which scales with pressure as $h_l \propto p^{-1}$. The way the collisionless, variable mobility, and constant diffusion solutions fit together to determine the edge-to-center density ratio h_l is illustrated in Figure 5.4. The heuristic scaling (5.3.16) is shown as the solid line, and the scalings of the collisionless flow, variable mobility diffusion, and constant diffusion coefficient models are indicated as dashed lines.

In Figure 5.5, the edge-to-center density ratio h_l determined by particle-in-cell (PIC) simulations is compared with the theoretical result (5.3.16) shown in Figure 5.4 for various inductively (ICP) and capacitively (CCP) coupled plasma configurations (Lafleur and Chabert, 2015a). There is good agreement for ICPs over the entire range of pressures $p \propto l/\lambda_i$ and also for CCPs at the lower pressures. However, the CCP and PIC results strongly deviate from the theory at the higher pressures, indicating that the theory strongly underestimates the particle losses. This issue is discussed in depth in Section 15.2, where it is shown that a transition occurs from an "active" to a "passive" bulk plasma regime at the higher pressures in CCPs.

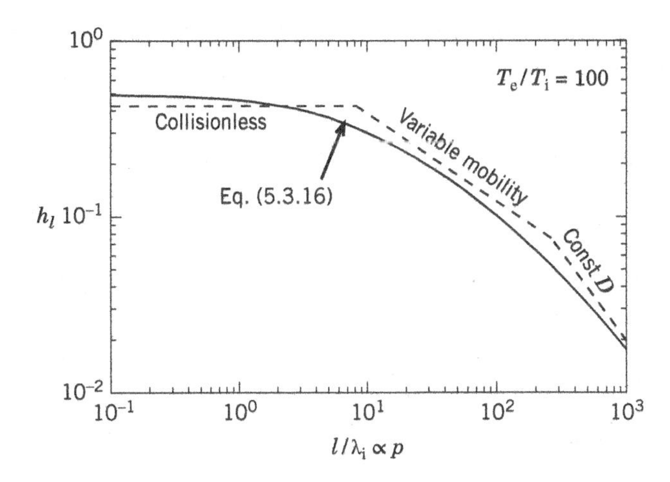

Figure 5.4 Edge-to-center density ratio h_l versus l/λ_i, illustrating the three regimes of collisionless flow, variable mobility diffusion, and constant diffusion coefficient models.

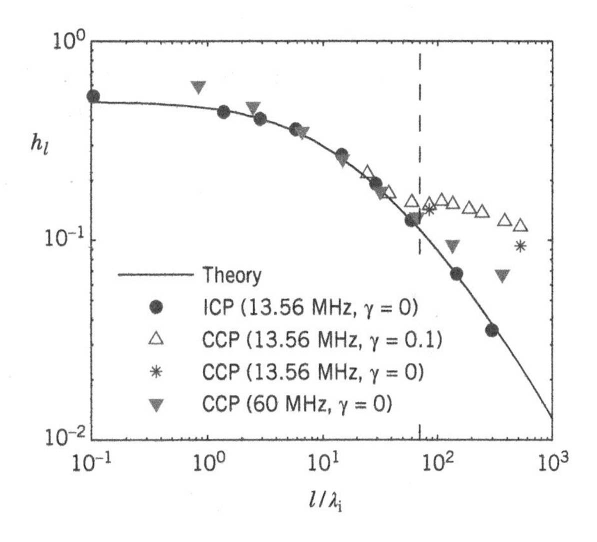

Figure 5.5 PIC simulation results showing the h_l factor versus $l/\lambda_i \propto$ pressure p for various inductively and capacitively coupled discharges, with the secondary electron emission coefficient γ indicated. Source: Lafleur and Chabert (2015a). ©IOP Publishing. Reproduced with permission. All rights reserved. https://doi.org/10.1088/0963-0252/24/2/025017.

5.4 Diffusion Across a Magnetic Field

We consider diffusion in the presence of magnetic fields, electric fields, and gradients, concentrating on the low temperature, weakly ionized plasma regime of interest to processing discharges. A review, including the differences between the low-temperature processing and hot fusion regimes, is given in Hagelaar and Oudini (2011). Generally the species for which the magnetic field is important, in weakly ionized plasmas, is the electrons that have small gyration orbits. To focus our attention, we first consider a long cylinder, with the magnetic field $\mathbf{B} = \hat{z}B_0$ taken along the cylinder. The density gradient points radially inward. When an electron gyrating around a line of force suffers a collision, it changes its direction, which would tend to move its center of gyration, on the average, by a gyration radius r_{ce}. This process is random, and therefore diffusive, with r_{ce} replacing λ_e as the diffusion mean free path when $r_{ce} \ll \lambda_e$.

To derive the perpendicular diffusion coefficient, we write the perpendicular component of the force equation for either species from (2.3.15):

$$0 = qn(\mathbf{E} + \mathbf{u}_\perp \times \mathbf{B}_0) - kT\nabla n - mn\nu_m \mathbf{u}_\perp$$

where we have again assumed an isothermal plasma and taken ν_m sufficiently large that the acceleration and inertial terms are negligible. It is convenient to express the vector equation in terms of the rectangular components (taken to be x and y):

$$mn\nu_m u_x = qnE_x - kT\frac{\partial n}{\partial x} + qnu_y B_0 \tag{5.4.1a}$$

and

$$mn\nu_m u_y = qnE_y - kT\frac{\partial n}{\partial y} - qnu_x B_0 \tag{5.4.1b}$$

Using the definitions of μ and D from (5.1.4) and (5.1.5), (5.4.1) can be rewritten:

$$u_x = \pm\mu E_x - \frac{D}{n}\frac{\partial n}{\partial x} + \frac{\omega_c}{\nu_m} u_y \tag{5.4.2a}$$

and

$$u_y = \pm\mu E_y - \frac{D}{n}\frac{\partial n}{\partial y} - \frac{\omega_c}{\nu_m} u_x \tag{5.4.2b}$$

where we have also used the definition of the gyration frequency $\omega_c = qB_0/m$. Equations (5.4.2) may be solved simultaneously for u_x and u_y to obtain

$$\left[1 + (\omega_c\tau_m)^2\right] u_x = \pm\mu E_x - \frac{D}{n}\frac{\partial n}{\partial x} + (\omega_c\tau_m)^2 \frac{E_y}{B_0} - (\omega_c\tau_m)^2 \frac{kT}{qB_0}\frac{1}{n}\frac{\partial n}{\partial y} \tag{5.4.3a}$$

$$\left[1 + (\omega_c\tau_m)^2\right] u_y = \pm\mu E_y - \frac{D}{n}\frac{\partial n}{\partial y} + (\omega_c\tau_m)^2 \frac{E_x}{B_0} + (\omega_c\tau_m)^2 \frac{kT}{qB_0}\frac{1}{n}\frac{\partial n}{\partial x} \tag{5.4.3b}$$

where we have defined $\tau_m \equiv 1/\nu_m$. Dividing by $1 + (\omega_c\tau_m)^2$, we define perpendicular mobility and diffusion coefficients,

$$\mu_\perp = \frac{\mu}{1 + (\omega_c\tau_m)^2} \tag{5.4.4}$$

$$D_\perp = \frac{D}{1 + (\omega_c\tau_m)^2} \tag{5.4.5}$$

and combining (5.4.3a) and (5.4.3b) in vector form, we find

$$\mathbf{u}_\perp = \pm\mu_\perp \mathbf{E} - D_\perp \frac{\nabla n}{n} + \frac{\mathbf{u}_E + \mathbf{u}_D}{1 + (\omega_c\tau_m)^{-2}} \tag{5.4.6}$$

Here, $\mathbf{u}_E$ and $\mathbf{u}_D$ are the $E \times B$ drift and the diamagnetic drift velocities, which are perpendicular to the field and the gradients:

$$\mathbf{u}_E = \frac{\mathbf{E} \times \mathbf{B}_0}{B_0^2} \tag{5.4.7}$$

$$\mathbf{u}_D = -\frac{kT}{qB_0^2}\frac{\nabla n \times \mathbf{B}_0}{n} \tag{5.4.8}$$

The drifts perpendicular to the field and gradients are slowed by the collisions, while the mobility and diffusion fluxes parallel to the gradients and perpendicular to the field exist only in the presence of collisions and are slowed by the presence of the magnetic field.

For some plasma discharges, the drifts can be important because they can lead to instabilities with a resulting anomalous transport, as we discuss below. The diamagnetic drift velocity depends on the species temperature and charge and leads to a plasma current flow dominated by electrons for $T_e \gg T_i$. For highly magnetized plasmas, the electron and ion $E \times B$ drift velocities are the same, yielding no net current flow. But in many weakly magnetized processing discharges, having magnetized electrons and unmagnetized ions, a strong electron $E \times B$ current can flow.

The factor $\omega_c \tau_m$, often called the *Hall parameter*, is an important quantity in magnetic confinement, with $\omega_c \tau_m \gg 1$ indicating strong retardation of diffusion. In this limit, dropping the 1, we have

$$D_\perp = \frac{kT}{m\nu_m}\frac{1}{(\omega_c \tau_m)^2} = \frac{kT\nu_m}{m\omega_c^2} \tag{5.4.9}$$

Comparing (5.4.9) with the diffusion coefficient without a magnetic field (or $D = D_\parallel$ parallel to B_0), from (5.1.5), we see that the position of the collision frequency is reversed, with $D_\perp \propto \nu_m$ while $D_\parallel \propto \nu_m^{-1}$. Since $\nu_m \propto m^{-1/2}$ at fixed energy and cross section, we also find $D_\perp \propto m^{1/2}$ and $D_\parallel \propto m^{-1/2}$. This is easily understood in that the lighter electrons move faster without a magnetic field but are strongly inhibited across the field. We can also understand these relations in terms of random walk distances. As in Section 5.1, we use $\bar{v}^2 = 8kT/\pi m$, and with the mean gyroradius $\bar{r}_c = \bar{v}/\omega_c$ substituted into (5.4.9), we have

$$D_\perp = \frac{\pi}{8}\bar{r}_c^2 \nu_m \tag{5.4.10}$$

Comparing (5.4.10) with (5.1.6), we see that the mean gyration radius has taken the place of the mean free path as the characteristic random walk step.

If plasma can be lost only across the magnetic field, then equating the electron and ion fluxes, as in Section 5.1, leads to a cross-field ambipolar diffusion coefficient as in (5.1.12), except that the quantities refer to the perpendicular mobility and diffusion

$$D_{\perp a} = \frac{\mu_{\perp i} D_{\perp e} + \mu_{\perp e} D_{\perp i}}{\mu_{\perp i} + \mu_{\perp e}} \tag{5.4.11}$$

If the magnetic field is sufficiently strong that $\mu_{\perp i} \gg \mu_{\perp e}$, reversing the inequality used in Section 5.1, then the simpler form, analogous to (5.1.14), is

$$D_{\perp a} = D_{\perp e}\left(1 + \frac{T_i}{T_e}\right) \tag{5.4.12}$$

where $D_{\perp e}$ is given by (5.4.9). Again the slower diffusion controls the behavior, but in the usual weakly ionized plasma with $T_i \ll T_e$, the ambipolar and electron diffusion coefficients perpendicular to B_0 are not significantly different.

A more thorough one-dimensional (r) fluid model of the transition from weakly to strongly magnetized radial diffusion in a plasma cylinder is given in Sternberg et al. (2006). They find the condition

$$\frac{R\lambda_e}{r_{ce} r_{ci}} \lesssim 1 \tag{5.4.13}$$

for weak magnetization, where the electric force dominates the magnetic force and electron Boltzmann equilibrium is obtained. Here R is the discharge radius, λ_e is the electron mean free path, and r_{ce} and r_{ci} are the electron and ion gyroradii given in (4.1.10) and (4.1.12), respectively.

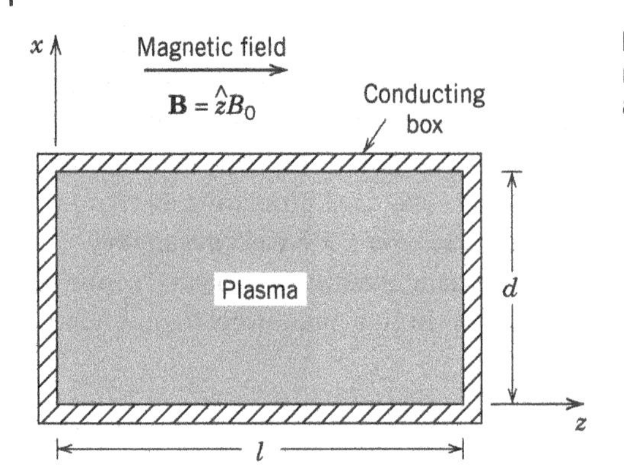

Figure 5.6 A plasma-filled conducting box in a dc magnetic field, illustrating the calculation of ambipolar diffusion in a magnetized plasma.

5.4.1 Nonambipolar Diffusion

The assumption that diffusion takes place only across the magnetic field is almost never satisfied in typical processing discharges. Even for finite length systems in which l (along B_0) $\gg d$ (across B_0), the more rapid diffusion along B_0 is usually important. We therefore consider the regime in which $l \sim d$, as shown in Figure 5.6. For simplicity, rectangular coordinates are used and the y direction is taken to be uniform and of infinite extent. Assuming that the walls are conducting, it is clear that the net (ion + electron) diffusion particle currents collected by each of the four walls are not necessarily zero. The only requirement is that the total electron and ion fluxes integrated over all four wall surfaces must be equal to maintain plasma quasineutrality. In fact, for magnetized electrons, $\omega_{ce}\tau_e \gg 1$, the electrons will preferentially flow out along the magnetic field, and the ions will flow out perpendicular to the field. Therefore, currents must flow in the walls; this is often called *the short circuit effect.*

The diffusion is obtained from the continuity equations for electrons and ions:

$$\frac{\partial n}{\partial t} = D_e \frac{\partial^2 n}{\partial z^2} + \mu_e \frac{\partial}{\partial z}(nE_z) + D_{\perp e} \frac{\partial^2 n}{\partial x^2} + \mu_{\perp e} \frac{\partial}{\partial x}(nE_x) \tag{5.4.14}$$

$$\frac{\partial n}{\partial t} = D_i \frac{\partial^2 n}{\partial z^2} - \mu_i \frac{\partial}{\partial z}(nE_z) + D_{\perp i} \frac{\partial^2 n}{\partial x^2} - \mu_{\perp i} \frac{\partial}{\partial x}(nE_x) \tag{5.4.15}$$

Exact two-dimensional solutions to these two coupled nonlinear diffusion equations have not been obtained. Let $V_{s\perp}$ and $V_{s\parallel}$ be the potential drops across the perpendicular and parallel sheaths, then because the plasma is surrounded by a conducting wall, the potential in the center can be estimated as

$$\Phi \sim V_{s\parallel} + \frac{1}{2}E_z l \sim V_{s\perp} + \frac{1}{2}E_x d$$

Two limiting cases can be considered depending on the size of E_x. For $E_x d \lesssim T_i$, the perpendicular mobility terms in (5.4.14) and (5.4.15) are small compared to the perpendicular diffusion terms. Dropping the perpendicular mobility terms, as done by Simon (1959), yields the diffusion equation (see Problem 5.11)

$$\frac{\partial n}{\partial t} = D_\parallel \frac{\partial^2 n}{\partial z^2} + D_\perp \frac{\partial^2 n}{\partial x^2} \tag{5.4.16}$$

with $D_\parallel \approx \mu_i \mathrm{T_e}$ the usual one-dimensional ambipolar diffusion coefficient, and $D_\perp \approx \mu_{\perp i}\mathrm{T_i}$. In this situation, $\mathbf{\Gamma}_i \neq \mathbf{\Gamma}_e$, a nonambipolar flow.

However, the assumption $E_x d \lesssim T_i$ is almost never satisfied for typical magnetized processing discharges. If electron flow along field lines is impeded by inertial or collisional effects or if the axial sheath voltage $V_{s\parallel}$ varies with x, then there can be a substantial ion acceleration potential $E_x d \gtrsim T_i$. In this case, the perpendicular ion diffusion term in (5.4.15) is smaller than the mobility term. There is experimental evidence (see Lieberman and Gottscho, 1994, section VIII.D.2) and also computer simulations (Porteous et al., 1994) that indicate the existence of these radial potentials in magnetized processing discharges such as ECRs (see Section 13.1). Measurements and simulations both show that ions are lost radially from the bulk plasma with a characteristic loss velocity of order the Bohm velocity $u_B = (eT_e/M)^{1/2}$. If an electric field exists across field lines with magnitude $E_x \sim T_e/d$, then we can estimate $\Gamma_{\perp i} \sim \mu_{\perp i} n T_e/d$. Then, defining $D_\perp$ through $\Gamma_{\perp i} \equiv -D_\perp dn/dx \sim D_\perp n/d$, we obtain

$$D_\perp \sim \mu_{\perp i} T_e \tag{5.4.17}$$

in the place of $D_\perp \sim \mu_{\perp i} T_i$. For $d \sim l$ and $T_e \gg T_i$, this can lead to substantial perpendicular ion losses in magnetized processing discharges. An experiment and model testing the use of various perpendicular diffusion coefficients in a high-frequency magnetized discharge (Vidal et al., 1999) finds the best agreement with (5.4.17).

Various models of classical diffusion in a magnetized, finite length plasma cylinder have been given. Notably, Fruchtman (2009) finds a linear analytical solution for constant diffusion and mobility coefficients and gives conditions for both ambipolar flows, and for nonambipolar flows having magnetized electrons and unmagnetized ions, exhibiting the "short circuit" effect with substantial wall currents. Curreli and Chen (2011, 2014) give a fluid model of these nonambipolar flows based on the assumptions of Maxwellian electrons and cross-field Boltzmann electrons, which they apply to helicon plasma equilibrium, obtaining good agreement with an experiment.

5.4.2 PIC Simulations

Lafleur and Boswell (2012a) performed two-dimensional particle-in-cell simulations of magnetized diffusion for an argon discharge in a rectangular conducting box with $d = 5$ cm and $l = 10$ cm. They simulated a range of pressures of 1–50 mTorr and magnetic fields of 0–200 G. Concentrating on a pressure of 1 mTorr, they examined the transition from no magnetic field to a 50 G field, obtaining the perpendicular and parallel fluxes to the conducting surfaces shown in Figure 5.7. The fluxes shown are normalized, with the total ion particle current to the parallel and perpendicular walls equal to that for electrons. As can be seen for $B_0 = 0$ in Figure 5.7*a* and *b*, the fluxes to the walls are nonambipolar, even in the absence of a magnetic field, as also seen by Lucken et al. (2018). However, there are significant ion and electron fluxes flowing to both perpendicular and parallel walls. This is not true for the 50 G magnetized diffusion fluxes shown in Figures 5.7*c* and *d*. The perpendicular electron flux $\Gamma_{\perp e}$ in Figure 5.7*c* is almost entirely absent, and there is a much larger electron flux than ion flux in the parallel direction, as seen in Figure 5.7*d*. There are large currents flowing in the walls; i.e., the flow is strongly nonambipolar. Significant wall currents, with or without confining magnetic fields, can lead to substrate damage during plasma processing, as described in Section 16.6.

Lafleur and Boswell give quantitative results for the total perpendicular and parallel particle currents for 50 G at 1, 10, and 50 mTorr. At this field, the electrons are strongly magnetized, $\omega_{ce}\tau_i \gg 1$. For 1 mTorr, the ions are weakly magnetized: $\omega_{ci}\tau_i \approx 1$ and the gyration radius $r_{ci} \approx 2$ cm. At the higher pressures, the ions are unmagnetized (see also (5.4.13)). From the currents, effective diffusion coefficients can be estimated. At 10 and 50 mTorr, the perpendicular diffusion coefficients are in reasonable agreement with $D_\perp \approx \mu_{\perp i} T_e$ given in (5.4.17). At 1 mTorr, the PIC result gives about 40% of $\mu_{\perp i} T_e$.

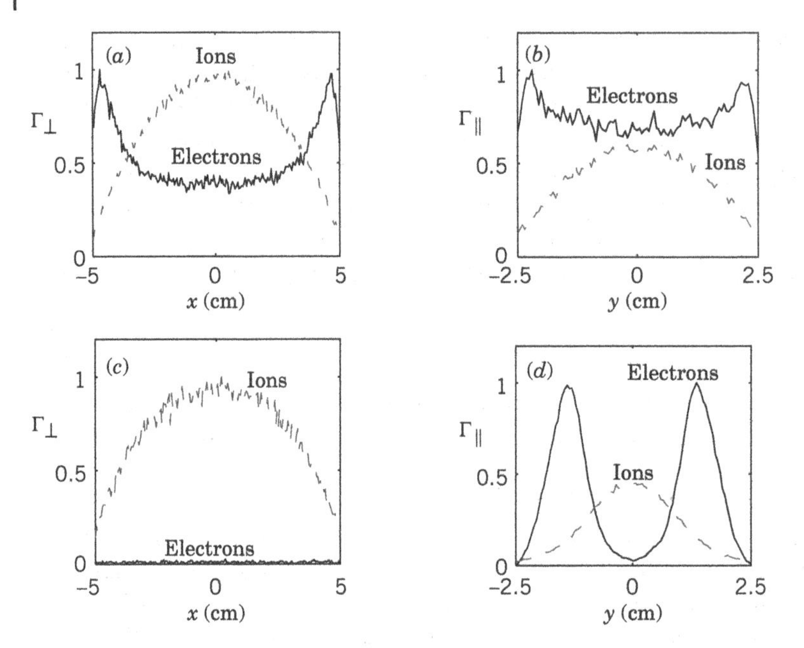

Figure 5.7 Two-dimensional simulation results for electron and ion fluxes to the rectangular conducting walls with $d = 5$ cm and $l = 10$ cm at 1 mTorr argon; (*a*) perpendicular flux $\Gamma_\perp$ versus x at $B_0 = 0$ G; (*b*) parallel flux $\Gamma_\parallel$ versus y at $B_0 = 0$ G; (*c*) perpendicular flux $\Gamma_\perp$ versus x at $B_0 = 50$ G; (*d*) parallel flux $\Gamma_\parallel$ versus y at $B_0 = 50$ G. Source: Reproduced from Lafleur and Boswell (2012a), with the permission of AIP Publishing. https://doi.org/10.1063/1.4719701.

In many processing discharges, the perpendicular or the parallel walls are insulating, e.g., quartz, ceramic, and anodized aluminum. In these systems, there is no short circuit effect; the perpendicular and parallel flows are not strongly coupled by currents flowing between the walls. The flows are then globally ambipolar but are usually still locally nonambipolar, i.e., $\mathbf{\Gamma}_\text{i} \neq \mathbf{\Gamma}_\text{e}$ within the plasma. For unmagnetized or weakly magnetized ions, a reasonable estimate is that $D_\parallel \approx D_\text{a} \approx \mu_\text{i}\text{T}_\text{e}$, and $D_\perp \approx D_{\perp a}$ given in (5.4.11). For high magnetic fields, the classical result is $D_\perp \approx D_{\perp e}$ given in (5.4.12), but this small diffusion coefficient may not hold due to saturated instabilities leading to turbulent diffusion, which can destroy the cross-field electron confinement. The transition of the diffusion with change of wall material from conducting to insulating has been observed experimentally in a 4 MHz rf discharge containing a magnetic filter (Lafleur and Aanesland, 2014).

It is well known that plasmas not in thermal equilibrium are subject to instabilities. This is a major subject of fully ionized, near collisionless plasmas and is treated in detail in most texts on plasma physics (see, e.g., Chen, 1984). Magnetic field confinement is an important source of such disequilibrium that leads to various instabilities which tend to destroy the confinement. A historical review of the resulting anomalous (nonclassical) transport is given in Curreli and Chen (2014). Boeuf (2014) gives examples from PIC simulations of anomalous transport induced by $E \times B$-driven instabilities. Generally, but not always, the diffusion has the upper limit of the *Bohm diffusion* coefficient,

$$D_\text{B} = \frac{1}{16}\frac{\text{T}_\text{e}}{B_0} \tag{5.4.18}$$

The scaling with B_0 makes Bohm diffusion increasingly important as a source of cross-field diffusion at high magnetic fields, since from (5.4.10), we see that classical cross-field diffusion scales as

$D_\perp \propto 1/B_0^2$. Bohm diffusion tends to be less important at high collisionality (low temperature and high pressure) both due to the comparative scaling of D_B to $D_\perp$ and also due to the fact that high collisionality tends to inhibit some of the instabilities.

An example of anomalous cross-field transport obtained from two-dimensional PIC simulations is given in Lucken et al. (2019). In the absence of the short circuit effect, they show that as the magnetic field is increased, the cross-field diffusion coefficient is approximately classical, (5.4.11), up to the point where an electron drift instability appears and saturates at a large amplitude. The resulting short-wavelength fluctuations destroy the cross-field electron confinement, giving a large anomalous electron collision frequency and a resulting diffusion coefficient of order (5.4.17).

5.5 Magnetic Multipole Confinement

In magnetic multipole confinement, a set of alternating rows of north and south pole permanent magnets is placed around the surface of a discharge chamber. A typical configuration, with the rows arranged around the circumference of a cylindrical chamber, is shown in Figure 5.8. In some cases, one or both cylindrical endwalls are also covered with rows of magnets. Commonly, each row is composed of a set of many permanent magnets (diameter ~ length ~2.5 cm, $B_0 \sim 1$ kG). The alternating rows of magnets generate a *line cusp* magnetic configuration in which the magnetic field strength B is a maximum near the magnets and decays with distance into the chamber, as shown in Figure 5.8. Hence, most of the plasma volume can be virtually magnetic field free, while a strong field can exist near the discharge chamber wall, inhibiting plasma loss and leading to an increase in plasma density and uniformity.

5.5.1 Magnetic Fields

The structure of the magnetic field can be understood by unwrapping the circumference to obtain the alternating periodic arrangement of magnet rows in rectangular geometry shown in Figure 5.9.

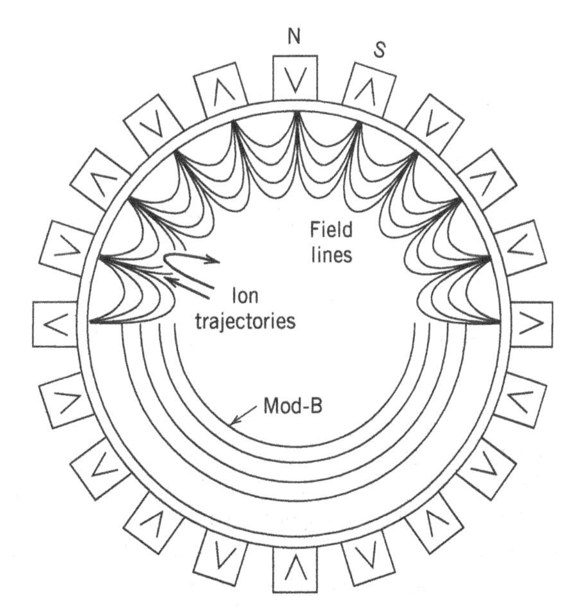

Figure 5.8 Magnetic multipole confinement in cylindrical geometry, illustrating the magnetic field lines and the $|\mathbf{B}|$ surfaces near the circumferential walls.

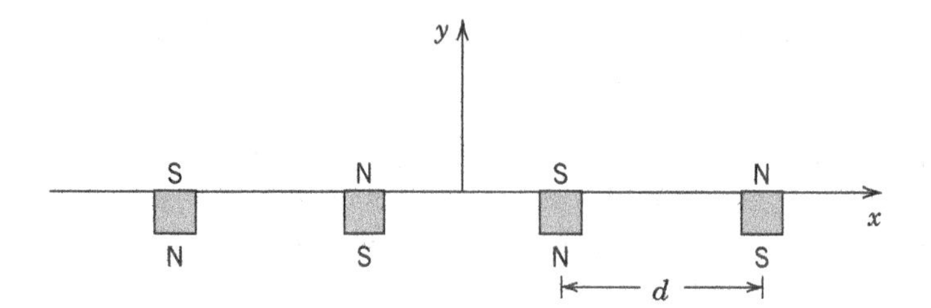

Figure 5.9 Schematic for determining multipole fields in rectangular geometry.

Assuming that each row of magnets has a width $\Delta \ll d$, the separation of the rows, then B_y at $y = 0$ can be approximated as

$$B_y(x,0) = B_0\Delta \sum_{i=-\infty}^{\infty} (-1)^i \delta\left(x - id - \frac{d}{2}\right) \tag{5.5.1}$$

where δ is the Dirac delta function. Introducing the Fourier transform,

$$B_y(x,0) = \sum_{m=1}^{\infty} A_m \sin\frac{m\pi}{d}x \tag{5.5.2}$$

and equating (5.5.1) and (5.5.2), then if we multiply by $\sin(\pi x/d)$ and integrate from 0 to d, we obtain the fundamental ($m = 1$) Fourier mode amplitude A_1, such that

$$B_{y1}(x,0) = \frac{2B_0\Delta}{d}\sin\frac{\pi x}{d} \tag{5.5.3}$$

Because $\nabla \cdot \mathbf{B} = 0$ and $\nabla \times \mathbf{B} = 0$ for $y > 0$, B_{y1} satisfies Laplace's equation:

$$\frac{\partial^2 B_{y1}}{\partial x^2} + \frac{\partial^2 B_{y1}}{\partial y^2} = 0 \tag{5.5.4}$$

The solution to (5.5.4) with boundary conditions that $B_{y1}(x,0)$ is given by (5.5.3) and that $B_{y1}(x, y \to \infty)$ is not infinite is

$$B_{y1}(x,y) = \frac{2B_0\Delta}{d}\sin\frac{\pi x}{d}\mathrm{e}^{-\pi y/d} \tag{5.5.5}$$

From the z component of $\nabla \times \mathbf{B} = 0$, we have

$$\frac{\partial B_{x1}}{\partial y} = \frac{\partial B_{y1}}{\partial x} \tag{5.5.6}$$

Using (5.5.5) in (5.5.6) and integrating with respect to y, we obtain

$$B_{x1}(x,y) = -\frac{2B_0\Delta}{d}\cos\frac{\pi x}{d}\mathrm{e}^{-\pi y/d} \tag{5.5.7}$$

The field amplitude is $B_1 = (B_{x1}^2 + B_{y1}^2)^{1/2}$. Using (5.5.5) and (5.5.7), we obtain

$$B_1(x,y) = \frac{2B_0\Delta}{d}\mathrm{e}^{-\pi y/d} \tag{5.5.8}$$

showing an exponential decay that is independent of x into the discharge column with decay length d/π. The smooth B_1 surfaces as well as the alternating B_{y1} and B_{x1} components can be clearly seen in Figure 5.8. The higher-order Fourier modes with nonzero coefficients ($m = 3, 5, \ldots$) have even shorter decay lengths ($d/3\pi, d/5\pi, \ldots$), and their effect is negligible a short distance from the chamber wall. Thus, we expect this picture to hold at distances significantly greater than d/π within the plasma chamber. Midway between the magnets (at $x = 0, \pm d, \ldots$), the magnetic field is zero at $y = 0$ and rises to a maximum value

$$B_{\mathrm{max}} = \frac{\pi^2}{8}\frac{\Delta^2}{d^2}B_0$$

at $y \approx 0.28\,d$, after which it decays exponentially with y. The diffusion across this region is important in determining the confinement properties of the multipoles.

5.5.2 Plasma Confinement

Experimentally (Leung et al., 1975, 1976), multipole fields have been found to have three important effects on low-pressure plasma confinement:

1. Hot electrons, having energies $\gtrsim$ dc sheath potential, can be efficiently confined, provided there is end confinement either with magnetic mirrors, multipoles, or negative electrostatic potentials. These electrons, if created and trapped at low pressures (large mean free path compared to the discharge size), can be the main ionization source for a discharge.
2. Significant (but not large) improvements can be obtained in the confinement of the bulk (low-temperature) plasma in a discharge.
3. Significant improvements in radial plasma uniformity can be obtained.

The effects can, at least partly, be understood in terms of magnetic mirroring in the cusps as governed by (4.3.15). The energetic electrons that are not lost by moving parallel to field lines are mirrored as they move into the higher field near the cusp. Their velocity vectors with respect to the magnetic field at the wall are randomized within the central plasma chamber, where (4.3.15) does not hold. The number of reflections from the cusp then depends on the size of the "loss cone" angle in velocity space compared to the possible solid angle of 4π within which the velocity vector can be found. At lower velocities (or higher pressures), the scattering can take place collisionally on the outward flight, greatly increasing the loss rate. Ambipolar fields also play a part, but in a complicated manner. The improvement in plasma uniformity follows because the diffusion is inhibited in the region of strong magnetic field, as described in Section 5.4. Thus, most of the density gradient occurs at the plasma edge, where the diffusion coefficient is small, leading to a relatively uniform central region.

As an example (Leung et al., 1975), a low-pressure dc argon discharge was created in a 30-cm-diameter, 33-cm-long chamber by primary energetic electrons emitted from a hot filament placed inside the chamber and biased at −60 V. With multipoles and at $p = 0.8$ mTorr, the energetic electrons were confined for up to 70 bounces within the chamber, and the plasma density was increased by approximately a factor of 100. Of this increase, roughly a factor of 30 was measured to be due to the increased confinement of the energetic electrons, and an additional factor of three increase was due to the improvement in confinement for the bulk plasma. However, in most processing discharges, the ionization is not produced by a class of very energetic electrons, and the second and third effects listed above are the most significant.

A useful concept to discuss confinement is the *effective leak width* w of a line cusp. If there are $\mathcal{N}$ cusps of width w, then the effective circumferential loss width is $\mathcal{N}w$ and the fraction f_{loss} of diffusing electron–ion pairs that will be lost to the wall is

$$f_{\mathrm{loss}} = \frac{\mathcal{N}w}{2\pi R}, \qquad \mathcal{N}w < 2\pi R \tag{5.5.9}$$

The boundary condition at the wall ($y \approx 0$) for the ambipolar diffusion of plasma within the field-free discharge volume is then

$$\Gamma_{\rm w} = f_{\rm loss} n_{\rm s} u_{\rm B} \tag{5.5.10}$$

We return to the example in Section 5.2 of steady-state diffusion in a plasma slab of length l with an ionization source proportional to the density. The density profile is given by (5.2.22). Equating $\Gamma(l/2)$ in (5.2.24) to $\Gamma_{\rm w}$ in (5.5.10), we obtain, for a thin sheath,

$$\frac{f_{\rm loss} u_{\rm B}}{D_{\rm a}\beta} = \tan\frac{\beta l}{2} \tag{5.5.11}$$

This transcendental equation for β must in general be solved numerically. However, if $f_{\rm loss}$ is not too small, such that the left-hand side of (5.5.11) still remains much greater than unity, then we can approximate $\beta \approx \pi/l$ on the left-hand side to obtain

$$\tan\frac{\beta l}{2} = \frac{f_{\rm loss} u_{\rm B} l}{\pi D_{\rm a}} \tag{5.5.12}$$

This is the usual regime for most processing discharges. Taking the ratio of $n_{\rm s} \equiv n(l/2)$ to $n_0 \equiv n(0)$, and using (5.2.22) to substitute for $\tan(\beta l/2)$ in terms of $n_{\rm s}$, we find

$$\frac{n_{\rm s}}{n_0} = \left[1 + \left(\frac{f_{\rm loss} u_{\rm B} l}{\pi D_{\rm a}}\right)^2\right]^{-1/2} \tag{5.5.13}$$

We see that the uniformity of the plasma improves as $f_{\rm loss}$ is reduced below unity by the presence of the multipoles. Since uniformity is often a critical issue in plasma processing, multipole confinement may offer a means to control this parameter. A measured density profile with and without multipole confinement is shown in Figure 12.13, where rf inductive discharges are discussed. As will be shown in Chapter 10 (see (10.2.15) and accompanying discussion), for a fixed absorbed power, the plasma density is inversely proportional to the loss area. Hence, we would expect $n_0 \propto f_{\rm loss}^{-1}$, when $f_{\rm loss} u_{\rm B} l/\pi D_{\rm a} \gg 1$.

5.5.3 Leak Width w

The size of the leak width w is not fully understood. At very low pressures, theoretical calculations, confirmed by measurements (see Hershkowitz et al., 1975), indicate that

$$w \approx 4(\bar{r}_{\rm ce}\bar{r}_{\rm ci})^{1/2} \tag{5.5.14}$$

where $\bar{r}_{\rm ce}$ and $\bar{r}_{\rm ci}$ are the mean electron and ion gyroradii at the location where the magnetic field lines enter the wall. However, the leak width is observed to increase with pressure and is much larger than indicated by (5.5.14) at typical process pressures ($\gtrsim 1$ mTorr). The mechanism for this increase in w is that ions and electrons collisionally diffuse across magnetic field lines, and diffuse or flow along the field lines to the wall. An estimate of the leak width for intermediate pressures is (Matthieussent and Pelletier, 1992)

$$w \approx \frac{2}{\pi}(\bar{r}_{\rm ce}\bar{r}_{\rm ci})^{1/2}\frac{d}{(\lambda_{\rm me}\lambda_{\rm mi})^{1/2}} \tag{5.5.15}$$

where $\lambda_{\rm me}$ and $\lambda_{\rm mi}$ are the electron and ion mean free paths. By comparing (5.5.14) and (5.5.15), a heuristic formula valid for low and intermediate pressures can be constructed. The general scalings have been observed experimentally. At some pressure where $w \approx 2\pi R/\mathcal{N}$, $f_{\rm loss}$ given by (5.5.9) rises to unity and the multipoles have little effect on the bulk plasma confinement. Other mechanisms, such as *Bohm diffusion* across magnetic fields due to fluctuating electric fields in the plasma, can also be present and are known to be important for particle losses, e.g., from weakly collisional cusp magnetic fields.

Problems

5.1 The Congruence Assumption The congruence assumption $\Gamma_e = \Gamma_i$ is used to derive the ambipolar relation (5.1.11).

(a) Show from particle conservation that $\nabla \cdot \Gamma_e = \nabla \cdot \Gamma_i$.

(b) Show that

$$\nabla \times \Gamma_e = -\mu_e \nabla\Phi \times \nabla n$$
$$\nabla \times \Gamma_i = \mu_i \nabla\Phi \times \nabla n$$

Hence, for $\nabla\Phi \times \nabla n = 0$, we find $\nabla \times \Gamma_e = \nabla \times \Gamma_i = 0$.

(c) For $\nabla\Phi \times \nabla n = 0$, from parts (a) and (b), show that $\Gamma_e = \Gamma_i + \text{const}$. The boundary conditions generally set the condition that the constant is zero, and, hence, $\Gamma_e = \Gamma_i$.

(d) Show that if $n(\mathbf{r}) = n[\Phi(\mathbf{r})]$, e.g., if n is given by the Boltzmann relation (2.4.16), with T_e a constant, then $\nabla\Phi \times \nabla n = 0$.

5.2 Ambipolar Diffusion Coefficient Making the assumptions of electric field-driven flux for ions and Boltzmann equilibrium for electrons, as in (5.1.15) and (5.1.16), solve to obtain the ambipolar diffusion coefficient D_a and compare with (5.1.14).

5.3 High-Pressure Diffusion with Specified Ionization Source A high-pressure, steady-state argon plasma discharge confined between two parallel plates located at $x = \pm l/2$ is created in argon gas at density n_g by uniformly illuminating the region within the plates with ultraviolet radiation. The radiation creates a uniform number G_0 of electron–ion pairs per unit volume per unit time everywhere within the plates. Assume that the electron and ion temperatures are uniform, with $T_e \gg T_i$. Electrons and ions are lost to the walls by ambipolar diffusion, with ambipolar diffusion coefficient $D_a \approx \mu_i T_e$ (T_e is in volts). Choose boundary conditions such that $n(x) \approx 0$ at the walls.

(a) Find the plasma density $n(x)$ and the peak density n_0 within the plates. Find the steady-state particle flux $\Gamma(x)$, ambipolar electric field $E(x)$, potential $\Phi(x)$, and total charge density $\rho(x)$. Sketch Γ, E and Φ for $|x| \leq l/2$.

(b) Plot $\rho(x)/e$ and $n(x)$ *on the same graph* for $|x| \leq l/2$. Are the ambipolar solutions valid for $\rho(x)/e > n(x)$? Explain your answer.

5.4 Ambipolar Diffusion with a Delta Function Source Consider ambipolar diffusion between two absorbing parallel plates separated by a distance l, with one plate located at $x = -l/4$ and the other plate located at $x = 3l/4$. Assume that $G = G_0$ electron–ion pairs per unit time per unit volume are created within a thin layer $-w < x < w$ within the plates and that $G = 0$ everywhere else within the plates. *You may assume that* $w \lll$. The ambipolar diffusion coefficient is $D_a = D_i(1 + T_e/T_i)$, where D_i is the ion diffusion coefficient, T_e is the electron temperature, and T_i is the ion temperature. Assume that D_i, T_e, and T_i are constants, with $T_i \ll T_e$.

(a) Find and sketch the electron density $n_e(x)$ everywhere between the plates. *You may assume that $n_e \approx 0$ at the surfaces of the two plates.*
(b) Give an expression for the ion flux $\Gamma_i(x)$ in terms of $n_e(x)$ and sketch $\Gamma_i(x)$ everywhere between the plates.
(c) Give an expression for the electric potential $\Phi(x)$ in terms of $n_e(x)$, and sketch $\Phi(x)$ everywhere between the plates.

5.5 Ambipolar Diffusion with an Ionization Source Near One Wall Consider ambipolar diffusion between two absorbing parallel plates separated by a distance l, with one plate located at $x = -l/2$ and the other plate located at $x = l/2$. Assume that $G = G_0$ electron–ion pairs per unit time per unit volume are created within the region to the left of the origin, $-l/2 < x < 0$, within the plates, and that $G = 0$ everywhere else within the plates. The ambipolar diffusion coefficient is $D_a = D_i(1 + T_e/T_i)$, where D_i is the ion diffusion coefficient, T_e is the electron temperature, and T_i is the ion temperature. Assume that D_i, T_e, and T_i are constants, with $T_i \ll T_e$.
(a) Find and sketch the electron density $n_e(x)$ everywhere between the plates. *You may assume that $n_e \approx 0$ at the surfaces of the two plates.*
(b) Give an expression for the ion flux $\Gamma_i(x)$ in terms of $n_e(x)$ and sketch $\Gamma_i(x)$ everywhere between the plates. What fraction of the created electron–ion pairs are lost to the right-hand wall $x = l/2$?
(c) Give an expression for the electric potential $\Phi(x)$ in terms of $n_e(x)$ and sketch $\Phi(x)$ everywhere between the plates.

5.6 Highly Collisional Ambipolar Diffusion
A highly collisional rf discharge is ignited between two parallel electrodes located at $x = \pm l/2$. Assume a constant ion-neutral collision frequency $\nu_{i,m}$. The steady-state diffusion equation

$$\frac{d^2n}{dx^2} + \beta^2 n = 0$$

with the boundary condition that $n(\pm l/2) = 0$ has the solution $n(x) = n_0 \cos(\pi x/l)$, where $\beta^2 = \nu_{iz}/D_a = (\pi/l)^2$, ν_{iz} is the electron–neutral ionization rate, and $D_a \approx \mu_i T_e$ is the ambipolar diffusion coefficient (T_e is in volts).
(a) Find the steady-state (dc) particle flux $\Gamma(x)$, ambipolar electric field $E(x)$, potential $\Phi(x)$, and total charge density $\rho(x)$. Sketch Γ, E, and Φ for $|x| \leq l/2$.
(b) Plot $\rho(x)/e$ and $n(x)$ *on the same graph* for $|x| \leq l/2$. Are the ambipolar solutions valid for $\rho(x)/e > n(x)$? Explain your answer.
(c) Taking the condition $\rho(x)/e = n(x)$ for the breakdown of quasineutrality and the onset of a sheath, with $x = l/2 - s$ and sheath width $s \ll l$, show that $s \approx (\lambda_{De}^2 l/\pi)^{1/3}$, where λ_{De} is the central Debye length.
(d) Show that $s \approx \lambda_{Ds}$, the Debye length at the sheath edge.
(e) Let the ion mobility at the sheath edge be the variable mobility relation (5.3.1), $\mu_{is} = 2e\lambda_{is}/\pi M|u_s|$, with λ_{is} the ion–neutral mean free path and u_s the ion speed at the sheath edge, show that $u_s = u_B(2\lambda_{is}/\pi\lambda_{Ds})^{1/2}$.
(f) Let the ion mobility be a constant, $\mu_i = e/M\nu_{mi}$, independent of ion speed, show that $u_s = u_B^2/\nu_{mi}\lambda_{Ds}$.

5.7 **Density at a Sheath Edge** For a constant ambipolar diffusion coefficient D_a and for a diffusion velocity equal to the Bohm velocity u_B at the sheath edge $x = l'/2$, show that the ratio of the density n_{sl} at the sheath edge to the density n_0 in the center of a plane-parallel discharge of length l' is

$$\frac{n_{sl}}{n_0} = \left[1 + \left(\frac{1}{\beta}\frac{u_B}{D_a}\right)^2\right]^{-1/2}$$

where β is given by (5.2.29).

5.8 **Diffusion in a Rectangular Box** Consider a high-pressure steady-state discharge confined inside of a rectangular box having edges of length a meters along x, b meters along y, and c meters along z. The center of the box is located at $x = 0$, $y = 0$, and $z = 0$. The plasma is created by a volume ionization $G = \nu_{iz} n_e$ and is lost to the walls by ambipolar diffusion with a constant ambipolar diffusion coefficient D_a. Here, ν_{iz} is the electron–neutral ionization frequency. Assume that the electron density n_e is n_0 in the center of the box and is zero on the walls.

(a) Find an expression for the density $n_e(x, y, z)$ inside the box.
(b) Find the relation between D_a, ν_{iz} and the dimensions of the box for your solution in (a) to be valid.
(c) Find the particle fluxes Γ flowing to each of the six walls.
(d) Find the total number of particles per second lost to the walls by integrating the particle fluxes Γ over the areas of the walls.
(e) Find the total number of particles per second created by ionization by integrating the volume generation rate G over the volume of the box. Your answer to parts (d) and (e) should be the same.

5.9 **Particle Balance for Diffusion in a Cylinder** Consider a high-pressure steady-state discharge confined inside of a cylindrical chamber of radius R and length l. The center of the chamber is located at $r = 0$ and $z = 0$. The plasma is created by a volume ionization $G = \nu_{iz} n_e$ and is lost to the walls by ambipolar diffusion with a constant ambipolar diffusion coefficient D_a. Assume that the electron density is n_0 in the center of the chamber and is zero on the walls. Then, the diffusion equation is given by (5.2.34), with the density $n(r, z)$ given by (5.2.37) and the fluxes Γ_{iz} and Γ_{ir} to the walls given by (5.2.39) and (5.2.40).

(a) Find the total number of particles per second created by ionization by integrating the volume generation rate G over the volume of the chamber.
(b) Find the total number of particles per second lost to the walls by integrating the particle fluxes Γ over the areas of the walls. Your answer to part (a) and (b) should be the same.

5.10 **Diffusive Decay in a Plasma Cylinder** Consider the diffusive decay of the plasma density in an infinite cylinder of radius R with a constant diffusion coefficient D. The density at time $t = 0$ is given by (5.2.33).

(a) Show that the time-dependent radial distribution can be expressed as a sum of Bessel functions and indicate how the amplitudes of the terms are determined.
(b) Show that late in time the decay is exponential in time and find the time constant τ for the decay in terms of R and D.

5.11 Cross-Field Diffusion with Negligible Cross-Field Potential Drop Neglecting the perpendicular mobility terms in (5.4.14) and (5.4.15), show that the diffusion equation reduces to (5.4.16) with $D_\perp = \mu_{\perp i} T_i$.

5.12 Diffusion in a Magnetic Field A plasma is generated in a cylindrical tube of radius R and length l in argon ($M_{Ar}/M_H = 40$) at $p = 3$ mTorr with a strong magnetic field $B_0 = 1$ kG along the axis of the tube.

(a) Assuming that the ambipolar diffusion coefficient along B_0 has been measured to be $D_a = C/p$(Torr), with $C = 10^4$ cm^2-Torr/s, and that the ambipolar ion drift velocity corresponds to an energy $\mathcal{E}_\parallel = 10$ V, calculate the mean free path of argon ions along B_0.

(b) Considering that the transverse ion velocity corresponds to a temperature $T_\perp = 1$ V, calculate the ion gyration radius and determine if the radial diffusion will be significant for $L = 30$ cm and $R = 10$ cm.

5.13 Random Walk Diffusion In a multiple-mirror device, which has been proposed for confining fusion plasmas, ions are injected into the central magnetic mirror and diffuse through a series of mirrors to the device ends. In the steady state, a flux Γ_0 flows out through each half of the machine. The density is a maximum in the center of the machine and falls linearly to n_{min} at each end. The axial diffusion mechanism is that an ion travels an axial distance $l_z = \lambda_i / R_m$, where λ_i is the ion mean free path and $R_m = B_{max}/B_{min}$ is the "mirror ratio." The ion remains trapped in the mirror for a time τ_i before again escaping axially in either direction. Assume that $l_z \gg l$, the length between mirrors, and that $\tau_i \gg \overline{v}_i / l_z$, the flight time between mirrors, and that the total device length $2L \gg l_z$.

(a) Derive an approximate one-dimensional diffusion equation for the ion transport in terms of the above parameters (electron effects are neglected) and find the axial diffusion coefficient D_z.

(b) The density falls to n_{min} at $z = \pm L$. Solve the diffusion equation for the central density n_0 as a function of Γ_0, n_{min}, D_z, and L.

5.14 Diffusion in a Magnetized Plasma Solve (5.4.16) in the steady state with a source term $\nu_{iz} n$ and boundary conditions that $n = 0$ at the rectangular walls $x = \pm d/2$ and $z = \pm l/2$. Find ν_{iz} as a function of D_a, $D_{\perp i}$, d, and l.

6

DC Sheaths

6.1 Basic Concepts and Equations

At the edge of a bounded plasma, a potential exists to contain the more mobile charged species. This allows the flow of positive and negative carriers to the wall to be balanced. In the usual situation of an electropositive plasma, consisting of equal numbers of positive ions and electrons, the electrons are far more mobile than the ions. The plasma will therefore charge positively with respect to a grounded wall. The nonneutral potential region between the plasma and the wall is called a *sheath.*

In a weakly ionized plasma, the energy to sustain the plasma is generally by heating of the electrons by the source, while the ions are at near equilibrium with the background gas. The electron temperature is then typically of few volts, while the ions are cold. In this situation, we may think of monoenergetic ions being accelerated through the sheath potential, while the electron density decreases according to a Boltzmann factor, as described in Section 2.4. The electron density would then decay on the order of a Debye length λ_{De}, to shield the electrons from the wall. However, we cannot linearize the Poisson equation, as we did in deriving λ_{De} in Section 2.4, if we wish to obtain the exact flux balance. Furthermore, we will show that a transition layer or *presheath* must exist between the neutral plasma and the nonneutral sheath in order to maintain the continuity of ion flux, giving rise to an ion velocity at the plasma–sheath edge known as the *Bohm velocity* u_B. The need for this presheath will arise naturally in our derivation in Section 6.2. If a potential is placed between bounding electrodes, then, while the overall flux balance is maintained, each electrode may separately draw current. The most straightforward analysis is of a boundary with a large negative potential with respect to the plasma. The simplest example is a uniform ion charge density, or *matrix sheath.* This occurs in the cathode sheath of a dc discharge, for example, considered in Section 14.3. A matrix sheath is also created transiently with a pulsed negative electrode voltage in which the electrons are expelled from a plasma region, leaving a uniform ion density behind. This occurs naturally in *plasma immersion ion implantation,* discussed in Chapter 17. We consider the matrix sheath in Section 6.3.

For a high-voltage sheath, the current to the electrode is almost all ion current. Provided the ion motion in the sheath is collisionless, then the steady self-consistent ion density is not uniform, but rather is described by the *Child–Langmuir law* of *space-charge-limited current* in a planar diode. We also discuss this situation in Section 6.3.

The idealized conditions described in Sections 6.2 and 6.3 are not always met. The temperature of the ions cannot always be ignored with respect to the electron temperature. This situation arises, for example, in highly ionized plasmas. In this case, more complicated kinetic treatments

Principles of Plasma Discharges and Materials Processing, Third Edition. Michael A. Lieberman and Allan J. Lichtenberg.

are required. In a similar vein, the electron distribution may not be Maxwellian. This may arise due to particular heating or loss mechanisms, which occur, for example, in low-pressure capacitive rf plasmas, discussed in Chapter 11. In this situation, the decrease in electron density in the sheath is not given by a Boltzmann factor but must be obtained kinetically. If the neutral gas is electronegative, such that electron attachment is significant, then the negative charges divide between electrons and negative ions. If the fraction of negative ions present becomes large, the mobility of the negative charges can be greatly reduced, changing the conditions at the sheath edge. We consider these various topics, which, in fact, have some unity of analysis, in Section 6.4. Electronegative plasmas are of considerable importance in processing applications, and their analysis is described in Chapter 10.

Other situations that differ from the basic theory arise due to collisional effects in the sheath region. In this case, the ion flow is impeded by collisional processes with neutrals, and the transport is mobility rather than inertia limited, similar to that already described in Chapter 5. We discuss two simple limiting collisional cases in Section 6.5. A full treatment, including both inertial and collisional effects, is very complicated, requiring numerical solution of the kinetic equations.

This chapter deals with sheaths that are constant in time. Two other interesting cases are sheaths formed in oscillating rf potentials and sheaths formed transiently by pulsed potentials. In both situations, approximate solutions can be obtained if there is a separation of time scales such that electrons respond rapidly to the time variation while ions respond slowly. This separation is characterized by the inequalities

$$f_{pe} \gg \frac{1}{\tau} \gg f_{pi} \tag{6.1.1}$$

where τ is the time scale of field variation ($\tau = 2\pi/\omega$ for an oscillatory variation) and f_{pe} and f_{pi} are the electron and ion plasma frequencies, respectively. An oscillatory potential applied to an electrode is characteristic of a capacitively excited rf discharge, and we consider this sheath in Chapter 11. The pulsed potential sheath is analyzed in Chapter 17. A review of sheath physics in laboratory and space plasmas is given in Robertson (2013).

6.1.1 The Collisionless Sheath

We use the assumptions (1) Maxwellian electrons at temperature $\mathrm{T_e}$, (2) cold ions ($\mathrm{T_i} = 0$), and (3) $n_e(0) = n_i(0)$ at the plasma–sheath interface (interface between essentially neutral and nonneutral regions) at $x = 0$. As shown in Figure 6.1, we define the zero of the potential Φ at $x = 0$ and we take the ions to have a velocity u_s there. Ion energy conservation (no collisions) then gives

$$\frac{1}{2}Mu^2(x) = \frac{1}{2}Mu_s^2 - e\Phi(x) \tag{6.1.2}$$

The continuity of ion flux (no ionization in the sheath) is

$$n_i(x)u(x) = n_{is}u_s \tag{6.1.3}$$

where n_{is} is the ion density at the sheath edge. Solving for u from (6.1.2) and substituting into (6.1.3), we have

$$n_i = n_{is}\left(1 - \frac{2e\Phi}{Mu_s^2}\right)^{-1/2} \tag{6.1.4}$$

The electron density is given by the Boltzmann relation

$$n_e(x) = n_{es}\mathrm{e}^{\Phi(x)/\mathrm{T_e}} \tag{6.1.5}$$

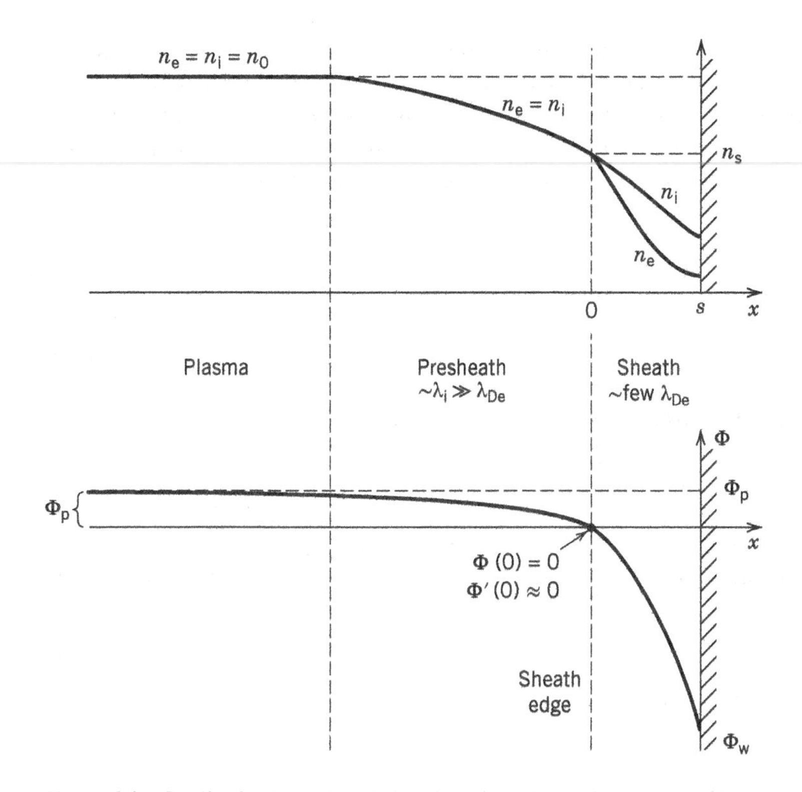

Figure 6.1 Qualitative behavior of sheath and presheath in contact with a wall.

Setting $n_{es} = n_{is} \equiv n_s$ at the sheath edge and substituting n_i and n_e into Poisson's equation

$$\frac{d^2\Phi}{dx^2} = \frac{e}{\epsilon_0}(n_e - n_i)$$

we obtain

$$\frac{d^2\Phi}{dx^2} = \frac{en_s}{\epsilon_0}\left[\exp\frac{\Phi}{T_e} - \left(1 - \frac{\Phi}{\mathcal{E}_s}\right)^{-1/2}\right] \tag{6.1.6}$$

where $e\mathcal{E}_s = \frac{1}{2}Mu_s^2$ is the initial ion energy. Equation (6.1.6) is the basic nonlinear equation governing the sheath potential and ion and electron densities. However, as we shall see in the next section, it has stable solutions only for sufficiently large u_s, created in an essentially neutral *presheath* region.

6.2 The Bohm Sheath Criterion

A first integral of (6.1.6) can be obtained by multiplying (6.1.6) by $d\Phi/dx$ and integrating over x:

$$\int_0^{\Phi} \frac{d\Phi}{dx}\frac{d}{dx}\left(\frac{d\Phi}{dx}\right)dx = \frac{en_s}{\epsilon_0}\int_0^{\Phi}\frac{d\Phi}{dx}\left[\exp\frac{\Phi}{T_e} - \left(1 - \frac{\Phi}{\mathcal{E}_s}\right)^{-1/2}\right]dx \tag{6.2.1}$$

Canceling the dxs and integrating with respect to Φ, we obtain

$$\frac{1}{2}\left(\frac{d\Phi}{dx}\right)^2 = \frac{en_s}{\epsilon_0}\left[T_e\exp\frac{\Phi}{T_e} - T_e + 2\mathcal{E}_s\left(1 - \frac{\Phi}{\mathcal{E}_s}\right)^{1/2} - 2\mathcal{E}_s\right] \tag{6.2.2}$$

where we have set $\Phi = 0$ and $\mathrm{d}\Phi/\mathrm{d}x = 0$ at $x = 0$ corresponding to a field free plasma. Equation (6.2.2) can be integrated numerically to obtain $\Phi(x)$. However, it is apparent that the RHS of (6.2.2) should be positive for a solution to exist. Physically this means that the electron density must always be less than the ion density in the sheath region. Since we expect this to be a problem only for small Φ, we expand the RHS of (6.2.2) to second order in a Taylor series to obtain the inequality

$$\frac{1}{2}\frac{\Phi^2}{\mathrm{T_e}} - \frac{1}{4}\frac{\Phi^2}{\mathcal{E}_\mathrm{s}} \geq 0 \tag{6.2.3}$$

We see that (6.2.3) is satisfied for $\mathcal{E}_\mathrm{s} \geq \mathrm{T_e}/2$ or, substituting for $\mathcal{E}_\mathrm{s}$,

$$u_\mathrm{s} \geq u_\mathrm{B} = \left(\frac{e\mathrm{T_e}}{M}\right)^{1/2} \tag{6.2.4}$$

This result is known as the *Bohm sheath criterion.*

6.2.1 Plasma Requirements

The condition (6.2.4) that $u_\mathrm{s} \geq u_\mathrm{B}$ for a collisionless sheath to form is complemented by an opposing condition that $u < u_\mathrm{B}$ in the quasineutral bulk plasma. To see this, we examine the quasineutral equilibrium

$$n_\mathrm{e} = n_\mathrm{i} \equiv n \tag{6.2.5}$$

in a plane-parallel discharge. We use ion conservation (2.3.7)

$$n\frac{\mathrm{d}u_\mathrm{i}}{\mathrm{d}x} + u_\mathrm{i}\frac{\mathrm{d}n}{\mathrm{d}x} = G \tag{6.2.6}$$

ion momentum conservation (2.3.9)

$$Mnu_\mathrm{i}\frac{\mathrm{d}u_\mathrm{i}}{\mathrm{d}x} = enE + \mathrm{f_c} \tag{6.2.7}$$

and the Boltzmann relation (2.4.13) for electrons

$$enE + kT_\mathrm{e}\frac{\mathrm{d}n}{\mathrm{d}x} = 0 \tag{6.2.8}$$

Here, G is the rate of production of electron–ion pairs per unit volume and $\mathrm{f_c}$ is the collisional force per unit volume acting on the ions.

Solving (6.2.8) for E and substituting into (6.2.7), we obtain

$$nu_\mathrm{i}\frac{\mathrm{d}u_\mathrm{i}}{\mathrm{d}x} + u_\mathrm{B}^2\frac{\mathrm{d}n}{\mathrm{d}x} = \frac{\mathrm{f_c}}{M} \tag{6.2.9}$$

Solving (6.2.6) and (6.2.9) for the derivatives $\mathrm{d}n/\mathrm{d}x$ and $\mathrm{d}u_\mathrm{i}/\mathrm{d}x$, we find

$$\frac{\mathrm{d}n}{\mathrm{d}x} = \frac{\mathrm{f_c}/M - Gu_\mathrm{i}}{u_\mathrm{B}^2 - u_\mathrm{i}^2} \tag{6.2.10}$$

and

$$n\frac{\mathrm{d}u_\mathrm{i}}{\mathrm{d}x} = \frac{Gu_\mathrm{B}^2 - u_\mathrm{i}\mathrm{f_c}/M}{u_\mathrm{B}^2 - u_\mathrm{i}^2} \tag{6.2.11}$$

Because $\mathrm{f_c}$ is always negative, we see that both derivatives become singular as $u_\mathrm{i} \to u_\mathrm{B}$. Since $u_\mathrm{i} = 0$ in the center of the discharge (by symmetry) and increases as we move toward the walls, we see that the quasineutral bulk solution can break down near the walls where $u_\mathrm{i} \to u_\mathrm{B}$. For the limiting case of a collisionless sheath (and bulk plasma), the quasineutral plasma (having $u_\mathrm{i} < u_\mathrm{B}$) joins the collisionless sheath (having $u_\mathrm{i} \geq u_\mathrm{B}$) exactly at $u_\mathrm{s} = u_\mathrm{B}$.

6.2.2 The Presheath

Although derived above for a plane-parallel equilibrium, the Bohm condition has a more general validity. To give the ions the directed velocity u_B, there must be a finite electric field in the plasma over some region, typically much wider than the sheath, called the *presheath* (see Figure 6.1). Hence, the presheath region is not strictly field free, although E is very small there. At the sheath–presheath Interface, there is a transition from subsonic ($u_i < u_B$) to supersonic ($u_i > u_B$) ion flow, where the condition of charge neutrality must break down. The transition can arise from geometric contraction of the plasma, from ion friction forces in the presheath, or from ionization in the bulk plasma (Riemann, 1991). Putting in specific values of momentum mean free path, ionization, or geometric contraction, the presheath equations can be solved analytically. This has been done, for example, for (a) a geometric presheath with current contraction onto a spherical probe, (b) a plane-parallel collisional presheath, and (c) an ionizing presheath with the ionization proportional to n_e. These solutions are plotted in Figure 6.2. They show quite different behavior in the plasma region: the geometric presheath (a) relaxes to the undisturbed (field free) plasma, the collisional presheath (b) tends to a logarithmic potential shape (see below), indicating that the ion transport requires a residual plasma field, and the ionizing presheath (c) ends with zero field at a finite point representing the midplane of a symmetric plasma. For (b) or (c), the presheath width is of order the mean free path for ion–neutral collisions or for electron–neutral ionization, respectively. Despite the differences, all solutions run quite similarly into the singularity $u_i = u_B$ at the sheath edge. The growing field inhomogeneity approaching this singularity indicates the formation of space charge and the breakdown of the quasineutral approximation.

The potential drop across a collisionless presheath, which accelerates the ions to the Bohm velocity, is given by

$$\frac{1}{2}Mu_B^2 = e\Phi_p$$

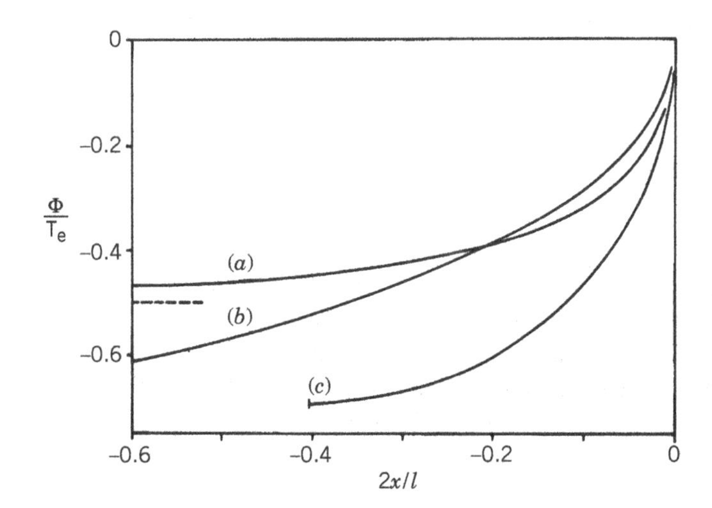

Figure 6.2 Φ/T_e versus position within the presheath, showing (a) the geometric presheath; (b) a planar collisional presheath; and (c) a planar ionization presheath. The sheath–presheath edge is at the right. Source: Riemann, 1991/IOP Publishing.

where Φ_p is the plasma potential with respect to the potential at the sheath–presheath edge. Substituting for the Bohm velocity from (6.2.4), we find

$$\Phi_p = \frac{T_e}{2} \tag{6.2.12}$$

This is shown as the dashed line in Figure 6.2. The spatial variation of the potential $\Phi_p(x)$ in a collisional presheath has been estimated by Riemann (1991) to be determined from

$$\frac{1}{2} - \frac{1}{2}\exp\left(-\frac{2\Phi_p}{T_e}\right) - \frac{\Phi_p}{T_e} = \frac{x}{\lambda_i}$$

where x is the distance from the Bohm point at the presheath–sheath edge and λ_i is the ion–neutral mean free path. The ratio of the density at the sheath edge to that in the plasma is then found from the Boltzmann relation

$$n_s = n_b e^{-\Phi_p/T_e} \approx 0.61 n_b \tag{6.2.13}$$

where n_b is the density where the presheath and bulk plasma join.

6.2.3 Sheath Potential at a Floating Wall

It is quite straightforward to determine the potential drop within the sheath between a plasma and a floating wall. We equate the ion flux (assumed constant through the sheath)

$$\Gamma_i = n_s u_B \tag{6.2.14}$$

to the electron flux at the wall

$$\Gamma_e = \frac{1}{4} n_s \bar{v}_e e^{\Phi_w/T_e} \tag{6.2.15}$$

where $\bar{v}_e = (8eT_e/\pi m)^{1/2}$ is the mean electron speed and Φ_w is the potential of the wall with respect to the sheath–presheath edge. We have, after substituting for the Bohm velocity from (6.2.4),

$$n_s\left(\frac{eT_e}{M}\right)^{1/2} = \frac{1}{4} n_s \left(\frac{8eT_e}{\pi m}\right)^{1/2} e^{\Phi_w/T_e} \tag{6.2.16}$$

Solving for Φ_w, we obtain

$$\Phi_w = -T_e \ln\left(\frac{M}{2\pi m}\right)^{1/2} \tag{6.2.17}$$

The wall potential Φ_w is negative and is related linearly to T_e with a factor proportional to the logarithm of the square root of the mass ratio. For hydrogen, for example, $\ln(M/2\pi m)^{1/2} \approx 2.8$, while for argon ($M = 40$ amu), the factor is 4.7. Thus, argon ions with initial energy $\mathcal{E}_s = T_e/2$ at the sheath–presheath edge that fall through a collisionless dc sheath to a floating wall would bombard the wall with an energy of $\mathcal{E}_i \approx 5.2T_e$. Of course, electrodes that have potentials on them, either dc or rf, can be bombarded with much higher energy, but these electrodes must draw a substantial net current, as we will show in Section 6.3.

The sheath width s is found by integrating (6.2.2) to obtain $\Phi(x)$ and setting $\Phi(s) = \Phi_w$, with Φ_w given by (6.2.17). The integral must be done numerically. Typical sheath widths are a few electron Debye lengths λ_{Ds}, where λ_{Ds} is the electron Debye length at the plasma–sheath edge.

6.2.4 Collisional Sheaths

As we have seen, for the collisionless case, a unique Bohm velocity can be defined at the position where the quasi-neutral presheath solution becomes singular. For weakly collisional plasmas, a unique edge position is not exactly defined, and approximate methods of separating the plasma and sheath regions become more subtle. The true behavior is quite complicated at this interface. For more details, including a kinetic treatment, the reader is referred to a review paper by Riemann (1991). Numerical solutions, including Poisson's equation in both the plasma and the sheath, have also been performed, e.g., Su and Lam (1963), Franklin and Ockendon (1970), Godyak and Sternberg (1990a), and Riemann (1997) but are not easy to apply to complete discharge calculations. For weakly collisional plasmas, the presheath scale length is the ion–neutral mean free path λ_i, and the sheath thickness is, as for the collisionless case, a few Debye lengths λ_{Ds}, with $\lambda_i \gg \lambda_{Ds}$. In this case, the presheath and sheath scale lengths are well-separated, and both theory and numerical calculations indicate that there is an intermediate length scale $\lambda_i^{1/5}\lambda_{De}^{4/5}$ over which the transition occurs from the presheath to the sheath. The ion drift speed in this region lies somewhat below the Bohm speed.

For highly collisional plasmas with $\lambda_i \lesssim \lambda_{Ds}$, the ion motion is mobility-limited, $u_i \approx \mu_i E$, the intermediate presheath region disappears, and the analysis of Problem 5.6 applies. As shown there, the bulk plasma quasineutrality breaks down at a sheath width

$$s \approx K(\lambda_{De}^2 l)^{1/3} \approx (\pi K^3)^{1/2} \lambda_{Ds} \tag{6.2.18}$$

with λ_{De} and λ_{Ds} the central and edge Debye lengths, l the plate separation, and K a coefficient of order unity (Blank, 1968). Franklin and Snell (2000c) give $K \approx 2.2 + 0.125\log_{10}(\nu_{iz}/\nu_{mi})$, with ν_{iz} the ionization frequency and ν_{mi} the ion–neutral momentum transfer frequency.

As shown in Problem 5.6(e), for the variable mobility relation $\mu_i = 2e\lambda_i/\pi M|u_i|$ given in (5.3.1), the ion speed at the sheath edge lies below the Bohm speed (see also Franklin, 2002)

$$u_s \approx u_B(C\lambda_i/\lambda_{Ds})^{1/2} \tag{6.2.19}$$

with C a coefficient of order unity. Using an electric field

$$E_s = \frac{T_e}{\lambda_{Ds}} \tag{6.2.20}$$

at the plasma–sheath edge for the breakdown of quasi-neutrality Godyak and Sternberg (1990a), find $C = 2/\pi$, and they give a heuristic expression for u_s over the range of mean free paths

$$u_s \approx \frac{u_B}{(1 + \pi\lambda_{Ds}/2\lambda_i)^{1/2}} \tag{6.2.21}$$

Using (6.2.20) for a constant mobility $\mu_i = e/(M\nu_{mi})$, independent of ion speed, the corresponding result, also lying below the Bohm speed, is

$$u_s = \frac{1}{2}\left[(\nu_{mi}^2\lambda_{Ds}^2 + 4u_B^2)^{1/2} - \nu_{mi}\lambda_{Ds}\right] \tag{6.2.22}$$

(see also Problem 5.6(f) for the highly collisional limit).

Since the bulk plasma and sheath regions merge, the exact position of the sheath is a matter of definition (Franklin, 2004). However, for the situations of most interest in this book, the exact position and ion drift speed for the plasma–sheath transition are not that important. For equilibrium calculations, the ion flux Γ_i is the main quantity of physical interest, and its decomposition into $\Gamma_i = n_s u_B$, the product of a density and a flow velocity at a "sheath edge," is not of great importance in determining the behavior in the "sheath" or the "bulk plasma." For the sheath, it is the ion flux entering the sheath that mainly determines the sheath properties. For the bulk plasma, as we saw in (5.2.3), the Bohm velocity is also used as part of the boundary conditions to determine the diffusion solution. However, for collisional plasmas, with $\lambda_i \ll l$, the plasma size, the diffusion solutions become quite insensitive to the edge ion drift speed, and the simpler boundary condition (5.2.1) can be used.

6.2.5 Simulation Results

Particle-in-cell (PIC) simulations can illustrate some of the phenomena we have described, as well as introduce some new features. Figure 6.3 shows a simulation of sheath formation during the decay of a warm, initially uniform density electron–proton plasma between short-circuited parallel plates (no source). The initial plasma parameters are $T_e = T_i = 1$ V and $n_0 = 10^8$ cm^{-3}, with $p =$ 50 mTorr, $l = 1$ cm, and an ion–neutral momentum transfer cross section $\sigma_{mi} = 5 \times 10^{-15}$ cm^2. For these parameters, $\lambda_{De} \approx 0.074$ cm, $f_{pe}^{-1} \approx 1.11 \times 10^{-8}$ s, $D_a \approx 1.5 \times 10^5$ cm^2/s, and the fundamental diffusion mode timescale is $\tau_0 \approx 0.68 \times 10^{-6}$ s. The density, field, and potential profiles are shown in (*a*), (*b*), and (*c*) at $t = 4 \times 10^{-8}$ s, after the sheaths have partially formed, but before the decay of the higher-order ($i > 1$) diffusion modes. Hence, the ion density in (*a*) is relatively uniform in the bulk plasma rather than the cosine variation given in (5.2.7), and the steady-state sheaths have not fully formed due to ion transit timescale effects. However, we clearly see the sheath formation. The midpotential variation with time is shown on a short timescale in (*d*), illustrating its formation with

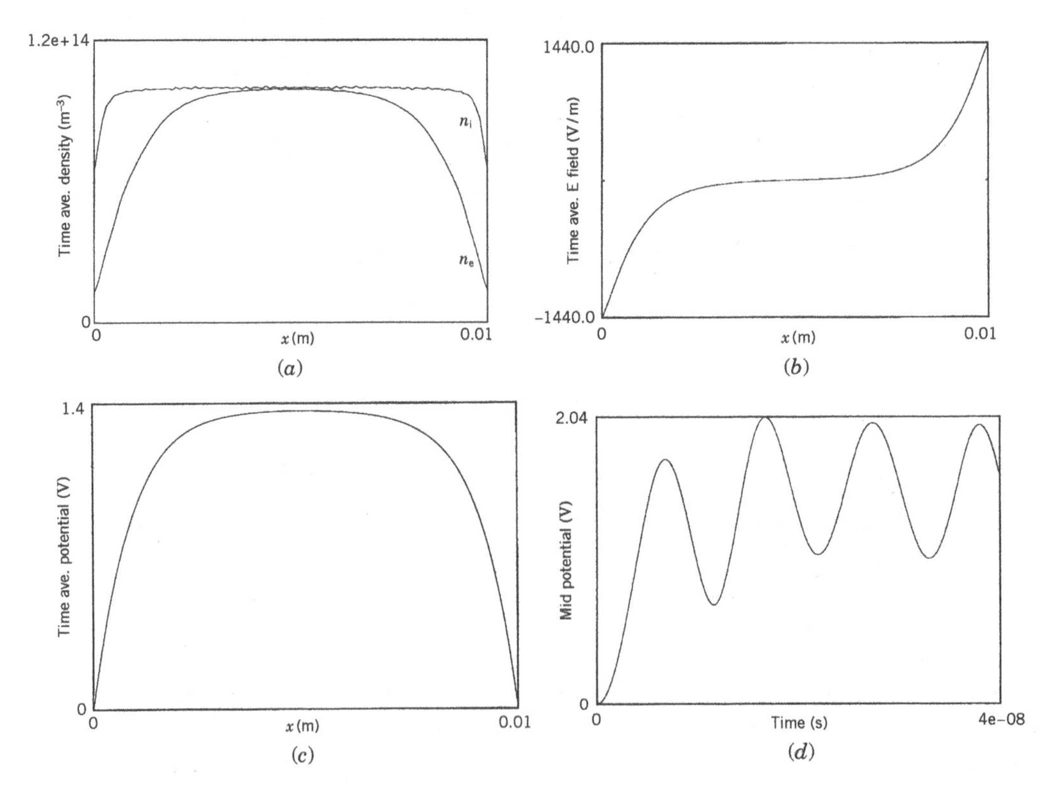

Figure 6.3 Particle-in-cell simulation showing sheath formation from warm, initially uniform electron–proton plasma between short-circuited parallel plates: (a) density profiles at time $t = 4 \times 10^{-8}$ s; (b) electric field profile; (c) potential profile; (d) midpotential versus time.

$\Phi_{max} \sim T_e$ as the sheaths form on the very fast electron timescale f_{pe}^{-1}, along with accompanying electron plasma oscillations, as noted previously for Figure 1.12*f*.

6.3 The High-Voltage Sheath

6.3.1 Matrix Sheath

Sheath voltages are often driven to be very large compared to T_e. The potential Φ in these sheaths is highly negative with respect to the plasma–sheath edge; hence, $n_e \sim n_s e^{\Phi/T_e} \to 0$ and only ions are present in the sheath. The simplest high-voltage sheath, with a uniform ion density, is known as a *matrix sheath*. Let $n_i = n_s$ = const within the sheath of thickness s and choosing $x = 0$ at the plasma–sheath edge, then from (2.2.3),

$$\frac{dE}{dx} = \frac{en_s}{\epsilon_0} \tag{6.3.1}$$

which yields a linear variation of E with x:

$$E = \frac{en_s}{\epsilon_0}x \tag{6.3.2}$$

Integrating $d\Phi/dx = -E$, we obtain a parabolic profile

$$\Phi = -\frac{en_s}{\epsilon_0}\frac{x^2}{2} \tag{6.3.3}$$

Setting $\Phi = -V_0$ at $x = s$, we obtain the matrix sheath thickness

$$s = \left(\frac{2\epsilon_0 V_0}{en_s}\right)^{1/2} \tag{6.3.4}$$

In terms of the electron Debye length $\lambda_{De} = (\epsilon_0 T_e/en_s)^{1/2}$ at the sheath edge, we see that

$$s = \lambda_{De}\left(\frac{2V_0}{T_e}\right)^{1/2} \tag{6.3.5}$$

Hence, the sheath thickness can be tens of Debye lengths.

6.3.2 Child Law Sheath

In the steady state, the matrix sheath is not self-consistent since it does not account for the decrease in ion density as the ions accelerate across the sheath. In the limit that the initial ion energy $\mathcal{E}_s$ is small compared to the potential, the ion energy and flux conservation equations (6.1.2) and (6.1.3) reduce to

$$\frac{1}{2}Mu^2(x) = -e\Phi(x) \tag{6.3.6}$$

$$en(x)u(x) = J_0 \tag{6.3.7}$$

where J_0 is the constant ion current. Solving for $n(x)$, we obtain

$$n(x) = \frac{J_0}{e}\left(-\frac{2e\Phi}{M}\right)^{-1/2} \tag{6.3.8}$$

Using this in Poisson's equation, we have

$$\frac{d^2\Phi}{dx^2} = -\frac{J_0}{\epsilon_0}\left(-\frac{2e\Phi}{M}\right)^{-1/2} \tag{6.3.9}$$

Multiplying (6.3.9) by $d\Phi/dx$ and integrating from 0 to x, we have

$$\frac{1}{2}\left(\frac{d\Phi}{dx}\right)^2 = 2\frac{J_0}{\epsilon_0}\left(\frac{2e}{M}\right)^{-1/2}(-\Phi)^{1/2} \tag{6.3.10}$$

where we have chosen $d\Phi/dx = -E = 0$ at $\Phi = 0$ ($x = 0$). Taking the (negative) square root (since $d\Phi/dx$ is negative) and integrating again, we obtain

$$(-\Phi)^{3/4} = \frac{3}{2}\left(\frac{J_0}{\epsilon_0}\right)^{1/2}\left(\frac{2e}{M}\right)^{-1/4} x \tag{6.3.11}$$

Let $\Phi = -V_0$ at $x = s$ and solving for J_0, we obtain

$$J_0 = \frac{4}{9}\epsilon_0\left(\frac{2e}{M}\right)^{1/2}\frac{V_0^{3/2}}{s^2} \tag{6.3.12}$$

Equation (6.3.12) is the well-known Child law of space–charge-limited current in a plane diode. With fixed spacing s, it gives the current between two electrodes as a function of the potential difference between them, and has been traditionally used for electron diodes. However, with J_0 given explicitly as

$$J_0 = en_s u_B \tag{6.3.13}$$

in (6.3.12), we have a relation between the sheath potential, the sheath thickness, and the plasma parameters, which can be used to determine the sheath thickness s. Substituting (6.3.13) into (6.3.12) and introducing the electron Debye length at the sheath edge, we obtain

$$s = \frac{\sqrt{2}}{3}\lambda_{De}\left(\frac{2V_0}{T_e}\right)^{3/4} \tag{6.3.14}$$

Comparing this to the matrix sheath width, we see that the Child law sheath is larger by a factor of order $(V_0/T_e)^{1/4}$. The Child law sheath can be of order of 100 Debye lengths (~1 cm) in a typical processing discharge. Since there are no electrons within the sheath to excite the gas, the sheath region appears dark when observed visually.

Inserting (6.3.12) into (6.3.11) yields the potential within the sheath as a function of position

$$\Phi = -V_0\left(\frac{x}{s}\right)^{4/3} \tag{6.3.15}$$

The electric field $E = -d\Phi/dx$ is

$$E = \frac{4}{3}\frac{V_0}{s}\left(\frac{x}{s}\right)^{1/3} \tag{6.3.16}$$

and the ion density $n = (\epsilon_0/e)\, dE/dx$ is

$$n = \frac{4}{9}\frac{\epsilon_0}{e}\frac{V_0}{s^2}\left(\frac{x}{s}\right)^{-2/3} \tag{6.3.17}$$

We see that n is singular as $x \to 0$, a consequence of the simplifying assumption in (6.3.6) that the initial ion energy $\mathcal{E}_s = 0$. The analysis can be carried through for a finite $e\mathcal{E}_s = \frac{1}{2}Mu_B^2$, using (6.1.2), resolving the singularity and yielding $n \to n_s$ as $x \to 0$ (Problem 6.1).

The ion motion within the sheath can be determined using conservation of energy (6.3.6). Assuming that an ion enters the sheath with initial velocity $u(0) = 0$, we insert (6.3.15) into (6.3.6) and solve for $u = \mathrm{d}x/\mathrm{d}t$ to obtain

$$\frac{\mathrm{d}x}{\mathrm{d}t} = v_0 \left(\frac{x}{s}\right)^{2/3} \tag{6.3.18}$$

with

$$v_0 = \left(\frac{2eV_0}{M}\right)^{1/2} \tag{6.3.19}$$

the characteristic ion velocity in the sheath. Integrating (6.3.18) yields

$$\frac{x(t)}{s} = \left(\frac{v_0 t}{3s}\right)^3 \tag{6.3.20}$$

Setting $x = s$ in (6.3.20), we obtain the ion transit time across the sheath:

$$\tau_\mathrm{i} = \frac{3s}{v_0} \tag{6.3.21}$$

The Child law solution is valid if the sheath potentials are large compared to the electron temperature. It is therefore not appropriate for use where the sheath potential is the potential between a plasma and a floating electrode. However, with some modification, we shall see in Chapter 12 that it is useful in determining the sheath width of an rf-driven discharge. Because the ion motion was assumed collisionless, it is also not appropriate for higher pressure discharges. We shall treat collisional formulations of the sheath region in Section 6.5.

6.4 Generalized Criteria for Sheath Formation

Using a kinetic treatment without ion collisions, the Bohm criterion for a stable sheath can be generalized to arbitrary ion and electron distributions. First formulated by Boyd and Thompson (1959), a more rigorous and complete treatment in the limit $\lambda_\mathrm{De} \to 0$ can be found in Riemann (1991). The result is

$$\frac{e\mathrm{T_e}}{M} \int_0^\infty \frac{1}{v^2} f(v)\,\mathrm{d}v \le \mathrm{T_e} \left.\frac{\mathrm{d}(n_\mathrm{e} + n_-)}{\mathrm{d}\Phi}\right|_{\Phi=0} \tag{6.4.1}$$

where $f(v)$ is the one-dimensional speed distribution of the positive ions, $n_\mathrm{e} + n_-$ is the sum of the densities of the negatively charged species, and Φ is the potential, with $\Phi = 0$ at the sheath–presheath edge. For our previous case of cold ions and Maxwellian electrons, (6.4.1) becomes

$$\frac{e\mathrm{T_e}}{M} \int_0^\infty \frac{1}{v^2} \delta(v - u_\mathrm{s})\,\mathrm{d}v \le \mathrm{T_e} \left.\frac{\mathrm{d}}{\mathrm{d}\Phi}\left(\mathrm{e}^{\Phi/\mathrm{T_e}}\right)\right|_{\Phi=0} \tag{6.4.2}$$

where $\delta(v - u_\mathrm{s})$ is the Dirac δ function. Evaluating the integral on the left and taking the derivative on the right, we have

$$\frac{e\mathrm{T_e}}{M} \frac{1}{u_\mathrm{s}^2} \le 1$$

or

$$u_\mathrm{s} \ge \left(\frac{e\mathrm{T_e}}{M}\right)^{1/2} = u_\mathrm{B}$$

which is the Bohm criterion from (6.2.4).

The more general form can be calculated for finite temperature ion distributions, but can lead to mathematical difficulties at low energies due to the average over $1/v^2$. Non-Maxwellian electron distributions, such as power-law distributions that can arise from stochastic rf heating (Chapter 11), can also lead to mathematical singularities. In physical devices, however, collisional processes at low energies generally allow nonsingular solutions to exist.

6.4.1 Electronegative Gases

A physical situation in which (6.4.1) is particularly useful is for electronegative gases in which electron attachment allows a significant number of negative ions to be present. This situation was treated by Boyd and Thompson (1959) and we follow their approach here. The Poisson equation for the potential is

$$\nabla^2\Phi = -\frac{e}{\epsilon_0}\left(n_+ - n_e - n_-\right) \tag{6.4.3}$$

where n_+, n_e, and n_- are the positive ion, electron, and negative ion densities, respectively. At the sheath edge, we use quasineutrality, $n_{s+} = n_{se} + n_{s-}$, and define the ratio of negative ions to electrons as $\alpha_s \equiv n_{s-}/n_{se}$. Then, the quasineutral condition becomes

$$n_{s+} = \left(1 + \alpha_s\right) n_{se} \tag{6.4.4}$$

If we further consider that the electron and negative ion distributions are Maxwellian, with a temperature ratio $T_e/T_i \equiv \gamma$, then for cold positive ions, we can directly repeat the calculation in Section 6.2 to obtain a new Bohm criterion (Problem 6.2). Here, we use the more general expression (6.4.1). The Boltzmann relation for electrons and negative ions gives

$$n_e + n_- = n_{se}e^{\Phi/T_e} + \alpha_s n_{se}e^{\gamma\Phi/T_e} \tag{6.4.5a}$$

which combined with (6.4.4) gives

$$n_e + n_- = \frac{n_{s+}}{1+\alpha_s}\left(e^{\Phi/T_e} + \alpha_s e^{\gamma\Phi/T_e}\right) \tag{6.4.5b}$$

Taking a derivative of (6.4.5b) with respect to Φ, and evaluating at $\Phi = 0$, on the RHS of (6.4.1), the equation becomes

$$\frac{eT_e}{M}\int_0^\infty \frac{1}{v^2} f(v)\,dv \le n_{s+}\left(\frac{1+\alpha_s\gamma}{1+\alpha_s}\right) \tag{6.4.6}$$

For cold positive ions, evaluating the integral as in (6.4.2) and taking the reciprocal, we have

$$u_s \ge \left[\frac{eT_e\left(1+\alpha_s\right)}{M\left(1+\alpha_s\gamma\right)}\right]^{1/2} \tag{6.4.7}$$

which is the generalization of the Bohm criterion (6.2.4) for an electronegative plasma. It is immediately apparent that, if γ is large and α_s not too small, the negative ions strongly reduce the velocity required at the sheath edge. However, in this situation, the positive ion temperature cannot be ignored and the left-hand side of (6.4.6) must be integrated over the ion distribution.

This is not the end of the story, because the potentials in the bulk plasma and presheath regions will repel the colder negative ions, thus reducing α_s at the sheath edge as compared to $\alpha_b \equiv n_{b-}/n_{be}$ where the presheath and bulk plasma join, thus increasing the importance of the electrons in the sheath region. If Φ_p is the potential at this position with respect to the sheath–presheath edge, then using the Boltzmann relation for both electrons and negative ions, $n_{se} = n_{be} \exp(-\Phi_p/T_e)$, $n_{s-} = n_{b-} \exp(-\gamma\Phi_p/T_e)$, we combine these expressions with the definition of α to obtain

$$\alpha_s = \alpha_b \exp\left[\frac{\Phi_p(1-\gamma)}{T_e}\right] \tag{6.4.8}$$

We have previously found that for $\alpha_s = 0$ (electrons only) that $\Phi_p/T_e = 1/2$. Using the same argument of conservation of ion energy, we obtain (Problem 6.3)

$$\frac{\Phi_p}{T_e} = \frac{1+\alpha_s}{2(1+\gamma\alpha_s)} \tag{6.4.9}$$

Substituting (6.4.9) into (6.4.8), we can solve explicitly for α_b:

$$\alpha_b = \alpha_s \exp\left[\frac{(1+\alpha_s)(\gamma-1)}{2(1+\gamma\alpha_s)}\right] \tag{6.4.10}$$

Considering α_b as the known quantity for an electronegative gas, then α_s must be determined numerically from (6.4.10), and Φ_p from (6.4.9). This was done by Boyd and Thompson with the result shown in Figure 6.4. The ratio Φ_p/T_e is seen to be very nearly $1/2$ for electronegative discharges if $\alpha_b < 2$ and $\gamma > 30$, which hold in weakly electronegative gases under typical discharge operating conditions. As will be seen in Chapter 10, α_b is, in turn, determined from a diffusion solution within the bulk plasma, in terms of $\alpha_0 = n_{0-}/n_{0e}$, the value at the center of the plasma.

The preceding calculation is not complete, because it ignores the possibility of double layers, where the quasineutrality condition breaks down. A calculation by Braithwaite and Allen (1988) indicates that the solutions $\Phi_p(\alpha_b)$ are triple-valued over a certain range of α_b for $\gamma > 5 + \sqrt{24}$. However, the proper physical solution is essentially as given in Figure 6.4. If the plasma is also collisional, then there are additional effects, which have been examined by Sheridan (1999). We explore some of these in Chapter 10, in the context of electronegative discharge equilibrium. In this context, the work of Kouznetsov et al. (1999) is also relevant. However, as also discussed in Chapter 10, the assumption of Boltzmann negative ions may not be valid in the bulk plasma.

The expression (6.4.5a), and all the following arguments, can be extended to multiple negative ion species, and also to more than one class of electrons (e.g., hot and cold), provided that all negatively charged species are individually Maxwellian. Restricting our attention to multiple negative ion species, (6.4.5a) generalizes to

$$n_e + \sum_j n_{-j} = n_{se}\, e^{\Phi/T_e} + n_{se} \sum_j \alpha_{sj}\, e^{\gamma_j\Phi/T_e} \tag{6.4.11}$$

where j runs over the negative ion species. The rest of the calculation is straightforward, provided that the left-hand side of (6.4.6) can be evaluated.

6.4.2 Multiple Positive Ion Species

If there are more than one positive ion species in the plasma, a condition often encountered with feedstock gases used in processing, then analysis of the sheath region becomes much more difficult. For simplicity, considering electrons as the only negative species, then the charge in Poisson's equation can be written

$$\rho = e\sum_j n_j - en_e \tag{6.4.12}$$

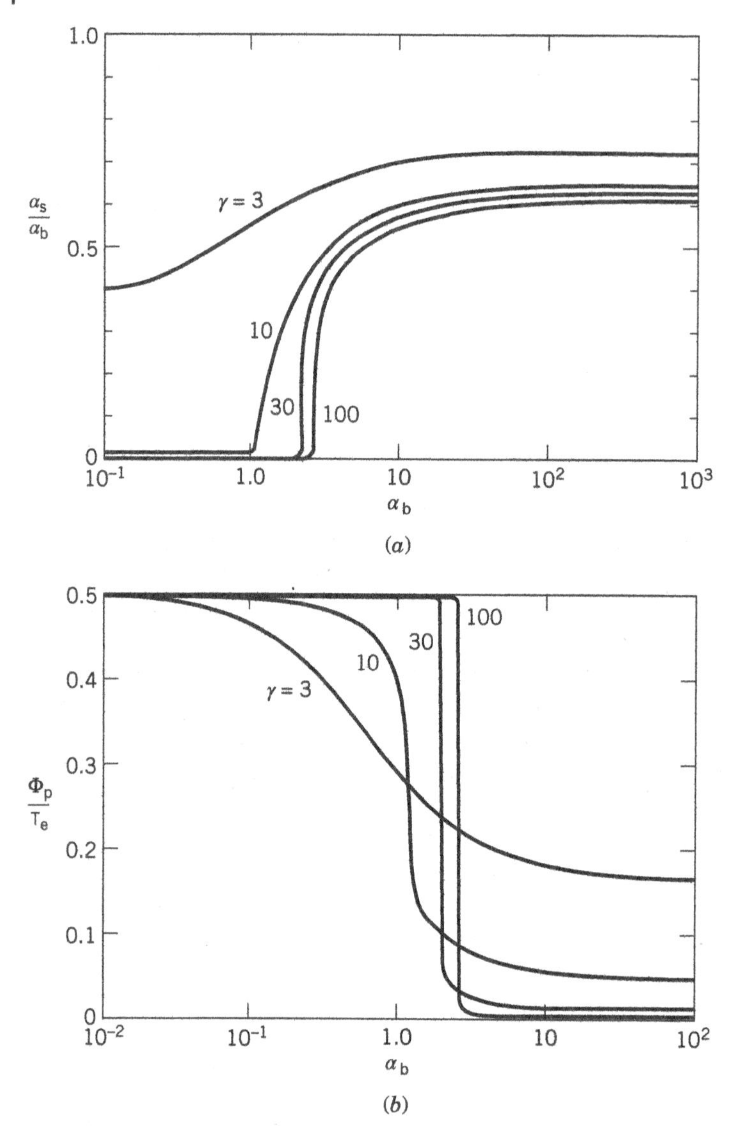

Figure 6.4 Negative ion sheath solutions; (*a*) α_s/α_b and (*b*) Φ_p/T_e versus α_b, with γ as a parameter. Source: Boyd and Thompson, 1959/The Royal Society.

where the summation is over the positive ion species with densities n_j per species. For cold ions, combining the continuity equation (2.3.7) in the steady state without sources or sinks,

$$n_j \frac{du_j}{dx} + u_j \frac{dn_j}{dx} = 0 \tag{6.4.13}$$

and the force equation (2.3.9) in the steady state without magnetic field, pressure, or collision terms,

$$m_j u_j \frac{du_j}{dx} = -e \frac{d\Phi}{dx} \tag{6.4.14}$$

we have for each species

$$\frac{dn_j}{d\Phi} = \frac{en_j}{m_j u_j^2} \tag{6.4.15}$$

Using the Boltzmann relation (2.4.16) for Maxwellian electrons, we have at the sheath edge

$$\frac{1}{n_e} \frac{dn_e}{d\Phi}\bigg|_s = \frac{1}{T_e} \tag{6.4.16}$$

Using the Bohm criterion, as in Section 6.2, that

$$\frac{d\rho}{d\Phi}\bigg|_s \leq 0 \tag{6.4.17}$$

taking the equality, and using (6.4.17) with (6.4.12), we have

$$\sum_j \frac{dn_j}{d\Phi}\bigg|_s = \frac{dn_e}{d\Phi}\bigg|_s \tag{6.4.18}$$

Substituting (6.4.15) and (6.4.16) into (6.4.18), we obtain the multispecies Bohm criterion

$$\sum_j \frac{en_{js}}{m_j u_{js}^2} = \frac{n_{es}}{T_e} \tag{6.4.19}$$

The work can be generalized to include finite temperature ions in both a fluid and a kinetic description, as has been done by Riemann (1995). However, (6.4.19) does not uniquely define a Bohm velocity for each species. For example, for two species, (6.4.19) becomes

$$\frac{en_{1s}}{m_1 u_{1s}^2} + \frac{en_{2s}}{m_2 u_{2s}^2} = \frac{n_{es}}{T_e} \tag{6.4.20}$$

Normalizing u_{1s} and u_{2s} to their individual Bohm velocities

$$u_{1n} = \left(\frac{m_1}{eT_e}\right)^{1/2} u_{1s} \qquad u_{2n} = \left(\frac{m_2}{eT_e}\right)^{1/2} u_{2s} \tag{6.4.21}$$

then (6.4.20) becomes

$$\frac{n_{1s}}{u_{1n}^2} + \frac{n_{2s}}{u_{2n}^2} = n_{1s} + n_{2s} \tag{6.4.22}$$

It is easy to see from (6.4.22) that either both ion species reach the sheath edge with their individual Bohm velocities ($u_{1n} = u_{2n} = 1$) or one will be subsonic and the other supersonic. If the ion flow across the presheath were purely collisionless, then each ion would indeed fall through the same potential ($T_e/2$) and acquire its individual Bohm velocity at the sheath edge. For a collisional presheath, each ion species can experience a different collisional force, depending on its mobility, which restricts the energy gain. One might then expect the most collisional ion species to have $u_n < 1$ and the less collisional one to have $u_n > 1$, as was observed in PIC simulations by Gozadinos (2001).

However, at low pressures, ions of different masses falling through the same potential have different velocities. This can give rise to a two-stream ion–ion instability, leading to an instability-enhanced friction that brings the two ion speeds together (Baalrud et al., 2009). The instability threshold, incorporating finite electron and ion temperature effects, was determined by Baalrud et al. (2015). The lowest threshold is for ions of widely disparate masses (e.g., He^+ and Xe^+). PIC simulations below (Gudmundsson and Lieberman, 2011) and above (Baalrud et al., 2015) the threshold confirm the results and show the transition from individual to common

Bohm speeds. The existence of instability-enhanced ion friction was confirmed experimentally in 0.05–0.4 mTorr xenon plasmas by Yip et al. (2015). The effects of gas pressure on these results have been examined using PIC simulations by Adrian et al. (2017).

6.5 High-Voltage Collisional Sheaths

If the mean free path for ion momentum transfer $\lambda_i < s$, the sheath width, then the assumption (6.3.6) of energy conservation, used to derive the Child law, fails. This modifies both the dynamics in the high potential sheath region and the ion velocity at the sheath edge. Consider first the high-voltage sheath region. If the ionization within the sheath is negligible, current continuity still holds, which is expressed as

$$n_i u_i = n_s u_s \tag{6.5.1}$$

where n_s and u_s are the values at the sheath edge. Considering the collisional case, we take

$$u_i = \mu_i E \approx \frac{2e\lambda_i}{\pi M |\mathbf{u}_i|} E \tag{6.5.2}$$

where μ_i is the mobility as defined in (5.3.1). Generally, both μ_i and λ_i are functions of the velocity. However, as we have discussed in Section 5.3, depending on the gas pressure and ion velocity, one or the other of these quantities may be relatively independent of velocity. For argon, for example, λ_i is relatively independent of velocity at intermediate pressures and with sheath voltages commonly used for plasma processing. With the assumption of constant λ_i, solving for $u_i > 0$ from (6.5.2) and substituting the result in (6.5.1), we have

$$n_i = \frac{n_s u_s}{(2e\lambda_i E/\pi M)^{1/2}} \tag{6.5.3}$$

Substituting this in Gauss' law (2.2.3), we have

$$\frac{\mathrm{d}E}{\mathrm{d}x} = \frac{e n_s u_s}{\epsilon_0 (2e\lambda_i E/\pi M)^{1/2}} \tag{6.5.4}$$

Separating variables, we can integrate and solve for E to obtain

$$E = \left[\frac{3 e n_s u_s}{2\epsilon_0 (2e\lambda_i/\pi M)^{1/2}}\right]^{2/3} x^{2/3} \tag{6.5.5}$$

where we have set $E(0) \approx 0$ at the sheath edge. A second integration gives the potential

$$\Phi = -\frac{3}{5}\left(\frac{3}{2\epsilon_0}\right)^{2/3} \frac{(e n_s u_s)^{2/3}}{(2e\lambda_i/\pi M)^{1/3}} x^{5/3} \tag{6.5.6}$$

where we have set $\Phi(0) = 0$. Noting that $e n_s u_s = J_0$, the constant current, we can take the $3/2$ power of (6.5.6), rearrange, and taking $\Phi = -V_0$ at the electrode position $x = s$, we obtain

$$J_0 = \left(\frac{2}{3}\right)\left(\frac{5}{3}\right)^{3/2} \epsilon_0 \left(\frac{2e\lambda_i}{\pi M}\right)^{1/2} \frac{V_0^{3/2}}{s^{5/2}} \tag{6.5.7}$$

Equation (6.5.7), first derived by Warren (1955), gives a collisional form of the Child law for the regime in which λ_i is independent of ion velocity. We note that the current scales the same with voltage, but differently with sheath spacing, than for the collisionless case. For a fixed J_0 and V_0, the sheath width scales as $s \propto \lambda_i^{1/5}$ and therefore weakly decreases as the gas pressure is increased.

Alternatively to our relation (6.5.7), we could equally well have chosen the higher pressure regime to make the calculation, taking ν_{mi} and, hence, μ_i independent of velocity. In this case, a similar integration procedure leads to the result (Problem 6.4)

$$J_0 = \frac{9}{8}\epsilon_0\mu_i\frac{V_0^2}{s^3} \tag{6.5.8}$$

We note here that the scalings of J_0 with both V_0 and s in (6.5.8) are different from (6.5.7). More detailed use of these various relations will be given in Chapter 11, where we use sheath physics in a complete description of capacitive discharges.

6.6 Electrostatic Probe Diagnostics

A metal probe, inserted in a discharge and biased positively or negatively to draw electron or ion current, is one of the earliest and still one of the most useful tools for diagnosing a plasma. These probes, introduced by Langmuir and analyzed in considerable detail by Mott-Smith and Langmuir (1926), are usually called *Langmuir probes*. As with any other electrode, the probe is surrounded by a sheath, such that its analysis naturally fits into the present chapter. However, unlike large electrode surfaces that are used to control a plasma, probes are usually quite small and, under suitable conditions, produce only minor local perturbations of the plasma.

The voltage and current of a probe defined in Figure 6.5 lead to a typical probe voltage–current characteristic as shown in Figure 6.6. The probe is biased to a voltage V_B with respect to ground, and the plasma is at a potential Φ_p with respect to ground. At the probe voltage $V_B = \Phi_p$, the probe

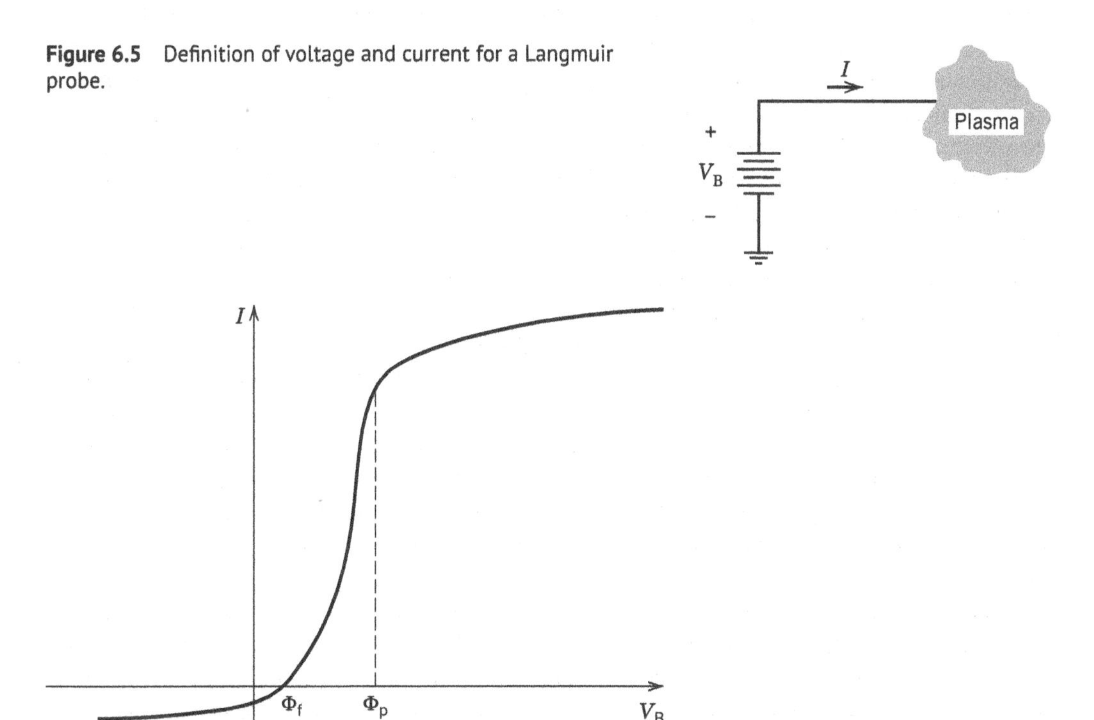

Figure 6.5 Definition of voltage and current for a Langmuir probe.

Figure 6.6 Typical $I-V_B$ characteristic for a Langmuir probe.

is at the same potential as the plasma and draws mainly current from the more mobile electrons, which is designated as positive current flowing from the probe into the plasma. For increasing V_B above this value, the current tends to saturate at the electron saturation current, but, depending on the probe geometry, can increase due to increasing effective collection area. For $V_B < \Phi_p$, electrons are repelled according to the Boltzmann relation, until at Φ_f, the probe is sufficiently negative with respect to the plasma that the electron and ion currents are equal such that $J = 0$. Φ_f is known as the *floating potential*, because it is the potential at which an insulated probe, which cannot draw current, will float. For $V_B < \Phi_f$, the current is increasingly ion current (negative into the plasma), tending to an ion saturation current that may also vary with voltage due to a change of the effective collection area. The magnitude of the ion saturation current is, of course, much smaller than the electron saturation current due to the much greater ion mass.

The basic theory for a plane collecting area, based on the sheath calculations of the previous sections, is quite simple. However, to minimally disturb the plasma and also for ease of construction, Langmuir probes are often thin wires with the wire radius $a < \lambda_{De}$. The trajectories of charged particles in the sheath then become important in determining the collected current, and the analysis becomes quite complicated. As the voltage is raised, either to large positive or large negative values with respect to the plasma, the sheath thickness s increases according to Child's law, and consequently the effective collecting area also increases. If $T_i \sim T_e$, then additional complications arise to make calculations very involved. There are also difficulties if the momentum transfer mean free paths λ_i, $\lambda_e \lesssim s$, which can occur in high-pressure discharges. A review of the analysis, including many of these complications, is given by Chen (1965). The extension to $T_i \sim T_e$, which is not usually of great interest in processing discharges, is given in a report by Laframboise (1966).

The story does not end here. More complicated probe configurations, such as double probes and emissive probes, have proven quite useful in various situations. These are also reviewed in Chen, and we consider them below. In an rf-driven plasma, an additional complication arises in that the potential of the plasma oscillates with respect to ground. Since we generally wish to use probes in a quasielectrostatic manner, the probe is usually made to oscillate with the plasma to eliminate the effect of the oscillating potential. Detailed consideration of oscillating plasma potentials and methods of using probes in their presence is found in reviews by Godyak (1990a) and Hershkowitz (1989). We summarize some of these considerations below.

Probe theory has generally been developed for plasmas in which the electron distribution is approximated by a Maxwellian. Various deviations from Maxwellian electrons exist in discharges. As described in Chapter 11, sheath heating in a capacitive discharge can result in a high-energy tail to the electron distribution, leaving the bulk electrons considerably colder than they would be in an equilibrium discharge with a Maxwellian distribution. These "two-temperature" distributions modify the results of Langmuir probes. Godyak et al. (1993) have critically examined this phenomenon and argue that the use of standard electron and ion saturation current techniques for analyzing probe data can lead to considerable error in the resulting plasma parameters. They present an alternative technique in which the electron energy distribution function (EEDF) is measured and used directly in calculating the plasma density. We review their arguments and techniques below.

Plasma densities obtained from the ion saturation current to probes have been compared with other measurement techniques such as microwaves (see Section 4.6). Generally, the comparisons have indicated that probe-predicted densities, using ion saturation current, are somewhat high when compared under conditions for which the microwave predictions are expected to be highly accurate. This result would generally agree with the arguments presented by Godyak and associates. However, in many situations, the densities obtained by probe and microwave techniques are quite close (e.g., see Figure 4.13). The accuracy of using the ion saturation current to measure the plasma density depends on the closeness of the electron distribution to an assumed Maxwellian at the probe sheath edge, and therefore to the type of plasma being diagnosed.

Finally, we shall briefly discuss practical probes and circuits for their use. Details of probes and probe circuitry are usually to be found in original articles, references to which can be found in the review articles cited here. Basic information on probes and circuits, beyond that given here, can also be found in the review articles by Chen (1965), Hershkowitz (1989), Godyak (1990a, 2021), Godyak and Alexandrovich (2015), and Godyak et al. (1992).

6.6.1 Planar Probe with Collisionless Sheath

Consider a flat plate probe with the (two-sided) physical probe area $A \gg s^2$, where s is the sheath thickness, such that the collecting area A is essentially independent of s. As we saw in Section 6.3, if a large voltage is applied to the probe, then $s \gg \lambda_{De}$, and we find that A is quite large to satisfy the above condition. For this reason, we expect that biasing the probe strongly positive to collect only electron current would strongly perturb the plasma. Consider therefore that the probe is biased sufficiently negatively to collect only ion current. From (6.3.13), the current "collected" (see direction in Figure 6.5) by the probe is

$$I = -I_i = -en_s u_B A \tag{6.6.1}$$

where, as in (6.2.4) with $T_i \ll T_e$, the Bohm velocity u_B is given by

$$u_B = \left(\frac{eT_e}{M}\right)^{1/2} \tag{6.6.2}$$

If we know T_e, then the density at the sheath edge n_s is determined from the measurement of I_i. As in (6.2.13), the plasma density in the probe neighborhood is then obtained as

$$n_0 \approx \frac{n_s}{0.61} \tag{6.6.3}$$

Since the electron temperature in most discharges is clamped in the range of 2–5 V by particle balance (see Section 10.1), a reasonable estimate of density can be obtained without knowing T_e. However, by varying the probe voltage, it is also straightforward to measure T_e. Considering that the probe potential is retarding with respect to the plasma potential, then, using Boltzmann's relation as in Section 6.2, the electron component of the probe current is

$$I + I_i = I_e = \frac{1}{4} e n_0 \bar{v}_e A \exp\left(\frac{V_B - \Phi_p}{T_e}\right) \tag{6.6.4}$$

where $\bar{v}_e = (8eT_e/\pi m)^{1/2}$, and $V_B - \Phi_p < 0$ is the potential between the probe and the plasma. There is an exponential increase in I_e with increasing V_B in this range. Defining an electron saturation current

$$I_{esat} = \frac{1}{4} e n_0 \bar{v}_e A \tag{6.6.5}$$

and taking the logarithm of (6.6.4), we have

$$\ln\left(\frac{I_e}{I_{esat}}\right) = \frac{V_B - \Phi_p}{T_e} \tag{6.6.6}$$

From (6.6.6), we see that the inverse slope of the logarithmic electron probe current with respect to V_B (in volts) gives T_e directly in volts.

The above simple interpretation is limited by the dynamic range over which (6.6.4) holds. For I_e too small, adding the measured I_i to I can introduce errors in the determination of I_e. For V_B too large, the Boltzmann exponential no longer is accurate, as electron saturation is approached. The nominal useful range of voltages over which the slope can be measured is then

$$\frac{|\Delta V_B|}{T_e} \approx \ln\left(\frac{\bar{v}_e}{4u_B}\right) = \ln\left(\frac{M}{2\pi m}\right)^{1/2} \tag{6.6.7}$$

which is approximately 4.7 for argon. This range is sufficient, provided there are no geometric complications.

The floating potential Φ_f and the plasma potential Φ_p are often of interest in discharge operation. The floating potential is the potential at which the probe draws equal electron and ion currents. If the plasma is mainly surrounded by a grounded conducting surface, then we would expect the floating potential to lie near this ground, as shown in Figure 6.6. This follows because the ground is usually not, itself, drawing significant net current, and thus at $V_B = \Phi_f$ the probe behaves as part of the ground. The plasma (space) potential, given by (6.6.7) with $\Phi_p - \Phi_f = \Delta V_B$, can be approximately determined from the knee (point of maximum first derivative) of the electron saturation portion of the I–V_B characteristic of Figure 6.6. For planar probes, the knee is easily recognizable, but the current drawn may be too large, either modifying the plasma or destroying the probe. For cylindrical probes, considered below, the measurement is usually possible, but its accuracy is reduced due to the variation of current with voltage in the electron saturation region.

6.6.2 Non-Maxwellian Electrons

A low-pressure discharge often has an electron energy distribution that departs significantly from a Maxwellian. For example, in Figure 11.9*a*, the electron distribution of a low-pressure rf discharge is given, which can be approximated by a two-temperature Maxwellian. At higher pressures, for which a two-temperature distribution is not evident, high accelerating fields may also result in a non-Maxwellian distribution. For an arbitrary distribution function, the electron current to a planar probe in the retarding potential region $\Phi_p - V_B > 0$ can be written as

$$I_e = eA \int_{-\infty}^{\infty} dv_x \int_{-\infty}^{\infty} dv_y \int_{v_{min}}^{\infty} dv_z\, v_z f_e(\mathbf{v}) \tag{6.6.8}$$

where

$$v_{min} = \left[\frac{2e(\Phi_p - V_B)}{m}\right]^{1/2} \tag{6.6.9}$$

is the minimum velocity along z for an electron at the plasma–sheath edge to reach the probe. For an isotropic distribution, we can introduce spherical polar coordinates in velocity to obtain

$$I_e = eA \int_{v_{min}}^{\infty} dv \int_0^{\theta_{min}} d\theta \int_0^{2\pi} d\phi\, v\cos\theta\, v^2 \sin\theta f_e(v) \tag{6.6.10}$$

where A is the physical collecting area of the probe and where

$$\theta_{min} = \cos^{-1}\frac{v_{min}}{v} \tag{6.6.11}$$

The ϕ and θ integrations are easily done, yielding

$$I_e = \pi eA \int_{v_{min}}^{\infty} dv\, v^3 \left(1 - \frac{v_{min}^2}{v^2}\right) f_e(v) \tag{6.6.12}$$

A transformation of (6.6.12) allows f_e to be obtained directly in terms of the second derivative of I_e with respect to $V = \Phi_p - V_B$. Introducing the change of variable $\mathcal{E} = \frac{1}{2}mv^2/e$, then (6.6.12) becomes

$$I_e = \frac{2\pi e^3}{m^2} A \int_V^\infty d\mathcal{E}\, \mathcal{E} \left\{ \left(1 - \frac{V}{\mathcal{E}}\right) f_e[v(\mathcal{E})] \right\} \tag{6.6.13}$$

where $v(\mathcal{E}) = (2e\mathcal{E}/m)^{1/2}$. Differentiating I_e, we obtain[1]

$$\frac{dI_e}{dV} = -\frac{2\pi e^3}{m^2} A \int_V^\infty d\mathcal{E} f_e[v(\mathcal{E})]$$

and a second differentiation yields

$$\frac{d^2 I_e}{dV^2} = \frac{2\pi e^3}{m^2} A f_e[v(V)] \tag{6.6.14}$$

It is usual to introduce the electron *energy distribution function* (EEDF) $g_e(\mathcal{E})$ by

$$g_e(\mathcal{E})\, d\mathcal{E} = 4\pi v^2 f_e(v)\, dv \tag{6.6.15}$$

Using the relation between $\mathcal{E}$ and v, we find

$$g_e(\mathcal{E}) = 2\pi \left(\frac{2e}{m}\right)^{3/2} \mathcal{E}^{1/2} f_e[v(\mathcal{E})] \tag{6.6.16}$$

Using this to eliminate f_e from (6.6.14), we obtain

$$g_e(V) = \frac{2m}{e^2 A} \left(\frac{2eV}{m}\right)^{1/2} \frac{d^2 I_e}{dV^2} \tag{6.6.17}$$

which gives $g_e(V)$ directly in terms of the measured value of d^2I_e/dV^2. The electron energy probability function (EEPF)

$$g_p(\mathcal{E}) = \mathcal{E}^{-1/2} g_e(\mathcal{E}) \tag{6.6.18}$$

is often introduced instead. For a Maxwellian distribution,

$$g_p(\mathcal{E}) = \frac{2}{\sqrt{\pi}} n_e \mathrm{T}_e^{-3/2} e^{-\mathcal{E}/\mathrm{T}_e} \tag{6.6.19}$$

such that $\ln g_p$ is linear with $\mathcal{E}$. The electron density n_e and the average energy $\langle \mathcal{E}_e \rangle$ can then be determined as

$$n_e = \int_0^\infty g_e(\mathcal{E})\, d\mathcal{E} \tag{6.6.20}$$

and

$$\langle \mathcal{E} \rangle = \frac{1}{n_e} \int_0^\infty \mathcal{E} g_e(\mathcal{E})\, d\mathcal{E} \tag{6.6.21}$$

The effective temperature is defined as $\mathrm{T}_{\mathrm{eff}} = \frac{2}{3}\langle \mathcal{E} \rangle$. The maximum in the first derivative dI_e/dV_B of the electron current is also a good indicator for the location of the plasma potential Φ_p. The use of (6.6.17), along with (6.6.20) and (6.6.21) to determine n_e and $\mathrm{T}_{\mathrm{eff}}$ from the probe characteristic, has a number of virtues. First, (6.6.20) can be shown to be valid for any isotropic electron velocity distribution. Second, (6.6.17) is valid for any convex probe geometry, planar, cylindrical or spherical (Kagan and Perel, 1964), e.g., $A = 2\pi a d$ for a cylindrical probe of radius a and length d. Third, non-Maxwellian distributions can be measured. Fourth, the result (6.6.17) does not depend on the ratio of probe dimension to Debye length or the ratio $\mathrm{T}_i/\mathrm{T}_e$ (Godyak, 1990a).

---

1 Note that if $G = \int_{x_1}^{x_2} g(x_1, x)\, dx$, then $\partial G/\partial x_1 = \int_{x_1}^{x_2} (\partial g/\partial x_1)\, dx - g(x_1, x_1)$.

6.6.3 Cylindrical Probe with a Collisionless Sheath

As we have seen in Section 6.3, the sheath thickness s can be quite significant, $s \gg \lambda_{De}$ such that one cannot routinely satisfy $A \gg s^2$. This recognition led to analysis of cylindrical and spherical probes (Mott-Smith and Langmuir, 1926). Because the cylindrical probe, consisting of a simple wire, is much more convenient and consequently almost exclusively used, we concentrate our attention on that geometry. The initial analysis and most subsequent improvements in analysis have concentrated on the pressure range for which the sheath is collisionless, $\lambda_i \gg s$, and we consider that pressure range here.

We consider first the case of a thin wire probe for which $s \gg a$, the probe radius, but take the probe tip length d (collecting part of the wire) to be sufficiently long, $s \ll d$, that an infinite cylinder approximation applies. In the saturation condition, where only a single species is collected, if all the electrons or ions entering the sheath were collected, then the collisionless Child's law would predict that $I \propto s \propto |\Phi_p - V_B|^{3/4}$. However, the collisionless trajectories preclude this happening, giving a weaker scaling which we now determine. The geometry is shown in Figure 6.7. A given incoming particle in the attractive central force of the probe has initial velocity components $-v_r$ and v_ϕ in the radial and azimuthal directions at the edge of the sheath $r = s$. At the probe radius $r = a$, the corresponding components are $-v_r'$ and v_ϕ'. For a collisionless sheath, we require conservation of energy,

$$\frac{1}{2}m(v_r^2 + v_\phi^2) + e|\Phi_p - V_B| = \frac{1}{2}m(v_r'^2 + v_\phi'^2) \tag{6.6.22}$$

and conservation of angular momentum,

$$sv_\phi = av_\phi' \tag{6.6.23}$$

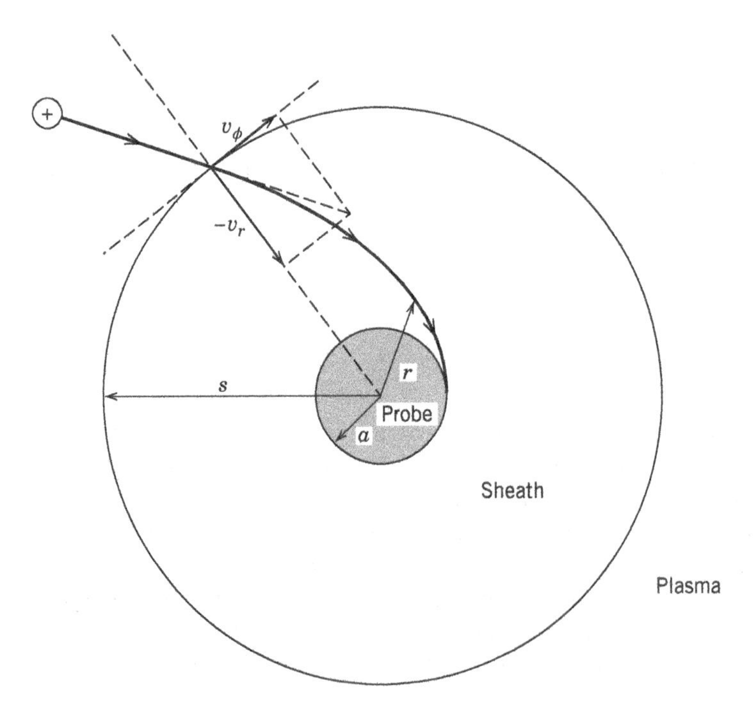

Figure 6.7 Ion orbital motion within the sheath of a cylindrical Langmuir probe.

where m is the mass of the attracted species, either electrons or ions. Solving, we obtain

$$v'_\phi = \frac{s}{a} v_\phi \tag{6.6.24}$$

$$v'^2_r = v_r^2 + v_\phi^2 + \frac{2e|\Phi_\text{p} - V_\text{B}|}{m} - \frac{s^2}{a^2} v_\phi^2 \tag{6.6.25}$$

For an ion to reach the probe, $v_r < 0$ and $v'^2_r > 0$. Setting $v'^2_r = 0$ in (6.6.25), we obtain

$$v_{\phi 0} = \left(\frac{v_r^2 + 2e|\Phi_\text{p} - V_\text{B}|/m}{s^2/a^2 - 1} \right)^{1/2} \tag{6.6.26}$$

such that particles only reach the probe if $|v_\phi| \leq v_{\phi 0}$.

The saturation current collected by the probe is found by integrating the radial flux $-n_\text{s} v_r$ over the distribution function at the plasma–sheath edge, for those particles that reach the probe:

$$I = -2\pi s d n_\text{s} e \int_{-\infty}^{0} v_r \, dv_r \int_{-v_{\phi 0}}^{v_{\phi 0}} dv_\phi f(v_r, v_\phi) \tag{6.6.27}$$

where f is the normalized distribution function of electrons or ions. Making the rather strong assumption that the distribution is an isotropic Maxwellian, averaged over the third velocity coordinate, we have

$$f = \frac{m}{2\pi e \text{T}_\text{s}} \exp\left[-\frac{m(v_r^2 + v_\phi^2)}{2e\text{T}_\text{s}} \right] \tag{6.6.28}$$

where T_s is the temperature of the collected species at the sheath edge. The integrations can be performed in terms of error functions, but the results, which can be found in the literature quoted above, are not particularly illuminating. However, for large probe voltages, we can simplify the evaluation of (6.6.27) by assuming that

$$\frac{a}{s} \ll 1 \tag{6.6.29a}$$

$$v_r^2 \ll \frac{e|\Phi_\text{p} - V_\text{B}|}{m} \tag{6.6.29b}$$

and

$$v_{\phi 0}^2 \ll \frac{e\text{T}_\text{s}}{m} \tag{6.6.29c}$$

Then, using (6.6.29a) and (6.6.29b) to evaluate (6.6.26), we obtain

$$v_{\phi 0} = \frac{a}{s} \left(\frac{2e|\Phi_\text{p} - V_\text{B}|}{m} \right)^{1/2}$$

We note that since $s \propto |\Phi_\text{p} - V_\text{B}|^{3/4}$ for Child's law, (6.6.29c) is well satisfied at high voltages. Using $v_{\phi 0}$ in (6.6.27), with (6.6.28) and the condition (6.6.29c), we integrate to find that

$$I = 2en_\text{s} a d \left(\frac{2e|\Phi_\text{p} - V_\text{B}|}{m} \right)^{1/2} \tag{6.6.30}$$

where I represents either electron or ion saturation current. We see that I is independent of T_s in this limit. Hence, a plot of I^2 versus $-V_\text{B}$ should be linear, with n_s^2 determined by the slope of this line, independent of T_e and T_i. Expression (6.6.30) is widely used to determine n_s in low-pressure discharges. However, the orbital ion motion is sensitive to ion collisions in the sheath, and orbital motion is destroyed at quite low pressures. In addition, the result (6.6.30) is sensitive to the isotropy of the distribution function at the sheath edge. From Figure 6.7, it is apparent that significant radial anisotropy will enhance the fraction of particles that are collected. For electrons, we might reasonably expect to find an isotropic distribution at the sheath edge, even if it is not Maxwellian. We have seen in Section 6.2 that ions, on the other hand, gain an energy $\mathrm{T_e}/2$ in a presheath, which may lead to significant anisotropy. Although we have assumed a collisionless sheath, the presheath is not necessarily so, and presheath collisions will tend to isotropize the distribution of ion velocities. For an alternative distribution at the sheath edge of monoenergetic ions on a cylindrical (isotropic) shell in velocity space, Hershkowitz finds that the coefficient 2, in (6.6.30), is replaced by $\pi/\sqrt{2}$, which is quite similar. A more extreme assumption of anisotropy of f_i, which might be approached at very low pressures, is that the radial ion velocity component is given by the Bohm velocity $u_\mathrm{B} = (e\mathrm{T_e}/M)^{1/2}$, while the azimuthal component remains Maxwellian at temperature $\mathrm{T_i}$,

$$f_\mathrm{i} = \delta(v_r + u_\mathrm{B})\left(\frac{M}{2\pi e\mathrm{T_i}}\right)^{1/2} \exp\left(-\frac{Mv_\phi^2}{2e\mathrm{T_i}}\right)$$

Using this in (6.6.27) along with the conditions (6.6.29), we integrate to obtain

$$I_\mathrm{i} = 2e\left(\frac{2\pi\mathrm{T_e}}{\mathrm{T_i}}\right)^{1/2} n_\mathrm{s} a d \left[\frac{2e(\Phi_\mathrm{p} - V_\mathrm{B})}{M}\right]^{1/2} \tag{6.6.31}$$

Comparing (6.6.31) to (6.6.30), we see that n_s is smaller by a factor of $(\mathrm{T_i}/2\pi\mathrm{T_e})^{1/2}$ for the same current. We do not expect to find such extreme overestimations of density from the measured orbital ion saturation current, but the sensitivity to the ion velocity distribution suggests that (6.6.30) provides only a semiquantitative estimate of the ion density. Similarly, in low-pressure discharges, the ion drift velocity tends to exceed the ion thermal velocity (see Section 5.3), leading to further modifications in the collected ion current. For a capacitive rf discharge at a pressure $p = 30$ mTorr argon, Godyak et al. (1993) found, by using the measured $I_\mathrm{e}(V)$ in the expression (6.6.17) for the energy distribution, a two-temperature distribution, as in Figure 11.9a, with $\mathrm{T_{ec}} = 0.50$ V and $\mathrm{T_{eh}} \approx 3.4$ V. Using the energy distribution (6.6.17) in (6.6.20) and (6.6.21), they found $n_\mathrm{e} \approx 4.4 \times 10^9\ \mathrm{cm^{-3}}$ and $\mathrm{T_{eff}} \approx 0.67$ V. From the standard Langmuir procedure (6.6.6) applied to the electron current I_e collected by the cylindrical probe, and using the measured electron saturation current (6.6.5) at the plasma potential to find the density, they found $\mathrm{T_{ec}} \approx 0.73$ V, $\mathrm{T_{eh}} \approx 4.2$ V, and $n_\mathrm{e} \approx 3.3 \times 10^9\ \mathrm{cm^{-3}}$, close to the values determined from the measured energy distribution, as expected for an isotropic distribution. The density determined from the orbital ion current I_i using (6.6.30) was $n_\mathrm{e} \approx 1.1 \times 10^{10}\ \mathrm{cm^{-3}}$, a factor of 2.5 larger than found from the measurement of the electron distribution, as might be expected if the ion distribution had significant anisotropy at the sheath edge.

It should be pointed out, however, that the more accurate calculational procedure, using $\mathrm{g_e}$ determined from (6.6.17), is considerably more difficult, experimentally. In particular, taking derivatives of measured quantities results in the introduction of system noise, much of which is intrinsic to the plasma. While averaging procedures can be employed to increase the signal-to-noise ratio, it is all too easy to substitute experimental uncertainty for the uncertainties of the ion orbital theory. Measuring electron saturation current, which does not suffer particularly from the above uncertainties, may be excluded by consideration of the power limits to the probe, unless the measurement system is pulsed, which introduces additional complexities. The experimenter must navigate carefully among these alternatives.

6.6.4 Double Probes and Emissive Probes

Other probe configurations have also been used to measure plasma parameters, with various claims as to accuracy, convenience, etc. Two of the most frequently used alternatives are double probes and emissive probes. Double probes are generally used if there is no well-defined ground electrode in the plasma. A schematic of a double probe is shown in Figure 6.8*a*, with a typical probe characteristic in Figure 6.8*b*. Since the two probes draw no net current, they will both be negative with respect to the plasma. Current flows between the probes if the differential potential $V \neq 0$. As V becomes large, the more negative probe (in this case probe 2) essentially draws ion saturation current, which is just balanced by the net electron current to probe 1. The probe system has the advantage that the net current never exceeds the ion saturation current, minimizing the disturbance to the discharge, but has a consequent disadvantage that only the high-energy tail of the electron distribution is collected by either probe. The distribution of these electrons may not be representative of the distribution of bulk electrons in the discharge.

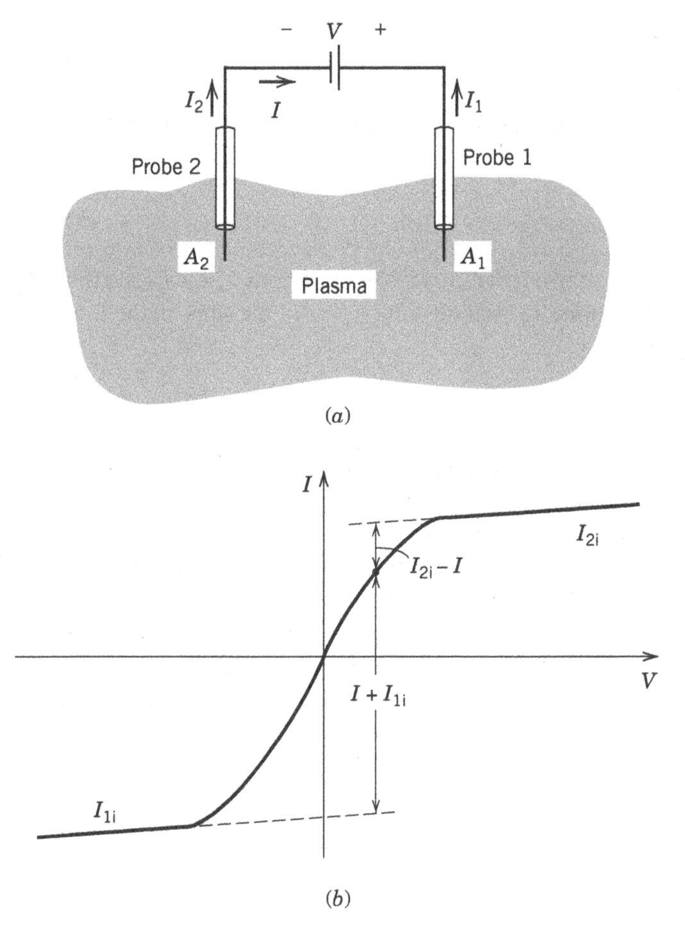

Figure 6.8 Schematic of double probe measurement: (*a*) definition of voltage and currents; (*b*) typical current–voltage characteristic. Source: Chen, 1965/with permission of Elsevier.

Defining the ion and electron currents to probes 1 and 2 as $I_{1i}, I_{1e}, I_{2i}, I_{2e}$, then the condition that the system float is

$$I_{1i} + I_{2i} - I_{1e} - I_{2e} = 0 \tag{6.6.32a}$$

The loop current is

$$I_{2i} - I_{2e} - (I_{1i} - I_{1e}) = 2I \tag{6.6.32b}$$

Combining (6.6.32a) with (6.6.32b), we obtain

$$I = I_{1e} - I_{1i} = I_{2i} - I_{2e} \tag{6.6.32c}$$

For the electron current, we have

$$I_{1e} = A_1 J_{esat} e^{V_1/T_e}, \quad I_{2e} = A_2 J_{esat} e^{V_2/T_e} \tag{6.6.33}$$

where J_{esat} is the electron random current density and V_1 and V_2 are the probe potentials with respect to the plasma potential. Using $V = V_1 - V_2$ and substituting (6.6.33) into (6.6.32c), we obtain

$$\frac{I + I_{1i}}{I_{2i} - I} = \frac{A_1}{A_2} e^{V/T_e} \tag{6.6.34}$$

which generally plots as shown in Figure 6.8*b*. For $A_1 = A_2$, then $I_{1i} = I_{2i} \equiv I_i$, such that (6.6.34) simplifies to

$$I = I_i \tanh\left(\frac{V}{2T_e}\right) \tag{6.6.35}$$

It is straightforward to fit (6.6.35) to the experimental curve, obtaining both T_e and I_i (and thus n). A simpler procedure can be used to determine T_e. Again taking $A_1 = A_2$, the slope of the I–V plot at the origin ($V = 0$) can be calculated to be

$$\left.\frac{dI}{dV}\right|_{V=0} = \frac{I_i}{2T_e} \tag{6.6.36}$$

The details are left to a problem. Note that for cylindrical probes, the I_i in either (6.6.35) or (6.6.36) is that obtained by extrapolation, as shown by the dashed lines in Figure 6.8*b*.

A hot wire electron-emitting (emissive) probe can be used for a simple measurement of the plasma space potential. Since it works with electron emission, it has the disadvantage of requiring a separate filament circuit carrying high currents, but because it is hot, it is less subject to contamination, which can be a serious problem with other probe measurements. The basic idea is very simple. Since the temperature T_w of the electrons emitted from the hot probe wire is related to the wire temperature, we have $T_w \ll T_e$. This results in a sharp change in probe current as the probe potential passes through the plasma potential. This is easily seen from the equations for the electron current. The plasma electron current is approximately (Hershkowitz, 1989)

$$\begin{aligned} I_{pe} &= I_{p0} e^{-(\Phi_p - V_B)/T_e} && V_B < \Phi_p \\ &= I_{p0}\left[1 + \frac{(V_B - \Phi_p)}{T_e}\right]^{1/2} && V_B > \Phi_p \end{aligned} \tag{6.6.37}$$

and the emission current is approximately

$$\begin{aligned} I_{we} &= I_{w0} e^{-(V_B - \Phi_p)/T_w} g_w(V_B - \Phi_p) \quad V_B > \Phi_p \\ &= I_{w0} \quad V_B < \Phi_p \end{aligned} \tag{6.6.38}$$

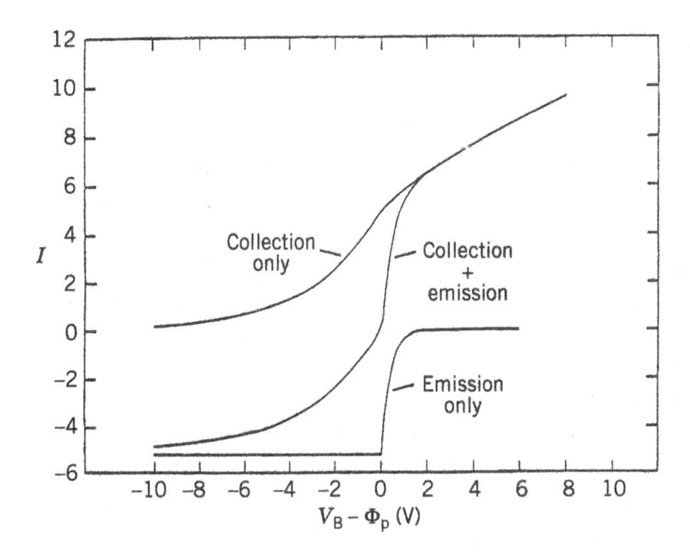

Figure 6.9 Typical collecting and emitting current–voltage characteristics for an emissive wire probe in a plasma; the electron and wire temperatures are $T_e = 3$ V and $T_w = 0.3$ V. Source: Hershkowitz, 1989/with permission of Elsevier.

It can be shown that $g_w \approx [1 + (V_B - \Phi_p)/T_w]^{1/2}$, but this result is not necessary for the argument. Neglecting the small ion current, the total probe current is given by

$$I = I_{pe} - I_{we} \tag{6.6.39}$$

Without detailed calculation, if we choose $I_{w0} \approx I_{p0}$, then for the case considered with $T_w \ll T_e$, there is a sharp change in I due to the exponential term in I_{we} at $(V_B - \Phi_p) \approx T_w$, which thus gives the plasma potential. The result for a typical case with $T_e = 3$ V and $T_w = 0.3$ V is shown in Figure 6.9. We note that measuring $V_B - \Phi_p$ also gives an estimate of T_e as obtained from (6.2.17) with $\Phi_w = V_B - \Phi_p$.

6.6.5 Effect of Collisions and DC Magnetic Fields

Collisions can significantly affect planar probe diagnostics when the mean free path λ_i becomes of the order of the sheath width s, or, for cylindrical or spherical probes, of order of the sum of the sheath width and probe radius, $s + r_p$. For planar probes with $\lambda_i \ll s$, we can directly use the collisional sheath theory in Section 6.5, just as we used collisionless sheath theory to describe collisionless planar probes. However, in the transition region, even the planar theory is complicated and difficult to use. For other geometries, the analysis becomes still more complicated and difficult to interpret. A good account of collisional effects can be found in Chen (1965). A fairly complete theory has been developed for large spherical probes by Su and Lam (1963). Sternberg and Godyak (2017), and Godyak and Sternberg (2024) give analyses of spherical and cylindrical probe depletion effects, respectively, in the collisional regime of a constant ion–neutral mean free path (see Section 5.3). These effects can be quite important for $\lambda_i \lesssim s + r_p$, leading to significant underestimation of densities.

One reason for studying collisional effects is that they also bear on the use of probes in the presence of an applied dc magnetic field. As we have seen in Section 5.4, the electron diffusion across a magnetic field is severely inhibited. For each species (without considering ambipolar effects), the diffusion across the field is related to the diffusion along the field by

$$D_{\perp} = \frac{D_{||}}{1 + \omega_c^2 \tau_c^2}$$

where $\omega_c = eB/m$ is the gyration frequency, and τ_c is the mean collision time. For electrons in a gas with $p = 10$ mTorr and $B = 100$ G, we find $\omega_c \tau_c \approx 10^2$. For ions, since ω_c is decreased by m/M and τ_c increased by $(M/m)^{1/2}$, $\omega_c \tau_c \propto (m/M)^{1/2}$ and therefore the ion diffusion is not severely limited. The result is that the probe, drawing electron current, behaves similarly to a plane probe without B but with an effective probe area equal to the probe cross section along the field lines. The ion orbital collection regime (6.6.30) may be used as previously, if the ions have gyroradii large compared to the sheath width. The above simple interpretation of a probe in a B field is limited by a phenomenon called shadowing. Because the probe collects electrons from a thin layer of plasma corresponding to the probe cross section, it acts similarly to a plane probe, as discussed in the first subsection. We indicated there that a large probe can deplete the nearby plasma, thus modifying the plasma it is supposed to measure. This probe shadowing can occur even for small-diameter probes with a magnetic field present. However, the depleted region can be refilled by diffusion across the magnetic field from the neighboring plasma. As one might expect, the calculations can become quite complicated, and the reader is again referred to the review by Chen (1965) for a summary and further references.

As mentioned above, shadowing can also occur for flowing plasma or when electron beams are present. If the plasma is flowing with a velocity of order of the Bohm velocity, then the ion collection can be distorted such that operation in the ion orbital motion region is modified. This is a common situation in low-pressure discharges in which the ion drift velocity typically exceeds the ion thermal velocity. Similarly, if the electrons are streaming through ions with beam velocities comparable to the electron thermal velocity, the electron collection will be distorted.

6.6.6 Probe Construction and Circuits

A basic cylindrical Langmuir probe consists of a thin wire surrounded by a thin insulator that, for dc discharges, may itself be encased in a thin grounded shield. The probe tip usually extends many wire diameters from the insulator. A typical probe, shown in Figure 6.10, has a tungsten wire probe tip 6.3 mm in length and 38 μm in radius, with a quartz or ceramic capillary sleeve preventing electrical contact between the probe and any conductive material on the probe holder. The insulating holder surrounding the capillary sleeve should have a radius smaller than an electron mean free path to prevent perturbation of the plasma by the probe. To construct other geometries, a small plate (plane probe geometry) or sphere (spherical probe geometry) may be attached to the probe tip. Complications include vacuum sealing the probe, allowing the probe tip to be replaceable (tip burnout can be a serious problem) and allowing the probe body to slide through a vacuum seal in order to scan the plasma. Details of various probe designs can be found in the literature; a typical design is shown in Figure 6.10.

For other types of probes, obvious constructional changes are made. The simplest emissive probe construction uses a high-resistivity refractory wire loop tip with the two sides of the loop returning with low-resistivity insulated wire through the probe body, where they can be connected to a power source for heating. The heating current is switched off during the measurement. For dense plasmas, a single probe can be made emissive by heating from electron current alone, but such probes are more subject to burnout. The simple Langmuir probe may also incorporate some means of heating to drive off impurities which can severely affect current measurements. Double probes are also

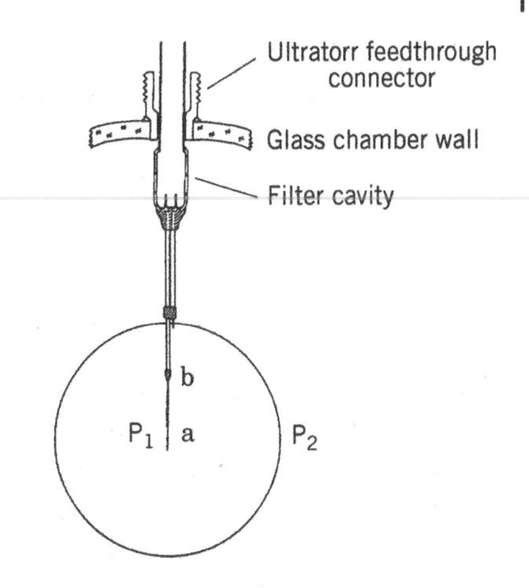

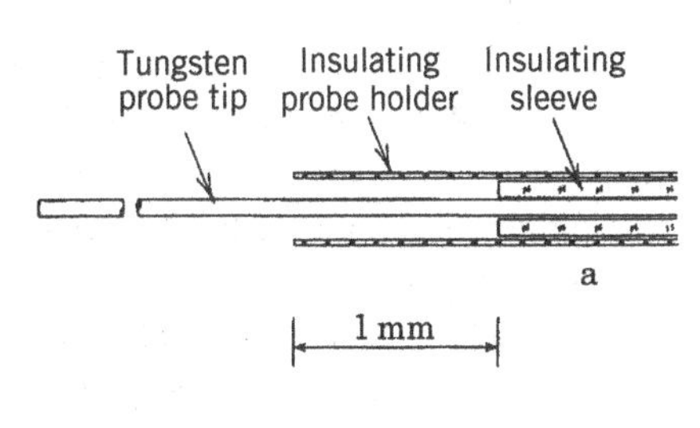

Figure 6.10 Construction of a cylindrical probe for rf discharge measurements. Source: Godyak et al., 1992/with permission of IOP Publishing.

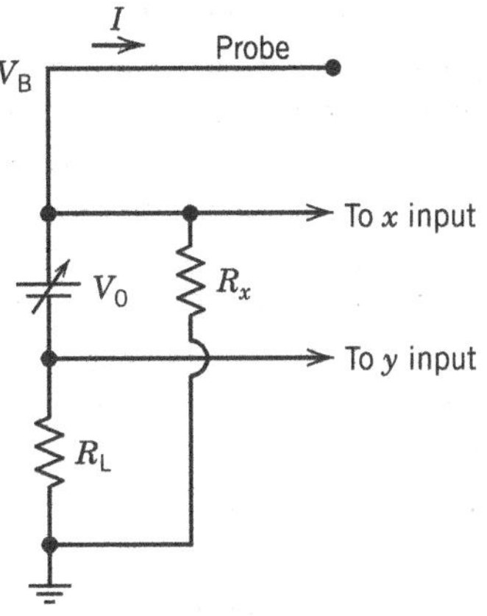

Figure 6.11 Simple Langmuir probe biasing circuit.

often constructed with the two probe tips emanating from a single probe body. In this case, the wires must be sufficiently far apart that the sheaths surrounding the wires do not interact. For expected plasma parameters, estimates of the sheath widths should be made before designing the probe separation.

A simple probe biasing circuit is shown in Figure 6.11, with the probe voltage V_B given by

$$V_B = V_0 - R_L I \tag{6.6.40}$$

where $R_X \gg R_L$ and the current through R_X is neglected. The current I is measured directly from the voltage across R_L, and V_B is measured either directly, as shown, or by measuring V_0 and subtracting $R_L I$. Clearly, $R_L \ll \partial V_B/\partial I$ for the measurement technique to work, that is, V_B must be able to be varied by varying V_0. The points labeled y input, measuring I, and x input, measuring V_B, may be the vertical and horizontal inputs on an oscilloscope, x, y recorder, or simply voltmeters. The circuit is usually a little more complicated, since V_0 is not only variable, but must be able to change signs. The voltage can also be swept at a slow rate. For a floating potential measurement $I = 0$, we take $V_0 = 0$. Amplifiers may also be used to adjust impedance levels in practical circuits.

6.6.7 Probes in Time-Varying Fields

A capacitive discharge driven between an rf excited electrode and a grounded electrode is widely used for plasma processing. We discuss this discharge in Chapter 11. The rf voltage capacitively divides across the system, and therefore part of the rf voltage appears between the plasma and the grounded electrode. The space potential Φ_p of the plasma with respect to the grounded electrode therefore oscillates in time. In this situation, the time-average current drawn to a probe biased to a constant voltage V_B through a low impedance is quite different than described in the preceding subsections.

The reason for this is illustrated in Figure 6.12, which shows the instantaneous I–V_B probe curves for various values of $\Phi_p(t)$. The "knee" of the probe curve, marked with a vertical dashed line, gives the value of V_B where $\Phi_p(t) = V_B$. As Φ_p oscillates in time as shown, the probe curve oscillates horizontally back and forth. The time average of this motion, indicated as the heavy line, gives the apparent "probe curve" $\bar{I}$ versus V_B. It is clear from the figure that the electron temperature determined from this curve will be much higher than the actual T_e.

Although it is possible to interpret the time-averaged current measurements (see Hershkowitz, 1989), it is also possible to modify the probe circuits so that the probe characteristic can be interpreted in the normal way. One common technique is to place an inductor L near the probe tip in series with the probe such that the probe reactance to ground $\omega L \gg 1/\omega C_s$, the reactance between the probe and the plasma, where ω is the radian rf driving frequency. This may be somewhat hard to achieve if ω is not too high, but can reasonably be obtained at $\omega/2\pi = 13.56$ MHz, a commonly used frequency.

The probe circuit elements, the additional series inductive "choke" element L, and a large bypass capacitor C_{bypass} are shown in Figure 6.13. Here, C_s is the effective capacitance of the probe sheath. The amplitude $\tilde{\Phi}_p - \tilde{V}_{\text{rf}}$ of the rf voltage across the probe sheath must satisfy $(\tilde{\Phi}_p - \tilde{V}_{\text{rf}})/T_e \ll 1$. In fact, fractional measurement errors appear to be $\lesssim 0.2$ if $(\tilde{\Phi}_p - \tilde{V}_{\text{rf}})/T_e \lesssim 1$. Using the voltage

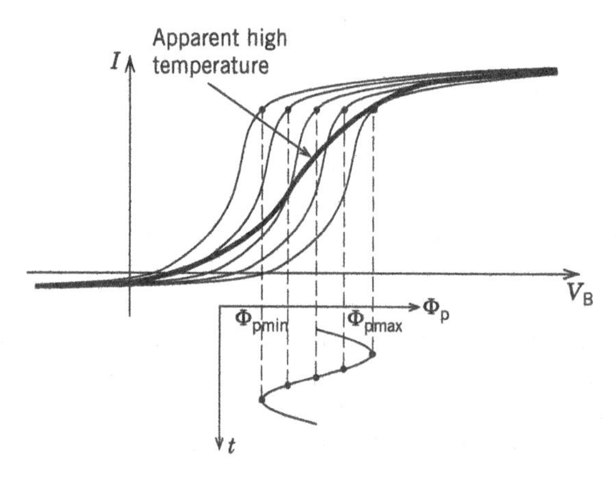

Figure 6.12 Probe characteristics I versus V_B in a plasma with an oscillating space potential $\Phi_p(t)$, showing (heavy solid line) a time-averaged probe characteristic having an apparent electron temperature much higher than the actual T_e.

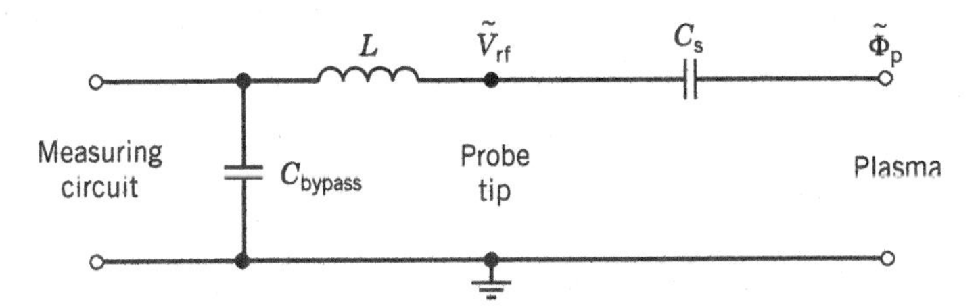

Figure 6.13 Probe circuit elements and blocking inductor used to measure the current–voltage characteristics in an rf discharge.

divider formula with the impedances $Z_s = (j\omega C_s)^{-1}$ and $Z_L = j\omega L$,

$$\tilde{\Phi}_p - \tilde{V}_{rf} = \tilde{\Phi}_p \frac{Z_s}{Z_L + Z_s}$$

we obtain the criterion

$$\frac{\tilde{\Phi}_p - \tilde{V}_{rf}}{T_e} = \frac{Z_s}{Z_L + Z_s} \frac{\tilde{\Phi}_p}{T_e} \lesssim 1 \tag{6.6.41}$$

A particular measurement of a probe with a 5 mH inductive choke, at 13.56 MHz, gave $|Z_L| = 450$ kΩ and $|Z_s| = 12$ kΩ ($C_s \approx 1$ pF), limiting $\tilde{\Phi}_p/T_e$ to less than $|Z_L/Z_s| \approx 37$. We shall see in Chapter 11 that this may limit the use of a simple blocking inductance in practical discharges. To overcome this limitation, one can include a capacitance C in parallel with L, such that the parallel LC circuit is in resonance at the desired frequency. If harmonics of the driving frequency are present, additional series resonant LC circuits can be used tuned to the second and third harmonics of the driving frequency.

For measurement of rf plasmas, the inductance required to allow the probe to follow the oscillating plasma space potential is usually incorporated into the probe body to minimize stray capacitance. The probe labeled P1 in Figure 6.10 can be used in this way. In this design, a large circular wire loop P2 is used to establish a ground reference for P1. Note that the probe does not have a grounded shield, which, if present, would greatly increase the stray capacitance of the probe tip to ground.

The above discussion, and that of the previous subsections, does not include all of the complications that can be encountered in probe diagnostics. The experimenter wishing to use probes as a diagnostic tool can proceed from the information given here, but may also wish to look further into the reviews referenced in this section, and also into the original literature referenced in those reviews.

Problems

6.1 Finite Density for Collisionless Child Law The Child law density (6.3.17) is singular at the sheath edge $x = 0$, while the potential (6.3.15) is not. Assuming that (6.3.15) still holds and that all ions enter the sheath with the Bohm velocity u_B, find a nonsingular expression for $n(x)$ as a function of J_0, u_B, $\Phi(x)$, and other constants. Plot n/n_s versus x/s for $V_0/T_e = 100$. Plot n/n_s given from (6.3.17) on the same graph to compare with your result.

6.2 Bohm Criterion for an Electronegative Plasma Derive the Bohm criterion (6.4.7) for an electronegative plasma with cold positive ions along with electrons and negative ions in

Boltzmann equilibrium at temperatures T_e and T_i, respectively, by repeating the calculation leading to (6.2.4) with three species rather than two.

6.3 Potential Across an Electronegative Presheath Show that the potential Φ_p across the presheath in an electronegative plasma is given by (6.4.9).

6.4 Collisional Sheath Law For a high-pressure, high-voltage, collisional sheath, the ion drift velocity can be written as $u_i = \mu_i E$, where $\mu_i = e/M\nu_{mi}$ is the constant ion mobility, with ν_{mi} a constant ion–neutral momentum transfer frequency.

(a) Using particle conservation and Poisson's equation, derive the high-pressure, collisional Child law for ions (6.5.8).

(b) For an argon discharge with $\lambda_i = (330p)^{-1}$ cm, with the pressure p in Torr and $p = 10$ Torr, calculate the sheath thickness s for $n_s = 10^9$ cm^{-3} at the sheath edge, $T_e = 2$ V, $T_i = 0.026$ V, and $V_0 = 100$ V across the sheath. Assume an average $\nu_{mi} = u_B/\lambda_i$ for ions within the sheath, with an ion speed entering the sheath given by (6.2.21). Compare this s to that obtained for the same discharge parameters from the collisionless Child law.

6.5 Langmuir Probe Calculation A probe whose collecting surface is a square tantalum foil 2×2 mm is found to give a saturation ion current of 100 μA in an argon plasma (atomic mass = 40). If $T_e = 2$ V, what is the approximate plasma density? (Assume that the probe can be considered as a plane collector with both sides collecting.) If a bias voltage of −20 V is applied between the probe and ground, calculate the sheath thickness, using the collisionless Child law, to determine if the plane collector assumption is justified.

6.6 Langmuir Probe Theory

(a) Referring to Figure 6.7, starting from (6.6.22) and (6.6.23), and using (6.6.27) and (6.6.28), fill in the steps to obtain (6.6.30).

(b) Starting from (6.6.32) and using (6.6.33), derive (6.6.34) and (6.6.35).

(c) Verify (6.6.36).

6.7 Analysis of Cylindrical Langmuir Probe Data A cylindrical Langmuir probe with radius $a = 50$ μm and length $d = 6.3$ mm is used to determine the plasma density n_s and electron temperature T_e in an argon discharge. The plasma potential Φ_p (with respect to ground) is measured to be 30 V. The Langmuir probe I versus V_B characteristic is measured to be (V_B is the probe voltage with respect to ground):

I(μA)	−25	−22	−19.3	−14.8	−8.7	15	50.5	131	313	733
V_B(V)	−20	−10	0	10	15	20	22.5	25	27.5	30

(a) According to (6.6.30), a plot of I^2 versus $\Phi_p - V_B$ should be a straight line in the ion saturation regime $\Phi_p - V_B \gg T_e$. Plot I^2 versus $\Phi_p - V_B$ on *linear scales* for $\Phi_p - V_B \gg T_e$. Extrapolate the linear part of this curve to determine the ion saturation current I_i over the *entire voltage range* $0 < \Phi_p - V_B < 50$ V. Then, apply (6.6.30) to I_i (where m in (6.6.30) is the ion mass) to determine n_s.

(b) Subtract I_i from I to determine the electron current I_e, and plot I_e (*log scale*) versus $\Phi_p - V_B$ (*linear scale*). You should obtain a straight line as in (6.6.6). Find T_e and n_s from your data. Compare the n_s value with the value you found in part (a), and comment briefly on any discrepancy.

6.8 Spherical Probe Theory Consider ion collection for a spherical probe of radius a. Use the collisionless analysis for sheath thickness $s \gg a$, as described in (6.6.22)–(6.6.26).

(a) Show that (6.6.27) is replaced by

$$I = -4\pi s^2 e n_s \cdot 2\pi \int_{-\infty}^{0} v^3 \, dv \int_{0}^{\theta_0} \sin\theta \, d\theta \, f(v)$$

where v is the spherical velocity coordinate, $v \sin\theta_0 = v_{\phi 0}$, with $v_{\phi 0}$ given by (6.6.26), and

$$f = \left(\frac{m}{2\pi e \mathrm{T_e}}\right)^{3/2} \mathrm{e}^{-mv^2/2e\mathrm{T_s}}$$

is the normalized Maxwellian distribution.

(b) Making the assumption of $\theta_0 \ll 1$ as in (6.6.29), show that

$$I = -4\pi a^2 \cdot \frac{1}{4} n_s \bar{v} \cdot \frac{\Phi_p - V_B}{\mathrm{T_s}}$$

where $\bar{v} = (8e\mathrm{T_s}/\pi m)^{1/2}$. A more accurate expression, valid for $\Phi_p - V_B > 0$, is (Laframboise, 1966; Laframboise and Parker, 1973)

$$I = -4\pi a^2 \cdot \frac{1}{4} n_s \bar{v} \cdot \left(1 + \frac{\Phi_p - V_B}{\mathrm{T_s}}\right)$$

6.9 Emissive Probes The relation between the floating potential and the probe potential for an emissive probe is found accurately by equating the emission current I_{we} to the plasma electron current I_{pe} to the probe. Taking $g_w = \left[1 + (\Phi_f - \Phi_p)/\mathrm{T_w}\right]^{1/2}$ and assuming that $\mathrm{T_e} \gg (\Phi_f - \Phi_p)$ in the emission current $I_{we} = I_{w0} \exp\left[-(\Phi_f - \Phi_p)/\mathrm{T_w}\right] g_w(\Phi_f - \Phi_p)$, show that

$$\frac{\Phi_f - \Phi_p}{\mathrm{T_w}} - \frac{1}{2} \ln\left(1 + \frac{\Phi_f - \Phi_p}{\mathrm{T_w}}\right) = \ln\left(\frac{I_{w0}}{I_{pe}}\right)$$

Plot $(\Phi_f - \Phi_p)/\mathrm{T_w}$ versus $\ln(I_{w0}/I_{p0})$ for $1 < I_{w0}/I_{p0} < 10$.

7

Chemical Reactions and Equilibrium

7.1 Introduction

Gas- and surface-phase chemical reactions play a critical role in plasma-assisted materials processing. To see why, consider the typical reactor, shown in Figure 7.1, that is used to etch a SiO_2 film. A CF_4/O_2 gas mixture is fed into the reactor and rf or microwave energy is applied to form a plasma. Electron impact ionization and dissociation create ions such as CF_3^+, CF_2^+, O_2^+, O^-, F^-, and free radicals such as CF_3, CF_2, O, and F. Chemical reactions in the gas phase and on the SiO_2 surface create additional molecules such as CO, CO_2, SiF_2, and SiF_4. The etch rates, anisotropies, and selectivities depend on the concentrations and energy distributions of all these species.

The concentrations are determined by general chemical reactions such as

$e + AB \rightarrow AB^+ + 2e$	(electron–ion pair production)
$e + AB \rightarrow e + A + B$	(radical production)
$e + AB \rightarrow A^- + B$	(negative ion production)
$A + B \rightarrow C + D$	(gas-phase chemical reactions)
$\Gamma_i = -D_a \nabla n_i$	(ion transport to surfaces)
$\Gamma_A = -D_A \nabla n_A$	(radical transport to surfaces)
$A(g) + B(s) \rightarrow C(g)$	(surface-phase reactions)

The net energy absorbed by these and other reactions must be supplied by the discharge power source. For example, electron energy is lost due to ionization, excitation, elastic scattering, and dissociation (the second reaction listed above). Hence, the discharge model must account for these energy losses. Further, the rates of these reactions depend critically on the energy distributions or temperatures of the reactants. Although thermodynamics determines the energy of reaction and can constrain the extent of reaction, most reactions occurring in typical reactors are far from thermodynamic equilibrium. Then collisions between pairs of species determine the reaction kinetics, including the reaction rates and the steady-state distribution of reactor species.

Another aspect seen in Figure 7.1 is the dual importance of homogeneous reactions in the gas-phase and heterogeneous reactions of gas-phase species with surfaces (the last reaction listed above). Hence, one must describe not only the properties of a given species but also possible changes in the phase of that species, e.g., from solid to gas, as well as changes in composition due to chemical reactions.

Principles of Plasma Discharges and Materials Processing, Third Edition. Michael A. Lieberman and Allan J. Lichtenberg.

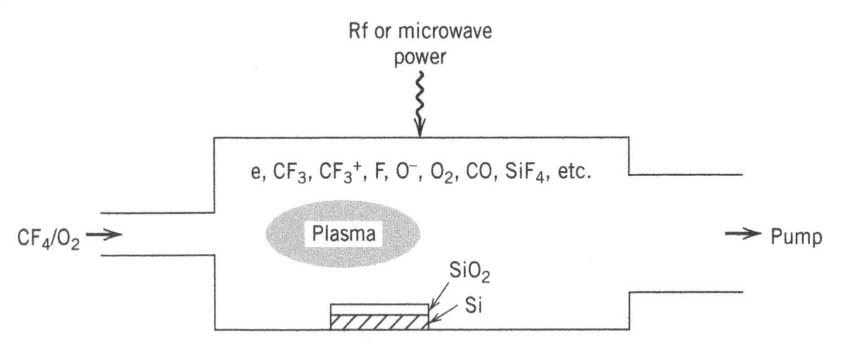

Figure 7.1 Typical materials processing reactor.

This and the following two chapters deal with the fundamentals of chemical dynamics. In this chapter, we describe the energetics of gas-phase and surface chemical reactions and chemical equilibrium. In Chapter 8, building on the study of atomic collisions in Chapter 3, we describe the fundamentals of molecular collisions, including such processes as dissociation, attachment, and recombination, and introduce appropriate rate constants. In Chapter 9, we introduce the principles of gas-phase and surface chemical kinetics, using the rate constants obtained in the previous chapter. We also describe the principles of surface interactions, including physical and chemical surface processes, the transport of species to surfaces, and surface reactions.

7.2 Energy and Enthalpy

The state of a system of $\mathcal{M}$ chemical species is uniquely determined by the temperature T, the total volume $\mathcal{V}$, and the number N_j of moles of each species (1 mol = 6.022×10^{23} molecules). This is illustrated for $\mathcal{M} = 3$ in Figure 7.2, for two states labeled 1 and 2, where the five axes shown in the figure are considered to be mutually perpendicular. State variables, such as the internal energy U, pressure p, entropy S, enthalpy H, and Gibbs free energy G, are then uniquely determined. For example, $U = \frac{3}{2}NRT$, and the equation of state determines $p = NRT/\mathcal{V}$ for a perfect gas. Often the equations for U, p, S, etc., can be inverted. Hence, other combinations of $\mathcal{M} + 2$ variables, such as (U, p, N_j), uniquely specify the state and thus determine $T, \mathcal{V}, S$, etc.

If a chemical system can exchange heat and work, but not matter with its surroundings, and undergo changes in chemical composition, then the first law of thermodynamics states that the

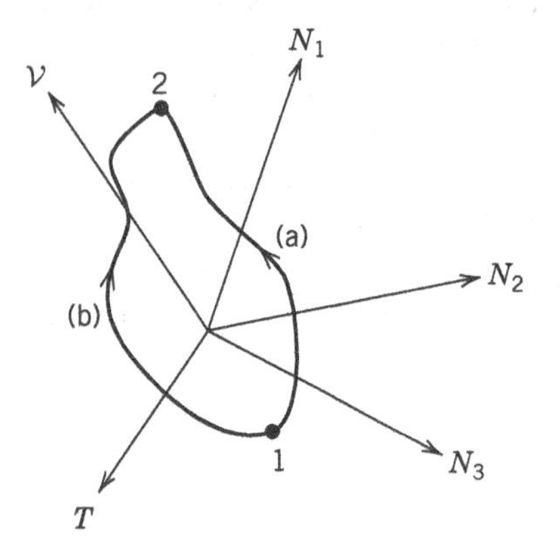

Figure 7.2 State space for a chemically reactive system.

increase $\mathrm{d}U$ in internal energy is equal to the sum of the heat flow $đQ$ into the system and the work done $đW$ on the system

$$\mathrm{d}U = đQ + đW \tag{7.2.1}$$

If neither heat nor work is exchanged with the surroundings, then U does not change. Equation (7.2.1) is the law of conservation of energy. Physically, U accounts for the random translational, vibrational, and rotational kinetic energy of the molecules in the system, the potential energies stored in the molecular chemical bonds, and the interaction energies between molecules.

The notation $đ$ is used for small changes of heat and work because $đQ$ and $đW$ are not, in general, exact differentials. Consider a process leading to a change from an initial state 1 to a final state 2 along two different paths (a) and (b), as shown in Figure 7.2. For exact differentials, such as $\mathrm{d}U$, the total change is independent of the path:

$$\Delta U_\mathrm{a} = \int_{\mathrm{path\ a}} \mathrm{d}U = \Delta U_\mathrm{b} = \int_{\mathrm{path\ b}} \mathrm{d}U = U_2 - U_1$$

The differentials of all state variables are exact, e.g., $\Delta p = p_2 - p_1$ and $\Delta \mathcal{V} = \mathcal{V}_2 - \mathcal{V}_1$. However, heat and work are not state variables. Hence,

$$\Delta Q_\mathrm{a} = \int_{\mathrm{path\ a}} đQ \neq \Delta Q_\mathrm{b} = \int_{\mathrm{path\ b}} đQ$$

and, similarly, $\Delta W_\mathrm{a} \neq \Delta W_\mathrm{b}$. Different kinds of work (mechanical, electrical, etc.) can be done on a system. We are considering here only $p\,\mathrm{d}\mathcal{V}$ work due to a change of volume $\mathcal{V}$. The work done on a system by its surroundings is found, from Newton's laws, to be

$$đW = -p_\mathrm{ext}\,\mathrm{d}\mathcal{V} \tag{7.2.2}$$

where p_ext is the pressure of the surroundings. In general, p_ext is not equal to the system pressure p. However, if the system is in near equilibrium with its surroundings, then $p_\mathrm{ext} \approx p$ and $T_\mathrm{ext} \approx T$. In this case,

$$đW = -p\,\mathrm{d}\mathcal{V} \tag{7.2.3}$$

If, during a process of change from state 1 to state 2, the system remains in near equilibrium with its surroundings, then the process is called *reversible*. Examples of reversible processes are the slow heating of a gas in a closed container ($\mathcal{V}$ = const) or in an open container capped by a piston exerting a constant pressure on the gas. The *reversible work* done on the system is found by integrating (7.2.3). Substituting (7.2.3) into (7.2.1), we see that

$$\mathrm{d}U = đQ - p\,\mathrm{d}\mathcal{V} \tag{7.2.4}$$

at every point along the path of a reversible process.

The work done is zero for a constant-volume reversible process. Integrating (7.2.4) shows that the increase in internal energy is equal to the total heat flow into the system:

$$U_2 - U_1 = \Delta Q \tag{7.2.5}$$

However, in plasma reactors, most processes occur at constant pressures, not constant volumes. It is useful to introduce a new state variable, the enthalpy

$$H = U + p\mathcal{V} \tag{7.2.6}$$

for constant-pressure processes. For example, for a perfect gas, $U = \frac{3}{2}NRT$ and $p\mathcal{V} = NRT$, so $H = \frac{5}{2}NRT$. Differentiating H and using (7.2.4), we obtain

$$dH = đQ + \mathcal{V}\,dp \tag{7.2.7}$$

Hence, the increase in enthalpy is equal to the total heat flow for constant-pressure processes:

$$H_2 - H_1 = \Delta Q \tag{7.2.8}$$

In general, there is a change of volume for a constant-pressure process. Integrating (7.2.3) yields the total work done on the system:

$$\Delta W = -p(\mathcal{V}_2 - \mathcal{V}_1) \tag{7.2.9}$$

Differentiating (7.2.6) at constant pressure, we obtain

$$\Delta H = \Delta U + p\Delta\mathcal{V} \tag{7.2.10}$$

Hence, the enthalpy change is equal to the sum of the internal energy change and the $p\,d\mathcal{V}$ work done *by* the system on its surroundings. Generally, $|\Delta W| \ll |\Delta U|$ for chemical reactions at the low pressures characteristic of plasma processing discharges; hence, $\Delta H \approx \Delta U$.

If a system containing N_j moles of each species undergoes a chemical reaction at constant temperature and pressure, then the N_js change, and the total enthalpy $H_2(T, p, N_j')$ after the reaction is not the same as the enthalpy $H_1(T, p, N_j)$ before the reaction. By (7.2.8), the excess enthalpy appears as heat. For $\Delta H > 0$, the reaction is called *en*dothermic and heat *en*ters the system. For $\Delta H < 0$, the reaction is called *ex*othermic and heat *ex*its. Although the enthalpy H_f for formation of a particular product species is a function of T, p, and the N_js, a *standard molar formation enthalpy* $H_f^\ominus(T_0)$ is tabulated in the thermodynamic literature for a standard temperature and pressure (STP) and for one mole (1 mol) of the product created by the reaction of the most stable natural forms of the elements. The standard pressure, denoted with a superscript $\ominus$, is usually taken to be either 1 bar = 10^5 Pa in the newer tables or 1 atm = 760 Torr = 1.013 bar in the older tables; the difference is not significant for our purposes. The standard temperature, denoted T_0, is taken to be 298.15 K = 25 °C. An example is the reaction for formation of the SiO_2: $Si(s) + O_2(g) \rightarrow SiO_2(s, \alpha)$; $H_f^\ominus(T_0) = -910.9$ kJ/mol, where s, l, and g denote solid, liquid, and gas, and α denotes the most stable (α) phase of SiO_2. In older tables, enthalpies are often specified in kcal/mol, where 1 kcal = 4.184 kJ. We also note that an energy equivalent voltage of 1 V/mol corresponds to 96.49 kJ/mol. When considering chemical reactions, only changes in enthalpies are significant. Hence, the standard enthalpies of formation of the elements in their most stable state are taken to be zero at all temperatures. Some standard enthalpies of formation are given in Tables 7.1 and 7.2.

The standard enthalpy $H_r^\ominus(T_0)$ for any chemical reaction can be calculated by subtracting the enthalpies of formation of the reactants from those of the products. For example, consider the etching of one mole of $SiO_2(s)$ by fluorine gas:

$$SiO_2(s) + 2F_2(g) \rightarrow SiF_4(g) + O_2(g) \tag{7.2.11}$$

From Table 7.1, $H_f^\ominus(T_0) = -910.9$ kJ/mol for one mole of $SiO_2(s)$ and $H_f^\ominus(T_0) = -1614.9$ kJ/mol for one mole of SiF_4. Hence,

$$H_r^\ominus(T_0) = (1)(-1614.9) - (1)(-910.9) = -704.0 \text{ kJ/mol}$$

and the reaction is exothermic.

For $SiO_2(s)$ etching by chlorine gas,

$$SiO_2(s) + 2Cl_2(g) \rightarrow SiCl_4(g) + O_2(g) \tag{7.2.12}$$

Table 7.1 Thermodynamic Properties

Substances	$H_f^\ominus(T_0)$ (kJ/mol)	$G_f^\ominus(T_0)$ (kJ/mol)
O	249.2	231.7
O_3	142.7	163.2
H	218.0	203.2
OH	39.0	34.2
H_2O(l)	−285.8	−237.1
H_2O	−241.8	−228.6
F	78.99	61.91
HF	−271.1	−273.2
Cl	121.7	105.7
HCl	−92.3	−95.3
Br	111.9	82.4
Br_2	30.9	3.11
S	278.8	238.3
SF_4	−774.9	−731.6
SF_6	−1209	−1105
N	472.7	455.6
C (graphite cr)	0	0
C (diamond cr)	1.90	2.90
CO	−110.5	−137.2
CO_2	−393.5	−394.4
CH_2	390.4	372.9
CH_3	145.7	147.9
CH_4	−74.8	−50.7
CF_3	−477	−464
CF_4	−925	−879
COF_2	−634.7	−619.2
CH_2F_2	−446.9	−419.2
CHF_3	688.3	−653.9
CCl_4	−102.9	−60.59
$COCl_2$	−218.8	−204.6
CH_3Cl	−80.8	−57.4
CH_2Cl_2	−92.5	−65.9
$CHCl_3$	−103.1	−70.3
C_2H_2	226.7	209.2
C_2H_4	52.3	68.2
C_2H_6	−84.7	−32.8

(Continued)

Table 7.1 (Continued)

Substances	$H^{\ominus}_{f}(T_0)$ **(kJ/mol)**	$G^{\ominus}_{f}(T_0)$ **(kJ/mol)**
C_2F_4	−650.6	−615.9
C_2F_6	−1297	−1213
Si (cr)	0	0
Si	455.6	411.3
SiO	−99.6	−126.4
SiO_2 (α quartz, cr)	−910.9	−856.6
SiO_2 (amorphous)	−903.5	−850.7
SiH_4	34.3	56.9
SiF	7.1	−24.3
SiF_2	−619	−628
SiF_4	−1614.9	−1572.7
$SiCl_2$	−165.6	−177.2
$SiCl_4$(l)	−687.0	−619.8
$SiCl_4$	−657	−617
Si_3N_4 (α, cr)	−743.5	−642.6
SiC (β, cubic)	−65.3	−62.8
Al_2O_3 (α)	−1675.7	−1582.3
AlF_3 (cr)	−1510.4	−1431.1
AlF_3	−1204.6	−1188.2
$AlCl_3$ (cr)	−704.2	−628.8
WF_6	−1721.7	−1632.1

Note. Substances are in gas phase unless otherwise specified.

we obtain

$$H^{\ominus}_{r}(T_0) = (1)(-657) - (1)(-910.9) = 253.9 \text{ kJ/mol}$$

and the reaction is endothermic.

The reactions in plasma processing do not necessarily take place at the standard temperature. To determine the temperature dependence of the enthalpy, we note that at constant pressure and composition, a small heat flow $đQ$ into the system produces a proportionate temperature rise,

$$đQ = C_p \, \mathrm{d}T \tag{7.2.13}$$

where the constant of proportionality C_p is called the *specific heat at constant pressure.* Since $\mathrm{d}H = đQ$ under these conditions, we find that

$$C_p = \left(\frac{\partial H}{\partial T}\right)_{p,N_j} \tag{7.2.14}$$

For a perfect gas, $H = \frac{5}{2}NRT$ and $C_p = \frac{5}{2}RN$. The specific heat for one mole of perfect gas is $C_{pm} \approx$ 20.8 J/K-mol. Most substances, including real gases, have $C_{pm} \sim$ 30–100 J/K-mol. The enthalpy at

Table 7.2 Enthalpies of Formation

Substances	$H_f^{\ominus}(T_0)$ **(kJ/mol)**
CH	595.8
CCl_3	59
CF_2	−194.1
CF_3	−467.4
SiH	377
SiH_2	269.0
SiH_3	194.1
SiF	−19.3
SiF_2	−587.9
SiF_3	−1025
SiCl	195.8
$SiCl_2$	−163.6
$SiCl_3$	−318
$AlCl_3$	−583.2
$Al(CH_3)_3$	−86.5

Note. All substances are in gas phase.

temperature T can be written as

$$H(T) = H(T_0) + \int_{T_0}^{T} C_p(T')\,dT' \tag{7.2.15}$$

Since reaction enthalpies are typically hundreds of kilojoules per mole, the integral in (7.2.15) is not too important for temperatures within a few hundred degrees of T_0, as is common in processing discharges.

Similarly, the enthalpy depends only weakly on the pressure. In fact, for a perfect gas, $H = \frac{5}{2}NRT$ and therefore is independent of p. At the low pressures of processing discharges, the pressure dependence is negligible.

The enthalpies associated with breaking chemical bonds to form neutral products are also of interest.

The dissociation reaction for the molecule AB,

$$AB(g) \rightarrow A(g) + B(g)$$

where both A and B may be groups of atoms, has a *dissociation enthalpy* $H_{diss}^{\ominus}(T_0)$ for breaking the AB bond. Some bond dissociation enthalpies are given in Table 7.3. A *mean bond dissociation enthalpy*, which is an average of $H_{diss}^{\ominus}(T_0)$ over many different types of molecules containing the bond, can also be defined. For example, $H_{diss}^{\ominus}(T_0) = 492$ kJ/mol for the HO–H bond and 428 kJ/mol for the O–H radical bond; the mean enthalpy of O–H bonds in many different molecules is 463 kJ/mol. The *enthalpy of phase transition* is also of interest, including *sublimation* s → g, *vaporization* l → g, and *melting* s → l, e.g., $H_2O(l) \rightarrow H_2O(g)$ has $H_{vap}^{\ominus}(100\,°C) = 40.66$ kJ/mol. Some enthalpies of formation of gaseous atoms are given in Table 7.4. The data in Tables 7.3 and 7.4

Table 7.3 Bond Dissociation Enthalpies

Bond	$H^{\ominus}_{\text{diss}}(T_0)$ **(kJ/mol)**
Al–Al	186.2
Al–Cl	511.3
Al–Cu	216.7
Al–F	663.6
Al–O	512.1
B–Cl	536
B–O	806
C–C	607
C–F	552
$CF_2{=}CF_2$	319.2
CF_3–CF_3	413.0
C–H	338.3
C–O	1076.5
C=O	749
C–Si	451.5
Cl–O	272
Cl–Si	381
Cl–Cl	243
F–F	158.75
F–Ni	435
F–O	222
F–S	342.7
F–SF_5	381.2
F–SF_4	222.2
F–SF_3	351.9
F–SF_2	264.0
F–SF	383.7
F–Si	552.7
F–W	548
F–Zn	368
H–O	427.5
H–Si	299.2
Hf–O	791
O–Si	799.6
Si–Si	326.8

Table 7.4 Enthalpies of Formation of Gaseous Atoms

Element	$H_f^{\ominus}(T_0)$ (kJ/mol)
Si	455.6
C	716.7
Br	111.9
Cl	121.7
F	79.4
H	218.0
Al	329.7
Mo	658.1
O	249.2
S	278.8
W	849.8
Zn	130.42
N	472.7
Cu	341
Ge	328
Ni	425

can be used to estimate the enthalpy of formation of various substances (see Problem 7.2). Other enthalpies include *ionization*,

$$A(g) \rightarrow A^{+}(g) + e$$

and *electron affinity*,

$$A(g) + e \rightarrow A^{-}(g)$$

For example, the enthalpy for ionization of Cl is 1251 kJ/mol, corresponding to 12.96 V/atom. The electron affinity enthalpy for Cl^- is −348.6 kJ/mol, corresponding to −3.61 V/atom. The affinity reaction is exothermic for Cl^- production.

7.3 Entropy and Gibbs Free Energy

We have seen in the previous section that for a reversible change, the system moves slowly through a succession of equilibrium states. There is no spontaneous tendency to move in one direction or the other. An example is the expansion of a gas as the volume of its container is slowly increased. But some things do happen spontaneously. Gas from a burst balloon expands to fill an available volume; it does not spontaneously contract to a smaller volume. A cold body absorbs heat from hotter surroundings; it does not supply heat to the surroundings and get colder. The second law of thermodynamics asserts that there is a state variable S, the entropy, that determines the direction of spontaneous change, which is defined by

$$dS = đQ/T \tag{7.3.1}$$

where $đQ$ is the heat injected into a system by a reversible process. The second law also asserts that, for a spontaneous process,

$$dS > đQ/T \tag{7.3.2}$$

The entropy is a measure of the disorder in the system.

Consider a thermally isolated system of chemical species that irreversibly (spontaneously) undergoes a chemical reaction, leading to a change in temperature, pressure, and species concentrations. Since $đQ = 0$ for a thermally isolated system, (7.3.2) shows that the system entropy must increase, i.e., the direction of spontaneous change in a thermally isolated system is to increase the system disorder.

Inserting (7.3.1) into the internal energy change (7.2.4), we find

$$dU = TdS - p\,d\mathcal{V} \tag{7.3.3}$$

Although (7.3.3) was derived for a reversible process, it applies to any process, reversible or irreversible. This is because the internal energy U depends only on the state of the system, so we may as well determine the change in energy from one state to another by using a reversible process. Although $đQ = T\,dS$ and $đW = -p\,d\mathcal{V}$ for a reversible process, and $đQ < T\,dS$ and $đW > -p d\mathcal{V}$ for an irreversible process, the sum $đQ + đW$ is always equal to $T\,dS - p\,d\mathcal{V}$. Similarly, inserting (7.3.1) into the enthalpy change (7.2.7), we find

$$dH = T\,dS + \mathcal{V}\,dp \tag{7.3.4}$$

for any process, reversible or irreversible.

Equation (7.3.1) can be used to determine the variation of S with temperature and pressure. Consider a constant pressure process for which the surroundings are heated slowly from T_0 to T_1. Then from the definition of specific heat (7.2.13), a reversible heat

$$đQ = dH(p, T) = C_p(p, T)\,dT \tag{7.3.5}$$

flows into the system. Inserting (7.3.5) into (7.3.1) and integrating, we obtain

$$S(p_0, T) - S(p_0, T_0) = \int_{T_0}^{T} \frac{C_p(p_0, T')}{T'} dT' \tag{7.3.6}$$

For a perfect gas, $C_p = \frac{5}{2}RN$ and

$$S(p_0, T) - S(p_0, T_0) = \frac{5}{2}RN \ln\left(\frac{T}{T_0}\right) \tag{7.3.7}$$

which gives the temperature variation of the entropy.

Similarly, the change in internal energy for a constant-volume reversible process is, from the internal energy change (7.2.4),

$$đQ = dU(\mathcal{V}, T) = C_{\mathcal{V}}(\mathcal{V}, T)\,dT \tag{7.3.8}$$

where

$$C_{\mathcal{V}} = \left(\frac{\partial U}{\partial T}\right)_{\mathcal{V}} \tag{7.3.9}$$

is the *specific heat at constant volume*. Inserting (7.3.8) into (7.3.1) and integrating, we find

$$S(\mathcal{V}, T) - S(\mathcal{V}, T_0) = \int_{T_0}^{T} \frac{C_{\mathcal{V}}(\mathcal{V}, T')}{T'} dT' \tag{7.3.10}$$

For a perfect gas, $C_{\mathcal{V}} = \frac{3}{2}RN$ and

$$S(\mathcal{V}, T) - S(\mathcal{V}, T_0) = \frac{3}{2}RN \ln\left(\frac{T}{T_0}\right) \tag{7.3.11}$$

Equations (7.3.6) and (7.3.10) can be used to determine the variation of entropy with pressure by considering the two-step reversible process

$$(p_0, T_0) \overrightarrow{p\ const} (p_0, T) \overrightarrow{\mathcal{V}\ const} (p, T_0)$$

For a perfect gas, using (7.3.7) and (7.3.11), and noting that $\mathcal{V} = NRT/p_0 = NRT_0/p$, we obtain

$$S(p, T_0) - S(p_0, T_0) = -RN \ln\left(\frac{p}{p_0}\right) \tag{7.3.12}$$

which gives the pressure variation of the entropy.

In general, the specific heats are continuous functions of temperature except at isolated values of T where the system undergoes a change of phase (first-order phase transition). At these temperatures, the specific heats are singular. An example is shown in Figure 7.3 for a change of phase of a pure substance from a solid to a liquid to a gas. The third law of thermodynamics states that the entropy of all perfect crystalline compounds may be taken to be zero at $T = 0$. Hence, integrating C_p from 0 to T, including the appropriate δ functions at T_{melt} and T_{vap}, yields the entropy. The *standard molar entropies* $S_{\text{m}}^{\ominus}(T_0)$ J/K-mol of various pure substances and compounds are tabulated in the thermodynamic literature. The *standard reaction entropies* $S_{\text{r}}^{\ominus}(T_0)$ for any reaction are found by subtracting the standard entropies of the reactants from those of the products.

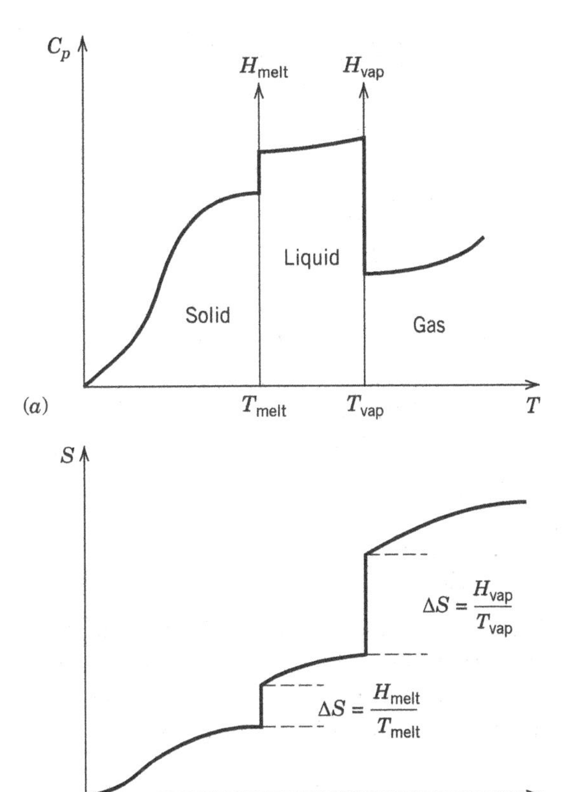

Figure 7.3 (*a*) Specific heat C_p at constant pressure and (*b*) entropy S versus temperature T.

7.3.1 Gibbs Free Energy

For a constant-pressure process $đQ = dH$ and the second law, (7.3.1) and (7.3.2) can be written

$$dH - T\,dS \leq 0 \tag{7.3.13}$$

where the equality applies for a reversible process. Introducing a new state variable, the *Gibbs free energy*

$$G = H - TS \tag{7.3.14}$$

such that

$$dG = dH - T\,dS - S\,dT \tag{7.3.15}$$

and comparing (7.3.13) and (7.3.15) at constant temperature, we see that

$$dG = dH - T\,dS \leq 0 \tag{7.3.16}$$

Hence, for a chemical reaction to proceed spontaneously at constant temperature and pressure, the Gibbs free energy must decrease. Inserting (7.3.4) into (7.3.15), we obtain

$$dG = \mathcal{V}\,dp - S\,dT \tag{7.3.17}$$

If we let $G = G(p, T, N_j)$, where (p, T, N_j), $j = 1, \ldots, \mathcal{M}$, specifies the state of the system, then the differential of G is

$$dG = \left(\frac{\partial G}{\partial p}\right)_{T,\{N_i\}} dp + \left(\frac{\partial G}{\partial T}\right)_{p,\{N_i\}} dT + \sum_{j=1}^{\mathcal{M}} \left(\frac{\partial G}{\partial N_j}\right)_{p,T,\{N_i \neq N_j\}} dN_j \tag{7.3.18}$$

Comparing (7.3.18) with (7.3.17), we see that

$$\mathcal{V} = \left(\frac{\partial G}{\partial p}\right)_{T,\{N_i\}} \tag{7.3.19}$$

$$S = -\left(\frac{\partial G}{\partial T}\right)_{p,\{N_i\}} \tag{7.3.20}$$

and, introducing the *chemical potential*

$$\mu_j = \left(\frac{\partial G}{\partial N_j}\right)_{p,T,\{N_i \neq N_j\}} \tag{7.3.21}$$

we see that

$$\sum_{j=1}^{\mathcal{M}} \mu_j\,dN_j = 0 \tag{7.3.22}$$

The chemical potential specifies how G changes as various substances j are added to the system. For a *closed system*, for which heat and work, but not matter, can be exchanged with the surroundings, (7.3.22) must hold, i.e., dG is independent of changes in composition. However, for an *open system*, for which matter can be exchanged with the surroundings, we must write

$$dG = \mathcal{V}\,dp - S\,dT + \sum_{j=1}^{\mathcal{M}} \mu_j\,dN_j \tag{7.3.23}$$

in place of (7.3.17). We note that $\mathcal{V}$, S, and μ_j in (7.3.23) are all functions of the state (p, T, N_j), $j = 1, \ldots, \mathcal{M}$. However, for a single substance

$$\mu = \left(\frac{\partial G}{\partial N}\right)_{p,T} \tag{7.3.24}$$

is independent of N. Hence, μ is equal to the *molar Gibbs free energy* $G_{\mathrm{m}}(p, T)$ for that substance.

From (7.3.14), the *standard molar Gibbs free energy of formation* of any substance from the elements in their most stable natural states is

$$\mu^{\ominus}(T_0) \equiv G_{\mathrm{f}}^{\ominus}(T_0) = H_{\mathrm{f}}^{\ominus}(T_0) - T_0 S_{\mathrm{m}}^{\ominus}(T_0) \tag{7.3.25}$$

These data are tabulated in the thermodynamic literature, and some selected values are given in Table 7.1. The standard Gibbs free energy $G_{\mathrm{r}}^{\ominus}(T_0)$ for any chemical reaction is found by subtracting the standard Gibbs free energies for formation of the reactants from those of the products. Again, $G_{\mathrm{f}}^{\ominus}(T_0)$ for the elements in their most stable natural state is taken to be zero.

As an example, consider reaction (7.2.12) for etching one mole of SiO_2(s) by chlorine gas. From Table 7.1, we find

$$G_{\mathrm{r}}^{\ominus}(T_0) = (1)(-617.0) - (1)(-856.6) = 239.6 \text{ kJ/mol} \tag{7.3.26}$$

The pressure and temperature variations of μ are found by integrating (7.3.19) and (7.3.20) for one mole of substance from STP at $(p^{\ominus}, T_0)$ to (p, T). First integrating (7.3.20) from $(p^{\ominus}, T_0)$ to $(p^{\ominus}, T)$ and assuming a perfect gas, such that $S(p^{\ominus}, T)$ is found from (7.3.7), we obtain

$$\mu^{\ominus}(T) = \mu^{\ominus}(T_0) + \left(T - T_0\right)\left[\frac{5}{2}R - S_{\mathrm{m}}^{\ominus}(T_0)\right] - \frac{5}{2}RT \ln\left(\frac{T}{T_0}\right) \tag{7.3.27}$$

To obtain the pressure variation, we integrate (7.3.19) from $(p^{\ominus}, T)$ to (p, T), using $\mathcal{V} = RT/p$ for one mole of a perfect gas, to obtain

$$\mu(p, T) = \mu^{\ominus}(T) + RT \ln\left(\frac{p}{p^{\ominus}}\right) \tag{7.3.28}$$

For a mixture of perfect gases, p is replaced by the partial pressure p_j in (7.3.28):

$$\mu_j(\mathrm{g}) = \mu_j^{\ominus}(T) + RT \ln\left(\frac{p_j}{p^{\ominus}}\right) \tag{7.3.29}$$

Introducing the *mole fractions* $x_j = p_j/p = N_j/\sum_{i=1}^{\mathcal{M}_{\mathrm{g}}} N_i$ for the $\mathcal{M}_{\mathrm{g}}$ gas-phase species, we have

$$\mu_j(\mathrm{g}) = \mu_j^{\ominus}(T) + RT \ln\left(\frac{x_j p}{p^{\ominus}}\right) \tag{7.3.30}$$

The x_js give the composition dependence. For typical processing discharges, most gases can be considered ideal. For solids or liquids, (7.3.28) is replaced by

$$\mu_j = \mu_j^{\ominus}(T) + RT \ln a_j \tag{7.3.31}$$

where $a_j = \gamma_j x_j$, a_j is the *activity*, γ_j is the *activity coefficient*, and x_j is the mole fraction in the solid or liquid phase. For a pure solid or liquid, $x_j = 1$ and γ_j is chosen to be unity at standard pressure $p^{\ominus}$. Hence, $a_j = 1$ and $\mu_j = \mu_j^{\ominus}$ for the pure substance at $p^{\ominus}$. Integrating (7.3.19) and (7.3.20) for one mole of solid or liquid substance shows that μ_j depends only weakly on p and T for typical values of the molar volume $\mathcal{V}_{\mathrm{m}}$ and molar entropy S_{m}. Assuming that the mutual solubilities of the constituents in the solid or liquid phases are small, then these phases are pure, and the a_js can be taken to be unity for the solid or liquid reactants and products at the pressures and temperatures in typical processing discharges:

$$\mu_j(\mathrm{s}\,,\mathrm{l}) = \mu_j^{\ominus}(T) \tag{7.3.32}$$

7.4 Chemical Equilibrium

Consider a chemical reaction in a closed system, such as

$$3\mathrm{A} + \mathrm{B} = 2\mathrm{C} + 4\mathrm{D} \tag{7.4.1}$$

Let $\mathcal{J}_1 = \mathrm{A}$, $\mathcal{J}_2 = \mathrm{B}$, $\mathcal{J}_3 = \mathrm{C}$, $\mathcal{J}_4 = \mathrm{D}$, etc. denote the species and introducing the *stoichiometric coefficients* $\alpha_1 = -3$, $\alpha_2 = -1$, $\alpha_3 = 2$, $\alpha_4 = 4$, etc., (the αs are negative for reactants and positive for products), the reaction can be written as

$$\sum_j \alpha_j \mathcal{J}_j = 0 \tag{7.4.2}$$

Let the number of moles of $\mathcal{J}_j$ change by $\mathrm{d}N_j = \alpha_j \mathrm{d}N$, where $\mathrm{d}N$ is the *extent* of the reaction. For reaction at constant pressure and temperature, (7.3.18) shows that

$$\mathrm{d}G = \sum_j \alpha_j \mu_j \,\mathrm{d}N \tag{7.4.3}$$

If the reaction (7.4.1) proceeds spontaneously, either to the right (C and D are formed) or to the left (A and B are formed), then the second law (7.3.16) shows that $\mathrm{d}G < 0$ and, hence, G must decrease. Eventually, the system attains a state of equilibrium in which the concentrations of the various species no longer change spontaneously; at this equilibrium state, $\mathrm{d}G = 0$. Hence, as shown in Figure 7.4, the equilibrium state is a minimum of G with respect to composition changes. Using (7.4.3), we see that

$$\sum_{j=1}^{\mathcal{M}} \alpha_j \mu_j = 0 \tag{7.4.4}$$

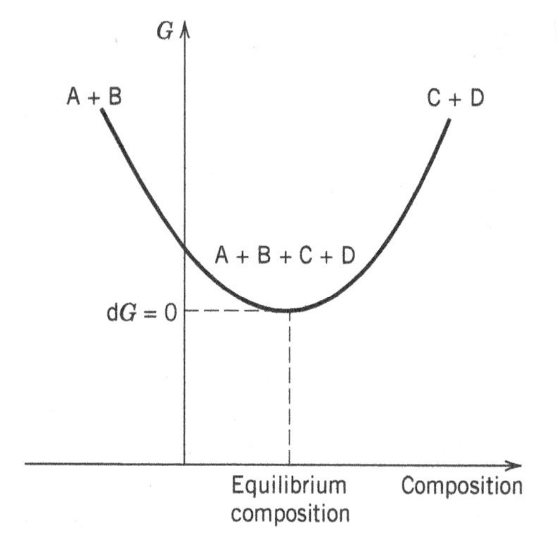

Figure 7.4 Gibbs free energy G versus composition.

at equilibrium. Inserting the chemical potentials (7.3.29) for the gas-phase constituents and (7.3.32) for the liquid- and solid-phase constituents into (7.4.4), we obtain

$$-RT\sum_{j=1}^{\mathcal{M}_g}\alpha_j \ln\left(\frac{\overline{p}_j}{p^\ominus}\right) = \sum_{j=1}^{\mathcal{M}}\alpha_j\mu_j^\ominus(T) \tag{7.4.5}$$

where $\overline{p}_j$ is the equilibrium partial pressure of the jth species, and the sum on the left is over the $\mathcal{M}_g$ gas-phase constituents only. The term on the right-hand side of (7.4.5) is the Gibbs free energy $G_r^\ominus(T)$ of the reaction. Using this and introducing the *equilibrium constant*

$$\mathcal{K} = \prod_{j=1}^{\mathcal{M}_g}\left(\frac{\overline{p}_j}{p^\ominus}\right)^{\alpha_j} \tag{7.4.6}$$

into (7.4.5), we obtain

$$\mathcal{K}(T) = \exp\left[-\frac{G_r^\ominus(T)}{RT}\right] \tag{7.4.7}$$

Species	Initial moles	Equilibrium moles
Cl_2	$x_{Cl_2}N_0$	$x_{Cl_2}N_0 - 2N$
O_2	$x_{O_2}N_0$	$x_{O_2}N_0 + N$
$SiCl_4$	$x_{SiCl_4}N_0$	$x_{SiCl_4}N_0 + N$

Equations (7.4.6) and (7.4.7) are the fundamental equations of chemical equilibrium. $\mathcal{K}$ can be written in terms of the equilibrium mole fractions $\overline{x}_j = \overline{p}_j/p$ as

$$\mathcal{K} = \mathcal{K}_x\left(\frac{p}{p^\ominus}\right)^{\alpha_g} \tag{7.4.8}$$

where

$$\mathcal{K}_x = \prod_{j=1}^{\mathcal{M}_g}\overline{x}_j^{\alpha_j} \tag{7.4.9}$$

and

$$\alpha_g = \sum_{j=1}^{\mathcal{M}_g}\alpha_j \tag{7.4.10}$$

is the sum of the gas-phase stoichiometric coefficients.

As an example, consider the reaction (7.2.12) for the etching of one mole of SiO_2 by Cl_2 gas at STP. The reaction Gibbs free energy is, from (7.3.26), $G_r^\ominus(T_0) = 239.6$ kJ/mol. Using (7.4.7) with $RT = 2.479$ kJ/mol, we find $\mathcal{K}(T_0) = 1.02\times10^{-42}$. Let x_{Cl_2}, x_{O_2}, and x_{SiCl_4} be the gas-phase mole fractions and N_0 be the total number of gas-phase moles in the initial state. Let N be the extent of the reaction to attain the equilibrium state. Then we obtain the following table based on conservation of Cl_2 and O_2 for the reaction (7.2.12): using $\alpha_{Cl_2} = -2$, $\alpha_{O_2} = 1$, $\alpha_{SiCl_4} = 1$, we obtain $\alpha_g = 0$ from (7.4.10) and, hence, $\mathcal{K} = \mathcal{K}_x$ from (7.4.9). Dividing each element in the third column of the table by the initial number of gas-phase moles, we obtain from (7.4.9) that

$$\mathcal{K} = \frac{(x_{O_2} + N/N_0)(x_{SiCl_4} + N/N_0)}{(x_{Cl_2} - 2N/N_0)^2} = 1.02\times10^{-42} \tag{7.4.11}$$

If the initial state contains only SiO_2 and Cl_2, then $x_{O_2} = x_{SiCl_4} = 0$ and $x_{Cl_2} = 1$. Then (7.4.11) becomes

$$\left(\frac{N/N_0}{1-2N/N_0}\right)^2 = 1.02 \times 10^{-42}$$

from which we obtain $N/N_0 = 1.01 \times 10^{-21} \ll 1$. Hence, only a negligible etching of SiO_2 occurs before equilibrium is obtained.

In contrast, consider reaction (7.2.11) for SiO_2 etching by fluorine gas, for which, using the data from Table 7.1, $G_r^\ominus(T_0) = -716.1$ kJ/mol. Using (7.4.7), we obtain $\mathcal{K} = 3.2 \times 10^{125} \gg 1$. Hence, almost the entire F_2 gas charge reacts to attain the equilibrium state.

It is necessary to emphasize at this point that thermodynamics has nothing to say about the rate of the reaction to attain the equilibrium state. The reaction timescale might be microseconds or centuries. Rates are typically fast for gas- or liquid-phase reactions due to the high mobilities of the reactants and products, but they can be very slow if one of the reactants or products is a solid. Catalysts can be used to increase the reaction rates without altering the thermodynamic equilibrium. Reaction rates are the provenance of chemical kinetics, which we consider in Chapter 9.

7.4.1 Pressure and Temperature Variations

Changing the reaction pressure and temperature can have a strong effect on the equilibrium. First considering pressure variations, we note from (7.4.7) that $\mathcal{K}$ is independent of pressure. However, the mole fractions $\bar{x}$ will generally change as p changes. Inserting (7.4.9) into (7.4.8), we obtain

$$\prod_{j=1}^{\mathcal{M}_g} \bar{x}_j^{\alpha_j} = \left(\frac{p}{p^\ominus}\right)^{-\alpha_g} \mathcal{K} \tag{7.4.12}$$

Recall that α_g, given by (7.4.10), is the difference between the number of gas-phase product and reactant molecules for the stoichiometric reaction. For $\alpha_g > 0$, there are more gas-phase product molecules than reactant molecules. If the pressure is decreased, then the right-hand side of (7.4.12) is increased, driving the reaction to the right, i.e., the $\bar{x}_j$s for the products increase and the $\bar{x}_j$s for the reactants decrease. Hence, at low pressures, it is desirable to seek reactions having $\alpha_g > 0$. For $\alpha_g < 0$, a decrease in pressure drives the reaction to the left (fewer products, more reactants). For $\alpha_g = 0$, the $\bar{x}_j$s are independent of pressure. These variations are summarized in the following table:

p Change	$\alpha_g > 0$	$\alpha_g = 0$	$\alpha_g < 0$
$p \downarrow$	Products ↑	No change	Products ↓
$p \uparrow$	Products ↓	No change	Products ↑

Equation (7.4.7) shows that the temperature variation of $\mathcal{K}$ is specified by the variation of $G_r^\ominus(T)/T$, which we can derive as follows: inserting the entropy (7.3.20) into the definition of G in (7.3.14), we obtain, at constant pressure,

$$G = H + T\frac{\partial G}{\partial T} \tag{7.4.13}$$

Dividing (7.4.13) by T^2 and rearranging, we obtain the *Gibbs–Helmholtz equation*

$$-\frac{H}{T^2} = -\frac{G}{T^2} + \frac{1}{T}\frac{\partial G}{\partial T} = \frac{\partial}{\partial T}\left(\frac{G}{T}\right) \tag{7.4.14}$$

Using (7.4.14) for each reaction species, we find

$$\left(\frac{\partial}{\partial T}\frac{G_r^{\ominus}(T)}{T}\right) = -\frac{H_r^{\ominus}(T)}{T^2} \tag{7.4.15}$$

Substituting (7.4.15) into the derivative of (7.4.7) and then dividing by (7.4.7), we obtain

$$\frac{d}{dT}\ln \mathcal{K}(T) = \frac{H_r^{\ominus}(T)}{RT^2} \tag{7.4.16}$$

We see from (7.4.16) that increasing the temperature for an exothermic reaction ($H_r < 0$) drives the reaction toward the left (fewer products, more reactants). Increasing the temperature drives an endothermic reaction toward the right (more products, fewer reactants). Integrating (7.4.16) over a temperature change from T_0 to T_1 and assuming that $H_r^{\ominus} \approx$ const, independent of temperature, we obtain

$$\mathcal{K}(T_1) = \mathcal{K}(T_0)\exp\left[\frac{H_r^{\ominus}}{R}\left(\frac{1}{T_0} - \frac{1}{T_1}\right)\right] \tag{7.4.17}$$

The following table summarizes the temperature variation:

T Change	$H_r < 0$	$H_r > 0$
$T\uparrow$	Products $\downarrow$	Products $\uparrow$
$T\downarrow$	Products $\uparrow$	Products $\downarrow$

7.5 Heterogeneous Equilibrium

7.5.1 Equilibrium Between Phases

We consider equilibrium between gas and liquid phases of a pure substance, e.g., H_2O, at constant temperature and pressure. Suppose that N_g moles of gas are in equilibrium with N_l moles of liquid. Let μ_g and μ_l be the chemical potentials of the gas and liquid. If dN moles are transferred from the gas to the liquid, then the Gibbs free energy changes by

$$dG = -\mu_g\, dN + \mu_l\, dN$$

If $\mu_g \neq \mu_l$, then dN can be chosen to make $dG < 0$; hence, the system is not in equilibrium. Therefore, in equilibrium,

$$\mu_g = \mu_l = \mu_s \equiv \mu \tag{7.5.1}$$

independent of phase.

Now suppose that T and p are changed slightly so as to remain in equilibrium with N_g and N_l constant. Using the Gibbs free energy change (7.3.17), we obtain

$$d\mu_g = -S_{gm}\, dT + \mathcal{V}_{gm}\, dp \tag{7.5.2}$$

$$d\mu_l = -S_{lm}\, dT + \mathcal{V}_{lm}\, dp \tag{7.5.3}$$

where S_{gm} and S_{lm} are the entropy per mole and $\mathcal{V}_{gm}$ and $\mathcal{V}_{lm}$ are the volume per mole of the gas and liquid phases. Using (7.5.1), we can equate the RHSs of (7.5.2) and (7.5.3) to obtain

$$\frac{dp}{dT} = \frac{\Delta S_m}{\Delta \mathcal{V}_m} \tag{7.5.4}$$

where

$$\Delta S_{\mathrm{m}} = S_{\mathrm{gm}} - S_{\mathrm{lm}} \tag{7.5.5}$$

$$\Delta \mathcal{V}_{\mathrm{m}} = \mathcal{V}_{\mathrm{gm}} - \mathcal{V}_{\mathrm{lm}} \tag{7.5.6}$$

From the entropy change (7.3.1) with $đQ = \mathrm{d}H$ (see also Figure 7.3), the change in the molar entropy is

$$\Delta S_{\mathrm{m}} = \frac{H_{\mathrm{vap}}}{T} \tag{7.5.7}$$

Assuming that $\mathcal{V}_{\mathrm{lm}} \ll \mathcal{V}_{\mathrm{gm}}$ and using the perfect gas law to determine $\mathcal{V}_{\mathrm{gm}}$, we have

$$\Delta \mathcal{V}_{\mathrm{m}} \approx \mathcal{V}_{\mathrm{gm}} = \frac{RT}{p} \tag{7.5.8}$$

Inserting (7.5.7) and (7.5.8) into (7.5.4) yields

$$\frac{\mathrm{d}p}{\mathrm{d}T} = \frac{H_{\mathrm{vap}}}{RT^2} p \tag{7.5.9}$$

which is known as the *Clausius–Clapeyron equation.* Assuming that H_{vap} varies only weakly with T, we can integrate this to find

$$p_j = p_{0j} \exp\left(-\frac{H_{\mathrm{vap}}}{RT}\right) \tag{7.5.10}$$

where the subscript j denotes a pure substance. Equation (7.5.10) specifies the *vapor pressure* p_j of the gas in equilibrium with the liquid at temperature T. For the two phases to coexist, p_j and T cannot be independently chosen. Conversely, if p_j and T do not satisfy (7.5.10), then one of the phases does not exist.

The preceding analysis can be applied similarly to equilibrium between the gas and solid phases, yielding

$$p_j = p'_{0j} \exp\left(-\frac{H_{\mathrm{subl}}}{RT}\right) \tag{7.5.11}$$

where H_{subl} is the sublimation enthalpy per mole. For most substances, $H \gg RT$, and thus p is a strong function of T. Plotting $\ln p_j$ versus $1/RT$ yields a straight line with slope $-H$. In the usual case, the curves (7.5.10) and (7.5.11) intersect at the *triple point* (p_3, T_3), leading to the phase diagram shown in Figure 7.5. All three phases can coexist only at the triple point. As an example, for H_2O, $H_{\mathrm{vap}} \approx 40.66$ kJ/mol and $p_j = 1$ atm at 100 °C. This determines p_{0j} in (7.5.10). Table 7.5 gives some vapor pressure data for various substances.

For a mixture of substances, (7.5.10) and (7.5.11) hold for the partial pressures p_j, where the total pressure is the sum of the partial pressures:

$$p = \sum_j p_j \tag{7.5.12}$$

Referring to Figure 7.5, we see that if $T > T_3$ and $p > p_j(T)$ for vaporization, then the liquid and gas phases of substance j can coexist; if $T < T_3$ and $p > p_j(T)$ for sublimation, then the solid and gas phases can coexist.

As an application of these ideas, consider an etching process in which the etch product forms on the substrate in the liquid form and in equilibrium with the gas phase. Then the product gas equilibrium density is $\overline{n} = p/kT$, where p, the vapor pressure, is given by (7.5.10). Now the flux of product molecules to and from the surface must balance in equilibrium. Using (2.4.10), the flux

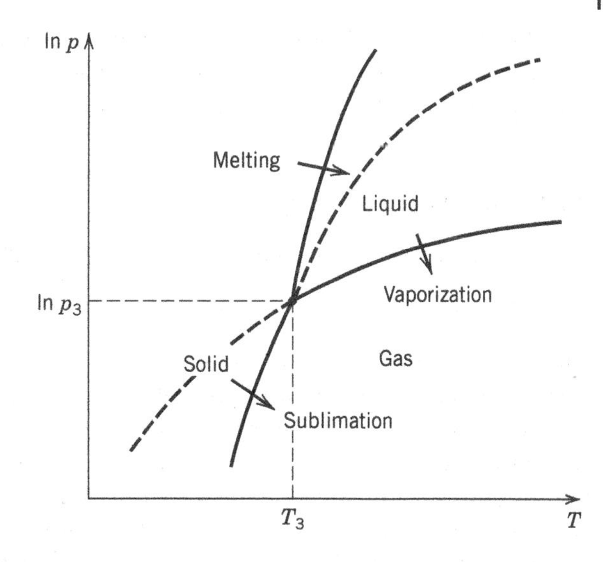

Figure 7.5 Phase diagram p versus T for a pure substance.

Table 7.5 Vapor Pressures

	Temperature (°C)		
Substance	**1 Torr**	**10 Torr**	**100 Torr**
$AlBr_3$	81.3 (s)	118.0	176.1
$AlCl_3$	100.0 (s)	123.8 (s)	152.0 (s)
AlF_3	1238	1324	1422
NH_3	−109.1 (s)	−91.9 (s)	−68.4
Br_2	−48.7 (s)	−25.0 (s)	9.3
Cl_2	−118.0 (s)	−101.6 (s)	−71.7
Cu_2Cl_2	546	702	960
$NiCl_2$	671 (s)	759 (s)	866 (s)
$SiCl_4$	−63.4	−34.4	5.4
SiF_4	−144.0 (s)	−130.4 (s)	−113.3 (s)
H_2O	−17.3 (s)	11.3	51.6
WF_6	−71.4 (s)	−49.2 (s)	−20.3 (s)

Note. s, solid phase.

to the surface is $\Gamma_{\text{in}} = \frac{1}{4}\overline{n}\overline{v}$, where $\overline{v} = (8kT/\pi M)^{1/2}$ is the mean speed of the product molecules. Hence, the flux from the surface is

$$\Gamma_{\text{out}} = \frac{1}{4}\overline{n}\overline{v} \tag{7.5.13}$$

Now consider the *nonequilibrium* situation in which the product gas is efficiently pumped away, such that the gas density $n \ll \overline{n}$. In this case, $\Gamma_{\text{in}} \ll \Gamma_{\text{out}}$. However, if the surface remains completely covered with the liquid etch product, then Γ_{out} is still given by (7.5.13). Hence, (7.5.13) determines a maximum etch product removal rate due to vapor pressure limitations. The removal rate can be less if the surface coverage is less than 100%, but it can never exceed this rate.

In this example, equilibrium thermodynamics (the vapor pressure p versus T) has been applied to determine an unknown kinetic rate (Γ_{out}) in terms of another known rate (Γ_{in}) for a system *that is not in equilibrium*. This important application of thermodynamics will be elaborated in Chapter 9.

7.5.2 Equilibrium at a Surface

We now consider thermal equilibrium for adsorption and desorption of gas molecules at a surface:

$$\text{A(g)} + \text{S} = \text{A:S} \tag{7.5.14}$$

where the notation A:S denotes an adsorbed molecule A on the surface S. In almost all cases, adsorption (the forward reaction) proceeds only if it is exothermic, $H_{ads} < 0$, because the entropy change S_{ads} is almost always negative, due to the binding of the gas molecule to the surface. Consequently, $G_{ads} = H_{ads} - TS_{ads} < 0$ only if $H_{ads} < 0$. Adsorption must be balanced by desorption (the reverse reaction, with $G_{desor} = -G_{ads}$) in thermal equilibrium. Let $\overline{n}_A$ (m^{-3}) be the equilibrium gas-phase volume density, n'_0 (m^{-2}) be the area density of surface sites, and $\overline{\theta}_A$ be the equilibrium fraction of sites on which molecules have adsorbed, such that the area densities covered and not covered with A molecules are $\overline{n}'_{S:A} = n'_0\overline{\theta}_A$ and $\overline{n}'_S = n'_0(1 - \overline{\theta}_A)$, respectively. Then as was done for pure gas-phase reactions, leading to an equilibrium constant $\mathcal{K}$ given by (7.4.6) and (7.4.7), we can write for reaction (7.5.14),

$$\frac{n'_0\overline{\theta}_A}{\overline{n}_A n'_0(1 - \overline{\theta}_A)} = \frac{\overline{\theta}_A}{\overline{n}_A(1 - \overline{\theta}_A)} = \mathcal{K}_{ads}(T) \tag{7.5.15}$$

where

$$\mathcal{K}_{ads}(T) = \frac{1}{n^\ominus}\exp\left(-\frac{G^\ominus_{ads}}{RT}\right) \tag{7.5.16}$$

and $n^\ominus \approx 2.46 \times 10^{19}$cm^{-3} is the gas-phase density at STP.

Solving (7.5.15) for $\overline{\theta}_A$, we obtain

$$\overline{\theta}_A = \frac{\mathcal{K}_{ads}\overline{n}_A}{1 + \mathcal{K}_{ads}\overline{n}_A} \tag{7.5.17}$$

which is known as the *Langmuir isotherm* because it specifies the equilibrium surface coverage as a function of pressure at fixed temperature. Plotting $\overline{\theta}_A$ versus $\overline{n}_A$ in Figure 7.6, we see that $\overline{\theta}_A \propto \mathcal{K}_{ads}\overline{n}_A$ for $\mathcal{K}_{ads}\overline{n}_A \ll 1$, $\overline{\theta}_A \to 1$ for $\mathcal{K}_{ads}\overline{n}_A \gg 1$, and $\overline{\theta}_A = 1/2$ at $\mathcal{K}_{ads}\overline{n}_A = 1$. At fixed gas density and for $G_{ads} < 0$, increasing T decreases $\mathcal{K}_{ads}$ and, hence, reduces $\overline{\theta}_A$. This behavior can be important in determining processing rates due to chemical reactions at surfaces. Although the rate of reaction for an adsorbed molecule A:S generally increases with temperature, the surface coverage decreases. Hence, the overall reaction rate can first increase with T up to some maximum value, due to an increase in the surface reaction rate, and then decrease as T is further increased, due to a decrease in the adsorbed reactant density on the surface. Such behavior has been observed, for example, for silicon etching using XeF_2 gas.

Now let us consider the desorption and adsorption of two kinds of gas molecules on a surface:

$$\text{A:S} = \text{A(g)} + \text{S}$$

$$\text{B:S} = \text{B(g)} + \text{S}$$

Figure 7.6 The Langmuir isotherm.

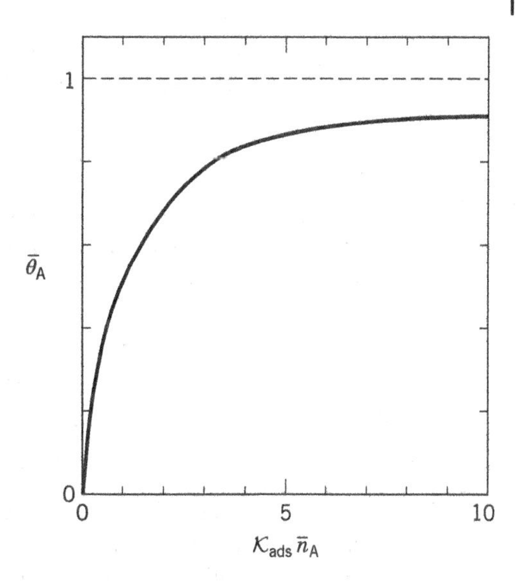

Let $\bar{\theta}_A$ and $\bar{\theta}_B$ be the surface fractions covered with A and B molecules in thermal equilibrium; hence, $1 - \bar{\theta}_A - \bar{\theta}_B$ is the surface fraction not covered. In thermal equilibrium, we must have

$$\frac{\bar{\theta}_A}{\bar{n}_A(1 - \bar{\theta}_A - \bar{\theta}_B)} = \mathcal{K}_A \tag{7.5.18a}$$

$$\frac{\bar{\theta}_B}{\bar{n}_B(1 - \bar{\theta}_A - \bar{\theta}_B)} = \mathcal{K}_B \tag{7.5.18b}$$

Solving for $\bar{\theta}_A$ and $\bar{\theta}_B$, we obtain

$$\bar{\theta}_A = \frac{\mathcal{K}_A \bar{n}_A}{1 + \mathcal{K}_A \bar{n}_A + \mathcal{K}_B \bar{n}_B} \tag{7.5.19a}$$

$$\bar{\theta}_B = \frac{\mathcal{K}_B \bar{n}_B}{1 + \mathcal{K}_A \bar{n}_A + \mathcal{K}_B \bar{n}_B} \tag{7.5.19b}$$

Comparing (7.5.19a) with (7.5.17), we see that the adsorption of B reduces the surface coverage of A. If A reacts at the surface and B does not, then B is an *inhibitor* for the reaction. Sidewalls in etching of silicon trenches are often protected by the use of inhibitors, which are cleared by ion bombardment at the bottom of the trench, thus yielding a low horizontal etch rate at the sidewall and a high vertical etch rate at the bottom.

Problems

7.1 **High-Temperature Equilibrium** A professor has suggested that hydrogen gas at a high temperature $T = 1100\,°\text{C}$ and pressure p can be used to convert a thin layer of a SiO_2 (quartz) wafer to silicon. The reaction is

$$2H_2(g) + SiO_2(s) \rightarrow Si(s) + 2H_2O(g)$$

At STP, $G_f = -228.6$ kJ/mol for $H_2O(g)$ and -856.6 kJ/mol for $SiO_2(s)$. Also, $H_f = -241.8$ kJ/mol for $H_2O(g)$ and -910.9 kJ/mol for $SiO_2(s)$. You may assume that H_f is independent of temperature.

(a) Show that the equilibrium constant for the reaction at 1100 °C is approximately 4.3×10^{-12}.

(b) Find the pressure p of H_2 gas necessary to convert a 1-nm-thick layer of SiO_2 to silicon. The SiO_2 wafer has an exposed area of 78.5 cm^2 and is placed in a reaction vessel having a volume of 10 L. Note that the density of SiO_2 is 2.65 g/cm^3.

7.2 Estimating Enthalpies of Formation The enthalpy of formation of $H_f^{\ominus}(AB)$ of the substance AB can be written in terms of the bond dissociation enthalpy $H_{diss}^{\ominus}(AB)$ and the enthalpies of formation $H_f^{\ominus}(A)$ and $H_f^{\ominus}(B)$ of the gaseous atoms A and B as

$$H_f^{\ominus}(AB) = H_f^{\ominus}(A) + H_f^{\ominus}(B) - H_{diss}^{\ominus}(AB)$$

This relation can be generalized to substances containing more than one bond.

(a) Using the data in Tables 7.3 and 7.4, estimate $H_f^{\ominus}(T_0)$ for CF_4, CF_3, CF_2, and CF. Compare your estimates with data given in Tables 7.1 and 7.2.

(b) Using the data in Tables 7.3 and 7.4, estimate $H_f^{\ominus}(T_0)$ for SiH_4, SiH_3, SiH_2, and SiH. Compare your estimates with data given in Tables 7.1 and 7.2.

(c) Using the data in Tables 7.3 and 7.4, estimate $H_f^{\ominus}(T_0)$ for TEOS [$Si(OC_2H_5)_4$] and compare your estimate to the measured value of −1397 kJ/mol.

(d) The enthalpies of formation at STP of $BF_3(g)$, $BF_2(g)$, BF(g), B(g), and F(g) are −1136 kJ/mol, −590 kJ/mol, −122.2 kJ/mol, 560 kJ/mol, and 79.4 kJ/mol, respectively. Using these data, find the bond dissociation energy (in equivalent voltage units) for dissociation of one molecule of $BF_3(g)$, $BF_2(g)$, and BF(g):

$$BF_3(g) \rightarrow BF_2(g) + F(g)$$

$$BF_2(g) \rightarrow BF(g) + F(g)$$

$$BF(g) \rightarrow B(g) + F(g)$$

7.3 The Triple Point Find p_3 and T_3 for H_2O by using the partial pressures for vaporization and sublimation (7.5.10) and (7.5.11) and compare to tabulated experimental data. Note that at standard pressure, the enthalpies of melting (at 273 K) and vaporization (at 373 K) are 6.01 and 40.66 kJ/mol, respectively. Assume that the heat capacity of liquid water is 1 cal/K-cm^3 and that the heat capacity of water vapor is given by the ideal gas formula.

7.4 Phase Equilibrium for a Mixture of Pure Substances A mixture of Cl_2 and $SiCl_4$ is in equilibrium at room temperature $T = 25$ °C and $p = 760$ Torr. Find all the phases that exist, and find the vapor pressures of the gas phases of the two substances.

7.5 Thermodynamics and Vapor Pressures Aluminum at $T = 298$ K (standard temperature) is etched reasonably fast in Cl_2 gas but not in F_2 gas, because the vapor pressure of AlF_3 is very low while that of $AlCl_3$ is reasonably high. The reactions are

$$Al(s) + \frac{3}{2}Cl_2(g) \rightarrow AlCl_3(s)$$

$$Al(s) + \frac{3}{2}F_2(g) \rightarrow AlF_3(s)$$

(a) Show that both reactions are thermodynamically strongly downhill (proceed far to the right) by finding the fraction x_{Cl_2} or x_{F_2} of unreacted Cl_2 or F_2 in equilibrium, given that the initial gas pressure of Cl_2 or F_2 is p_0. (Assume that there is a very large initial supply of aluminum to be etched.) Note that $G_f = -628.8$ kJ/mol for $AlCl_3$(s) and -1431 kJ/mol for AlF_3(s) at STP.

(b) Estimate the maximum etch rate (Å/min) at 298 K that can be achieved for Cl_2 and F_2 etching of aluminum due to vapor pressure limitations. Note that $H_{vap} = 116$ kJ/mol for $AlCl_3$ and 531 kJ/mol for AlF_3 at STP; the vapor pressure is 760 Torr at $T = 453.2$ K for $AlCl_3$ and 1810 K for AlF_3. The density of solid aluminum is 2.70g/cm^3. (Industrial processes generally require etch rates exceeding 2000 Å/min.)

7.6 Vapor Pressure Data The vapor pressure data for $NiCl_2(s) \rightarrow NiCl_2(g)$ is given below:

p (Torr)	1	10	100	760
T (°C)	671	759	866	987

Plot $\log p$ versus $1000/T$ (T in kelvins, not degrees Centigrade!) and use this plot to show that the sublimation enthalpy per mole at STP is $\approx$210 kJ/mol.

7.7 Equilibrium for Dissociation on a Surface For dissociative adsorption in thermal equilibrium with associative desorption,

$$\text{A:S} + \text{A:S} = \text{A}_2(\text{g}) + 2\text{S}$$

show that the equilibrium surface coverage is

$$\bar{\theta}_A = \frac{(\mathcal{K}\bar{n}_{A_2})^{1/2}}{1 + (\mathcal{K}\bar{n}_{A_2})^{1/2}}$$

where $\bar{n}_{A_2}$ is the equilibrium gas-phase density and $\mathcal{K}$ is the equilibrium constant for the reaction.

8

Molecular Collisions

8.1 Introduction

Basic concepts of gas-phase collisions were introduced in Chapter 3, where we described only those processes needed to model the simplest noble gas discharges: electron–atom ionization, excitation, and elastic scattering; and ion–atom elastic scattering and resonant charge transfer. In this chapter, we introduce other collisional processes that are central to the description of chemically reactive discharges. These include the dissociation of molecules, the generation and destruction of negative ions, and gas-phase chemical reactions.

Whereas the cross sections have been measured reasonably well for the noble gases, with measurements in reasonable agreement with theory, this is not the case for collisions in molecular gases. Hundreds of potentially significant collisional reactions must be examined in simple diatomic gas discharges such as oxygen. For feedstocks such as CF_4/O_2 and SiH_4/O_2, the complexity can be overwhelming. Furthermore, even when the significant processes have been identified, most of the cross sections have been neither measured nor calculated. Hence, one must often rely on estimates based on semiempirical or semiclassical methods, or on measurements made on molecules analogous to those of interest. As might be expected, data are most readily available for simple diatomic and polyatomic gases.

8.2 Molecular Structure

The energy levels for the electronic states of a single atom were described in Chapter 3. The energy levels of molecules are more complicated for two reasons. First, molecules have additional vibrational and rotational degrees of freedom due to the motions of their nuclei, with corresponding quantized energies $\mathcal{E}_v$ and $\mathcal{E}_J$. Second, the energy $\mathcal{E}_e$ of each electronic state depends on the instantaneous configuration of the nuclei. For a diatomic molecule, $\mathcal{E}_e$ depends on a single coordinate R, the spacing between the two nuclei. Since the nuclear motions are slow compared to the electronic motions, the electronic state can be determined for any fixed spacing. We can therefore represent each quantized electronic level for a frozen set of nuclear positions as a graph of $\mathcal{E}_e$ versus R, as shown in Figure 8.1. For a molecule to be stable, the ground (minimum energy) electronic state must have a minimum at some value $\overline{R}_1$ corresponding to the mean intermolecular separation (curve 1). In this case, energy must be supplied in order to separate the atoms ($R \to \infty$). An excited

Principles of Plasma Discharges and Materials Processing, Third Edition. Michael A. Lieberman and Allan J. Lichtenberg.

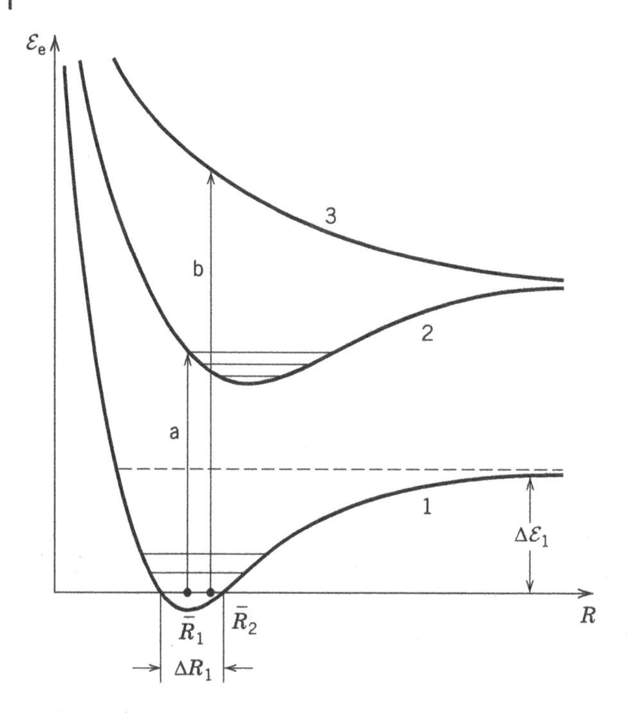

Figure 8.1 Potential energy curves for the electronic states of a diatomic molecule.

electronic state can either have a minimum ($\bar{R}_2$ for curve 2) or not (curve 3). Note that $\bar{R}_2$ and $\bar{R}_1$ do not generally coincide. As for atoms, excited states may be short-lived (unstable to electric dipole radiation) or may be metastable. Various electronic levels may tend to the same energy in the unbound ($R \to \infty$) limit.

For diatomic molecules, the electronic states are specified first by the component (in units of $\hbar$) Λ of the total orbital angular momentum along the internuclear axis, with the symbols Σ, Π, Δ, and Φ corresponding to $\Lambda = 0, \pm 1, \pm 2$, and ± 3, in analogy with atomic nomenclature. All but the Σ states are doubly degenerate in Λ. For Σ states, + and − superscripts are often used to denote whether the wave function is symmetric or antisymmetric with respect to reflection at any plane through the internuclear axis. The total electron spin angular momentum S (in units of $\hbar$) is also specified, with the multiplicity $2S + 1$ written as a prefixed superscript, as for atomic states. Finally, for homonuclear molecules (H_2, N_2, O_2, etc.), the subscripts g or u are written to denote whether the wave function is symmetric or antisymmetric with respect to interchange of the nuclei. Summing the orbital angular momenta l_i over all the electrons, then one obtains "g" if $\sum_i l_i$ is even, and "u" if $\sum_i l_i$ is odd. In this notation, the ground states of H_2 and N_2 are both singlets, ${}^1\Sigma_g^+$, and that of O_2 is a triplet, ${}^3\Sigma_g^-$.

In addition to the preceding nomenclature, it is customary to label the ground state with the letter X. Excited states with the same spin quantum number S (same multiplicity) are labeled with A, B, C, etc., in order of increasing energy. All other excited states are labeled with a, b, c, etc., in order of increasing energy.

For polyatomic molecules, the electronic energy levels depend on more than one nuclear coordinate, so Figure 8.1 must be generalized. Furthermore, since there is generally no axis of symmetry, the states cannot be characterized by the quantum number Λ, and other naming conventions are used. Such states are often specified empirically through characterization of measured optical emission spectra. Typical spacings of low-lying electronic energy levels range from a few to tens of volts, as for atoms.

8.2.1 Vibrational and Rotational Motion

Unfreezing the nuclear vibrational and rotational motions leads to additional quantized structure on smaller energy scales, as illustrated in Figure 8.2. The simplest (harmonic oscillator) model for the vibration of diatomic molecules leads to equally spaced quantized, nondegenerate energy levels

$$e\mathcal{E}_v = \hbar\omega_{\text{vib}}\left(v + \frac{1}{2}\right) \tag{8.2.1}$$

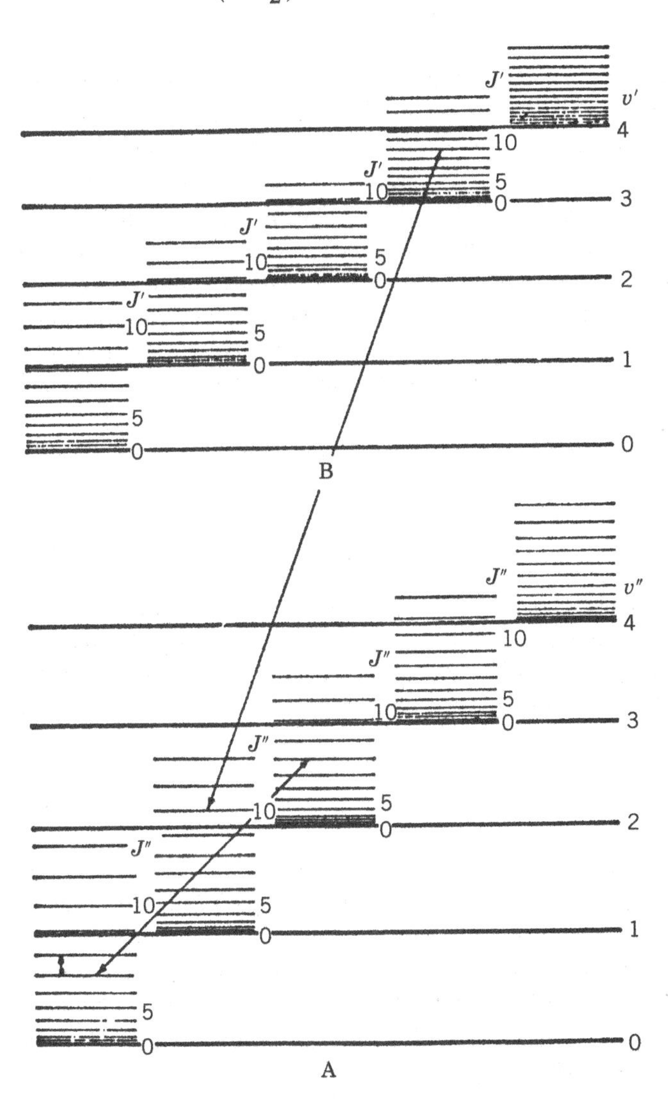

Figure 8.2 Vibrational and rotational levels of two electronic states A and B of a molecule; the three double arrows indicate examples of transitions in the pure rotation spectrum, the rotation–vibration spectrum, and the electronic spectrum (Herzberg, G. (1971)/with permission of Dover Publications).

where $\upsilon = 0,1,2,\ldots$ is the vibrational quantum number and ω_{vib} is the linearized vibration frequency. Fitting a quadratic function

$$e\mathcal{E}_\upsilon = \frac{1}{2}k_{\text{vib}}(R - \overline{R})^2 \tag{8.2.2}$$

near the minimum of a stable energy level curve such as those shown in Figure 8.1, we can estimate

$$\omega_{\text{vib}} \approx \left(\frac{k_{\text{vib}}}{m_{\text{Rmol}}}\right)^{1/2} \tag{8.2.3}$$

where k_{vib} is the "spring constant" and m_{Rmol} is the reduced mass of the AB molecule. The spacing $\hbar\omega_{\text{vib}}$ between vibrational energy levels for a low-lying stable electronic state is typically a few tenths of a volt. Hence, for molecules in equilibrium at room temperature (0.026 V), only the $\upsilon = 0$ level is significantly populated. However, collisional processes can excite strongly nonequilibrium vibrational energy levels.

We indicate by the short horizontal line segments in Figure 8.1 a few of the vibrational energy levels for the stable electronic states. The length of each segment gives the range of classically allowed vibrational motions. Note that even the ground state ($\upsilon = 0$) has a finite width ΔR_1 as shown, because from (8.2.1), the $\upsilon = 0$ state has a nonzero vibrational energy $\frac{1}{2}\hbar\omega_{\text{vib}}$. The actual separation ΔR about $\overline{R}$ for the ground state has a Gaussian distribution and tends toward a distribution peaked at the classical turning points for the vibrational motion as $\upsilon \to \infty$. The vibrational motion becomes anharmonic and the level spacings tend to zero as the unbound vibrational energy is approached ($\mathcal{E}_\upsilon \to \Delta\mathcal{E}_1$). For $\mathcal{E}_\upsilon > \Delta\mathcal{E}_1$, the vibrational states form a continuum, corresponding to unbound classical motion of the nuclei (breakup of the molecule). For a polyatomic molecule, there are many degrees of freedom for vibrational motion, leading to a very complicated structure for the vibrational levels.

The simplest (dumbbell) model for the rotation of diatomic molecules leads to the nonuniform quantized energy levels

$$e\mathcal{E}_J = \frac{\hbar^2}{2I_{\text{mol}}}J(J+1) \tag{8.2.4}$$

where $I_{\text{mol}} = m_{\text{Rmol}}\overline{R}^2$ is the moment of inertia and $J = 0,1,2,\ldots$ is the rotational quantum number. The levels are degenerate, with $2J + 1$ states for the Jth level. The spacing between rotational levels increases with J (see Figure 8.2). The spacing between the lowest ($J = 0$ to $J = 1$) levels typically corresponds to an energy of 0.001–0.01 V; hence, many low-lying levels are populated in thermal equilibrium at room temperature.

8.2.2 Optical Emission

An excited molecular state can decay to a lower energy state by emission of a photon or by breakup of the molecule. As shown in Figure 8.2, the radiation can be emitted by a transition between electronic levels, between vibrational levels of the same electronic state, or between rotational levels of the same electronic and vibrational state; the radiation typically lies within the optical, infrared, or microwave frequency range, respectively. Electric dipole radiation is the strongest mechanism for photon emission, having typical transition times of $t_{\text{rad}} \sim 10^{-9}$ s, as obtained in (3.4.13). The selection rules for electric dipole radiation are

$$\Delta\Lambda = 0, \pm 1 \tag{8.2.5a}$$

$$\Delta S = 0 \tag{8.2.5b}$$

In addition, for transitions between Σ states, the only allowed transitions are

$$\Sigma^+ \rightarrow \Sigma^+ \quad \text{and} \quad \Sigma^- \rightarrow \Sigma^- \tag{8.2.6}$$

and for homonuclear molecules, the only allowed transitions are

$$\mathrm{g} \rightarrow \mathrm{u} \quad \text{and} \quad \mathrm{u} \rightarrow \mathrm{g} \tag{8.2.7}$$

Hence, homonuclear diatomic molecules do not have a pure vibrational or rotational spectrum. Radiative transitions between electronic levels having many different vibrational and rotational initial and final states give rise to a structure of emission and absorption bands within which a set of closely spaced frequencies appear. These give rise to characteristic molecular emission and absorption bands when observed using low-resolution optical spectrometers. As for atoms, metastable molecular states having no electric dipole transitions to lower levels also exist. These have lifetimes much exceeding 10^{-6} s; they can give rise to weak optical band structures due to magnetic dipole or electric quadrupole radiation.

Electric dipole radiation between vibrational levels of the same electronic state is permitted for molecules having permanent dipole moments. In the harmonic oscillator approximation, the selection rule is $\Delta v = \pm 1$; weaker transitions $\Delta v = \pm 2, \pm 3, \ldots$ are permitted for anharmonic vibrational motion. Further details on vibrational and rotational electric dipole transitions can be found in Ricard (1996) and in many other books on plasma spectroscopy.

The preceding description of molecular structure and nomenclature applies to diatomic molecules having arbitrary electronic charge. This includes neutral molecules AB, positive molecular ions AB^+, AB^{2+}, etc., and negative molecular ions AB^-. The potential energy curves for the various electronic states, regardless of molecular charge, are commonly plotted on the same diagram. Figures 8.3 and 8.4 give these for some important electronic states of H_2^-, H_2, and H_2^+, and of O_2^-, O_2, and O_2^+. Examples of both attractive (having a potential energy minimum) and repulsive (having no minimum) states can be seen. The vibrational levels are labeled with the quantum number v for the attractive levels. The ground states of both H_2^+ and O_2^+ are attractive; hence, these molecular ions are stable against *autodissociation* ($AB^+ \rightarrow A + B^+$ or $A^+ + B$). Similarly, the ground states of H_2 and O_2 are attractive and lie below those of H_2^+ and O_2^+; hence, they are stable against autodissociation and *autoionization* ($AB \rightarrow AB^+ + e$). For all molecules, the AB ground state lies below the AB^+ ground state and is stable against autoionization. Excited states can be attractive or repulsive. A few of the attractive states may be metastable; some examples are the $^3\Pi_u$ state of H_2 and the $^1\Delta_g$, $^1\Sigma_g^+$, $^3\Delta_u$ states of O_2.

For some molecules, e.g., the diatomic noble gases, the molecular ion and some of the excited neutral molecular states are stable, but the bimolecular ground state neutral is not stable. The neutral excited states are often called excimers (from "excited dimers"). The physics of excimers is reviewed in Smirnov (1983). Some excimers are metastable and therefore can have quite high densities in a discharge. Noble gas molecular ions and excimers are typically created by three-body collisional processes, such as $He^+ + He + He \rightarrow He_2^+ + He$ and $He^* + He + He \rightarrow He_2^* + He$, respectively. These processes can be very important in high-pressure discharges, as we describe in Chapter 15. Table 15.1 gives a set of rate coefficients for the He/N_2 system, including these processes. A simplified set of potential energy curves for diatomic helium, useful in modeling high-pressure helium-containing discharges, is given in Figure 8.5 (Golubovskii et al., 2003). More complete sets of curves are given in Ginter and Battino (1970) and Brutschy and Haberland (1979). Some diatomic argon potential energy curves are given in Hutchinson (1980), Smirnov (1983), and Yates et al. (1983).

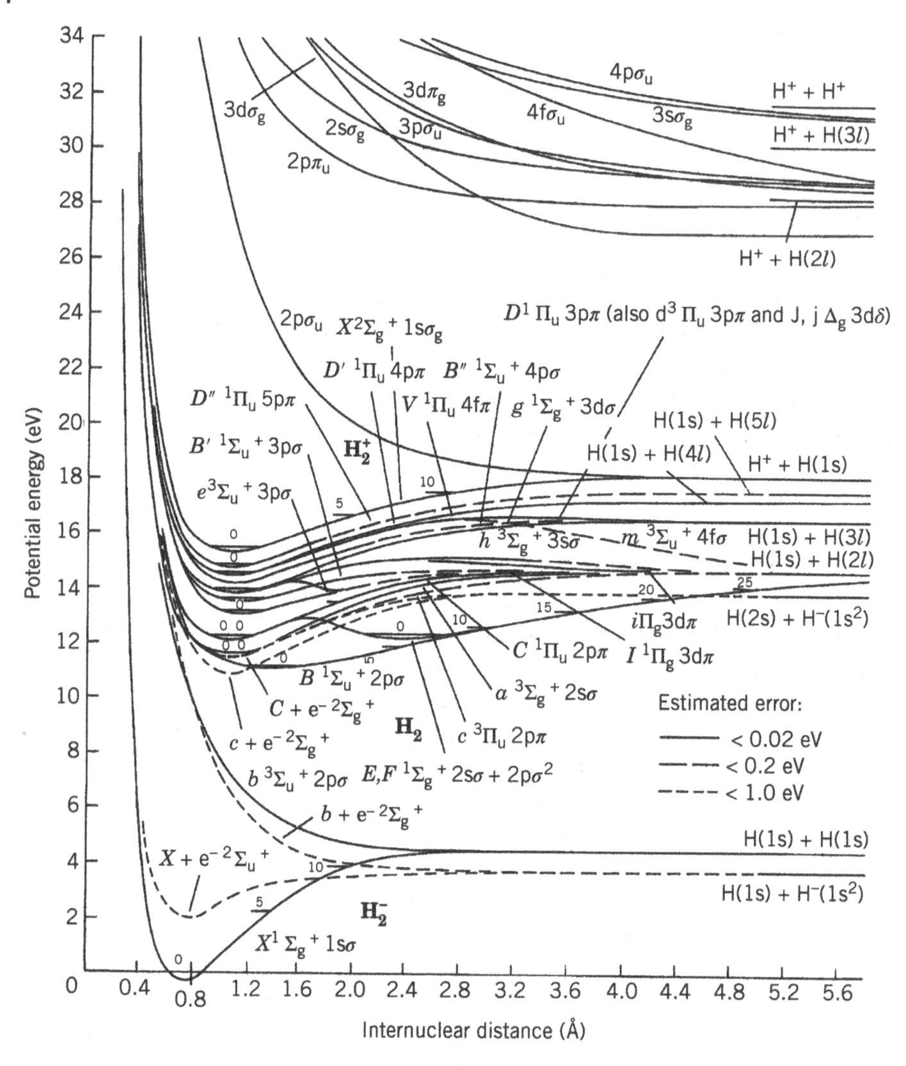

Figure 8.3 Potential energy curves for H_2^-, H_2, and H_2^+. Source: Steinfeld (1985)/with permission of Dover Publications.

8.2.3 Negative Ions

Recall from Section 7.2 that many neutral atoms have a positive electron affinity $\mathcal{E}_{aff}$, i.e., the reaction

$$A + e \rightarrow A^-$$

is exothermic with energy $\mathcal{E}_{aff}$ (in volts). If $\mathcal{E}_{aff}$ is negative, then A^- is unstable to *autodetachment*, $A^- \rightarrow A + e$. A similar phenomenon is found for negative molecular ions. A stable AB^- ion exists if its ground (lowest energy) state has a potential minimum that lies below the ground state of AB. This is generally true only for strongly electronegative gases having large electron affinities, such as O_2 ($\mathcal{E}_{aff} \approx 1.463$ V for O atoms) and the halogens ($\mathcal{E}_{aff} > 3$ V for the atoms). For example, Figure 8.4 shows that the $^2\Pi_g$ ground state of O_2^- is stable, with $\mathcal{E}_{aff} \approx 0.43$ V for O_2. For weakly electronegative or for electropositive gases, the minimum of the ground state of AB^- generally lies

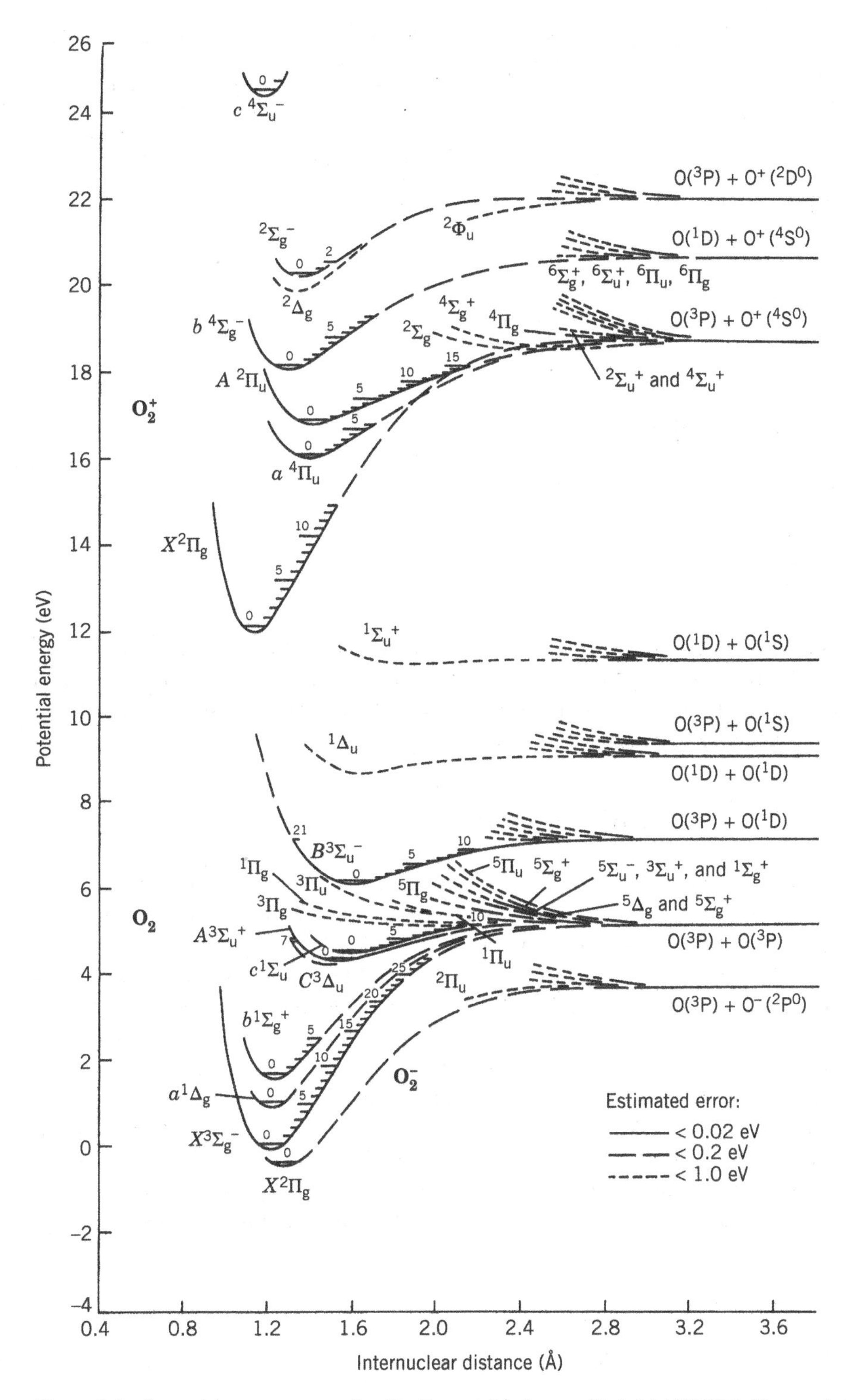

Figure 8.4 Potential energy curves for O_2^-, O_2, and O_2^+. Source: Steinfeld (1985)/with permission of Dover Publications.

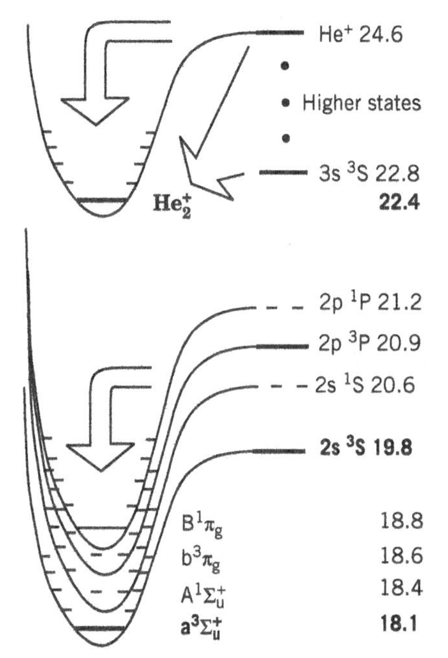

Figure 8.5 Simplified potential energy curves for $He_2^+(X^2\Sigma_u^+)$ and the four lowest-energy excimers of He_2; the energies (in volts) above the He ground state are given at the right. The repulsive $He_2(X^1\Sigma_g^+)$ ground state is not shown. Source: Golubovskii et al. (2002). ©IOP Publishing. Reproduced with permission. All rights reserved. https://doi.org/10.1088/0022-3727/36/1/306.

above the ground state of AB, and AB^- is unstable to autodetachment. An example is hydrogen, which is weakly electronegative ($\mathcal{E}_{aff} \approx 0.754$ V for H atoms). Figure 8.3 shows that the $^2\Sigma_u^+$ ground state of H_2^- is unstable, although the H^- ion itself is stable. In an electropositive gas such as N_2 ($\mathcal{E}_{aff} \lesssim 0$), both N_2^- and N^- are unstable.

8.3 Electron Collisions with Molecules

The interaction time for the collision of a typical (1–10 V) electron with a molecule is short, $t_c \sim 2a_0/v_e \sim 10^{-16}$–$10^{-15}$ s, compared to the typical time for a molecule to vibrate, $\tau_{vib} \sim 10^{-14}$–10^{-13} s. Hence, for electron collisional excitation of a molecule to an excited electronic state, the new vibrational (and rotational) state can be determined by freezing the nuclear motions during the collision. This is known as the *Franck–Condon principle* and is illustrated in Figure 8.1 by the vertical line a, showing the collisional excitation at fixed R to a high quantum number bound vibrational state and by the vertical line b, showing excitation at fixed R to a vibrationally unbound state, in which breakup of the molecule is energetically permitted. Since the typical transition time for electric dipole radiation ($\tau_{rad} \sim 10^{-9}$–10^{-8} s) is long compared to the dissociation (~vibrational) time τ_{diss}, excitation to an excited state will generally lead to dissociation when it is energetically permitted. Finally, we note that the time between collisions $\tau_c \gg \tau_{rad}$ in typical low-pressure processing discharges. Summarizing the ordering of timescales for electron–molecule collisions, we have

$$t_{at} \sim t_c \ll t_{vib} \sim t_{diss} \ll \tau_{rad} \ll \tau_c$$

8.3.1 Dissociation

Electron impact dissociation,

$$e + AB \rightarrow A + B + e$$

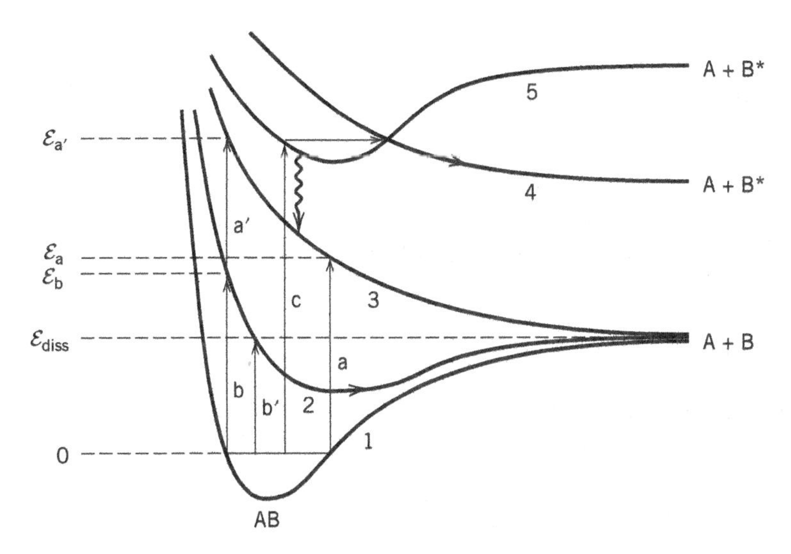

Figure 8.6 Illustrating the variety of dissociation processes for electron collisions with molecules.

of feedstock gases plays a central role in the chemistry of low-pressure reactive discharges. The variety of possible dissociation processes is illustrated in Figure 8.6. In collisions a or a′, the $v = 0$ ground state of AB is excited to a repulsive state of AB. The required threshold energy $\mathcal{E}_{thr}$ is $\mathcal{E}_a$ for collision a and $\mathcal{E}_{a'}$ for collision a′, and it leads to an energy after dissociation lying between $\mathcal{E}_a - \mathcal{E}_{diss}$ and $\mathcal{E}_{a'} - \mathcal{E}_{diss}$ that is shared among the dissociation products (here, A and B). Typically, $\mathcal{E}_a - \mathcal{E}_{diss} \sim$ few volts; consequently, hot neutral fragments are typically generated by dissociation processes. If these hot fragments hit the substrate surface, they can profoundly affect the process chemistry. In collision b, the ground state AB is excited to an attractive state of AB at an energy $\mathcal{E}_b$ that exceeds the binding energy $\mathcal{E}_{diss}$ of the AB molecule, resulting in dissociation of AB with fragment energy $\mathcal{E}_b - \mathcal{E}_{diss}$. In collision b′, the excitation energy, $\mathcal{E}_{b'} = \mathcal{E}_{diss}$, and the fragments have low energies; hence, this process creates fragments having energies ranging from essentially thermal energies up to $\mathcal{E}_b - \mathcal{E}_{diss} \sim$ few volts. In collision c, the AB molecule is excited to the bound excited state AB* (labeled 5), which subsequently radiates to the unbound AB state (labeled 3), which then dissociates. The threshold energy required is large, and the fragments are hot. Collision c can also lead to dissociation of an excited state by a radiationless transfer from state 5 to state 4 near the point where the two states cross:

$$AB^*(\text{bound}) \rightarrow AB^*(\text{unbound}) \rightarrow A + B^*$$

The fragments can be both hot and in excited states. We discuss such radiationless electronic transitions in the next section. This phenomenon is known as *predissociation.* Finally, a collision (not labeled in the figure) to state 4 can lead to dissociation of AB*, again resulting in hot excited fragments.

The process of electron impact excitation of a molecule is similar to that of an atom, and, consequently, the cross sections have a similar form. A simple classical estimate of the dissociation cross section for a level having excitation energy U_1 can be found by requiring that an incident electron having energy W transfer an energy W_L lying between U_1 and U_2 to a valence electron. Here, U_2 is the energy of the next higher level. Then, integrating the differential cross section $d\sigma$ (given in (3.4.20) and repeated here),

$$d\sigma = \pi \left(\frac{e^2}{4\pi\epsilon_0} \right)^2 \frac{1}{W} \frac{dW_L}{W_L^2} \tag{3.4.20}$$

over W_L, we obtain

$$\sigma_{diss} = \begin{cases} 0 & W < U_1 \\ \pi \left(\frac{e^2}{4\pi\epsilon_0} \right)^2 \frac{1}{W} \left(\frac{1}{U_1} - \frac{1}{W} \right) & U_1 < W < U_2 \\ \pi \left(\frac{e^2}{4\pi\epsilon_0} \right)^2 \frac{1}{W} \left(\frac{1}{U_1} - \frac{1}{U_2} \right) & W > U_2 \end{cases} \tag{8.3.1}$$

Let $U_2 - U_1 \ll U_1$ and introducing voltage units $W = e\mathcal{E}$, $U_1 = e\mathcal{E}_1$, and $U_2 = e\mathcal{E}_2$, we have

$$\sigma_{diss} = \begin{cases} 0 & \mathcal{E} < \mathcal{E}_1 \\ \sigma_0 \frac{\mathcal{E} - \mathcal{E}_1}{\mathcal{E}_1} & \mathcal{E}_1 < \mathcal{E} < \mathcal{E}_2 \\ \sigma_0 \frac{\mathcal{E}_2 - \mathcal{E}_1}{\mathcal{E}} & \mathcal{E} > \mathcal{E}_2 \end{cases} \tag{8.3.2}$$

where

$$\sigma_0 = \pi \left(\frac{e}{4\pi\epsilon_0 \mathcal{E}_1} \right)^2 \tag{8.3.3}$$

We see that the dissociation cross section rises linearly from the threshold energy $\mathcal{E}_{thr} \approx \mathcal{E}_1$ to a maximum value $\sigma_0(\mathcal{E}_2 - \mathcal{E}_1)/\mathcal{E}_{thr}$ at $\mathcal{E}_2$ and then falls off as $1/\mathcal{E}$. Actually, $\mathcal{E}_1$ and $\mathcal{E}_2$ can depend on the nuclear separation R. In this case, (8.3.2) should be averaged over the range of Rs corresponding to the ground state vibrational energy, leading to a broadened dependence of the average cross section on energy $\mathcal{E}$. The maximum cross section is typically of order 10^{-15} cm^2. Typical rate constants for a single dissociation process with $\mathcal{E}_{thr} \gtrsim T_e$ have an Arrhenius form (3.5.7)

$$K_{diss} \propto K_{diss0} \exp\left(-\frac{\mathcal{E}_{thr}}{T_e} \right) \tag{8.3.4}$$

where $K_{diss0} \sim 10^{-7}$ cm^3/s. However, in some cases, $\mathcal{E}_{thr} \lesssim T_e$. For excitation to an attractive state, an appropriate average over the fraction of the ground-state vibration that leads to dissociation must be taken.

8.3.2 Dissociative Ionization

In addition to normal ionization,

$$e + AB \rightarrow AB^+ + 2e$$

electron–molecule collisions can lead to dissociative ionization

$$e + AB \rightarrow A + B^+ + 2e$$

These processes, common for polyatomic molecules, are illustrated in Figure 8.7. In collision a having threshold energy $\mathcal{E}_{iz}$, the molecular ion AB^+ is formed. Collisions b and c occur at higher threshold energies $\mathcal{E}_{diz}$ and result in dissociative ionization, leading to the formation of fast, positively charged ions and neutrals. These cross sections have a similar form to the Thompson ionization cross section for atoms.

8.3.3 Dissociative Recombination

The electron collision,

$$e + AB^+ \rightarrow A + B^*$$

illustrated as d and d′ in Figure 8.7, destroys an electron–ion pair and leads to the production of fast excited neutral fragments. Since the electron is captured, it is not available to carry away a part of the reaction energy. Consequently, the collision cross section has a resonant character, falling to very low values for $\mathcal{E} < \mathcal{E}_d$ and $\mathcal{E} > \mathcal{E}_{d'}$. However, a large number of excited states A^* and B^* having increasing principal quantum numbers n and energies can be among the reaction products. Consequently, the rate constants can be large, of order 10^{-7}–10^{-6} cm^3/s. Dissociative recombination to the ground states of A and B cannot occur because the potential energy curve for AB^+ is always greater than the potential energy curve for the repulsive state of AB. Some dissociative recombination rate constants for molecular ions are given in Mitchell (1990).

Two-body recombination for atomic ions or for molecular ions that do not subsequently dissociate can only occur with emission of a photon:

$$e + A^+ \rightarrow A + h\nu.$$

As shown in Section 9.2, the rate constants are typically three to five orders of magnitude lower than for dissociative recombination.

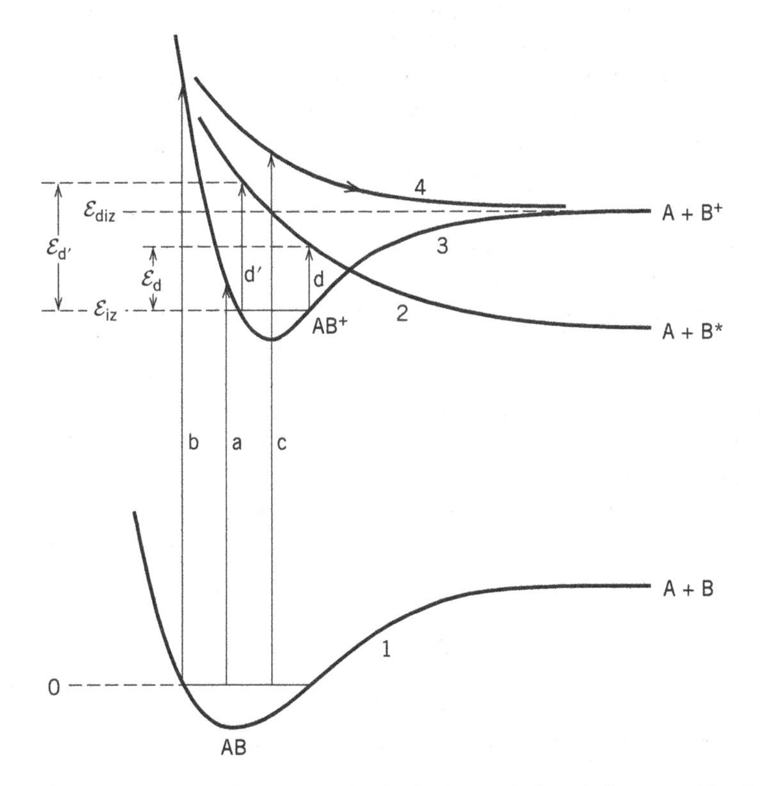

Figure 8.7 Illustrating dissociative ionization and dissociative recombination for electron collisions with molecules.

8.3.4 Example of Hydrogen

The example of H_2 illustrates some of the inelastic electron collision phenomena we have discussed. In order of increasing electron impact energy, at a threshold energy of ~8.8 V, there is excitation to the repulsive $^3\Sigma_u^+$ state followed by dissociation into two fast H fragments carrying ~2.2 V/atom. At 11.5 V, the $^1\Sigma_u^+$ bound state is excited, with subsequent electric dipole radiation in the UV region to the $^1\Sigma_g^+$ ground state. At 11.8 V, there is excitation to the $^3\Sigma_g^+$ bound state, followed by electric dipole radiation to the $^3\Sigma_u^+$ repulsive state, followed by dissociation with ~2.2 V/atom. At 12.6 V, the $^1\Pi_u$ bound state is excited, with UV emission to the ground state. At 15.4 V, the $^2\Sigma_g^+$ ground state of H_2^+ is excited, leading to the production of H_2^+ ions. At 28 V, excitation of the repulsive $^2\Sigma_u^+$ state of H_2^+ leads to the dissociative ionization of H_2, with ~5 V each for the H and H^+ fragments.

8.3.5 Dissociative Electron Attachment

The processes,

$$e + AB \rightarrow A + B^-$$

produce negative ion fragments as well as neutrals. They are important in discharges containing atoms having positive electron affinities, not only because of the production of negative ions but because the threshold energy for production of negative ion fragments is usually lower than for pure dissociation processes. A variety of processes are possible, as shown in Figure 8.8. Since the impacting electron is captured and is not available to carry excess collision energy away, dissociative attachment is a resonant process that is important only within a narrow energy range. The maximum cross sections are generally much smaller than the hard-sphere cross section of the molecule. Attachment generally proceeds by collisional excitation from the ground AB state to a repulsive AB^- state, which subsequently either autodetaches or dissociates. The attachment cross section is determined by the balance between these processes. For most molecules, the dissociation energy $\mathcal{E}_{diss}$ of AB is greater than the electron affinity $\mathcal{E}_{affB}$ of B, leading to the potential energy curves shown in Figure 8.8*a*. In this case, the cross section is large only for impact energies lying between a minimum value $\mathcal{E}_{thr}$, for collision a, and a maximum value $\mathcal{E}'_{thr}$, for collision a′. The fragments are hot, having energies lying between minimum and maximum values $\mathcal{E}_{min} = \mathcal{E}_{thr} + \mathcal{E}_{affB} - \mathcal{E}_{diss}$ and $\mathcal{E}_{max} = \mathcal{E}'_{thr} + \mathcal{E}_{affB} - \mathcal{E}_{diss}$. Since the AB^- state lies above the AB state for $R < R_x$, autodetachment can occur as the molecules begin to separate: $AB^- \rightarrow AB + e$. Hence, the cross section for production of negative ions can be much smaller than that for excitation of the AB^- repulsive state. As a crude estimate, for the same energy, the autodetachment rate is $\sqrt{M_R/m} \sim 100$ times the dissociation rate of the repulsive AB^- molecule, where M_R is the reduced mass. Hence, only one out of 100 excitations lead to dissociative attachment.

Excitation to the AB^- bound state can also lead to dissociative attachment, as shown in Figure 8.8*b*. Here the cross section is significant only for $\mathcal{E}_{thr} < \mathcal{E} < \mathcal{E}'_{thr}$, but the fragments can have low energies, with a minimum energy of zero and a maximum energy of $\mathcal{E}'_{thr} + \mathcal{E}_{affB} - \mathcal{E}_{diss}$. Collision b,

$$e + AB \rightarrow AB^{-*}$$

does not lead to production of AB^- ions because energy and momentum are not generally conserved when two bodies collide elastically to form one body (see Problem 3.13). Hence, the excited AB^{-*} ion separates,

$$AB^{-*} \rightarrow e + AB$$

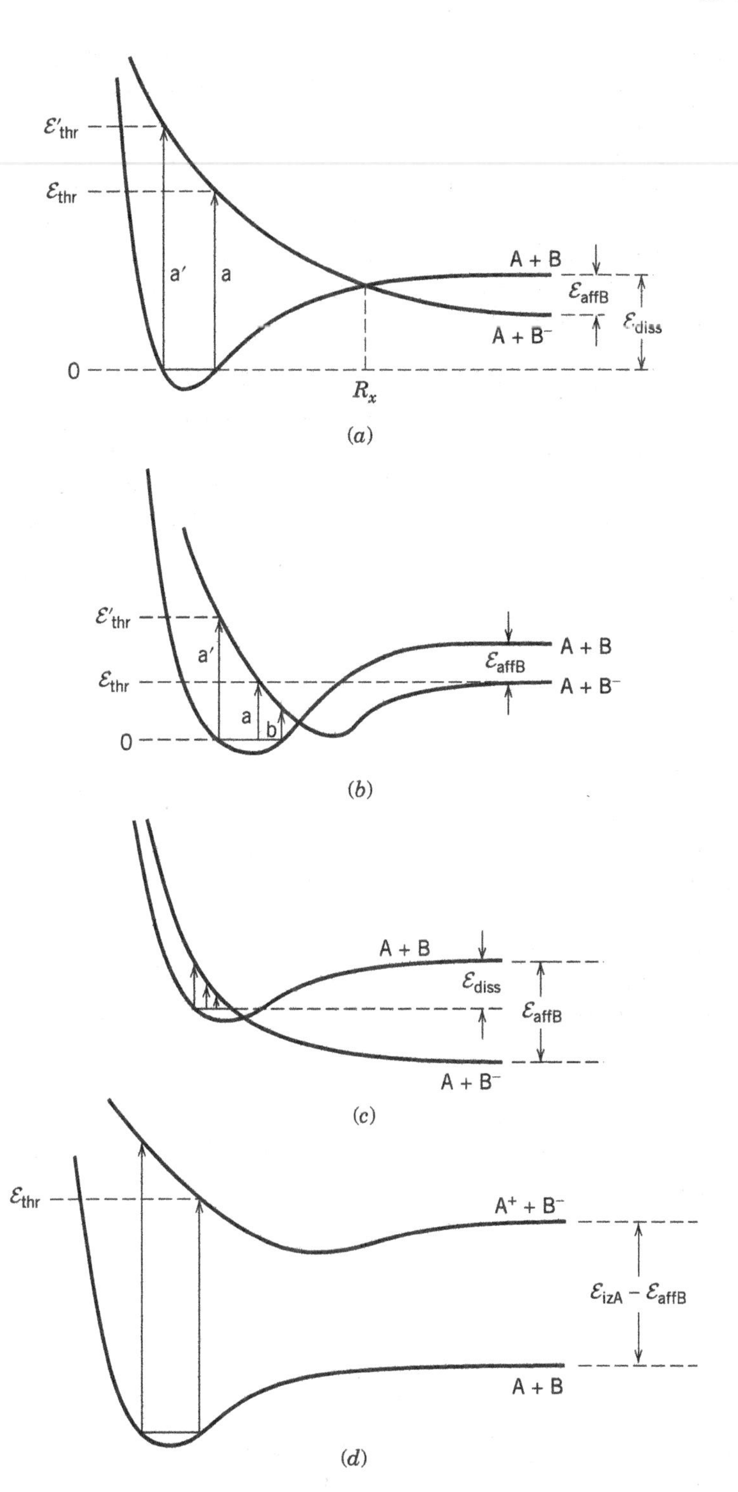

Figure 8.8 Illustrating a variety of electron attachment processes for electron collisions with molecules; (*a*) capture into a repulsive state; (*b*) capture into an attractive state; (*c*) capture of slow electrons into a repulsive state; (*d*) polar dissociation.

unless vibrational radiation or collision with a third body carries off the excess energy. These processes are both slow in low-pressure discharges (see Section 9.2). At high pressures (say, atmospheric), three-body attachment to form AB^- can be very important.

For a few molecules, such as some halogens, the electron affinity of the atom exceeds the dissociation energy of the neutral molecule, leading to the potential energy curves shown in Figure 8.8*c*. In this case, the range of electron impact energies $\mathcal{E}$ for excitation of the AB^- repulsive state includes $\mathcal{E} = 0$. Consequently, there is no threshold energy, and very slow electrons can produce dissociative attachment, resulting in hot neutral and negative ion fragments. The range of Rs over which autodetachment can occur is small; hence, the maximum cross sections for dissociative attachment can be as high as 10^{-16} cm^2.

A simple classical estimate of electron capture can be made using the differential scattering cross section for energy loss (3.4.20), in a manner similar to that done for dissociation. For electron capture to an energy level $\mathcal{E}_1$ that is unstable to autodetachment, and with the additional constraint for capture that the incident electron energy lie within $\mathcal{E}_1$ and $\mathcal{E}_2 = \mathcal{E}_1 + \Delta\mathcal{E}$, where $\Delta\mathcal{E}$ is a small energy difference characteristic of the dissociative attachment timescale, we obtain, in place of (8.3.2),

$$\sigma_{\text{att}} = \begin{cases} 0 & \mathcal{E} < \mathcal{E}_1 \\ \sigma_0 \frac{\mathcal{E} - \mathcal{E}_1}{\mathcal{E}_1} & \mathcal{E}_1 < \mathcal{E} < \mathcal{E}_2 \\ 0 & \mathcal{E} > \mathcal{E}_2 \end{cases} \tag{8.3.5}$$

where

$$\sigma_0 \approx \pi \left(\frac{m}{M_{\text{R}}}\right)^{1/2} \left(\frac{e}{4\pi\epsilon_0 \mathcal{E}_1}\right)^2 \tag{8.3.6}$$

The factor of $(m/M_{\text{R}})^{1/2}$ roughly gives the fraction of excited states that do not autodetach. We see that the dissociative attachment cross section rises linearly at $\mathcal{E}_1$ to a maximum value $\sigma_0 \Delta\mathcal{E}/\mathcal{E}_1$ and then falls abruptly to zero.

As for dissociation, $\mathcal{E}_1$ can depend strongly on the nuclear separation R, and (8.3.5) must be averaged over the range of $\mathcal{E}_1$s corresponding to the ground state vibrational motion, e.g., from $\sim \mathcal{E}_{\text{thr}}$ to $\sim \mathcal{E}'_{\text{thr}}$ in Figure 8.8*a*. Because generally $\Delta\mathcal{E} \ll \mathcal{E}'_{\text{thr}} - \mathcal{E}_{\text{thr}}$, we can write (8.3.5) in the form

$$\sigma_{\text{att}} \approx \pi \left(\frac{m}{M_{\text{R}}}\right)^{1/2} \left(\frac{e}{4\pi\epsilon_0}\right)^2 \frac{(\Delta\mathcal{E})^2}{2\mathcal{E}_1^3} \delta(\mathcal{E} - \mathcal{E}_1) \tag{8.3.7}$$

where δ is the Dirac delta function. Using (8.3.7), the average over the vibrational motion can be performed, leading to a cross section that is strongly peaked lying between $\mathcal{E}_{\text{thr}}$ and $\mathcal{E}'_{\text{thr}}$. We leave the details of the calculation to a problem.

8.3.6 Polar Dissociation

The process,

$$e + AB \rightarrow A^+ + B^- + e$$

produces negative ions without electron capture. As shown in Figure 8.8*d*, the process proceeds by excitation of a polar state A^+B^- of AB^* that has a separate atom limit of A^+ and B^-. Hence, at large R, this state lies above the A + B ground state by the difference between the ionization potential of A and the electron affinity of B. The polar state is weakly bound at large R by the Coulomb attraction force, but is repulsive at small R. The maximum cross section and the dependence of

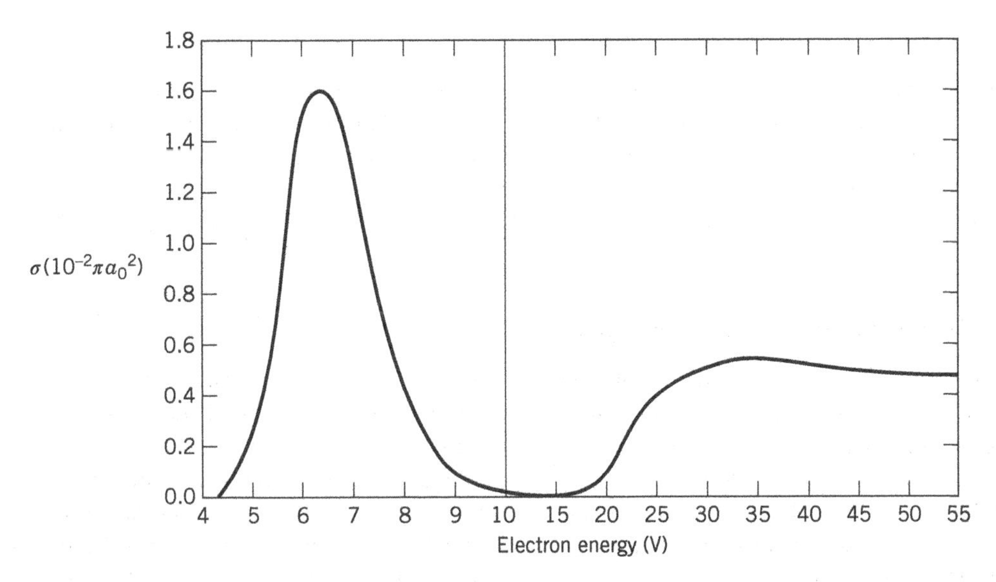

Figure 8.9 Cross section for production of negative ions by electron impact in O_2. Source: Rapp and Briglia, 1965/with permission of AIP Publishing.

the cross section on electron impact energy are similar to that of pure dissociation. The threshold energy $\mathcal{E}_{\rm thr}$ for polar dissociation is generally large.

The measured cross section for negative ion production by electron impact in O_2 is shown in Figure 8.9. The sharp peak at 6.5 V is due to dissociative attachment. The variation of the cross section with energy is typical of a resonant capture process. The maximum cross section of $\sim 10^{-18}$ cm^2 is quite low because autodetachment from the repulsive O_2^- state is strong, inhibiting dissociative attachment. The second gradual maximum near 35 V is due to polar dissociation; the variation of the cross section with energy is typical of a nonresonant process.

8.3.7 Metastable Negative Ions

In some complex molecules, a negative ion state lies at an energy very close to but just above the ground state. In this case, pure attachment of electrons having nearly zero energy can occur at low pressures. A good example is SF_6 where the SF_6^- state lies about 0.1 V above the SF_6 state, leading to the process

$$e + SF_6 \rightarrow SF_6^-$$

The negative ion is unstable to autodetachment and may also be unstable to autodissociation, but in some complex molecules, such as SF_6, these processes are weak, leading to lifetimes for the SF_6^- metastable ion in excess of 10^{-6} s. The cross section is sharply resonant with a maximum value of order 10^{-15} cm^2. For very low electron energies, this process might be important in low-pressure SF_6 discharges.

8.3.8 Electron Impact Detachment

The processes

$$e + A^- \rightarrow A + 2e$$

$$e + AB^- \rightarrow AB + 2e$$

can be important in destroying atomic or molecular negative ions. The process is similar to electron neutral ionization, with the electron affinity $\mathcal{E}_{aff}$ of A or AB playing the role of the ionization potential. However, the peak in the cross section is shifted to energies of order 10–20 $\mathcal{E}_{aff}$ due to the repulsive Coulomb force between the incident electron and the negative ion. The maximum cross section per valence electron is smaller than the Thomson result (3.4.21), with $\mathcal{E}_{aff}$ replacing $\mathcal{E}_{iz}$, due to the same effect.

8.3.9 Vibrational and Rotational Excitations

Vibrational and rotational energy levels are separated by energies of order $\mathcal{E}_v$~0.2 V and $\mathcal{E}_J$~0.01 V, respectively. Classically, slow electrons are unlikely to excite ground state molecules to higher vibrational or rotational levels because an electron having energy $\mathcal{E}$ transfers an energy $\sim (2m/M)\mathcal{E} \ll \mathcal{E}_v, \mathcal{E}_J$ in an elastic collision with a heavy particle. However, it is found experimentally that there can be significant excitations when $\mathcal{E} \sim \mathcal{E}_v$ or $\mathcal{E}_J$, respectively. For vibrational excitations, the cross sections are generally sharply peaked, indicating that a resonant (electron capture) process is involved. A common mechanism is a two-step process in which the electron is first captured by the $v = 0$ AB ground state to form an unstable negative molecular ion:

$$e + AB(v = 0) \rightarrow AB^-$$

The AB^- ion is unstable, but its lifetime for decay (typically 10^{-15}–10^{-10} s) can be comparable to or larger than its vibrational (or autodissociation) timescale (10^{-14} s). Eventually, the unstable negative ion undergoes autodetachment to an excited vibrational state of AB:

$$AB^- \rightarrow AB(v > 0) + e$$

For N_2, the N_2^- ground state is attractive (has a potential energy minimum), lies about 2.3 V above the ground state, and has a lifetime of about 10^{-14} s. Hence, the cross section for vibrational excitation of N_2 is strongly peaked about 2.3 V. The maximum cross section is large, about 5×10^{-16} cm^2. For O_2, the $v' = 0$ to $v' = 3$ states of the $^2\Pi_g$ ground state of O_2^- lie below the $v = 0$ ground state of O_2 and do not autodetach. The set of O_2^- states with $v' > 3$ lie above the O_2 ground state and can autodetach. The lifetimes of these states are long, e.g., 10^{-10} s for $v' = 4$. Excitation of these states by electron impact leads to a series of 8–10 peaks for the total vibrational cross section lying between 0.3 and 2.5 V, with the energy-integrated cross section associated with each peak in the range 10^{-19}–10^{-18} cm^2.

Direct excitation of vibrational levels due to electron interaction with the dipole moment of the vibrating molecule is also possible. The excitation cross section generally increases sharply for energies approaching the vibrational excitation threshold. A notable example is vibrational excitation of the asymmetric stretch mode of CF_4 (Christophorou et al., 1996), which is the dominant electron energy loss process for all energies below the threshold for electronic excitations.

Pure rotational excitation by electron impact can be a resonant process as for vibrational excitation or can be a nonresonant process in which the electron interacts with the permanent dipole moment of the molecule (or with the quadrupole moment for a homonuclear diatomic molecule). Angular momentum is transferred to the molecule; hence, the angular momentum of the electron must change. The cross sections for $J \rightarrow J'$ are of order 10^{-18}–10^{-16} cm^2 at energies a few times the rotational energy level difference.

Neither vibrational nor rotational cross sections have been especially well measured or calculated for most molecules. This is unfortunate because electron impact excitations to higher vibrational (and, to a lesser extent, rotational) levels can be an important source of electron energy loss in low-pressure discharges, particularly for the lower range of electron temperatures ($\lesssim$2 V) in these discharges. We consider these energy losses further in Section 8.5.

8.3.10 Elastic Scattering

Elastic scattering of electrons by atoms was described in Section 3.3. For slow electrons, polarization scattering dominates, and the cross sections typically vary as $1/v$, with v the incident electron velocity, as described by the Langevin cross section (3.3.13). In some cases, however, a relatively constant cross section is found at low energies (see Figure 3.9). For molecules having a permanent dipole moment, scattering by the resulting $1/r^2$ potential can also be significant, and the Langevin cross section is increased (Su and Bowers, 1973). We consider this process in Section 8.4.

8.4 Heavy-Particle Collisions

Heavy-particle energies in a discharge range from room temperature (~0.026 V) for most ions and neutrals in the bulk plasma, to a few volts for ion and neutral fragments newly created by dissociation processes, to hundreds of volts for ions in rf discharge sheaths. In all cases, however, the heavy particle velocities are much smaller than the characteristic velocities of orbital electron motion in an atom or molecule. The time $t_c \sim 2a_0/v_i$ for a collision between two slowly moving heavy particles is $\sim 10^{-13}$ s for room-temperature energies and is 10^{-15}–10^{-14} s for fast moving particles. These times are comparable to the molecular vibration timescale and are much longer than the timescale $t_{at} \sim 10^{-16}$–10^{-15} s for electron motion in the molecule. Hence, we have the ordering for heavy-particle collisions,

$$t_{at} \ll t_c \sim \tau_{vib} \ll \tau_{rad} \ll \tau_c$$

where, as previously, τ_{rad} is the timescale for electric dipole radiation and τ_c is the mean free time between collisions. Because $t_{at} \ll t_c$, we expect that as two heavy particles approach each other, the electronic states and their corresponding energy levels will adiabatically vary, in a manner described by the variation of the potential energy with nuclear separation R shown in Figure 8.1 and in succeeding figures. During a collision, two heavy particles move toward smaller separations along the potential energy curve, reflect at some minimum radius R_{min} corresponding to their center of mass energy, and retrace the incoming trajectory along the same curve to larger separations. This corresponds to an elastic scattering between heavy particles without a change of electronic state.

If two potential energy curves cross or nearly touch at some separation R_x, then a change of electronic state can occur with a very small energy transfer as the collision passes through R_x. A small energy transfer is required classically because the energy transferred by a heavy particle of energy $\mathcal{E}$ to an orbital electron is $\sim(2m/M)\mathcal{E}$, which is much less than the typical energy (1–10 V) required for electronic excitations of the molecule. The condition for a change of state between two electronic energy levels separated by an energy $\Delta\mathcal{E}$ during a heavy-particle collision can be estimated by requiring that the collision time $\sim R_x/v_i$ be shorter than the characteristic time $\sim\hbar/e\Delta\mathcal{E}$ for the orbital electron to change its state:

$$\frac{R_x}{v_i} \lesssim \frac{\hbar}{e\Delta\mathcal{E}} \tag{8.4.1}$$

which yields

$$\Delta\mathcal{E} \lesssim \frac{\hbar v_{\mathrm{i}}}{eR_x} \tag{8.4.2}$$

This is known as the *adiabatic Massey criterion.* In practical units, we find

$$\Delta\mathcal{E} \lesssim \frac{1}{6R_x'}\left(\frac{\mathcal{E}}{A_{\mathrm{R}}}\right)^{1/2} \tag{8.4.3}$$

where $\Delta\mathcal{E}$ is in volts, $\mathcal{E}$ is the center of mass energy in volts, R_x' is the nuclear separation in units of the Bohr radius, and A_{R} is the reduced mass in atomic mass units (amu). For example, let $\mathcal{E} = 1$ V, $A_{\mathrm{R}} = 8$, and $R_x' = 2$, we find that $\Delta\mathcal{E} \lesssim 0.03$ V for a nonadiabatic transition to occur. Hence, the states must cross or nearly touch.

8.4.1 Resonant and Nonresonant Charge Transfer

For some processes, such as resonant charge transfer,

$$\mathrm{A}^+ + \mathrm{A} \rightarrow \mathrm{A} + \mathrm{A}^+$$

which was described in Section 3.4, the two states have exactly the same energy, such that $\Delta\mathcal{E} \equiv 0$ for all separations. From the present point of view, the transition is very likely even at large separations, leading to a large cross section of the high- or low-energy form (3.4.33) or (3.4.37).

Nonresonant charge transfer between atoms,

$$\mathrm{A}^+ + \mathrm{B} \rightarrow \mathrm{A} + \mathrm{B}^+$$

is illustrated in Figure 8.10 for the reactions between N^+ and O and between O^+ and N. Since the ionization potentials of N and O are 14.53 and 13.61 V, respectively, the separated $\mathrm{N}^+ + \mathrm{O}$ level is 0.92 V higher than the $\mathrm{N} + \mathrm{O}^+$ level. At the crossing separation R_x between the attractive $\mathrm{N}^+ + \mathrm{O}$ and the repulsive $\mathrm{O}^+ + \mathrm{N}$ level, a change of state corresponding to a transfer of charge can occur. Collision a–x–b in Figure 8.10 for the exothermic reaction

$$\mathrm{N}^+ + \mathrm{O} \rightarrow \mathrm{N} + \mathrm{O}^+$$

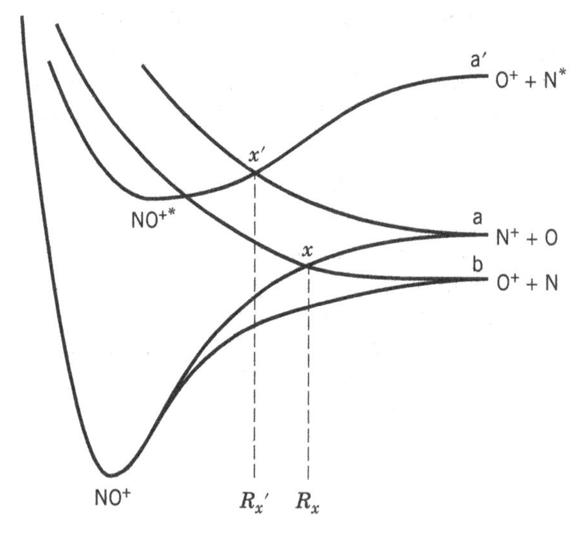

Figure 8.10 Illustrating nonresonant charge transfer processes for heavy-particle collisions.

does not have a threshold energy, and the N and O^+ products share an increase in kinetic energies of 0.92 V; hence, for slow (thermal) collisions of N^+ and O, the charge transfer products are fast. The cross section is of order the resonant cross section (3.4.33) or (3.4.37). The inverse reaction

$$O^+ + N \rightarrow O + N^+$$

is endothermic with a threshold energy of 0.92 V; hence, the rate constant for charge transfer collisions of O^+ and N at thermal energies is very small. However, if either the O^+ ion or the N atom is in an excited atomic state, then the reaction a′–x'–a, e.g.,

$$O^+ + N^* \rightarrow O + N^+$$

has no threshold, and the cross section can be large at thermal energies. Consequently, excited atoms and molecules (particularly metastables) can be important in charge transfer processes.

Similar collisions can occur between atoms and molecules. The ionization potential of O_2 is 12.2 V, so the cross section for the reaction

$$O^+ + O_2 \rightarrow O + O_2^+$$

does not have a threshold and can be expected to be large, while the cross section for the endothermic reverse reaction,

$$O_2^+ + O \rightarrow O_2 + O^+$$

has a threshold energy of 1.4 V; hence, it is very unlikely for collisions between thermal particles. As for collisions between atoms, excited O_2^+ and/or O atom charge transfer collisions can have no threshold. In fact, a proper combination of excited electronic and vibrational states can have $\Delta\mathcal{E} \approx 0$, leading to a large (resonant) cross section.

The charge transfer cross section between O_2 molecules

$$O_2^+ + O_2 \rightarrow O_2 + O_2^+$$

is resonant if the molecules have the same vibrational and rotational states after the collision, but this is not very likely. However, we may expect any energy change due to the change in vibrational and rotational quantum numbers to be small, leading to a near resonant cross section.

Charge transfer processes between negative ions and neutrals can be important in electronegative discharges. For example, in oxygen discharges, we have

$$O_2^- + O \rightarrow O_2 + O^-$$
$$O^- + O_2 \rightarrow O + O_2^-$$
$$O^- + O \rightarrow O + O^-$$
$$O_2^- + O_2 \rightarrow O_2 + O_2^-$$

Since the electron affinities of O_2 and O are 0.43 and 1.463 V, respectively, the first reaction has no threshold energy, while the second reaction has a threshold energy of 1.03 V. Hence, we expect a large cross section for the first reaction, but the second reaction is very unlikely for thermal particles. The last two processes are resonant or near resonant and have large cross sections.

8.4.2 Positive–Negative Ion Recombination

This process

$$A^- + B^+ \rightarrow A + B^*$$

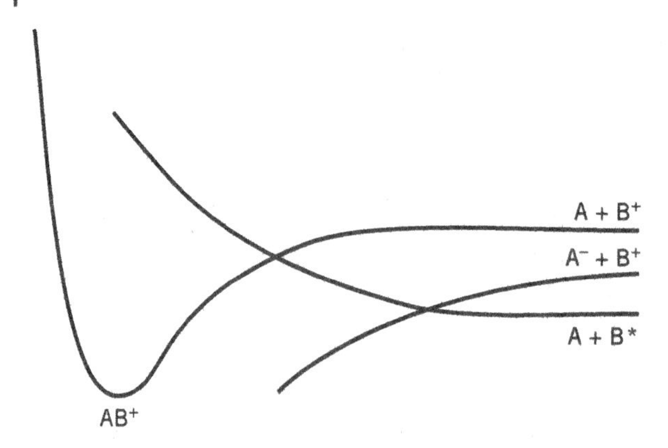

Figure 8.11 Illustrating positive–negative ion recombination for heavy-particle collisions.

is a type of charge transfer and can be the dominant mechanism for the loss of negative ions in a low-pressure discharge. The potential energy diagram is shown in Figure 8.11. The separated $A^- + B^+$ state lies below the separated $A + B^+$ state by the electron affinity $\mathcal{E}_{affA}$ of A and lies above the separated $A + B^*$ state. The $A^- + B^+$ potential energy falls as the nuclear separation decreases because of the attractive Coulomb force between the A^- and the B^+ ions. The energy level difference between the separated $A^- + B^+$ and $A + B^*$ states is of order

$$\Delta\mathcal{E} \sim \frac{\mathcal{E}_{izB}}{n^2} - \mathcal{E}_{affA} \tag{8.4.4}$$

where $\mathcal{E}_{izB}$ is the ionization potential of B and n is the principal quantum number of the excited state B^*. For $\mathcal{E}_{affA} \approx 1$ V and $\mathcal{E}_{izB} \approx 14$ V, we find that $\Delta\mathcal{E}$ is small for $n \approx 3$–4. Since $\Delta\mathcal{E}$ can be quite small, the separation R_x at the crossing can be large, and positive–negative ion recombination can have a large near-resonant cross section. A crude classical estimate of σ_{rec} can be found by putting $n \sim 3$–4 in (3.4.28), to obtain

$$\sigma_{rec} \sim 3000 - 10{,}000\, \pi a_0^2 \tag{8.4.5}$$

However, this does not expose the energy dependence. To estimate this for attractive Coulomb collisions with $\mathcal{E} \ll \mathcal{E}_{affA}$, we write conservation of angular momentum and energy during a collision as

$$v_i b = v_{max} b_0 \tag{8.4.6}$$

$$\frac{1}{2} m_R v_{max}^2 \approx \frac{e^2}{4\pi\epsilon_0 b_0} \tag{8.4.7}$$

where v_i and b are the initial velocity and impact parameter in the center of mass system, respectively, v_{max} is the velocity at the distance of closest approach b_0, and m_R is the reduced mass. Solving (8.4.6) and (8.4.7) for b, we obtain an estimate for the cross section

$$\sigma_{rec} \approx \pi b^2 = \pi \frac{e}{4\pi\epsilon_0 \mathcal{E}} b_0 \tag{8.4.8}$$

where $e\mathcal{E} = \frac{1}{2} m_R v_i^2$. We can crudely estimate the value of b_0 for a significant probability of transition to be $b_0 \approx R_x$, where for $\Delta\mathcal{E} \approx \mathcal{E}_{affA}$,

$$b_0 \approx R_x \approx \frac{e}{4\pi\epsilon_0 \mathcal{E}_{affA}} \tag{8.4.9}$$

Substituting (8.4.9) into (8.4.8), we obtain

$$\sigma_{rec} \approx \pi \left(\frac{e}{4\pi\epsilon_0}\right)^2 \frac{1}{\mathcal{E}\mathcal{E}_{affA}} \tag{8.4.10}$$

We see that $\sigma_{rec} \propto 1/\mathcal{E}$, where $\mathcal{E}$ is the collision energy in the center of mass system. Hence, for collisions between heavy particles at thermal energies, the cross sections are very large. If we put $\mathcal{E} \approx 0.026$ V and $\mathcal{E}_{affA} \approx 1$ V, then (8.4.10) yields a value of σ_{rec} in the range given by (8.4.5).

Actually, b_0 is more properly determined from a consideration of quantum mechanical electron tunneling. This was done in Section 3.4, to obtain the result (3.4.36) for b_0, which scales as $b_0 \propto \mathcal{E}_{affA}^{-1/2}$; hence, $\sigma_{rec} \propto \mathcal{E}_{affA}^{-1/2}$, not $\propto \mathcal{E}_{affA}^{-1}$, as in (8.4.10). The reader should consult Smirnov (1982) for further details.

8.4.3 Associative Detachment

This process

$$A^- + B \rightarrow AB + e$$

proceeds by formation of an unstable AB^- state that autodetaches. Figure 8.12*a* gives a potential energy diagram illustrating this process. At low energies, the collision partners move along path a–b–c of the attractive AB^- state 2, which autodetaches at c to the AB ground state 1, often falling into a highly excited vibrational state. If the collision partners follow path a–b′–a along the repulsive AB^- state 3, then there is mainly elastic scattering with little detachment. If the two AB^- states have equal statistical weight, then roughly half the collisions will lead to associative detachment.

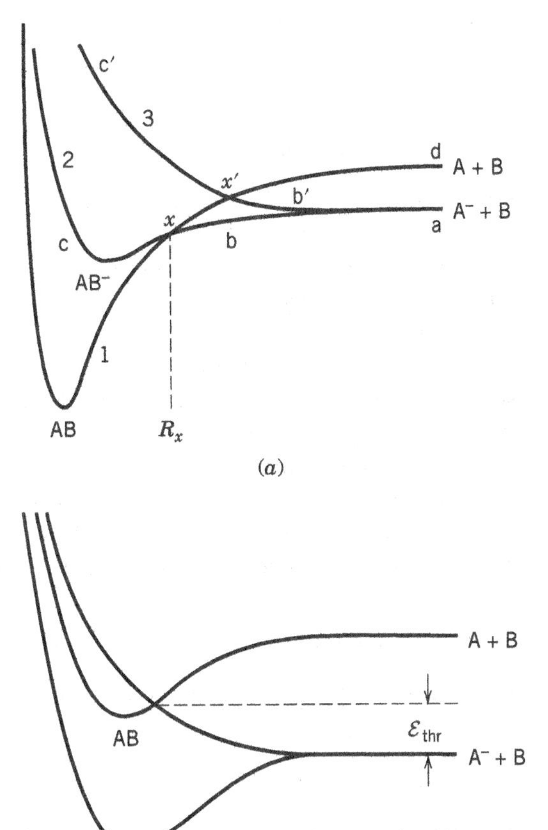

Figure 8.12 Illustrating associative detachment processes for heavy particle collisions; (*a*) the AB^- ground state lies above the AB ground state; (*b*) the AB^- ground state lies below the AB ground state.

At higher energies, the path a–b′–c′ can result in autodetachment from the repulsive state 3, instead of elastic scattering.

At thermal energies, the interaction between the negative ion and neutral is dominated by the polarization force, and the cross section for associative detachment will tend toward half the Langevin value (for a statistical weight of $\frac{1}{2}$):

$$\sigma_{\text{adet}} \approx \frac{1}{2}\sigma_{\text{L}} \tag{8.4.11}$$

where σ_{L} is given in (3.3.13). At higher energies, where the trajectories are practically straight lines, the cross section will be of order $\frac{1}{2}\pi R_x^2$. Finally, at energies higher than the electron affinity $\mathcal{E}_{\text{affA}}$ of A, the process

$$\text{A}^- + \text{B} \to \text{A} + \text{B} + \text{e}$$

can occur, as shown by the path a–b′–x'–d, leading to detachment of the electron from A^- by collision with B.

If the AB^- attractive ground state lies below the AB ground state, as shown in Figure 8.12*b*, then associative detachment from the ground state of AB^- cannot occur. However, at high energies, $\mathcal{E} > \mathcal{E}_{\text{thr}}$, associative detachment from the repulsive AB^- state is possible, and, at still higher energies, detachment from A^- due to collision with B can occur.

Associative detachment reactions in an oxygen discharge include

$$\text{O}^- + \text{O} \to \text{O}_2 + \text{e}$$

$$\text{O}^- + \text{O}_2 \to \text{O}_3 + \text{e}$$

$$\text{O}_2^- + \text{O} \to \text{O}_3 + \text{e}$$

$$\text{O}_2^- + \text{O}_2 \to \text{O}_4 + \text{e} \to 2\text{O}_2 + \text{e}$$

For oxygen, the O_2^- ground state lies below the O_2 ground state and is stable against autodetachment. However, there are a large number of shallow attractive O_2^- electronic states that lie above the O_2 ground state, and, hence, are subject to autodetachment. Consequently, there is a large rate constant for associative detachment of O^- on O (the first reaction listed above); at thermal energies, $K_{\text{adet}} \sim 5 \times 10^{-10}\ \text{cm}^3/\text{s}$. The importance of the second and third reactions listed above can be understood by noting that ozone (O_3) has a dissociation energy of only 1.04 V. Because the electron affinity of O is 1.463 V, the potential energy diagram for the second reaction is similar to that shown in Figure 8.12*b*, and the reaction has a very small rate constant at thermal energies, of order $5 \times 10^{-15}\ \text{cm}^3/\text{s}$. Since $\mathcal{E}_{\text{aff}} \approx 0.43$ V for O_2, the third reaction has a potential energy diagram similar to that shown in Figure 8.12*a*, and the rate constant is large at thermal energies, of order $1.5 \times 10^{-10}\ \text{cm}^3/\text{s}$. The fourth reaction requires a threshold energy equal to the electron affinity of O_2, $\mathcal{E}_{\text{aff}} \approx 0.43$ V, and is not very likely at thermal energies.

8.4.4 Transfer of Excitation

Ionization or excitation by the impact of ground-state atoms or molecules,

$$\text{A} + \text{B} \to \text{A}^+ + \text{B} + \text{e}$$

$$\text{A} + \text{B} \to \text{A}^* + \text{B}$$

is improbable because, as we have already seen, the potential energy curve for the A+B state is widely separated from the potential energy curves of the A^++B and A^*+B states. Classically, as noted earlier, only a very small fraction, $\sim 2m/M$, of the initial kinetic energy can be transferred

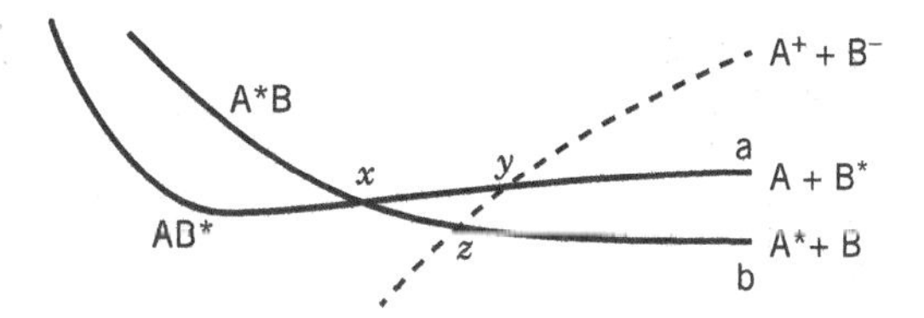

Figure 8.13 Illustrating transfer of excitation for heavy-particle collisions.

to an orbital electron. However, transfer of energy from an excited electronic state to another excited state

$$A + B^* \rightarrow A^* + B$$

can be accomplished if the potential energy curves cross or nearly touch at some nuclear separation R_x. In discharges, the excited state B^* is usually metastable. Some potential energy curves for this reaction are illustrated in Figure 8.13. The transfer of excitation can proceed along the path a–x–b. Alternately, the temporary A^+B^- state can be involved, which has a Coulomb-type potential energy curve (dashed), with the transition a–y–z–b. An example of the latter process is $Ar(^3P) + N_2 \rightarrow Ar + N_2(c^3\Pi_u)$ (Ricard, 1996).

Because there is no emitted electron to carry away any excess energy, these processes are resonant near the crossings. The energy uncertainty of the A+B* and A*+B levels is of order $\hbar/\tau_{rad}$, where τ_{rad} is the lifetime of the excited states, and the excitation energies of the A^* and B^* states must coincide to within this uncertainty. An important example of transfer occurs in the He–Ne gas laser, where the transfers

$$He(^1S) + Ne \rightarrow He + Ne(5s)$$

$$He(^3S) + Ne \rightarrow He + Ne(4s)$$

are near-resonant, resulting in a population inversion for the 4s and 5s levels of neon and subsequent laser action.

8.4.5 Penning, Associative, and Pooling Ionization

Chemi-ionization, the transfer of energy from an excited state to an ionized state, can be an important process in discharges, especially at high pressures. Some examples are described in Chapter 15. These reactions include

$A + B^* \rightarrow A^+ + e + B$	Penning ionization
$A + B^* \rightarrow A^+B + e$	Associative ionization
$B^* + B^* \rightarrow B^+ + e + B$	Metastable pooling

The Penning and associative ionization reactions can occur at all internuclear separations where the AB^* potential energy curve exceeds the A^+B curve. For the pooling reaction, the B^*B^* energy must exceed the B^+B energy. Let us consider Penning and associative ionization as shown in Figure 8.14 (Manus, 1976; Ricard, 1996). First, it is important to realize that when the AB^* curve lies above the A^+B curve, the AB^* state actually lies within a continuum of states of the ionized type $A^+B + e$. The AB^* state is therefore *autoionizing*. At some point 2 within the entrance channel during the collision of A and B^*, the autoionization occurs, with a vertical transition to point 4, and an electron with energy $\mathcal{E}(2) - \mathcal{E}(4)$ is released. The relative kinetic energy $\mathcal{E}_{kin}$ of the nuclei is conserved, so $\mathcal{E}_{kin} = \mathcal{E}(1) - \mathcal{E}(2) = \mathcal{E}(3) - \mathcal{E}(4)$, which allows curve 3 ($\mathcal{E}_{final}$) to be constructed from the known curves 1 ($\mathcal{E}_{init}$), 2 (AB^*), and 4 (A^+B). If point 3 lies above the dissociation limit

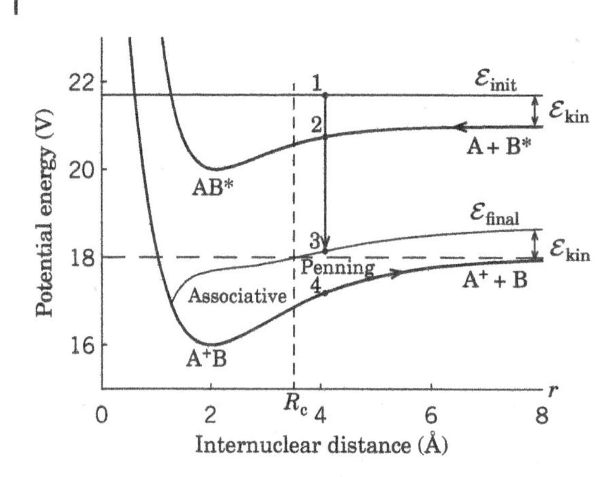

Figure 8.14 Illustrating Penning and associative ionization for heavy-particle collisions.

of A^+B (dashed horizontal line), then the two nuclei separate, resulting in Penning ionization. If point 3 lies below the dissociation limit, the nuclei are bound and the result (classically) is associative ionization. The vertical dashed line in Figure 8.14 shows the critical separation R_c; Penning ionization occurs to the right and associative ionization to the left.

The upper AB^* and/or lower A^+B molecular states can be repulsive. For example, for He/N_2 Penning ionization, the upper He^*N_2 state is repulsive, and the lower HeN_2^+ state is attractive. For He/O_2, the situation is reversed (Ricard, 1996).

As illustrated in Figure 8.15, there are actually two electron emission processes that can lead to autoionization. For the direct process (*a*), the excited electron of B^* makes a transition to the ground-state B, and an electron from the ground-state A is ejected into the continuum. For this process, the separation of A and B^* can be significantly larger than the sum of their orbital electron radii. In the exchange process (*b*), an electron from A tunnels into B^*, after which the excited electron of B^* is ejected into the continuum. The exchange process is similar to ion-induced Auger emission from surfaces, which we discuss in Section 9.4, and requires that the A and B^* electron orbitals are close to each other.

Where does the transition occur? The transition frequency at every point on the entrance channel depends on the internuclear distance r, roughly having the form $\nu_{trans} \propto e^{-r/R_t}$. The transition frequency can become quite significant even when the outer electron orbitals of A and B^* do not overlap much, leading to large cross sections. The most important process is generally Penning ionization. The most effective metastable atom is helium with 19.82 V for the 2^3S state and 20.6 V for the 2^1S state. Because the valence electrons in excited ($n > 1$) states have large radii, $a \sim a_0 n^2$ from (3.4.3), the maximum cross sections can be very large: $\sigma_{max} \sim 10^{-15}$ cm^2 for He(2^3S) ionization

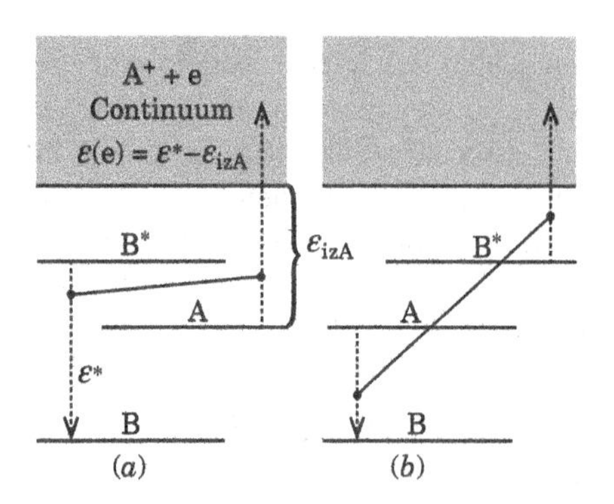

Figure 8.15 Schematic representation of Penning ionization; (*a*) direct and (*b*) exchange processes.

of Ar, and 1.4×10^{-14} cm^2 for Hg. Argon metastables are used for Penning ionization of gases such as Xe, CO_2, and CH_4, which have ionization energies lower than the 11.55 and 11.72 V energies of the $^3P_2$ and $^3P_0$ metastables (Şahin et al., 2010). Some rate coefficients for Penning ionization by He, Ne, and Ar metastables are given in Ricard (1996). Associative ionization can also have a large maximum cross section $\sim 10^{-15}$ cm^2. The metastable pooling ionization proceeds in a manner similar to Penning ionization, with the upper (entrance) molecular state B^*B^* generally repulsive.

8.4.6 Rearrangement of Chemical Bonds

Exothermic chemical reactions between ions and neutrals of the form

$$\begin{aligned} AB^+ + CD &\rightarrow AC^+ + BD \\ &\rightarrow ABC^+ + D \\ &\rightarrow \text{etc.} \end{aligned}$$

result in rearrangements of chemical bonds. For thermal collisions, the collision is dominated by the polarization force, and the maximum rate constant for reactions of this type might be expected to be the Langevin value (3.3.17). However, the thermal rate constants are often considerably smaller than this, indicating that the collision complex does not live long enough to allow for efficient bond rearrangement. An exception occurs for exothermic *proton abstraction* processes,

$$AH^+ + B \rightarrow BH^+ + A$$

which have rate constants close to the Langevin value for thermal collisions. An example of bond rearrangement in oxygen discharges is the exothermic reaction

$$O^+ + O_3 \rightarrow O_2^+ + O_2$$

which has a rate constant of $\sim 10^{-10}$ cm^3/s.

Exothermic neutral–neutral bond rearrangements

$$\begin{aligned} AB + CD &\rightarrow AC + BD \\ &\rightarrow ABC + D \\ &\rightarrow \text{etc.} \end{aligned}$$

generally have rate constants $\sim 10^{-11}$ cm^3/s, one or two orders of magnitude smaller than the Langevin value. The maximum cross sections are of order the gas kinetic value $\pi(a_1 + a_2)^2$, where a_1 and a_2 are the mean radii of the reactants. Generally, even exothermic reactions are impeded by energy barriers, such that many such reactions have an Arrhenius form

$$K(T) = K_0 \exp\left(-\frac{\mathcal{E}_a}{T}\right) \tag{8.4.12}$$

with the pre-exponential factor K_0 and the activation energy $\mathcal{E}_a$ roughly independent of temperature T. An example of oxygen discharges is

$$O + O_3 \rightarrow 2O_2$$

with $K \approx 2 \times 10^{-11} \exp(-0.2/T)$ cm^3/s.

8.4.7 Ion–Neutral Elastic Scattering

If the molecular ion has a permanent dipole moment p_d, then the polarization scattering and the Langevin capture cross section are increased due to the additional interaction potential $U \propto p_d/r^2$. The increase in the Langevin rate constant has been calculated by Su and Bowers (1973) for thermal collisions of ions and neutrals, with the result

$$K = \left(\frac{\pi q^2}{\epsilon_0 m_R}\right)^{1/2} \left[\alpha_p^{1/2} + Cp_d \left(\frac{1}{2\pi^2 \epsilon_0 kT}\right)^{1/2}\right] \tag{8.4.13}$$

where the first term in square brackets gives the Langevin rate constant (3.3.15) and the second term gives the increase due to the permanent dipole moment. The quantity C is a parameter between 0 and 1 that describes the effectiveness of the charge "locking" in the dipole and is a function of T and $p_d/\alpha_p^{1/2}$ alone. At $T = 300$ K, C is plotted against $p_d/\alpha_p^{1/2}$ in Figure 8.16.

8.4.8 Three-Body Processes

We have said little in this and the previous section about three-body reactions such as electron–ion recombination

$$e + A^+ \ (+e) \rightarrow A \ (+e)$$

attachment

$$e + A \ (+M) \rightarrow A^- \ (+M)$$

association

$$A^+ + B \ (+M) \rightarrow AB^+ \ (+M)$$

and positive–negative ion recombination

$$A^- + B^+ \ (+M) \rightarrow AB \ (+M)$$

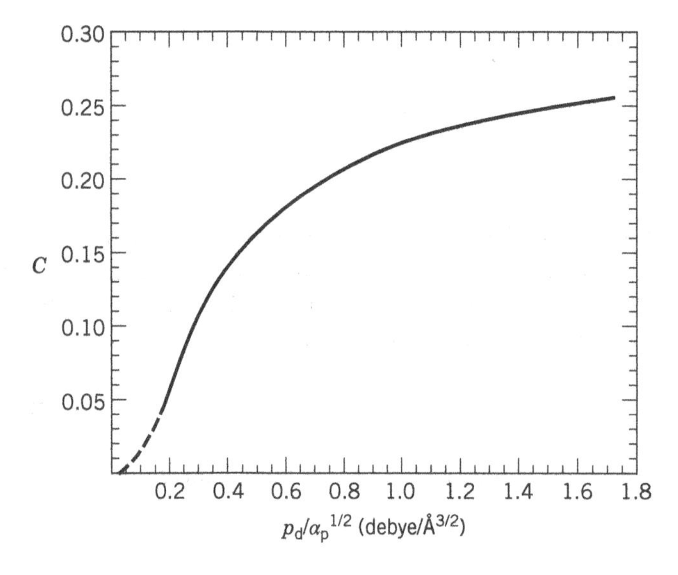

Figure 8.16 A plot of the dipole locking constant C; 1 debye $\approx 3.34 \times 10^{-30}$ C-m. Source: Su and Bowers, 1973/with permission of Elsevier.

Here, A or B can be any atom or molecule, and M can be any atom or molecule including A or B. The rate constants in the forward and reverse directions for elementary reactions involving three bodies are discussed in the next section. However, in most cases, for the densities of interest in low-pressure materials processing discharges, three-body reactions proceed by a series of two-body elementary reactions in which the third body (shown in parentheses for the reactions listed above) absorbs the excess reaction energy. For low densities of the third body, the equivalent two-body rate constants (cm^3/s) for three-body processes are proportional to the density of the third body and are generally smaller than the rate constants for two-body processes. We will show this and consider other aspects of three-body processes in Chapter 9.

8.5 Reaction Rates and Detailed Balancing

As described in Section 3.5, the cross sections must be averaged over the energy distributions of the colliding particles in order to determine the reaction rates. For a general reaction of A and B particles,

$$\mathrm{A} + \mathrm{B} \rightarrow \text{products}$$

the number of A and B particles reacting per unit volume per unit time is

$$\frac{dn_A}{dt} = \frac{dn_B}{dt} = -K_{AB} n_A n_B \tag{8.5.1}$$

where the two-body rate constant K_{AB} is a function of the particle energy distributions but is independent of their densities.

We described the averaging required for electron collisions with heavy particles in Section 3.5. Here, we consider the case of heavy-particle collisions. If A and B are unlike particles (of different species) that each have a Maxwellian distribution with a common temperature T, then the averaging yields

$$K_{AB}(T) = \langle \sigma_{AB} v_R \rangle = \int_0^\infty f_m v_R \sigma_{AB}(v_R) 4\pi v_R^2 \, dv_R \tag{8.5.2}$$

where

$$f_m = \left(\frac{m_R}{2\pi kT} \right)^{3/2} \exp\left(-\frac{m_R v_R^2}{2kT} \right) \tag{8.5.3}$$

and m_R is the reduced mass. If A and B are like particles, e.g., for the collision of two ground-state oxygen atoms, then

$$K_{AA}(T) = \frac{1}{2} \langle \sigma_{AA} v_R \rangle = \frac{1}{2} \int_0^\infty f_m v_R \sigma_{AA}(v_R) 4\pi v_R^2 \, dv_R \tag{8.5.4}$$

The reason for the factor of $\frac{1}{2}$ in (8.5.4) can be understood by numbering the A and B particles within a unit volume. For unlike particles, the collisions of A1 with B2 and A2 with B1 are different collisions, whereas for like particles, the collisions of A1 with A2 and A2 with A1 are the same collision and must not be counted twice.

8.5.1 Temperature Dependence

For thermal collisions ($T \sim 300$ K) with a constant cross section σ_0 near zero velocity, as for hard-sphere collisions, the averages in (8.5.2) and (8.5.4) are easily done, yielding

$$K_{AB} = \sigma_0 \bar{v}_R \tag{8.5.5}$$

$$K_{AA} = \frac{1}{2}\sigma_0 \bar{v}_R \tag{8.5.6}$$

where $\bar{v}_R = (8kT/\pi m_R)^{1/2}$. Hence, K_{AB} and K_{AA} vary weakly as $\sqrt{T}$. For the polarization interaction, with $\sigma \propto 1/v_R$, we have already seen for the Langevin rate constant (3.3.15) that K is independent of T.

Consider now a process that has a threshold energy $\mathcal{E}_{thr}$. The variation of the cross section with energy near the threshold can be estimated from conservation of angular momentum and energy,

$$v_R b = v_\theta b_0 \tag{8.5.7}$$

$$e\mathcal{E} = \frac{1}{2}m_R v_R^2 \approx \frac{1}{2}m_R v_\theta^2 + e\mathcal{E}_{thr} \tag{8.5.8}$$

where b_0 is the effective radius for the reaction and v_θ is the angular component of the velocity. The influence of the interaction potential has been neglected in (8.5.8). Solving (8.5.7) for v_θ, substituting this into (8.5.8), and solving for $\sigma \approx \pi b^2$, we obtain

$$\sigma = \begin{cases} 0 & \mathcal{E} < \mathcal{E}_{thr} \\ \sigma_0\left(1 - \dfrac{\mathcal{E}_{thr}}{\mathcal{E}}\right) & \mathcal{E} > \mathcal{E}_{thr} \end{cases} \tag{8.5.9}$$

where $\sigma_0 = \pi b_0^2$. We see that the cross section rises linearly just above the threshold energy and tends to a maximum value σ_0 for large $\mathcal{E}$. The rise is linear rather than abrupt because the centrifugal energy $\frac{1}{2}m_R v_\theta^2$ is not available to excite the reaction. Many cross sections display this linear rise.

Inserting (8.5.9) into either (8.5.2) or (8.5.4) and integrating, we obtain

$$K_{AB} = \sigma_0 \bar{v}_R \, e^{-\mathcal{E}_{thr}/T} \tag{8.5.10}$$

$$K_{AA} = \frac{1}{2}\sigma_0 \bar{v}_R \, e^{-\mathcal{E}_{thr}/T} \tag{8.5.11}$$

respectively, which have an Arrhenius form, with the preexponential factor varying weakly as $\sqrt{T}$.

8.5.2 The Principle of Detailed Balancing

The cross sections and rate constants for forward and reverse reactions are related by the principle of detailed balancing, which expresses the time reversibility of the equations of motion for a collision. Hence, knowledge of the cross section for a two-body reaction allows one to determine the properties of the reverse reaction. The cross section $\sigma(v_R)$ for the inelastic reaction (endothermic with threshold energy $\mathcal{E}_a$),

$$A + B \rightarrow C + D$$

is related to the cross section $\sigma'(v_R')$ for the reverse reaction,

$$C + D \rightarrow A + B$$

by (Smirnov, 1981, Appendix A2)

$$m_R^2 g_A g_B v_R^2 \sigma(v_R) = m_R'^2 g_C g_D v_R'^2 \sigma'(v_R') \tag{8.5.12}$$

where

$$\frac{1}{2} m_R v_R^2 = \frac{1}{2} m_R' v_R'^2 + e\mathcal{E}_a \tag{8.5.13}$$

m_R and m_R' are the reduced masses for particles A and B, and C and D, respectively, and the gs are the degeneracies of the energy levels of the particles, for example, $g_e = 2$ for a free electron (the two spin states have the same energy), and $g_O = 5$ for the $O(^3P_2)$ ground state (the five m_J values 2,1, 0, −1, −2, have the same energy). We can integrate (8.5.12) over a Maxwellian distribution of v_R to obtain (Problem 8.9)

$$\frac{K(T)}{K'(T)} = \left(\frac{m_R'}{m_R}\right)^{3/2} \frac{\bar{g}_C \bar{g}_D}{\bar{g}_A \bar{g}_B} e^{-\mathcal{E}_a/T} \tag{8.5.14}$$

which expresses the ratio of the rate constants for the forward and reverse reactions in terms of a ratio of reduced masses and energy level degeneracies times a Boltzmann factor. As shown in Problem 8.10, the reduced mass factor is actually the ratio of *translational statistical weights per unit volume*, $\hat{g}$, of the reaction products, divided by those of the reactants

$$\left(\frac{m_R'}{m_R}\right)^{3/2} = \frac{\hat{g}_C \hat{g}_D}{\hat{g}_A \hat{g}_B} \tag{8.5.15}$$

where

$$\hat{g} = \left(\frac{2\pi m e T}{h^2}\right)^{3/2} \quad (m^{-3}) \tag{8.5.16}$$

with h Planck's constant.

For the internal energy level degeneracies in (8.5.14), we have written $\bar{g}$ rather than g because we are generally more interested in the rate constants for a group of closely spaced energy levels for each particle, rather than for a single state. For example, we specify the ground state of an oxygen atom as $O(^3P)$, which comprises three closely spaced levels: the $^3P_1$ and $^3P_0$ levels lie 0.020 and 0.028 V above the $^3P_2$ level, respectively. We can apply (8.5.14) to this case if we interpret the gs as $\bar{g}$s, the *statistical weights* of the internal degrees of freedom, i.e., the mean number of occupied states for the group of levels. The ratio of statistical weights can be evaluated by assuming that the A, B, C, and D particles are all in thermal equilibrium at temperature T. Generally, for an atom or atomic ion somewhat above room temperature, the electronic states within the fine structure of a group of energy levels are all occupied; consequently, $\bar{g}_{at}$ is equal to the total degeneracy g_{at} of the group of levels. For example, the $O(^3P)$ ground-state triplet has five states for $^3P_2$, three states for $^3P_1$, and one state for $^3P_0$, for a total degeneracy $\bar{g}_O = 9$. At room temperature and below, $\bar{g}_O < g_O$ because the upper levels do not have a high probability of being occupied (Problem 8.12). Typically, $\bar{g}_{at} \sim 1$–10 for ground-state atoms or atomic ions.

For molecules at thermal energies (0.026 V) and above, in addition to the electronic degeneracy $\bar{g}_{at}$ of the molecular level, many rotational states and some vibrational states can be occupied. The energy of a molecule in a vibrational–rotational state (v, J) above the (0, 0) ground state is, summing the vibrational and rotational energies in (8.2.1) and (8.2.4),

$$e\mathcal{E} = \hbar\omega_{vib}\left(v + \frac{1}{2}\right) + eB_{rot}J(J+1)$$

where ω_{vib} is the vibrational frequency and $B_{rot} = \hbar^2/2eI_{mol}$ is the rotational energy constant of the molecule. In thermal equilibrium, the mean number of levels occupied by a heteronuclear diatomic molecule can be shown to be (Problem 8.13)

$$\bar{g}_{rot}\bar{g}_{vib} = \frac{T}{B_{rot}} \frac{1}{1 - e^{-\hbar\omega_{vib}/eT}} \tag{8.5.17}$$

For a homonuclear diatomic molecule, $\bar{g}_{rot}$ must be divided by two because the two states with the molecule rotated by 180° are identical. For polyatomic molecules, $\bar{g}_{vib}$ consists of a product of factors, one for each vibrational degree of freedom. The statistical weight of the molecule is then $\bar{g}_{mol} = \bar{g}_{at}\bar{g}_{vib}\bar{g}_{rot}$. For typical diatomic molecules at room temperature, $\bar{g}_{mol} \sim 10^2$–10^3.

Although the statistical weights in (8.5.14) are determined for thermal equilibrium, the ratio of statistical weights is the same for a system that is not in thermal equilibrium. The only assumption required is that the distribution of v_R (and, consequently, v'_R) be Maxwellian. This is because each rate constant in (8.5.14) depends only on the collision dynamics (the cross section) and the assumed velocity distribution (a Maxwellian). Consequently, the ratio of rate constants must be the same whether or not the particles are in thermal equilibrium.

As will be shown in Section 9.1, (see (9.1.13)), the right-hand side of (8.5.14) is the equilibrium constant $\mathcal{K}(T)$, as given in (7.4.7), for the reaction of A + B to form C + D. Writing the Gibbs free energy of reaction, $G_r^\ominus$, in terms of the enthalpy and entropy of reaction using the definition of G (7.3.14) and substituting this into the expression (7.4.7) for $\mathcal{K}$, we obtain

$$\frac{K(T)}{K'(T)} = \mathcal{K}(T) = e^{-G_r^\ominus/RT} = e^{S_r^\ominus/R} e^{-H_r^\ominus/RT} \tag{8.5.18}$$

The terms exponential in $S_r^\ominus$ and $H_r^\ominus$ on the right-hand side of (8.5.18) is equal, in (8.5.14), to the product of mass and statistical weight factors, and to the exponential energy factor, respectively. If $G_r^\ominus$ is known, then K' can be determined if K is known, and vice versa. This relationship will be elaborated in Chapter 9. Let us note some examples where (8.5.14) can be applied. The rate constant for de-excitation of an excited state

$$A + B^* \rightarrow A + B$$

can be determined from the rate constant for collisional excitation of that state:

$$A + B \rightarrow A + B^*$$

Here, A can be an electron, atom, or molecule, and B can be an atom or molecule. The rate constant for associative ionization

$$A + B^* \rightarrow AB^+ + e$$

can be determined from the rate constant for dissociative recombination

$$e + AB^+ \rightarrow A + B^*$$

Relations similar to (8.5.14) can be found for reactions that change the number of particles, such as

$$e + A \underset{K_{rec}}{\overset{K_{iz}}{\rightleftharpoons}} e + e + A^+$$

$$AB + M \underset{K_{rec}}{\overset{K_{diss}}{\rightleftharpoons}} A + B + M$$

These relations connect the two-body rate constants to the three-body rate constants for the reverse reactions. For example, analogous to equations (8.5.14)–(8.5.16)

$$\frac{K_{\mathrm{iz}}}{K_{\mathrm{rec}}} = \frac{\hat{g}_{\mathrm{e}}\hat{g}_{\mathrm{A}^+}}{\hat{g}_{\mathrm{A}}} \frac{\bar{g}_{\mathrm{e}}\bar{g}_{\mathrm{A}^+}}{\bar{g}_{\mathrm{A}}} \mathrm{e}^{-\mathcal{E}_{\mathrm{iz}}/\mathrm{T}_{\mathrm{e}}} \tag{8.5.19}$$

Assuming that the forward and reverse reactions are totally balanced in thermal equilibrium, with a common temperature $\mathrm{T_e} = \mathrm{T}$ for all species, then (8.5.19) gives the *Saha equation* for a particular gas, which specifies the fractional ionization $n_{\mathrm{e}}/n_{\mathrm{g}}$ as a function of n_{g} and T. We leave the calculation to Problem 8.14.

Finally, let us note that detailed balancing is not as useful to determine rate constants as might first be imagined, because the "forward" and "reverse" reactions of interest may not actually be inverses. For example, electron excitation to B*

$$\mathrm{e} + \mathrm{B} \rightarrow \mathrm{B}^* + \mathrm{e}$$

often proceeds by a compound process of excitation to a higher level or set of levels, followed by radiative decay:

$$\mathrm{e} + \mathrm{B} \rightarrow \mathrm{B}^{*2} + \mathrm{e}$$

$$\mathrm{B}^{*2} \rightarrow \mathrm{B}^* + \hbar\omega$$

The reverse reaction of interest might be direct de-excitation of B* to the ground state:

$$\mathrm{e} + \mathrm{B}^* \rightarrow \mathrm{B} + \mathrm{e}.$$

These two processes are not inverses and are not connected by detailed balancing. Similarly, excitation of a molecule

$$\mathrm{e} + \mathrm{AB} \rightarrow \mathrm{AB}^* + \mathrm{e}$$

may be to a high vibrational state $v' \gg 0$, while de-excitation

$$\mathrm{e} + \mathrm{AB}^* \rightarrow \mathrm{AB} + \mathrm{e}$$

is from the ground vibrational state $v' = 0$. The reader should consult other sources (e.g., Smirnov, 1981) for further discussion of these methods for the determination of rate constants.

8.5.3 A Data Set for Oxygen

To illustrate the complexity of molecular processes, we give some data for oxygen, which is a simple diatomic gas that has been particularly well studied. This data set will be used throughout this book to illustrate various features of chemically reactive discharges. In an oxygen discharge, there can be significant ground-state concentrations of O, O_2, O_3, O^+, O_2^+, O_4^+, O_3^-, O_2^-, O^-, and electrons, as well as metastable states such as the ^{1}D and ^{1}S states of O and the $^1\Delta_g$ and $^1\Sigma_g^+$ states of O_2. Some basic constants for some of these species are given in Table 8.1. The cross sections for binary processes among these species have mostly not been carefully measured or calculated. To give an example of some of the best data, some cross sections for electron impact excitation of O_2, useful for determining the energy losses, are given in Figure 8.17. These include momentum transfer, rotational and vibrational excitation, two- and three-body attachment, $^1\Delta_g$ and $^1\Sigma_g^+$ metastable excitation, excitations to states involving energy losses of approximately 4.5, 6.0, 8.4, 10.0, and 14.7 V, and ionization with an energy loss of 12.06 V. The momentum transfer cross section is also given. The identification of the energy losses with specific processes such as dissociation and attachment is uncertain. Using

Table 8.1 Basic Constants for Oxygen Discharges

State	$\mathcal{E}_{diss}$ (V)	$\mathcal{E}_{iz}$ (V)	Lifetime (s)	α_p (a_0^3)
$O(^3P)$	—	13.61	—	5.4
$O^-(^2P)$	—	1.463	—	—
$O^*(^1D)$	—	11.64	147.1	—
$O_2(^3\Sigma_g^-)$	5.12	12.14	—	10.6
$O_2^+(^2\Pi_g)$	6.59	—	—	
$O_2^-(^2\Pi_g)$	4.06	0.44	—	—
$O_2^*(^1\Delta_g)$	4.14	11.16	4400[a)]	—
O_3	1.05	12.67	—	
O_3^-	1.69	2.10	—	

a) Newman et al. (2000).

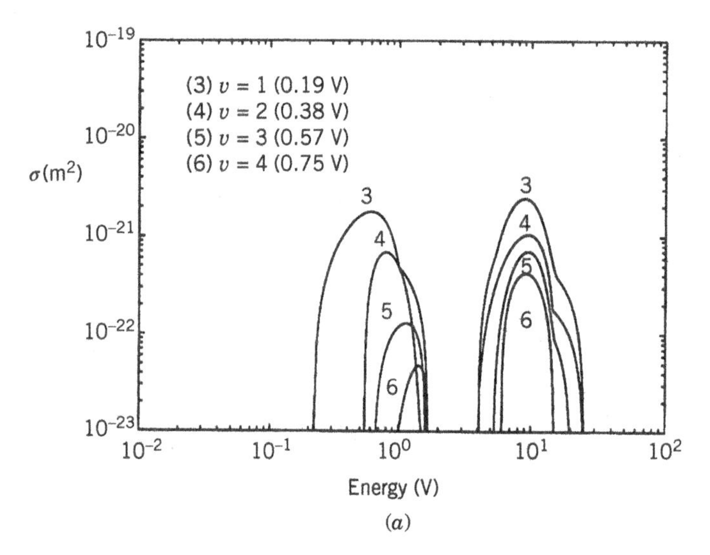

Figure 8.17 Cross sections for electron excitation of O_2 (Lawton and Phelps, 1978; Phelps, 1985; compiled by Vahedi, 1993).

these data, the energy loss $\mathcal{E}_c$ per e–O_2^+ pair created in oxygen has been determined and plotted in Figure 3.17. Similar cross-section sets have been compiled for electron collisions in many reactive gases of interest for materials processing by Hayashi (1987). Up-to-date cross-sectional data for many gases is available on the open-access LXCat website https://www.lxcat.net/home (see Carbone et al., 2021, for a review and tutorial), which is described briefly at the end of Section 19.2.

Table 8.2 gives some rate constants for a restricted set of two-body reactions of interest in modeling low-pressure oxygen discharges. These include reactions among ground states O, O_2, O_3, O^+, O_2^+, O^-, O_2^-, O_3^-, and electrons, and metastable states $O^*(^1D)$ and $O_2^*(^1\Delta_g)$. Electrons are assumed to have a Maxwellian distribution in the range $1 < T_e < 7$ V, and the heavy particles are assumed to be Maxwellian at a common temperature T near room temperature 0.026 V. A first set of reactions is given involving just the species O, O_2, O_2^+, O^-, and electrons, because these often suffice for the

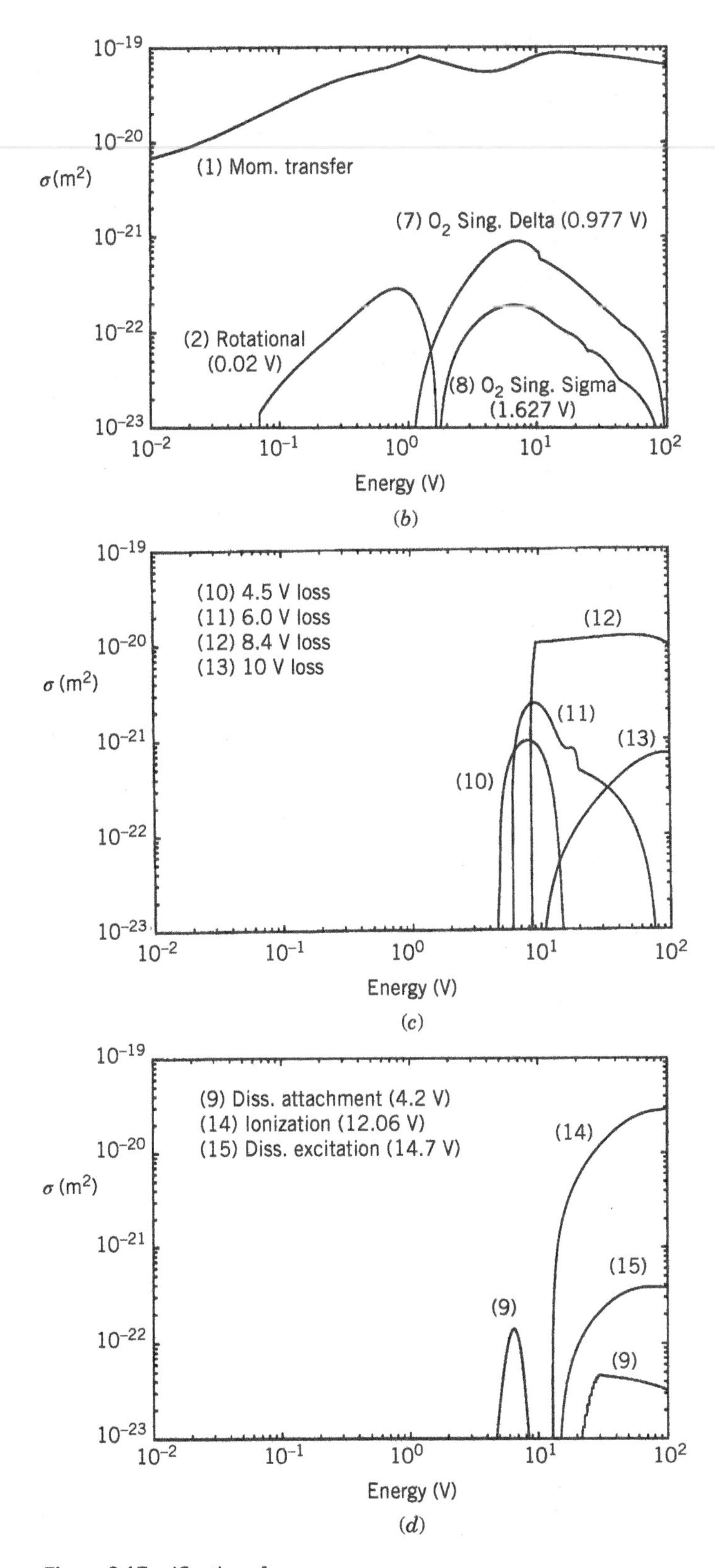

Figure 8.17 *(Continued)*

Table 8.2 Selected Second-Order Reaction Rate Constants for Oxygen Discharges

Number	Reaction	Rate Constant (cm^3/s)	Source
Reactions among e, O_2, O_2^+, O, *and* O^-			
1	$e + O_2$ momentum transfer	$4.7E{-}8T_e^{0.5}$	a
2	$e + O_2 \rightarrow O^- + O$	$1.07E{-}9T_e^{-1.391}\exp(-6.26/T_e)$	j
3	$e + O_2 \rightarrow 2O + e$	$6.86E{-}9\exp(-6.29/T_e)$	g2
4	$e + O_2 \rightarrow O_2^+ + 2e$	$2.34E{-}9T_e^{1.03}\exp(-12.29/T_e)$	n
5	$e + O^- \rightarrow O + 2e$	$5.47E{-}8\,T_e^{0.324}\exp(-2.98/T_e)$	v
6	$e + O_2^+ \rightarrow 2O$	$2.2E{-}8/T_e^{1/2}$	g3
7	$O^- + O_2^+ \rightarrow O + O_2$	$2.6E{-}8(300/T)^{0.44}$	g3
8	$O^- + O \rightarrow O_2 + e$	(1.9, 3, 5)E−10	h,m,k
9	$O^- + O_2^+ \rightarrow 3O$	$2.6E{-}8(300/T)^{0.44}$	g3
Addition of O^+			
10	$e + O_2 \rightarrow O^- + O^+ + e$	$7.1E{-}11T_e^{0.5}\exp(-17/T_e)$	r
11	$e + O_2 \rightarrow O + O^+ + 2e$	$1.88E{-}10T_e^{1.699}\exp(-16.81/T_e)$	n
12	$e + O \rightarrow O^+ + 2e$	$9.0E{-}9T_e^{0.7}\exp(-13.6/T_e)$	d
13	$O^- + O^+ \rightarrow 2O$	$4.0E{-}8(300/T)^{0.44}$	g3
14	$O^+ + O_2 \rightarrow O + O_2^+$	$2.0E{-}11(300/T)^{0.5}$	e
Addition of metastable $O_2^*(^1\Delta_g)$ *; see note f below*			
15	$e + O_2 \rightarrow O_2^* + e$	$1.37E{-}9\exp(-2.14/T_e)$	g2
16	$e + O_2^* \rightarrow e + O_2$	$2.06E{-}9\exp(-1.163/T_e)$	b
17	$e + O_2^* \rightarrow O + O^-$	$4.19E{-}9T_e^{-1.376}\exp(-5.19/T_e)$	j
18	$O_2^* + O_2 \rightarrow 2O_2$	$2.2E{-}18(T/300)^{0.8}$	e,k
19	$O_2^* + O \rightarrow O_2 + O$	(1.0, 7)E−16	e,k
20	$O^- + O_2^* \rightarrow O_3 + e$	2.2E−11	g0
21	$O^- + O_2^* \rightarrow O_2^- + O$	1.1E−11	g0
Addition of metastable $O(^1D)$			
22	$e + O_2 \rightarrow O + O^* + e$	$3.49E{-}8\exp(-5.92/T_e)$	g2
23	$e + O \rightarrow O^* + e$	$4.54E{-}9\exp(-2.36/T_e)$	g2
24	$e + O^* \rightarrow e + O$	$8.17E{-}9\exp(-0.4/T_e)$	b
25	$e + O^* \rightarrow O^+ + 2e$	$9.0E{-}9T_e^{0.7}\exp(-11.6/T_e)$	d
26	$O^* + O \rightarrow 2O$	8.0E−12	e
27	$O^* + O_2 \rightarrow O + O_2$	$(6.4, 7.0)E{-}12\exp(67/T)$	k,e
28	$O^* + O_2 \rightarrow O + O_2^*$	1.0E−12	e

Table 8.2 (Continued)

Number	Reaction	Rate Constant (cm³/s)	Source
Addition of selected reactions for O_2^- and O_3			
29	$O^- + O_2 \rightarrow O_3 + e$	5E−15	k
30	$e + O_3 \rightarrow O_2^- + O$	$9.75E{-}8\, T_e^{-1.309} \exp(-1.007/T_e)$	c
31	$e + O_3 \rightarrow O^- + O_2$	$2.12E{-}9\, T_e^{-1.058} \exp(-0.93/T_e)$	s
32	$O_2^- + O_2^+ \rightarrow 2O_2$	$2E{-}7(300/T)^{0.5}$	k
33	$O_2^- + O^+ \rightarrow O_2 + O$	$(1, 2)E{-}7(300/T)^{0.5}$	e,k
34	$O_3 + O_2 \rightarrow O_2 + O + O_2$	$7.3E{-}10 \exp(-11{,}400/T)$	e
35	$O_3 + O \rightarrow 2O_2$	$1.8E{-}11 \exp(-2300/T)$	e

Note. T_e in volts and T in kelvins. Two values from different sources are sometimes given in parentheses. The notation E−8 means 10^{-8}.
[a]Based on Phelps (1985). [b]Based on detailed balance. [c]Based on Rangwala et al. (1999).
[d]Based on Lee et al. (1994). [e]Eliasson and Kogelschatz (1986).
[f]Reactions 1, 3, 4, 10, and 11 for O_2^* have activation energies reduced by ~1 V.
[g0]Gudmundsson et al. (2000); [g1]Gudmundsson et al. (2001); [g2]Gudmundsson (2002a).
[g3]Gudmundsson (2002b). [h]Fehsenfeld (1967).
[j]Based on Jaffke et al. (1992). [k]Kossyi et al. (1992). [m]Sommerer and Kushner (1992).
[n]Based on Krishnakumar and Srivastava (1992). [r]Based on Rapp and Briglia (1965).
[s]Based on Senn et al. (1999). [v]Based on Vejby-Christensen et al. (1996).

simplest discharge models. Additional sets of reactions give added complexity as additional species are added to the model. A key task of the modeler is to choose the set of reactions appropriate to the parameter range of interest. Finally, Table 8.3 gives some rate constants for three-body reactions. These processes are described in Section 9.2.

A more complete set of Maxwellian rate constants for oxygen is given in Gudmundsson (2002b, 2004). Maxwellian rate constants, as well as non-Maxwellian rate constants, can be calculated for many gases from data available on the open-access LXCat website https://www.lxcat.net/home, using the on-line program Bolsig+ (Hagelaar and Pitchford, 2005); this is briefly described at the end of Section 19.2.

8.6 Optical Emission and Actinometry

Optical diagnostics are powerful tools for the noninvasive measurement of the properties of chemically complex discharges. A wide variety of optical diagnostic techniques are currently in use. A relatively simple technique is that in which the wavelength-resolved optical emission is measured. More complex and expensive schemes, such as laser-induced fluorescence (LIF) and optogalvanic techniques, in which laser beam probes are used to excite specific optical transitions whose subsequent emission or other response is measured, have also been widely used. Infrared emission and absorption techniques are also receiving increasing attention. We refer the reader to review articles by Donnelly (1989), Manos and Dylla (1989), and Selwyn (1993), and research articles referenced therein, for a detailed exposition of the subject.

In this section, we first discuss the simplest technique of optical emission and actinometry (defined below) to illustrate the usefulness of optical diagnostics. Small variations in discharge

Table 8.3 Selected Third-Order Reaction Rate Constants for Oxygen Discharges

Number	Reaction	Rate Constant (cm^6/s)	Source
Reactions among e, O_2, O_2^+, and O^-			
1	$e + e + O_2^+ \rightarrow e + O_2$	$1E{-}19(0.026/T_e)^{4.5}$	ke
2	$e + O_2^+ + O_2 \rightarrow O_2 + O_2$	$6E{-}27(0.026/T_e)^{1.5}$, 1E−26	k,e
3	$e + O + O_2 \rightarrow O^- + O_2$	1E−31	ke
4	$O^- + O_2^+ + O_2 \rightarrow O + O_2 + O_2$	$2E{-}25(300/T)^{2.5}$	k
5	$O + O + O_2 \rightarrow O_2 + O_2$	$2.45E{-}31T^{-0.63}$	k
		$1.3E{-}32(300/T)\exp(-170/T)$	e
6	$O + O + O \rightarrow O_2 + O$	$6.2E{-}32\exp(-750/T)$	e
Addition of O^+			
7	$e + e + O^+ \rightarrow e + O$	$1E{-}19(0.026/T_e)^{4.5}$	ke
8	$e + O^+ + O_2 \rightarrow O + O_2$	$6E{-}27(0.026/T_e)^{1.5}$, 1E−26	k,e
9	$O^- + O^+ + O_2 \rightarrow O_2 + O_2$	$2E{-}25(300/T)^{2.5}$, 2E−25	k,e
10	$O^- + O^+ + M \rightarrow O + O + M$	$2E{-}25(300/T)^{2.5}$	k
11	$O^+ + O + O_2 \rightarrow O_2^+ + O_2$	1E−29	ke
Addition of metastable $O(^1D)$			
12	$O + O^* + O_2 \rightarrow O_2 + O_2$	9.9E−33	e
Addition of selected reactions for metastable $O_2^*(^1\Delta_g)$, O_2^-, *and* O_3			
13	$e + O_2 + O_2 \rightarrow O_2^- + O_2$	$1.4E{-}29(0.026/T_e)$	
		$\times\exp(100/T - 0.061/T_e)$	k
14	$e + O_2 + O \rightarrow O_2^- + O$	1E−31	k
15	$O^- + O_2^+ + O_2 \rightarrow O_3 + O_2$	$2E{-}25(300/T)^{2.5}$	k,e
16	$O + O_2 + O_2 \rightarrow O_3 + O_2$	$6.9E{-}34(300/T)^{1.25}$,	
		$6.4E{-}35\exp(663/T)$	k,e
17	$O + O_2 + O \rightarrow O_3 + O$	$2.15E{-}34\exp(345/T)$	e
18	$e + O_2^* + O_2 \rightarrow O_2^- + O_2$	1.9E−30	e
19	$e + O_2^* + O \rightarrow O_2^- + O$	1E−31	e
20	$O_2^- + O^+ + M \rightarrow O_3 + M$	$2E{-}25(300/T)^{2.5}$	e
21	$O_2^- + O_2^+ + O_2 \rightarrow O_2 + O_2 + O_2$	$2E{-}25(300/T)^{2.5}$	e

Note. T_e in volts and T in kelvins; M denotes either O_2 or O. Two values from different sources are sometimes given. The notation E−19 means 10^{-19}.
[e] Eliasson and Kogelschatz (1986).
[k] Kossyi et al. (1992).

operation due to contamination, aging, vacuum leaks, etc., can produce large changes in emission. Hence, process reproducibility is often monitored, and even actively controlled, by measurement of emission. Detection of the endpoint for a materials process, particularly an etch, is also conveniently accomplished using optical emission. In this case, an emission line associated with an etch product can be monitored; a sharp decrease in the emission intensity versus time generally signals the completion of the etch process. Optical emission can be spatially resolved perpendicular to the line of sight, but generally is a spatial average along the line of sight. As will be shown below, the emission intensity is a convolution of the species density for the optical wavelength being monitored, the electron distribution function, and the cross section for electron impact excitation of the optical level. For example, a qualitative measure of the importance of F atoms can be obtained by monitoring the 7037-Å F-atom emission line as the discharge power and pressure are varied.

A quantitative measure of relative F-atom density can be found by using a tracer gas of known density, such as argon, and measuring the intensity of both an F-atom and an Ar-atom emission line. This widely used comparison technique is called *optical actinometry*. More sophisticated measurements, at finer wavelength resolution, can be used to determine ion and neutral energies. Time-resolved emission measurements can be used to determine both volume and surface rate constants.

The excitation of the tracer gas emission line is very sensitive to the high-energy part of the electron distribution, which controls the time- and space-varying ionization within an rf discharge. Therefore, time-resolved (phase-resolved within an rf period) tracer gas emission can be used as a powerful diagnostic to understand the energy deposition and ionization dynamics. We discuss this *phase-resolved optical emission spectroscopy* (PROES) technique in the last subsection.

8.6.1 Optical Emission

Figure 8.18 illustrates the electron impact excitation of the ground state of atom A to an excited state A*, followed by subsequent emission at frequency ω to some lower energy state A_f. The emission wavelength is

$$\lambda = \frac{2\pi c}{\omega} \tag{8.6.1}$$

where $\hbar\omega = e(\mathcal{E}_{A^*} - \mathcal{E}_f)$. The usual wavelengths are in the optical band, $\lambda \sim 2000 - 8000$ Å. The emission is sharply peaked about λ, with a small intrinsic linewidth due to the spontaneous emission rate from level A*, $\Delta\lambda^* \sim 10^{-3}$ Å. The Doppler-broadened linewidth due to a velocity distribution of ions or neutrals is wider,

$$\frac{\Delta\lambda}{\lambda} = \frac{v_{thi}}{c} \tag{8.6.2}$$

For 1 V argon atoms, $\Delta\lambda \sim 0.025$ Å.

Let n_A be the concentration of the free radical A and let I_λ (in watts) be the optical emission intensity, integrated over the linewidth. The emission due to excitation from the ground-state A can be written as

$$I_\lambda = \alpha_{\lambda A} n_A \tag{8.6.3}$$

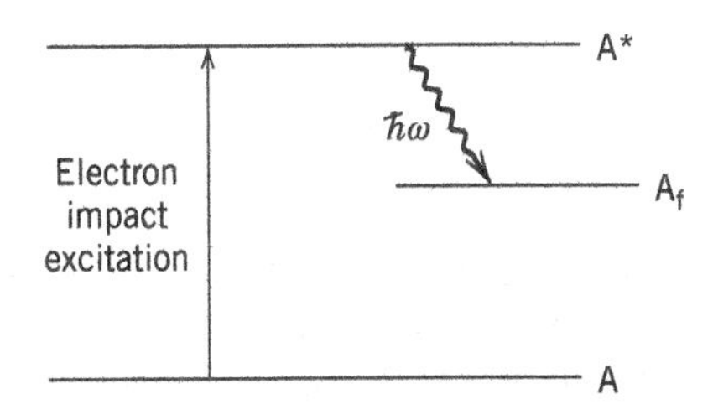

Figure 8.18 Energy level diagram for emission of radiation from an excited state.

where

$$\alpha_{\lambda A} = k_D(\lambda) \int_0^\infty 4\pi v^2 dv\, Q_{A^*}(p,\ n_e)\sigma_{\lambda A}(v) v f_e(v) \tag{8.6.4}$$

Here, f_e is the (assumed isotropic) electron distribution function, $\sigma_{\lambda A}$ is the cross section for emission of a photon of wavelength λ due to electron impact excitation of A, Q_{A^*} is the quantum yield for photon emission from the excited state ($0 \leq Q_{A^*} \leq 1$), and k_D is the detector response constant. For low-pressure discharges and excited states having short lifetimes, $Q_{A^*} \approx 1$. Q_{A^*} is generally less than unity for metastable states, due to collisional or electric field de-excitation, ionization, or other processes that depopulate the state without emission of a photon. We note that the cross section $\sigma_{\lambda A}$ differs from the cross section σ_{A^*} for excitation of A to level A*, because spontaneous emission to more than one lower lying level can occur. The two cross sections are related by

$$\sigma_{\lambda A} = b_\lambda \sigma_{A^*} \tag{8.6.5}$$

where b_λ is the *branching ratio* for emission of a photon of wavelength λ from the excited state A*.

Typically, $\sigma_{\lambda A}$ is known but f_e is not, i.e., f_e is not generally a single-temperature Maxwellian. As discharge parameters (pressure, power, driving frequency, and length) are varied, f_e changes shape, as shown in Figure 11.9. In particular, the high-energy tail of the distribution, near the excitation energy E_{A^*}, can vary strongly as discharge parameters are changed. Consequently, $\alpha_{\lambda A}$ changes and I_λ given by (8.6.3) is not proportional to n_A. This limits the usefulness of a measurement of I_λ, which provides only qualitative information on the radical density n_A.

8.6.2 Optical Actinometry

An inert tracer gas of known concentration n_T can be added to the feedstock to provide quantitative information on the radical density n_A (Coburn and Chen, 1980). We choose an excited state T* of the tracer T that has nearly the same excitation threshold energy, $\mathcal{E}_{T^*} \approx \mathcal{E}_{A^*} \approx \mathcal{E}_*$. The cross sections $\sigma_{\lambda A}(v)$ and $\sigma_{\lambda' T}(v)$ for photon emission of λ (from A) and λ' (from T) are sketched in Figure 8.19. A typical form for the multiplicative factor $v^3 f_e(v)$ in the integrand of (8.6.4) is also shown, with the overlap shown as the shaded area. For the tracer gas,

$$I_{\lambda'} = \alpha_{\lambda' T} n_T \tag{8.6.6}$$

with

$$\alpha_{\lambda' T} = k_D(\lambda') \int_0^\infty 4\pi v^2 dv\, Q_{T^*}(p, n_e)\sigma_{\lambda' T}(v) v f_e(v) \tag{8.6.7}$$

Since, from Figure 8.19, there is only a small range of overlap of f_e with σ, we can replace the cross sections with values near the threshold: $\sigma_{\lambda' T} \approx C_{\lambda' T}(v - v_{thr})$ and $\sigma_{\lambda A} \approx C_{\lambda A}(v - v_{thr})$, where the Cs are proportionality constants. We then take the ratio of (8.6.3) and (8.6.6) to obtain

$$n_A = C_{AT}\, n_T \frac{I_\lambda}{I_{\lambda'}} \tag{8.6.8}$$

where

$$C_{AT} = \frac{k_D(\lambda')\ Q_{T^*}\ C_{\lambda' T}}{k_D(\lambda)\ Q_{A^*}\ C_{\lambda A}} \tag{8.6.9}$$

It is often possible to choose $\lambda' \approx \lambda$ such that $k_D(\lambda) \approx k_D(\lambda')$ and also to choose $Q_{A^*} \approx Q_{T^*}$. Hence, the constant of proportionality $C_{AT} \approx C_{\lambda A}/C_{\lambda' T}$ is related to the threshold behavior of the two cross sections. If n_T is known and I_λ and $I_{\lambda'}$ are measured, an absolute value of n_A can be determined.

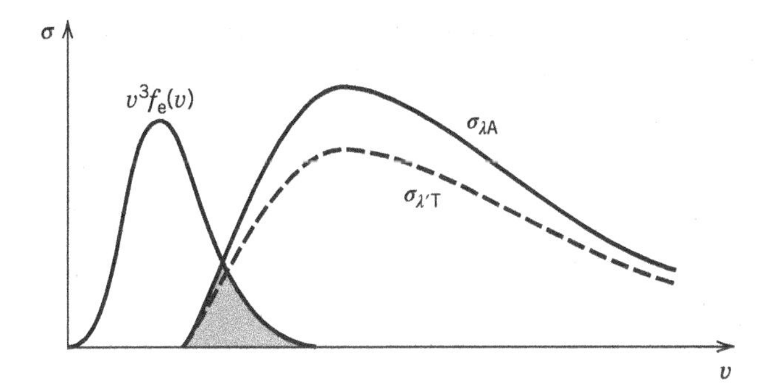

Figure 8.19 Overlap of excitation cross sections and electron velocity distribution.

Even if C_{AT} is not known, the relative variation of n_A with variation of discharge parameters can be found. For F-atom actinometry, a common choice for the tracer gas is argon with $\lambda' = 7504$ Å; the cross section has a threshold energy of 13.5 V. For F atoms, $\lambda = 7037$Å is commonly chosen, with a threshold energy of 14.5 V. Typically, n_T is chosen to be 1–5%, of the feedstock gas density.

8.6.3 O Atom Actinometry

To illustrate both the utility and the pitfalls of optical actinometry, we consider O atoms with argon as the tracer gas. Figure 8.20 shows data (Walkup et al., 1986) for n_O for an O_2/CF_4 feedstock mix with 2–3% argon added as a tracer gas. The data were taken in a 13.56 MHz capacitive rf discharge. The oxygen radical density n_O was determined actinometrically using O atom emission at two different wavelengths, $\lambda = 7774$ Å ($3p^5P \rightarrow 3s^5S$ transition) and $\lambda = 8446$ Å ($3p^3P \rightarrow 3s^3S$ transition), each ratioed to the argon emission at wavelength $\lambda' = 7504$ Å. The actinometric measurements were compared with a more accurate (and much more expensive) determination of n_O using two-photon absorption laser-induced fluorescence (TALIF) (see Walkup et al., 1986). It can be seen that the 8446/7504-Å actinometric measurement tracks the TALIF measurement fairly well as the %CF_4 is varied. However, the 7774/7504-Å measurement yields a saturation of n_O rather than a decrease as the CF_4 concentration is lowered below 20%, contrary to the TALIF measurement. Similar results have been obtained by Katsch et al. (2000).

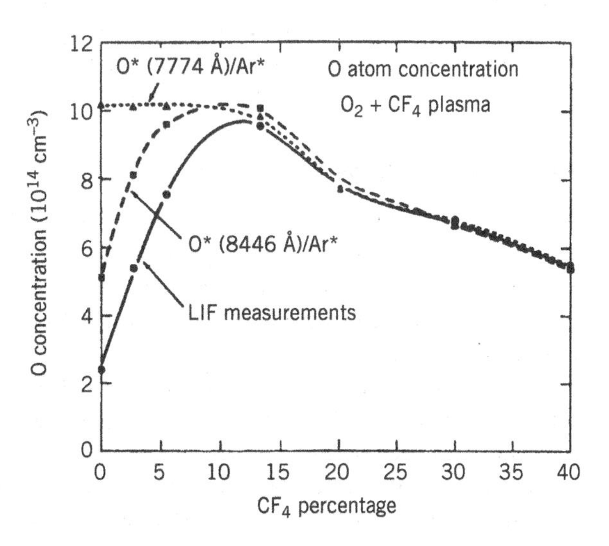

Figure 8.20 Comparison of actinometric measurements with a two-photon absorption laser-induced fluorescence (TALIF) measurement of oxygen atom density in an O_2/CF_4 discharge. Source: Walkup et al., 1986/with permission of AIP Publishing.

To understand this behavior, we first note that emission of a photon of wavelength λ can occur due to processes other than excitation from the ground-state A. For example, the *dissociative excitation* process

$$e + O_2 \rightarrow O + O^* + e \rightarrow 2O + e + \hbar\omega \tag{8.6.10}$$

can compete with the direct excitation process

$$e + O \rightarrow O^* + e \rightarrow O + e + \hbar\omega \tag{8.6.11}$$

such that the measured emission intensity

$$I_\lambda = \alpha_{\lambda O} n_O + \alpha_{\lambda O_2} n_{O_2}, \tag{8.6.12}$$

has a component proportional to the feedstock density n_{O_2} as well as the radical density n_O. The actinometric measurement of n_O will fail if $\alpha_{\lambda O} n_O \lesssim \alpha_{\lambda O_2} n_{O_2}$, which is the case for the 7774 Å measurement.

Using a high-resolution monochromator or spectrometer, the radiation due to direct and dissociative excitation can be distinguished. Because dissociative excitation generally results in excited neutral fragments having many volts of energy, the radiation is Doppler broadened according to (8.6.2) and can therefore be distinguished from the much sharper linewidth for radiation produced by direct excitation of a room temperature atom. Subtracting the emission intensity in the broadened tail from the total intensity allows the intensity due to direct excitation alone to be determined. However, other processes can also increase or decrease I_λ. These include cascade processes, which are radiative transitions from higher-energy excited states to A*, electron impact excitation of metastable states to A*, and collisional and electric field quenching of A*. These can invalidate an actinometric measurement unless the optical transition and discharge operating regime have been selected to minimize their effects.

Multiple small concentrations of noble gases can be added to the feedstock gas, enabling the use of *trace rare gases optical emission spectroscopy* to determine the high-energy tail electron temperature and, under some conditions, the electron density in low-pressure discharges. The method is based on a comparison of measured atomic emission intensities from known densities of trace amounts of He, Ne, Ar, Kr, and Xe added to the plasma, with intensities calculated from a model. The analysis procedure is described in Malyshev and Donnelly (1997, 1999). Measurements of T_e and n_e in a chlorine planar inductive discharge, of the type described in Section 12.3, are given in Malyshev and Donnelly (2000b). These measurement results are discussed briefly in Section 11.3. A variation of this method has been applied to determine the time and space variation of the energetic part of the EEDF using PROES (Gans et al., 2004a).

8.6.4 Phase-Resolved Optical Emission Spectroscopy (PROES)

There are generally strong time and space variations of optical emission in rf discharges, on time scales that are smaller than the rf period, and space scales that are smaller than the discharge size. Measurements of these variations using PROES can give highly useful information on the discharge physics and chemistry. PROES is sensitive to high-energy electrons ($\gtrsim$10 V), which mainly determine the ionization dynamics.

Time-dependent excitation processes using PROES were observed by de Rosny et al. (1983) in a 150 mTorr, 13.56 MHz silane capacitive discharge. Gans et al. (2001a,b) applied the method for H_α line emission to determine excited state quenching coefficients and rotational ground-state populations in a 20–400 Pa hydrogen capacitive discharge. Reviews with some experimental results are given in Gans et al. (2004b, 2006a) and Schulze et al. (2010).

In the simplest case, a noble gas discharge is used, or a small concentration of a noble gas is added to the feed gas mix. As illustrated in Figure 8.18, the electron excitation rate $R_{gi}(t) = K_{gi}n_e n_g$ from the noble gas ground-state g to an excited state i is determined from measurements of the number of photons per second, $\dot{N}_{if}$, emitted at wavelength λ due to radiative decay from the excited state to a chosen final state f. The measurement is usually done with an intensified charge-coupled device (ICCD) camera, which provides spatial resolution within the discharge perpendicular to the optical line-of-sight.

The density $n_i(t)$ of the excited state at time t is a convolution of the excitation rate over all previous times

$$n_i(t) = \int_{-\infty}^{t} dt' \, R_{gi}(t') \, e^{-A_i(t-t')} \tag{8.6.13}$$

where A_i (s^{-1}) is the effective decay frequency of state i

$$A_i = \sum_j A_{ij} g_{ij} + \sum_q k_{iq} n_q \tag{8.6.14}$$

The effective decay frequency accounts for spontaneous emission with frequency A_{ij} (s^{-1}) from state i to all lower states j, to radiation readsorption by the plasma with escape factor g_{ij} (Walsh, 1959), and to radiationless collisional de-excitation (quenching) of state i by collisions with species states q, with k_{iq} the quenching coefficient and n_q the corresponding density. Generally, the escape factors are very nearly unity and the quenching coefficients are small for a properly chosen emission line, such that $A_i \approx \sum_j A_{ij} \approx A_{if}/b_\lambda$, with b_λ the branching ratio for emission from state i to state f.

Taking the time derivative of (8.6.13) and rearranging, we obtain

$$R_{gi}(t) = \frac{dn_i(t)}{dt} + A_i n_i(t) \tag{8.6.15}$$

For $n_i(t)$ periodic in time with period T, (8.6.13) reduces to (see Problem 8.16)

$$n_i(t) = \left[\frac{\int_0^T dt' \, R_{gi}(t') \, e^{A_i t'}}{e^{A_i T} - 1} + \int_0^t dt' \, R_{gi}(t') \, e^{A_i t'} \right] e^{-A_i t} \tag{8.6.16}$$

where here $0 \leq t < T$. Equation (8.6.15) directly determines $R_{gi}(t)$ from $n_i(t)$. Alternately, the integral equation (8.6.16) can be solved. The former method is more susceptible to corruption by noise, since the derivative must be evaluated. In any case, for good time resolution in the presence of noise, the effective decay time A_i^{-1} should be significantly less than the rf period T.

The density n_i of state i is related to the number of photon emitted per second, $\dot{N}_{if}$, by

$$\dot{N}_{if}(t) = A_{if} n_i(t) \tag{8.6.17}$$

As for actinometry, $I_{if}(t) = k_D \dot{N}_{if}(t)$, where I_{if} is the measured optical intensity and k_D is the measurement apparatus response constant, determined by optical system calibration, measurement averaging time, etc. In this way, measurements of $I_{if}(t)$ determine $\dot{N}_{if}(t)$. Then, $n_i(t)$ is found from (8.6.17), and the excitation rate $R_{gi}(t)$ at any particular position within the discharge is found from (8.6.15) or (8.6.16).

A typical measurement system is shown in Figure 8.21. A 13.56 MHz rf generator (period $T = 73.75$ ns) drives the discharge through a matching network. To choose a particular wavelength emission line for examination, the light is passed through an optical filter into an ICCD camera. The camera provides spatial resolution perpendicular to the optical line-of-sight. A variable delay, synchronizing signal from the generator triggers the camera over a short (few ns) window, to measure the photon emission at that delay time. To obtain a good signal-to-noise ratio, the

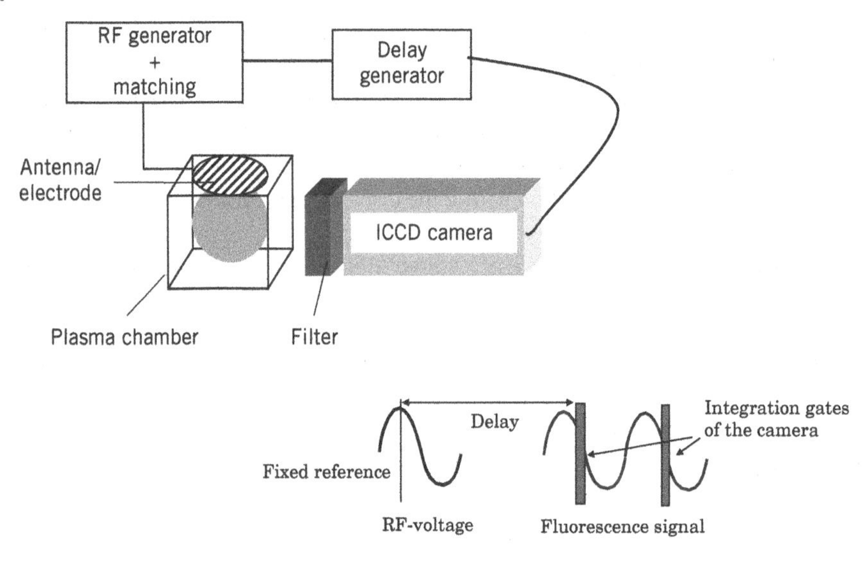

Figure 8.21 Schematic of a typical PROES measurement system. Source: Reproduced from Gans et al. (2006a), with the permission of AIP Publishing. https://doi.org/10.1063/1.2406035.

emission at a given delay time is averaged over many thousands of rf cycles. The delay time is then advanced to the next window interval, and the procedure is repeated. The time required to scan the entire rf period is typically from seconds to hours, depending on the emission intensities and the read-out capabilities of the ICCD camera. Typical time and space resolutions of the measurement are 5 ns and 0.5 mm.

The noble gas emission line must be chosen carefully. Cascades from higher excited levels should be small; otherwise, they modify the basic rate equation dynamics (Gans et al., 2006a; Schulze et al., 2010). Similarly, excitations from metastable levels should be small, or they need to be taken into account. Fortunately, most of the optical transition rates and quenching coefficients are reasonably well known for the noble gases.

An example PROES measurement is shown in Figure 8.22 for a 13.56 MHz, 5 cm gap, 2 Pa neon capacitive discharge (Schulze et al., 2010). Excitation to the 18.97 V Paschen $2p_1$ level $(3p)^1S_0$ is

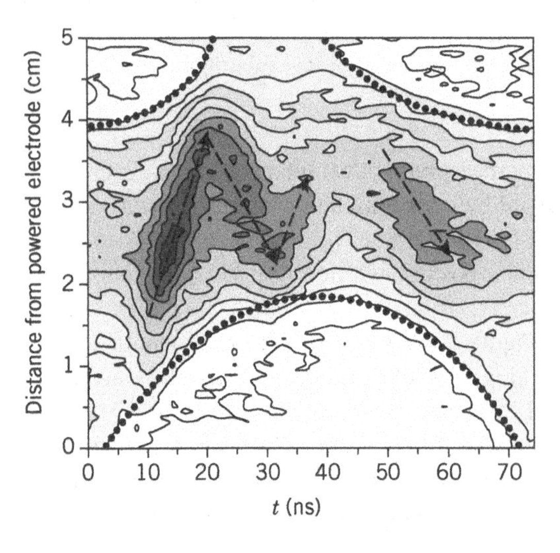

Figure 8.22 PROES measurement of the time- and space-resolved excitation to the Paschen $2p_1$ level of neon, in a 13.56 MHz, 2 Pa neon discharge with a 5 cm gap; the dotted lines indicate the sheath motions, and the arrows indicate the trajectories of energetic electron groups generated by the expanding sheaths. Source: Schulze et al. (2010). ©IOP Publishing. Reproduced with permission. All rights reserved. https://doi.org/10.1088/0022-3727/43/12/124016.

determined using photon emission to the Paschen $1s_2$ level $(3s)^1P_1$. The emission wavelength is $\lambda = 585.2$ nm, and the $2p_1$ lifetime is short, $A_i^{-1} = 14.5$ ns. The powered (lower dotted line) and grounded (upper dotted line) sheath motions are indicated. As discussed in Section 11.4, the powered sheath amplitude is larger due to capacitive coupling to the grounded discharge sidewalls. The time (horizontal) and space (vertical) contours of the excitation rate $R_{gi}(x, t)$ are shown, with the darker shade indicating a higher excitation rate. As can be seen (arrows), a high-energy electron group is generated near the expanding sheath at the bottom electrode. It propagates through the plasma bulk and reflects from the top electrode sheath. A smaller amplitude energetic electron group is also generated near the smaller amplitude expanding sheath at the top electrode. These features are seen in particle-in-cell (PIC) simulations (Wood, 1991; Vender and Boswell, 1990), as shown in Figures 11.11 and 11.12 of Chapter 11. As another example, Figure 11.25 from Gans et al. (2006a) shows a PROES measurement for a dual-frequency driven, 490 mTorr He/O_2 capacitive discharge.

For chemically complex feed gas mixes, small concentrations of various noble gases can be added. Applying the method of trace gases optical emission spectroscopy (Malyshev and Donnelly, 1997, 1999), the time and space variations of an assumed drifting Maxwellian electron energy distribution can be determined, e.g., in a 148 Pa hydrogen capacitive discharge (Gans et al., 2004a).

PROES has also been applied to atmospheric pressure discharges. Figure 15.18 from Waskoenig et al. (2010) shows a measurement for a 13.56 MHz $He/0.5\%O_2$ capacitive discharge. Bischoff et al. (2018) measured the 706.5 nm He $(3s)^3S_1 \rightarrow (2p)^3P_0$ emission, as well as the emission from molecular bands $N_2^+(B) \rightarrow N_2^+(A)$ and $N_2(B) \rightarrow N_2(A)$ in a He/N_2 atmospheric pressure discharge.

Problems

8.1 Vibration and Dissociation of H_2

(a) By fitting $\mathcal{E}_e$ for the $H_2(^1\Sigma_g^+)$ ground state in Figure 8.3 to a parabolic function of $R - \bar{R}$ and using (8.2.2) and (8.2.3), estimate the spring constant k_{vib} and the vibration period $\tau_{vib} = 2\pi/\omega_{vib}$.

(b) From the potential energy curve for the $H_2(^3\Sigma_u^+)$ repulsive state in Figure 8.3, estimate the timescale τ_{diss} for dissociation of the molecule after electron impact excitation to this excited state.

(c) For excitation of $H_2(^3\Sigma_u^+)$ from the ground vibrational state of $H_2(^1\Sigma_g^+)$, estimate the threshold energy for dissociation and the minimum and maximum energies of the dissociated H atoms.

8.2 Metastable Molecular States

(a) In order of increasing energy, the five lowest attractive states of O_2 are $^3\Sigma_g^-$, $^1\Delta_g$, $^1\Sigma_g^+$, $^3\Delta_u$, and $^3\Sigma_u^+$ (see Figure 8.4). Which of these states are metastable? (give the reasons). Give the total (orbital + spin) electronic degeneracy of these states.

(b) Which of the He_2 excimer states in Figure 8.5 are metastable?

8.3 Dissociation Cross Section for O_2

(a) Using (8.3.2) and the potential energy curves in Figure 8.4, estimate the cross section $\sigma_{diss}(\mathcal{E})$ for electron impact dissociation of O_2 at the equilibrium nuclear separation $\bar{R}$ to form ground-state O atoms. Assume that the dissociation results from direct excitation of the repulsive $^1\Pi$ and $^3\Pi$ energy level curves and does not average over the vibrational motion. Plot $\sigma_{diss}(\mathcal{E})$ versus $\mathcal{E}$ using linear scales.

(b) Approximating $\sigma_{diss}(\mathcal{E})$ by

$$\sigma_{diss} = \begin{cases} 0 & \mathcal{E} < \mathcal{E}_{thr} \\ \sigma_{max}\mathcal{E}_{thr}/\mathcal{E} & \mathcal{E} > \mathcal{E}_{thr} \end{cases}$$

then integrate $\sigma_{diss}(\mathcal{E})$ over a Maxwellian electron distribution (T_e in the range 2–7 V) to determine the rate constant $K_{diss}(T_e)$. Compare your result to that given in Table 8.2.

8.4 Dissociative Attachment of O_2

(a) For dissociative attachment to a single molecular level having $\mathcal{E}_{att}$~4 V, estimate the rate constant $K_{att}(T_e)$ for T_es in the range of 2–7 V by integrating (8.3.7) over a Maxwellian electron distribution.

(b) Suppose $\mathcal{E}_{att}$ varies linearly with nuclear separation R over the range of ground-state vibrational motions

$$\mathcal{E}_{att}(R) = \overline{\mathcal{E}}_{att} + \Delta\mathcal{E}_{thr}x$$

where $x = (R - \overline{R})/\Delta R$ has a Gaussian distribution

$$f(x) = \frac{e^{-x^2}}{\sqrt{\pi}}$$

Average (8.3.7) over the vibrational motion and plot your result for $\overline{\sigma}_{att}$ versus $\mathcal{E}$ for $\overline{\mathcal{E}}_{att} = 4$ V, $\Delta\mathcal{E}_{thr} = 1$ V, and $\Delta\mathcal{E}_{att} = 0.2$ V. On the same graph, plot σ_{att} from (8.3.5) with $\mathcal{E}_{att} = 4$ V.

(c) Using detailed balancing (8.5.14), estimate the rate constant for associative detachment

$$O^- + O \rightarrow O_2 + e$$

using your result in (a). You will need to use (8.5.17) to estimate the statistical weight of O_2; $\hbar\omega_{vib}/e \approx 0.192$ V and $B_{rot} \approx 1.79 \times 10^{-4}$ V for O_2.

8.5 Polar Ionization of O_2 Interpreting the second (higher energy) peak in Figure 8.9 as the cross section for polar ionization of O_2, estimate the rate constant for this process for T_es in the range 2–7 V by fitting the cross section in the energy region above threshold to the form (8.5.9) and then using (8.5.10). Compare your result to that given in Table 8.2.

8.6 Positive Charge Transfer in O_2 Discharges For thermal (T ~ room temperature) ground-state particles:

(a) Estimate the reaction rate constant for the resonant reaction

$$O^+ + O \rightarrow O + O^+$$

using (3.4.37) and the data in Table 3.2.

(b) Estimate the reaction rate constant for the near-resonant reaction

$$O_2^+ + O_2 \rightarrow O_2 + O_2^+$$

using (3.4.37) and the data in Table 3.2.

(c) Estimate the reaction rate constant for the exothermic (1.4 V) reaction

$$O^+ + O_2 \rightarrow O + O_2^+$$

using (3.4.37) and the data in Table 3.2.

(d) The reaction

$$O_2^+ + O \rightarrow O_2 + O^+$$

has a threshold energy of 1.4 V. Estimate the reaction rate constant using detailed balancing (8.5.14) and your result in (c). To simplify the calculations, you may assume that ω_{vib} and B_{rot} are the same for both molecules and that the fine structure of the atoms is equally occupied. Note that the ground state of O^+ is ^4S.

8.7 Negative Charge Transfer in O_2 Discharges For thermal ($T \sim$ room temperature) ground-state particles:

(a) Estimate the reaction rate constant for the resonant reaction

$$O^- + O \rightarrow O + O^-$$

using (3.4.37) and the data in Table 3.2.

(b) Estimate the reaction rate constant for the near-resonant reaction

$$O_2^- + O_2 \rightarrow O_2 + O_2^-$$

using (3.4.37) and the data in Table 3.2.

(c) Estimate the reaction rate constant for the exothermic (1.0 V) reaction

$$O_2^- + O \rightarrow O_2 + O^-$$

using (3.4.37) and the data in Table 3.2.

(d) The reaction

$$O^- + O_2 \rightarrow O + O_2^-$$

has a threshold energy of 1.0 V. Estimate the reaction rate constant using detailed balancing (8.5.14) and your result in (c). To simplify the calculations, you may assume that ω_{vib} and B_{rot} are the same for both molecules and that the fine structure of the atoms is equally occupied.

8.8 Positive–Negative Ion Recombination For thermal particles at temperature T (near room temperature), estimate the rate constant for the reaction

$$O_2^+ + O^- \rightarrow O_2^* + O$$

by integrating the classical cross section (8.4.10) over a Maxwellian distribution of relative velocities. Compare your answer (both magnitude and scaling with T) with that given in Table 8.2.

8.9 Detailed Balancing For a Maxwellian distribution of relative velocities v_R, integrate the relation (8.5.12) for detailed balancing of the cross sections for forward and reverse reactions using the energy conservation relation (8.5.13), to obtain the relation (8.5.14) for detailed balancing between the rate constants.

8.10 Translational Statistical Weight Per Unit Volume According to the uncertainty principle, the number N of distinguishable translational states for a free particle of mass m within a velocity interval d^3v around $\mathbf{v}$ and a position interval d^3x around $\mathbf{x}$ is $\hat{N} = m^3 d^3v\, d^3x/h^3$, where h is Planck's constant. The probability that such a state is occupied is given by the Boltzmann factor, $f_B = \exp[-mv^2/(2e\mathrm{T})]$. The translational statistical weight per unit volume $\hat{g}$ is defined as the number of occupied states per unit volume d^3x, summed (integrated) over all velocities. Show that

$$\hat{g} = \left(\frac{2\pi m e\mathrm{T}}{h^2}\right)^{3/2} \quad (m^{-3})$$

8.11 Application of Detailed Balancing

(a) For a Maxwellian electron distribution at temperature $\mathrm{T_e}$, the direct electron collisional excitation of an atom B having statistical weight $\bar{g}_B$ to an excited state having energy $\mathcal{E}$ and statistical weight $\bar{g}_*$ is measured to have an Arrhenius form $K_{ex} = K_0 \exp(-\mathcal{E}_a/\mathrm{T_e})$, where $\mathcal{E}_a \neq \mathcal{E}$ is the activation energy. Using detailed balancing, find the rate constant K_q for quenching (electron collisional de-excitation) of B* to the ground-state B.

(b) Apply your formula to determine the rate constant for

$$e + O(^1D) \rightarrow O(^3P) + e$$

using the data in Tables 8.1 and 8.2. Compare your result to that given in Table 8.2.

(c) If $\mathcal{E}_a$ is markedly different from $\mathcal{E}$, then is your result in (a) correct? Explain your answer.

8.12 Statistical Weights

(a) The $^3P_1$ and $^3P_0$ levels of an oxygen atom lie at energies 0.020 and 0.028 V above the $^3P_2$ ground-state level. Assuming that the probability that a level is occupied is given by a Boltzmann factor $e^{-\mathcal{E}/\mathrm{T_e}}$, find the statistical weight of $O(^3P)$ at room temperature (0.026 V) and at twice room temperature.

(b) The ground-state of N and O^+ is ^4S. Find the statistical weight if all levels in the fine structure are equally occupied.

(c) The ground-state level of fluorine and chlorine atoms is $^2P_{3/2}$; the $^2P_{1/2}$ levels lie 0.050 and 0.109 V above the ground state, respectively. Find the statistical weights of $F(^2P)$ and $Cl(^2P)$ at room temperature.

(d) The ground-state level of an argon atom is $^1S_0$. Find its statistical weight.

(e) The vibrational and rotational energy constants for $O_2(^3\Sigma_g^-)$, $O_2^+(^2\Pi_g)$, and $O_2^-(^2\Pi_g)$ are $\hbar\omega_{vib}/e = 0.196$, 0.236, and 0.136 V and $B_{rot} = 1.79 \times 10^{-4}$, 2.09×10^{-4}, and 1.45×10^{-4} V, respectively. Find the statistical weights of these molecules at room temperature (0.026 V).

8.13 Statistical Weight for Molecules

(a) Show that

$$\bar{g}_{vib} = \frac{1}{1 - \exp(-\hbar\omega_{vib}/e\mathrm{T})}$$

by summing the probability $\exp(-\hbar\omega_{vib}v/e\mathrm{T})$ over the $v = 0$ to $v = \infty$ vibrational levels.

(b) Show that at temperatures $\mathrm{T} \gg B_{rot}$, the mean number of rotational states occupied is $\bar{g}_{rot} = \mathrm{T}/B_{rot}$ by summing the probability $\exp[-B_{rot}J(J+1)/\mathrm{T}]$ over the $J = 0$ to $J = \infty$ levels. *Hint*: Convert the sum over J to an integral over dJ and recall that the degeneracy of level J is $2J + 1$.

8.14 **Saha Equation for Argon** Consider an argon discharge with gas density n_g in which the only reactions that create and destroy charged particles are

$$\mathrm{e} + \mathrm{Ar} \underset{K_{\mathrm{rec}}}{\overset{K_{\mathrm{iz}}}{\rightleftharpoons}} \mathrm{e} + \mathrm{e} + \mathrm{Ar}^+$$

Assuming complete thermal equilibrium, with a common temperature $\mathrm{T_e} = \mathrm{T}$ for all species (including the neutral gas), the creation and destruction rates must balance,

$$K_{\mathrm{iz}}\overline{n}_{\mathrm{e}}n_{\mathrm{Ar}} = K_{\mathrm{rec}}\overline{n}_{\mathrm{e}}^{2}\overline{n}_{\mathrm{Ar^+}}$$

Here, $\overline{n}_{\mathrm{e}} = \overline{n}_{\mathrm{Ar^+}}$ are the equilibrium charged particle densities. Note that the ground-state atomic term for Ar is $^1\mathrm{S}_0$ and that Ar^+ has two closely spaced "ground" terms, $^2\mathrm{P}_{3/2}$ and $^2\mathrm{P}_{1/2}$. Using (8.5.19) to evaluate $K_{\mathrm{iz}}/K_{\mathrm{rec}}$, find the argon Saha equation, which specifies the fractional ionization $\overline{n}_{\mathrm{e}}/n_g$, as a function of n_g, T and the ionization energy $\mathcal{E}_{\mathrm{iz}}$. Plot (logarithmic scales) the fractional ionization for $0.01\,\mathcal{E}_{\mathrm{iz}} < \mathrm{T} \le \mathcal{E}_{\mathrm{iz}}$, for gas pressures $p_g = n_g e\mathrm{T}$ of 1 mTorr, 1 Torr, and 760 Torr. For what temperature is there 50% ionization fraction in each of these cases?

8.15 **Negative Ions in an O_2 Discharge** Negative ions in a discharge are generally created and lost only through processes in the plasma volume because the plasma potential is positive with respect to all wall surfaces; hence, the negative ions are electrostatically trapped. Use the rate constants given in Table 8.2 to perform the following:

(a) For an oxygen discharge containing room temperature O_2, O_2^+ and O^-, and electrons at temperature $\mathrm{T_e}$, obtain the condition on $\mathrm{T_e}$ for dissociative attachment (reaction 2) to dominate over polar ionization (reaction 10) for production of O^- by electron impact on O_2.

(b) Obtain a condition on $\mathrm{T_e}$ such that O_2^+–O^- recombination (sum of reactions 7 and 9) dominates over electron detachment (reaction 5) for destruction of O^-.

8.16 **Phase-Resolved Optical Emission for a Periodic Discharge** Show, either by solving (8.6.15) with appropriate periodic boundary conditions or by direct summations of (8.6.13) over all previous full periods T that for $n_i(t)$ periodic in time with period T, the convolution integral (8.6.13) reduces to (8.6.16), with $0 \le t < T$.

9

Chemical Kinetics and Surface Processes

9.1 Elementary Reactions

In this chapter, we describe aspects of gas-phase and surface chemical kinetics that are important to materials processing. We first introduce the concept of *elementary reactions*, give the definition of the appropriate rate constants, and show their connection to the equilibrium constants for the reactions. Section 9.2 deals with gas-phase kinetics. We introduce the first-, second-, and third-order kinetics, and the concept of a rate-limiting step. Although some examples of time-varying kinetics are given, the main applications are to the steady state. The third-order kinetics are described with emphasis on three-body recombination and three-body chemical reactions which, at the low pressures of interest, can often be considered to be a series of two or more one- or two-body reactions. In Sections 9.3 and 9.4, we turn to surface processes and reaction kinetics. The various physical and chemical processes of interest for processing are described in Section 9.3. Section 9.4 deals with heterogeneous reactions on the surface and between the surface and the gas phase. The surface reaction mechanisms for most plasma processes are not well understood or characterized experimentally. Some simple models of surface reactions are introduced, but these, for the most part, should not be regarded as correctly representing the actual plasma-induced reactions at substrate surfaces. Rather, they are intended to provide some insight into the more complicated processes that go on in actual surface processing. Consider stoichiometric reactions such as

$$3\mathrm{A} + 2\mathrm{B} \rightarrow \mathrm{C} + 2\mathrm{D} \tag{9.1.1}$$

$$\mathrm{A} + \mathrm{B} \rightarrow \mathrm{C} + \mathrm{D} \tag{9.1.2}$$

$$\mathrm{A} \rightarrow \mathrm{B} + \mathrm{C} \tag{9.1.3}$$

where A, B, C, and D are molecules. A reaction is called *elementary* if it proceeds in one step directly as written, i.e., in a simultaneous "collision" of all the reactant molecules for (9.1.1) and (9.1.2), or by a single "decomposition" for (9.1.3). The first reaction is not elementary because it is very unlikely for five particles to simultaneously collide. The second and third reactions might or might not be elementary. If two reactant molecules A and B collide to immediately ($\Delta t \sim t_c$) form two product molecules C and D, then the reaction is elementary. An example from Chapter 8 is

$$\mathrm{O}^+ + \mathrm{O}_2 \rightarrow \mathrm{O} + \mathrm{O}_2^+$$

Similarly, if an A molecule suddenly decomposes, then the reaction is elementary. An example is

$$\mathrm{A}^* \rightarrow \mathrm{A} + \hbar\omega$$

On the other hand, the reaction

$$\mathrm{Cl}_2 + \mathrm{H}_2 \rightarrow 2\,\mathrm{HCl}$$

Principles of Plasma Discharges and Materials Processing, Third Edition. Michael A. Lieberman and Allan J. Lichtenberg.

having the form (9.1.2), is known not to be elementary. There is no way of knowing from the stoichiometric equations (9.1.2) or (9.1.3) whether a reaction is elementary; additional information is needed. A significant effort in chemical kinetics has been to determine the set of elementary reactions into which a given stoichiometric reaction can be decomposed.

The most important elementary reactions are *unimolecular*

$$\mathrm{A} \rightarrow \text{products}$$

and *bimolecular*

$$\mathrm{A} + \mathrm{B} \rightarrow \text{products}$$

At high pressures, some *termolecular* gas-phase reactions

$$\mathrm{A} + \mathrm{B} + \mathrm{C} \rightarrow \text{products}$$

are elementary; however, in low-pressure discharges, almost all gas-phase termolecular reactions with significant reaction rates are complex. The reaction rate R for a gas-phase reaction is defined in terms of the stoichiometric coefficients α_j for the reaction introduced in Section 7.4. Recall that these are negative for reactants and positive for products. We define R as

$$R = \frac{1}{\alpha_j}\frac{dn_j}{dt}, \qquad \text{for all } j \tag{9.1.4}$$

where n_j is the volume density ($\mathrm{m^{-3}}$) of molecules of the jth substance. For (9.1.1), for example, this yields

$$R = -\frac{1}{3}\frac{dn_\mathrm{A}}{dt} = -\frac{1}{2}\frac{dn_\mathrm{B}}{dt} = \frac{dn_\mathrm{C}}{dt} = \frac{1}{2}\frac{dn_\mathrm{D}}{dt}$$

For surface reactions, n_j is replaced by the area density n'_j ($\mathrm{m^{-2}}$) on the surface. In general, R is a complicated function of the n_js of the reactants. However, for elementary reactions, R has the following simple forms:

$$\mathrm{A} \rightarrow \text{products}$$

$$R = -\frac{dn_\mathrm{A}}{dt} = K_1 n_\mathrm{A} \tag{9.1.5}$$

$$\mathrm{A} + \mathrm{A} \rightarrow \text{products}$$

$$R = -\frac{1}{2}\frac{dn_\mathrm{A}}{dt} = K_2 n_\mathrm{A}^2 \tag{9.1.6}$$

$$\mathrm{A} + \mathrm{B} \rightarrow \text{products}$$

$$R = -\frac{dn_\mathrm{A}}{dt} = -\frac{dn_\mathrm{B}}{dt} = K_2 n_\mathrm{A} n_\mathrm{B} \tag{9.1.7}$$

$$\mathrm{A} + \mathrm{A} + \mathrm{A} \rightarrow \text{products}$$

$$R = -\frac{1}{3}\frac{dn_\mathrm{A}}{dt} = K_3 n_\mathrm{A}^3 \tag{9.1.8}$$

$$\mathrm{A} + \mathrm{A} + \mathrm{B} \rightarrow \text{products}$$

$$R = -\frac{1}{2}\frac{dn_\mathrm{A}}{dt} = -\frac{dn_\mathrm{B}}{dt} = K_3 n_\mathrm{A}^2 n_\mathrm{B} \tag{9.1.9}$$

$$A + B + C \rightarrow \text{products}$$

$$R = -\frac{dn_A}{dt} = -\frac{dn_B}{dt} = -\frac{dn_C}{dt} = K_3 n_A n_B n_C \tag{9.1.10}$$

The quantities K_1 (s^{-1}), K_2 (m^3/s), and K_3 (m^6/s) are the *first-* , *second-* , and *third-order rate constants.* They are functions of temperature but are independent of the densities.

9.1.1 Relation to Equilibrium Constant

Let us consider the two opposing elementary reactions

$$A + B \underset{K_{-2}}{\overset{K_2}{\rightleftarrows}} C + D$$

The rate at which C is created by the forward reaction is $K_2 n_A n_B$, and the rate at which C is destroyed by the reverse reaction is $K_{-2} n_C n_D$. In thermal equilibrium (reactants and products at temperature T), the rates must balance:

$$K_2 \bar{n}_A \bar{n}_B = K_{-2} \bar{n}_C \bar{n}_D$$

or

$$\frac{K_2(T)}{K_{-2}(T)} = \frac{\bar{n}_C \bar{n}_D}{\bar{n}_A \bar{n}_B} \tag{9.1.11}$$

But from the condition for thermal equilibrium (7.4.6), we find

$$\frac{\bar{n}_C \bar{n}_D}{\bar{n}_A \bar{n}_B} = \mathcal{K}(T) \tag{9.1.12}$$

Substituting this into (9.1.11), we obtain

$$\frac{K_2(T)}{K_{-2}(T)} = \mathcal{K}(T) \tag{9.1.13}$$

Although (9.1.13) was derived for thermal equilibrium between A, B, C, and D, it is also true for a system that is not in thermal equilibrium. The only requirement is that the distribution of relative velocities of the colliding particles be Maxwellian at temperature T. As was noted in Section 8.5, this is because the rate constants K_2 and K_{-2} depend only on the reactant particle collision dynamics and the relative velocity distribution. Therefore, (9.1.13) gives an important relation between the rate constants for the forward and reverse reactions. If the equilibrium constant is known, then K_{-2} can be determined if K_2 is known, and vice versa.

The relation (9.1.13) is just another form of detailed balancing (8.5.14), which was described in Section 8.5 from the point of view of microscopic two-body collision dynamics. However, detailed balancing holds for all opposing pairs of elementary reactions, as is obvious from the derivation presented here. Thus, for the opposing reactions

$$A \underset{K_{-2}}{\overset{K_1}{\rightleftarrows}} B + C$$

we find

$$\frac{K_1(T)}{K_{-2}(T)} = \frac{\bar{n}_B \bar{n}_C}{\bar{n}_A} = \mathcal{K}(T) \tag{9.1.14}$$

and for

$$A + B \underset{K_{-3}}{\overset{K_2}{\rightleftharpoons}} C + D + E$$

we find

$$\frac{K_2(T)}{K_{-3}(T)} = \frac{\bar{n}_C \bar{n}_D \bar{n}_E}{\bar{n}_A \bar{n}_B} = \mathcal{K}(T) \tag{9.1.15}$$

etc.

9.2 Gas-Phase Kinetics

Materials processing reactions in the gas phase are almost never elementary, but consist of a complex set of opposing, consecutive, and parallel reactions. For example, for F-atom etching of silicon in a CF_4 discharge, F atoms are created and destroyed by consecutive opposing reactions of the form

$$e + CF_x \rightleftharpoons CF_{x-1} + F + e, \qquad x = 1,\ 2,\ 3,\ 4$$

Most processing is done in steady state, i.e., the processing time is long compared to the reaction or transport times for the gas-phase species of interest. In steady state, there is a constant flow of feedstock gas and a constant discharge power, and the gas-phase species are continuously pumped away or deposited on surfaces. In steady state, all gas-phase densities are constant, independent of time. However, these densities cannot be determined from equilibrium thermodynamics because the system is not in thermal equilibrium. If the reaction rate constants (Ks) are known, then the densities can be found by solving the rate equations for particle conservation for each species. Since the reaction set is often very complex, the set of rate equations must generally be solved numerically. However, insight can be developed by considering simplified reaction sets under both time-varying and steady-state conditions, which we do here.

A complete self-consistent discharge model cannot be developed without considering the full set of particle and energy conservation equations. As will be shown in Chapter 10, the full set of equations determines not only the particle densities but also the electron temperature, and, hence, the self-consistent rate constants for the discharge equilibrium, which are, in many cases, functions of the electron temperature.

9.2.1 First-Order Consecutive Reactions

Consider the consecutive time-varying first-order reactions

$$A \xrightarrow{K_A} B \xrightarrow{K_B} C \tag{9.2.1}$$

with no sources or sinks. The rate equations are

$$\frac{dn_A}{dt} = -K_A n_A \tag{9.2.2}$$

$$\frac{dn_B}{dt} = K_A n_A - K_B n_B \tag{9.2.3}$$

$$\frac{dn_C}{dt} = K_B n_B \tag{9.2.4}$$

We let $n_A = n_{A0}$ and $n_B = n_C = 0$ at $t = 0$. Then, (9.2.2) can be integrated to obtain

$$n_A = n_{A0}\, e^{-K_A t} \tag{9.2.5}$$

Substituting this into (9.2.3) and integrating, we obtain

$$n_B = n_{A0} \frac{K_A}{K_B - K_A} \left(e^{-K_A t} - e^{-K_B t} \right) \tag{9.2.6}$$

This procedure can be repeated to find $n_C(t)$ by substituting (9.2.6) into (9.2.4) and integrating. However, summing (9.2.2)–(9.2.4) yields

$$n_A(t) + n_B(t) + n_C(t) = \text{const} = n_{A0} \tag{9.2.7}$$

Substituting (9.2.5) and (9.2.6) into (9.2.7), we obtain

$$n_C = n_{A0} \left[1 + \frac{1}{K_A - K_B} \left(K_B e^{-K_A t} - K_A e^{-K_B t} \right) \right] \tag{9.2.8}$$

The time variation of the densities is sketched in Figure 9.1 for the two cases of (*a*) $K_A \ll K_B$ and (*b*) $K_B \ll K_A$. For (*a*), we obtain the approximate variation

$$n_C = n_{A0} \left(1 - e^{-K_A t} \right) \tag{9.2.9}$$

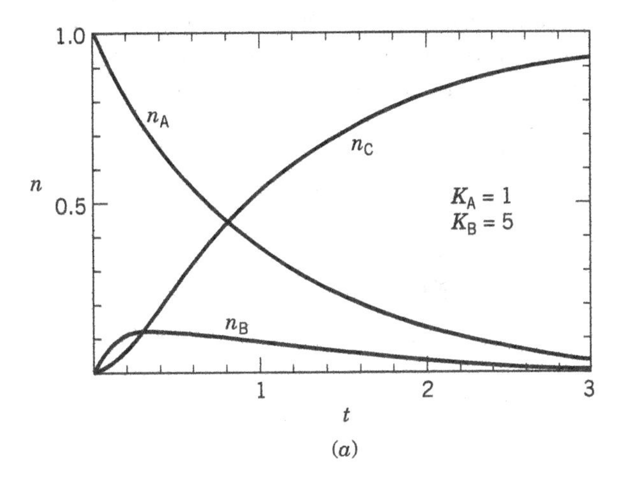

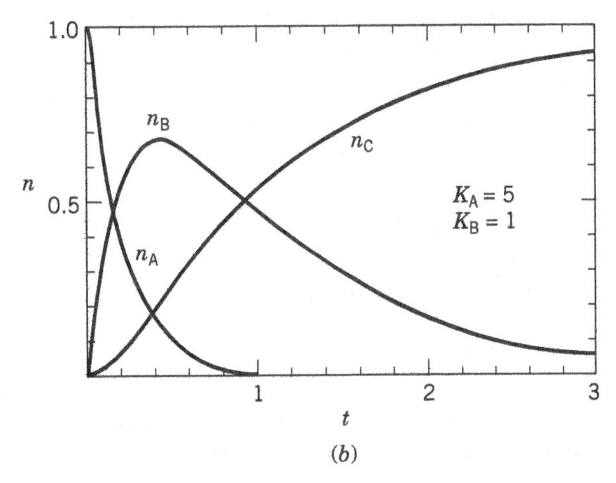

Figure 9.1 Transient kinetics for gas-phase reaction A → B → C; (*a*) $K_A = 1$, $K_B = 5$; (*b*) $K_A = 5$, $K_B = 1$.

For case (*b*), after a short initial transient time $t \sim K_A^{-1}$, we obtain the approximate variation

$$n_C = n_{A0}\left(1 - e^{-K_B t}\right) \tag{9.2.10}$$

In both cases, the rate of formation of the product species C is governed by the *smallest* rate constant. In general, for a series of many consecutive elementary reactions, the reaction with the smallest rate constant limits the overall rate of product formation. The consecutive reaction with the smallest rate constant is called the *rate-limiting step.*

In case (*a*) ($K_B \gg K_A$), species B is created from A at a slow rate K_A and is immediately converted into C. Hence, we should expect that after a short transient time, n_B decays with t at a rate K_A, such that $n_B \ll n_A$ and that $dn_B/dt \sim K_A n_B \ll K_A n_A$ at all times. Therefore, dn_B/dt can be set to zero in (9.2.3) to obtain the approximate solution

$$\begin{aligned} n_A &= n_{A0}\, e^{-K_A t} \\ n_B &\approx \frac{K_A}{K_B} n_{A0}\, e^{-K_A t} \\ n_C &\approx n_{A0}\left(1 - e^{-K_A t}\right) \end{aligned} \tag{9.2.11}$$

Species B is known as a *reactive intermediate,* and setting $dn_B/dt \approx 0$ is known as the *steady-state approximation for reactive intermediates.*

In case (*b*) ($K_A \gg K_B$), A creates B before B creates C. Hence, there are, approximately, two *uncoupled* first-order reactions having solutions

$$\begin{aligned} n_A &= n_{A0}\, e^{-K_A t} \\ n_B &\approx n_{A0}\left(1 - e^{-K_A t}\right) \\ n_C &\approx 0 \end{aligned} \tag{9.2.12}$$

for $0 < t < \bar{t}$, and

$$\begin{aligned} n_A &\approx 0 \\ n_B &\approx n_{A0}\, e^{-K_B t} \\ n_C &\approx n_{A0}\left(1 - e^{-K_B t}\right) \end{aligned} \tag{9.2.13}$$

for $t > \bar{t}$, where $\bar{t} = (K_A K_B)^{-1/2}$ is the characteristic time that divides the fast and slow timescales. The fast reaction in which A is first converted to B is known as a *pre-equilibrium reaction* for the formation of the product C.

For reaction (9.2.1) in the steady state with a source G (m^{-3}-s^{-1}) for A, and adding a loss term $-K_C n_C$ for C, the rate equations become

$$\begin{aligned} \frac{dn_A}{dt} &= G - K_A n_A = 0 \\ \frac{dn_B}{dt} &= K_A n_A - K_B n_B = 0 \\ \frac{dn_C}{dt} &= K_B n_B - K_C n_C = 0 \end{aligned} \tag{9.2.14}$$

Here, K_C could represent a first-order rate constant for loss of C to the surfaces or to the vacuum pump. Solving these equations yields $n_A = G/K_A$, $n_B = G/K_B$, and $n_C = G/K_C$.

9.2.2 Opposing Reactions

Consider the two opposing steady-state reactions

$$\mathrm{A} \underset{K_{-\mathrm{A}}}{\overset{K_{\mathrm{A}}}{\rightleftarrows}} \mathrm{B} \underset{K_{-\mathrm{B}}}{\overset{K_{\mathrm{B}}}{\rightleftarrows}} \mathrm{C} \tag{9.2.15}$$

with $n_{\mathrm{A}} = n_{\mathrm{A0}}$ and no sources or sinks. Then,

$$\frac{dn_{\mathrm{A}}}{dt} = -K_{\mathrm{A}}n_{\mathrm{A}} + K_{-\mathrm{A}}n_{\mathrm{B}} = 0 \tag{9.2.16}$$

$$\frac{dn_{\mathrm{B}}}{dt} = K_{\mathrm{A}}n_{\mathrm{A}} - K_{-\mathrm{A}}n_{\mathrm{B}} - K_{\mathrm{B}}n_{\mathrm{B}} + K_{-\mathrm{B}}n_{\mathrm{C}} = 0 \tag{9.2.17}$$

$$\frac{dn_{\mathrm{C}}}{dt} = K_{\mathrm{B}}n_{\mathrm{B}} - K_{-\mathrm{B}}n_{\mathrm{C}} = 0 \tag{9.2.18}$$

with the solution

$$n_{\mathrm{B}} = \frac{K_{\mathrm{A}}}{K_{-\mathrm{A}}}n_{\mathrm{A0}} = \overline{n}_{\mathrm{B}}$$

and

$$n_{\mathrm{C}} = \frac{K_{\mathrm{B}}}{K_{-B}}n_{\mathrm{B}} = \overline{n}_{\mathrm{C}}$$

which are the solutions in thermal equilibrium. For opposing elementary reactions with no sources or sinks, the thermal equilibrium solutions must be obtained. However, now consider (9.2.15) with a source G for A and an added first-order loss $-K_{\mathrm{C}}n_{\mathrm{C}}$ for C. Solving (9.2.16)–(9.2.18) under these conditions, we find

$$\frac{n_{\mathrm{B}}}{n_{\mathrm{A}}} = \frac{K_{\mathrm{A}}}{K_{-\mathrm{A}} + \dfrac{K_{\mathrm{B}}K_{\mathrm{C}}}{K_{-B} + K_{\mathrm{C}}}}$$

$$\frac{n_{\mathrm{C}}}{n_{\mathrm{B}}} = \frac{K_{\mathrm{B}}}{K_{-\mathrm{B}} + K_{\mathrm{C}}} \tag{9.2.19}$$

with $n_{\mathrm{C}} = G/K_{\mathrm{C}}$. We see that $n_{\mathrm{B}}/n_{\mathrm{A}}$ and $n_{\mathrm{C}}/n_{\mathrm{B}}$ are both depressed below their thermal equilibrium values ($K_{\mathrm{A}}/K_{-\mathrm{A}}$ and $K_{\mathrm{B}}/K_{-\mathrm{B}}$, respectively) by the presence of the source and sink. This situation holds for most low-pressure processing discharges, i.e., the species densities are not in thermal equilibrium.

9.2.3 Bimolecular Association with Photon Emission

Consider the association reaction

$$\mathrm{A} + \mathrm{B} \longrightarrow \mathrm{AB} \tag{9.2.20}$$

On a molecular level, this reaction cannot occur because energy and momentum cannot be simultaneously conserved in the collision (see Problem 3.16). However, there are many examples known of such stoichiometric reactions, e.g., the associative attachment

$$\mathrm{e} + \mathrm{SF_6} \longrightarrow \mathrm{SF_6^-}$$

mentioned in Section 8.3.

To understand how a reaction like (9.2.20) can arise, let us note that the molecular reaction

$$\mathrm{A} + \mathrm{B} \xrightarrow{K_2} \mathrm{AB}^* \tag{9.2.21}$$

can occur, leading to an unstable molecular state. If energy is not taken from AB*, then it immediately dissociates:

$$AB^* \xrightarrow{K_{-1}} A + B \tag{9.2.22}$$

One possible mechanism for loss of energy from AB* is photon emission. This suggests the complex reaction

$$A + B \underset{K_{-1}}{\overset{K_2}{\rightleftarrows}} AB^* \xrightarrow{K_1} AB + \hbar\omega \tag{9.2.23}$$

for production of AB. The steady-state rate equations are

$$\begin{aligned}
\frac{dn_A}{dt} &= \frac{dn_B}{dt} = -K_2 n_A n_B + K_{-1} n_{AB^*} + G = 0 \\
\frac{dn_{AB^*}}{dt} &= K_2 n_A n_B - K_{-1} n_{AB^*} - K_1 n_{AB^*} = 0 \\
\frac{dn_{AB}}{dt} &= K_1 n_{AB^*} - K_{1w} n_{AB} = 0
\end{aligned} \tag{9.2.24}$$

where, to obtain a steady state, a net input source G and a first-order loss term for AB having rate constant K_{1w} have been added, with $G = K_{1w} n_{AB} = K_1 n_{AB^*}$. The solution of (9.2.24) is

$$n_{AB^*} = \frac{K_2}{K_{-1} + K_1} n_A n_B \tag{9.2.25}$$

Hence, the rate of production of n_{AB} has the form

$$R = K_1 n_{AB^*} = \frac{K_1 K_2}{K_{-1} + K_1} n_A n_B \tag{9.2.26}$$

of a second-order elementary reaction (9.1.7) with rate constant

$$K_2' = \frac{K_1 K_2}{K_{-1} + K_1} \tag{9.2.27}$$

However, this reaction is not elementary.

We can estimate K_2' from the rate constants in (9.2.24). The characteristic time for dissociation of an unstable AB* molecular state is found in Chapter 8 to be 10^{-13}–10^{-12} s, so that $K_{-1} \approx 10^{12}$–10^{13} s^{-1}. The radiative lifetime for electric dipole radiation was found to be 10^{-9}–10^{-8} s, so that $K_1 \approx 10^8$–10^9 s^{-1}. Using these estimates in (9.2.27), we find $K_2' \approx 10^{-5}$–$10^{-3}\, K_2$. Therefore, the rate constant for the association reaction (9.2.20) due to photon emission is small; consequently, such reactions are usually not important in low-pressure discharges.

9.2.4 Three-Body Association

A second mechanism for the association reaction (9.2.20) is collision with a third body,

$$A + B + M \rightarrow AB + M \tag{9.2.28}$$

Here, M can be A or B or any other molecule in the system. However, simultaneous collisions of three bodies are very rare at low pressures. This suggests the complex reaction

$$A + B \underset{K_{-1}}{\overset{K_2}{\rightleftarrows}} AB^* \tag{9.2.29}$$

$$AB^* + M \xrightarrow{K_{2M}} AB + M \tag{9.2.30}$$

The rate equations are

$$\frac{dn_A}{dt} = -K_2 n_A n_B + K_{-1} n_{AB^*} + G = 0$$
$$\frac{dn_{AB^*}}{dt} = K_2 n_A n_B - K_{-1} n_{AB^*} - K_{2M} n_{AB^*} n_M = 0$$
$$\frac{dn_{AB}}{dt} = K_{2M} n_{AB^*} n_M - K_{1w} n_{AB} = 0 \quad (9.2.31)$$

with the solution

$$G = K_{1w} n_{AB} = K_{2M} n_{AB^*} n_M$$
$$n_{AB^*} = \frac{K_2}{K_{-1} + K_{2M} n_M} n_A n_B \quad (9.2.32)$$

Hence, the rate of production of n_{AB} has the form

$$R = K_{2M} n_{AB^*} n_M = \frac{K_2 K_{2M} n_M}{K_{-1} + K_{2M} n_M} n_A n_B \quad (9.2.33)$$

This rate depends in a complicated way on the third-body density. In the low- and high-pressure limits, we find

$$R = \begin{cases} \dfrac{K_2 K_{2M}}{K_{-1}} n_A n_B n_M & K_{2M} n_M \ll K_{-1} \\ K_2 n_A n_B & K_{2M} n_M \gg K_{-1} \end{cases} \quad (9.2.34)$$

Therefore, at low pressure, reaction (9.2.28) looks like an elementary three-body reaction:

$$A + B + M \xrightarrow{K_3'} AB + M \quad (9.2.35)$$

with rate constant

$$K_3' = \frac{K_2 K_{2M}}{K_{-1}} \quad (9.2.36)$$

The equivalent second-order rate constant K_2' for the reaction

$$A + B \xrightarrow{K_2'} AB \quad (9.2.37)$$

at low pressures is then

$$K_2' = K_2 \frac{K_{2M} n_M}{K_{-1}} \quad (9.2.38)$$

Consider the ratio K_2'/K_2 for neutral particle collisions at thermal energies (300 K). From Section 8.4, we have the estimate $K_{2M} \approx \sigma_{el}\bar{v} \approx 10^{-11}$–$10^{-10}$ cm^3/s, and $K_{-1} \approx 10^{12}$–10^{13} s^{-1} for dissociation of the unstable AB* molecule. Hence, from (9.2.38), we have

$$K_2' \approx (10^{-24} - 10^{-22})\, n_M K_2$$

where n_M is in cm^{-3}. At $p = 1$ Torr, $K_2' \approx (10^{-7} - 10^{-6}) K_2$; consequently, three-body processes involving neutrals are weak in low-pressure discharges. However, let us note that neutral three-body processes can be important for other applications, particularly near or at atmospheric pressures. For example, the three-body reaction

$$O + O_2 + O_2 \rightarrow O_3 + O_2$$

can be the most significant source of ozone in high-pressure discharges ($p \gtrsim 1$ atm) and is known to be the most important source in the earth's ionosphere. Similarly, three-body ion-molecule association reactions such as

$$O_2^+ + O_2 + O_2 \rightarrow O_4^+ + O_2$$

$$Ar^+ + Ar + Ar \rightarrow Ar_2^+ + Ar$$

can be important in generating higher mass molecular ions in high-pressure discharges. One example is the generation of Ar_2^+ in high-pressure argon, rf and dc discharges. Generation of higher mass ions by association reactions an atmospheric pressure helium discharge is described in Section 15.4. Some three-body rate constants in oxygen discharges are given in Table 8.3.

9.2.5 Three-Body Positive–Negative Ion Recombination

Three-body processes involving charged particles can have rate constants that are much higher than three-body processes involving only neutrals. Consider positive–negative ion recombination at thermal energies as an example:

$$A^+ + B^- + M \rightarrow AB + M$$

The basic theory of this process was first developed by Thomson (1924) and can be understood as follows. Let the positive and negative ions approach each other within a critical radius b_0 such that the Coulomb interaction energy is equal to the mean kinetic energy

$$\frac{e^2}{4\pi\epsilon_0 b_0} = \frac{3}{2}kT \tag{9.2.39}$$

If, during the time the ions are within the critical radius, one of them collides with a neutral molecule M, then energy is transferred with high probability from the ion to the neutral, and the ions become bound to each other. From this description, estimates of the rate constants for the elementary reactions in (9.2.29) and (9.2.30) are

$$K_2 \approx \pi b_0^2 \overline{v}_i \tag{9.2.40}$$

$$K_{-1} \approx \frac{\overline{v}_i}{b_0} \tag{9.2.41}$$

$$K_{2M} \approx (\sigma_{M+}\overline{v}_{M+} + \sigma_{M-}\overline{v}_{M-}) \tag{9.2.42}$$

where $\overline{v}_i$ and $\overline{v} \approx \overline{v}_i$ are the mean speeds of relative motion of the ion pair and the ion–neutral pairs, respectively, and σ_{M+} and σ_{M-} are the cross sections for energy transfer from ions to neutrals. Then from (9.2.38), the equivalent two-body rate constant is

$$K_2' = K_3' n_M \approx \pi b_0^3 K_{2M} n_M \tag{9.2.43}$$

where b_0 is found from (9.2.39):

$$b_0 = \frac{2}{3}\frac{e^2}{4\pi\epsilon_0 kT} \tag{9.2.44}$$

We note from (9.2.42) and (9.2.43) that $K_2' \propto T^{-5/2}$. At room temperature (300 K), we find $b_0 \approx$ 550 Å, a very large critical radius. Consider the example of an estimate of K_2' for the reaction

$$O_2^+ + O^- + O_2 \rightarrow O_3 + O_2$$

In this case, near-resonant charge transfer of O_2^+ on O_2 dominates in (9.2.42), and we estimate from the polarization rate constant (3.3.17) with $\alpha_R \approx 10.6$ and $A_R = 16$ that $K_{2M} \approx 7 \times 10^{-10}$ cm^3/s. Then, (9.2.43) yields $K_2' \approx 3.7 \times 10^{-25} n_M$. At 1 Torr, $n_M \approx 3.3 \times 10^{16}$ cm^{-3}, such that $K_2' \approx 1.2 \times 10^{-8}$ cm^3/s, a very respectable rate constant. Consequently, three-body positive–negative ion recombination can be quite important for processing discharges at pressures $p \gtrsim 1$ Torr. Some rate constants for this process in oxygen discharges are given in Table 8.3.

At very high pressures, the ion–neutral mean free path becomes smaller than the critical radius, leading to multiple ion–neutral collisions within the critical sphere, and the preceding analysis of the mechanism is not correct. This regime is not of interest for low-pressure processing. The reader is referred to Smirnov (1982) for further information.

9.2.6 Three-Body Electron–Ion Recombination

For this process,

$$e + A^+ + e \rightarrow A + e \tag{9.2.45}$$

with $T_e \gg T$, we have, in place of (9.2.40)–(9.2.42), the rate constants

$$K_2 \approx \pi b_0^2 \bar{v}_e \tag{9.2.46}$$

$$K_{-1} \approx \frac{\bar{v}_e}{b_0} \tag{9.2.47}$$

$$K_{2M} \approx \sigma_1 \bar{v}_e \tag{9.2.48}$$

where

$$b_0 = \frac{2}{3} \frac{e^2}{4\pi\epsilon_0 k T_e} \tag{9.2.49}$$

and

$$\pi b_0^2 \approx \sigma \tag{9.2.50}$$

is the cross section for a single electron–electron Coulomb collision (see Section 3.3) that transfers an energy $\sim \frac{3}{2} T_e$. Substituting (9.2.46)–(9.2.48) into (9.2.38), we obtain

$$K_2' \approx \pi^2 b_0^5 \bar{v}_e n_e \tag{9.2.51}$$

which scales as $K_2' \propto T_e^{-9/2} n_e$. A calculation shows that this process is not important in processing discharges with $T_e \gtrsim 1$ V and $n_e \lesssim 10^{13}$ cm^{-3}.

9.3 Surface Processes

Physical and chemical surface processes are central to plasma processing. For example, in F-atom etching of silicon in a CF_4 discharge, the F atoms created in the gas phase are transported to and successively fluorinate the surface through reactions such as

$$F(g) + SiF_x(s) \rightleftharpoons SiF_{x+1}(s), \quad x = 0,\ 1,\ 2,\ 3$$

with production of etch products,

$$SiF_x(s) \rightleftharpoons SiF_x(g), \quad x = 2,\ 4$$

which are transported back into the gas phase. In addition, adsorption–desorption reactions such as

$$\mathrm{F(g) + S \rightleftharpoons F{:}S}$$

play a critical role in determining gas-phase species concentrations. Finally, the discharge equilibrium itself is affected by surface processes such as positive ion neutralization and secondary electron emission at surfaces. For these reasons, the gas-phase and surface reaction sets are coupled, with the coupling being strong at low gas pressures. In this section, we describe some important physical processes, primarily involving positive ions, and some important physical and chemical processes involving neutrals.

9.3.1 Positive Ion Neutralization and Secondary Electron Emission

The strongly exothermic neutralization reaction

$$\mathrm{e + A^+ \rightarrow A}$$

is forbidden in the gas phase because energy and momentum cannot be conserved for the formation of one body from two. However, at the surface, the three-body neutralization reaction

$$\mathrm{e + A^+ + S \rightarrow A + S}$$

is fast. For positive ion energies at the surface in the range 10–1000 V, typical of processing discharges, essentially all positive ions are immediately neutralized at the surface.

To understand the neutralization mechanism, the confinement of electrons in a solid must be briefly described. Figure 9.2*a* shows the energy versus position near a metal surface. The electrons in the metal fill a set of closely spaced energy levels (conduction band) up to a maximum energy (from the bottom of the conduction band) called the *Fermi energy* $\mathcal{E}_F$. The Fermi energy lies below the $\Phi = 0$ potential energy level for a free electron by an energy equal to the *work function* $\mathcal{E}_\phi$. Hence, electrons at $\mathcal{E}_F$ are confined within the solid by a potential barrier of height $\mathcal{E}_\phi$.

A simple classical estimate of $\mathcal{E}_\phi$ for a metal is that it is the work done in moving an electron initially located at a distance $x = a_{\mathrm{eff}}$ from a perfectly conducting surface to $x = \infty$, where a_{eff} is a distance of order an atomic radius. As shown in Figure 9.3*a*, the force F_x acting on an electron −e located at x can be found using the method of images (Ramo et al., 1994, chapter 1), with the image charge +e located at $-x$. From Coulomb's law, we have

$$F_x = -\frac{e^2}{4\pi\epsilon_0 (2x)^2} \tag{9.3.1}$$

The work done in moving the electron from a_{eff} to ∞ is then

$$e\mathcal{E}_\phi = -\int_{a_{\mathrm{eff}}}^{\infty} F_x \, \mathrm{d}x$$

which yields, upon integration,

$$\mathcal{E}_\phi = \frac{e}{16\pi\epsilon_0 a_{\mathrm{eff}}} \tag{9.3.2}$$

Let $a_{\mathrm{eff}} = a_0$, the Bohr radius, we obtain $\mathcal{E}_\phi \approx 6.8$ V. Work functions for most materials are in the range 4–6 V, although the alkalis and alkali earths are lower. There is a rough correlation $\mathcal{E}_\phi \propto \mathcal{E}_{\mathrm{iz}}^{1/2}$, where $\mathcal{E}_{\mathrm{iz}}$ is the ionization potential of the metal atoms. This can be understood from (9.3.2) because the atomic radius $a_{\mathrm{eff}} \propto \mathcal{E}_{\mathrm{iz}}^{-1/2}$, as given in (3.4.9).

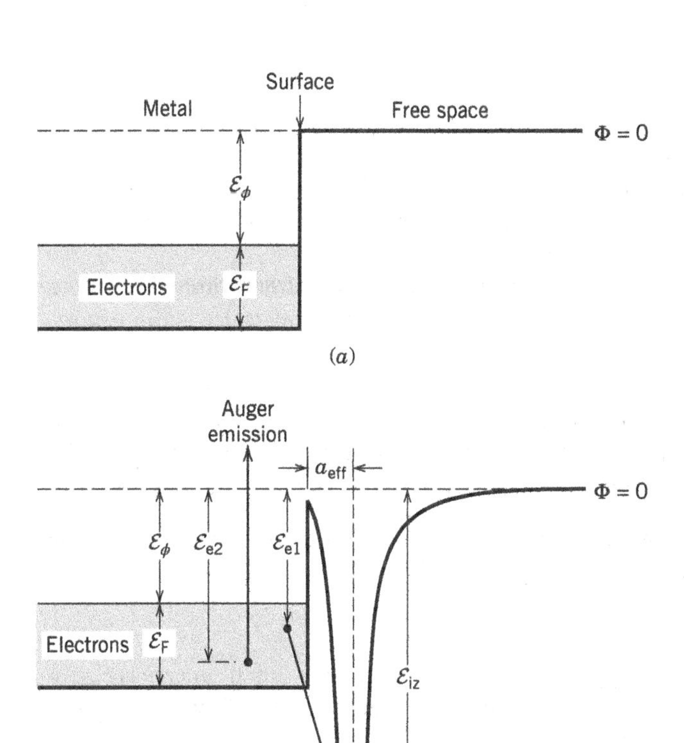

Figure 9.2 Illustrating ion neutralization and secondary emission at a metal surface; (*a*) the work function $\mathcal{E}_\phi$ and the Fermi energy $\mathcal{E}_F$; (*b*) Auger emission due to electron tunneling.

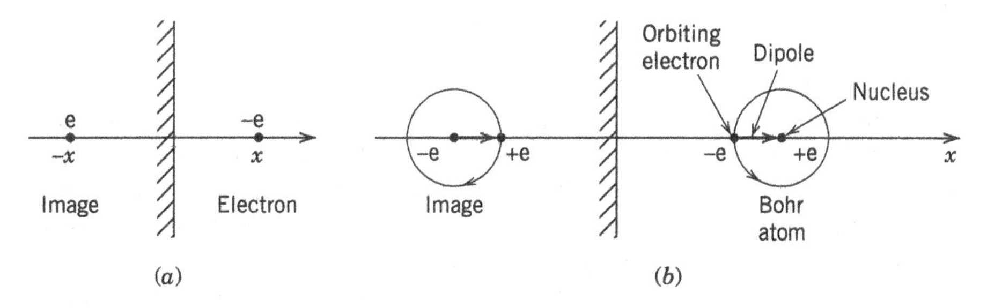

Figure 9.3 Illustrating the method of images for a metal surface to determine (*a*) the work function and (*b*) the van der Waals force.

Now consider the approach of a positive ion to within an atomic radius a_{eff} of the surface. As shown in Figure 9.2*b*, this creates a deep potential well very near the surface that is separated from the surface by a narrow potential barrier of width $\sim a_{eff}$. An electron with energy $\mathcal{E}_e$ from within the conduction band can tunnel through the barrier into the positive ion to neutralize it. There are two possibilities:

1. The electron enters an excited state

$$e + A^+ + S \rightarrow A^* + S$$

where $\mathcal{E}_* \approx \mathcal{E}_{iz} - \mathcal{E}_e$. If the excited state is not metastable, it radiates a photon in a transition to the ground state or to a metastable state. Hence, positive ion neutralization at the surface can create metastables as well as *recombination radiation.*

2. The electron enters the ground state of the atom, and a *second* electron from within the conduction band absorbs the excess energy of neutralization. This mechanism, called *Auger neutralization,* is a nonradiative transition involving two electrons. The electron that enters the ground state of the atom loses an energy $\Delta\mathcal{E} = \mathcal{E}_{iz} - \mathcal{E}_{e1}$, which the second electron gains. If $\Delta\mathcal{E} < \mathcal{E}_{e2}$ for the second electron, then it remains trapped within the solid. However, if $\Delta\mathcal{E} > \mathcal{E}_{e2}$, then the second electron is released from the solid and is free to move away from the surface. This process is called *Auger emission,* or, more commonly, *secondary emission.*

From Figure 9.2*b*, the condition for release of the second electron is most easily met if both electrons come from the top of the conduction band: $\mathcal{E}_{e1} = \mathcal{E}_{e2} = \mathcal{E}_\phi$. The condition for emission is then

$$\mathcal{E}_{iz} \geq 2\mathcal{E}_\phi \tag{9.3.3}$$

The released electron has kinetic energy $\mathcal{E}_{max} = \mathcal{E}_{iz} - 2\mathcal{E}_\phi$. The minimum kinetic energy is $\mathcal{E}_{min} = \mathcal{E}_{iz} - 2\mathcal{E}_\phi - 2\mathcal{E}_F$ (both electrons come from the bottom of the conduction band) or zero if this is negative. Equation (9.3.3) shows that secondary emission is favored for noble gas ions ($\mathcal{E}_{iz}$ is high) and for alkali or alkali earth solids ($\mathcal{E}_\phi$ is low). Because the electron tunneling time (see Section 3.4) is short compared to the ion collision time with the surface, the secondary emission process is practically independent of ion kinetic energy and depends only on the atomic ion species and the near-surface composition of the solid.

Although neutralization and secondary emission have been described for metals, essentially the same processes occur for ions incident on semiconducting and insulating surfaces. As mentioned in Section 3.5, secondary emission is usually characterized by the *secondary emission coefficient* γ_{se}, which is the number of secondary electrons created per incident ion. An empirical expression is (Raizer, 1991)

$$\gamma_{se} \approx 0.016\,(\mathcal{E}_{iz} - 2\mathcal{E}_\phi) \tag{9.3.4}$$

provided $\mathcal{E}_{iz} > 2\mathcal{E}_\phi$. Metastables and other excited atoms produce Auger electron emission very efficiently; the condition for emission is $\mathcal{E}_* > \mathcal{E}_\phi$. Some data for argon metastables and 4p excited states are given in Schohl et al. (1992); see also Gudmundsson et al. (2021) for additional references. In addition to Auger emission, secondary electrons can be created by kinetic ejection for ion (or neutral) impact energies $\gtrsim 1$ kV. These heavy particle energies are not common in processing discharges except for ion implantation applications (see Chapter 17).

Although (9.3.4) provides a rough estimate, the actual value of γ_{se} depends sensitively on surface conditions, morphology, impurities, and contamination. Some measured values of $\mathcal{E}_\phi$ and γ_{se} for ions incident on atomically clean surfaces are given in Table 9.1. However, surfaces are never atomically clean in processing applications. Dirty, contaminated surfaces typically have lower secondary emission coefficients than clean surfaces at low ion energies, and higher coefficients, at higher ion energies, where kinetic emission processes are important (see review by Phelps and Petrović, 1999; Phelps, 1999).

Vacuum ultraviolet (VUV) photon-induced secondary emission can be significant under low field conditions near the surface; some data is given in Phelps and Petrović (1999) and in Raizer

Table 9.1 Work Functions and Secondary Emission Coefficients

Solid	Work Function (V)	Ion	Energy (V)	γ_{se}
Si(100)	4.90	He^+	100	0.168
		Ar^+	10	0.024
			100	0.027
Ni(111)	4.5	He^+	100	0.170
		Ar^+	10	0.034
			100	0.036
Mo	4.3	He^+	100	0.274
		Ar^+	100	0.115
		N_2^+	100	0.032
		O_2^+	100	0.026
W	4.54	He^+	100	0.263
		Ar^+	10	0.096
			100	0.095
		H_2^+	100	0.029
		N_2^+	100	0.025
		O_2^+	100	0.015

Source: Konuma (1992)/with permission of Springer Nature.

(1991, Section 4.6). Secondary emission is an important process in dc and high-pressure rf capacitive discharges, which are described in Chapters 14 and 15, respectively. Apart from neutralization and Auger emission, heavy particles (ions and neutrals) have much the same behavior when they impact surfaces. At low (thermal) energies, physi- and chemisorption and desorption can occur. At higher energies (tens of volts), molecules can fragment into atoms. At still higher energies (hundreds of volts), atoms can be sputtered from the surface, and at still higher energies (thousands of volts), implantation is important.

9.3.2 Adsorption and Desorption

Adsorption and desorption are very important for plasma processing because, in many cases, one or the other of these reactions is the rate-limiting step for a surface process. Adsorption

$$A + S \rightarrow A{:}S$$

is the reaction of a molecule with a surface. Desorption is the reverse reaction. Adsorption is due to the attractive force between an incoming molecule and a surface. There are two kinds of adsorption. *Physisorption* is due to the weak attractive van der Waals force between a molecule and a surface. We can understand how this force arises by considering the example of a Bohr atom near a metal surface. As shown in Figure 9.3*b*, the Bohr model gives rise to an oscillating dipole moment $p_{dx}(t) \approx a_0 e \cos \omega_{at} t$ normal to the surface as the electron orbits the nucleus. The force F_x acting on the dipole can be found using the method of images. For a dipole $+p_{dx}(t)$ at x, there is an image dipole $+p_{dx}(t)$ at $-x$, and the force is attractive (Problem 9.6),

$$F_x = -\frac{6\langle p_{\mathrm{dx}}^2(t)\rangle}{4\pi\epsilon_0(2x)^4} \approx -\frac{3a_0^2e^2}{4\pi\epsilon_0(2x)^4}$$

The van der Waals interaction potential is found from $F_x = -e\,\mathrm{d}V/\mathrm{d}x$ to be

$$V(x) = -\frac{a_0^2 e}{64\pi\epsilon_0 x^3} \tag{9.3.5}$$

When the atom comes to within a distance of order $d \sim$ 1–3 Å from the surface, then the Coulomb clouds of the atom and surface interact, leading to a repulsive force. Hence, a shallow potential well is formed near the surface. Let $d \sim$ 1–3 Å at equilibrium, the well depth is estimated from (9.3.5) to be $\mathcal{E}_{\mathrm{physi}} \sim$ 0.01–0.25 V. Hence, physisorption is exothermic with $|\Delta H| \sim$ 1–25 kJ/mol. The vibration frequency ω_{vib} for a molecule trapped in the well can be estimated assuming a harmonic oscillator potential,

$$\frac{1}{2}M\omega_{\mathrm{vib}}^2 d^2 \sim e\mathcal{E}_{\mathrm{physi}} \tag{9.3.6}$$

which yields $\omega_{\mathrm{vib}} \sim 10^{12}$–$10^{13}$ s^{-1}. Physisorbed molecules are often so weakly bound to the surface that they can diffuse rapidly along the surface.

Chemisorption is due to the formation of a chemical bond between the atom or molecule and the surface. The reaction is strongly exothermic with $|\Delta H| \sim$ 40–400 kJ/mol, corresponding to a potential well depth $\mathcal{E}_{\mathrm{chemi}} \sim$ 0.4–4 V. The minimum of the well is typically located at a distance $d \sim$ 1–1.5 Å from the surface. Chemisorption of a molecule having multiple (double, triple, etc.) bonds can occur with the breaking of one bond as the molecule bonds to the surface,

$$\mathrm{A{=}B + S \rightarrow AB{:}S}$$

Molecules that are single bonded are often torn apart as they bond to the surface,

$$\mathrm{AB + 2S \rightarrow A{:}S + B{:}S}$$

This process is called *dissociative chemisorption* and requires two adsorption sites. Physi- and chemisorption are often found in the same system, with different regimes favored depending on the surface temperature and the form of the potential energy curves. Figure 9.4 gives three examples. In (*a*), the A+B dissociated chemisorbed state combines with the AB physisorbed state to give a minimum potential energy curve (solid line) that is everywhere negative. AB molecules at low energies incident on the surface can easily pass through the physisorbed region and enter the dissociated chemisorbed state. In (*b*), there is a potential barrier $\mathcal{E}_{\mathrm{ads}}$ to chemisorption, but incident AB molecules can be trapped in the physisorbed state. If the barrier is low, then thermal molecules can be first physisorbed and later pass into the lower energy, dissociated chemisorbed state. In (*c*), there is molecular chemisorption but not dissociative adsorption because the A+B chemisorbed state lies everywhere above the AB state.

Molecules that impinge on a surface cannot be adsorbed unless they lose energy in the collision with the surface. The normal component of the energy loss must be sufficient to trap the molecule in the adsorption well. Let $\Gamma_{\mathrm{A}} = \frac{1}{4}\bar{v}_{\mathrm{A}} n_{\mathrm{AS}}$ be the flux of molecules incident on the surface, where $\bar{v}_{\mathrm{A}}$ is the mean speed of the molecule and n_{AS} is the gas-phase volume density of molecules at the surface. Then, the flux of molecules that are chemisorbed can be written as

$$\Gamma_{\mathrm{ads}} = s\Gamma_{\mathrm{A}} = \frac{1}{4}s\bar{v}_{\mathrm{A}} n_{\mathrm{AS}} \tag{9.3.7}$$

which defines the *sticking coefficient* s. In general, s is a function of the surface coverage θ (fraction of sites covered with adsorbate) and the gas and surface temperatures. If the gas and surface are

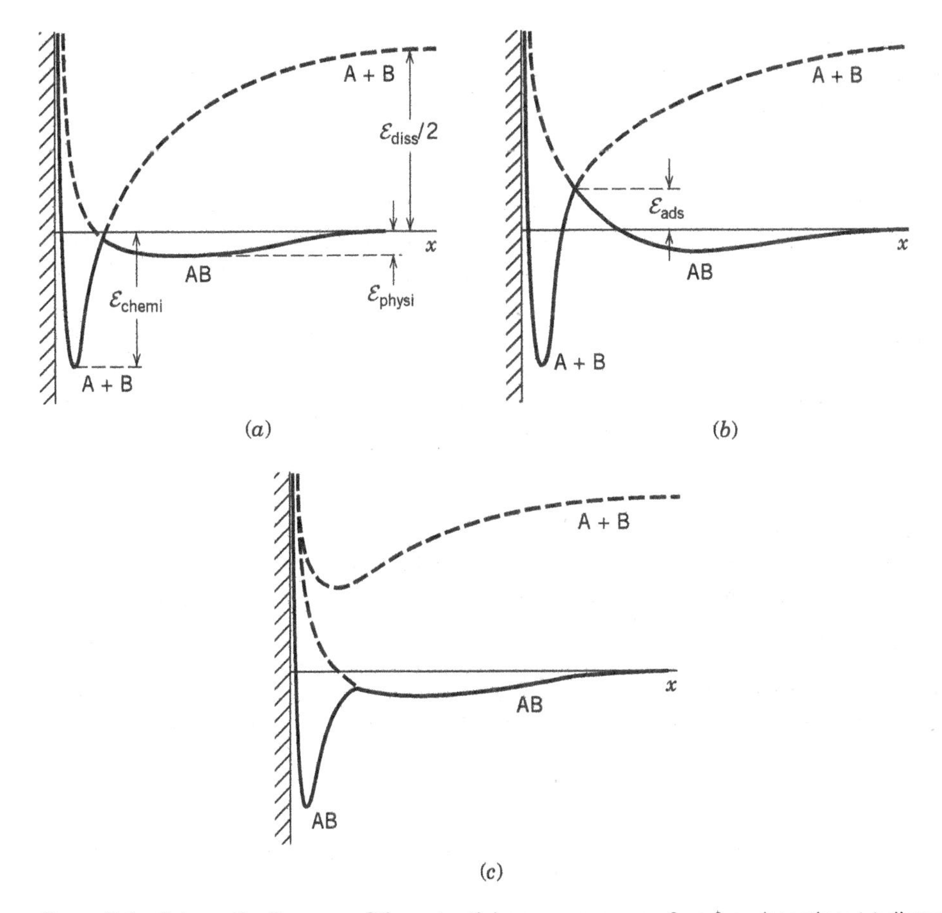

Figure 9.4 Schematic diagrams of the potential energy near a surface for adsorption: (*a*) dissociative chemisorption; (*b*) physisorption; and (*c*) molecular chemisorption.

in thermal equilibrium at temperature T, then the surface coverage $\bar{\theta}(\mathrm{T})$ is determined and $s = \bar{s}(\mathrm{T})$, the equilibrium *thermal sticking coefficient*. A common assumption for s for nondissociative adsorption for systems not in thermal equilibrium is Langmuir kinetics,

$$s(\theta, \mathrm{T}) = s_0(\mathrm{T})(1 - \theta) \tag{9.3.8}$$

where s_0 is the initial or *zero coverage sticking coefficient,* and $1 - \theta$ is the fraction of the surface not covered with adsorbate. Langmuir kinetics is often found to underestimate the sticking coefficient for chemisorption at intermediate values of θ, because molecules that impact sites already filled with adsorbate can be trapped by physisorption and diffuse along the surface to vacant sites, where they chemisorb. Generally, chemisorption ceases after all active sites have been filled; this roughly corresponds to a monolayer of coverage. Continued adsorption is only by the much less tightly bonded physisorption mechanism. Many monolayers can be physisorbed and, in fact, continuous condensation of adsorbate can occur. Usually, however, nonactive surfaces, e.g., reactor walls, come to an equilibrium where physisorption and desorption balance; hence, the *net* flux of molecules to these surfaces is zero. The kinetics of physi- and chemisorption are treated in Section 9.4.

The temperature variation of s_0 depends on whether there is an energy barrier to chemisorption (Figure 9.4*b*) or not (Figure 9.4*a*). If there is no barrier, then s_0 can be near unity at low temperatures and decreases with increasing T, because the fraction of incident molecules that lose sufficient energy to trap decreases as T increases. If there is an activation barrier of height $\mathcal{E}_{ads}$, then very little sticking can occur until T $\sim \mathcal{E}_{ads}$. Then, s_0 has an Arrhenius form,

$$s_0 = s_{00}(\mathrm{T})\,\mathrm{e}^{-\mathcal{E}_{ads}/\mathrm{T}} \tag{9.3.9}$$

where the preexponential factor s_{00} decreases as T increases, as for the case with no barrier. Measured sticking coefficients at T = 0.026 V (300 K) vary over a wide range 10^{-6}–1 and strongly depend on crystal orientation and surface roughness, with s_0 increasing as the roughness increases (Morris et al., 1984). For many surfaces, the active sites for sticking are at surface imperfections such as steps, kinks, vacancies, and dislocations. Chemically reactive gases, and especially radicals, usually stick with a high probability of $s_0 \sim 0.1$–1 on transition metals (Fe, Ni, etc.). Sticking probabilities can be lower for other surfaces. For example, $s_0 \sim 1$ for H on Si, but s_0 is a few percent for H_2 on Si, and $s_0 \sim 10^{-4}$–10^{-3} for O_2 on Si (Joyce and Foxon, 1984).

Desorption,

$$\mathrm{A{:}S} \rightarrow \mathrm{A} + \mathrm{S}$$

is the reverse reaction to adsorption. In thermal equilibrium, the two reactions must balance. The (first-order) desorption rate constant can be shown to have an Arrhenius form (Zangwill, 1988)

$$K_{desor} = K_0\,\mathrm{e}^{-\mathcal{E}_{desor}/\mathrm{T}} \tag{9.3.10}$$

where K_0 (s^{-1}) is the pre-exponential factor and $\mathcal{E}_{desor}$ is the depth ($\mathcal{E}_{chemi}$ or $\mathcal{E}_{physi}$) of the potential well from the zero of potential energy. A crude classical estimate is that K_0 is the number of attempted escapes per second from the adsorption well; hence,

$$K_0 \sim \omega_{vib}/2\pi \tag{9.3.11}$$

where ω_{vib} is the vibration frequency of the adsorbed molecule, as estimated in (9.3.6). A more precise estimate from transition rate theory (Zangwill, 1988) is that

$$\begin{aligned} K_0 &\approx \bar{s}(\mathrm{T})\frac{e\mathrm{T}}{h}\frac{\bar{g}_{esc}}{\bar{g}_{ads}} \\ &\approx 6\times 10^{12}\,\bar{s}(\mathrm{T})\frac{\bar{g}_{esc}}{\bar{g}_{ads}} \quad \mathrm{s}^{-1} \end{aligned} \tag{9.3.12}$$

where $\bar{g}_{esc}/\bar{g}_{ads}$ is a ratio of statistical weights for escaping and trapped molecules. For physisorption, $K_0 \sim 0.2$–$1\times 10^{12}\ \mathrm{s}^{-1}$. For chemisorption, $K_0 \sim 0.2$–$0.5\times 10^{13}\ \mathrm{s}^{-1}$. For activated adsorption, $\bar{s}$ also has an Arrhenius dependence (see (9.3.9)); therefore,

$$K_{desor} \propto \mathrm{e}^{-(\mathcal{E}_{ads}+\mathcal{E}_{desor})/\mathrm{T}}$$

Associative desorption,

$$2\,\mathrm{A{:}S} \rightarrow \mathrm{A}_2 + 2\mathrm{S}$$

the reverse of dissociative adsorption, also has a (second order) rate constant on the surface with the Arrhenius form (9.3.10). The classical estimate of the preexponential factor is that it is the number of collisions per second per unit area on the surface between two adsorbed atoms:

$$K_0 \sim \bar{d}\left(\frac{\pi e\mathrm{T}}{M_R}\right)^{1/2} \tag{9.3.13}$$

where $\bar{d}$ is the mean diameter for a collision, and $(\pi e\mathrm{T}/M_R)^{1/2}$ is the characteristic collision velocity. Typically, $K_0 \sim 10^{-3}$–$1\ \mathrm{cm}^2/\mathrm{s}$.

9.3.3 Fragmentation

Ionic and neutral molecules that have sufficient impact energy can fragment into atoms that are reflected or adsorbed when they hit a surface. The threshold energy for fragmentation is of the order of the energy of the molecular bond. At energies four or five times the threshold energy, over half of the molecules typically fragment. Since molecular bond energies are in the range of 1–10 V, and ion-bombarding energies at surfaces are often considerably higher (particularly at capacitively driven electrodes), molecular ions often fragment when they hit surfaces.

9.3.4 Sputtering

At energies above a threshold of $\mathcal{E}_{\text{thr}} = 20\text{–}50$ V, heavy particles can sputter atoms from a surface. Usually, ions are the impacting species. The sputtering yield γ_{sput} (atoms sputtered per incident ion) increases rapidly with energy up to a few hundred volts, where the yield becomes significant for processing applications, with 200–1000 V argon ions the usual projectile for physical sputtering. For these energies, the bombarding ion transfers energy to many target atoms, which in turn collide with other atoms in the solid. The final distribution of atom energies is isotropic with mean energy $\mathcal{E}_{\text{t}}$, the surface binding energy (roughly, the enthalpy of vaporization in units of volts, see Table 7.4). Most of the atoms in this collision cascade are trapped in the solid, but one or several can escape from the surface (Sigmund, 1981; Smith, 1995, Chapter 8; Mahan, 2000, Chapter 7). When the atomic numbers of the target and incident ion are both large and not too different ($0.2 \lesssim Z_{\text{t}}/Z_{\text{i}} \lesssim 5$ with Z_{t}, $Z_{\text{i}} \gg 1$), then a reasonable estimate for the sputtering yield is (Zalm, 1984)

$$\gamma_{\text{sput}} \approx \frac{0.06}{\mathcal{E}_{\text{t}}}\sqrt{\overline{Z}_{\text{t}}}\left(\sqrt{\mathcal{E}_{\text{i}}} - \sqrt{\mathcal{E}_{\text{thr}}}\right) \tag{9.3.14}$$

where

$$\overline{Z}_{\text{t}} = \frac{2Z_{\text{t}}}{(Z_{\text{i}}/Z_{\text{t}})^{2/3} + (Z_{\text{t}}/Z_{\text{i}})^{2/3}} \tag{9.3.15}$$

For mass ratios $M_{\text{i}}/M_{\text{t}} \gtrsim 0.3$, a reasonable estimate of the threshold energy is (Bohdansky et al., 1980)

$$\mathcal{E}_{\text{thr}} \approx 8\,\mathcal{E}_{\text{t}}(M_{\text{i}}/M_{\text{t}})^{2/5} \tag{9.3.16}$$

These semi-empirical formulas encompass the main ions, targets, and energy regimes of interest in plasma processing.

We can understand the $\sqrt{\mathcal{E}_{\text{i}}}$ energy dependence in (9.3.14) as follows: the incoming ion and the cascade of energetic atoms partially penetrate the electronic cores of the target atoms during their collisions. The collision dynamics in this energy regime is reasonably well described by a Thomas–Fermi interaction potential (Wilson et al., 1977), which scales as $U(r) \propto 1/r^4$ for large r, yielding a collision cross section $\sigma(\mathcal{E}) \propto 1/\sqrt{\mathcal{E}}$ (see Table 3.1). The range of ion penetration into the target can be estimated as $\lambda(\mathcal{E}_{\text{i}}) \sim (n_{\text{t}}\sigma)^{-1}$, where n_{t} is the target atom density in the solid. From energy conservation, the number of atoms in the collision cascade having average energy $\mathcal{E}_{\text{t}}$ is $\mathcal{N} \approx \mathcal{E}_{\text{i}}/\mathcal{E}_{\text{t}}$. Of this number, only those atoms within a distance $\lambda(\mathcal{E}_{\text{t}})$ of the surface can escape. The sputtering yield then scales as

$$\gamma_{\text{sput}} \sim \mathcal{N}\lambda(\mathcal{E}_{\text{t}})/\lambda(\mathcal{E}_{\text{i}}) \propto \sqrt{\mathcal{E}_{\text{i}}} \tag{9.3.17}$$

as in (9.3.14). The threshold energy in (9.3.16) is about an order of magnitude greater than $\mathcal{E}_{\text{t}}$ because multiple (at least three) energy-transferring collisions are necessary to finally eject one backward-traveling atom having energy $\geq \mathcal{E}_{\text{t}}$ from the surface.

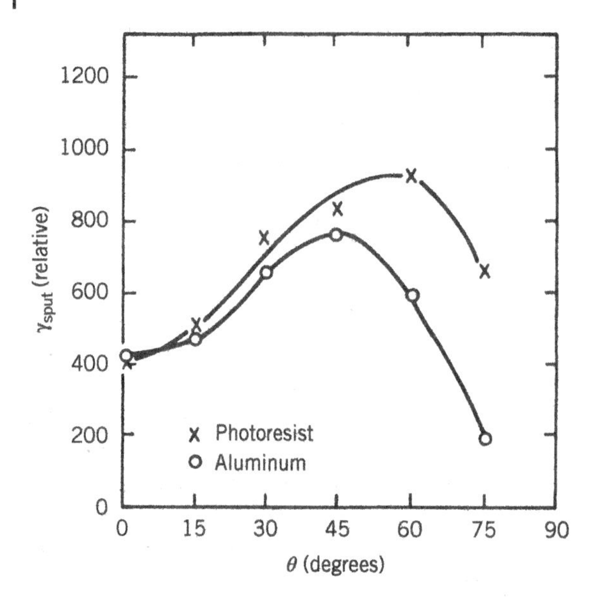

Figure 9.5 Relative sputtering yields for photoresist and aluminum versus angle of incidence θ. Source: Flamm and Herb, 1989/with permission Elsevier.

For $\mathcal{E}_i \gg \mathcal{E}_{thr}$, the sputtered atoms are emitted with a cascade-type energy distribution and with a cosine law in the emission angle χ (Sigmund, 1981; Winters and Coburn, 1992)

$$f(\mathcal{E}, \chi) \propto \frac{\mathcal{E}}{(\mathcal{E}_t + \mathcal{E})^3} \cos \chi \tag{9.3.18}$$

The maximum of this distribution occurs for $\mathcal{E} = \mathcal{E}_t/2$. Since $\mathcal{E}_t \sim$ 3–6 V, the characteristic sputtered atom energies are 1.5–3 V, much hotter than room temperature.

The sputtering yield depends on the angle of incidence of the ion. Figure 9.5 shows typical angular dependences for argon ions incident on aluminum and photoresist. In both cases, the yield rises from its normal (0°) incidence value to some maximum value γ_{max} at θ_{max}, and then falls to zero at grazing incidence (90°). The increase in γ_{sput} with increasing θ is due to the shortening of the range of ion penetration normal to the surface. The range can be estimated as $\lambda(\mathcal{E}_i) \cos \theta$. Using this rather than $\lambda(\mathcal{E}_i)$ in (9.3.17) yields $\gamma_{sput} \propto \sec \theta$, as roughly seen in the figure. However, as $\theta \to 90°$, the incoming ion is increasingly deflected by its first few collisions and emerges from the surface without transferring most of its energy, thus reducing the number of atoms in the collision cascade. Hence, $\gamma_{sput} \to 0$ as $\theta \to 90°$.

In addition to the dependence (9.3.14), measured high fluence sputtering yields have a periodic variation of peaks and valleys versus projectile atomic number, which are not seen in low fluence measurements. These are due to changes in the surface layer due to implantation or deposition of the projectile ion. The peaks are for sputtering by the noble gases and are believed to be due to gas agglomeration and bubble formation in the target material. The valleys are due to buildup of a surface layer which blocks sputtering of the target, e.g., for carbon or calcium projectiles. Some measured sputtering yields for argon ion bombardment at 600 V are given in Table 9.2 (Konuma, 1992). The role of sputtering in dc discharges is described in Sections 14.4–14.6. The application of (9.3.14) to plasma-assisted etch processes is considered in Sections 16.1–16.2. We examine the sputtering deposition and reactive sputtering deposition of thin films in Section 17.4.

Above a few hundred volts, there is a significant chance that ions will be implanted in the solid (Feldman and Mayer, 1986). This process becomes increasingly important above 1 kV. These energies are not common in processing discharges, but can be accessed by applying extremely high dc

Table 9.2 Measured Sputtering Yields for Ar^+ at 600 V

Target	γ_{sput}
Al	0.83
Si	0.54
Fe	0.97
Co	0.99
Ni	1.34
Cu	2.00
Ge	0.82
W	0.32
Au	1.18
Al_2O_3	0.18
SiO_2	1.34
GaAs	0.9
SiC	1.8
SnO_2	0.96

Source: Konuma (1992)/with permission of Springer Nature.

or pulsed voltages to an electrode immersed in a plasma. This application, plasma-immersion ion implantation, is described in Section 17.5.

9.4 Surface Kinetics

A general reaction set for a surface process is illustrated in Figure 9.6. This might apply to the etching of a carbon substrate in an oxygen discharge, with O the etchant and CO the etch product. The etchant atoms diffuse or flow to the surface (rate constant K_a), where they are adsorbed (K_b) and react (K_c) to form the product, which then desorbs (K_d) and diffuses or flows into the gas phase (K_e). In addition, etchants can desorb without reaction (K_f and K_g for normal or associative desorption),

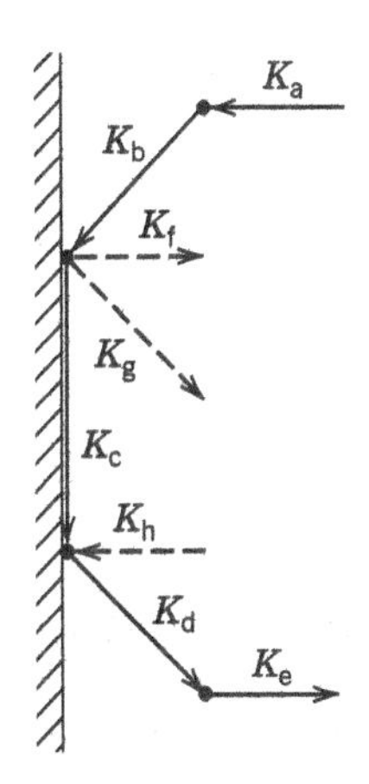

Figure 9.6 Illustrating the processes that can occur for reaction of an etchant with a surface.

and etch products in the gas phase can adsorb back onto the surface (K_h). More complicated reactions can also occur, e.g., to form CO_2 product.

9.4.1 Diffusion of Neutral Species

Charged particle diffusion was the subject of Chapter 5. Neutral species also diffuse. The diffusion coefficient for A molecules due to collisions with B molecules is

$$D_{AB} = \frac{e\mathrm{T}}{M_R \nu_{AB}} \tag{9.4.1}$$

where M_R is the reduced mass and

$$\nu_{AB} = n_B \sigma_{AB} \overline{v}_{AB}$$

is the collision frequency for a constant cross-section (hard-sphere) process, with $\overline{v}_{AB} = (8e\mathrm{T}/\pi M_R)^{1/2}$ the mean speed of relative motion. Inserting ν_{AB} into D_{AB} yields [1]

$$D_{AB} = \frac{\pi}{8} \lambda_{AB} \overline{v}_{AB} \tag{9.4.2}$$

where $\lambda_{AB} = 1/n_B \sigma_{AB}$ is the mean free path. The cross section can be estimated from

$$\sigma_{AB} \approx \pi (r_A + r_B)^2 \tag{9.4.3}$$

where r_A and r_B are the mean radii of the molecules. Some gas kinetic cross sections are given in Table 9.3 (Smirnov, 1977). Cross sections are typically in the range $2\text{–}6 \times 10^{-15}$ cm^2. For self-diffusion of A molecules due to collisions with A molecules, $M_R = M_A/2$ in (9.4.1). A simple and useful model for binary gas-phase diffusion coefficients in terms of atomic or molecular "diffusion volumes" is given by Fuller et al. (1966).

9.4.2 Loss Rate for Diffusion

Let us consider the transport and loss of gas-phase molecules to surfaces by diffusion. We consider a simple steady-state plane-parallel discharge model in which molecules (e.g., etchant atoms, density n_A) are created by electron-impact dissociation of the parent neutral gas (density $n_g \equiv n_B$), and molecules incident on the walls are lost with some probability γ. We assume a uniform

Table 9.3 Gas Kinetic Cross Sections in Units of 10^{-15} cm^2

	He	Ar	H_2	N_2	O_2	CO	CO_2
He	1.6	2.9	2.2	3.1	2.9	3.0	3.6
Ar		5.0	3.7	5.4	5.2	5.3	5.7
H_2			2.7	3.8	3.7	3.9	4.5
N_2				5.2	4.1	5.1	6.8
O_2					4.9	4.8	5.9
CO						5.0	6.3
CO_2							7.8

Source: Smirnov (1977)/with permission of Mir Publishers.

1 The result from kinetic theory is 3/4 of this value (McDaniel, 1964, p. 50).

profile $n_e = n_{e0}$ for the electron density, appropriate for a low-pressure discharge (see Section 10.2), and leave the higher pressure case of a cosine density profile to Problem 9.7. Then, the diffusion equation (5.1.8) becomes

$$-D_{AB}\frac{d^2 n_A}{dx^2} = K_{diss} n_{e0} n_B \tag{9.4.4}$$

with K_{diss} the dissociation rate coefficient. Integrating (9.4.4) yields the symmetric solution

$$n_A = \frac{G_0 l^2}{8D_{AB}}\left(1 - \frac{4x^2}{l^2}\right) + n_{AS} \tag{9.4.5}$$

where $G_0 = K_{diss} n_{e0} n_B$ and the constant of integration n_{AS} is the gas-phase density at the surface. To evaluate this, we use the boundary condition (5.2.4) at $x = l/2$ that

$$\Gamma_A(l/2) = -D_{AB}\left.\frac{dn_A}{dx}\right|_{l/2} = \frac{\gamma}{2(2-\gamma)} n_{AS}\bar{v}_{AB} \tag{9.4.6}$$

The incident flux is obtained by differentiating (9.4.5)

$$\Gamma_A = -D_{AB}\frac{dn_A}{dx} = G_0\frac{l}{2} \tag{9.4.7}$$

Applying the boundary condition (9.4.6) by evaluating (9.4.7) at $x = l/2$, we obtain

$$n_{AS} = \frac{(2-\gamma)}{\gamma}\frac{G_0 l}{\bar{v}_{AB}} \tag{9.4.8}$$

Substituting (9.4.8) into (9.4.5) yields the central density

$$n_A(0) = G_0\left(\frac{l^2}{8D_{AB}} + \frac{(2-\gamma)l}{\gamma\bar{v}_{AB}}\right) \tag{9.4.9}$$

Integrating (9.4.5) over $(-l/2,\ l/2)$ and dividing the result by l yields the average density

$$\bar{n}_A = G_0\left(\frac{l^2}{12D_{AB}} + \frac{(2-\gamma)l}{\gamma\bar{v}_{AB}}\right) \tag{9.4.10}$$

To determine the first-order rate coefficient K_{loss} for loss of particles to the walls, we note from (9.4.7) that the total particle flux lost to both walls is $2\Gamma_A(l/2) = G_0 l$. Hence, we can write

$$K_{loss} = \frac{2\Gamma_A(l/2)}{\bar{n}_A l} = \left(\frac{l^2}{12D_{AB}} + \frac{(2-\gamma)l}{\gamma\bar{v}_{AB}}\right)^{-1} \tag{9.4.11}$$

If we substitute for $l^2/12$ the square of the diffusion length, Λ_0^2 given by the equation after (5.2.10), into the first term in (9.4.11), and substitute $l = 2\mathcal{V}/S$ into the second term, with $\mathcal{V}$ the discharge volume and $S = 2A$ the surface area for loss, then K_{loss} can be written in the form

$$K_{loss} = \left(\frac{\Lambda_0^2}{D_{AB}} + \frac{2(2-\gamma)}{\gamma\bar{v}_{AB}}\frac{\mathcal{V}}{S}\right)^{-1} \tag{9.4.12}$$

Chantry (1987) has shown that the generalization (9.4.12) gives a good heuristic approximation of K_{loss} for all (non-reentrant) discharge wall shapes (e.g., cylinders and rectangular boxes). The loss rate in (9.4.12) can be written in the form

$$\frac{1}{K_{loss}} = \frac{1}{K_D} + \frac{1}{K_w} \tag{9.4.13}$$

where

$$K_D = \frac{D_{AB}}{\Lambda_0^2} \equiv \tau_0^{-1} \tag{9.4.14}$$

and

$$K_w = \frac{\gamma}{2(2-\gamma)} \frac{\overline{v}_{AB} S}{\mathcal{V}} \tag{9.4.15}$$

with τ_0 the fundamental diffusion loss time given by (5.2.11). K_D gives the diffusion rate for transport of molecules to the walls, and K_w gives the loss rate at the walls. Since D_{AB} varies inversely with the neutral gas pressure p, for high pressures and γ not too small, the diffusion is the rate-limiting step for loss, such that n_A has a diffusion profile (cosine profile for a plane-parallel system) and $K_{loss} \approx K_D$. At low pressures or for γ near unity, the surface loss term dominates, such that n_A is nearly uniform and $K_{loss} \approx K_w$. Chantry (1987) has also shown that (9.4.12) gives a reasonable estimate for K_{loss} even in the low-pressure regime $\lambda_{AB} \gtrsim \Lambda_0$. In this limit, the volume loss to the walls is no longer diffusive; the molecules flow freely to the walls. Their characteristic rate of loss is determined by their mean speed, the distance they travel, and their probability of loss to the surfaces.

Let us estimate K_{loss} for O atoms diffusing through O_2 molecules in a plane-parallel reactor with $l = 10$ cm, $\gamma = 10^{-2}$, and $T_B = 300$ K. Let $\sigma_{AB} \approx 3 \times 10^{-15}$ cm^2 and $p = 10$ mTorr ($n_B \approx 3.3 \times 10^{14}$ cm^{-3}). Then, $\lambda_{AB} \approx 1$ cm and $\overline{v}_{AB} \approx 7.7 \times 10^4$ cm/s. This yields $D_{AB} \approx 3.0 \times 10^4$ cm^2/s from (9.4.2). Substituting these values into (9.4.14) and (9.4.15), we find that $K_D \approx 3600$ s^{-1} and $K_w \approx 77$ s^{-1}. Because the diffusion is fast compared to the wall loss, the rate-limiting step is the wall loss: $K_{loss} \approx K_w$. This is typical for low-pressure plasma processing systems.

The loss probability γ may be known from measurements, but it can also be inferred from the kinetics of adsorption, desorption, and reaction on the surface. We discuss the relation between γ and these fundamental surface processes at the end of this section.

9.4.3 Adsorption and Desorption

Consider the opposing reactions for nondissociative adsorption and desorption of A molecules on a surface,

$$A(g) + S \underset{K_d}{\overset{K_a}{\rightleftharpoons}} A{:}S$$

Let n'_0 be the area density of adsorption sites and $n'_{A:S} = n'_0\theta$ be the density of sites covered with adsorbed molecules. Assuming Langmuir kinetics, such that the flux of A adsorbing on the surface is proportional to $1 - \theta$, the fraction of sites not covered with adsorbate, we can write

$$\Gamma_{ads} = K_a n_{AS} n'_0 (1-\theta) \tag{9.4.16}$$

where n_{AS} is the gas-phase density at the surface. The adsorption rate coefficient is given in terms of fundamental quantities by equating (9.4.16) to (9.3.7) and eliminating θ by using (9.3.8), yielding

$$K_a = \frac{1}{4} s_0 \frac{\overline{v}_{AS}}{n'_0} \tag{9.4.17}$$

with s_0 the zero-coverage ($\theta = 0$) sticking coefficient. The flux of desorbing molecules is

$$\Gamma_{desor} = K_d n'_0 \theta \tag{9.4.18}$$

Equating the adsorption and desorption fluxes, we can solve for θ to obtain

$$\theta = \frac{\mathcal{K}_{ads} n_{AS}}{1 + \mathcal{K}_{ads} n_{AS}} \tag{9.4.19}$$

where

$$\mathcal{K}_{ads} = \frac{K_a}{K_d} \tag{9.4.20}$$

This is the Langmuir isotherm for thermal equilibrium (7.5.17). In addition to these direct surface processes, adsorption and desorption can proceed via an intermediate precursor state. For example, chemical adsorption can be from a physisorbed precursor state, leading to an isotherm different from the simple Langmuir isotherm (see Problem 9.10).

9.4.4 Dissociative Adsorption and Associative Desorption

Consider now the opposing reactions

$$A_2(g) + 2S \underset{K_d}{\overset{K_a}{\rightleftharpoons}} 2\,A{:}S$$

Because two sites are required for adsorption, the molecular flux adsorbed is

$$\Gamma_{ads} = K_a n_{A_2S} n'_0 (1-\theta)^2 \tag{9.4.21}$$

and the molecular flux desorbed is

$$\Gamma_{desor} = K_d n'_0 \theta^2 \tag{9.4.22}$$

Here, K_a (m^3/s) in (9.4.17) and K_d (s^{-1}) are the adsorption and desorption rate constants, respectively. Equating fluxes and solving for θ, we obtain the isotherm

$$\theta = \frac{(\mathcal{K} n_{A_2S})^{1/2}}{1 + (\mathcal{K} n_{A_2S})^{1/2}} \tag{9.4.23}$$

where $\mathcal{K} = K_a/K_d$. For θ small, we see that $\theta \propto n_{A_2S}^{1/2}$ for dissociative adsorption, a slower variation than for normal adsorption.

9.4.5 Physical Adsorption

While the density of available sites is usually fixed at some n'_0 for chemisorption, many monolayers can be physisorbed. Let n'_i be the area density of sites having a thickness of i physisorbed atoms. Then, equating the adsorption to desorption flux for these sites,

$$K_a n'_i n_{AS} = K_d n'_{i+1} \tag{9.4.24}$$

we obtain

$$n'_{i+1} = \beta n'_i \tag{9.4.25}$$

where $\beta = K_a n_{AS}/K_d = \mathcal{K} n_{AS}$. Hence, by induction,

$$n'_i = n'_0 \beta^i \tag{9.4.26}$$

The total number of physisorbed molecules per unit area is given by

$$\begin{aligned} n'_T &= \sum_{i=1}^{\infty} i n'_i = n'_0 \sum_{i=1}^{\infty} i\beta^i \\ &= n'_0 \frac{\beta}{(1-\beta)^2} \end{aligned} \tag{9.4.27}$$

and the number of sites covered per unit area is

$$n'_C = \sum_{i=1}^{\infty} n'_i = n'_0 \frac{1}{1-\beta} \tag{9.4.28}$$

For $\beta \lesssim 1$, many monolayers can be adsorbed. The condition $\beta = 1$ signals the onset of continuous condensation. The combination of physi- and chemisorption can also be analyzed, leading to the so-called BET isotherm (see Atkins, 1986, p. 779).

9.4.6 Reaction with a Surface

Consider the reaction set where A is adsorbed on the surface S ($\equiv$ B) and reacts directly with the surface to form the gas-phase product AS ($\equiv$ AB):

$$A(g) + S \underset{K_d}{\overset{K_a}{\rightleftharpoons}} A{:}S$$

$$A{:}S \xrightarrow{K_r} AS\ (g)$$

The surface coverage θ is found from the conservation of adsorbed sites,

$$\frac{dn'_{A:S}}{dt} = K_a n_{AS} n'_0(1-\theta) - K_d n'_0 \theta - K_r n'_0 \theta = 0 \tag{9.4.29}$$

Solving for θ, we obtain

$$\theta = \frac{1}{1 + (K_d + K_r)/K_a n_{AS}} \tag{9.4.30}$$

The reaction rate (m^{-2}-s^{-1}) for the production of AB is then

$$R_{AB} = K_r n'_0 \theta \tag{9.4.31}$$

9.4.7 Reactions on a Surface

A common reaction mechanism on the surface, called *Langmuir–Hinshelwood kinetics,* involves the reaction of two adsorbed species:

$$A(g) + S \underset{K_{d1}}{\overset{K_{a1}}{\rightleftharpoons}} A{:}S$$

$$B(g) + S \underset{K_{d2}}{\overset{K_{a2}}{\rightleftharpoons}} B{:}S$$

$$A{:}S + B{:}S \xrightarrow{K_r} AB\ (g) + 2S$$

For ease of analysis let the reaction itself be the rate-limiting step. Then, the surface concentrations of A and B are the thermal equilibrium values, from (7.5.19),

$$\theta_A = \frac{\mathcal{K}_A n_{AS}}{1 + \mathcal{K}_A n_{AS} + \mathcal{K}_B n_{BS}} \tag{9.4.32a}$$

$$\theta_B = \frac{\mathcal{K}_B n_{BS}}{1 + \mathcal{K}_A n_{AS} + \mathcal{K}_B n_{BS}} \tag{9.4.32b}$$

where $\mathcal{K}_A = K_{a1}/K_{d1}$ and $\mathcal{K}_B = K_{a2}/K_{d2}$. The rate of production of AB(g) is then

$$R_{AB} = K_r n_0'^2 \theta_A \theta_B \tag{9.4.33}$$

At low pressures, $\mathcal{K}_A n_{AS} \ll 1$ and $\mathcal{K}_B n_{BS} \ll 1$, the kinetics is the second order,

$$R_{AB} = K_r \frac{K_{a1} K_{a2}}{K_{d1} K_{d2}} n_0'^2 n_{AS} n_{BS} \tag{9.4.34}$$

As previously, n_{AS} and n_{BS} are related to G_{A0} and G_{B0} by using (9.4.8).

A second reaction mechanism, called *Eley–Rideal kinetics*, involves the reaction of adsorbed A directly with an impinging gas-phase molecule B:

$$\mathrm{A(g)} + \mathrm{S} \underset{K_{\mathrm{d1}}}{\overset{K_{\mathrm{a1}}}{\rightleftarrows}} \mathrm{A{:}S}$$

$$\mathrm{B(g)} + \mathrm{S} \underset{K_{\mathrm{d2}}}{\overset{K_{\mathrm{a2}}}{\rightleftarrows}} \mathrm{B{:}S}$$

$$\mathrm{A{:}S} + \mathrm{B(g)} \xrightarrow{K_{\mathrm{r}}} \mathrm{AB\ (g)} + \mathrm{S}$$

Again assuming that the reaction itself is the rate-limiting step, then

$$R_{\mathrm{AB}} = K_{\mathrm{r}} n_0' \theta_{\mathrm{A}} n_{\mathrm{BS}} \tag{9.4.35}$$

which, at low pressures, reduces again to the second-order kinetics,

$$R_{\mathrm{AB}} = K_{\mathrm{r}} \frac{K_{\mathrm{a1}}}{K_{\mathrm{d1}}} n_0' n_{\mathrm{AS}} n_{\mathrm{BS}} \tag{9.4.36}$$

9.4.8 Surface Kinetics and Loss Probability

Let us consider the coupling of the surface kinetics to the transport and loss of a diffusing species A. From (9.4.6), the loss flux is given in terms of the loss probability γ as

$$\Gamma_{\mathrm{A}}(l/2) = \frac{\gamma}{2(2-\gamma)} n_{\mathrm{AS}} \bar{v}_{\mathrm{AB}} \tag{9.4.37}$$

We consider the simplest kinetics of adsorption and desorption of A on the surface and reaction of A with the surface. Inserting (9.4.30) into (9.4.31), the loss flux is given in terms of the surface rate coefficients as

$$\Gamma_{\mathrm{A}}(l/2) = \frac{K_{\mathrm{r}} n_0' K_{\mathrm{a}} n_{\mathrm{AS}}}{K_{\mathrm{a}} n_{\mathrm{AS}} + K_{\mathrm{d}} + K_{\mathrm{r}}} \tag{9.4.38}$$

Equating (9.4.37) and (9.4.38), we obtain

$$\frac{\gamma \bar{v}_{\mathrm{AB}}}{2(2-\gamma)} = \frac{K_{\mathrm{r}} n_0' K_{\mathrm{a}}}{K_{\mathrm{a}} n_{\mathrm{AS}} + K_{\mathrm{d}} + K_{\mathrm{r}}} \tag{9.4.39}$$

For a small surface coverage $\theta \ll 1$ in (9.4.30), we have $K_{\mathrm{a}} n_{\mathrm{AS}} \ll K_{\mathrm{d}} + K_{\mathrm{r}}$. Then, (9.4.39) can be solved to obtain γ as a function of the surface rate coefficients, independent of n_{AS}. For the usual case of a small loss probability, $\gamma \ll 1$, the result is

$$\gamma = \frac{4n_0'}{\bar{v}_{\mathrm{AB}}} \frac{K_{\mathrm{r}} K_{\mathrm{a}}}{K_{\mathrm{d}} + K_{\mathrm{r}}} \tag{9.4.40}$$

Substituting K_{a} from (9.4.17) into (9.4.40), we obtain

$$\gamma = \frac{s_0 K_{\mathrm{r}}}{K_{\mathrm{d}} + K_{\mathrm{r}}} \tag{9.4.41}$$

which gives γ in terms of the fundamental surface quantities. With (9.4.41) for γ, the loss flux given by (9.4.6) depends linearly on n_{AS}, a first-order loss kinetics. The first-order kinetics are typical for surface reactions at low pressures, such as are found in processing discharges.

In the opposite limit $K_{\mathrm{a}} n_{\mathrm{AS}} \gg K_{\mathrm{d}} + K_{\mathrm{r}}$, (9.4.39) reduces to

$$\frac{\gamma \bar{v}_{\mathrm{AB}}}{2(2-\gamma)} = \frac{K_{\mathrm{r}} n_0'}{n_{\mathrm{AS}}} \tag{9.4.42}$$

With (9.4.42), the loss flux (9.4.6) is then independent of n_{AS}, a zeroth-order surface kinetics, which is not uncommon at high pressures. This regime is generally not of interest for low-pressure processes.

As one application of these kinetics, Kota et al. (1998, 1999) measured γ for halogen atom recombination to form halogen molecules on various surfaces and compared their measurements to surface kinetic models. They found that the measured γ was independent of the incident halogen flux and that the surface reaction rate was the first order in the incident flux. As the simplest model, they considered the reaction of physisorbed chlorine atoms with a surface saturated with chemisorbed chlorine atoms, using (9.4.41) to describe the recombination kinetics. More elaborate models including the kinetics of chemical adsorption and reactions on the surface are given in Kota et al. (1999).

9.5 Showerhead Gas Flow

The flows of low-pressure feed gases, along with deposition and etching precursors and products from surface reactions, are important issues in materials processing of substrates over large areas. Typically, the substrate rests on the bottom electrode, and feed gas is introduced from an upper "showerhead" gas diffuser surface, often an rf-powered electrode. The showerhead contains numerous, closely spaced small holes through which the feedgas flows. This enables gas flow within the reactor chamber without turbulence, in contrast to the use of a simple single gas feed orifice. An idealized, two-dimensional representation of a showerhead reactor is shown in Figure 9.7. The gas impinges on the substrate electrode and then flows laterally to the reactor periphery or sidewalls, where it is pumped away. This is a type of *stagnation flow*, with a *stagnation point* at $x = z = 0$ having zero fluid velocity. For deposition reactors, typical dimensions are $l \sim 2.5$ cm, $w \sim 0.25$–1.5 m, and the gas pressures are $p \sim 0.5$–10 Torr (67–1330 Pa). As might be imagined, the flow of gas feedstock, and the removal of deposition or etch products, can be nonuniform along the lateral dimension x, leading to processing nonuniformities.

Howling et al. (2012) have given an analytical description of the gas flow in such a reactor, which gives considerable insight into the basic phenomena and flow patterns, and we summarize their treatment here. At these pressures with gas mean free path $\lambda_g \ll l$, the flow can be described by fluid equations. Because the gas densities and flow velocities are small, the flow is laminar. Furthermore, the small flow speeds require only a very small pressure drop to drive them; this means that to a good approximation, the flow is incompressible

$$\nabla \cdot \mathbf{u} = 0 \tag{9.5.1}$$

Steady incompressible gas flow is described using the momentum balance equation (2.3.9), by dropping the time-varying term $\partial \mathbf{u}/\partial t$, the field term $\mathbf{E} + \mathbf{u} \times \mathbf{B}$, and the friction term $\mathbf{f}_c$, and by

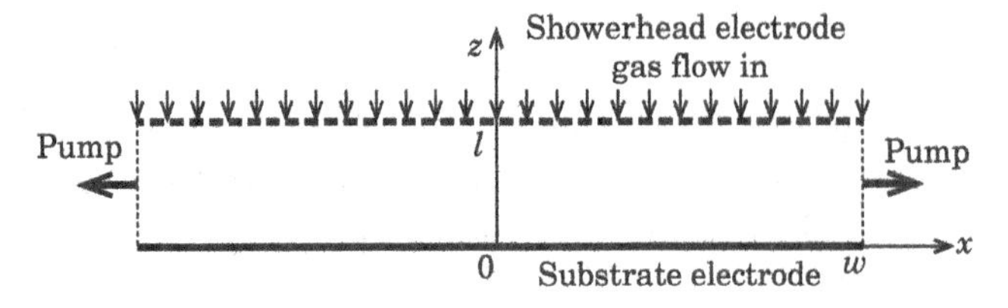

Figure 9.7 Idealized, two-dimensional reactor configuration for showerhead gas flow calculation, with no variation along the y-coordinate.

expanding the pressure tensor $\mathbf{\Pi}$ into isotropic and anisotropic parts

$$\rho_g(\mathbf{u}\cdot\nabla)\mathbf{u} = -\nabla p + \nabla\cdot\mathbf{T}_g \tag{9.5.2}$$

Here, $\rho_g = M_g n_g$ is the mass density and $\mathbf{T}_g$ is the gas stress tensor. For an incompressible, isothermal, low-pressure gas

$$\nabla\cdot\mathbf{T}_g = \mu_g\nabla^2\mathbf{u} \tag{9.5.3}$$

(Bird et al., 2002, chapter 3), where μ_g (Pa-s) is the gas *absolute viscosity*. The right-hand side of (9.5.3) accounts for the shear stress forces per unit volume on a fluid element due to the diffusion of molecules of faster flowing fluid into slower flowing fluid, and vice versa.

Absolute viscosities are almost independent of pressure, with typical values 0.5–2 × 10^{-5} Pa-s, depending on the gas and its temperature. Sometimes the *kinematic viscosity* or *viscous diffusivity* $D_g = \mu_g/\rho_g$ (m^2/s) is introduced. From mean free path theory, $D_g \approx \frac{1}{3}\bar{v}_g\lambda_g$, with $\bar{v}_g = (8kT_g/\pi M_g)^{1/2}$ the gas atom mean speed. Then, $\mu_g \approx \frac{1}{3}M_g n_g \bar{v}_g \lambda_g$.

Substituting (9.5.3) into (9.5.2) yields the steady-state *Navier–Stokes equation*

$$\rho_g(\mathbf{u}\cdot\nabla)\mathbf{u} = -\nabla p + \mu_g\nabla^2\mathbf{u} \tag{9.5.4}$$

Due to the slow flow velocity $\mathbf{u} = \hat{x}u_x + \hat{z}u_z$, the inertial term in (9.5.4) is small, and the pressure and sheer stress forces approximately balance; this is the limiting case of *creeping flow*. The x-component of the Navier–Stokes equation then gives

$$\frac{\partial p}{\partial x} = \mu_g\left(\frac{\partial^2 u_x}{\partial x^2} + \frac{\partial^2 u_x}{\partial z^2}\right) \tag{9.5.5}$$

9.5.1 Approximate Solution

Let us consider a uniform showerhead gas flow rate $u_z(x, z = l) = -u_0$ into the reactor, independent of x. Then, an approximate solution for this stagnation flow is found with the two assumptions (1) that u_z is independent of x everywhere in the reactor

$$u_z(x, z) = u_z(z) \tag{9.5.6}$$

and (2) that the pressure $p(x, z) = p(x)$, independent of z. Substituting (9.5.6) into (9.5.1) and integrating with respect to x, with the boundary condition that $u_x = 0$ at $x = 0$, gives

$$u_x(x, z) = -x\frac{\mathrm{d}u_z}{\mathrm{d}z} \tag{9.5.7}$$

Substituting (9.5.7) into (9.5.5), taking the derivatives, and using assumption (2) gives

$$\frac{\mathrm{d}^3u_z}{\mathrm{d}z^3} = -\frac{1}{\mu_g x}\frac{\mathrm{d}p}{\mathrm{d}x} \tag{9.5.8}$$

The left-hand side of (9.5.8) is only a function of z, and the right-hand side is only a function of x. Therefore both the left- and right-hand sides are equal to a constant value K.

To determine u_z and K, we use the boundary conditions (a) $u_z(x, 0) = 0$, (b) $u_z(x, l) = -u_0$, and the no-slip boundary conditions at the showerhead and substrate surfaces that (c) $u_x(x, 0) = 0$ and $u_x(x, l) = 0$. Note that (a) implies no gas absorption (deposition, etc.) on the substrate. Integrating $\mathrm{d}^3u_z/\mathrm{d}z^3 = K$ three times and using the boundary conditions (a)–(d), we obtain $K = 12u_0/l^3$ and

$$u_z = \frac{u_0}{l^3}\left(2z^3 - 3lz^2\right) \tag{9.5.9}$$

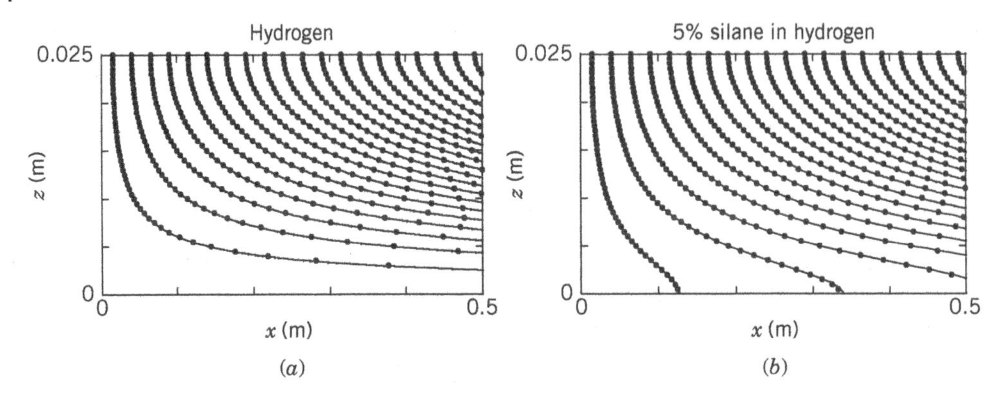

Figure 9.8 Showerhead streamlines for the idealized, two-dimensional reactor configuration; (*a*) pure H_2, no substrate deposition; (*b*) 5%/95% SiH_4/H_2, with a loss probability $\gamma = 0.107$ on the electrodes. Source: Howling et al. (2012). ©IOP Publishing. Reproduced with permission. All rights reserved. https://doi.org/10.1088/0963-0252/21/1/015005.

Evaluating u_x from (9.5.7) gives

$$u_x = \frac{6u_0}{l^3} x\left(lz - z^2\right) \tag{9.5.10}$$

Finally, the pressure is found by integrating $\mathrm{d}p/\mathrm{d}x = -K\mu_g x$, with the boundary condition at the pump location $x = \pm w$ that $p = p_0$, to obtain

$$p = p_0 + \frac{6u_0\mu_g}{l^3}(w^2 - x^2) \tag{9.5.11}$$

For typical deposition conditions, we leave the confirmation that the gas flow is nearly incompressible, that the inertial terms in (9.5.4) are negligible, and that the flow is laminar, as Problem 9.12 for the student.

The streamlines of the flow with initial values $x = x_s$ and $z = l$ are given by (see Problem 9.13)

$$x(z) = x_s \frac{l^3}{3lz^2 - 2z^3} \tag{9.5.12}$$

Figure 9.8*a* shows typical streamlines for a hydrogen feedgas with $l = 2.5$ cm and $w = 0.5$ m. The lines are the approximate model, and the dots are the results of a numerical calculation (Howling et al., 2012), which incorporates lateral nonuniformity of the vertical flow velocity, inertial terms in the Navier–Stokes equation, and non-vertical diffusion.

The preceding analysis has zero net mass deposition on the showerhead and substrate electrodes. For deposition, there is a finite flow to these surfaces; this case is called *Stefan flow.* With a finite fraction γ of the incident gas deposited on the electrodes, the boundary conditions become (a′) $u_z(x, 0) = -\gamma u_0$ and (b′) $u_z(x, l) = -u_0(1 - \gamma)$. This gives (Howling et al., 2012)

$$u_z = \frac{u_0}{l^3}\left[(1 - 2\gamma)\left(2z^3 - 3lz^2\right) - \gamma l^3\right] \tag{9.5.13}$$

with u_x again determined from (9.5.7). Let us note that $u_z = -\gamma u_0$ at the substrate surface is a constant, independent of x. However, the pressure p (and gas density n_g) varies with x, as given by (9.5.11). This indicates that the deposition is nonuniform, varying as $p(x)$. Of course, other nonuniformities also arise due to lateral plasma variations, which affect precursor generation rates.

Streamlines for this finite deposition are shown for a 5%/95% SiH_4/H_2 mixture in Figure 9.8*b*. Note that the streamlines from the showerhead originating from $x = 0$ to $x \approx 0.05$ m terminate on the substrate surface with a non-zero perpendicular component. This represents the finite deposition rate on the substrate.

Problems

9.1 Complex Reaction for Ozone Consider the loss of ozone in a dilute, low-pressure O_3/O_2 gas mixture in the steady state at standard (room) temperature due to the reactions

$$O_3 + O_2 \underset{K_3}{\overset{K_2'}{\rightleftharpoons}} O + O_2 + O_2$$

$$O + O_3 \xrightarrow{K_2} O_2 + O_2$$

(a) Find the reaction rate R (cm^{-3}-s^{-1}) for the destruction of ozone based on the above reaction set. Estimate R using the data in Tables 8.2 and 8.3 for $n_{O_2} = 3.3 \times 10^{16}$ cm^{-3} and $n_{O_3} = 3.3 \times 10^{14}$ cm^{-3}.

(b) The reverse reaction,

$$O_2 + O_2 \xrightarrow{K_{-2}} O_3 + O$$

is not listed in Table 8.2. Find the rate constant K_{-2} for this reaction using (9.1.13) and the data in Table 8.2. The standard Gibbs free energies for the formation of O and O_3 are 231.75 and 163.16 kJ/mol respectively, and the standard enthalpies of formation of O and O_3 are 249.17 and 142.7 kJ/mol, respectively.

9.2 Reaction Rate Calculations

(a) Consider the kinetics of a stable molecule A that "spontaneously" decomposes into molecules B and C,

$$A \xrightarrow{K_1} B + C$$

Determine the conditions for this to happen and obtain the first-order rate constant K_1 by considering the elementary reactions

$$A + A \underset{K_{-2}}{\overset{K_2}{\rightleftharpoons}} A^* + A$$

$$A^* \xrightarrow{K_{1*}} B + C$$

Assume that the last reaction is rate limiting.

(b) Consider the first-order reaction chain

$$A \xrightarrow{K_{AB}} B \xrightarrow{K_{BC}} C \xrightarrow{K_{CD}} D \xrightarrow{K_{DE}} E$$

Assuming that the concentration $n_A = n_{A0}$, that all other ns are zero at time $t = 0$, and that $C \rightarrow D$ is the rate-limiting reaction, then find an approximate expression for $n_E(t)$. Sketch on the same graph the time-varying behavior of n_A, n_B, n_C, n_D, and n_E.

9.3 Stepwise Ionization Ionization can occur as a two-step process involving excited atoms:

$$e + A \xrightarrow{K_{ex}} e + A^*$$

$$e + A^* \xrightarrow{K_{iz*}} 2e + A^+$$

Competing reactions for loss of A* are collisional de-excitation

$$e + A^* \xrightarrow{K_{dex}} e + A$$

and the first-order losses

$$A^* \xrightarrow{K_1} A$$

where K_1 is the total first-order rate constant for loss of A* due to radiative emission and de-excitation at the reactor walls. Let $\mathcal{E}_{iz}$ and $\mathcal{E}_{iz*}$ be the ionization potentials of A and A*, respectively, and let the statistical weights of A and A* be the same. Assume that $T_e \ll \mathcal{E}_{iz*}$, $\mathcal{E}_{iz}$ and that $K_{iz*} \ll K_{dex}$.

(a) From detailed balance (8.5.14), show that

$$K_{ex} = K_{dex} \exp\left(-\frac{\mathcal{E}_{iz} - \mathcal{E}_{iz*}}{T_e}\right)$$

(b) Find n_{A^*} as a function of n_e, n_A, and the rate constants.

(c) Using the Thomson ionization rate constants (3.5.5), show that the ratio of two-step to single-step ionization rates is

$$\frac{R_{iz*}}{R_{iz}} = \frac{n_e K_{dex}}{n_e K_{dex} + K_1} \frac{\mathcal{E}_{iz}^2}{\mathcal{E}_{iz*}^2}$$

Hence, two-step ionization is *always* more important than single-step ionization for thermal equilibrium $K_1 \ll n_e K_{dex}$.

(d) Estimate the ratio R_{iz*}/R_{iz} for a typical low-pressure processing discharge with $K_1 = K_{loss}$ given by (9.4.12) with $\gamma = 1$. Is two-step ionization important or not?

9.4 Ionization Rate Due to Argon Metastables The rate coefficient K_4 for electron impact excitation of the 3p ($^1S_0$) ground state of argon to the 4s metastable levels is given in Table 3.3.

(a) By applying (8.5.14), find the rate coefficient K_5 for the inverse process of electron impact de-excitation ("quenching") of the 4s metastable levels to the 3p ground state. (There are actually two 4s metastable levels, $^3P_0$ and $^3P_2$, which are separated by a small energy gap $\Delta\mathcal{E} = 0.17$ V. You may assume that $T_e \gg \Delta\mathcal{E}$, i.e., you may assume that the two metastable levels have essentially the same energy and can be treated as a single metastable level having a total of six states.) Compare your result to K_5 in Table 3.3.

(b) Assuming that argon metastables are created only by electron impact excitation with rate coefficient K_4 and are lost only by electron impact de-excitation to the ground state with rate coefficient K_5, find the ratio of argon metastables to ground state atoms at an electron temperature $T_e = 3$ V.

(c) The rate coefficients for ionization of ground state and metastable argon atoms are given as K_2 and K_{10}, respectively, in Table 3.3. Accounting only for these two processes and the two processes in part (b), find the fraction of the total ionizations per second that are due to metastable argon. Are metastable argon atoms important in this discharge?

9.5 Three-Body Recombination

(a) Estimate the rate constant (9.2.38) for K_2' at low pressures for the three-body recombination reaction

$$\mathrm{e} + \mathrm{A}^+ + \mathrm{M} \rightarrow \mathrm{A} + \mathrm{M}$$

by modifying the analysis done for positive–negative ion recombination leading to (9.2.40)–(9.2.42). You should obtain the scaling $K_2' = n_\mathrm{M} K_3' \propto \mathrm{T}_\mathrm{e}^{-3/2}$

(b) Compare your result in (a) with the tabulated data in Table 8.3.

9.6 Dipole–Dipole Force Consider two electric dipoles $p_{\mathrm{d}1}$ and $p_{\mathrm{d}2}$ oriented along x and separated by a distance r. Each dipole can be regarded as a pair of point charges $+q$ and $-q$ separated by a small distance $d \ll r$:$p_{\mathrm{d}1} = p_{\mathrm{d}2} = qd$. Using Coulomb's law for the electrostatic force on a point charge due to another point charge, show that the net force on dipole $p_{\mathrm{d}1}$ due to dipole $p_{\mathrm{d}2}$ is attractive and has a magnitude

$$F_x = \frac{6 p_{\mathrm{d}1} p_{\mathrm{d}2}}{4\pi\epsilon_0 r^4}$$

9.7 Diffusion Loss with a Nonuniform Source Consider a steady-state plane-parallel discharge model in which a neutral species A is created within the discharge region $-l/2 < x < l/2$ at a rate $G_0 \cos \pi x/l$ (m^{-3}-s^{-1}) and is lost to the walls with a loss probability γ. Show that the rate coefficient K_loss for loss of particles to the walls is given by (9.4.11) with 12 replaced by π^2. (Recall that (9.4.11) was obtained for a uniform creation rate.)

9.8 Diffusion Loss in an Asymmetric Discharge Consider the creation, diffusion, and loss of a species A in an *asymmetric* one-dimensional slab geometry, with a uniform rate of production G_0 (m^{-3}-s^{-1}) within the slab. Assume that one of the two electrode surfaces is *inactive*, such that the net flux of A to this surface is zero. The other electrode is *active*, such that a fraction γ of the flux incident on the surface is lost to the surface.

(a) Find $n_\mathrm{A}(x)$ within the slab in terms of G_0 and γ.

(b) Find K_loss, the first-order rate coefficient for loss of A to the walls, and compare your expression with (9.4.11).

9.9 Diffusion and Reaction in the Volume Consider a steady-state plane-parallel discharge model in which a neutral species A is created *uniformly* within the discharge region $-l/2 < x < l/2$ at a constant rate G_0 (m^{-3}-s^{-1}). A fraction γ of the flux of A incident on the walls is lost to the walls. Species A is also lost inside the discharge region by reaction with the background neutral gas (density n_B) at a rate $-K_\mathrm{rec} n_\mathrm{A} n_\mathrm{B}$ (m^{-3}-s^{-1}).

(a) Give the diffusion equation that determines $n_\mathrm{A}(x)$.

(b) Give the boundary conditions necessary to solve the diffusion equation of part (a) and then solve the diffusion equation to determine $n_\mathrm{A}(x)$.

(c) Evaluate the rate coefficient K_loss (surface) for loss of A to the surfaces, and evaluate the overall rate coefficient K_loss (total) for loss of A both to the surfaces and by reaction with the background neutral gas.

9.10 Chemical Adsorption via a Physical Adsorption State Consider chemical adsorption and desorption kinetics in which the adsorption is via a physisorbed precursor state, but the desorption is directly from the chemisorbed state into the gas phase. The kinetics is described by the reaction chain

$$A(g) + S \underset{K_d^*}{\overset{K_a^*}{\rightleftharpoons}} A{:}S\,(physi) \xrightarrow{K_a} A{:}S\,(chemi) \xrightarrow{K_d} A(g) + S$$

where K_a^* (m^3/s) is the second-order rate coefficient for adsorption of a gas-phase atom into the physisorbed surface state, K_d^* (s^{-1}) is the first-order rate coefficient for desorption of a physisorbed atom into the gas phase, K_a' (m^2/s) is the second-order rate coefficient for adsorption of a physisorbed atom into the chemisorbed surface state, and K_d (s^{-1}) is the first-order rate coefficient for desorption of a chemisorbed atom directly into the gas phase. Assume a surface density of n_0' (m^{-2}) for both physisorption and chemisorption sites, and let θ^* be the surface coverage for physisorption and θ be the surface coverage for chemisorption. You may assume that because the physisorption sites lie above the chemisorption sites on the surface, a physisorption site can be located over either an empty or an occupied chemisorption site.

(a) Write the site balance equation for physisorbed atoms in the steady state. From this, show that the rate of chemisorption is

$$\Gamma_{ads} = \frac{K_a' K_a^* n_{AS} n_0'^2 (1-\theta)}{K_a^* n_{AS} + K_d^* + K_a' n_0' (1-\theta)}$$

(b) Write the site balance equation for chemisorbed atoms in the steady state. Using this and your result in (a), find the equation to determine θ.

9.11 Normal and Dissociative Adsorption

(a) Consider the steady-state *chemisorption* and desorption reactions at a reactor wall at room temperature (300 K):

$$A(g) + S \underset{K_d}{\overset{K_a}{\rightleftharpoons}} A{:}S$$

$$A{:}S + A{:}S \xrightarrow{K_{d2}} A_2(g) + 2S$$

Make the following assumptions: $s_0 \approx 1$, the chemisorption well depth for atoms is $\mathcal{E}_{desor} = 3$ V, and the molecular dissociation energy is $\mathcal{E}_{diss} = 5$ V. Note that the activation energies for desorption of atoms and molecules are $\mathcal{E}_{desor}$ and $2\mathcal{E}_{desor} - \mathcal{E}_{diss}$, respectively (see Figure 9.4). Also use as typical parameters: $\overline{v}_A \approx 8 \times 10^4$ cm/s, $n_0' \approx 10^{15}$ cm^{-2}, $n_{AS} \approx 10^{13}$ cm^{-3}, and use the preexponential factors for normal and associative desorption of 10^{14} s^{-1} and 0.1 cm^2/s, respectively. For these parameters, show that the reactor walls are completely passivated; i.e., $\theta \approx 1$.

(b) Show that for chemisorption with these parameters, virtually all atoms desorb as molecules rather than as atoms.

(c) Find the ratio Γ_{A_2}/Γ_A of the desorbing molecular flux Γ_{A_2} to the flux Γ_A of atoms incident on the surface, and show that this ratio is very small.

(d) Now consider *physisorption* of A along with desorption of A and A_2 on a completely passivated wall (no chemisorption). Make the same assumptions as in (a), except let the physisorption well depth be $\mathcal{E}_{desor} \approx 0.2$ V; hence, the activation energy for desorption of atoms is $\mathcal{E}_{desor}$, but the activation energy for desorption of molecules is zero (this reaction is now exothermic). Use a preexponential factor for normal desorption of 10^{13} s^{-1} and an associative desorption rate constant of 0.1 cm^2/s. For these parameters, show that the surface coverage for physisorption is *very small*; $\theta \ll 1$.

(e) Show that for physisorption with these parameters virtually all atoms desorb as atoms rather than as molecules.

(f) Find the ratio Γ_{A_2}/Γ_A of the desorbing molecular flux Γ_{A_2} to the flux Γ_A of atoms incident on the surface, and show that this ratio is very small.

Note that in view of your results in (c) and (f), the surface recycles most reactive atoms back into the discharge as atoms. This is typical for fluorine atoms.

9.12 Showerhead Gas Flow Assumptions Typical amorphous silicon deposition conditions for a 5%/95% silane/hydrogen feedgas might be gas pressure $p \sim 200$ Pa, temperature $T_g \sim 500$ K, and showerhead gas injection velocity $u_0 \sim 0.2$ m/s. Consider the idealized two-dimensional reactor shown in Figure 9.7 with $l = 2.5$ cm and $w = 0.5$ m. Use an absolute viscosity $\mu_g \sim 10^{-5}$ Pa-s.

(a) Find the gas density n_g and the gas mass density ρ_g.

(b) Determine the lateral pressure drop Δp from $x = 0$ to $x = w$, and show that it is much less than the deposition pressure. Therefore, to a good approximation, the gas flow is incompressible.

(c) From (9.5.10), estimate an average lateral flow velocity $\bar{u}_x$. Using this, show that the inertial terms in (9.5.4) are negligible compared to the pressure term. Therefore, the flow is creeping.

(d) The flow is laminar if the Reynolds number $\text{Re} = \rho_g \bar{u}_x l/\mu_g \lesssim 1500$ (Bird et al., 2002, chapter 5). Evaluate the Reynolds number and show that the flow is laminar.

9.13 Showerhead Gas Flow Streamlines By integrating the streamline equation

$$\frac{dz}{dx} = \frac{u_z}{u_x}$$

with u_z and u_x given by (9.5.9) and (9.5.10), respectively, show that the streamlines of the flow with initial values $x = x_s$ and $z = l$ are given by (9.5.12).

10

Particle and Energy Balance in Discharges

10.1 Introduction

For low-pressure discharges, the plasma is not in thermal equilibrium and the electrical power is coupled most efficiently to plasma electrons. In the bulk plasma, energy is transferred inefficiently from electrons to ions and neutrals by weak collisional processes. The fraction of energy transferred by elastic collision of an electron with a heavy ion or neutral is of order $2m/M \sim 10^{-4}$, where m and M are the electrons and heavy particle masses. Hence, the electron temperature T_e greatly exceeds the ion and neutral temperatures, T_i and T_g, respectively, in the bulk; typically $T_e \sim 2$–5 V whereas T_i and T_g are at most a few times room temperature (0.026 V). However, dissociation and excitation processes (see Section 8.3) can create a subgroup of relatively high-energy heavy particles. Also, the ambipolar electric fields accelerate positive ions toward the sheath edge, where they typically acquire a directed energy of order $T_e/2$.

At low pressures, the energy relaxation mean free path $\lambda_\mathcal{E}$ for ionizing electrons, with energies $\gtrsim$ 10–15 V, can be comparable to the discharge dimensions. Here, $\lambda_\mathcal{E} = (\lambda_m \lambda_{inel}/3)^{1/2}$, with λ_m the momentum transfer mean free path, and λ_{inel} the mean free path accounting for all collisional electron energy losses. (See Section 15.2 for further discussion about the energy relaxation length.) Hence, even if the electrical power is deposited in a small volume, within an unmagnetized discharge, the electron–neutral ionization rate ν_{iz} can be relatively uniform, since the ionization occurs on the distance scale of $\lambda_\mathcal{E}$. In magnetized plasmas, on the other hand, the ionization may be highly nonuniform as the magnetized electrons have trouble crossing field lines, so ionization along a magnetic flux tube might be uniform but significant cross-field nonuniformities may persist. In addition, the propagation and absorption of exciting electromagnetic fields can depend on the electron density profile. In some instances, this can steer power into regions of higher or lower density and make the plasma more or less uniform, respectively (see Chapter 13). At higher pressures, $\lambda_\mathcal{E}$ is often smaller than the discharge dimensions. Hence, for a nonuniform electron power deposition, the ionization frequency within the discharge can be nonuniform. Some examples are discussed in Chapter 15.

The electron distribution function f_e need not be Maxwellian. However, insightful estimates of source operation can be obtained by approximating f_e to be Maxwellian, with uniform temperature T_e, and with the various electron rate constants assumed to be uniform within the bulk plasma. The detailed distribution depends on the collisional processes, the gas pressure, and the heating mechanism (see Chapter 19).

Principles of Plasma Discharges and Materials Processing, Third Edition. Michael A. Lieberman and Allan J. Lichtenberg.

Electron–neutral collisions are important not only for particle production (ionization, dissociation) but also for other collisional energy losses (excitation, elastic scattering). Ion–neutral collisions (charge transfer, elastic scattering) are also important in determining particle production, plasma transport, and ion energy distributions at a substrate surface. The myriad of collisional processes that can occur in molecular feedstock gas mixtures can obscure the fundamental principles of particle and energy balance. Consequently, a noble gas, such as argon, is often used as a reference for describing discharge operation. Although this provides some understanding of plasma properties, it provides little understanding of gas and surface chemistry, which are critical to most processing applications. Furthermore, most process gases are molecular and electronegative (containing negative ions), leading to significant differences in plasma properties compared to argon. To obtain insight into the more complicated plasma and chemical phenomena that occur in typical materials processing discharges, we must examine properties of electronegative discharges.

In electropositive discharges, there are only two species normally considered, electrons and one positive ion species. The diffusion analysis of Sections 5.2 and 5.3 is usually adequate to treat the particle transport. If magnetic fields are present, then the methods of Sections 5.4 and 5.5 can be used. Similarly, sheath dynamics is treated as in Sections 6.2–6.3. In Section 6.4, we also included a negative ion species, in preparation for a treatment of particle transport in electronegative plasma. Although charge neutrality still holds within the bulk plasma, the low mobility, low temperature, negative ions may constitute most of the negative charge, thus profoundly influencing the equilibrium. In addition, it is generally not possible to neglect recombination of negative and positive ions because this process has a large rate constant (see Section 8.4). Recombination makes the diffusion equation nonlinear, and therefore much more difficult to solve.

In Section 10.2, we treat electropositive plasmas, where the equations for particle balance and energy balance decouple, the former giving the electron temperature, and the latter giving the electron and ion densities. In Section 10.3, we extend the treatment to electronegative plasmas. Simplifying assumptions must be made to specify an ambipolar diffusion coefficient for the positive ions, and the resulting equations are fundamentally nonlinear. The particle and energy balances are also coupled, further complicating the analysis. In Section 10.4, we present approximate analytic solutions for electronegative plasmas, which may include an electropositive edge region. In Section 10.5, calculations are performed for oxygen and chlorine. Particle-in-cell (PIC) simulations are also given and compared with the analytically obtained equilibria.

There are advantages to operating a discharge repetitively pulsed, rather than steady state. In this mode of operation, there is no true equilibrium, but rather a repeated transient build-up and decay. However, nonlinearities make the time evolution more complicated than exponential build-ups and decays. In Section 10.6, we treat some simple models of this operation.

In all cases, the models for the plasma equilibrium that we discuss in this chapter are not complete. The voltage across a plasma sheath cannot be specified independently of the heating mechanism and the power absorbed by the plasma. To obtain a complete heating model, we must specify the method of sustaining the plasma from an external energy source and determine how that source transfers energy to the electrons and (indirectly) to the ions. We consider various ways of transferring energy from fields to plasma discharges in Chapters 11–15. The resulting electron heating mechanisms are of the following types:

- Ohmic heating
- Stochastic heating
- Resonant wave–particle interaction heating
- Secondary electron emission heating

Ohmic heating is present in all discharges due to the transfer of energy gained from the acceleration of electrons in electric fields to thermal electron energy through local collisional processes. Ohmic heating is particularly important at high pressures at which the collision frequency is high, where it can be the dominant heating mechanism.

Stochastic electron heating (sometimes called collisionless heating) has been found to be a powerful mechanism in capacitive rf discharges. Here, electrons impinging on the oscillating sheath edge suffer a change of velocity upon reflection back into the bulk plasma. As the sheath moves into the bulk, the reflected electrons gain energy; as the sheath moves away, the electrons lose energy. However, averaging over an oscillation period, there is a net energy gain. Since the electric fields in the sheath are much larger than the fields inside the plasma, stochastic heating is often the dominant heating mechanism in low pressure capacitive discharges. Sheath heating can also preferentially heat the higher energy electrons, leading to bi-Maxwellian distributions at low pressure. We treat this heating mechanism in Chapter 11. Collisionless heating can also be important in low-pressure inductive discharges, which we treat in Chapter 12.

Wave–particle interactions are a fundamental method of transferring energy from fields to electrons and are important in high-density discharges such as electron cyclotron resonance, helicon, and surface wave sources. The heating can involve both collisional (ohmic) and collisionless energy transfers. We consider these processes in Chapter 13.

Secondary emission heating does not play a central role in most low-pressure discharges. At high pressures, especially in dc and capacitive rf discharges, secondary emission can play a crucial role in plasma production and can also contribute substantially to electron heating. It is fundamental to the operation of dc glow discharges. We consider some of these effects in Chapters 11, 14, and 15. The plasma heating mechanism often defines the type of plasma that is generated, as will be seen in the following chapters.

10.2 Electropositive Plasma Equilibrium

10.2.1 Basic Properties

We consider the example of argon discharges. The most important rate constants for electron collisions in argon are K_{iz}, K_{ex}, and K_{el} for electron–neutral ionization, excitation, and momentum transfer, respectively. These are given in Figure 3.16 as a function of T_e. The most important cross sections for ion–neutral collisions in argon are for resonant charge transfer and elastic scattering. As shown in Figure 3.15, the cross section for resonant charge transfer of Ar^+ on Ar somewhat exceeds that for elastic scattering. The combined ionic momentum transfer cross section σ_i for these two processes is large ($\sigma_i \approx 10^{-14}\,cm^2$) and relatively constant for the (thermal) ion energies of interest. The corresponding ion–neutral mean free path is given in (3.5.10): $\lambda_i = 1/n_g\sigma_i$, where n_g is the neutral argon density.

The overall discharge particle losses for a cylindrical plasma having radius R and length l depend on the particle fluxes to the walls. These fluxes can be written as a product $n_s u_B$, where n_s is the ion density at the plasma–sheath edge and u_B is the Bohm (ion loss) velocity. The relation between the density n_s at the sheath edge and the density n_0 at the plasma center is complicated because the ambipolar transport of ions and electrons spans the regime $\lambda_i \sim (R,\ l)$, depending on the pressure and the values for R and l. As discussed in Chapter 5, there are three regimes.

10.2.1.1 Low Pressure: $\lambda_i \gtrsim (R,\ l)$

This *Langmuir* regime was described in Section 5.3. The ion transport is collisionless and well described by an ion free-fall profile (Figure 5.3 in plane-parallel geometry) within the bulk plasma. This profile is relatively flat near the plasma center and dips near the sheath edge, with $n_s/n_0 \approx 0.5$ for $R \gg l$ (planar geometry) and $n_s/n_0 \approx 0.4$ for $l \gg R$ (infinite cylinder geometry).

10.2.1.2 Intermediate Pressures: $(R,\ l) \gtrsim \lambda_i \gtrsim (T_i/T_e)(R,\ l)$

In this regime, also described in Section 5.3, the transport is diffusive and ambipolar. However, the ion drift velocity u_i much exceeds the ion thermal velocity within most of the bulk plasma, leading to a nonlinear diffusion equation with the solution (5.3.7) for the density profile in plane-parallel geometry shown in Figure 5.2*b*. Again the profile is relatively flat in the center and steep near the sheath edge. As discussed in Section 5.3, joining the collisionless (low pressure) and collisional (intermediate pressure) results leads to the estimates (5.3.13) and (5.3.15), repeated here:

$$h_l \equiv \frac{n_{sl}}{n_0} \approx 0.86\left(3 + \frac{l}{2\lambda_i}\right)^{-1/2} \tag{10.2.1}$$

at the axial sheath edge, and

$$h_R \equiv \frac{n_{sR}}{n_0} \approx 0.80\left(4 + \frac{R}{\lambda_i}\right)^{-1/2} \tag{10.2.2}$$

at the radial sheath edge.

10.2.1.3 High Pressures: $\lambda_i \lesssim (T_i/T_e)(R,\ l)$

In this regime, described in Section 5.2, the transport is diffusive and ambipolar, and the density profile is well described by a J_0 Bessel function variation along r and a cosine variation along z. For this highly collisional regime, the assumption of a relatively uniform density within the plasma bulk, falling sharply near the sheath edge, is not good. This regime is relevant, for example, to the higher pressure planar rf capacitive discharge analyses of Chapters 11 and 15. As shown in Problem 5.7,

$$h_l = \frac{n_{sl}}{n_0} = \left[1 + \left(\frac{l}{\pi}\frac{u_B}{D_a}\right)^2\right]^{-1/2} \tag{10.2.3}$$

where u_B is the Bohm velocity and D_a is the ambipolar diffusion coefficient. However, for these higher pressures, it is often adequate to use a solution in which the edge density $n_{sl} \simeq 0$ and the wall flux is found from the density gradient at the wall, $\Gamma_i = -D_a dn_i/dx$ (in parallel-plane geometry).

The overall discharge energy losses depend on the collisional energy losses, $\mathcal{E}_c$, as well as on the kinetic energy carried by electrons and ions to the walls. Using the rate constants in Figure 3.16, the collisional energy $\mathcal{E}_c$ lost per electron–ion pair created was defined in (3.5.11) and is plotted versus T_e in Figure 3.17. For Maxwellian electrons, the mean kinetic energy lost per electron lost was shown in (2.4.11) to be $\mathcal{E}_e = 2T_e$. The mean kinetic energy lost per ion lost, $\mathcal{E}_i$, is the sum of the ion energy entering the sheath and the energy gained by the ion as it traverses the sheath. The ion velocity entering the sheath is u_B, corresponding to a directed energy of $T_e/2$. The dc sheath voltage V_s takes various forms depending on whether significant rf or dc currents are drawn to the surface, as described in Sections 6.2 and 6.3. If there are no significant voltages applied across the sheath, then for an insulating wall, the ion and electron fluxes must balance in the steady state, leading to (6.2.17), and setting $V_s \equiv -\Phi_w$ we have

$$V_s = T_e \ln\left(\frac{M}{2\pi m}\right)^{1/2} \tag{10.2.4}$$

or $V_s \approx 4.7\,\mathrm{T_e}$ for argon. Accounting for the initial ion energy, we obtain $\mathcal{E}_i \approx 5.2\,\mathrm{T_e}$. At an undriven conducting wall, the fluxes need not balance. However, if the fluxes are not too dissimilar, then (10.2.4) remains a good estimate due to the logarithmic dependence of V_s on the ratio of fluxes.

Let us note that the separation of kinetic energies lost into ion and electron components depends on position within the sheath. Ions and electrons crossing the sheath to the wall gain and lose an energy V_s, respectively. Hence, $\mathcal{E}_e = 2\mathrm{T_e} + V_s$ and $\mathcal{E}_i = \frac{1}{2}\mathrm{T_e}$ at the plasma–sheath edge, but $\mathcal{E}_e = 2\mathrm{T_e}$ and $\mathcal{E}_i = \frac{1}{2}\mathrm{T_e} + V_s$ at the wall. The sum $\mathcal{E}_e + \mathcal{E}_i$ is independent of position.

A *high-voltage sheath* exists at the negatively driven electrode (cathode) surface of a dc discharge, as described in Section 6.3, with the sheath voltage

$$V_s \approx V_{dc} \tag{10.2.5}$$

where V_{dc} is the anode–cathode voltage. Similar high-voltage sheaths exist near capacitively driven electrode surfaces. For a symmetrically driven capacitive rf discharge, with $V_{rf} \gg \mathrm{T_e}$

$$V_s \approx 0.4\,V_{rf} \tag{10.2.6}$$

where V_{rf} is the driving voltage across the electrodes. For a strongly asymmetrically driven discharge, all the voltage appears across a single sheath at the driven (powered) electrode

$$V_s \approx 0.8\,V_{rf} \tag{10.2.7}$$

More precise calculations of the coefficients in (10.2.5)–(10.2.7) are given in Chapter 14 for dc discharges and in Chapter 11 for capacitive rf discharges. The ion kinetic energy lost at a surface is then

$$\mathcal{E}_i = V_s + \frac{1}{2}\mathrm{T_e} \tag{10.2.8}$$

where V_s is given by one of (10.2.4)–(10.2.7). We see from the above discussion that estimating ion energy is not so simple, as it depends not only on electron temperature but also on source geometry and the application of bias voltages. Summing the three contributions yields the total energy lost per electron–ion pair lost from the system:

$$\mathcal{E}_T = \mathcal{E}_c + \mathcal{E}_e + \mathcal{E}_i \tag{10.2.9}$$

The discharge equilibrium generally depends only weakly on the sheath thickness. Undriven sheath thicknesses s rarely exceed a few Debye lengths λ_{De}; hence, such sheaths are less than a millimeter thick in typical discharges. The thickness of a high-voltage sheath follows that of a Child law, with s given by (6.3.14) with $V_0 \sim V_s$. For typical dc or capacitive rf discharges, s is of the order of 0.5 cm.

10.2.2 Uniform Density Discharge Model

We consider a simple cylindrical discharge model in the low-to-intermediate ion mean free path regime to estimate the plasma parameters and their variation with power, pressure, and source geometry. The electron temperature $\mathrm{T_e}$, the ion bombarding energy $\mathcal{E}_i$, the plasma density n_0, and the ion current density J_i are the most significant quantities for plasma-processing applications. We approximate the density to be nearly uniform in the bulk cylindrical plasma, with the density falling sharply near the sheath edges, with (10.2.1) and (10.2.2) giving the ratios of sheath to bulk density. This approximation is one form of a *global model*, in which the profile is assumed. We assume Maxwellian electrons absorbing an electrical power P_{abs}.

We determine the electron temperature T_e from particle balance by equating the total surface particle loss to the total volume ionization,

$$n_0 u_B A_{\text{eff}} = K_{iz} n_g n_0 \pi R^2 l \tag{10.2.10}$$

where

$$A_{\text{eff}} = 2\pi R^2 h_l + 2\pi R l h_R \tag{10.2.11}$$

is the *effective area* for particle loss. Since the ionization and loss terms are both proportional to the plasma density, n_0 cancels and (10.2.10) can be rewritten as

$$\frac{K_{iz}(T_e)}{u_B(T_e)} = \frac{1}{n_g d_{\text{eff}}} \tag{10.2.12}$$

where

$$d_{\text{eff}} = \frac{\pi R^2 l}{A_{\text{eff}}} = \frac{1}{2}\frac{Rl}{Rh_l + lh_R} \tag{10.2.13}$$

is an *effective plasma size* for particle loss, and the explicit T_e dependences of K_{iz} and u_B are assumed known. Given $n_g d_{\text{eff}}$, we can solve (10.2.12) for T_e. For argon with K_{iz} from Table 3.3, and for typical plasma pressures and sizes, we obtain values of T_e shown in Figure 10.1. We see that T_e varies over a narrow range between 2 and 5 V, because the exponential variation of K_{iz} with T_e allows wide variations of K_{iz} for small variations of T_e. We note that T_e is determined by particle conservation alone and is *independent of the plasma density* and therefore the input power.

We determine the central plasma density n_0 from energy balance by equating the total power absorbed, P_{abs}, to the total power lost

$$P_{\text{abs}} = e n_0 u_B A_{\text{eff}} \mathcal{E}_T \tag{10.2.14}$$

Solving for n_0, we obtain

$$n_0 = \frac{P_{\text{abs}}}{e u_B A_{\text{eff}} \mathcal{E}_T} \tag{10.2.15}$$

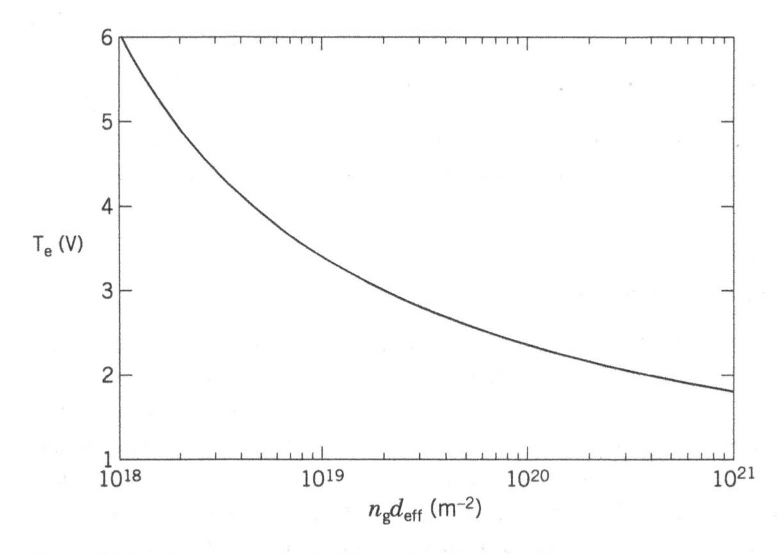

Figure 10.1 T_e versus $n_g d_{\text{eff}}$ for Maxwellian electrons in argon.

which yields n_0 for a specified P_{abs} and T_e determined from (10.2.12) or Figure 10.1. Note that n_0 is determined by the total power balance in the discharge and is a function of pressure only through the dependence of h_l and h_R on p and through the dependence of T_e on p.

Let us consider a discharge ohmically heated by a uniform rf electric field $\tilde{E}$. From (4.2.27), the ohmic heating power is

$$P_{abs} = \pi R^2 l \cdot \frac{e^2 n_0}{m \nu_m} \frac{\nu_m^2}{\omega^2 + \nu_m^2} |\tilde{E}|^2 \tag{10.2.16}$$

where ν_m is the electron–neutral momentum transfer collision frequency. Inserting (10.2.16) into (10.2.14), the densities cancel and we can solve for the electric field magnitude

$$|\tilde{E}| = \left(\frac{m}{e} \frac{u_B \mathcal{E}_T}{d_{eff}} \frac{\omega^2 + \nu_m^2}{\nu_m} \right)^{1/2} \tag{10.2.17}$$

Similar to T_e, we see that $|\tilde{E}|$ is *independent of the plasma density* and therefore the input power. For high frequencies and/or low collisionality (low pressure), $\nu_m/\omega \ll 1$, and $|\tilde{E}| \propto \omega$. In the opposite limit $\nu_m/\omega \gg 1$, $|\tilde{E}|$ is independent of ω. For more complex discharges having nonuniform densities and other heating mechanisms in addition to ohmic heating, the "average" plasma electric field is also more or less restricted to a narrow range of values at a given pressure, independent of the discharge power.

We have assumed in (10.2.14) and (10.2.15) that the same energy loss $\mathcal{E}_T$ occurs at all surfaces. If this is not the case, then these equations must be modified, e.g., for effective areas A_{eff1} and A_{eff2} with energy losses $\mathcal{E}_{T1}$ and $\mathcal{E}_{T2}$, (10.2.14) becomes

$$P_{abs} = e n_0 u_B (A_{eff1} \mathcal{E}_{T1} + A_{eff2} \mathcal{E}_{T2})$$

Example 10.1 Consider a cylindrical discharge having low-voltage sheaths at all surfaces, with V_s given by (10.2.4). Let $R = 0.15$ m, $l = 0.3$ m, $n_g = 3.3 \times 10^{19}$ m^{-3} ($p = 1$ mTorr at 298 K), and $P_{abs} = 800$ W. At 1 mTorr, $\lambda_i \approx 0.03$ m from (3.5.10). Then, from (10.2.1) and (10.2.2), $h_l \approx 0.31$, $h_R \approx 0.27$, and from (10.2.13), $d_{eff} \approx 0.18$ m. From Figure 10.1, $T_e \approx 3.8$ V, and from Figure 3.17, $\mathcal{E}_c \approx 47$ V. Using (10.2.9) with $\mathcal{E}_i \approx 5.2 T_e \approx 20$ V and $\mathcal{E}_e = 2T_e \approx 7.6$ V, we find $\mathcal{E}_T \approx 74$ V. The Bohm velocity is $u_B \approx 3.0 \times 10^3$ m/s, and $A_{eff} \approx 0.12$ m^2 from (10.2.11). Substituting these values into the energy balance (10.2.15) yields $n_0 \approx 1.9 \times 10^{17}$ m^{-3}, corresponding to a flux at the axial boundary $\Gamma_{il} = n_0 h_l u_B \approx 1.7 \times 10^{20}$ m^{-2}-s^{-1} or an ion current density of $J_{il} \approx 2.8$ mA/cm^2.

Example 10.2 If a strong dc magnetic field is applied along the cylinder axis, then particle loss to the circumferential wall is inhibited. For the parameters of Example 10.1, in the limit of no radial loss, a calculation similar to that in Example 10.1 yields $n_0 \approx 5.2 \times 10^{17}$ m^{-3}, and $J_{il} \approx 7.0$ mA/cm^2. There is a significant increase in charge density and ion flux due to the magnetic field confinement. The details of the calculation for this example and for Example 10.3 are left to Problem 10.1.

Example 10.3 Consider the parameters of Example 10.2 for a symmetrically driven rf discharge with high-voltage sheaths, e.g., $V_s \approx 500$ V at each of the cylinder endwalls. There is a large increase in $\mathcal{E}_i \approx 520$ V and therefore in $\mathcal{E}_T \approx 570$ V at the endwalls, which leads to a significant reduction in n_0 and J_{il}; $n_0 \approx 7.5 \times 10^{16}$ m^{-3} and $J_{il} \approx 1.0$ mA/cm^2.

A comparison of Examples 10.2 and 10.3 illustrates an important difference between discharges having high-voltage sheaths over a significant fraction of the surface area and discharges having low-voltage sheaths at all surfaces. The densities are significantly lower and the ion bombarding energies are significantly higher for the same input power and geometry for the high-voltage case than for the low-voltage case. Consequently, in practical applications, low-pressure discharges tend to be divided into two types:

10.2.2.1 Low-Density Discharges

These discharges have high-voltage sheaths over a significant surface area. We treat the important cases of capacitive rf discharges in Chapters 11 and 15, and dc discharges in Chapter 14.

10.2.2.2 High-Density Discharges

These discharges have low-voltage sheaths near almost all surfaces. We treat the cases of rf-driven inductive and helical resonator discharges in Chapter 12 and helicon, ECR, and surface wave discharges in Chapter 13. The ion bombarding energy $\mathcal{E}_i$ in high-density discharges is often too low for the materials process of interest. In this case, the substrate surface is often capacitively driven by an additional rf power supply to increase $\mathcal{E}_i$. In this way, the desired ion bombarding energy at an rf-powered substrate holder can be obtained. The additional ion energy flux $en_s u_B \mathcal{E}_i$ striking the wafer holder is supplied by an rf power source driving the holder. The independent control of the ion energy and the ion flux hitting the substrate is a highly desirable feature of high-density (low sheath voltage) discharges.

It should be noted that V_s was arbitrarily chosen to be 500 V in Example 10.3. In general, as mentioned in Section 10.1, it is not possible to choose the power absorbed P_{abs} and the discharge voltage V_{rf} (or V_{dc}) independently, as was done in Example 10.3. Therefore, for capacitive rf and dc discharges, the preceding analysis is not complete. We elaborate this in Chapter 11, where we determine the I–V characteristic for capacitive rf discharges and complete the analysis presented here.

10.2.3 Nonuniform Discharge Model

At relatively high pressures, $\lambda_i \lesssim (T_i/T_e)l$, the uniform global model cannot be used. The ambipolar diffusion profile in a one-dimensional slab geometry was obtained, in Section 5.2, by solving the ion conservation equation,

$$\frac{d\Gamma_i}{dx} = K_{iz} n_g n_e \tag{10.2.18}$$

where $n_e = n_i$ and

$$\Gamma_i = -D_a \frac{dn_i}{dx} \tag{10.2.19}$$

to obtain the density n_i and particle flux Γ_i. The results (5.2.22) and (5.2.24), repeated here, are

$$n_i(x) = n_0 \cos \beta x \tag{10.2.20}$$

$$\Gamma_i(x) = D_a \beta n_0 \sin \beta x \tag{10.2.21}$$

with D_a the constant ambipolar diffusion coefficient. The simplest assumption made to obtain a solution is that $n_i \approx 0$ at $x = \pm l/2$ which gives $\beta = \pi/l$. This is reasonable because $\lambda_i \ll l$ (see discussion following (5.2.25)). Integrating (10.2.18) from $x = 0$ to $x = l/2$, we obtain

$$\Gamma_i(l/2) = K_{iz} n_g \int_0^{l/2} n_i(x)\, dx \tag{10.2.22}$$

Equation (10.2.22) expresses the overall particle conservation in a nonuniform plasma slab, in analogy to (10.2.10), which expresses this same conservation for a finite cylinder of plasma with uniform bulk density except near the edges. Substituting (10.2.20) and (10.2.21) with $x = l/2$ into (10.2.22) and performing the integration, we obtain

$$\frac{\pi}{l} D_a = \frac{l}{\pi} K_{iz} n_g \tag{10.2.23}$$

Since

$$D_a = \frac{e\mathrm{T}_e}{M n_g K_{mi}} \tag{10.2.24}$$

where $K_{mi}(\mathrm{T}_i)$ is the ion–neutral momentum transfer rate constant, and substituting $u_B = (e\mathrm{T}_e/M)^{1/2}$, (10.2.23) can be rewritten

$$\frac{[K_{mi}K_{iz}(\mathrm{T}_e)]^{1/2}}{u_B(\mathrm{T}_e)} = \frac{\pi}{n_g l} \tag{10.2.25}$$

Equation (10.2.25) is analogous to (10.2.12) in that it determines T_e for a given $n_g l$.

Similarly, equating the total power absorbed by a unit area of the discharge, S_{abs}, to the total power lost, we have

$$S_{abs} = 2\Gamma_i(l/2)\, e(\mathcal{E}_e + \mathcal{E}_i) + 2e\mathcal{E}_c \int_0^{l/2} K_{iz} n_g n_e(x)\, dx \tag{10.2.26}$$

Using (10.2.22) to eliminate the integral in (10.2.26), we find

$$S_{abs} = 2\Gamma_i(l/2)\, e\mathcal{E}_T \tag{10.2.27}$$

Substituting (10.2.21) with $\sin \beta l/2 \approx 1$ into (10.2.27), and solving for n_0, we obtain

$$n_0 = \frac{S_{abs} l}{2\pi D_a e\mathcal{E}_T} \tag{10.2.28}$$

which is analogous to (10.2.15).

This procedure can be generalized to a finite cylinder nonuniform discharge with a constant D_a, which can be solved, as done in Section 5.2. Using the solution found there, e.g., with the approximations of zero densities at the plasma edge with a thin sheath, then the particle conservation equation is

$$2\pi R \int \Gamma_{iR}(R, z)\, dz + 4\pi \int \Gamma_{il}(r, l/2)\, r\, dr = K_{iz} n_g \int n_i(r, z) 2\pi r\, dr\, dz \tag{10.2.29}$$

where Γ_{iR}, Γ_{il}, and n_i are evaluated using the product solution (5.2.37). Equation (10.2.29), analogous to (10.2.22), determines T_e for a given n_g, R, and l. In a similar manner, an energy balance relation analogous to (10.2.26) can be obtained, which can be solved for n_0 (see Problem 10.2).

As discussed in Section 5.2, for some plasmas, the edge density may not be small compared to the central density. In such cases, the boundary conditions can be modified, as in (5.2.28), to specify that $\Gamma_i = n_s u_B$ at the plasma boundaries. The calculation is straightforward, but the algebra becomes considerably more complicated. In all cases, the plasma parameters can be determined analogously to the procedure used for a plasma cylinder with $n \approx 0$ on the boundaries. The particle and power balance relations in this section can also be extended to describe magnetized plasmas. This has been done by Margot et al. (2001) for a finite length cylindrical high-density argon plasma at low pressures, using the cross-field ambipolar diffusion coefficient for $D_{\perp a}$ given in (5.4.17), showing a good fit to experimental data.

10.2.4 Neutral Radical Generation and Loss

For the feedstock gases used in processing applications, dissociation into neutral products occurs. In many applications, only the dissociated neutrals and the ions of the primary neutral species are important. Oxygen, which we consider as an example in the following sections, has this property in some ranges of pressure and power. To illustrate the dissociation process and its scaling with discharge parameters, we consider the simple model of a one-dimensional, uniform plasma slab in the low-pressure regime. The scaling in the high-pressure regime is left to Problem 10.10. For electrode separation l and area A, the particle balance yields, from (10.2.12) and (10.2.13) in slab geometry (with $R \gg l$)

$$\frac{K_{\mathrm{iz}}(\mathrm{T_e})}{u_{\mathrm{B}}(\mathrm{T_e})} = \frac{2h_l}{n_{\mathrm{g}}l} \tag{10.2.30}$$

which determines $\mathrm{T_e}$. The overall discharge power balance yields the ion density, from (10.2.15)

$$n_{\mathrm{is}} = h_l n_{\mathrm{i}} = \frac{P_{\mathrm{abs}}}{2e\mathcal{E}_{\mathrm{T}} u_{\mathrm{B}} A} \tag{10.2.31}$$

with the corresponding ion flux

$$\Gamma_{\mathrm{is}} = n_{\mathrm{is}} u_{\mathrm{B}} \tag{10.2.32}$$

Consider now the production of oxygen atoms by dissociation of the feedstock oxygen molecules

$$\mathrm{e} + \mathrm{O_2} \xrightarrow{K_{\mathrm{diss}}} 2\mathrm{O} + \mathrm{e}$$

where K_{diss} has an Arrhenius form:

$$K_{\mathrm{diss}} = K_{\mathrm{diss0}}\, \mathrm{e}^{-\mathcal{E}_{\mathrm{diss}}/\mathrm{T_e}} \tag{10.2.33}$$

The ionization rate constant can also be fitted to a similar form:

$$K_{\mathrm{iz}} = K_{\mathrm{iz0}}\, \mathrm{e}^{-\mathcal{E}_{\mathrm{iz}}/\mathrm{T_e}} \tag{10.2.34}$$

Raising (10.2.34) to the power $\mathcal{E}_{\mathrm{diss}}/\mathcal{E}_{\mathrm{iz}}$, we obtain

$$\mathrm{e}^{-\mathcal{E}_{\mathrm{diss}}/\mathrm{T_e}} = \left(\frac{K_{\mathrm{iz}}}{K_{\mathrm{iz0}}}\right)^{\mathcal{E}_{\mathrm{diss}}/\mathcal{E}_{\mathrm{iz}}} \tag{10.2.35}$$

Substituting this into (10.2.33), we obtain

$$K_{\mathrm{diss}} = C_0 K_{\mathrm{iz}}^{\mathcal{E}_{\mathrm{diss}}/\mathcal{E}_{\mathrm{iz}}} \tag{10.2.36}$$

where $C_0 = K_{\mathrm{diss0}}/K_{\mathrm{iz0}}^{\mathcal{E}_{\mathrm{diss}}/\mathcal{E}_{\mathrm{iz}}}$. Substituting (10.2.30) into (10.2.36) to eliminate the temperature-sensitive K_{iz}, we obtain

$$K_{\mathrm{diss}} = C_0 \left(\frac{2h_l u_{\mathrm{B}}}{n_{\mathrm{g}} l}\right)^{\mathcal{E}_{\mathrm{diss}}/\mathcal{E}_{\mathrm{iz}}} \tag{10.2.37}$$

In this form, it can be seen that K_{diss} depends only weakly on the temperature $\mathrm{T_e}$.

Assume now that the net flux of O atoms to the electrodes is zero (passivated electrodes) such that the only loss of O atoms is due to the vacuum pump. We will discuss loading effects due to etching and nonpassivated walls in Section 16.2. We also assume low pressures for which the O-atom diffusion rate is fast compared to the pumping rate. In this regime, the O-atom density is nearly uniform, $n_{OS} \approx n_O$ as described in Section 9.4 (see discussion following (9.4.15)). Assuming no other significant sources or sinks of O atoms, then the steady-state rate equation is

$$Al\frac{dn_O}{dt} = 2AlK_{diss}n_g n_i - S_p n_{OS} = 0 \tag{10.2.38}$$

where S_p (m^3/s) is the pumping speed, and we take $n_i \approx n_e$ (negligible negative ion density) as a simplifying approximation. Solving (10.2.38) for n_{OS}, we obtain

$$n_{OS} = \frac{2Aln_g}{S_p}K_{diss}n_i \tag{10.2.39}$$

Substituting for n_i given by (10.2.31) and K_{diss} given by (10.2.37) into (10.2.39), we obtain

$$n_{OS} = \frac{2P_{abs}}{e\mathcal{E}_T S_p}C_0\left(\frac{n_g l}{2h_l u_B}\right)^{1-\mathcal{E}_{diss}/\mathcal{E}_{iz}} \tag{10.2.40}$$

Typically, $\mathcal{E}_{diss}/\mathcal{E}_{iz} \approx 0.3$–$0.5$. The flux of O atoms incident on an electrode is then

$$\Gamma_{OS} = \frac{1}{4}n_{OS}\bar{v}_O \tag{10.2.41}$$

where $\bar{v}_O = (8kT_O/\pi M_O)^{1/2}$. Equations (10.2.32) and (10.2.41) give the ion flux at the plasma–sheath edge and the neutral atom flux at the surface, respectively. We see from (10.2.32) that Γ_i increases linearly with power and is almost independent of pressure. (There is a weak pressure dependence given by the variation of $\mathcal{E}_T u_B$ with pressure.) The neutral atom flux, from (10.2.41), also increases linearly with power and increases with the pressure ($\Gamma_O \propto p^{0.5-0.7}$). These scalings are important in determining etch and deposition rates, as discussed in Chapters 16 and 17.

10.3 Electronegative Plasma Equilibrium

The addition of a negative ion species greatly complicates the equilibrium plasma structure in a discharge. As shown in Figure 10.2, the plasma tends to stratify into an electronegative core and an electropositive edge (Tsendin, 1989). The stratification occurs because the ambipolar field required to confine the more energetic mobile electrons pushes the negative ions into the discharge center. Since the negative ions generally have a low temperature compared to the electrons, only a very small field is required to confine them to the core. The higher temperature electrons, in Boltzmann equilibrium with this field, have a nearly uniform density in the presence of the negative ions, but then form a more usual electropositive plasma in the edge regions, as shown in the figure.

The analysis of particle and energy balance in low-pressure electronegative discharges is difficult for the following reasons:

- An additional particle conservation equation is required for the negative ions.
- Negative ions are confined by the ambipolar potentials and are not lost to the walls, so various volume loss processes must be considered. These processes result in nonlinear equations for the particle balances.
- The Bohm velocity, which signals the end of the plasma and the beginning of the sheath, is modified by the presence of negative ions (see (6.4.7)).

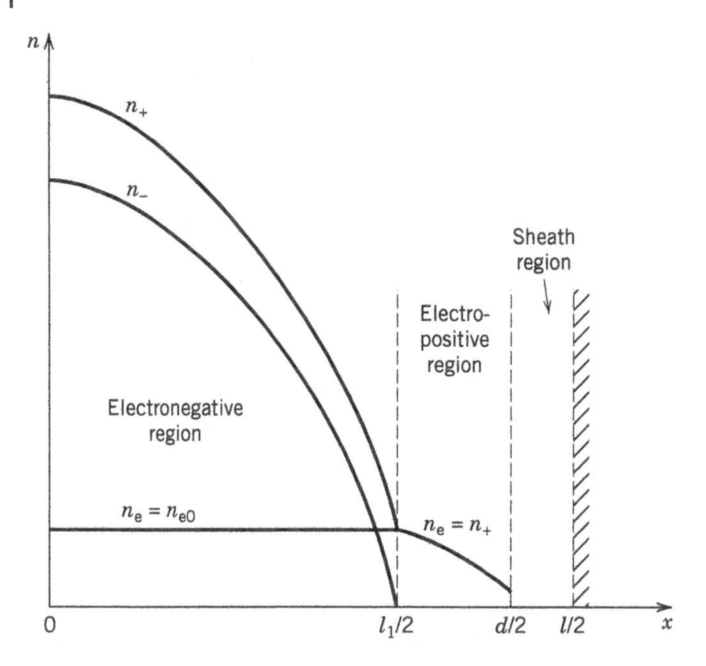

Figure 10.2 Positive ion, negative ion, and electron densities versus position for a plane-parallel electronegative discharge, showing the electronegative, electropositive, and sheath regions.

- At low pressures, different diffusion models may be required in the electronegative core and the electropositive edge.
- In general, a set of nonlinear diffusion equations for the various species must be solved simultaneously.

In spite of these complications, the near-constancy of n_e in the electronegative core over large parameter ranges allows simple approximate solutions to be obtained for the equilibria. These are described in Section 10.4. There is a trade-off between the more approximate analytic solutions, which expose the scaling of the plasma parameters with external parameters, and the more accurate numerical solutions. In this and the following section, we follow the treatment of Kaganovich and Tsendin (1993), Lichtenberg et al. (1994, 1997, 2000), Kouznetsov et al. (1996, 1999), Kolobov and Economou (1998), Berezhnoj et al. (2000), Franklin and Snell (1994, 2000a,b), Franklin (2001), and Kim et al. (2006a).

We will mainly consider the simplest case in which there is one positive and one negative ion species in addition to electrons, and one excited neutral detaching species for negative ions. However, the plasma chemistry can be quite complicated, and other species can play significant roles. We will return to this point, briefly, after treating the case of three charged species.

10.3.1 Differential Equations

As in electropositive plasmas, for each charged species, we can write a particle balance equation (2.3.7) and a drift–diffusion equation (5.1.3) for the flux Γ. A simplified set of particle balance equations is

$$\nabla \cdot \Gamma_+ = K_{iz} n_g n_e - K_{rec} n_+ n_- \tag{10.3.1a}$$

$$\nabla \cdot \Gamma_- = K_{att} n_g n_e - K_{rec} n_+ n_- - K_{det}\, n_* n_- \tag{10.3.1b}$$

$$\nabla \cdot \Gamma_e = (K_{iz} - K_{att})n_g n_e + K_{det} n_* n_- \qquad (10.3.1c)$$

$$\nabla \cdot \Gamma_* = K_{ex*} n_e n_g - K_{det} n_* n_- - K_{dex*} n_e n_* \qquad (10.3.1d)$$

The subscripts +, −, e denote positive ions, negative ions, and electrons, respectively, n_g is the neutral gas density, K_{iz} is the ionization rate constant, K_{rec} is the recombination rate constant, K_{att} is the dissociative attachment rate constant, n_* is the density of an excited neutral species, K_{det} is the rate constant for detachment of negative ions by collision with the excited neutrals, and K_{dex*} is the rate constant for electron impact de-excitation of the excited neutral species. We neglect the electron detachment term $-K_{edet} n_e n_-$ in (10.3.1b), although it can be important for some gases, e.g., O_2. Dissociative attachment is usually mainly from the ground vibrational molecular state, but it can also be mainly from vibrationally excited states, as has been measured for H_2; we assume the former here. The dominant excited species for detachment can be a dissociation product of the feedstock gas, such as O atoms for O_2 feedstock, but is commonly a metastable molecule or atom, e.g., $O_2^*(^1\Delta_g)$ (see Table 8.1). The relative importance of negative ion recombination versus detachment losses depends on the gas pressure and ratio of n_-/n_e. At low pressures and/or high n_-/n_e ratios, recombination losses exceed detachment losses, and (10.3.1a)–(10.3.1c) form an essentially complete set of particle balance relations (with $n_* \approx 0$). At high pressures and/or low n_-/n_e ratios, detachment losses can exceed recombination losses. In this case, the particle balance relation (10.3.1d) for n_* must be solved.

The drift–diffusion equations for the charged particles and Fick's law for the excited neutral species are

$$\Gamma_+ = -D_+ \nabla n_+ + n_+ \mu_+ E \qquad (10.3.2a)$$

$$\Gamma_- = -D_- \nabla n_- - n_- \mu_- E \qquad (10.3.2b)$$

$$\Gamma_e = -D_e \nabla n_e - n_e \mu_e E \qquad (10.3.2c)$$

$$\Gamma_* = -D_* \nabla n_* \qquad (10.3.2d)$$

where the Ds and μs are taken to be constants. As will be shown below, the negative ions in the core reduce the ambipolar electric fields there to low values. Consequently, except at very low pressures, ion drift velocities are small compared to ion thermal velocities, such that a constant diffusion coefficient model of the ion transport can be used. We also make the ambipolar assumption that the sum of the fluxes must balance

$$\Gamma_+ = \Gamma_- + \Gamma_e \qquad (10.3.3)$$

and we have the quasineutrality condition

$$n_+ = n_- + n_e \qquad (10.3.4)$$

Depending on plasma conditions, we can determine n_* from a simple model. In particular, if the wall losses dominate detachment and electron impact de-excitation losses for the excited species, then, inserting (10.3.2d) into (10.3.1d), we obtain the simple diffusion equation

$$-D_* \nabla^2 n_* = K_{ex*} n_e n_g \qquad (10.3.5)$$

For a uniform $n_e = \bar{n}_e$, (10.3.5) can be easily solved (see Section 5.2). The boundary condition for loss of n_* at the walls is

$$-D_* \nabla n_* = \frac{1}{4} \gamma_* n_{*s} \bar{v}_* \qquad (10.3.6)$$

where $\bar{v}_*$ is the mean speed of the excited species, and γ_* is a loss probability for the excited species on the wall (see (9.4.37)). Typically $\gamma_* = \gamma_q$, the wall quenching probability for de-excitation of the excited species. For a typical value of γ_*, e.g., 10^{-3} for metastable oxygen molecules, the solution for the density n_* is practically uniform (see the discussion following (9.4.15)). Assuming this and integrating (10.3.5) over the volume and using the boundary condition (10.3.6), we find

$$n_* \approx \frac{4K_{ex*}\mathcal{V}}{\gamma_*\bar{v}_*A} n_g \bar{n}_e \tag{10.3.7}$$

where $\mathcal{V}$ is the discharge volume, A is the surface area for loss, and $\bar{n}_e \approx n_{e0}$ is the mean electron density (Problem 10.4). The effects of the volume loss terms in (10.3.1d) are explored in Problems 10.5 and 10.6; see also Franklin (2001).

Assuming that wall losses dominate, we substitute (10.3.7) into the negative ion balance (10.3.1b) to obtain

$$\nabla \cdot \Gamma_- = K_{att} n_g n_e - K_{rec} n_+ n_- - K_* n_g \bar{n}_e n_- \tag{10.3.8}$$

with K_* a third-order rate coefficient

$$K_* = \frac{4K_{det}K_{ex*}\mathcal{V}}{\gamma_*\bar{v}_*A} \tag{10.3.9}$$

Equation (10.3.8) then replaces (10.3.1b) in determining the discharge equilibrium.

The equilibrium naturally divides into two regimes depending on whether recombination or detachment dominates the negative ion loss. From (10.3.8), we see that recombination dominates detachment for

$$n_+ > \frac{K_* n_g}{K_{rec}} \bar{n}_e \tag{10.3.10}$$

Introducing the *electronegativity* $\alpha = n_-/n_e$, then for large α, we will find that $n_e \approx n_{e0} =$ const and $n_+ \approx n_-$, as will be shown in the following subsection. Then, (10.3.10) yields the condition for recombination-dominated negative ion loss

$$\alpha > \frac{K_* n_g}{K_{rec}} \tag{10.3.11}$$

Hence, highly electronegative discharges at low pressures are recombination-dominated, while moderately electronegative discharges at higher pressures are detachment-dominated. The transition between these regimes depends on the generation rate of excited species and their surface loss probability.

For both regimes, since the electrons are very mobile, we can eliminate the electric field by use of a Boltzmann assumption for the electrons. Setting $\Gamma_e \approx 0$ in (10.3.2c) and using $D_e = \mu_e \mathrm{T}_e$ from the Einstein relation (5.1.9), we obtain

$$\mathrm{T}_e \nabla n_e + n_e E = 0 \tag{10.3.12a}$$

yielding

$$n_e = n_{e0}\, \mathrm{e}^{\Phi/\mathrm{T}_e} \tag{10.3.12b}$$

with n_{e0} the central electron density and Φ the potential. On the other hand, the negative ions are not necessarily in Boltzmann equilibrium with the potential (see (10.3.36)). To treat this general case, we combine the particle balance and drift–diffusion equations for positive and negative ions to obtain a pair of differential equations which in plane-parallel geometry are

$$\frac{\mathrm{d}}{\mathrm{d}x}\left(-D_+\frac{\mathrm{d}n_+}{\mathrm{d}x} + n_+\mu_+ E\right) = K_{iz} n_g n_e - K_{rec} n_+ n_- \tag{10.3.13}$$

and

$$\frac{\mathrm{d}}{\mathrm{d}x}\left(-D_-\frac{\mathrm{d}n_-}{\mathrm{d}x} - n_-\mu_-E\right) = K_{\mathrm{att}}n_\mathrm{g}n_\mathrm{e} - K_{\mathrm{rec}}n_+n_- - K_*n_\mathrm{g}\overline{n}_\mathrm{e}n_- \tag{10.3.14}$$

The electric field and the positive ion density may be eliminated from (10.3.13) and (10.3.14) using the Boltzmann relation (10.3.12a) for electrons and the quasineutrality condition (10.3.4). Making these substitutions, and taking $\mu_- = \mu_+$ and $D_- = D_+$ ($\mathrm{T}_- = \mathrm{T}_+ \equiv \mathrm{T}_\mathrm{i}$, a common ion temperature) for simplicity, we obtain

$$-\frac{\mathrm{d}}{\mathrm{d}x}\left(D_+\frac{\mathrm{d}}{\mathrm{d}x}(n_- + n_\mathrm{e}) + \mu_+\mathrm{T}_\mathrm{e}\frac{n_- + n_\mathrm{e}}{n_\mathrm{e}}\frac{\mathrm{d}n_\mathrm{e}}{\mathrm{d}x}\right) = K_{\mathrm{iz}}n_\mathrm{g}n_\mathrm{e} - K_{\mathrm{rec}}(n_- + n_\mathrm{e})n_- \tag{10.3.15}$$

and

$$\frac{\mathrm{d}}{\mathrm{d}x}\left(-D_+\frac{\mathrm{d}n_-}{\mathrm{d}x} + \mu_+\mathrm{T}_\mathrm{e}\frac{n_-}{n_\mathrm{e}}\frac{\mathrm{d}n_\mathrm{e}}{\mathrm{d}x}\right) = K_{\mathrm{att}}n_\mathrm{g}n_\mathrm{e} - K_{\mathrm{rec}}(n_- + n_\mathrm{e})n_- - K_*n_\mathrm{g}\overline{n}_\mathrm{e}n_- \tag{10.3.16}$$

Equations (10.3.15) and (10.3.16) can be solved simultaneously, together with the appropriate boundary conditions, to obtain the density profiles.

10.3.2 Boltzmann Equilibrium for Negative Ions

If we make the more restrictive assumption that the negative ion species is also in Boltzmann equilibrium, then setting $\Gamma_- \approx 0$ in (10.3.2b) and using $D_- = \mu_-\mathrm{T}_\mathrm{i}$ from (5.1.9), we obtain

$$\mathrm{T}_\mathrm{i}\nabla n_- + n_-E = 0 \tag{10.3.17a}$$

yielding

$$n_- = n_{-0}\,\mathrm{e}^{\Phi/\mathrm{T}_\mathrm{i}} \tag{10.3.17b}$$

Eliminating E from (10.3.12a) and (10.3.17a), we obtain

$$\frac{\nabla n_-}{n_-} = \gamma\frac{\nabla n_\mathrm{e}}{n_\mathrm{e}} \tag{10.3.18}$$

where $\gamma = \mathrm{T}_\mathrm{e}/\mathrm{T}_\mathrm{i}$. Using (10.3.18) together with

$$\nabla n_+ = \nabla n_- + \nabla n_\mathrm{e}$$

obtained from quasineutrality (10.3.4), we find

$$\nabla n_\mathrm{e} = \frac{1}{1+\gamma\alpha}\nabla n_+ \qquad \nabla n_- = \frac{\gamma\alpha}{1+\gamma\alpha}\nabla n_+ \tag{10.3.19}$$

We now show that the positive ion flux can be written in the form of Fick's law

$$\Gamma_+ = -D_{\mathrm{a}+}\nabla n_+ \tag{10.3.20}$$

where $D_{\mathrm{a}+}$ depends on the electronegativity α. Substituting $n_+ = n_- + n_\mathrm{e}$ and $\mu_+ = D_+/\mathrm{T}_\mathrm{i}$ into the second term in (10.3.2a), we obtain

$$\Gamma_+ = -D_+(\nabla n_+ - n_\mathrm{e}E/\mathrm{T}_\mathrm{i} - n_-E/\mathrm{T}_\mathrm{i})$$

Substituting $n_\mathrm{e}E$ from (10.3.12a) and n_-E from (10.3.17a) into the preceding equation, and using the gradients (10.3.19), we obtain, analogous to electropositive plasmas, (10.3.20) with an ambipolar diffusion coefficient (see Problem 10.7)

$$D_{\mathrm{a}+} = D_+\frac{1+\gamma+2\gamma\alpha}{1+\gamma\alpha} \tag{10.3.21}$$

Thompson (1959) gives a form similar to (10.3.21), but including corrections of order $\alpha\mu_+/\mu_e$, which are much less than unity except at very high α. The variation of D_{a+} with α is easily seen from (10.3.21). For $\alpha \gg 1$, γ cancels out such that $D_{a+} \approx 2D_+$. When α decreases below unity but $\gamma\alpha \gg 1$, then $D_{a+} \approx D_+/\alpha$ such that D_{a+} decreases with increasing α. For $\gamma\alpha < 1$, we find $D_{a+} \approx \gamma D_+ \equiv D_a$, the usual ambipolar diffusion coefficient without negative ions. For plasmas in which $\alpha \gg 1$ in the center of the discharge, the entire transition region takes place over a small range of $1/\gamma < \alpha < 1$ near the edge of the electronegative region, such that the simpler value of

$$D_{a+} = 2D_+ \tag{10.3.22}$$

holds over most of the electronegative core.

Since $2D_+ \ll \gamma D_+$, the presence of negative ions greatly increases the plasma confinement. The discontinuous slope of dn_+/dx near the boundary between electronegative and electropositive regions shown in Figure 10.2 is due to the sharp change in D_{a+} from $2D_+$ to γD_+. Because the ion flux near the boundary is the product of the diffusion coefficient and density gradient, a sharp change in diffusion coefficient for the same flux results in a sharp change in density gradient.

Although D_{a+} is a function of α given by (10.3.21), α is implicitly given as a function of n_+ through the Boltzmann relations. Eliminating Φ from (10.3.12b) and (10.3.17b) yields

$$n_e = n_{e0}\left(\frac{n_-}{n_{-0}}\right)^{1/\gamma} \tag{10.3.23}$$

Inserting this into quasineutrality (10.3.4) yields

$$n_+ = n_- + n_{e0}\left(\frac{n_-}{n_{-0}}\right)^{1/\gamma} \tag{10.3.24}$$

which implicitly gives n_- as a function of n_+ (and the central densities n_{e0} and n_{-0}). Similarly, solving (10.3.23) for n_- and inserting this into (10.3.4) yields

$$n_+ = n_{-0}\left(\frac{n_e}{n_{e0}}\right)^{\gamma} + n_e \tag{10.3.25}$$

which implicitly gives n_e as a function of n_+. The electronegativity $\alpha = n_-/n_e$ therefore also implicitly depends on n_+. Inserting (10.3.20) into (10.3.1a) and using (10.3.24) for n_- and (10.3.25) for n_e, we obtain a single nonlinear diffusion equation for n_+

$$-\frac{d}{dx}\left(D_{a+}\frac{dn_+}{dx}\right) = K_{iz}n_g n_e - K_{rec}n_+n_- \tag{10.3.26}$$

where D_{a+}, n_e, and n_- are known functions of n_+ that depend on n_{e0} and n_{-0}. Although (10.3.26) does not appear to depend on the detachment process, n_{e0} and n_{-0} depend on the detachment, as will be seen below.

Equation (10.3.26) has as a boundary condition at the sheath edge $x = d/2$ that the ion flow cannot exceed the Bohm velocity. Stating this condition as an equality, it becomes the Bohm flux condition

$$-D_{a+}\left.\frac{dn_+}{dx}\right|_{x=d/2} = n_+(d/2)\,u_{B\alpha} \tag{10.3.27}$$

Since negative ions may be present when (10.3.27) is satisfied, the Bohm velocity has the general form from (6.4.7)

$$u_s \equiv u_{B\alpha} = \left[\frac{eT_e(1+\alpha_s)}{M_+(1+\gamma\alpha_s)}\right]^{1/2} \tag{10.3.28}$$

where $\alpha_s = \alpha(d/2) = n_-(d/2)/n_e(d/2)$. The generalized Bohm velocity $u_{B\alpha}$ reduces to the usual expression $u_B = (e\mathrm{T}_e/M_+)^{1/2}$ when $\alpha_s = 0$. For $\alpha_s > 1/\gamma$, the negative ion density at the sheath edge significantly reduces the Bohm velocity.

In the electronegative core of the discharge, where a significant negative ion density exists, (10.3.26) can be simplified. Since $\gamma \gg 1$, (10.3.23) implies that $n_e \approx n_{e0}$ in the core, as shown in Figure 10.2. Hence, we can write $n_e = n_{e0}$ and $n_- = n_+ - n_{e0}$. Therefore the diffusion equation in the core can be written as

$$-\frac{d}{dx}\left(D_{a+}\frac{dn_+}{dx}\right) = K_{iz}n_g n_{e0} - K_{rec}n_+(n_+ - n_{e0}) \tag{10.3.29}$$

We also have $D_{a+} \approx 2D_+ = \text{const}$ for $\alpha \gg 1$. Then, except in the transition layer of a highly electronegative core, (10.3.29) reduces to a relatively simple diffusion equation with a constant diffusion coefficient.

In the electropositive edge regions of a low-pressure stratified discharge, the diffusion equation (10.3.26) may not be valid. This is because the constant diffusion coefficient model applies only at high pressures. At low pressures, a variable mobility model must be used (see Section 5.3 and Appendix C). Such a model is described in Section 10.4.

10.3.3 Conservation Equations

Equation (10.3.26) can be characterized by three parameters: $\alpha_0 = n_{-0}/n_{e0}$ (the ratio of n_- to n_e at the plasma center), n_{e0}, and T_e. We can determine these by solving (10.3.26) together with two particle conservation equations, which are the integrated forms of (10.3.26) and (10.3.14), and an energy conservation equation. These are positive ion particle balance

$$-D_{a+}\frac{dn_+}{dx}\bigg|_{x=d/2} = \int_0^{d/2} K_{iz}n_g n_e\, dx - \int_0^{d/2} K_{rec}n_+n_-\, dx \tag{10.3.30}$$

negative ion particle balance (negligible negative ion wall flux)

$$\int_0^{d/2} K_{att}n_g n_e\, dx - \int_0^{d/2} K_{rec}n_+n_-\, dx - \int_0^{d/2} K_* n_g \overline{n}_e n_-\, dx = 0 \tag{10.3.31}$$

and energy balance for the discharge

$$S_{abs} = 2e\mathcal{E}_c \int_0^{d/2} K_{iz}n_g n_e\, dx + 2e(\mathcal{E}_e + \mathcal{E}_i)n_+(d/2)u_{B\alpha} \tag{10.3.32}$$

Here, $\mathcal{E}_c(\mathrm{T}_e)$ is the collisional energy lost per electron–positive ion pair created, and $\mathcal{E}_e + \mathcal{E}_i$ is the kinetic energy lost to the wall per electron–ion pair lost to the wall. Given the neutral density n_g and the power per unit area deposited in the discharge, S_{abs}, the three equations can be simultaneously solved for the three unknowns T_e, α_0, and n_{e0}, provided the plasma half-width $d/2$ is known. A common assumption (sometimes not satisfied in capacitive rf discharges) is that the sheath width $s \ll d/2$, such that $d = l$, the gap width.

At relatively low pressure and low electronegativity, there may be a significant electropositive edge region in which the positive ion mobility is not constant, such that the basic equations (10.3.13) and (10.3.14) have nonconstant coefficients. We examine this in Section 10.4. Also, in regions where the negative ion density is very small, large electric fields can sweep them into the electronegative core, creating additional friction forces (Deutsch and Räuchle, 1992).

10.3.4 Validity of Reduced Equations

We examine the condition for validity of the Boltzmann equilibrium for negative ions, from which we have derived a single ambipolar diffusion equation (10.3.26) for the positive ions. From (10.3.2b), we have

$$\Gamma_- = -D_- \frac{dn_-}{dx} - n_- \mu_- E \tag{10.3.33}$$

with the condition for Boltzmann equilibrium being that

$$\eta_B(x) = \left| \Gamma_- \Big/ D_- \frac{dn_-}{dx} \right| \ll 1 \tag{10.3.34}$$

everywhere. Integrating the negative ion balance equation (10.3.8) from 0 to x, Γ_- can be written as

$$\Gamma_- = \int_0^x K_{att} n_g n_e \, dx' - \int_0^x K_{rec} n_+ n_- \, dx' - \int_0^x K_* n_g \bar{n}_e n_- \, dx' \tag{10.3.35}$$

If we have profiles for n_e, n_-, and n_+, (10.3.34) can be explicitly evaluated. We have obtained the profiles in Section 10.4, finding $n_e \approx n_{e0}$, a parabolic solution for $n_-(x)$, and $n_+ = n_{e0} + n_-(x)$, with parabolic scale length $l_1/2$ (see Figure 10.2). For these profiles, (10.3.34) has its maximum value at $x = 0$, giving the condition for Boltzmann equilibrium of negative ions

$$\eta_B \equiv \frac{n_{e0} l_1^2}{8D_-} \left(\frac{7}{15} K_{rec} \alpha_0 + \frac{1}{3} K_* n_g \right) < 1 \tag{10.3.36}$$

where we use a simple inequality since we have taken the maximum value of the ratio.

If (10.3.36) is not satisfied, the negative ions are not in Boltzmann equilibrium and (10.3.26) is not valid, but the electron profile may still be quite flat, which also allows the reduction to a single differential equation for the profile. Adding (10.3.15) and (10.3.16) and dropping small terms, we obtain

$$-\frac{d}{dx}\left(2D_+ \frac{dn_+}{dx} + \gamma D_+ \frac{dn_e}{dx} \right) = (K_{iz} + K_{att}) n_g n_e - 2K_{rec} n_+ (n_+ - n_e) - K_* n_g \bar{n}_e (n_+ - n_e) \tag{10.3.37}$$

where we have used the Einstein relation to write $\mu_+ T_e = \gamma D_+$. Equation (10.3.37) is still a function of two variables n_e and n_+, so that a known form of n_e is required to obtain a general solution. However, because $T_e \gg T_i$, there is a large parameter range in which (10.3.36) is not satisfied but n_e is still essentially flat, as determined by the Boltzmann relation. Assuming such a flat solution with $\gamma dn_e/dx \ll dn_+/dx$ in (10.3.37), we obtain

$$2D_+ \frac{d^2 n_+}{dx^2} + (K_{iz} + K_{att}) n_g n_{e0} - 2K_{rec} n_+ (n_+ - n_{e0}) - K_* n_g n_{e0} (n_+ - n_{e0}) = 0 \tag{10.3.38}$$

We would expect that, with increasing η_B, at sufficiently high pressure and α_0, the ionization and attachment are increasingly balanced locally by the recombination and detachment, leading to a relatively flat positive ion profile. In this regime, the LHS of (10.3.37) is a perturbation to the RHS. The RHS by itself gives, for large α

$$n_e(x) = \frac{2K_{rec} n_+^2(x) + K_* n_g \bar{n}_e n_+(x)}{(K_{iz} + K_{att}) n_g} \tag{10.3.39}$$

Dropping the excitation term for clarity, we find $n_e \propto n_+^2$. This is quite different from the regime $\eta_B < 1$ for which $n_e \approx n_{e0}$, a constant, with n_+ varying with position. However, because the LHS of (10.3.37) is small, both n_e and n_+ are nearly constant over the central part of the discharge.

10.4 Approximate Electronegative Equilibria

The stratification of the discharge into a parabolic electronegative core and an electropositive edge at moderate electronegativity α_0 allows approximate equilibrium solutions to be obtained. However, other factors arise in which different solutions hold. At higher α_0, the positive ion drift velocity near the edge of the core can become equal to the local ion sound velocity. If this occurs, local field build-up rather abruptly cuts off the negative ions, initiating an electropositive edge. A further increase in α_0 leads to the disappearance of the electropositive edge altogether.

At higher gas densities, the integrated loss of positive ions by recombination becomes large compared to the diffusion loss. One consequence is that the negative ions are no longer in Boltzmann equilibrium, which leads to a flattening of the ion profile in the electronegative core region. A qualitative criterion for the transition to a flattened core is that (10.3.36) is not satisfied (Lichtenberg et al., 1997). As α_0 is increased at a fixed higher pressure (by decreasing n_{e0}), the flattened profile goes through the same sequence of variations as for the parabolic profile at lower pressure. A classification of these regimes in terms of the two important parameters $n_{e0}d$ and $n_g d$ is given in Lichtenberg et al. (2000).

10.4.1 Global Models

The complexity of electronegative equilibria motivates us to consider *global models* in which the plasma spatial variations are assumed, rather than calculated. The simplest model of this type, also called a *zero-dimensional model* because all spatial variations are ignored, is useful to provide a first estimate of the plasma parameters and their scaling in complicated discharges, and to study the effects of a large number of reactions, and the effects of negative ions, and more than one positive ion species, which occur in real gases (e.g., Lee et al., 1994, 1997a; Lee and Lieberman, 1995; Meeks and Shon, 1995; Stoffels et al., 1995; Kimura and Ohe, 1999; Kimura et al., 2001; Kim et al., 2006a; Monahan and Turner, 2008). A review of capabilities and limitations is given in Hurlbatt et al. (2017). Although global models are usually employed to treat multi-species systems, we confine ourselves here to the principal reactions and a single positive ion species, as in Section 10.3.

10.4.1.1 Particle and Energy Balance

To put our equations in the form used in Section 10.2 to model uniform density electropositive plasmas, but allow for profiles characteristic of electronegative plasmas, we take volume-averaged quantities. Using (10.3.30)–(10.3.32), the equations for conservation of positive ions, negative ions and energy within the volume are written in the form

$$K_{iz} n_{e0} n_g \mathcal{V} - K_{rec} \overline{n}_+ \overline{n}_- \mathcal{V}_{rec} - \Gamma_{+s} A = 0 \tag{10.4.1}$$

$$K_{att} n_{e0} n_g \mathcal{V} - K_{rec} \overline{n}_+ \overline{n}_- \mathcal{V}_{rec} - K_* n_g n_{e0} \overline{n}_- \mathcal{V} = 0 \tag{10.4.2}$$

$$P_{abs} = e\mathcal{E}_c K_{iz} n_{e0} n_g \mathcal{V} + \Gamma_{+s} A e(\mathcal{E}_e + \mathcal{E}_i) \tag{10.4.3}$$

Here, $\mathcal{V}$ and A are the volume and surface area of the plasma (sheath thicknesses are assumed to be small), n_{e0} is the assumed uniform electron density, and $\overline{n}_+$ and $\overline{n}_-$ are volume averaged positive and negative ion densities. K_* is the third-order rate constant (10.3.9) for detachment losses, which assumes that wall losses dominate volume losses for the excited species. The quantities $\mathcal{V}_{\text{rec}}$ and $\Gamma_{+\text{s}}$ are the effective volume for recombination and the average positive ion flux normal to the surface, respectively, which must be defined for a given problem. The ability to approximate $\mathcal{V}_{\text{rec}}$ and $\Gamma_{+\text{s}}$ from the plasma parameters and dimensions is the essence of a global model.

Equations (10.4.1)–(10.4.3) along with the quasineutrality condition $\overline{n}_+ = \overline{n}_- + n_{\text{e0}}$ are four equations that must be simultaneously solved to determine $\overline{n}_+$, $\overline{n}_-$, n_{e0}, and T_e, for the specified neutral density n_g, total absorbed power P_{abs} and geometry ($\mathcal{V}$, A, $\mathcal{V}_{\text{rec}}$, and $\Gamma_{+\text{s}}$). The general solution must be found numerically. The electron temperature is determined from positive ion balance, as for the electropositive case. However, (10.4.1) depends on both n_{e0} and $\overline{n}_+$. Therefore, T_e is not a function of $n_\text{g} d_{\text{eff}}$ alone, independent of plasma density, as it is for the electropositive case. Now T_e depends as well on the density, or, equivalently, on the discharge power per unit area, P_{abs}/A. We illustrate the solution procedure with two examples.

Example 10.4 For $\alpha_0 \gtrsim 3$, we estimate $\Gamma_{+\text{s}}$ using the simplest form of the diffusion equation (10.3.20) with $D_{\text{a}+} = 2D_+$ from (10.3.22),

$$\Gamma_{+\text{s}} \approx -2D_+ \nabla n_{+\text{s}} \approx \frac{2D_+ \overline{n}_+}{\Lambda_0} \tag{10.4.4}$$

where Λ_0 is an effective diffusion length in the bulk plasma. For a slab geometry, we shall see in the following subsection that when surface (diffusive) losses dominate volume losses, there is an approximate parabolic solution (Lee et al., 1997a)

$$n_+(x) \approx n_-(x) = n_{+0}\left(1 - \frac{4x^2}{d^2}\right) \tag{10.4.5}$$

with d the bulk plasma width. In this case, $\overline{n}_+ = \frac{2}{3} n_{+0}$ and, from (10.4.4), $\Lambda_0 = d/6$. Furthermore, averaging $n_+ n_-$ over the profile, we find $\mathcal{V}_{\text{rec}} = \frac{6}{5}\mathcal{V}$.

Example 10.5 For $\alpha_0 \gg 1$ and when volume losses dominate surface losses, the profile becomes nearly uniform, such that $\mathcal{V}_{\text{rec}} = \mathcal{V}$. Even though the edge gradient steepens, the recombination loss dominates the diffusion loss, such that an approximate solution can be obtained by setting $\Gamma_{+\text{s}} \approx 0$.

An important scaling follows from the negative ion balance. Using quasineutrality (10.3.4) to eliminate $\overline{n}_+$ from (10.4.2) and introducing the average electronegativity $\overline{\alpha} = \overline{n}_-/n_{\text{e0}}$, we can solve for $\overline{\alpha}$ to obtain

$$\overline{\alpha} = \frac{K_{\text{att}} n_\text{g} \mathcal{V}}{K_{\text{rec}} \overline{n}_- \mathcal{V}_{\text{rec}}} - \frac{K_* n_\text{g} \mathcal{V}}{K_{\text{rec}} \mathcal{V}_{\text{rec}}} - 1 \tag{10.4.6}$$

There are two limiting regimes.

(1) For recombination-dominated negative ion loss with $K_* n_\text{g} \mathcal{V} \ll \overline{\alpha} K_{\text{rec}} \mathcal{V}_{\text{rec}}$, (10.4.6) can be solved to obtain

$$\overline{\alpha} = \frac{K_{\text{att}} n_\text{g} \mathcal{V}}{K_{\text{rec}} \overline{n}_+ \mathcal{V}_{\text{rec}}} \tag{10.4.7a}$$

For large $\overline{\alpha}$, substituting $\overline{n}_+ \approx \overline{\alpha} n_{\text{e0}}$ into (10.4.7a), we obtain

$$\overline{\alpha} = \left(\frac{K_{\text{att}} n_\text{g} \mathcal{V}}{K_{\text{rec}} n_{\text{e0}} \mathcal{V}_{\text{rec}}}\right)^{1/2} \tag{10.4.7b}$$

(2) For detachment-dominated loss with $\overline{\alpha} K_{\text{rec}} \mathcal{V}_{\text{rec}} \ll K_* n_{\text{g}} \mathcal{V}$, (10.4.6) yields the simple result

$$\overline{n}_- = \frac{K_{\text{att}}}{K_*} \tag{10.4.8}$$

which is independent of n_{g} (gas pressure). For a given gas, (10.4.8) sets an upper limit on $\overline{n}_-$ as n_{e0} increases.

To get a further feeling for the behavior of the global model, we examine two limiting cases for recombination-dominated negative ion loss with $\overline{\alpha} \gg 1$.

(a) *Surface losses dominate.* For the case of Example 10.4, when the ratio of surface to volume loss for the positive ions is large, (10.4.1) gives

$$K_{\text{iz}} n_{\text{e0}} n_{\text{g}} \mathcal{V} = \Gamma_{+\text{s}} S \tag{10.4.9}$$

Inserting (10.4.9) into the power balance (10.4.3) to eliminate K_{iz} and using (10.4.4) for $\Gamma_{+\text{s}}$, we obtain $\overline{n}_+$ as a function of n_{g} and P_{abs}. Then, (10.4.7) determines $\overline{\alpha}$, with $n_{\text{e0}} = \overline{n}_+/(1+\overline{\alpha})$ and $\overline{n}_- = \overline{\alpha} n_{\text{e0}}$. Solving (10.4.9) then yields T_{e}.

(b) *Volume loss dominates.* For Example 10.5, when the ratio of volume to surface loss for the positive ions is large, dropping $\Gamma_{+\text{s}} A$ in (10.4.1) and subtracting (10.4.2), we obtain

$$K_{\text{iz}} = K_{\text{att}} \tag{10.4.10}$$

which determines T_{e}. Setting $\mathcal{V}_{\text{rec}} = \mathcal{V}$ and eliminating the ionization term from the power balance (10.4.3) and the positive ion balance (10.4.1) gives

$$P_{\text{abs}} \approx \mathcal{V} \mathcal{E}_{\text{c}} K_{\text{rec}} \overline{n}_+^2 \tag{10.4.11}$$

which determines $\overline{n}_+ \approx \overline{n}_-$. Since (10.4.11) determines n_+, we substitute this into (10.4.7a) to determine

$$n_{\text{e0}} \approx \frac{K_{\text{rec}} \overline{n}_+^2}{K_{\text{att}} n_{\text{g}}} \tag{10.4.12}$$

Summarizing the conditions for (10.4.7) and (10.4.10)–(10.4.12), this solution holds for $K_* n_{\text{g}} \ll \overline{\alpha} K_{\text{rec}}$, $n_{\text{e0}} \ll \overline{n}_+$, and $\Gamma_{+\text{s}} A \ll K_{\text{rec}} \overline{n}_+^2 \mathcal{V}$.

10.4.1.2 Ion Loss Flux $\Gamma_{+\text{s}}$ and Recombination Volume $\mathcal{V}_{\text{rec}}$

A model for the ion loss flux and recombination volume for electronegative discharges over the entire range of density profile changes was developed by Kim et al. (2006a). The loss flux is written as

$$\Gamma_{+\text{s}} = h_l n_{+0} u_{\text{B}} \tag{10.4.13}$$

with h_l the heuristic sum of three h-factors

$$h_l \approx h_{\text{a}} + h_{\text{b}} + h_{\text{c}} \tag{10.4.14}$$

These "stitch together" the h-factors from (a) a parabolic model with an electropositive edge, (b) a one-region parabolic model, and (c) a one-region flat-topped model, respectively. They also incorporate possibly different properties (temperature, mass, and mean free path) of the positive and negative ions. For the parabolic model with electropositive edge

$$h_{\text{a}} = \frac{0.86}{(3 + \eta d/2\lambda_+)^{1/2}} \frac{1}{1+\alpha_0} \equiv h_{l0} \frac{1}{1+\alpha_0} \tag{10.4.15}$$

with $\eta = 2T_+/(T_+ + T_-)$, d the bulk plasma width, and λ_+ the positive ion mean free path. As electronegativity $\alpha_0 \to 0$, h_a approaches the pressure-dependent electropositive result h_{l0} given in (10.2.1). For the one-region parabolic model

$$h_b = \frac{(2\pi)^{1/2}}{(2\pi\gamma_-)^{1/2} + \gamma_+^{1/2}\eta d/2\lambda_+}\frac{\alpha_0}{1+\alpha_0} \equiv h_{l\infty}\frac{\alpha_0}{1+\alpha_0} \tag{10.4.16}$$

with $\gamma_+ = T_e/T_+$ and $\gamma_- = T_e/T_-$. As $\alpha_0 \to \infty$, h_b gives a result $h_{l\infty}$ that is again only pressure-dependent. For the one-region flat-topped model

$$h_c = \frac{1}{\gamma_-^{1/2} + \gamma_+^{1/2} n_*^{1/2} n_{+0}/n_{-0}^{3/2}} \approx \frac{1}{\gamma_-^{1/2} + \gamma_+^{1/2} n_*^{1/2}/n_{-0}^{1/2}} \tag{10.4.17}$$

where

$$n_* = \frac{15}{64}\frac{\eta^2 \overline{v}_+}{K_{rec}\lambda_+}$$

and $\overline{v}_+ = (8eT_+/\pi M_+)^{1/2}$ is the mean positive ion speed. For large α_0, h_c depends on the product $n_{-0}K_{rec}$, as well as the pressure.

The recombination volume $\mathcal{V}_{rec}$ is defined by the width l_1 of the electronegative core. For a one-dimensional parabolic model with an electropositive edge, $\mathcal{V}_{rec} = l_1 A/2$, with A the sum of the two electrode areas. Kim et al. (2006a) give a heuristic expression

$$\frac{l_1}{d} \approx 1 - \frac{1}{(\beta_1^{-3} + \beta_2^{-3})^{1/3}}, \qquad 0 < \alpha_0 R_\lambda < 1 \tag{10.4.18}$$

where

$$\beta_1 = 1 - \left(\frac{\alpha_0 R_\lambda}{h_{eff}}\right)^{1/2}, \qquad \beta_2 = \frac{h_{eff} - \alpha_0 R_\lambda}{R_{rec} F(\alpha_0)} \tag{10.4.19}$$

Here,

$$R_\lambda = \left(\frac{2\pi}{\gamma_+}\right)^{1/2}\frac{2\lambda_+}{\eta d}, \qquad R_{rec} = \frac{K_{rec} n_{e0} d}{2u_B}$$
$$h_{eff} = h_{l0} + (1 - h_{l0})\alpha_0 R_\lambda, \qquad F(\alpha_0) = \frac{8}{15}\alpha_0^2 + \frac{2}{3}\alpha_0 \tag{10.4.20}$$

As can be seen, the factors β_1 and β_2 are "stitched together" in (10.4.18) and individually give $1 - l_1/d$ for $l_1 \ll d$ and $d - l_1 \ll d$, respectively. Note that for $\alpha_0 R_\lambda \geq 1$, there is no electropositive edge and $l_1 = d$.

This model was applied to both steady-state and pulsed oxygen discharges in a finite (radius R, height L) cylindrical geometry (Kim et al., 2006a). Comparisons were made to steady-state experimental measurements of n_{e0} and T_e in an inductive discharge with $R = 8$ cm and $L = 7.5$ cm, over the pressure range of 3–70 mTorr, and for absorbed powers of 120 and 180 W. The model and measured T_es were in good agreement, except at the lowest pressures, where the model was about 25% higher. The n_{e0}s agreed at the lower pressures, but the model was 50% lower at the higher pressures. This latter effect was attributed to the enhanced T_e near the inductive coil, with T_e in the bulk plasma falling to lower values.

Monahan and Turner (2008) compared the model to 1D PIC simulations in argon and argon/oxygen discharges over the pressure range of 1–100 mTorr (see Section 10.5). They found better agreement using a heuristic sum-of-squares, $h_l^2 = h_a^2 + h_b^2 + h_c^2$, rather than the linear sum of (10.4.14). They also point out the important influence of the electron energy distribution function (EEDF) on the comparisons. A significantly better agreement was found by using the

EEDF extracted from the simulations in the model, rather than using an assumed Maxwellian distribution.

Using the results of fluid simulations as a guide, Chabert (2016) determined

$$h_a = 0.86\left(3 + \frac{d}{2\lambda_+} + \frac{1}{5}\frac{(1+\alpha_0)^{1/2}}{\gamma_+}\frac{d^2}{\lambda_+^2}\right)^{-1/2}\left(\frac{\gamma_- - 1}{\gamma_-(1+\alpha_0)^2} + \frac{1}{\gamma_-}\right)^{1/2} \tag{10.4.21}$$

when the negative ions are in Boltzmann equilibrium and recombination losses are small. This expression incorporates both the lower pressure, constant mean free path (variable mobility) and higher pressure, constant collision frequency regimes, as in (5.3.16), and it approximates well the lower-pressure regime.

Global models can be criticized because the profiles and fluxes of the charged particles are not calculated from first principles. For example, the approximation of (10.4.4) is good only for $\alpha \gg 1$, as seen from (10.3.21). For $\alpha_0 \lesssim 1$, the negative ions are localized in the center of the discharge, such that $D_{a+} \approx D_a$ over most of the discharge region, and a parabolic model clearly cannot be used to determine n_-, as in (10.4.5). A nonuniform description based on a solution to the appropriate diffusion equation is required. It is clearly desirable to calculate the profiles and fluxes from the basic equations or to determine them from simulations and/or measurements. However, these can be burdensome for multiple ion species and realistic processing gas mixtures and (two- and three-dimensional) discharge geometries. In such cases, global models can be a good first step in understanding the discharge equilibrium.

10.4.2 Parabolic Approximation for Low Pressures

We now give an example of the complexities of the electronegative spatial profiles by examining a spatially varying model for low pressures when the positive ion wall loss is larger than or comparable to the volume recombination loss. We assume a slab geometry with bulk plasma width d in which α is sufficiently large that $n_e \approx n_{e0}$ and $D_{a+} \approx 2D_+$, but the effect of recombination can be neglected in determining the spatial distribution (but not necessarily the plasma parameters). The diffusion equation (10.3.26) then takes the simple form

$$-2D_+\frac{d^2 n_+}{dx^2} = K_{iz} n_g n_{e0} \tag{10.4.22}$$

In this approximation, $n_+(x)$ has a parabolic solution of the form (see (5.2.20))

$$\frac{n_+}{n_{e0}} = \alpha_0\left(1 - \frac{4x^2}{l_1^2}\right) + 1, \quad -l_1/2 < x < l_1/2 \tag{10.4.23}$$

where $l_1/2$ is the position where $\alpha = 0$.

We would not necessarily expect the Bohm flux condition to be met within the range of this solution, so the $\alpha \gg 1$ solution must be matched to an $\alpha = 0$ electropositive solution which extends from $x = l_1/2$ to $x = d/2$. This determines the position of the plasma–sheath boundary satisfying the Bohm flux condition (10.3.27), which for $\alpha_s = 0$ reduces to

$$-D_a \left.\frac{dn_+}{dx}\right|_{d/2} = n_+(d/2)\,u_B \tag{10.4.24}$$

The complete structure is illustrated in Figure 10.2. We further simplify our analysis by assuming that n_{e0} is known. The absorbed power per unit area, S_{abs}, is then obtained *a posteriori* from (10.3.32).

Substituting (10.4.23) into positive ion balance (10.3.26) and integrating *only* over the electronegative core $0 < x < l_1/2$, we obtain

$$K_{iz} n_g n_{e0} \frac{l_1}{2} = K_{rec} n_{e0}^2 \left(\frac{8}{15}\alpha_0^2 + \frac{2}{3}\alpha_0 \right) \frac{l_1}{2} + \frac{8D_+ \alpha_0 n_{e0}}{l_1} \tag{10.4.25}$$

Substituting (10.4.23) into negative ion balance (10.3.31) and integrating over the *entire* bulk plasma $0 < x < d/2$, we obtain

$$K_{att} n_g n_{e0} \frac{d}{2} = n_{e0}^2 \alpha_0 \left[\frac{8}{15} K_{rec} \alpha_0 + \frac{2}{3}(K_{rec} + K_* n_g) \right] \frac{l_1}{2} \tag{10.4.26}$$

Note in (10.4.26) that attachment occurs over the entire volume while recombination and detachment occur only in the electronegative core.

At $x = l_1/2$, this electronegative solution must be matched to an electropositive solution. Various electropositive solutions can be used. For pressures not too high, the variable mobility solution of Section 5.3 and Appendix C is appropriate. To determine the total positive ion balance, we equate the sum of the flux leaving the electronegative core and the ionization in the electropositive edge region to the Bohm flux at $x = d/2$. This is approximately given by

$$\frac{8D_+ \alpha_0 n_{e0}}{l_1} + \frac{K_{iz} n_g n_{e0}(d - l_1)}{2} = h_{le} u_B n_{e0} \tag{10.4.27}$$

For simplicity, we have taken $n_e = n_{e0}$ for calculating the ionization in the electropositive edge region. The factor $h_{le} = n_s/n_{e0}$ gives the ratio of the density at the sheath edge $x = d/2$ to the density at $x = l_1/2$. The variable mobility model gives the result (C.16)

$$h_{le} = \left[\frac{2\nu_{iz}\lambda_+/\pi u_B + (u_1/u_B)^3}{1 + 2\nu_{iz}\lambda_+/\pi u_B} \right]^{1/3} \tag{10.4.28}$$

where $\nu_{iz} = K_{iz} n_g$, λ_+ is the ion mean free path, and u_1 is the positive ion flow velocity at $x = l_1/2$, $u_1 = \Gamma_+(l_1/2)/n_{e0}$, where $\Gamma_+(l_1/2)$ is given by the last term in (10.4.25). For given input variables n_g, n_{e0}, and d, the unknown quantities α_0, l_1, h_{le}, and T_e can be determined from (10.4.25) to (10.4.28).

The above equations are readily solved by noting that K_{iz} is a strong (exponential) function of T_e, such that the temperature is essentially clamped by the particle balance for positive ions. Substituting K_{iz} from (10.4.25) into (10.4.27) and (10.4.28), we obtain a set of three equations (10.4.26)–(10.4.28) to determine α_0, l_1, and h_{le} that depend only weakly on T_e. We can therefore take the temperature as given (say $T_e = 2.5$ V) and solve for α_0, l_1, and h_{le}. T_e can then be obtained from (10.4.25) and, if necessary, all parameters improved by iteration.

Solving (10.4.26) for α_0, we obtain

$$\alpha_0 = -\frac{5}{8}\left(1 + \frac{K_* n_g}{K_{rec}}\right) + \frac{5}{8}\left[\left(1 + \frac{K_* n_g}{K_{rec}}\right)^2 + \frac{24}{5}\frac{K_{att} n_g d}{K_{rec} n_{e0} l_1}\right]^{1/2} \tag{10.4.29}$$

For large α_0 and recombination-dominated negative ion loss, (10.4.29) reduces to

$$\alpha_0 \approx \left(\frac{15}{8} \frac{K_{att}}{K_{rec}} \frac{n_g}{n_{e0}} \frac{d}{l_1} \right)^{1/2} \tag{10.4.30}$$

In the regime where the electropositive edge is thin, setting $d/l_1, \approx 1$, substituting $n_{e0} \approx n_{+0}/\alpha_0$ into (10.4.30), and solving for α_0, we obtain

$$\alpha_0 = \frac{15}{8} \frac{K_{att}}{K_{rec}} \frac{n_g}{n_{+0}} \tag{10.4.31}$$

which agrees with the result (10.4.7) for the global model with $\mathcal{V}_{\rm rec} = \frac{6}{5}\mathcal{V}$ and $\overline{n}_+ = \frac{2}{3}n_{+0}$ as given below (10.4.5). From (10.4.30) or (10.4.31), we see the scaling of α_0 with $n_{\rm g}$ for fixed $n_{\rm e0}$ or n_{+0}, respectively.

For large α_0 and detachment-dominated loss, (10.4.29) reduces to

$$\alpha_0 = \frac{3}{2}\frac{K_{\rm att}}{K_* n_{\rm e0}}\frac{d}{l_1} \tag{10.4.32}$$

which agrees with (10.4.8) for $d/l_1 \approx 1$.

To determine l_1, we eliminate $K_{\rm iz}$ from (10.4.25) and (10.4.27) to obtain

$$8D_+\alpha_0\frac{d}{l_1^2} + K_{\rm rec}n_{\rm e0}\left(\frac{8}{15}\alpha_0^2 + \frac{2}{3}\alpha_0\right)\left(\frac{d-l_1}{2}\right) = h_{l\rm e}u_{\rm B} \tag{10.4.33}$$

Solving (10.4.26) for $n_{\rm e0}$ and inserting this into (10.4.33), we can obtain l_1/d in terms of α_0. The result is complicated and not particularly illuminating. In the limit that diffusion loss dominates recombination loss, then (10.4.33) reduces to

$$\frac{l_1}{d} = \left(\frac{8D_+\alpha_0}{h_{l\rm e}u_{\rm B}d}\right)^{1/2} \tag{10.4.34}$$

Equations (10.4.30) or (10.4.32) and (10.4.34) can be solved simultaneously for α_0 and l_1/d within various approximations (Problem 10.8) for recombination- or detachment-dominated discharges.

The condition that the negative ions are in Boltzmann equilibrium can be verified, as described in Section 10.3, by comparing the total negative ion flux Γ_- to the negative ion diffusion flux $-D_-{\rm d}n_-/{\rm d}x$. Using the model profile $n_- = n_{\rm e0}\alpha_0(1 - 4x^2/l_1^2)$ from (10.4.23) and performing the integration in (10.3.14), we obtain for $\alpha_0 \gg 1$ and $x \ll l_1 \approx d$ that

$$|\Gamma_-| \approx \left(\frac{7}{15}K_{\rm rec}\alpha_0 + \frac{1}{3}K_* n_{\rm g}\right) n_{\rm e0}^2\alpha_0 x \tag{10.4.35}$$

Similarly evaluating the diffusion flux yields

$$-D_-\frac{{\rm d}n_-}{{\rm d}x} = \frac{8D_- n_{\rm e0}\alpha_0}{l_1^2}x \tag{10.4.36}$$

The ratio of (10.4.35) and (10.4.36) yields $\eta_{\rm B}$ in (10.3.34), and the condition (10.3.36) for Boltzmann equilibrium of negative ions.

Depending on plasma parameters, it is possible to reach the local ion sound speed in the electronegative core. In this case, a nonlinear potential structure forms that confines negative ions (an ion-acoustic shock, as described at the end of Section 4.2). The core then terminates abruptly, over a few Debye lengths, at a position $< l_1/2$ (Kolobov and Economou, 1998; Kouznetsov et al., 1999; see Monahan and Turner, 2008 for a PIC simulation example).

For high electronegativity, the electropositive edge disappears. The condition is that the flux leaving the electronegative core is equal to the Bohm flux out of the electropositive region

$$n_{\rm e0}(1+\alpha_{\rm s})u_{{\rm B}\alpha} = n_{\rm e0}u_{\rm B} \tag{10.4.37}$$

where we put $\alpha = \alpha_{\rm s}$ since the electropositive edge disappears. Approximating $u_{{\rm B}\alpha}$ from (10.3.28) by

$$u_{{\rm B}\alpha} = v_{\rm th+}\left(\frac{1+\alpha_{\rm s}}{\alpha_{\rm s}}\right)^{1/2} \tag{10.4.38}$$

where $v_{th+} = (e\mathrm{T_i}/M_+)^{1/2}$, we obtain

$$\left(1+\alpha_s\right)\left(\frac{1+\alpha_s}{\alpha_s}\right)^{1/2} = \gamma^{1/2} \tag{10.4.39}$$

For a nominal value of $\gamma = \mathrm{T_e}/\mathrm{T_i} = 100$, we obtain a transition at $\alpha_s \approx 8.5$.

At still higher pressures, (10.3.36) is not satisfied, the central region flattens and the edge steepens, so that a parabolic approximation is not adequate. The increased flattening of the central density at increasing α suggests a model in which all of the variation of the ion density occurs in a transition edge region. A heuristic model that captures this profile is described in Lichtenberg et al. (1997). An estimate of the transition layer thickness is also given in Berezhnoj et al. (2000).

10.5 Electronegative Discharge Experiments and Simulations

The usefulness of models must be validated with more complete models, simulations, and experiments. Various gases have differing characteristics that must be treated separately. Two commonly used gases that illustrate this variety are oxygen and chlorine, with chlorine being considerably more electronegative than oxygen. We consider some examples for these gases in the following.

10.5.1 Oxygen Discharges

Oxygen is a widely used feedstock for thin film processing. It is the primary gas for photoresist ashing and is a common additive in halogen gas mixes for metal and polysilicon etching (see Sections 16.3 and 16.4). It is also the primary feedstock (with silicon-containing gases) for plasma-assisted silicon dioxide and oxynitride depositions (see Section 17.2).

Oxygen is weakly electronegative with a dissociative attachment rate constant K_{att} (reaction 2 in Table 8.2) having a threshold energy of about 4.7 V. There are two low-lying, easily excited metastable molecular states (see Figure 8.4). Metastable collisions with O^- ions can result in electron detachment, an important negative ion loss process. The $O_2(a^1\Delta_g)$ metastable has a high rate constant K_{ex*} for excitation (reaction 15 in Table 8.2) and a low loss probability γ_q on most wall surfaces through the wall quenching reaction $O_2(a) + \text{wall} \rightarrow O_2$. A nominal value for a room temperature aluminum wall is 10^{-3} (Gudmundsson et al., 2001; Franklin, 2001). Some $O_2(a)$ quenching coefficients are given in Proto and Gudmundsson (2018). The $O_2(b^1\Sigma_g^+)$ metastable has approximately one-fourth the excitation rate coefficient of $O_2(a)$. Its detachment rate coefficient K_{det} is somewhat uncertain but is believed to be about a factor of 20 larger than that of $O_2(a)$. Therefore, it can play a very important role in oxygen discharge equilibrium (Gudmundsson, 2004).

Because γ_q is so small, the vacuum pumping speed S_p (m^3/s) can contribute to metastable surface loss. Therefore (neglecting electron impact de-excitation processes) the metastable loss flux (10.3.6) can be written more generally as

$$-D_*\nabla n_* = \frac{1}{4}\gamma_q n_{*s}\bar{v}_* + \frac{S_p}{A}n_{*s}$$

which yields an effective metastable loss probability

$$\gamma_* = \gamma_q + \frac{4S_p}{\bar{v}_* A} \tag{10.5.1}$$

Hence, varying the pumping speed (or, equivalently, the gas flow rate at a fixed pressure) varies γ_*, as seen experimentally below.

Because γ_* is small and the metastables are easy to excite, the metastable densities can be high, and the detachment loss of negative ions can exceed the positive–negative ion recombination loss (reaction 7 in Table 8.2) at quite modest pressures. The transition pressure depends also on the associative detachment rate constants K_{det} (processes 20 and 21 in Table 8.2 for $O_2(a^1\Delta_g)$; see Gudmundsson, 2004 for $O_2(b^1\Sigma_g^+)$). In early work, the best fit of a global discharge model to one set of oxygen discharge data gave K_{det} a factor of three higher than that given in Table 8.2 for $O_2(a)$ alone, suggesting the importance of $O_2(b)$ (Stoffels et al., 1995).

10.5.1.1 Measurements and Global Models

A number of measurements of negative ion densities have been reported in low-to-moderate pressure (below 100 mTorr) oxygen discharges, and comparisons have been made to global and spatially varying models. The dominant negative and positive ions have been measured to be O^- and O_2^+. Stoffels et al. (1995) measured the O^- and electron densities in an asymmetrically driven rf capacitive discharge and developed a global model to describe their experimental results. They found that O^- is mainly produced by dissociative attachment of O_2 and is mainly lost by positive–negative ion recombination at pressures below 20–30 mTorr and by detachment against $O_2(a^1\Delta_g)$ metastables at higher pressures. At fixed input power, the measured negative ion density was found to increase with pressure for low pressures and then to decrease with pressure at higher pressures, with the maximum negative ion density at the recombination–detachment transition. The negative ion density was found to increase with oxygen feedstock flow rate at both 25 and 100 mTorr. This result is expected from (10.5.1) if detachment is important because higher flow rates (higher chamber pumping speeds) at fixed pressure lead to a reduced metastable density.

Vender et al. (1995) measured the negative ion density profile in the system of Stoffels et al. (1995) at 10 W rf power, with the results shown in Figure 10.3*a* and *b*. In Figure 10.3*a* at 10 mTorr, $n_{e0} \approx 5 \times 10^8\ \mathrm{cm}^{-3}$ and a classic parabolic negative ion profile was measured, which is well described by (10.4.23). The recombination loss was found to be small compared to the diffusion loss and recombination loss exceeded detachment loss. In Figure 10.3*b* at 40 mTorr, detachment by O^- collisions with $O_2(a^1\Delta_g)$ was found to exceed recombination, leading to roughly constant (but somewhat decreasing) negative ion density with increasing pressure, expected from (10.4.8). The scaling of n_- with pressure seen experimentally by both Vender et al. (1995) and Stoffels et al. (1995) was fitted reasonably well with a global model incorporating both the recombination and detachment losses

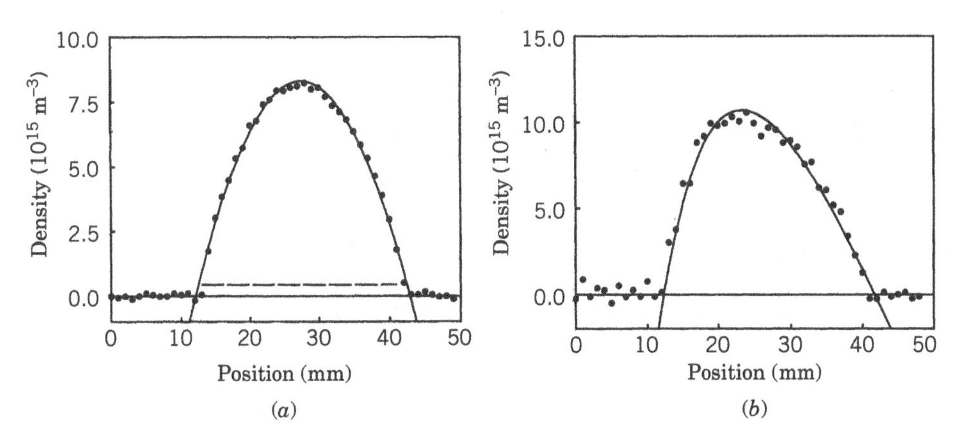

Figure 10.3 Density profiles versus position in oxygen; negative ions (dots) measured at 10 W input power at (*a*) 10 mTorr, with the electron density shown by the dashed line; and (*b*) at 40 mTorr. Source: Vender et al., 1995/with permission of American Physical Society.

along with the particle balance equation for the $O_2(a)$ metastable and several other less important species. As in the modeling presented in Sections 10.3 and 10.4, the electron density was taken as the input, so that sheath physics and power absorption were not considered. The assumed Maxwellian temperature of 3 eV also required a rescaling of the peak ion density.

The asymmetry in the profile at 40 mTorr in Figure 10.3*b* is due to the nonuniform ionization rate in the discharge gap. The electrode area ratio was about three, so most of the applied rf voltage appeared across the (smaller) powered electrode sheath (on the left), leading to a large stochastic sheath and local electron heating there. At the higher pressures, the mean free path for ionizing electrons becomes sufficiently short that an enhanced ionization occurs near the powered sheath, producing the observed density asymmetry. These phenomena will be discussed in Chapter 11.

Berezhnoj et al. (2000) studied the spatially varying charged particle densities in a symmetrically driven capacitively coupled rf discharge in oxygen over a range of pressures from 21.5 to 215 mTorr and for interelectrode gaps of 2 to 10 cm. At low powers for a 3 cm gap, they obtained the results shown in Figure 10.4*a* and *b*. The solid squares and hollow circles denote two different types of negative ion density measurements, and the triangles denote the positive ion density measurements. The solid and dashed lines represent calculated positive and negative ion densities determined from a fluid model (essentially integrating (10.3.1a) and (10.3.8)) with a metastable quenching probability $\gamma_q = 10^{-3}$. In Figure 10.4*a* at 45 mTorr, they measured $n_{+0} \approx 4.0 \times 10^9$ cm^{-3} and $n_{e0} \approx 1.1 \times 10^8$ cm^{-3}, with a parabolic profile corresponding to (10.4.23). In Figure 10.4*b* at 150 mTorr, $n_{+0} \approx 3.5 \times 10^9$ cm^{-3} and $n_{e0} \approx 1.6 \times 10^8$ cm^{-3}. At the higher pressure, we see a flat-topped profile corresponding to non-Boltzmann negative ions. They also found that n_{-0} varied very little with pressure, which is consistent with the prediction of (10.4.8), as the negative ion recombination losses were calculated to be small compared with the detachment losses. A similar transition from parabolic to flat-topped profiles was measured at 75 mTorr as the gap length was increased from 2 to 4 cm. As described above, at the higher pressures, the nonuniform electron temperature leads to increased edge ionization. Together with radial loss, not incorporated in the model, this resulted in the higher ion densities near the plasma edge.

At higher powers than for the results shown in Figures 10.3 and 10.4, the electronegativity α decreases for the data of Stoffels et al. (1995), $\alpha \approx 10$ at 10 W rf power and $\alpha \approx 1$ at 40 W. A number

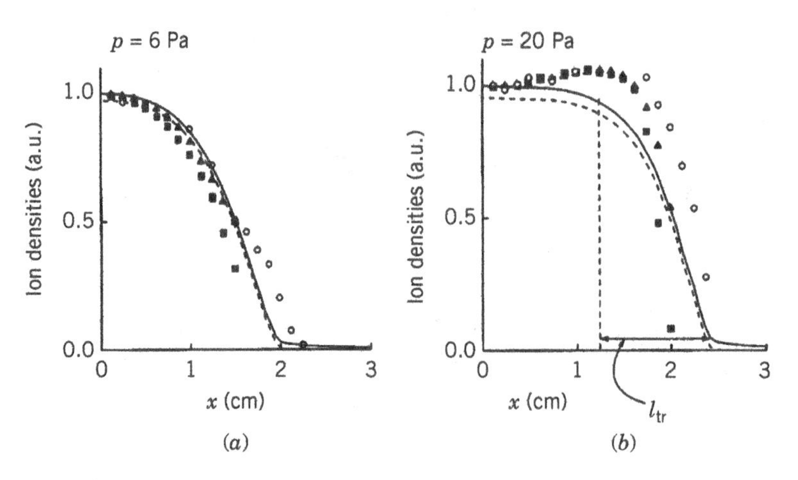

Figure 10.4 Density profiles versus position in oxygen; measured negative (squares, open circles) and positive (triangles) ions at (*a*) 45 mTorr, and (*b*) 150 mTorr; solid and dashed lines are calculated positive and negative ion densities. Source: Berezhnoj et al., 2000/with permission of AIP Publishing.

of global model studies have been done in the higher power range, which is more characteristic of high-density plasmas used in processing (see Lee et al., 1994; Lee and Lieberman, 1995; Kimura and Ohe, 1999; Kimura et al., 2001; Gudmundsson et al., 2001; Gudmundsson, 2004; Kim et al., 2006a). At these higher powers, if the O-atom wall recombination probability γ_O is not too high, the O-atom density, which is important for processing applications, may be dominant, and must be included in a calculation. The main generation is by electron impact dissociation of O_2 (processes 3 and 22 in Table 8.2) and the main loss is by O-atom recombination on the walls. The O-atom loss probability γ_{rec} has been measured to be 0.2–0.5 on stainless steel walls, leading to relatively low fractional dissociations, even in high power discharges, as seen experimentally (Fuller et al., 2000), and in reasonable agreement with global model predictions described below. To obtain higher O-atom dissociation fractions, wall materials such as quartz or anodized aluminum can be used, which have a lower γ_{rec} for O-atoms (Problem 10.11).

Figure 10.5 shows one example of predictions from a global model at high powers for a cylindrical stainless steel plasma chamber with $l = 7.6$ cm and $R = 15$ cm, in which the electronegativity α and the electron density n_e are plotted versus pressure for various discharge powers (Gudmundsson, 2004). The positive ion flux loss model of Lee and Lieberman (1995) was used. The metastable wall loss probability was taken to be $\gamma_q = 0.007$ for both $O_2(a)$ and $O_2(b)$. We see from Figure 10.5*a* that the electronegativity decreases with increasing power and is low except at the highest pressures, $\alpha < 1$ for $p < 100$ mTorr at 500 W discharge power. The electron density in Figure 10.5*b* increases with power. It decreases with pressure at the higher pressures, due to the increasing energy losses (see Figure 3.17). At the lower pressures, it decreases due to the larger effective loss areas (10.2.7) for positive ions. The fractional dissociation at 10 mTorr and 500 W was about 0.12. The critical gas pressure p_{crit} where recombination losses equal detachment losses varies roughly linearly with power, $p_{crit} \approx 5$ mTorr at 300 W and 12 mTorr at 2000 W. Hence, at high powers O^- loss is dominated by metastable detachment at all but the lowest pressures. Such global model results are consistent with measurements, e.g., see Tuszewski (1996).

A global model of the argon/oxygen discharge at low pressures, accounting for the variation of the argon and oxygen excited states, has been given by Gudmundsson and Thorsteinsson (2007) and compared to various experiments. They also examined the production and loss of O atoms and

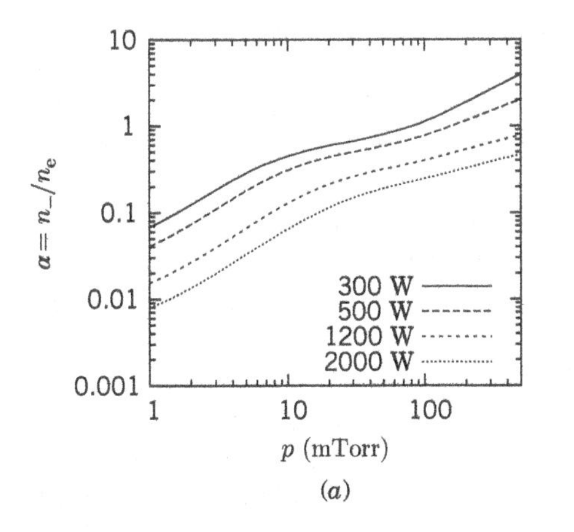

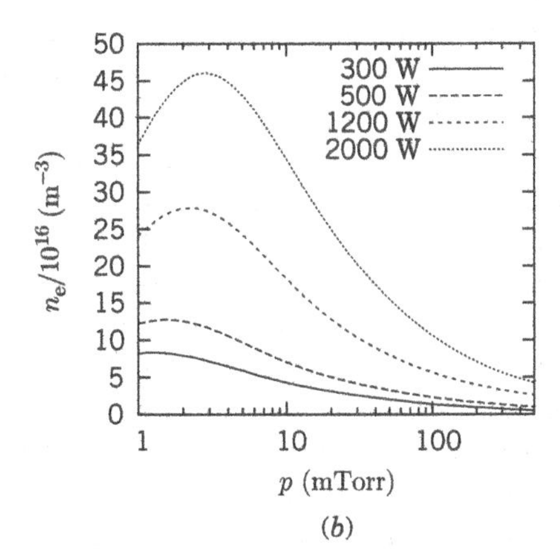

Figure 10.5 (*a*) Electronegativity α and (*b*) electron density n_e versus discharge pressure p in oxygen at various discharge powers, in a cylindrical stainless steel chamber with $l = 7.6$ cm and $R = 15.2$ cm; 50 sccm flow rate, $\gamma_q = 0.007$, $\gamma_O = 0.5$, $T_g = 600$ K. Source: Gudmundsson (2004). ©IOP Publishing. Reproduced with permission. All rights reserved. https://doi.org/10.1088/0022-3727/37/15/005.

O^- ions in a stainless steel chamber with $R = L = 10$ cm and powers from 300 to 2000 W. They note, importantly, that the majority of ground state $O(^3P)$ atoms originate from the metastable $O(^1D)$ atomic state, as also seen by comparing the rate constants for reactions 3 and 22 in Table 8.2. The fractional oxygen dissociation was found to decrease with increasing pressure and to increase with increasing power, as expected. Both n_e and, for high argon fractions, T_e, were found to increase with increasing argon fraction. The former may be due to the decreasing energy losses as argon is added to the mixture, while the latter may be due to the higher ionization threshold for argon. These combined increases produce an increasing dissociation rate for O_2. Therefore, somewhat surprisingly, at all powers, the oxygen fractional dissociation was found to increase with increasing argon fraction.

10.5.1.2 PIC Simulations

In addition to the global and fluid simulations described above, PIC computer simulations have been compared with basic space-varying models, such as described in Section 10.4. Such simulations allow for tests of some of the basic idealizations incorporated into analytical, fluid, and global models, for example, the assumption of a Maxwellian electron distribution. On the other hand, the number of species that can be handled in a PIC simulation is limited, and species with slow timescales for production and loss are not easily incorporated into the simulation, e.g., metastable O_2^*.

An early PIC simulation (Lichtenberg et al., 1994) used a 13.56 MHz plane-parallel capacitive discharge with a plate separation of 4.5 cm at $p = 50$ mTorr and low power ($n_{e0} = 2.4 \times 10^9$ cm^{-3}), with the following dynamics: O_2^+ and electrons are created by electron impact ionization of O_2, and O^- is created by dissociative attachment of electrons on O_2. Negative ions are lost only by recombination with positive ions in the volume. Positive ions are lost to the walls by diffusion and in the volume by recombination with negative ions. O^- detachment losses and O-atom dissociation were omitted from the calculations. The simulation results showed the general features of an electronegative core plasma surrounded by an electropositive halo, as seen in Figure 10.2. There was a large sheath at these parameters with u_B reached at $x \approx 1.2$ cm from the discharge center (see Section 11.2). The approximate parabolic variation of n_- and n_+ and the flat profile for n_e within the electronegative core were seen for this relatively low α_0 case ($\alpha_0 \approx 8$).

The importance of the $O_2(a)$ metastable to oxygen capacitive discharge equilibrium was examined using 1D PIC kinetic simulations of both the three charged species and the $O_2(a)$ and $O(^1D)$ metastables (Gudmundsson and Lieberman, 2015). The 50 mTorr discharge had a 4.5 cm gap and was driven by 222 V at 13.56 MHz. $O_2(a)$ detachment was found to have a significant influence on the discharge properties such as electronegativity, effective electron temperature and electron heating processes. The electron heating changed from having contributions from both bulk and sheath heating to being dominated by sheath heating when $O_2(a)$ is included.

The importance of the $O_2(a)$ wall quenching coefficient γ_q in determining the discharge properties was investigated by Proto and Gudmundsson (2018) using 1D PIC simulations for the charged particles, along with global models of the neutral densities. While mainly a study of electron heating mechanisms, Figure 10.6 shows the central electronegativity α_0 versus the $O_2(a)$ wall quenching coefficient γ_q at 10, 25, and 50 mTorr. At 10 mTorr, the discharge is recombination-dominated at all values of γ_q. At 25 and 50 mTorr, it is recombination-dominated for $\gamma_q > 10^{-2}$ and detachment-dominated for $\gamma_q < 10^{-3}$. These experiments and simulations indicate that the presence of the $O_2(a)$ metastable turns the oxygen discharge from highly electronegative, recombination-dominated O^- loss at low pressures to weakly electronegative, attachment-dominated O^- loss at higher pressures.

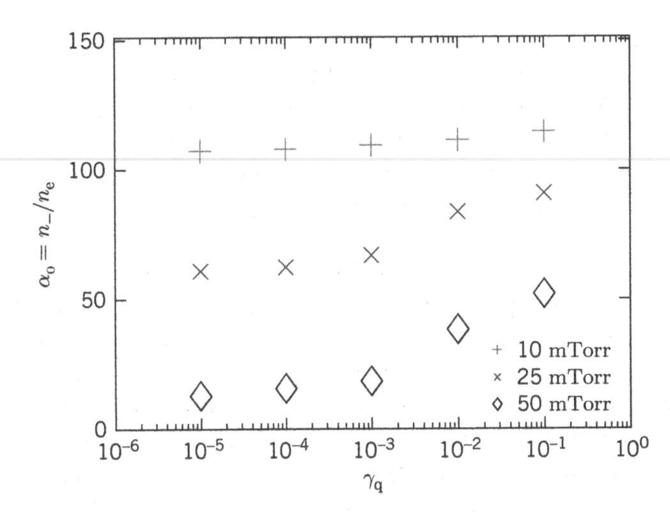

Figure 10.6 Central electronegativity α_0 versus $O_2(a)$ wall quenching coefficient γ_q at pressures of 10 (pluses), 25 (crosses), and 50 (diamonds) mTorr, for a 4.5 cm gap, 222 V, 13.56 MHz plane-parallel electronegative discharge in oxygen. Source: Proto and Gudmundsson (2018). ©IOP Publishing. Reproduced with permission. All rights reserved. https://doi.org/10.1088/1361-6595/aaca06.

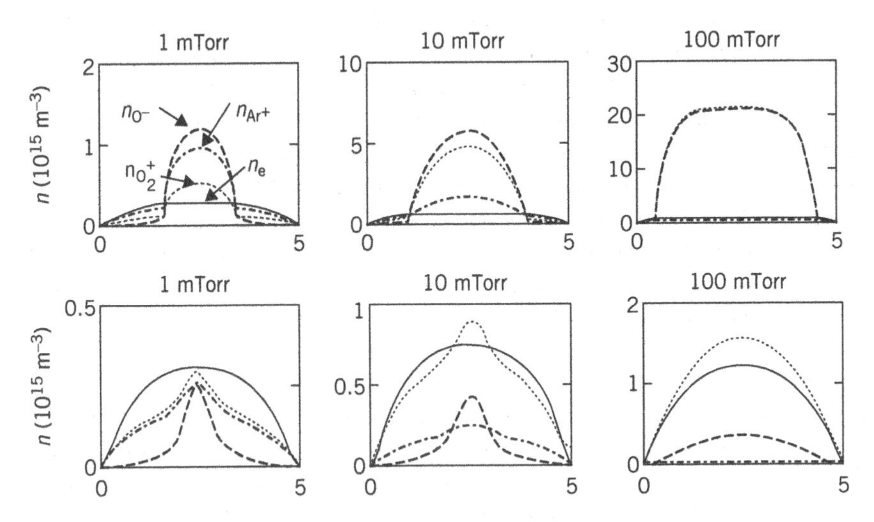

Figure 10.7 Simulation results for recombination-dominated 0.5/0.5 Ar/O_2 (top panel) and detachment-dominated 0.4/0.4/0.2 $Ar/O_2/O_2(a^1\Delta_g)$ (bottom panel) for 1, 10, and 100 mTorr, for a 5 cm gap, low power, 13.56 MHz plane-parallel inductively driven discharge. Source: Monahan and Turner (2008). ©IOP Publishing. Reproduced with permission. All rights reserved. https://doi.org/10.1088/0963-0252/17/4/045003.

A wide range of PIC simulations of argon/oxygen mixtures were used by Monahan and Turner (2008) to evaluate the utility of global modeling. They investigated low powers (low densities) over the pressure range of 1–1000 mTorr, with and without a fixed fraction of $O_2(a)$ metastable background density. The plasma was maintained by a *spatially uniform, 13.56 MHz electric field parallel to the plates.* This corresponds best to a low-density, low-pressure inductively driven discharge. The simulation results are shown in Figure 10.7. In the top panel, we see the transition from parabolic to flat-topped model as the pressure is increased, for recombination-dominated O^- loss. The bottom panel shows the profile changes with pressure for detachment-dominated O^- loss. They compared the pressure variation of the PIC results to the global model of Kim et al. (2006a), as described in Section 10.4, using both a Maxwellian EEDF, and an EEDF extracted from the PIC simulations and fit to the non-Maxwellian global model form (19.6.1) described in Section 19.6. A major conclusion was that the evolution of the EEDF with pressure, and the resulting changes in the global model rate coefficients were very important in obtaining good agreement between simulation and global model results.

Proto and Gudmundsson (2020) used 1D PIC simulations and Boltzmann term analysis (see Section 11.3 for a description) to determine the space- and time-varying decompositions of the electric field and the electron power absorption in a 4.5 cm gap, oxygen discharge driven by 400 V at 13.56 MHz. Pressures of 10 and 100 mTorr were studied. The discharge was weakly electronegative, $\alpha_0 \approx 3.6$, at 100 mTorr, and quite strongly electronegative, $\alpha_0 \approx 94$, at 10 mTorr. Overall, at 100 mTorr, about 70 kW/m^3 of the space–time-averaged heating was ohmic, and about 35 kW/m^3 was identified as ∇p_e pressure ("stochastic") heating, with large but nearly canceling terms due to the ∇n_e and ∇T_e components. At 10 mTorr, about 60 kW/m^3 was found to be ohmic heating and 40 kW/m^3 was pressure heating, with a small (−10 kW/m^3) negative inertial contribution.

10.5.2 Chlorine Discharges

Chlorine is a strongly electronegative gas that is widely used for thin film etching, e.g., polysilicon. Cl_2 has an ionization energy of 11.47 V, a low dissociation energy (2.5 V) and a high electron affinity (2.45 V), with a near-zero threshold energy for dissociative attachment (see Figure 8.8c). Consequently, it has high dissociation and dissociative attachment rate constants. All electronic excitations appear to be dissociative; hence, there are no metastable molecular states. Ground-state Cl has an electron affinity of 3.61 V, such that the detachment reaction $Cl + Cl^- \rightarrow e + Cl_2$ has a high threshold energy of 1.13 V. Therefore detachment losses of negative ions are not significant. Data on electron interactions with Cl_2 have been reviewed by Christophorou and Olthoff (1999a), Maxwellian rate constants are given in Kemaneci et al. (2014), and the potential energy diagram of Cl_2 is given in Peyerimhoff and Buenker (1981).

10.5.2.1 Measurements

A fairly complete set of measurements of charged and neutral particle densities, electron temperature, and gas temperature has been reported (Malyshev et al., 1999a; Malyshev and Donnelly, 2000a,b, 2001) for an inductively coupled chlorine plasma. Their discharge was excited at 13.56 MHz by a planar coil through a quartz vacuum window at one end of a stainless steel chamber having a radius $R = 18.5$ cm and a length $l = 20$ cm (see Section 12.3 for a description of this type of discharge). Below an incident discharge power of about 150 W, the discharge operates in a low-density capacitively coupled mode, while above 150 W, the discharge makes a transition to a high-density inductively coupled mode (see Section 12.2). The measurements were made over a pressure range of 1–20 mTorr for various powers.

Figure 10.8 shows the on-axis positive ion and electron densities 9.5 cm below the quartz window and 3 cm above the substrate, determined from Langmuir probe measurements at 20 mTorr. The capacitive and inductive modes are apparent. In the low-power capacitive mode, the electronegativity α is about 60 at the lowest power ($n_e \approx 3 \times 10^7$ cm^{-3}), decreasing to about 20 at a higher power ($n_e \approx 2 \times 10^8$ cm^{-3}). This variation is consistent with the predicted scaling of $\alpha \propto n_e^{-1/2}$ given from (10.4.7b). The α also was observed to increase somewhat with increasing pressure (gas density n_g); however, the measured variation of α with n_g is weaker than the scaling $\alpha \propto n_g^{1/2}$ of (10.4.7b). The

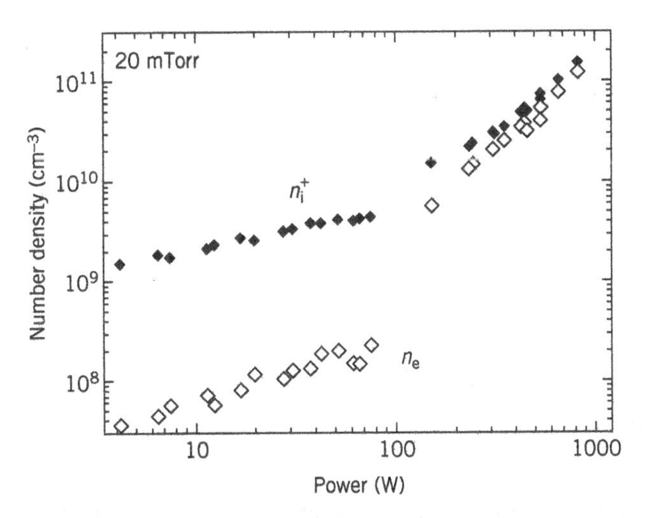

Figure 10.8 Positive ion and electron density versus discharge incident power in chlorine at 20 mTorr, in a cylindrical stainless steel chamber with $l = 20$ cm and $R = 18.5$ cm. Source: Malyshev and Donnelly, 2001/with permission of AIP Publishing.

scaling $n_+ \propto P_{\text{abs}}^{1/2}$ of (10.4.11), roughly independent of pressure, is also seen in the data, consistent with a high-α discharge dominated by positive ion volume recombination loss. Even at the highest power in capacitive mode, the dissociation fraction was measured to be low (Malyshev and Donnelly, 2000a); the neutral density consisted mainly of Cl_2.

In contrast, the measurements in the high-density inductive mode showed completely different scalings. As seen in Figure 10.8, the maximum α was about 2.5 just above the transition to inductive mode. The transition from the low-density capacitive mode to the high-density inductive mode is considered in detail in Section 12.2. At lower pressures, α was typically less than unity. Furthermore, the density in the higher power regime scales roughly as $n_+ \propto P_{\text{abs}}$, typical of the scaling (10.2.15) expected for high-density electropositive discharges. This scaling is also seen at lower pressures. Hence, a low-pressure high-power chlorine discharge is mainly electropositive. We can understand this result from the global model scaling in (10.4.7a), and from the measurements of the Cl_2 and Cl densities. In inductive mode, the chlorine feedstock is strongly dissociated; the Cl density exceeds the Cl_2 density. The low Cl_2 density implies that the dissociative attachment rate ($\propto n_{\text{Cl}_2}$) for production of Cl^- is low. Furthermore, the recombination losses for Cl^- are high because n_+ is high. This results in a low negative ion density. We should expect the dominant positive ion in the high-density inductive mode to be Cl^+, not Cl_2^+, and this was measured to be the case.

10.5.2.2 Global Models and Simulations

In addition to experiments, direct numerical solution of the fundamental diffusion equations (10.3.13) and (10.3.14), global models, fluid simulations, and 1D PIC simulations can be used to examine some assumptions of the approximate analytic models given in Section 10.4. Lichtenberg et al. (1997) and Lee et al. (1997a) considered a chlorine feedstock gas with a bulk plasma width (excluding sheath widths) $d = 0.9$ cm, $n_{e0} \approx 10^{16}$ m^{-3}, and $p = 200$–600 mTorr. This corresponds to parameters of a capacitive discharge operating at reasonably high power but at pressures for which α is relatively high and negative ion loss is dominated by recombination. At 300 mTorr with the detachment term in (10.3.14) equal to zero and with $T_- = T_+ = T_i = 300$ K, the coupled equations (10.3.13) and (10.3.14) were solved, with the positive ion flux limited to the Bohm flux

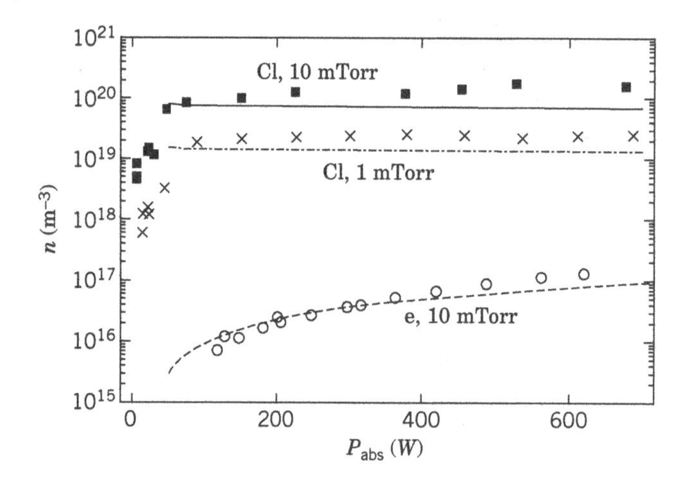

Figure 10.9 Global model (lines) and measured (symbols) atomic chlorine density at 1 and 10 mTorr, and electron density at 10 mTorr, versus discharge power. Source: Thorsteinsson and Gudmundsson (2010a). ©IOP Publishing. Reproduced with permission. All rights reserved. https://doi.org/10.1088/0963-0252/19/1/015001.

at the plasma edge, $\Gamma_{+s} = n_+ u_{B\alpha}$, with $u_{B\alpha}$ given by (10.3.28). The numerically determined positive ion density was found to display a flattened profile (Lichtenberg et al., 1997), with $\eta_B \approx 5$ from (10.3.36), indicating non-Boltzmann negative ions. The numerical result was compared with that obtained from a heuristic flat-topped model, which was found to work quite well in the range of α_0 where the simple parabolic model is not a good approximation.

Some global model comparisons to the data of Malyshev and Donnelly (2000a, 2001) are given by Thorsteinsson and Gudmundsson (2010a). Figure 10.9 shows their calculated Cl densities at 1 and 10 mTorr, and n_es at 10 mTorr, versus discharge power. As expected, the Cl densities were found to be relatively independent of power above about 100 W, when the chlorine feedstock is nearly completely dissociated. The measured and global model n_es agree well for this high-power regime. The measured T_es (not shown), however, were always higher than the model at low pressures (1–10 mTorr). This may be due to measurement errors, and/or to the use of Maxwellian rate constants in the model. The effects of non-Maxwellian EEDFs on chlorine discharge modeling have been examined by Gudmundsson et al. (2012b).

Global models of chlorine/argon discharges for continuous (cw) and pulsed power are described in Thorsteinsson and Gudmundsson (2010b,c), respectively, and Cl_2/O_2 in Thorsteinsson and Gudmundsson (2010d). A thorough global study of cw and pulsed-power chlorine discharges, with comparisons to experimental results, was done by Kemaneci et al. (2014). We examine some pulsed-power results in Section 10.6.

Two-dimensional fluid simulations of inductively coupled, high-density chlorine discharges were performed by Lymberopoulos and Economou (1995). The plasma was highly dissociated and n_e was found to scale linearly with P_{abs}, consistent with (10.4.11) and (10.4.12). Significant loading effects for the etching of polysilicon wafers were seen (see Section 16.2).

Low-power 1D PIC simulations of capacitively coupled chlorine discharges, along with comparisons to measurements, were performed by Huang and Gudmundsson (2013) over the pressure range of 1–100 mTorr. The gap and electrode diameter were 4.5 cm and 14.36 cm, respectively. The discharge was driven by a 222 V rf source at 13.56 MHz through a large blocking capacitor. Some results for the midplane plasma parameters as a function of pressure are given in the table.

p (mTorr)	P_{abs} (W)	T_e (V)	n_e (m^{-3})	α_0	f_{diss}
5	0.66	3.75	2.04×10^{14}	67	0.053
10	0.87	3.35	2.81×10^{14}	93	0.064
50	2.57	3.21	4.74×10^{14}	138	0.089
100	7.80	3.04	6.51×10^{14}	157	0.138

Here, f_{diss} is the chlorine fractional dissociation. As in the top panel in Figure 10.7, the transition from parabolic (5 and 10 mTorr) to flat-topped (50 and 100 mTorr) ion density profiles was clearly seen. At all pressures, the electron energy distributions were found to be non-Maxwellian, depleted of electrons below the Cl_2 ionization energy of 11.47 V. In later work, Huang and Gudmundsson (2014) examined a 10 mTorr current-driven chlorine discharge over a range of driving currents, frequencies (13.56–60 MHz), and secondary emission coefficients. They also added a second low-frequency current to examine dual-frequency operation (see Section 11.6). The electron energy distribution was found to evolve from a Druyvesteyn to a Maxwellian form as the current density was increased from 20 to 80 A/m^2. The effects of non-Maxwellian distributions are discussed in Section 19.6 and, specifically for chlorine, by Gudmundsson et al. (2012a).

Proto and Gudmundsson (2021) did a Boltzmann term analysis (see a description in Section 11.3) of their 1D PIC simulation results to determine the space- and time-varying decompositions of the electric field and the electron power absorption in a chlorine capacitive discharge with a 2.54 cm gap, driven by 222 V at 13.56 MHz. The pressures investigated were 1, 10, 25, 35, and 50 Pa. The discharge was highly electronegative, $\alpha_0 \approx 100$–150, at all pressures. The fractional dissociation f_{diss} for this low-power discharge, determined using a global model, was low, in the range 0.02–0.06, except at 50 Pa, where $f_{diss} \approx 0.18$. The space–time averaged electron heating power was found to be almost entirely ohmic at all pressures except at 1 Pa, where it was about two-thirds ohmic and one-third ∇p_e pressure ("stochastic") heating. The powers increased from about 5.3 kW/m^3 at 1 Pa to almost 60 kW/m^3 at 50 Pa. The conclusion is that the low-power chlorine discharge is strongly ohmic, except at pressures below of order 1 Pa.

10.6 Pulsed Discharges

Discharges operated using single-frequency modulated power are of strong interest for materials processing. They can have higher average charged particle densities at the same average power and significantly lower wafer damage. Both effects can be attributed to a lower electron temperature during the off-time, as described below. In addition, the negative ions in electronegative plasmas may be able to escape during the off-time, which can be useful in processing.

Discharges can be driven with dual frequencies (see Section 11.6), with the high-frequency source providing most of the power deposition, and the low-frequency source providing control of the bias voltage across the wafer-containing electrode sheath. Either or both sources can be pulsed to control discharge characteristics. Bias voltage pulsing is discussed briefly at the end of Section 11.6, and its application to atomic layer etching (ALE) is described at the beginning of Section 16.5. Some reviews of pulsed-power and pulsed bias-voltage discharge physics, and applications to materials processing, are given in Banna et al. (2012) and Economou (2014).

Discharges can be self-pulsing. An example is given in Section 12.2 for an electronegative discharge, in which the absorbed power oscillates between a capacitively driven and an inductively driven mode for certain discharge parameters.

Feedstock flows in discharges can be pulsed on slow (0.1–10 s) timescales. Alternating flows of various feedstocks can enable exquisite control of etching and deposition processes, in which films are etched or deposited on the atomic layer scale length. Feedstock pulsing for ALE and atomic layer deposition (ALD) processes is described in Sections 16.5 and 17.3, respectively.

Here, we describe mainly pulsed-power single frequency discharges. Ashida et al. (1995) and Lieberman and Ashida (1996) investigated the behavior of argon plasmas driven by time-modulated power in high-density plasma reactors using a spatially averaged (global) model. The time evolution of the electron temperature and the plasma density was calculated by solving the particle and energy balance equations. In their calculation, the species included ground state Ar, 4s (resonance and metastable) excited Ar, 4p excited Ar, and Ar^+ ions. However, for typical pressures and absorbed powers, the excited Ar states affect the calculated plasma density by at most 25% and have practically no effect on the electron temperature. We therefore describe a simplified global model (without excited states) to emphasize the physical ideas.

Although pulsed power argon discharges are useful benchmarks, electronegative discharges are used for most materials processing. A simplified model for pulsed power electronegative discharges such as O_2 or Cl_2 is also presented, and the model predictions are compared with experiments. More complete global models of high-density pulsed Cl_2 discharges are given by Meyyappan (1996), Ashida and Lieberman (1997), Thorsteinsson and Gudmundsson (2010c), and Kemaneci et al. (2014). Low-density capacitively coupled pulsed discharges can be described in a similar manner, for early studies, see Overzet and Leong-Rousey (1995) and the references cited therein. For electronegative plasmas, the spatial variation of the negative ions during the turn-on and turn-off times can significantly change the dynamics from the results of a global model. These phenomena are described by Kaganovich and Tsendin (1993) and more completely by Kaganovich (2001). We shall introduce these ideas briefly.

Controlling the power flow to the substrate can be a major concern for plasma etching and deposition processes. In many applications, the process is driven mainly by the density (or flux) of a neutral etchant or deposition precursor. To conclude this section, we show using a simple model that pulsed discharges can have much lower average power flows than cw discharges for the same neutral etchant or precursor density. This is a widely used application of pulsed discharges.

10.6.1 Pulsed Electropositive Discharges

We consider a cylindrical argon discharge of radius R and length l, with uniform spatial distributions of plasma parameters over the bulk plasma volume, with the plasma density n_e in the bulk dropping sharply to edge values n_{sl} and n_{sR} at thin sheaths close to the axial and circumferential walls. Electron–ion pairs are assumed to be created by electron-impact ionization of the background gas and to be lost by diffusive flow to the walls. Including the time derivative in the particle balance equation (10.2.10), we have

$$\mathcal{V}\frac{dn_e}{dt} = K_{iz}n_e n_g \mathcal{V} - n_e u_B A_{eff} \tag{10.6.1}$$

where $\mathcal{V} = \pi R^2 l$ is the plasma volume, n_g is the argon gas density, and A_{eff} given by (10.2.11) is the effective area for particle loss given by low-pressure diffusion theory.

As in (3.5.11), the rate of collisional energy loss within the discharge volume can be expressed as

$$P_c = en_e n_g \mathcal{V} \sum K_i \mathcal{E}_i = en_e n_g \mathcal{V}(K_{iz}\mathcal{E}_{iz} + K_{ex}\mathcal{E}_{ex} + K_{el}\mathcal{E}_{el}) \tag{10.6.2}$$

where K_{iz}, K_{ex}, and K_{el} are the rate constants, and $\mathcal{E}_{iz}$, $\mathcal{E}_{ex}$, and $\mathcal{E}_{el}$ (in volts) are the energies lost per ionization, excitation, and elastic collision, respectively. Similarly, the part of the input power lost as kinetic energy of particles to the walls has the components, for ions,

$$P_i = e\left(V_s + \frac{1}{2}T_e\right) n_e u_B A_{eff} \tag{10.6.3}$$

and for electrons,

$$P_e = 2eT_e n_e u_B A_{eff} \tag{10.6.4}$$

where V_s given by (10.2.4) is the sheath voltage drop. Therefore, including the time derivative term, the entire power balance equation (10.2.14) is written as

$$P_{abs}(t) = \left[\frac{d}{dt}\left(\frac{3}{2}en_eT_e\right) + en_en_g\sum K_i\mathcal{E}_i\right]\mathcal{V} + e\left(V_s + \frac{5}{2}T_e\right)n_eu_BA_{eff} \tag{10.6.5}$$

where P_{abs} is the total power absorbed, assumed known. By numerically solving ordinary differential equations (10.6.1) and (10.6.5) simultaneously, we obtain $n_e(t)$ and $T_e(t)$.

We assume that the power in (10.6.5) is modulated by an ideal rectangular waveform

$$\begin{aligned} P_{abs} &= P_{max}, \quad 0 \le t < \eta\tau, \\ &= 0, \quad \eta\tau \le t < \tau, \end{aligned} \tag{10.6.6}$$

where η is the *duty ratio* and τ is the *period.* We can then solve the global equations to find the approximate transient behavior. Equation (10.6.1) can be written as

$$\frac{1}{n_e}\frac{dn_e}{dt} = \nu_{iz} - \nu_{loss} \tag{10.6.7}$$

where $\nu_{iz} = K_{iz}n_g$ is the ionization rate and $\nu_{loss} = u_B/d_{eff}$ is a characteristic low-pressure particle loss rate, with $d_{eff} = \mathcal{V}/A_{eff}$ given by (10.2.13); ν_{iz} depends strongly (exponentially) on T_e, while $\nu_{loss} \propto u_B \propto T_e^{1/2}$ depends only weakly on T_e. Using (10.6.7) to eliminate dn_e/dt in (10.6.5) yields

$$\frac{1}{T_e}\frac{dT_e}{dt} = \frac{P_{abs}(t)}{W_e} - \left(\frac{2}{3}\frac{\mathcal{E}_c}{T_e} + 1\right)\nu_{iz} - \left(\frac{2}{3}\frac{V_s + \frac{5}{2}T_e}{T_e} - 1\right)\nu_{loss} \tag{10.6.8}$$

where $W_e = \frac{3}{2}en_eT_e\mathcal{V}$ is the plasma energy, and we have used the usual definition (3.5.11) that the sum of the collisional energy losses can be combined into a single term with energy loss $\mathcal{E}_c$.

Consider times just after the pulse turns on. Initially n_e and T_e must build up, so we approximate (10.6.7) and (10.6.8) by

$$\frac{1}{n_e}\frac{dn_e}{dt} \approx \nu_{iz} \tag{10.6.9}$$

$$\frac{1}{T_e}\frac{dT_e}{dt} \approx \frac{P_{max}}{W_e} - \left(\frac{2}{3}\frac{\mathcal{E}_c}{T_e} + 1\right)\nu_{iz} \tag{10.6.10}$$

Since T_e is low initially, ν_{iz} is small and the second term on the RHS of (10.6.10) is small. This gives a very sharp rise in T_e at a rate P_{max}/W_e, up to some maximum value $T_{e\,max}$. We can estimate $T_{e\,max}$ by setting $dT_e/dt = 0$ in (10.6.10). From Figure 3.17, we recognize that $\mathcal{E}_c/T_e \gg 1$. Using this and substituting for W_e, we have

$$\nu_{iz} = n_gK_{iz}(T_e) \approx \frac{P_{max}}{n_ee\mathcal{E}_c\mathcal{V}} \tag{10.6.11}$$

Since (10.6.9) and (10.6.10) imply that T_e increases much faster than n_e, we can set n_e equal to the initial density, $n_e = n_{e\,min}$, in (10.6.11), yielding $T_{e\,max}$.

Beyond this time, (10.6.10) remains in quasi-steady state with $d/dt \approx 0$, and T_e falls toward its equilibrium steady-state value $T_{e\infty}$ as n_e increases. Since ν_{loss} varies slowly with T_e, we assume a constant value $\nu_\infty = u_{B\infty}/d_{\text{eff}}$ for ν_{loss}, where $u_{B\infty} = (e\mathrm{T}_{e\infty}/M)^{1/2}$. Substituting (10.6.11) into (10.6.7) and multiplying by n_e, we obtain

$$\frac{dn_e}{dt} \approx (n_{e\infty} - n_e)\nu_\infty \tag{10.6.12}$$

where

$$n_{e\infty} = \frac{P_{\max}}{e\mathcal{E}_c \nu_\infty \mathcal{V}} \tag{10.6.13}$$

the equilibrium value for an infinitely long pulse. The solution to (10.6.12) is

$$n_e(t) \approx n_{e\infty}(1 - e^{-\nu_\infty t}) + n_{e\min} e^{-\nu_\infty t}, \qquad 0 < t < \eta\tau \tag{10.6.14}$$

Substituting (10.6.14) into (10.6.11) yields $\nu_{iz}(t)$, from which $\mathrm{T}_e(t)$ can be determined.

Consider now times after the pulse is turned off (the "afterglow"). Then T_e falls such that $\nu_{iz} \ll \nu_{\text{loss}}$ in (10.6.7) and (10.6.8), yielding

$$\frac{1}{n_e}\frac{dn_e}{dt} \approx -\nu_{\text{loss}}(t) \tag{10.6.15}$$

and

$$\frac{1}{\mathrm{T}_e}\frac{d\mathrm{T}_e}{dt} \approx -\left(\frac{2}{3}\frac{V_s + \frac{5}{2}\mathrm{T}_e}{\mathrm{T}_e} - 1\right)\nu_{\text{loss}}(t) \tag{10.6.16}$$

Assuming a low-voltage sheath (10.2.4) for V_s, we find that the term in parentheses in (10.6.16) has the numerical value of $3.8 \approx 4$ for argon. We see that both T_e and n_e decay with time, but that the decay rate for T_e is faster than the decay rate of n_e by a factor of approximately four. Using $\nu_{\text{loss}}(t) = u_B(\mathrm{T}_e)/d_{\text{eff}} \propto \mathrm{T}_e^{1/2}$, we can solve (10.6.16) to obtain the temperature decay

$$\mathrm{T}_e(t) = \mathrm{T}_{e\infty}[1 + 2\nu_\infty(t - \eta\tau)]^{-2}, \qquad \eta\tau < t < \tau \tag{10.6.17}$$

where we have assumed that the pulse is sufficiently long that $\mathrm{T}_e \approx \mathrm{T}_{e\infty}$ at the end of the on-time. The density decay follows immediately by eliminating $\nu_{\text{loss}}(t)$ from (10.6.15) and (10.6.16):

$$n_e(t) \approx n_{e\max}[1 + 2\nu_\infty(t - \eta\tau)]^{-1/2}, \qquad \eta\tau < t < \tau \tag{10.6.18}$$

The character of the solutions for different pulse lengths can be seen from the results in Figure 10.10*a*–*c* (Ashida et al., 1995) for the complete argon model (with excited states). They show the time evolution of n_e, the excited atom densities, and T_e for a discharge with $p = 5$ mTorr (600 K gas temperature), $R = 15.25$ cm, and $l = 7.5$ cm, with three different periods τ for $P_{\text{abs}}(t)$. The time-averaged power was fixed at 500 W. Each of these graphs shows one cycle of the power on-off duration. During the on-time, 2000 W power is applied and during the off-time the power is 0 W; the duty ratio is 25%. Results representing the 500 W continuous wave (cw) case are also shown.

For a modulation period much less than 10 μs, the electron temperature responds weakly to the modulated power, while the plasma density hardly changes. Therefore both the electron density and the electron temperature are very close to those for the continuous 500 W case. For periods much greater than 10 μs, both the electron temperature and the plasma density respond to the applied modulated power. For all cases, the electron temperatures first rise sharply to peak values larger than those for the cw case, while the densities hardly change. After this, the temperatures fall and the densities rise, approaching quasi-steady values during the pulse on-times. After the pulse is turned off, the temperatures and densities decay toward zero; the temperature decays at a considerably faster rate than the density.

From the numerical results, we see that the analytic model best applies in the long pulse-length regime, where near-asymptotic values are obtained both during the on-time and the off-time. Comparing the analytic results to Figure 10.10*c*, and taking values of $T_{e\infty} = 3.2$ V and $d_{eff} = 8.5$ cm, we obtain $\nu_\infty \approx 3.3 \times 10^4\ s^{-1}$, in reasonable agreement with the numerical result from Figure 10.10*c*. Using this value in (10.6.17) and (10.6.18), the decay times from the analytic model of 16 and 61 μs for T_e and n_e are in good agreement with the decay times of 10 and 50 μs, respectively, from the more complete numerical model of Figure 10.10*c*.

It is worth noting that, for the same time-averaged power, the average plasma density for pulsed operation can be higher than the density for cw operation. In fact, in Figure 10.10*a* and *b*, the time-varying densities are higher than the cw density for *all* times. These results have been confirmed experimentally (Ashida et al., 1996; Tang and Manos, 1999). Physically, average pulsed densities are higher than steady-state densities because the electron temperature decreases rapidly after the power is turned off. This leads to a decrease of the loss rate of charged particles because the Bohm velocity, which accounts for the particle loss process, is proportional to the square root of the electron temperature. If the period is long compared to the time constants, as in Figure 10.10*c*, the electron temperature drops rapidly in the afterglow while the electron density drops more slowly. Consequently, there is a time during the afterglow when the plasma has a low T_e, which is particularly favored for some wafer etching processes, to reduce charging damage and distortions in pattern transfer during etching (see Section 16.6).

Processes that have threshold energies, e.g., ionization, excitation, and chemical reactions, often have rate constants of Arrhenius form that are approximately proportional to $\exp(-\mathcal{E}_a/T_e)$, with $\mathcal{E}_a$ the activation energy. The time-averaged production rates are then $\langle n_e(t)\exp(-\mathcal{E}_a/T_e(t))\rangle$. Higher $\mathcal{E}_a$s represent reactions such as ionization and high-energy electronic excitation. Lower $\mathcal{E}_a$s represents processes such as attachment and dissociation in molecular gases. The average production rate depends sensitively on the time variation of $T_e(t)$ and on $n_e(t)$, and, hence, on the pulse period and duty ratio. For example, if these parameters are chosen to yield a large initial T_e "spike" as shown in Figure 10.10*c*, then processes such as ionization and electronic excitation with high $\mathcal{E}_a$s can have higher production rates, compared to a steady-state discharge with the same time-averaged power. Because $\langle n_e\rangle$ for a pulsed discharge can be greater than n_e for a steady state-discharge, processes with very low activation energies ($\mathcal{E}_a \lesssim T_e$) can also have higher production rates. In contrast, for intermediate $\mathcal{E}_a$s, the time-averaged production rates of processes, such as low-energy dissociation, can be reduced.

We have used low-pressure diffusion theory to write $\nu_{loss} = u_B/d_{eff} \propto T_e^{1/2}$. At higher pressures, the particle losses are given from high-pressure diffusion theory (10.2.21) to be proportional to the ambipolar diffusion coefficient D_a, where $D_a \propto T_e$, i.e., $\nu_{loss} = \nu_\infty T_e/T_{e\infty}$, with ν_∞ the appropriate loss rate at the end of the on-time given from high-pressure diffusion theory. Using this, ν_{loss} in (10.6.15) and (10.6.16) yields (Problem 10.13)

$$\begin{aligned} T_e(t) &\approx T_{e\infty}[1 + 4\nu_\infty(t - \eta\tau)]^{-1} \\ n_e(t) &\approx n_{e\max}[1 + 4\nu_\infty(t - \eta\tau)]^{-1/4} \end{aligned} \qquad \eta\tau < t < \tau \tag{10.6.19}$$

These decays have been found to be in reasonable agreement with PIC simulations (Smith, 1998).

The approximation of an ideal rectangular power waveform $P_{abs}(t)$ is quite simplified from a real discharge for which P_{abs} may vary as the electron density builds up and may not be exactly zero during the off period if there is a continuous bias voltage in addition to the pulsed power. It also does not account for the time response of the matching network (see Sections 11.9 and 12.1) used to couple the source power to the plasma. Furthermore, for long off-times, a pulsed discharge can extinguish or can enter a different operating mode during the initial phase of the on-time, e.g.,

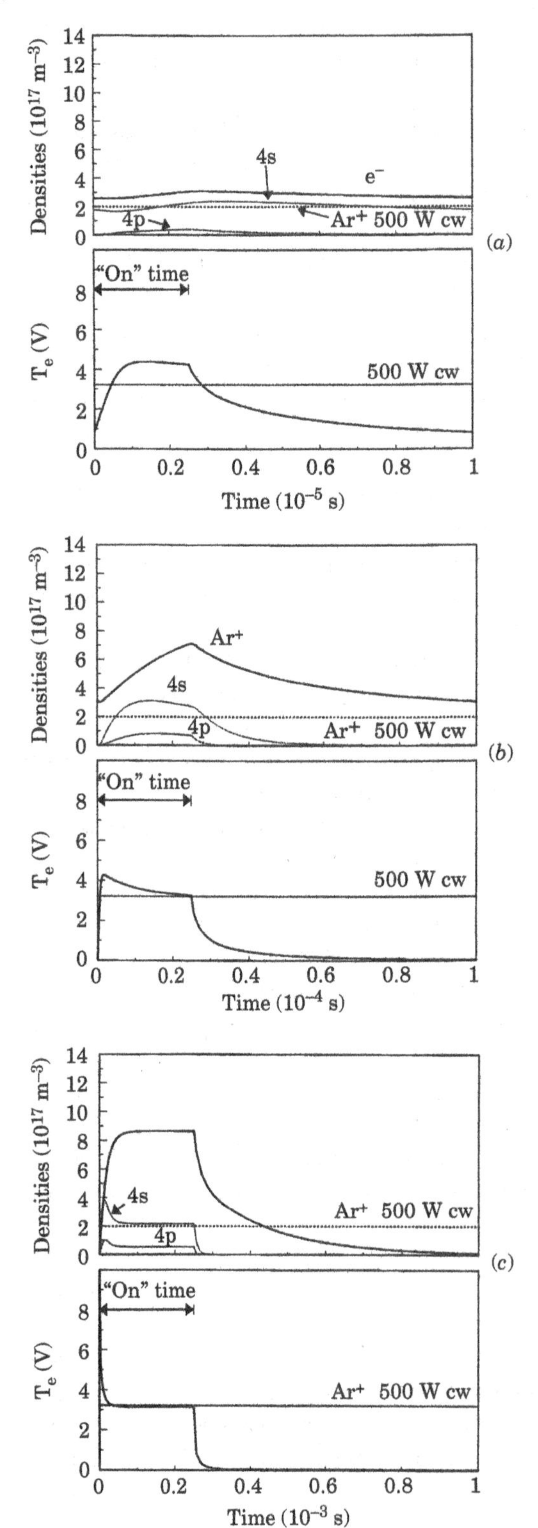

Figure 10.10 Time evolution of the plasma density n_e, the electron temperature T_e, and the excited atom (4s and 4p) densities for different periods τ, for a time-averaged power of 500 W and a duty ratio of 0.25: (*a*) $\tau = 10$ μs; (*b*) $\tau = 100$ μs; (*c*) $\tau = 1$ ms. Source: Ashida et al., 1995/with permission of AIP Publishing.

a capacitively coupled mode for an inductive or helicon discharge, or a "low mode" (Carl et al., 1991) for an electron cyclotron discharge. In such cases, new physical phenomena can arise, such as weak power absorption (Carl et al., 1991) or multipacting (Boswell and Vender, 1995), which are not described well by a rectangular power absorption waveform.

10.6.2 Pulsed Electronegative Discharges

The use of an electronegative molecular gas greatly complicates the analysis of particle and energy balance in high-density, low-pressure discharges, as seen in Sections 10.3 and 10.4. Even for steady state power, the high fractional dissociation of the molecular gas implies that neutral particle balance equations for the dissociation fragments are important. These in turn may depend on poorly known rate constants at the chamber walls for recombination, reaction, etc. Furthermore, there can be multiple positive and negative ions, such that the usual assumption of ambipolar diffusion for the charged particle fluxes may not be valid. It is also seen that the discharge can stratify into an electronegative core region, surrounded by an electropositive halo region. Hence, the assumption of relatively uniform particle density profiles in volume-averaged models may not be adequate.

In typical high-density, low-pressure, cw processing discharges, the ratio n_-/n_e of negative ion to electron density can be less than or of order unity, even with highly electronegative feedstocks such as Cl_2 (see Section 10.5). Hence, the major issue may not be the negative ion dynamics, but the dissociation of the gas into multiple neutral and positive ion species. This can also be the situation during the on-time of a pulsed discharge. However, for low pulsing frequencies, the situation changes markedly when the power is turned off. During the off-time, T_e rapidly decreases due to energy loss processes such as ionization, electronic and vibrational excitation, dissociation, and elastic scattering, and n_e decreases due to dissociative attachment to the molecular gas and diffusive losses to the walls. Because negative ions are confined within the discharge by the positive space charge potential there, they are not initially lost, and their density can initially increase for some attaching gases (those having an attachment rate constant that increases with decreasing T_e). At some point in time, when n_e has dropped to a low enough value compared to n_-, the potential collapses to near-zero, resulting in a diffusive flux of negative ions to the wall. This flux can have important effects on materials processing at surfaces. For example, it has been studied as a way to prevent notching in narrow trenches (Hwang and Giapis, 1998).

The condition for potential collapse can be estimated from charge conservation at the walls:

$$\Gamma_e + \Gamma_- = \Gamma_+ \tag{10.6.20}$$

or, setting edge densities approximately equal to center densities for a strongly electronegative plasma,

$$\frac{1}{4} n_e \bar{v}_e \mathrm{e}^{-\Phi/T_e} + \frac{1}{4} n_- \bar{v}_- \mathrm{e}^{-\Phi/T_i} = n_+ u_B \tag{10.6.21}$$

For $\Phi \sim T_e \gg T_i$, the flux of negative ions in (10.6.21) is essentially zero. Neglecting this term and solving for Φ yields

$$\Phi \approx T_e \ln\left[\frac{n_e}{n_+}\left(\frac{M_+}{2\pi m}\right)^{1/2}\right] \tag{10.6.22}$$

Hence, Φ collapses to near-zero at a time $t = t_0$ when

$$n_+(t_0) \approx n_-(t_0) \approx \left(\frac{M_+}{2\pi m}\right)^{1/2} n_e(t_0) \tag{10.6.23}$$

After this time, a significant flux of negative ions can escape to the walls.

Let us examine the negative ion dynamics in the afterglow of a recombination-dominated discharge. Although the initial negative ion density may be small compared to the electron density, we consider that $\alpha(t) \equiv n_-(t)/n_e(t) > 1$ during most of the decay. Furthermore, because of the strong energy loss processes that operate at higher electron energy, we assume that the electron temperature falls to some relatively small value T_{e0}, which holds during most of the decay. For notational simplicity, taking $t = 0$ to be the time when the power pulse is turned off, from (10.6.7) with $\nu_{iz} = 0$, we have

$$n_+(t) = n_+(0)e^{-\nu_{loss}t} \tag{10.6.24}$$

At sufficiently low pressures that the recombination loss is small compared to the wall flux, we have $\nu_{loss} = u_B(T_{e0})A_{eff}/\mathcal{V}$. A_{eff} is taken appropriately for a highly electronegative plasma, where we can often use the simple approximation that $A_{eff} \approx A = 2\pi R^2 + 2\pi Rl\,(h_l = h_R = 1)$. The evolution of the negative ion density is described by

$$\frac{dn_-}{dt} = K_{att}n_e(t)n_g - K_{rec}n_+(t)n_-(t) \tag{10.6.25}$$

where K_{att} and K_{rec} are, as in previous sections, the dissociative attachment and positive–negative ion recombination rate constants, and n_g is the molecular (e.g., O_2 or Cl_2) gas density. Since the temperature dependence of K_{att} is not usually strong, we can assume it to be constant, and using quasineutrality ($n_e = n_+ - n_-$) and (10.6.24), (10.6.25) can be integrated to obtain $n_-(t)$. The general behavior of the result can be seen by making the quasistatic assumption $dn_-/dt \approx 0$ in (10.6.25). This is justified because n_e in (10.6.25) varies on a fast timescale compared to n_-. Substituting $n_e = n_+ - n_-$ into (10.6.25) with $dn_-/dt = 0$, we obtain

$$n_-(t) = \frac{K_{att}n_gn_+(t)}{K_{rec}n_+(t) + K_{att}n_g} \tag{10.6.26}$$

and substituting from (10.6.24), we find

$$n_-(t) = \frac{K_{att}n_gn_+(0)\,e^{-\nu_{loss}t}}{K_{rec}n_+(0)\,e^{-\nu_{loss}t} + K_{att}n_g} \tag{10.6.27}$$

Calculating $n_e(t) = n_+(t) - n_-(t)$ using (10.6.24) for n_e and (10.6.27) for n_-, we then form the ratio $\alpha(t) \equiv n_-(t)/n_e(t)$ to obtain

$$\alpha(t) = \alpha_0\,e^{\nu_{loss}t} \tag{10.6.28}$$

where $\alpha_0 \equiv n_-(0)/n_e(0) = K_{att}n_g/K_{rec}n_+(0)$ is the initial electronegativity. Setting $\alpha(t) \approx (M_+/2\pi m)^{1/2}$ according to (10.6.23) and substituting for ν_{loss} gives an estimate for the time to potential collapse,

$$t_0 \approx \frac{\mathcal{V}}{2u_BA_{eff}}\ln\left(\frac{M_+}{2\pi m\alpha_0^2}\right) \tag{10.6.29}$$

Since $u_B \propto T_{e0}^{1/2}$ depends only weakly on T_{e0}, its value need not be known very accurately, but may be estimated either from the lowest-lying important excitation energy, or taken as the ion temperature itself.

The preceding description of positive and negative ion dynamics must be modified in detachment-dominated plasmas (Kaganovich et al., 2000) or in higher pressure regimes. For low initial electronegativities, the global model obscures the physics of the expansion of the negative ion core as the electron density decays. In addition, when the space charge potential V_s no longer confines the negative ions, there is a significant negative ion flux to the walls, and the resulting loss

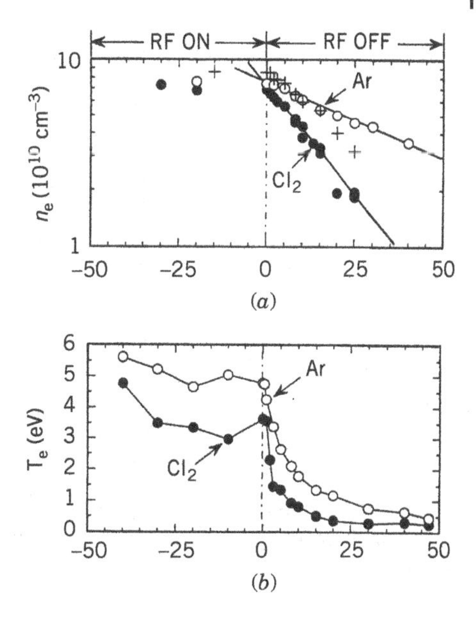

Figure 10.11 Time variation of (*a*) electron density n_e and (*b*) electron temperature T_e for 100 μs period and 0.50 duty ratio in chlorine (8 mTorr, 400 W) and in argon (6 mTorr, 200 W); the open and closed circles indicate the data for Ar and Cl_2, respectively; the crosses in (*a*) indicate the data obtained after photodetachment. Source: Ahn et al., 1995/with permission of IOP Publishing.

term must be included. The subsequent decay of this positive–negative ion plasma is not governed by the low-pressure diffusion solutions (10.2.1) and (10.2.2) for an electropositive plasma. Some of these effects have been incorporated into a one-dimensional model (Kaganovich, 2001).

Ahn et al. (1995) studied experimentally the afterglow in an inductively excited pulsed chlorine discharge ($\tau = 100$ μs, $\eta = 0.5$). They measured n_+ and T_e using a planar Langmuir probe, n_e using an electron-beam excited plasma oscillation detection method, and n_{Cl^-} using a laser photodetachment method, in which the electrons were detached from the negative ions by an 0.3 μs ultraviolet XeCl excimer laser pulse and the sudden increase in n_e was measured. Figure 10.11 shows n_e and T_e in chlorine ($p = 8$ mTorr, P_{in}(on) = 400 W), and, for comparison, in argon ($p = 6$ mTorr, P_{in}(on) = 200 W). We see the characteristic feature in the afterglow that T_e falls more rapidly than n_e for both chlorine and argon, as predicted for argon by (10.6.17) and (10.6.18). The decay of n_e for chlorine is faster than the decay for argon for two reasons: (1) T_e falls faster for chlorine due to the higher collisional energy losses in the molecular component of the neutral gas, and (2) n_e falls faster than n_{Cl^+} because n_{Cl^-} increases or remains relatively constant during the decay of n_e. The crosses in Figure 10.11*a* give n_e after photodetachment (sum of $n_e + n_{Cl^-} = n_{Cl^+}$ before photodetachment) and the solid dots give n_e. By subtraction, one obtains the negative ion density variation. In fact, these data show that n_{Cl^-} increases from approximately 1×10^{10} cm^{-3} at the beginning of the afterglow $t = 0$, to a maximum of approximately 2×10^{10} cm^{-3} at $t = 25$ μs, after which n_{Cl^-} slowly decreases. This type of behavior is predicted by (10.6.26) if the attachment rate ν_{att} increases from its initial value to a larger value within the afterglow. This is indeed the case for chlorine; ν_{att} increases by about a factor of six as T_e varies from 3 to 0.05 V in the afterglow (Ashida and Lieberman, 1997).

Malyshev et al. (1999b) measured the time dependences of electron, positive ion, and negative ion densities and electron temperature in a 13.56 MHz inductively coupled chlorine plasma for pressures between 3 and 20 mTorr. An on-time of 50 μs and off-times from 30 to 100 μs were used. They found good agreement to the global model results, with the T_e time variation during the off-time well described by (10.6.17).

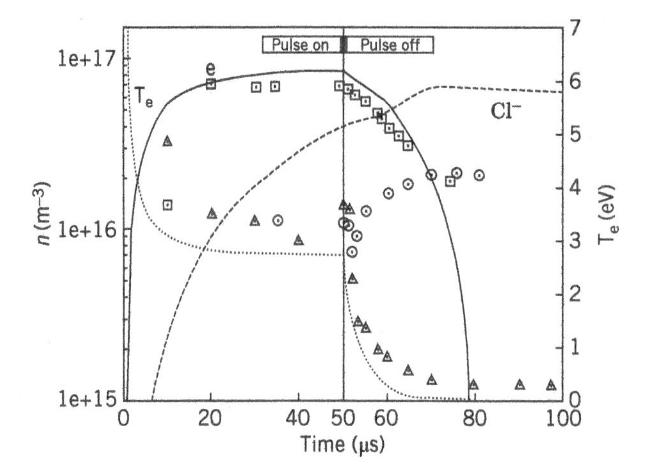

Figure 10.12 Global simulation results for $n_e(t)$, $n_-(t)$, and $T_e(t)$ (lines) for the chlorine measurements (symbols) of Ahn et al. (1995) shown in Figure 10.11. Source: Reproduced with permission from Kemaneci et al. (2014)/IOP Publishing.

Thorsteinsson and Gudmundsson (2010c) and Kemaneci et al. (2014) have compared the chlorine measurements of Ahn et al. (1995) to a pulsed power global model. Figure 10.12 shows the comparison of Kemaneci et al. (2014) for $n_e(t)$, $n_-(t)$, and $T_e(t)$; the symbols are the measurements and the lines give the model results. The model agrees well with the measurements, although in the afterglow the model decay time for T_e is somewhat shorter than that measured. Also, the n_- model results are larger than the photodetachment-measured values (open circles), but both show an increasing Cl^- density over the first 25 microseconds of the afterglow, followed by a slow decay. The large spread in measured n_- values just after discharge power turn-off may be due to measurement error.

Thorsteinsson and Gudmundsson (2010c) compared the global model $T_e(t)$s with those measured by Ahn et al. (1995), finding good agreement. They also did a comparison to the measurements of Malyshev et al. (1999b), for a 100 μs, 50% duty ratio-pulsed chlorine discharge, finding reasonable agreement for $T_e(t)$ and the electronegativity $\alpha(t) = n_-(t)/n_e(t)$, but also some discrepancies in $n_e(t)$ and $n_-(t)$, some of which could be due to changes in the power coupling efficiency, especially near discharge turn-on.

Ramamurthi and Economou (2002) developed a two-dimensional $(r,\ z)$ fluid model of a 100 μs period, 50% duty ratio, square-wave modulated chlorine discharge. The discharge was inductively driven by a planar coil across an upper quartz window of radius 5.7 cm, with a cylinder height of 3.9 cm (see Figure 12.1*b* and Section 12.3 for the configuration). The base case pressure was 20 mTorr and the peak power was 320 W. Their results showed the separation of the plasma into an electronegative core and an electropositive edge during the 50 μs power-on time, and the formation of an ion–ion plasma 15 μs into the afterglow (power off), as suggested by (10.6.29). They observed strong electric-field-driven negative ion flows during the early power-on stage. The time-averaged positive ion flux was fairly uniform radially over the first 3 cm of the substrate electrode, and then dipped slightly. The Cl-atom flux at the substrate electrode was found to be strongly center high. They made comparisons to both continuous (cw) and pulsed experiments, finding reasonable agreement for cw discharges, but some discrepancies for the variation of n_e with duty ratio for pulsed discharges.

10.6.3 Neutral Radical Dynamics

The generation of neutral etchant or deposition precursors at greatly reduced power levels has been an important application of pulsed discharges. To understand this, we consider a simple model of the neutral dynamics

$$\frac{dn}{dt} = 2K_{diss}n_e n_g - \nu_{loss}n \tag{10.6.30}$$

where K_{diss} is the rate constant for electron impact dissociation of a diatomic gas (density n_g) to produce the radicals or precursors (density n), and ν_{loss} is the radical loss frequency to the walls. Because n varies with time, n_g does also. These variations are generally fast compared to the loss frequency $S_p/\mathcal{V}_p$ for the vacuum pump, where S_p is the pumping speed and $\mathcal{V}_p$ is the chamber volume. Hence, we can write

$$2n_g(t) + n(t) = n_{a0} \approx \text{const} \tag{10.6.31}$$

where n_{a0} is the total density of atoms in both atomic and molecular forms.

Eliminating n_g from (10.6.30) using (10.6.31), we obtain

$$\frac{dn}{dt} = K_{diss}n_e n_{a0} - (K_{diss}n_e + \nu_{loss})n \tag{10.6.32}$$

For a given n_e, setting $dn/dt = 0$, we obtain the corresponding steady-state density

$$n_\infty = \frac{K_{diss}n_e n_{a0}}{K_{diss}n_e + \nu_{loss}} \tag{10.6.33}$$

which is also the on-time density for a long pulse. Let us consider now a time-varying density $n_e(t)$. Assuming that n is initially small and that n_e turns on, then from (10.6.32) and (10.6.33), we find that n rises to its steady-state value on a timescale

$$\tau_{rise} \approx \frac{1}{K_{diss}n_e + \nu_{loss}} \tag{10.6.34}$$

Also from (10.6.32), when n_e suddenly turns off, we obtain a characteristic decay time $\tau_{loss} \approx 1/\nu_{loss}$ for n. A typical case given by Lieberman and Ashida (1996) for a chlorine discharge has $\tau_{rise} \approx$ 0.4 ms and $\tau_{loss} \approx 1.7$ ms, which are both long compared to the characteristic electron density and temperature rise and decay times.

Equation (10.6.32) can be solved to determine $n(t)$ for a given $n_e(t)$ and $T_e(t)$. We can assume rectangular waveforms for $n_e(t)$ and $T_e(t)$, because their rise and decay times are much shorter than those of n. The solution (Problem 10.14) has various regimes depending on whether the radical rise time τ_{rise} is greater than or less than the on-time τ_{on} of the pulse, and on whether the radical decay time τ_{loss} is greater than or less than the off-time τ_{off} of the pulse. The interesting regime is for $\tau_{on} > \tau_{rise}$ and $\tau_{loss} > \tau_{off}$. In this regime, the radical density n builds up to and remains nearly at its steady-state value and varies only weakly with time: $n(t) \approx$ const. Since n is determined by the on-time value of n_e, it depends on the on-time power P_{max}. Therefore, the *time-averaged* radical density $\bar{n}$ depends only on the *peak* power P_{max}, and not on the average power P_{abs}. Holding P_{max} fixed, we can then decrease the on-time to of order τ_{rise} and increase the off-time to of order τ_{loss}. This reduces the average power to the walls (and substrate) from P_{max} (for cw operation) to approximately $P_{max}\tau_{rise}/\tau_{loss}$, without much affecting the neutral radical flux. In the example above with $\tau_{rise} = 0.4$ ms and $\tau_{loss} = 1.7$ ms, this corresponds to roughly 25% of P_{max}.

Charles et al. (1995) and Charles and Boswell (1998) studied silicon dioxide deposition from a 1 to 2 mTorr oxygen/silane feedstock in a pulsed helicon discharge with $P_{max} = 800$–900 W. In one set of experiments, they fixed the duty ratio at 50% ($\tau_{on} = \tau_{off}$) and measured the average deposition rate from low pulse frequencies up to 1 kHz. They observed that deposition continues long after the plasma is extinguished, obtaining a time constant for the process of about 200 ms. The corresponding plasma decay time constant was about 1 ms. In another set of experiments in a different reactor, they fixed the pulse length at 500 μs and varied the duty ratio. The deposition rate increased by a factor of 2.5 as the duty ratio was varied from 10 to 100%. They found that deposition continued in the post-discharge with a time constant of 1 ms. The corresponding plasma decay time was 130 μs. These experiments imply that by properly pulsing the discharge, the power flux to the substrate can be significantly reduced, without reducing significantly the deposition rate.

Problems

10.1 Low-Pressure Equilibrium

(a) Using the method outlined in Example 10.1, calculate d_{eff}, A_{eff}, $\mathrm{T_e}$, $\mathcal{E}_c$, $\mathcal{E}_i$, $\mathcal{E}_T$, and u_B, for Example 10.2. Confirm that $n_0 \approx 5.2 \times 10^{17}\ \mathrm{m^{-3}}$ and $J_{il} \approx 7.0\ \mathrm{mA/cm^2}$.

(b) Repeat for Example 10.3, confirming that $n_0 \approx 7.5 \times 10^{16}\ \mathrm{m^{-3}}$ and $J_{il} \approx 1.0\ \mathrm{mA/cm^2}$.

10.2 High-Pressure Argon Discharge Consider a cylindrical argon plasma of radius $R = 5$ cm, length $l = 30$ cm, pressure $p = 20$ mTorr, and absorbed power $P_{\text{abs}} = 500$ W. Assume that the ionization rate is $\nu_{iz} n(r, z)$, with $\nu_{iz} =$ const, and that there is diffusive loss to the cylinder side and end walls with a constant axial and radial (ambipolar) diffusion coefficient D_a.

(a) Assuming that the ion neutral mean free path $\lambda_i \ll R, l$ such that the plasma density $n \approx 0$ at the cylinder side and end walls, show that

$$n(r, z) \approx n_0 \mathrm{J_0}(\chi_{01} r/R) \cos(\pi z/l)$$

where $\chi_{01} \approx 2.405$ is the first zero of the zero-order Bessel function $\mathrm{J_0}(\chi)$.

(b) Determine $\mathrm{T_e}$ (V) by equating the total (axial + radial) particle loss rate to the total particle creation rate. (Integrate the particle flux $-D_a \nabla n$ over the wall area to obtain the former, and integrate $\nu_{iz} n(r, z)$ over the cylinder volume to obtain the latter.)

(c) At high pressures, the ion bombarding energy is due to the sheath voltage V_s, given by (10.2.4), that develops at the walls. Assuming that the sheath thickness $s \ll \lambda_i$, equate ion and electron fluxes at the walls to show that $\mathcal{E}_i \approx 5.2\mathrm{T_e}$.

(d) From energy balance, estimate the central density n_0 ($\mathrm{cm^{-3}}$) and the total current I_z (amperes) incident on one end wall. Assume low-voltage sheaths at all surfaces.

10.3 High-Pressure Argon Discharge with Local Ionization Consider a one-dimensional slab model $0 < x < l$ of a high-pressure argon discharge in which the ionization rate is localized near the left-hand plate: $G(x) = \nu_{iz} n_e(x)$ for $0 < x < \lambda < l$ and $G = 0$ otherwise. Here, ν_{iz} is a constant ionization frequency and $\lambda \propto p^{-1}$ is an energy diffusion mean free path for ionizing electrons heated locally near the left hand plate. Let D_a be the ambipolar diffusion coefficient and assume boundary conditions that $n \approx 0$ at both walls.

(a) Show that the solution for $n(x)$ can be written as

$$\begin{aligned} n &= n_0 \sin \beta x, && 0 < x < \lambda \\ &= n_0 \frac{l - x}{l - \lambda} \sin \beta\lambda, && \lambda < x < l \end{aligned}$$

where $\beta^2 = \nu_{iz}/D_a$.

(b) Find the equation that determines β (and hence ν_{iz}). The equation can only be solved numerically.

(c) Find β (numerically) for $\lambda = 1$ cm and $l = 5$ cm. Sketch the corresponding solution for $n(x)$.

(d) For $\lambda = 1$ cm, $l = 5$ cm, plate diameter = 20 cm, and an input power absorbed of 50 W, find the peak density n_0. Assume low-voltage sheaths at both plates with $T_e = 3$ V and $T_i = 0.026$ V.

10.4 Detaching Neutral Species Density Derive expression (10.3.7) for the detaching neutral species density n_* using (10.3.5) and (10.3.6).

10.5 Detaching Neutral Species Wall and Volume Losses For an oxygen discharge with a given n_-, n_e, $\mathcal{V}/A$, and $\overline{v}_*$, and with K_{dex*} and K_{det} for reactions 16, and the sum of 20 and 21, respectively, in Table 8.2, find the condition on the recombination coefficient γ_* for surface losses of n_* to be larger than volume losses.

10.6 De-excitation Losses for Detaching Neutral Species Consider the effect of de-excitation losses on the density n_* of an excited neutral detaching species in a one-dimensional slab model. Then, (10.3.5) is modified to

$$-D_*\nabla^2 n_* = K_{ex*} n_e n_g - K_{dex*} n_e n_*$$

Assume a uniform density distribution for n_* and n_e.

(a) Find an expression for n_* analogous to (10.3.7), but including both de-excitation and wall losses.

(b) For nominal values in oxygen of $T_e = 3$ V, $T_* = 0.026$ V, $\gamma_* = 10^{-3}$, $l = 5$ cm, and K_{dex*} given by reaction 16 in Table 8.2, find the condition on $\overline{n}_e$ for wall losses to dominate de-excitation losses.

10.7 Ambipolar Diffusion Coefficient in Electronegative Plasmas Assuming that both electrons and negative ions are in Boltzmann equilibrium, derive the ambipolar diffusion coefficient (10.3.21).

10.8 Electronegative Discharge at Low-Pressure and High n_{e0} Assume that α_0 is in the range where the parabolic solution (10.4.23) without an ion sound limitation holds, l_1/d can be approximated by (10.4.34), and h_{le} in (10.4.28) is given approximately by $h_{le} \approx (2\nu_{iz}\lambda_i/\pi u_B)^{1/3}$.

(a) For a plasma with negative ion loss dominated by recombination, such that α_0 can be approximated by (10.4.30), find the scaling of α_0 and l_1/d, separately, as functions of the input parameters $n_g d$ and $n_{e0} d$.

(b) For a plasma with negative ion loss dominated by detachment, such that α_0 can be approximated by (10.4.32), find the scaling of α_0 and l_1/d, separately, as functions of the input parameters $n_g d$ and $n_{e0} d$.

10.9 Parabolic Solution for Electronegative Equilibrium For an oxygen discharge with negative ion loss dominated by recombination, with $p = 50$ mTorr, $n_{e0} = 2.4 \times 10^{15}$ m^{-3}, and $d = 4.5$ cm, use the equations for the approximate parabolic solution in the electronegative core and equations (10.4.27) and (10.4.28) in the electropositive region to find α_0

and l_1. Assume an initial value of $\mathrm{T_e} = 3$ V and iterate your solution once. Sketch your results and compare them with Figure 10.2.

10.10 Ion and Neutral Radical Densities in a High-Pressure Discharge Repeat the analysis leading to the scalings (10.2.31) and (10.2.40) of n_{is} and n_{OS} with discharge parameters for a high-pressure slab model of a high-density discharge. Assume that the discharge is electropositive and that the ion flux to the wall is determined by an ambipolar diffusion coefficient $D_{\mathrm{a}} \propto n_{\mathrm{g}}^{-1}$.

10.11 High-Density Oxygen Discharge Model The recombination probability for O atoms on quartz walls is very low. Consider a high-density oxygen discharge slab model (thickness $l = 10$ cm) in a quartz chamber at low pressures. Assume that the only volume reactions are 3, 4, 11, and 12 in Table 8.2 for generation of O, $\mathrm{O_2^+}$, and $\mathrm{O^+}$ due to electron impact. Assume further that K_{O}, $K_{\mathrm{O_2^+}}$, and $K_{\mathrm{O^+}}$ are the first-order rate constants for loss of O, $\mathrm{O_2^+}$ and $\mathrm{O^+}$ to the vacuum pump and/or to the walls. Let $K_{\mathrm{O}} = S_{\mathrm{p}}/\mathcal{V} = 30\ \mathrm{s^{-1}}$, where S_{p} is the pumping speed and $\mathcal{V}$ is the discharge volume. Let $K_{\mathrm{O_2^+}} = 2u_{\mathrm{BO_2^+}}/l$ and $K_{\mathrm{O^+}} = 2u_{\mathrm{BO^+}}/l$. Assume that all heavy particles are at 300 K, that there are no other sources for generation or loss of O, $\mathrm{O_2^+}$, and $\mathrm{O^+}$, and that $\mathrm{O^-}$ generation is negligible.

(a) Estimate the first-order rate constant for loss of O atoms to the walls due to recombination for a recombination probability on quartz of 10^{-4}, and compare this with the value of K_{O} given above due to the vacuum pump.

(b) Write the steady-state rate equations for n_{O}, $n_{\mathrm{O_2^+}}$, and $n_{\mathrm{O^+}}$.

(c) Find an expression for $n_{\mathrm{O}}/n_{\mathrm{O_2}}$ as a function of n_{e} and the rate constants K_3, K_4, K_{11}, K_{12}, and K_{O}. For $K_{12}n_{\mathrm{e}} \gg K_{\mathrm{O}}$, show that

$$\frac{n_{\mathrm{O}}}{n_{\mathrm{O_2}}} = \frac{2K_3 + K_{11}}{K_{12}}$$

independent of n_{e}. Evaluate $n_{\mathrm{O}}/n_{\mathrm{O_2}}$ and the condition on n_{e} to achieve this high-density limit for an electron temperature $\mathrm{T_e} = 3$ V. Show that $n_{\mathrm{O}} \gg n_{\mathrm{O_2}}$.

(d) Find an expression for $n_{\mathrm{O^+}}/n_{\mathrm{O_2^+}}$ in terms of n_{e} and the rate constants. In the high-density limit, show that

$$\frac{n_{\mathrm{O^+}}}{n_{\mathrm{O_2^+}}} = \frac{2K_{\mathrm{O_2^+}}}{K_{\mathrm{O^+}}}\frac{K_3 + K_{11}}{K_4}$$

Evaluate this in the high-density limit for $\mathrm{T_e} = 3$ V, and show that $n_{\mathrm{O^+}} \gg n_{\mathrm{O_2^+}}$.

(e) Consider now the volume reactions 2, 8, and 13 in Table 8.2 for $\mathrm{O^-}$ generation and loss. Find $n_{\mathrm{O^-}}$ in the high-density limit and show that $n_{\mathrm{O^-}} \ll n_{\mathrm{e}}$.

10.12 Electronegative Equilibrium in Cylindrical Coordinates Using the form of the solution (5.2.33) for high-pressure diffusion in an infinite cylinder with a uniform source of ionization, obtain algebraic equations for the electronegative core plasma, analogous to (10.4.25) and (10.4.26) for the plane parallel case. Comment on the difference between these equations and those of the plane-parallel case.

10.13 Afterglow of a High-Pressure, Pulsed Power Discharge Using $\nu_{\mathrm{loss}} = D_{\mathrm{a}}/l_{\mathrm{eff}}^2$, as in (5.2.11), where $D_{\mathrm{a}} \approx e\mathrm{T_e}/M_+\nu_{\mathrm{mi}}$ is the ambipolar diffusion coefficient and $l_{\mathrm{eff}} \approx l/\pi$, then find the time dependence (10.6.19) for the decay of $n_{\mathrm{e}}(t)$ and $\mathrm{T_e}(t)$ for a high-pressure argon discharge model after the power has been turned off. Assume that ν_{mi} is a constant.

10.14 **Neutral Radical Dynamics in a Pulsed Power Discharge** Assume a rectangular waveform model for electron density and temperature in a pulsed electropositive discharge. During the on-time, $n_e = n_{e\infty}$ and $T_e = T_{e\infty}$; during the off-time, n_e and T_e are approximately zero.

(a) Show that the solution of (10.6.32) for the neutral radical dynamics $n(t)$ during the on-time is

$$n(t) = n_\infty - (n_\infty - n_{min})\, e^{-\nu_{rise} t}$$

(b) Show that the solution of (10.6.32) during the off-time is

$$n(t) = n_{max}\, e^{-\nu_{loss}(t-\eta\tau)}$$

where η is the duty ratio.

(c) Setting $n(t) = n_{max}$ at $t = \eta\tau$ in (a) and $n(t) = n_{min}$ at $t = \tau$ in (b), solve the resulting two equations to obtain n_{min} and n_{max} in terms of n_∞.

11

Low-Pressure Capacitive Discharges

As discussed in the previous chapter, a complete description of a plasma discharge requires a choice of heating mechanisms to sustain it. These mechanisms play essential roles in determining the plasma density, the voltages between the plasma and the surfaces, and the bombarding ion energies. In this and the next two chapters, we discuss the main types of low-pressure processing discharges. One of the most widely used is sustained by radio-frequency (rf) currents and voltages applied directly to an electrode immersed in the plasma. This creates a high-voltage capacitive sheath between the electrode and the bulk plasma. The rf currents flowing across the sheath and through the bulk plasma lead to *stochastic* or *collisionless* heating in the sheath (see Section 19.4 for a kinetic description) and ohmic heating in the bulk. The complete self-consistent model is quite complicated, even in the simplest plane-parallel geometry. This leads to simplifying assumptions in order to obtain analytic solutions in which the scalings of plasma parameters with control parameters are explicit. The heating mechanisms and resulting plasma parameters are the subject of this chapter.

An important feature distinguishing low- and higher-pressure regimes is the size of the energy relaxation length $\lambda_\mathcal{E} = (\lambda_\mathrm{m}\lambda_\mathrm{inel}/3)^{1/2}$ for ionizing electrons, compared with the discharge gap length l. Here, λ_m is the momentum transfer mean free path, and λ_inel is the mean free path accounting for all collisional electron energy losses. (See Section 15.2 for a discussion of the energy relaxation length.) For low pressures, $\lambda_\mathcal{E} \gtrsim l$, and no matter where the electrons are heated, they flow everywhere, leading to a relatively uniform ionization rate constant K_iz within the gap. As we will see, in capacitive discharges, electrons are mainly heated within or near the sheath edges. The global models introduced in this chapter are well-suited to treating this situation. In contrast, at the higher pressures, $\lambda_\mathcal{E} \ll l$, the electrons can be locally heated, producing local ionization, with K_iz a strong function of position within the gap. High-pressure capacitive discharges are described in Chapter 15.

In the 1970s, Godyak and collaborators, finding clear experimental evidence for collisionless heating (see Figure 11.7), developed a simple model, by approximating the plasma and the sheath as having homogenous densities and the electron distribution as Maxwellian (see Godyak, 1986). Considerable insight into the behavior of capacitive discharges can be obtained from the homogeneous model, which we describe in Section 11.1. However, because simplifying assumptions are made, the model only partially predicts the quantitative behavior of "real" discharges. In Section 11.2, we consider sheath and plasma nonuniformities in symmetric discharges and develop formulae from which more realistic calculations can be made. We also describe various model limitations and alternate explanations. In Section 11.3, we give comparisons to symmetric experiments and computer simulations. Most discharges are asymmetric because more electrode surfaces are naturally

Principles of Plasma Discharges and Materials Processing, Third Edition. Michael A. Lieberman and Allan J. Lichtenberg.

grounded rather than driven. This leads to a dc bias voltage on the driven electrode with respect to ground. We describe asymmetric discharges in Section 11.4.

As shown in Sections 11.1 and 11.2, the voltage across a single sheath is nonlinear for a sinusoidal current. For a symmetric discharge, the two sheath nonlinearities cancel or nearly cancel, and it little matters whether the discharge is current or voltage driven. For an asymmetric discharge, with one dominant sheath, this is no longer true. The nonlinear current of a sinusoidally voltage-driven sheath can excite a high harmonic series resonance between the sheath capacitance and the bulk plasma inductance. Square-wave modulated voltage bursts of rf can also be useful as an ion flux diagnostic. We consider voltage-driven rf sheaths in Section 11.5.

Multiple-frequency-driven capacitive discharges provide additional control of critical processing parameters. For example, two widely separated incommensurate frequencies can be used to independently control the ion flux to the substrate and the ion-bombarding energy. The first and second harmonic frequencies can be used to electrically tune the ion-bombarding energy by varying the second harmonic phase. More than two harmonics can further control discharge characteristics. These topics are discussed in Section 11.6.

For high frequencies and/or large areas, electromagnetic effects such as standing waves and skin effects can arise in capacitive discharges, which cannot be described using conventional electrostatic analysis. These effects are due to surface waves that propagate radially in the discharge, and they can strongly modify the transverse (radial) discharge uniformity in narrow gap discharges. We describe these phenomena in Section 11.7.

In Sections 11.1–11.7, the applied frequency is sufficiently high and the plasma density is sufficiently low that the ion transit time across the sheath is long compared to the rf period. This is not the case for all capacitive discharges; lower frequencies are often used, for practical considerations or desirability in some applications. Further, in the high-density discharges that we describe in Chapters 12 and 13, the substrate holder is often capacitively driven at a lower frequency. We treat these lower-frequency and/or higher-density rf sheaths in Section 11.8. The ion energy distribution is strongly affected by the transit time effects. We examine these distributions for long and short transit time regimes in Section 11.9.

One approach to improve the performance of capacitive discharges involves application of a dc magnetic field lying in the plane of the driven electrode. These discharges, *magnetically enhanced reactive ion etchers* (MERIEs) or *rf magnetrons*, are described in Section 11.10. Capacitive discharges are commonly driven by 50 Ω rf power sources, typically at 13.56 MHz, although lower and higher frequencies are also used. For efficient power transfer, the power source must drive the discharge through a matching network. We describe matching network operation and rf power measurement techniques in Section 11.11.

11.1 Homogeneous Model

Figure 11.1*a* shows the basic symmetric discharge model. A sinusoidal current $I_{rf}(t)$, having complex representation $I_{rf} = \mathrm{Re}\,\tilde{I}_{rf}e^{j\omega t}$, flows across discharge plates a and b. Here, we take $\tilde{I}_{rf} = I_1$, a real number. The plates are separated by a distance l and each has a cross-sectional area A. A gas having neutral density n_g is present between the plates. In response to the current flow, a discharge plasma forms between the plates, accompanied by a voltage $V(t)$ across the plates and a power flow $P(t)$ into the plasma. The plasma has an ion density $n_i(\mathbf{r}, t)$ and an electron temperature $\mathrm{T_e}(\mathbf{r}, t)$. Because of quasineutrality, $n_e \approx n_i$ almost everywhere except within the oscillating sheaths near

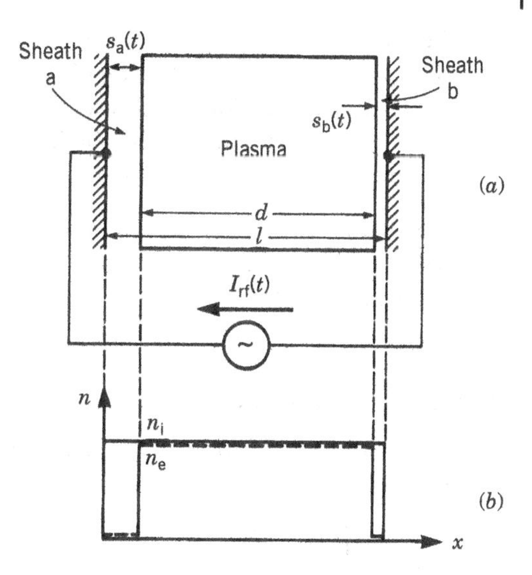

Figure 11.1 The basic rf discharge model: (*a*) sheath and plasma thicknesses; (*b*) electron and ion densities.

the plates, where $n_e < n_i$. The instantaneous sheath thickness is $s(t)$ and its time-averaged value is $\bar{s}$. Typically, $\bar{s} \ll l$.

The state of the discharge is specified once a complete set of control parameters is given. The remaining plasma and circuit parameters are then specified as functions of the control parameters. A convenient choice for the control parameters is I_{rf}, ω, n_g, and l. Given these, we develop the basic model to determine n_e, T_e, s, $\bar{s}$, V, and P. The choice of control parameters is not unique. We choose I rather than V or P, for ease of analysis.

In general, the discharge parameters n_e, n_i, and T_e are complicated functions of position and time. We assume the following to simplify the analysis:

(a) The ions respond only to the time-averaged potentials. This is a good approximation provided

$$\omega_{pi}^2 \ll \omega^2$$

where ω_{pi} is the ion plasma frequency.

(b) The electrons respond to the instantaneous potentials and carry the rf discharge current. This is a good approximation provided

$$\omega_{pe}^2 \gg \omega^2 \left(1 + \frac{\nu_m^2}{\omega^2}\right)^{1/2}$$

where ω_{pe} is the electron plasma frequency and ν_m is the electron–neutral collision frequency for momentum transfer.

(c) The electron density is zero within the sheath regions. This is a good approximation provided $\lambda_{De} \ll \bar{s}$, where λ_{De} is the electron Debye length. This holds if $T_e \ll \overline{V}$, where $\overline{V}$ is the dc voltage across the sheath.

(d) There is no transverse variation (along the plates). This is a good approximation provided $l \ll \sqrt{A}$ and provided that there are no electromagnetic wave and skin effects (Raizer et al., 1995; Lieberman et al., 2002; see Section 11.7). With these assumptions, a one-dimensional (along x) electrostatic solution of Maxwell's equations determines the fields. Since the divergence of Maxwell's equation $\nabla \times \mathbf{H} = \mathbf{J} + \epsilon_0 \partial \mathbf{E}/\partial t$ is zero, we see that, at any instant of time, the sum of the conduction current $\mathbf{J}$ and the displacement current $\epsilon_0 \partial \mathbf{E}/\partial t$ within the discharge is independent of x.

These assumptions hold both for the uniform model of this section and for the inhomogeneous model of Section 11.2. For the simplified model in this section, we also assume the following:

(e) The ion density is uniform and constant in time everywhere in the plasma and sheath regions: $n_i(\mathbf{r}, t) = n =$ const. The electron and ion density profiles for the simplified model are shown in Figure 11.1*b*, corresponding to the position of the plasma as shown in Figure 11.1*a*.

As we shall see in Section 11.2, the variation of the ion density in the sheath, which we obtain from a Child law calculation as in Section 6.3, considerably modifies the results obtained here using the approximation (e).

11.1.1 Plasma Admittance

The admittance of a bulk plasma slab of thickness d and cross-sectional area A is $Y_p = j\omega\epsilon_p A/d$, where

$$\epsilon_p = \epsilon_0 \left[1 - \frac{\omega_{pe}{}^2}{\omega(\omega - j\nu_m)}\right] \tag{11.1.1}$$

is the plasma dielectric constant given by (4.2.18). We show below that, within the uniform ion density approximation

$$d = l - 2\bar{s} = \text{const} \tag{11.1.2}$$

independent of time. We then find the plasma admittance (see Problem 11.1)

$$Y_p = j\omega C_0 + \frac{1}{j\omega L_p + R_p} \tag{11.1.3}$$

where $C_0 = \epsilon_0 A/d$ is the vacuum capacitance, $L_p = \omega_{pe}^{-2} C_0^{-1}$ is the plasma inductance, and $R_p = \nu_m L_p$ is the plasma resistance. This form for Y_p represents the series combination of L_p and R_p in parallel with C_0. By assumption (b), the displacement current that flows through C_0 is much smaller than the conduction current that flows through L_p and R_p. The sinusoidal current

$$I_{rf}(t) = \text{Re}\, \tilde{I}_{rf} e^{j\omega t} \tag{11.1.4}$$

that flows through the plasma bulk produces a voltage across the plasma

$$V_p(t) = \text{Re}\, \tilde{V}_p e^{j\omega t} \tag{11.1.5}$$

where $\tilde{V}_p = \tilde{I}_{rf}/Y_p$ is the complex voltage amplitude. We see that the plasma voltage is linear in the applied current and that there is no harmonic generation (multiples of ω) or dc component of V_p. For low pressures and $\omega \ll \omega_{pe}$, $\tilde{V}_p$ and the corresponding plasma electric field $\tilde{E}_p$ are small.

11.1.2 Sheath Admittance

In contrast to the plasma, the current that flows through the two sheaths is almost entirely displacement current, i.e., it is due to a time-varying electric field. This is true because the conduction current in a discharge is carried mainly by electrons, and the electron density is approximately zero within the time-varying sheath. We will see that the conduction current carried by the steady flow of ions across the sheath to the plates is much smaller than the displacement current.

11.1.2.1 Displacement Current

The electric field $\mathbf{E} = \hat{x}E$ within sheath a (see Figure 11.1) is given by Poisson's equation

$$\frac{\mathrm{d}E}{\mathrm{d}x} = \frac{en}{\epsilon_0}, \qquad x \leq s_\mathrm{a}(t) \tag{11.1.6}$$

which on integration yields

$$E(x,t) = \frac{en}{\epsilon_0}[x - s_\mathrm{a}(t)] \tag{11.1.7}$$

The boundary condition is $E \approx 0$ at $x = s_\mathrm{a}$ because E is continuous across the plasma–sheath interface (no surface charge) and the electric field is small in the plasma. The displacement current flowing through sheath a into the plasma is

$$I_\mathrm{ap}(t) = \epsilon_0 A \frac{\partial E}{\partial t} \tag{11.1.8}$$

Substituting (11.1.7) into (11.1.8), we obtain

$$I_\mathrm{ap}(t) = -enA\frac{\mathrm{d}s_\mathrm{a}}{\mathrm{d}t} \tag{11.1.9}$$

From (11.1.9), the sheath boundary s_a oscillates linearly with the applied current. Setting $I_\mathrm{ap}(t) = I_\mathrm{rf}(t)$, where $I_\mathrm{rf} = I_1 \cos\omega t$, we integrate (11.1.9) to obtain

$$s_\mathrm{a} = \bar{s} - s_0 \sin\omega t \tag{11.1.10}$$

where

$$s_0 = \frac{I_1}{en\omega A} \tag{11.1.11}$$

is the sinusoidal oscillation amplitude about the dc value $\bar{s}$. The voltage across the sheath is given by

$$V_\mathrm{ap}(t) = \int_0^{s_\mathrm{a}} E\,\mathrm{d}x = -\frac{en}{\epsilon_0}\frac{s_\mathrm{a}^2}{2} \tag{11.1.12}$$

From (11.1.12), the sheath voltage is a nonlinear function of s_a and therefore of the applied current. Substituting (11.1.10) into (11.1.12), we obtain

$$V_\mathrm{ap} = -\frac{en}{2\epsilon_0}\left(\bar{s}^2 + \frac{1}{2}s_0^2 - 2\bar{s}s_0\sin\omega t - \frac{1}{2}s_0^2\cos 2\omega t\right) \tag{11.1.13}$$

We see that the nonlinearity leads to both second-harmonic voltage generation and a constant average value.

Similarly, for sheath b, we obtain

$$I_\mathrm{bp} = -enA\frac{\mathrm{d}s_\mathrm{b}}{\mathrm{d}t} \tag{11.1.14}$$

and the voltage across this sheath is

$$V_\mathrm{bp} = -\frac{en}{\epsilon_0}\frac{s_\mathrm{b}^2}{2} \tag{11.1.15}$$

By continuity of current, $I_\mathrm{bp} = -I_\mathrm{ap}$, so that adding (11.1.9) and (11.1.14), we find

$$\frac{\mathrm{d}}{\mathrm{d}t}(s_\mathrm{a} + s_\mathrm{b}) = 0$$

Integrating, we obtain

$$s_\mathrm{a} + s_\mathrm{b} = 2\bar{s}, \text{ a constant} \tag{11.1.16}$$

so that $d = l - 2\bar{s} = \text{const}$, as previously stated. For sheath b,

$$s_{\mathrm{b}} = \bar{s} + s_0 \sin \omega t \tag{11.1.17}$$

with the nonlinear voltage response, using (11.1.15),

$$V_{\mathrm{bp}} = -\frac{en}{2\epsilon_0}\left(\bar{s}^2 + \frac{1}{2}s_0^2 + 2\bar{s}s_0 \sin \omega t - \frac{1}{2}s_0^2 \cos 2\omega t\right) \tag{11.1.18}$$

Although V_{ap} and V_{bp} are nonlinear, the combined voltage $V_{\mathrm{ab}} = V_{\mathrm{ap}} - V_{\mathrm{bp}}$ across both sheaths, obtained by subtracting (11.1.18) from (11.1.13), is

$$V_{\mathrm{ab}} = \frac{en\bar{s}}{\epsilon_0}(s_{\mathrm{b}} - s_{\mathrm{a}})$$

Substituting for s_{b} and s_{a} from (11.1.10) and (11.1.17), we find

$$V_{\mathrm{ab}} = \frac{2en\bar{s}s_0}{\epsilon_0} \sin \omega t \tag{11.1.19}$$

which is a linear voltage response. We obtain the surprising result that although each sheath is nonlinear, the combined effect of both sheaths is linear. This is true only for the simplified model assumptions of a symmetric, homogeneous (constant ion density) discharge. The total voltage $\tilde{V}_{\mathrm{rf}}$ across the discharge is the sum of $\tilde{V}_{\mathrm{ab}}$ and $\tilde{V}_{\mathrm{p}}$. However, for typical low-pressure discharge conditions, we usually have $|\tilde{V}_{\mathrm{p}}| \ll |\tilde{V}_{\mathrm{ab}}|$, and we often approximate $\tilde{V}_{\mathrm{rf}} \approx \tilde{V}_{\mathrm{ab}}$.

11.1.2.2 Conduction Current

Although the conduction current in each sheath is small, the average sheath thickness $\bar{s}$ is determined by the balance between ion and electron conduction currents. By assumption (a), there is a steady flow of ions from the plasma through sheath a, carrying a steady current

$$\bar{I}_{\mathrm{i}} = enu_{\mathrm{B}}A \tag{11.1.20}$$

where the loss velocity is taken to be the Bohm velocity u_{B}.

By symmetry, the time-averaged conduction current flowing to plate a is zero. There is a steady flow of ions to the plate. For the basic model, the electron density is assumed zero in the sheath. The sheath thickness $s_{\mathrm{a}}(t)$ must therefore collapse to zero at some time during the rf cycle in order to transfer electrons from the plasma to the plate. It follows from (11.1.10) and (11.1.11) that

$$\bar{s} = s_0 = \frac{I_1}{en\omega A} \tag{11.1.21}$$

and from (11.1.13) that

$$V_{\mathrm{pa}} = \frac{en}{2\epsilon_0}s_0^2(1 - \sin \omega t)^2 \tag{11.1.22}$$

Since the sheath voltage collapses to zero at the time that the electrons are transferred to the plate, this acts like an ideal diode across the sheath whose preferred direction of current flow is into the plasma. A similar result holds for sheath b.

We can define a linear sheath capacitance C_{s} because the voltage (11.1.19) across both sheaths is sinusoidal. Differentiating (11.1.19) and substituting for I_{rf} using (11.1.21), we obtain the simple result

$$I_{\mathrm{rf}} = C_{\mathrm{s}}\frac{\mathrm{d}V_{\mathrm{ab}}}{\mathrm{d}t}$$

where

$$C_s = \frac{\epsilon_0 A}{2s_0} \tag{11.1.23}$$

is a linear capacitance. Physically, this capacitance is the series combination of the two nonlinear capacitances $C_a = \epsilon_0 A/s_a(t)$ and $C_b = \epsilon_0 A/s_b(t)$. We see that ignoring the impedance of the bulk plasma, the homogeneous sheath model gives a linear discharge rf voltage–current characteristic. This is also seen to a good approximation in many experiments (Godyak, 2020), particularly at lower driving voltages. Therefore, it can be highly useful in understanding and modeling the discharge physics.

The dc sheath voltage is found by time averaging (11.1.22)

$$\overline{V} \equiv \overline{V}_{pa} = \frac{3}{4}\frac{en}{\epsilon_0}s_0^2 \tag{11.1.24}$$

Let us note from (11.1.22) and (11.1.24) that if $V_1 = ens_0^2/\epsilon_0$ is the fundamental component of the rf voltage amplitude across one sheath, then the time-maximum sheath voltage is $2V_1$ and the dc voltage is

$$\overline{V} = \frac{3}{4}V_1 = \frac{3}{8}V_{ab} \tag{11.1.25}$$

The voltages $V_{ap}(t)$, $V_{pb}(t)$, and their sum $V_{ab}(t)$ are plotted versus t in Figure 11.2. The manner in which the sum of the two nonsinusoidal voltages yields the $V_{ab} = 2V_1$ amplitude sinusoid is clearly seen. The time-averaged value $\overline{V}$ for V_{pb} is also shown as the horizontal dashed line.

The spatial variation of the total potential at various times within the rf cycle is shown (solid lines) in Figure 11.3. It is assumed that the right-hand electrode is grounded (held at $V = 0$ at all times). The dashed curve shows the spatial variation of the time-averaged potential.

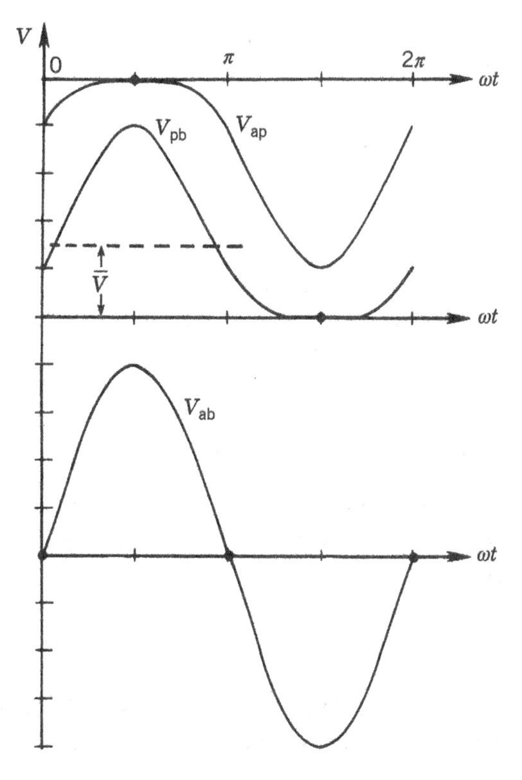

Figure 11.2 Sheath voltages V_{ap}, V_{pb}, and their sum V_{ab} versus time; the time-average value $\overline{V}$ of V_{pb} is also shown.

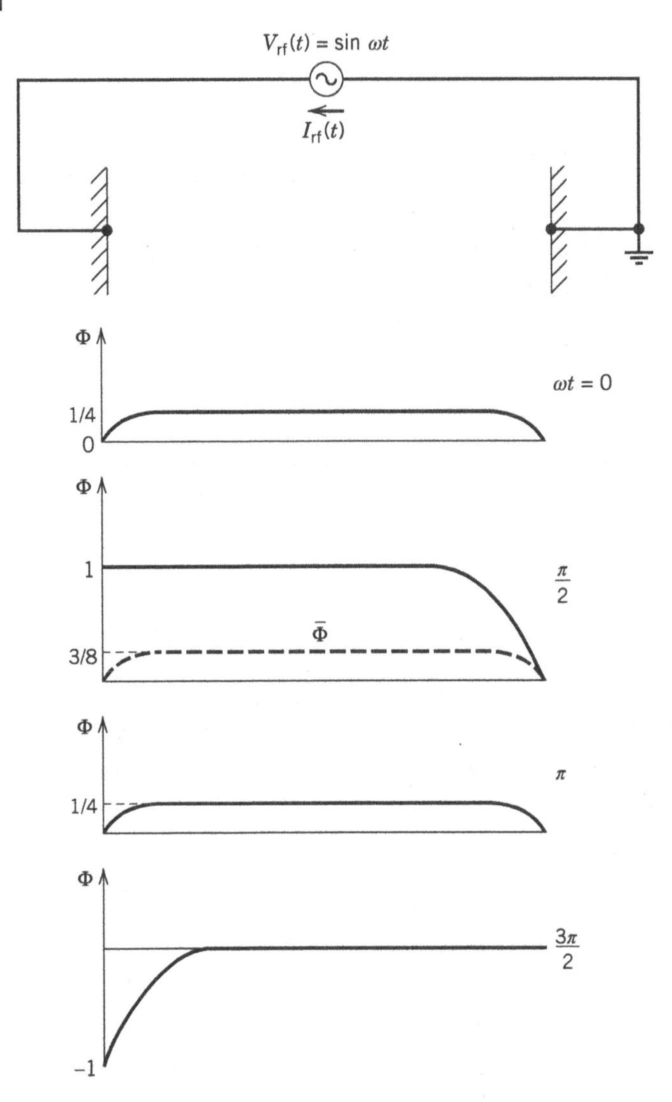

Figure 11.3 Spatial variation of the total potential Φ (solid curves) for the homogeneous model of Section 11.1, at four different times during the rf cycle. The dashed curve shows the spatial variation of the time-averaged potential $\bar{\Phi}$.

11.1.3 Particle and Energy Balance

To complete the analysis, we need to evaluate expressions for particle and energy balance as developed in Chapter 10. Particle balance per unit area, for a uniform plasma, is straightforwardly given by

$$nK_{iz}n_g d = 2nu_B \tag{11.1.26}$$

as in (10.2.12), with $d_{eff} = d/2$. If the sheaths are thin, such that $d \approx l$, we can evaluate the temperature from (11.1.26) alone.

To calculate the plasma density, we must evaluate the time-averaged power per unit area absorbed by the electrons, S_e, which involves the rf currents and voltages, and the sheath oscillations.

11.1.3.1 Ohmic Heating

The time-averaged power per unit area deposited by ohmic heating in the bulk plasma, $\overline{S}_{ohm}$, is due to collisional momentum transfer between the oscillating electrons and the neutrals. Integrating (4.2.30) over the bulk plasma length d, we obtain

$$\overline{S}_{ohm} = \frac{1}{2} J_1^2 \frac{d}{\sigma_{dc}} \tag{11.1.27}$$

where $J_1 = I_1/A$ and σ_{dc} is the dc plasma conductivity. Substituting (4.2.22) for σ_{dc} into (11.1.27), we find

$$\overline{S}_{ohm} = \frac{1}{2} J_1^2 \frac{m\nu_m d}{e^2 n} \tag{11.1.28}$$

11.1.3.2 Stochastic Heating

Electrons reflecting from the large decelerating fields of a moving high-voltage sheath can be approximated by assuming the reflected velocity is that which occurs in an elastic collision of a ball with a moving wall

$$u_r = -u + 2u_{es} \tag{11.1.29}$$

where u and u_r are the incident and reflected electron velocities parallel to the time-varying electron sheath velocity u_{es}. If the parallel electron velocity distribution at the sheath edge is $f_{es}(u, t)$, then in a time interval dt and for a speed interval du, the number of electrons per unit area that collide with the sheath is given by $(u - u_{es})f_{es}(u, t)du dt$. This results in a power transfer per unit area,

$$dS_{stoc} = \frac{1}{2} m(u_r^2 - u^2)(u - u_{es})f_{es}(u, t)\, du \tag{11.1.30}$$

Using $u_r = -u + 2u_{es}$ and integrating overall incident velocities, we obtain

$$S_{stoc} = -2m \int_{u_{es}}^{\infty} u_{es}(u - u_{es})^2 f_{es}(u, t)\, du \tag{11.1.31}$$

In the physical problem, f_{es} varies with time, as the sheath oscillates, and the problem becomes quite complicated. For our uniform density model, we note that

$$\int_{-\infty}^{\infty} f_{es}(u, t)\, du = n_{es}(t) = n, \text{ a constant} \tag{11.1.32}$$

Furthermore, for the purpose of understanding the heating mechanism, we make the simplifying approximation that $f_{es}(u, t)$ can be approximated by a Maxwellian, ignoring the bulk plasma rf oscillation velocity u_{es}. Then, we can set the lower limit in (11.1.31) to zero. If the oscillation of the bulk plasma is self-consistently included in the above calculation, then the homogeneous model does not predict collisionless (stochastic) heating. Nevertheless, the above simple picture allows us to better understand the more self-consistent results of the inhomogeneous model, described in the next section, for which the calculation is considerably more complicated. The inhomogeneous model is also not fully self-consistent, and we return to this more subtle question at the end of Section 11.2. Before performing the average over the distribution function, we substitute

$$u_{es} = u_0 \cos \omega t \tag{11.1.33}$$

in (11.1.31) and average over time. Only the term in $\sin^2\omega t$ survives giving

$$\overline{S}_{\text{stoc}} = 2mu_0^2 \int_0^\infty u f_{\text{es}}(u)\,du \tag{11.1.34}$$

Now, consistent with our approximation that f_{es} is Maxwellian, we note that the integral gives the usual random flux $\Gamma_{\text{e}} = \frac{1}{4}n\overline{v}_{\text{e}}$, and (11.1.34) becomes

$$\overline{S}_{\text{stoc}} = \frac{1}{2}mu_0^2 n\overline{v}_{\text{e}} \tag{11.1.35}$$

Inside the plasma the rf current is almost entirely conduction current, such that

$$I_1 = J_1 A = -enu_0 A \tag{11.1.36}$$

Substituting (11.1.36) into (11.1.35) yields the stochastic electron power in terms of the (assumed) known current. Since we are calculating the power per unit area, we use the current density, to obtain, for a single sheath,

$$\overline{S}_{\text{stoc}} = \frac{1}{2}\frac{m\overline{v}_{\text{e}}}{e^2 n}J_1^2 \tag{11.1.37}$$

11.1.4 Discharge Parameters

Adding (11.1.37) (for two sheaths) and (11.1.28), the total time-averaged electron power per unit area is

$$S_{\text{e}} = \frac{1}{2}\frac{m\nu_{\text{eff}} d}{e^2 n}J_1^2 \tag{11.1.38}$$

where

$$\nu_{\text{eff}} = \nu_{\text{m}} + \frac{2\overline{v}_{\text{e}}}{d} \tag{11.1.39}$$

is an effective collision frequency for the sum of ohmic and stochastic heating. Assuming J_{rf}, ω, n_{g}, A, and l are the specified control parameters, we equate the electron energy deposited in the plasma to the electron energy lost from the plasma:

$$S_{\text{e}} = 2enu_{\text{B}}(\mathcal{E}_{\text{c}} + \mathcal{E}_{\text{e}}') \tag{11.1.40}$$

In (11.1.40), the kinetic energy $\mathcal{E}_{\text{e}}'$ lost per electron lost from the plasma is not the same as the kinetic energy $\mathcal{E}_{\text{e}} = 2\text{T}_{\text{e}}$ lost per electron hitting the wall. This is because (see discussion following (10.2.4)) an electron crossing the presheath and sheath fields loses an energy $V_{\text{s}} + \frac{1}{2}\text{T}_{\text{e}}$, with V_{s} given by (10.2.4). Hence, we have $\mathcal{E}_{\text{e}}' = \mathcal{E}_{\text{e}} + V_{\text{s}} + \frac{1}{2}\text{T}_{\text{e}} \approx 7.2\text{T}_{\text{e}}$ for argon. Setting (11.1.38) equal to (11.1.40) and solving for n, we obtain

$$n = \frac{1}{2}\left[\frac{m\nu_{\text{eff}} d}{e^3 u_{\text{B}}(\mathcal{E}_{\text{c}} + \mathcal{E}_{\text{e}}')}\right]^{1/2} J_1 \tag{11.1.41}$$

With the temperature assumed known from (11.1.26) and if we again let $d \approx l$, the density can be calculated. With n known, the sheath thickness is calculated from (11.1.21). If $2\overline{s}$ is a significant fraction of l, then we determine $d \approx l - 2\overline{s}$, and the equations can be iterated to determine more accurate values for T_{e}, n, and $d = l - 2\overline{s}$. However, this iteration compromises the simplicity of the model.

Finally, to obtain the total power dissipated, we must calculate the power lost by the ions. To do this, we need the average voltage across each sheath, which is given in (11.1.24):

$$\overline{V} = \frac{3}{4}\frac{en}{\epsilon_0}s_0^2 = \frac{3}{4}\frac{J_1^2}{e\epsilon_0 n\omega^2} \tag{11.1.42}$$

where the second equality comes from substituting for s_0 from (11.1.21). Using (10.2.14), the power per unit area lost by the ions is

$$S_i = 2enu_B\overline{V} = \frac{3}{2}u_B\frac{J_1^2}{\epsilon_0\omega^2} \tag{11.1.43}$$

where the factor of two is for two sheaths. The total power absorbed per unit area, S_{abs}, is found by adding (11.1.38) and (11.1.43).

The stochastic heating $\overline{S}_{stoc}$ leads to equivalent sheath resistances R_a and R_b defined by $\overline{S}_{stoc} = (1/2)J_1^2AR_{a,b}$. These resistances are in series with the sheath capacitances, as shown in Figure 11.4. The ion heating S_i can be modeled as equivalent dc current sources $\overline{I}_i = \overline{J}_iA$, as shown in the figure. Because this dc current flows across a dc sheath voltage $\overline{V}$, it represents a power dissipation within the sheath. Note that R_a and $\overline{I}_i$ are not constants, but are functions of the rf voltage. For low-pressure discharges, the inductive and resistive impedances of the bulk plasma are small compared to the capacitive impedance of the sheaths, such that almost all of the applied rf voltage appears across the two sheath capacitors. This situation is described in Chapter 4 (see Figure 4.4, along with accompanying discussion in the text). Although the voltage drops across the resistors are generally small for an electropositive plasma, the power dissipation due to the flow of current through these resistors is important, as we have described. At very high frequencies (typically much exceeding 13.56 MHz), the bulk plasma inductance and the sheath capacitance can resonate, leading to a *series resonance discharge* regime in which the overall rf voltage across the discharge drops to a very low value, of order a few volts (see Section 11.5, and Problems 4.7 and 11.4). In a high-pressure or strongly electronegative plasma with $n_e \ll n_i$, the voltage drops across the resistors can be comparable to, or even exceed, the voltage drops across the sheath capacitors, and the discharge can enter a resistive regime (see Chapter 15).

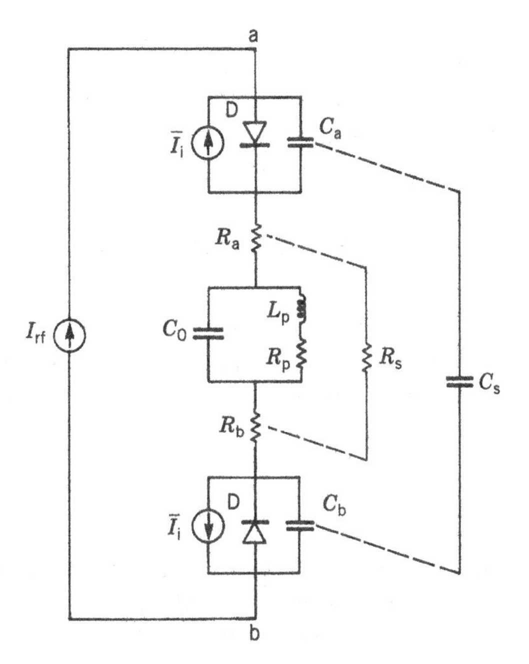

Figure 11.4 Nonlinear circuit model of the homogeneous rf plasma discharge. The dashed lines indicate that the series connection of the nonlinear elements C_a and C_b, and R_a and R_b, yield the corresponding linear elements C_s and R_s, respectively.

In real devices, the control parameter is usually V_{rf} or S_{abs}, rather than J_{rf}. This would make the above calculations more cumbersome. We examine this in the next section, and in Section 11.5, where we make more quantitatively correct calculations and give examples of calculating the parameters in real discharges.

11.2 Inhomogeneous Model

In this section, we describe a realistic inhomogeneous model for a high-voltage capacitive discharge and give the set of equations that are required for a quantitative calculation of the discharge parameters. For the inhomogeneous model, we retain approximations (a)–(d) in Section 11.1, but allow the plasma and the sheath to be inhomogeneous. The inhomogeneity in the plasma is not critical, taking different forms depending on the pressure, as discussed in detail in Chapters 5 and 10. The inhomogeneous sheath, however, strongly modifies the results, and the consequences of this are the main subject of this section. The basic processes are the following. The decreasing ion density within the sheath between the plasma–sheath edge and the collecting boundary leads to a Child law variation of the density and an increased sheath width compared to the matrix sheath width in Section 11.1. It also leads to an increase in the sheath velocity in the regions of decreasing ion density. This follows because the rf current must be continuous, while the electron density is decreasing to preserve charge neutrality. The result is a substantial increase in the stochastic sheath heating. Due to the partial shielding of the ion space charge by the oscillating electrons, the Child law (6.3.12) for the ions is also modified. The increase in sheath width decreases the total sheath capacitance. A self-consistent analysis must consider all of these effects together. The analysis for a collisionless sheath is given somewhat briefly in the first part of this section; a more detailed calculation can be found in Lieberman (1988). The results required to make a quantitative calculation of the discharge parameters are summarized in (11.2.32)–(11.2.38), and their use is illustrated in several following examples. The reader who wishes to calculate parameters for a given discharge can skip to these equations without following the preceding analysis.

At higher pressures where the ion mean free path $\lambda_i < s_m$, the sheath width, collisional models similar to those described in Section 6.5 must be used to describe the self-consistent sheath dynamics. We summarize the results for these models in this section. We also briefly describe nonideal effects for the self-consistent sheath, including low to moderate rf driving voltages, ohmic heating in the sheaths, and self-consistency conditions for collisionless heating.

11.2.1 Collisionless Sheath Dynamics

The structure of the rf sheath is shown in Figure 11.5. Ions crossing the ion sheath boundary at $x = 0$ accelerate within the sheath and strike the electrode at $x = s_m$ with high energies. Since the ion flux $n_i u_i$ is conserved and u_i increases as ions transit the sheath, n_i drops. This is sketched as the heavy solid line in Figure 11.5. The ion particle and energy conservation equations are, respectively,

$$n_i u_i = n_s u_B \tag{11.2.1}$$

$$\frac{1}{2} M u_i^2 = \frac{1}{2} M u_B^2 - e\overline{\Phi} \tag{11.2.2}$$

where n_s is the plasma density at the plasma–sheath edge at $x = 0$ and $\overline{\Phi}$ is the time-averaged potential within the sheath; $\overline{\Phi}$, n_i, and u_i are functions of x. The Poisson equation for the instantaneous electric field $E(x, t)$ within the sheath is

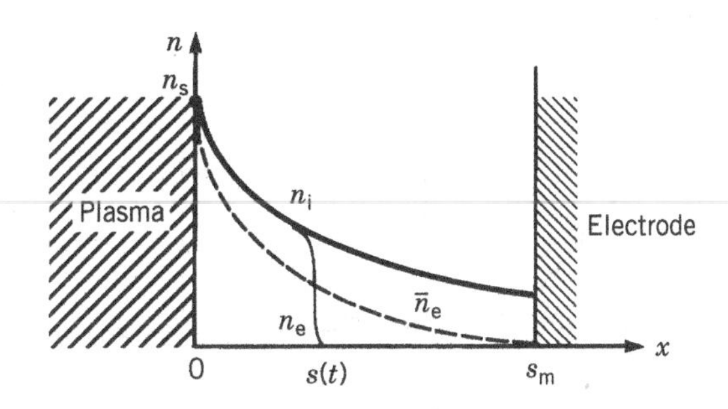

Figure 11.5 Schematic plot of the densities in a high-voltage, capacitive rf sheath.

$$\frac{\partial E}{\partial x} = \frac{e}{\epsilon_0} n_i(x) \qquad s(t) < x$$
$$= 0 \qquad s(t) > x \tag{11.2.3}$$

Here, $s(t)$ is the distance from the ion sheath boundary at $x = 0$ to the electron sheath edge.

Time averaging (11.2.3) over an rf cycle, we obtain the equations for the time-averaged electric field $\overline{E}(x)$:

$$\frac{d\overline{E}}{dx} = \frac{e}{\epsilon_0}\left(n_i(x) - \overline{n}_e(x)\right) \tag{11.2.4}$$

$$\frac{d\overline{\Phi}}{dx} = -\overline{E} \tag{11.2.5}$$

where $\overline{n}_e(x)$ is the time-averaged electron density within the sheath. We determine $\overline{E}$ and $\overline{n}_e$ from $s(t)$ as follows. We note that $n_e(x, t) = 0$ during the part of the rf cycle where $s(t) < x$; otherwise, $n_e(x, t) = n_i(x)$. We therefore have

$$\overline{n}_e(x) = \left(1 - \frac{2\phi}{2\pi}\right) n_i(x) \tag{11.2.6}$$

where $2\phi(x) = 2\omega t$ is the phase interval during which $s(t) < x$. Qualitatively, we sketch $\overline{n}_e$ as the dashed line in Figure 11.5. For x near zero, $s(t) < x$ during only a small part of the rf cycle; therefore, $2\phi \approx 0$ and $\overline{n}_e \approx n_i(x)$. For x near s_m, $s(t) < x$ during most of the rf cycle; therefore, $2\phi \approx 2\pi$ and $\overline{n}_e \approx 0$. To determine the time averages quantitatively, we assume that a sinusoidal rf current density passes through the sheath, which, equated to the conduction current at the electron sheath boundary, gives the equation for the electron sheath motion:

$$-en_i(s)\frac{ds}{dt} = -J_1 \sin \omega t \tag{11.2.7}$$

The solutions to these equations are rather involved, and we present only a few results. Combining (11.2.1)–(11.2.7), we obtain (see Lieberman, 1988 for details)

$$\frac{x}{s_0} = (1 - \cos\phi) + \frac{H}{8}\left(\frac{3}{2}\sin\phi + \frac{11}{18}\sin 3\phi - 3\phi\cos\phi - \frac{1}{3}\phi\cos 3\phi\right) \tag{11.2.8}$$

for $0 \leq \phi \leq \pi$, as sketched in Figure 11.6; and, at the electron sheath edge, the ion density is determined to be

$$\frac{n_i(x)}{n_s} = \left[1 - H\left(\frac{3}{8}\sin 2\phi - \frac{1}{4}\phi\cos 2\phi - \frac{1}{2}\phi\right)\right]^{-1} \tag{11.2.9}$$

Here,

$$s_0 = \frac{J_1}{e\omega n_s} \tag{11.2.10}$$

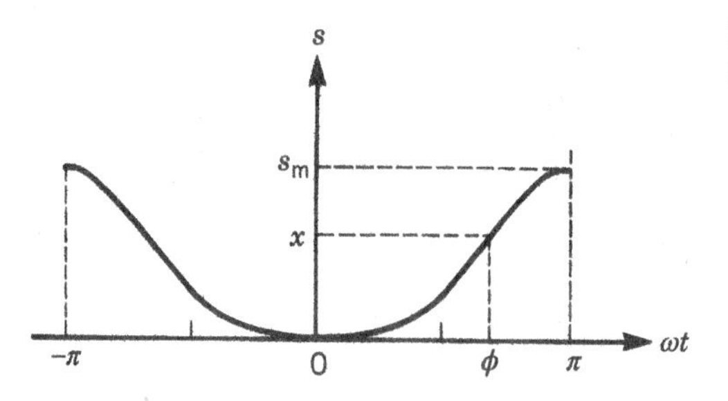

Figure 11.6 Sketch of the electron sheath thickness s versus ωt, showing the definition of the phase $\phi(x)$.

is an effective oscillation amplitude, and

$$H = \frac{J_1^2}{\pi e \epsilon_0 \mathrm{T_e} \omega^2 n_s} = \frac{1}{\pi} \frac{s_0^2}{\lambda_{De}^2} \tag{11.2.11}$$

with $\lambda_{De} = (\epsilon_0 \mathrm{T_e}/en_s)^{1/2}$ the electron Debye length at the ion sheath edge ($x = 0$, $n_i = n_s$). The ion density and average electron density are as sketched in Figure 11.5 in the usual regime of a high-voltage sheath with $V_{rf} \gg \mathrm{T_e}$.

11.2.2 Child Law

The Child law for the self-consistent ion sheath is obtained by integrating (11.2.4) with $n_i(x)$ and $\overline{n}_e(x)$ given by (11.2.9) and (11.2.6). Performing the integrations, we find

$$\frac{\overline{\Phi}}{\mathrm{T_e}} = \frac{1}{2} - \frac{1}{2}\left[1 - H\left(\frac{3}{8}\sin 2\phi - \frac{1}{4}\phi\cos 2\phi - \frac{1}{2}\phi\right)\right]^2 \tag{11.2.12}$$

The ion sheath voltage $\overline{V}$ is then found by putting $\phi = \pi$ at $\overline{\Phi} = -\overline{V}$ in (11.2.12) to obtain, for $H \gg 1$,

$$\frac{\overline{V}}{\mathrm{T_e}} = \frac{9\pi^2 H^2}{32} \tag{11.2.13}$$

Similarly putting $\phi = \pi$ at $x = s_m$ in (11.2.8), we obtain for $H \gg 1$ that

$$\frac{s_m}{s_0} = \frac{5\pi H}{12} \tag{11.2.14}$$

At low pressures, the ion current is obtained from the Bohm flux at the plasma edge where $n_i = n_s$. Substituting for H from (11.2.11), we use (11.2.13) and (11.2.14) to construct the Bohm flux, finding

$$\overline{J}_i = en_s u_B = K_i \epsilon_0 \left(\frac{2e}{M}\right)^{1/2} \frac{\overline{V}^{3/2}}{s_m^2} \tag{11.2.15}$$

where $K_i = 200/243 \approx 0.82$. This has the same scaling with $\overline{V}$ and s_m as the normal Child law (6.3.12) without electron shielding, which has $K_i = 4/9 \approx 0.44$. For a fixed current density and sheath voltage, the self-consistent rf ion sheath thickness s_m is larger than the Child law sheath thickness by the factor $\sqrt{50/27} \approx 1.36$. This increase is produced by the reduction in space charge within the sheath due to the nonzero, time-averaged electron density.

11.2.3 Sheath Capacitance

To obtain a complete self-consistent model, we need a relationship between the rf voltage and rf current, which involves the total capacitance of both sheaths. Unlike the uniform model in

Section 11.1, the sum of the two sheath capacitances is no longer a constant, producing weak harmonics of the rf driving frequency. In the model, the current has been taken to be sinusoidal; hence, we Fourier decompose the voltage to obtain a capacitance associated with the fundamental component of the voltage

$$I_{ab}(t) \equiv C_{ab}\frac{d}{dt}V_{ab1}(t) \tag{11.2.16}$$

Using (11.2.8) and integrating Poisson's equation twice, to obtain the time-varying total voltage, we find

$$\begin{aligned} V_{ab} = -\frac{\pi}{4}H\mathrm{T}_e\Big\{8\cos\omega t + H\Big[\frac{10}{3}\pi\cos\omega t - \frac{5}{9}\sin 2\omega t \\ - \frac{25}{288}\sin 4\omega t + (2\omega t - \pi)\left(\frac{3}{8} + \frac{1}{3}\cos 2\omega t + \frac{1}{48}\cos 4\omega t\right)\Big]\Big\} \end{aligned} \tag{11.2.17}$$

for $0 \le \omega t \le \pi$. The peak-to-peak value of V_{ab} is $2V(0)$, with $V(0)$ given by

$$V(0) = \frac{\pi}{4}H\mathrm{T}_e\left[8 + H\left(\frac{125\pi}{48}\right)\right] \tag{11.2.18}$$

The amplitude of the fundamental voltage harmonic is

$$V_{ab1} = \frac{\pi}{4}H\mathrm{T}_e\left[8 + H\left(\frac{10\pi}{3} - \frac{4096}{675\pi}\right)\right] \tag{11.2.19}$$

Evaluating (11.2.19) and substituting into (11.2.16), we find

$$C_{ab} \approx \frac{0.613\,\epsilon_0 A}{s_m} \tag{11.2.20}$$

There is no second harmonic, and the third harmonic of the voltage is only 4.2% of the fundamental. Hence, to a good approximation, a sinusoidal sheath current produces a sinusoidal voltage across the sum of the two sheaths in a geometrically symmetric rf discharge.

From (11.2.16) and (11.2.20), we obtain

$$J_1 \approx 1.23\frac{\omega\epsilon_0}{s_m}V_1 \tag{11.2.21}$$

where $V_1 = V_{ab1}/2$ is the fundamental rf voltage amplitude across a single sheath. From (11.2.13) and (11.2.19) with $H \gg 1$, we also find

$$\overline{V} \approx 0.83\,V_1 \tag{11.2.22}$$

The relation between V_1, J_1, and n_s is found by eliminating $\overline{V}$ and s_m from (11.2.15), (11.2.21), and (11.2.22) to obtain

$$\frac{J_1^2}{n_s} \approx 1.73\,e\epsilon_0\omega^2\mathrm{T}_e^{1/2}V_1^{1/2} \tag{11.2.23}$$

11.2.4 Bulk Ohmic Heating

The ohmic heating is obtained straightforwardly as in Section 11.1, except that the density and therefore the resistivity is a function of position. The time-averaged ohmic power per unit area can therefore be written

$$\overline{S}_{ohm} \approx \frac{1}{2}J_1^2\int_{-l/2+s_m}^{l/2-s_m}\frac{m\nu_m}{e^2n(x)}\,dx \tag{11.2.24}$$

where $n(x)$ is the only function of position, depending on the equilibrium solution as calculated in Section 10.2, and the approximate equality is due to the integration limits, which neglect the heating in the oscillating sheath regions. At low pressures, $\lambda_i/d > T_i/T_e$, for which the density profile is rather flat, the central density can be substituted for $n(x)$, without significant error. At low pressures, the ohmic heating is small compared to the stochastic heating, such that the errors are negligible. At higher pressures, $\lambda_i/d < T_i/T_e$, most of the ohmic heating occurs at the plasma edge and the energy relaxation mean free path $\lambda_\mathcal{E}$ of the energetic (ionizing) electrons is generally less than the discharge length. This can lead to a flattening of the cosine solution $n = n_0 \cos \beta x$ of (10.2.20). However, we ignore this effect here and use (10.2.20) to integrate $1/n(x)$ to incorporate the density variation. Thus, we have

$$\bar{S}_{\text{ohm}} = \frac{1}{2} J_1^2 \frac{m\nu_{\text{m}}}{e^2 n_0} d, \qquad \lambda_i > \left(\frac{T_i}{T_e}\right) d \tag{11.2.25a}$$

$$\bar{S}_{\text{ohm}} = \frac{1}{2} J_1^2 \frac{m\nu_{\text{m}}}{e^2 n_0} \frac{2}{\beta} \ln \tan\left(\frac{\pi}{4} + \frac{\beta d}{4}\right), \qquad \lambda_i < \left(\frac{T_i}{T_e}\right) d \tag{11.2.25b}$$

where $d \approx l - 2s_{\text{m}}$ is the plasma length and $\cos(\beta d/2) = n_s/n_0$.

11.2.5 Stochastic Heating

The power transferred to the electrons by the sheath is found from (11.1.31) as in Section 11.1, but now f_{es} is not a fixed Maxwellian but is a time-varying function with a time-varying density $n_{\text{es}}(t)$ at the electron sheath edge $s(t)$. To determine f_{es}, we first note that the sheath is oscillating because the electrons in the bulk plasma are oscillating in response to a time-varying electric field. If the velocity distribution function within the plasma at the ion sheath edge $x = 0$ in the absence of the electric field is a Maxwellian $f_{\text{m}}(u)$ having density n_s, then the distribution within the plasma at the ion sheath edge is $f_s(u, t) = f_{\text{m}}(u - u_s)$, where $u_s(t) = -u_0 \sin \omega t$ is the time-varying oscillation velocity of the plasma electrons. At the moving electron sheath edge, because $n_{\text{es}} < n_s$, not all electrons having $u > 0$ at $x = 0$ collide with the sheath at s. Many electrons are reflected within the region $0 < x < s$ where the ion density drops from n_s to n_{es}. This reflection is produced by an ambipolar electric field whose value maintains quasineutrality $n_e \approx n_i$ at all times. The transformation of f_s across this region to obtain f_{es} is complicated. However, the essential features to determine the stochastic heating are seen if we approximate

$$f_{\text{es}} = \frac{n_{\text{es}}}{n_s} f_{\text{m}}(u - u_s) \tag{11.2.26}$$

As with the homogeneous model, this expression for f_{es} is not fully self-consistent with the flow of rf current across the moving sheath; we discuss this issue further at the end of this section. Inserting (11.2.26) into (11.1.31) and transforming to a new variable $u' = u - u_s$, we obtain

$$S_{\text{stoc}}(t) = -\frac{2m}{n_s} \int_{u_{\text{es}}-u_s}^{\infty} u_{\text{es}} n_{\text{es}} [u'^2 - 2u'(u_{\text{es}} - u_s) + (u_{\text{es}} - u_s)^2] f_{\text{m}}(u')\, du' \tag{11.2.27}$$

From (11.2.7), we note that

$$n_{\text{es}} u_{\text{es}} = -u_0 n_s \sin \phi \tag{11.2.28}$$

and differentiating (11.2.8), we obtain

$$v_{\text{R}}(\phi) \equiv u_{\text{es}} - u_s = \pm \frac{u_0 H}{8} \left(-\frac{3}{2} \cos \phi + 3\phi \sin \phi + \frac{3}{2} \cos 3\phi + \phi \sin 3\phi\right) \tag{11.2.29}$$

where the minus sign is used for the integration from 0 to π and the plus sign for the integration from $-\pi$ to 0. Substituting (11.2.28) and (11.2.29) into (11.2.27), we find the average stochastic power for a single sheath to be

$$\begin{aligned}\overline{S}_{\text{stoc}} = & -\frac{mu_0}{\pi}\int_{-\pi}^{\pi}\sin\phi\,\mathrm{d}\phi\int_{v_{\mathrm{R}}(\phi)}^{\infty}u'^2 f_{\mathrm{m}}(u')\,\mathrm{d}u' \\ & +\frac{2mu_0}{\pi}\int_{-\pi}^{\pi}v_{\mathrm{R}}(\phi)\sin\phi\,\mathrm{d}\phi\int_{v_{\mathrm{R}}(\phi)}^{\infty}u' f_{\mathrm{m}}(u')\,\mathrm{d}u' \\ & -\frac{mu_0}{\pi}\int_{-\pi}^{\pi}v_{\mathrm{R}}^2(\phi)\sin\phi\,\mathrm{d}\phi\int_{v_{\mathrm{R}}(\phi)}^{\infty}f_{\mathrm{m}}(u')\,\mathrm{d}u' \end{aligned} \tag{11.2.30}$$

or, for notational convenience, $\overline{S}_{\text{stoc}} = \overline{S}_1 + \overline{S}_2 + \overline{S}_3$.

If the assumption is made that the sheath motion is much slower than the electron thermal velocity, as in Section 11.1, then $v_{\mathrm{R}}(\phi)$ is small, and we can make the lower limit of the u' integrals equal to zero. Since $v_{\mathrm{R}}(\phi)$ is an odd function, the $\overline{S}_1$ and $\overline{S}_3$ integrands integrate to zero, with the $\overline{S}_2$ integral yielding

$$\overline{S}_{\text{stoc}} = \frac{3\pi}{32}Hmn_{\mathrm{s}}\overline{v}_{\mathrm{e}}u_0^2 \tag{11.2.31}$$

For $u_{\mathrm{es}} \gtrsim \overline{v}_{\mathrm{e}}$ in (11.2.30), the stochastic heating result (11.2.31) is not correct. An analytic calculation in the limit $u_{\mathrm{es}} \gg \overline{v}_{\mathrm{e}}$, and a numerical calculation using the complete expression for the stochastic heating from (11.2.30) have been made by Wood et al. (1995). The calculations give a somewhat larger power dissipation at the higher voltages. However, a fast sheath strongly perturbs the distribution of electrons within the sheath (Surendra and Vender, 1994).

11.2.6 Self-Consistent Model Equations

We summarize the complete set of equations used to calculate the parameters for an electropositive plasma, given a set of control parameters for a symmetric plane-parallel geometry. In addition to $f = \omega/2\pi$, l, A, and p, we have assumed that J_{rf} is known for deriving the self-consistent set. However, usually V_{rf} or the total absorbed power P_{abs} is the specified control parameter. The model includes assumptions that are only approximately satisfied, so we should not expect very close quantitative agreement with more detailed numerical simulations, or with actual experiments. In addition, for experiments, it is very difficult to control the transverse uniformity of the plasma, as implied in the plane-parallel assumption. However, reasonably accurate scaling of plasma parameters with control parameters can still be determined. In this subsection, we use the basic set of equations for sample calculations of plasma parameters. We then indicate the scaling that can be employed to estimate a wider set of plasma parameters. In Section 11.3, we compare analytic results to simulations and experiments, with the symmetric plane-parallel assumption. Then, in Section 11.4, we model asymmetric discharges.

The approximate self-consistent model equations are summarized here. We assume $d \approx l - 2s_{\mathrm{m}}$, with an initial estimate $s_{\mathrm{m}} \approx 1$ cm for numerical computations, which is a nominal value for low-pressure capacitive discharges. We can iterate on this value if we believe it will improve overall accuracy. From particle conservation (10.2.12) at intermediate and low pressures, we have

$$\frac{K_{\mathrm{iz}}}{u_{\mathrm{B}}} = \frac{1}{n_{\mathrm{g}}d_{\mathrm{eff}}} = \frac{2}{n_{\mathrm{g}}d}\frac{n_{\mathrm{s}}}{n_0}, \qquad \lambda_{\mathrm{i}} \gtrsim \left(\frac{\mathrm{T_i}}{\mathrm{T_e}}\right)d \tag{11.2.32a}$$

where n_{s}/n_0 is given by (10.2.1). At higher pressures, from (10.2.25), we have

$$\frac{(K_{\mathrm{mi}}K_{\mathrm{iz}})^{1/2}}{u_{\mathrm{B}}} = \frac{\pi}{n_{\mathrm{g}}d}, \qquad \lambda_{\mathrm{i}} \lesssim \left(\frac{\mathrm{T_i}}{\mathrm{T_e}}\right)d \tag{11.2.32b}$$

These equations determine $\mathrm{T_e}$ given n_g and d. Substituting (11.2.23) into (11.2.25), we obtain the electron ohmic heating power per unit area,

$$\overline{S}_{\text{ohm}} \approx 1.73 \frac{m}{2e} \frac{n_s}{n_0} \epsilon_0 \omega^2 \nu_m \mathrm{T_e}^{1/2} V_1^{1/2} d, \qquad \lambda_i \gtrsim \left(\frac{\mathrm{T_i}}{\mathrm{T_e}}\right) d \tag{11.2.33a}$$

$$\overline{S}_{\text{ohm}} \approx 1.73 \frac{m}{2e} \frac{n_s}{n_0} \epsilon_0 \omega^2 \nu_m \mathrm{T_e}^{1/2} V_1^{1/2} \frac{2}{\beta} \ln \tan\left(\frac{\pi}{4} + \frac{\beta d}{4}\right), \quad \lambda_i \lesssim \left(\frac{\mathrm{T_i}}{\mathrm{T_e}}\right) d \tag{11.2.33b}$$

where $\cos(\beta d/2) = n_s/n_0$. Substituting (11.2.23) into (11.2.31) with $u_0 = J_1/en_s$ and using (11.2.10) and (11.2.11) for a single sheath in the slow sheath limit, we obtain

$$\overline{S}_{\text{stoc}} \approx 0.45 \left(\frac{m}{e}\right)^{1/2} \epsilon_0 \omega^2 \mathrm{T_e}^{1/2} V_1, \qquad \omega s_m \lesssim \overline{v}_e \tag{11.2.34}$$

We also have, from (11.2.22), that the ion kinetic energy per ion hitting the electrode is

$$\mathcal{E}_i = \overline{V} \approx 0.83\, V_1 \tag{11.2.35}$$

The electron power balance equation is

$$S_e = \overline{S}_{\text{ohm}} + 2\overline{S}_{\text{stoc}} = 2en_s u_B (\mathcal{E}_c + \mathcal{E}_e') \tag{11.2.36}$$

where, as in (11.1.40), $\mathcal{E}_e' = \mathcal{E}_e + V_s + \frac{1}{2}\mathrm{T_e} \approx 7.2\,\mathrm{T_e}$ for argon. Since $\overline{S}_{\text{ohm}}$ and $\overline{S}_{\text{stoc}}$ are both functions of V_1 alone, independent of n_s and J_1, (11.2.36) explicitly determines n_s if V_1 is the specified electrical control parameter. The total power absorbed per unit area is then found as

$$S_{\text{abs}} = 2en_s u_B (\overline{V} + \mathcal{E}_c + \mathcal{E}_e') \tag{11.2.37}$$

Eliminating n_s from these two equations and using (11.2.35) for $\overline{V}$, we obtain

$$S_{\text{abs}} \approx \left(\overline{S}_{\text{ohm}} + 2\overline{S}_{\text{stoc}}\right) \left(1 + \frac{0.83 V_1}{\mathcal{E}_c + \mathcal{E}_e'}\right) \tag{11.2.38}$$

If S_{abs} is the specified control parameter, then (11.2.38) implicitly determines V_1 by substituting for $\overline{S}_{\text{ohm}}$ and $\overline{S}_{\text{stoc}}$ from (11.2.33) and (11.2.34). In this case, (11.2.36) or (11.2.37) can then be used to find n_s. The center density n_0 is then found using (10.2.3) or (10.2.1), and $\overline{V}$ is found from (11.2.35). To complete the summary, s_m and J_1 are found from (11.2.15) and (11.2.21), respectively.

Example 11.1 We take the following parameters:

- $p = 3$ mTorr argon at 300 K
- $l = 10$ cm
- $A = 1000\,\text{cm}^2$
- $f = 13.56$ MHz ($\omega = 8.52 \times 10^7\,\text{s}^{-1}$)
- $V_{\text{rf}} = 500$ V

Starting with an estimate $s_m \approx 1$ cm, and using $\lambda_i = 1/n_g\sigma_i$, we find from (3.5.10) with $n_g = 1.0 \times 10^{20}\,\text{m}^{-3}$ at 300 K, that $\lambda_i = 1.0$ cm $= 0.01$ m. Thus, with $d = l - 2s_m = 0.08$ m, $\lambda_i/d \approx 0.125$, which is in the intermediate mean free path regime, in which the plasma is relatively flat in the center. The ratio between the edge density and center density is given in (10.2.1), with d replacing l, as $n_s/n_0 = 0.326$. Then, $n_g d_{\text{eff}} \approx 1.21 \times 10^{19}\,\text{m}^{-2}$ from (11.2.32a). Solving the particle balance using Figure 10.1, or numerically, using the rate coefficient of reaction 2 in Table 3.3, we find $\mathrm{T_e} \approx 3.3$ V. This gives $u_B = (e\mathrm{T_e}/M)^{1/2} \approx 2.8 \times 10^3$ m/s. From Figure 3.17, $\mathcal{E}_c \approx 55$ V and $\mathcal{E}_c + \mathcal{E}_e' \approx 79$ V. Estimating $\nu_m \approx K_{\text{el}} n_g$ with K_{el} given by the rate coefficient of reaction 1 in Table 3.3, we obtain $\nu_m \approx 8.9 \times 10^6\text{s}^{-1}$. Then, (11.2.33a) can be evaluated to obtain

$$\overline{S}_{\text{ohm}} \approx 0.132\, V_1^{1/2}\ \text{W/m}^2 \tag{11.2.39}$$

Similarly evaluating (11.2.34) yields

$$\overline{S}_{\text{stoc}} \approx 0.125\, V_1 \text{ W/m}^2 \tag{11.2.40}$$

Neglecting the voltage drop across the bulk plasma, and let $V_1 \approx V_{\text{rf}}/2 = 250$ V in (11.2.39) and (11.2.40), we find $\overline{S}_{\text{ohm}} \approx 2.09$ W/m^2 and $\overline{S}_{\text{stoc}} \approx 31.2$ W/m^2. We see from this example that $\overline{S}_{\text{stoc}}$ considerably exceeds $\overline{S}_{\text{ohm}}$. Using these values in the electron power balance (11.2.36), we obtain $n_s \approx 9.1 \times 10^{14}$ m^{-3}. Since $n_s/n_0 \approx 0.326$, we have $n_0 \approx 2.8 \times 10^{15}$ m^{-3}. From (11.2.35), we find $\overline{V} = \mathcal{E}_i \approx 208$ V; from the two equations in (11.2.15), $\overline{J}_i \approx 0.41$ A/m^2 and $s_m \approx 1.1 \times 10^{-2}$ m, and from (11.2.21), $J_1 \approx 23.2$ A/m^2. The total power absorbed per unit area is then obtained from (11.2.37) to be $S_{\text{abs}} \approx 235$ W/m^2. For $A = 0.1$ m^2, the discharge power is 23.5 W. Since s_m is close to our initial estimate, the plasma parameters are probably calculated within the accuracy of the calculation, and therefore an iteration is not useful.

Example 11.2 We take the following parameters, with the absorbed power as the specified electrical parameter:

- $p = 3$ mTorr argon at 300 K
- $l = 10$ cm
- $A = 1000\,\text{cm}^2$
- $f = 13.56$ MHz ($\omega = 8.52 \times 10^7\,\text{s}^{-1}$)
- $P_{\text{abs}} = 200$ W

As in Example 11.1, $n_s/n_0 \approx 0.326$, $\text{T}_e \approx 3.3$ V, $u_B \approx 2.8 \times 10^3$ m/s, and $\mathcal{E}_c + \mathcal{E}_e' \approx 79$ V. Because n_g and T_e are the same as in Example 11.1, $\overline{S}_{\text{ohm}}$ is given by (11.2.39) and $\overline{S}_{\text{stoc}}$ is given by (11.2.40). Substituting these into (11.2.38) with $S_{\text{abs}} = P_{\text{abs}}/A = 2000$ W/m^2, we obtain

$$2000 = \left(0.132 V_1^{1/2} + 0.25 V_1\right)\left(1 + \frac{0.83 V_1}{79}\right) \tag{11.2.41}$$

Dropping the first (small) terms in each parenthesis yields an approximate solution $V_1 \approx 873$ V. A numerical solution of (11.2.41) gives a more exact result $V_1 = 817$ V. Then $V_{\text{ab1}} = 2V_1 \approx V_{\text{rf}} \approx 1634$ V, and (11.2.35) yields $\mathcal{E}_i \approx 678$ V. Using this in (11.2.37), we obtain $n_s \approx 2.9 \times 10^{15}$ m^{-3} and, with $n_s/n_0 \approx 0.326$, we find $n_0 \approx 9.0 \times 10^{15}$ m^{-3}. The ion current density and the sheath width are found from (11.2.15) to be $\overline{J}_i \approx 1.32$ A/m^2 and $s_m \approx 1.46 \times 10^{-2}$ m, and the rf current density is found from (11.2.21) to be $J_1 \approx 75.8$ A/m^2. Since the new s_m is about 50% larger than the old, an iteration with a new $d \approx 7$ cm would give somewhat more accurate estimates of the plasma parameters.

11.2.7 Scaling

We can use the basic equations to obtain the most important scalings of the plasma parameters with control parameters. These scalings can also be compared to the scalings obtained from simulations and experiments to investigate the validity of the various approximations. We assume that $d \approx l - 2s_m$ is essentially constant as the voltage and pressure are varied over reasonable ranges. We can then combine the model equations to obtain the scalings in various limiting cases. We assume that the pressure is sufficiently low that ohmic heating can be neglected. We leave the ohmic heating scalings to Problem 11.8. From (11.2.34), we have

$$\overline{S}_{\text{stoc}} \propto \omega^2 \text{T}_e^{1/2} V_{\text{rf}} \tag{11.2.42}$$

Dropping the ohmic term in (11.2.36), such that $S_e = 2\overline{S}_{stoc}$, assuming $\mathcal{E}_c \gg T_e$, and substituting for $\overline{S}_{stoc}$ from (11.2.34), we obtain

$$n_s \propto \frac{\omega^2 V_{rf}}{\mathcal{E}_c} \tag{11.2.43}$$

For low sheath voltages, taking $\mathcal{E}_i \ll \mathcal{E}_c$ in (11.2.37), we obtain

$$S_{abs} \propto \omega^2 T_e^{1/2} V_{rf} \tag{11.2.44}$$

For the more common situation of high sheath voltages, $\mathcal{E}_i \gg \mathcal{E}_c$, with $\overline{V} \propto V_{rf}$, we obtain

$$S_{abs} \propto \frac{\omega^2 T_e^{1/2} V_{rf}^2}{\mathcal{E}_c} \tag{11.2.45}$$

The weak dependence of s_m is found by substituting n_s from (11.2.43) into (11.2.15) to obtain

$$s_m \propto \frac{V_{rf}^{1/4} \mathcal{E}_c^{1/2}}{\omega T_e^{1/4}} \tag{11.2.46}$$

and, using this scaling in (11.2.21), we find

$$J_{rf} \propto \frac{\omega^2 V_{rf}^{3/4} T_e^{1/4}}{\mathcal{E}_c^{1/2}} \tag{11.2.47}$$

If J_{rf} is the control parameter, we can invert (11.2.47) and substitute for V_{rf}, in the other proportionalities, in terms of J_{rf}. Note that in the low-pressure regime, where stochastic heating dominates, variations in the temperature only enter logarithmically through the change in pressure. The generally strong frequency dependences should be noted. We can equally well consider the total absorbed power as the independent variable and solve for V_{rf}, n_s, s_m, and J_{rf}. Using the same approximations as above, we find, for high voltages, $\mathcal{E}_i \gg \mathcal{E}_c$, that

$$V_{rf} \propto S_{abs}^{1/2} \mathcal{E}_c^{1/2} / \omega T_e^{1/4} \tag{11.2.48}$$

$$n_s \propto S_{abs}^{1/2} \omega / \mathcal{E}_c^{1/2} T_e^{1/4} \tag{11.2.49}$$

$$s_m \propto S_{abs}^{1/8} \mathcal{E}_c^{5/8} / \omega^{5/4} T_e^{5/16} \tag{11.2.50}$$

$$J_{rf} \propto S_{abs}^{3/8} \omega^{5/4} T_e^{1/16} / \mathcal{E}_c^{1/8} \tag{11.2.51}$$

The above scalings are independent of pressure, except implicitly through the weak dependence of T_e on pressure. These scalings can be easily compared to experimental results. Since T_e only varies logarithmically with change in pressure, it can usually be held constant in comparing scalings. However, $\mathcal{E}_c$ can vary significantly with pressure, especially at high pressures where T_e is low.

11.2.8 Collisional Sheaths

If $\lambda_i \lesssim s_m$, then the ions suffer one or more collisions as they cross the sheath and the collisionless analysis is not valid. For argon with λ_i given by (3.5.10) and with $s_m \sim 1$ cm, we find $p \lesssim 3$ mTorr for a collisionless sheath, at the low end of typical processing discharges. At higher pressures, a self-consistent analysis of the collisional sheath is required, which has been given by Lieberman (1989a) and, over a wider range of collisionality, by Godyak and Sternberg (1990b). These authors assume λ_i = const, independent of velocity. The basic ion dynamical equations (11.2.1) and (11.2.2) are then modified, as in (6.5.1) and (6.5.2), to

$$n_i u_i = n_s u_B \tag{11.2.52}$$

and

$$u_i = \frac{2e\lambda_i}{\pi M u_i}\overline{E} \tag{11.2.53}$$

Carrying out the analysis as in the first part of this section, the dc ion current density is found to be

$$\overline{J}_i = e n_s u_B \approx 1.68\,\epsilon_0\left(\frac{2e}{M}\right)^{1/2}\frac{\overline{V}^{3/2}\lambda_i^{1/2}}{s_m^{5/2}} \tag{11.2.54}$$

where the coefficient is 1.68 for the self-consistent calculation rather than 1.43 as given in (6.5.7). Note that (11.2.54) differs from the collisionless Child law (11.2.15) because $\overline{J}_i$ now scales with λ_i and scales differently with s_m. For a fixed n_s and $\overline{V}$ (and T_e), the sheath thickness s_m decreases weakly with increasing n_g. The collisional sheath capacitance is found to be $0.76\,\epsilon_0 A/s_m$, leading to

$$J_1 \approx 1.52\frac{\omega\epsilon_0}{s_m}V_1 \tag{11.2.55}$$

in place of (11.2.21). We also find

$$\overline{V} \approx 0.78\,V_1 \tag{11.2.56}$$

in place of (11.2.22).

The average ion-bombarding energy $\mathcal{E}_{ic}$ is reduced below $\overline{V} \equiv V_s$ because ion energy is lost during charge transfer and elastic collisions in the sheath, creating fast neutrals there. The ion-bombarding energy is found to be

$$\mathcal{E}_{ic} = \frac{1}{2}Mu_i^2(s_m) \approx 0.62\,\frac{\lambda_i}{s_m}\overline{V} \tag{11.2.57}$$

Note, however, that the total kinetic energy lost per ion transiting the sheath is still $\overline{V}$, as for the collisionless sheath, and as used in (11.2.37). Thus, the effect of collisions in the sheath is to reduce the ion-bombarding energy but to proportionally increase the total energetic particle flux (ions + fast neutrals) to the electrode.

The stochastic heating is found to be

$$\overline{S}_{stoc} \approx 0.49\left(\frac{2\lambda_i s_0}{\pi^2\lambda_{De}^2}\right)^{1/2} m n_s \overline{v}_e u_0^2 \tag{11.2.58}$$

in place of (11.2.31). Substituting (11.2.54)–(11.2.56) into (11.2.58) with u_0 given by (11.2.28), we obtain (Problem 11.3)

$$\overline{S}_{stoc} \approx 0.61\left(\frac{m}{e}\right)^{1/2}\epsilon_0\omega^2 T_e^{1/2} V_1 \tag{11.2.59}$$

in place of (11.2.34). We see that, except for the numerical coefficients, (11.2.59) and (11.2.34) for the collisional and collisionless sheaths have the same form. However, $\overline{J}_i$ has somewhat different scaling between (11.2.54) and (11.2.15). The procedure for calculating the discharge parameters for the collisionless sheath can therefore be applied to the collisional sheath, with minor modifications.

Consider, for example, the scaling of discharge parameters with absorbed power for a plasma with collisional (constant mean-free path) sheaths in the regime where ohmic heating dominates stochastic heating and where ion energy losses dominate electron energy losses. Using (11.2.33a), (11.2.36), (11.2.37), (11.2.54), and (11.2.56), we find the scalings

$$\begin{aligned} V_{\mathrm{rf}} &\propto S_{\mathrm{abs}}^{2/3} \\ n_{\mathrm{s}} &\propto S_{\mathrm{abs}}^{1/3} \\ s_{\mathrm{m}} &\propto S_{\mathrm{abs}}^{4/15} \\ \overline{S}_{\mathrm{ohm}} &\propto S_{\mathrm{abs}}^{1/3} \end{aligned} \tag{11.2.60}$$

We leave the details to Problem 11.8.

11.2.9 Low and Moderate Voltages

Godyak and Sternberg (1990b) have treated the regimes from $V_{\mathrm{rf}} \ll \mathrm{T_e}$ to $V_{\mathrm{rf}} \gg \mathrm{T_e}$ in a unified manner. For $V_{\mathrm{rf}} \ll \mathrm{T_e}$, their results reduce to that of an undriven dc sheath, as in (6.2.17). At high voltages, $V_1/\mathrm{T_e} \gtrsim 200$, their numerical results asymptotically approach the analytic results $\overline{V} \propto V_1$, but these voltages are at the upper end of typical processing discharge regimes. At more moderate voltages, $50 \lesssim V_1/\mathrm{T_e} \lesssim 200$, $\overline{V}$ is seen to have a weaker scaling with V_1, such that $\overline{V} \propto V_1^{\beta}$, with $\beta \approx 2/3$–$3/4$. With the weaker scaling, s_{m} in (11.2.46) is found to be nearly independent of V_{rf}, and from (11.2.21), the J_{rf} versus V_{rf} discharge characteristics are nearly linear, as observed in many experiments (see Godyak and Sternberg (1990b) and Godyak (2020) for further discussion).

11.2.10 Ohmic Heating in the Sheath

Ohmic heating due to electron–neutral collisions of the oscillating electrons within the sheaths can be an important additional electron heating mechanism. For a Child law scaling, the density within the sheath is, from (6.3.8), of order

$$n_{\mathrm{sh}} \sim n_{\mathrm{s}} \left(\frac{\mathrm{T_e}}{V_{\mathrm{rf}}} \right)^{1/2} \tag{11.2.61}$$

Because the ohmic power density p_{ohm} scales as $J_{\mathrm{rf}}^2/n_{\mathrm{sh}}$ and J_{rf} is not spatially varying, we see that p_{ohm} is a factor of $(V_{\mathrm{rf}}/\mathrm{T_e})^{1/2}$ larger within the sheath than at the sheath edge. Hence, for a uniform bulk plasma, the ohmic heating within the sheath exceeds the bulk heating when

$$s_{\mathrm{m}} \left(\frac{V_{\mathrm{rf}}}{\mathrm{T_e}} \right)^{1/2} \gtrsim d \tag{11.2.62}$$

This condition can be met in a high-voltage discharge. To determine the sheath ohmic power per unit area for a sinusoidal current density $J_{\mathrm{rf}}(t) = J_1 \sin \omega t$, we must time-average $J_{\mathrm{rf}}^2(t)$ over the oscillating sheath width $s(t)$

$$\overline{S}_{\mathrm{ohm,sh}} = \frac{m\nu_{\mathrm{m}}}{e^2} \left\langle J_1^2 \sin^2 \omega t \int_{s(t)}^{s_{\mathrm{m}}} \frac{\mathrm{d}x}{n_{\mathrm{i}}(x)} \right\rangle_{\mathrm{rf}} \tag{11.2.63}$$

For a collisionless ion sheath, Misium et al. (1989) give the expression

$$\overline{S}_{\mathrm{ohm,sh}} \approx 1.73 \frac{m}{2\pi e} \epsilon_0 \omega^2 \nu_{\mathrm{m}} s_{\mathrm{m}} (\mathrm{T_e} V_1)^{1/2} \left[0.235 \left(\frac{V_1}{\mathrm{T_e}} \right)^{1/2} + 1.16 + 4.39 \left(\frac{\mathrm{T_e}}{V_1} \right)^{1/2} \right] \tag{11.2.64a}$$

and for high-voltage collisional sheaths, Chabert et al. (2004a) give

$$\overline{S}_{\mathrm{ohm,sh}} \approx 0.236 \frac{m}{2e} \epsilon_0 \omega^2 \nu_{\mathrm{m}} s_{\mathrm{m}} V_1 \tag{11.2.64b}$$

This should be added to (11.2.33) for each sheath. Although $\overline{S}_{\text{ohm,sh}}$ depends on s_m, explicitly, a nominal value $s_m = 1$ cm can be assumed initially, and the equations can be iterated if greater accuracy is required.

11.2.11 Self-Consistent Collisionless Heating Models

Although the Fermi model is physically appealing, the heating rates (11.1.37) and (11.2.31) for the homogeneous and inhomogeneous sheaths have not been obtained self-consistently. For the homogeneous model, the electron distribution $f_{es}(u, t)$ was approximated to be a Maxwellian *without a superimposed rf oscillation velocity.* For the inhomogeneous model, f_{es} was approximated by (11.2.26), with u_s the oscillation velocity *in the bulk plasma.* However, the form chosen for f_{es} should be consistent with conservation of rf current at the moving sheath edge

$$n_s(t)u_{es}(t) = \int_{-\infty}^{\infty} du\, u f_{es}(u, t) \tag{11.2.65}$$

Consider, for example, the physically appealing choice of a distribution function with a time-varying density $n_s(t)$ and drift velocity $u_{ed}(t)$

$$f_{es} = n_s(t) f_0\left(u - u_{ed}(t)\right) \tag{11.2.66}$$

Inserting (11.2.66) into (11.2.65) and changing variables to $u' = u - u_{ed}$, we obtain $u_{ed} = u_{es} - u_{e0}$, where $u_{e0} = \int_{-\infty}^{\infty} du'\, u' f_0(u')$ is a time-independent velocity. Substituting (11.2.66) with $u_{ed} = u_{es} - u_{e0}$ into the fundamental expression (11.1.31) for stochastic heating, we have

$$S_{stoc} = -2mn_s(t)u_{es}(t) \int_{u_{e0}}^{\infty} du'\, (u' - u_{e0})^2 f_0(u') \tag{11.2.67}$$

The time-average of (11.2.67) is zero because the rf current $J_{rf} = -en_s(t)u_{es}(t)$ has a time average of zero, producing no heating. The actual motion of the electrons, however, gives a more complicated distribution than that postulated in (11.2.66).

For the homogeneous model ($n_s(t) = \text{const}$), it can be shown that there is no heating in the self-consistent model (Lieberman, 1988; Kaganovich and Tsendin, 1992a), independent of the form of f_{es}. This can be seen physically by transforming to the inertial frame of the bulk plasma oscillations. In this frame, which oscillates with a sinusoidal velocity $u_s(t)$, the net (electric field + inertial) force acting on an electron is zero everywhere in the plasma; hence, there is no heating.

For collisionless heating in the inhomogeneous rf sheath, Kaganovich (2002) and Kaganovich et al. (2004) develop a kinetic model consistent with rf current conservation at the sheath to determine f_{es} analytically for a two-step ion density profile with $n_i = n_{sh} = \text{const}$ in the sheath and $n_i = n_s = \text{const}$ in the bulk plasma, with $n_{sh} \leq n_s$. This profile is meant to model the Child law sheath which has a lower density in the sheath region than in the bulk plasma, as given in (11.2.61). For the homogeneous model with $n_{sh} = n_s$, there is no heating. For $n_{sh} < n_s$, the heating consistent with rf current conservation is found to be $\overline{S}_{stocK} = G_K \overline{S}_{stoc}$, with $\overline{S}_{stoc}$ given by (11.2.31)

$$G_K \approx \frac{H}{H + 1.1} \tag{11.2.68}$$

Here, $H \gtrsim 2$ is given by (11.2.11). For high-voltage sheaths, this reduction in heating is small, but for low voltages, it can be important.

Gozadinos et al. (2001a), following earlier work (Surendra and Dalvie, 1993; Turner, 1995), give an alternate model of collisionless heating that associates the heating with acoustic disturbances in the electron fluid. They develop an analytic model based on moments of the Vlasov equation

(2.3.2) in which the electron distribution at the sheath edge is characterized by separate densities and temperatures for electrons entering and leaving the sheath. Their model gives results for H not too large that scale in the same way with parameters as the Fermi result (11.2.31), but with a coefficient about 40% of the Fermi result. The fluid and Fermi theories are compared with various particle-in-cell (PIC) simulation results in Kawamura et al. (2006) and Lafleur et al. (2014a,b). In the latter work, it is shown that for a self-consistent sheath rf current, the stochastic heating is reduced to half the result given in (11.2.31).

Various kinetic approaches have been developed to determine the ohmic and non-ohmic contributions to the total electron heating. Aliev et al. (1997) developed a kinetic treatment in which the collisionless heating is considered to arise from a resonant wave–particle interaction. We introduce this method in Section 19.4. Surendra and Dalvie (1993), Lafleur et al. (2014a), and Schulze et al. (2018) calculated the ohmic and non-ohmic heating from an analysis of the various terms for electron mechanical energy conservation, obtained from the Boltzmann equation. We describe this *Boltzmann term analysis* and give a summary of the results in Section 11.3.

It should be noted that for the electron mean free path $\lambda_e < l$, the phase randomization that is required to produce thermal electron energy from the stochastic energy gain is due to electron interparticle collisions (Kaganovich et al., 1996; Lafleur and Chabert, 2015b). This is generally the case at all but the lowest discharge pressures, where dynamical phase randomization could occur (Lichtenberg and Lieberman, 1992).

Other significant issues are heating due to electron inertia effects during sheath contraction (Vender and Boswell, 1992; Turner and Hopkins, 1992) and energy losses of electrons escaping to the electrode (Wendt and Hitchon, 1992; Gozadinos et al., 2001a,b). We will see an example of heating during sheath contraction in the PIC calculations shown in Figure 11.12. Energy losses occur preferentially when the sheath edge is near the electrode (see Figure 11.13). While electrons are lost, the motion of the sheath edge is not symmetrical around the time when the sheath edge lies closest to the electrode. This gives an additional heating effect (Gozadinos et al., 2001a). For this reason, comparisons of the analytically determined collisionless heating rates (11.2.31) or (11.2.34) with PIC simulations including the electron energy losses give good agreement for macroscopic quantities, although details of microscopic predictions agree significantly less well with simulations (Gozadinos et al., 2001b). At low pressures where collisionless heating is dominant, experiments and simulations indicate that the electron distribution is approximately bi-Maxwellian; the assumption of Maxwellian electrons used in fluid models can be considerably in error. Stochastic sheath heating with a self-consistent bi-Maxwellian distribution gives results more in agreement with PIC simulations (see Wang et al., 1999, and Section 19.6).

In summary, it appears that collisionless heating of electrons occurs around the plasma–sheath interface, but the detailed mechanisms and exact value are not accurately known. The calculations of collisionless heating given in this section appear to give reasonable estimates that can be used to determine discharge equilibrium parameters. For low-pressure plasmas, the dominant heating is clearly in the sheath region, as we see in the following section.

11.2.12 Electronegative Plasmas

Although the discussion in this and the preceding sections has been for electropositive plasmas, much of it can also be applied to electronegative plasmas. However, some care must be taken to understand the assumptions to see if they hold without change or must be modified. As we saw in Sections 10.3–10.5, there are various regions in parameter space, each requiring some modifications of the analysis. Here, we discuss some of the general considerations, without specific calculations.

Most of the detailed calculations for the inhomogeneous sheath in Section 11.2 can be carried over to an electronegative plasma. The calculation of the stochastic and ohmic heating powers, given by (11.2.34) and (11.2.33), is of prime importance. If the plasma–sheath transition density is such that the electron and positive ion densities can both be taken to be n_s, then provided there is an electropositive edge, (11.2.34) is unchanged, and (11.2.33a) is modified only by taking $n_0 = n_s$ in the bulk plasma. The electron power balance (11.2.36) must be changed to take into account the electronegative equilibria which we have analyzed in Sections 10.3 and 10.4. This can be done by estimating the general range of the expected n_{e0} for a given pressure and power. For example, if we take parameters with a moderate central value of electronegativity (e.g., $5 < \alpha_0 < 20$) and at not too high a pressure (e.g., 5 mTorr $< p <$ 20 mTorr), then a reasonable approximation is a flat electron profile and a parabolic negative ion profile which goes to zero at the plasma–sheath interface. For these conditions, we replace (11.2.36) by

$$\bar{S}_{\text{ohm}} + 2\bar{S}_{\text{stoc}} = \frac{8}{15} K_{\text{rec}} n_{+0}^2 de\mathcal{E}_{\text{c}} + 2D_+ \frac{4n_{+0}}{d} e(\mathcal{E}_{\text{c}} + \mathcal{E}_{\text{e}}') \tag{11.2.69}$$

where n_{+0} is the central ion density $n_{+0} = n_{-0} + n_{e0}$, D_+ is the ion diffusion coefficient with $T_+ = T_-$, and K_{rec} is the recombination rate of positive and negative ions. We obtain n_{+0} using α_0 in (10.4.29) with $d/l_1 = 1$.

Because n_{+0} appears with different powers in the two terms on the right-hand side of (11.2.69), we no longer have simple scalings in electronegative plasmas. This is also true for electropositive plasmas if the ohmic and stochastic heating are comparable. At higher pressures, there is increasing flattening of the center of the electronegative core with accompanying steepening of the profile near the edge. A calculation can still be performed by use of a somewhat more complicated flat-topped model, as described in Section 10.4.

As shown in Section 10.5, much of the understanding of experiments can be obtained without a complete model of the heating. As examples, in both Figures 10.3 and 10.4, the parabolic structure of the electronegative region at low pressure is evident. The deviation from the theoretical equilibrium structures at higher pressures in both experiments is due in part to the nonuniform electron temperature. This effect, related to the short electron mean free path at the higher pressures, combined with the primary heating at the plasma edge, has not been treated in our modeling calculations.

11.3 Experiments and Simulations

Models are based on a particular set of assumptions that must be tested by experimentally determining if the observable consequences of those assumptions are in agreement with the experiment that the model is designed to represent. In recent years, the tool of computer simulation, added to analytic modeling and experiments, has improved our understanding. The particular simulations described here for modeling low-pressure discharges, called particle-in-cell (PIC) simulations, follow large numbers of representative particles acted upon by the basic forces. Many of the assumptions of the analytic models need not be used. It is also possible to determine various microscopic quantities that are not observable experimentally. In these ways, the simulations serve as an intermediary between the models and the experiments.

In this section, we shall first give some experimental observations and relate the results to the model of Section 11.2. We then present the results of simulations to obtain further understanding of the plasma behavior. We restrict our attention to symmetric plane-parallel geometry. Simulations can be performed in more complicated geometries, but the calculations become more involved.

11.3.1 Experimental Results

An early experiment to investigate stochastic sheath heating is described in Godyak's (1986) review. In an apparatus designed to approximate a plane-parallel discharge, an effective collision frequency ν_{eff} was measured versus pressure, using the relationship for the power absorbed per unit area,

$$S_{\text{abs}} = \frac{1}{2}\frac{|\tilde{J}_{\text{rf}}|^2}{e^2 n} m\nu_{\text{eff}} d \tag{11.3.1}$$

S_{abs}, $\tilde{J}_{\text{rf}}$, and n being simultaneously measured. The measurements were done at relatively low voltages, such that $\overline{V} \lesssim \mathcal{E}_{\text{c}}$; consequently, $S_{\text{abs}} \sim S_{\text{e}}$, the power per unit area absorbed by the electrons. The result is shown in Figure 11.7. Both the asymptotic leveling off of ν_{eff} at low pressure p, characteristic of stochastic heating which is independent of p, and the linear increase of ν_{eff} with p at high p, characteristic of ohmic heating, are clearly visible. The good agreement of the measurements with ν_{eff} calculated from the stochastic heating formula is somewhat fortuitous; however, as a uniform sheath rather than a self-consistent sheath was used in the calculation, and the ion power loss S_{i} was neglected in determining ν_{eff} from the measurements. We further examine the non-ohmic heating later in this section.

A thorough experimental study of symmetric rf discharge characteristics in argon at 13.56 MHz has been performed by Godyak and collaborators (Godyak and Piejak, 1990b; Godyak et al., 1991, 1992). The discharge length and diameter were 6.7 and 14.3 cm, respectively, approximating a uniform plane-parallel configuration. Measurements were made of rf voltage, rf current, total power absorbed, dc bias voltage, central plasma density n_0, mean electron energy $\langle\mathcal{E}_{\text{e}}\rangle$, and electron energy probability function g_{p} given in (6.6.18). The time-averaged power was determined by averaging $V_{\text{rf}}(t)I_{\text{rf}}(t)$ over an rf cycle (see Section 11.11), and n_0, $\langle\mathcal{E}_{\text{e}}\rangle$, and g_{p} were determined using Langmuir probes (see Section 6.6). Measurements were performed over a wide range of pressures from

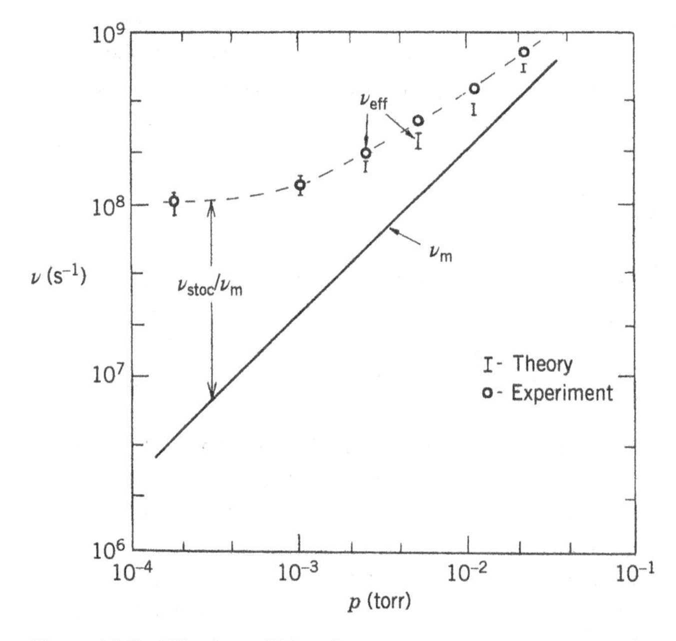

Figure 11.7 Effective collision frequency ν_{eff} versus pressure p, for a mercury discharge driven at 40.8 MHz. The solid line shows the collision frequency due to ohmic dissipation alone. Source: Popov and Godyak, 1985/with permission of AIP Publishing.

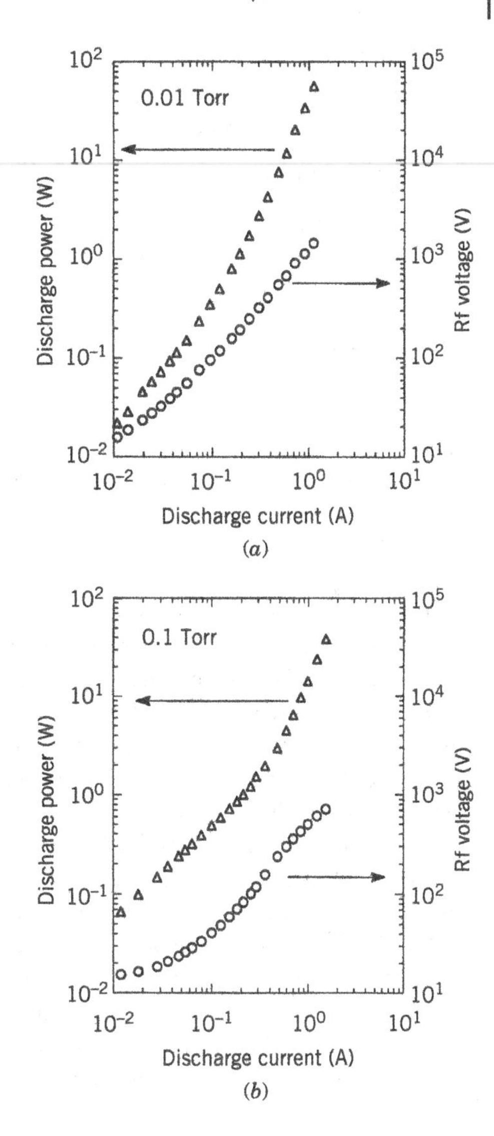

Figure 11.8 Discharge power absorbed P_{abs}, and rf voltage V_{rf} versus discharge current I_{rf} at (*a*) $p = 10$ mTorr and (*b*) $p = 100$ mTorr in argon. Source: Reproduced from Godyak et al. (1991)/IEEE.

3 mTorr to 3 Torr and for powers up to 100 W. The corresponding rf voltage amplitudes were up to 1500 V, and the rf current amplitudes were up to 2 A. Figure 11.8 shows V_{rf} and P_{abs} versus I_{rf} at relatively low (10 mTorr) and relatively high (100 mTorr) pressures. At 10 mTorr, where ohmic heating is small, and at low to moderate voltages, the voltage scales roughly linearly with the current, with a transition to the scaling $V_{rf} \propto I_{rf}^{4/3}$ predicted from (11.2.47) at the higher voltages. The power scales as $P_{abs} \propto I_{rf} \propto V_{rf}$ at low voltages, with a transition to $P_{abs} \propto V_{rf}^2$ at higher voltages, in agreement with (11.2.45). The density n_0, however, was found to scale more strongly with the voltage than the linear scaling predicted by (11.2.43), and the mean electron energy $\langle \mathcal{E}_e \rangle$, which corresponds to $\frac{3}{2}\mathrm{T_e}$ for a Maxwellian distribution, fell significantly at the higher voltages, contrary to the analytic model in which $\mathrm{T_e}$ depends only on the pressure and is independent of the voltage.

Generally, the experimental density is higher than the model predictions, indicating somewhat more efficient electron power absorption at a given applied voltage, which may be due to the effect of bi-Maxwellian distributions and to secondary electron emission. The discrepancy can be partly understood by examining the measured electron energy probability function g_p (see (6.6.19)),

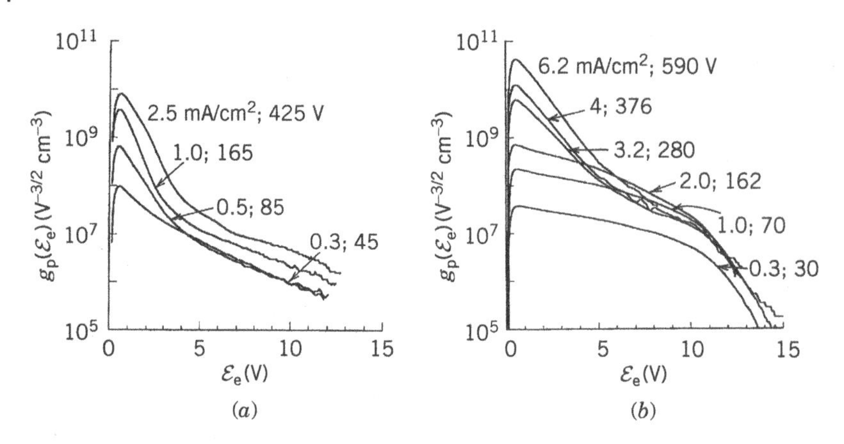

Figure 11.9 Electron energy probability function g_p versus $\mathcal{E}_e$ for various discharge currents for argon gas with $f = 13.56$ MHz and $l = 6.7$ cm: (*a*) $p = 10$ mTorr and (*b*) $p = 100$ mTorr. Source: Godyak, 1990b/with permission of American Physical Society.

which is plotted versus $\mathcal{E} = mv^2/2e$ in Figure 11.9. We see a transition from a single Maxwellian for $V_{rf} \lesssim 100$ V to a two-temperature distribution at higher voltages, with most of the electrons in the lower energy class, which therefore determines $\langle \mathcal{E}_e \rangle$. The high-temperature tail maintains the ionization balance required by (11.2.32a), allowing $\langle \mathcal{E}_e \rangle$ to drop to low values. As we will see from simulations, a two-temperature distribution is characteristic of stochastic heating. A similar behavior is seen at 3 and 30 mTorr. At 100 mTorr, ohmic heating dominates the electron power absorption below approximately 300 V, leading to a single temperature Maxwellian, as seen in Figure 11.9*b* with $\langle \mathcal{E}_e \rangle \approx 4$–5 V. From Figure 11.8*b*, we see a near-linear scaling of P_{abs} with V_{rf} at low voltages, with a transition to a steeper scaling of power with voltage at higher voltages. At higher V_{rf} there is a transition to a two-temperature distribution, as seen in Figure 11.9*b*, with $\langle \mathcal{E}_e \rangle$ falling to 1.5–2 V. These results indicate a transition from ohmic heating at low voltage to stochastic heating at high voltage. At still higher current densities than shown in Figure 11.9, the discharge can undergo an α-to-γ transition, in which ionization by secondary electrons emitted from the electrodes sustains the plasma. The midplane electron energy distribution function (EEDF) can then further evolve to a single, low-temperature Maxwellian distribution (Godyak et al., 1992), due to strong electron–electron collisions. We examine this transition in Section 15.3.

Figure 11.10 gives the measured (open circles) central plasma density (*a*) and average electron energy (*b*) versus argon pressure from 70 mTorr to 3 Torr, for a 2 cm gap discharge with 0.3 A rms rf current at 13.56 MHz (Godyak and Piejak, 1990b). Also shown are corresponding PIC simulation results (closed triangles), and a measurement in a somewhat larger gap (2.5 cm) discharge (Lafleur et al., 2014a). The simulation densities are somewhat lower than those measured, and the simulation temperatures are somewhat higher, with about a factor of two discrepancies at the lowest pressure of 70 mTorr. The simulation results here are consistent with those of Vahedi et al. (1993a). We return to these results in the next subsection.

The experimental results of Godyak et al. (1991) and Godyak and Piejak (1990b) for the total power as a function of the applied rf voltage can be compared with the nonuniform density model results for a 3 mTorr argon discharge. At this low pressure, stochastic heating is the principal heating mechanism. For the modeling results, $\overline{S}_{stoc}$ was computed numerically from (11.2.30), and other quantities were computed using these numerical results. The self-consistent single Maxwellian temperature was 3.2 V. The total power in the experiments was then normalized to the electrode

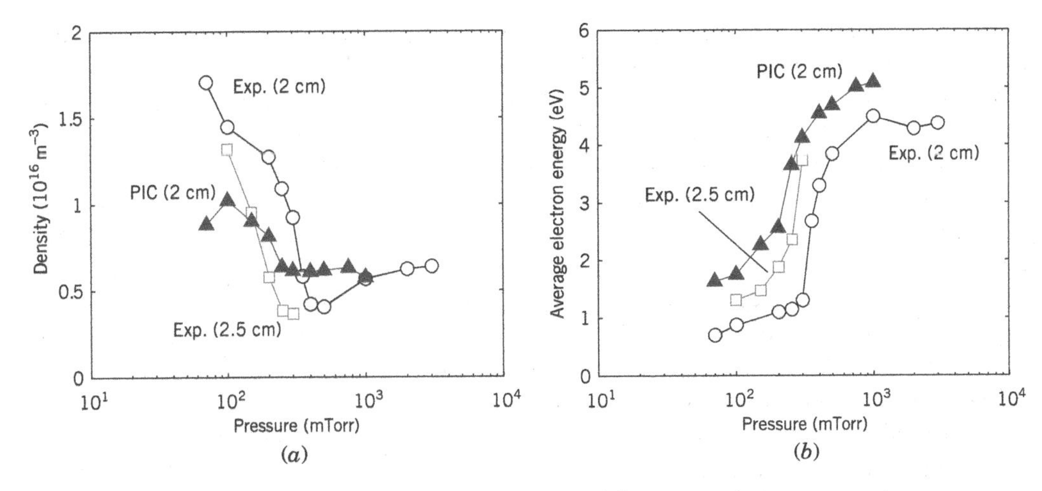

Figure 11.10 (*a*) Central plasma density versus pressure and (*b*) average electron temperature versus pressure, for an rms discharge current of 0.3 A; open circles are measurements from Godyak and Piejak (1990a) for a 2 cm gap; closed triangles are corresponding PIC simulations from Lafleur et al. (2014a); open squares are measurements from Lafleur et al. (2014a) for a 2.5 cm gap. Source: Reproduced from Lafleur et al. (2014a)/with permision of IOP Publishing.

area. The model results are in qualitative agreement with the experiments. However, there are also some significant disagreements. Generally, the experimental density is more steeply varying with rf voltage than the model predictions, which is at least in part related to the changing electron energy probability function.

Experimentally, the dc voltage $\overline{V}$ across a single sheath is found to track the rf voltage V_{ab} across both sheaths, with $\overline{V} \approx 0.4\, V_{ab} = 0.8\, V_1$ at high voltages, as predicted. For lower voltages, we find a weaker variation $\overline{V} \propto V_{ab}^{\beta}$, with $\beta \sim 2/3$–$3/4$, as described in Section 11.2. At 10 mTorr, we find that $V_{ab} \approx V_{rf}$ over the entire voltage range. However, at 100 mTorr, V_{ab} falls below V_{rf} at low V_{rf}, due to the additional rf voltage V_p dropped across the bulk plasma.

Some discrepancy exists for the sheath width, with the experimental widths being somewhat larger, but scaling more weakly with V_{rf}, than the high-voltage model. This may be a consequence of the somewhat weaker-than-linear scaling of $\overline{V}$ with V_{rf} at moderate discharge voltages. Despite these differences, the model has reasonable predictive power. We shall discuss the discrepancies further after giving simulation results.

11.3.2 Particle-in-Cell Simulations

The symmetric measurements of Godyak and Piejak (1990a) shown in Figure 11.10 were compared to PIC simulations by Vahedi et al. (1993a). The gas pressure was varied between 70 and 500 mTorr to observe the transition from stochastically to ohmically dominated electron heating. Except for the normalization, the g_ps obtained from the simulations agree well with the measured g_ps, showing the transition from a two-temperature distribution at 70 mTorr to a single-temperature distribution at 500 mTorr. The simulation temperatures are in good agreement with the measured temperatures over the entire range of pressures. Two sets of simulation results were examined, with and without secondary emission due to ion impact on the electrodes (see discussion of secondaries below). The plasma density showed a better agreement with measurements when secondaries were included, but the density was lower than the measurements by roughly a factor of 1.5 at low gas pressures. Possible explanations include incomplete modeling of the atomic collision processes, e.g., neglect of energetic ion–neutral ionization processes within the sheaths, and neglect of metastable atom production, electron impact ionization, and quenching.

Another simulation of discharge behavior (Wood, 1991) was performed at $p = 3$ mTorr (argon) with a spacing of 10 cm between parallel plates, and over a range of rf voltages between 100 and 1000 V. A two-temperature distribution was found, as in the experiments, and the distribution varied in both space and time. It is clear that a deeper understanding of the discharge behavior involves the space and time variations of f_e. Figure 11.11 shows the one-dimensional electron distribution function $f_e(x, v_x, t)$ versus v_x at 15 positions near the sheath region ($x = 0$–3 cm) and at four different times during the rf cycle. Each plot covers $\frac{1}{32}$ of a cycle temporally, and each line in a plot covers a 2-mm-thick region spatially. The units on the vertical axis are proportional to f_e. At time $\frac{0}{32}$, the sheath is fully expanded, and the two-temperature nature of the discharge near the sheath can be seen as the wide "base" and narrow "peak" of the distribution. As the rf cycle progresses to time $\frac{8}{32}$ (not shown), the distributions in the sheath region at each position display a drift toward the electrode (negative velocity) that is approximately equal to the sheath velocity. By time $\frac{12}{32}$, fast electrons have arrived from the opposite electrode, moving at a velocity of about 4×10^6 m/s (small peak at the extreme left of the figure). At time $\frac{16}{32}$ (not shown), the sheath is fully collapsed, the

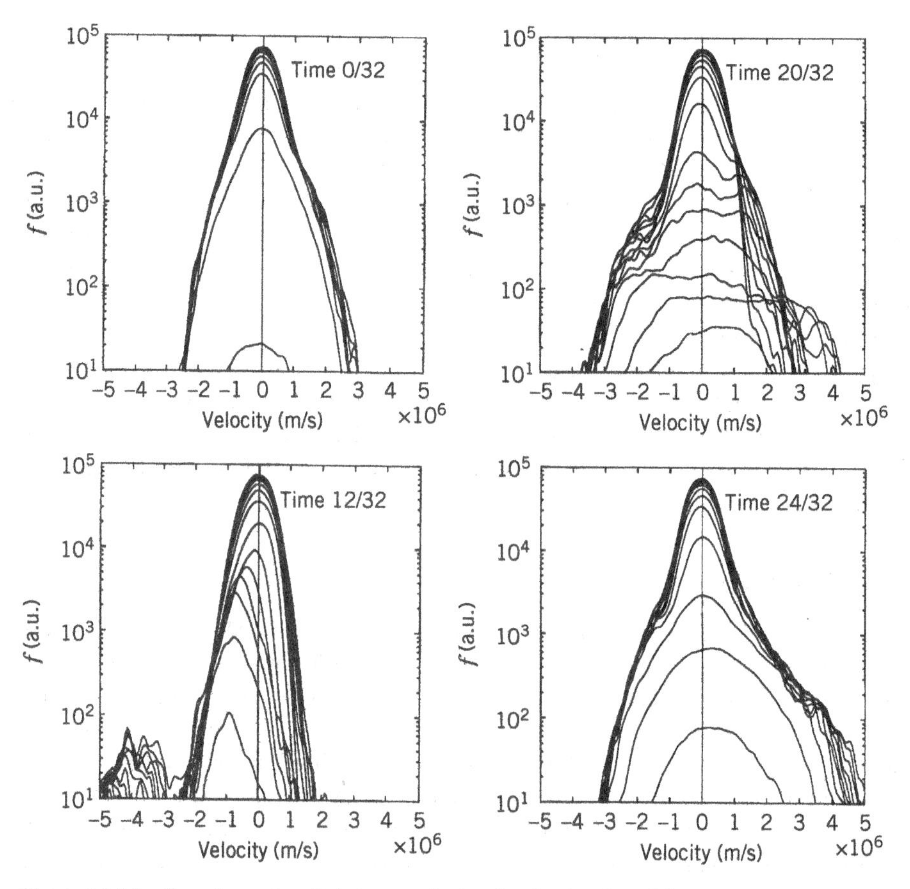

Figure 11.11 One-dimensional electron velocity distribution function $f_e(x, v_x, t)$ for a 10 cm electrode spacing; each plot covers a time window of $\frac{1}{32}$ of an rf cycle. Each line on a plot represents a spatial window of 2 mm. Source: Wood, 1991/with permission of Wood, B. P.

drift in the sheath has disappeared, and the fast electron group moving toward the electrode shows a lower velocity as slower electrons arrive from the opposite electrode. As the sheath expands, as shown here at time $\frac{20}{32}$, the electrons in the sheath region are strongly heated, and the beginning of an electron beam produced by this expansion can be seen moving away at a positive velocity. As the sheath continues to expand, the drift of the distribution in the sheath away from the electrode initially matches the sheath velocity but then decays (time $\frac{24}{32}$) to a velocity much slower than when the sheath was collapsing.

The existence of more energetic electrons near the plasma edge due to stochastic heating increases the ionization there, tending to flatten the plasma profile. Furthermore, the ionization is not constant, but follows the density variations in space and time of the more energetic electrons. This is shown for a PIC simulation by Vender and Boswell (1990) in the plot of Figure 11.12, in which the darkness of each square is proportional to the number of ionizing collisions within that square of position and time intervals. Most of the ionization is seen to occur along a path of fastest electrons that are reflected off of the sheath at the phase at which it is most rapidly expanding. There is also somewhat more ionization near the sheaths, an effect that becomes more pronounced at higher pressures where the ionization mean-free-path is shorter, which has been observed in various experiments.

In Figure 11.13, the time-dependent ion and electron conduction currents are given as a function of time, with V_{rf} and V_{pb} also indicated. We see that $J_i(t)$ is nearly constant, as assumed,

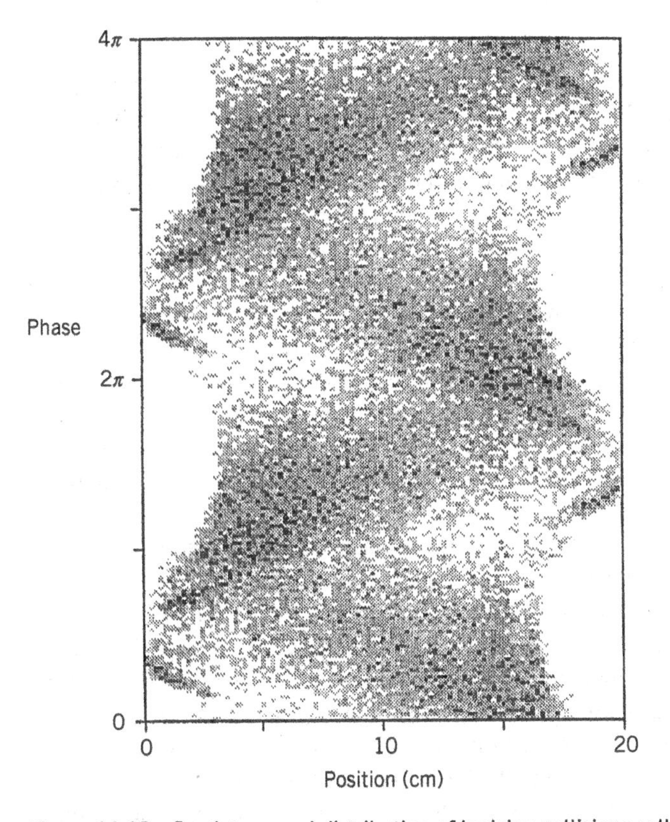

Figure 11.12 Spatiotemporal distribution of ionizing collisions collected over 20 rf cycles, for a 10 MHz, 1 kV, 20 mTorr hydrogen discharge. Source: Reproduced from Vender and Boswell (1990)/with permission of IEEE.

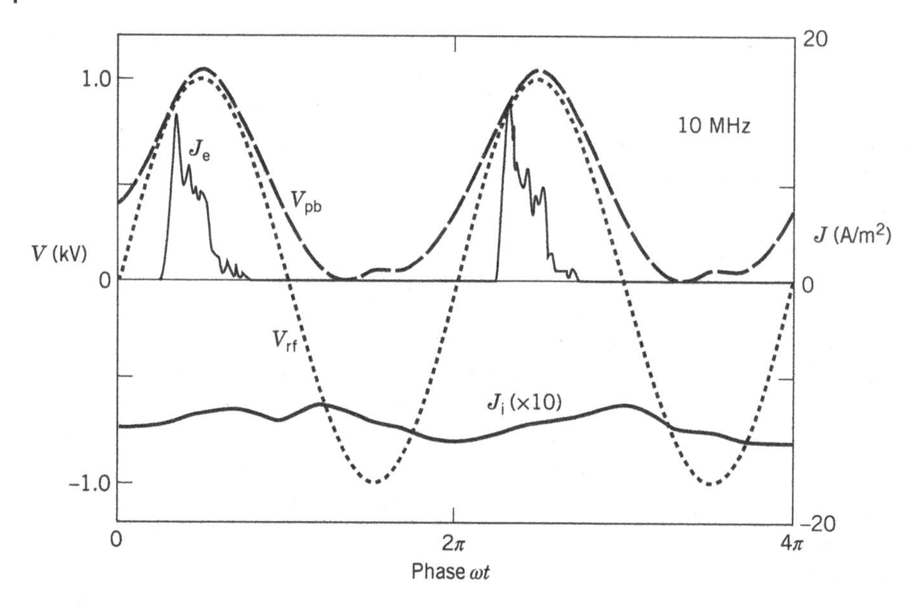

Figure 11.13 Central plasma potential V_{pb} (dashed), driving voltage V_{rf} (dotted), and electron (positive) and ion (negative) currents to the electrode. The ion current is plotted 10 times enlarged to show modulation within the rf cycle. Source: Reproduced from Vender and Boswell (1990)/with permission of IEEE.

but also contains some ripple which is not important for the modeling. $J_e(t)$, on the other hand, is spread over a significant fraction of the rf cycle, when significant voltages exist between the plasma and the electrode. This is possible because of the distribution of electron energies. Because the time-averaged electron and ion fluxes must balance, the average potential of the plasma with respect to the electrode must decrease slightly, as will be seen in (11.5.10). From this equation, we see that the zero-order value of $\overline{V} = V_1$ is increased by the usual thermal term but is decreased because the electrons reach the electrode over a finite time interval. For example, with $V_{rf} = 500$ V ($V_1 = 250$ V) and $T_e = 3.4$ V in argon gas, we find that the thermal enhancement to the voltage is the usual 4.8 T_e while the finite electron loss effect reduces $\overline{V}$ by 3.2 T_e, leading to a net increase in $\overline{V}$ over V_1 of 1.6 $T_e = 5.4$ V. Recall, however, for our model in Sections 11.1 and 11.2 with sinusoidal current and nonsinusoidal single-sheath voltage, that the zero-order result is $\overline{V} = 0.83\, V_1$. In this case, for $V_1 \gg T_e$, the correction is not significant compared to other approximations in the model. The relation between $\overline{V}$ and V_1 for sinusoidal current drive has been obtained over the entire range of V_1/T_e by Godyak and Sternberg (1990b).

The effect of a group of fast beamlike particles, traversing the plasma from one sheath to the other, indicates that the discharge length may enter into the dynamics of stochastic heating in a more sensitive manner than in the model equations. This has been demonstrated by following a class of representative electrons, with energy greater than the ionization energy, over several rf cycles, for $l = 13$, 10, and 7.5 cm. One finds a tendency to have a resonant increase of energy near $l = 10$ cm, producing a higher-energy tail on the distribution, because, for this length, electrons heated at one sheath arrive at the opposite sheath at a phase resulting in further heating. The average electron energy lost from the discharge also varies but in all cases is increased by a factor of 2–3 over the value of $2T_e$ for a single Maxwellian. Other interesting effects have been observed in simulations and sometimes confirmed in experiments. For example, high-harmonic components of the driving frequency have been observed, both in simulations and in experiments, that can be much larger than predicted from the sheath nonlinearity. This has been shown to result from a series resonance of the bulk plasma inductance L_p with the sheath capacitance C_s, occurring below the electron plasma frequency (see Section 11.5, and Problems 4.7 and 11.4).

11.3.3 Secondaries, Gas Heating, and Excited Neutral States

The incoming fluxes Γ_{in} of ions, electrons, excited neutrals, and photons striking the electrodes can generate outgoing secondary electron fluxes $\Gamma_e = \gamma\Gamma_{in}$ that are accelerated back into the plasma. Typically, $\gamma \sim 0.05$–0.2 for these various processes on metal electrodes (see Section 9.3). Depending on the phase of the rf voltage, the secondaries gain various energies up to the maximum sheath voltage. The effect of ion-induced secondaries does not generally dominate the particle and energy balances for argon at pressures below 100 mTorr at the usual operating frequency of 13.56 MHz. Unlike dc discharges, in which continuity of current requires secondaries, the rf current at 13.56 MHz can be sustained by the sheath capacitance. However, high-energy secondaries do produce some ionization. At very low pressures, most of the secondary electrons are lost from the discharge before significant ionization occurs (Problem 11.10).

Another effect of secondaries is to increase the power dissipation. Some of this additional energy loss goes into ionization and other collisional processes. Part of the energy may be lost directly to surfaces at a lower potential than the emitting surface, acting as a power drain. Both increased ionization and increased power loss can be included in a self-consistent model (Misium et al., 1989). For $\gamma_{se} = 0.1$ at $p = 10$ mTorr, they found little effect on most plasma parameters, except for an increase of up to 30% in the total power absorbed.

At higher pressures, the effect of secondaries becomes greater, as more of their energy is captured by the plasma. As the rf frequency is reduced, I_{rf} decreases with ω at fixed V_{rf}, and secondaries also play a more important role in sustaining the plasma. In addition, a transition to a different mode can take place with increasing voltage, in which the plasma is sustained by ionization from secondaries. The bulk electron temperature then falls to prevent additional ionization. This transition to the *γ-mode* has been observed experimentally and predicted theoretically (see Section 15.3). At these higher pressures, secondary electron multiplication within the sheath typically occurs, leading to a mechanism that sustains the discharge similar to that which sustains a dc glow discharge. In Chapter 14, we consider dc discharges in which the entire current in the sheath is sustained by secondaries and electrons that are created by ionization in the sheath. The important role of secondaries in the physics of high-pressure capacitive discharges is treated in Chapter 15.

Gas heating by electron–neutral, ion–neutral, and charged particle recombination collisions can have a significant effect on the discharge equilibrium. The gas temperature T_g in the discharge midplane may be increased considerably above the electrode wall temperature T_w. At a fixed gas pressure $p = n_g k T_g$, this decreases the midplane gas density n_g, altering the particle and energy balances. The simplest modeling approach in a narrow gap discharge is to solve the Fourier heat conduction equation (Bird et al., 2002, Chapters 9 and 10)

$$-\kappa_T \frac{d^2 T_g}{dx^2} = S_g(x) \tag{11.3.2}$$

where κ_T is the gas thermal conductivity (see (2.3.21)) and $S_g(x)$ is the power per unit area absorbed by the gas. For a uniform power S_{g0} across the gap, with the midplane at $x = 0$, the solution of (11.3.2) is

$$T_g(x) = T_w + \frac{S_{g0} l^2}{8\kappa_T}\left(1 - \frac{4x^2}{l^2}\right) \tag{11.3.3}$$

with l the gap length. Note that except for ion, electron, and possibly fast neutral and energetic photon powers hitting the electrodes, all the power absorbed by the discharge produces gas heating. For a fixed rf driving voltage, S_g increases with n_g, and $\kappa_T \propto m_g^{-1/2}$, with m_g the molecular mass, independent of n_g. This indicates that the midplane T_g will increase with increasing feed gas pressure, gap length, and molecular mass. Further discussion of gas heating issues, especially those relevant to high-density, low-pressure discharges, is given at the end of Section 12.3.

In addition to altering the particle and energy balances, gas heating can induce discharge instability at the higher pressures. This issue is described in Chapter 15 and motivates the use of low mass (helium) feed gas mixtures and narrow gaps in high-pressure capacitive discharges.

Excited neutral states, such as the argon metastable Ar_m, resonance Ar_r, and Ar(4p), can play important roles in the discharge particle balance due to processes such as multistep ionization

$$e + Ar_m \rightarrow 2e + Ar^+$$

and metastable pooling ionization

$$Ar_m + Ar_m \rightarrow e + Ar^+ + Ar$$

These processes are very important in high-pressure capacitive discharges, and are described throughout Chapter 15, and, in particular, in Section 15.2. At the lower pressures considered in this chapter, they do produce some ionization, but do not dominate the particle balance.

Figure 11.14 shows comparisons between measurements and 1D PIC simulation results of Schulenberg et al. (2021), and later simulations by Wen et al. (2023). The symmetric argon discharge had 12 cm diameter stainless steel electrodes with a 4 cm gap and was driven at 13.56 MHz rf voltages between 150 and 350 V. The pressures were varied from 1 to 100 Pa. Langmuir probe measurements were used to determine the central electron density n_e, and tunable diode laser absorption was used to determine the gas temperature T_g. Both simulations used an ion-induced secondary electron emission coefficient $\gamma_{se} = 0.7$. Schulenberg et al. (2021) considered various choices of an "effective" electron reflection coefficient γ_e and gas temperature T_g. The later simulations by Wen et al. (2023) added a more physical model of electron-induced secondary emission (Vaughan, 1989), and secondary emission by excited argon neutrals and by photons.

The comparisons of the measured and simulation values of n_e by Schulenberg et al. (2021), at $V_{rf} = 250$ V and 1–20 Pa pressures, are shown in Figure 11.14*a*. The importance of the gas heating is apparent. Choosing a "best-fit" value of γ_e, along with the measured gas temperature, gave a good agreement between measured and simulation values of n_e. The comparisons of Wen et al. (2023) in Figure 11.14*b* replace the best-fit γ_e with the Vaughan model and also incorporate the excited state and photon secondary emission processes. The measured and simulation densities agree well, except at the highest pressure (20 Pa in Figure 11.14*b*), where there is a significant disagreement. This could be due to 2D effects, as the discharge is more local radially at the higher pressures. Also, an additional discharge around the Langmuir probe began to appear at pressures above 20 Pa, which might have affected the density measurement.

11.3.4 Boltzmann Term Analysis for PIC Simulations

PIC simulations of rf discharges offer unique opportunities to determine fundamental discharge mechanisms, since the electron distribution function $f_e(\mathbf{x}, \mathbf{v}, t)$ is determined on space scales much shorter than the sheath and discharge scale lengths, and on timescales much shorter than the rf period. One important procedure is *Boltzmann term analysis*, in which the electron heating is associated with various terms in the Boltzmann equation, rather than being divided into the ohmic

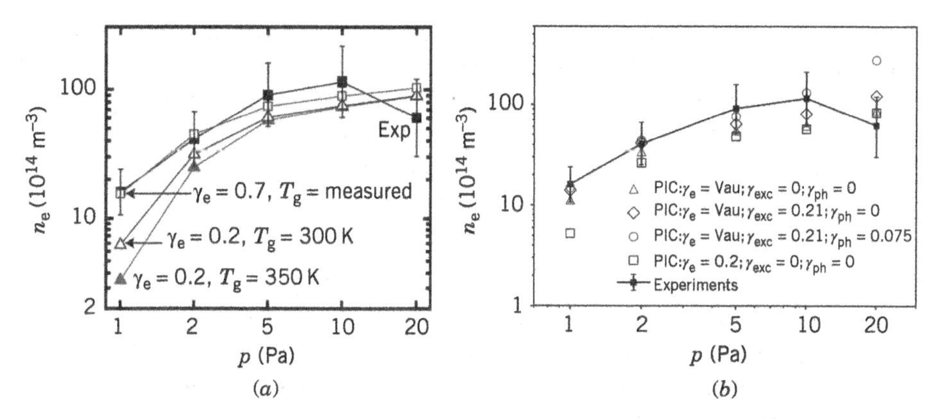

Figure 11.14 Symmetric discharge experiments compared to 1D PIC simulations, for central electron density n_e versus argon pressure p; (*a*) Langmuir probe measurements (closed squares with error bars), PIC with "effective" electron reflection coefficient γ_e, ion-induced secondary emission coefficient $\gamma_{se} = 0.7$, and gas heating (Schulenberg et al., 2021); (*b*) electron-induced γ_e (Vaughan model), ion-induced $\gamma_{se} = 0.7$, excited neutral-induced γ_{exc}, photon-induced γ_{ph} (Wen et al., 2023), and measured gas heating; 12 cm diameter stainless steel electrodes, 4 cm gap, 13.56 MHz, 250 V rf amplitude. Source: (*a*) Schulenberg et al. (2021)/IOP Publishing/CC BY 4.0. https://doi.org/10.1088/1361-6595/ac2222. (*b*) Wen et al. (2023)/IOP Publishing/CC BY 4.0.

and stochastic heating contributions of Sections 11.1 and 11.2. This analysis has been applied to categorize electron heating mechanisms in one-dimensional (1D) PIC simulations of rf capacitive discharges (Surendra and Dalvie, 1993; Lafleur et al., 2014a; Schulze et al., 2018). We summarize the method and some results below.

In 1D, the electron momentum balance (2.3.9) is (see Problem 11.12)

$$\frac{\partial}{\partial t}(mn_e u_e) + \frac{\partial}{\partial x}(mn_e u_e^2) = -\frac{\partial p_e}{\partial x} - en_e E_x - mu_e G_{iz} - C_m \tag{11.3.4}$$

where $n_e = \int d^3v f_e$ is the electron density, $u_e = \int d^3v\, v_x f_e$ is the x-component of the electron mean velocity (2.3.5), $p_e = m\int d^3v\, v_x^2 f_e - mn_e u_e^2$ is the xx-component of the electron pressure tensor (2.3.10), $G_{iz} = \int d^3v\,(\partial f_e/\partial t)_c$ is the net electron production rate per unit volume due to collisions, and

$$C_m = m\int d^3v\, v_x(\partial f_e/\partial t)_c = -mn_g\int d^3v\,\sigma_m v v_x f_e \tag{11.3.5}$$

is the electron momentum transfer rate per unit volume due to elastic electron–neutral collisions, with $\sigma_m(v)$ the momentum transfer cross section given in (3.1.15). Multiplying (11.3.4) by u_e and re-arranging, with $J_e = -en_e u_e$, we obtain the space- and time-varying electron power absorbed per unit volume, $p_{abs} = J_e E_x$, as

$$p_{abs}(x,t) = \underbrace{u_e\frac{\partial}{\partial t}(mn_e u_e) + u_e\frac{\partial}{\partial x}(mn_e u_e^2)}_{\text{inertial}} + \underbrace{eu_e\frac{\partial}{\partial x}(n_e \mathrm{T}_e)}_{\text{pressure}} + \underbrace{mu_e^2 G_{iz} + u_e C_m}_{\text{ohmic}} \tag{11.3.6}$$

where we note that $p_e = en_e\mathrm{T}_e$ with $\mathrm{T}_e = p_e/(en_e)$. The time-averaged power is found by averaging over an rf cycle, $\overline{p}_{abs}(x) = \langle p_{abs}(x,t)\rangle_{\omega t}$. Integrating $\overline{p}_{abs}(x)$ over the discharge length gives the total electron power absorbed per unit area, $S_e = S_{e,in} + S_{e,press} + S_{e,ohm}$. Generally, the inertial heating terms are small, and the power is absorbed by pressure and ohmic heating, with a negligible ohmic contribution from G_{iz}.

If the compression and rarefaction of the electrons is done reversibly (without entropy change), then the time-averaged pressure heating would vanish. This would be true in the absence of all collisions, $(\partial f_e/\partial t)_c \equiv 0$, since the Boltzmann H-theorem (B.10) then guarantees an unchanging entropy. However, there are energy, momentum, and particle losses during an rf cycle due to the various collisional processes, so in general there is significant pressure heating. This can be decomposed into density gradient and temperature gradient-driven terms

$$p_{\text{press}} = eu_e \mathrm{T}_e \frac{\partial n_e}{\partial x} + eu_e n_e \frac{\partial \mathrm{T}_e}{\partial x} \tag{11.3.7}$$

Typically, the temperature gradient and density gradient terms partially cancel.

The Boltzmann term analysis can be compared with various models of ohmic and stochastic heating. This has been done by Lafleur et al. (2014a) for the argon discharge measurements of Godyak and Piejak (1990b) shown in Figure 11.10 over the pressure range from 70 mTorr to 1 Torr. In Figure 11.15a, the experimental data and PIC simulations were analyzed using the simple homogeneous model expressions (11.1.28) with average bulk density $\overline{n}$, and (11.1.37) with sheath edge density n_s

$$\frac{S_{\text{stoc}}}{S_{\text{ohmic}}} = \frac{\overline{n}}{n_s} \frac{2\overline{v}_e}{\nu_m d} \tag{11.3.8}$$

Here, $\nu_m = n_g \sigma_m \overline{v}_e$ and d is the oscillating bulk width. We see that stochastic heating dominates ohmic heating at the lowest pressures. Figure 11.15*b* shows the Boltzmann term analysis of the PIC simulations. We see that at the lowest pressures, ohmic and pressure heating contribute roughly equally to the total heating, with negligible inertial heating.

The differences in the results reflect the different ways in which the total heating is calculated and the different ways in which the heating is categorized. The stochastic heating for the self-consistent collisionless or collisional Child law sheath, (11.2.34) or (11.2.59), respectively, would give a different contribution than the homogeneous expression used to evaluate (11.3.8). Similarly, the collisional momentum transfer per unit volume, $C_m = -mn_e \nu_m u_e = -mn_e n_g \sigma_m \overline{v}_e u_e$, used in (11.3.8), is only an estimate of the exact expression (11.3.5) obtained from kinetic theory. The exact expression

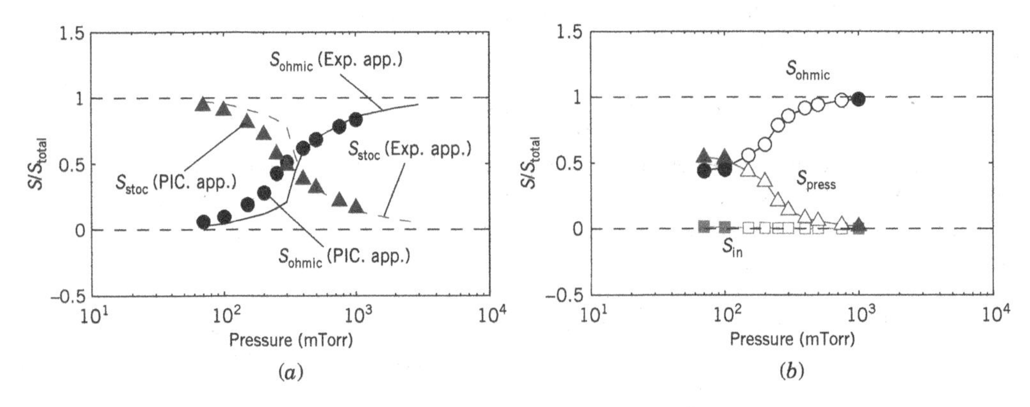

Figure 11.15 Electron heating fractions S/S_{tot} as a function of argon pressure; (*a*) apparent $S_{\text{ohmic}}/S_{\text{tot}}$ and $S_{\text{stoc}}/S_{\text{tot}}$, from the data in Godyak and Piejak (1990b) (closed symbols), and from the PIC simulations (dashed curves), both analyzed using the homogeneous model of Section 11.1; and (*b*) $S_{\text{in}}/S_{\text{tot}}$, $S_{\text{press}}/S_{\text{tot}}$ and $S_{\text{ohmic}}/S_{\text{tot}}$ (symbols) from the PIC simulations using Boltzmann term analysis expression (11.3.6); the open and closed symbols correspond to two different, but equivalent, models of electron scattering. Source: Lafleur et al. (2014a)/with permission of IOP Publishing.

is accessible from PIC simulations, but would be very difficult to evaluate from discharge measurements. Other more accurate, but approximate, expressions for C_m from two-term kinetic theory are described in Section 19.1. Both pressure and ohmic heating scale with collision rates, and the distribution function itself depends on all terms in the Boltzmann equation.

Boltzmann term analysis of PIC simulations gives access to space- and time-varying decompositions of the heating. Schulze et al. (2018) give the decomposition of the spatial variation of the time-averaged heating into seven electric field-driven terms (three for inertial heating, two for pressure heating, and two for ohmic heating). Their 1D PIC simulation was done for a 400 V, 13.56 MHz, 5 cm gap argon discharge over the pressure range from 1 to 50 Pa. Secondary electron emission was omitted, and a 25% electron reflection coefficient was used. The spatio-temporal, time-averaged spatial, and time-averaged total heating were obtained for each of the seven terms. The (non-ohmic) pressure heating was shown to be concentrated near the plasma–sheath edge, with the decomposition (11.3.7) into a larger positive ∇n_e and a smaller negative ∇T_e term seen over the entire range of pressures. Proto and Gudmundsson (2020, 2021) give Boltzmann term analyses of the electron power deposition for low-power oxygen and chlorine capacitive discharges, respectively, determined by 1D PIC simulation results. We have discussed these results briefly in Section 10.5.

11.4 Asymmetric Discharges

11.4.1 Capacitive Voltage Divider

Most capacitive discharges are asymmetric, because more conducting surfaces are naturally grounded than driven. The dc voltage between the plasma and the driven electrode is then larger than the dc voltage between the plasma and the grounded electrode. This is easily seen from a model of the rf voltage drops across the two sheaths connecting the driven electrode to ground, as shown in Figure 11.16, along with the linear relation (11.2.22) between the rf and dc voltages, which lead to

$$V_{ab1} = V_{a1} + V_{b1} = \frac{1}{0.83}(\overline{V}_a + \overline{V}_b) \tag{11.4.1}$$

where the dc voltages are taken between the plasma and the electrodes. Here, the subscript "1" denotes the fundamental rf magnitude. An easily measurable dc *bias voltage* is set up at the driven electrode with respect to ground,

$$V_{bias} = -(\overline{V}_a - \overline{V}_b) \tag{11.4.2}$$

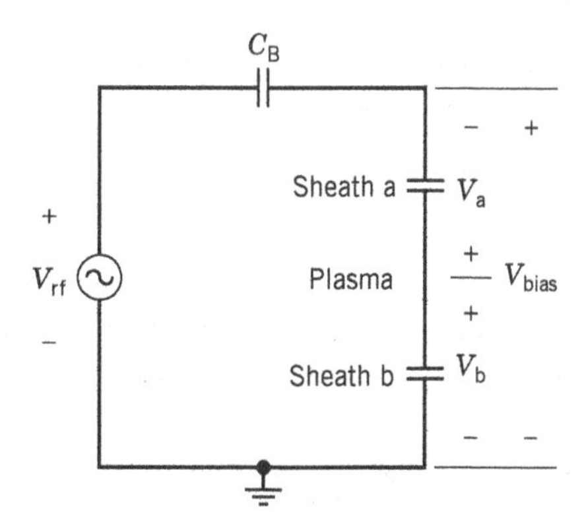

Figure 11.16 Capacitive voltage divider model of bias voltage formation in an asymmetric discharge.

which is negative in the usual case that $\overline{V}_a > \overline{V}_b$. Since the voltage drops across the sheaths are inversely proportional to the sheath capacitances, the sheath with the smaller area has a smaller capacitance and therefore a larger voltage drop. The situation is more complicated, because the sheath thickness also depends on the voltage across it, through Child's law, which must be solved self-consistently to obtain the voltage. This is relatively easy to do within various simplifying assumptions.

Consider arbitrary electrodes a and b having areas A_a and A_b and dc voltage drops $\overline{V}_a$ and $\overline{V}_b$, independent of the vector position $\mathbf{x}$ along (in the plane of) the sheath. We then have the proportionalities

$$J_{a1}(\mathbf{x}) \propto \frac{\overline{V}_a}{s_a(\mathbf{x})} \tag{11.4.3}$$

for the capacitive sheath, and

$$n_a(\mathbf{x}) \propto \frac{\overline{V}_a^{3/2}}{s_a^2(\mathbf{x})} \tag{11.4.4}$$

for the collisionless Child law. In terms of the total rf current

$$I_{a1} = \int_{A_a} J_{a1}(\mathbf{x})\,\mathrm{d}^2x \tag{11.4.5}$$

we can eliminate s_a in favor of n_a by substituting (11.4.4) into (11.4.3) to get

$$I_{a1} \propto \overline{V}_a^{1/4} \int_{A_a} n_a^{1/2}(\mathbf{x})\,\mathrm{d}^2x$$

and similarly

$$I_{b1} \propto \overline{V}_b^{1/4} \int_{A_b} n_b^{1/2}(\mathbf{x})\,\mathrm{d}^2x$$

For rf current continuity, we can equate $I_{a1} = I_{b1}$ to obtain

$$\frac{\overline{V}_a}{\overline{V}_b} = \left(\frac{\int_{A_b} n_b^{1/2}\mathrm{d}^2x}{\int_{A_a} n_a^{1/2}\mathrm{d}^2x} \right)^4 \tag{11.4.6}$$

In the simplest plasma model, we set $n_a = n_b$, independent of $\mathbf{x}$, to find the scaling

$$\frac{\overline{V}_a}{\overline{V}_b} = \left(\frac{A_b}{A_a} \right)^4 \tag{11.4.7}$$

This very strong scaling with area is not in accordance with most experimental observations which have typically found

$$\frac{\overline{V}_a}{\overline{V}_b} \approx \left(\frac{A_b}{A_a} \right)^q \tag{11.4.8}$$

with $q \lesssim 2.5$. The experiments were mainly done at higher pressure, where the sheath dynamics do not follow the collisionless Child–Langmuir law. If, for example, we consider a collisional, constant λ_i sheath, as described in Section 6.4, the proportionality

$$n_a(\mathbf{x}) \propto \frac{\overline{V}_a^{3/2}}{s_a^{5/2}(\mathbf{x})} \tag{11.4.9}$$

leads to

$$\frac{\overline{V}_{\rm a}}{\overline{V}_{\rm b}} = \left[\frac{\int_{A_{\rm b}} n_{\rm b}^{2/5} {\rm d}^2x}{\int_{A_{\rm a}} n_{\rm a}^{2/5} {\rm d}^2x}\right]^{5/2} \tag{11.4.10}$$

For unequal edge densities, we obtain

$$\frac{\overline{V}_{\rm a}}{\overline{V}_{\rm b}} = \frac{n_{\rm b}}{n_{\rm a}}\left(\frac{A_{\rm b}}{A_{\rm a}}\right)^{5/2} \tag{11.4.11}$$

which for $n_{\rm b} = n_{\rm a}$ is much closer to the experimental range. Similarly, for homogeneous (uniform density) sheaths, which may be appropriate at the lower sheath voltages, we find

$$\frac{\overline{V}_{\rm a}}{\overline{V}_{\rm b}} = \frac{n_{\rm b}}{n_{\rm a}}\left(\frac{A_{\rm b}}{A_{\rm a}}\right)^{2} \tag{11.4.12}$$

We leave the derivation of these scalings to Problem 11.13. However, we should note that some experiments had high-voltage sheaths that were more collisionless than collisional, so the sheath scaling is not the only factor involved.

11.4.2 Spherical Shell Model

It is clear that geometric factors alone, even at low density, will make $n_{\rm a} \neq n_{\rm b}$. Additionally, for higher pressure, factors such as local ionization can further increase the density at the higher-voltage electrode.

We now consider these effects using the one-dimensional spherical shell model shown in Figure 11.17 (Lieberman, 1989b). The powered electrode is the inner sphere a having radius $r_{\rm a}$, and the grounded electrode is the outer sphere b having radius $r_{\rm b}$. The electrode separation l, plasma thickness d, and sheath thicknesses $s_{\rm a}$ and $s_{\rm b}$ are defined in the figure. The discharge is driven by an rf current source through a blocking capacitor $C_{\rm B}$ having negligible impedance at the driving frequency. Since the system is spherically symmetric, the model is purely one-dimensional (along r). The freedom to choose not only the discharge length $l = r_{\rm b} - r_{\rm a}$ but also the powered-to-grounded electrode area ratio $A_{\rm a}/A_{\rm b} = r_{\rm a}^2/r_{\rm b}^2 < 1$ allows us to model an asymmetric discharge. We consider the intermediate mean free path regime, where the ion drift velocity is much greater than the ion thermal velocity, $u_{\rm i} \gg v_{\rm thi}$, and assume that the dominant

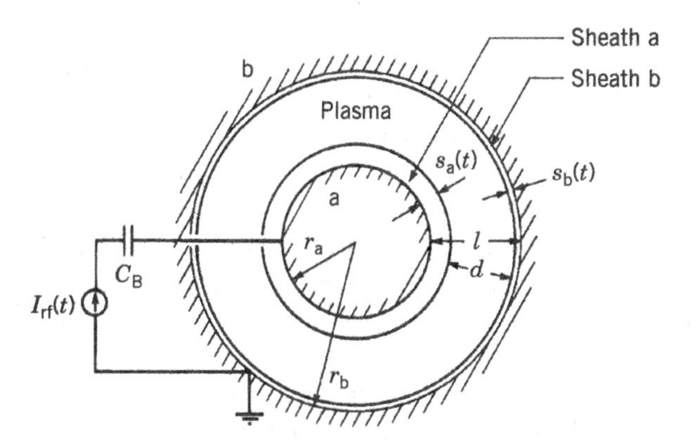

Figure 11.17 Spherical shell model of an asymmetric rf discharge. Source: Lieberman, 1989b/with permission of AIP Publishing.

ion collisional process is charge exchange of the ion with the parent neutral gas atom, such that the mean free path $\lambda_i = (n_g \sigma_i)^{-1}$ is nearly constant.

For this intermediate pressure regime, we found the drift velocity in (5.3.5) to be

$$u_i^2 = \left| \frac{eT_e}{M} \frac{2}{\pi} \frac{\lambda_i}{n} \frac{dn}{dr} \right| \tag{11.4.13}$$

Substituting this into the continuity equation (5.3.4), in spherical coordinates, we obtain an equation for the density,

$$\frac{1}{r^2} \frac{d}{dr} \left[r^2 n \left| \frac{eT_e}{M} \frac{2}{\pi} \frac{\lambda_i}{n} \frac{dn}{dr} \right|^{1/2} \right] = \nu_{iz} n \tag{11.4.14}$$

Using various transformations to simplify (11.4.14), Lieberman (1989b) numerically solved this differential equation to obtain a simple expression for the ratio of densities at the grounded and powered electrodes. In the usual regime for processing discharges for which $(2\lambda_i \nu_{iz}/\pi u_B)(A_a/A_b)^{3/8} \ll 1$, this gives

$$\frac{n_b}{n_a} \approx \left(\frac{A_a}{A_b} \right)^{0.29} \tag{11.4.15}$$

With this geometric scaling of density, we can recompute the voltage ratios from (11.4.6) and (11.4.9) to obtain, for a collisionless Child law sheath,

$$\frac{\overline{V}_a}{\overline{V}_b} = \left(\frac{A_b}{A_a} \right)^{3.42} \tag{11.4.16}$$

and, for a collisional (λ_i = const) sheath,

$$\frac{\overline{V}_a}{\overline{V}_b} = \left(\frac{A_b}{A_a} \right)^{2.21} \tag{11.4.17}$$

The above results do not exhaust the assumptions that can be made to describe the plasma glow region and the sheath region. Spatially uniform ionization by secondaries, and edge ionization by stochastically heated electrons can also be introduced as important ionization processes. For the sheath dynamics, a collisional constant mobility sheath law (6.5.8) or a homogeneous sheath can be introduced. The scaling results for all these cases (except the homogeneous sheath) are shown in Table 11.1. We note that considerably lower values of q can be found with the appropriate combination of glow and sheath dynamics, but we hasten to add that many of these combinations are mutually incompatible. Generally, the higher-pressure assumptions, to the right and down in the table, produce the lower values of q.

The results in the table do not give the whole story. For large-area ratios, the voltage at the large-area electrode saturates at its dc value given by (6.2.17) for an undriven sheath, and the scaling laws in the table must be modified. Geometries that would more closely resemble physical reactors, on which experimental measurements have been made, have also been investigated using more complicated models. One such study investigated various cylindrical and coaxial systems and compared the results with similar experimental configurations, obtaining reasonable agreement (Lieberman and Savas, 1990). For two-dimensional geometries, such as a cylinder, the voltage ratio does not simply scale as a power of the area ratio but depends in a complicated way on the cylinder length and radius. For details, the reader should consult the original paper.

Table 11.1 Scaling Exponent q for the Equation $\overline{V}_a/\overline{V}_b = (A_b/A_a)^q$

Glow Physics	Sheath Physics Scaling Law	Child's Law $J \propto V^{3/2}/s^2$	Constant λ_i Law $J \propto V^{3/2}/s^{5/2}$	Constant μ_i Law $J \propto V^2/s^3$
Homogeneous	$n = \text{const}$	4.0	2.5	3.0
Thermal electron	$n \propto A^{-7/24}$	3.42	2.21	2.71
Secondary electron	$n \propto A^{-1/2}$	3.0	2.0	2.5
Local ionization	$n \propto V$	1.33	1.25	1.5

Source: Lieberman (1989b)/with permission of AIP Publishing.

11.5 Voltage-Driven Sheaths and Series Resonance

In Sections 11.1 and 11.2, for ease of analysis, we treated the case of sheaths driven by a sinusoidal rf current. For the homogeneous sheath in Section 11.1, we found that the voltage (11.1.13) across a single sheath is nonlinear, having a significant second harmonic component. However, for a symmetric (equal electrode area) discharge, the sum of the voltages (11.1.19) across both sheaths is linear, with only a fundamental frequency component. Similarly, for the self-consistent Child law sheath in Section 11.2, there is a strong voltage nonlinearity for a single sheath, but the sum of the sheath voltages (11.2.17) for a symmetric discharge is nearly linear, with no second harmonic and a small (4.2%) third harmonic. These results suggest that it little matters whether a symmetric discharge is current- or voltage-driven, but that there will be significant differences for highly asymmetric discharges, where the rf voltage drops across the powered sheath only.

The sheath dynamics driven by a sinusoidal voltage source was first examined by Klick (1996). Klick et al. (1997) applied this to determine the current harmonics and resulting excitation of the series resonance in a voltage-driven asymmetric discharge. Czarnetzki (2013) developed a near-self-consistent analytic model for the sinusoidal voltage-driven sheath.

Let us examine a single homogeneous (constant density) sheath driven by

$$V_s = V_B - V_0 \cos \omega t \tag{11.5.1}$$

where V_s is the voltage at the plasma–sheath edge with respect to the electrode, V_B is the edge-to-electrode dc bias voltage, and V_0 is the rf voltage amplitude. The sheath charge $Q_s(t) = ens(t)A$ is found in terms of the voltage from (11.1.12) as

$$Q_s^2 = 2e\epsilon_0 nA^2 V_s \tag{11.5.2}$$

where n is the sheath ion density, $s(t)$ is the oscillating sheath width, and A is the sheath area. The current flowing toward the electrode is

$$I_s = \frac{dQ_s}{dt} \tag{11.5.3}$$

For the simple case with $V_B = V_0$, where the oscillating sheath collapses onto the electrode at one instant per rf cycle, the voltage, charge, and current waveforms are shown versus $\phi = \omega t$ as the solid lines in Figure 11.18. Here, $V_{norm} = V_s/V_0$, $Q_{norm} = Q_s/Q_0$, with $Q_0 = (2e\epsilon_0 nV_0)^{1/2}A$, and $I_{norm} = I_s/(\omega Q_0)$. As can be seen, the sheath charge has a discontinuous slope, which leads to a discontinuous current. The discontinuity in the current generates a rich set of high harmonics of the rf frequency. Expanding I_{norm} in a Fourier sine series gives (Problem 11.14)

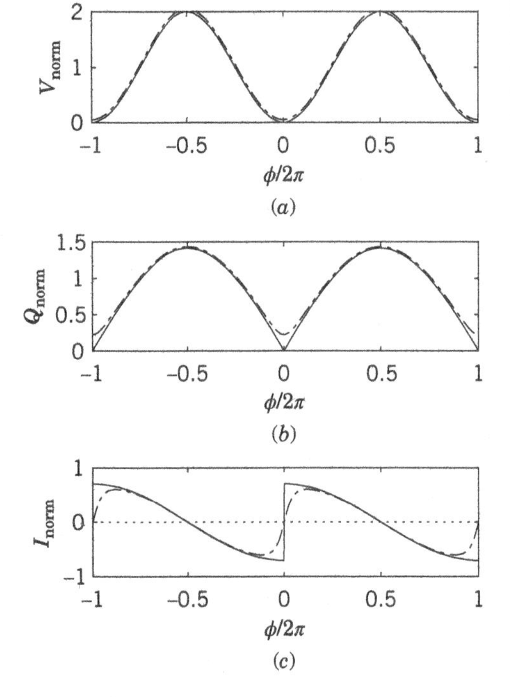

Figure 11.18 Voltage-driven rf sheath, showing (*a*) normalized voltage V_{norm}, (*b*) normalized charge Q_{norm}, and (*c*) normalized current I_{norm}, over two rf cycles; the solid lines show $V_B = V_0$, and the dot-dashed lines show $V_B = 1.05\,V_0$.

$$I_{norm} = \frac{4}{\pi}\sum_{m=1}^{\infty}\frac{m}{4m^2-1}\sin m\omega t \tag{11.5.4}$$

As can be seen from (11.5.4), the harmonic amplitudes fall off slowly, $\propto m^{-1}$, for large m.

For $V_B > V_0$, the discontinuity in the current I_s in (11.5.3) is replaced by a steeply increasing current near $\phi = 0$ (mod 2π). An example for $V_B = 1.05\,V_0$ is shown as the dot-dashed lines in Figure 11.18. This still gives a strong set of high harmonics of the current waveform.

11.5.1 Bias Voltage

The choice of $V_B = V_0$ does not account for the effect of a finite dc floating potential (10.2.4) in the absence of an rf voltage across the sheath. In addition, there is a correction to V_B due to the sinusoidal voltage amplitude V_0 itself. Accounting for both effects, V_B is found by equating the time-averaged electron and ion sheath fluxes. The average electron flux $\overline{\Gamma}_e$ is

$$\overline{\Gamma}_e = \frac{1}{4}\langle n_e(t)\rangle\overline{v}_e \tag{11.5.5}$$

where from Boltzmann's law

$$n_e(t) = n_s \exp\left(-\frac{V_B - V_0\cos\omega t}{\mathrm{T_e}}\right) \tag{11.5.6}$$

with n_s is the sheath edge density. Substituting (11.5.6) into (11.5.5), we have

$$\overline{\Gamma}_e = \frac{1}{4}n_s\overline{v}_e\,\mathrm{e}^{-V_B/\mathrm{T_e}}\frac{1}{2\pi}\int_0^{2\pi}\exp\left(\frac{V_0\cos\omega t}{\mathrm{T_e}}\right)d(\omega t) \tag{11.5.7}$$

The averaged integral yields the modified Bessel function

$$I_0(V_0/T_e) \approx \left(\frac{T_e}{2\pi V_0}\right)^{1/2} e^{V_0/T_e} \tag{11.5.8}$$

where the approximate equality holds for $V_0 \gg T_e$. Using this and equating $\overline{\Gamma}_e$ to the Bohm ion flux $\Gamma_i = n_s u_B$, we have

$$e^{(-V_B+V_0)/T_e} = \left(\frac{2\pi V_0}{T_e}\frac{2\pi m}{M}\right)^{1/2} \tag{11.5.9}$$

Taking the logarithm of both sides and solving for V_B, we obtain

$$V_B = V_0 + \frac{T_e}{2}\left(\ln\frac{M}{2\pi m} - \ln\frac{2\pi V_0}{T_e}\right) \tag{11.5.10}$$

We see that the zero-order value of $V_B = V_0$ is increased by the usual floating potential term but is decreased because the electrons reach the electrode over a finite time interval.

The result (11.5.10) applies to both sheaths of a high-frequency capacitive discharge. It also applies to the powered (lower area) sheath for a low-frequency highly asymmetric discharge, which we examine in Section 11.8.

11.5.2 Wall Ion Flux Probe

As one application, consider a small planar disk of area $A_d \sim 1\text{–}10\ \text{cm}^2$, embedded in, and insulated from, a grounded rf discharge reactor wall. The disk is driven by periodic square wave bursts of an rf voltage with amplitude $V_0 \gg T_e$ through a capacitor C_d. A frequency ω_d is used that differs from the discharge driving frequency ω. The rf voltage is on for the first half-period $\tau_d/2$ and is off for the second half-period. During the on-time, C_d charges up to a bias voltage $V_d(0) = V_B$ given by (11.5.10). During the off-time, the capacitor discharges due to the plasma ion and electron currents flowing to the disk

$$\frac{dV_d(t)}{dt} = -\frac{eA_d}{C_d}\left(\Gamma_i - \Gamma_e(V_d)\right) \tag{11.5.11}$$

Note that V_d decreases linearly with time due to the steady ion current, until $V_d \sim 3\text{–}5\,T_e$, when the electron current becomes non-negligible. This gives a constant capacitor current $I_d = eA_d\Gamma_i$ during the first part of the off-time. Measuring this current determines the ion flux Γ_i.

The measurement is insensitive to thin dielectric layers deposited on the probe (and reactor walls), provided $C_s \ll C_d < C_f$, with $C_s \sim \epsilon_0 A/\lambda_{Ds}$ the low-voltage sheath capacitance and C_f the film capacitance (Braithwaite et al., 1996). Additionally, the voltage V_0 and square wave period τ must be chosen properly, e.g., $\tau \gg \tau_i$, the ion transit time across the sheath. In practice, a guard ring also driven by V_0 can be used to reduce measurement edge effects. These and other measurement issues, such as finite rf source impedance, negative ions, and electron temperature effects, are discussed in Braithwaite et al. (1996, 2003) and Booth et al. (2000).

11.5.3 Nonlinear Excitation of Series Resonance

The high harmonics ($m \gg 1$) of the current in (11.5.4) can strongly excite an asymmetric rf discharge near its series resonance frequencies. Recall from Problem 11.4 that for a symmetric homogeneous discharge of gap length l, density n, with the (constant) sum of the two oscillating sheath widths $s_a(t) + s_b(t) = s_m$, the series resonance frequency is

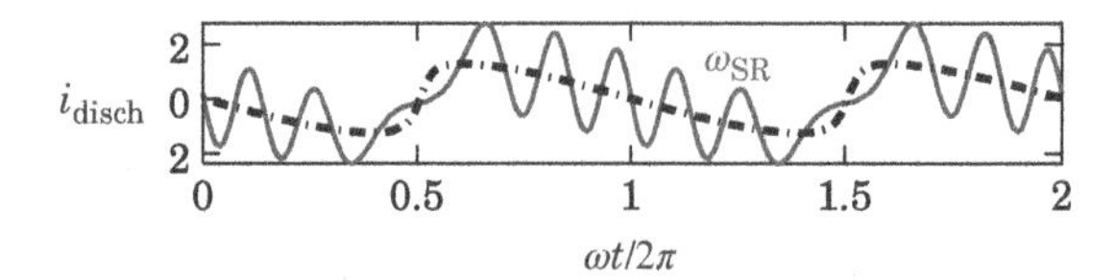

Figure 11.19 Normalized discharge current over two rf cycles; the dot-dashed line shows the current for the nonlinear sheath alone, in the absence of the bulk plasma; the solid line shows the series resonance oscillations that are excited in an asymmetric discharge; $V_{rf} = 200$ V, $f = 13.56$ MHz, $l = 5.7$ cm, $n = 10^{15}$ m^{-3}, $p = 3$ mTorr argon. Source: Reproduced from Lieberman et al. (2008), with the permission of AIP Publishing. https://doi.org/10.1063/1.2928847.

$$\omega_{res} = \omega_{pe}\left(\frac{s_m}{l}\right)^{1/2} \tag{11.5.12}$$

with ω_{pe} the plasma frequency. However, the sheath width for the single-powered sheath of an asymmetric discharge oscillates between 0 and s_m, leading to a time-varying sheath capacitance, and, hence, to a wide range of series resonance frequencies that can be excited.

For a homogeneous asymmetric discharge, neglecting the small vacuum capacitance, the voltage across the bulk plasma is

$$V_p = L_p\frac{dI_s}{dt} + I_sR_p \tag{11.5.13}$$

Equating the sum of the plasma voltage (11.5.13) and the sheath voltage V_s from (11.5.2) to the discharge voltage $V_B - V_0\cos\omega t$, we obtain the nonlinear differential equation

$$L_p\frac{dI_s}{dt} + I_sR_p + \frac{Q_s^2}{2e\epsilon_0 nA^2} = V_B - V_0\cos\omega t \tag{11.5.14}$$

Solving the two differential equations (11.5.3) and (11.5.14) gives the excitation of the series resonances due to the nonlinear sheath motion. This has been done by Klick et al. (1997), Mussenbrock and Brinkmann (2006, 2007), Lieberman et al. (2008), and Mussenbrock et al. (2008), taking various additional physics into account. An approximate analytic calculation is given in Czarnetzki et al. (2006).

A typical solution at low pressure (3 mTorr argon) is shown in Figure 11.19. The series resonances are essentially excited near the moment of sheath collapse, where the current discontinuity occurs. At this low pressure, the plasma resistance R_p is relatively small and the resulting oscillations are relatively undamped.

Figure 11.20 shows measurements at pressures of 2 Pa (upper curve) and 10 Pa (lower curve) for an argon asymmetric discharge excited by a 13.56 MHz sinusoidal voltage (Klick et al., 1997). The damped oscillations are excited near the current discontinuity, with greater damping at the higher pressure where the plasma bulk resistance is larger.

11.6 Multi-frequency Capacitive Discharges

Multi-frequency-driven capacitive discharges are of increasing importance for materials processing. As shown in Figure 11.21, with two or more rf sources, both diode and triode configurations are possible. For the diode, all rf current entering an electrode exits at the opposite electrode. The two voltage sources are simply in series, with $V_{rf} = V_l + V_h$. For the triode, some current entering each electrode exits at the radial sidewall (ground). The current flows to the three electrodes must

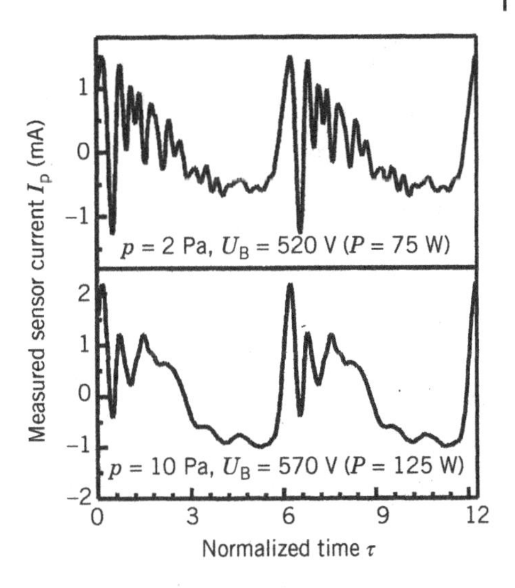

Figure 11.20 Measured sensor current (proportional to the discharge current), showing the nonlinearly excited series resonance oscillations over two rf cycles, for a 13.56 MHz asymmetric discharge with a gap length of 6.7 cm; (upper curve) argon pressure 2 Pa, dc bias voltage 520 V, rf power 75 W; (lower curve) argon pressure 10 Pa, dc bias voltage 570 V, rf power 125 W. Source: Klick et al. (1997). ©IOP Publishing. Reproduced with permission. All rights reserved. https://doi.org/10.1143/JJAP.36.4625.

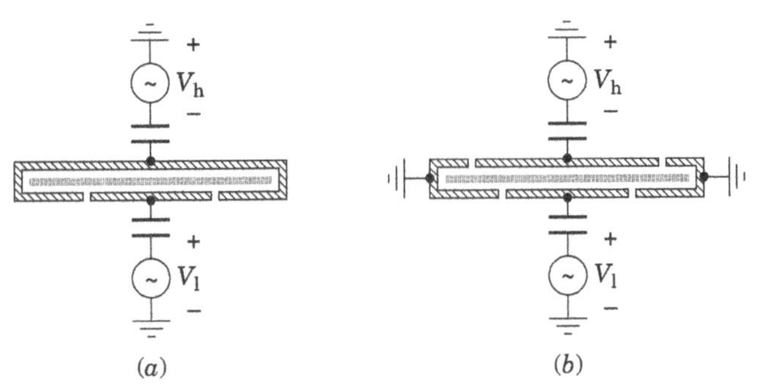

Figure 11.21 Diode (*a*) and triode (*b*) configurations of a dual-frequency capacitive discharge.

be separately considered. Here, we describe only the diode configuration. Some elements of triode discharge analysis are considered in Kawamura et al. (2007).

Dual-frequency discharges with one high- and one low-frequency source

$$V_{rf} = V_l \cos\omega_l t + V_h \cos(\omega_h t + \chi) \tag{11.6.1}$$

are in wide use for materials processing. (Here, V_l and V_h are taken to be positive, with $\omega_h \neq \omega_l$.) The dual-frequency sheath motions are complicated. We will give a simple description of homogeneous (constant density) sheaths.

Discharges with widely separated, usually incommensurate, frequencies are often used for critical anisotropic etching applications. The dual-frequency sources can give separate control of the ion flux and ion-bombarding energy to a substrate (Goto et al., 1992; Kitajima et al., 2000; Lieberman et al., 2003; Kim et al., 2003; Boyle et al., 2004a). From (11.2.43), higher frequency produces a reduced ion-bombarding energy and thinner sheaths for a given ion flux to the substrate. It also permits the addition of a second low-frequency driving voltage, which strongly affects the dc sheath voltage drop, and therefore the ion-bombarding energy. A typical dual-frequency waveform is shown in Figure 11.22*a*.

Discharges synchronously driven at both fundamental (first) and second harmonics, e.g., $\omega_h = 2\omega_l$, can independently control the ion-bombarding energy (Heil et al., 2008). The control parameter is the phase χ between the first and second harmonic. This is true even in geometrically

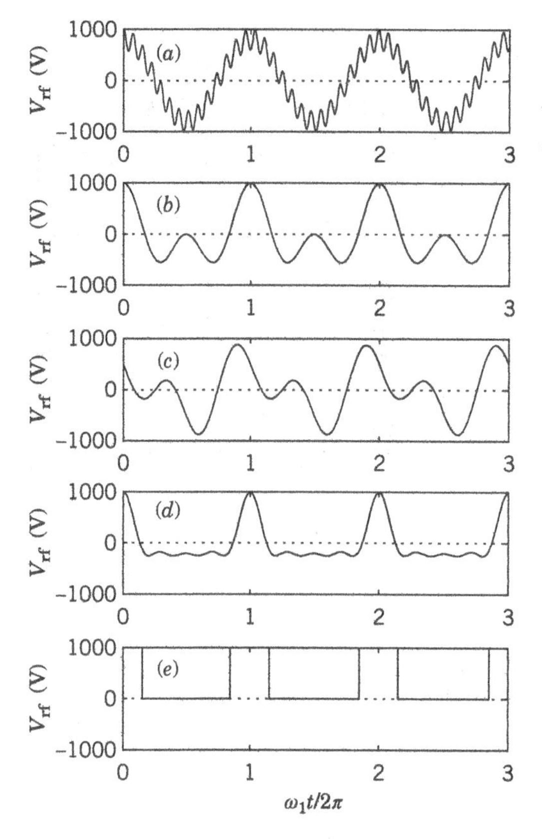

Figure 11.22 Typical multi-frequency-driven discharge waveforms over three low-frequency periods; (*a*) widely separated incommensurate frequencies, with $V_l = 800$ V, $V_h = 200$ V, $f_l = 2$ MHz, $f_h = 27.12$ MHz, $\chi = 0$; (*b*) fundamental and second harmonic, with $V_l = V_h = 500$ V, $\chi = 0$; (c) fundamental and second harmonic, with $V_l = V_h = 500$ V, $\chi = \pi/2$; (*d*) tailored voltage, with $V_{rf} = 400\cos\omega_l t + 300\cos 2\omega_l t + 200\cos 3\omega_l t + 100\cos 4\omega_l t$; (*e*) rectangular waveform with 30% duty ratio.

symmetric discharges with equal electrode areas. Typical voltage waveforms for this *electrical asymmetry effect* (EAE) are shown in Figure 11.22*b* and *c* for $\chi = 0$ and $\pi/2$, respectively. An important variation is to drive the discharge with multiple synchronized harmonics of the driving frequency. These *tailored voltage waveforms* allow more precise control of discharge and processing characteristics (Johnson et al., 2010; Lafleur and Booth, 2012). A typical waveform is shown in Figure 11.22*d*. Various types of multi-frequency-driven, low-pressure capacitive discharges are reviewed in Donkó et al. (2012).

11.6.1 Dual-Frequency Sheaths

As will be seen below, the voltage maximum and minimum of (11.6.1) play critical roles in understanding dual-frequency sheath physics. The extrema of (11.6.1) are found from $\mathrm{d}V_{rf}/\mathrm{d}t = 0$

$$\omega_l V_l \sin\omega_l t + \omega_h V_h \sin(\omega_h t + \chi) = 0 \tag{11.6.2}$$

which specifies the values of t at the extrema of V_{rf}, and which in general has to be solved numerically. Substituting these into (11.6.1) gives the rf voltages at the extrema. In general, there are one or more maxima and one or more minima; the largest maximum $V_{rf\,max}$ and the smallest minimum $V_{rf\,min}$ must be chosen.

Let us consider a low-pressure asymmetric discharge driven by a dual-frequency voltage through the usual large blocking capacitor, as in Figures 11.16 and 11.17. We use a homogeneous model with constant sheath density n, and powered and grounded electrode areas A_a and A_b, respectively.

As the quasineutral discharge bulk oscillates, the sum of the time-varying charges within each of the two sheaths is a constant

$$A_a s_a(t) + A_b s_b(t) = (A_a + A_b) s_m \tag{11.6.3}$$

Here, $s_a(t)$ and $s_b(t)$ are the time-varying powered and grounded sheath widths, measured from their respective electrode surfaces, and s_m is a maximum sheath width, determined below. Introducing an area ratio $\eta = A_a/A_b$, we obtain

$$\eta s_a(t) + s_b(t) = (1+\eta) s_m \tag{11.6.4}$$

From (11.1.12) and (11.1.15), the plasma-to-electrode sheath voltages are

$$V_a(t) = \frac{1}{2}\frac{en}{\epsilon_0} s_a^2(t) \tag{11.6.5}$$

and

$$V_b(t) = \frac{1}{2}\frac{en}{\epsilon_0} s_b^2(t) \tag{11.6.6}$$

In addition to $V_{rf}(t)$, a bias voltage V_B will generally appear across the electrodes. Then, ignoring any small voltage drop across the bulk plasma, the total powered-to-grounded voltage across the electrodes can be written in terms of the individual sheath voltages as $V_b(t) - V_a(t)$

$$V_B + V_{rf}(t) = \frac{1}{2}\frac{en}{\epsilon_0}\left(s_b^2(t) - s_a^2(t)\right) \tag{11.6.7}$$

To determine V_B, we note that at the voltage maximum $V_{rf\,max}$, s_a has collapsed to zero width, and from (11.6.4), $s_b = (1+\eta)s_m$. Substituting (11.6.6) into (11.6.7), we obtain

$$V_B + V_{rf\,max} = \frac{1}{2}\frac{en}{\epsilon_0} s_m^2 (1+\eta)^2 \tag{11.6.8}$$

Similarly, at the voltage minimum $V_{rf\,min}$, sheath b has collapsed to zero and

$$V_B + V_{rf\,min} = -\frac{1}{2}\frac{en}{\epsilon_0} s_m^2 \frac{(1+\eta)^2}{\eta^2} \tag{11.6.9}$$

Substituting the right-hand side of (11.6.8) into (11.6.9), we obtain

$$V_B = -\frac{V_{rf\,max} + \eta^2 V_{rf\,min}}{1+\eta^2} \tag{11.6.10}$$

which gives the bias voltage. Finally, substituting (11.6.10) into (11.6.8), we obtain

$$\frac{1}{2}\frac{en}{\epsilon_0} s_m^2 (1+\eta)^2 = \frac{\eta^2}{1+\eta^2}\left(V_{rf\,max} - V_{rf\,min}\right) \tag{11.6.11}$$

which gives s_m.

The individual sheath motions are now determined. For example, substituting $s_b(t)$ from (11.6.4) into (11.6.7), we obtain the equation for $s_a(t)$ in terms of the known voltages

$$V_B + V_{rf}(t) = \frac{1}{2}\frac{en}{\epsilon_0}(1+\eta)^2\left(s_m - s_a(t)\right)\left(s_m + \frac{1-\eta}{1+\eta} s_a(t)\right) \tag{11.6.12}$$

with s_m given from (11.6.11). Solving this quadratic equation gives $s_a(t)$. From (11.1.9), the rf discharge current entering electrode a is

$$I_{rf} = -enA_a \frac{ds_a}{dt} \tag{11.6.13}$$

Finally, time-averaging (11.6.5) gives the dc plasma-to-sheath voltage across sheath a

$$\overline{V}_{\rm a} = \frac{1}{2}\frac{en}{\epsilon_0}\langle s_{\rm a}^2(t)\rangle_{\rm rf} \tag{11.6.14}$$

and hence the ion acceleration voltage.

The preceding derivation is modified if the sheath edge densities are not equal. In this case, the bias voltage expression for $V_{\rm B}$ in (11.6.10) still holds, but with η^2 replaced by the *voltage asymmetry factor* ϵ

$$\epsilon \equiv \frac{\overline{V}_{\rm b}}{\overline{V}_{\rm a}} = \frac{n_{\rm a}A_{\rm a}^2}{n_{\rm b}A_{\rm b}^2} \tag{11.6.15}$$

Note that ϵ is the inverse of (11.4.12). We leave the calculation to Problem 11.15. Additional effects accounting for Child law sheaths, the dc floating potentials of the sheaths, the bulk plasma voltage drop, and possible differences between the maximum charges in the two sheaths are described in Heil et al. (2008) and in an extended model by Schulze et al. (2011c).

For a symmetric discharge ($\eta = 1$), (11.6.12) can be solved to obtain

$$s_{\rm a}(t) = -\frac{1}{en}C_{\rm d}'\left(V_{\rm rf}(t) - V_{\rm rf\,max}\right) \tag{11.6.16}$$

where $C_{\rm d}' = \epsilon_0/(2s_{\rm m})$ is a constant discharge capacitance per unit area. Using (11.6.13), we find $I_{\rm rf} = C_{\rm d}\,{\rm d}V_{\rm rf}/{\rm d}t$. We see that the symmetric voltage-driven discharge (11.6.1) is equivalent to a symmetric current-driven discharge

$$J_{\rm rf} = J_{\rm l}\sin\omega_{\rm l}t + J_{\rm h}\sin(\omega_{\rm h}t + \chi) \tag{11.6.17}$$

with

$$J_{\rm l} = -\frac{\omega_{\rm l}\epsilon_0}{2s_{\rm m}}V_{\rm l}, \qquad J_{\rm h} = -\frac{\omega_{\rm h}\epsilon_0}{2s_{\rm m}}V_{\rm h} \tag{11.6.18}$$

Finally, substituting (11.6.16) into (11.6.14) and time-averaging, we obtain the ion acceleration voltage across sheath a, in terms of the voltages alone

$$\overline{V}_{\rm a} = \frac{1}{4}\frac{V_{\rm l}^2 + V_{\rm h}^2 + 2V_{\rm rf\,max}^2}{V_{\rm rf\,max} - V_{\rm rf\,min}} \tag{11.6.19}$$

11.6.2 Dual-Frequency Discharges

A common choice for dual-frequency operation is widely separated, incommensurate frequencies, $\omega_{\rm h} \gg \omega_{\rm l}$. Examples are 27.1/2 MHz, 60/13.56 MHz, and 160/13.56 MHz. In this situation, the phase shift χ is unimportant, since the voltage waveforms for the two frequencies rapidly drift through each other, giving $V_{\rm rf\,max} = -V_{\rm rf\,min} \approx V_{\rm l} + V_{\rm h}$. We should note that although 2 MHz has been used commercially, the ion plasma frequency $f_{\rm pi}$ might significantly exceed 2 MHz at high plasma densities, resulting in significant ion rf motions within the discharge.

11.6.2.1 Homogeneous Model

An analytic model of a current-driven, homogeneous symmetric discharge was developed by Kim et al. (2003). As shown above, current- and voltage-driven symmetric discharges are equivalent. For equal electrode areas, $V_{\rm B} = 0$ from (11.6.10). The dc ion acceleration voltage across the sheath is found from (11.6.19) to be

$$\overline{V}_{\rm a} = \frac{3}{8}(V_{\rm l} + V_{\rm h}) - \frac{1}{4}\frac{V_{\rm l}V_{\rm h}}{V_{\rm l} + V_{\rm h}} \tag{11.6.20}$$

The cross-term, due to the sheath nonlinearity, gives at most a 17% correction to $\overline{V}_a$ for the case of equal high- and low-frequency voltages.

The particle and energy balance relations determine the electron temperature T_e and the density n, in the same manner as for a single-frequency, low-pressure homogeneous discharge, as described in Section 11.1. The particle balance (11.1.26) is the same for single or dual frequency and determines T_e

$$K_{iz}(T_e)n_g d = 2eu_B \tag{11.6.21}$$

with d the oscillating bulk width. The time-averaged electron power per unit area is given by (11.1.38) with $\nu_{eff} = \nu_m + 2\overline{v}_e/d$ from (11.1.39) the effective collision frequency due to the sum of the bulk ohmic heating and the stochastic heating at the two sheaths. For the dual-frequency case, time-averaging $J_{rf}^2(t)$ in (11.6.17) gives

$$S_e = S_{el} + S_{eh} = \frac{1}{2}\frac{m\nu_{eff}d}{e^2 n}(J_l^2 + J_h^2) \tag{11.6.22}$$

Equating S_e to the electron power per unit area lost from the plasma, given by (11.1.40), we obtain the plasma density in terms of the driving current magnitudes

$$n = \left[\frac{m\nu_{eff}d}{4e^3 u_B}(J_l^2 + J_h^2)\right]^{1/2} \tag{11.6.23}$$

In terms of the voltages, from (11.6.18), and substituting for ns_m^2 using (11.6.11), we obtain

$$S_e = \frac{1}{4}\frac{\epsilon_0 m\nu_{eff}d}{e} \cdot \frac{\omega_l^2 V_l^2 + \omega_h^2 V_h^2}{V_l + V_h} \tag{11.6.24}$$

Equating (11.6.24) to (11.1.40), we obtain n in terms of the two voltage magnitudes

$$n = \frac{8\epsilon_0 m\nu_{eff}d}{e^2 u_B(\mathcal{E}_c + \mathcal{E}_e')} \cdot \frac{\omega_l^2 V_l^2 + \omega_h^2 V_h^2}{V_l + V_h} \tag{11.6.25}$$

The ion power flux is $S_i = 2enu_B\overline{V}_a$.

Note that for the homogeneous model, the condition $J_h^2 \gg J_l^2$, or equivalently, $\omega_h^2 V_h^2 \gg \omega_l^2 V_l^2$ ensures that the high-frequency current J_h controls S_e, and, therefore, the plasma density n and the ion fluxes Γ_i to each electrode. Similarly, for $V_l \gg V_h$, the low-frequency voltage controls the ion acceleration voltage $\overline{V}_a$. For the homogeneous model, these two conditions are

$$\frac{J_h^2}{J_l^2} = \frac{\omega_h^2 V_h^2}{\omega_l^2 V_l^2} \gg \frac{V_l}{V_h} \gg 1 \tag{11.6.26}$$

with the natural control parameters J_h and V_l.

However, for two specified currents J_h and J_l, V_l depends on J_h. At a fixed J_l, an increase in J_h increases n, which reduces the sheath width s_m in (11.6.18), and therefore reduces V_l. Similarly, for two specified voltages V_h and V_l, we note from (11.6.25) that S_e and therefore n depend not only on V_h but also inversely on V_l. The reason is that increasing V_l increases s_m. This reduces J_h, and, hence, n, at a fixed V_l. In practice, the conditions (11.6.26) are often only marginally met, giving further couplings between the high and low frequencies, as the following example shows.

Example 11.3 We take the following dual-frequency parameters:

- $p = 100$ mTorr argon at 300 K ($n_g = 3.3 \times 10^{21}$ m^{-3})
- $l = 4.5$ cm
- $f_l = 2$ MHz and $f_h = 27.12$ MHz

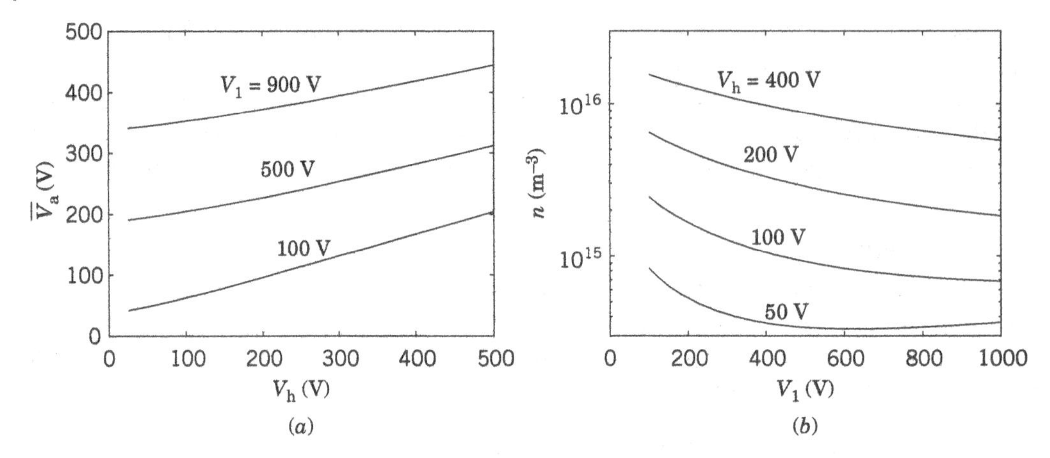

Figure 11.23 Homogeneous model solution for a 2 MHz/27.12 MHz, 100 mTorr argon, dual-frequency discharge with 4.5 cm gap; (*a*) ion acceleration voltage $\overline{V}_a$ versus high-frequency voltage amplitude V_h, for various low-frequency voltages V_l; and (*b*) density n versus low-frequency voltage V_l, for various high-frequency voltages V_h.

Assuming the sheath width is small compared to the gap width l, we solve (11.6.21) with bulk width $d \approx l$ as in Figure 10.1 to obtain $T_e = 2.5$ V. Then, $u_B = 2450$ m/s and $\overline{v}_e = 1.06 \times 10^6$ m/s. Evaluating the rate coefficients in Figure 3.16 gives $K_{iz} = 3.8 \times 10^{-17}$ m^3/s, $K_{ex} = 2.0 \times 10^{-16}$ m^3/s, and $K_m = 6.4 \times 10^{-14}$ m^3/s, with momentum transfer frequency $\nu_m = K_m n_g = 2.1 \times 10^8$ s^{-1}. The collisional energy loss per electron–ion pair created is $\mathcal{E}_c = 81$ V from Figure 3.17. As in (11.1.40), $\mathcal{E}_e' = 7.2\,T_e = 18$ V for argon. The effective collision frequency is $\nu_{eff} = \nu_m + 2\overline{v}_e/d = 2.3 \times 10^8$ s^{-1}. We see that at 100 mTorr, the discharge is mostly ohmically heated. Examining a range of high- and low-frequency voltages to determine the density n and the ion acceleration voltage $\overline{V}_a$, we obtain the results shown in Figure 11.23*a* and *b*. In Figure 11.23*a*, the ion acceleration voltage $\overline{V}_a$ is plotted versus V_h for three values of V_l. We see a strong variation of $\overline{V}_a$ with V_l, but also that $\overline{V}_a$ increases significantly with V_h, especially at the lower value of $V_l = 100$ V. In an ideal decoupled case, with $V_h \ll V_l$, the three plotted curves should be horizontal lines. In Figure 11.23*b*, the density n is plotted versus V_l for four different values of V_h. We see a strong variation of n with V_h, but also that the density decreases significantly with increasing V_l, due to the increase in the sheath width, leading to a reduced J_h and high-frequency heating. Similar results are seen for current-driven discharges, with specified values of J_l and J_h, which we leave as Problem 11.16 for the student.

11.6.2.2 Child Law Models

More consistent Child law sheaths, having spatially varying density profiles, lead to somewhat different scalings than homogeneous sheaths. The sheath motion for dual-frequency excitation is complicated (Robiche et al., 2003; Franklin, 2003; Boyle et al., 2004b; Kawamura et al., 2006). Useful approximate forms are given in Turner and Chabert (2014) for collisionless sheaths, and in Chabert and Turner (2017) for collisional sheaths. For independent control of dc sheath voltage and ion flux, it suffices to examine the conditions from the single-frequency scaling. For low-pressure discharges, with stochastic sheath heating much greater than ohmic bulk heating, we see from (11.2.43) that for $\omega_h^2 V_h \gg \omega_l^2 V_l$, the high-frequency source produces a much higher density than the low-frequency source. On the other hand, the ion-bombarding energy is controlled by the total voltage (high + low) across the sheath. Hence, with a wide separation of frequencies, it is possible to meet both conditions simultaneously

$$\frac{\omega_\mathrm{h}^2 V_\mathrm{h}}{\omega_\mathrm{l}^2 V_\mathrm{l}} \gg \frac{V_\mathrm{l}}{V_\mathrm{h}} \gg 1 \tag{11.6.27}$$

However, as for the homogeneous model, V_h does not independently control the ion flux. For a fixed V_h, an increase in V_l increases the sheath width and reduces the high-frequency electron power absorbed, due to the reduced J_h (Lieberman and Lichtenberg, 2010).

In addition, for a given J_h, the density and ion flux are also functions of J_l, even when the condition (11.6.27) is met (Kawamura et al., 2006; Turner and Chabert, 2006a, 2007). A larger J_l, while not directly heating the electrons, reduces the electron density within the sheaths, and, hence, increases the stochastic and ohmic sheath heating produced by J_h. This in turn increases n and the ion flux. This situation is illustrated in Figure 11.24. As shown there, the small-amplitude, high-frequency sheath motion is reasonably well described by a homogeneous sheath model, which oscillates back and forth within a large Child law sheath on the low-frequency timescale. Increasing J_l (and V_l) increases the low-frequency sheath width, leading to a reduced average electron density in the sheath. The sheath density scales as $n_\mathrm{sh} \sim n_\mathrm{a}(\mathrm{T_e}/V_\mathrm{l})^{1/2}$ from (11.2.61). Since $S_\mathrm{eh} \propto J_\mathrm{h}^2/n_\mathrm{sh}$, we expect $S_\mathrm{eh} \propto (V_\mathrm{l}/\mathrm{T_e})^{1/2}$ for $V_\mathrm{l} \gg V_\mathrm{h}$. Also, changing the ratio of stochastic to ohmic power by varying V_l can lead to various heating mode transitions (Turner and Chabert, 2006b).

For the high-frequency stochastic heating power per unit area, Kawamura et al. (2006) find, analogous to the single-frequency result (11.2.31)

$$\overline{S}_\mathrm{stoc} = 0.5m\overline{v}_\mathrm{e} n_\mathrm{s} u_\mathrm{sh}^2 F(H_\mathrm{l}) \tag{11.6.28}$$

where n_s is the bulk plasma density at the low-frequency plasma–sheath edge, $u_\mathrm{sh} = J_\mathrm{h}/(en_\mathrm{s})$ is the high-frequency oscillation velocity at the sheath edge, and

$$F(H_\mathrm{l}) \approx \left(1 + \frac{\pi}{4} H_\mathrm{l}\right)\left(\frac{H_\mathrm{l}}{H_\mathrm{l} + 2.2}\right) \tag{11.6.29}$$

Similar to (11.2.11), the low-frequency H_l factor is

$$H_\mathrm{l} = \frac{1}{\pi} \frac{u_\mathrm{sl}^2}{\omega_\mathrm{l}^2 \lambda_\mathrm{Ds}^2} \tag{11.6.30}$$

with $u_\mathrm{sl} = J_\mathrm{l}/(en_\mathrm{s})$ and λ_Ds the low-frequency oscillation velocity and the Debye length at the sheath edge. Since $H_\mathrm{l} \propto (V_\mathrm{l}/\mathrm{T_e})^{1/2}$ from (11.2.13), we see that the first factor on the right-hand side in (11.6.29) gives the expected scaling for a fixed J_h that $\overline{S}_\mathrm{stoc} \propto (V_\mathrm{l}/\mathrm{T_e})^{1/2}$. Similar to the single-frequency result (11.2.68), the second factor accounts for a finite but small low-frequency bulk oscillation velocity, compared to the average oscillation velocity in the sheath. The ohmic

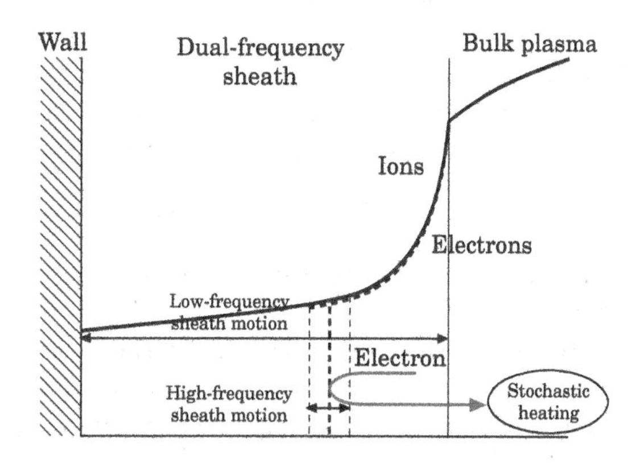

Figure 11.24 Illustrating the high- and low-frequency sheath motions in a dual-frequency-driven Child law sheath.

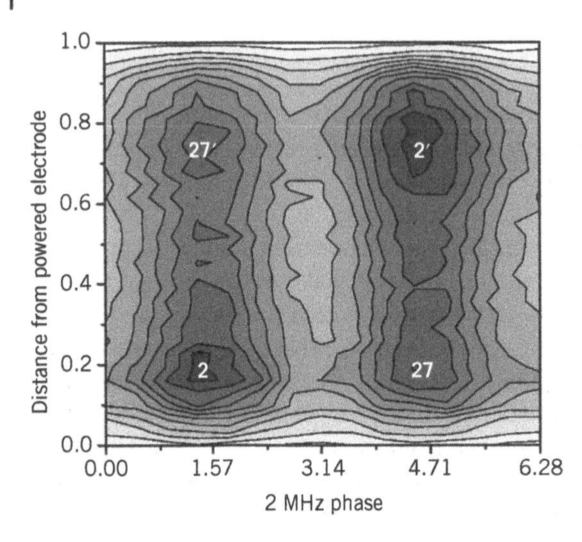

Figure 11.25 Contour plot of space- and time-resolved optical emission for a dual-frequency discharge, illustrating the electron dynamics within the low-frequency rf cycle; only the 2 MHz motion is resolved; 490 mTorr He/O_2, 2 MHz at 800 W, 27 MHz at 200 W. Source: Reproduced from Gans et al. (2006b), with the permission of AIP Publishing. https://doi.org/10.1063/1.2425044.

heating in the sheath is given by a simple generalization of (11.2.63), in which $J_1^2 \sin^2 \omega t$ is replaced by $J_1^2 \sin^2 \omega_1 t + J_h^2 \sin^2 \omega_h t$. Some analytic expressions are given in Turner and Chabert (2014) for collisionless sheaths and in Chabert and Turner (2017) for collisional sheaths.

Simulations (Kawamura et al., 2006; Schulze et al., 2009a) and measurements (Gans et al., 2006a; Schulze et al., 2007) using phase-resolved optical emission spectroscopy (PROES) reasonably confirm the preceding picture of the dual-frequency sheath dynamics. Figure 11.25 shows a contour plot of the space- and time-resolved 706.5 nm 3^3S helium optical emission. The 490 mTorr He/O_2 discharge was driven with 800 W at 2 MHz and 200 W at 27 MHz. The large peaks, labeled 2 and 2′, correspond to the time-maxima of the 2 MHz oscillating sheath voltages, where the low-frequency motion enhances the 27 MHz heating by widening the sheath width and transporting the 27 MHz electron oscillations to regions of lower plasma density, and, hence, higher oscillating sheath velocity. Increasing the 2 MHz power level increases the size of these peaks. Smaller peaks, labeled 27 and 27′, also appear near the time-minima of the 2 MHz sheath voltages. These 27 MHz heating peaks are relatively unaffected by the 2 MHz power level.

11.6.3 Electrical Asymmetry Effect (EAE)

As shown by Heil et al. (2008), for multi-frequency harmonic excitation of a geometrically symmetric discharge, the no-bias voltage condition is that the plasma-to-electrode voltages satisfy $V_a(\omega_1 t) = -V_b(\omega_1 t + \pi)$. This implies that only odd harmonics of the fundamental frequency ω_1 can appear in the applied waveform. Then, $V_{\text{rf max}} = -V_{\text{rf min}}$ and (11.6.10) gives $V_B = 0$ for $\eta^2 = 1$. Therefore, even harmonics of ω_1 must be present to obtain a nonzero bias voltage. An important case is for the first and second harmonics, $\omega_h = 2\omega_1$, with equal voltage amplitudes $V_h = V_1$

$$V_{\text{rf}} = V_1 \left(\cos \omega_1 t + \cos(2\omega_1 t + \chi) \right) \tag{11.6.31}$$

Then varying the phase χ in (11.6.31) causes V_B and the ion acceleration voltage $\overline{V}_a$ to vary, an *EAE*.

Two examples of the voltage waveforms used are given in Figure 11.22*b* and *c*. For $\chi = 0$, we clearly see that $V_{\text{rf max}} > -V_{\text{rf min}}$, producing $V_B < 0$. For $\chi = \pi/2$, we see that $V_{\text{rf max}} = -V_{\text{rf min}}$ and, consequently, $V_B = 0$. Figure 11.26 shows $V_{\text{rf max}}$ versus χ, which varies from a maximum of

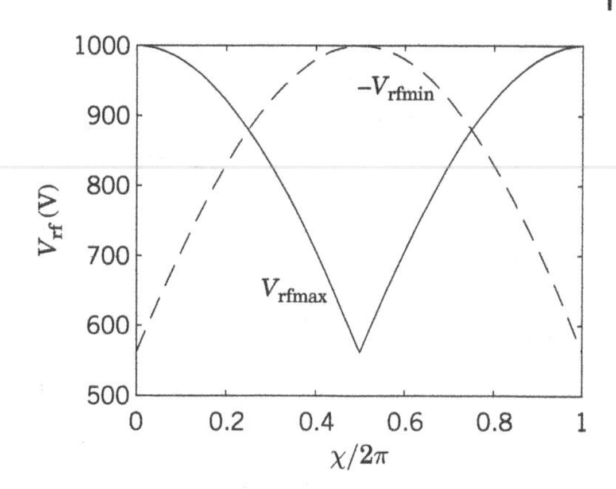

Figure 11.26 Electrical asymmetry effect for $V_h = V_l = 500$ V and $\omega_h = 2\omega_l$, showing $V_{rf\,max}$ and $-V_{rf\,min}$ versus the second harmonic phase χ; note that $V_{rf\,min}(\chi) = -V_{rf\,max}(\chi + \pi)$.

1000 V at $\chi = 0$ to a minimum of 566 V at $\chi = \pi$. From the applied voltage waveform symmetry, note that $V_{rf\,min}(\chi) = -V_{rf\,max}(\chi + \pi)$.

The resulting variations of $V_B(\chi)$ and $\overline{V}_a(\chi)$ are shown in Figure 11.27*a* and *b*, respectively. To a good approximation, the shapes are triangular (Heil et al., 2008), with $V_{B\,min} = \frac{7}{16}V_l = -219$ V and $\overline{V}_{a\,max} = \frac{4}{5}V_l = 400$ V at $\chi = 0$, and $V_{B\,max} = 219$ V and $\overline{V}_{a\,min} = V_{a\,max} - V_{B\,max} = 181$ V at $\chi = \pi$. We see that indeed the ion acceleration voltage $\overline{V}_a$ can be varied by over a factor of two by varying the second harmonic phase χ. This procedure can even be used to make an asymmetric discharge electrically symmetric (Schulze et al., 2011b), which we leave as Problem 11.17 for the student.

Simulations (Donkó et al., 2009) and experiments (Schulze et al., 2009a,b) verify the essential correctness of the preceding analyses. Figure 11.28 shows measurements of the mean energy $\langle \epsilon_i \rangle$ bombarding the grounded electrode, and the relative ion flux, as functions of the 27.12 MHz phase angle χ. At 20 Pa, the gap size was 1 cm, giving a reasonably symmetric discharge. At 4 and 10 Pa, the gap had to be increased to 2.5 cm, which gave a significant geometric asymmetry due to the enlarged grounded sidewall area. At the lowest pressure of 4 Pa, $\langle \epsilon_i \rangle$ could be varied by almost a factor of two, with little change in the ion flux. At the higher pressures, there was considerable ion–neutral scattering and charge transfer in the sheath, leading to the reduced mean ion energies

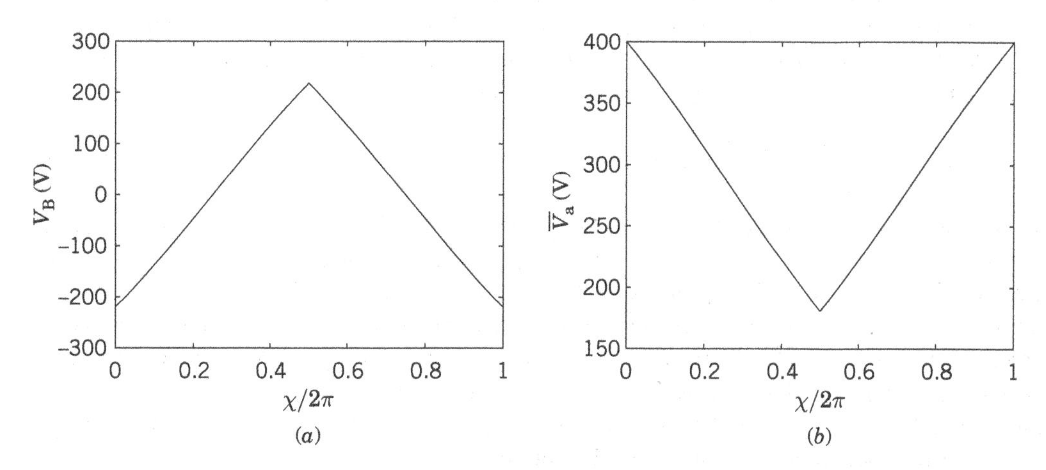

Figure 11.27 Electrical asymmetry effect for $V_h = V_l = 500$ V and $\omega_h = 2\omega_l$, showing (*a*) bias voltage V_B versus the second harmonic phase χ; (*b*) ion acceleration voltage $\overline{V}_a$ versus χ.

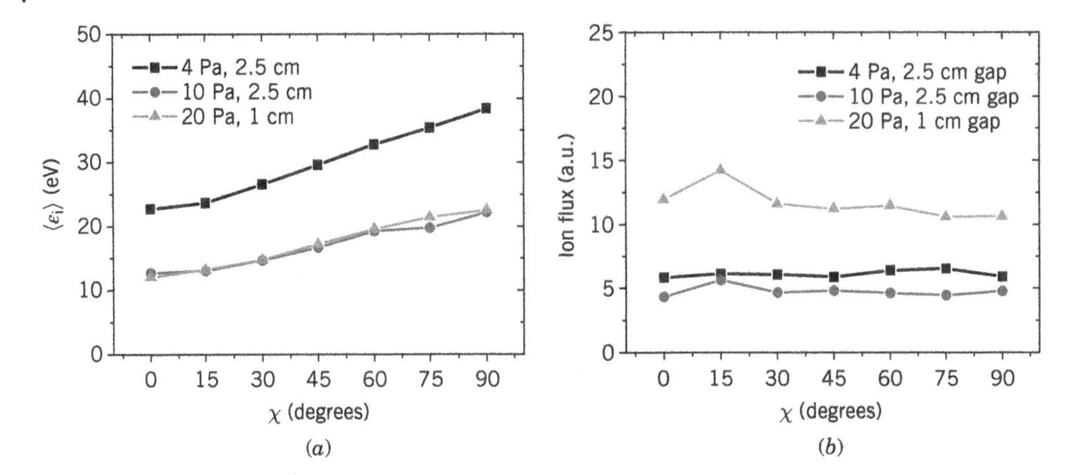

Figure 11.28 (*a*) Measured mean energy $\langle \epsilon_i \rangle$ with the gap size indicated, and (*b*) relative ion flux at the grounded electrode versus phase angle χ in degrees; 10 cm diameter argon discharge driven at 13.56 and 27.12 MHz with equal amplitude voltages. Source: Schulze et al. (2009b). ©IOP Publishing. Reproduced with permission. All rights reserved. https://doi.org/10.1088/0022-3727/42/9/092005.

seen in Figure 11.28*a*. Collisionally reduced ion energies are briefly described in Section 11.9, see in particular Figure 11.46.

11.6.4 Tailored Voltage Waveforms

Additional control over discharge properties can be obtained using three or more frequencies. For example, with widely separated dual frequencies, $\omega_h \gg \omega_l$, adding a second harmonic of ω_l can provide increased control of the bias voltage and ion-bombarding energy via the EAE (Czarnetzki et al., 2011; Schulze et al., 2011c). Tailoring the applied voltage waveform can also modify the electron energy deposition mechanisms, increasing the sheath heating (Lafleur et al., 2012a) and modifying the EEDF. Control of the EEDF can be beneficial for fine-tuning materials etching and deposition processes through its effects on neutral and radical precursor generation and the ion energy distribution bombarding the substrate (Johnson et al., 2010; Lafleur et al., 2012b).

One voltage waveform that has been extensively investigated (Czarnetzki et al., 2011; Lafleur et al., 2012a; Lafleur and Booth, 2012) has a Gaussian-like profile, being composed of a finite number N of harmonics having gradually decreasing amplitudes

$$V_{\mathrm{rf}}(t) = V_0 \sum_{\mathrm{m}=1}^{N} \frac{N-m+1}{N} \cos(m\omega_l t + \chi_m) \tag{11.6.32}$$

where the χ_ms are N tunable phases. An example is shown in Figure 11.22*d* for four harmonics with all the χ_ms equal to zero. This gives a sharply-peaked waveform, with a larger maximum bias voltage $V_{\mathrm{B\,max}}$ then for dual frequencies. By tuning the χ_ms, the bias voltage can again be varied from $-V_{\mathrm{B\,max}}$ to $V_{\mathrm{B\,max}}$ in a symmetric discharge. Measurements and simulations for low-pressure argon (Lafleur et al., 2012a,2012b; Lafleur and Booth, 2012) and oxygen (Derzsi et al., 2016) discharges give results similar to those shown in Figure 11.28, with an increased range of control of the ion-bombarding voltage, and with the simulations giving insight into spatiotemporal discharge characteristics such as the electron power absorption.

One should note that the higher harmonics of a high driving frequency $f_h \gtrsim 13.56$ MHz can give rise to electromagnetic standing wave and skin effects, which can affect the radial discharge uniformity. These phenomena, which are not described by the usual electrostatic discharge models, are considered in Section 11.7. However, tailored voltage waveforms can be very useful for low-frequency bias voltages. For example, consider an f_l = 500 kHz *rectangular* bias voltage waveform, as shown in Figure 11.22*e*, with a plasma-to-substrate voltage $V_{B\,max}$ = 5 kV during the "on" time of the waveform cycle, and a very low plasma-to-sheath voltage during the "off" time. Such a waveform, which contains many higher harmonics of f_l (see Problem 11.18), produces a nearly monoenergetic ion-bombarding energy distribution $\mathcal{E}_i \approx 5$ kV on the substrate. This can be highly useful for etching high aspect ratio trenches and holes, where the presence of a significant component of moderate energy ions leads to undesired sidewall etching. We consider the ion-bombarding energy distribution in Section 11.9, and high aspect ratio etching, at the end of Section 16.2.

11.7 Standing Wave and Skin Effects

At high frequencies and/or large areas, electromagnetic effects such as standing waves and skin effects arise, which cannot be described using conventional electrostatic analysis (Lieberman et al., 2002; Chabert et al., 2004a). To see this, consider the excitation of a capacitive discharge shown in Figure 11.29. The driving voltage source V_h is connected to the center of the upper powered electrode through a hole in the grounded metal chamber. Because the fields cannot pass through the conducting electrode, the discharge must be excited at its outer radius R. The source launches an outward traveling wave along the radial transmission line comprising the upper grounded chamber and the powered electrode. This wave "turns 180°" and propagates radially inward into the discharge between the powered and grounded electrodes. The wave reflects at $r = 0$ and then travels back to the source. The result is a standing wave in the discharge region, with radial wavenumber k and corresponding wavelength $\lambda = 2\pi/k$.

Let us consider the symmetric, low-pressure, narrow gap cylindrical discharge shown in Figure 11.30, having radius R and gap length $l = 2\,l'$ (Lieberman et al., 2002). A uniform bulk plasma of density n and thickness $d = 2\,d'$ is separated from the electrodes by two vacuum sheaths, each of width s. The plasma is a low loss dielectric, $\nu_m \ll \omega$, having a constant relative permittivity $\kappa_p \approx 1 - \omega_{pe}^2/\omega^2$ given by (4.2.18), with $\omega_{pe} = (e^2 n/\epsilon_0 m)^{1/2}$ the plasma frequency.

We assume that the excitation voltage at the outer radius is axisymmetric around the circumference, with all system properties independent of the ϕ-coordinate. For a transverse magnetic

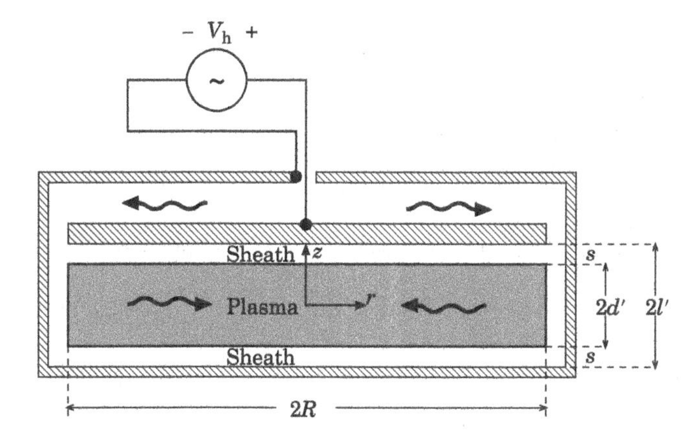

Figure 11.29 Schematic of high-frequency excitation of a capacitive discharge.

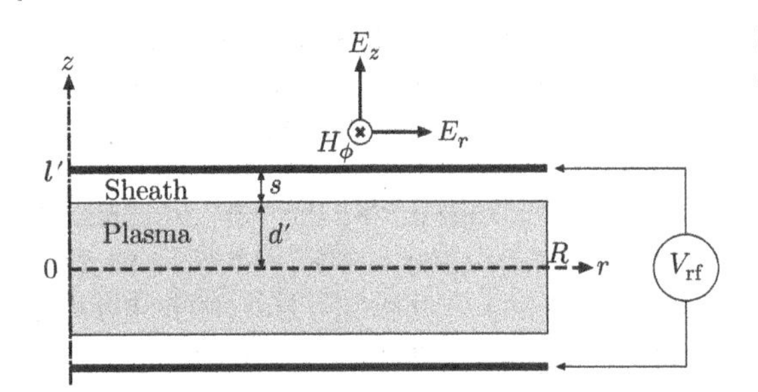

Figure 11.30 Simple model for surface waves in a symmetrically driven capacitive discharge.

(TM) mode having only the magnetic field component $H_\phi \propto e^{j\omega t}$, the Maxwell equations in the bulk plasma are

$$\frac{\partial H_\phi}{\partial z} = -j\omega\epsilon_0\kappa_p E_r \tag{11.7.1}$$

$$\frac{1}{r}\frac{\partial(rH_\phi)}{\partial r} = j\omega\epsilon_0\kappa_p E_z \tag{11.7.2}$$

$$\frac{\partial E_r}{\partial z} - \frac{\partial E_z}{\partial r} = -j\omega\mu_0 H_\phi \tag{11.7.3}$$

Here, E_z is a "capacitive" electric field (perpendicular to the discharge plates) and E_r is an "inductive" field (parallel to the plates). Substituting for E_r and E_z from (11.7.1) and (11.7.2) into (11.7.3) yields the propagation equation for H_ϕ

$$\frac{\partial}{\partial r}\left(\frac{1}{r}\frac{\partial(rH_\phi)}{\partial r}\right) + \frac{\partial^2 H_\phi}{\partial z^2} + k_0^2\kappa_p H_\phi = 0 \tag{11.7.4}$$

where $k_0 = \omega/c$ is the free space wavenumber, with c the speed of light. The same equations hold in the two vacuum sheath regions, with $\kappa_p \to 1$.

In general, the two conducting surfaces and the central bulk plasma form a three-electrode radial transmission line system. Therefore, for radial wavelength λ large compared to the gap width l, (11.7.4) has two radially propagating modes. All other modes are radially evanescent, decaying exponentially into the plasma from their excitation radius R. The field symmetries for the propagating modes are shown in Figure 11.31. The two modes have symmetric and antisymmetric $E_z(z)$ variations about the midplane. Only the symmetric mode is excited for the symmetric discharge of Figure 11.30, while both modes can be excited in an asymmetric capacitive discharge, having unequal powered and grounded electrode areas (Lieberman et al., 2016).

Here, we consider the symmetric mode. The boundary conditions are

1. $E_r = 0$ and $H_\phi = 0$ at $r = 0$ (by symmetry)
2. $E_r = 0$ at the electrode surfaces $z = \pm l'$ (for perfectly conducting plates)
3. H_ϕ is continuous at the plasma–sheath interfaces $z = \pm d'$
4. E_r is continuous at $z = \pm d'$

Using boundary conditions 1–3, the solution of (11.7.4) in the plasma has field components

$$E_r = -A\frac{\alpha_p \cosh\alpha_0 s}{j\omega\epsilon_0\kappa_p}\sinh\alpha_p z\, J_1(kr) \tag{11.7.5}$$

$$H_\phi = A\cosh\alpha_0 s\cosh\alpha_p z\, J_1(kr) \tag{11.7.6}$$

$$E_z = A\frac{k\cosh\alpha_0 s}{j\omega\epsilon_0\kappa_p}\cosh\alpha_p z\, J_0(kr) \tag{11.7.7}$$

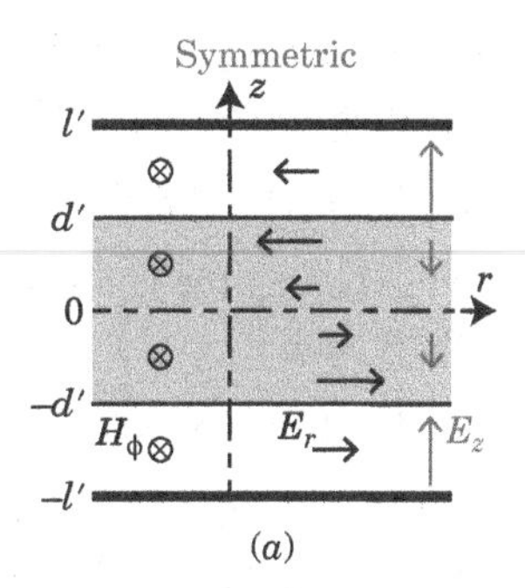

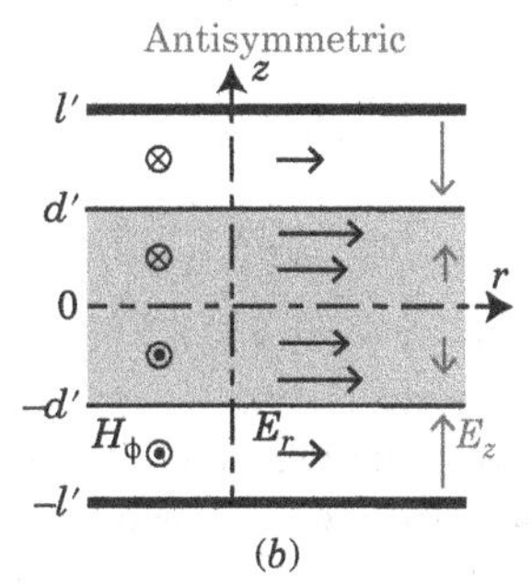

Figure 11.31 Symmetry properties of the symmetric (*a*) and antisymmetric (*b*) surface wave modes.

Similarly, the fields in the sheath region $d' < z < l'$ are

$$E_r = A\frac{\alpha_0 \cosh \alpha_p d'}{j\omega\epsilon_0} \sinh \alpha_0(l' - z)\,J_1(kr) \tag{11.7.8}$$

$$H_\phi = A \cosh \alpha_p d' \cosh \alpha_0(l' - z)\,J_1(kr) \tag{11.7.9}$$

$$E_z = A\frac{k \cosh \alpha_p d'}{j\omega\epsilon_0} \cosh \alpha_0(l' - z)\,J_0(kr) \tag{11.7.10}$$

The fields in the lower sheath region $-l' < z < -d'$ are given from the symmetry properties in Figure 11.31. In (11.7.5)–(11.7.10), J_0 and J_1 are the zero- and first-order Bessel functions. Equations (11.7.5)–(11.7.10) represent a surface wave that propagates into the discharge center from the outer radial periphery and has a radial wavenumber k, and axial decay constants $\alpha_{p,0}$ in the plasma and sheath, respectively, related by

$$k^2 - \alpha_p^2 = k_0^2 \kappa_p \tag{11.7.11}$$

$$k^2 - \alpha_0^2 = k_0^2 \tag{11.7.12}$$

The quantity α_p^{-1} is the characteristic field penetration distance (in the z-direction) of the surface wave into the plasma. See Section 13.3 for another application of surface waves: the generation of a surface-wave sustained plasma column.

Applying boundary condition 4 yields the dispersion relation

$$\alpha_0 \kappa_p \tanh \alpha_0 s + \alpha_p \tanh \alpha_p d' = 0 \tag{11.7.13}$$

Equations (11.7.11)–(11.7.13) determine α_p, α_0, and k for the symmetric surface wave mode.

Let us consider the inequalities that the axial field decay length in the sheath is large compared to the sheath width, $\alpha_0^{-1} \gg s$, and the frequency is low compared to the plasma frequency, $\omega \ll \omega_{pe}$, which are usually satisfied in conventional capacitive discharges. At low densities, the condition $\alpha_p^{-1} \gg d'$ is also satisfied. Then, (11.7.11)–(11.7.13) can be solved to obtain

$$k \approx k_0(1 + d'/s)^{1/2}, \qquad \alpha_0 \approx k_0(d'/s)^{1/2}, \qquad \alpha_p \approx \omega_{pe}/c \equiv \delta_p^{-1} \tag{11.7.14}$$

with $\delta_p = c/\omega_{pe}$ the collisionless skin depth given in (12.1.3). At higher densities, $\alpha_p^{-1} \ll d'$, we similarly find

$$k \approx k_0(1 + \delta_p/s)^{1/2}, \qquad \alpha_0 \approx k_0(\delta_p/s)^{1/2}, \qquad \alpha_p \approx \delta_p^{-1} \tag{11.7.15}$$

In both cases, radial wavelength is reduced below the free space wavelength $\lambda_0 = c/f$, and the radial phase velocity $v_{ph} = \omega/k$ is reduced below the speed of light

$$\lambda \approx \frac{\lambda_0}{(1 + \Delta/s)^{1/2}} \tag{11.7.16}$$

$$v_{ph} \approx \frac{c}{(1 + \Delta/s)^{1/2}} \tag{11.7.17}$$

with $\Delta = \min(d', \delta_p)$ giving the magnetic field penetration distance in the two cases (Problem 11.19).

We can understand the slowing of the wave velocity physically from a simple transmission line model as follows: considering by symmetry a half-thickness l' of the system, a strong wave electric field E_z exists only in the sheath region. Hence, the capacitance per unit length (along r) can be estimated as $C' \approx \epsilon_0 2\pi r/s$. On the other hand, the wave magnetic field H_ϕ penetrates through the sheath thickness s into the plasma a distance Δ. Hence, the inductance per unit length can be estimated as $L' \approx \mu_0(s + \Delta)/2\pi r$. Substituting L' and C' into the usual expression (Ramo et al., 1994, chapter 5) for the phase velocity of a wave on a transmission line, $v_{ph} = 1/\sqrt{L'C'}$, we obtain (11.7.17).

As seen from (11.7.16) at low densities $\Delta \equiv d'$, the local wavelength depends on the radially varying sheath width, $\lambda(r) \approx \lambda_0[s(r)/l']^{1/2}$. The sheath width is radially varying because the transmission line voltage $V(r)$ is radially varying. For a narrow gap discharge, $l \ll R$, $s(r)$ is determined from an appropriate Child law by $V(r)$ and the local (in r) particle balance and power balances, which set n_e and T_e. Therefore, the transmission line capacitance C' depends on V, yielding a nonlinear transmission line equation. This has been solved for various sheath and electron power deposition models by Chabert et al. (2004a, b). For a constant mean free path Child law sheath (11.2.54) and a dominant stochastic heating process (11.2.59), the nonlinear wavelength is found to be

$$\lambda_{NL} = 1.28\,\lambda_0 \left(\frac{e u_B^2 \mathcal{E}_T \lambda_i V_0}{m T_e^2 l^5 \omega^4} \right)^{1/10} \tag{11.7.18}$$

where u_B is the Bohm speed, $\mathcal{E}_T$ is the energy loss per electron–ion pair created, $\lambda_i \lesssim s$ is the ion–neutral mean free path, and V_0 is the rf voltage amplitude at $r = 0$. In (11.7.18), λ_{NL} is defined to be four times the radius r_0 of the first transmission line voltage node $V(r_0) = 0$. This result is insensitive to V_0, the gas pressure, T_e, and $\mathcal{E}_T$; it mainly depends on the gap width l and radian frequency ω. A practical formula for argon (MKS units) is $\lambda_{NL} \approx 40\,\lambda_0 V_0^{1/10} l^{-1/2} f^{-2/5}$, with f the rf frequency. This expression gives a good agreement with the experimental results at 81.36 MHz shown in Figure 11.34.

A similar electromagnetic analysis gives the wave dispersion for the antisymmetric mode, which is also excited in an asymmetric discharge geometry, having unequal electrode areas (Sansonnens et al., 2006; Lieberman et al., 2016)

$$\alpha_0 \kappa_p \tanh \alpha_0 s + \alpha_p \coth \alpha_p d' = 0 \tag{11.7.19}$$

For a low-density plasma with $\alpha_p^{-1} \gg d'$, the approximate solutions are

$$k \approx \left(k_0^2 - \frac{1}{\kappa_p s d'} \right)^{1/2}, \qquad \alpha_0 \approx (k^2 - k_0^2)^{1/2}, \qquad \alpha_p \approx (k^2 - k_0^2 \kappa_p)^{1/2} \tag{11.7.20}$$

The surface currents flowing in the electrode plates are $\mathbf{K} = \hat{\mathbf{n}} \times \mathbf{H}$ (A/m), with $\hat{\mathbf{n}}$ the inward-pointing surface normal unit vector and $\mathbf{H}$ the magnetic field vector. From the symmetry of H_ϕ shown in Figure 11.31, let us note that for the symmetric mode, the upper and lower electrode plates carry equal radial currents that flow in opposite directions. For the antisymmetric mode, both electrode currents flow in the same direction, and they both return by flowing radially through the bulk plasma in the opposite direction. Therefore, the antisymmetric mode exists only in the presence of the plasma and is analogous to the propagating mode in a coaxial cable, with the plasma as the inner conductor and the two electrodes as the outer conductor.

At high pressures, the plasma permittivity κ_p has both real and imaginary parts, due to its dependence on the electron collision frequency ν_m. Then, all wavenumbers have both real and imaginary parts; i.e., due to collisions the incident and reflected radial waves are both attenuating as they propagate. This leads to somewhat complicated expressions for the fields. For the antisymmetric wave mode, resistive losses due to the plasma current flowing in the bulk can cause strong wave attenuation. At low densities, the wave propagates along r as $e^{-jkr-k_i r}$, with a decay constant $k_i \approx \frac{1}{2} k \nu_m / \omega$. We leave the calculation of k_i for the student as Problem 11.20.

The radial distributions of electron power absorbed per unit area due to the capacitive field E_z and the inductive field E_r are proportional to the square of the field components: $P_{\text{cap}} \propto \int_0^d dz\, E_z^2$ and $P_{\text{ind}} \propto \int_0^d dz\, E_z^2$, respectively. Figure 11.32 shows calculated normalized radial power distributions for the symmetric mode, including evanescent radial edge effects, for a low-density ($n = 10^{15}$ m^{-3}) large-area, symmetric (equal plate area) discharge driven at (*a*) 13.56 MHz and (*b*) 40.12 MHz (Lieberman et al., 2002). In both cases, the inductive power deposition is negligible. At the lower frequency, the power distribution is flat, except near the edge where the evanescent radial modes

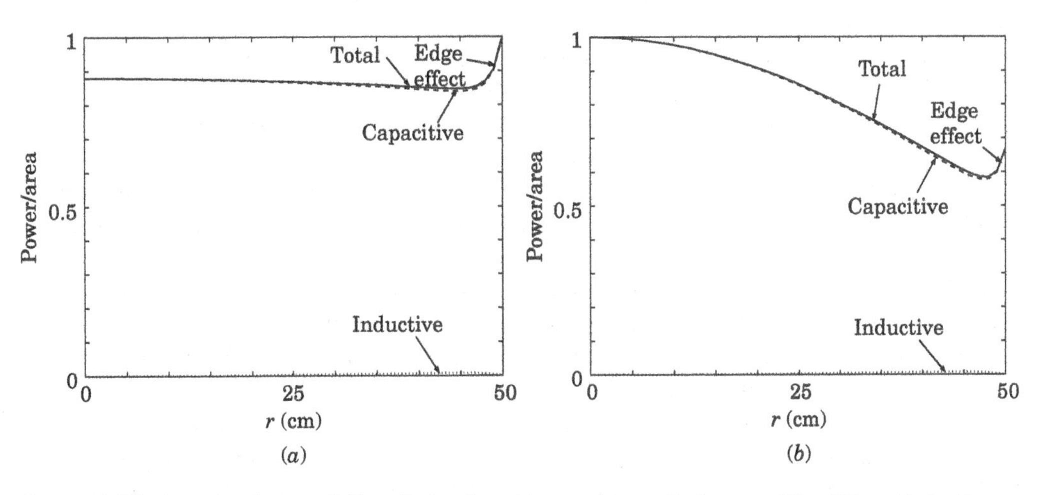

Figure 11.32 Low-density ($n = 10^{15}$ m^{-3}) standing wave model results for capacitive (E_z) and inductive (E_r) normalized power deposition versus radius r; (*a*) $f = 13.56$ MHz; (*b*) $f = 40.12$ MHz; the edge effects result from evanescent modes excited at the outer radius $R = 50$ cm; $s = 0.4$ cm and bulk width $d = 2d' = 4$ cm. Source: Adapted from Lieberman et al. (2002) ©IOP Publishing.

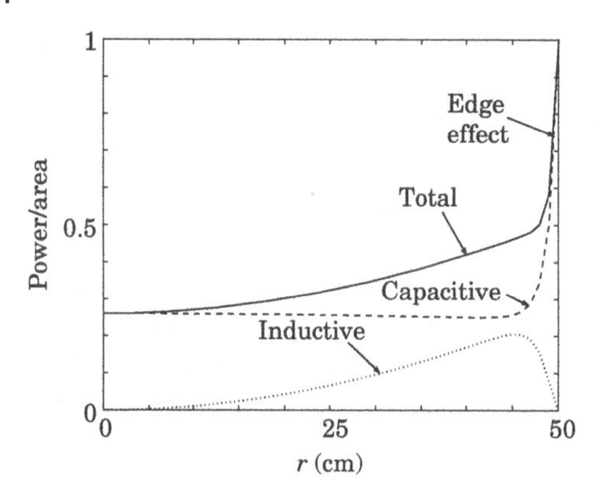

Figure 11.33 High-density ($n = 10^{16}$ m^{-3}) standing wave model results for capacitive (E_z) and inductive (E_r) normalized power deposition versus radius r; $f = 13.56$ MHz; the edge effects result from evanescent modes excited at the outer radius $R = 50$ cm; $s = 0.4$ cm and bulk width $d = 2\,d' = 4$ cm. Source: Adapted from Lieberman et al. (2002) ©IOP Publishing.

are important. At the higher frequency, a pronounced center-high standing wave is seen, and the power deposition is nonuniform.

Skin effects are found at high densities where the plasma interior shields itself from the applied fields. In this case, the inductive fields become important. The combined power deposition can have a center-low radial profile. An example is shown in Figure 11.33. Skin effects are small if $\delta_p \gg \sqrt{d'R}$. For high frequencies (and concomitant high densities), and/or large areas, one or both of the conditions for negligible standing wave or skin effects may be difficult to meet.

11.7.1 Experiments and Simulations

Some experimental results for a 40 cm × 40 cm, 4.5 cm gap, 150 mTorr argon discharge driven at 50 W rf power are shown in the left panel of Figure 11.34. The square electrodes are surrounded by a 4 cm thick teflon barrier, so the discharge is symmetrically driven. The ion saturation currents to a set of 8 × 8 planar probes in the top electrode were measured. At 150 mTorr, the plasma diffusion length is small, so the ion current profile is expected to depend locally on the power deposition profile. As seen in the figure, there is very little standing wave effect at 13.56 MHz, but a center-high standing wave is seen at 60 MHz and is even more pronounced at 81.36 MHz. At the latter frequency, the discharge approaches the first (quarter-wave) spatial resonance of the zero-order Bessel function, $kR \to 2.405$, where $E_z(\text{edge})/E_z(\text{center}) \to 0$.

The right panel of Figure 11.34 shows the measured variation with rf power at 60 MHz. At 50 W power, the measured maximum ion density was $n_i = 4.5 \times 10^{16}$ m^{-3}. At 265 W, $n_i = 2.1 \times 10^{17}$ m^{-3}, and strong plasma production was measured near the sides and corners of the discharge, with an overall center-low profile. This is due to a combination of skin and electrostatic edge effects, as shown in Figure 11.33.

Two-dimensional electromagnetic fluid simulations of the wave effects were performed by Lee et al. (2008) and Rauf et al. (2008). For the simulations of Lee et al. (2008), the fluid equations in the bulk plasma were solved, along with an analytic Child law sheath model, with the electromagnetics determined by a frequency domain analysis (fundamental frequency only) in the entire (plasma + sheaths) region. The discharge is symmetric, with $p = 150$ mTorr argon, $f = 80$ MHz, $R = 20$ cm, and $l = 2\,l' = 4.8$ cm. The transition at low powers (low densities) to a significant standing wave effect with increasing frequency was found to agree quite well with that seen experimentally in the left panel of Figure 11.34.

Results for the spatial profiles of electron density n_e with increasing power are shown in Figure 11.35, for rf powers of (*a*) 40 W, (*b*) 110 W, and (*c*) 190 W. Similar to that seen for the ion

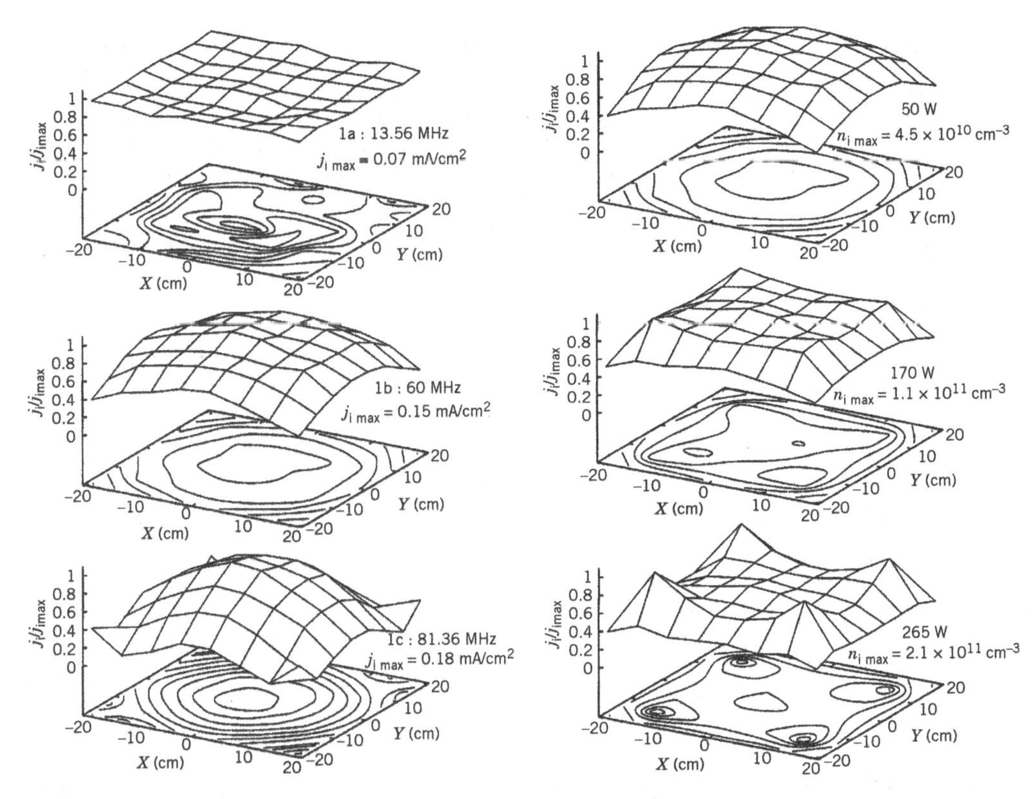

Figure 11.34 Experimental results in a 40 cm × 40 cm, 4.5 cm gap, 150 mTorr, symmetrically driven argon discharge, showing the two-dimensional normalized ion fluxes to the upper electrode; (left panel) 13.56, 60, and 81.36 MHz at 50 W rf power; (right panel) 50, 170, and 265 W rf power at 60 MHz. Source: Reproduced from Perret et al. (2003), with the permission of AIP Publishing. https://doi.org/10.1063/1.1592617.

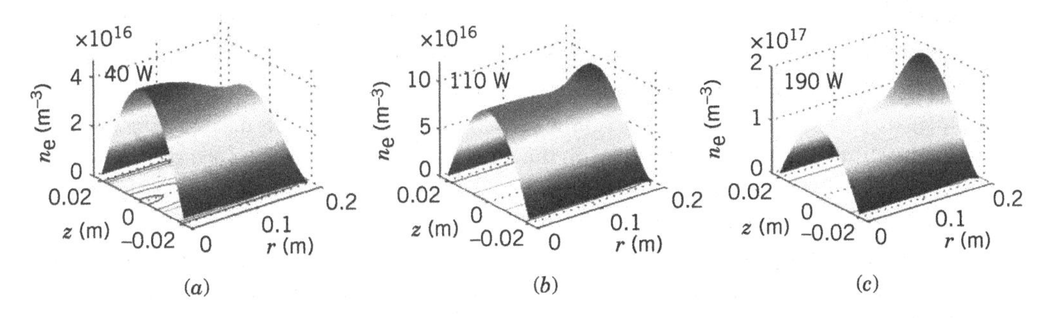

Figure 11.35 Two-dimensional fluid simulation results for the spatial distribution of the electron density in a symmetric cylindrical discharge ($R = 20$ cm and $l = 2l' = 4.8$ cm) at 80 MHz and 150 mTorr argon, for rf powers of (*a*) 40 W, (*b*) 110 W, and (*c*) 190 W. Source: Adapted from Lee et al. (2008) ©IOP Publishing.

current in the right panel of Figure 11.34, the transition from a standing wave at the lowest power to strong skin/edge effects at the highest power is seen in the simulated density profiles. Similar effects are seen for an asymmetrically driven case where both symmetric and antisymmetric modes are excited.

Rauf et al. (2008) also saw the transition from standing wave to skin effects as the power was increased. These results are shown in Figure 11.36 for a 100 mTorr argon asymmetric discharge

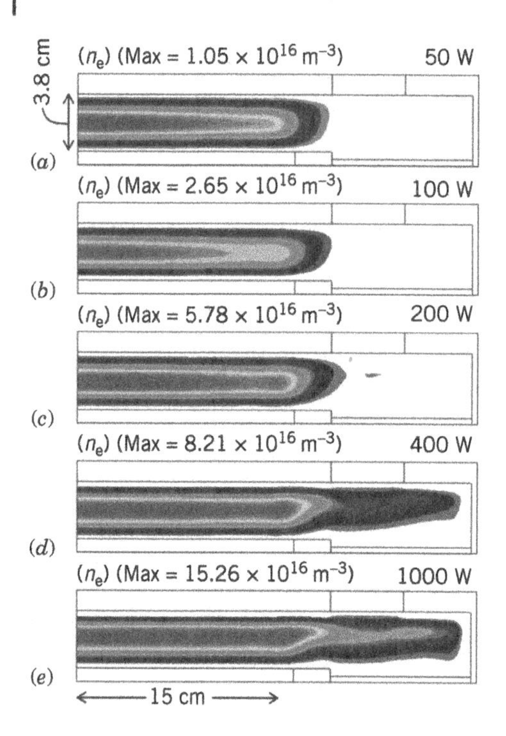

Figure 11.36 Electron density contours for various rf powers in an asymmetric, 100 mTorr argon capacitive discharge driven at 180 MHz; 3.8 cm gap, 15 cm lower electrode radius, and 17.8 cm upper electrode radius. Source: Rauf et al. (2008). ©IOP Publishing. Reproduced with permission. All rights reserved. https://doi.org/10.1088/0963-0252/17/3/035003.

driven at 180 MHz, with a 3.8 cm gap, 15 cm lower electrode radius, and 17.8 cm upper electrode radius. Above 800 W, there is significant inductive (E_r) power deposition near the radial edge due to the skin effect, which increases the density there and leads to a center-low profile at the higher powers. The simulations of Rauf et al. (2008) in Ar/SF_6 and Kawamura et al. (2016) in chlorine also show the effects of electronegative gas chemistry, often important in processing applications.

As shown in Problem 11.19, the lossless symmetric surface wave is radially evanescent for $\omega_{pe}/\sqrt{2} < \omega < \omega_{pe}$; i.e., there is damping rather than propagation from the discharge radial edge into the center. For collisional plasmas, there is also a frequency region below $\omega_{pe}/\sqrt{2}$ where the propagating wave is highly damped and cannot penetrate into the discharge center. The low-frequency limit of this region depends on the ratio ν_m/ω. This highly damped region has been observed in simulations with $\nu_m/\omega \sim 0.23$ to appear at $\omega/\omega_{pe} \approx 0.1$ (Lee et al., 2008); it may be of practical importance at high operating frequencies.

Electromagnetic simulations in asymmetrically driven discharges show the importance of both symmetric and antisymmetric wave modes and resonances. Another important issue is nonlinearities, which arise when the sheath physics is more correctly modeled, e.g., by a Child law sheath width. Then, $s(r)$ depends nonlinearly on the radially varying voltage. This can also lead to strong generation of higher harmonics of the fundamental driving frequency, e.g., via nonlinear series resonance excitation (see Section 11.5). This has been examined using various transmission line models coupled to various sheath laws and to particle and energy balance to obtain more self-consistent radial variations of discharge properties. Numerical simulations of the nonlinear discharge dynamics in the time-domain allows for understanding the harmonic generation and have also been performed in various approximations.

Various methods have been examined to reduce the processing nonuniformities due to the standing wave effects. These include reduced gap widths, dielectric lenses and shaped electrodes, segmented and graded conductivity electrodes, and dual-frequency operation. Reviews of these phenomena, with relevant references, are given in Chabert (2007) and Liu et al. (2015).

11.8 Low-Frequency Sheaths

At low frequencies $\omega \lesssim \omega_i$, where $\tau_i = 2\pi/\omega_i$ is the ion transit time across the sheath, the ions respond to the time-varying fields within the sheath, rather than to their average value. In this case, the sheath analysis of Sections 11.1–11.4 is invalid. The ion transit time for a collisionless Child law sheath is given by (6.3.21)

$$\tau_i = \frac{3s}{v_0} \tag{11.8.1}$$

where $v_0 = (2eV_0/M)^{1/2}$ is the characteristic ion velocity in the sheath and V_0 is the dc component of the voltage across the sheath. Substituting (6.3.12) for s into (11.8.1) with $J_0 = en_s u_B$ from (6.3.13), we obtain

$$\omega_i = \pi\omega_{pi}\left(\frac{2\mathrm{T_e}}{V_0}\right)^{1/4} \tag{11.8.2}$$

where $\omega_{pi} = (e^2 n_s/\epsilon_0 M)^{1/2}$ is the ion plasma frequency at the sheath edge. For typical operating conditions $V_0 \sim 100$ V, $\mathrm{T_e} \sim 3$ V, we find that $\omega_i \sim \omega_{pi}$. However, strictly speaking, it is ω/ω_i that determines the ion behavior in the sheath, and not ω/ω_{pi}.

At both high and low frequencies, the current density everywhere within the sheath is the sum of the ion and electron conduction currents J_i and J_e and the displacement current J_d. To examine the importance of J_d, we use an estimate based on a parallel-plate vacuum model,

$$J_d \approx \frac{\omega\epsilon_0 V_1}{s} \tag{11.8.3}$$

where V_1 is the rf voltage amplitude. Using the Child law (6.3.12) for J_i, we form the ratio

$$\frac{J_d}{J_i} = \frac{3\pi}{2}\frac{V_1}{V_0}\frac{\omega}{\omega_i} \tag{11.8.4}$$

For a high-voltage sheath with $V_1 \sim V_0$, we see that in the low frequency or thin sheath (high-density) limit, $J_d \ll J_i$. In this limit, the displacement current is small and the sheath is resistive. This is in contrast to high-frequency rf sheaths that are capacitive. In contrast to capacitive sheaths, the time-varying voltages across resistive sheaths have a rectifying character and can be strongly non-sinusoidal. In addition, for asymmetric discharges, resistive sheaths give a quite different scaling with electrode size than capacitive sheaths give. We examine these issues below.

To determine the sheath characteristics and also the effects of two resistive sheaths at the powered and grounded electrodes of an asymmetric capacitive discharge, we consider a low-pressure voltage-driven system with sheath areas A_a and A_b and corresponding edge densities n_{sa} and n_{sb}, as shown in Figure 11.37 (Song et al., 1990; Kawamura et al., 1999). The bulk plasma resistance and inductance are assumed to be negligible. The current flowing from the bulk plasma to electrode a is the sum of a steady ion current (characteristic of the bulk plasma) and a time-varying electron current:

$$I_a(t) = en_{sa}u_B A_a - \frac{1}{4}en_{sa}\bar{v}_e A_a\, \mathrm{e}^{-V_a(t)/\mathrm{T_e}} \tag{11.8.5}$$

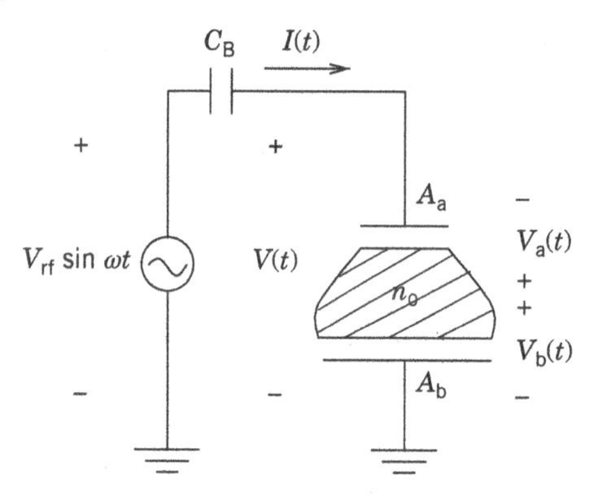

Figure 11.37 Asymmetric low-frequency capacitive discharge.

with V_a the plasma voltage with respect to electrode a. Similarly,

$$I_b(t) = en_{sb}u_B A_b - \frac{1}{4}en_{sb}\overline{v}_e A_b\, e^{-V_b(t)/T_e} \tag{11.8.6}$$

with V_b the plasma voltage with respect to electrode b. We have ignored the displacement current at low frequencies. By current continuity, we have $I_b(t) = -I_a(t)$. A circuit model of the discharge is shown in Figure 11.38.

Because we are assuming that $V_{rf} \gg T_e$, we can model each sheath to be an ideal diode, corresponding to the time-varying electron current, in parallel with an ideal current source, corresponding to the steady ion current. The currents carried by the sheath capacitances, shown as dashed lines in the figure, are assumed to be small. Typical variations of $V_a(t)$ and $V_b(t)$ are shown in Figure 11.39. Because one or the other sheath alternately limits the current to that of the ions alone, the total current has the square wave shape shown in Figure 11.39, with $I_{b0} = en_{sb}u_B A_b$ and $I_{a0} = en_{sa}u_B A_a$. Because neither V_a nor V_b can ever be significantly negative, V_a and V_b are alternately positive and clamped near zero volts; i.e., the sheaths are rectifying. The electron currents are nonzero only when the sheath voltages are clamped near zero.

The circuit in Figure 11.38 gives a quite different voltage divider action than that of the high-frequency capacitive voltage divider in Figure 11.16. Although the diodes (representing electron conduction current) still appear at high frequencies, as seen in the general circuit model for a high-frequency discharge of Figure 11.4, the capacitive displacement current is the dominant current at high frequencies, so the effect of the diodes was neglected in Figure 11.16. At low frequencies, in contrast, we neglect the displacement currents through the sheath capacitors.

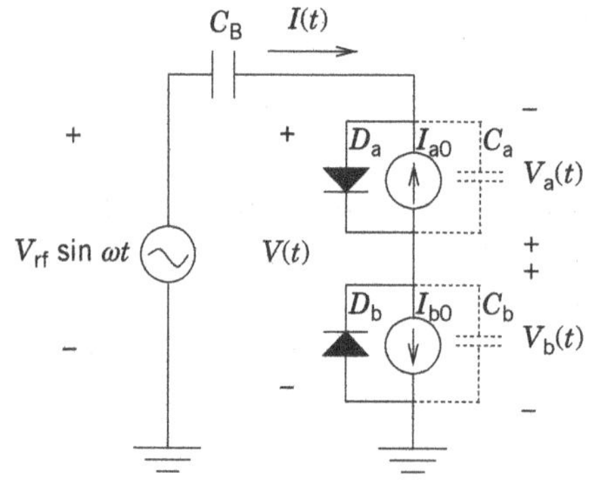

Figure 11.38 Model of low-frequency asymmetric capacitive discharge.

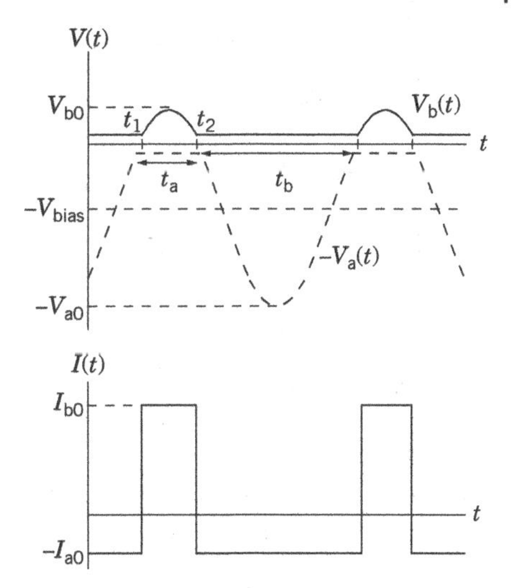

Figure 11.39 Time-varying sheath voltages and currents.

Due to the blocking capacitor (C_B in Figure 11.38), a dc self-bias voltage V_B builds up, such that the voltage $V(t) = V_b(t) - V_a(t)$ across the discharge can be written as

$$V(t) = V_{rf} \sin \omega t - V_B \tag{11.8.7}$$

To determine V_B, we note first that the total electron charge Q_{ea} collected by electrode a over one rf period must equal the total ion charge collected:

$$Q_{ea} = en_{sa}u_B A_a(t_a + t_b) \tag{11.8.8}$$

Here, t_a and t_b, as shown in Figure 11.39, are the time intervals for electron collection by electrodes a and b, respectively, with

$$\omega(t_a + t_b) = 2\pi \tag{11.8.9}$$

Because the plasma must remain quasineutral during t_a, the total electron charge lost to electrode a must equal the total ion charge lost to both electrodes:

$$Q_{ea} = eu_B(n_{sa}A_a + n_{sb}A_b)t_a \tag{11.8.10}$$

Substituting Q_{ea} from (11.8.8) into (11.8.10), we obtain

$$\frac{t_a}{t_b} = \frac{n_{sa}A_a}{n_{sb}A_b} \tag{11.8.11}$$

Using (11.8.9) to eliminate t_b from (11.8.11), we obtain

$$\omega t_a = 2\pi \frac{n_{sa}A_a}{n_{sa}A_a + n_{sb}A_b} \tag{11.8.12}$$

Referring to Figure 11.39 with $V(t)$ given as a shifted sinusoid by (11.8.7), we see that $\omega t_1 = \sin^{-1}(V_B/V_{rf})$, $\omega t_2 = \pi - \omega t_1$, and, hence, that

$$\omega t_a = \omega(t_2 - t_1) = \pi - 2\sin^{-1}(V_B/V_{rf}) \tag{11.8.13}$$

Equating (11.8.12) to (11.8.13) and solving for V_B, we obtain

$$V_B = V_{rf} \sin\left(\frac{\pi}{2}\frac{n_{sb}A_b - n_{sa}A_a}{n_{sb}A_b + n_{sa}A_a}\right) \tag{11.8.14}$$

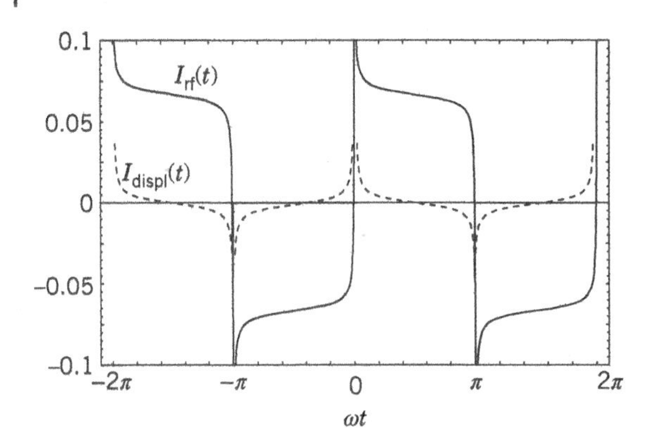

Figure 11.40 Symmetric low-frequency capacitive discharge showing total current I_{rf} and displacement current I_{displ} versus time. Source: Kawamura et al., 1999/with permission of IOP Publishing.

The maximum accelerating potentials at the electrodes are (see Figure 11.39) $V_{a0} = V_{rf} + V_B$ and $V_{b0} = V_{rf} - V_B$. A reasonable fit to (11.8.14) for $n_{sb}A_b \lesssim 5\,n_{sa}A_a$ is

$$\frac{V_{a0}}{V_{b0}} \approx \left(\frac{n_{sb}A_b}{n_{sa}A_a}\right)^{\pi/2} \tag{11.8.15}$$

Setting $n_{sa} = n_{sb}$, we contrast the low-frequency scaling exponent of $\pi/2$ to the high-frequency exponent of 4 in (11.4.7) for the same conditions of collisionless Child law sheaths with equal sheath edge densities.

For a highly asymmetric discharge, $n_{sa}A_a \ll n_{sb}A_b$ in (11.8.14), we find that $V_B \to V_{rf}$, $V_{a0} \to 2\,V_{rf}$, and $V_{b0} \to 0$. As described in Section 11.5, the former limit for V_B at the powered sheath has a small correction, to account for both the effects of a finite dc floating potential and the sinusoidal voltage amplitude V_{rf} itself. The latter limit for V_{b0} at the grounded sheath should also be corrected, as V_{b0} should tend to the dc floating potential (10.2.4).

The effect of displacement currents on these solutions for various discharge asymmetries and with a finite blocking capacitor C_B has been examined by Metze et al. (1986) by numerically integrating the equations for the circuit model of Figure 11.38, including the effects of the nonlinear sheath capacitances, shown as dashed lines in the figure. Displacement current effects were also examined analytically by Kawamura et al. (1999).

Summing the displacement and conduction currents yields the total current, as shown in Figure 11.40. More accurate models of the transition from low-frequency/high-density to high-frequency/low-density sheath dynamics must incorporate the ion inertial effects (finite du_i/dt). This has been done by Miller and Riley (1997) using an ion relaxation time model and by Sobolewski (2000) using the full ion dynamics described by the fluid equations. The latter model spans the entire range of frequencies and densities and is in excellent agreement with experimental measurements.

11.9 Ion-Bombarding Energy at Electrodes

The energy distribution $g_i(\mathcal{E})$ of ions bombarding substrate surfaces is critical to the plasma processing of materials. At low pressures for which the ion transport across the sheath is collisionless, g_i depends on the time-varying flux Γ_i of ions entering the sheath (assumed constant in previous sections) and on the time-varying sheath potential V_s. At higher pressures, ion–neutral collisions within the sheath strongly affect the energy distribution. Considering first the collisionless case, then cold ions that enter the sheath at time t_0 will strike the surface at time t_f with energy $\mathcal{E}$.

The fraction of ions per unit area that enter the sheath during a fractional time interval $\omega\,\mathrm{d}t_0/2\pi$ must equal the fraction of ions per unit area within an energy interval $\mathrm{d}\mathcal{E}$:

$$\frac{\omega}{2\pi}\Gamma_{\mathrm{i}}(\omega t_0)\,\mathrm{d}t_0 = \mathrm{g}_{\mathrm{i}}(\mathcal{E})\,\mathrm{d}\mathcal{E} \tag{11.9.1}$$

Summing overall times t_{0j} during one rf cycle that give energy $\mathcal{E}$, we obtain

$$\mathrm{g}_{\mathrm{i}}(\mathcal{E}) = \frac{1}{2\pi}\sum_j \Gamma_{\mathrm{i}}(\omega t_{0j})\left|\frac{\mathrm{d}\mathcal{E}}{\mathrm{d}(\omega t_{0j})}\right|^{-1} \tag{11.9.2}$$

For low frequencies $\omega \lesssim \omega_{\mathrm{i}}(\mathrm{T_e}/V_0)^{1/2}$, the sheath velocity is small compared to the Bohm velocity, and $\Gamma_{\mathrm{i}}(\omega t_0) \approx n_{\mathrm{s}}u_{\mathrm{B}}$ = const. The ion flight time across the sheath is also small compared to the rf period, and, hence, $\mathcal{E}(\omega t_{\mathrm{f}}) \approx \mathcal{E}(\omega t_0) \approx V_{\mathrm{s}}(\omega t_0)$. This yields the low-frequency energy distribution

$$\mathrm{g}_{\mathrm{i}}(\mathcal{E}) = \frac{1}{2\pi}\sum_j n_{\mathrm{s}}u_{\mathrm{B}}\left|\frac{\mathrm{d}V_{\mathrm{s}}}{\mathrm{d}(\omega t_{0j})}\right|^{-1}_{V_{\mathrm{s}}=\mathcal{E}} \tag{11.9.3}$$

The physical explanation for the bimodal distribution of $\mathrm{g}_{\mathrm{i}}(\mathcal{E})$ is illustrated in Figure 11.41 for a sinusoidal sheath voltage $V_{\mathrm{s}}(\phi)$, with $\phi = \omega t$. For an energy interval $\Delta\mathcal{E}$, there is a large phase interval $\Delta\phi$ near the maximum and minimum of $V_{\mathrm{s}}(\phi)$, and a small phase interval near the maximum and minimum of $\mathrm{d}V_{\mathrm{s}}/\mathrm{d}\phi$. As shown in the figure, $\Delta\phi = \Delta\mathcal{E}|\mathrm{d}V_{\mathrm{s}}/\mathrm{d}\phi|^{-1}$. The larger and smaller phase intervals give corresponding larger and smaller contributions to $\mathrm{g}_{\mathrm{i}}(\mathcal{E})$, as seen in (11.9.3).

Let us determine g_{i} at the powered electrode a in Figure 11.37. There are two parts to the distribution depending on the voltage variation during the two time intervals t_{a} and t_{b} (see Figure 11.39). During the time interval t_{b}, using $V_{\mathrm{s}} = -V(t)$ in (11.8.7), we have (see Figure 11.39) that

$$V_{\mathrm{s}} = -V_{\mathrm{rf}}\sin\omega t_0 + V_{\mathrm{bias}} \tag{11.9.4}$$

Differentiating V_{s} yields

$$\frac{\mathrm{d}V_{\mathrm{s}}}{\mathrm{d}(\omega t_0)} = -V_{\mathrm{rf}}\cos\omega t_0 = -V_{\mathrm{rf}}(1-\sin^2\omega t_0)^{1/2} \tag{11.9.5}$$

Substituting $\sin\omega t_0$ from (11.9.4) with $V_{\mathrm{s}} \equiv \mathcal{E}$ into (11.9.5), inserting this into (11.9.3), and noting that there are two values of t_0 during one rf cycle for each value of $\mathcal{E}$, we obtain

$$\mathrm{g}_{\mathrm{i}}(\mathcal{E}) = \frac{n_{\mathrm{sa}}u_{\mathrm{B}}}{\pi}\left[V_{\mathrm{rf}}^2 - (V_{\mathrm{bias}} - \mathcal{E})^2\right]^{-1/2}, \qquad 0 < \mathcal{E} < V_{\mathrm{rf}} + V_{\mathrm{bias}} \tag{11.9.6}$$

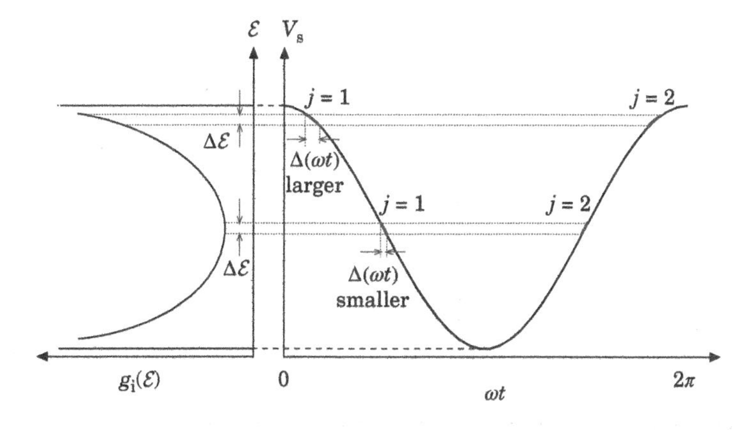

Figure 11.41 Illustrating the formation of the ion energy distribution $\mathrm{g}_{\mathrm{i}}(\mathcal{E})$ on an electrode.

The distribution at the maximum value $\mathcal{E} = V_{rf} + V_{bias}$ is singular but integrable. The preceding analysis does not account for the finite dc floating potential $\overline{V}_s$ across the sheath. Taking this into account, the lowest value of $\mathcal{E}$ is $\overline{V}_s$ rather than zero.

During the time interval t_a (see Figure 11.39), the voltage collapses to $\overline{V}_s$ across sheath a and (11.9.3) yields a mono-energetic (δ-function) contribution to the distribution

$$g_i(\mathcal{E}) = n_{sa} u_B \frac{\omega t_a}{2\pi} \delta(\mathcal{E} - \overline{V}_s) \tag{11.9.7}$$

The total distribution is the sum of (11.9.6) and (11.9.7). This is sketched in Figure 11.42 for a low-frequency and/or high-density sheath in a symmetrically driven capacitive discharge ($V_{bias} = 0$ and $V_{a0} = V_{rf}$). The spectrum is broad and independent of ion mass because ions of any mass respond to the full range of the slowly varying sheath voltage. The ions have maximum energy $V_{rf} + V_{bias}$, and there is a considerable population of low-energy ions.

Let us now consider the opposite limit of high frequencies, $\omega \gtrsim \omega_i$, for which the ion transit time τ_i is long compared to the rf period $\tau_{rf} = 2\pi/\omega$. This regime was described analytically for a Child law sheath by Benoit-Cattin and Bernard (1968) for a sinusoidal sheath voltage (11.9.4) and a constant (ion) sheath width s. To zero order in the ratio τ_{rf}/τ_i, the ion motion within the sheath is determined by the dc fields alone, independent of the rf modulation. For an ion entering the sheath at time t_0 with an initial velocity $u(t_0) = 0$, the unperturbed motion is given by (6.3.20) as

$$\frac{x(t)}{s} = \left[\frac{v_0(t - t_0)}{3s}\right]^3 \tag{11.9.8}$$

where $v_0 = (2eV_{bias}/M)^{1/2}$ with V_{bias} the dc voltage across sheath a.

To the first order, the ion motion is found from the force equation

$$M\frac{d^2x}{dt^2} = eE = e(-V_{rf} \sin \omega t + V_{bias})\frac{4}{3s}\left(\frac{x}{s}\right)^{1/3} \tag{11.9.9}$$

where we have used (6.3.16) for the electric field E. We have assumed a sinusoidal voltage V_{rf} across sheath a, with V_{bias} and V_{rf} related by (11.5.10). Substituting the zero-order solution (11.9.8) into the right-hand side of (11.9.9) and integrating once, we obtain

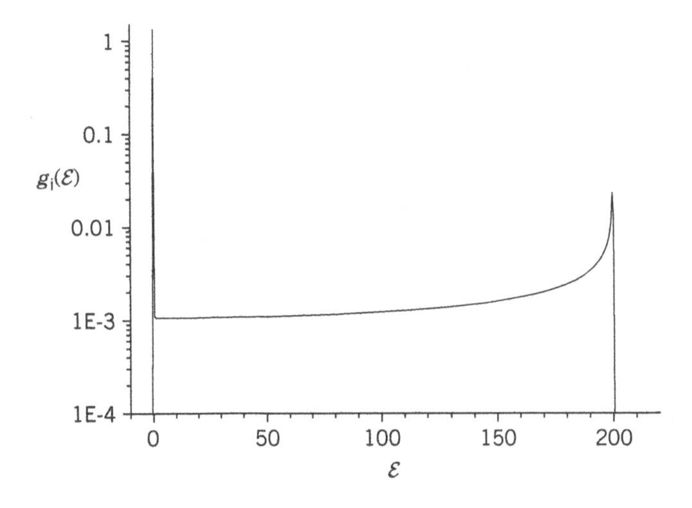

Figure 11.42 Ion energy distribution $g_i(\mathcal{E})$ for a symmetrically driven capacitive discharge with a low-frequency and/or high-density sheath; $V_{rf} = 200$ V.

$$M\frac{\mathrm{d}x}{\mathrm{d}t} = \frac{4}{9}\frac{v_0}{s^2}\left\{\frac{eV_{\text{bias}}}{2}(t-t_0)^2 - eV_{\text{rf}}\left[\frac{\sin\omega t - \sin\omega t_0}{\omega^2} - (t-t_0)\frac{\cos\omega t}{\omega}\right]\right. \tag{11.9.10}$$

Evaluating (11.9.10) at $t = t_{\text{f}}$ yields the ion-bombarding velocity $u_{\text{f}} = (\mathrm{d}x/\mathrm{d}t)_{t_{\text{f}}}$. To the first order, we can use the unperturbed value of $t_{\text{f}} - t_0 = \tau_{\text{i}} = 3s/v_0$ in (11.9.10) and retain only the two largest terms to obtain

$$Mu(t_{\text{f}}) \approx \frac{4}{3s}\left[\frac{eV_{\text{bias}}}{2}\frac{3s}{v_0} + \frac{eV_{\text{rf}}}{\omega}\cos\omega\left(t_0 + \frac{3s}{v_0}\right)\right] \tag{11.9.11}$$

Assuming that the second term in (11.9.11) is small compared to the first term, we form the energy, again retaining the two largest terms:

$$\mathcal{E}(\omega t_{\text{f}}) = \frac{M}{2e}u_{\text{f}}^2 \approx \frac{1}{2eM}\frac{16}{9s^2}\left[\frac{e^2V_{\text{bias}}^2}{4}\frac{9s^2}{v_0^2} + e^2V_{\text{bias}}V_{\text{rf}}\frac{3s}{\omega v_0}\cos\omega\left(t_0 + \frac{3s}{v_0}\right)\right]$$

Using $v_0 = (2eV_{\text{bias}}/M)^{1/2}$ in the above, we obtain the final result

$$\mathcal{E}(\omega t_{\text{f}}) \approx V_{\text{bias}} + \frac{4}{3}\frac{v_0}{\omega s}V_{\text{rf}}\cos\omega\left(t_0 + \frac{3s}{v_0}\right) \tag{11.9.12}$$

The width $\Delta\mathcal{E}$ of the modulation in energy is

$$\Delta\mathcal{E} = \frac{8}{3}\frac{v_0}{\omega s}V_{\text{rf}} = \frac{4}{\pi}\frac{\tau_{\text{rf}}}{\tau_{\text{i}}}V_{\text{rf}} \tag{11.9.13}$$

Assuming that the ion flux entering a high-frequency sheath at $x = 0$ is constant, the energy distribution is, from (11.9.2),

$$g_{\text{i}}(\mathcal{E}) = \frac{n_{\text{s}}u_{\text{B}}}{2\pi}\sum_j\left|\frac{\mathrm{d}\mathcal{E}}{\mathrm{d}(\omega t_{0j})}\right|^{-1} \tag{11.9.14}$$

Using (11.9.12) to evaluate the derivative and noting that there are two values of t_0 during one rf cycle for a given energy, we obtain

$$g_{\text{i}}(\mathcal{E}) = \frac{2n_{\text{s}}u_{\text{B}}}{\pi\Delta\mathcal{E}}\left[1 - 4\left(\frac{\mathcal{E} - V_{\text{bias}}}{\Delta\mathcal{E}}\right)^2\right]^{-1/2} \tag{11.9.15}$$

which yields a characteristic bimodal distribution with two peaks symmetric about V_{bias}. The two peaks are singular because of the assumed monoenergetic initial ion velocity, but the peaks are integrable.

The intermediate frequency regime $\omega\tau_{\text{i}} \sim 1$ is difficult to treat analytically. Let us consider a sheath voltage $V_{\text{s}}(t) = V_{\text{bias}} - V_{\text{rf}}(t)$, as in (11.9.4), with $V_{\text{rf}}(t) = V_{\text{rf}}\sin\omega t$ and $V_{\text{rf}} \leq V_{\text{bias}}$. We can introduce a simple relaxation model to give the approximate energy spread $\Delta\mathcal{E}$ (distance between the energy peaks) about $\mathcal{E} = V_{\text{bias}}$ (Miller and Riley, 1997; Panagopoulos and Economou, 1999)

$$\frac{\mathrm{d}V_{\text{i}}}{\mathrm{d}t} = -4\frac{V_{\text{i}} - V_{\text{rf}}}{\tau_{\text{i}}} \tag{11.9.16}$$

having a relaxation time $\tau_{\text{i}}/4$. Equation (11.9.16) has the same form as (8.6.15), yielding a general convolution solution for V_{i} of the form (8.6.13) or (8.6.16).

Alternately, a Fourier transform procedure can be used, $\tilde{V}_{\text{i}}(\omega) = \tilde{A}(\omega)\tilde{V}_{\text{rf}}(\omega)$, where $\tilde{V}_{\text{i}}(\omega)$ and $\tilde{V}_{\text{rf}}(\omega)$ are the (complex) Fourier transforms of $V_{\text{i}}(t)$ and $V_{\text{rf}}(t)$, and

$$\tilde{A}(\omega) = \frac{4}{4 + j\omega\tau_{\text{i}}} \tag{11.9.17}$$

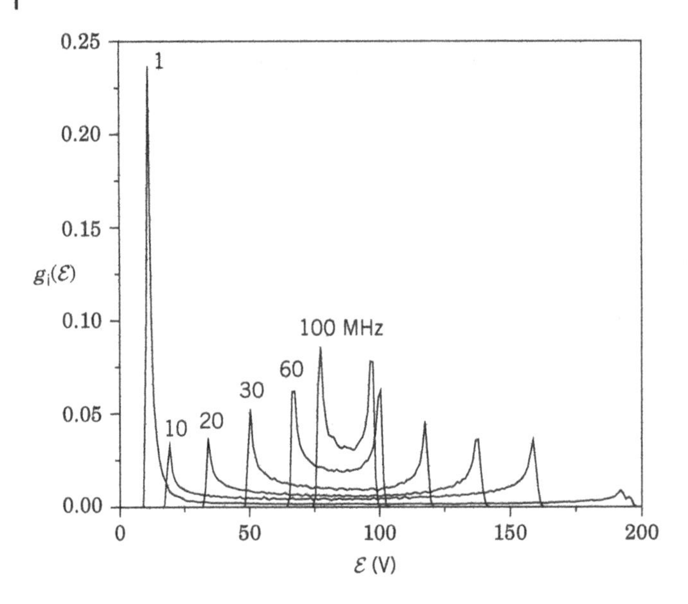

Figure 11.43 Simulation results showing ion energy distributions $g_i(\mathcal{E})$ for a single sheath in a current-driven helium discharge at frequencies from 1 to 100 MHz; the maximum sheath voltage drop was about 200 V in every case.

is the voltage response function. This corresponds to a low-pass filter having $|\tilde{V}_i| = |\tilde{V}_{rf}|$ for $\omega\tau_i/4 \ll 1$ and $|\tilde{V}_i| = 4|\tilde{V}_{rf}|/(\omega\tau_i)$ for $\omega\tau_i/4 \gg 1$. Taking the inverse Fourier transform of $\tilde{V}_i(\omega)$ yields $V_i(t)$. Using $V_s(t) = V_{bias} - |V_i(t)| \sin\omega t$ rather than (11.9.4) in (11.9.3) then approximately determines the ion-bombarding energy distribution. Note that $\Delta\mathcal{E} = 2|V_i|$, with $\Delta\mathcal{E} = 2|V_{rf}|$ for $\omega\tau_i/4 \ll 1$, and with $\Delta\mathcal{E}$ given by (11.9.13) for $\omega\tau_i/4 \gg 1$.

Figure 11.43 shows the energy distribution obtained from a one-dimensional PIC simulation of a single collisionless sheath in a helium discharge driven by a sinusoidal current source (Kawamura et al., 1999). A current-driven sheath was used in order to avoid arbitrarily setting the dc bias voltage. The plasma parameters were chosen to fix the ion transit time at $\tau_i \approx 77$ ns, and the rf frequency was varied from 1 to 100 MHz. As expected, we see bimodal distributions that become narrower as the frequency increases.

In the high-frequency regime $\omega \gtrsim \omega_i$, (11.9.13) yields the result that the energy width $\Delta\mathcal{E}$ scales as the inverse square root of the ion mass. For a plasma containing ions of various masses, we expect the total energy distribution g_i to have a series of peaks due to density-weighted sums of bimodal distributions of the form (11.9.15), with peaks for lighter mass ions showing a larger width than heavier mass ions. An extreme example of this energy separation is shown in the experimental data of Figure 11.44, taken in a 75 mTorr capacitive discharge driven at 13.56 MHz. The Eu ions (mass = 152 amu) display a single unresolved energy peak; the H_2O^+ ions (mass = 18 amu) show the characteristic bimodal shape of the high-frequency regime, and the H_3^+ ions show the wide separation and unsymmetrical shape that characterizes the low-frequency regime. The preciseness with which ions of different masses can be distinguished in the measured energy distribution in the frequency regime $\omega \gtrsim \omega_i$ is shown in Figure 11.45, taken in a CF_4 capacitive discharge at 3 mTorr.

Ion–neutral collisions within the sheath can strongly modify the bombarding energy distribution and can lead to an additional structure of peaks within the energy distribution. These features have been described by Wild and Koidl (1991) as a consequence of the rf modulation of the sheath in

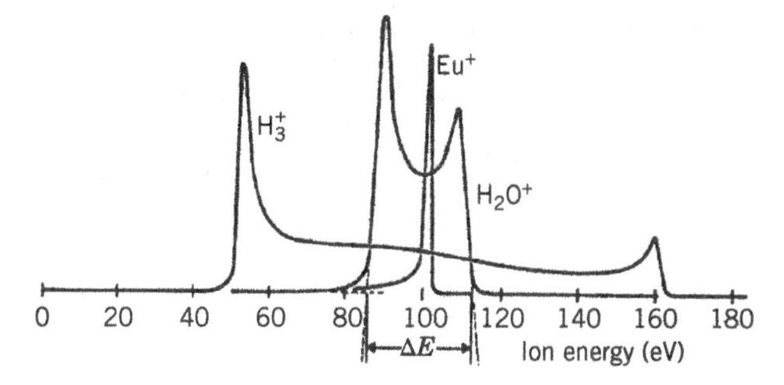

Figure 11.44 Measured ion energy distributions $g_i(\mathcal{E})$ for H_3^+, H_2O^+, and Eu^+ ions at the grounded electrode of a 75 mTorr argon rf discharge driven at 13.56 MHz. Source: Coburn and Kay, 1972/with permission of AIP Publishing.

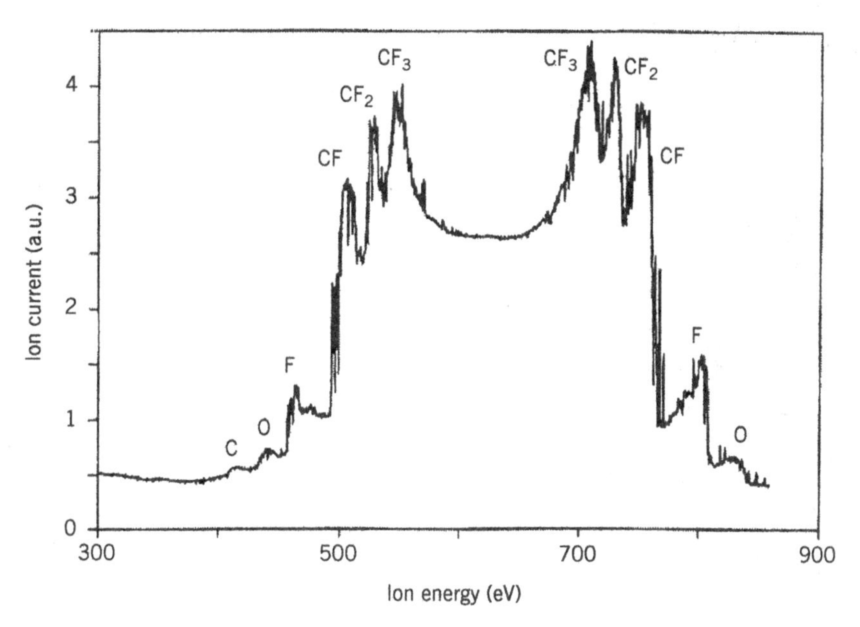

Figure 11.45 Measured ion energy distributions $g_i(\mathcal{E})$ at the powered electrode of a CF_4 discharge driven at 13.56 MHz. Source: Kuypers and Hopman, 1990/with permission of AIP Publishing.

combination with the creation of "secondary" cold ions by charge exchange processes within the sheath.

Cold ions created at some position x_0 within the sheath during a finite time interval $t_2 - t_1$, when the electron sheath edge oscillates from position x_0 at time t_1 to the electrode surface and back to position x_0 at time t_2, all see a quasineutral plasma ($n_e \approx n_i$) with a local electric field $E \approx 0$. All of these ions start their journey to the surface at the same time t_2, just after n_e falls to zero at x_0, yielding the additional peaks in the distribution.

Figure 11.46 shows a quantitative comparison of theory and experiment in a capacitively coupled argon discharge with 500 V across the sheath. At 3 μbar, the high-frequency bimodal distribution of the primary ions is clearly seen centered around this voltage, but the lower energy peaks due to the secondary ions created by primary ion charge exchange in the sheath are also prominent. At higher pressures, the bimodal distribution of the primary ions is much diminished.

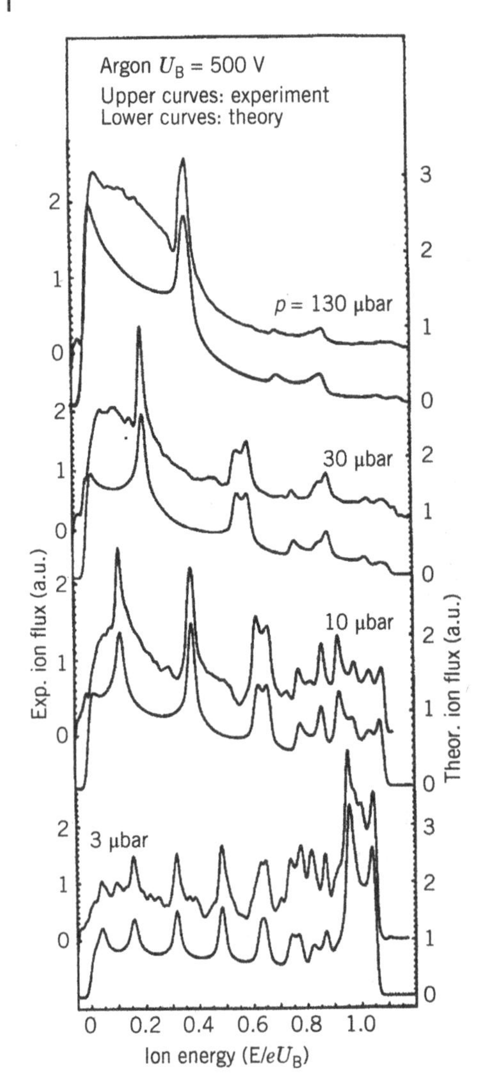

Figure 11.46 Comparison of experimental and theoretical ion energy distributions $g_i(\mathcal{E})$ in an argon discharge driven at 13.56 MHz at various pressures (1 μbar = 0.76 mTorr). Source: Wild and Koidl, 1991/with permission of AIP Publishing.

The ion-bombarding energy distribution $g_i(\mathcal{E})$ for the multiple-frequency-driven discharges described in Section 11.6 can be more complicated than for single-frequency discharges because (1) there can typically be multiple maxima and minima of the driving voltage, and (2) the different frequencies lie in different $\omega\tau_i$ regimes. Here, recall that τ_i given by (11.8.1) is the ion transit time across the sheath. Effect (1) leads to multiple peaks in the ion energy distribution, and effect (2) determines an overall ion energy distribution width.

For low ion collisionality in the sheath, the relaxation model of (11.9.16) can be used.

The application of the Fourier transform relaxation model and comparisons to PIC simulations are given in Wu et al. (2007). A somewhat different voltage response function than (11.9.17) was used. An example is shown in Figure 11.47. The model predicts the 64 MHz/2 MHz ion-bombarding energy distribution reasonably well, including the 64 MHz-induced substructure. In usual industrial practice, the dual frequencies are incommensurate, and the substructure is averaged away, leaving a relatively smooth ion energy distribution.

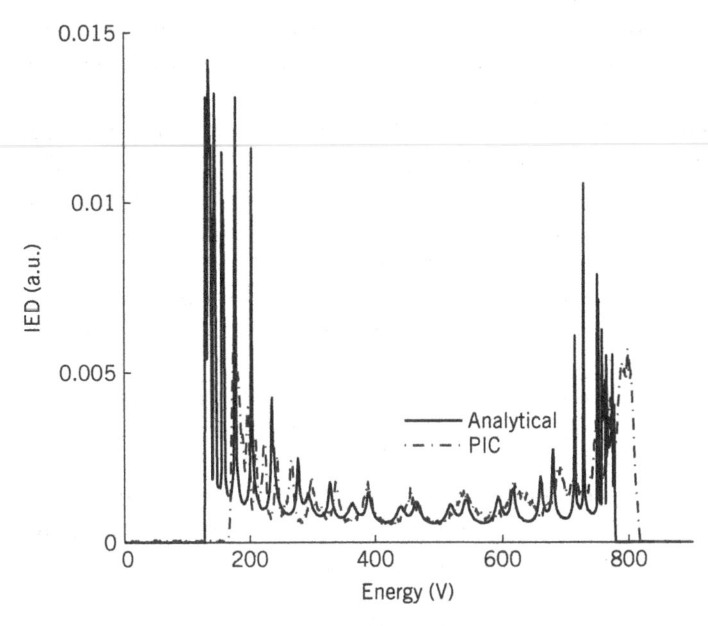

Figure 11.47 Ion-bombarding energy distribution versus ion energy; comparison of Fourier transform relaxation model and PIC simulations for a 3 cm gap, 30 mTorr argon discharge driven by 800 V at 2 MHz and 400 V at 64 MHz. Source: Reproduced from Wu et al. (2007), with the permission of AIP Publishing. https://doi.org/10.1063/1.2435975.

11.10 Magnetically Enhanced Discharges

Capacitive discharges have been the most widely used source for low-pressure materials processing. However, they suffer from the disadvantages of high sheath voltages with consequent low ion density (ion flux) and high ion-bombarding energy at a given power level. The ion-bombarding energy also cannot be varied independently of the ion flux in these devices. To circumvent these disadvantages, other sources have been employed and various attempts have been made to improve the performance of capacitive discharges. In the following two chapters, we consider alternative methods of producing and heating plasmas, particularly to achieve lower sheath voltages, higher densities, and independent control of both. In this section, we consider one modification of the capacitive discharge that can also achieve these goals, the *magnetically enhanced reactive ion etcher* (MERIE) or *rf magnetron*. In this discharge, a controllable, relatively weak (50–200 G) dc magnetic field is imposed parallel to the surface of the rf powered electrode.

The magnetic field introduces a number of effects that act to increase the density and reduce the sheath voltage at a fixed absorbed power. (1) As described in Section 5.4, the electron motion can be strongly inhibited across the field, leading to a reduced nonambipolar flux in the cross-field directions. If most of the collecting surface is across field lines, then the power loss can be significantly reduced, resulting in higher density at a given power. (2) The magnetic field can confine the energetic (ionizing) electrons to a small volume near the electrode. This reduces the overall effective loss area of the discharge and directs a greater fraction of the escaping plasma to the powered electrode, increasing the density and ion flux there. (3) A transverse field can increase the efficiency of stochastic heating due to multiple correlated collisions of electrons with the oscillating sheaths, increasing the density. (4) There is an increase in the efficiency of ohmic heating due to reduced electron cross-field mobilities, and consequently, increased electric fields in the plasma. In a simple model presented below, we examine effects (3) and (4), the increased efficiency of stochastic heating and ohmic heating.

Although application of the magnetic field results in a reduced sheath voltage and increased plasma density, the plasma generated can be strongly nonuniform both radially and azimuthally due to $\mathbf{E} \times \mathbf{B}$ drifts, where $\mathbf{E}$ and $\mathbf{B}$ are the local dc electric and magnetic fields, respectively. To increase process uniformity (at least azimuthally), the magnetic field can be rotated in the plane of the wafer at a low frequency (~ 0.5 Hz). While this is an improvement, MERIE systems may not have good uniformity, which may limit their performance. A strongly nonuniform plasma over the wafer can also give rise to lateral dc currents within a film on the wafer that can damage the film (see Section 16.6).

In early work, Yeom et al. (1989a, b) investigated magnetic enhancement in a 3 mTorr argon cylindrical rf magnetron discharge with inner radius 5 cm and outer radius 9.7 cm, which has azimuthally closed magnetic field lines. At 13.56 MHz, they measured a decrease from 85° to 18° of the rf current–voltage phase angle as the magnetic field was increased from 0 to 200 G, indicating a reduced electron mobility and a more resistive plasma. The bias voltage also decreased, suggesting an increased ion flux to the inner radius (powered) electrode due to nearby increased ionization, a result of increased magnetic confinement of electrons. As will be seen, these features are common in parallel plate systems having a uniform (open) magnetic field parallel to the plates, which we examine here.

In the following analysis, we return to the homogeneous model of Section 11.1 for the plasma and the sheath in order to calculate the heating. Effect 1, nonambipolar ion losses, could be included in the model if desired, but the effect may not be significant at the typical 50–200 G fields. Effect 2, energetic electron confinement, may be quite important but cannot be treated in a homogeneous model. We later summarize two-dimensional simulation results from Kushner (2003) describing this effect.

11.10.1 Parallel Plate Homogeneous Model

The model is the same as given in Figure 11.1, except that a uniform magnetic field B_0 is oriented parallel to the electrode surfaces in the z direction. As in Section 11.1, a uniform sinusoidal current density $J_x(t) = \mathrm{Re}\, J_1 \mathrm{e}^{j\omega t}$ flows between the plates. In the plasma, the current density is related to the electric field vector through the dielectric tensor. Let $J_\alpha(t) = \mathrm{Re}\, \tilde{J}_\alpha \mathrm{e}^{j\omega t}$ and $E_\alpha(t) = \mathrm{Re}\, \tilde{E}_\alpha \mathrm{e}^{j\omega t}$, where $\alpha = x, y$, or z, we have

$$\begin{pmatrix} \tilde{J}_x \\ \tilde{J}_y \\ \tilde{J}_z \end{pmatrix} = j\omega\epsilon_0 \begin{pmatrix} \kappa_\perp & -\kappa_\times & 0 \\ \kappa_\times & \kappa_\perp & 0 \\ 0 & 0 & \kappa_\parallel \end{pmatrix} \begin{pmatrix} \tilde{E}_x \\ \tilde{E}_y \\ \tilde{E}_z \end{pmatrix} \tag{11.10.1}$$

where the tensor elements are given in Section 4.4. Since $\tilde{J}_x = J_1$, and $\tilde{J}_y = \tilde{J}_z = 0$, we can solve (11.10.1) to obtain $\tilde{E}_z = 0$,

$$\tilde{E}_y = -\frac{\tilde{E}_x \kappa_\times}{\kappa_\perp} \tag{11.10.2}$$

and

$$\tilde{E}_x = \left[j\omega\epsilon_0 \left(\kappa_\perp + \frac{\kappa_\times^2}{\kappa_\perp} \right) \right]^{-1} J_1 \tag{11.10.3}$$

In the sheath region a (see Figure 11.1), the x component of the electric field is found by integrating Poisson's equation to obtain

$$E_{xa}(x,t) = en\frac{x - s_a(t)}{\epsilon_0} + E_x(t) \tag{11.10.4}$$

where $E_x(t)$ is the field in the plasma, and we have chosen $E_{xa} = E_x$ at the instantaneous position of the sheath edge $x = s_a$. Although the usual assumption is that $|E_x| \ll |E_{xa}|$, the field in the plasma can be significant for sufficiently large magnetic fields (Lieberman et al., 1991). For 50–200 G fields, it suffices to set $E_x(t) \equiv 0$ in (11.10.4), as in (11.1.7). The analysis of the sheath then proceeds as in Section 11.1, with the total rf voltage drop across both sheaths $V_{ab}(t)$ in this symmetric discharge given by (11.1.19) with $\bar{s} = s_0$. The complex amplitude of $V_{ab}(t)$ is

$$\tilde{V}_{ab} = -\frac{2jens_0^2}{\epsilon_0} \tag{11.10.5}$$

where s_0 is given by (11.1.11). Adding to this the voltage drop $\tilde{E}_x d$ across the plasma, where $d = l - 2s_0$ is the bulk plasma thickness, we obtain the complex amplitude of the discharge voltage

$$\tilde{V}_{rf} = -\frac{2jens_0^2}{\epsilon_0} + \tilde{E}_x d \tag{11.10.6}$$

The dc voltage across a single sheath is given by (11.1.42).

The dynamics of the sheath heating can be changed by the addition of magnetic fields. For weak magnetic fields, we assume that the sheath motion remains unchanged, but that the particle interaction is modified due to multiple correlated collisions of electrons with the moving sheath. A gyrating electron that collides once with the moving sheath collides again in a time interval of approximately half a gyroperiod. The electron trajectory can be coherent over many such sheath collisions, leading to large energy gains. The coherent motion is destroyed on the timescale for electron collisions with neutral gas atoms.

To determine the heating, one starts with the basic sheath heating equation (11.2.27). For the homogeneous model, this simplifies to

$$\bar{S}_{stoc} = 2m\Gamma_e \langle \Delta u(\Delta u - u_{es})\rangle_\phi \tag{11.10.7}$$

where Δu is half the change in electron velocity for a set of multiple collisions. To determine Δu for the multiple sheath collisions, we let $2u_{es}(\omega t)$ be the change in electron velocity for a single collision with the sheath at time t. For a slowly moving sheath, successive collisions take place at time intervals of $\Delta t = \pi/\omega_{ce}$, where $\omega_{ce} = eB_0/m$ is the electron gyration frequency. These collisions result in coherent energy gain. However, the coherent energy gain is terminated by electron collisions with neutral gas atoms. Hence, we write

$$\Delta u = \sum_{i=0}^{\infty} u_{es}(\omega t + i\omega\Delta t)\,\mathrm{e}^{-i\nu_{el}\Delta t} \tag{11.10.8}$$

where ν_{el} is the electron–neutral elastic scattering frequency. The exponential factor in (11.10.8) gives the fraction of electrons that have not collided with neutral gas atoms after a time $i\Delta t$. For the regime of interest ω, $\nu_{el} \ll \omega_{ce}$, we can convert the sum to an integral, obtaining (Lieberman et al., 1991) the time-averaged power per unit area delivered to the electrons by the oscillating sheath

$$\bar{S}_{stoc} = \frac{1}{4}mn\bar{v}_e|\tilde{u}_{es}|^2 \frac{\omega_{ce}}{\pi(\nu_{el}^2 + \omega^2)}\left(\nu_{el} + \frac{\omega_{ce}}{\pi}\right) \tag{11.10.9}$$

To complete the model, we add the usual equilibrium conditions of flux balance from (11.2.32a) or (11.2.32b), and electron and total power balance from (11.2.36) and (11.2.37).

The modification of the heating due to the magnetic field has two main effects on the discharge equilibrium. (1) The stochastic heating increases with increasing B_0, provided most of the rf voltage appears across the sheaths; and (2) a significant fraction of the total rf discharge voltage can be dropped across the bulk plasma at high magnetic fields. If the bulk plasma voltage is small, then we can estimate the scaling of the discharge equilibrium with S_{abs} and B_0, in various regimes, as follows. We first note by current continuity that $J_{rf} \propto ns_0$. Since the sheaths are capacitive, $J_{rf} \propto V_{rf}/s_0$. Hence, it follows that $n \propto V_{rf}/s_0^2$. Using this result in (11.10.9), we obtain the scaling of the stochastic heating power $\overline{S}_{stoc} \propto B_0^2 V_{rf}$, for $\omega_{ce} \gg \omega$, ν_{el}. Similarly, scaling the ohmic power yields $\overline{S}_{ohm} \propto V_{rf}$. The power balance equations can then be evaluated in various limiting cases, depending on whether stochastic heating or ohmic heating is the dominant heating mechanism and on whether ion energy losses or electron energy losses are the dominant loss mechanism. Considering the high-voltage case for which ion losses are dominant, then at low pressures where stochastic heating dominates, we obtain the scaling

$$\begin{aligned} V_{rf} &\propto \frac{S_{abs}^{1/2}}{B_0} \\ n &\propto S_{abs}^{1/2} B_0 \\ s_0 &\propto B_0^{-1} \\ \overline{S}_{stoc} &\propto S_{abs}^{1/2} B_0 \end{aligned} \tag{11.10.10}$$

We leave the details to Problem 11.22. At very high B_0, the bulk plasma voltage drop can dominate the sheath drop, in which case the scaling can become very different, but this is not the usual regime for rf magnetrons. We caution the reader that the analysis leading to (11.10.9) and the scaling in (11.10.10) is illustrative rather than rigorous.

11.10.2 Measurements and Simulations

Measurements were made in a commercial, strongly asymmetric etch chamber in argon at 13.56 MHz to compare with the model (Lieberman et al., 1991). The plasma density n was measured with a Langmuir probe approximately 3 cm in front of the 200 cm^2 powered electrode. The experiment and model results for the density were compared for three magnetic fields of 10, 30, and 100 G, and three power densities of 0.25, 0.5, and 1.0 W/cm^2, with $p = 10$ mTorr. The experimental magnetic field dependence was somewhat weaker than predicted, and the experimental values were about 30% lower than the theory. Park and Kang (1997a) also compared experiments with a model incorporating multiple bounce stochastic heating and a Child law sheath, obtaining reasonable agreement. The increased plasma resistance and resulting transition from stochastic to ohmic heating was also observed with increasing magnetic field in one-dimensional PIC simulations by Hutchinson et al. (1995). You et al. (2011) used an rf-compensated Langmuir probe to measure the electron energy distributions in a 4 cm gap, 1.33–40 Pa argon MERIE discharge. The discharge was asymmetrically driven at 13.56 MHz, and the magnetic field was varied from 0 to 80 G. They determined ion-to-electron heating transitions, electron energy cooling, and $E \times B$-induced spatial density variations as the field was increased.

Kushner (2003) performed hybrid-fluid simulations at 40 mTorr argon in a two-dimensional asymmetric geometry with a 2.5 cm gap, with the magnetic field modeled as purely radial and parallel to the electrode surface. As seen in Figure 11.48, there is a pronounced shift of the argon ion production rate, and the resulting argon ion density, toward the powered (smaller area) electrode as the magnetic field is increased from 0 to 220 G, with a large increase in the powered-to-grounded

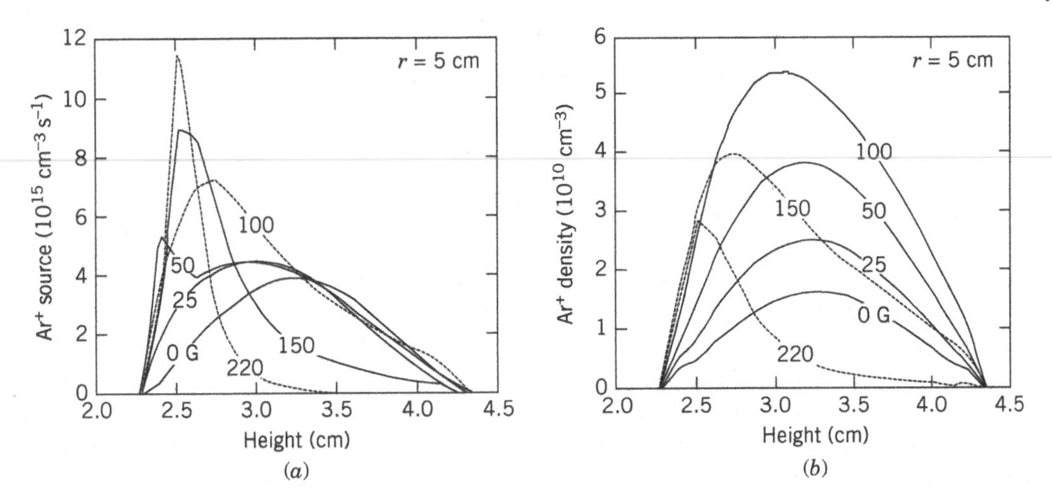

Figure 11.48 Two-dimensional hybrid-fluid simulation of a magnetically enhanced capacitive discharge, showing (*a*) Ar^+ production rate versus height (distance from powered electrode) and (*b*) Ar^+ ion density versus height, for magnetic field strengths varied from 0 to 220 G; rf frequency 10 MHz, pressure 40 mTorr argon, powered electrode radius 10 cm, grounded electrode radius ≈ 20 cm, gap height 4 cm. Source: Reproduced from Kushner (2003), with the permission of AIP Publishing. https://doi.org/10.1063/1.1587887.

edge densities. These results are attributed to the reduced electron mobility in the presence of the magnetic field, leading to enhanced electron confinement near the powered electrode.

The bias voltage on the (smaller) powered electrode (*a*) with respect to the (larger) grounded electrode (*b*) is $V_B = \overline{V}_b - \overline{V}_a$. Substituting for $\overline{V}_b$ from (11.4.12) for the homogeneous sheath model, we obtain

$$V_B = -\left(1 - \frac{n_a A_a^2}{n_b A_b^2}\right)\overline{V}_a \tag{11.10.11}$$

From (11.1.25), the rf voltage drop across both sheaths for the homogenous model is

$$V_{rf} = \frac{4}{3}\left(1 + \frac{n_a A_a^2}{n_b A_b^2}\right)\overline{V}_a \tag{11.10.12}$$

For a fixed V_{rf}, these two equations imply that the field-induced increase in the ratio of powered-to-grounded edge densities n_a/n_b shown in Figure 11.48 decreases the magnitude of the (negative) bias voltage and increases the magnitude of the flux $\Gamma_i = n u_B$, as observed in both experiments and simulations.

Rauf (2005) used a two-dimensional hybrid-fluid simulation to examine the effects of negative ions on MERIE operation in a 37.5 mTorr Ar/C_2F_6 plasma. For both constant voltage and constant power conditions, the charged and neutral particle densities increased with increasing frequency from 13.56 to 60 MHz. At constant voltage, these densities also increased by at least a factor of two with a magnetic field increase from 0 to 50 G. However, at constant power, density decreases were observed, contrary to that found in pure argon discharges. At the higher magnetic fields, the electron cross-field motion was significantly inhibited, allowing negative ions to play important roles in the sheath and plasma dynamics. This led to increased power absorption by negative ions. Rauf (2003) also used the 2D hybrid-fluid model to investigate the effects of magnetic enhancement on the electrical characteristics of dual-frequency discharges. The similar trends of theory and experiments and simulations indicate that the basic modeling approach includes much of the essential physics.

11.11 Matching Networks and Power Measurements

Although this text is mainly concerned with the internal dynamics of the plasma, some knowledge of the external circuit is necessary. If the discharge is driven directly by an rf power source, then generally power is not transferred efficiently from the source to the discharge. To understand this, consider a discharge modeled as a load having impedance $Z_D = R_D + jX_D$, where R_D is the discharge resistance and X_D is the discharge reactance. The power source connected to Z_D is modeled by its Thevenin-equivalent circuit, consisting of a voltage source with complex amplitude $\tilde{V}_T$ in series with a source resistance R_T. The time-averaged power flowing into the discharge is

$$\overline{P} = \frac{1}{2}\mathrm{Re}\,(\tilde{V}_{rf}\tilde{I}_{rf}^*) \tag{11.11.1}$$

where $\tilde{V}_{rf}$ is the complex voltage across Z_D. Solving for $\tilde{I}_{rf}$ and $\tilde{V}_{rf}$ for these series elements, we obtain

$$\tilde{I}_{rf} = \frac{\tilde{V}_T}{R_T + R_D + jX_D} \tag{11.11.2}$$

$$\tilde{V}_{rf} = \tilde{I}_{rf}(R_D + jX_D) \tag{11.11.3}$$

Substituting (11.11.2) and (11.11.3) into (11.11.1), we obtain

$$\overline{P} = \frac{1}{2}|\tilde{V}_T|^2 \frac{R_D}{(R_T + R_D)^2 + X_D^2} \tag{11.11.4}$$

For fixed source parameters $\tilde{V}_T$ and R_T, maximum power transfer is obtained by setting $\partial\overline{P}/\partial X_D = 0$ and $\partial\overline{P}/\partial R_D = 0$, which gives $X_D = 0$ and $R_D = R_T$. The maximum power supplied by the source to the load is then

$$\overline{P}_{max} = \frac{1}{8}\frac{|\tilde{V}_T|^2}{R_T} \tag{11.11.5}$$

If maximum power transfer is obtained, then we say that the source and load are matched.

Since X_D is not zero, and, typically, $R_D \ll R_T$, the power $\overline{P}$ is generally much less than $\overline{P}_{max}$. To increase $\overline{P}$ to $\overline{P}_{max}$, thus matching the source to the load, a lossless *matching network* can be placed between them. Because R_D and X_D are two independent components of Z_D, the simplest matching network consists of two independent components. The most common configuration, called an "L-network," is shown inserted between the source and the load in Figure 11.49. It consists of a shunt capacitor having susceptance $B_M = \omega C_M$ and a series inductor having a reactance $X_M = \omega L_M$.

To determine X_M and B_M, we write the admittance looking toward the right at location 2 in Figure 11.49 as the inverse of the impedance:

$$Y_2 \equiv G_2 + jB_2 \equiv Z_2^{-1} = (R_D + jX_2)^{-1} \tag{11.11.6}$$

where $X_2 = X_M + X_D$. Separating real and imaginary parts and solving for G_2 and B_2, we obtain

$$G_2 = \frac{R_D}{R_D^2 + X_2^2} \tag{11.11.7}$$

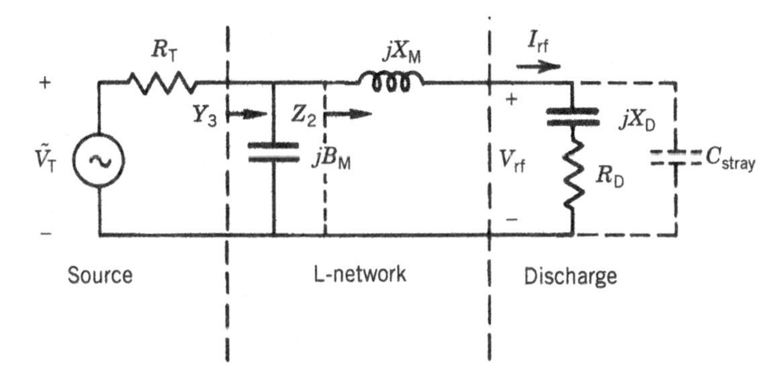

Figure 11.49 Equivalent circuit for matching the rf power source to the discharge using an L-network.

$$B_2 = -\frac{X_2}{R_D^2 + X_2^2} \tag{11.11.8}$$

Next, we note from Figure 11.49 that $Y_3 = G_3 + jB_3$, with $G_3 = G_2$ and $B_3 = B_2 + B_M$. The matched condition of maximum power transfer is $G_3 = 1/R_T$ and $B_3 = 0$. Setting $G_3 = 1/R_T$ in (11.11.7), we solve for X_2 to obtain

$$X_2 = (R_D R_T - R_D^2)^{1/2} \tag{11.11.9}$$

Since $X_2 = X_M + X_D$, the required X_M is

$$X_M = (R_D R_T - R_D^2)^{1/2} - X_D \tag{11.11.10}$$

Since X_D is negative, X_M must be positive; i.e., a matching inductor $L_M = X_M/\omega$ must be used. Using (11.11.9) in (11.11.8) and setting $B_3 = 0$ ($B_M = -B_2$), we obtain

$$B_M = \left(\frac{1}{R_T R_D} - \frac{1}{R_T^2}\right)^{1/2} \tag{11.11.11}$$

Since B_M is positive, a matching capacitor $C_M = B_M/\omega$ is required. Because there must be real solutions for B_M and X_M, we see from (11.11.10) or (11.11.11) that R_D must be less than R_T for a match to be achieved with an L-network. This is the usual regime at the higher power levels used for typical processing discharges. For low powers, $R_D > R_T$ and a different form of matching network must be used (Problem 11.24). A three-element (T or Π) network can be used to match any discharge; hence, such networks are commonly used to provide added flexibility. The three elements are not uniquely determined by the maximum power condition, but the inductive element is usually a fixed value, and the two capacitors can be varied to achieve the match.

Because R_D and X_D are actually functions of the discharge voltage or the absorbed power, we must specify these to determine the matched condition. For a specified voltage or absorbed power, we can determine $\tilde{I} \equiv I_1$ and s_m as in Example 11.1 or 11.2 of Section 11.2. Then, R_D and X_D are determined from

$$P_{abs} = \frac{1}{2} I_1^2 R_D \tag{11.11.12}$$

and

$$X_D = -\frac{1}{\omega C_{ab}} \tag{11.11.13}$$

where C_{ab} is given by (11.2.20).

Because typically $X_D \gg R_D$, the voltage and current across the discharge are nearly 90° out of phase. Setting the current and voltage across the discharge to be

$$I_{rf}(t) = I_1 \cos \omega t \tag{11.11.14}$$

$$V_{rf}(t) = V_{rf} \cos\left(\omega t - \frac{\pi}{2} + \psi\right) \tag{11.11.15}$$

we find that

$$V_{rf} = (R_D^2 + X_D^2)^{1/2} I_1 \approx -X_D I_1 \tag{11.11.16}$$

$$\psi = -\tan^{-1}\frac{R_D}{X_D} \approx -\frac{R_D}{X_D} \tag{11.11.17}$$

The time-averaged power absorbed by the discharge is

$$P_{abs} = \frac{1}{\tau}\int_0^{\tau} V_{rf}(t) I_{rf}(t)\,dt \tag{11.11.18}$$

$$= \frac{1}{2} I_1 V_{rf} \sin \psi \tag{11.11.19}$$

where $\tau = 2\pi/\omega$. Under matched conditions, the voltage and current at the source are in phase with each other, $\tilde{V}_T = 2R_T\tilde{I}_T$, with the power supplied by the source,

$$P_T = \frac{1}{8}\frac{|\tilde{V}_T|^2}{R_T} \tag{11.11.20}$$

For a lossless matching network, $P_{abs} = P_T$. Equating (11.11.19) and (11.11.20), using (11.11.16), and solving for $|\tilde{V}_T|$, we obtain

$$|\tilde{V}_T| = -2\frac{(R_D R_T)^{1/2}}{X_D} V_{rf} \tag{11.11.21}$$

For the usual discharge conditions, $|\tilde{V}_T| \ll V_{rf}$.

Does a matched capacitive discharge appear as voltage- or current-driven? For symmetric discharges, there is little harmonic generation due to the sheath nonlinearities, which cancel or nearly cancel. But for asymmetric voltage-driven discharges (see Section 11.4), nonlinearly generated current harmonics from the dominant sheath motion, including the series resonance harmonics described in Section 11.5, would be blocked from flowing through the large inductive impedance magnitude $m\omega L_M$, with $m = 2, 3, \ldots$ the harmonic number. With only the fundamental component of the current, the matched system would appear as sinusoidally current-driven.

However, as shown in Figure 11.49, the harmonic currents can return to ground in the presence of a significant stray capacitance C_{stray}, having a low impedance magnitude $(m\omega C_{stray})^{-1}$, and therefore a low-voltage drop. In this case, the voltage across the discharge is mainly at the fundamental frequency. Hence, a matched discharge with a significant stray capacitance would appear as sinusoidally voltage-driven. This is a common case seen in experiments, e.g., Klick et al. (1997).

11.11.1 Power Measurements

An rf wattmeter placed between the source and the matching network is conventionally used to measure the time-averaged power P_T supplied by the source. This instrument is often an integral part of the rf power supply. For sinusoidal voltages and currents, the time-averaged powers P_f and P_b flowing in the forward and backward directions are then measured, with

$$P_T = P_f - P_b \tag{11.11.22}$$

For the voltage and current nearly in phase at the measurement location, we have $P_b \ll P_f$, and the measurements accurately determine P_T. Under strongly out-of-phase conditions, $P_b \approx P_f \gg P_T$, such that subtracting P_b from P_f does not determine P_T accurately. Hence, rf wattmeters cannot be placed between the matching network and the discharge to determine the power P_{abs} absorbed by the discharge. For a lossless matching network, $P_{abs} = P_T$, but the nonideal matching networks used in typical processing systems often absorb a considerable fraction of the source power, such that $P_{abs} < P_T$. The usual source of loss is the finite resistance of the wire with which the matching inductor is wound. This nonideal inductor can be modeled as an ideal inductor L_M in series with a resistor R_M, through which the discharge current I_1 flows.

Equation (11.11.19) can be used to determine P_{abs} if the amplitudes I_1 and V_{rf}, and the phase difference ψ, can be accurately measured. The discharge current $I_{rf}(t)$ is conventionally measured with a calibrated, miniature current transformer or *Rogowski coil* (Huddlestone and Leonard, 1965). This often consists of a helical coil of wire wound around an insulating toroid, which surrounds the center conductor of the coaxial cable at the matchbox feeding the discharge. If $I_{rf}(t)$ and $V_{rf}(t)$ are displayed on an oscilloscope or measured using a dual channel vector voltmeter, then V_{rf}, I_1, and ψ can be determined. However, because ψ can be as small as 2–3°, phase shifts between the measured voltage and current signals due to nonideal instrumental and cabling effects can render the measurement meaningless. Accurate calibration of the phase shift for known calibration loads is essential. This can be a difficult measurement to make at 13.56 MHz in a practical processing discharge, where, for example, a 4 cm length of coaxial cable has a phase shift of approximately 1°. If the measured voltage and current waveforms are not approximately sinusoidal, then the power must be determined by direct averaging of the $I_{rf}(t)V_{rf}(t)$ product using (11.11.18).

For the circuit shown in Figure 11.49, if the voltage and current waveforms are reasonably sinusoidal in shape, the power absorbed by the discharge alone can be determined, as follows:

1. Measure I_1 and the source power P_T. This is the power absorbed by the lossy matching inductor and by the discharge.
2. Extinguish the discharge by raising the pressure in the chamber to 1 atm or by reducing the pressure to a very low value. The measured I_1 will in general change.
3. Readjust the source voltage V_T so that I_1 is the same as that measured in (1) and measure the source power $P_T^{(0)}$. The matching capacitor can be retuned if desired, but a perfect match is not necessary. This measurement yields the power absorbed by the lossy inductor only.

Because I_1 is the same with and without the plasma present, the same power is lost in the inductor. Therefore, the power absorbed by the plasma is

$$P_{abs} = P_T - P_T^{(0)} \tag{11.11.23}$$

In many discharges, there is a large stray capacitance C_{stray} in parallel with the discharge impedance $R_D + jX_D$. C_{stray} represents the capacitance to ground of the powered electrode and center conductor of its coaxial cable feed, typically $C_{stray} \sim$ 100–200 pF in processing discharges. Then, $I_1 \approx \omega C_{stray} V_{rf}$, with or without the plasma present. In this case, following the same procedure as above using the measured V_{rf}, rather than I_1, determines P_{abs}. For further discussion of power measurements in rf discharges, the reader is referred to Godyak and Piejak (1990b) and Godyak (2021).

Problems

11.1 Plasma Admittance Derive expression (11.1.3) for the bulk plasma admittance Y_p and show, using assumption (b), that the displacement current that flows through C_0 is much smaller than the conduction current that flows through L_p and R_p.

11.2 Low-Pressure Homogeneous Discharge Equilibrium Consider a symmetric, capacitively coupled rf discharge in argon gas having a total applied rf voltage of amplitude 800 V at a frequency $f = \omega/2\pi = 13.56$ MHz. The plate separation is $l = 0.1$ m, and the gas pressure is 5 mTorr. Use the low-pressure homogeneous sheath/bulk plasma discharge model of Section 11.1, with a uniform ion density n_0 everywhere in the discharge.

(a) From (11.1.26), determine the electron temperature T_e in the discharge.
(b) From (11.1.21) and (11.1.42), determine the maximum sheath width $s_m = 2s_0$.
(c) From (11.1.23), determine the discharge rf current density J_1 (A/m^2).
(d) Find the plasma density n_0 using (11.1.21).
(e) Determine the dc potential of the plasma with respect to the plates and the ion-bombarding energy (in volts).
(f) Determine the total rf power per unit area required to sustain the discharge.

11.3 Stochastic Heating Derive the stochastic heating result (11.2.59), showing that $\bar{S}_{stoc}$ is proportional to the rf voltage, independent of the rf discharge current and plasma density.

11.4 Series Resonant Discharge Consider a uniform density bulk plasma of thickness d with two vacuum sheaths, each of thickness $s_m/2$, such that $l = d + s_m$. Show that the frequency ω_{res} of the series resonance between the bulk plasma inductance L_p and the parallel combination of the vacuum capacitance C_0 (see (11.1.3)) and the overall sheath capacitance C_s is

$$\omega_{res} = \omega_{pe}\left(\frac{s_m}{l}\right)^{1/2}$$

Evaluate ω_{res} for $d = 8$ cm, $s_m = 2$ cm, and $n_0 = 2 \times 10^{10}$ cm^{-3}, and compare ω_{res} to the driving frequency ω for a 13.56 MHz discharge.

11.5 Low-Pressure RF Discharge Equilibrium with Voltage Specified Consider a symmetric, capacitively coupled rf discharge in argon gas having a total applied rf voltage of amplitude 800 V at a frequency $f = \omega/2\pi = 13.56$ MHz. The plate separation is $l = 0.1$ m, and the gas pressure is 5 mTorr. Use a low-pressure discharge model, as in Example 11.1 of Section 11.2.

(a) Determine the electron temperature T_e in the discharge.
(b) Determine the plasma density n_0 and the ion flux.
(c) Determine the dc potential of the plasma with respect to the plates and the ion-bombarding energy (in volts).
(d) Determine the sheath thickness s_m.
(e) Determine the total rf power per unit area required to sustain the discharge.
(f) Determine the rf current amplitude drawn by the discharge.
(g) Use the scaling formulas to check your answers against Example 11.1.

11.6 Low-Pressure RF Discharge Equilibrium with Power Specified Verify the results of Example 11.2.

11.7 Intermediate Pressure RF Discharge Equilibrium with Voltage Specified Consider a symmetric, capacitively coupled rf discharge in argon gas having a total applied rf voltage of amplitude 800 V at a frequency $f = \omega/2\pi = 13.56$ MHz. The plate separation is $l = 0.1$ m, and the gas pressure is 30 mTorr. Use a low-pressure discharge model, as in Example 11.1 of Section 11.2, but use the collisional sheath results (11.2.54)–(11.2.59) instead of the collisionless sheath results.

(a) Determine the electron temperature T_e in the discharge.
(b) Determine the plasma density n_0 and the ion flux.
(c) Determine the dc potential of the plasma with respect to the plates and the ion-bombarding energy (in volts).
(d) Determine the sheath thickness s_m.
(e) Determine the total rf power per unit area required to sustain the discharge.
(f) Determine the rf current amplitude drawn by the discharge.
(g) For each plate of the discharge having a cross-sectional area of 1000 cm^2, find, using your results above, the effective resistance R_D (in ohms) and capacitance C_D (in farads) for a series *RC* model of the discharge.
(h) Design a matching network to match the discharge to a 50 Ω rf generator.

11.8 Scaling with Power for an Ohmically Heated Discharge Derive the scalings (11.2.60) for a low-pressure symmetric capacitive discharge with collisional (constant mean free path) sheaths in the regime where ohmic heating dominates stochastic heating and where ion energy losses dominate electron energy losses. Ignore ohmic heating in the sheaths.

11.9 Ion and Neutral Radical Densities in a Capacitive Rf Discharge Repeat the analysis leading to the scalings (10.2.31) and (10.2.40) of n_{is} and n_{OS} with P_{abs}, n_g, l, T_e, ω, and A for a low-pressure slab model of a symmetric capacitive rf discharge. Assume that the discharge is electropositive, that stochastic heating dominates ohmic heating, and that ion energy losses dominate electron collisional losses.

11.10 Secondary Electrons in a Low-Pressure Discharge Consider a symmetric, capacitively coupled rf discharge in argon gas having a total applied rf voltage of amplitude 800 V at a frequency $f = \omega/2\pi = 13.56$ MHz. The plate separation is $l = 0.1$ m, and the gas pressure is 5 mTorr. Use a low-pressure discharge model, as in Example 11.1 of Section 11.2.

(a) Sketch the total (rf + dc) potential $\Phi(x, t)$ inside the discharge versus the distance x between the plates, at four times $\omega t = 0, \pi/2, \pi$, and $3\pi/2$.
(b) Because energetic ions bombard the plates, secondary electrons are released which can accelerate to high energies through the sheaths back into the discharge, where they may become electrostatically trapped. Assume that these "hot electrons" have a temperature $T_h \sim 100$ V and are weakly collisional ($\lambda_e \gg l$). Estimate the range of times within an rf cycle $\omega t = 2\pi$ over which secondary electrons will be accelerated to high energies ($T_h \gg T_e$) and become trapped. Estimate the timescale for these hot electrons to be lost from the discharge. To do this, consider the time-varying behavior of the trapping potential $\Phi(x, t)$, as shown by your sketches in (a).
(c) From your answers in (b), and the hot-electron particle conservation law, show that the steady-state hot-electron density can be estimated as $n_h \approx \gamma_{se} n_0 u_B/(\omega l)$, where u_B is the Bohm velocity and γ_{se} is the secondary emission coefficient.

(d) These hot electrons will ionize argon atoms, producing electron–ion pairs. For $\gamma_{se} = 0.1$, over what range of discharge frequencies ω will this mechanism be important compared to ionization by the thermal ($T_e \sim 3$ V) electrons in the discharge? (You will need to use the data shown in Figure 3.16.)

11.11 **Electronegative Discharge** A voltage of $V_{rf} = 500$ V is applied across a plane-parallel device, with plates separated by 6 cm, with an oxygen feedstock gas at $p = 5$ mTorr. Assuming $T_e = 3$ V, stochastic heating dominates ohmic heating, and using (11.2.69) and a value of K_{rec} from Table 8.2, with $d = 4$ cm, calculate n_{e0}, n_{+0}, P_{abs}, and s_m for a discharge with recombination-dominated negative ion loss. Using the value of s_m, obtain a new value of d and recalculate all quantities.

11.12 **Particle and Momentum Balance from Boltzmann's Equation** One spatial dimension (x) and three velocity dimensions $\mathbf{v}$ are commonly used for PIC capacitive discharge simulations. In one spatial dimensional, the Boltzmann equation (2.3.3) for unmagnetized electrons is

$$\frac{\partial f_e}{\partial t} + v_x \frac{\partial f_e}{\partial x} - \frac{e}{m} E_x \frac{\partial f_e}{\partial v_x} = \left(\frac{\partial f_e}{\partial t}\right)_c$$

(a) By integrating the Boltzmann equation overall velocities, show that the macroscopic particle balance is

$$\frac{\partial n_e}{\partial t} + \frac{\partial}{\partial x}(n_e u_e) = G_{iz}$$

where $n_e = \int d^3v f_e$ is the electron density, $u_e = \int d^3v\, v_x f_e$ is the x-component of the electron mean velocity (2.3.5), and $G_{iz} = \int d^3v\, (\partial f_e/\partial t)_c$ is the net electron production rate per unit volume due to collisions.

(b) Multiplying the Boltzmann equation by v_x and integrating overall velocities, show that the macroscopic momentum balance is

$$\frac{\partial}{\partial t}(m n_e u_e) + \frac{\partial}{\partial x}(m n_e u_e^2) = -\frac{\partial p_e}{\partial x} - e n_e E_x - m u_e G_{iz} - C_m$$

where $p_e = m \int d^3v\, v_x^2 f_e - m n_e u_e^2$ is the xx-component of the electron pressure tensor (2.3.10), and $C_m = m \int d^3v\, v_x (\partial f_e/\partial t)_c$ is the electron momentum transfer rate per unit volume due to collisions.

11.13 **Asymmetric High-Frequency Capacitive Discharge** Consider a high-frequency capacitive discharge as in Section 11.4.

(a) Derive the scaling (11.4.11) for an asymmetric high-frequency capacitive discharge with two collisional (constant mean free path) sheaths.

(b) Derive the scaling (11.4.12) for an asymmetric high-frequency capacitive discharge with two homogeneous sheaths.

11.14 **Fourier Series of Charge and Current for Voltage-Driven Sheath** Find the Fourier series for the sheath charge $Q_s(t)$ and the sheath current $I_s(t)$ for a voltage-driven homogeneous sheath with density n, area A, and $V_s(t) = V_0(1 + \cos \omega t)$.

11.15 Dual-Frequency Sheaths with Unequal Edge Densities For homogeneous sheaths with unequal edge densities $n_a \neq n_b$, and assuming that the sum of the time-varying charges within each of the two sheaths is a constant, show that the bias voltage is given by (11.6.10), but with η^2 replaced by the voltage asymmetry ratio ϵ given in (11.6.15).

11.16 Current-Driven Dual-Frequency Discharge Equilibrium Repeat Example 11.3 in Section 11.6 for the current-driven case, with specified values $2 < J_l < 20$ A/m^2, and $20 < J_h < 500$ A/m^2. Plot $\overline{V}_a$ versus J_h for $J_l = 2$, 5, 10, and 20 A/m^2, and plot n versus J_l for $J_h = 20$, 50, 100, 200, and 500 A/m^2. Compare your plots with those in Figure 11.23 and comment on the differences.

11.17 Electrically Symmetric, Geometrically Asymmetric Discharge The EAE can be used to make a modestly asymmetric discharge electrically symmetric.

(a) For the waveform (11.6.31) driving an asymmetric discharge with $\eta = A_a/A_b = 0.85$, show numerically that the bias voltage $V_B = 0$ at a second harmonic phase $\chi = 2.5$.
(b) For $V_B = 0$ and $V_l = V_h = 500$ V, find numerically the ion acceleration energy $\overline{V}_a$, by solving the quadratic equation for $s_a(t)$ in (11.6.12) and then evaluating $\overline{V}_a$ from (11.6.14).

11.18 Rectangular Waveform Frequency Analysis For a rectangular waveform $V_{rf}(t)$ as shown in Figure 11.22e, with a duty ratio of on-time-to-period η, show that

$$V_{rf}(t) = \sum_{m=0}^{\infty} a_m \cos m\omega_1 t$$

where the Fourier coefficients are $a_0 = \eta V_{rf0}$ and

$$a_m = \frac{2V_{rf0}}{m\pi} \sin m\pi\eta, \qquad m \geq 1$$

11.19 Electromagnetic Surface Wave Dispersion Consider the dispersion equations for the symmetric and antisymmetric modes (11.7.13) and (11.7.19), respectively.

(a) For the asymptotic limit $k \to 0$ (assume also, and verify, that $\omega \to 0$ as $k \to 0$), show that $\tanh \alpha_0 s \to \alpha_0 s$ and $\alpha_p \to \omega_{pe}/c$.
(b) Show that the wave phase velocities $v_{ph} = \omega/k$ for $k \to 0$ are

$$v_{ph} = c\left(1 + \frac{c}{s\omega_{pe}} \tanh \frac{\omega_{pe} d'}{c}\right)^{-1/2}$$

and

$$v_{ph} = c\left(1 + \frac{c}{s\omega_{pe}} \coth \frac{\omega_{pe} d'}{c}\right)^{-1/2}$$

for the symmetric and antisymmetric modes, respectively. Compare with (11.7.14), (11.7.15), and (11.7.20).
(c) Consider the asymptotic limit that $k \to \infty$. Show that $\alpha_0 \to k$ and $\alpha_p \to k$ in this limit. Then, show that both the symmetric and antisymmetric modes have the limiting frequency

$$\omega \to \frac{\omega_{pe}}{\sqrt{2}}$$

(d) Numerically solve and plot (ω/ω_{pe} versus kc/ω_{pe}) the symmetric and antisymmetric dispersion equations for sheath width $s = 0.2$ cm, plasma half-width $d' = 1.8$ cm, and density $n = 2 \times 10^{16}$ m^{-3}.

(e) Show that for $\omega_{pe}/\sqrt{2} < \omega < \omega_{pe}$, the symmetric wave is evanescent (non-propagating), and that for $\omega > \omega_{pe}$, it propagates again as a fast wave ($v_{ph} > c$) with $k \approx k_0 (1 - \omega_{pe}^2/\omega^2)^{1/2}$. For $\omega \gg \omega_{pe}$, this corresponds to a conventional transverse electromagnetic wave on a parallel plate transmission line.

11.20 Attenuation of the Low-Density Antisymmetric Wave Mode From transmission line theory (Ramo et al., 1994, chapter 5), for a half-thickness of the low-density antisymmetric wave mode, the capacitance per unit length is $C' = 2\pi r\epsilon_0/s$, the same as for the symmetric wave mode. However, the inductance per unit length, L', is mainly the internal inductance of the bulk plasma, defined as in (11.1.3).

(a) Show that $L' = (2\pi r d' \epsilon_0 \omega_{pe}^2)^{-1}$.

(b) Show that the resistance per unit length of the plasma is $R' = (2\pi r d' \sigma)^{-1}$, with $\sigma = e^2 n_e/m\nu_m$ the dc bulk conductivity.

(c) Show that $R' = \nu_m L'$, as in (11.1.3).

(d) From transmission line theory, the complex propagation constant of the wave is

$$k_i + jk = \pm[(R' + j\omega L')(j\omega C')]^{1/2}$$

Here, k is the wave propagation constant (wavenumber) and k_i is the wave attenuation constant. For $\nu_m/\omega \ll 1$, show that $k \approx \pm\omega/(\omega_{pe} s d')^{1/2}$ and $k_i \approx \pm\frac{1}{2}k\nu_m/\omega$.

11.21 Asymmetric Low-Frequency Capacitive Discharge Consider a low-frequency capacitive discharge as in Section 11.8.

(a) Find V_{a0} and V_{b0} (see Figure 11.39) for a low frequency capacitive discharge with $V_{rf} = 200$ V, $A_b = 2A_a$, and $n_{sb} = n_{sa}$. Use the model of Figure 11.38 and neglect displacement currents.

(b) Determine and plot the ion energy distribution $g_i(\mathcal{E})$ for a collisionless sheath at the powered electrode a.

(c) Determine and plot the ion energy distribution $g_i(\mathcal{E})$ for a collisionless sheath at the grounded electrode b.

11.22 Magnetically Enhanced RF Discharge Derive the scaling results quoted in (11.10.10) for an rf magnetron discharge.

11.23 Design of a Matching Network

(a) For the discharge of Example 11.1, find the effective resistance R_D (in ohms) and capacitance C_D (in farads) for a series RC model of the discharge.

(b) Design an L-type matching network to match the discharge to a 50 Ω rf generator.

11.24 Low-Power Matching Networks Design an L-type matching network to match a discharge impedance $Z_D = R_D + jX_D$ to an rf generator having characteristic impedance R_T, for the low-power case $R_D > R_T$.

12

Inductive Discharges

The limitations of capacitive rf discharges and their magnetically enhanced variants have led to the development of various low-pressure, high-density plasma discharges. The distinction between low- and high-density discharges is described in Sections 10.1 and 10.2. A few examples are shown schematically in Figure 1.16, and typical parameters are given in Table 1.1. In addition to high density and low pressure, a common feature is that the rf or microwave power is coupled to the plasma across a dielectric window or wall, rather than by direct connection to an electrode in the plasma, as for a capacitive discharge. This noncapacitive power transfer is the key to achieving low voltages across all plasma sheaths at electrode and wall surfaces. The dc plasma potential, and, hence, the ion acceleration energy, is then typically 20–40 V at all surfaces. To control the ion energy, the electrode on which the substrate is placed can be independently driven by a capacitively coupled rf source. Hence, independent control of the ion/radical fluxes (through the source power) and the ion-bombarding energy (through the substrate electrode power) is possible. The relation between substrate electrode power and ion-bombarding energy at the substrate is described in Chapter 11. High-density inductive rf discharges are described in this chapter, and high-density wave heated discharges are described in Chapter 13. Inductive discharges operated at high densities and low pressures, which are driven at frequencies below the self-resonant frequency of the exciting coil, are described in Section 12.1. Other operating regimes, power transfer considerations, and the resonant coil *helical resonator* configuration are described in Section 12.2. The planar coil configuration, which is commonly used for materials processing, is described in Section 12.3, with examples of experimental measurements and two-dimensional simulations. Finally, low rf frequency planar discharges, and means for improving the coupling of the planar coil to the plasma, are described in Section 12.4. Some other aspects of high-density discharges, including issues of plasma transport and substrate damage, are described in a review article by Lieberman and Gottscho (1994), from which some of the materials in Chapters 12 and 13 are drawn.

12.1 High-Density, Low-Pressure Discharges

Inductive discharges are nearly as old as the invention of electric power, with the first report of an "electrodeless ring discharge" by Hittorf (1884). He wrapped a coil around an evacuated tube and observed a discharge when the coil was excited with a Leyden jar. A subsequent 50-year controversy developed as to whether these discharges were capacitively driven by plasma coupling to the low- and high-voltage ends of the cylindrical coil, as in a capacitive discharge (see Chapter 11), or were driven by the induced electric field inside the coil. This issue was resolved with the recognition

Principles of Plasma Discharges and Materials Processing, Third Edition. Michael A. Lieberman and Allan J. Lichtenberg.

that the discharge was capacitively driven at low plasma densities, with a transition to an inductive mode of operation at high densities. Succeeding developments, which focused on pressures exceeding 20 mTorr in a cylindrical coil geometry, are described in a review article by Eckert (1986). The high-pressure regime was intensively developed in the 1970s with the invention of the open air induction torch and its use for spectroscopy. In the late 1980s, the planar coil configuration was developed, renewing interest in the use of high-density inductive discharges for materials processing at low pressures (< 50 mTorr) and in low aspect ratio geometries ($l/R \lesssim 1$ for a cylindrical discharge). Such discharges can be driven with either planar or cylindrical coils. It is this regime that is the primary focus here.

Plasma in an inductive discharge is created by the application of rf power to a nonresonant inductive coil. Inductive sources have potential advantages over high-density wave-heated sources, including simplicity of concept, no requirement for dc magnetic fields (as required for ECRs and helicons, discussed in Chapter 13), and rf rather than microwave source power.

In contrast to ECRs and helicons, which can be configured to achieve densities $n_0 \gtrsim 10^{13}$ cm^{-3}, as we will see in Chapter 13, inductive discharges may have natural density limits, $n_0 \lesssim 10^{13}$ cm^{-3}, for efficient power transfer to the plasma. However, the density regime $10^{11} \lesssim n_0 \lesssim 10^{12}$ cm^{-3} for efficient inductive discharge operation, typically a factor of 10 times higher than for capacitive discharges, is of considerable interest for low-pressure processing. Inductive discharges for materials processing are usually referred to as ICPs (inductively coupled plasmas); sometimes the names TCPs (transformer-coupled plasmas) or RFI plasmas (rf inductive plasmas) are used.

12.1.1 Inductive Source Configurations

The two coil configurations, cylindrical and planar, are shown in Figure 12.1 for a low aspect ratio ($l/R \lesssim 1$) discharge. The planar coil is a set of concentric circular conductors, or a flat helix wound from near the axis to near the outer radius of the discharge chamber ("electric stovetop" coil shape). Operating with $l \sim R$ can provide reasonable radial uniformity. Inner and outer coils separately powered, or nonuniformly wound coils, allow additional uniformity control. Multipole permanent magnets (see Section 5.5) can also be used around the process chamber circumference, as shown in Figure 12.1*b*, to increase radial uniformity.

Inductive coils are commonly driven at 13.56 MHz or below, using a 50-Ω rf supply through a capacitive matching network, which we describe later in this section. The coil can also be driven push–pull using a balanced transformer, which places a virtual ground in the middle of the coil and reduces the maximum coil-to-plasma voltage by a factor of two. This reduces the undesired capacitively coupled rf current flowing from coil to plasma by a factor of two. An electrostatic shield placed between the coil and the plasma further reduces the capacitive coupling if desired, while allowing the inductive field to couple unhindered to the plasma. For the dc plasma potential to be clamped to a low value, 20–40 V, in the presence of stray capacitive coupling from the exciting coil and from the capacitively driven substrate holder, it is essential that the plasma be in contact with or strongly capacitively coupled to a grounded metal surface of substantial area (see Figure 12.1 and Section 11.4).

12.1.2 Power Absorption and Operating Regimes

In an inductively coupled plasma, power is transferred from the electric fields to the plasma electrons within a skin depth layer of thickness δ near the plasma surface by collisional (ohmic) dissipation and by a collisionless heating process in which bulk plasma electrons "collide" with the

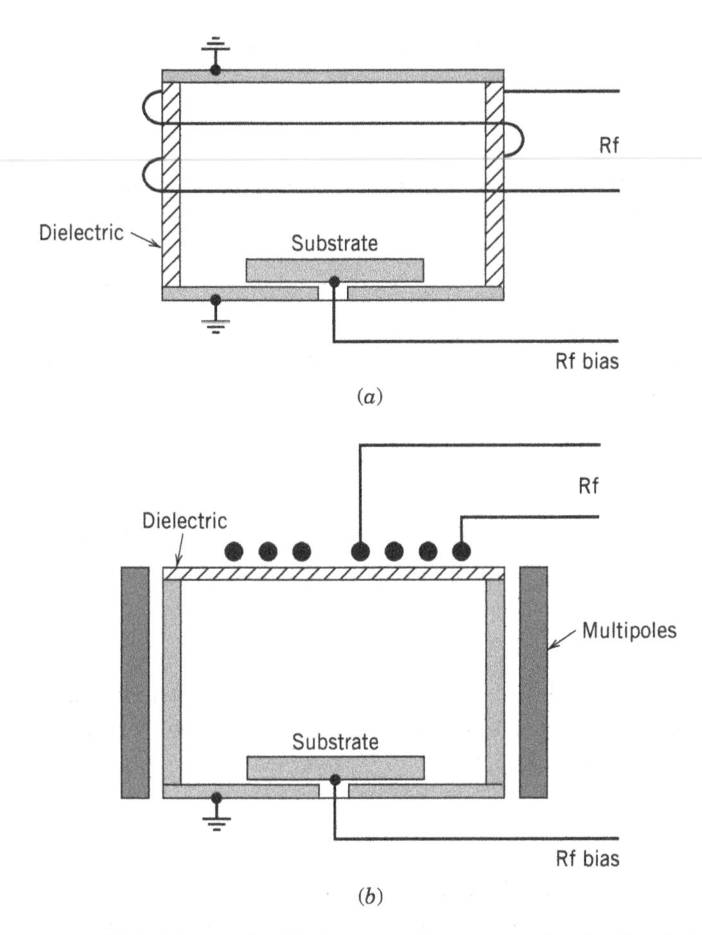

Figure 12.1 Schematic of inductively driven sources in (*a*) cylindrical and (*b*) planar geometries.

oscillating inductive electric fields within the layer. In the latter situation, electrons are accelerated and subsequently thermalized much like stochastic heating in capacitive rf sheaths, which we discussed in Section 11.1. We first consider the ohmic heating process.

The spatial decay constant α within a plasma for an electromagnetic wave normally incident on the boundary of a uniform density plasma can be calculated as discussed in Section 4.2 and is (Problem 12.1)

$$\alpha = -\frac{\omega}{c}\mathrm{Im}\left(\kappa_{\mathrm{p}}^{1/2}\right) \equiv \delta^{-1} \tag{12.1.1}$$

From (4.2.18), the relative plasma dielectric constant is

$$\kappa_{\mathrm{p}} = 1 - \frac{\omega_{\mathrm{pe}}^2}{\omega(\omega - j\nu_{\mathrm{m}})} \approx -\frac{\omega_{\mathrm{pe}}^2}{\omega^2(1 - j\nu_{\mathrm{m}}/\omega)} \tag{12.1.2}$$

with ω_{pe} the plasma frequency near the boundary, and ν_{m} the electron–neutral momentum transfer frequency. There are two collisionality regimes.

(a) For $\nu_{\mathrm{m}} \ll \omega$, we drop ν_{m}/ω in (12.1.2) to obtain

$$\alpha = \frac{\omega_{\mathrm{pe}}}{c} \equiv \frac{1}{\delta_{\mathrm{p}}} \tag{12.1.3}$$

where δ_p is the collisionless skin depth. Substituting for ω_{pe} into (12.1.3), we find

$$\delta_p = \left(\frac{m}{e^2\mu_0 n_s}\right)^{1/2} \tag{12.1.4}$$

(b) For $\nu_m \gg \omega$, we drop the 1 in the parentheses of (12.1.2), expanding for $\nu_m \gg \omega$, and substituting the imaginary part of $\kappa_p^{1/2}$ into (12.1.1), we obtain

$$\alpha = \frac{1}{\sqrt{2}}\frac{\omega_{pe}}{c}\left(\frac{\omega}{\nu_m}\right)^{1/2} \equiv \frac{1}{\delta_c} \tag{12.1.5}$$

where δ_c is the collisional skin depth. Substituting for the dc conductivity $\sigma_{dc} = e^2 n_s/m\nu_m$ from (4.2.22), δ_c can be written in the forms

$$\delta_c = \delta_p\left(\frac{2\nu_m}{\omega}\right)^{1/2} = \left(\frac{2}{\omega\mu_0\sigma_{dc}}\right)^{1/2} \tag{12.1.6}$$

(c) There is a third situation (Weibel, 1967; Turner, 1993) for which electrons incident on a skin layer of thickness δ_e satisfy the condition

$$\frac{\overline{v}_e}{2\delta_e} \gg \omega, \nu_m \tag{12.1.7}$$

where δ_e is determined below. In this case, the interaction time of the electrons with the skin layer is short compared to the rf period or the collision time. In analogy to collisionless heating at a capacitive sheath, a stochastic collision frequency (19.5.12) can be defined

$$\nu_{stoc} = \frac{C_e\overline{v}_e}{4\delta_e} \tag{12.1.8}$$

where C_e is a quantity of order unity that depends weakly on $\overline{v}_e$, δ_e, and ω, provided the ordering (12.1.7) is satisfied (see Section 19.5). We then substitute ν_{stoc} for ν_m in (12.1.2) and expand for $\nu_{stoc} \gg \omega$ as in (b), to obtain

$$\delta_e = \frac{c}{\omega_{pe}}\left(\frac{C_e\overline{v}_e}{2\omega\delta_e}\right)^{1/2}$$

Solving for δ_e, we find

$$\delta_e = \left(\frac{C_e c^2\overline{v}_e}{2\omega\omega_{pe}^2}\right)^{1/3} = \left(\frac{C_e\overline{v}_e}{2\omega\delta_p}\right)^{1/3}\delta_p \tag{12.1.9}$$

where δ_e is the *anomalous skin depth* (see Alexandrov et al., 1984).

At 13.56 MHz in argon, we find $\nu_m = \omega$ for $p^* \approx 25$ mTorr. We are interested primarily in the low-pressure regimes with $p \ll p^*$, which we consider first. For each pressure regime, we also distinguish two density regimes:

(a) High density, $\delta \ll R, l$
(b) Low density, $\delta \gtrsim R, l$

For typical low-pressure processing discharges, we are generally in the regime for which the frequency ordering is $\omega \sim \overline{v}_e/2\delta \gtrsim \nu_m$ and such that the skin depth is approximately δ_p. For typical plasma dimensions $R, l \sim 10$ cm, we are in the high-density regime. We shall discuss the high-pressure ($\nu_m \gg \omega$) and low-density ($\delta \gg R, l$) regimes later when we consider the minimum current and power necessary to generate an inductively coupled plasma.

12.1.3 Discharge Operation and Coupling

Although many systems are operated with planar coils (see Figure 12.1*b*), finite geometry effects make these configurations difficult to analyze. To illustrate the general principles of inductive source operation, we examine a uniform density cylindrical discharge (Figure 12.1*a*) in the geometry $l \gtrsim R$. We take the coil to have $\mathcal{N}$ turns at radius $b > R$. To determine the equivalent circuit elements, we integrate the time-averaged complex power (see Ramo et al., 1994, chapter 3) deposited by the field into the plasma, with $\delta_p \ll R$, obtaining (Problem 12.2)

$$\tilde{P}_{cmplx} = P_{abs} + jP_{reac} = \frac{1}{2}\frac{J_{\theta 0}^2}{\sigma_p}\pi R l \delta_p \tag{12.1.10}$$

where P_{abs} and P_{reac} are the real and imaginary parts of $\tilde{P}_{cmplx}$, respectively, and $J_{\theta 0}$ is the amplitude of the induced rf azimuthal current density at the plasma edge $r = R$ (opposite in direction to the applied azimuthal current in the coil). In analogy to the complex plasma conductivity in (4.2.20),

$$\sigma_p = \frac{e^2 n_s}{m(\nu_{eff} + j\omega)} \tag{12.1.11}$$

with $\nu_{eff} = \nu_m + \nu_{stoc}$ as in (19.5.15), a sum of collisional and stochastic heating. An estimate for ν_{stoc} is given in (19.5.14) or Figure 19.6. The real part of (12.1.10) is equivalently the resistive power deposited for a uniform sinusoidal current density $J_{\theta 0}$ flowing within a skin thickness δ_p. Let $I_p = J_{\theta 0} l \delta_p$ be the total induced rf current amplitude in the plasma skin and defining the plasma resistance through $P_{abs} = \frac{1}{2}I_p^2 R_p$, we obtain

$$R_p = \frac{\pi R}{\sigma_{eff} l \delta_p} \tag{12.1.12}$$

where

$$\sigma_{eff} = \frac{e^2 n_s}{m\nu_{eff}} \tag{12.1.13}$$

is the effective dc plasma conductivity. Note that the factor πR rather than $2\pi R$ appears in the numerator of (12.1.12) because $J_\theta^2(r)$ has a radial decay length into the plasma of $\delta_p/2$, not δ_p (see Problem 12.2).

As in (11.1.3) (see also Figure 11.4), there is also a plasma inductance L_p, which accounts for the phase lag between the rf electric field and the rf conduction current due to the finite electron inertia. Defining L_p through the imaginary part of (12.1.10), $P_{reac} = \frac{1}{2}I_p^2\omega L_p$, we obtain

$$L_p = \frac{R_p}{\nu_{eff}} \tag{12.1.14}$$

In addition to L_p, there is the usual magnetic energy storage inductance L_{22}, because the rf plasma current creates a magnetic flux $\Phi_p = L_{22} I_p$ linked by the current. Using $\Phi_p = \mu_0 \pi R^2 H_z$, where $H_z = J_\theta \delta_p$ is the magnetic field produced by the skin current, we obtain

$$L_{22} = \frac{\mu_0 \pi R^2}{l} \tag{12.1.15}$$

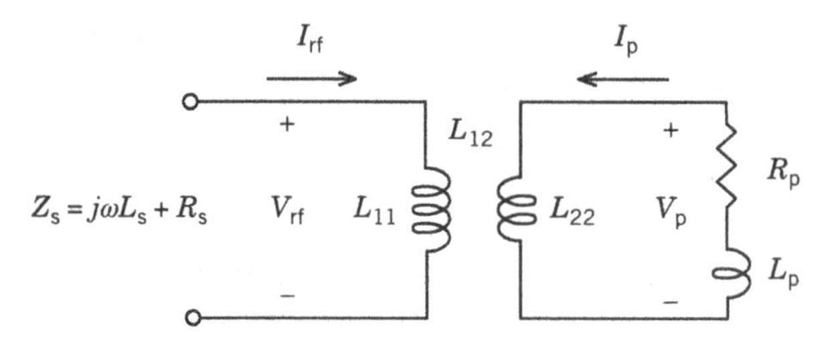

Figure 12.2 Equivalent transformer-coupled circuit model of an inductive discharge.

Let the coil have $\mathcal{N}$ turns at a radius $b \gtrsim R$, where $b - R$ is the "thickness" of the dielectric interface separating coil and plasma, then we can model the source as the transformer shown in Figure 12.2. Evaluating the inductance matrix for this transformer, defined by (Schwarz and Oldham, 1984)

$$\tilde{V}_{rf} = j\omega L_{11}\tilde{I}_{rf} + j\omega L_{12}\tilde{I}_p \tag{12.1.16}$$

$$\tilde{V}_p = j\omega L_{21}\tilde{I}_{rf} + j\omega L_{22}\tilde{I}_p \tag{12.1.17}$$

where the tildes denote the complex amplitudes, e.g., $V_{rf}(t) = \text{Re}\,\tilde{V}_{rf}\,e^{j\omega t}$, we obtain (Problem 12.3)

$$L_{11} = \frac{\mu_0 \pi b^2 \mathcal{N}^2}{l} \tag{12.1.18}$$

$$L_{12} = L_{21} = \frac{\mu_0 \pi R^2 \mathcal{N}}{l} \tag{12.1.19}$$

Using $\tilde{V}_p = -\tilde{I}_p(R_p + j\omega L_p)$ (see Figure 12.2) in (12.1.17) and inserting into (12.1.16), we can solve for the impedance seen at the coil terminals:

$$Z_s = \frac{\tilde{V}_{rf}}{\tilde{I}_{rf}} = j\omega L_{11} + \frac{\omega^2 L_{12}^2}{R_p + j\omega(L_{22} + L_p)} \tag{12.1.20}$$

We will assume the usual high-density ordering $\delta_p \sim \delta_c \ll R$ for the validity of (12.1.12)–(12.1.15). With this ordering, it can easily be seen from (12.1.12)–(12.1.15) that $R_p^2 + \omega^2 L_p^2 \ll \omega^2 L_{22}^2$. Hence, expanding the denominator in (12.1.20), we obtain

$$L_s \approx \frac{\mu_0 \pi R^2 \mathcal{N}^2}{l}\left(\frac{b^2}{R^2} - 1\right) \tag{12.1.21}$$

$$R_s \approx \mathcal{N}^2 \frac{\pi R}{\sigma_{eff} l \delta_p} \tag{12.1.22}$$

where $Z_s = R_s + j\omega L_s$. The power balance,

$$P_{abs} = \frac{1}{2}|\tilde{I}_{rf}|^2 R_s \tag{12.1.23}$$

then yields the required rf source current, and the rf voltage is determined from

$$\tilde{V}_{rf} = \tilde{I}_{rf}|Z_s| \tag{12.1.24}$$

Example 12.1 We let $R = 10$ cm, $b = 15$ cm, $l = 20$ cm, $\mathcal{N} = 3$ turns, $n_g = 1.7 \times 10^{14}$ cm^{-3} (5 mTorr argon at 298 K), $\omega = 8.5 \times 10^7$ s^{-1} (13.56 MHz), and $P_{abs} = 600$ W. At 5 mTorr, $\lambda_i \approx 0.6$ cm. Then from (10.2.1) and (10.2.2) $h_l \approx 0.20$, $h_R \approx 0.18$, and from (10.2.13), $d_{eff} \approx 18.2$ cm. For argon, we then obtain from Figure 10.1 that $\mathrm{T_e} \approx 2.8$ V, and from Figure 3.17, that $\mathcal{E}_c \approx 68$ V. From (10.2.4), we obtain $\mathcal{E}_i + 2\mathrm{T_e} \approx 20$ V, and using (10.2.9), we find $\mathcal{E}_T \approx 88$ V. The Bohm velocity is

$u_B \approx 2.6 \times 10^5$ cm/s, and from (10.2.11), $A_{\text{eff}} \approx 340$ cm^2. Then from (10.2.15), we obtain $n_0 \approx 4.8 \times 10^{11}$ cm^{-3} and $n_s = h_R n_0 \approx 9.3 \times 10^{10}$ cm^{-3}. Estimating ν_m for argon from Figure 3.16, we find $\nu_m \approx 1.2 \times 10^7$ s^{-1}. Using (12.1.4), we find $\delta_p \approx 1.7$ cm. Evaluating ν_{stoc}, we first find $\alpha \approx 2.2$ from (19.5.7). Then with $\overline{v}_e \approx 1.1 \times 10^8$ cm/s and $\delta = \delta_p$, we obtain $\nu_{\text{stoc}} \approx 2.8 \times 10^7$ s^{-1} from Figure 19.6 or (19.5.14), such that $\nu_{\text{eff}} \approx 4.0 \times 10^7$ s^{-1}. Using this in (12.1.13), we find $\sigma_{\text{eff}} \approx 66$ mho/m. Evaluating (12.1.22) and (12.1.21), we find $R_s \approx 12.3$ Ω and $L_s \approx 2.2$ μH, such that $\omega L_s \approx 190$ Ω. Equations (12.1.23) and (12.1.24) then yield $I_{\text{rf}} \approx 9.9$ A and $V_{\text{rf}} \approx 1870$ V.

We note that $\omega > \nu_{\text{eff}}$ for this example, such that $\delta \approx \delta_p$, the collisionless skin depth, verifying our assumed ordering. We also note that stochastic heating somewhat dominates at this pressure: $\nu_{\text{stoc}} \sim 2\nu_m$. Godyak et al. (1993) have measured ν_{eff} in an inductive discharge, finding that ν_{eff} is independent of pressure at low pressures, indicating the dominance of stochastic over collisional heating. The measured ν_{eff} was $\sim 2\overline{v}_e/\delta_e$. For this experiment, $\delta_e \sim \delta_p \sim \delta_c$, so the scale length dependence could not be distinguished.

Because stochastically heated electrons are "kicked" in the skin layer and flow back into the bulk plasma with a characteristic thermal speed v_{th}, which is generally small compared to the phase velocity v_{ph} of the wave, the electron current (carried by v_{th}) can become out-of-phase with the electric field (carried by v_{ph}) downstream from the skin layer, giving rise to regions of negative $\tilde{\mathbf{J}} \cdot \tilde{\mathbf{E}}$. For a phase change an odd multiple of π, $\omega t - kz = (2i-1)\pi$, we would expect a series of negative $\tilde{\mathbf{J}} \cdot \tilde{\mathbf{E}}$ regions centered about positions

$$z_i = \frac{\pi(2i-1)}{\dfrac{\omega}{v_{\text{th}}} - \dfrac{\omega}{v_{\text{ph}}}}, \qquad i = 1, 2, \ldots$$

as has been observed experimentally (Godyak and Kolobov, 1997).

The rf magnetic field scales as $\tilde{\mathbf{B}} \sim \tilde{\mathbf{E}}/\omega\delta$ from (2.2.1), and the rf electron velocity scales as $\tilde{\mathbf{u}}_e \sim e\tilde{\mathbf{E}}/m\omega$ from (2.3.9). At low frequencies, $\tilde{\mathbf{B}}$ and $\tilde{\mathbf{u}}_e$ become large, leading to significant nonlinear inertial and Lorentz forces ($\tilde{\mathbf{u}}_e \cdot \nabla\tilde{\mathbf{u}}_e$ and $\tilde{\mathbf{u}}_e \times \tilde{\mathbf{B}}$ terms in (2.3.9)). These give rise to second harmonic (2ω) rf currents and to ponderomotive forces (Smolyakov et al., 2003); the latter can expel low-energy electrons from the skin layer. We consider these phenomena further in Section 12.4.

Under some conditions, the electron drift velocity associated with the induced rf plasma current within the skin depth layer can be larger than the electron thermal velocity. This drifting Maxwellian distribution with large mean energy can produce an increased ionization, leading to a lowering of T_e and significant changes in the density profile over that found for the global (constant T_e) ionization model (10.2.12) (see Problem 12.4b).

12.1.4 Matching Network and Power Measurements

The high inductive voltage required for a three-turn coil can be supplied from a 50-Ω rf power source through a capacitive matching network, as shown in Figure 12.3. The admittance looking to the right at the terminals A–A′ is

$$Y_A \equiv G_A + jB_A = \frac{1}{R_s + j(X_1 + X_s)} \tag{12.1.25}$$

where the conductance is

$$G_A = \frac{R_s}{R_s^2 + (X_1 + X_s)^2} \tag{12.1.26}$$

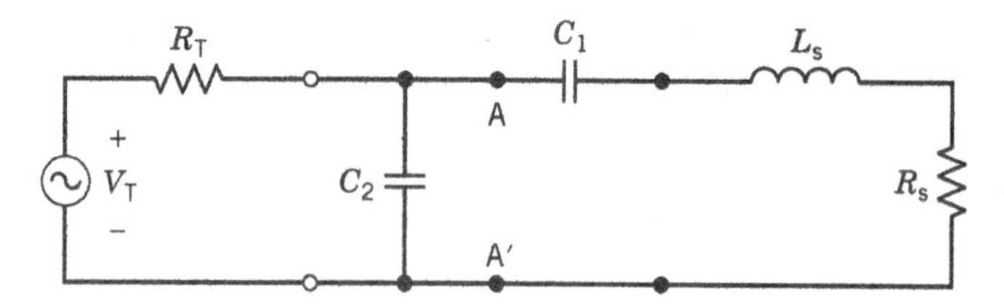

Figure 12.3 Equivalent circuit for matching an inductive discharge to a power source.

and the susceptance is

$$B_A = -\frac{X_1 + X_s}{R_s^2 + (X_1 + X_s)^2} \tag{12.1.27}$$

and where $X_1 = -(\omega C_1)^{-1}$. As described in Section 11.11, we must choose G_A to be equal to $1/R_T$, for maximum power transfer, where $R_T = 50\ \Omega$ is the Thevenin-equivalent source resistance. For $R_s \approx 12.3\ \Omega$ and $X_s \approx 190\ \Omega$, we obtain from (12.1.26) that $X_1 \approx -168\ \Omega$. Hence, $C_1 \approx 70$ pF. Evaluating B_A for this value of X_1, we obtain $B_A \approx -0.035\ \Omega^{-1}$. We must choose C_2 to cancel this susceptance, i.e., $B_2 = \omega C_2 = -B_A$, which determines $C_2 = 410$ pF to achieve the matched condition. In practice, C_1 and C_2 are variable capacitors that are tuned to achieve the match. The power absorption, $P_{abs} = \frac{1}{2}I_T^2 R_T$, then determines $I_T \approx 4.9$ A and $V_T = 2I_T R_T \approx 490$ V.

12.1.4.1 Power Measurements

As for capacitive discharges, as described in Section 11.11, an effective procedure to determine the power absorbed by the discharge alone if the voltage and current waveforms are reasonably sinusoidal in shape is as follows:

1. With the discharge ignited, measure the coil current I_{rf} and the source power P_T. This is the power absorbed by the lossy coil, the matching network and the discharge.
2. Extinguish the discharge by raising the pressure in the chamber to 1 atm or by reducing the pressure to a very low value. The current measured will in general change.
3. Readjust the rf source so that I_{rf} is the same as that measured in (1) and measure the source power $P_T^{(0)}$. The matching capacitor can be retuned if desired, but a perfect match is not necessary. This measurement yields the power absorbed by the lossy coil and lossy matching network only.

Because I_{rf} is the same with and without the plasma present, the same power is lost in the coil and matching network with and without the plasma present. Therefore, the power absorbed by the plasma is given by $P_T - P_T^{(0)}$, as in (11.11.23). For further information about power measurements in rf inductive discharges, see Hwang et al. (2013), Godyak and Alexandrovich (2017), and Godyak (2021).

12.2 Other Operating Regimes

12.2.1 Low-Density Operation

Since the effective conductivity $\sigma_{eff} \propto n_0$ and $\delta_p \propto n_0^{-1/2}$, it follows from (12.1.22) and (12.1.23) that at high densities:

$$P_{abs} \propto n_0^{-1/2} I_{rf}^2 \tag{12.2.1}$$

Hence, at fixed I_{rf}, we have that $P_{abs} \propto n_0^{-1/2}$. However, at low densities, such that $\delta_p \gg R$, the conductivity is low and the fields fully penetrate the plasma. In this case, expressions (12.1.12)–(12.1.15) are no longer correct. To find the absorbed power for this case, we apply Faraday's law to determine the induced electric field E_θ within the coil

$$E_\theta(r) = \frac{12j\omega r\mu_0 \mathcal{N} I_{rf}}{l} \tag{12.2.2}$$

and, writing $J_\theta = j\omega\epsilon_0\kappa_p E_\theta$ for $\nu_m \ll \omega$, we have $J_\theta \propto n_0 r I_{rf}$. Evaluating the power absorbed, we have

$$\begin{aligned} P_{abs} &= \frac{1}{2}\int_0^R \frac{J_\theta^2(r)}{\sigma_{eff}} 2\pi r l\,dr \\ &= \frac{1}{2} I_{rf}^2 \frac{\pi e^2 n_0 \nu_{eff} \mu_0^2 \mathcal{N}^2 R^4}{8ml} \end{aligned} \tag{12.2.3}$$

such that

$$P_{abs} \propto n_0 I_{rf}^2 \tag{12.2.4}$$

In this low-density regime where the fields fully penetrate the plasma, the power absorbed is simply proportional to the number (density) of electrons in the discharge. Comparing (12.2.1) with (12.2.4) and holding I_{rf} fixed, we see that P_{abs} versus n_0 has a maximum near $\delta_p \sim R$. This corresponds to a variation with density as sketched as the solid curves in Figure 12.4 for two different values of I_{rf}. Now consider the power balance requirement (10.2.14), which is plotted as a straight line in the figure. The intersection of this line with each of the solid curves defines the equilibrium point for inductive discharge operation for that particular value of I_{rf}. The intersection shown at $I_{rf} > I_{min}$ gives an inductive mode equilibrium. We see that inductive source operation is impossible if the source current I_{rf} lies below some minimum value I_{min}. However, a weak capacitive discharge can exist for $I_{rf} < I_{min}$, as we describe in the following subsection.

12.2.2 Capacitive Coupling

At this point, the reader might ask: since a high voltage $V_{rf} \approx 1870$ volts exists at the high-voltage end of the coil, what is the effect of capacitive coupling on the discharge? We will see below that for high densities, only a small fraction of V_{rf} appears across the sheath, such that the additional

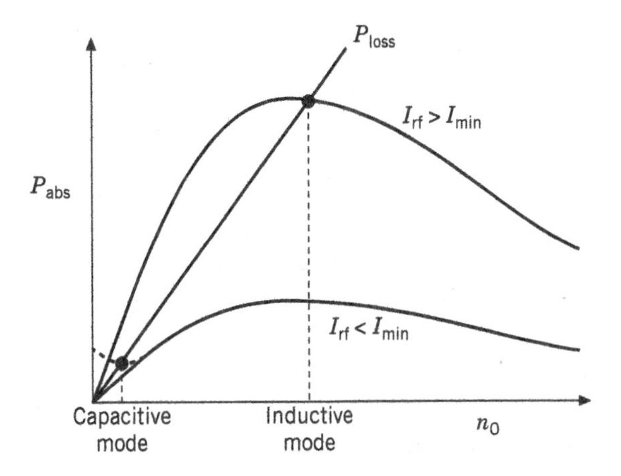

Figure 12.4 Absorbed power versus density from the inductive source characteristics (curves) for two different values of the driving current I_{rf}, and power lost versus density (straight line); the dotted curve includes the additional capacitive power at low density for $I_{rf} < I_{min}$.

ion (and electron) energy loss is small. However, at low densities, the capacitive coupling can be the major source of power deposition.

To estimate the rf voltage across the sheath, $\tilde{V}_{sh}$, at the high-voltage end of the coil, we note that the sheath capacitance per unit area is $\sim \epsilon_0/s_m$ and the capacitance per unit area of the dielectric cylinder is $\sim \epsilon_0/(b-R)$. Assuming that the plasma is at ground potential, then the voltage across the sheath is found from the capacitive voltage divider formula,

$$\tilde{V}_{sh} = V_{rf}\frac{s_m}{b-R+s_m} \tag{12.2.5}$$

Using the modified Child law (11.2.15), we calculate the sheath thickness from

$$en_s u_B = 0.82\,\epsilon_0\left(\frac{2e}{M}\right)^{1/2} V_{rf}^{3/2}\left(\frac{s_m}{b-R+s_m}\right)^{3/2}\frac{1}{s_m^2} \tag{12.2.6}$$

which is a quartic equation in s_m. However, for high densities for which $s_m \ll b-R$, (12.2.6) simplifies to

$$s_m \approx \left(\frac{0.82\,\epsilon_0}{en_s u_B}\right)^2\left(\frac{2e}{M}\right)\frac{V_{rf}^3}{(b-R)^3} \tag{12.2.7}$$

The right-hand side is generally small for the usual voltages of inductive discharges, so that s_m is much smaller than in a capacitive discharge. In our example, we find $s_m \approx 6.4\times10^{-4}$ cm, so that, from (12.2.5), $\tilde{V}_{sh} \approx 0.22$ V. Actually, for a sheath this thin, the high-voltage sheath relation (12.2.6) is not valid. From (2.4.23), the Debye length is $\lambda_{De} \approx 3.8\times10^{-3}$ cm. The sheath is a few Debye lengths thick. Using calculations for capacitive discharges (see Godyak and Sternberg, 1990b), we estimate $s_m \sim 2\times10^{-2}$ cm, such that (12.2.5) yields $\tilde{V}_{sh} \sim 9$ V, which contributes only a small correction to the dc sheath voltage.

From relations (12.1.21)–(12.1.24), we can see one reason why the designer wants to keep the number of turns of the exciting coil small in a high-density discharge. From (12.1.21) and (12.1.22), we see that $Z_s \propto \mathcal{N}^2$, and at fixed P_{abs}, from (12.1.23), we find $\tilde{I}_{rf} \propto 1/\mathcal{N}$. Then, (12.1.24) gives $\tilde{V}_{rf} \propto \mathcal{N}$. From (12.2.7), we see that $s_m \propto \mathcal{N}^3$ at a fixed density, such that doubling $\mathcal{N}$ would increase s_m by almost an order of magnitude. The increased ion energy loss across this larger sheath (increased $\mathcal{E}_T$) leads to lower density and generally less favorable discharge parameters. We examine the frequency dependence of the high-density capacitive coupling in Section 12.4.

In contrast to the high-density case, at low densities, the sheath width s_m becomes comparable to or larger than the vacuum (or dielectric) window gap width $b-R$, and from (12.2.5) most of V_{rf} can be dropped across the sheath. From (11.2.33) or (11.2.34), the capacitive power increases with increasing $\tilde{V}_{sh}$, and from (12.2.5), $\tilde{V}_{sh}$ increases with decreasing density n_0 (increasing sheath thickness s_m). Therefore, the capacitive power absorbed increases with decreasing n_0. In this regime, any discharge must be capacitively driven. Including the additional capacitively coupled power due to the sheath voltage (12.2.5) gives the dotted curve in Figure 12.4 at low densities instead of the solid curve for $I_{rf} < I_{min}$. There is an intersection with the P_{loss} line and, therefore, a capacitive mode equilibrium at low densities. Increasing I_{rf} from below I_{min}, this low-density capacitive (E) mode plasma makes a relatively abrupt "E-to-H" transition to a high-density inductive (H) mode when I_{rf} exceeds I_{min}. Decreasing I_{rf} in inductive mode results in a similar inductive-to-capacitive transition when I_{rf} falls below I_{min}. Two-dimensional fluid simulations of this mode transition, including the full electromagnetics, are given by Zhao et al. (2009) and Kawamura et al. (2011).

The capacitive coupling can be convenient for start-up of an inductive discharge, as the ignition can rely on a high voltage in the discharge chamber, before the high-density inductive plasma is formed. However, it prevents inductive operation at low densities and can give rise to instabilities, as described in the next subsection. Some possible means to overcome the low-density and instability limitations are described in Section 12.4.

12.2.3 Hysteresis and Instabilities

Various additional plasma and circuit effects can produce a *hysteresis* in the discharge characteristics, in which the capacitive-to-inductive transition occurs at a higher rf coil current $I_{\text{min}\,2}$ than the inductive-to-capacitive transition at $I_{\text{min}\,1}$. For the range of currents between $I_{\text{min}\,1}$ and $I_{\text{min}\,2}$, the mode actually present depends on the history of the system. Hysteresis of discharge characteristics is often seen experimentally, and an explanation in terms of power balance arguments has been given by Turner and Lieberman (1999). For example, considering again the power balance curves shown in Figure 12.4, with capacitive coupling present, it can be seen (Problem 12.5) that there is only a single intersection of the P_{abs} and P_{loss} curves as the current is varied, yielding a discharge characteristic without hysteresis. However, if the P_{loss} versus n_{e} curve is not linear but has a convex curvature, or if the curve is linear but is displaced upward from the origin ($P_{\text{loss}} > 0$ for $n_{\text{e}} = 0$), then there is a range of currents where there are three intersections, such that the discharge characteristic has hysteresis. The low-density intersection in a stable capacitive equilibrium, the high-density intersection is a stable inductive equilibrium, and the intermediate-density intersection is an unstable equilibrium. Mechanisms that can produce a convex curvature (nonlinearity) for the P_{loss} curve include multi-step ionization, electron distribution function changes due to electron–electron collisions, and a reduction in inductive coupling due to a capacitive rf sheath (Turner and Lieberman, 1999). An extensive review of mode transitions and hysteresis is given in Lee (2018).

A modified P_{loss} curve can also be produced by the presence of negative ions in the discharge. In this latter case, the additional dynamics of negative ion generation and destruction can result in an "E-to-H transition" instability, in which there is no stable discharge equilibrium. Experimentally, it is found that if the plasma contains negative ions, e.g., from feedstock gases such as O_2, SF_6, Cl_2, and CF_4, that over a significant power range around the transition between lower power capacitive operation and higher power inductive operation, there is a relaxation oscillation between high- and low-density modes. For example, an experiment to investigate these instabilities was performed in an Ar/SF_6 (1:1) gas mixture in a device 30 cm in diameter and 19 cm long, with a three turn planar coil driven at 13.56 MHz (see Section 12.3). At a pressure of 5 mTorr, with an average absorbed power of 550 W, the relaxation oscillation shown in Figure 12.5 was found (Chabert et al., 2001). Varying the power at this pressure, a range of oscillatory (unstable) behavior was found between $P_{\text{abs}} = 400$ W and $P_{\text{abs}} = 700$ W. Above 700 W, the plasma was stable in the inductive mode with ion and electron densities in the higher-density range, and, below 400 W, the ion and electron densities were more than a factor of 10 lower, characteristic of capacitively driven operation.

To analyze the process, three time varying equations of an electronegative plasma for positive ion, negative ion, and electron energy balance were solved, using a global model (see Chapter 10), together with the conditions of quasi-neutrality and Boltzmann electrons. However, we can understand the instability mechanism in a rather straightforward way from the electron energy balance alone. As for pulsed power discharges in (10.6.5), we have

$$\frac{\mathrm{d}}{\mathrm{d}t}\left(\frac{3}{2}en_{\text{e}}\mathrm{T}_{\text{e}}\right) = P_{\text{abs}} - P_{\text{loss}} \tag{12.2.8}$$

For the losses, in keeping with a global approximation, we use the RHS of (11.2.69), but we substitute for $\frac{8}{15}K_{\text{rec}}n_+^2$, using (10.4.31), and consider $\alpha_0 = n_-/n_{\text{e}} \gg 1$, such that $n_- \approx n_+$, to obtain

$$P_{\text{loss}} = K_{\text{iz}}e\mathcal{E}_{\text{c}}dAn_{\text{g}}n_{\text{e}} + h_l u_{\text{B}\alpha}e(4.8\,\mathrm{T}_{\text{e}} + 2\mathrm{T}_{\text{e}})An_- \tag{12.2.9}$$

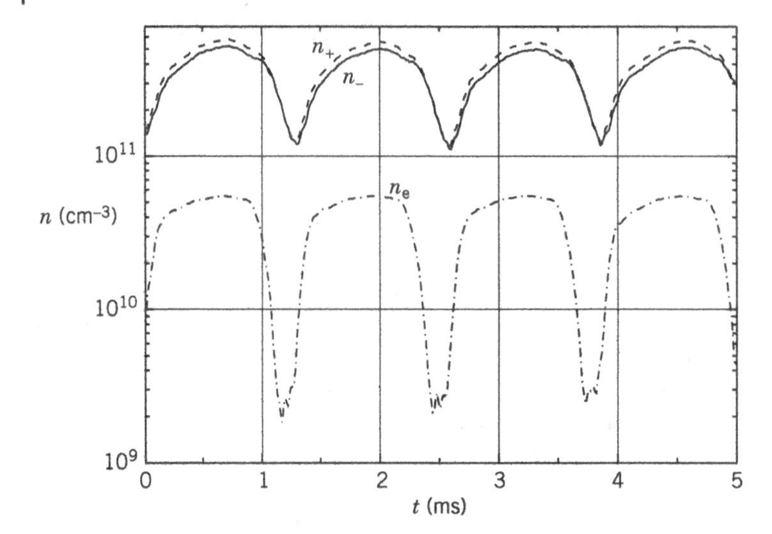

Figure 12.5 Positive ion, negative ion, and electron densities as a function of time for 1:1 Ar/SF$_6$ mixture; the total pressure is 5 mTorr, the average power absorbed is 550 W.

For the absorbed power (solid curve in Figure 12.4), we take a sum of inductive and capacitive powers

$$P_{abs} = \frac{1}{2} I_{rf}^2 R_{abs} \left(\frac{n_{ind} n_e}{n_{ind}^2 + n_e^2} + \frac{n_{cap}}{n_{cap} + n_e} \right) \tag{12.2.10}$$

where the first and second terms approximate the inductive and capacitive powers, with parameters R_{abs} a resistance chosen to give the correct power, at the power maximum, n_{ind} chosen to give the correct maximum of the inductive power versus n_e, as in Figure 12.4, and n_{cap} chosen to give the correct ratio of capacitive-to-inductive power at some low density, falling off with n_e at higher densities.

With the further observation that the electron density can build up and decay much more rapidly than the negative ion density, particularly for high α_0, we obtain the physical instability mechanism illustrated in Figure 12.6. From (12.2.9), with T$_e$ nearly constant, the power loss has the form $P_{loss} = K_e n_e + K_- n_-$, with constants K_e and K_-. This gives a linear variation of P_{loss} with n_e whose intercept at $n_e = 0$ is proportional to the slowly varying negative ion density n_-. The two loss curves, P_{loss1} and P_{loss3}, have been chosen at the two tangencies with the P_{abs} curve. At the end of phase 4, the loss curve decreases below the P_{loss1} curve, the quasi-capacitive equilibrium is lost, and the discharge enters phase 1, with n_e increasing rapidly due to ionization. Similarly, at the end of phase 2, during which the negative ion density builds up, the loss curve increases above P_{loss3}. The quasi-inductive equilibrium is lost, and the discharge enters phase 3, with the loss of positive ions in the escaping flux being matched by a rapid loss of the lower-density electrons. The decay of the negative ions toward a lower density equilibrium, in phase 4, then repeats the relaxation oscillation cycle.

The rapid rise and fall of the electron density, seen in Figure 12.5, are consistent with this physical mechanism. The time scale of the relaxation oscillation is set by the build-up and decay of the negative ions, with the scaling determined from

$$\frac{1}{n_-} \left(\frac{dn_-}{dt} \right)_{decay} \approx K_{rec} n_- \approx (K_{att} K_{rec} n_g n_e)^{1/2} \tag{12.2.11}$$

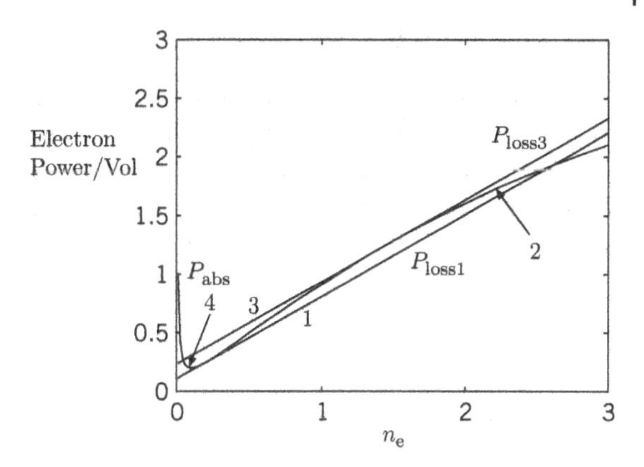

Figure 12.6 Absorbed electron power P_{abs} versus electron density n_e and two different curves of electron power lost versus n_e (P_{loss1} at a low negative ion density n_- and P_{loss3} at a high n_-).

where the second equality follows by use of (10.4.7b). The n_g and n_e scalings have been qualitatively seen experimentally.

Measurements and global modeling in a low pressure, inductively driven chlorine discharge, accounting for the gas chemistry, also show the E-to-H transition instability (Corr et al., 2005; Soberón et al., 2006; Despiau-Pujo and Chabert, 2009). At the border between capacitive and inductive modes, Cl_2 was found from global modeling to be only moderately dissociated, $f_{diss} \approx 0.16$. Kawamura et al. (2012) used a two-dimensional hybrid fluid-analytic simulation to determine the space- and time-varying densities of electrons, positive and negative ions, and neutral species, and electron and neutral gas temperatures, for the E-to-H transition instability.

12.2.4 Power Transfer Efficiency

Let us note that the driving coil (primary of the transformer shown in Figure 12.2) has some resistance R_{coil}. Hence, even if the discharge is extinguished ($n_0 = 0$), there is a minimum power $P_{T\min} = \frac{1}{2}I^2_{\min} R_{coil}$ supplied by the source. Because $P_{abs} \propto n_0^{-1/2}$ at high densities, we see from Figure 12.4 that the power transfer efficiency P_{abs}/P_T falls continually as n_0 is increased, hence, limiting source operation at high densities because of power supply limitations. Although $\delta_p \sim R$ is the preferred operating regime for maximum power efficiency, other considerations often indicate operation at lower or higher densities. The poor power transfer to the plasma at very low and at very high densities is analogous to the well-known property of an ordinary transformer with an open and a shorted secondary winding. In both cases, no power is dissipated in the load (here the plasma), but in both cases, there is power dissipated in the primary winding (here the coil) due to its inherent resistance. Piejak et al. (1992) have given a complete analysis of an inductive discharge in terms of measurable source voltages and currents, based on this analogy.

For completeness, we note that at very high densities, the electron–ion collision frequency may be larger than the electron–neutral collision frequency. In this collisional regime, ν_{90} from the equation preceding (3.3.7) replaces ν_m in determining σ_{dc}. Since $\nu_{90} \propto n_0$ (the Spitzer conductivity is independent of n_0), the scaling (12.2.1) is replaced by

$$P_{abs} \propto I_{rf}^2$$

independent of n_0 in this regime. However, low-pressure inductive discharges for materials processing are rarely operated at such high densities.

12.2.5 Exact Solutions

One-dimensional solutions over the entire range of densities can be given for the case where a uniform density plasma fills a long cylindrical coil ($b = R$ and $l \gg R$). These were first obtained by Thomson (1927) in the collisional (high-pressure) regime $\nu_m \gg \omega$, where the penetration of the rf fields into the discharge is governed by the collisional skin depth (12.1.6). Here, we extend Thomson's treatment to the entire range of collisionalities from $\nu_m \ll \omega$ to $\nu_m \gg \omega$. Maxwell's equations (2.2.1) and (2.2.2) for the $\tilde{E}_\theta$ and $\tilde{H}_z$ field components are

$$\frac{d}{dr}(r\tilde{E}_\theta) = -j\omega\mu_0 r\tilde{H}_z \tag{12.2.12}$$

$$-r\frac{d\tilde{H}_z}{dr} = j\omega\epsilon_0\kappa_p(r\tilde{E}_\theta) \tag{12.2.13}$$

with κ_p given by (12.1.2). Eliminating $r\tilde{E}_\theta$ from these equations, we obtain

$$\frac{d^2\tilde{H}_z}{dr^2} + \frac{1}{r}\frac{d\tilde{H}_z}{dr} + k_0^2\kappa_p\tilde{H}_z = 0 \tag{12.2.14}$$

which is Bessel's equation with $k_0 = \omega/c$. With the boundary condition that $\tilde{H}_z(R) = H_{z0}$, the solution is

$$\tilde{H}_z = H_{z0}\frac{J_0(kr)}{J_0(kR)} \tag{12.2.15}$$

where

$$k = k_0\sqrt{\kappa_p} \tag{12.2.16}$$

is the complex propagation constant. We see that the Bessel functions have complex argument. Using (12.2.15) to evaluate the left-hand side of (12.2.13), and solving for $\tilde{E}_\theta$, we obtain

$$\tilde{E}_\theta = H_{z0}\frac{k}{j\omega\epsilon_0\kappa_p}\frac{J_1(kr)}{J_0(kR)} \tag{12.2.17}$$

The time-averaged power flowing into the discharge is found in terms of the field amplitudes at the plasma surface $r = R$ using the complex Poynting theorem (Ramo et al., 1994, chapter 3),

$$P_{abs} = 2\pi R l S_{abs} = 2\pi R l\,\mathrm{Re}\left(-\frac{1}{2}\tilde{E}_{\theta 0}H_{z0}\right) \tag{12.2.18}$$

As the plasma density n is increased from zero at fixed ω, R, and H_{z0} (equivalent to holding the coil current fixed), then one finds from (12.2.18) that P_{abs} rises from zero to a maximum and then falls to zero. The variation is similar to that shown in the solid curves of Figure 12.4. For a nearly collisionless plasma $\nu_m \ll \omega$, one finds a maximum power at a density such that $\delta_p \approx 0.37\,R$. For a collisional plasma $\nu_m \gg \omega$, one finds a maximum power at $\delta_c \approx 0.57\,R$. Hence, in both cases, the maximum power efficiency (for a coil having a finite resistance) occurs when the appropriate skin depth is of order of the plasma radius.

Other issues of inductive discharge operation include finite geometry effects ($l \sim R$), planar coil source operation, startup, and resonant effects due to stray coil capacitances. We address some of these issues in the following sections, and refer the reader to the literature (Piejak et al., 1992; Eckert, 1986; Hopwood et al., 1993a,b) for further information.

12.2.6 Helical Resonator Discharges

Helices have long been used to propagate electromagnetic waves with phase velocity $v_{\mathrm{ph}} \ll c$, the velocity of light. This property allows a helix to resonate in the MHz range such that it can be used for efficient plasma generation at low pressures (Cook et al., 1990). Helical resonator plasmas operate at radio frequencies (3–30 MHz), do not require a dc magnetic field (as do ECRs and helicons; see Chapter 13), exhibit high Q (600–1500 typically without the plasma present) and high characteristic impedance, and can be operated without a matching network. As shown in Figure 12.7, the source consists of a coil surrounded by a grounded coaxial cylinder. The composite structure becomes resonant when an integral number of quarter waves of the rf field fit between the two ends. When this condition is satisfied, the electromagnetic fields within the helix can sustain a plasma with low matching loss at low gas pressure. Typically, an electrostatic shield is added between the helix and the plasma column to reduce the capacitive coupling to a negligible value. The shield is typically a metal cylinder slotted along z that allows the inductive field $\tilde{E}_\theta$ to penetrate into the plasma, while shorting out the capacitive $\tilde{E}_r$ and $\tilde{E}_z$ fields.

The basic design parameters for a helical resonator discharge consist of pressure, rf power, source length, plasma radius, helix radius, outer cylinder radius, winding pitch angle, and excitation frequency (Niazi et al., 1994, 1995; Park and Kang, 1997b). A thorough review of the physics, along with new theoretical calculations and experimental data, is given in Martines et al. (2021). For an unshielded plasma column, the wave dispersion, k_z versus ω, and the relation among the field quantities are determined in the approximation of a uniform, collisionless ($\omega \gg \nu_{\mathrm{m}}$) plasma having relative dielectric constant $\kappa_{\mathrm{p}} = 1 - \omega_{\mathrm{pe}}^2/\omega^2$ (see (4.2.24)) by using a *sheath helix model*, in which the rf current in the helical wires is replaced by a continuous current sheet ("sheath"). With an outer conductor, there are generally two modes, a helix mode, whose axial wavenumber is associated with the helix pitch angle Ψ,

$$k_{z\mathrm{h}} \approx \frac{\omega}{c \tan \Psi} \tag{12.2.19}$$

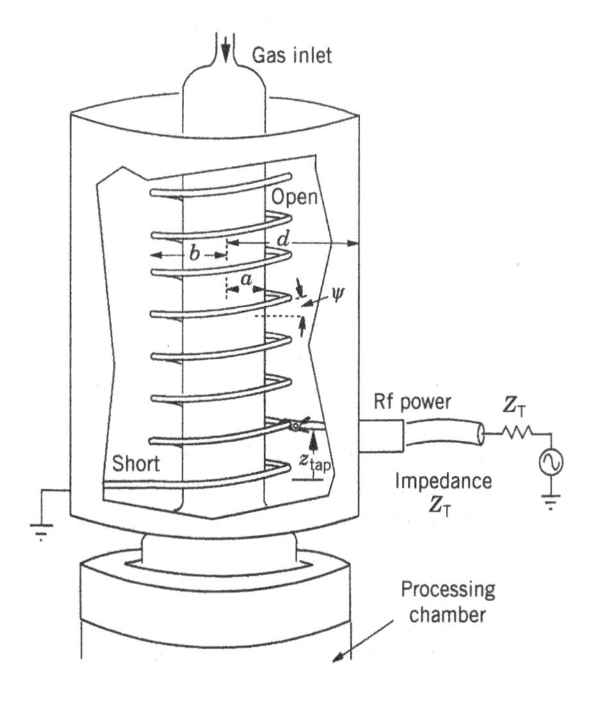

Figure 12.7 Schematic of a helical resonator plasma source.

and a coax mode, associated with a transverse electromagnetic wave propagating near the speed of light, $k_{z0} \approx \omega/c$. Note from the figure that the coil length is $l = 2\pi b\mathcal{N} \tan\Psi$.

The useful mode is the helix mode. The mode has a resonance $k_z \to \infty$ at relatively low density, such that $\omega_{pe} = \omega$, and exists above that density. At large density, the plasma and outer cylinder are at nearly the same voltage, and the helix is at a high voltage with respect to them both. In this high-density limit the plasma acts like a conducting cylinder. During typical source operation, only the helix mode is resonant, and it dominates the source operation.

The approximate resonance condition, $k_{zh}l = \mathrm{m}\pi/2$, with m an integer, can also be written, using (12.2.19) with $\tan\Psi = l/(2\pi b\mathcal{N})$, as

$$2\pi b\mathcal{N} = \mathrm{m}\frac{\lambda}{4} \tag{12.2.20}$$

where b is the helix radius, $\mathcal{N}$ is the number of turns, and λ is the free space wavelength. A simple estimate of the resonant frequency from (12.2.19) gives $f \approx 5$ MHz at $l = \lambda/4 = 30$ cm ($k_z = 5.2\ \mathrm{m}^{-1}$) and $\Psi = 0.02$ rad.

The helical resonator propagation and matching characteristics are conveniently obtained from a circuit model (Park and Kang, 1997b). By measuring the input impedance characteristics as the tap position is varied, they determined the propagation constant. By matching the input resistance of the experiment to the model results, they also determined the power absorption in the plasma, and the slight decay of the propagating signal away from the source. Theoretical results for helical resonator operation were compared with measurements on four different helical resonator configurations, with and without a conducting shield, by Martines et al. (2021), showing reasonable agreement.

A somewhat different resonator configuration was used by Vinogradov et al. (1998). They made the helical resonator a full wavelength long, which they called the *lambda-resonator.* The resonator had shorted ends and the power was injected in a capacitively compensated mode, in which the voltages on the two halves of the resonator are out of phase, such that the plasma remains near rf ground potential. This minimizes capacitive current to the grounded surfaces, reducing various undesirable effects.

12.3 Planar Coil Configuration

The planar coil discharge shown in Figure 12.1*b*, usually without multipole magnets, is a commonly used configuration for materials processing, typically generating relatively uniform low aspect ratio plasmas with densities between 10^{11} and $10^{12}\ \mathrm{cm}^{-3}$ over substrate diameters of 30 cm or more. In axisymmetric geometry, the coil generates an inductive field having magnetic components $\tilde{H}_r(r,z)$ and $\tilde{H}_z(r,z)$, and an electric component $\tilde{E}_\theta(r,z)$. As shown in Figure 12.8*a*, the rf magnetic field lines in the absence of a plasma encircle the coil and are symmetric with respect to the plane of the coil. If a plasma is formed below the coil, as shown in Figure 12.8*b*, then from Faraday's law (2.2.1), an azimuthal electric field $\tilde{E}_\theta$ and an associated current density $\tilde{J}_\theta$ are induced within the plasma. The plasma current, opposite in direction to the coil current, is confined to a layer near the surface having a thickness of order the skin depth δ. The total magnetic field, which is the sum of the fields due to the $\mathcal{N}$ turn exciting coil current and the "single-turn" induced plasma current, is shown in Figure 12.8*b*. The dominant magnetic field components within the plasma are $\tilde{H}_z$ near the axis and $\tilde{H}_r$ away from the axis, as shown. Near the axis, Faraday's law implies that both $\tilde{E}_\theta$

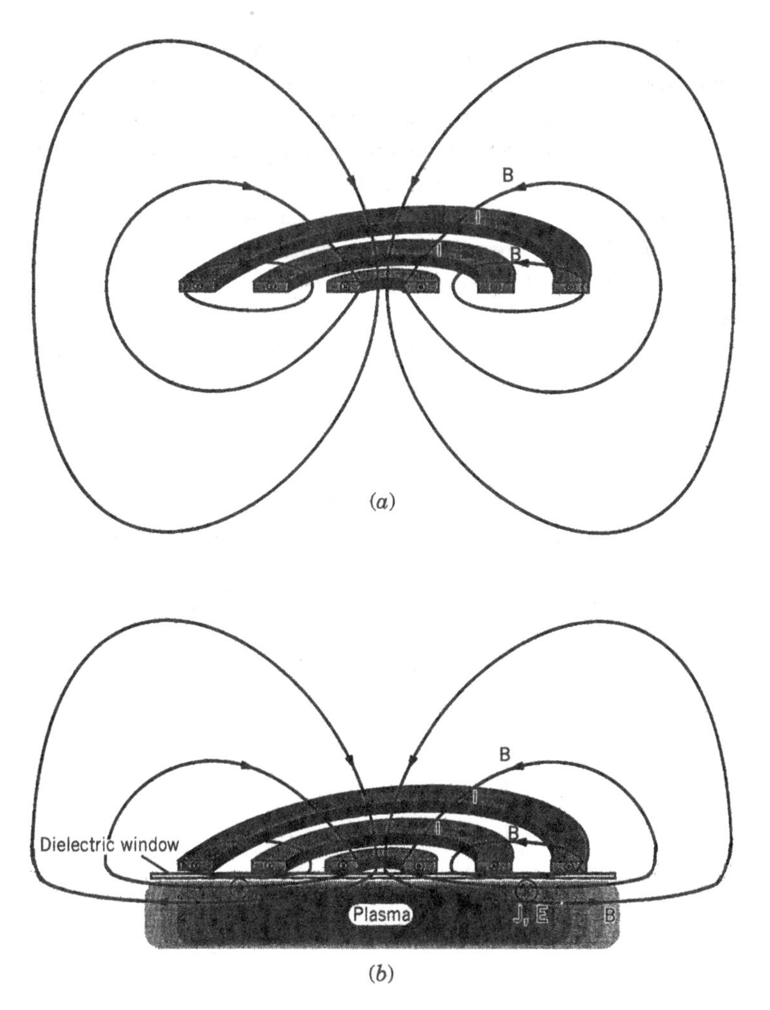

Figure 12.8 Schematic of the rf magnetic field lines near a planar inductive coil: (*a*) without nearby plasma and (*b*) with nearby plasma. Source: Wendt, 1993/with permission of IOP Publishing.

and $\tilde{J}_\theta$ vanish as $\tilde{E}_\theta$, $\tilde{J}_\theta \propto r$. This implies that the absorbed power density,

$$p_{\text{abs}} = \frac{1}{2}\text{Re}\,\tilde{J}_\theta \tilde{E}_\theta^* \tag{12.3.1}$$

vanishes on axis, leading to a ring-shaped profile for the absorbed power. The transformer model of the previous section can be applied to the planar configuration, but the inductance matrix elements are difficult to determine from simple electromagnetic models (Gudmundsson and Lieberman, 1998).

The rf magnetic fields within the plasma have been measured by Hopwood et al. (1993a) for an inductive discharge excited by a planar square coil, which was separated from a rectangular aluminum plasma chamber 27 cm on a side and 13 cm high by a 2.54-cm-thick quartz window. Although this system is not axisymmetric, the general structure of the fields and the absorbed power profile are similar to those in an axisymmetric system. We use Hopwood's results to illustrate the general features observed in planar inductive discharges.

Figure 12.9 shows the measured variation of $B_r \equiv |\tilde{B}_r|$ with z at $r = 6.3$ cm in a 5 mTorr oxygen discharge. The field decreases exponentially with distance from the window, with a maximum of 2.7–5.1 G, depending on the incident power P_{inc}, and with a skin depth δ (characteristic length for the exponential decay) varying from 2.1 to 2.7 cm, and scaling roughly as $P_{\mathrm{inc}}^{-1/2}$, in agreement with (12.1.4) or (12.1.6) with $n_0 \propto P_{\mathrm{inc}}$. In general, the skin depth lies between the values δ_{p} and δ_{c} given by (12.1.4) and (12.1.6) and is fairly close to both. Figure 12.10 shows the measured variation of B_r with r (along the diagonal of the chamber) at three different positions below the window in a 5 mTorr, 500 W, argon discharge. We see that B_r falls to zero on the axis and has a maximum at approximately 9.5 cm off the axis.

The rf electric field $\tilde{E}_\theta$ can be related to $\tilde{B}_r \equiv \mu_0 \tilde{H}_r$ by using the r component of Faraday's law (2.2.1),

$$\frac{\partial \tilde{E}_\theta}{\partial z} = j\omega\mu_0 \tilde{H}_r \tag{12.3.2}$$

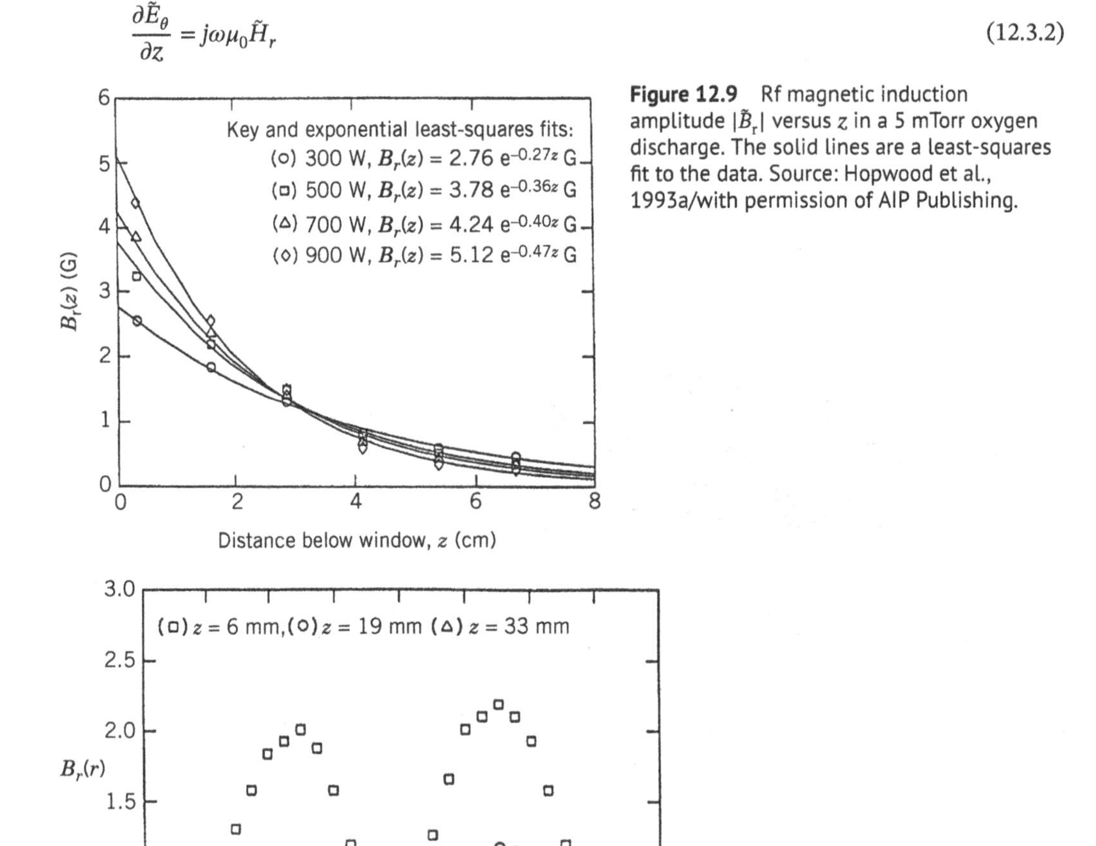

Figure 12.9 Rf magnetic induction amplitude $|\tilde{B}_r|$ versus z in a 5 mTorr oxygen discharge. The solid lines are a least-squares fit to the data. Source: Hopwood et al., 1993a/with permission of AIP Publishing.

Figure 12.10 Rf magnetic induction amplitude $|\tilde{B}_r|$ versus diagonal radius r at three different distances below the window as measured in a 5 mTorr, 500 W argon discharge. Source: Hopwood et al., 1993a/with permission of AIP Publishing.

Assuming that $\tilde{E}_\theta = \tilde{E}_{\theta 0}\, e^{-z/\delta}$ and inserting this into (12.3.2), we obtain

$$\tilde{E}_\theta = -j\omega\mu_0\delta\tilde{H}_r \tag{12.3.3}$$

Hence, $\tilde{E}_\theta$ has the same axial and radial variation as $\tilde{B}_r$ given in Figures 12.9 and 12.10.

In addition to the field measurements, Langmuir probes (see Section 6.6) were used to determine the ion density n_i, electron temperature T_e, and plasma potential V_s (Hopwood et al., 1993b). The ion density measurement was confirmed by comparison to a 35 GHz microwave interferometer measurement (see Section 4.6). Figure 12.11 shows n_i versus incident power P_{inc} at a location on-axis and 5.7 cm ($\sim$ 3 skin depths) below the window, for pressures between 0.5 and 15 mTorr in argon. We see that n_i varies linearly with P_{inc}, but that $n_i \approx 0$ (on a 10^{11}-cm^{-3} scale) at approximately 100 W. Below this incident power, an inductive discharge cannot be sustained (see Section 12.2), and a low-density plasma is sustained by capacitive coupling between the coil and the plasma.

Figure 12.12 shows the measured variation of n_i, T_e, and V_s with argon pressure for $P_{inc} = 500$ W. We see that T_e falls slowly as p increases, as determined from an ion particle balance relation, e.g., (10.2.12) plotted in Figure 10.1. The ion density is seen to increase with increasing pressure. This is consistent with the power balance relation, which indicates that the density varies inversely with the effective plasma area, as follows. For this discharge, with $l = 13$ cm and $R \gg l$, we can estimate

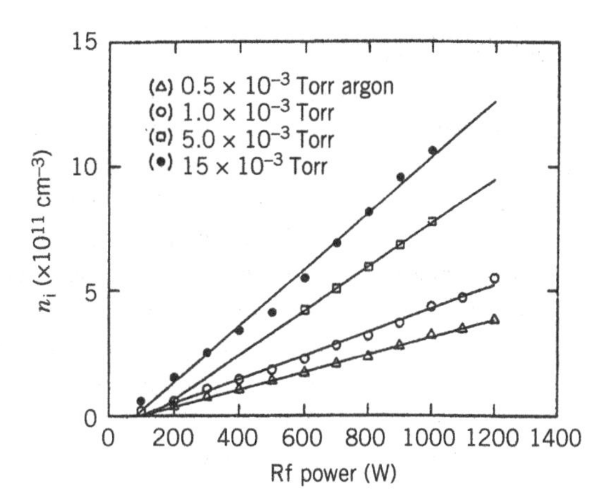

Figure 12.11 Ion density versus rf power at various argon pressures. Source: Hopwood et al., 1993b/with permission of AIP Publishing.

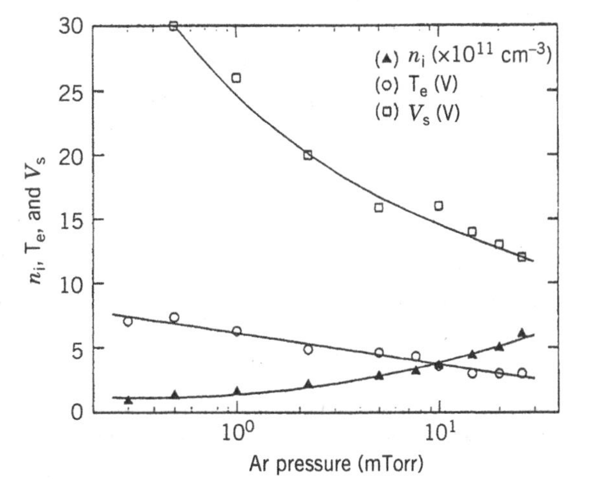

Figure 12.12 Ion density, electron temperature, and plasma potential versus argon pressure in a 500 W discharge with magnetic multipole confinement. Source: Hopwood et al., 1993b/with permission of AIP Publishing.

that A_{eff} in (10.2.11) scales as

$$A_{\text{eff}} \propto h_l \propto \left(3 + \frac{l}{2\lambda_{\text{i}}}\right)^{-1/2} \tag{12.3.4}$$

Using (3.5.10) to determine the ion–neutral mean free path λ_{i} in argon, we find $\lambda_{\text{i}} \approx 3$ cm at 1 mTorr and $\lambda_{\text{i}} \approx 0.15$ cm at 20 mTorr. Hence, from (10.2.15), the predicted density ratio is

$$\frac{n_{\text{i}}(20\ \text{mTorr})}{n_{\text{i}}(1\ \text{mTorr})} \approx \frac{A_{\text{eff}}(1\ \text{mTorr})}{A_{\text{eff}}(20\ \text{mTorr})} \approx \frac{0.44}{0.15} \approx 3.0$$

which is in reasonable agreement with the measured ratio of ~ 3.3 obtained from Figure 12.12. The plasma potential V_{s} is seen to lie between 12 and 30 V, roughly consistent with (10.2.4). The potential increases as the pressure decreases, in qualitative agreement with the scaling predicted from (10.2.4).

All preceding measurements were performed with multipole magnets placed along the four 27 cm × 13 cm sidewall areas (see Section 5.5). In Figure 12.13, the normalized ion saturation current (proportional to the density) is plotted along a diagonal within the chamber with and without the multipole magnets in place, for a 5 mTorr oxygen discharge. We see that the multipole magnets greatly increase the uniformity of the density. The ratio of the standard deviation to the average density across the central 20 cm of the discharge with multipole magnets was measured to be 2.5%. This result is qualitatively consistent with the dominant losses being axial, when quadrupoles are present.

A relatively complete set of characterization measurements of a planar inductive argon discharge has been given by Godyak et al. (1994, 1999, 2002) and Godyak and Piejak (1997). External electrical characteristics such as voltage, current phase angle, resistance, reactance, and coupling efficiency were measured over a wide range of discharge powers, driving frequencies, and gas pressures. Magnetic probes were used to determine the internal rf electric and magnetic fields and currents, and Langmuir probes were used to determine the electron energy distribution function and plasma parameters such as density, average energy, and effective collision frequency. The transition from collisional to stochastic heating was observed as the pressure was lowered. In the stochastic (nonlocal) regime, the expected non-exponential decays of the field profiles were observed.

Some global and two-dimensional fluid simulations of planar inductive discharges in chlorine were described briefly in Section 10.5. Two-dimensional fluid simulations in Ar, Ar/O_2, and

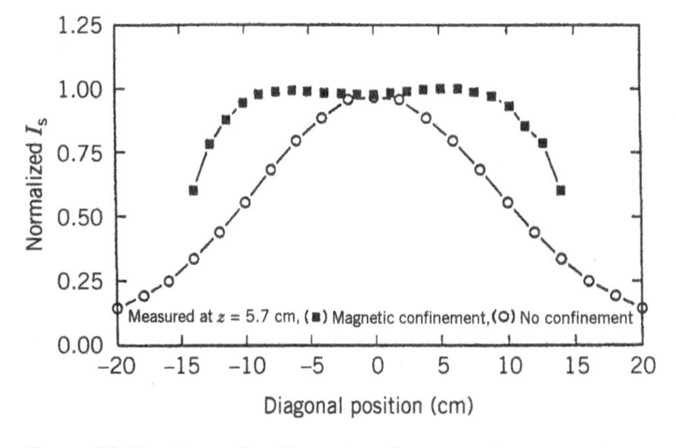

Figure 12.13 Normalized ion saturation current measured across the diagonal of the plasma chamber with and without magnetic multipole confinement. Source: Hopwood et al., 1993b/with permission of AIP Publishing.

$Ar/O_2/Cl_2$, along with comparisons to an experiment, were reported by Hsu et al. (2006). They found good agreement for Ar and Ar/O_2, but only partial agreement for the more complex $Ar/O_2/Cl_2$ gas mixture, indicating some deficiencies in their chemical reaction database.

12.3.1 Neutral Gas Depletion and Heating

At low pressures and high plasma densities (high powers), neutral gas depletion and heating within the discharge region can be important, even if the discharge is weakly ionized. Summing the force equations (2.3.15) for all the species, the dominant terms in the sum are the pressure gradient forces. Assuming a fixed wall gas density n_w and temperature T_w, the equilibrium total pressure balance is then

$$n_g k T_g + n_e k T_e = n_w k T_w \tag{12.3.5}$$

where the small ion pressure has been neglected since $T_i \sim T_g \ll T_e$. For a typical value $T_e/T_g \sim 50$, and with $n_e/n_g \sim 0.02$, (12.3.5) gives a large (50%) neutral depletion within the discharge region.

Neutral gas depletion also results from gas heating, even if $n_e k T_e$ is negligible. Gas heating is mainly due to electron–neutral collisions. For elastic collisions, the gas heating power per unit volume is $w_g \approx (3m/M_g) K_{el} n_e n_g k(T_e - T_g)$. Rotational, vibrational, and electronic excitations can also contribute to w_g. As described in Section 11.3, w_g increases the gas temperature within the discharge region according to the heat equation coupled with Fourier's law (Bird et al., 2002, chapters 9 and 10)

$$-\nabla \cdot (\kappa_T \nabla T_g) = w_g \tag{12.3.6}$$

with κ_T (W/m-K) the gas thermal conductivity. Integrating (12.3.6) over the reactor volume and approximating $|\nabla T_g| \approx T_g/\Lambda_0$, with Λ_0 the characteristic reactor diffusion length, we obtain (Abada et al., 2002)

$$\kappa_T S \frac{T_g - T_w}{\Lambda_0} \approx W_g \tag{12.3.7}$$

with S the reactor surface area and with $W_g = \int d\mathcal{V}\, w_g$ the total gas heating power in the reactor volume $\mathcal{V}$. Equation (12.3.7) determines the globally averaged T_g for the given heating power W_g. For a cylindrical discharge, $S = 2\pi R(R + l)$, $\mathcal{V} = \pi R^2 l$, and Λ_0 is given by (5.2.38). Hebner (1996) measured T_g as high as 900 K in a 300 W, 10 mTorr argon inductive discharge. At 50 mTorr CF_4 and $w_g \approx 0.15$ W/cm^3, Abada et al. (2002) measured T_g in the discharge center as high as 900 K, and they found good agreement to the global heating model (12.3.7).

The gas depletion is, of course, coupled to the plasma equilibrium particle and energy balances. For example, according to the particle balance (10.2.12), the equilibrium T_e increases with increasing n_e (increasing w_g), since n_g decreases. The decreasing n_g also increases the edge-to-center density ratios h_l and h_R in (10.2.1) and (10.2.2) (Raimbault and Chabert, 2009). According to the energy balance (10.2.14), with (10.2.11) for A_{eff}, this increases the diffusion-dominated plasma loss. In the absence of gas heating, planar geometry analytical models of the coupling are given by Raimbault et al. (2007). The effects of gas heating on these models are described in Liard et al. (2007). An extensive review of gas depletion and heating, and the resulting plasma-depletion coupling, is given in Fruchtman (2017).

12.4 High-Efficiency Planar Discharges

Highly efficient inductive discharges using ferrite cores at frequencies as low as 50–60 Hz have long found applications in lighting as electrodeless (long-life) fluorescent lamps (Godyak, 2013; Kolobov and Godyak, 2017). A similar technology can be applied to improve the performance of planar materials processing discharges driven in the frequency range of 450 kHz–2 MHz. This can enable reasonably efficient reactor operation with low capacitive coupling, giving access to lower plasma densities (to the left of the peak in P_{abs} shown in Figure 12.4). In this section, we first examine some issues of inductive discharge operation at low frequencies. We then consider means of improving the coil-to-plasma coupling, such as reduced coil-to-plasma loop spacing and ferrite enhancement, to increase the reactor efficiency. Finally, we describe measurements on a close-coupled, ferrite-enhanced, 2 MHz planar inductive discharge.

12.4.1 Low Frequencies

For low frequencies $f \lesssim 2$ MHz and not-so-low gas pressures, $2\pi f = \omega \ll \nu_m$, a significant rf magnetic field can exist in the skin layer. The Lorentz force due to this field can produce strong nonlinear effects, such as second harmonic currents and ponderomotive forces. To see this, let us consider the rf magnetic fields in the skin layer of a low-frequency planar discharge. As shown in (10.2.17), the rf electric field magnitude E_θ that maintains the discharge can be pressure-dependent but is independent of the frequency. From (12.3.3), the main component of the rf magnetic field magnitude is

$$B_r \approx \frac{E_\theta}{\omega\delta} \tag{12.4.1}$$

which varies inversely with frequency for relatively fixed skin depth δ. Let us estimate the ratio of rf magnetic (Lorentz) to electric forces on an electron. The electric force magnitude is $F_E = eE_\theta$. The magnetic force magnitude is $F_L = ev_\theta B_r$, with $v_\theta \approx eE_\theta/(m\nu_m)$. This gives the ratio

$$\frac{F_L}{F_E} \approx \frac{e}{m}\frac{E_\theta}{\omega\delta\nu_m} \tag{12.4.2}$$

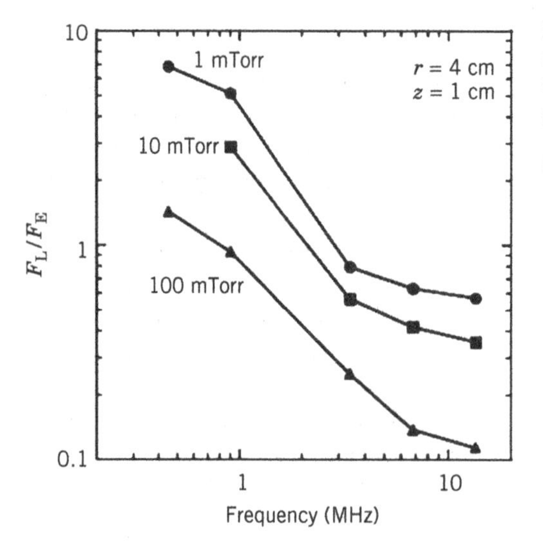

Figure 12.14 Measured ratios (symbols) of Lorentz-to-electric-field force, F_L/F_E at $r = 4$ cm and $z = 1$ cm, versus frequency, at various argon pressures; 19.8 cm diameter, 10.5 cm length planar ICP. Source: Godyak, 2003/with permission of IOP Publishing.

At low frequencies, $F_L/F_E \gg 1$. This ratio also varies with gas pressure, since E_θ, δ, and ν_m all vary with pressure.

Figure 12.14 shows the measured ratio F_L/F_E versus frequency for various pressures in a 19.5 cm diameter, 10.5 cm length planar argon discharge (Godyak, 2003). The measurements were taken in the middle of the skin layer at a radius of 4 cm using magnetic and Langmuir probes to determine the fields, densities and plasma potentials. At low frequencies, we see roughly the frequency scaling $F_L/F_E \propto f^{-1}$, and a weaker scaling $F_L/F_E \propto p^{-1/2}$ with pressure. These results indicate that we can expect strong nonlinear effects due to the rf magnetic fields at low frequencies and low pressures. However, even at high pressures, the Lorentz force can be important at frequencies below 2 MHz.

The low pressure, low-frequency regime has been examined by Godyak (2003). In this regime, the skin depth is anomalous, corresponding mainly to stochastic heating in the skin layer. For sinusoidal rf fields, the electrons are alternately magnetized and de-magnetized in time during the rf period, leading to strong nonlinear effects. Since v_θ and B_r both vary sinusoidally, their product F_L lies along z and has both a second harmonic and a constant value. This gives rise to a second harmonic plasma potential and resulting electric field $E_{z2\omega}$, which decays sharply within the skin layer, as seen experimentally. $E_{z2\omega}$ is a polarization (Hall effect) field that can be larger than the fundamental E_θ field, but it does not drive axial ohmic currents that heat the plasma.

On the other hand, the interaction of the v_θ motion with B_r and the smaller B_z rf fields in a planar inductive discharge can produce a Lorentz force along the poloidal, closed rf magnetic field lines, producing substantial circulating second harmonic current flows. The currents flow within the skin layer and plasma along closed loops, similar to those shown for the rf magnetic field lines in Figure 12.8*b*.

The dc component of F_L gives rise to a ponderomotive potential energy W_P whose gradient gives a force that can have measurable effects on the discharge equilibrium at low frequencies, being comparable or larger than the gradient of thermal potential energy $\nabla p_e/n_e$ (Godyak, 2003). The classical ponderomotive potential energy is $W_P = \frac{1}{2}m\langle v_\theta^2\rangle_{\omega t}$ (Gapanov and Miller, 1958), with a somewhat different expression in the anomalous skin effect regime (Smolyakov et al., 2001).

Finally, we should realize that in some inductive geometries, a substantial induced E field can exist in a region with a very weak time-varying B field (Kolobov and Godyak, 2017). An example is the outer region $r > R$ of a long solenoid of radius R and length $\ell \gg R$, carrying a time-varying surface current density K_θ. Since the (tangential) E_θ field must be continuous across the surface $r = R$, the induced E_θ field within the solenoid extends into the outer region $r > R$, where the B_z field is very small. Note, however, that if the solenoid actually consists of $\mathcal{N} \gg 1$ conducting wire turns, each carrying a current I, then an inductive voltage $V = \omega LI$ exists across the solenoid ends, corresponding to an E_z field in the exterior region. In this case, the electrostatic interaction of the induced E_θ field with the conducting wire produces an exterior E_z field.

12.4.2 Close Coupling

As described in the previous subsection, the discharge requires a certain E_θ field generated by the time-varying magnetic flux φ linking the area A of the plasma loop, according to the integral form of Faraday's law (2.2.1)

$$\oint_C \mathbf{E} \cdot d\ell = -j\omega \int_A \mathbf{B} \cdot d\mathbf{A} \tag{12.4.3}$$

where $\mathbf{B} = \mu_0\mathbf{H}$ is the magnetic induction vector. For a plasma loop of mean radius R, $2\pi RE_\theta \approx -j\omega\varphi$, with $\varphi \approx \pi R^2 B_z$, and with B_z the magnetic induction produced by the coil at the plasma loop. However, the usual inductive discharge with a thick dielectric window has a low coupling efficiency, that is, much of the magnetic flux generated by the coil is "wasted" because it does not link the plasma loop. To see this, let us consider a planar coil with 10 and 15 cm inner and outer radius mounted at the top of a 2.5-cm thick window. This thickness is typically set by the need to hold atmospheric pressure across the window.

Approximating the coil as a single loop of mean radius R carrying a total current I within a conductor radius a, then the flux linking the primary coil is $\varphi_{11} = L_{11}I$, with

$$L_{11} \approx \mu_0 R \left(\ln \frac{8R}{a} - 2\right), \qquad a \ll R \tag{12.4.4}$$

the self-inductance of the one-turn coil (Ramo et al., 1994, chapter 4). Correspondingly, the flux linking the one-turn plasma loop is

$$L_{12} \approx \mu_0 R \left(\ln \frac{8R}{w+a+b} - 2\right), \qquad w+a+b \ll R \tag{12.4.5}$$

with L_{12} the mutual inductance and $b \approx \delta/2$ the average plasma loop radius. As an example, let $R = 12.5$ cm, $w = 2.5$ cm, $a = b = 0.25$ cm, one finds $L_{12}/L_{11} \approx 0.36$, i.e., only a fraction 0.36 of the coil flux links the plasma loop. If the coil is embedded within the window, closer to the plasma loop, than the coupling can be increased. For example, let $w = 0.25$ cm, then $L_{12}/L_{11} \approx 0.65$. The required coil current is then reduced by a factor of $0.65/0.36 = 1.8$, with a corresponding reduction in resistive losses. This also reduces the coil voltage and therefore the coil–plasma capacitive coupling.

The dependence of the capacitive coupling on frequency, primary coil turns, and discharge geometry can be estimated from the basic transformer relations. Since $E_\theta \approx$ const for a fixed pressure, the plasma loop voltage magnitude V_p is roughly constant

$$V_p \approx 2\pi R E_\theta \approx \omega L_{22} I_p \approx \omega L_{12} I_{rf} \approx \text{const} \tag{12.4.6}$$

From the first and third terms, this yields an inductive current magnitude

$$I_p \approx V_p/\omega L_{22} \tag{12.4.7}$$

For not-too-strong coupling (see next subsection), the coil voltage and current magnitudes are related by $V_{rf} \approx \omega L_{11} I_{rf}$. From the first and fourth terms in (12.4.6), we find $V_{rf} \approx (L_{11}/L_{12})V_p$. The capacitively coupled current magnitude is therefore

$$I_{cap} \approx \omega C V_{rf} \approx \frac{L_{11}}{L_{12}} \omega C V_p \tag{12.4.8}$$

with C the coil-to-plasma capacitance (mainly, due to the coil capacitance across the dielectric window into the plasma). The ratio of capacitive to inductive currents is

$$\frac{I_{cap}}{I_p} = \omega^2 L_{22} C \frac{L_{11}}{L_{12}} \tag{12.4.9}$$

Note that the ratio L_{11}/L_{12} is proportional to the coil-to-plasma turns ratio $\mathcal{N}$. Equation (12.4.9) indicates that the capacitive coupling can be strongly reduced by lowering the frequency, and, as also described in Section 12.2, by reducing $\mathcal{N}$.

12.4.3 Ferrite Enhancement

Further inductive coupling increases of the primary coil to the plasma loop can be achieved by surrounding the upper part of the primary coil with a π-shaped, low-coercivity, low conductivity

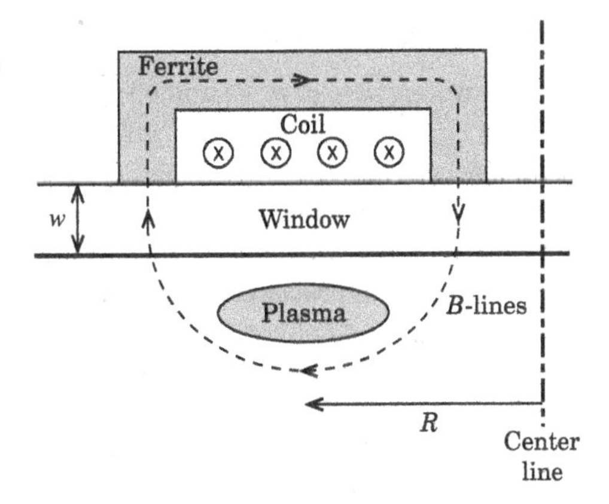

Figure 12.15 Illustrating the use of ferrite materials to increase the coupling efficiency of a planar inductive discharge.

ferrite material. For a finite coercivity, the B versus H characteristics have a hysteresis loop. The energy lost per rf period during one traversal of the hysteresis loop is equal to the hysteresis loop area (Ramo et al., 1994, chapter 2). Since the power loss scales with the frequency, low loss requires low frequencies. For a finite ferrite conductivity, eddy currents also contribute to the power loss. In the low rf frequency range, typically below a few megahertz, ferrites can have high magnetic permeabilities $\mu \gtrsim 100\,\mu_0$, and quite low resistive and hysteresis losses.

The basic configuration is illustrated in Figure 12.15. For an interface between a magnetic and nonmagnetic material, the Maxwell equation (2.2.4) implies that the normal component of **B** across the ferrite/window interface is continuous. That is, $B_\mathrm{f} = |\mathbf{B}_\mathrm{f}|$ in the ferrite is comparable to $B_\mathrm{wp} = |\mathbf{B}_\mathrm{wp}|$ in the dielectric and plasma region of the figure.

Neglecting the small displacement current term $\epsilon_0 \partial \mathbf{E}/\partial t$, the integral form of Ampere's law (2.2.2) is

$$\oint_C \mathbf{H} \cdot \mathrm{d}\boldsymbol{\ell} = I \tag{12.4.10}$$

For the dashed field line contour C shown in Figure 12.15, we obtain

$$\frac{B_\mathrm{f}}{\mu}\ell_\mathrm{f} + \frac{B_\mathrm{wp}}{\mu_0}\ell_\mathrm{wp} = I \tag{12.4.11}$$

with I the total coil current. With comparable Bs and $\mu \gtrsim 100\,\mu_0$, the first term in (12.4.11) is negligible compared to the second term, giving $B_\mathrm{wp} \approx \mu_0 I/\ell_\mathrm{wp}$. Without the ferrite, $\mu \to \mu_0$, and $B_\mathrm{wp} \approx \mu_0 I/(\ell_\mathrm{f} + \ell_\mathrm{wp})$. Therefore, for the same B_wp in the plasma, the coil current I (and the coil voltage) is reduced by a factor of $1 + \ell_\mathrm{f}/\ell_\mathrm{wp}$ by adding the ferrite.

Combining an embedded window coil with a π-shaped ferrite can give a more closely coupled, efficient coil-loop system. However, this still may not approximate an ideal transformer, which has a coupling coefficient $k_\mathrm{T} = L_{12}^2/(L_{11}L_{22}) \to 1$, ensuring that all the primary coil flux links the secondary coil, and vice versa. Making this assumption in (12.1.20), we obtain the input impedance Z_s at the coil terminals

$$\frac{1}{Z_\mathrm{s}} = \frac{1}{j\omega L_{11}} + \frac{L_{22}}{L_{11}}\frac{1}{Z_\mathrm{p}} \tag{12.4.12}$$

with $Z_\mathrm{p} = R_\mathrm{p} + j\omega L_\mathrm{p}$ the plasma impedance. An ideal transformer also has $L_{11} \to \infty$. From (12.4.12), this gives the ideal transformer relation $Z_\mathrm{s} = \mathcal{N}^2 Z_\mathrm{p}$, with $L_{11}/L_{22} = \mathcal{N}^2$, with $\mathcal{N}$ the number of coil turns.

12.4.4 Experimental Results

Various researchers have examined ferrite-enhanced processing discharges (Lloyd et al., 1999; Meziani et al., 2001; Colpo et al., 2005; Kim et al., 2010; Godyak, 2011), these studies have been reviewed by Godyak (2013). Lloyd et al. (1999) investigated a cylindrical discharge that used a two-turn coil covered with ferrite, placed outside of a 10-cm diameter quartz tube, to excite a 13.56 MHz argon discharge. This system did not show much improved efficiency, probably because of the high hysteresis losses at 13.56 MHz. In a planar inductive geometry at 13.56 MHz, Meziani et al. (2001) obtained a factor of three increase in plasma density and halving of the coil current using ferrite enhancement and a thin dielectric window. A large rectangular area (0.75 × 0.72 m) discharge based on this technology, but driven at 2 MHz, is described by Colpo et al. (2005). At 2 MHz, the losses were considerably reduced, compared to 13.56 MHz. Similar results were obtained by Kim et al. (2010) using an internal, ferrite-enhanced antenna operated at 2 and 13.56 MHz. A factor of four increase in plasma density was found at 2 MHz, compared to 13.56 MHz. Due to the greater coupling efficiency at 2 MHz, the antenna voltage for the same delivered power was also significantly decreased.

Let us examine in detail the close-coupled, 20 cm diameter planar inductive discharge driven at 2 MHz developed and characterized by Godyak (2011, 2013). An 8-turn coil surrounded at the top by π-shaped ferrite pieces was used to excite the discharge across a 2-mm thick dielectric window having inner and outer diameters of 10.7 and 19.7 cm. The coil was driven by a push–pull arrangement with a grounded center tap. This reduces the maximum coil voltage by a factor of two over that of a coil grounded at one end, decreasing the capacitive coupling to the plasma.

Measurements were made in argon at 1, 10, 100, and 1000 mTorr over a power range of roughly 20–500 W. The capacitive coupling was measured to be small, with an rf potential less than a few volts at 100 W over the entire range of pressures. Figure 12.16 shows the coil voltage (left-hand scale) and coil current (right-hand scale) versus absorbed power. As can be seen, the voltage and current track each other at all pressures, implying a constant coil impedance magnitude $|Z_s| = |V/I|$. The measured coil voltages were relatively constant over the entire power range: approximately 1600 V at 1 mTorr, 1200 V at 10 mTorr, 700 V at 100 mTorr, and 500 V at 1000 mTorr. The constant coil voltage versus power implies a constant E_θ to maintain the plasma loop, as predicted

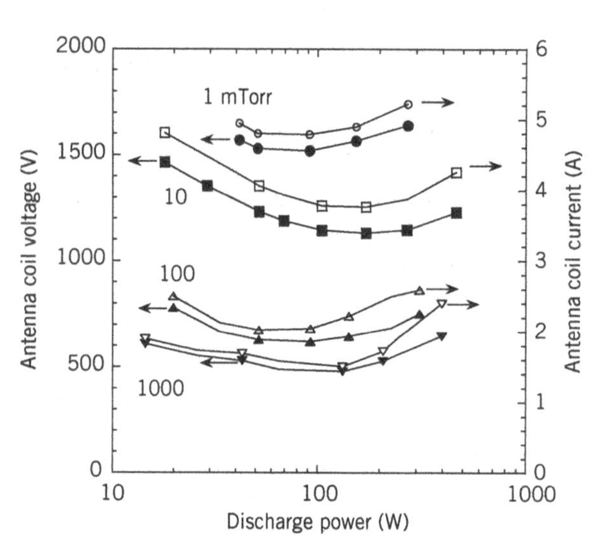

Figure 12.16 Measured coil voltage and current versus absorbed discharge power at 1, 10, 100, and 1000 mTorr argon in a 2 MHz, 20 cm diameter, close-coupled, ferrite-enhanced planar inductive discharge. Source: Godyak (2011). ©IOP Publishing. Reproduced with permission. All rights reserved. https://doi.org/10.1088/0963-0252/20/2/025004.

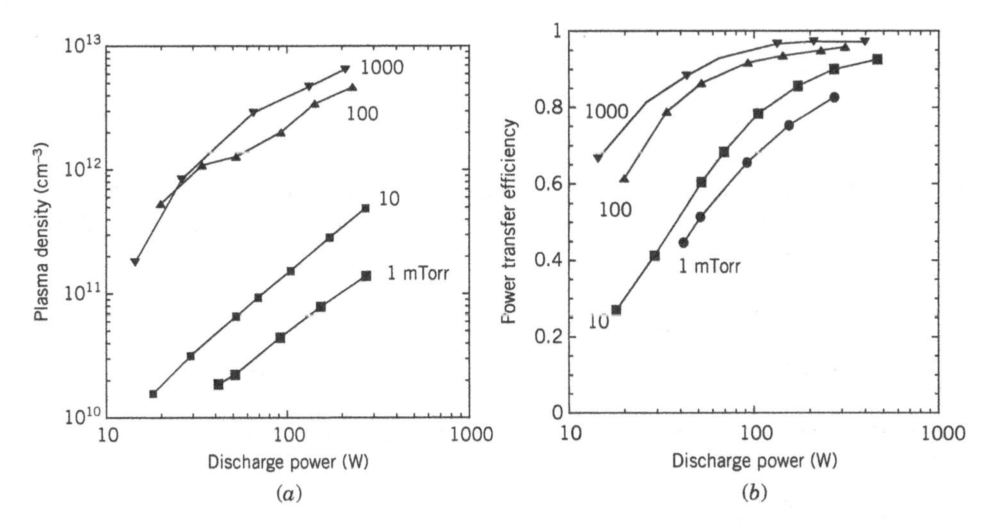

Figure 12.17 Measured central plasma densities (*a*) and discharge power efficiencies (*b*) versus absorbed discharge power at 1, 10, 100, and 1000 mTorr argon in a 2 MHz, 20 cm diameter, close-coupled, ferrite-enhanced planar inductive discharge. Source: Godyak (2011). ©IOP Publishing. Reproduced with permission. All rights reserved. https://doi.org/10.1088/0963-0252/20/2/025004.

by (10.2.17). $|Z_\mathrm{s}|$ was about 300 Ω without the plasma present, corresponding to a coil inductance $L_{11} \approx 24$ μH. $|Z_\mathrm{s}|$ was only a few percent larger with the plasma present, indicating that the magnitude of the loop impedance transformed to the coil, $\mathcal{N}^2 Z_\mathrm{p}$, was still significantly smaller than ωL_{11}, the magnitude of the unloaded primary coil impedance. Hence, this system, while having improved coupling, still does not well-approximate an ideal transformer.

However, Figure 12.17 shows that low-density inductive operation was achieved, with significant gains in discharge efficiency. In Figure 12.17*a* and *b*, respectively, the measured central plasma density and the power transfer efficiency are plotted versus the absorbed discharge power. At the lower pressures, the densities scale roughly linearly with the power, in agreement with (10.2.15). The lowest central density achieved in inductive mode was of order 2×10^{16} m^{-3}. At the highest pressures and powers, densities of order 5×10^{12} m^{-3} were obtained. The measured power transfer efficiencies were typically above 0.8 at the higher powers, significantly higher than for conventional weakly coupled discharges (Godyak, 2011). This improvement is at least partially attributed to the use of 2 MHz, rather than 13.56 MHz, to drive the discharge, resulting in significantly reduced ferrite hysteresis losses.

Various other schemes to use ferrites are described in the Godyak (2013) review. These include the use of closed ferrite cores and multiple numbers of coil antennas spread over a large area.

Problems

12.1 Skin Depth Consider a uniform electric field,

$$E_z(x,t) = \mathrm{Re}\,\tilde{E}_z(x)\,\mathrm{e}^{j\omega t}$$

at the surface of a half-space $x > 0$ of plasma having dielectric constant $\epsilon_\mathrm{p} = \epsilon_0 \kappa_\mathrm{p}$ given by (12.1.2).

(a) Using Maxwell's equations (2.2.1) and (2.2.2) with $\mathbf{J}_\mathrm{T} = \mathbf{J} + j\omega\epsilon_0\mathbf{E} = j\omega\epsilon_\mathrm{p}\mathbf{E}$ and with $\mathbf{E} = \hat{z}E_z$ in the form given above, show that

$$\frac{\mathrm{d}^2\tilde{E}_z}{\mathrm{d}x^2} = -\frac{\omega^2}{c^2}\kappa_\mathrm{p}\tilde{E}_z.$$

(b) Obtain the solution for $\tilde{E}_z(x)$ with the boundary conditions that $\tilde{E}_z = E_0$ at $x = 0$ and that $\tilde{E}_z$ is noninfinite as $x \to \infty$, and show that the electric field magnitude $|\tilde{E}_z(x)|$ decays exponentially into the plasma with a decay constant

$$\alpha = -\frac{\omega}{c}\,\mathrm{Im}\left(\kappa_\mathrm{p}^{1/2}\right)$$

(c) Evaluate α in the two limits $\nu_\mathrm{m} \ll \omega$ and $\nu_\mathrm{m} \gg \omega$, thus verifying (12.1.3) and (12.1.5).

12.2 Complex Power Deposition Starting from the basic expression for the time-averaged complex power deposition,

$$\tilde{P}_\mathrm{cmplx} = \frac{1}{2}\int J_\theta^* E_\theta\,\mathrm{d}\mathcal{V}$$

where $E_\theta(r) = J_\theta(r)/\sigma_\mathrm{p}$ with σ_p the (complex) plasma conductivity, and with $J_\theta(r)$ having an initial amplitude $J_{\theta 0}$ and decaying exponentially into the plasma with a decay constant δ_p^{-1} given by (12.1.3), with skin depth $\delta_\mathrm{p} \ll R$, obtain expression (12.1.10) for $\tilde{P}_\mathrm{cmplx}$. (Note that for σ_p complex, $\tilde{P}_\mathrm{cmplx}$ has both real and imaginary parts. This calculation neglects the small displacement current contribution $j\omega\epsilon_0 E_\theta$ flowing in the plasma.)

12.3 Self- and Mutual Inductance of Concentric Solenoids Consider two concentric solenoids of length l. The outer solenoid has $\mathcal{N}_1$ turns at radius b, and the inner solenoid has $\mathcal{N}_2$ turns at radius R. The elements of the inductance matrix are defined as

$$\Phi_1 = L_{11}I_1 + L_{12}I_2$$
$$\Phi_2 = L_{21}I_1 + L_{22}I_2$$

where Φ_i is the total magnetic flux linking the $\mathcal{N}_i$ turns of solenoid i and I_i is the feed current. The magnetic induction inside a solenoid having $\mathcal{N}_i$ turns each carrying a current I_i is uniform and given by $B_{zi} = \mu_0\mathcal{N}_i I_i/l$. Using this and the above definition, for $\mathcal{N}_1 = \mathcal{N}$ and $\mathcal{N}_2 = 1$, obtain (12.1.15), (12.1.18), and (12.1.19) for the elements of the inductance matrix.

12.4 Inductive Discharge Equilibrium

(a) Verify all calculations for the example of inductive discharge equilibrium given in Section 12.1.

(b) Estimate the electron drift velocity v_e within the skin depth layer, compare $\mathcal{E}_\mathrm{e} = \frac{1}{2}mv_\mathrm{e}^2$ to T_e, and comment on the validity of the global ionization model (10.2.12) for these discharge parameters.

12.5 Hysteresis and Stability in an Inductive Discharge Consider a low pressure, electropositive inductive discharge (no negative ions). The electron power absorption P_abs is given by (12.2.10), which includes both inductive and capacitive power deposition.

(a) Assume a linear electron power loss $P_\mathrm{loss} = K_\mathrm{e}n_\mathrm{e}$. Show that there is one and only one intersection of the $P_\mathrm{abs}(n_\mathrm{e})$ and $P_\mathrm{loss}(n_\mathrm{e})$ curves for any given value of I_rf. Hence, show that there is no hysteresis in the capacitive-to-inductive transition.

(b) Consider now a nonlinear electron loss curve of the form $P_{loss} = K_e(n_e - an_e^2)$ with $an_e < 1$ over the density range of interest. Sketch the $P_{abs}(n_e)$ and $P_{loss}(n_e)$ curves for the cases of (i) one intersection and low n_e (capacitive mode); (ii) one intersection and high n_e (inductive mode); and (iii) three intersections (region of hysteresis).

(c) The stability of the intersections can be examined from the time-varying electron particle and energy conservation equations (10.6.1) and (12.2.8) as follows: consider a small displacement $\Delta n_e > 0$ from the equilibrium value n_{e0} at an intersection. If $P_{loss}(n_{e0} + \Delta n_e) > P_{abs}(n_{e0} + \Delta n_e)$, then from (12.2.8), we find that T_e decreases. Hence, K_{iz} decreases and (10.6.1) shows that n_e decreases, i.e., n_e will be restored to its equilibrium value n_{e0}. On the other hand, if $P_{loss}(n_{e0} + \Delta n_e) < P_{abs}(n_{e0} + \Delta n_e)$, then from (12.2.8), we find that T_e increases. Hence, K_{iz} increases and (10.6.1) shows that n_e increases, a run-away situation that yields an unstable equilibrium. Using this simple picture of stability of an equilibrium, investigate the stability of the intersections in (a) and (b).

12.6 Discharge Equilibrium at High Pressure For the same R, b, l, $\mathcal{N}$, f, P_{abs}, as in the example, but with a higher pressure $p = 50$ mTorr, find all the equilibrium discharge parameters.

12.7 Discharge Equilibrium with Anomalous Skin Depth For the same R, b, l, $\mathcal{N}$, p, and P_{abs} as in the example, but with a lower frequency $f = 2$ MHz, find all the equilibrium discharge parameters. Assume that $\overline{v}_e/2\delta_e \gg \omega, \nu_m$.

12.8 Discharge Equilibrium and Matching Network

(a) Verify all calculations for the values of the matching network capacitors C_1 and C_2 given at the end of Section 12.1.

(b) Suppose P_{abs} is increased from 600 to 1200 W in the example given in Section 12.1, with R, b, l, $\mathcal{N}$, p, and f remaining the same. Find all the equilibrium discharge parameters.

(c) For part (b), determine values of C_1 and C_2 to match the discharge to a 50-Ω rf power source, using the procedure given at the end of Section 12.1.

12.9 Minimum Current for an Inductive Discharge For the same R, b, l, $\mathcal{N}$, p, and f as given in the example of Section 12.1, use (12.2.3) to determine the minimum rf current amplitude I_{min} to sustain an inductive discharge.

13

Wave-Heated Discharges

Waves generated near a plasma surface can propagate into the plasma or along the surface where they can be subsequently absorbed, leading to heating of plasma electrons and excitation of a discharge. For electron cyclotron resonance (ECR) discharges, described in Section 13.1, a right circularly polarized wave propagates along dc magnetic field lines to a resonance zone, where the wave energy is absorbed by a collisionless heating mechanism. ECR discharges are generally excited at microwave frequencies (e.g., 2450 MHz), and the wave absorption requires application of a strong dc magnetic field (875 G at resonance). The aspect ratio of these discharges, l/R for a plasma cylinder, can range from $l/R \ll 1$ to $l/R \gg 1$. For helicon discharges, described in Section 13.2, a *whistler wave* launched by an antenna propagates along a plasma column and is subsequently absorbed by a collisional or collisionless mechanism, resulting in heating of the bulk plasma electrons. Helicon discharges are usually excited at rf frequencies (e.g., 13.56 MHz), and a magnetic field of order 100 G or greater is required for wave propagation and absorption. The aspect ratio ranges from $l/R \sim 1$ to $l/R \gg 1$. For surface wave discharges, described in Section 13.3, a wave launched along the surface of the plasma propagates and is absorbed by collisional heating of the plasma electrons near the surface. The heated electrons subsequently diffuse into the bulk plasma. Surface wave discharges can be excited by either rf or microwave sources and do not require dc magnetic fields, but generally a long propagation distance is needed for efficient wave absorption, leading to discharges with high aspect ratios, $l/R \gg 1$. In contrast to capacitive rf discharges (see Chapter 11), wave-heated discharges share with inductive discharges (see Chapter 12) the characteristic that the potential of the plasma with respect to all wall surfaces is low, of order $5\,T_e$. As shown in Section 10.1, this leads to high-density plasmas at reasonably absorbed power levels. A brief description of some of the characteristics of wave-heated discharges is given in Chapter 1.

13.1 Electron Cyclotron Resonance Discharges

13.1.1 Characteristics and Configurations

Microwave generation of plasmas has been employed since the invention of high-power microwave sources in World War II. At low plasma densities, the high electric fields obtainable in a resonant microwave cavity can break down a low-pressure gas and sustain a discharge. For good field penetration in the absence of a magnetic field, $\omega_{pe} \lesssim \omega$, which sets a critical density limit $n_c \lesssim \omega^2 \epsilon_0 m/e^2$, or, in practical units, $n_c(\mathrm{m}^{-3}) \lesssim 0.012 f^2$, with f in Hz. More restrictively, for the high fields required, the cavity Q must be high, further limiting the range of operation.

The introduction of a steady magnetic field $\mathbf{B}$, in which there is a resonance between the applied frequency ω and the electron cyclotron frequency $\omega_{ce} = eB/m$ somewhere within the discharge, allows operation at high density and without a cavity resonance. Because of the cyclotron resonance, the gyrating electrons rotate in phase with the right-hand circularly polarized (RHP) wave, seeing a steady electric field over many gyro-orbits. Thus the high field of the cavity resonance, acting over a short time, is replaced by a much lower field, but acting over a much longer time. The net result is to produce sufficient energy gain of the electrons to allow ionization of the background gas. Furthermore, the injection of the microwaves along the magnetic field, with $\omega_{ce} > \omega$ at the entry into the discharge region, allows wave propagation to the absorption zone $\omega_{ce} \approx \omega$, even in a dense plasma with $\omega_{pe} > \omega$ ($n_0 > n_c$). Various aspects of ECR discharge heating, equilibrium, and applications to materials processing have been reviewed by Popov (1994) and Asmussen et al. (1997).

Figure 13.1*a* shows a typical high aspect ratio, i.e., $l > R$, ECR system, with the microwave power injected along the magnetic field lines. The power at frequency $f = \omega/2\pi$ is coupled through a vacuum end window into a cylindrical metal source chamber, which is often lined with a dielectric to minimize metal contamination resulting from wall sputtering. One or several magnetic field coils are used to generate a nonuniform, axial magnetic field $B(z)$ within the chamber. The magnetic field strength is chosen to achieve the ECR condition, $\omega_{ce}(z_{res}) \approx \omega$, where z_{res} is the axial resonance position. When a low-pressure gas is introduced, the gas breaks down and a discharge forms inside the chamber. The plasma streams or diffuses along the magnetic field lines into a process chamber toward a wafer holder. Energetic ions and free radicals generated within the entire discharge region

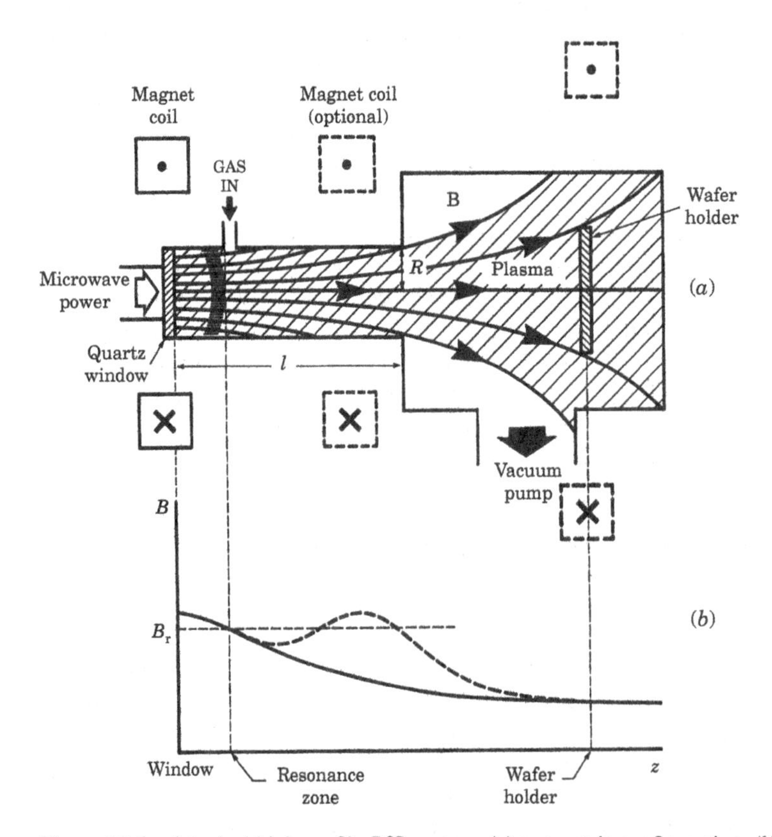

Figure 13.1 A typical high-profile ECR system: (*a*) geometric configuration; (*b*) axial magnetic field variation, showing one or more resonance zones. Source: Lieberman and Gottscho (1994)/with permission of Elsevier.

(source and process chambers) impinge on the wafer. A magnetic field coil at the wafer holder is often used to modify the uniformity of the etch or deposition process.

Typical parameters for ECR discharges used for semiconductor materials processing are shown in the last column of Table 1.1. The electron cyclotron frequency f_{ce}(MHz) $\approx 2.8B$, with B in gauss. For $f_{ce} = f = 2450$ MHz, we obtain a resonant magnetic field $B_{res} \approx 875$ G. A typical source diameter is 15 cm. In some cases, there are multiple resonance positions, as shown by the heavy dashed line in Figure 13.1*b*. A uniform profile can be used only for a low aspect ratio system ($l \lesssim R$), where the substrate is located near the point of microwave power injection, because of the difficulty of maintaining exact resonance and the possibility of overheating the electrons. The monotonically decreasing profile $\mathrm{d}B/\mathrm{d}z < 0$ shown as the solid line in Figure 13.1*b*, with one resonant zone near the window, is often used. The mirror profile shown as the heavy dashed line in Figure 13.1*b* has one resonant zone near the window and two additional zones under the second magnet. This profile can yield higher ionization efficiencies, due to enhanced confinement of hot (superthermal) electrons that are magnetically trapped between the two mirror (high-field) positions. However, the longer length of a two-mirror system leads to enhanced radial diffusion at high pressures and consequently may reduce the plasma density at the substrate.

A typical microwave power system is shown in Figure 13.2. A dc power supply drives a magnetron or klystron source coupled to the discharge by means of a TE_{10} waveguide transmission system. This consists of a circulator, to divert reflected power to a water-cooled, matched load; a directional coupler, to monitor the transmitted and reflected power; a multiscrew tuner, to match the source to the load through the dielectric window, achieving a condition of low reflected power; and, often, a mode converter, to convert the TE_{10} linear polarized, rectangular waveguide mode to an appropriate mode in the cylindrical source chamber.

The simplest mode converter (Figure 13.3*a*) is from TE_{10} rectangular to TE_{11} circular mode. At 2450 MHz, the minimum source chamber diameter for TE_{11} mode propagation (in vacuum) is 7.18 cm. However, the electric field profile and corresponding power flux are peaked on axis and are not azimuthally symmetric for this mode, leading to possible nonaxisymmetric processing profiles on the wafer. A common converter to an axisymmetric mode configuration (Figure 13.3*b*) is from TE_{10} rectangular to TM_{01} circular mode, having a minimum diameter for mode propagation of 9.38 cm at 2450 MHz. The profile is ringlike, with a vanishing on-axis power flux. The electric field

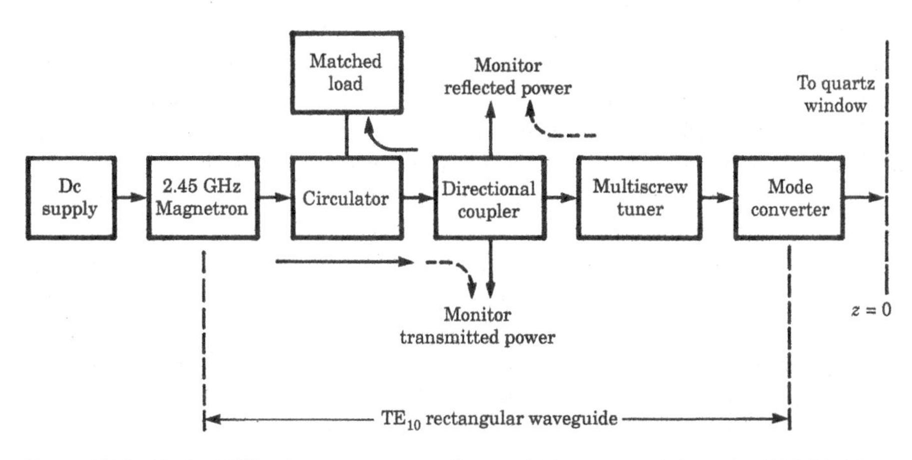

Figure 13.2 Typical ECR microwave system. Source: Lieberman and Gottscho (1994)/with permission of Elsevier.

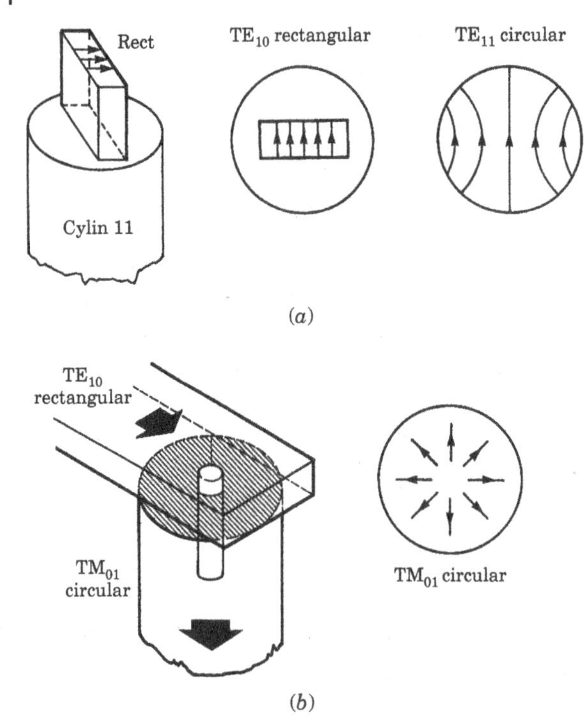

Figure 13.3 Microwave field patterns for ECR excitation; (*a*) TE_{10} rectangular to TE_{11} circular mode; (*b*) TE_{10} rectangular to TM_{01} circular mode. Source: Lieberman and Gottscho (1994)/with permission of Elsevier.

for both modes is linearly polarized, consisting of equal admixtures of RHP and left-hand circularly polarized (LHP) waves. The basic power-absorption mechanism is the absorption of the RHP wave on a *magnetic beach,* where the wave propagates from higher to lower magnetic field to the resonance $\omega_{ce}(B) \approx \omega$. The fate of the LHP wave is unclear, but it is probably inefficiently converted to a RHP wave due to multiple reflections from waveguide feed or source surfaces, or, more efficiently, from a critical density layer in the source (Musil and Zacek, 1970, 1971). An efficient scheme uses a microwave polarizer to convert from TE_{10} rectangular to a TE_{11} circular mode structure that rotates in the right-hand sense at frequency ω. This yields a time-averaged azimuthally symmetric power profile peaked on axis and having an on-axis electric field that is right-hand polarized. Hence, most of the power can be delivered to the plasma in the form of the RHP wave alone.

There are a variety of ECR processing discharges, with somewhat different coupling of the microwave power to the resonance zone. Three categories are (1) traveling wave propagation mainly along $\mathbf{B}$ (wave vector $\mathbf{k}||\mathbf{B}$), (2) propagation mainly across $\mathbf{B}$ ($\mathbf{k} \perp \mathbf{B}$), and (3) standing wave excitation (mainly cavity coupled). While these distinctions are significant, most of these ECR sources rely on the magnetic beach absorption of the RHP wave. Additionally, the sources are not neatly broken into these categories, e.g., wave propagation is at an angle to $\mathbf{B}$, and absorption can involve standing waves. In addition to 2450 MHz, lower frequencies are used in materials processing applications, e.g., 915 and 450 MHz, with corresponding resonance fields of 330 and 160 G, respectively.

Various ECR configurations are shown in Figure 13.4. A high aspect ratio system with the source plasma far from the wafer and with microwave injection along $\mathbf{B}$ is shown in Figure 13.4*a*. The resonance (heating) zone can be ring- or disk-shaped (the latter is shown) and may be as much as 50 cm from the wafer. Expansion of the plasma from the resonance zone to the wafer reduces the ion flux and increases the ion impact energy at the wafer. Hence, high aspect ratio systems have

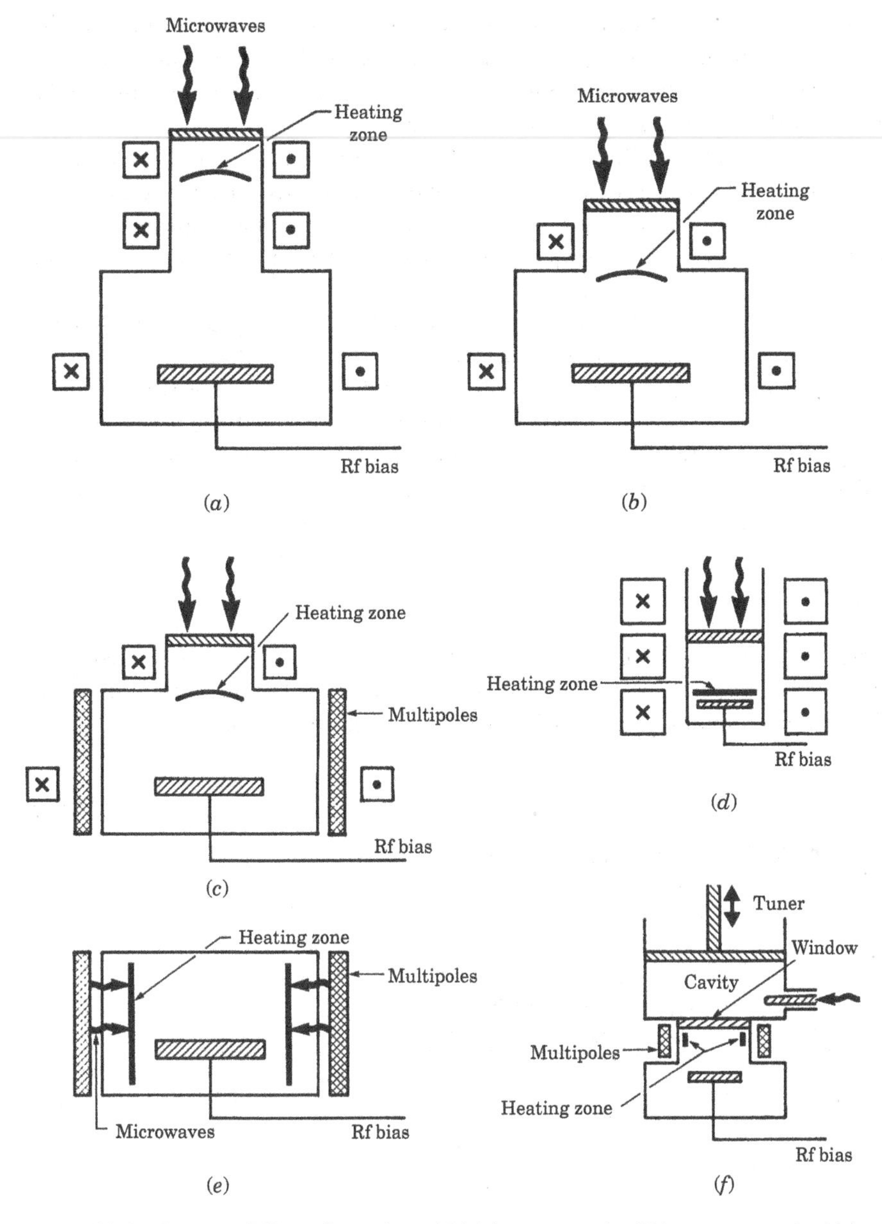

Figure 13.4 Common ECR configurations: (*a*) high aspect ratio; (*b*) low aspect ratio; (*c*) low aspect ratio with multipoles; (*d*) close-coupled; (*e*) distributed (DECR); (*f*) microwave cavity excited. Source: Lieberman and Gottscho (1994)/with permission of Elsevier.

given way to low aspect ratio systems, as shown in Figure 13.4*b*, where only a single high-field magnet is used and where the resonance zone is placed within the process chamber and may be only 10–20 cm from the wafer. Uniformity is controlled at least in part by shaping the axial magnetic field. Uniformity can be further improved and density increased by adding 6–12 linear multipole permanent magnets around the circumference of the process chamber, as shown in Figure 13.4*c*. Multipole magnetic confinement is described in Section 5.5. As another variation, a strong (rare earth) permanent magnet that generates a diverging axial magnetic field can also replace the source coil. Another approach to achieving adequate uniformity and density is to combine the source and process chambers and place the resonance zone close to the wafer, leading to a low aspect ratio *close-coupled* configuration, shown in Figure 13.4*d*. Uniformity requirements can be met by using a relatively flat, radially uniform resonance zone.

The multipole, distributed ECR (DECR) system shown in Figure 13.4*e* is powered by microwave injection perpendicular to the strong, permanent magnet, multipole magnetic fields. Typically, four or more microwave applicators are arranged around the circumference to achieve adequate uniformity. Each applicator creates an approximately linear resonance zone near the process chamber wall, as shown.

A microwave cavity source is shown in Figure 13.4*f*. The coaxial feed is tuned using a sliding short on top and a stub tuner from the side. In earlier, lower-density versions, a grid was used below the plasma generation region providing microwave containment while allowing the plasma to diffuse out. The linear resonance zones, similar to those in the DECR (Figure 13.4*e*), are generated by a set of 8–12 strong permanent magnets arranged around the circumference of the source chamber as shown. More details of the configurations in Figure 13.4 are given in review articles by Popov (1994) and Asmussen et al. (1997).

13.1.2 Electron Heating

The basic principle of ECR heating is illustrated in Figure 13.5. A linearly polarized microwave field launched into the source chamber can be decomposed into the sum of two counter-rotating circularly polarized waves. Assuming a sinusoidal steady state with the incident wave polarized along $\hat{x}$,

$$\mathbf{E}(\mathbf{r}, t) = \mathrm{Re}\,\hat{x}E_x(\mathbf{r})\mathrm{e}^{j\omega t} \tag{13.1.1}$$

where the complex amplitude E_x is here taken to be pure real, we have

$$\hat{x}E_x = (\hat{x} - j\hat{y})E_\mathrm{r} + (\hat{x} + j\hat{y})E_\mathrm{l} \tag{13.1.2}$$

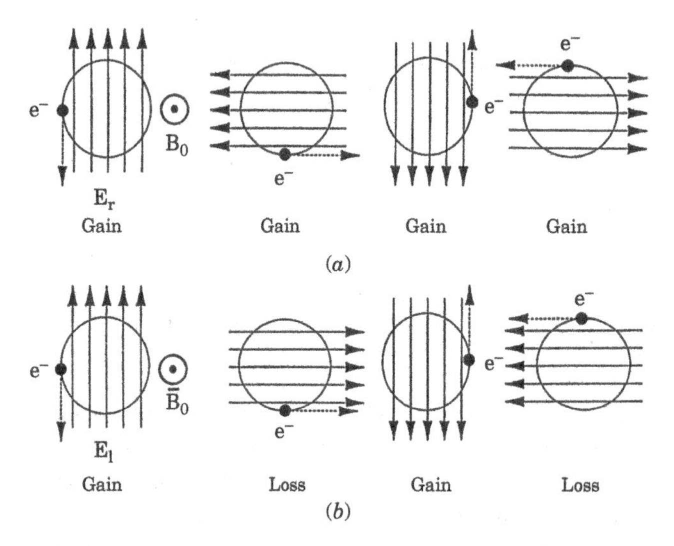

Figure 13.5 Basic principle of ECR heating: (*a*) continuous energy gain for right-hand polarization; (*b*) oscillating energy for left-hand polarization. Source: Lieberman and Gottscho (1994)/with permission of Elsevier.

where $\hat{x}$ and $\hat{y}$ are unit vectors along x and y and where E_{r} and E_{l} are the amplitudes of the RHP and LHP waves, with $E_{\mathrm{r}} = E_{\mathrm{l}} = E_x/2$. The electric field vector of the RHP wave rotates in the right-hand sense around the magnetic field at frequency ω while an electron in a uniform magnetic field B_0 also gyrates in a right-hand sense at frequency ω_{ce}. Consequently, as shown in Figure 13.5*a*, for $\omega_{\mathrm{ce}} = \omega$, the force $-eE$ accelerates the electron along its circular orbit, resulting in a continuous transverse energy gain. In contrast, as shown in Figure 13.5*b*, the LHP wave field produces an oscillating force whose time average is zero, resulting in no energy gain.

13.1.2.1 Collisionless Heating Calculation

To determine the overall heating power, the nonuniformity in the magnetic field profile $B(z)$ must be considered. For $\omega_{\mathrm{ce}} \neq \omega$, an electron does not continuously gain energy, but rather its energy oscillates at the difference frequency $\omega_{\mathrm{ce}} - \omega$. As an electron moving along z passes through resonance, its energy oscillates as shown in Figure 13.6, leading to a transverse energy gained (or lost) in one pass. For low-power absorption, where the electric field at the resonance zone is known, the heating can be estimated as follows. We expand the magnetic field near resonance as

$$\omega_{\mathrm{ce}}(z') = \omega(1 + \alpha z') \tag{13.1.3}$$

where $z' = z - z_{\mathrm{res}}$ is the distance from exact resonance, $\alpha = \partial\omega_{\mathrm{ce}}/\partial z'$ is proportional to the gradient in $B(z)$ near the resonant zone, and we approximate $z'(t) \approx v_{\mathrm{res}}t$, where v_{res} is the parallel speed at resonance.

The complex force equation for the right-hand component of the transverse velocity, $v_{\mathrm{r}} = v_x + jv_y$, can be written in the form

$$\frac{dv_{\mathrm{r}}}{dt} - j\omega_{\mathrm{ce}}(z)v_{\mathrm{r}} = -\frac{e}{m}E_{\mathrm{r}}\mathrm{e}^{j\omega t} \tag{13.1.4}$$

where E_{r} is the amplitude of the RHP wave with

$$\mathbf{E} = \mathrm{Re}\left[(\hat{x} - j\hat{y})E_{\mathrm{r}}\mathrm{e}^{j\omega t}\right] \tag{13.1.5}$$

Using (13.1.3) and substituting $v_{\mathrm{r}} = \tilde{v}_{\mathrm{r}}\exp(j\omega t)$ into (13.1.4), we obtain

$$\frac{d\tilde{v}_{\mathrm{r}}}{dt} - j\omega\alpha v_{\mathrm{res}}t\tilde{v}_{\mathrm{r}} = -\frac{e}{m}E_{\mathrm{r}} \tag{13.1.6}$$

Multiplying by the integrating factor $\mathrm{e}^{-j\theta(t)}$ and integrating (13.1.6) from $t = -T$ to $t = T$, we obtain

$$\tilde{v}_{\mathrm{r}}(T)\mathrm{e}^{-j\theta(T)} = \tilde{v}_{\mathrm{r}}(-T)\mathrm{e}^{-j\theta(-T)} - \frac{eE_{\mathrm{r}}}{m}\int_{-T}^{T} dt'\,\mathrm{e}^{-j\theta(t')} \tag{13.1.7}$$

where

$$\theta(t) = \omega\alpha v_{\mathrm{res}}t^2/2 \tag{13.1.8}$$

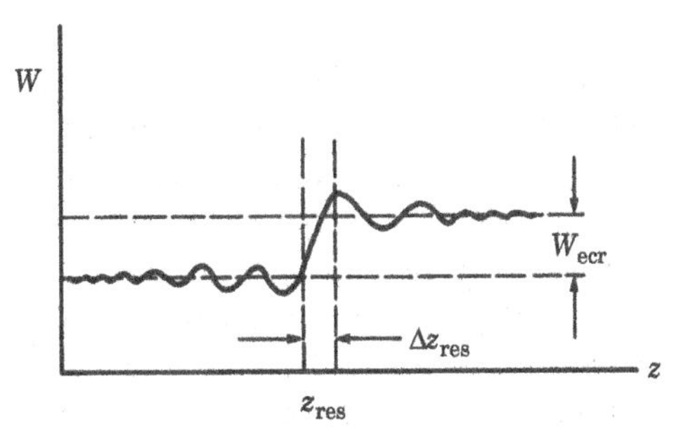

Figure 13.6 Energy change in one pass through a resonance zone. Source: Lieberman and Gottscho (1994)/with permission of Elsevier.

In the limit $T \gg (2\pi/\omega|\alpha|v_{res})^{1/2}$, the integral in (13.1.7) is the integral of a Gaussian of complex argument, which has the standard form

$$\int_{-T}^{T} dt' \, e^{-j\theta(t')} = (1-j)\left(\frac{\pi}{\omega|\alpha|v_{res}}\right)^{1/2} \tag{13.1.9}$$

Substituting (13.1.9) into (13.1.7), multiplying (13.1.7) by its complex conjugate, and averaging over the initial "random" phase $\theta(-T)$, we obtain

$$|\tilde{v}_r(T)|^2 = |\tilde{v}_r(-T)|^2 + \left(\frac{eE_r}{m}\right)^2 \left(\frac{2\pi}{\omega|\alpha|v_{res}}\right) \tag{13.1.10}$$

The average energy gain per pass is thus

$$W_{ecr} = \frac{\pi e^2 E_r^2}{m\omega|\alpha|v_{res}} \tag{13.1.11}$$

This can also be written as

$$W_{ecr} = \frac{1}{2}m(\Delta v)^2 \tag{13.1.12}$$

where $\Delta v = (eE_r/m)\Delta t_{res}$, and

$$\Delta t_{res} = \left(\frac{2\pi}{\omega|\alpha|v_{res}}\right)^{1/2} \tag{13.1.13}$$

is the effective time in resonance. The effective resonance zone width (see Figure 13.6) is

$$\Delta z_{res} \equiv v_{res}\Delta t_{res} = \left(\frac{2\pi v_{res}}{\omega|\alpha|}\right)^{1/2} \tag{13.1.14}$$

which, for typical ECR parameters, gives $\Delta z_{res} \sim 0.5$ cm.

The absorbed power per unit area, or energy flux, is found by integrating (13.1.11) over the flux nv_{res} of electrons incident on the zone, yielding

$$S_{ecr} = \frac{\pi n e^2 E_r^2}{m\omega|\alpha|} \tag{13.1.15}$$

We can understand the form of Δt_{res} as follows. An electron passing through the zone coherently gains energy for a time Δt_{res} such that

$$\left[\omega - \omega_{ce}(v_{res}\Delta t_{res})\right] \Delta t_{res} \approx 2\pi \tag{13.1.16}$$

Inserting (13.1.3) into (13.1.16) and solving for Δt_{res}, we obtain (13.1.13). The rf magnetic force was neglected in calculating (13.1.15) from (13.1.4). Because a magnetic force does no work on a moving charged particle, it does not contribute to the total power absorbed (see Section 19.5). A more careful derivation of the absorbed power (13.1.15), including the effect of non-constant v_{res} during passage through resonance, is presented by Jaeger et al. (1972), giving similar results. For both calculations, S_{ecr} is proportional to the density and the square of the RHP electric field amplitude at the resonance. The calculation here also gives S_{ecr} proportional to α^{-1} and independent of the axial electron velocity, which are not true for nonconstant v_{res}.

13.1.2.2 Collisional Heating Calculation

The fact that (13.1.15) is independent of v_{res} suggests that we can examine the $v_{\mathrm{res}} \to 0$ limit in considering the effects of electron collisions. Adding collisional (friction) terms $-\nu_{\mathrm{m}}\tilde{v}_x$ and $-\nu_{\mathrm{m}}\tilde{v}_y$ to the force equations (4.4.1*a*) and (4.4.1*b*), respectively, and solving for the transverse velocity amplitudes, we obtain

$$\tilde{v}_x + j\tilde{v}_y = -\frac{2eE_{\mathrm{r}}}{m}\frac{1}{\nu_{\mathrm{m}} + j(\omega - \omega_{\mathrm{ce}})} \tag{13.1.17}$$

where ν_{m} is the electron momentum transfer frequency. The time-averaged power absorbed per electron is

$$\overline{p}_{\mathrm{ecr}} = \frac{1}{2}\mathrm{Re}\left(-e\tilde{E}_x\tilde{v}_x^* - e\tilde{E}_y\tilde{v}_y^*\right) \tag{13.1.18}$$

Substituting (13.1.17) into (13.1.18) with $\tilde{E}_x = E_{\mathrm{r}}$, $\tilde{E}_y = -jE_{\mathrm{r}}$, we find

$$\overline{p}_{\mathrm{ecr}} = m\left(\frac{eE_{\mathrm{r}}}{m}\right)^2\frac{\nu_{\mathrm{m}}}{\nu_{\mathrm{m}}^2 + (\omega - \omega_{\mathrm{ce}})^2} \tag{13.1.19}$$

For $\omega_{\mathrm{ce}} \to \omega$, we see that

$$\overline{p}_{\mathrm{ecr}} \to \frac{e^2E_{\mathrm{r}}^2}{m\nu_{\mathrm{m}}}$$

This implies a singular behavior $\overline{p}_{\mathrm{ecr}} \to \infty$ as $\nu_{\mathrm{m}} \to 0$. However, this behavior is found only at exact resonance. To obtain the total heating power, we average (13.1.19) over the distribution of electrons near the resonance zone. Substituting the linear expansion (13.1.3) into (13.1.19), we obtain

$$\overline{p}_{\mathrm{ecr}} = \frac{e^2E_{\mathrm{r}}^2}{m}\frac{\nu_{\mathrm{m}}}{\nu_{\mathrm{m}}^2 + \omega^2\alpha^2z'^2} \tag{13.1.20}$$

Multiplying (13.1.20) by $n\,\mathrm{d}z$ and integrating from $z = -z_0$ to $z = z_0$, we obtain

$$\overline{S}_{\mathrm{ecr}} = \frac{2e^2E_{\mathrm{r}}^2n}{m\omega|\alpha|}\tan^{-1}\left(\frac{\omega|\alpha|z_0}{\nu_{\mathrm{m}}}\right) \tag{13.1.21}$$

The total power absorbed is obtained by let $z_0 \to \infty$ such that $\tan^{-1} \to \pi/2$ and (13.1.21) reduces to (13.1.15). We see that the power absorbed is independent of ν_{m} for constant electric field, and the nonlinear collisionless and the collisional power absorption calculations correspond. If we insert $z_0 \equiv \Delta z_{\mathrm{res}}$, from (13.1.14) into (13.1.21), we find that, since $\nu_{\mathrm{m}} \ll \omega|\alpha|\Delta z_{\mathrm{res}}$, almost all of the power is absorbed by collisionless heating within the resonance zone. This is the usual regime for ECR processing discharges.

13.1.3 Resonant Wave Absorption

A serious limitation on the result (13.1.15) is that it assumes that the electric field within the resonance zone is constant and known from the input power. That this cannot be true in the case of strong absorption is clear, since the absorbed power cannot exceed the incident power. The resolution of this difficulty lies in the attenuation of the wave in the resonance zone, so that the resonant value of E_{r} is in fact much smaller than the value of the incident E_{r}.

The propagation and absorption of microwave power in ECR sources are not fully understood. For excitation at an end window (Figure 13.4*a*–*d*), the waves in a cylindrical magnetized plasma are neither exactly RHP nor propagating exactly along **B**. The waves are not simple plane waves and the mode structure in a magnetized plasma of finite dimension must be considered. Nevertheless, the essence of the wave coupling, and transformation and absorption at the resonance zone, can be understood by considering the one-dimensional problem of an RHP plane wave propagating strictly along **B** in a plasma that varies spatially only along z. For right-hand polarization (13.1.5), where now $E_r(z)$ is the spatially varying electric field amplitude, the wave equation for plane waves propagating along **B** parallel to z can be written as (Problem 13.1)

$$\frac{d^2E_r}{dz^2} + k_0^2\kappa_r E_r = 0 \tag{13.1.22}$$

Far from resonance such that $\omega - \omega_{ce} \gg \nu_m$, we have the relative dielectric constant (4.5.5a),

$$\kappa_r = 1 - \frac{\omega_{pe}^2(z)}{\omega\left[\omega - \omega_{ce}(z)\right]} \tag{13.1.23}$$

with $k_0 = \omega/c$ and c the velocity of light. κ_r varies with z due to the dependence of ω_{pe}^2 on the density $n(z)$ and of ω_{ce} on the magnetic field $B(z)$. If the variation of κ_r with z is weak,

$$\frac{d\lambda}{dz} \ll 1 \tag{13.1.24}$$

with $\lambda = 2\pi/k$, and

$$k(z) = k_0\kappa_r^{1/2}(z) \tag{13.1.25}$$

then a Wentzel–Kramers–Brillouin (WKB) wave expansion can be made (Stix, 1992):

$$E_r(z) = E_{r1}(z)\exp\left[-j\int^z k(z')\,dz'\right] \tag{13.1.26}$$

where $E_{r1}(z) = E_{r0}k_0^{1/2}/k^{1/2}(z)$ is the spatially varying amplitude of the wave. The WKB wave propagates without absorption for $\kappa_r > 0$, where k is real, and the wave is evanescent for $\kappa_r < 0$, where k is imaginary. The WKB result can be understood from a calculation of the time-averaged power per unit area carried by the wave,

$$\overline{S}_r = \frac{1}{2}Z_0^{-1}\kappa_r^{1/2}E_{r1}^2 = \text{const} \tag{13.1.27}$$

where $Z_0 = (\mu_0/\epsilon_0)^{1/2} \approx 377\ \Omega$ is the impedance of free space. This indicates that the propagation is without reflection. The result is characteristic of slowly varying solutions called *adiabatic*. We note that for propagation close to cyclotron resonance, from (13.1.23), κ_r becomes large, as does k, and from (13.1.26), E_{r1} becomes small. However, as resonance is approached from the propagating side ($\omega_{ce} > \omega$), the condition of slow spatial variation (13.1.24) is no longer satisfied, and the WKB approximation breaks down.

What happens as a wave propagates through the resonance into an evanescent region where $\omega_{ce} < \omega$? The answer was obtained analytically by Budden (1966), for the approximation of constant density and linear magnetic field. Reintroducing collisions into the plasma dielectric constant, (13.1.22) becomes

$$\frac{d^2E_r}{dz^2} + k_0^2\left\{1 - \frac{\omega_{pe}^2(z)}{\omega\left[\omega - \omega_{ce}(z) - j\nu_m(z)\right]}\right\}E_r = 0 \tag{13.1.28}$$

Taking ω_{pe} and ν_m to be constants independent of z, and linearizing ω_{ce} about the resonance point, (13.1.28) reduces to

$$\frac{d^2E_r}{ds^2} + \left(1 + \frac{\eta}{s + j\gamma}\right)E_r = 0 \tag{13.1.29}$$

where we have normalized z by $s = k_0(z - z_{res})$, ω_{pe}^2 by $\eta = \omega_{pe}^2/(\omega c|\alpha|)$, and ν_m by $\gamma = \nu_m/(c|\alpha|)$. The dielectric function has both a pole and a zero, with the pole, in the absence of collisions, occurring at $s = 0\,(z = z_{res})$ and the zero at $s = -\eta$. In this approximation, Budden has obtained a solution to (13.1.29) in the limit of $\gamma \to 0$. For a wave traveling into a decreasing magnetic field (the magnetic beach), he obtained

$$\frac{S_{abs}}{S_{inc}} = 1 - e^{-\pi\eta} \tag{13.1.30}$$

$$\frac{S_{trans}}{S_{inc}} = e^{-\pi\eta} \tag{13.1.31}$$

$$\frac{S_{refl}}{S_{inc}} = 0 \tag{13.1.32}$$

Hence, some of the wave power is absorbed at the resonance while some tunnels through to the other side, but no power is reflected. Taking a typical case for which $\alpha = 0.1\ \text{cm}^{-1}$ and $k_0 = 0.5\ \text{cm}^{-1}$, we find that $\eta > 1$ corresponds to $\omega_{pe}^2/\omega^2 > 0.2$. Thus at 2450 MHz, we expect that most of the incident power will be absorbed for a density $n_0 \gtrsim 1.5 \times 10^{10}\ \text{cm}^{-3}$.

The Budden result can be qualitatively understood in terms of the dispersion diagrams of k versus ω_{ce}/ω, as shown in Figure 13.7, with $\omega_{pe}^2 = \text{const}$. In Figure 13.7*a*, for low density ($\omega_{pe} < \omega$), the wave is evanescent downstream of the resonance in the region

$$1 - \frac{\omega_{pe}^2}{\omega^2} < \frac{\omega_{ce}}{\omega} < 1$$

and is propagating otherwise. For $\omega_{pe} \ll \omega$, the region of evanescence is thin (in z), and the wave can tunnel through this region to propagate again further downstream. In Figure 13.7*b*, for high density ($\omega_{pe} > \omega$), the wave is always evanescent downstream such that the tunneling fields fall off exponentially. Within Budden's approximation, we have the nonintuitive result that there is no reflected power for a wave incident on the resonance zone from the high-field side.

Since from (10.2.14) and (10.2.15), with the pressure dependence of h_l and h_R determined from (10.2.1) and (10.2.2), the bulk density scales as $n_0 \propto S_{abs}$ at low pressures and as $n_0 \propto p^{1/2} S_{abs}$ at high pressures, we obtain from (13.1.30) the region of good power absorption $\eta \geq 1$, as sketched in Figure 13.8. For parameters well within this region, the incident microwave power is efficiently absorbed over the entire cross section of the resonance zone. For operation outside this region, considerable microwave power can impinge on the substrate.

The minimum S_{inc} to sustain an ECR discharge can similarly be found. Expanding (13.1.30) for small η (n_0 small) yields $S_{abs} = \pi\eta S_{inc}$. Substituting this into (10.2.14), we obtain the minimum value of S_{inc} to sustain the discharge. At a given pressure, this minimum is found to be a factor of two below the $\eta = 1$ condition for good power absorption, as illustrated in Figure 13.8. The situation is analogous to the case for an inductive discharge (see Figure 12.4 and accompanying discussion), in which there is a tangency between power absorbed and power lost versus density, at low density. As in the inductive discharge, the plasma does not turn off, but has a transition to a considerably lower density state (sometimes referred to as a "low mode"). This can be qualitatively understood in terms of cavity resonance effects (Williamson et al., 1992). We should also note that the discharge cannot be sustained if the pressure drops below some minimum value p_{min}, because the particle balance equation (10.2.12) has no solution for T_e. This limit is also illustrated in Figure 13.8.

In a number of respects, the Budden theory is rather idealized for direct application to a physical system. The reflections in a plasma chamber generate interference of waves that can significantly affect the absorption. The variation of axial density causes initial upstream power reflection. The

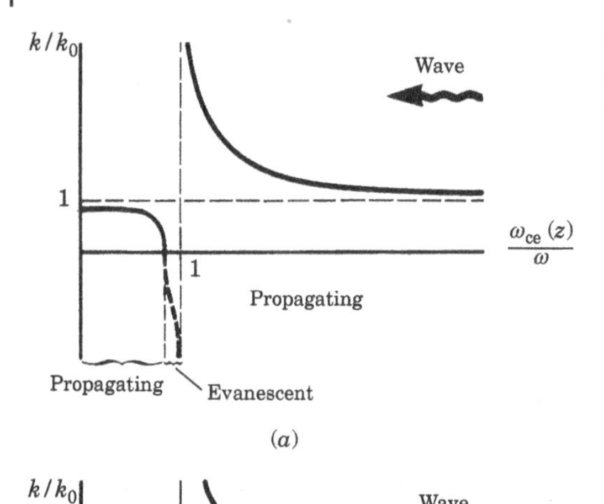

Figure 13.7 k/k_0 versus ω_{ce}/ω for (a) low density $\omega_{pe}/\omega \ll 1$ and (b) high density $\omega_{pe}/\omega \gg 1$. The heavy dashed curves denote imaginary values for k.

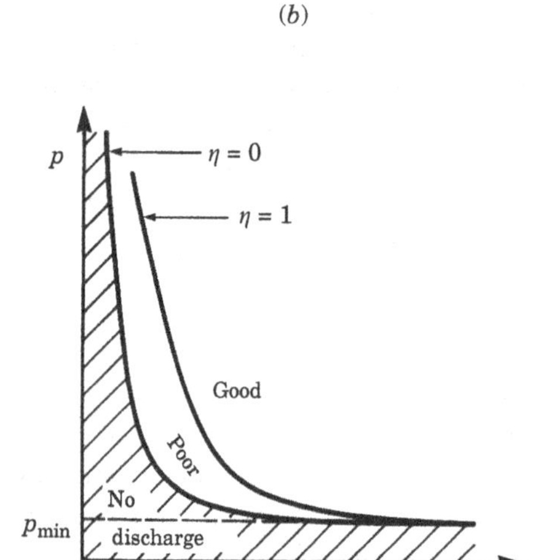

Figure 13.8 Parameters for good ECR source operation: pressure p versus incident power S_{inc}. Source: Lieberman and Gottscho (1994)/with permission of Elsevier.

collisionality is locally enhanced by nonlinear absorption of power in the resonance zone. These effects can be taken into account in a one-dimensional model by numerical integration of the fundamental equation (13.1.28) with boundary conditions imposed at each end of the region of interest (Williamson et al., 1992). The results indicate that the Budden theory holds reasonably well for strong absorption.

The size, shape, and location of the resonant zone are set by the magnet coil configuration and the magnet currents. The zone shape and location are also modified by the Doppler effect for electrons incident on the zone. The actual resonance position is determined by the Doppler-shifted frequency,

$$\omega + kv_{\text{res}} = \omega_{\text{ce}}(z_{\text{res}}) \tag{13.1.33}$$

At high densities, from (13.1.25), k can be large near the zone, leading to a large Doppler shift. For example, for $k = 6.3\ \text{cm}^{-1}$ ($\lambda = 1$ cm), a typical value at the edge of the resonance zone, and $v_{\text{res}} = 10^8$ cm/s (a 3 V electron), we obtain $kv_{\text{res}}/\omega \approx 0.094$. Hence, the resonant magnetic field is 910 G for this electron and not 875 G. For $\alpha = 0.1\ \text{cm}^{-1}$, this leads to a broadening of the zone of ± 0.4 cm. By using a coaxial electrostatic probe to sample, the microwave field in an ECR discharge and beating that signal against a reference signal from the incident microwaves, Stevens et al. (1992) have measured the microwave field amplitude as a function of position and verified that the resonant zone is Doppler broadened, with the absorption beginning at ~975 G for their case.

Radial density and magnetic field variations can lead to wave refraction effects that are significant. The radial gradients are generally much larger than the axial gradients in high aspect ratio ($l/R \gg 1$) sources. A radial density profile that is peaked on axis leads to a dielectric constant κ_{r} that is peaked on axis. This in turn can lead to a self-focusing effect that can increase the sharpness of the microwave power profile as the wave propagates to the zone, adversely affecting uniformity. The mechanism is analogous to the use of a graded dielectric constant optical fiber to guide an optical wave. However, the ECR refraction problem is much more complicated because the density profile is not known a priori and the magnetized plasma medium cannot be represented as an isotropic dielectric. A simplified picture of the refraction is obtained in the geometrical optics limit by examining the trajectories of optical rays as they propagate. The ray dynamics are derivable from the dispersion equation and have a Hamiltonian form (Born and Wolf, 1980, Appendix II), with $(k_\perp, r)$ and (k_z, z) canonically conjugate variable pairs and with $\omega(k_\perp, k_z, r, z)$ the Hamiltonian. For high densities and magnetic fields ($\omega_{\text{pe}}, \omega_{\text{ce}} \gg \omega$) and propagation at an angle to the magnetic field, the dispersion equation reduces to that of whistler waves (see Problem 4.14), with

$$\omega = \frac{k_0^2 \omega_{\text{pe}}^2}{k k_z \omega_{\text{ce}}} \tag{13.1.34}$$

where $k = (k_\perp^2 + k_z^2)^{1/2}$ is the wave-vector magnitude and $k_\perp$ and k_z are the radial and axial components, respectively. Choosing $\omega_{\text{pe}}^2/\omega_{\text{ce}}$ to have radial variation only, independent of z, Hamilton's equations are

$$\frac{\text{d}k_\perp}{\text{d}t} = -\frac{\partial \omega}{\partial r}, \qquad \frac{\text{d}r}{\text{d}t} = \frac{\partial \omega}{\partial k_\perp}, \qquad \frac{\text{d}k_z}{\text{d}t} = -\frac{\partial \omega}{\partial z} \equiv 0, \qquad \frac{\text{d}z}{\text{d}t} = \frac{\partial \omega}{\partial k_z}$$

They show that k_z is conserved along the path of a ray and that the ray propagates in the direction of the group velocity $\mathbf{v}_{\text{g}} = \nabla\omega(\mathbf{k})$. If $\omega_{\text{pe}}^2/\omega_{\text{ce}}$ is a decreasing function of r, then (13.1.34) shows that $k_\perp$ decreases with increasing r, implying that the ray bends toward the axis, a focusing action. On the other hand, for some parameter choices, e.g., $\omega_{\text{pe}} \sim \omega \sim \omega_{\text{ce}}$, a refraction of the wave away from the axis has been found by numerical integration of the ray equations, leading, for this particular case, to an increased uniformity of the power flux profile (Stevens et al., 1992). For some source concepts (e.g., DECR in Figure 13.4e), the microwave power is injected perpendicular to the magnetic field, and not parallel to the field. In this case, the feed structure excites the extraordinary (x) wave (see Section 4.5), which in the WKB limit has a resonance at the upper hybrid frequency (4.5.15), $\omega_{\text{UH}} = (\omega_{\text{pe}}^2 + \omega_{\text{ce}}^2)^{1/2}$, where the wave power is absorbed. Since ω_{UH} depends on both ω_{pe} and ω_{ce}, we see that the shape and location of the resonance zone depend on the density as well as the magnetic configuration. Furthermore, the x wave is evanescent for frequencies such that

$\omega_{UH} < \omega < \omega_R$, where ω_R is given by (4.5.11). For a fixed driving frequency $\omega > \omega_{ce}$, the x wave must tunnel through this evanescent layer on its journey from the feed structure to the zone. For $\omega_{pe} > \omega$, the tunneling becomes rapidly small with increasing density and the wave cannot propagate to the zone. This can limit the density obtainable in these sources to the order of 2×10^{12} cm^{-3} at 2450 MHz. However, the limitation is not severe for typical processing applications. Microwave cavity sources (Figure 13.4*f*) can suffer from similar limitations. On the other hand, densities as high as 3×10^{13} cm^{-3} have been generated using 2450 MHz RHP wave injection along **B**.

A number of other power absorption issues and mechanisms can play a role in ECR sources. The electron oscillation velocity under the application of the microwave field at resonance may become large compared to the electron thermal speed. Higher harmonic resonances ($\omega = 2\omega_{ce}$, $3\omega_{ce}$), the upper hybrid resonance (4.5.15), and the left-hand polarized wave can give rise to significant power absorption. These and other heating issues have been reviewed in the context of materials processing discharges by Popov (1994).

13.1.4 Model and Simulations

The spatially averaged model of the discharge equilibrium described in Section 10.2 can be applied to determine the plasma parameters for a given geometry, magnetic field, pressure, and absorbed power. For a cylindrical plasma of radius R and length l with a strong axial magnetic field, particle balance (10.2.10) can be used to determine T_e, and power balance (10.2.14) then determines n_0. The procedure was described in Example 10.2 of Section 10.2 for the case appropriate to ECR discharges where particle loss to the walls is inhibited by the strong field. In this case, d_{eff} in (10.2.13) is $l/2h_l$, A_{eff} in (10.2.11) is $2\pi R^2 h_l$, and the ion-bombarding energy is $\mathcal{E}_i \approx V_s + \frac{1}{2}T_e$, with V_s given by (10.2.4).

Porteous et al. (1994) have compared results from a two-dimensional simulation of plasma transport in a low aspect ratio ($l/R \lesssim 1$) ECR source with predictions from the spatially averaged model. A two-dimensional hybrid simulation was used. The electrons were treated as a fluid, with the particle, momentum, and energy conservation equations (2.3.7), (2.3.9), and (2.3.21) coupled by the electric field in two dimensions (r, z) to the motion of the ions. These were treated as a collection of particles acted on by the Lorentz force (2.2.12), along with Monte Carlo collisions against the background neutral gas. Argon gas with a simplified set of cross sections, similar to that described in Chapter 3, was used. The source geometry and magnetic field lines are shown in Figure 13.9. At $P_{abs} = 850$ W, the spatially averaged electron temperature, ion-bombarding energy at the endwall $z = 21.5$ cm, and density are shown versus the pressure p for the simulation and the model in Figure 13.10*a*, *b*, and *c*, respectively. The model and simulation agree to within about 10% over the pressure range 0.5–10 mTorr and power range 850–1500 W. The model provides insight into the discharge behavior and scaling with control parameters, while the simulation provides spatial profile information on the plasma density and the ion-bombarding energy and flux at the substrate surface. Comparisons of measured plasma densities and electron temperatures with a global model have been given by Vidal et al. (1999) (see discussion in Section 5.4).

13.1.5 Plasma Expansion

In a high aspect ratio system, $l \gg R$, where the plasma flows from a small-diameter source chamber into a larger-diameter process chamber along the diverging magnetic field lines, the plasma density n_0 in the source chamber can considerably exceed the density n_s in the process chamber where the

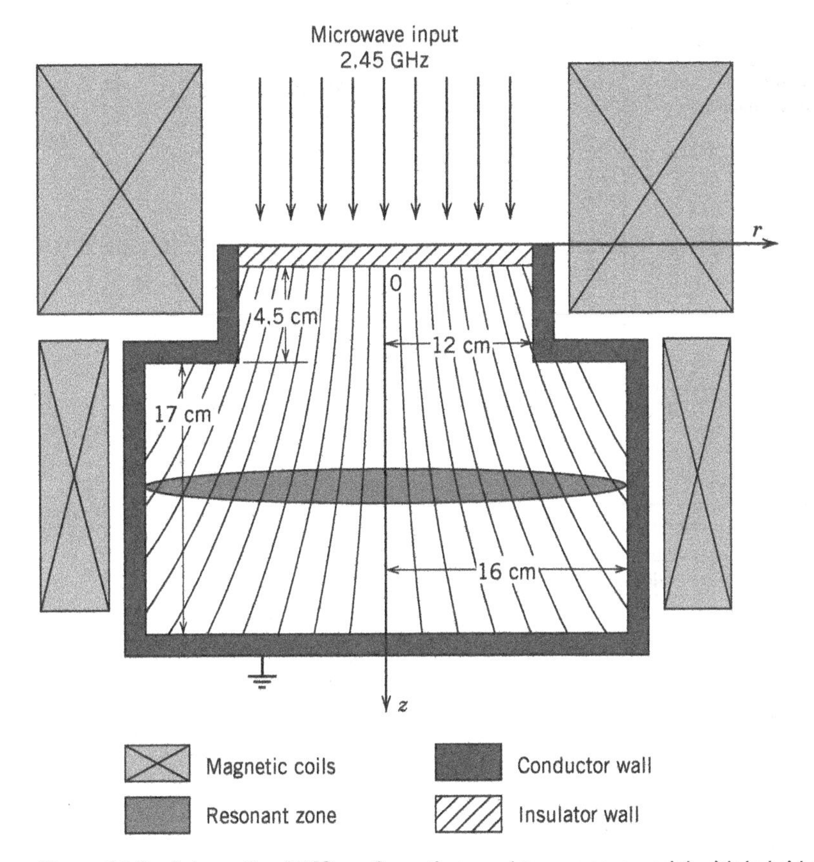

Figure 13.9 Schematic of ECR configuration used to compare model with hybrid simulation.

substrate is located. In this case, illustrated in Figure 13.11, a *distributed potential* V_d exists between the source and process chambers, with V_d related to n_0 and n_s by the Boltzmann factor,

$$n_s = n_0\, e^{-V_d/T_e} \tag{13.1.35}$$

This dc potential acts to accelerate ions from the source exit to the plasma–sheath edge near the substrate. As the ions cross the sheath, they are further accelerated by the wall sheath potential V_s. The drop in density in a high aspect ratio system is due to the expansion in the area of the plasma as it flows along field lines, to the increase in ion velocity at a fixed flux, and to particle loss by radial diffusion to the walls.

An estimate of the potentials can be made in the collisionless (very low pressure) limit, ignoring radial diffusion, using the model in Figure 13.11. The assumptions are that ions are generated only within the source chamber and flow out of the source with a characteristic velocity $u_i \sim u_B \ln(n_{max}/n_0) \sim u_B$, the Bohm velocity, where the magnetic field is B_0 and the cross-sectional area is A_0. The ions flow along the magnetic field lines as the magnetic field decreases, such that the cross-sectional area expands. By conservation of magnetic flux, we have

$$A_s = A_0 \frac{B_0}{B_s} \tag{13.1.36}$$

The ion particle and energy balance equations are

$$n_s u_s A_s = n_0 u_B A_0 \tag{13.1.37}$$

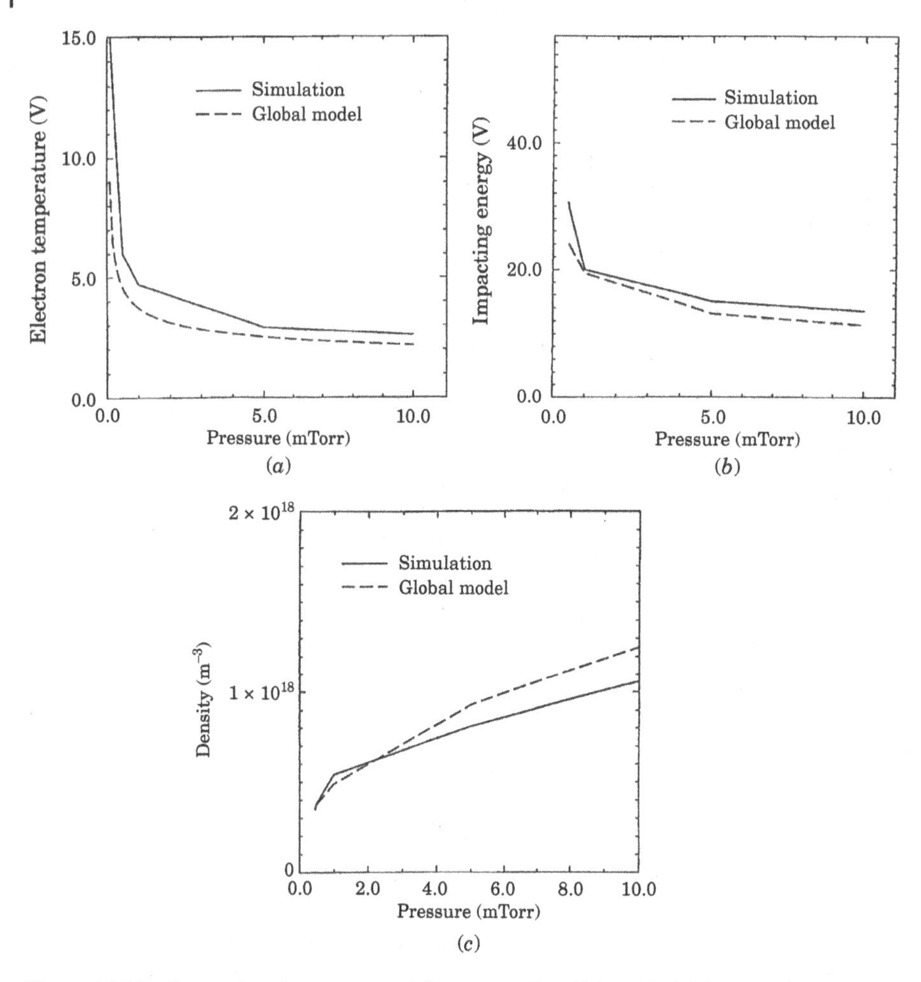

Figure 13.10 Comparison between spatially averaged model and hybrid simulation predictions of (*a*) electron temperature, (*b*) ion impact energy, and (*c*) plasma density, versus neutral gas pressure, for $P_{abs} = 850$ W. Source: Adapted from Porteous et al. (1994).

$$\frac{1}{2}Mu_s^2 = \frac{1}{2}Mu_B^2 + eV_d \tag{13.1.38}$$

Equations (13.1.35)–(13.1.38) can be solved numerically to determine V_d/T_e, u_s/u_B, and n_s/n_0 as functions of the area expansion ratio A_s/A_0. The results for V_d/T_e are shown in Figure 13.12. The further potential drop across the wall sheath can then be found by equating ion and electron fluxes at the wall. With

$$\Gamma_i = n_s u_s \tag{13.1.39}$$

$$\Gamma_e = \frac{1}{4}n_s\bar{v}_e\, e^{-V_s/T_e} \tag{13.1.40}$$

and setting $\Gamma_i = \Gamma_e$, we obtain V_s/T_e as shown in Figure 13.12. The total potential drop $V_T = V_d + V_s$ from the source to the wall is also shown in the figure.

The ion-bombarding energy for a collisionless ion flow from the source to the wall is

$$\mathcal{E}_i = V_d + V_s + \frac{1}{2}T_e \tag{13.1.41}$$

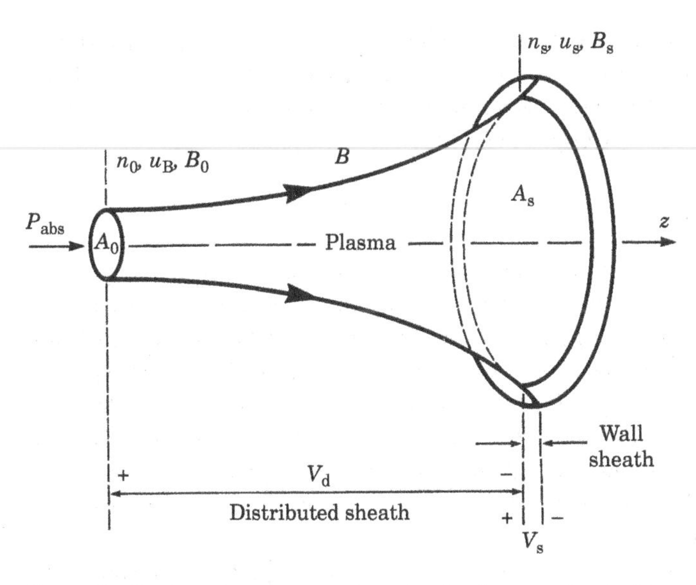

Figure 13.11 Model used to calculate the distributed potential V_d and the sheath potential V_s in a diverging field ECR system.

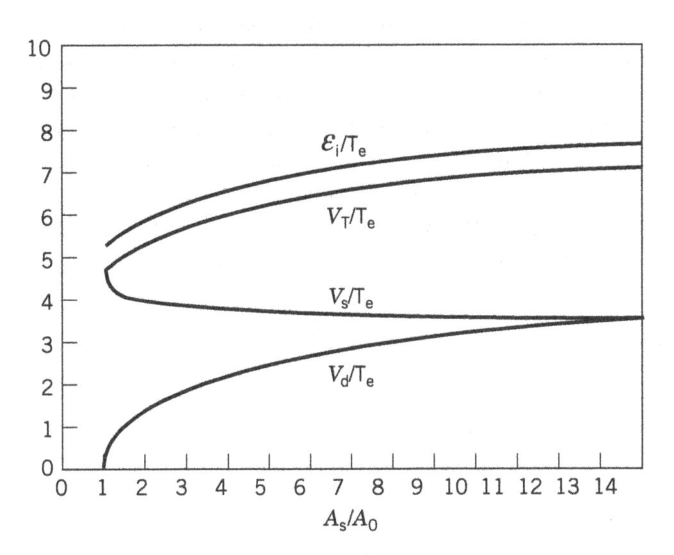

Figure 13.12 Potential drops V_T, V_s, and V_d versus A_s/A_0 for a diverging field ECR system.

where the last term is the initial ion energy at the source exit. Figure 13.12 shows that $\mathcal{E}_i \approx 5$–$8\,T_e$ over a wide range of area expansion ratios. However, the ion flow across the distributed potential is collisionless only at very low pressures, such that $\lambda_i \gtrsim l_d$, where l_d is the length of the distributed potential region (see Figure 13.11). For $\lambda_i < l_d$, the more usual pressure regime, the ion energy is modified to

$$\mathcal{E}_i \approx \frac{\lambda_i}{l_d} V_d + V_s + \frac{1}{2} T_e \tag{13.1.42}$$

The first term in (13.1.42) is reduced below V_d because ion energy is lost during charge transfer and elastic collisions in the expansion region. The situation is similar to that described in Section 11.2

for the ion-bombarding energy of the collisionless and collisional sheaths in a capacitive rf discharge. For $\lambda_i \ll l_d$, V_s reduces to the usual sheath voltage (10.2.4), and the first term in (13.1.42) is negligible. Hussein and Emmert (1990) have given a more complete description of the potential drops and ion-bombarding energies in a diverging magnetic field ECR system.

13.1.6 Measurements

Most measurements of ion energy distributions in high-density sources have been done for diverging field systems. The work of Matsuoka and Ono (1988) is typical. Microwaves are launched from a cavity into a high magnetic field region so that the RHP wave propagates and then is absorbed, heating electrons in the process. Because the magnetic field continues to decrease, the plasma expands, the plasma density decreases, and an ambipolar field is created that accelerates ions along the magnetic field gradient. At some point downstream, ions are sampled through a 50 μm pinhole and energy analyzed using two grids and a collector. Although the relatively large orifice diameter and the use of arbitrary units for spatial distance makes the work of mostly qualitative value, the trends are notable and are borne out in many other experiments.

Matsuoka and Ono focused primarily on the effects of magnetic field configuration and pressure. By varying the dc current i_m in an electromagnet located near the sampling plane, they modified the divergence of the magnetic field, varying the field from a mirror at high current to a cusp at negative current ($i_c = -i_m$). Figure 13.13 shows their ion energy distributions for different currents i_m. As the field is made uniform (large i_m), the parallel ion temperature T_i (spread in $N(\mathcal{E}_i)$) decreases and the average energy $\overline{\mathcal{E}}_i$ shifts to lower values. At the same time, the ion current density increases, the plasma potential (deduced from Langmuir probe current–voltage characteristics) decreases, and the plasma potential gradient or electric field decreases. These effects are all consistent with reduced plasma expansion. By contrast, the largest $\overline{\mathcal{E}}_i$ and T_i are obtained when the subcoil magnet is used to produce a cusp (highly diverging field) before the sampling orifice. Under these conditions, the plasma expansion is largest as the magnetic field decreases to zero and then reverses on

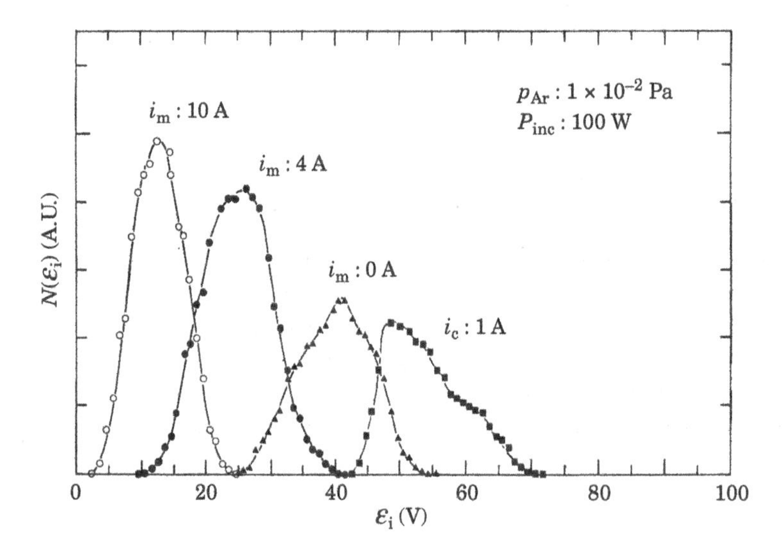

Figure 13.13 Change in the bombarding ion energy distribution as the wafer-level coil current i_m is varied. Source: Matsuoka and Ono (1988)/with permission of AIP Publishing.

the other side of the cusp. Note that ions and electrons do not follow field lines through a cusp since the field decreases to zero. Regardless of the magnetic field configuration, both $\overline{\mathcal{E}}_i$ and T_i decrease as the pressure is increased and charge exchange cools the ions. Many other experimental studies in research reactors have been made and compared to various model results to determine heating mechanisms and global plasma parameters such as density and electron temperature and their scaling with discharge parameters, as well as radial and axial profiles. The reader is referred to the reviews of Popov (1994) and Asmussen et al. (1997) for further information.

On the other hand, due to very limited access, there have been few reported measurements in commercial plasma processing reactors. As one example, phase-resolved optical emission spectroscopy (PROES, see Section 8.6) was performed on the commercial system shown in Figure 13.14 (Milosavljević et al., 2013; Milosavljević and Cullen, 2015). This is a close-coupled configuration as shown in Figure 13.4*d*. The discharge was excited in 8–30 mTorr argon by both 2.45 GHz microwave power and 2 MHz substrate bias power, with a silicon dioxide wafer on the substrate holder. The time-varying optical emission was observed from both longitudinal and transverse viewports. Unlike the usual dual frequency measurement, the PROES signal was synchronized to the low-frequency bias supply period (500 ns), because synchronization to the very short (0.4 ns) microwave period was beyond the PROES measurement capabilities. Both 750 nm argon and 777.2 nm oxygen emission lines were monitored. One conclusion from this work is that the bias power can increase the argon emission, and therefore, the ionization rate, by up to 15%. Thus the microwave power alone does not exclusively control the density and flux to the substrate. Similar results for dual frequency capacitive discharges are described in Section 11.6. However, in practice, independent control of ion flux and ion bombarding energy is not a serious issue in commercial use.

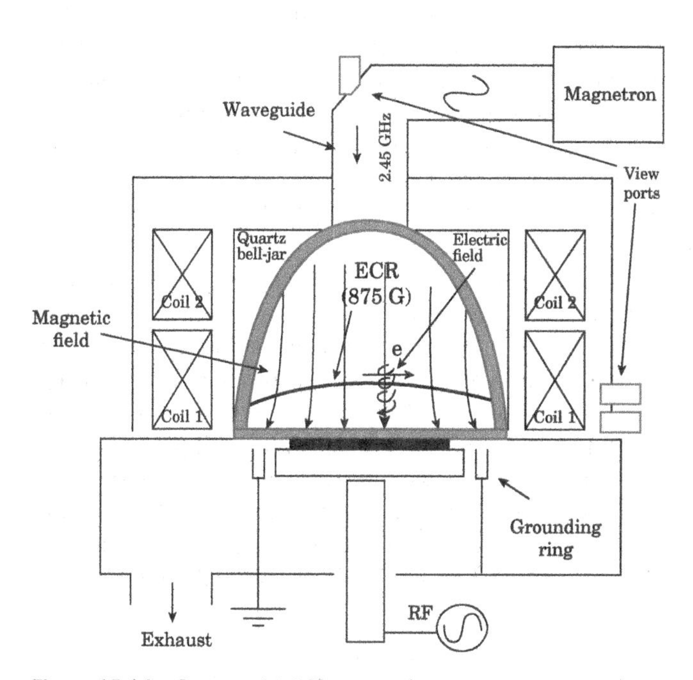

Figure 13.14 Commercial ECR reactor for phase-resolved optical emission spectroscopy. Source: Milosavljević et al. (2013)/with permission of AIP Publishing.

13.2 Helicon Discharges

Helicon generation of plasmas was first employed by Boswell (1970), following a 10-year history of helicon propagation studies, first in solid-state and then in gaseous plasmas. The early history is described in a review article by Boswell and Chen (1997). A detailed theory of helicon propagation and absorption is given by Chen (1991). Helicon discharge physics is reviewed by Chen and Boswell (1997), and, along with newer work, by Curreli and Chen (2011), Tarey et al. (2012), and Li et al. (2023). Helicons are propagating whistler wave modes in a finite diameter, axially magnetized plasma column, with dispersion as given in (13.1.34) (see also Problem 4.14). The electric and magnetic fields of the modes have radial, axial, and, usually, azimuthal variation, and they propagate in a low-frequency, low magnetic field, high-density regime characterized by

$$\omega_{\rm LH} \ll \omega \ll \omega_{\rm ce} \tag{13.2.1}$$

$$\omega_{\rm pe}^2 \gg \omega\omega_{\rm ce} \tag{13.2.2}$$

where $\omega_{\rm LH}$ is the lower hybrid frequency given by (4.5.20), with $\omega_{\rm pi}$ and $\omega_{\rm ci}$ the ion plasma frequency and ion gyrofrequency, respectively. The driving frequency is typically 1–50 MHz, with 13.56 MHz commonly used for processing discharges. The magnetic fields vary from 100 G for typical processing discharges up to 1000 G for some fundamental plasma studies. Plasma densities range from 10^{11}–10^{14} cm^{-3}, with 10^{11}–10^{12} cm^{-3} typical for processing.

Helicons are excited by an rf-driven antenna that couples to the transverse mode structure across an insulating chamber wall. The mode then propagates along the column, and the mode energy is absorbed by plasma electrons due to collisional or collisionless damping. All helicon applications to materials processing to date have utilized a process chamber downstream from the source. A typical helicon system is shown in Figure 1.16*c*. The plasma potential in helicon discharges is typically low, of order 15–20 V, as in ECRs. However, the magnetic field is much lower than the 875 G required for ECRs, and the helicon power is supplied by rf rather than microwave sources. The smaller magnetic field, in particular, may lead to lower cost for helicon sources when compared to the ECR sources. However, as we will see, the resonant coupling of the helicon mode to the antenna can lead to nonsmooth variation of density with source parameters, known as "mode jumps," restricting the operating regime for a given source design.

The rf power system driving the helicon antenna can be of conventional design (as for rf capacitive discharges; see Section 11.11). A 500–5000 W, 50 Ω, 13.56 MHz supply can be used to drive the antenna through a matching network to minimize the reflected power seen by the supply. The matching network can be an L-design with two variable capacitors, as for inductive discharges. The antenna can also be driven through a balanced transformer so that the antenna coil is isolated from ground. This reduces the maximum antenna–plasma voltage by a factor of two, thus also reducing the undesired capacitive current coupled to the plasma by a factor of two. Since low aspect ratio geometries have not been developed for helicons, as they have for ECRs (see Figure 13.4*d*), the transport and diffusion of the source plasma into the process chamber may be a significant limitation. The process chamber can have multipole confinement magnets to increase uniformity (see Section 5.5) or can have a wafer-level magnet coil (e.g., as in Figure 13.1*a*) to keep the source plasma more tightly focused, thus increasing the etch rate but with some reduction in uniformity.

13.2.1 Helicon Modes

Helicon modes are a superposition of low-frequency *whistler waves* propagating at a common (fixed) angle to $\mathbf{B}_0$. Hence, although helicons have a complex transverse mode structure, they have the same dispersion equation as whistler waves, which is, from (13.1.34),

$$\frac{kk_z}{k_0^2} = \frac{\omega_{\text{pe}}^2}{\omega\omega_{\text{ce}}} \tag{13.2.3}$$

where

$$k = (k_\perp^2 + k_z^2)^{1/2} \tag{13.2.4}$$

is the wave-vector magnitude, $k_\perp$ and k_z are the radial and axial components, respectively, and $k_0 = \omega/c$. The helicon modes are mixtures of electromagnetic ($\nabla \cdot \mathbf{E} \approx 0$) and quasistatic ($\nabla \times \mathbf{E} \approx 0$) fields having the form

$$\mathbf{E}, \mathbf{H} \sim \exp j(\omega t - k_z z - m\theta)$$

where the integer m specifies the azimuthal mode. For an insulating (or conducting) wall at $r = R$ and assuming a uniform plasma density, the boundary condition on the total radial current density amplitude $\tilde{J}_r = 0$ (or $\tilde{E}_\theta = 0$) leads to (Chen, 1991)

$$mk\mathrm{J}_m(k_\perp R) + k_z k_\perp R\,\mathrm{J}_m'(k_\perp R) = 0 \tag{13.2.5}$$

where the prime denotes a derivative of the Bessel function, J_m, with respect to its argument. For a given frequency ω, density n_0, and magnetic field B_0, (13.2.3)–(13.2.5) can be solved to obtain $k_\perp$, k_z, and k.

Helicon sources based on excitation of the $m = 0$ mode and the $m = 1$ mode have been developed. Since the $m = 0$ mode is axisymmetric and the $m = 1$ mode has a helical variation, both modes generate time-averaged, axisymmetric field intensities. The transverse electric field patterns and the way these propagate along z are shown in Figure 13.15*a* for the $m = 0$ mode and in Figure 13.15*b* for the $m = 1$ mode. Undamped helicon modes have $\tilde{E}_z = 0$ (i.e., the quasistatic and electromagnetic components of $\tilde{E}_z$ exactly cancel). The antenna couples to the transverse electric or magnetic fields to excite the modes.

Equation (13.2.5) can be solved for $k_\perp R$ as a function of k_z/k. There are an infinite number of solutions corresponding to different radial field variations, and, in any real system, a mixture of modes

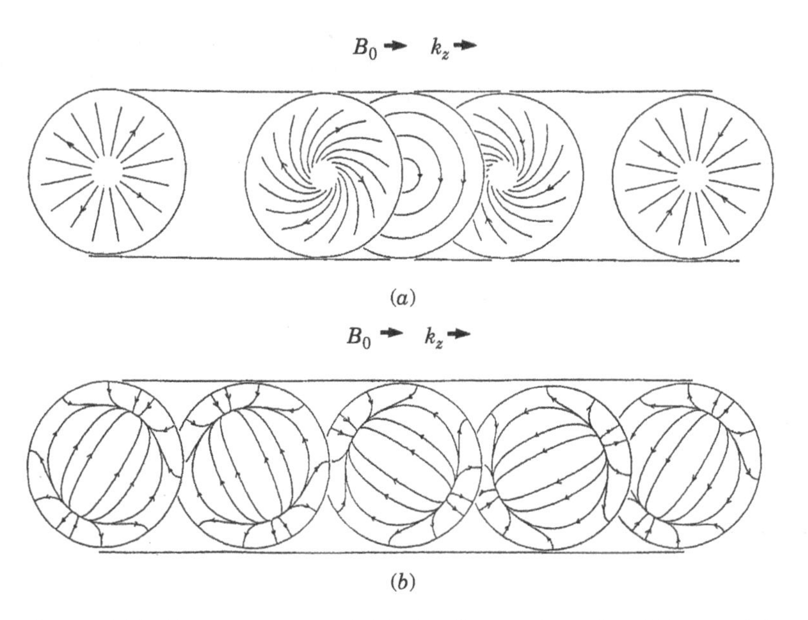

Figure 13.15 Transverse electric fields of helicon modes at five different axial positions: (a) $m = 0$; (b) $m = 1$. Source: Chen (1991)/with permission of IOP Publishing.

is very likely excited. For simplicity, let us consider the first radial mode, shown in Figure 13.15. For $m = 0$, from (13.2.5), $J_0'(k_\perp R) = 0$, which gives

$$k_\perp R = 3.83$$

for any k_z/k. For $m = 1$, we solve (13.2.5) numerically to obtain the graph shown in Figure 13.16, with the limiting values

$$k_\perp R = 3.83 \qquad (k_z \ll k_\perp)$$

$$k_\perp R = 2.41 \qquad (k_z \gg k_\perp)$$

To design an antenna for efficient power coupling, we must solve (13.2.3)–(13.2.5) and determine $k_\perp$ and k_z. Rewriting (13.2.3) in more physical terms,

$$kk_z = \frac{e\mu_0 n_0 \omega}{B_0} \tag{13.2.6}$$

we see that the $k_z \ll k_\perp$ limit corresponds to low density, and the $k_z \gg k_\perp$ limit corresponds to high density. These two limits are treated analytically below. We distinguish them by setting the condition $n = n_0^*$ for which $k_z = k_\perp$ for the $m = 1$ mode; this separates the low- and high-density limits. We have $k = \sqrt{2}k_z$ and, from Figure 13.16, $k_z = k_\perp \approx 2.5/R$. Choosing typical source parameters of $R = 5$ cm, $f = 13.56$ MHz, and $B_0 = 200$ G, we obtain $n_0^* \approx 4.0 \times 10^{12}$ cm^{-3}. Hence, for this source with $n_0 \ll n_0^*$, we have $k_\perp \gg k_z$ and, from (13.2.4), $k \approx k_\perp$. For this case, (13.2.6) yields the axial wavelength of the helicon mode for low-density operation:

$$\lambda_z = \frac{2\pi}{k_z} = \frac{3.83}{R}\frac{B_0}{e\mu_0 n_0 f} \tag{13.2.7}$$

This regime is of limited interest for materials processing because, setting the antenna length $l_a \sim \lambda_z$ (see the next subsection), the condition on $k_\perp R$ requires $R \ll l_a < l$ for a cylindrical discharge of radius R and length l. Hence, the source would be long and thin, and uniformity over a large area

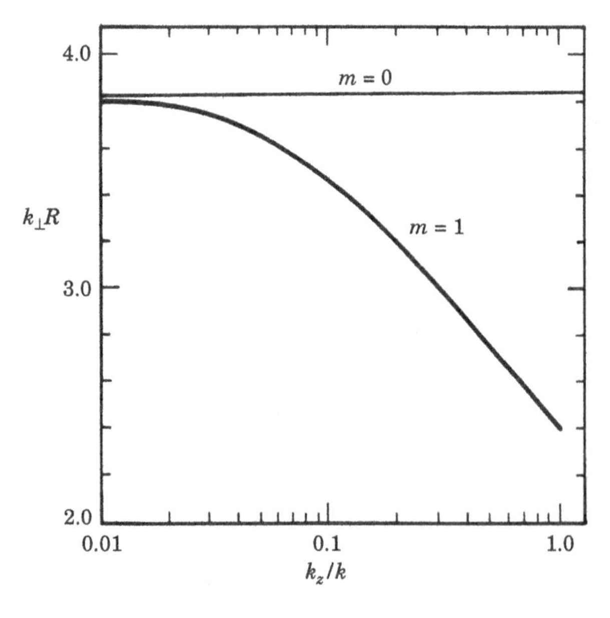

Figure 13.16 $k_\perp R$ versus k_z/k for helicon modes. Source: Lieberman and Gottscho (1994)/with permission of Elsevier.

would be compromised. However, (13.2.7) is useful in understanding source operation, as described below.

For $n_0 \gg n_0^*$, we have $k_z \gg k_\perp$ and $k \approx k_z$. In this high-density regime, we find

$$\lambda_z = \left(\frac{2\pi B_0}{e\mu_0 n_0 f}\right)^{1/2} \tag{13.2.8}$$

This regime is also of limited interest because it requires the antenna length $l_a \sim \lambda_z \ll R$, which leads to inefficient coupling of power from the antenna to the plasma. For a given current, only a small axial voltage is induced, leading to a small axial charge separation to drive the helicon mode. The regime of most interest for materials processing sources is $n_0 \sim n_0^*$, for which $k_z \sim k_\perp$; hence we have $R \sim l_a \sim l$, yielding an aspect ratio of order unity. This regime is not easy to analyze. For $m = 1$, the solution must be found numerically. One usually chooses $k_\perp$ somewhat larger than k_z; hence, we can use (13.2.7) for simple estimates of source operation. Komori et al. (1991) have measured the helicon wave magnetic field using a magnetic pick-up coil. The dependence of λ_z on B_0/n_0 was found to roughly follow (13.2.7) at densities below n_0^*.

Recall from power balance (10.2.14) that the bulk density n_0 is determined by the absorbed power P_{abs} and the pressure p. Once B_0, f, and R (for low density) are chosen, then (13.2.7) or (13.2.8) determine λ_z. Ideally, the antenna must be designed to excite modes having that particular λ_z. At first sight, this seems to limit source operation to one particular density unless B_0 or f can be conveniently varied. Fortunately, antennas excite a range of λ_zs, thus allowing source operation over a range of n_0s.

13.2.2 Antenna Coupling

A typical saddle-type antenna used to excite the $m = 1$ mode is shown in Figure 13.17. Other antennas are described by Chen (1992). Looking at the x–y transverse coördinates shown in the figure, we see that this antenna generates a $\tilde{B}_x$ field over an axial antenna length l_a, which can couple to the transverse magnetic field of the helicon mode. The antenna also induces a current within the plasma column just beneath each horizontal wire, in a direction opposite to the currents shown. This current produces charge of opposite signs at the two ends of the antenna, which in turn generates a transverse quasistatic field $\tilde{E}_y$, which can couple to the transverse quasistatic fields of the helicon mode (see Figure 13.15*b*). The conditions for which each form of coupling dominates are not well understood.

To illustrate the wavelength-matching condition for helicon excitation, we consider an ideal antenna field for quasistatic coupling:

$$\tilde{E}_y(z) \sim \tilde{E}_{y1}\Delta z\left[\delta\left(z + \frac{l_a}{2}\right) - \delta\left(z - \frac{l_a}{2}\right)\right] \tag{13.2.9}$$

where δ is the Dirac delta function. This ideal field is sharply peaked in Δz near the two antenna ends, as shown schematically in Figure 13.18*a*. Taking the Fourier transform,

$$E_y(k_z) = \int_{-\infty}^{\infty} dz\, \tilde{E}_y(z) \exp(-jk_z z)$$

and squaring this to obtain the spatial power spectrum of the antenna, we obtain

$$E_y^2(k_z) = 4\tilde{E}_{y1}^2(\Delta z)^2 \sin^2\frac{k_z l_a}{2} \tag{13.2.10}$$

which is plotted in Figure 13.18*b*. We see from (13.2.10) that the antenna couples well to the helicon mode ($E_y^2(k_z)$ is a maximum) for $k_z \approx \pi/l_a$, $3\pi/l_a$, etc., corresponding to $\lambda_z \approx 2l_a$, $2l_a/3$, etc.

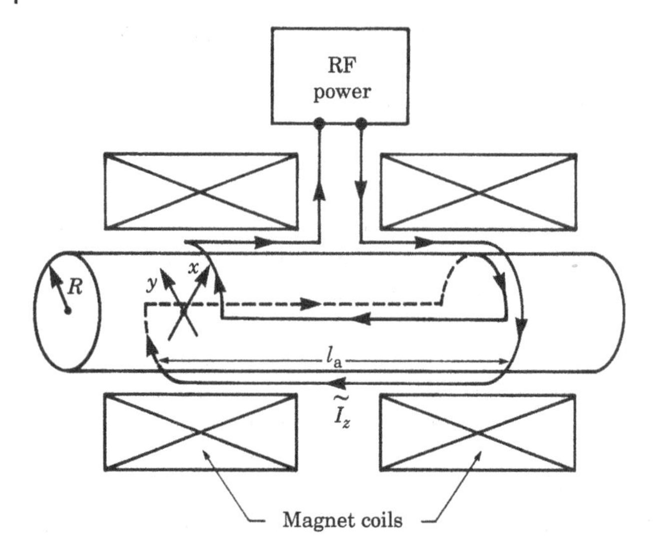

Figure 13.17 The antenna for $m = 1$ helicon mode excitation. Source: Lieberman and Gottscho (1994)/with permission of Elsevier.

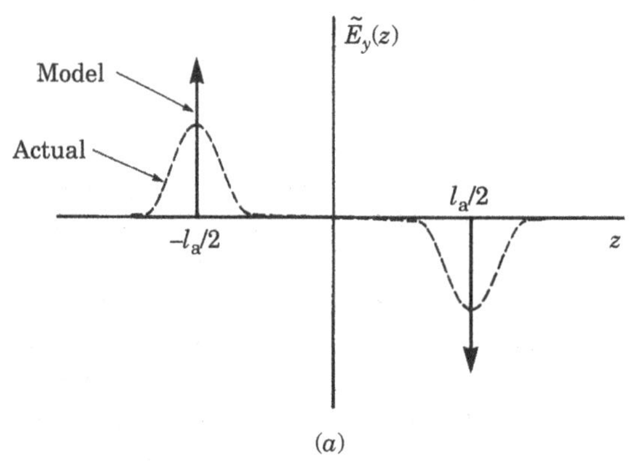

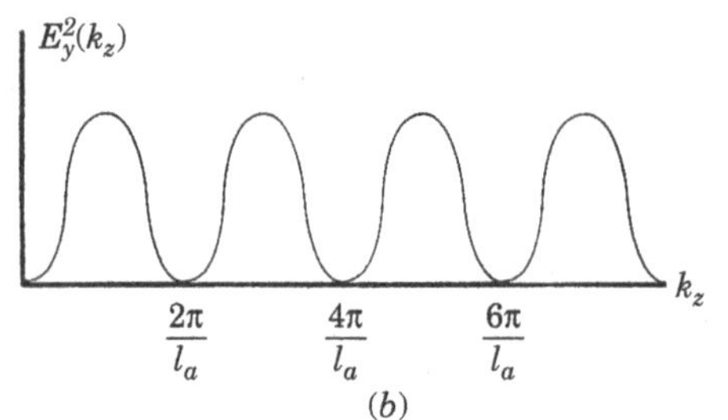

Figure 13.18 The quasistatic antenna coupling field $\tilde{E}_y$: (*a*) ideal and actual field; (*b*) spatial power spectrum of a typical field. Source: Lieberman and Gottscho (1994)/with permission of Elsevier.

The coupling is poor ($E_y^2(k_z) \approx 0$) for $k_z \approx 0$, $2\pi/l_a$, $4\pi/l_a$, etc., corresponding to $\lambda_z \to \infty$, $\lambda_z \approx l_a$, $\lambda_z \approx l_a/2$, etc.

Figure 13.19*a* shows the effect of the antenna coupling on the density n_0 as the power P_{inc} supplied to the antenna is increased, using a 36 GHz microwave interferometer to measure n_0 (see Section 4.6 for details of the measurement technique). For $P_{inc} < 350$ W, n_0 determined from the power balance (10.2.14) is low, leading to $k_z \ll \pi/l_a$ and, from (13.2.10), poor coupling to the

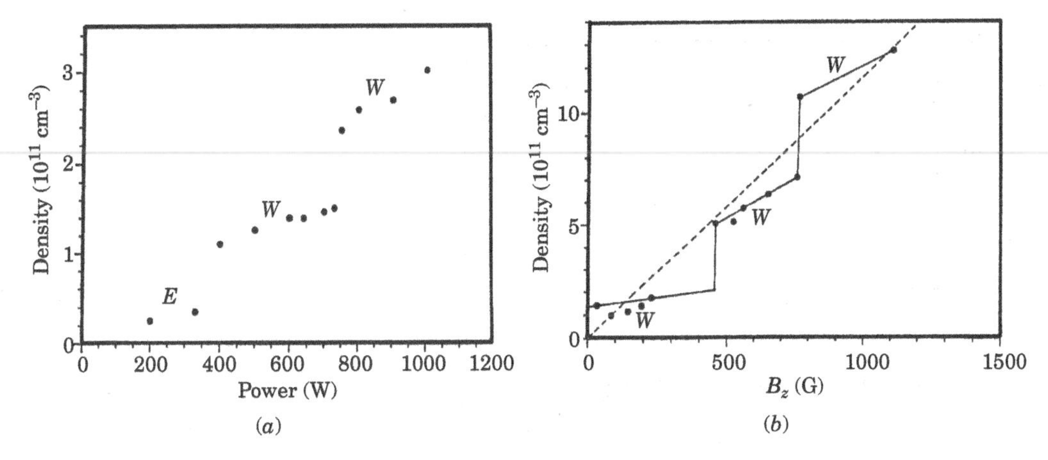

Figure 13.19 Saddle-antenna excited helicon discharges; (*a*) measured density (dots) as a function of 13.56 MHz input power for $B_0 = 80$ G in a 15-cm diameter chamber at 5 mTorr argon, showing transitions from *E*-mode to two different helicon wave (*W*) modes; (*b*) measured density (dots) as a function of magnetic field at 180 W of 8.8 MHz input power in a 10 cm diameter chamber at 1.5 mTorr argon, showing three different helicon wave mode transitions; the dashed line represents the resonance condition imposed by the antenna. Source: Perry et al. (1991)/with permission of AIP Publishing.

helicon mode. The discharge in this regime is probably capacitively driven, with a relatively high antenna voltage ($\sim$2 kV) and plasma potential (> 30 V). The transition to helicon mode operation with $k_z \approx \pi/l_a \approx 0.4k_\perp$ for $P_{\text{inc}} \approx 400$ W and $n_0 \approx 1.4 \times 10^{11}$ cm^{-3} is clearly seen. A further increase in power is not reflected in a proportional density increase, as the antenna coupling becomes increasingly inefficient. A second transition is seen to $k_z \approx 3\pi/l_a \approx k_\perp$ with $n_0 \approx 2.7 \times 10^{11}$ cm^{-3}. Standing helicon wave effects may also play a role in this transition. Figure 13.19*b* shows the roughly linear scaling of n_0 with B_0 predicted from (13.2.7) or (13.2.8), for a different source than that of Figure 13.19*a*. Again we see the density steps imposed by the antenna coupling condition. Depending on the specific experimental configuration, for example, the distance between the antenna and the outer surface of the source dielectric cylinder, the density steps are not always as evident as shown in these data. There may also be large relaxation oscillations as the discharge "hunts" between the various excitation modes. Similar effects are seen for $m = 0$ mode helicons. This mode is excited by an antenna consisting of two circular coils of radius R, separated by a length l_a, carrying oppositely directed currents.

The antenna can also be designed to couple efficiently to a wide range of k_zs, reducing the importance of mode jumps in the density range of interest. An example is shown in Figure 13.20, using the double half-turn antenna shown schematically in the upper left of the figure. This antenna, located at $z = 0$, has an excitation $E_y(z) \sim \tilde{E}_{y1}\Delta z\,\delta(z)$, which gives a broad Fourier spectrum $E_y^2(k_z) \approx$ const. Using this antenna to drive a large helicon source at 13.56 MHz, with $B_0 = 50$ G and 3 mTorr argon gas pressure, Degeling et al. (1996) measured the density downstream of the source, obtaining the results shown in the figure. We clearly see the transitions from capacitive (*E*) to inductive (*H*) to helicon wave (*W*) modes as the power increases. In this case, there is a large (order of magnitude) range of densities where the discharge remains in the helicon wave mode.

13.2.3 Power Absorption

The power absorption is believed to be due to a combination of helicon mode and Trivelpiece–Gould (TG) mode heating (Curreli and Chen, 2011). We first consider the helicon mode heating. The

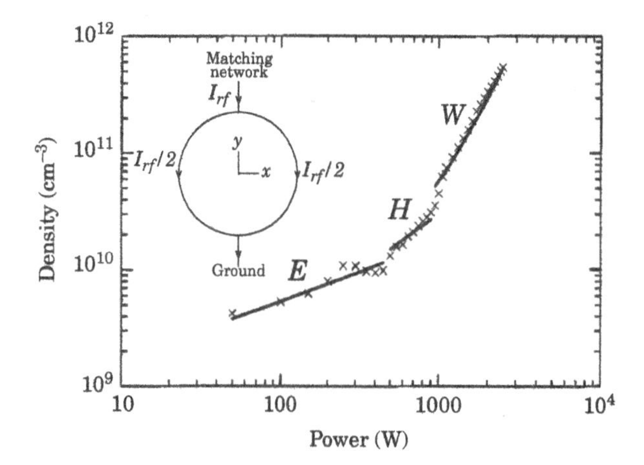

Figure 13.20 Langmuir probe measurements (×) of the central downstream density versus power, in an 18-cm diameter helicon discharge excited at 13.56 MHz by a double half-turn antenna (shown upper left); $B_0 = 50$ G at 3 mTorr argon pressure; the transition from capacitive (E) to inductive (H) to helicon wave (W) modes is clearly seen. Source: Adapted from Degeling et al. (1996).

helicon mode energy is believed to be transferred to the plasma electrons as the mode propagates along the column by collisional or collisionless (Landau) damping. The former mechanism transfers the energy to the thermal (bulk) electron population, while the latter mechanism can act to preferentially heat a nonthermal electron population to energies greatly exceeding the bulk electron temperature. There is considerable evidence that collisional absorption is too weak to account for energy deposition at low pressures (< 10 mTorr argon), although this mechanism may dominate at higher pressures. Landau damping is a process by which a wave transfers energy to electrons having velocities near the phase velocity $v_{ph} = \omega/k_z$ of the wave. (See, e.g., Chen (1984) for an exposition of the phenomenon.) Chen (1991) has estimated the effective collision frequency ν_{LD} for Landau damping of the helicon mode as

$$\nu_{LD} \approx 2\sqrt{\pi}\omega\zeta^3 \exp(-\zeta^2) \qquad \zeta \gg 1 \tag{13.2.11}$$

$$\nu_{LD}(\max) \approx 1.45\,\omega \qquad \zeta \approx 1.2 \tag{13.2.12}$$

where $\zeta = \omega/(k_z\sqrt{2}v_{th})$, with $v_{th} = (eT_e/m)^{1/2}$ (here m is the electron mass). From (13.2.7) or (13.2.8), we see that ζ decreases with increasing density. Thus for $\zeta \gg 1$, ν_{LD} increases with increasing electron density at constant magnetic field. However, in typical helicon sources where ζ may be less than or of order unity, ν_{LD} can decrease with increasing n. The total effective collision frequency can be written as

$$\nu_T = \nu_c + \nu_{LD}$$

where here ν_c is the sum of the electron–neutral and electron–ion collision rates. The axial decay length α_z^{-1} for helicon mode damping is (see Problem 13.2)

$$\alpha_z^{-1} \approx \frac{\omega_{ce}}{k_\perp \nu_T} \tag{13.2.13}$$

for low density ($k_z \ll k_\perp$); and

$$\alpha_z^{-1} \approx \frac{2\omega_{ce}}{k_z \nu_T} \tag{13.2.14}$$

for high density ($k_z \gg k_\perp$). For efficient power transfer to the plasma electrons, we require that $\alpha_z^{-1} \lesssim l$, where l is the helicon chamber source length.

By choosing the antenna length l_a such that $k_z \approx \pi/l_a$, it is possible to heat electrons, by Landau damping, whose energies are near that corresponding to the wave phase velocity

$$e\mathcal{E} = \frac{1}{2}m\left(\frac{\omega}{k_z}\right)^2 \tag{13.2.15}$$

If $\mathcal{E}$ is chosen near the peak of the ionization cross section (~50 V in argon), then the collisional energy $\mathcal{E}_c$ lost per electron–ion pair created can be reduced to a low value, of order of twice the ionization energy $\mathcal{E}_{iz}$. It follows from (10.2.14) that this can lead to a significant increase in density for the same absorbed power. However, the effective collision frequency ν_{LD} falls precipitously for $\omega/k_z \gg v_{th}$, leading to a low spatial decay rate which is not compatible with materials processing sources having $l \sim R$.

Example 13.1 As an example of helicon design, let $R = 5$ cm, $l = 20$ cm, $B_0 = 200$ G, $n_g = 6.6 \times 10^{13}$ cm^{-3} (2 mTorr), $\omega = 85 \times 10^6$ s^{-1} (13.56 MHz), and $P_{abs} = 400$ W. Due to the magnetic confinement (see Section 5.5), we assume that radial losses are small compared to axial losses. At 2 mTorr, $\lambda_i \approx 1.5$ cm. Then, from (10.2.1), we find $h_l \approx 0.28$, we choose $h_R = 0$, and from (10.2.13), $d_{eff} \approx 36$ cm. For argon, we then obtain from Figure 10.1 that $T_e \approx 2.9$ V, and from Figure 3.17, that $\mathcal{E}_c \approx 64$ V. Using (10.2.9), we find $\mathcal{E}_T \approx 85$ V. The Bohm velocity is $u_B \approx 2.6 \times 10^5$ cm/s, and from (10.2.11), $A_{eff} \approx 43$ cm^2. Then, from (10.2.15), we obtain $n_0 \approx 2.5 \times 10^{12}$ cm^{-3}. Using the value of $n_0^* \approx 4.0 \times 10^{12}$ cm^{-3} at $B_0 = 200$ G from our previous discussion, we see that $n_0 < n_0^*$. From (13.2.7), we find $\lambda_z = 22$ cm, and, hence, we choose an antenna length $l_a = \lambda_z/2 = 11$ cm to optimize power coupling. We note that $\omega/k_z = 3.0 \times 10^8$ m/s, compared with the electron thermal velocity $v_{th} = 7.2 \times 10^7$ cm/s. Hence, $\zeta \approx 2.9$, not too far from the peak of the Landau damping rate for thermal electrons.

Experimental evidence of Landau damping has been reported (Chen and Boswell, 1997; Celik et al., 2009), but electron heating due to excitation of Trivelpiece–Gould modes can also play a role in energy transfer. The TG mode is a short wavelength, electron cyclotron wave that can be excited near the radial plasma edge. It propagates radially inward with a slow group velocity and can deposit energy collisionally into the electrons. This mode can be excited along with the helicon mode by an antenna that excites waves with a characteristic axial wavenumber $k_z = N_z\omega/c$, where N_z is the axial index of refraction. The helicon and TG modes are two branches of the same electromagnetic wave, labeled as the dumbbell-shaped phase velocity surface "r" lying between the $u_\times = 0$ and $u_R = 0$ lines on the right-hand edge of the CMA diagram in Figure 4.11. To exhibit the two branches, we first note using (4.4.5), (4.5.4), and (4.5.5) the identity

$$\kappa_\perp^2 - \kappa_\times^2 = \kappa_r\kappa_l = \kappa_\parallel\kappa_\perp + \kappa_\perp - \kappa_\parallel \tag{13.2.16}$$

In the low frequency, high-density helicon discharge regime, we have the ordering

$$\omega \lesssim \omega_{ce} \ll \omega_{pe} \tag{13.2.17}$$

Using this in (4.4.5a) and (4.4.5c), (13.2.16) reduces to $\kappa_\perp^2 - \kappa_\times^2 = \kappa_\parallel\kappa_\perp$. Using this expression in (4.5.2*b*,*c*), we obtain $b = 2\kappa_\perp\kappa_\parallel$ and $c = \kappa_\perp\kappa_\parallel^2$. Then, the wave dispersion equation (4.5.1) that determines the index of refraction $N = kc/\omega$ becomes

$$N^4[\kappa_\perp(1 - \cos^2\theta) + \kappa_\parallel\cos^2\theta] - 2\kappa_\perp\kappa_\parallel N^2 + \kappa_\perp\kappa_\parallel^2 = 0 \tag{13.2.18}$$

Let $N_z = N\cos\theta$, we can re-write this in the form

$$\kappa_\perp(N^2 - \kappa_\|)^2 + N^2 N_z^2(\kappa_\| - \kappa_\perp) = 0 \tag{13.2.19}$$

with the positive solution for N^2,

$$N^2 - \kappa_\| = NN_z\left(\frac{\kappa_\perp - \kappa_\|}{\kappa_\perp}\right)^{1/2} \tag{13.2.20}$$

In the frequency regime (13.2.17), we use (4.4.5a) and (4.4.5c) to evaluate

$$\left(\frac{\kappa_\perp - \kappa_\|}{\kappa_\perp}\right)^{1/2} = \frac{\omega_{ce}}{\omega} \tag{13.2.21}$$

in (13.2.20).

For $\omega < \omega_{ce}/2$, there are two propagating mode solutions to the quadratic equation (13.2.20) at a fixed N_z:

$$N = N_z\frac{\omega_{ce}}{2\omega}\left[1 \pm \left(1 - \frac{4\omega_{pe}^2}{N_z^2\omega_{ce}^2}\right)^{1/2}\right] \tag{13.2.22}$$

A plot of N_z versus $N_\perp = (N^2 - N_z^2)^{1/2}$ at a fixed frequency, density, and magnetic field is given in Figure 13.21, with the two modes at a fixed N_z (horizontal dashed line) indicated in the figure. Such a plot is often called an *index surface*. For large N_z, the negative sign gives the helicon mode (13.2.6), and the positive sign gives the TG mode. For $N \gg 1$, the TG mode has the limit

$$N \approx \frac{\omega_{ce}}{\omega}N_z \quad \text{or} \quad k \approx \frac{\omega_{ce}}{\omega}k_z \tag{13.2.23}$$

which is plotted as the dashed line in the figure. We note from the figure that $N_\perp$(TG mode) $\gg N_\perp$ (W mode), indicating that the TG mode has a short radial wavelength compared to the helicon mode. The direction of the group (energy propagation) velocity v_g is perpendicular to the index surface, as indicated in the figure (see Problem 13.4). In addition, the TG mode is heavily damped compared to the helicon mode at a given collisionality ν_m. The radially varying plasma density profile couples the two modes together. Because the coupling is mainly near the edge, the highly damped TG mode preferentially deposits its energy in the edge region. The coupling of the two modes leads to enhanced energy deposition, as has been seen in some experiments (Chen and Boswell, 1997; Curreli and Chen, 2011).

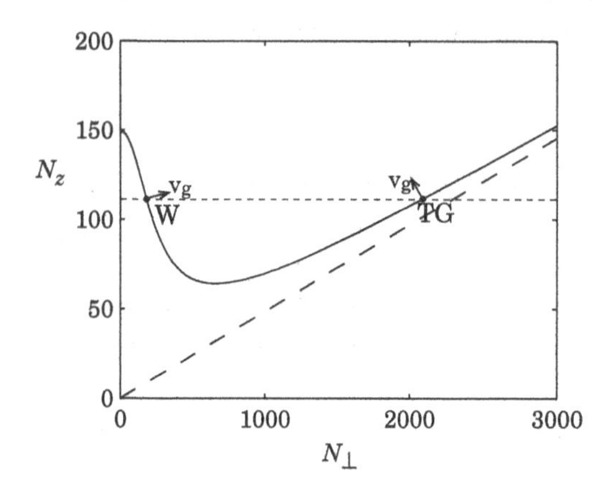

Figure 13.21 Index surface N_z versus $N_\perp$ for the helicon-TG mode system with $f = 13.56$ MHz, $B_0 = 100$ G, and $n_e = 10^{12}$ cm^{-3}; the group velocity vector directions at a given N_z are indicated; the dashed line shows the TG mode limit for $N \gg 1$.

Nonlinear absorption mechanisms, such as electron trapping in large amplitude helicon waves, may also play a role in electron heating. This can produce pulses of high-energy electrons appearing downstream of the source, leading to the pulses of ionization seen downstream in some experiments (Degeling and Boswell, 1997).

13.2.4 Neutral Gas Depletion

The high plasma densities at low pressures achievable in helicon (and ECR) discharges can give rise to significantly reduced gas densities within the discharge volume compared to the gas density within the chamber in the region surrounding the discharge volume, due to removal of neutrals from the plasma by ionization. This *ionization pumping* and other gas depletion effects have been found to play a significant role in helicon discharges (Boswell, 1984; Sudit and Chen, 1996; Gilland et al., 1998; Curreli and Chen, 2011). A global model of depletion including finite gas feeds through the discharge was developed by Cho (1999). Instabilities leading to relaxation oscillations have been observed in some helicon discharges, and gas depletion effects are thought to play a critical role in modeling these oscillations (Degeling et al., 1999). Gas depletion and heating effects for high-density, low-pressure inductive discharges are quite relevant to helicon discharges and are described at the end of Section 12.3. Strong depletion effects exist for pulsed planar magnetron discharges, as described in Section 14.5.

13.3 Surface Wave Discharges

Electromagnetic surface waves that propagate along a cylindrical plasma column can be efficiently absorbed by the plasma, hence, sustaining a discharge. Surface waves, which are propagating modes having strong fields only near the plasma surface, were described by Smullin and Chorney (1958) and Trivelpiece and Gould (1959). As described in Section 11.7, these waves play an important role in describing electromagnetic effects in large-area, high-frequency capacitive discharges. Moisan and his group at the Université de Montréal have extensively analyzed surface-wave-driven discharges and developed high-power wave-launching systems over a wide frequency range (1 MHz–10 GHz). Cylindrical surface wave sources have been reviewed by Moisan and Zakrzewski (1991). Discharges having diameters as large as 15 cm have been operated, although diameters of 3–10 cm are more commonly used. The simplest sources operate without an imposed axial magnetic field. At the high densities of interest here, the sources must be driven at microwave frequencies in the range of 1–10 GHz. Although there have been some applications to materials processing, the absorption length α_z^{-1} for the surface modes tends to be long, such that $l \gg R$ for these discharges. Hence, the cylindrical configuration cannot be operated as a low aspect ratio source. However, planar (rectangular) configurations have been developed (Komachi, 1993) that may be suitable for large-area processing applications.

13.3.1 Planar Surface Waves

Two types of configurations can support electromagnetic surface waves at an interface between a dielectric and a plasma. (1) At an interface between a semi-infinite plasma and a dielectric, a solution can be found for which the wave amplitude decays in both directions away from the plasma–dielectric interface. Maxwell's equations admit solutions of the form ($x > 0$ in the plasma)

$$\tilde{H}_{yd} = H_{y0}\, e^{\alpha_d x - jk_z z} \tag{13.3.1}$$

$$\tilde{H}_{yp} = H_{y0}\, e^{-\alpha_p x - jk_z z} \tag{13.3.2}$$

where we have assumed that H_y is continuous across the interface at $x = 0$. From the wave equation, the transverse decay constants are related to the propagation constant k_z by

$$-\alpha_d^2 + k_z^2 = \kappa_d \frac{\omega^2}{c^2} \tag{13.3.3}$$

and

$$-\alpha_p^2 + k_z^2 = \kappa_p \frac{\omega^2}{c^2} \tag{13.3.4}$$

where κ_p, given by (4.2.24), is the lossless plasma relative dielectric constant. From Maxwell's equations (2.2.2), we obtain the electric field components (e.g., see Ramo et al., 1994, chapter 8)

$$\tilde{E}_{zd} = H_{y0} \frac{\alpha_d}{j\omega\epsilon_0\kappa_d} e^{\alpha_d x - jk_z z} \tag{13.3.5}$$

and

$$\tilde{E}_{zp} = -H_{y0} \frac{\alpha_p}{j\omega\epsilon_0\kappa_p} e^{-\alpha_p x - jk_z z} \tag{13.3.6}$$

Using continuity of $\tilde{E}_z$ at the interface $x = 0$, we can eliminate the arbitrary constant H_{y0} by equating (13.3.5) to (13.3.6) to obtain

$$\frac{\alpha_p}{\kappa_p} = -\frac{\alpha_d}{\kappa_d} \tag{13.3.7}$$

Substituting (13.3.3) and (13.3.4) into (13.3.7), we obtain

$$\kappa_d^2 \left(k_z^2 - \kappa_p \frac{\omega^2}{c^2} \right) = \kappa_p^2 \left(k_z^2 - \kappa_d \frac{\omega^2}{c^2} \right) \tag{13.3.8}$$

which can be solved for k_z to determine the wave dispersion,

$$k_z = \kappa_d^{1/2} \frac{\omega}{c} \left[\frac{\omega_{pe}^2 - \omega^2}{\omega_{pe}^2 - (1 + \kappa_d)\omega^2} \right]^{1/2} \tag{13.3.9}$$

Figure 13.22 shows k_z versus ω for the lossless case. We see that k_z is real for $\omega \leq \omega_{res}$, where $\omega_{res} = \omega_{pe}/(1 + \kappa_d)^{1/2}$ gives the resonance $k_z \to \infty$ of the surface wave. For $\omega \ll \omega_{res}$, we see that $k_z \approx \kappa_d^{1/2}\omega/c$. The region of interest for surface wave sources is ω near but just below ω_{res}. Hence, for high-density sources, the frequencies of interest are above 1 GHz, i.e., microwave frequencies. Fixing ω for the source, we introduce the resonance value of the density $n_{res} = \epsilon_0 m\omega^2(1 + \kappa_d)/e^2$ (here, m is the electron mass). Then, the surface wave propagates for densities $n_0 \geq n_{res}$.

(2) A configuration, in which the plasma is separated from a conducting plane by a dielectric slab of thickness d, also admits a wave that decays into the plasma region. However, this wave does not decay into the dielectric, but is confined within the dielectric by the conducting plane. This type of surface wave, similar to that used for optical wave guiding, has also been used for surface wave discharges (Komachi, 1993). We will not consider this configuration here.

13.3.2 Cylindrical Surface Waves

A surface wave also propagates on a nonmagnetized plasma column of radius R confined by a thick dielectric tube of outer radius b. In analogy to the wave in slab geometry, assuming $b \gg R$, so that interaction with external surfaces can be neglected, the azimuthally symmetric mode has $\tilde{H}_z = 0$ and

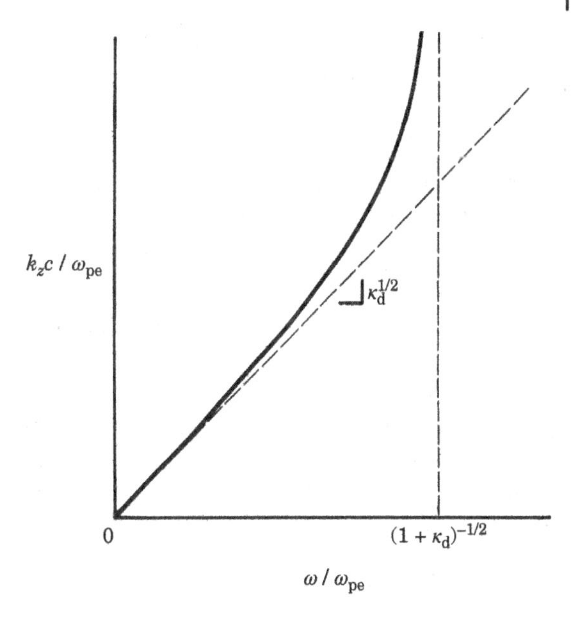

Figure 13.22 Surface wave dispersion k_z versus ω. Source: Lieberman and Gottscho (1994)/with permission of Elsevier.

$$\tilde{E}_{zp} = \tilde{E}_{z1} \frac{\mathrm{I}_0(\alpha_p r)}{\mathrm{I}_0(\alpha_p R)} \exp j(\omega t - k_z z) \qquad r < R \tag{13.3.10}$$

$$\tilde{E}_{zd} = \tilde{E}_{z1} \frac{\mathrm{K}_0(\alpha_d r)}{\mathrm{K}_0(\alpha_d R)} \exp j(\omega t - k_z z) \qquad r > R \tag{13.3.11}$$

where α_d and α_p are related to k_z by (13.3.3) and (13.3.4), and I_0 and K_0 are the modified Bessel functions of the first and second kinds. We note from the form of the Bessel functions that the fields decay away from the surface of the plasma in both directions. The transverse fields are obtained from $\tilde{E}_z$ using Maxwell's equations. In particular, we find

$$\tilde{H}_\theta = -\frac{j\omega\epsilon_0\kappa}{\alpha^2} \frac{\partial \tilde{E}_z}{\partial r}$$

in the two regions. The continuity of the tangential magnetic field $\tilde{H}_\theta$ then yields the dispersion equation

$$\frac{\kappa_p}{\alpha_p R} \frac{\mathrm{I}_0'(\alpha_p R)}{\mathrm{I}_0(\alpha_p R)} = \frac{\kappa_d}{\alpha_d R} \frac{\mathrm{K}_0'(\alpha_d R)}{\mathrm{K}_0(\alpha_d R)} \tag{13.3.12}$$

From (13.3.10) and (13.3.11), if $\alpha R \gg 1$, then the surface modes decay rapidly, which greatly simplifies the analysis. Using the asymptotic expansions of the Bessel functions $\mathrm{I}_0'/\mathrm{I}_0 = 1$ and $\mathrm{K}_0'/\mathrm{K}_0 = -1$, we then obtain the result (13.3.7), i.e., the cylinder looks like a plane in this approximation. The dispersion is the same as in (13.3.9) and as illustrated in Figure 13.22. However, in the cylinder, at low frequencies, the ordering $\alpha R \gg 1$ is not valid, and the complete dispersion equation (13.3.12) must be solved numerically. The result is similar to that shown in Figure 13.22.

13.3.3 Power Balance

We treat the power balance in the geometrically simple case of a long, thin source, $l \gg R$, using the general principles described in Section 10.2. In particular, the local power balance along z determines the density n_0 for a given absorbed power P'_{abs} per unit length along the column, as in the derivation leading to (10.2.14). Let P_w be the power carried by the wave along the column at the position z, at which the density is n_0, then

$$P'_\mathrm{abs}(n_0) = 2\alpha_z(n_0)P_\mathrm{w} \tag{13.3.13}$$

where α_z is the axial attenuation constant of the wave fields at the density n_0. Equating P'_abs to the power loss per unit length,

$$P'_\mathrm{loss}(n_0) = en_0u_\mathrm{B}A'_\mathrm{eff}\mathcal{E}_\mathrm{T} \tag{13.3.14}$$

where $A'_\mathrm{eff} = 2\pi Rh_R$ is the effective (radial) loss area per unit length, we obtain $n_0(z)$ for a given wave power $P_\mathrm{w}(z)$.

The mode attenuates as it propagates along z due to a nonzero electron–neutral momentum transfer frequency ν_m. Let $\nu_\mathrm{m} \ll \omega$ in (4.2.18), substituting this into (13.3.9), solving for the complex propagation constant k_z, and taking the imaginary part, we obtain the attenuation constant $\alpha_z(n_0) = -\mathrm{Im}\, k_z$ at a fixed ω. The expression is complicated and we give only the scaling for n_0 greater than, but not too near, resonance:

$$\alpha_z \propto \frac{n_0\nu_\mathrm{m}}{(n_0 - n_\mathrm{res})^{3/2}} \tag{13.3.15}$$

At resonance, there is a finite α_z, while for $n_0 < n_\mathrm{res}$, the wave does not propagate. For the variation of α_z in (13.3.15), P'_abs given by (13.3.13) is plotted versus n_0 for several different values of P_w in Figure 13.23. The linear variation of P'_loss given by (13.3.14) is also plotted in the figure. The intersection of P'_abs with P'_loss determines the equilibrium density along the column. It can be seen that there is a minimum value $P_{\mathrm{w\,min}}(z)$, below which a discharge at that value of z cannot be sustained. A discharge forms near the position of surface wave excitation $z = 0$ for $P_{\mathrm{w\,max}} > P_{\mathrm{w\,min}}$. As the wave propagates, P_w attenuates along z due to wave absorption. A discharge cannot be sustained when P_w falls below $P_{\mathrm{w\,min}}$ at $z = z_\mathrm{max}$. Hence, the discharge exists as a finite length plasma column over $0 < z < z_\mathrm{max}$. Typical plasma column variations of n_0 and P_w are shown in Figure 13.24. We note in Figure 13.23 that there are generally two intersections of $P'_\mathrm{abs}(n_0)$ with $P'_\mathrm{loss}(n_0)$. The lower-density intersection is an unstable equilibrium because a fluctuation that decreases n_0 leads to $P'_\mathrm{abs} < P'_\mathrm{loss}$, thus further decreasing n_0. The higher-density intersection is stable by similar reasoning.

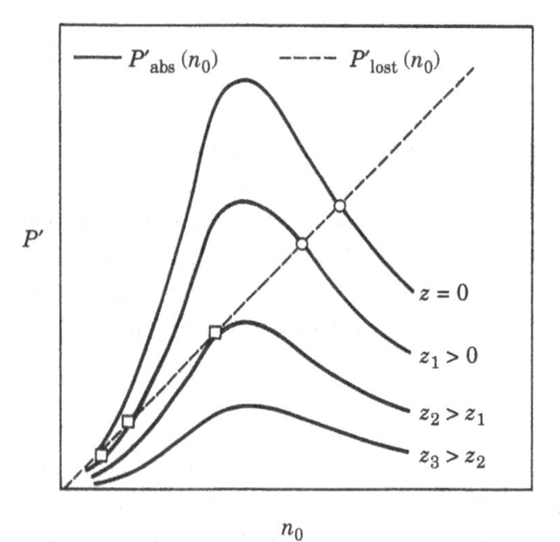

Figure 13.23 Determination of the equilibrium density in a surface wave discharge. The high-density intersection of P'_abs and P'_loss gives the equilibrium density. Source: Moisan and Zakrzewski (1991)/with permission of IOP Publishing.

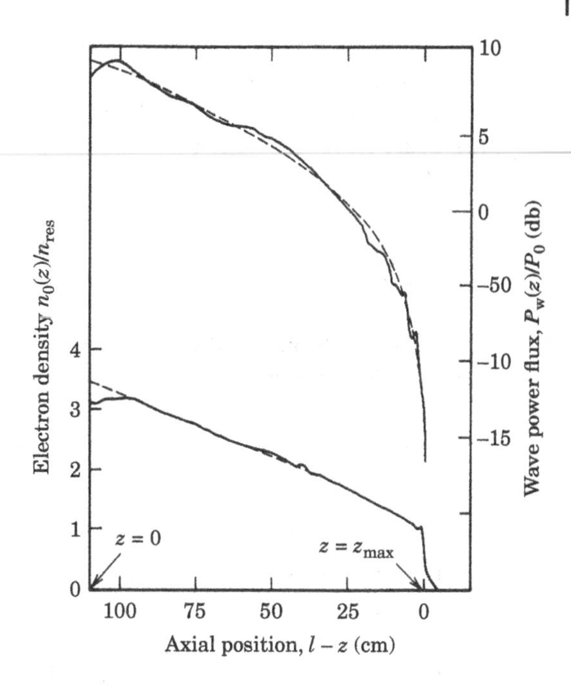

Figure 13.24 Comparison of theory (dashed) and experiment (solid) of density n_0 and wave power P_w versus z for a typical surface wave source. Source: Moisan and Zakrzewski (1991)/with permission of IOP Publishing.

Problems

13.1 Wave Equation for Right Circularly Polarized Wave Starting from Maxwell's equations (2.2.1)–(2.2.4), with variation in z only, with J given by (4.4.3) and following, derive the wave equation (13.1.22), with κ_r given by (13.1.23).

13.2 Helicon Mode Decay Constants Starting from (13.2.3), perform the following:

(a) Re-derive the right-hand side to include collisions.

(b) Introducing $k_z = \beta_z - j\alpha_z$ on the right-hand side, expand in the two limits $k_z \ll k_\perp$ and $k_z \gg k_\perp$ to obtain (13.2.13) and (13.2.14), respectively.

13.3 Helicon Discharge Equilibrium

(a) Taking Example 13.1 in Section 13.2, obtain all of the values given there by making the appropriate calculations.

(b) Repeat for $p = 10$ mTorr and $P_{abs} = 600$ W. Discuss the modification of the results.

13.4 Group Velocities of Helicon and Trivelpiece–Gould Modes Given an index surface $\omega(\mathbf{N}) = \omega(N_\perp, N_z) = \text{const}$, show that the group velocity vector $\mathbf{v}_g = \nabla_{\mathbf{N}}\, \omega$ is perpendicular to the index surface. Hint: Compare the slope of a vector **ds** lying within the surface to the slope of $\mathbf{v}_g$.

13.5 ECR Discharge Equilibrium An ECR discharge is excited in argon gas at a pressure of 1 mTorr by a 2.45 GHz, right circularly polarized wave carrying $P_{inc} = 1000$ W of incident microwave power through a quartz window at one end of a cylindrical discharge chamber of radius $R = 10$ cm and length $l = 50$ cm. The magnetic field monotonically decreases from the window into the chamber, and the logarithmic field gradient at the resonance zone is $\alpha = 4\text{m}^{-1}$.

(a) Assuming that all of the incident power is uniformly absorbed over the cross-sectional area of the plasma and that there is axial loss, but no radial loss of plasma, find the electron temperature T_e and the central plasma density n_0.
(b) Using (13.1.30), verify that essentially all of the incident power is absorbed by the plasma.
(c) Find the minimum incident power that will sustain the discharge.
(d) Using your results in (a), (b), and (c), sketch n_0 versus P_{inc} (linear scales) for $0 < P_{inc} <$ 1000 W.

13.6 ECR Wave at Angle to Magnetic Field in Overdense Plasma

(a) Show from the general wave dispersion equation (4.5.1) in a magnetized plasma that a wave propagating at an angle θ to a uniform magnetic field B_0 in a uniform high-density plasma, $\omega \lesssim \omega_{ce} \ll \omega_{pe}$, has an index of refraction

$$N^2 = \frac{\omega_{pe}^2}{\omega(\omega_{ce}\cos\theta - \omega)}$$

Hint: This is the same wave as the helicon-TG wave found in (13.2.22), but it is written in a different form.

(b) For waves launched through a window into a plasma at $z = 0$ at a larger magnetic field than for ECR resonance (see Figure 13.1*b*), find the range of angles and the range of wavelengths $\lambda(\theta) = 2\pi/k(\theta)$, with $k(\theta) = N(\theta)\omega/c$, for a propagating wave to exist. Take $f_{ce} = 1.1f$ and $n_e = 10^{12}$ cm^{-3}, with $f = 2.45$ GHz. Plot $\lambda(\theta)$ (θ in degrees) and compare λ to the free space wavelength $\lambda_0 = c/f$.

14

DC Discharges

14.1 Qualitative Characteristics of Glow Discharges

The dc glow discharge has been historically important, both in applications of weakly ionized plasmas and in studying the properties of the plasma medium. A dc discharge has one obvious feature, its macroscopic time independence, that is simpler than rf discharges. However, the need for the current, which provides the power for the discharge, to be continuous through the dc sheath provides an additional complication to the operation. This complication is not present in rf or microwave discharges where displacement current provides current continuity through the sheath. To understand the glow discharge, we consider the usual configuration of a long glass cylinder with the positive anode at one end and a negative cathode at the other. Although not necessarily the configuration used in processing applications, it has the advantage of symmetry and has been well studied. The usual pressure range of operation is between 10 mTorr and 10 Torr. Typically, a few hundred volts between cathode and anode is required to maintain the discharge. The approximate characteristics of the discharge are shown in Figure 14.1. It is clear from the many light and dark regions identified in Figure 14.1*a* that the behavior is quite complicated. The length of the positive column region can be varied by changing the distance between electrodes at a constant pressure and approximately constant voltage drop, while the other regions maintain their lengths. It is therefore apparent that the positive column can be analyzed per unit length, while the other features must be analyzed in their entirety. All of the regions are gas, pressure, and voltage dependent in their size and intensity, with some of the smaller features being essentially absent over various parameter ranges.

We now describe qualitatively the essential operation of the various regions in maintaining the discharge. The treatment follows most closely that in Cobine (1958) where additional material and references can be found.

14.1.1 Positive Column

The axially uniform plasma is maintained by the $\mathbf{J} \cdot \mathbf{E}$ power integrated over the cross section, which balances the loss of energy per electron–ion pair created, which, in the axially uniform model, is assumed to be radial. The dynamics are very similar to that of the bulk rf discharge, with the power lost per electron–ion pair created going to excitation (the glow), ionization, electron–neutral elastic scattering energy losses, and kinetic energy of the electrons and ions striking the walls. The normal glow discharge tends to have a negative voltage–current characteristic

Principles of Plasma Discharges and Materials Processing, Third Edition. Michael A. Lieberman and Allan J. Lichtenberg.

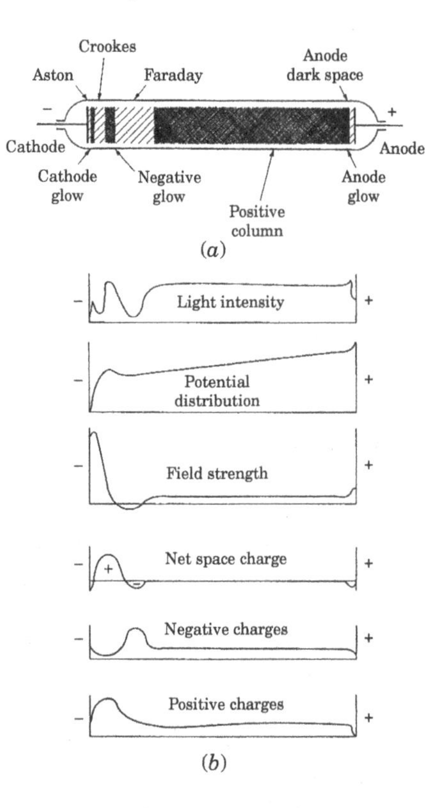

Figure 14.1 Qualitative characteristics of a dc glow discharge. Source: Brown (1959)/with permission of Massachusetts Institute of Technology.

(negative differential resistance dV/dI) that is stabilized by an external resistor, which is varied to adjust the current to the desired value. The power balance determines the (weak) axial E field required to maintain the positive column. Once E is known, the drift velocity of the electrons along the column can be found using the dc electron mobility and then, from J, the density can be determined. We use this prescription in Section 14.2 to calculate the characteristics of the positive column.

14.1.2 Cathode Sheath

This region, known also as the *cathode fall* or *Crookes dark space*, is the region over which most of the voltage drop occurs. The electrons, which carry most of the current in the positive column, are, of course, prevented from reaching the cathode. The massive ions, however, are incapable of carrying the full current. The discharge is maintained by secondary electrons produced at the cathode by the impact of the energetic ions. This process, which is incidental (although often important) in rf discharges, is essential for the operation of the dc discharge. The current is built up by ionization within the sheath, which is generated by the secondary electrons accelerating in the large electric fields of this region. The electron density and flux grow exponentially from the cathode, with the exponent known as the *first Townsend coefficient*. This mechanism is important, not only for the steady-state discharge but also for understanding the *breakdown* that initiates the discharge. In breakdown, the entire region between the cathode and the anode participates in the process, which requires a much higher voltage and therefore leads to hysteresis in the voltage–current characteristic. We analyze this dynamics in Section 14.3.

14.1.3 Negative Glow and Faraday Dark Space

The exponentially increasing density of high-velocity electrons near the cathode leads rapidly to a bright cathode glow in which intense ionization and excitation occur. The electric field must decrease rapidly at the end of this region, where the transition to the positive column occurs. However, the high electron velocities must be dissipated by elastic and inelastic collisions before the equilibrium conditions of the positive column can be established. This is done in a rather complicated process in which the electrons first lose almost all of their energy and then are reaccelerated in a weak field over approximately a mean free path (the Faraday dark space). We give a simple approximate analysis of this behavior at the end of Section 14.3.

14.1.4 Anode Fall

The drift velocity of the electrons in the weak electric field of the positive column is typically less than their thermal velocity. This requires a retarding electric field in the neighborhood of the anode to prevent the full thermal electron current from reaching the anode. However, the anode itself must clearly be positive with respect to the positive column to maintain the current. The result is a *double layer*, which is also seen in various other types of discharges, for essentially the same reason. Since the total voltage drop in this region is small and plays little role in the overall dynamics, we will not analyze it quantitatively.

14.1.5 Other Effects

The various other regions indicated in Figure 14.1 are not of particular significance for an overall understanding of the discharge behavior. In addition to the axial variations there are, of course, radial variations. In a long cylindrical discharge, we shall obtain the usual Bessel function radial variation as part of our solution for the positive column given in Section 14.2. We may assume qualitatively similar radial variations of density in other regions, but quantitative calculations are very difficult. Additional radial features exist, such as an incomplete coverage of the cathode surface by the discharge, as we discuss in Section 14.3.

In the previous discussion, we have considered the typical characteristics in the normal glow, which occurs over a range of current densities, typically between 10^{-5} and 10^{-3} A/cm^2. Considering current density as the controlling variable, the voltage–current characteristic of a dc discharge is shown in Figure 14.2. The flat region with slightly negative slope $\mathrm{d}V/\mathrm{d}I$ is that of the normal glow. From low currents, the region below and around I_A is called a dark or *Townsend discharge*. The electron and ion charge densities are low and do not much perturb the applied electric field. The glow gradually builds up until a "Townsend-to-subnormal glow" transition is reached, often with hysteresis, entering the *normal glow* at a voltage V_S. The voltage remains constant as the current increases until I_B, at which point there is an increasing voltage–current characteristic called the *abnormal glow*. A further increase in current results in a rather abrupt transition at I_C, again characterized by hysteresis, to a considerably lower-voltage discharge known as an *arc discharge*. The voltage continues to decrease with increasing current, approaching an asymptote. For a typical pressure (say 1 Torr) and a typical discharge tube of a few centimeters cross section, the transitions might occur at $I_\mathrm{A} \approx 10^{-6}$ A, $I_\mathrm{B} \approx 10^{-2}$ A, and $I_\mathrm{C} \approx 10^{-1}$ A, but these currents depend on various other factors such as gas and electrode surfaces. There are applications of these various regions, particularly for high current arc discharges, which we do not consider. The reader can find further descriptions of the behavior and the applications in various monographs, e.g., in Cobine (1958) and Roth (1994).

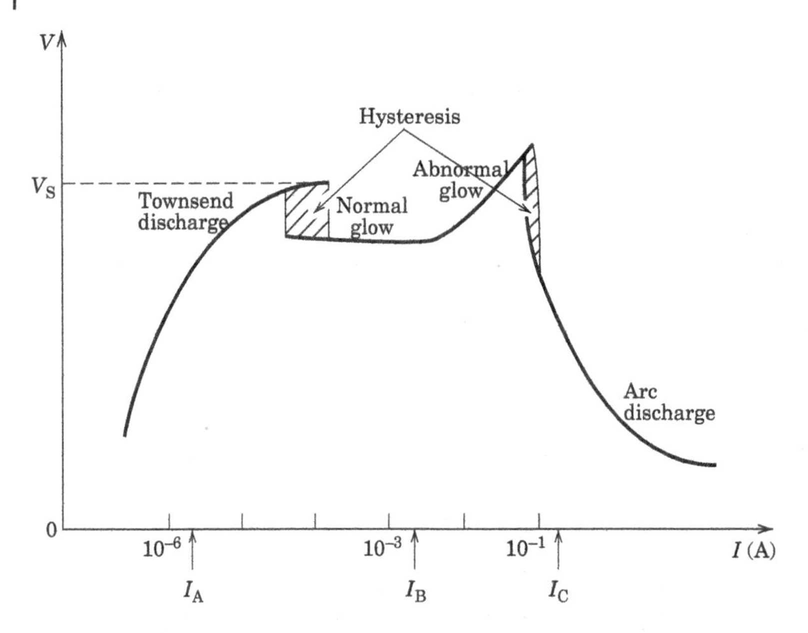

Figure 14.2 Typical voltage–current characteristic of a dc glow discharge.

In some pressure and voltage ranges, there are also interesting time-varying phenomena, such as moving transverse striations and longitudinal filaments. At high pressures, arc spots can form at the cathode, which correspond to an entirely different range of operation, not considered here, in which the secondary emission process is thermionic. For further study, the interested reader is referred to the literature (Cobine, 1958; Franklin, 1976; Raizer, 1991; Roth, 1994).

14.1.6 Sputtering and Other Configurations

A phenomenon that is not part of the discharge dynamics, but is important both for applications and in limiting the use of glow discharges, is cathode sputtering. The potential drop across a cathode sheath is typically several hundred volts. These ion-bombarding voltages lead to severe sputtering of the cathode surface and consequently deposit material on other surfaces. We describe physical sputtering in Section 9.3 and its application to the deposition of thin films in Section 17.4. Since there is little control over the large voltage drop in the cathode sheath, the existence of sputtering is important in defining appropriate applications. Low aspect ratio dc discharges have been used for sputtering. To enhance sputtering efficiency, other configurations of dc discharges have been employed. One configuration that has proved to be important for optical radiation sources and for metal-ion lasers is *hollow cathode discharges*. We treat this configuration in Section 14.4. Another method of enhancing sputtering, used primarily for depositing metallic films on substrates, employs a nonuniform dc magnetic field. This configuration is called a *dc planar magnetron discharge* and is analyzed in Section 14.5.

14.2 Analysis of the Positive Column

As in the analysis of rf and microwave discharges, there are various pressure regimes for which different dynamics apply. We will assume the following: (1) The pressure is sufficiently high,

$\lambda_i \leq (T_i/T_e)R$, that a diffusion equation with a constant diffusion coefficient D_a applies. The low-pressure (collisionless) limit with freely falling ions, $\lambda_i \gtrsim R$, was described very early by Tonks and Langmuir (1929); and the intermediate pressure regime, $R \geq \lambda_i \geq (T_i/T_e)R$, is discussed in Godyak (1986). In fact, as described in Section 5.3, the radial distributions in the low and intermediate regimes tend to look quite similar. Franklin (1976) describes these various solutions and relations between them. (2) As discussed in Section 14.1, it is often adequate to assume only radial variation, which we do here.

14.2.1 Calculation of T_e

The calculation of T_e follows from the particle balance as described in Section 10.2. Ion particle balance is obtained from the diffusion equation (5.2.21)

$$-\nabla \cdot D_a \nabla n = \nu_{iz} n \tag{14.2.1}$$

where $n = n_e = n_i$ is the plasma density, D_a is the ambipolar diffusion coefficient, and $\nu_{iz} = K_{iz} n_g$ is the ionization rate as defined in (3.5.1). In cylindrical coordinates, (14.2.1) becomes

$$\frac{d^2 n}{dr^2} + \frac{1}{r}\frac{dn}{dr} + \frac{\nu_{iz}}{D_a} n = 0 \tag{14.2.2}$$

Equation (14.2.2) is Bessel's equation with the solution given by (5.2.35)

$$n = n_0 J_0(\beta r) \tag{14.2.3}$$

where $\beta = (\nu_{iz}/D_a)^{1/2}$ and J_0 is the usual zero-order Bessel function. If the ion mean free path λ_i and the sheath thickness s ($s \approx$ few λ_{De}) are both small compared to the column radius R, then the boundary condition $n(R) \approx 0$ can be used, with the solution approximately given by (5.2.36)

$$\beta = \left(\frac{\nu_{iz}}{D_a}\right)^{1/2} = \frac{\chi_{01}}{R} \tag{14.2.4}$$

where $\chi_{01} \approx 2.405$ is the first zero of the zero-order Bessel function. Although (14.2.4) does not give a completely self-consistent solution, since the finite ion flux at the wall implies infinite velocity at zero density (see Section 5.2), it can give a reasonably accurate value of T_e. The reason is that ν_{iz} is a very sensitive function of T_e of the form (see Chapter 3)

$$\nu_{iz} \propto p \exp\left(-\frac{\mathcal{E}_{iz}}{T_e}\right) \tag{14.2.5}$$

with p the pressure and with the ionization voltage $\mathcal{E}_{iz} \gg T_e$. Thus T_e depends only weakly on all parameters except for $\mathcal{E}_{iz}$. A more accurate solution is obtained by setting the radial particle flux Γ_r equal to $n_s u_B$, where, as previously, n_s is the density at the sheath edge and $u_B = (eT_e/M)^{1/2}$ is the Bohm velocity. For this case, since $\Gamma_r = -D_a dn/dr$, we can take a derivative of (14.2.3) to obtain a transcendental equation for the electron and ion flux to the wall (see also Section 10.2):

$$-(D_a \nu_{iz})^{1/2} J_1(\beta R) = J_0(\beta R)\, u_B \tag{14.2.6}$$

Because $\lambda_i \ll R$ for this constant D_a solution, (14.2.6) essentially reduces to (14.2.4).

In the intermediate- and low-pressure regimes, $\lambda_i \gtrsim (T_i/T_e)R$, the radial profile becomes relatively uniform, and the estimate for ν_{iz} (5.3.14) applies,

$$\nu_{iz} \approx 2.2\frac{u_B}{R}\left(4 + \frac{R}{\lambda_i}\right)^{-1/2} \tag{14.2.7}$$

An additional issue at low pressures is the deviation of the electron distribution from a Maxwellian. In using (14.2.5), we have assumed a Maxwellian distribution, thus ignoring the electron drift motion u_e. This motion can readily be included (see Franklin, 1976), with $u_e \ll (e\mathrm{T}_e/m)^{1/2}$ this does not appreciably change the results. More important, particularly at low densities, there are various kinetic effects and particle losses, that can affect the distribution at high velocities. We discuss these qualitatively at the end of this section.

14.2.2 Calculation of E and n_0

The electric field E along the z-axis of the discharge is calculated by equating the input power absorbed to the power lost. In the rf discharge, this was used to determine the density. Here, the density cancels, leaving an expression for the electric field. However, once the field is known, a subsidiary condition immediately gives the density. Equating the ohmic power absorbed

$$P_{\mathrm{abs}} = 2\pi \int_0^R \mathbf{J} \cdot \mathbf{E}\, r\, \mathrm{d}r \tag{14.2.8}$$

to the power lost

$$P_{\mathrm{loss}} = 2\pi R \Gamma_r e \mathcal{E}_{\mathrm{T}} \tag{14.2.9}$$

where $e\mathcal{E}_{\mathrm{T}}$ is the total energy lost per electron–ion pair created, and substituting our radial density solution (14.2.3), we have

$$en_0\mu_e E^2 2\pi \int_0^R \mathrm{J}_0(\beta r) r\, \mathrm{d}r = 2\pi R (D_a \nu_{iz})^{1/2} n_0 \mathrm{J}_1(\beta R) e\mathcal{E}_{\mathrm{T}} \tag{14.2.10}$$

where we have assumed a constant mobility μ_e, substituted for the current density J along z using

$$J = en\mu_e E \tag{14.2.11}$$

and have taken E out of the integral by assuming that it is a constant in the long thin approximation. We see that n_0 cancels from (14.2.10) giving an equation for E alone. Performing the integration we find that J_1 cancels, and we can solve for E to obtain

$$E = \left(\frac{\nu_{iz}\mathcal{E}_{\mathrm{T}}}{\mu_e} \right)^{1/2} \tag{14.2.12}$$

Substituting $\mu_e = e/m\nu_m$, from (5.1.4), then (14.2.12) can also be written in the form

$$E = \left(\frac{m}{e} \nu_{iz} \nu_m \mathcal{E}_{\mathrm{T}} \right)^{1/2} \tag{14.2.13}$$

We note that ν_{iz} and ν_m are both linearly dependent on pressure and that the only other dependence on the right-hand side is T_e. Although (14.2.12) gives E as a function of p and as an exponentially sensitive function of T_e through its dependence on ν_{iz}, we can eliminate ν_{iz} using (14.2.4) to obtain

$$E = \frac{\chi_{01}}{R} \left(\frac{D_a \mathcal{E}_{\mathrm{T}}}{\mu_e} \right)^{1/2} = \frac{\chi_{01}}{R} \left(\frac{mK_{\mathrm{me}}}{MK_{\mathrm{mi}}} \mathrm{T}_e \mathcal{E}_{\mathrm{T}} \right)^{1/2} \tag{14.2.14}$$

which shows that E depends only on T_e, independent of p. Integrating (14.2.11) over the discharge cross section yields

$$I = 2\pi e n_0 \left(\frac{R^2}{\chi_{01}} \right) \mathrm{J}_1(\chi_{01}) \mu_e E \tag{14.2.15}$$

which can be solved to determine n_0 for a given discharge current I, with E given by (14.2.14).

14.2.3 Kinetic Effects

Although the preceding subsections give a qualitative description of the positive column, various quantitative discrepancies, particularly at lower pressures, have led to more sophisticated treatments. Particular phenomena to be explained are significantly higher average temperatures than predicted from (14.2.7) (with ν_{iz} calculated for a Maxwellian distribution), higher average energies near the column edge, an excess of local ohmic heating near the column edge compared to the local power dissipated in collisional processes, and a somewhat higher axial electric field.

A full kinetic theory including the radial density variation is very complicated, so that various approximate kinetic methods have been employed. One important method is the nonlocal approximation, which we describe in Chapter 19. The basic idea is that if the pressure is sufficiently low that $\lambda_\mathcal{E}/R > 1$, where $\lambda_\mathcal{E}$ is the electron energy loss mean free path, then the total energy $e\mathcal{E} = \frac{1}{2}mv^2 + e\Phi(r)$ can be taken to be a constant. For a Maxwellian electron distribution, the conservation of total energy is equivalent to the Boltzmann assumption that the temperature is constant and the potential and density are related in the usual logarithmic manner $\Phi(r) = -\mathrm{T_e} \ln{(n(r)/n(0))}$, with $\Phi(0) = 0$ at the plasma center. In this case, a local macroscopic theory applies, as it does at high pressure for any distribution. However, we will see in Chapter 19 that the electron energy distribution in the positive column tends to be Druyvesteyn-like, falling more rapidly at high energies than a Maxwellian, with the high-energy electrons further truncated by the inelastic processes.

Because of the non-Maxwellian distribution, the average energy is significantly higher near the plasma edge than in the discharge center, since the lower energy electrons are confined by the potential, while the higher energy electrons can overcome the potential hill. The average energy is significantly higher than predicted by a Maxwellian because overall there are fewer high-energy (ionizing) electrons. These effects have been confirmed by comparison with a more complete kinetic theory by Busch and Kortshagen (1995). Because the non-local method is limited to low pressures, other methods valid at higher pressure have been proposed (see Ingold, 1997 for another method of analysis and comparison among various methods).

14.3 Analysis of the Cathode Region

Considering the analysis of the previous section, we take as an example an argon glow discharge at $p = 100$ mTorr and $\mathrm{T_e} = 4$ V. The current density carried by the electrons in the glow is calculated from (14.2.11)

$$J(r) = en(r)\mu_\mathrm{e}E$$

with $\mu_\mathrm{e} \approx 10^3$ m^2/V-s and $E = 60$ V/m. Continuity of current requires the same current at the edge of the cathode sheath region, where the current is carried only by the ions. This can be approximated by

$$J_\mathrm{i}(r) = en_\mathrm{s}(r)u_\mathrm{B}$$

where for argon at $\mathrm{T_e} = 4$ V, we calculate $u_\mathrm{B} = (e\mathrm{T_e}/M)^{1/2} \approx 3 \times 10^3$ m/s. This is considerably less than the electron drift velocity $u_\mathrm{e} = \mu_\mathrm{e}E = 6 \times 10^4$ m/s, and thus, even ignoring the difference between n_s and n, it is not possible for the ions to carry the current in the cathode sheath. The resolution of this contradiction is that secondary electrons, created by ion impact at the cathode, are required to sustain the discharge. The process is similar to that involved in the breakdown of a gas-filled gap and was first analyzed in that context. We first consider this case and then discuss the modifications required to treat the cathode sheath.

14.3.1 Breakdown of a Gas-Filled Gap

Consider electrons emitted from a cathode at $z = 0$ being accelerated by an electric field and ionizing a neutral background. For a flux Γ_e in the z direction (the direction of the field), a differential equation for the increase in flux can be written

$$d\Gamma_e = \alpha(z)\Gamma_e \, dz \tag{14.3.1}$$

with the solution

$$\Gamma_e(z) = \Gamma_e(0)\exp\left[\int_0^z \alpha(z')\,dz'\right] \tag{14.3.2}$$

where $\alpha(z) \equiv 1/\lambda_{iz}(z)$ is the inverse of an "ionization" mean free path, analogous to the collisional mean free path defined in a similar way in Section 3.1. By continuity of total charge (creation of equal numbers of electron–ion pairs), the electron flux leaving the sheath edge at $z = d$, minus the electron flux emitted at $z = 0$, must be equal to the ion flux striking the cathode at $z = 0$, minus the ion flux that enters at $z = d$:

$$\Gamma_i(0) - \Gamma_i(d) = \Gamma_e(0)\left\{\exp\left[\int_0^d \alpha(z')\,dz'\right] - 1\right\} \tag{14.3.3}$$

where we have substituted for $\Gamma_e(d)$ from (14.3.2). For breakdown, the discharge must be self-sustaining. That is, set $\Gamma_e(0) = \gamma_{se}\Gamma_i(0)$ where γ_{se} is the secondary electron emission coefficient at the cathode $z = 0$, then (14.3.3) must be satisfied with $\Gamma_i(d) = 0$. Solving for the exponential, we obtain

$$\exp\left(\int_0^d \alpha(z')\,dz'\right) = 1 + \frac{1}{\gamma_{se}} \tag{14.3.4}$$

as the self-sustaining condition. For an uncharged region, E is a constant and the electron drift velocity $u_e(z) = \mu_e E = \text{const}$. Hence, the electron energy is a constant, allowing us to set $\alpha = \text{const}$ in (14.3.4). Taking the logarithm of both sides, we have

$$\alpha d = \ln\left(1 + \frac{1}{\gamma_{se}}\right) \tag{14.3.5}$$

the usual form for the breakdown condition of a dc discharge. The quantity α is known as the *first Townsend coefficient*. As might be expected from our knowledge of cross sections, α is a complicated function of the pressure and the accelerating field, which is very difficult to calculate. However, we might expect α to be expressed in the form

$$\alpha = \frac{\text{const}}{\lambda_e}\exp\left(-\frac{\mathcal{E}_{iz}}{E\lambda_e}\right) \tag{14.3.6}$$

where λ_e is the mean free path for inelastic (mainly ionization) electron–neutral collisions, $E\lambda_e$ is a typical electron energy gain in the field between collisions, and $\mathcal{E}_{iz}$ is an energy (in volts) for ionization. Here, $E\lambda_e$ plays the role that T_e plays in (14.2.5). Recognizing that $\lambda_e \propto n_g^{-1}$, or that $\lambda_e \propto p^{-1}$ at room temperature (here 20 °C), then (14.3.6) can be written in the form

$$\frac{\alpha}{p} = A\exp\left(-\frac{Bp}{E}\right) \tag{14.3.7}$$

where A and B are determined experimentally and found to be roughly constant over a restricted range of E/p for any given gas. Some experimental values of α/n_g versus E/n_g are shown in Figure 14.3. Here, the gas density $n_g(\text{m}^{-3}) = 3.25 \times 10^{22}\,p(\text{Torr})$ at room temperature from

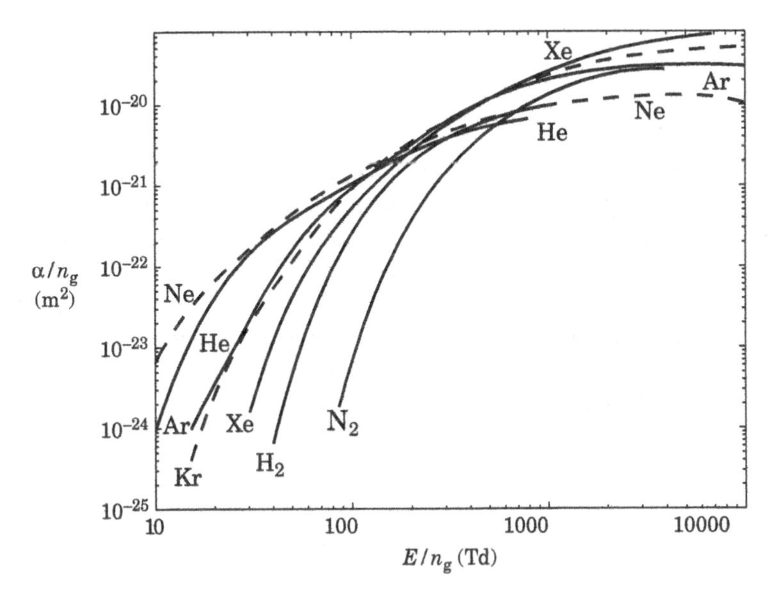

Figure 14.3 Field-intensified ionization cross section α/n_g versus reduced field E/n_g (1 Td $\equiv 10^{-21}$ V-m^2). Source: (Data provided by Petrović and Marić, 2004).

Table 14.1 Constants of the Equation $\alpha/p = A\exp(-Bp/E)$

Gas	A (cm^{-1}-Torr^{-1})	B (V/cm-Torr)	Range of E/p (V/cm-Torr)
He	2.8	77	30–250
Ne	4.4	111	100–400
Ar	11.5	176	100–600
Kr	15.6	220	100–1000
Xe	24	330	200–800
H_2	4.8	136	15–600
N_2	11.8	325	100–600
O_2	6.5	190	50–130
CH_4	17	300	150–1000
CF_4	11	213	25–200

Source: Fits to data at 20 °C supplied by Petrović and Marić (2004).

(2.3.18). The quantity α/n_g is a field-intensified ionization cross section. The reduced field E/n_g is often specified in units of townsends (1 Td $\equiv 10^{-21}$ V-m^2). Fitting the form (14.3.7) to data such as shown in Figure 14.3, the coefficients in Table 14.1 are constructed.

Combining (14.3.7) with (14.3.5), and setting the breakdown voltage $V_b = Ed$, we have the relation

$$Apd\exp\left(-\frac{Bpd}{V_b}\right) = \ln\left(1 + \frac{1}{\gamma_{se}}\right) \tag{14.3.8}$$

Solving (14.3.8) for V_b, we obtain

$$V_b = \frac{Bpd}{\ln Apd - \ln\left[\ln\left(1 + 1/\gamma_{se}\right)\right]} \tag{14.3.9}$$

We see that the breakdown voltage is a function of the product *pd*. For large values of *pd*, V_b increases essentially linearly with *pd*. For small *pd*, $V_b \to \infty$ at $pd = (pd)_{asym} = A^{-1}\ln(1 + 1/\gamma_{se})$, below which breakdown cannot occur. The breakdown voltage has a minimum value $V_{min} = (eB/A)\ln(1 + 1/\gamma_{se})$ at $pd = (pd)_{min} = eA^{-1}\ln(1 + 1/\gamma_{se})$, with $e \approx 2.72$ the natural base (Problem 14.2). The curve $V_b(pd)$ is called the *Paschen curve* and is a function of the gas and weakly a function of the electrode material. Typical breakdown curves for plane-parallel electrodes are shown in Figure 14.4. As we shall see, the values of V_{min} and $(pd)_{min}$ play important roles in the more complicated problem of the cathode sheath.

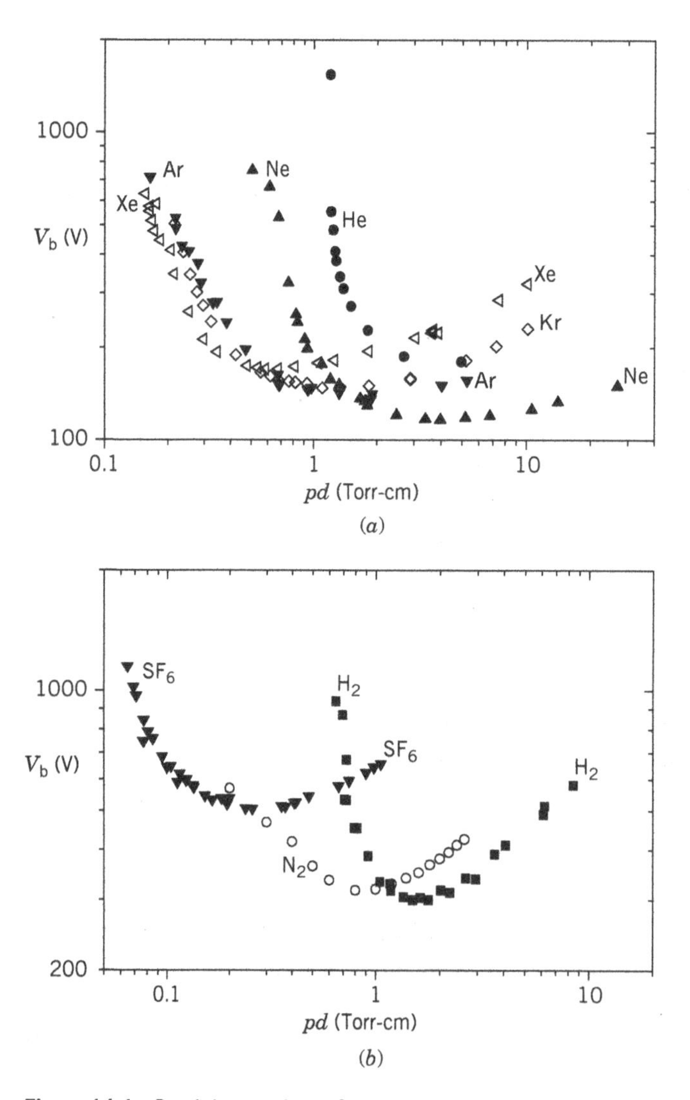

Figure 14.4 Breakdown voltage for plane-parallel electrodes at 20 °C: (*a*) noble gases; (*b*) molecular gases. Source: (Data supplied by Petrović and Marić, 2004).

In attaching (electronegative) gases, breakdown will not occur for low values of E/p, where the electron attachment coefficient β exceeds the ionization coefficient α. The multiplication mechanism (14.3.1) is replaced by

$$d\Gamma_e = [\alpha(z) - \beta(z)]\Gamma_e \, dz \tag{14.3.10}$$

leading to an effective first Townsend coefficient $\alpha_{eff} = \alpha(E/p) - \beta(E/p)$. Typically $\beta > \alpha$ at a critical E/p, below which multiplication does not occur. For example, $(E/p)_{crit} \approx$ 40–42 V/cm-Torr for oxygen and air, which are weakly attaching, and $\approx$ 100 and 117 V/cm-Torr for chlorine and SF_6 respectively, which are highly attaching (Raizer, 1991, Table 7.1). Therefore, breakdown voltages for highly attaching gases are high, as seen in Figure 14.4*b* for SF_6. It is for this reason that attaching gases are used as insulators in various applications such as gas-filled circuit breakers.

14.3.2 Cathode Sheath

We now consider the cathode sheath region of a discharge for which the electric field, and consequently α, is not a constant with position. For a large sheath multiplication, we can still take $\Gamma_i(d) = 0$ in (14.3.3). Taking the logarithm of (14.3.4), we have

$$\int_0^d \alpha(z) \, dz = \ln\left(1 + \frac{1}{\gamma_{se}}\right) \tag{14.3.11}$$

An exact solution for $\alpha(z)$ would involve an integral equation for the field and be very difficult to solve. A simpler alternative is to measure the electric field distribution, which then becomes a known variation in determining $\alpha(z)$. Somewhat surprisingly (Cobine, 1958), it is found that the matrix sheath (constant ion space charge density, see Section 6.3) well approximates the region, giving a linear field variation

$$E \approx E_0 \left(1 - \frac{z}{d}\right) \tag{14.3.12}$$

with $z = 0$ at the cathode and $z = d$ at the sheath edge. Substituting (14.3.12) into (14.3.7), we have

$$\frac{\alpha}{p} = A \exp\left[-\frac{Bp}{E_0(1 - z/d)}\right] \tag{14.3.13}$$

and substituting (14.3.13) into (14.3.11), we obtain

$$\int_0^d Ap \exp\left[-\frac{Bp}{E_0(1 - z/d)}\right] dz = \ln\left(1 + \frac{1}{\gamma_{se}}\right) \tag{14.3.14}$$

which can be evaluated to give E_0 as a function of d. Integrating E into (14.3.12) from 0 to d, we can express E_0 in terms of the cathode sheath (cathode fall) voltage V_c as $E_0 = 2V_c/d$, which when substituted into (14.3.14) gives

$$\frac{AB(pd)^2}{2V_c} S\left(\frac{2V_c}{Bpd}\right) = \ln\left(1 + \frac{1}{\gamma_{se}}\right) \tag{14.3.15}$$

where

$$S(\zeta) = \int_0^\zeta e^{-1/y} \, dy \tag{14.3.16}$$

is a known tabulated integral. If one plots $V_c(pd)$ for a given gas (given A and B) and given electrode material (given γ_{se}), we find, as expected, curves that have a minimum $V_c = V_{cmin}$ at some $(pd)_{min}$. We might expect the discharge to adjust itself to this stable value of d, and this is indeed the case

Table 14.2 Normal Cathode Fall in Volts

Cathode	Air	Ar	H_2	He	Hg	N_2	Ne	O_2
Al	229	100	170	140	245	180	120	311
Ag	280	130	216	162	318	233	150	
C			240		475			
Cu	370	130	214	177	447	208	220	
Fe	269	165	250	150	298	215	150	290
Hg				142	340	226		
K	180	64	94	59		170	68	
Mg	224	119	153	125		188	94	310
Na	200		185	80		178	75	
Ni	226	131	211	158	275	197	140	
Pb	207	124	223	177		210	172	
Pt	277	131	276	165	340	216	152	364
Zn	277	119	184	143		216		354

Source: Cobine (1958)/with permission of Dover Publications.

Table 14.3 Normal Cathode Fall Thickness *pd* in Torr-cm

Cathode	Air	Ar	H_2	He	Hg	N_2	Ne	O_2
Al	0.25	0.29	0.72	1.32	0.33	0.31	0.64	0.24
C			0.9		0.69			
Cu	0.23		0.8		0.6			
Fe	0.52	0.33	0.9	1.30	0.34	0.42	0.72	0.31
Hg			0.9					
Mg			0.61	1.45		0.35		0.25
Ni			0.9		0.4			
Pb			0.84					
Pt			1.0					

Source: Cobine (1958)/with permission of Dover Publications.

in the normal glow region (see Figure 14.2). Some values of the cathode fall voltage are given in Table 14.2, and some corresponding normal glow cathode fall thicknesses are given in Table 14.3. These values are similar to the values for breakdown.

We have not quite reached the end of the story. It is also possible to eliminate d in favor of the current density and gain both new insight into the operation of the normal glow region and also understand the abnormal glow operation. The total current density at the cathode is given by

$$J(0) = en_{\rm i}(0)v_{\rm i}(0)(1 + \gamma_{\rm se}) \tag{14.3.17}$$

where $n_{\rm i}$ is the ion density, $v_{\rm i}$ is the ion velocity, and $\gamma_{\rm se}$ gives the fraction of the current due to secondary electrons. Using Poisson's equation with the assumption of constant charge density, we

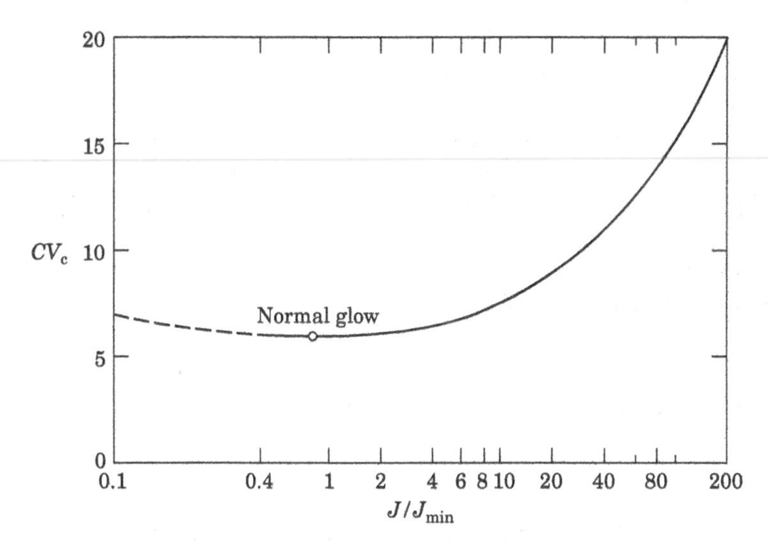

Figure 14.5 Cathode voltage drop versus discharge current, illustrating the normal and abnormal glow; $C = 2A/B\ln[1 + (1/\gamma_{se})]$. Source: Cobine (1958)/with permission of Dover Publications.

can write en_i in terms of the cathode fall potential $en_i(0) = \epsilon_0 2V_c/d^2$. Similarly, assuming a collisional sheath, we have $v_i(0) = \mu_i 2V_c/d$, where μ_i is the ion mobility. Substituting these values into (14.3.17), we obtain

$$J(0) = \frac{4\epsilon_0\mu_i V_c^2(1+\gamma_{se})}{d^3} \tag{14.3.18}$$

from which we can eliminate d in favor of $J(0)$. Hence, we can determine a Paschen-type curve of V_c versus $J(0)$. This is shown in Figure 14.5 in terms of normalized parameters.

It is clear that with a fixed external voltage source V_T and resistance R_T, the dashed curve is unstable, such that if $J = I/A < J_{min}$, where A is the effective cathode area, that is, if

$$\frac{V_T - V_{cmin}}{R_T A} < J_{min} \tag{14.3.19}$$

then the cathode fall area will constrict to a smaller value. This is the normal glow region. On the other hand, for

$$\frac{V_T - V_{cmin}}{R_T A} > J_{min} \tag{14.3.20}$$

the solution is stable, and V_c will increase with increasing current density. It is this region that is called the *abnormal glow*, but as we can see, it is just as normal as the normal glow.

14.3.3 The Negative Glow and Faraday Dark Space

As discussed qualitatively in Section 14.1, when the electrons have multiplied sufficiently that they can carry the current in the cathode sheath, the high electric field must decrease to create plasma-like, rather than sheath-like conditions. However, the majority of electrons have been accelerated by a high field and are thus far from equilibrium. A local region of high ionization and excitation must therefore exist while the electrons are slowing down. This is characterized by a mean free path for the combination of scattering and energy loss processes, sometimes referred to as the range of the energetic electrons. In fact, the electric field can actually reverse in this region to keep the electron current in balance, contributing to the slowing down process. Overall, the visual region of the negative glow has been correlated with the measured range of fast electrons in various gases, obtaining good agreement.

The reversal of the field tends to exclude ions from a region of the column, as shown in Figure 14.1. This region of low ion density prevents the negative glow from joining directly onto the essentially neutral positive column, requiring one more transition region. Although an exact analysis is difficult, a simple calculation produces the correct scaling and surprisingly good quantitative agreement with experiment. If electrons are assumed to start from rest and accelerate through a mean free path, the kinetic energy gained is

$$\frac{1}{2}mv_{\text{res}}^2 = eE\lambda_{\text{e}} = eV_{\text{res}} \tag{14.3.21}$$

where V_{res} is known as the resonance voltage. Set the current density as

$$J = env_{\text{res}} = en\left(\frac{2eE\lambda_{\text{e}}}{m}\right)^{1/2} \tag{14.3.22}$$

where we have substituted for v_{res} from (14.3.21), then n can be used in Poisson's equation to obtain

$$\frac{\text{d}E}{\text{d}z} = \frac{J}{\epsilon_0}\left(\frac{m}{2eE\lambda_{\text{e}}}\right)^{1/2} \tag{14.3.23}$$

Integrating (14.3.23) and substituting $E = V_{\text{res}}/\lambda_{\text{e}}$ from (14.3.21), we obtain

$$V_{\text{res}} = \left(\frac{2e}{m}\right)^{-1/3}\left(\frac{3}{2}\frac{J\lambda_{\text{e}}}{\epsilon_0}z\right)^{2/3} \tag{14.3.24}$$

Assuming that V_{res} is a constant, then we find that the length of the Faraday dark space scales as

$$z \propto \frac{1}{\lambda_{\text{e}}J} \propto \frac{p}{J} \tag{14.3.25}$$

which is found to hold experimentally provided the pressure is sufficiently high that the Faraday dark space is collisional.

14.4 Hollow Cathode Discharges

Hollow cathodes were first used as thermionic emitters to produce electron beams. The large area of the hollow cathode emitter prolonged the life of the delicate emitting surface at a given beam current. These early hollow cathode devices were operated in high vacuum. Plasma discharges employing hollow cathodes are operated in a quite different manner, akin to the glow discharges described in the previous sections. Nevertheless, the large interior surface area of the cathode surface for a given discharge current plays a similar role.

The basic configuration of the hollow cathode is shown in Figure 14.6, with an internal cathode cylinder of length l and radius R, and a front side anode disk of radius R. The back side disk can be either a cathode or an another anode, depending on the application, but, providing $l \gg 2R$, the operation is much the same. This follows because the cathode sheath is confined to a narrow layer between the cathode cylinder and the slightly smaller plasma cylinder. The anode can also have a central hole or be remote, depending on the application. Hollow cathode discharges have been operated at pressures of 0.1–10 Torr and current densities in the range of 0.01–1 A/cm^2 in tubes with $2R \lesssim 1$ cm for use in atom and ion lasers (see, e.g., Warner et al., 1979; van Veldhuizen and de Hoog, 1984). More recently, they have been operated at lower pressures for plasma processing, particularly for ionized physical vapor deposition (I-PVD) (see Section 14.6).

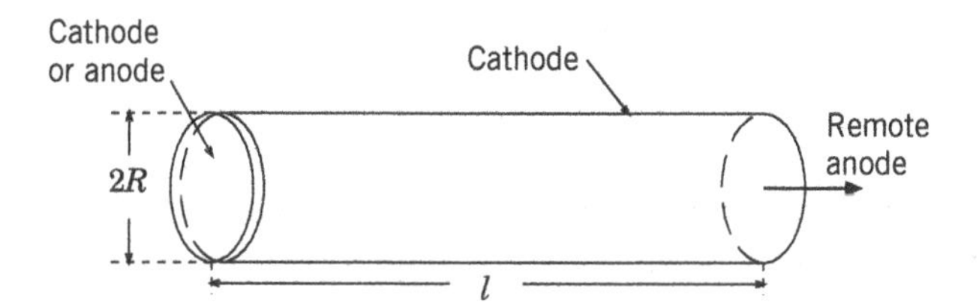

Figure 14.6 Cylindrical configuration of a hollow cathode discharge.

There is also a higher current density range of 1–10 A/cm^2 that is being investigated for plasma propulsion. In this range, the intense ion bombardment of the cathode leads to very hot cathode surfaces and consequent thermionic emission. This mode of operation is similar to the operation of original electron beam hollow cathodes and quite different from the types of plasma discharges treated in this book. We do not consider them further, but refer the interested reader to the literature (see, e.g., Siegfried and Wilbur, 1984).

Although the basic hollow cathode configuration is efficient and works well in many high current applications, various other configurations have been investigated that have advantages for particular applications. A particular variation that has been analyzed in some detail by Arslanbekov et al. (1997) is the *segmented hollow cathode discharge*. In the quadrupole configuration, for example, the cylinder is broken into two facing cathodes and two facing anodes. In this configuration, the basic trapping mechanism of secondary electrons between sheaths is still preserved, but operating voltages are larger at a given current, due to the large internal anodes, leading to more efficient excitation of metal-ion laser lines and also improved stability. The interested reader is referred to the literature for details, e.g., Arslanbekov et al. (1997) and references therein.

14.4.1 Simple Discharge Model

To model the basic cylindrical equilibrium of Figure 14.6, we consider a discharge containing hot electrons, cold electrons, and positive ions with a cylindrical cathode having length l and radius R carrying a (radial) current I_{dc}. The population of hot electrons is created by secondary emission due to ion bombardment of the cathode surface. The electrons accelerate across the sheath potential V_{dc} and are trapped within the discharge by the confining space potential. It is this population that creates the ionization required to sustain the discharge. We assume that the pressure is low, such that the hot electron energy relaxation length $\lambda_{\varepsilon\text{h}} > 2R$. Hence, the hot electrons traverse the plasma before losing their energy through inelastic collisions with the background gas. If we consider a single positive ion species, then the plasma diffusion equation is of the form (5.2.30), which for our system be written in the form

$$-D_{\text{a}}\frac{1}{r}\frac{\text{d}}{\text{d}r}\left(r\frac{\text{d}n_{\text{i}}}{\text{d}r}\right) = K_{\text{izh}}n_{\text{g}}n_{\text{h0}} \tag{14.4.1}$$

where n_{h0} is the density, assumed uniform, of hot electrons generated by secondary emission from the cathode, K_{izh} is the ionization rate coefficient for hot electrons with average energy $\sim V_{\text{dc}}/2$, and D_{a} is the ambipolar diffusion coefficient. D_{a} depends on the cold electron temperature, which is determined by a power balance relation for cold electrons. We do not consider this relation here but instead take reasonable values $\text{T}_{\text{e}} \sim 0.25$–$1$ V to determine D_{a}. Integrating (14.4.1) once yields

$$\frac{\text{d}n_{\text{i}}}{\text{d}r} = -\frac{1}{D_{\text{a}}}K_{\text{izh}}n_{\text{g}}n_{\text{h0}}\frac{r}{2} \tag{14.4.2}$$

A second integration yields, as in (5.2.33), the parabolic ion density profile

$$n_i = n_{h0} + n_{e0} - \frac{K_{izh} n_g n_{h0}}{4D_a} r^2 \tag{14.4.3}$$

where the constant of integration n_{e0} is the density of low-temperature electrons in the discharge center. Using the boundary condition in (14.4.3) that $n_i = n_{h0}$ at $r = R$ sets the relation between n_{e0} and n_{h0}

$$n_{e0} = \frac{K_{izh} n_g R^2}{4D_a} n_{h0} \tag{14.4.4}$$

such that $n_i = n_{e0}(1 - r^2/R^2) + n_{h0}$. The hot electron fraction is generally small in these discharges, $n_{h0} \ll n_{e0}$. We note that this parabolic equilibrium profile has also been found in an electronegative core plasma in Section 10.4.

Using (14.4.2) and (14.4.4) to evaluate the flux $-D_a dn_i/dr$ at the cathode surface, we find

$$\Gamma_i(R) = 2D_a n_{e0}/R \tag{14.4.5}$$

The current $I_{dc} = 2\pi Rl \cdot e\Gamma_i$, assumed fixed externally by a voltage source and resistor, sets the plasma density n_{e0} by (14.4.5)

$$I_{dc} = 2\pi l \cdot 2eD_a n_{e0} \tag{14.4.6}$$

For the hot electrons, the current balance in the sheath requires that

$$\Gamma_i(R) = \Gamma_h(R)/\gamma_{se} \tag{14.4.7}$$

where γ_{se} is the secondary emission coefficient creating the hot electron return flux $\Gamma_h(R)$. Unlike the simple glow discharge (see Section 14.3), because here the hot electrons are confined by the space potential for a number of bounces, no sheath multiplication is required. The model is completed by assuming that all hot electrons lose energy in the discharge by inelastic collisions, including ionization. For a given dc sheath voltage V_{dc} and collisional energy lost per electron–ion pair created, $\mathcal{E}_c$, then each hot electron creates $\mathcal{M} = V_{dc}/\mathcal{E}_c$ electron–ion pairs; $\mathcal{M}$ is the *multiplication factor*. In steady state, for ion balance, we require

$$\frac{\gamma_{se} V_{dc}}{\mathcal{E}_c} = 1 \tag{14.4.8}$$

which sets the voltage V_{dc}. Note that V_{dc} is independent of I_{dc} (and n_{e0}) in this simple model.

For a hot electron mean free path $\lambda_{mh} > 2R$, the hot electrons bounce between thin sheaths, giving an approximately uniform density, as assumed in (14.4.2). Even if $\lambda_{mh} < 2R$, the electrons can diffusively traverse the plasma, again leading to a uniform density. A condition for this can be estimated from the hot electron diffusion equation, using rectangular coordinates ($r \to x$) for simplicity

$$-D_h \frac{d^2 n_h}{dx^2} \approx -K_{inel} \frac{\mathcal{E}_c}{V_{dc}} n_h n_g \tag{14.4.9}$$

where D_h is the hot electron diffusion coefficient, K_{inel} is the hot electron inelastic rate coefficient, and the factor $\mathcal{E}_c/V_{dc}$ gives the inverse number of inelastic collisions for a hot electron to lose its energy. Substituting $D_h \approx \frac{1}{3}\lambda_{mh}\bar{v}_h$ from (5.1.6) (see comment below this equation) and $K_{inel} n_g \approx \bar{v}_h/\lambda_{inel}$ into (14.4.9), we solve to obtain an exponential decay of n_h from the sheath edge, with a decay length

$$\lambda_{\mathcal{E}h} \approx \left(\frac{\lambda_{mh}\lambda_{inel}}{3}\frac{V_{dc}}{\mathcal{E}_c}\right)^{1/2} \tag{14.4.10}$$

The condition for a relatively uniform hot electron density is therefore $\lambda_{\varepsilon h} > 2R$. This is considerably less stringent than the condition $\lambda_{mh} > 2R$ for collisionless bouncing of the hot electrons.

The current density $J_a(x)$ flowing from the anode into the exit plane of the source has a parabolic profile in r. The total anode current must equal the total cathode current

$$I_{dc} = \pi R^2 J_{a0}/2 \tag{14.4.11}$$

where J_{a0} is the on-axis anode current density at the source exit. The axial electric field E_0 at this point is given by

$$J_{a0} = en_{e0}\mu_e E_0 \tag{14.4.12}$$

where μ_e is the cold electron mobility.

Example 14.1 Let $R = 0.25$ cm, $l = 5$ cm, $p = 0.1$ Torr in argon, and $I_{dc} = 0.1$ A. Then, $\lambda_i \approx 0.03$ cm from (3.5.10). We assume an ion temperature $T_i = 0.05$ V and a cold electron temperature $T_e = 0.5$ V. Then, $D_a \approx \mu_i T_e \approx (e\lambda_i/M\bar{v}_i)T_e \approx 6600$ cm^2/s from (5.1.14) and $\mu_e \approx e/mn_g K_{el} \approx 9.3 \times 10^7$ cm^2/V-s from (5.1.5), where K_{el} is taken from Figure 3.16. From Figure 3.17 for $V_{dc} > 100$ V, we estimate $\mathcal{E}_c \approx 20$ V, and we assume a secondary emission coefficient $\gamma_{se} = 0.1$. Using $\mathcal{E}_c$ and γ_{se} in (14.4.8) yields $V_{dc} \approx 200$ V. This value of the cathode sheath potential is in a typical range for these discharges. We can then estimate $K_{iz} \approx 3 \times 10^{-13}$ m^3/s from Figure 3.16. Solving for n_{e0} from (14.4.6) yields $n_{e0} \approx 1.5 \times 10^{12}$ cm^{-3}. Then, (14.4.4) yields $n_{h0} \approx 6.5 \times 10^8$ cm^{-3}. We find that $\lambda_{\varepsilon h} \approx 1.6$ cm from (14.4.10), which is large compared to R. From (14.4.11), we find $J_{a0} \approx 1.0$ A/cm^2, and from (14.4.12), we obtain $E_0 \approx 4.5$ V/m.

At very high or very low pressures, the hot electron density may not be radially uniform. If, for the hot electrons, the mean free path for momentum transfer $\lambda_{mh} < R$, but the energy loss mean free path $\lambda_{\varepsilon h} > R$, we might expect a reasonably uniform density. The simultaneous satisfaction of these inequalities depends on the gas density as well as the hot electron energy. At high values of pR the second inequality is not satisfied, and the density decays exponentially into the plasma. At lower pR, if the first inequality is not satisfied, then the fast electron distribution is anisotropic, with a density that geometrically peaks on the axis. These effects have been experimentally observed and calculated, approximately, for a helium discharge by observation of helium atom and helium ion emission lines in the pressure range of 1–10 Torr (Arslanbekov et al., 1992).

The high plasma densities in hollow cathode discharges can lead to considerable sputtering of cathode material into the plasma, where it can be subsequently ionized. This introduces a second ion species and also metal atoms. One application has been to metal-ion lasers, pumped by charge transfer from noble gas ions to sputtered metal atoms (see, e.g., McNeil et al., 1976). It is considerably more complicated to analyze than the simple hollow cathode discharge. The basic process is that the ions of a noble gas (usually called the buffer gas) striking the cathode produce secondary electrons, and, along with the metal ions, also produce sputtered metal atoms. The secondary electrons, accelerated across the sheath, produce buffer gas ions, which in turn are the dominant species for producing metal ions by charge exchange from metal atoms in the volume. Other metal ion production processes, such as direct ionization by the hot electrons, are less important at the high densities in this application. Charge exchange between the thermal buffer gas ions and atoms is very rapid but without consequence in the bulk plasma. However, in the sheath, the very short charge transfer mean free path reduces the energy of the buffer gas ions at the cathode surface, such that a smaller number of metal ions can produce most of the sputtering. Global models of these effects are given in Warner et al. (1979) and Lichtenberg and Lieberman (2000). Hybrid particle-in-cell (PIC)/fluid simulations for 0.3–1 Torr argon sputtering of copper atoms are given in Baguer and Bogaerts (2005) for an $R = 5$ mm, $l = 3.5$ cm hollow cathode.

The preceding discharge model is considerably simplified, giving no hint of the electron energy distribution, which has both a hot secondary electron component and a cold bulk electron component resulting from the ionization. These distributions can significantly modify the overall plasma behavior, particularly if there are reactions that are sensitive to lower energy electrons. These kinetic effects have been treated by various groups over a number of years, with increasing sophistication, for example, Kagan et al. (1975), Arslanbekov et al. (1992), and Kolobov and Tsendin (1995). We describe one example of kinetic analysis below.

14.4.2 Finite Sheath Effects

For the simple model, the sheath width s was assumed to be much less than the radius R, such that there is no electron multiplication in the sheath. This is the case at high currents (high plasma densities). Estimating the sheath width s for Example 14.1 gives $s \approx 0.014$ cm from (6.3.12), which is much less than R. However, at low currents, this would not be the case.

For a finite sheath width, the effective bulk radius shrinks, $R_{\text{eff}} = R - s$. This increases the radial diffusion loss rate, decreasing the discharge efficiency, requiring a higher voltage for the same current. At the same time, there is secondary electron multiplication in the high-field sheath region, increasing the discharge efficiency. For low pressures, $\lambda_{\mathcal{E}\text{h}} \gg 2R$, the hot secondary electrons radially oscillate ("pendulum effect") producing both ionization in the bulk, and secondary electron multiplication and consequent ionization in the sheath. At the same time, hot electrons can be reabsorbed at the opposite cathode surface and can be lost to the anode, reducing efficiency. At high pressures, $\lambda_{\mathcal{E}\text{h}} \lesssim R$, there is no "pendulum effect," but increased ionization within the sheath still exists. The sheath multiplication leads to a higher fraction of electrons created for each ion striking the cathode, and consequently a lower voltage for the same current. The totality of these phenomena is sometimes called the *hollow cathode effect* (HCE).

For the simple model, $V = V_{\text{dc}}$ is a constant given by (14.4.8), independent of I_{dc}. The HCE leads to a weak dependence of V_{dc} on I_{dc}. As s increases, the increased ionization in the sheath and the increased hot electron and diffusive losses can lead to an optimum choice of s/R, which minimizes the voltage V_{dc} as s/R is varied. Kolobov and Tsendin (1995) give a parallel plate model (plate separation L) based on a kinetic, nonlocal discharge theory (see Section 19.3). At the higher pressures having no pendulum effect, they find a minimum V_{dc} for $s/L \approx 0.2$–0.25, depending weakly on $\gamma_{\text{se}} \lesssim 0.1$. At the lower pressures, V_{dc} decreases by about a factor of two as s/L increases from 0 to 0.5. However, the low-pressure, one-dimensional slab model does not account for hot electron reabsorption on the cathode surfaces, or hot electron losses to the anode surfaces, which again might give an optimum s/L.

Hagelaar et al. (2010) give a cylindrical hollow cathode model including finite sheath width effects and axial transport. In the high-pressure regime $\lambda_{\text{he}} < R$, they determine the hot electron energy balance as

$$2\,\frac{\exp(\alpha s) - \alpha s - 1}{\alpha^2 s^2}\gamma_{\text{se}} V = \mathcal{E}_{\text{c}} \tag{14.4.13}$$

with α the (hot electron) first Townsend coefficient. Here $\mathcal{E}_{\text{c}}$ accounts for collisional losses and small additional axial electron losses. Equation (14.4.13) gives the hollow cathode voltage V for a given sheath width $s < R$. Note that V is a monotonically decreasing function of s, with a maximum value of $V = V_{\text{dc}} = \mathcal{E}_{\text{c}}/\gamma_{\text{se}}$ at $s = 0$. Inverting this gives, approximately

$$s \approx \frac{6}{\alpha}\left[\left(1 - \ln\frac{V}{V_{dc}}\right)^{1/2} - 1\right] \tag{14.4.14}$$

At lower pressures, $\lambda_{he} \gg 2R$, they find that these relations are modified by the pendulum effect, with $\alpha \to 2V/(5\mathcal{E}_c R)$.

14.4.3 RF-Driven Hollow Cathodes

Hollow cathode discharges operated at rf frequencies were first studied by Horwitz (1983). A review of early work is given in Bárdoš (1996). Their use has mainly been motivated by (1) increasing the density, and therefore the etching or deposition rates, in capacitive discharge reactors; and (2) improving processing uniformity over large areas (see Choi and White, 2009, for an example). As described in Section 9.5, these reactors are often rf-driven by an upper showerhead electrode, having many regularly spaced small holes through which the feedgas flows into the reactor volume. Hollow cathode discharges can form in these holes, either inadvertently or by design.

The hollow cathode discharge properties can be controlled by properly choosing the hole radius R and the axial hole length h at the showerhead surface. Each cylindrical hole surface acts as a hollow cathode, and the discharge bulk plasma intruding into the hole acts as the anode. The plasma-to-showerhead voltage V_s contains both dc and rf components, e.g., $V_s = V_B - V_0 \cos\omega t$, as in (11.5.1). As for dc discharges, rf hollow cathode plasmas are often sustained by ion-induced, high-energy secondary electrons, mostly created at the time of maximum $V_s(t)$. However, under some conditions, rf sheath heating (see Sections 11.1 and 11.2) can be an additional mechanism to create hot (ionizing) electrons.

The maximum rf sheath width s_m at the showerhead electrode is an important parameter, with $s_m < R$ for hollow cathode operation, allowing the bulk plasma to penetrate into the hole. The sheath width is defined by a Child law, for example (11.2.15), at low pressures, and (11.2.54) at high pressures. The sheath width scales weakly with the discharge power, see (11.2.50) or (11.2.60) for examples.

14.4.3.1 Experiments

Measurements of the density enhancement versus upper electrode dc sheath voltage, for various upper electrode hole radii, are shown in Figure 14.7 (Lee et al., 2011a; see also Lee et al., 2010a,b). The 129 mTorr argon discharge with 14 cm diameter electrodes and a 4 cm gap was driven at 13.56 MHz. The five exchangeable upper electrodes each contain a multi-hole hexagonal pattern with hole length $h = 10$ mm, hole radii R varying from 0 to 5 mm, and hole spacing $4R$. As can be seen in the figure, as R increases at the higher dc sheath voltages, n_e first increases, and then decreases. At 400 V, the maximum density $n_e \approx 2.5 \times 10^{11}$ cm^{-3} is for a hole radius of 2.5–3.5 mm and is about five times larger than that for an upper electrode with no holes. At this pressure, the argon energy relaxation length (14.4.10) is $\lambda_{\mathcal{E}h} \approx 10$ mm.

The sheath widths were calculated from the measured dc voltage and electron density using the collisionless rf Child law (11.2.15). At 400 V for the 3.5 mm radius hole, $s_m \approx 1$ mm, giving $s_m/R \approx 0.3$ as an optimum hole size for argon at this pressure. Similar measurements were made over the pressure range of 64.5–645 mTorr for argon and for neon and krypton discharges. These indicate an optimum hole radius of about 1.5–2.5 times the sheath length for the most efficient multihole hollow cathode operation (Lee et al., 2010b).

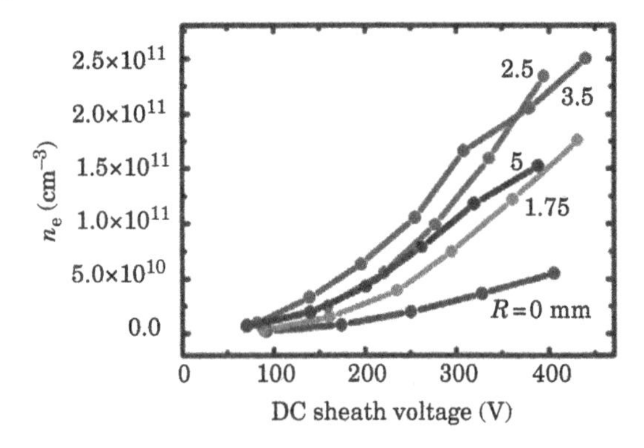

Figure 14.7 Hollow cathode-enhanced rf capacitive discharge, showing the measured electron density n_e versus dc cathode sheath voltage, in a 129 mTorr argon discharge driven at 13.56 MHz, for various hole radii R; the upper electrode contains a hexagonal pattern of uniformly spaced holes of length $h = 10$ mm and hole spacing $4R$. Source: Lee et al. (2011a)/with permission of Elsevier.

14.4.3.2 Simulations

Simulations are challenging because of the three-dimensional geometry and kinetic nature of hollow cathode discharges. Two-dimensional PIC simulations in single-slit geometry have been used to examine some aspects of multi-hole rf hollow cathode physics. Lafleur and Boswell (2012b) did such simulations and compared them to a multi-hole measurement of Lee et al. (2011a). The simulation was limited to low voltages (low densities), as run times would be excessive for higher densities. For a 258 mTorr argon plasma with 5 mm diameter holes in the upper electrode, the 5 mm slit simulation with $\gamma_{se} = 0.2$ and 100–120 V sheath bias showed a significant density enhancement compared to a no-hole electrode, and it was in reasonable agreement with the experimental results. The enhanced density was found to be due to the formation of hollow cathode discharges within the slit, resulting mainly from secondary emission on the slit sidewalls due to ion bombardment. The ionization was found to peak near the maximum of $V_s(t)$, where the secondary electron multiplication is the strongest, as expected. A hollow cathode discharge did not form for $\gamma_{se} = 0$.

He et al. (2020) used a 2D cylindrical PIC simulation to investigate the effects of the electrode gap on a 1 Torr argon, 13.56 MHz hollow cathode discharge, with hole diameter and depth both 5 mm. The electrode gap was varied from 4 to 12 mm. At small gaps of 4 and 6 mm, the rf sheath was found to occupy the entire hole, bulk plasma did not intrude into the hole, and a hollow cathode discharge did not form.

Park et al. (2023) examined hollow cathode plasma formation in a slit with a curved boundary using 2D PIC simulations, in the absence of secondary emission ($\gamma_{se} = 0$). Hole depths of 5–15 mm were investigated, with 2–8 mm slit widths at the gas inlet surface, widening toward the showerhead surface facing the plasma. For argon in the pressure range of 0.4–1 Torr, hollow cathode discharges were found when the sheath length was shorter than the hole size. When the sheath heating was strong enough to generate hot (ionizing) electrons, then plasma density increases were observed; weaker sheath heating led to density decreases.

14.5 Planar Magnetron Discharges

Dc planar magnetron discharges are widely used for sputter deposition of metallic thin films such as aluminum, copper, tungsten, gold, and various alloys, e.g., Al/2%Cu and Ti/W. When powered by an rf source, or mid-frequency pulsed sources, these discharges are also used for sputter deposition of insulating films such as oxides, nitrides, and ceramics. Physical sputtering is described in Section 9.3, and its application to thin-film deposition is described in Section 17.4. In this section, we first discuss the limitations of glow discharges as sputtering sources. We then describe the important planar magnetron configuration and present a simple equilibrium model that can be used to estimate discharge parameters and sputtering efficiency. Finally, we examine some issues for high-power impulse magnetron sputtering (HiPIMS). For a more thorough understanding, the early review articles by Thornton and Penfold (1978) and Waits (1978), and more recent reviews by Gudmundsson (2020) and Gudmundsson et al. (2012a, 2022) should be consulted.

14.5.1 Limitations of Glow Discharge Sputtering Source

Low aspect ratio ($l/R < 1$ for a cylindrical plasma) dc glow discharges have long been used as sputtering sources for metallic materials and are still used in some specialized applications. These are illustrated in Figure 14.8*a* for a planar discharge in argon gas driven by a constant current dc source. The upper aluminum electrode is the cathode, which serves as the target for ion impact sputtering of aluminum atoms. The substrates, on which the sputtered atoms are deposited, are placed on the lower electrode, which is the anode. The cathode–anode gap is typically $l \sim 5$ cm. Almost all of the anode–cathode voltage appears across the cathode sheath (dark space or cathode fall). The negative glow extends almost to the anode, and the positive column is absent in these short discharges. High ion current densities, $J_{\text{dc}} \gtrsim 1\ \text{mA/cm}^2$, are required in order to achieve, at best, commercially viable deposition rates of $\approx$350 Å/min. Hence, the discharge is operated in the abnormal glow regime with a high discharge voltage, $V_{\text{dc}} \sim 2$–5 kV. The sputtering power efficiency (sputtered atoms/ion-volt) is relatively low at these high energies and decreases with increasing energy (see Section 9.3).

As described in Section 14.3, the discharge is maintained in the usual manner by secondary electron emission from the cathode, with the energetic secondary electrons providing the ionization required to maintain the discharge. However, operating pressures must be high enough,

$$p \gtrsim 30\,\text{mTorr} \tag{14.5.1}$$

so that secondary electrons are not lost to the anode or side walls. These pressures are higher than optimum for deposition of sputtered atoms onto the substrates due to scattering of sputtered atoms by argon atoms. This results in sputtered atom redeposition on the cathode, deposition on the side walls, and, in some cases, poor adhesion of the sputtered film (see Section 17.4). For a neutral–neutral scattering cross section $\sigma \sim 2 \times 10^{-16}$ cm^2, setting the mean free path $\lambda = (n_{\text{g}}\sigma_{\text{el}})^{-1} \gtrsim l$, we obtain

$$p \lesssim 30\,\text{mTorr} \tag{14.5.2}$$

for acceptably low sputtered atom scattering. Equations (14.5.1) and (14.5.2) indicate that there is a narrow pressure range around 30 mTorr for dc glow discharge sputtering. As we saw in Section 14.4, the sputtering efficiency of a dc discharge can be improved by operation with a hollow cathode. However, the geometry also becomes a limiting factor for deposition, leading to the addition of a magnetic field, as described below.

14.5.2 Magnetron Configuration

It is clearly desirable to operate a sputtering discharge at higher current densities, lower voltages, and lower pressures than can be obtained in a conventional glow discharge. This has led to the use of a dc magnetic field at the cathode to confine the electrons. An axisymmetric dc magnetron configuration is shown in Figure 14.8*b*. The permanent magnet placed at the back of the cathode

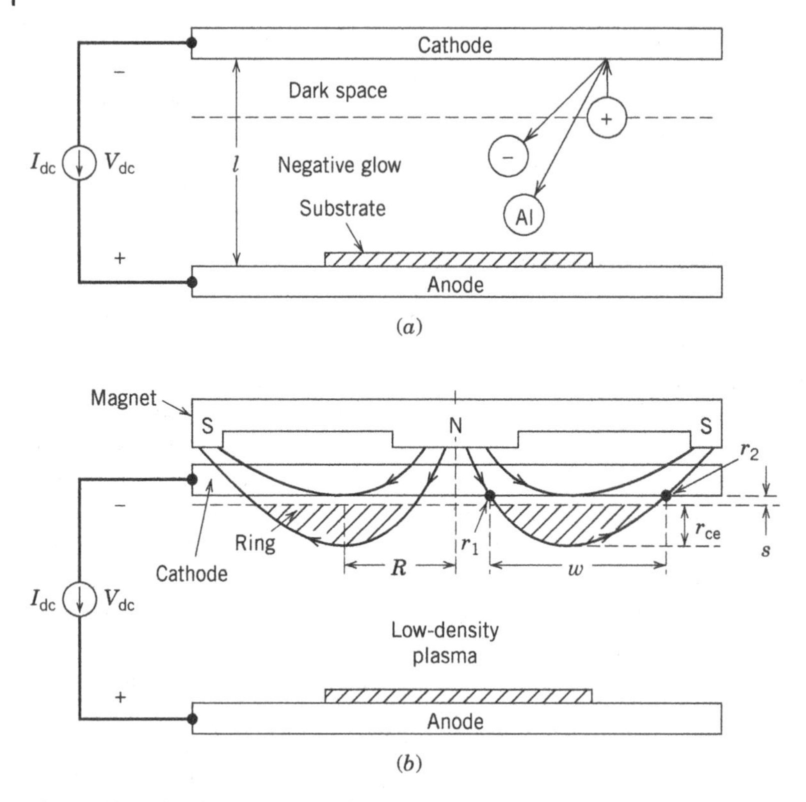

Figure 14.8 Dc discharges used for sputtering: (*a*) low aspect ratio dc glow discharge; (*b*) planar magnetron discharge.

target generates magnetic field lines that enter and leave through the cathode plate as shown. A discharge is formed when a negative voltage of 200 V or more is applied to the cathode. The discharge appears in the form of a high-density brightly glowing circular plasma ring of width w and mean radius R that hovers below the cathode, as illustrated in the figure, with sputtering occurring in a corresponding track on the cathode. The ring contains significant numbers of trapped secondary electrons, in addition to primary electrons, and is embedded in a lower density, low electron temperature bulk plasma (Sheridan et al., 1991). The plasma shields the electric field through most of the chamber, and a cathode sheath of thickness $s \sim 1$ mm develops, which sustains most of the externally applied voltage. Argon ions in the plasma, unconfined by the magnetic field, are accelerated toward the cathode and strike it at high energy. In addition to sputtering target material, the ion impact produces secondary electron emission. These electrons are accelerated back into the plasma and are confined near the cathode by the magnetic field. They can often undergo a sufficient number of ionizing collisions to maintain the discharge before being lost to a surface. Typical planar magnetron characteristics are

- $B_0 \sim 200$ G
- $p \sim 2$–5 mTorr argon
- $\bar{J}_\mathrm{i} \sim 40$ mA/cm^2
- $V_\mathrm{dc} \sim 800$ V
- Deposition rate ~ 2000 Å/min

Here, B_0 is the magnetic field strength at the radius R where the magnetic field line is tangential to the cathode target surface, and $\overline{J}_i$ is the average ion current density to the target area *under the ring*. The target area under the ring is typically about half the total target area in well-designed planar magnetron discharges.

14.5.3 Discharge Model

Because the magnetic field and discharge structure are highly nonuniform, a complete quantitative model of the discharge has not been developed. We present a qualitative model to indicate some issues that arise when determining the equilibrium properties of the discharge. The given discharge control parameters are I_{dc}, p, B_0, and R.

14.5.3.1 Magnetron Voltage V_{dc}

Almost all of the applied voltage is dropped across the cathode sheath. The secondary emission coefficient for argon ions on aluminum is $\gamma_{se} \sim 0.1$ for 200–1000 V argon ions. If $\mathcal{M}$ is the number of electron–ion pairs created by each secondary electron that is trapped within the ring, then an estimate for $\mathcal{M}$ is

$$\mathcal{M} \approx \frac{V_{dc}}{\mathcal{E}_c} \tag{14.5.3}$$

where $\mathcal{E}_c$ is the energy lost per electron–ion pair created by secondary electrons. For 200–1000 V secondaries, we take $\mathcal{E}_c \approx 30$ V (see Figure 3.17). Because of the tangential magnetic field, not all secondary electrons emitted at the cathode are trapped in the ring. Some electrons execute one or more gyro orbits and are reabsorbed at the cathode, and some ions in the ring do not return to the cathode. This leads to an effective secondary emission coefficient γ_{eff} that is less than γ_{se}. Thornton and Penfold (1978) have estimated

$$\gamma_{eff} \approx \frac{1}{2}\gamma_{se} \tag{14.5.4}$$

In steady state, for ion particle balance, we require

$$\gamma_{eff}\mathcal{M} = 1 \tag{14.5.5}$$

Inserting (14.5.3) and (14.5.4) into (14.5.5), we obtain

$$V_{dc} \approx \frac{2\mathcal{E}_c}{\gamma_{se}} \tag{14.5.6}$$

which is sometimes called the Thornton equation. For $\mathcal{E}_c = 30$ V and $\gamma_{se} = 0.1$, we find $V_{dc} \approx 600$ V.

The energetic electron gyroradius is found from (4.1.8) to be

$$r_{ce} = \frac{v_e}{\omega_{ce}} = \frac{1}{B_0}\left(\frac{2mV_{dc}}{e}\right)^{1/2} \tag{14.5.7}$$

where $v_e = (2eV_{dc}/m)^{1/2}$. For $B_0 = 200$ G and $V_{dc} = 600$ V, we obtain $r_{ce} \approx 0.5$ cm. The energetic ion gyroradius is similarly found from

$$r_{ci} = \frac{1}{B_0}\left(\frac{2MV_{dc}}{e}\right)^{1/2} \tag{14.5.8}$$

to give $r_{ci} \approx 1.3$ m for argon ions. Therefore, the energetic ions are not magnetized by the weak magnetic field in this discharge.

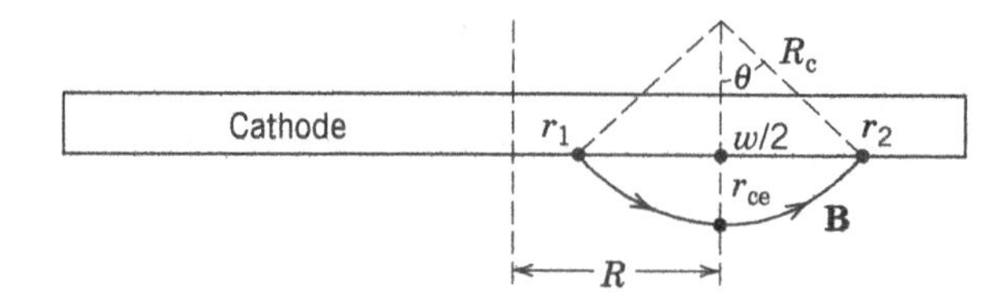

Figure 14.9 Calculation of planar magnetron ring width.

14.5.3.2 Ring Width w

Referring to Figure 14.9, we estimate that the ring has mean height (from the cathode) equal to the gyration radius r_{ce}. We assume that the sheath width $s \ll r_{ce}$ and will show this below. Hence, energetic secondary electrons are trapped on a magnetic field line and can oscillate back and forth between radii r_1 and r_2. The main force that reflects the electrons at r_1 and r_2 is the electrostatic sheath potential; there can also be some mirroring, due to the nonuniform magnetic field, which results in a parallel force as given by (4.3.9). For the field line having a radius of curvature R_c and height r_{ce} from the cathode, given $w \approx r_2 - r_1$, then, as shown in Figure 14.9,

$$\frac{w/2}{R_c} = \sin\theta \qquad (14.5.9)$$

and

$$r_{ce} + R_c \cos\theta = R_c \qquad (14.5.10)$$

Eliminating θ from these two equations yields w as a function of r_{ce} and R_c. For simplicity, assuming that $w/2 \ll R_c$, then (14.5.9) and (14.5.10) become

$$\frac{w}{2R_c} \approx \theta$$

$$\frac{2r_{ce}}{R_c} \approx \theta^2$$

Eliminating θ, we obtain an estimate of the ring width (Wendt and Lieberman, 1990)

$$w \approx 2(2r_{ce}R_c)^{1/2} \qquad (14.5.11)$$

For $r_{ce} \approx 0.5$ cm and choosing $R_c = 4$ cm, we obtain $w \approx 4$ cm. This is not fully consistent with the simplifying assumption $w/2 \ll R_c$, but a more accurate result can be found by solving (14.5.9) and (14.5.10) (see Problem 14.7).

14.5.3.3 Ion Current Density $\bar{J}_i$ and Sheath Thickness s

The ions are unmagnetized and the gas pressure is low; therefore the collisionless Child law (6.3.12) can be used to describe the flow of ions from the surface of the ring to the cathode target area under the ring

$$\bar{J}_i = \frac{4}{9}\epsilon_0\left(\frac{2e}{M}\right)^{1/2}\frac{V_{dc}^{3/2}}{s^2} \qquad (14.5.12)$$

Assuming for simplicity that the ring is thin, $w \ll R$, then we find

$$\bar{J}_i = \frac{I_{dc}}{2\pi R w} \qquad (14.5.13)$$

Taking the typical parameters $I_{dc} = 5$ A, $R = 5$ cm, $w = 4$ cm, and $V_{dc} = 600$ V, then we obtain $\bar{J}_i \approx 40$ mA/cm^2 from (14.5.13) and $s \approx 0.56$ mm from (14.5.12). For a typical target area under the ring that is half the total target area, this gives a current density averaged over the total target area of 20 mA/cm^2.

14.5.3.4 Ring Plasma Density n_i

We use the Bohm flux at the edge of a collisionless plasma to estimate n_i in the ring from

$$0.61\, e n_i u_B = \bar{J}_i \qquad (14.5.14)$$

where the electron temperature T_e enters only weakly. Using a typical value for low-pressure discharges of $T_e \approx 3$ V, then from (14.5.14) with $\bar{J}_i \approx 40$ mA/cm^2, we obtain $n_i \approx 1.5 \times 10^{12}$ cm^{-3}. We note that the density n_{es} of energetic electrons within the ring is generally much smaller than n_i (see Problem 14.6).

14.5.3.5 Sputtering Rate R_{sput} and Absorbed Power P_{abs}

Let γ_{sput} be the yield of sputtered atoms per incident ion, we have the sputtering rate

$$R_{sput} = \gamma_{sput} \frac{\bar{J}_i}{e} \frac{1}{n_{Al}} \text{ cm/s} \qquad (14.5.15)$$

Taking $n_{Al} \approx 6 \times 10^{22}$ cm^{-3} to be the atomic density of the aluminum target and evaluating (14.5.15) for $\gamma_{sput} \sim 1$ and $\bar{J}_i \approx 40$ mA/cm^2, we obtain $R_{sput} \approx 4.1 \times 10^{-6}$ cm/s. After 24 hours of operation, a target thickness of 3.6 mm has been sputtered. Thus the discharge digs an erosion track into the cathode material beneath the ring. The cathode must be replaced when the erosion track becomes comparable to the cathode thickness, which is usually an expensive proposition for the ultrapure materials commonly used as targets.

The discharge power is $P_{abs} = V_{dc} I_{dc}$. For our example with $I_{dc} = 5$ A and $V_{dc} = 600$ V, $P_{abs} =$ 3 kW. Almost all of this power is absorbed at the cathode; hence, the cathode must be cooled. Uniformity of the sputtered film deposited on the substrate is an important issue. For a ring-shaped source of sputtered atoms, the aspect ratio l/R controls the radial uniformity. As expected, for $l \ll R$, the radial distribution of deposited atoms has a maximum off-axis, while for $l \gg R$, the radial distribution is (mildly) peaked on-axis. Hence, an optimum ratio of l/R exists that maximizes the deposition uniformity. Measurements and analysis show that $l/R \approx 4/3$ (Thornton and Penfold, 1978).

The confining magnetic field of a planar magnetron produces a dense plasma near the cathode and a much weaker plasma near the substrate, thus minimizing ion bombardment flux to the substrate. This is desirable for some applications, but for others, an increased primary gas ion flux is desirable. This can be accomplished by using an *unbalanced magnetic configuration* where some of the field lines from the outer radius of the cathode flow to the substrate surface rather than to the inner radius of the cathode.

14.5.3.6 Ohmic Heating in the Plasma

The Thornton equation (14.5.6) for secondary electron discharge maintenance gives $V_{dc} \to \infty$ as $\gamma_{se} \to 0$. However, measurements show a finite limit $V_{dc} \to V_{ohm}$. This can be ascribed to plasma sustainment by bulk ohmic heating, due essentially to the low electron cross-field mobility. See Sections 5.4 and 11.10 for discussions of cross-field mobilities and the resulting conductivities.

To distinguish the two mechanisms, Brenning et al. (2016) plotted $1/V_{dc}$ measured by Depla et al. (2009) for 18 different target materials versus γ_{se} at a fixed magnetron current, e.g., $I_{dc} = 0.4$ A, obtaining the result

$$\frac{I_{dc}}{V_{dc}} = G_a \gamma_{se} + G_b \qquad (14.5.16)$$

The conductances $G_a \gamma_{se}$ and G_b give the contributions to V_{dc}, and the power $V_{dc} I_{dc}$, due to the secondary electron and bulk ohmic discharge maintenance, respectively. A simple parallel circuit

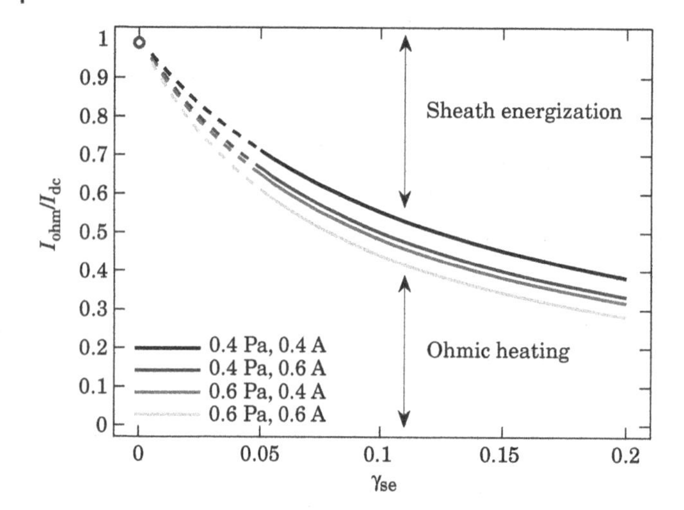

Figure 14.10 The relative contributions to the total ionization due to ohmic heating and sheath energization. The curves show (14.5.17) using the measured G_a and G_b from the four combinations of pressure and discharge current in the dc magnetron sputtering discharge studied by Depla et al. (2009); the lines are solid only in the range of γ_{se} where they are supported by the measurements; the circle in the upper left-hand corner marks the HiPIMS study by Huo et al. (2013). Source: Brenning et al. (2016)/with permission of IOP Publishing.

model of the two conductances, corresponding to the power dissipated by secondary and bulk electrons, gives the form (14.5.16). From the parallel circuit model, the ratio of ohmic to total current is

$$\frac{I_{ohm}}{I_{dc}} = \frac{G_b}{G_a\gamma_{se} + G_b} \tag{14.5.17}$$

A graph of I_{ohm}/I_{dc} versus γ_{se} is shown in Figure 14.10. We see that over the range of typical γ_{se}s plotted, ohmic heating makes an important contribution to the discharge maintenance. Note that this also implies that not all the applied voltage is dropped across the cathode sheath; there is also a voltage drop across an extended pre-sheath within the bulk plasma.

14.5.4 Time-Varying Power Sources

In addition to dc power, magnetron discharges can be powered using a wide variety of time-varying sources. As already mentioned, for sputtering ceramic or oxide targets, rf sources can be used. Deposition of dielectric films can be done in this way, as well as with dc reactive sputtering, as described in Section 17.4. Magnetrons can be pulsed, enabling operation in a different regime than dc- or rf-driven systems. An important example is *high-power impulse magnetron sputtering* (HiPIMS). The negative voltage pulses on the cathode are short, 10–400 μs, and the current densities are high, $\bar{J}_i \approx 2$–$20\ \text{A/cm}^2$. The peak power fluxes are high, typically 0.5–5 kW/cm^2. With such high-power fluxes, the pulse repetition frequency must be low, 50–5000 Hz, to remain below the average power flux threshold for target damage of about 0.05 kW/cm^2 (Gudmundsson et al., 2012a). The high-peak power density produces a high plasma density within the ring, and therefore a highly ionized flux of sputtered material, similar to that for I-PVD as described in Section 14.6. The large fraction of ionized sputtered material can enable deposition with favorable film properties for many applications. See Lundin et al. (2020) for a thorough description of the fundamentals, technologies, challenges, and applications of HiPIMS discharges.

Sputtered atom ionization. Measurements indicate (Gudmundsson, 2020) that the region between the ring and the substrate has an average plasma density

$$\overline{n}_e(\mathrm{m}^{-3}) \approx 10^{15}\,\overline{J}_i\,(\mathrm{mA/cm^2}) \tag{14.5.18}$$

The sputtered atoms typically enter the discharge with a kinetic energy of a few volts. The sputtered atom ionization mean free path due to this plasma can be estimated as

$$\lambda_{iz,s} = \frac{\overline{v}_s}{K_{iz,s}\overline{n}_e} \tag{14.5.19}$$

with $\overline{v}_s$ the atom mean speed and $K_{iz,s}$ the atom ionization rate coefficient. For typical dc magnetron current densities, $\overline{J}_i \sim 1$–$100\ \mathrm{mA/cm^2}$, one finds $\overline{n}_e \sim 10^{15}$–$10^{17}\ \mathrm{m^{-3}}$ and $\lambda_{iz,s} \gtrsim 1$ m, significantly greater than the target-to-substrate separation l. Hence, in dc magnetrons most sputtered atoms are deposited on the substrate as atoms, not ions. However, in HiPIMS discharges, with current densities of order 1–10 A/cm^2, $\overline{n}_e \sim 10^{18}$–$10^{19}\ \mathrm{m^{-3}}$, and $\lambda_{iz,s} \lesssim 10$ cm, significantly less than l, indicating that most sputtered atoms are ionized as they flow to the substrate, depositing as ions. We leave these calculations of $\lambda_{iz,s}$ as Problem 14.8 for the student.

14.5.4.1 Gas and Sputtered Atom Depletion and Recycling

Gas depletion effects can be significant near the high ring current density (high power) limit $\overline{J}_{imax} \sim 200\ \mathrm{mA/cm^2}$ of dc magnetron operation and can be very strong at the high-power fluxes of HiPIMS. This gas depletion is due to collisions by energetic sputtered species and by significant ionization of the argon atoms. In the steady state, the maximum ion current density above the ring (see Figure 14.8*b*) that can be drawn is determined by the primary working gas refill flux to the cathode target area under the ring (Gudmundsson, 2020)

$$\overline{J}^{(0)}_{i,max} \sim e \cdot \frac{1}{4} n_g \overline{v}_g \tag{14.5.20}$$

At 300 K and 1 Pa argon, $\overline{J}^{(0)}_{i,max} \sim 400\ \mathrm{mA/cm^2}$. This indicates that gas depletion effects might be important even for dc magnetrons.

However, $\overline{J}_i \gg 200\ \mathrm{mA/cm^2}$ during the pulse in a HiPIMS discharge, leading to severe gas atom depletion and recycling effects during the pulse. The discharge can generate much more ion current density than given in (14.5.20) by recycling primary gas atoms. These are generated from ions of the working gas that are ionized for the first time and then drawn to the cathode. Primary gas atoms are recycled when a primary gas ion hits the target, generating a "recycled" gas atom. Let p_g be the fraction (< 1) of gas atoms recycled per each recycling step, then summing over all successive recyclings[1] and adding the initial primary current density give

$$\overline{J}_{i,g} = \overline{J}^{(0)}_{i,max}\left(\frac{1}{1 - p_g}\right) \tag{14.5.21}$$

Equation (14.5.21) shows that the working gas recycling can give far larger current densities than that limited by the gas refill flux over the target area under the ring; this mechanism is essential for HiPIMS operation. In some cases where the self-sputter yield is high (e.g., argon ion sputtering of a copper target), sputtered atoms can also be ionized, drawn to the target, and recycled, leading to an additional current density (Brenning et al., 2017; Gudmundsson, 2020).

1 Note that $1 + p_g + p_g^2 + \cdots = 1/(1 - p_g)$ for $0 \le p_g < 1$.

14.5.4.2 PIC Simulations

In contrast to 1D PIC simulations, which are quite useful for understanding the physics of narrow gap capacitive discharges, planar magnetron PIC simulations are difficult because of the essential 3D nature of the discharge. Even in axisymmetric (r, z) discharge geometry, the θ-directed particle drifts can excite instabilities that saturate, giving a θ-dependence of the discharge properties (see Section 5.4). Highly parallelized codes, advanced computer hardware, and long run times are needed for each simulation. As an example (Jo et al., 2022), Figure 14.11 shows the 3D spatial profiles of (a) the electron density in a range greater than 3×10^{16} m^{-3} and (b) the electron temperature in a range greater than 1 V. The 5 cm × 5 cm square cathode (target) and anode (substrate) electrodes are separated by 2 cm, and the cathode magnets are placed off-axis to increase the utilization of the target material. The peak magnetic field is of order 1000 G near the target surface. Figure 14.11a and b clearly shows the ring structure, with a ring radius of about 1.6 cm. They also show strong azimuthal fluctuations, perhaps excited by the grad-B and diamagnetic drift velocities, as described briefly in Section 5.4.

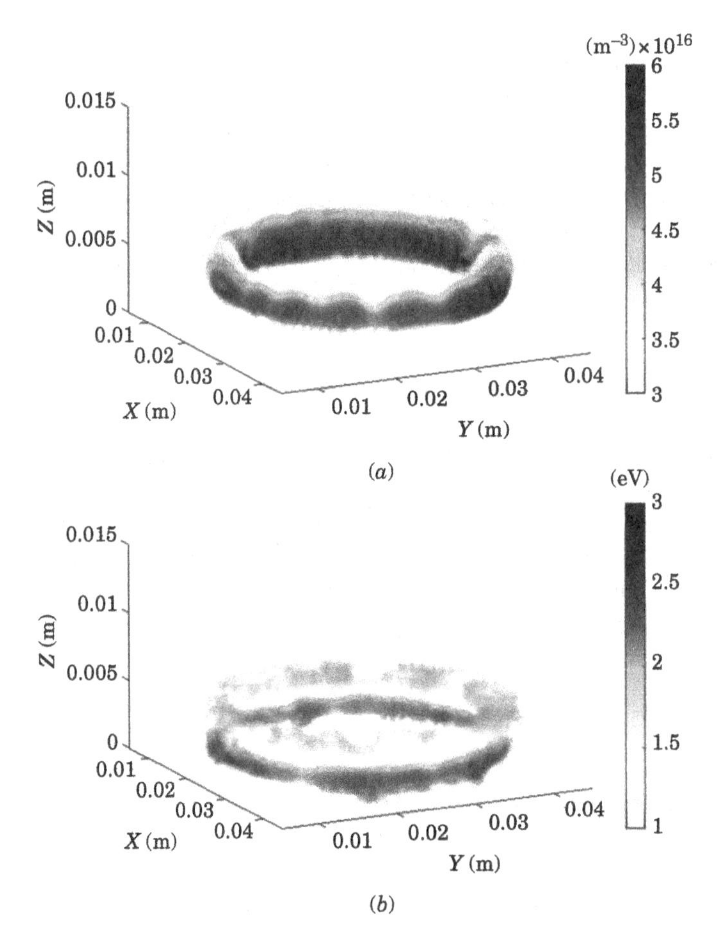

Figure 14.11 3D spatial profiles of (a) the electron density in a range greater than 3×10^{16} m^{-3} and (b) the electron temperature in a range greater than 1 V. Source: Jo et al. (2022)/with permission of Springer Nature.

14.6 Ionized Physical Vapor Deposition

For the dc planar magnetron discharge in Section 14.5, physical vapor deposition (PVD) directly deposits metals on the substrate. Low-pressure operation is desirable to minimize scattering in the intervening plasma region. However, even with minimal scattering, the angular spread at the substrate is large, preventing uniform deposition in deep trenches, as required for many processing applications. This situation can be greatly improved by ionizing the metal atoms in the discharge. The ions are then naturally accelerated across the plasma-to-substrate sheath, giving good collimation. The process is known as *ionized physical vapor deposition* (I-PVD). The same basic dc magnetron discharge, or some variant, is typically used, but operating in a somewhat different parameter regime. In particular, the gas pressure is larger by about a factor of ten, $p \sim 10$–50 mTorr, and the plasma region is also larger, so that the mean free path for scattering of the metal atoms is small compared to the cathode–substrate length. The multiple scattering, together with a high-density plasma generated in the background gas, results in a high degree of ionization of the metal atoms. This is facilitated by using a noble gas such as argon, with a higher ionization energy ($\mathcal{E}_{iz} = 15.76$ V) than the metal atoms (e.g., $\mathcal{E}_{iz} = 7.23$ V for aluminum and 8.77 V for copper). The latter are then efficiently ionized by the energetic electrons required to maintain the argon plasma.

The efficiency of ionizing the argon background gas is enhanced by using a supplementary source, such as an electron cyclotron resonance, inductive, or hollow cathode discharge, to produce a high density of electrons everywhere in the chamber. Alternately, the magnetron can be pulsed (e.g., HiPIMS, as described in Section 14.5), dispensing with a supplementary source. Hopwood (2000) gives a description of a dc magnetron sputtering source with a supplementary rf-driven inductively-coupled plasma (ICP) source that is placed between the cathode target and the substrate, as shown in Figure 14.12. These are commonly used for deposition of metal films in the microelectronics industry. Reviews of I-PVD, with descriptions of the various supplementary sources used, are given in Helmersson et al. (2006), Gudmundsson (2020), and Gudmundsson et al. (2022).

As with most complicated processing devices, complete analysis is not possible. For some specific I-PVD plasmas, there has been numerical modeling (Li et al., 2000). Simple global models can often be used to describe the basic physical principles and understand the parameter scaling, but

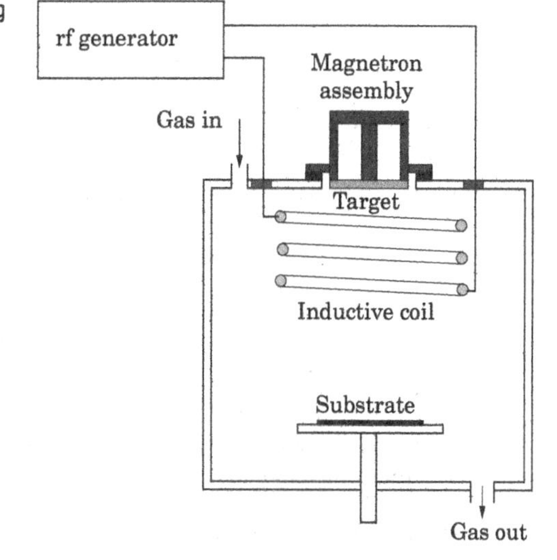

Figure 14.12 Schematic of an I-PVD system, showing a planar magnetron source and a supplementary inductively coupled rf source that is embedded between the magnetron cathode target and the substrate. Source: Gudmundsson (2020)/IOP Publishing/CC BY 4.0.

assume certain given quantities. For a global analysis, we assume a low-pressure cylindrical argon discharge (radius R, length l) with metal atoms (density $n_{\rm m} \ll n_{\rm Ar}$) created by sputtering at one end wall (the magnetron cathode) due to energetic (mainly argon) ions having energy $V_{\rm dc} \sim 600$ V. The central electron density is the sum of the average density $n_{\rm e0}$ produced by the magnetron and the density produced by the supplementary discharge power $P_{\rm rf}$

$$n_{\rm e} = n_{\rm e0} + \frac{P_{\rm rf}}{eA_{\rm eff}u_{\rm B}\mathcal{E}_{\rm T}} \tag{14.6.1}$$

where $A_{\rm eff} = 2\pi R^2 h_l + 2\pi R l h_R$ from (10.2.11) is the effective area for particle loss, h_l and h_R given by (10.2.1) and (10.2.2) are the edge-to-center density ratios, $u_{\rm B} = (e{\rm T}_{\rm e}/M)^{1/2}$ is the Bohm velocity, and $\mathcal{E}_{\rm T}$ given by (10.2.9) is the total energy lost per electron–ion pair created. The low-pressure approximation that the electron density is nearly uniform, except near the sheaths, is used. The total ion current to the target is then

$$I_{\rm T} = e n_{\rm e} u_{\rm B} h_l \pi R^2 \tag{14.6.2}$$

and the metal atom current entering the plasma is $I_{\rm m} = \gamma_{\rm sput} I_{\rm T}$. We assume a loss time $\tau_{\rm m}$ for fully collisional metal atoms given by

$$\frac{1}{\tau_{\rm m}} = D_{\rm m}\left(\frac{\pi^2}{l^2} + \frac{\chi_{01}^2}{R^2}\right) \tag{14.6.3}$$

with $D_{\rm m}$ the metal atom diffusion coefficient (see Section 5.2) and $\chi_{01} \approx 2.405$. Then, the particle balance for metal atoms is

$$\gamma_{\rm sput} h_l u_{\rm B} n_{\rm e} \pi R^2 \approx \frac{n_{\rm m}}{\tau_{\rm m}} \pi R^2 l \tag{14.6.4}$$

which yields an average metal atom density

$$n_{\rm m} = \frac{\gamma_{\rm sput} h_l u_{\rm B} \tau_{\rm m}}{l} n_{\rm e} \tag{14.6.5}$$

Let us assume that the dominant process producing metal ions is direct ionization of metal atoms. Then, the particle balance of metal ions is

$$K_{\rm izm} n_{\rm e} n_{\rm m} \pi R^2 l \approx n_{\rm m+} u_{\rm Bm} A_{\rm eff} \tag{14.6.6}$$

with $u_{\rm Bm}$ the Bohm speed for metal ions. Solving for the ratio $n_{\rm m+}/n_{\rm m}$, we obtain

$$\frac{n_{\rm m+}}{n_{\rm m}} = \frac{K_{\rm izm}\pi R^2 l}{A_{\rm eff} u_{\rm Bm}} n_{\rm e} \tag{14.6.7}$$

Substituting (14.6.5) into (14.6.7) to determine the metal ion density yields

$$n_{\rm m+} = \frac{K_{\rm izm}\gamma_{\rm sput}\tau_{\rm m}\pi R^2}{A_{\rm eff}} \frac{u_{\rm B}}{u_{\rm Bm}} n_{\rm e}^2 \tag{14.6.8}$$

We see that $n_{\rm m+}/n_{\rm m}$ scales as $n_{\rm e}$ and $n_{\rm m+}$ scales as $n_{\rm e}^2$, with $n_{\rm e} \propto P_{\rm rf}/\mathcal{E}_{\rm T} + I_{\rm dc0}$ from (14.6.1) and (14.6.2).

The electron temperature (needed to evaluate $K_{\rm izm}$ and $u_{\rm B}$) is found from the overall particle balance for ions. Assuming that most of the ionization is due to the supplementary source and that $n_{\rm m} \ll n_{\rm Ar}$, then the balance is for argon ions only

$$K_{\rm iz} n_{\rm e} n_{\rm Ar} \pi R^2 l = n_{\rm e} u_{\rm B} A_{\rm eff} \tag{14.6.9}$$

which yields T_e as given in Figure 10.1. If the ionization due to the supplementary source becomes small compared to the ionization produced by the hot electrons of the magnetron, then the bulk electron temperature drops, leading to reduced ionization of the metal atoms. Also, since many metals have a low ionization energy compared to argon, (14.6.9) must be modified to include metal ion production and loss when n_m is greater than a few percent of n_{Ar} (Problem 14.9).

To convert to the physically most significant quantities of surface fluxes, we note that the ion flux is

$$\Gamma_{m+} = h_l n_{m+} u_B \tag{14.6.10}$$

and a thermalized neutral metal flux is

$$\Gamma_m = \frac{1}{4} n_m \bar{v}_m \tag{14.6.11}$$

with $\bar{v}_m = (8e\mathrm{T}_m/\pi M_m)^{1/2}$. The ratio of ionized-to-neutral flux Γ_{m+}/Γ_m is enhanced over the ionized density fraction in (14.6.7) because typically $h_l u_B/\bar{v}_m > 1$.

For the low pressures used in I-PVD, charge transfer from argon ions to metal atoms does not contribute significantly to the production of metal ions, as it does in the high-pressure hollow cathode discharge. Hopwood (2000) has examined the additional production of metal ions due to Penning ionization by metastable argon atoms

$$\mathrm{Ar}^* + \mathrm{M} \rightarrow \mathrm{e} + \mathrm{M}^+ + \mathrm{Ar} \tag{14.6.12}$$

with the result for the ionized flux fraction $\Gamma_{m+}/(\Gamma_m + \gamma_{m+})$ for aluminum plotted versus electron density in Figure 14.13. Penning ionization dominates at low electron densities but becomes relatively unimportant at the higher electron densities that would normally be employed. Hopwood also compared the results to an experiment, with good agreement, considering the many simplifying assumptions.

The global model does not capture important information concerning spatial variations and can lead to quantitative errors. Various *ad hoc* methods can be employed to introduce spatial variations. The spatial variations may be rather insensitive to the exact global plasma parameters, so that useful information on these variations can be obtained. The reader is referred to Hopwood (2000), Helmersson et al. (2006), Gudmundsson (2020), and Gudmundsson et al. (2022) for discussions and references for various higher-order effects, including spatial variations.

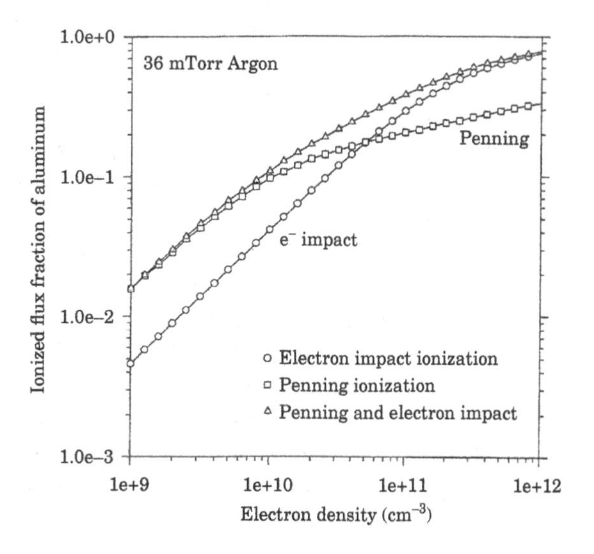

Figure 14.13 Electron impact ionization is the primary path for metal ion production in a high electron density plasma; Penning ionization dominates under conditions of low electron density. Source: Hopwood (2000)/with permission of Elsevier.

Problems

14.1 Positive Column of a DC Glow Discharge

(a) A glow discharge in argon with $R = 2$ cm, $l = 25$ cm is operated at $p = 100$ mTorr. This gives $T_e \approx 4$ V, $\mathcal{E}_T \approx 40$ V, $\lambda_e \approx 0.4$ cm, and $D_a p \approx 10^4$ cm^2-Torr/s. Assuming the solution in Section 14.2, and using (14.2.14), find the electric field strength E. Assuming that most of the discharge is positive column, what is the voltage drop in the positive column?

(b) If the discharge current in part (a) is 10 mA, what is the plasma density?

14.2 Paschen Minimum Voltage Show that the Paschen minimum voltage is $V_{\min} = (eB/A)\ln(1 + 1/\gamma_{se})$ and occurs at $pd = (pd)_{\min} = eA^{-1}\ln(1 + 1/\gamma_{se})$; here e ≈ 2.72 is the natural base.

14.3 Breakdown of a DC Discharge For the previous discharge, but at $p = 1$ Torr and a cathode secondary emission constant $\gamma_{se} = 0.12$, using values from Table 14.1, calculate the breakdown voltage for the discharge.

14.4 Operation of a DC Discharge For the parameters of Problem 14.3, take $T_e \approx 3$ V and $\mathcal{E}_T \approx 50$ V, and $D_a p \approx 10^4$ cm^2-Torr/s.

(a) Recompute the results for the positive column of Problem 14.1.

(b) Using Figure 14.5, calculate the voltage drop across the cathode sheath.

(c) For an applied voltage $V_B = 1500$ V ($V_B > V_b$ in Problem 14.3) what resistance should be put in series with the applied voltage to supply the 10 mA required.

14.5 High-Pressure Hollow Cathode Discharge Equilibrium
Redo the example in Section 14.4 with a new (higher) pressure of 1 Torr.

14.6 Planar Magnetron Discharge An axially symmetric planar magnetron discharge in argon with an aluminum cathode has a magnetic field strength $B_0 = 200$ G at a radius $R = 10$ cm, where the field line is tangent to the cathode surface. The field line radius of curvature is $R_c = 3$ cm. The discharge current is $I_{dc} = 2$ A and the pressure is $p = 2$ mTorr.

(a) Assuming that the effective secondary emission coefficient $\gamma_{eff} \approx 0.05$ for Ar^+ ions on Al and that 20% of the secondary electrons are lost by diffusive transport to the anode before creating electron–ion pairs in the plasma ring, estimate the dc voltage V_{dc} across the discharge.

(b) Estimate the mean width w (in the r direction) of the ring (erosion track) and the ion current density $\bar{J}_i$ (mA/cm^2) incident on the aluminum cathode over the erosion track area $2\pi Rw$.

(c) Assuming that the sputtering coefficient at the erosion track is unity (1 sputtered Al atom per incident Ar ion) and that sputtered atoms are deposited uniformly on the anode surface over an area of πR_a^2, where $R_a \approx 15$ cm, estimate the deposition rate (Å/min) for the aluminum film deposited on wafers located at the anode surface.

(d) Estimate the (low-temperature) plasma density n_i (cm^{-3}) within the ring and the secondary electron density n_{se} within the ring. (Use the data given in Figure 3.13 for the argon ionization cross section σ_{iz}(m^2) for secondary electrons having energy eV_{dc}, and note that secondary electrons lose about 30 V per ionization. From this information, the secondary electron lifetime can be determined.) Note that you should find $n_{se} \ll n_i$.

14.7 Planar Magnetron Ring Width Use the exact equations (14.5.9) and (14.5.10) to determine w for $r_{ce} \approx 0.5$ cm and $R_c = 4$ cm, and compare to the result $w \approx 4$ cm obtained in (14.5.11) for the simplified analysis.

14.8 Sputtered Atom Ionization in Planar Magnetron Discharges Consider the sputtering of copper atoms in the planar magnetron of Figure 14.8*b* with a target-to-substrate separation of $l = 30$ cm. Copper atoms have a mass $M_{Cu} = 63.5$ amu, an ionization energy $\mathcal{E}_{iz} = 7.73$ V, and an initial sputtered atom energy (see Section 9.3) $\mathcal{E}_s \approx 2$ V.

(a) For a sputtered atom with a uniform velocity $\overline{v}_s$, flowing from the target to the substrate through a plasma of uniform density $\overline{n}_e$ and electron temperature $\mathrm{T_e} = 3$ V, show that its ionization mean free path is given by (14.5.19).

(b) Estimate the ionization rate coefficient K_{iz} of copper atoms for $\mathrm{T_e} = 3$ V using the Thomson expression (3.5.5).

(c) At very low argon gas pressures in a dc magnetron with $\overline{J}_i = 40$ mA/cm^2, the copper atom speed as it flows to the substrate is $\overline{v}_s \approx (2e\mathcal{E}_s/M_{Cu})^{1/2}$. Using this and $\overline{n}_e$ given by (14.5.18), determine $\lambda_{iz,s}$ for the copper atoms, and compare this to l.

(d) At higher gas pressure, the copper atoms will thermalize, with $\overline{v}_s \approx (8e\mathrm{T_g}/\pi M_{Cu})^{1/2}$. For room temperature $\mathrm{T_g} = 0.026$ V, repeat part (c).

(e) Consider now a typical HiPIMS current density $\overline{J}_i \approx 3$ A/cm^2. At very low gas pressures, repeat part (c).

14.9 Electron Temperature for Ionized Physical Vapor Deposition Consider the generation and loss of electron–ion pairs in a two-species plasma

$$K_{izAr} n_{Ar} n_e \pi R^2 l = n_{Ar+} u_{BAr+} A_{eff}$$

$$K_{izAl} n_{Al} n_e \pi R^2 l = n_{Al+} u_{BAl+} A_{eff}$$

Let n_g be the total gas density (argon + aluminum atoms) and $f = n_{Al}/n_g$ be the fraction of the gas that is aluminum.

(a) Find the equation that determines the electron temperature $\mathrm{T_e}$.

(b) For $R = l = 15$ cm and 50 mTorr total (room temperature) gas pressure, solve the equation in (a) numerically to find $\mathrm{T_e}$ for $f = 0.0$, 0.005, 0.1, and 0.2. You may use the same edge-to-center density ratio factors h_l and h_R to determine A_{eff} as for a pure argon discharge. Use the argon ionization rate coefficient given in Table 3.3 and use $K_{izAl} \approx 1.23 \times 10^{-8}\, e^{-7.23/\mathrm{T_e}}$ cm^3/s for aluminum.

15

High-Pressure Capacitive Discharges

15.1 Introduction

Narrow gap, radio frequency (rf) capacitive discharges at intermediate pressures ($p \sim 0.2$–5 Torr, $l \sim 1$–10 cm) are widely used for thin film deposition in the semiconductor, flat panel display, and solar panel industries. Some examples are given in Chapter 17. For flat panel displays, the (rectangular) substrate sizes can be large, for example, 2.94 m × 3.37 m, imposing severe uniformity requirements. At still higher pressures and shorter gap lengths, and particularly including atmospheric pressure, there are important applications such as nanoparticle production, soft (plastic) surface film deposition and modification, and *in vitro* and *in vivo* biological treatments. Particulate production is described in Chapter 18.

Many, but not all, capacitive discharge properties scale as the ps_m product, with s_m the sheath width. We consider "intermediate pressures" to lie not far from the Paschen minimum $(ps_m)_{min} \sim$ 0.5–5 Torr-cm for breakdown of a gas-filled gap (see Figure 14.4), with "high pressures" lying well above 5 Torr-cm. There is also scaling with the pl product for energy relaxation and transport phenomena, so there are some similarities between intermediate and high-pressure regimes, as well as differences. We consider intermediate pressures in Sections 15.2 and 15.3, and atmospheric pressures in Sections 15.4 and 15.5.

Much new and complicating physics and chemistry appear at the higher pressures. An important distinguishing feature that separates the low and higher pressure regimes is the size of the energy relaxation length $\lambda_\mathcal{E}$ for ionizing electrons, compared with the discharge gap length l. For low pressures, $\lambda_\mathcal{E} \gtrsim l$, and no matter where the electrons are heated, they flow everywhere, leading to a relatively uniform ionization rate constant K_{iz} within the gap. The global models introduced in Chapters 10 and 11 are well-suited to treating this situation.

In contrast, at the higher pressures, $\lambda_\mathcal{E} \ll l$, and the electrons can be locally heated, producing local ionization, with K_{iz} a strong function of position within the gap. We have already seen in Chapter 11 that even at low pressures, electrons are mainly heated within or near the capacitive sheath edges; the former is due to stochastic sheath heating, and the latter is due to the increased bulk ohmic heating near the low electron density sheath edges. At higher pressures, almost all electron heating, *and the resultant ionization,* usually takes place within or near the plasma–sheath edges, and the bulk plasma can become "passive," i.e., devoid of ionization. The particle and energy balances for a passive bulk capacitive discharge are described in Section 15.2.

A second important feature is the role of secondary electron emission due to ion and excited state neutral bombardment of the electrodes. At low rf source voltages, the secondary emission is usually a perturbation of the basic particle and energy balances described in Chapters 10 and 11. At higher voltages, the secondary emission can dominate, leading to a new mode of discharge operation,

Principles of Plasma Discharges and Materials Processing, Third Edition. Michael A. Lieberman and Allan J. Lichtenberg.

the γ-mode. At intermediate pressures, this is characterized by a near-breakdown of the capacitive sheaths, similar to the dc breakdown of a gas-filled gap, which is examined in Section 14.3. At lower pressures, the sheaths may be relatively far from breakdown, but the ionization due to secondary emission can still dominate. The transition from the α-mode, where the secondary emission is not important, to the γ-mode is described in Section 15.3, along with a γ-mode model of the particle and energy balances.

At atmospheric pressures, it has been found that diffuse, transversely uniform, narrow gap rf discharges can be excited in the 6.78–40.68 MHz rf frequency range. The gap sizes are typically 0.5–4 mm, with $pl \sim 40$–300 Torr-cm, putting these discharges well into the high-pressure side of the Paschen curve. Helium, argon, or nitrogen gas feeds are commonly used, often with a small ($<1\%$) admixture of a molecular gas. Penning ionization of the molecular gas often dominates in determining the charged particle balances. At atmospheric pressures, in addition to $\lambda_{\mathcal{E}} \ll l$, the relaxation time for the electron temperature T_e is smaller than the rf period, such that $T_e(t)$ oscillates at twice the driving frequency. Also, the electron momentum transfer frequency is very large, $\nu_m \gg \omega_{pe} \gg \omega$, usually leading to large resistive voltage drops across the bulk plasma. As at intermediate pressures, an α–γ transition can occur, almost always accompanied by a discharge contraction or filamentation. An α-mode model of an atmospheric pressure, rf-driven Penning discharge is described in Section 15.4.

At atmospheric pressures but lower driving frequencies, 50 Hz–1 MHz, stable discharges can be excited using dielectric barriers on one or both electrodes. Transient charge accumulation on the dielectric surfaces creates an opposing potential that limits the current buildup after breakdown, ensuring that the transition to an arc discharge cannot occur. For source voltages just exceeding the Paschen breakdown limit, relatively transversely uniform glow or Townsend discharges can be generated, typically two short 1–20 μs discharges per low-frequency cycle. At higher source voltages, exceeding the streamer breakdown voltage, huge numbers of 1–10 ns microdischarge filaments, each about 0.1 mm in diameter, can be generated per unit area per second, leading to operation as a *dielectric barrier discharge.* In Section 15.5, we examine the atmospheric pressure low-frequency regime. We describe the Paschen to streamer breakdown transition and examine simple models of the relatively uniform glow discharge regime, and the locally nonuniform dielectric barrier discharge regime.

15.2 Intermediate Pressure RF Discharges

As discussed in Section 5.3, the heuristic edge-to-center density ratio h_l factor (5.3.16) shown in Figure 5.4 in narrow gap rf discharges has been confirmed by one-dimensional (1D) PIC simulations at the lower pressures in capacitive discharges (see Figure 5.5). However, the PIC results strongly deviate from the theory at the higher pressures, typically above 300 mTorr in argon, indicating that the theory strongly underestimates the particle losses and that the smaller values of $h_l \lesssim 0.1$ are not valid. The reason for this can be seen in Figure 15.1, which shows one-dimensional PIC results for time-averaged normalized ionization rate and normalized electron density versus x at three different argon gas pressures. At the lower pressure, the ionization rate in Figure 15.1*a* is quite uniform in the bulk plasma, even though there is significantly enhanced edge heating within and near the sheath edges. Hence, the global (volume-averaged) models of Chapters 10 and 11 are appropriate to describe the discharge equilibrium. In contrast, at the highest pressure, there is no ionization in the bulk plasma, and all the ionization occurs near the sheath edges due to strong stochastic and ohmic heating there. Peaked ionization near the sheath edges was also seen in previous fluid simulations at 1 Torr (Gogolides and Sawin, 1992;

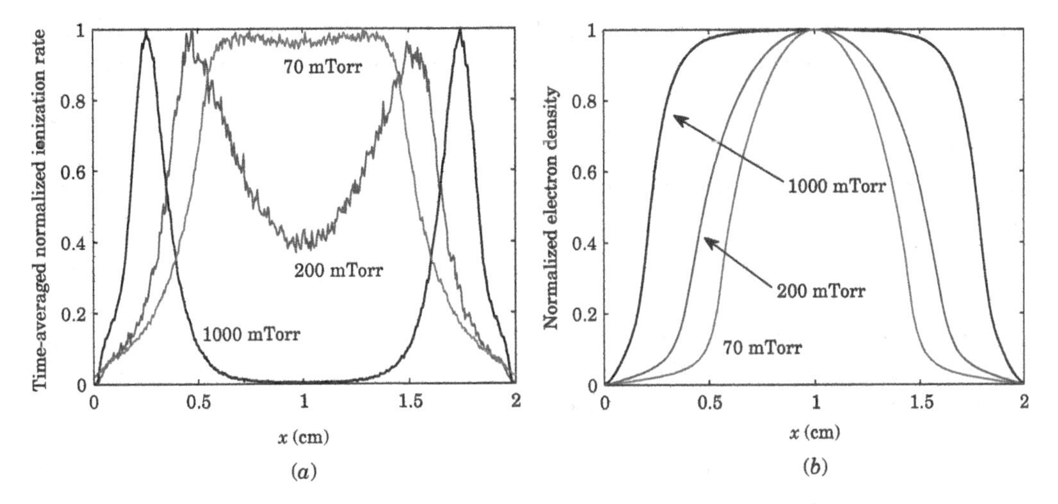

Figure 15.1 One-dimensional PIC simulation of (*a*) time-averaged normalized ionization rate and (*b*) normalized electron density versus x for a capacitive discharge at various argon gas pressures; the driving current density is $J = 2.56$ A/m^2 at 13.56 MHz, and the secondary electron emission coefficient is $\gamma = 0.1$. Source: Lafleur and Chabert (2015a)/with permission of IOP Publishing.

Gogolides et al., 1992; Lymberopoulos and Economou, 1993). This leads to the flat bulk density profile shown in Figure 15.1*b*, along with a greatly enhanced plasma density gradient near the sheath edges. This gradient drives a much larger ion loss flux to the walls, leading to the strongly increased simulation values of h_l seen at the higher pressures in Figure 5.5. Hence there is an "active" ionization process in the bulk plasma at the lower pressures, but the bulk plasma becomes "passive" at the higher pressures. Similarly, there can be an active-to-passive bulk transition as the driving rf current is increased at fixed pressure. These transitions can be abrupt. This was observed in hybrid PIC/fluid simulations of a 13.56 MHz, 6 cm gap, 75 mTorr CH_4 discharge by Schweigert (2004). He found an abrupt active-to-passive transition in the discharge equilibrium as the rf current density was increased from 1.0 to 1.1 mA/cm^2. Similar phenomena were seen in PIC simulations of electronegative (CF_4) discharges by Schulze et al. (2011c).

The rf "passive bulk" regime is quite similar to the situation in short dc discharges, in which the positive column is absent (Raizer et al., 1995, section 2.5.5), but with one important difference. In a dc discharge, the strong ionization within and near the sheath edge is due to secondary electrons emitted from the cathode, while for the passive bulk rf discharge regime, the low-density plasma electrons within and near the sheath edge provide the ionization. To understand this, let us consider the energy relaxation length for ionizing electrons in the discharge.

15.2.1 Energy Relaxation Length $\lambda_{\mathcal{E}}$

Ionizing electrons are hot, they lie in the tail of the electron energy distribution, and they have typical energies $\mathcal{E} \approx \mathcal{E}_{iz} + 2\,T_e$ (see Problem 3.12). At high pressures, they are mainly created within or near the sheath edges by ohmic heating. They then flow into the plasma bulk by a diffusive process, continually scattering against the neutral gas atoms, until they lose a significant energy by inelastic collisions. Assuming monoenergetic electrons with energy $\mathcal{E} = \frac{1}{2}mv^2/e$, the hot electron diffusion coefficient is (see (5.1.6) and the subsequent comment).

$$D = \frac{1}{3}\lambda_m v \tag{15.2.1}$$

with λ_m the momentum transfer mean free path for scattering of electrons against neutral gas atoms. From (5.1.8) in the steady state and with $G = 0$, the hot electron diffusion equation is

$$D\frac{d^2n}{dx^2} = \nu_{inel}n \tag{15.2.2}$$

where n is the hot electron density, x lies along the discharge gap length (here $x = 0$ at the left hand sheath edge), and

$$\nu_{inel} = \frac{v}{\lambda_{inel}} \tag{15.2.3}$$

is the energy-loss collision frequency of electrons against neutral gas atoms, accounting for all energy losses due to elastic and inelastic collisions. The corresponding mean free path λ_{inel} also accounts for all collisional energy loss processes, i.e.

$$\lambda_{inel}^{-1} = n_g[(2m/M)\sigma_{el} + \sigma_{ex} + \sigma_{iz} + \cdots] \tag{15.2.4}$$

The solution of (15.2.2) is

$$n(x) = n_0\,e^{-x/\lambda_\mathcal{E}} \tag{15.2.5}$$

where

$$\lambda_\mathcal{E}^2 = \frac{D}{\nu_{inel}} \tag{15.2.6}$$

Substituting D from (15.2.1) and ν_{inel} from (15.2.3) into (15.2.6), we obtain

$$\lambda_\mathcal{E} = \left(\frac{\lambda_m\lambda_{inel}}{3}\right)^{1/2} \tag{15.2.7}$$

From Figure 3.13, the excitation and ionization cross sections for argon at $\mathcal{E} \approx 20$ V are about 10^{-20} m^2, and the elastic scattering cross section is about 1.4×10^{-19} m^2. Using these to evaluate λ_m and λ_{inel}, we obtain $\lambda_\mathcal{E} = 1.1 \times 10^{19}/n_g$ m, with n_g the gas density. At 1 Torr, this gives $\lambda_\mathcal{E} \approx 0.33$ mm. Hence, electrons heated near the sheath edge do not diffusively flow into the plasma, giving rise to a "passive" bulk.

15.2.2 Passive Bulk Plasma Model

We assume an argon gas density n_g and bulk width d, driven by a sinusoidal rf current density of radian frequency ω and amplitude J, with no secondary emission, and Maxwellian electrons with temperature T_e. All ionization is assumed to be produced within and near the sheath edge by the tail of this Maxwellian EEPF. Within the plasma bulk, the tail is assumed to be depleted and therefore to not result in ionization. In most of the bulk plasma, except possibly very near the sheath edge, the E-field is low, with ion drift velocities small compared to ion thermal velocities, so we use a constant mobility bulk transport model $\mu_i = e/(M\nu_{mi})$, with ν_{mi} the constant ion–neutral collision frequency and M the ion mass. In the sheath, the ion drift and thermal velocities are comparable, so we could use either a constant mobility or a constant mean free path model. We choose the latter.

Setting the coordinate system origin $x = 0$ at the midplane, the particle balance in the bulk plasma is

$$\frac{d\Gamma}{dx} = G_{iz} \tag{15.2.8}$$

with Γ the particle flux. In view of the exponential decay (15.2.5) of the ionizing electrons from each sheath edge into the midplane, the volume generation rate taken to be

$$G_{iz} = \Gamma_s \beta \frac{2\cosh \beta x}{\exp(\beta d/2)} \tag{15.2.9}$$

where Γ_s is the particle flux at the plasma–sheath edge and $\beta = \lambda_\mathcal{E}^{-1}$ is the inverse of the energy relaxation length $\lambda_\mathcal{E}$ of the ionizing electrons. The assumed $G_{iz} \propto \cosh \beta x$ spatial variation is the same as for the usual models of secondary electron ionization (Godyak and Khanneh, 1986; Misium et al., 1989). Integrating (15.2.8), we obtain the spatially varying bulk plasma flux

$$\Gamma(x) = 2\Gamma_s \frac{\sinh \beta x}{\exp(\beta d/2)} \tag{15.2.10}$$

The flux $\Gamma(x)$ and density $n(x)$ are related by

$$\Gamma = -D_a \frac{dn}{dx} \tag{15.2.11}$$

with the ambipolar diffusion coefficient (for $T_e \gg T_i$)

$$D_a = \mu_i T_e \tag{15.2.12}$$

Substituting (15.2.10) into (15.2.11) and integrating to determine the density variation, we obtain

$$n(x) = n_s + \frac{2\Gamma_s}{\beta D_a} \frac{\cosh(\beta d/2) - \cosh \beta x}{\exp(\beta d/2)} \tag{15.2.13}$$

with n_s the density at the plasma–sheath edge. The edge flux and density are related by

$$\Gamma_s = n_s u_s \tag{15.2.14}$$

Substituting (15.2.14) into (15.2.13) with $x = 0$, and assuming $\lambda_\mathcal{E} \ll d$, i.e., $\beta d \gg 1$, we obtain the passive bulk edge-to-center density ratio

$$h_l = \frac{n_s}{n_0} = \left(1 + \frac{u_s}{\beta D_a}\right)^{-1} \tag{15.2.15}$$

The ion loss speed, obtained in Section 6.2 from the procedure described in Godyak and Sternberg (1990a), is

$$u_s = \frac{1}{2}\left[(\nu_{mi}^2 \lambda_{Ds}^2 + 4u_B^2)^{1/2} - \nu_{mi}\lambda_{Ds}\right]$$

with $\lambda_{Ds} = (\epsilon_0 T_e / e n_s)^{1/2}$ the Debye length at the plasma–sheath edge and $u_B = (eT_e/M)^{1/2}$ the Bohm speed. For $u_s \ll u_B$, this reduces to the simpler expression

$$u_s = \frac{u_B^2}{\nu_{mi}\lambda_{Ds}} \tag{15.2.16}$$

The energy balance for the sheath-heated electrons, which produce all the ionization, is

$$S_{ohm,sh} + S_{stoc} = e\mathcal{E}_c \Gamma_s \tag{15.2.17}$$

where $S_{sh} = S_{ohm,sh} + S_{stoc}$ is the time-averaged electron power flux deposited due to the sheath ohmic and stochastic heating, and $\mathcal{E}_c$ defined in (3.5.11) is the energy loss per electron–ion pair created.

The bulk electrons are ohmically heated, and we assume that they lose energy due to excitation processes only. The bulk power loss per unit volume is

$$\frac{S_{bulk}}{d} = \frac{1}{2}\frac{J^2 m \nu_m}{e^2 n} = K_{ex} n n_g e \mathcal{E}_{ex} \tag{15.2.18}$$

with $K_{ex}(T_e)$ the excitation rate constant. The elastic scattering power loss $K_m n n_g(3m/M_g)eT_e$ is small in the pressure regime considered here, where $T_e > 1$ V. Equation (15.2.18) gives the relation between T_e and n for a given J. The total electron energy lost $\mathcal{E}_{lost}$ per electron lost to the walls, due to the sum of sheath and bulk heating, is

$$\mathcal{E}_{lost} = \frac{2S_{sh} + S_{bulk}}{2e\Gamma_s} \tag{15.2.19}$$

Using (15.2.17) to eliminate Γ_s, we obtain

$$\mathcal{E}_{lost} = \mathcal{E}_c\left(1 + \frac{S_{bulk}}{2S_{sh}}\right) \tag{15.2.20}$$

Due to the relatively high pressures and low electron densities, the bulk plasma is quite resistive, and the second term, due to the bulk energy losses, is typically much larger than the first term, leading to a large value of $\mathcal{E}_{lost}$ (~300–1000 V).

The sheath heating power S_{sh} is a function of J, Γ_s, and T_e, obtained from the Child law and the capacitive nature of the sheath. For a constant ion–neutral mean free path, the Child law (11.2.54) gives the sheath width

$$s_m = K_{CL}\frac{\epsilon_0}{e\Gamma_s}\left(\frac{2e\lambda_i}{M}\right)^{1/2}\left(\frac{J}{\omega\epsilon_0}\right)^{3/2} \tag{15.2.21}$$

with s_m the maximum sheath width and $K_{CL} = 0.618$. The voltage–current relation for the capacitive sheath is

$$V_1 = K_1\frac{Js_m}{\omega\epsilon_0} \tag{15.2.22}$$

with V_1 is the fundamental frequency voltage amplitude and $K_1 = 0.658$.

The sheath stochastic and ohmic heating expressions are given in (11.2.59) and (11.2.64b). Substituting (15.2.21) and (15.2.22) into these expressions, we obtain S_{stoc} and $S_{ohm,sh}$ as functions of J, Γ_s, and T_e

$$S_{stoc} = K_{stoc}K_1K_{CL}\left(\frac{mT_e}{e}\right)^{1/2}\left(\frac{2e\lambda_i}{M\omega\epsilon_0}\right)^{1/2}\frac{J^{5/2}}{e\Gamma_s} \tag{15.2.23}$$

with $K_{stoc} = 0.61$, and

$$S_{ohm,sh} = K_{ohm,sh}K_1K_{CL}^2\frac{m}{2e}\nu_m\frac{2e\lambda_i}{M\omega^2\epsilon_0}\frac{J^4}{e^2\Gamma_s^2} \tag{15.2.24}$$

with $K_{ohm,sh} = 0.236$.

To estimate the passive bulk equilibrium parameters for $\lambda_{\mathcal{E}} \ll d$, i.e., $\beta d \gg 1$, we note that because K_{ex} in (15.2.18) is a strongly varying function of T_e, we can use a nominal value of T_e in the other equations, e.g., $T_e = 2$ V. After obtaining the equilibrium solution, the procedure can be iterated by solving the bulk power balance (15.2.18) for a new T_e.

Example 15.1 We take the following parameters:

- $p = 1.6$ Torr argon at 300 K ($n_g = 5.28 \times 10^{22}$ m^{-3})
- $l = 0.025$ m
- $f = 13.56$ MHz ($\omega = 8.52 \times 10^7$ s^{-1})
- $J = 20$ A/m^2

The "reduced" ion mobility μ_{i0} is defined as the ion mobility at 760 Torr and 300 K. At the low bulk plasma fields for argon ions in argon gas, $\mu_{i0} \approx 1.4 \times 10^{-4}$ m^2/V-s (Ellis et al., 1976). The ion mobility μ_i is related to the "reduced" mobility by $\mu_i = (T_g/273.16)(760/p)\mu_{i0} = 2.69 \times 10^{25}\mu_{i0}/n_g$. Rescaling μ_{i0} to 1.6 Torr at 300 K gives $\mu_i \approx 0.07$ m^2/V-s and $\nu_{mi} = 3.4 \times 10^7$ s^{-1}. From (15.2.7) at this gas density, $\lambda_{\mathcal{E}} = 1.93 \times 10^{-4}$ m, so we are clearly in a passive bulk regime. From (3.5.10), the ion–neutral mean free path $\lambda_i = 1.9 \times 10^{-5}$ m. Starting with an estimate $T_e = 2$ V, we evaluate the collision frequencies $\nu_{iz} = 3.0 \times 10^5$, $\nu_{ex} = 2.8 \times 10^6$, and $\nu_m = 2.5 \times 10^9$ s^{-1}, and we evaluate $\mathcal{E}_c = 127$ V from Figure 3.17. The Bohm speed is $u_B = 2.2 \times 10^3$ m/s. Substituting S_{sh} from the sum of (15.2.23) and (15.2.24) into (15.2.17), we solve numerically to find $\Gamma_s = 6.6 \times 10^{17}$ m^{-2}-s^{-1}. Using (6.2.22) for u_s with this Γ_s, we solve numerically to obtain $n_s = 1.4 \times 10^{15}$ m^{-3} and $u_s =$ 470 m/s. Then from (15.2.15), we find $n_0 = 1.6 \times 10^{15}$ m^{-3}. The sheath power flux from the sum of (15.2.23) and (15.2.24) is $S_{sh} = 13.4$ W/m^2. From (15.2.18), the bulk power flux is $S_{bulk} = 102$ W/m^2. From (15.2.20), this gives $\mathcal{E}_{lost} = 1095$ V. Neglecting any voltage drop across the bulk plasma, the discharge voltage from (15.2.22) is $V_{rf} \approx 2V_1 = 74$ V, and the sheath width is $s_m = 2.1$ mm from (15.2.21). From (11.2.56), the dc voltage across one sheath is 29 V, and, from (11.2.57), the ion bombarding energy is $\mathcal{E}_{ic} = 0.16$ V. The new $T_e = 2.1$ V from (15.2.18) is close to the initial estimate, so the plasma parameters are probably found within the accuracy of the calculation, and therefore an iteration is not useful. We note *post hoc* that $S_{stoc}/S_{ohm,sh} = 0.6$ and $u_s/u_B = 0.2$.

15.2.3 Simulation Results

Kawamura et al. (2020) performed 1D PIC simulations of a 2.5-cm gap argon discharge with gas temperature 300 K, driven at 13.56 MHz by an rf current source, and compared these to various models of the passive bulk regime. The argon-excited states were not considered. They first conducted simulations at 1.6 Torr and $J = 25.6$, 50, and 100 A/m^2, with the ion-induced secondary electron emission coefficient set to zero or 0.15 at the electrodes, finding that secondary emission had no significant effects on the simulation results at 25.6 A/m^2 and marginal effects at 50 A/m^2. They then fixed J to 50 A/m^2 and conducted simulations without secondary emission over a pressure range of 0.04–20 Torr, which encompasses the passive bulk pressure regime of interest.

A significant result from the simulations is that over the pressure range from 0.4 to 6 Torr, the bulk electron energy probability function (EEPF) was not Maxwellian, but was "Druyvesteyn-like," $g_p \sim \exp(-0.565\mathcal{E}^2/T_{eb}^2)$, with a strongly depressed tail above $\mathcal{E}_{ex} = 11.55$ V. Here, T_{eb} is two-thirds of the average energy per electron. The genesis of a bulk Druyvesteyn distribution is described in Section 19.1, and some more details are given in Section 19.5. Druyvesteyn distributions have also been observed in experiments at these intermediate pressures (Godyak et al., 1992), and an example is shown in Figure 11.9*b*. In contrast, the EEPF in the sheath region was found to be "Maxwellian-like" with a tail that extends above $\mathcal{E} > \mathcal{E}_{iz}$. So, the tail of the sheath EEPF produces practically all of the ionization in the discharge. Since only the sheath-heated electrons contribute to the ionization while most of the total electron heating is dissipated in the bulk via excitation, the ionization efficiency of the discharge is low, for example, $\mathcal{E}_{lost} \approx 532$ V at 1.6 Torr.

In contrast, at the low pressures of 0.04 and 0.16 Torr, both the sheath and bulk EEPFs were Maxwellian, and there was significant ionization in the bulk plasma as well as enhanced ionization within and near the sheaths. At 0.16 Torr, the bulk plasma density profile conforms reasonably well to typical lower-pressure models which predict a parabolic-like decrease in the density from the discharge center to the sheath edge. The ionization peaks near the sheath edge, but unlike the "passive" bulk base case, it is not sharply localized. Instead, there is significant ionization throughout the plasma bulk. This is expected as at this pressure, $\lambda_{\mathcal{E}} \approx 3.6$ mm, which is only somewhat smaller than the bulk half-width $d/2 \approx 6.6$ mm. Therefore, at 0.16 Torr, the discharge is in a transition between a low-pressure active bulk regime and an intermediate pressure passive bulk regime.

At the highest simulation pressure of 20 Torr, both sheath and bulk plasma EEPFs were Druyvesteyn-like, with the ionization strongly peaked near the plasma–sheath edges and negligible in the bulk. The electron heating was also peaked at the plasma–sheath edges but was significant and fairly uniform in the bulk. However, the electron heating was in equilibrium with the electron power loss throughout the discharge. Thus, at this high pressure, there is a local equilibrium within the sheaths as well as within the bulk. These results indicate that at 20 Torr, the entire discharge was in a local regime, which is not considered here. At 4 Torr, the discharge was in the transition between passive bulk and purely local regimes.

The simulation results were compared to various passive bulk models. All models incorporated a Druyvesteyn EEPF in the plasma bulk, and also used a variable mobility transport model in the plasma bulk, rather than a constant mobility model, as this gave somewhat better agreement with the simulations. Various sheath models were examined; we discuss here only the constant mean free path model used in the previous subsection.

The PIC central densities n_0 were found to increase from 0.04 to 0.16 Torr, as predicted by low pressure (active bulk) diffusion theory. However, the n_0 values decreased at the higher pressures, indicating a transition to the passive bulk regime. At 1.6 Torr, $n_0 \approx 1.2 \times 10^{16}$ m^{-3} for the simulation and $n_0 \approx 0.7 \times 10^{16}$ m^{-3} for the model. Over the range of passive bulk pressures, the n_0 values were almost a factor of two lower for the model. The PIC bulk temperatures $\mathrm{T_{eb}}$ increased from ≈ 1 V at the lowest pressure 0.04 Torr to ≈ 3 V at the highest two pressures, 6 and 20 Torr, and they showed reasonable agreement with the model in the passive bulk pressure regime: $\mathrm{T_{eb}} \approx 2.7$ V at 1.6 Torr for both PIC and model results. There was poorer agreement in the higher and lower pressure ranges because $\mathrm{T_{eb}}$ was calculated under the assumption that the EEPF is Druyvesteyn in the bulk and Maxwellian in the sheaths. However, this is only true at intermediate pressures in the passive bulk regime. As shown in Section 19.6, at the higher pressures, the Druyvesteyn $\mathrm{T_{eb}}$s are about a factor of 1.7 larger than the Maxwellian $\mathrm{T_{eb}}$s.

Figure 15.2 shows some comparisons of the sheath quantities. As seen in Figure 15.2*a* for the sheath edge density n_s, and Figure 15.2*b* for the rf sheath voltage V_1, the model tracks the PIC results quite well in the pressure range of the passive bulk regime, being only slightly lower. At the two lower pressures for n_s, and the highest pressure for V_1, the disagreement is stronger, due to the active-to-passive transition at the lower pressures, and the passive-to-local transition at the highest pressure. Figure 15.2*c* for the ratio g_l of the ion loss speed u_s to the Bohm speed u_B shows the model and PIC results in good agreement over the entire range of passive bulk pressures. Figure 15.2*d* for the edge-to-center density ratio $h_l = n_s/n_0$ gives slightly higher values for the model than for the simulations, except at the lowest and highest pressures, where the discharge transitions into a low-pressure active bulk and a high pressure fully local regime, respectively.

Not shown in the figure are the ion flux Γ_s and the sheath width s_m, with the model also in good agreement with the simulations in the passive bulk regime. The model and PIC bulk ohmic heating and sheath heating also agree well, except that above the transition pressure (~4 Torr) to a fully local discharge physics, the model significantly underestimates the sheath heating.

15.2.4 Metastables and Secondary Electrons

The influence of the argon-excited neutral states on the discharge equilibrium has been examined using 1D PIC simulations for the charged particles, self-consistently coupled to fluid simulations of a three-level model of the excited neutral states (Wen et al., 2021). Secondary electron emission

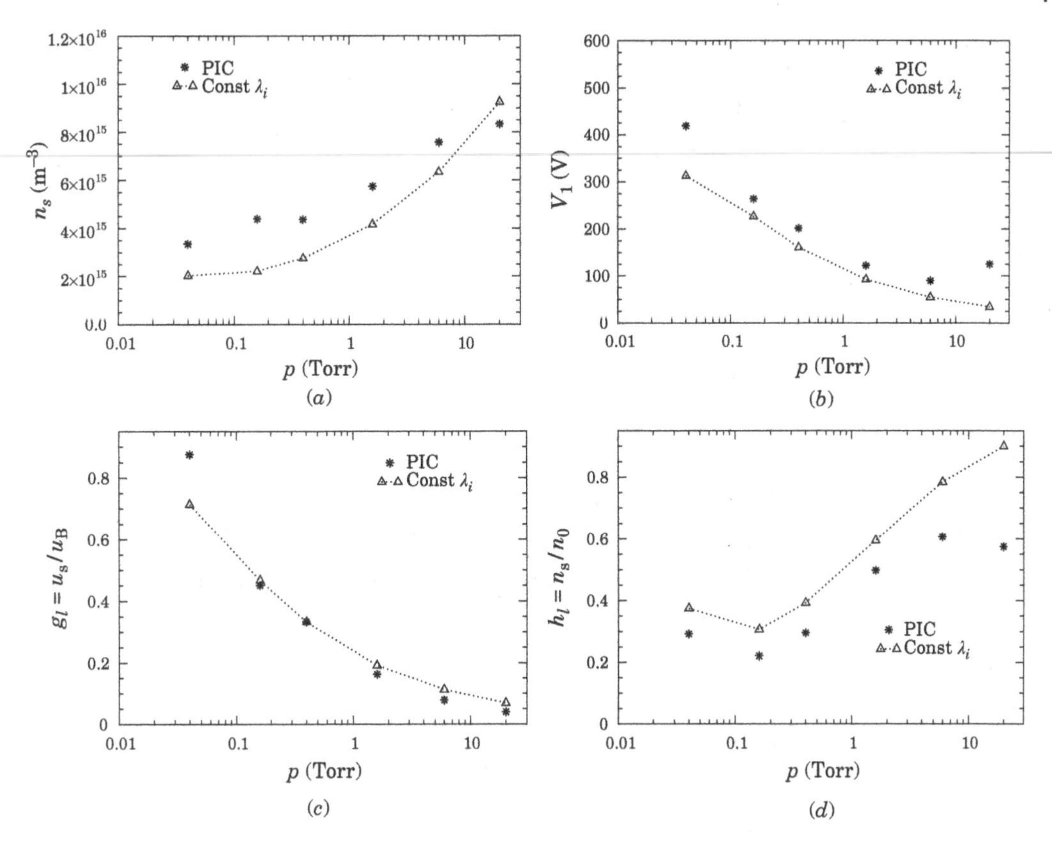

Figure 15.2 Comparison of one-dimensional PIC simulation results (stars) with a passive bulk model (triangles), versus pressure p, for a 2.5 cm-gap argon discharge, showing (*a*) plasma–sheath edge density n_s; (*b*) fundamental component of rf sheath voltage V_1; (*c*) normalized sheath edge ion speed $g_l = u_s/u_B$; and (*d*) edge-to-center density ratio $h_l = n_s/n_0$; the driving current density is $J = 50$ A/m² at 13.56 MHz. Source: Adapted from Kawamura et al. (2020).

was neglected. The diffusion fluids include metastable atoms Ar_m, resonance radiation atoms Ar_r, and Ar(4p) atoms excited to the 4p-manifold (see Figure 3.12). The three levels are treated as time- and space-evolving fluids, which diffuse against the ground state argon gas. The 2.5 cm discharge gap was excited by a 13.56 MHz current source at 50 A/m². The pressure was varied from 0.03 to 15 Torr. A major goal was to determine the relative importance of the various ionization reactions

$$
\begin{aligned}
e + Ar &\rightarrow 2e + Ar^+ && \text{(direct ionization)} \\
e + Ar_m &\rightarrow 2e + Ar^+ && \text{(Ar}_\text{m}\text{ stepwise ionization)} \\
Ar_m + Ar_m &\rightarrow e + Ar^+ + Ar && \text{(Ar}_\text{m}\text{–Ar}_\text{m}\text{ pooling ionization)} \\
Ar_m + Ar_r &\rightarrow e + Ar^+ + Ar && \text{(Ar}_\text{m}\text{–Ar}_\text{r}\text{ pooling ionization)} \\
Ar_r + Ar_r &\rightarrow e + Ar^+ + Ar && \text{(Ar}_\text{r}\text{–Ar}_\text{r}\text{ pooling ionization)}
\end{aligned}
$$

The results are shown in Figure 15.3. At low pressures, direct ionization dominates, as expected, with the order of a 10% contribution from Ar_m–Ar_m pooling ionization, and a smaller contribution from Ar_m stepwise ionization. However, there is a transition to a discharge dominated by Ar_m–Ar_m pooling (shaded region of the figure) above about 1.6 Torr. The much smaller contributions of the Ar_r pooling ionizations are due to the strong radiative emission losses of Ar_r to the ground state, leading to low Ar_r densities. Ar(4p) ionization reactions were also found not to be important, due to strong radiative transitions to the metastable and resonance levels.

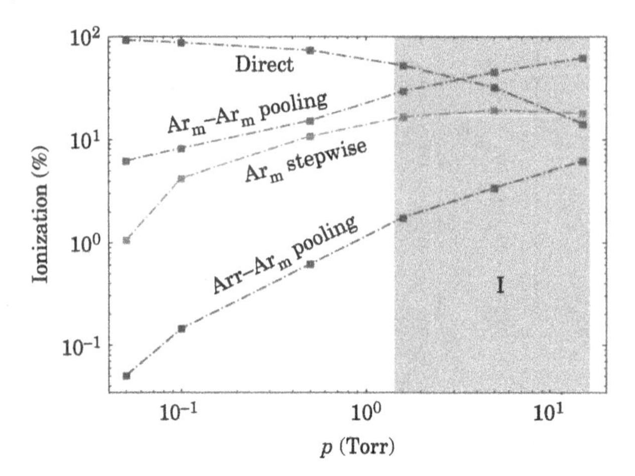

Figure 15.3 Percentage of each ionization reaction versus gas pressure p for a 2.5-cm gap argon discharge driven at 13.56 MHz by a 50 A/m^2 current source; the shaded region labeled I shows a transition from direct ionization to Ar_m−Ar_m pooling ionization with increasing pressure; at the highest pressure, Ar_m stepwise ionization is also significant. Source: Wen et al. (2021)/with permission of IOP Publishing.

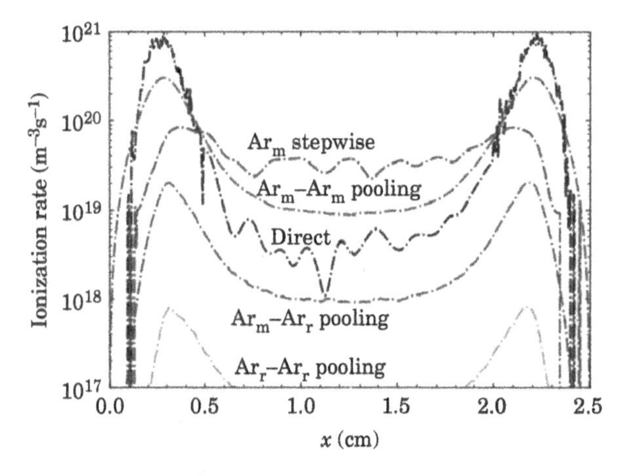

Figure 15.4 Ionization rates with no secondary emission versus gap position x for direct electron–neutral ionization, stepwise ionization, and metastable pooling ionization between excited state atoms, Ar_m, Ar_r and Ar(4p), for a pressure of 1.6 Torr and current density amplitude of $J = 50$ A/m^2. Source: Wen et al. (2021). ©IOP Publishing. Reproduced with permission. All rights reserved. https://doi.org/10.1088/1361-6595/ac1b22.

The ionization rates versus x within the gap at 1.6 Torr are shown in Figure 15.4. Near the sheath edges, Ar_m–Ar_m pooling ionization dominates, with a smaller contribution from direct ionization. Within the bulk plasma, the Ar_m stepwise ionization is most important, with a smaller contribution from direct ionization.

The combined effects of secondary electron emission due to ion *and* excited neutral atom bombardment of the electrodes can profoundly alter the discharge equilibrium, leading to a new mode of discharge operation. This has been examined by Gudmundsson et al. (2021) for a 2.5-cm gap, 1.6 Torr argon discharge driven at 13.56 MHz by a 50 A/m^2 current source. Secondary electrons, created at the electrode surfaces, and primary electrons, created in the discharge volume, were distinguished and separately tracked in the PIC simulation. The secondary emission coefficient for ions bombarding a clean surface was used (Phelps and Petrović, 1999; Phelps, 1999). For ion

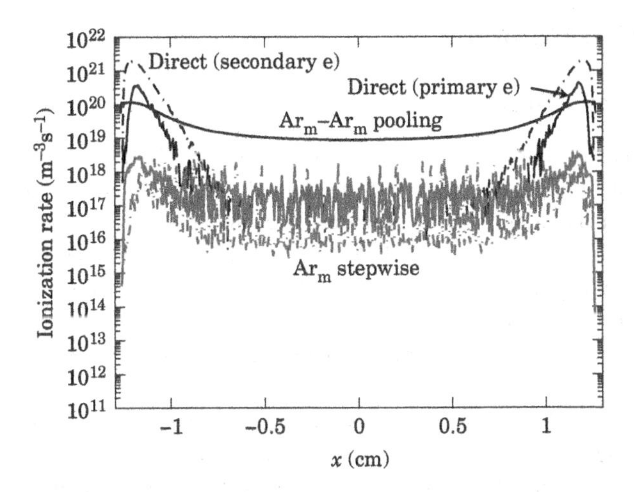

Figure 15.5 Ionization rates including excited state kinetics, energy-dependent secondary electron emission due to ion and excited atom bombardment of the electrodes, and electron reflection, for a 1.6 Torr argon discharge with a gap separation of 2.54 cm driven by a 50 A/m^2 rf current source at 13.56 MHz. Source: Gudmundsson et al. (2022)/with permission of IOP Publishing.

bombarding energies $\lesssim 1$ V, typical at 1.6 Torr, this reduces to $\gamma_{se} \approx 0.07$. For Ar_m, Ar_r, and Ar(4p), the values $\gamma_{se} = 0.21$, 0.21, and 0.27 were used, respectively, as given in Schohl et al. (1992). Additionally, a 20% electron reflection coefficient at the electrodes was assumed. The simulation results showed that bombardment of the electrodes by excited argon atoms was strongly effective in creating secondary electrons. The secondary emission flux due to the excited atoms was found to be 14.7 times greater than that due to the ions. Hence, referenced to the ion flux alone, the "effective" ion secondary emission coefficient was $(1 + 14.7) \times 0.07 = 1.1$, a very large value.

As previously described in the absence of secondary electrons, at 1.6 Torr, the bulk electron temperature was 2.5–3 V, the bulk plasma density was of order 1.2×10^{16} m^{-3}, and the density profile was strongly flattened in the discharge midplane (Gogolides and Sawin, 1992, Kawamura et al., 2020; Wen et al., 2021; Gudmundsson et al., 2021). This is consistent with the passive bulk model. Including the ion and excited atom secondary emission, the PIC results showed that the profile remained flattened, but the midplane density increased to about 8×10^{16} m^{-3}, and the bulk electron temperature dropped sharply to about 0.7 V.

To understand this, Figure 15.5 shows the PIC ionization rates versus x within the gap. Direct ionization by secondary electrons within and near the sheath edge was found to dominate, contributing 76% of the total ionization rate. Primary electron ionization, again within and near the sheath edge, contributed only 11% of the total. The most important ionization source in the bulk plasma was Ar_m–Ar_m pooling, contributing 13%. Ar_m stepwise ionization and the other pooling reactions contributed negligibly. The suppression of the significant bulk stepwise ionization seen in Figure 15.4 in the absence of secondary emission, compared with that seen in Figure 15.5 with secondary emission, is due to the sharply reduced temperature ($T_{eb} \approx 0.7$ V), since most bulk electrons have energies lower than the energy threshold for stepwise ionization (4.21 V).

These PIC results show that the greatly increased secondary emission due to the excited atoms strongly alters the discharge equilibrium. The very large ionization percentage (76%) due to secondary electrons, the factor of 6.7 increase in bulk plasma density, and the sharp decrease in bulk plasma temperature, from 2.7 to 0.7 V, indicate that the discharge has entered a new mode of operation, the *γ-mode*. We consider this in the next section.

15.3 Alpha-to-Gamma ($\alpha-\gamma$) Transition

15.3.1 Qualitative Description of α and γ Modes

An important feature of rf capacitive discharges is the existence of two distinct discharge modes: the *α-mode* and the *γ-mode*. In the α-mode, the discharge is mainly sustained by volume ionization due to electrons heated within the bulk plasma and/or near the discharge sheaths, as described by the basic particle and energy balances in Chapters 10 and 11. Electron–ion pairs created by multiplication within the sheaths of secondary electrons emitted from the electrode surfaces play a minor role. In contrast, in the γ-mode, ionization due to secondary electron emission sustains the discharge. At low rf source voltages, the discharge is in the α-mode. With increasing source voltage, an $\alpha-\gamma$ transition typically occurs.

The existence of two operating modes at higher pressures was identified by Levitskii (1957), who also posited that the γ-mode was due to a breakdown or near-breakdown of the rf sheath and was similar to the dc breakdown of a gas-filled gap or the formation of a dc cathode sheath (see Section 14.3). Yatsenko (1980, 1981) studied the $\alpha-\gamma$ transition in narrow gap capacitive discharges at intermediate and high pressures for various gases, gap lengths, and electrode surface materials. Lisovskii (1998) and Lisovskiy et al. (2006) investigated the $\alpha-\gamma$ transition in the low-pressure regime. A model of the γ-mode was given in Godyak and Khanneh (1986). Godyak et al. (1992) measured the electron energy distribution functions in argon and helium discharges during the $\alpha-\gamma$ transition, observing the transition to a low-temperature Maxwellian in the γ-mode. An extensive description of some experimental results and physical models for the $\alpha-\gamma$ transition is given in Raizer et al. (1995).

Two possible $\alpha-\gamma$ transitions at intermediate pressures are illustrated schematically in Figure 15.6 for a discharge having electrode area $A_0 = \pi R_0^2$ and maximum sheath width s_m driven by a high impedance rf source (e.g., an rf current source). At intermediate pressures, the $\alpha-\gamma$ transition voltage lies near, but always below, the Paschen breakdown voltage curve for a gas-filled gap of thickness s_m. The Paschen curve is used in the figure in place of the actual near-breakdown transition curve, for illustrative purposes. The Paschen breakdown voltages V_b for argon (solid) and oxygen (dashed) from (14.3.9) are shown as a function of ps_m, with V_b the maximum of the oscillating sheath voltage and s_m the maximum of the rf oscillating sheath width. As described in Section 14.3 and seen in Figure 14.4, the Paschen curve has a minimum voltage V_{min} at some $(ps_m)_{min}$. Using the A and B coefficients from Table 14.1, $(ps_m)_{min} = eA^{-1}\ln(1 + 1/\gamma_{se})$ and $V_{min} = B(ps_m)_{min}$, with e

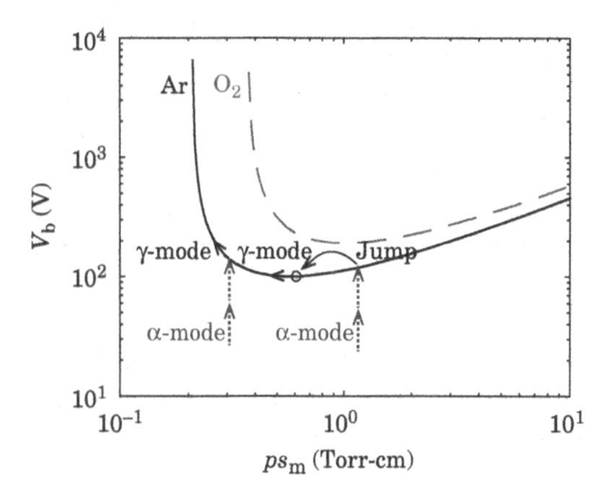

Figure 15.6 Sheath breakdown voltage (Paschen voltage) V_b versus ps_m, with p the gas pressure at 20 °C and s_m the maximum sheath thickness, for argon (solid line) and oxygen (dashed line), showing the $\alpha-\gamma$ transition; at low ps_m, the transition is smooth; at high ps_m, there is a discontinuous jump to the Paschen minimum voltage.

the natural base. At lower pressures, there is also an asymptote $V_b \to \infty$ as $ps_m \to A^{-1} \ln(1 + 1/\gamma_{se})$, below which breakdown cannot occur.

The dotted lines with arrows show V_b versus ps_m in the α-mode as the voltage is increased. For all but the lowest rf voltages, the α-mode discharge fills the entire area A_0 of the electrodes, similar to the "abnormal" regime of a dc discharge. As the voltage is increased, the sheath width changes only slightly, leading to a near-vertical $V_b(ps_m)$ variation. For intermediate pressures ps_m to the *left* of the Paschen minimum, the α–γ transition is found to be smooth, following the Paschen curve with a decreasing ps_m as V_b increases after the α–γ transition. The resulting γ-mode discharge continues to entirely fill the electrode area.

In contrast, at the intermediate pressures to the *right* of the Paschen minimum, the α–γ transition typically has a jump to the minimum voltage point, with the sheath width s_m dropping sharply. Since the discharge current density is set by the sheath capacitance per unit area, $J \propto \omega\epsilon_0 V_b/s_m$, the current density also increases sharply. Because the total current is set by the high impedance rf source to not change much, there must be a radial discharge contraction to an area $A < A_0$. Similar to the "normal" current density regime of a dc discharge, the γ-mode discharge no longer fills the entire electrode area. The α–γ transition in this case is discontinuous, passing from an abnormal glow α-regime to a normal glow γ-regime. For an ideal current source, the contraction is $A/A_0 = J_{min}/J_{\alpha\gamma}$, with $J_{\alpha\gamma}$ the current density at the α–γ transition.

With further increases of the rf source current, the discharge area A expands, while the sheath voltage, current density, and sheath width remain at their normal values V_{min}, J_{min}, and s_{min}, respectively. When A increases to A_0, then the discharge equilibrium enters the abnormal γ-mode. With further current increases, J increases above J_{min}, the sheath voltage increases and the sheath width decreases along the left-hand side of the Paschen curve.

The stability of the exceptional point at the Paschen minimum voltage for the normal γ-mode in rf discharges is also seen in dc glow discharges and has been the subject of much research (Raizer, 1991, sections 8.4.8–8.4.9; Raizer et al., 1995, section 2.3.1). The Paschen curve region to the right of $(ps_m)_{min}$ is unstable because it has a negative differential impedance; $dV_b/dJ < 0$. A small perturbation in this region that increases J and decreases s_m leads to a decreased voltage which further increases J, a "positive" feedback leading to instability. In the region to the left of $(ps_m)_{min}$, it is found that a discharge having current within $A < A_0$, and no current outside of A, is unstable at its current/currentless boundary. Hence, for $A < A_0$, the only stable operating point for the normal γ-mode is at the voltage minimum.

For a source impedance that is small compared to the discharge impedance (e.g., an ideal voltage source), the discharge voltage cannot decrease during the α–γ transition. The abnormal α-mode discharge to the right of the Paschen minimum can jump into the abnormal γ-mode to the left, at the same voltage, with a sudden increase in the current density J, a sudden decrease in the sheath width, and a sudden increase in the total discharge current.

A second phenomenon that can occur as the discharge current is increased is discharge filamentation. At a certain (typically large) current, a discharge instability can develop, the radial uniformity is destroyed, and the discharge contracts to a thin filament, or a number of "dancing" filaments. An important filamentation mechanism is local gas heating (Raizer et al., 1995, section 2.5.6; Raizer, 1991, section 9.4). An increasing local plasma density n_e increases the local $J_e \cdot E$, which increases the gas temperature T_g, which reduces the gas density n_g, which increases the E/n_g, which increases T_e, which further increases n_e, a "positive" feedback that leads to instability. This instability is inhibited at the low pressures (low electric fields E) and small discharge gaps l (large gas conduction heat removal).

Dielectric layers can be inserted into the discharge on one or both electrodes. These layers can drop additional voltage, reducing the Paschen minimum current density, and they can increase the propensity for filamentation, by inhibiting the discharge heat removal (Raizer et al., 1995, section 2.9.2).

15.3.2 Gamma Mode Model

A model of the γ-mode equilibrium in a cylindrical capacitive discharge of radius R and gap length l was developed by Godyak and Khanneh (1986). They account for both radial and axial diffusion and use a homogeneous rf sheath description (see Section 11.1). All ionization is assumed to be due to secondary electrons created in the sheaths, which diffuse into the bulk over a characteristic energy relaxation length $\lambda_{\mathcal{E}}$. The ionization due to the bulk plasma electrons is assumed to be zero. The sheath physics in this model bears a significant resemblance to the physics of dc gas-filled gap breakdown and dc cathode sheath formation developed in Section 14.3. Similarly, the γ-mode bulk plasma physics resembles the passive bulk model developed in Section 15.2. However, there are some differences, which are described below. Here, we give a simplified description of the γ-mode equilibrium for intermediate pressures, for a one-dimensional narrow gap discharge, $l \ll R$, neglecting radial losses.

As for dc breakdown described in Section 14.3, an incoming ion flux Γ_w at the electrode wall produces an outgoing electron flux $\Gamma_{we} = \gamma_{se}\Gamma_w$ of secondary electrons. The electron flux grows exponentially within the high-field region of the sheath, reaching a value of $\Gamma_{se} = \mathcal{M}\Gamma_{we}$ at the plasma–sheath edge, with $\mathcal{M}$ the multiplication factor. For dc breakdown, the multiplication factor was found to be $\mathcal{M} = 1 + 1/\gamma_{se}$, as given in (14.3.4). For large values of $\mathcal{M}$, most secondary electron ionization would occur at the time phase of the maximum sheath voltage, $V_s = 2V_1$, with V_1 the fundamental component rf sheath voltage amplitude. For modest values of $\mathcal{M}$, the average sheath voltage V_1 is more appropriate, as the secondary ionization is spread over a significant sheath extent and rf time-phase interval (Godyak and Khanneh, 1986; Kawamura et al., 2021). Typically $V_1 \sim$ 50–200 V, so each secondary electron creates $\sim V_1/\mathcal{E}_c$ ionizations, with $\mathcal{E}_c \sim 1$–$2\,\mathcal{E}_{iz} \sim 20$–$40$ V the energy lost per ionization due to all collision processes. Therefore, only $\mathcal{E}_c/(V_1 + \mathcal{E}_c)$ secondary electrons are needed to replenish each ion lost to the sheath from the bulk plasma. This reduces the required multiplication factor for sheath breakdown to

$$\mathcal{M}_\gamma \approx \frac{\mathcal{E}_c}{V_1 + \mathcal{E}_c}\left(1 + \frac{1}{\gamma_{se}}\right) \tag{15.3.1}$$

The secondary electron flux Γ_{se} entering the bulk plasma at each sheath is determined from the equal production of electron–ion pairs in the sheath, as was done in (14.3.3),

$$\Gamma_w - \Gamma_s = \Gamma_{se} - \Gamma_{we} \tag{15.3.2}$$

Substituting $\Gamma_w = \Gamma_{se}/(\mathcal{M}_\gamma\gamma_{se})$ and $\Gamma_{we} = \Gamma_{se}/\gamma_{se}$ into this balance equation, we obtain

$$\Gamma_{se} = \Gamma_s \frac{\gamma_{se}\mathcal{M}_\gamma}{1 + \gamma_{se} - \gamma_{se}\mathcal{M}_\gamma} \tag{15.3.3}$$

or $\Gamma_{se} \approx \gamma_{se}\mathcal{M}_\gamma\Gamma_s$ for $\gamma_{se}\mathcal{M}_\gamma \ll 1$, with $\mathcal{M}_\gamma$ given in (15.3.1).

For the passive bulk model of Section 15.2, an energy relaxation length for approximately 20 V ionizing electrons diffusing into the bulk plasma was determined in (15.2.7). However, for the γ-mode, each secondary electron creates multiple ionizations. Therefore, the disappearance of the secondary electrons on the right-hand side of (15.2.2) is not at the frequency ν_{inel}, but at the reduced frequency $\nu_{\text{inel}}\mathcal{E}_c/(V_1 + \mathcal{E}_c)$. Carrying through the energy relaxation length analysis, we obtain

$$\lambda_\mathcal{E} \approx \left(\frac{V_1 + \mathcal{E}_c}{\mathcal{E}_c} \frac{\lambda_m \lambda_{\text{inel}}}{3} \right)^{1/2} \tag{15.3.4}$$

for secondary electrons. In the following, we assume that $\lambda_\mathcal{E} \ll d$, the bulk plasma width.

As for the passive bulk model, the ionization rate G_{iz} in the bulk plasma due to the secondary electrons is taken to decay exponentially from both sheath edges into the discharge midplane, yielding the density $n(x)$ from (15.2.13), the ion flux Γ_s entering the sheath edge from (15.2.14), and the edge-to-center density ratio h_l from (15.2.15), with $\beta = \lambda_\mathcal{E}^{-1}$ here given in (15.3.4).

For the homogeneous sheath model, the ion density everywhere in the sheath is assumed to be n_s. The rf current density J is related to V_s by the capacitive sheath

$$J = \frac{\omega\epsilon_0}{s_m} V_s \tag{15.3.5}$$

with s_m the maximum sheath width. In the γ-mode, as a simplification, we will use the Paschen curve (14.3.9) to relate s_m to V_s

$$V_s = \frac{Bps_m}{\ln Aps_m - \ln\left[\ln\left(1 + 1/\gamma_{\text{se}}\right)\right]} \tag{15.3.6}$$

A modified α–γ transition curve is described at the end of this section.

Inverting (15.3.6) at given pressure p yields $s_m(V_s)$. However, we should note (see Figure 14.4 or Figure 15.6) that the Paschen curve is double-valued, having low and high values of ps_m at any voltage V_s above the Paschen minimum $V_{\min}$. The current–voltage relation in the γ-mode is

$$J = \frac{\omega\epsilon_0}{s_m(V_s)} V_s \tag{15.3.7}$$

When determining the (large) sheath electric field (11.1.7) for the homogeneous sheath model in Section 11.1, the assumption $E \approx 0$ at the plasma–sheath edge was made. There is actually a small oscillating rf electric field $\tilde{E}_s$, which drives the electron oscillation $\tilde{s}$. Solving the electron force equation in the sinusoidal steady state,

$$-m\omega^2\tilde{s} = -e\tilde{E}_s - j\omega m\nu_m\tilde{s} \tag{15.3.8}$$

we obtain

$$\tilde{s} = \frac{e}{m} \frac{1}{\omega(j\omega + \nu_m)} \tilde{E}_s \tag{15.3.9}$$

Considering intermediate pressures with $\nu_m^2 \gg \omega^2$ and taking the magnitudes of the left- and right-hand sides of (15.3.9), we obtain the relation between the magnitude of the field $E_s = |\tilde{E}_s|$ at the plasma–sheath edge and $s_m = 2\,|\tilde{s}|$

$$E_s = \frac{m\omega\nu_m s_m(V_s)}{2e} \tag{15.3.10}$$

By current continuity, again considering $\nu_m \gg \omega$, the capacitive displacement current (15.3.7) in the sheath must equal the ohmic current in the bulk plasma

$$J = \frac{\omega\epsilon_0}{s_m(V_s)} V_s = \frac{e^2 n_s}{m\nu_m} E_s \tag{15.3.11}$$

Substituting (15.3.10) for E_s into the last equality in (15.3.11), and solving for the edge density n_s, we obtain

$$n_s = \frac{2\epsilon_0 V_s}{e s_m^2(V_s)} \tag{15.3.12}$$

The center density, as a function of n_s and the ion loss speed u_s, is found from (15.2.15)

$$n_0 = n_s\left(1 + \frac{u_s}{\beta D_a}\right) \tag{15.3.13}$$

To determine u_s, we substitute n_s into (6.2.22). Then substituting u_s into the above equation, we obtain the center density n_0. The ion flux Γ_s entering the sheath is found from (15.2.14).

The rf source power flux supplied to the discharge is the sum of secondary, ohmic, and ion powers

$$S_{rf} = S_{se} + S_{ohm} + S_i \tag{15.3.14}$$

The secondary electron power flux is

$$S_{se} \approx 2e\Gamma_{se}V_1 \tag{15.3.15}$$

with Γ_{se} given by (15.3.3), and V_1 the fundamental voltage component across a single sheath. The ion power flux for the homogeneous sheath is, similarly

$$S_i \approx \frac{3}{4}e\left(\Gamma_s V_s + \Gamma_{se}V_1\right) \tag{15.3.16}$$

The second term assumes that on average, the sheath-generated ions fall across a dc potential of $\frac{3}{8}V_1$. The bulk ohmic power flux is found by integration over the inverse plasma density profile as

$$S_{ohm} = \frac{1}{2}J^2\frac{m\nu_m}{e^2}\int_{-d/2}^{d/2}\frac{dx}{n(x)} \tag{15.3.17}$$

with J given by the first equality in (15.3.11) and $n(x)$ given in (15.2.13). Neglecting the bulk edges, an approximate expression for the bulk power balance is

$$S_{ohm} \approx \frac{1}{2}J^2\frac{m\nu_m}{e^2 n_0}\cdot d = \nu_{iz}n_0 e\mathcal{E}_c(T_e)\cdot d \tag{15.3.18}$$

with $\mathcal{E}_c$ for the low temperature bulk electrons given in (3.5.11).

Similarly, the ohmic component of the voltage across the plasma is given by

$$V_p = J\frac{m\nu_m}{e^2}\int_{-d/2}^{d/2}\frac{dx}{n(x)} \tag{15.3.19}$$

or approximately

$$V_p = J\frac{m\nu_m}{e^2 n_0}\cdot d \tag{15.3.20}$$

Assuming a negligible reactive plasma voltage, $\nu_m \gg \omega$, the total discharge voltage amplitude is $V_{rf} = (V_s^2 + V_p^2)^{1/2}$. Typically, $V_p \ll V_s$ at intermediate pressures.

Example 15.2 We take the following parameters:

- $p = 1.6$ Torr argon at 300 K ($n_g = 5.28 \times 10^{22}$ m^{-3})
- Paschen parameters $A = 11.5$ cm^{-1}-Torr^{-1}. $B = 176$ V/cm-Torr
- $l \approx d = 0.025$ m
- $f = 13.56$ MHz ($\omega = 8.52 \times 10^7$ s^{-1})
- $V_s = 100$ V
- $\gamma_{se} = 0.15$

As for Example 15.1 in Section 15.2, the ion mobility is $\mu_i \approx 0.07$ m^2/V-s and $\nu_{mi} = 3.4 \times 10^7$ s^{-1}. From (15.3.4) at this gas density, $\lambda_{\mathcal{E}} = 3.7 \times 10^{-4}$ m. From (3.5.10), the ion–neutral mean free path $\lambda_i = 1.9 \times 10^{-5}$ m. From the Paschen curve (15.3.6), we find $s_{min} = 3.0$ mm and $V_{min} = 85$ V at 1.6 Torr. The sheath width for the α-mode of Example 15.1 at $V_s = 100$ V was less than s_{min}, so we use the low ps_m value $s_m = 1.9$ mm of the Paschen sheath width for the γ-mode. Then from (15.3.7), we obtain $J = 40$ A/m^2. Starting with an estimate for the low bulk plasma electron temperature $T_e = 1.5$ V in the γ-mode, we evaluate the electron momentum transfer frequency $\nu_m = 1.6 \times 10^9$ s^{-1}, the Bohm speed $u_B = 1890$ m/s, and the ambipolar diffusion coefficient $D_a = 0.105$ m^2/s. From (15.3.10), the field at the plasma–sheath edge is $E_s = 730$ V/m. The edge density is $n_s = 3.1 \times 10^{15}$ m^{-3} from (15.3.12). Substituting this into the right-hand side of (6.2.22), we obtain $u_s = 580$ m/s. Then from (15.3.13), we find $n_0 = 9.6 \times 10^{15}$ m^{-3}. For 100 volt secondary electrons, we estimate $\mathcal{E}_c = 18$ V from the cross sections for ionization and excitation in Figure 3.13. Then from (15.3.15), the secondary electron power flux is $S_{se} = 10.6$ W/m^2. The ion power flux is 26 W/m^2 from (15.3.16), and the ohmic power flux is 123 W/m^2 from (15.3.17). From (15.3.20), the ohmic plasma voltage component is 5.9 V, which is much smaller than the sheath voltage, so $V_{rf} \approx V_s$. Note that the required multiplication factor from (15.3.1) is not large, $\mathcal{M}_\gamma \approx 2.0$, with $\gamma_{se}\mathcal{M}_\gamma \approx 0.3$. Equation (15.3.18) can be used to determine an iterated value of the assumed bulk temperature T_e, if required. The new $T_e = 1.46$ V from this iteration is close to the initial estimate, so a T_e iteration is not useful.

Intermediate pressure deposition discharges often use gas temperatures T_g exceeding 293.15 °K (20 °C). As seen from (14.3.6), breakdown phenomena are actually a function of $n_g s_m$, not ps_m, so an effective pressure $p_{eff} = 293.15p/T_g$ should be used with the A and B coefficients given in Table 14.1. At higher pressures, the α–γ transition can take place within the high ps_m portion of the Paschen curve. In this case, discharge contraction may occur. This regime is explored for a 6 Torr argon discharge in Problems 15.2 and 15.3.

15.3.3 Experimental Results

Figure 15.7 shows a comparison of some experimental results, for a 3.2 MHz helium discharge between two titanium electrodes with the full two-dimensional diffusion theory at 3 Torr, for a 3.2 MHz helium discharge with a gap length $l = 7.8$ cm inside a glass vessel of radius $R = 3$ cm (Godyak and Khanneh, 1986). The solid lines with circles give the measurements; the dashed lines show the theory. At 3 Torr, the α–γ transition occurs on the low ps_m side of the Paschen curve. The transition into the γ-mode gives rise to an increased slope of the I–V_{rf} characteristic seen in Figure 15.7*a* due to the rapidly decreasing sheath width $s_m \propto dV_{rf}/dI$ in the γ-mode. The central density n_0 shown in Figure 15.7*b* increases markedly with discharge voltage, and the central electron temperature T_e in Figure 15.7*c* rapidly drops. Since the bulk plasma conductivity is proportional to n_0, the α-mode bulk plasma electric field strength $E_0 \approx 30$ V/cm seen in Figure 15.7*d* also drops rapidly after entering the γ-mode. The γ-mode theory is seen to give a good agreement with the experimental results.

Figure 15.8 shows the measured dependences of the α-mode minimum maintenance voltage V_{main} and the α–γ transition voltage $V_{\alpha\gamma}$, versus pressure p (Torr) (after Godyak and Khanneh, 1986). The α-mode lies between the two curves, and the transition to the γ-mode occurs above the top curve. At low pressures, there is a large voltage range for the α-mode, but this narrows as the pressure is increased. The increase of V_{main} with pressure p is ascribed to the increasing resistive bulk voltage component V_p as the pressure is increased.

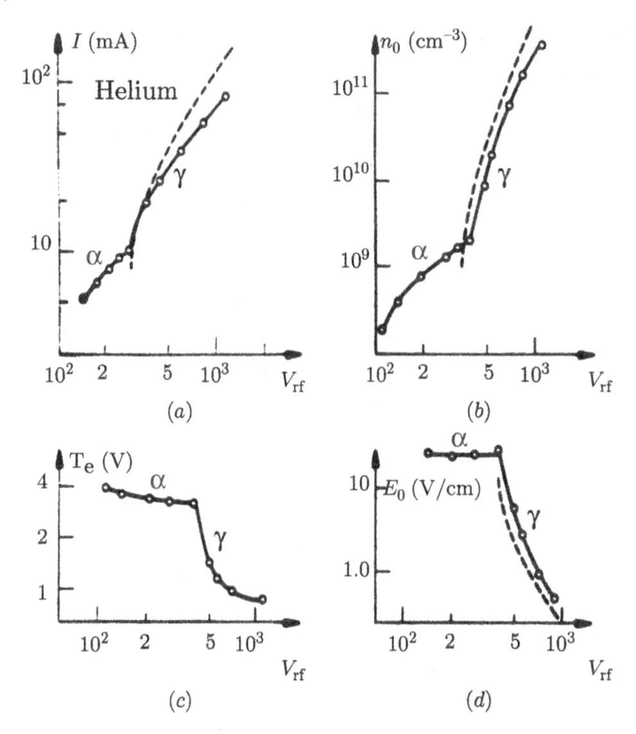

Figure 15.7 Evolution of (*a*) discharge rf current I (mA), (*b*) central plasma density n_0 (cm^{-3}), (*c*) central bulk electron temperature T_e (V), and (*d*) central electric field E_0 (V/cm), versus discharge voltage V_{rf} (V) during the $\alpha-\gamma$ transition, for a 3 Torr helium discharge of radius $R = 3$ cm and gap length $l = 7.8$ cm driven at 3.2 MHz; solid lines with circles give the measurements; dashed lines show the γ-mode theory. Source: Godyak and Khanneh (1986)/with permission of IEEE.

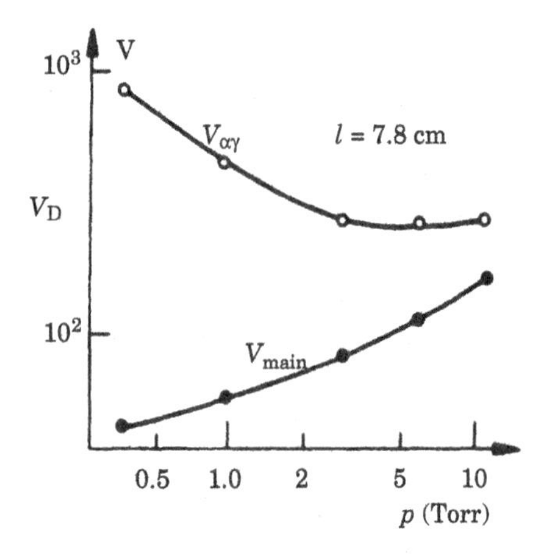

Figure 15.8 Measured dependences of the α-mode minimum maintenance voltage V_{main} (bottom curve) and the $\alpha-\gamma$ transition voltage $V_{\alpha\gamma}$ (top curve), versus pressure p (Torr), for a helium discharge of radius $R = 3$ cm and gap length $l = 7.8$ cm driven at 3.2 MHz. Source: Godyak and Khanneh (1986)/with permission of IEEE.

Since the total discharge voltage V_D is the sum of the capacitive sheath voltage V_{rf} and V_p, with a 90° phase shift between these, $V_D = (V_{rf}^2 + V_p^2)^{1/2}$. As seen in Figure 15.7*d* for the bulk electric field, V_p drops precipitously in the γ-mode, allowing the full discharge voltage V_D to appear across the sheaths. If V_D exceeds the Paschen voltage, the discharge can no longer operate in the α-mode. Experiments indicate that if pl exceeds a critical value of $(pl)_{crit}$, the α-mode discharge either makes a transition to the γ-mode, or it dies out (Raizer et al., 1995; Godyak and Khanneh, 1986). In helium, $(pl)_{crit} \approx 150$ Torr-cm, and in air, $(pl)_{crit} \approx 40$ Torr-cm (Raizer et al., 1995, Section 2.2).

With an improved Langmuir probe system, Godyak et al. (1992) measured the electron energy distribution function (EEDF) during the α–γ transition in argon and helium discharges as the rf current density was increased. The discharge was driven at 13.56 MHz in a symmetric system with an electrode diameter 14.3 cm and a discharge gap of 6.7 cm. At 0.3 Torr, they found a sharp decrease in electron temperature at ≈ 0.4 mA/cm^2 in argon and ≈ 2 mA/cm^2 in helium. The plasma density increased strongly through the transition. In argon, the plasma EEDF transitioned from Druyvesteyn-like to Maxwellian, in helium from Maxwellian to two-temperature Maxwellian. However, the high-energy part of the EEDF due to the γ-electrons could not be measured, due to their low density.

Figure 15.9 shows measured root-mean-square (rms) V–I characteristics for a 13.56 MHz discharge with 10 cm diameter electrodes, showing the α–γ transition for various gases, pressures, gap lengths, and electrode materials. The helium curve 1 at $pl = 27$ Torr-cm shows a smooth α–γ transition, consistent with Figure 15.7 at about the same pl product. The air curve 2 at the same pl value shows a discontinuous transition from the α-mode at low currents to the normal current density γ-mode at V_{min}. At the larger value of $pl = 90$ Torr-cm for the air curve 3, the α-mode cannot be excited, and the discharge operates only in the γ-mode. The CO_2 curves 4 and 5 are similar to the air curve 2. Finally, curves 6 and 7 for air show the strong effects of electrodes coated with dielectric materials.

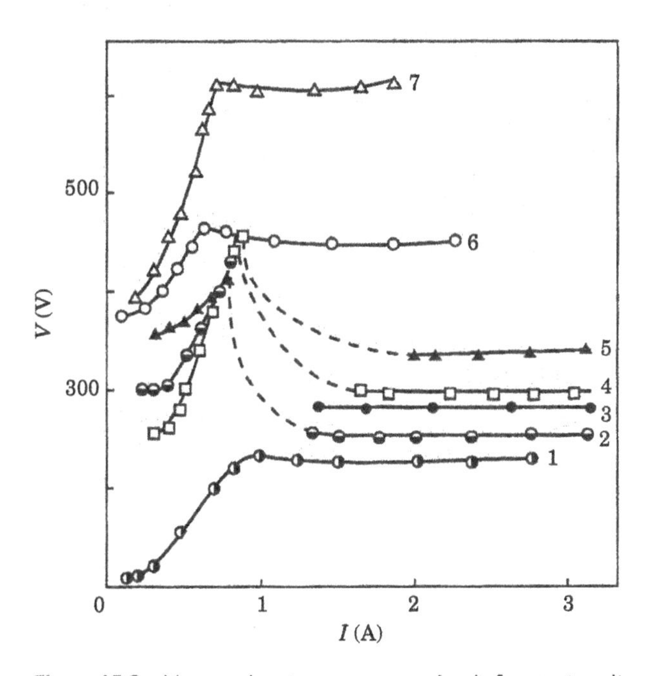

Figure 15.9 Measured root-mean-square (rms) rf current–voltage (I–V) characteristics for discharges driven at 13.6 MHz, with 10 cm electrode diameter; (1) helium, 30 Torr, 0.9 cm gap; (2) air, 30 Torr, 0.9 cm gap; (3) air, 30 Torr, 3 cm gap; (4) CO_2, 30 Torr, 0.9 cm gap; (5) CO_2, 15 Torr, 3 cm gap; (6) air, 7.5 Torr, 1 cm gap, glass-coated electrodes; (7) air, 7.5 Torr, 1 cm gap, teflon-coated electrodes. Source: Raizer et al. (1995)/with permission of Taylor & Francis.

15.3.4 Particle-in-Cell Simulations

Kawamura et al. (2021) performed one-dimensional (1D) particle-in-cell (PIC) simulations of the α–γ transition in a 1.6 Torr, 635 K, 2.5 cm gap capacitive nitrogen discharge driven at 13.56 MHz, with current density amplitudes $J = 7$–$150\ \text{A/m}^2$. They found that for low-current densities in the α-mode, ion-induced secondary electron emission had small effects on the discharge equilibrium, but became increasingly important near the α–γ transition. In the γ-mode, secondary emission was found to be essential for sustaining the discharge. The α-mode discharge was described using the passive bulk model of Section 15.2, in which the ionization is negligible in the central bulk region and is due solely to electron sheath heating, with no secondary electron emission. The discharge was found to undergo an α–γ transition in the applied J range, characterized by an increase in density and a decrease in electron temperature and sheath width. Figure 15.10 shows the PIC results (stars) and the theoretical passive bulk α, and Paschen γ, models for (a) the single-sheath rf voltage amplitude V_{sh} versus the maximum sheath width s_{m}, and (b) the corresponding curve V_{sh} versus the discharge current density J. The correspondence is because s_{m} and J vary inversely for a capacitive sheath at a given voltage

$$J = 1.52\,\omega_0\varepsilon_0 V_{\text{sh}}/s_{\text{m}} \tag{15.3.21}$$

with the coefficient 1.52 for a high-voltage, constant ion–neutral mean free path sheath, as used in the simulations. One should note that due to the high gas temperature, the effective pressure p_{eff} for the Paschen model in this PIC simulation is $p_{\text{eff}} = 273.15\,p/635 = 0.7$ Torr.

The PIC results clearly show a relatively constant bulk electron temperature of about 0.9 V for $J \leq 50\ \text{A/m}^2$ in the α-mode, falling to as low as 0.3 V in the γ-mode at the highest current density 150 A/m^2. Similarly, the electron temperature within and near the plasma–sheath edge is constant at 2 V in the α-mode and falls to 0.4 V in the γ-mode at 150 A/m^2.

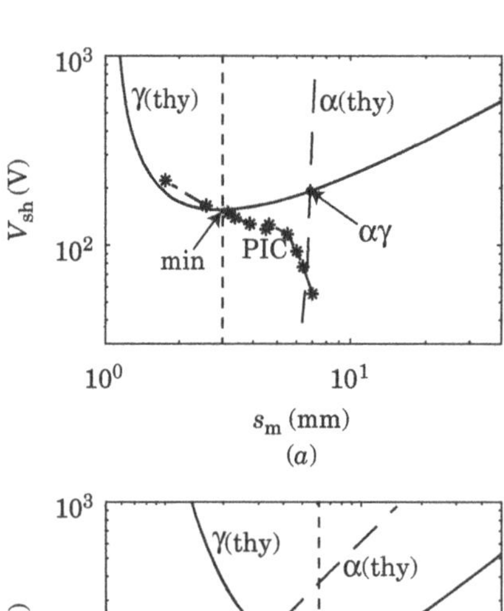

Figure 15.10 α–γ transition for a 2.5-cm gap, 635 K, 1.6 Torr nitrogen discharge driven at 13.56 MHz from the passive bulk α-mode model without secondary emission (dashed line) and the Paschen curve γ-mode model (solid line), along with the PIC simulations with $\gamma_{\text{se}} = 0.15$ (stars), showing (a) the single-sheath rf voltage amplitude V_{sh} versus the maximum sheath width s_{m}, and (b) the corresponding curve V_{sh} versus the discharge current density J.

It is important to realize that the 1D PIC simulations give a description of the actual discharge equilibrium for a "unit area" of the discharge. They do not describe transverse phenomena such as radial discharge contraction or radial filamentation. At first sight from the theoretical α–γ transition point in Figure 15.10*a*, it might appear that a 1.6 Torr N_2 discharge should contract radially during the α–γ transition, since the theoretical $V_{sh}(s_m)$ α-mode curve (dashed line) does intersect the theoretical $V_{sh}(s_m)$ γ-mode curve (solid line) to the right of the Paschen minimum. However, if this were true, the PIC simulation sheath voltages V_{sh} (stars) would show a voltage maximum at the value of J corresponding to the α–γ transition, which is not seen in the figure. Instead, the PIC V_{sh} increases monotonically with decreasing s_m, indicating that there can be no discharge contraction. As can be seen in Figure 15.10*a* and *b*, there is a significant discrepancy between the current-density-driven PIC V_{sh} and the theoretical α-mode V_{sh}. At the higher current densities, this is due to neglect of secondary emission in the theoretical α-mode model, which should give an ionization contribution, resulting in a reduced sheath voltage. At the lower current densities, there may be deviations from the high-voltage sheath model (15.3.21) used in the theory. In fact, theory suggests that the discharge does not contract, operating in the low-pressure abnormal glow regime in both the α-mode and γ-mode. For an N_2 rf discharge, Raizer et al. (1995, Section 2.6), give the condition for this low-pressure operation as $p \lesssim 2.5$ Torr, consistent with the PIC simulation results.

15.3.5 Low Pressures and α–γ Transition Curve

One might think that below the low-pressure Paschen asymptote, $ps_m = A^{-1}\ln(1 + 1/\gamma_{se})$, there could be no γ-mode, as there can be no sheath breakdown. At low discharge voltages, this is indeed the case, and the discharge is in the α-mode, with ionization mainly due to bulk electrons. However, at higher voltages, the ionization due to secondary electrons can still dominate the discharge equilibrium, even though the electron multiplication in the sheath may be low and the sheath is relatively far from the breakdown condition. This leads to an α–γ transition, as has been seen experimentally in argon (Lisovskii, 1998) and N_2O discharges (Lisovskiy et al., 2006). To understand this, let us note from (15.3.1) that the γ-mode multiplication factor $\mathcal{M}_\gamma$ is reduced from the gas-filled gap multiplication $\mathcal{M} = 1 + 1/\gamma_{se}$ by the factor $\mathcal{E}_c/(V_1 + \mathcal{E}_c)$. The Paschen curve (15.3.6) is accordingly modified to yield an α–γ transition curve

$$V_{rf} = \frac{Bps_m}{\ln Aps_m - \ln\left\{\ln\left[\frac{\mathcal{E}_c}{V_1+\mathcal{E}_c}\left(1 + \frac{1}{\gamma_{se}}\right)\right]\right\}} \tag{15.3.22}$$

Note that this transition voltage must always lie below $V_{rf,max} = 2\mathcal{E}_c/\gamma_{se}$. Figure 15.11 shows the argon Paschen breakdown curve (dashed) and α–γ transition curve (solid) for $\mathcal{E}_c = 18$ V and $\gamma_{se} = 0.15$. At the very low and very high pressures, well away from the Paschen minimum, the transition voltage is significantly lower than the Paschen breakdown voltage. This signals an α–γ transition with the sheaths relatively far from breakdown.

Figure 15.12 shows the measured extinguishing voltage (1) and α–γ transition voltage (2) as a function of pressure in a 2.2 cm gap argon discharge driven at 13.56 MHz. The transition voltage drops at low pressures, similarly to that seen in Figure 15.11. Note that γ_{se} is not known for this experiment. As described in Section 15.2, the "effective" ion secondary emission coefficient can be quite large in argon due to metastable-produced secondaries.

The Paschen breakdown and α–γ transition curves in Figure 15.11 assume a uniform electric field in the sheath. A better assumption would be a linear field variation, as for the cathode sheath of a dc glow discharge. We leave this as Problem 15.4 for the student.

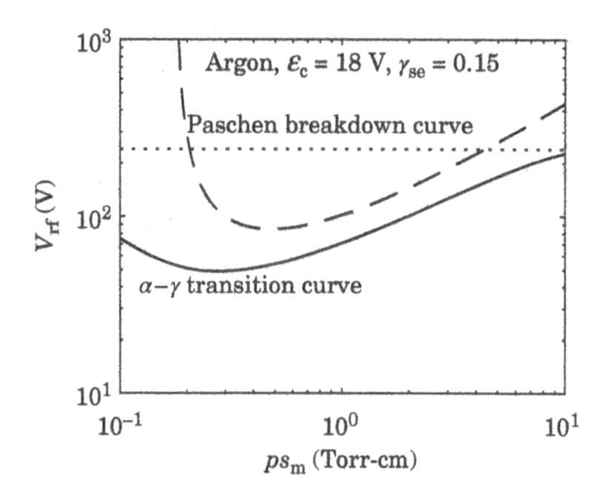

Figure 15.11 Paschen breakdown curve (dashed) from (15.3.6) and $\alpha-\gamma$ transition curve (solid) from (15.3.22), with p the argon gas pressure at 20 °C; the dotted line shows $V_{rf,max} = \mathcal{E}_c/\gamma_{se}$.

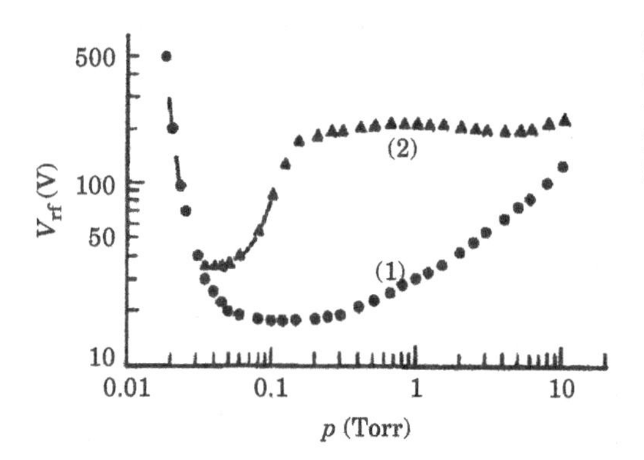

Figure 15.12 Discharge extinguishing voltage (1) and $\alpha-\gamma$ transition voltage (2), versus pressure p, for a 10-cm diameter, 2.2 cm-gap argon discharge. Source: Lisovskii (1998)/with permission of Springer Nature.

15.4 Atmospheric Pressure RF Discharges

15.4.1 The Atmospheric Pressure RF Regime

It is perhaps surprising that diffuse, transversely uniform, narrow gap rf discharges can be excited at atmospheric pressures in the 6.78–40.68 MHz common range of rf frequencies (Park et al., 2000). The usual gas feeds are helium or argon with a small ($< 1\%$) admixture of a molecular gas such as air or its constituents nitrogen, oxygen, and/or water vapor. The gap sizes are small, $l \sim 0.5$–4 mm, but even so, the $pl \sim 40$–300 Torr-cm regime puts these discharges well into the high-pressure side of the Paschen curve. The high thermal conductivity of helium gas promotes stability against discharge constriction and filamentation at the higher currents, particularly during and after the $\alpha-\gamma$ transition (see Section 15.3). In argon, gas heating can more easily lead to discharge contraction, instability and filamentation, particularly in the γ-mode (Moon et al., 2006; Balcon et al., 2008). Placing dielectric barriers on the electrodes can promote stability (Moon et al., 2004; Shi et al., 2007).

While the basic principles of particle and energy balance still apply, there are significant differences from low-pressure global discharge dynamics.

- The electron–neutral momentum transfer collision frequency $\nu_m \gg \omega_{pe} \gg \omega$, where ω_{pe} is the electron plasma frequency and ω is the rf radian frequency. This different ordering is illustrated in Figure 15.13a and b for a low-pressure argon discharge and an atmospheric pressure helium discharge, respectively.

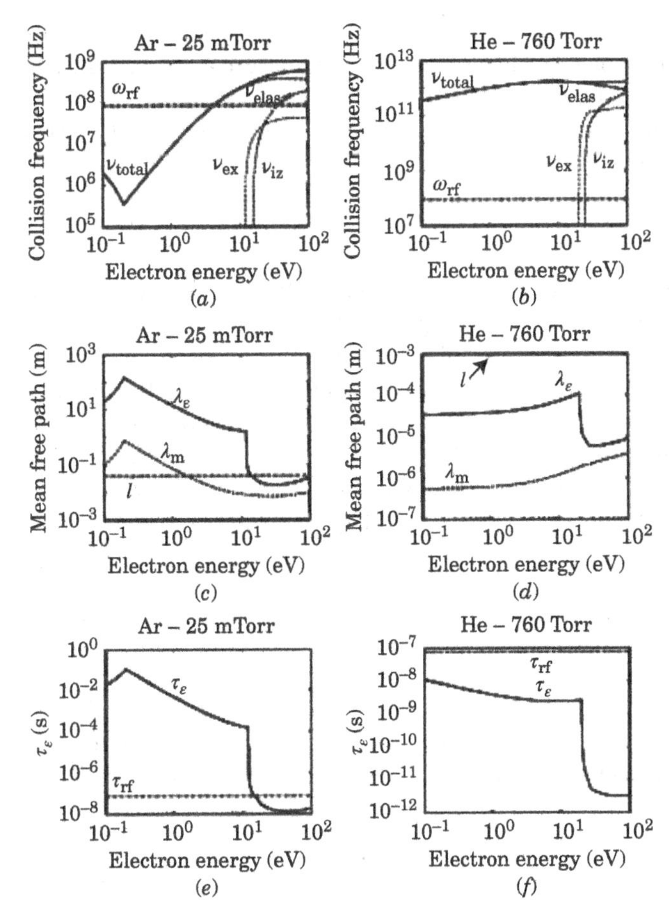

Figure 15.13 Comparison between a low-pressure argon rf discharge (first column) and an atmospheric pressure helium discharge (second column) driven at 13.56 MHz; (*a*, *b*) electron–neutral collision frequencies ν_{elas}, ν_{ex}, and ν_{iz} and driving frequency ω; (*c*, *d*) electron mean free path λ_m, energy relaxation length λ_ε and gap length l; (*e*,*f*) electron energy relaxation time τ_ε and rf period $\tau_{rf} = 2\pi/\omega$. Source: Iza et al. (2008)/with permission of John Wiley & Sons.

- The thermal electron energy relaxation length λ_ε is small compared to the discharge gap l. This different ordering is shown in Figure 15.13*c* and *d* for low-pressure argon with a 2.5 cm gap, and atmospheric pressure helium with a 1 mm gap, respectively.
- The electron energy relaxation time τ_ε is short compared to the rf period τ_{rf}. This implies that the electron temperature oscillates with time at twice the rf timescale. The different orderings are shown in Figure 15.13*e* and *f* for low-pressure argon and atmospheric pressure helium, respectively.
- The electron–neutral elastic collision energy losses are usually significantly larger than the inelastic (excitation, ionization, etc.) energy losses.
- There is a significant resistive electric field in the bulk plasma region, with the resistive voltage drop across the bulk plasma comparable to the sum of the capacitive voltage drops across the two sheaths.
- Penning processes involving noble gas metastables (He^*, Ar_m), such as $He^* + N_2 \rightarrow e + N_2^+ + He$, are often the dominant source of ionization.

- Three-body collisions, molecular ions such as Ar_2^+ and He_2^+, and excimers such as He_2^* can be important in the particle balance dynamics. Noble gas molecular ions and excimers are described in Section 8.2, see also Figure 8.5 for helium.
- High molecular weight ion and neutral clusters can be generated with significant densities. This is particularly true for H_2O-containing discharges.

15.4.2 Homogeneous Model

We use the uniform homogeneous model of Section 11.1 to describe the dynamics of the α-mode in an rf current-driven atmospheric pressure discharge with gap length l (Lazzaroni et al., 2012a,b). A more elaborate model including the α–γ transition is given in Lieberman (2015). The net heavy particle charge density $\rho = e(n_+ - n_-) = en_e$ is assumed to be constant everywhere within the gap, with n_+ and n_- the positive and negative ion densities in the bulk plasma, respectively. An electron cloud of uniform density n_e and fixed width $d < l$ oscillates within the gap due to the rf excitation $J(t) = \mathrm{Re}(\tilde{J}\,\mathrm{e}^{j\omega t})$, leading to the appearance of oscillating rf sheaths near each electrode. The complex amplitude of the total current density flowing in the bulk plasma is

$$\tilde{J} = j\omega\epsilon_0(\kappa_R + j\kappa_I)\tilde{E}_p \tag{15.4.1}$$

where $\tilde{J}$ is the sum of the displacement current density $\tilde{J}_d = j\omega\epsilon_0\tilde{E}_p$ and the reactive and resistive (ohmic) components of the electron conduction current density

$$\tilde{J}_c = j\omega\epsilon_0(\kappa_R - 1)\tilde{E}_p - \omega\epsilon_0\kappa_I\tilde{E}_p \tag{15.4.2}$$

Here, $\tilde{E}_p$ is the complex amplitude of the bulk plasma electric field, and κ_R and κ_I are the real and imaginary parts of the relative dielectric constant given in square brackets in (11.1.1). For rf atmospheric pressure discharges, $\nu_m \gg \omega_{pe} \gg \omega$, and the dielectric coefficients reduce to

$$\kappa_R = 1, \qquad \kappa_I = -\frac{\omega_{pe}^2}{\omega\nu_m} \tag{15.4.3}$$

and (15.4.2) reduces to the simpler resistive form

$$\tilde{J}_c = -\omega\epsilon_0\kappa_I\tilde{E}_p \tag{15.4.4}$$

Eliminating $\tilde{E}_p$ from (15.4.1) and (15.4.4), we obtain

$$\tilde{J}_c = \frac{j\kappa_I}{1 + j\kappa_I}\tilde{J} \tag{15.4.5}$$

The electron sheath width near the left-hand electrode at $x = 0$ is assumed to oscillate from $x = 0$ to $x = 2|\tilde{s}| \equiv s_m$ as $s(t) = |\tilde{s}| + \mathrm{Re}(\tilde{s}\,\mathrm{e}^{j\omega t})$, where the complex oscillation amplitude $\tilde{s}$ is determined by equating the sheath edge motion to the bulk conduction current

$$-j\omega en_e\tilde{s} = \tilde{J}_c \tag{15.4.6}$$

Substituting (15.4.5) into (15.4.6) and solving for $\tilde{s}$ yields

$$\tilde{s} = -\frac{\tilde{J}}{en_e\omega}\frac{\kappa_I}{1 + j\kappa_I} \tag{15.4.7}$$

The complex amplitude $\tilde{V}_D$ of the voltage of the electrode at $x = 0$, with respect to the electrode at $x = l$, is the sum of the capacitive voltage across the two sheaths and the voltage across the bulk plasma. Since the sum of the two oscillating sheath widths is $2|\tilde{s}| = s_m$, this yields

$$\tilde{V}_D = \tilde{V}_s + \tilde{V}_p = \frac{s_m}{j\omega\epsilon_0}\tilde{J} + d\tilde{E}_p \tag{15.4.8}$$

Substituting $\tilde{E}_p$ from (15.4.1) into (15.4.8), we obtain the discharge voltage–current characteristic

$$\tilde{V}_D = \frac{\tilde{J}}{j\omega\epsilon_0}\left(s_m + \frac{d}{1 + j\kappa_I}\right) \tag{15.4.9}$$

In contrast to low pressures where the bulk plasma voltage is small and inductive, at atmospheric pressure, the typical bulk plasma voltage is resistive and can be comparable to the capacitive voltage across the two sheaths.

15.4.2.1 Energy Balance

The time-varying ohmic power density due to the oscillating electron cloud is

$$p_{ohm}(t) = \mathrm{Re}\,(\tilde{J}_c\, e^{j\omega t}) \cdot \mathrm{Re}\,(\tilde{E}_p\, e^{j\omega t}) \tag{15.4.10}$$

Substituting $\tilde{J}_c$ from (15.4.4) into (15.4.10), we obtain

$$p_{ohm}(t) = -\omega\epsilon_0\kappa_I|\tilde{E}_p|^2 \cos^2 \omega t \tag{15.4.11}$$

Substituting the magnitude of $\tilde{E}_p$ from (15.4.1) into (15.4.11) yields

$$p_{ohm}(t) = -\frac{J^2}{\omega\epsilon_0}\frac{\kappa_I}{1 + \kappa_I^2}\cos^2 \omega t \tag{15.4.12}$$

where $J = |\tilde{J}|$ is the magnitude of the current density. The time average of $p_{ohm}(t)$ is

$$\overline{p}_{ohm} = -\frac{1}{2}\frac{J^2}{\omega\epsilon_0}\frac{\kappa_I}{1 + \kappa_I^2} \tag{15.4.13}$$

such that

$$p_{ohm}(t) = \overline{p}_{ohm}(1 + \cos 2\omega t) \tag{15.4.14}$$

As shown in Figure 15.13*f* at atmospheric pressure, the electron energy relaxation time $\tau_\mathcal{E}$ is short compared to the rf period τ_{rf}, so the electron temperature T_e varies with time as the bulk rf field oscillates (Makabe and Petrović, 2006, section 6.3). Assuming a uniform T_e, then the time-varying electron energy balance equation is

$$\frac{dT_e}{dt} = \frac{2}{3}\frac{p_{ohm}(t)}{en_e} - \frac{2}{3}\frac{3m}{M_g}\nu_m T_e - \frac{2}{3}\sum_j \nu_j \mathcal{E}_j \tag{15.4.15}$$

where M_g is the neutral gas mass, the ν_j are the (time-varying) inelastic electron–neutral collision frequencies, and the $\mathcal{E}_j$ are inelastic energy losses per collision. The second term on the right-hand side gives the energy losses due to elastic collisions. As seen in Figure 15.13*b* at atmospheric pressures in helium, the elastic collision frequency is approximately independent of T_e, $\nu_m = \overline{\nu}_m$. Also, for the low T_es ($\lesssim 1.5$ V) of typical atmospheric pressure discharges, the elastic energy losses are about an order of magnitude larger than the inelastic losses. Neglecting the inelastic losses, the electron energy relaxation frequency is then

$$\overline{\nu}_\mathcal{E} = \frac{2m}{M_g}\overline{\nu}_m \tag{15.4.16}$$

Substituting p_{ohm} from (15.4.14) and $\overline{\nu}_\mathcal{E}$ from (15.4.16) into (15.4.15), we obtain

$$\frac{dT_e}{dt} = \frac{2}{3}\frac{\overline{p}_{ohm}}{en_e}(1 + \cos 2\omega t) - \overline{\nu}_\mathcal{E} T_e \tag{15.4.17}$$

The solution is

$$T_e(t) = \overline{T}_e + \widetilde{T}_e \cos(2\omega t + \phi_0) \tag{15.4.18}$$

with average and oscillating components

$$\overline{\mathrm{T}}_{\mathrm{e}} = \frac{2}{3}\frac{\overline{p}_{\mathrm{ohm}}}{en_{\mathrm{e}}\overline{\nu}_{\mathcal{E}}}, \qquad \widetilde{\mathrm{T}}_{\mathrm{e}} = \frac{2}{3}\frac{\overline{p}_{\mathrm{ohm}}}{en_{\mathrm{e}}(4\omega^2 + \overline{\nu}_{\mathcal{E}}^2)^{1/2}} \tag{15.4.19}$$

and $\phi_0 = \tan^{-1}(2\omega/\overline{\nu}_{\mathrm{m}})$. The temperature oscillates between a minimum value $\mathrm{T}_{\mathrm{e,min}} = \overline{\mathrm{T}}_{\mathrm{e}} - \widetilde{\mathrm{T}}_{\mathrm{e}}$ and a maximum value $\mathrm{T}_{\mathrm{e,max}} = \overline{\mathrm{T}}_{\mathrm{e}} + \widetilde{\mathrm{T}}_{\mathrm{e}}$ at twice the rf frequency. For typical discharges, $\overline{\nu}_{\mathcal{E}}$ is about an order of magnitude larger than 2ω (see Figure 15.13*f*), so $\mathrm{T}_{\mathrm{e,min}}$ is significantly less than $\overline{\mathrm{T}}_{\mathrm{e}}$. Then, (15.4.18) reduces to

$$\mathrm{T}_{\mathrm{e}}(t) \approx \mathrm{T}_{\mathrm{e,max}} \cos^2(\omega t - \phi_0/2) \tag{15.4.20}$$

where $\mathrm{T}_{\mathrm{e,max}} \approx 2\overline{\mathrm{T}}_{\mathrm{e}}$.

Substituting (15.4.13) into the time average of (15.4.17), we obtain the discharge electron energy balance relation

$$-\frac{1}{2}\frac{J^2}{\omega\epsilon_0}\frac{\kappa_{\mathrm{I}}}{1+\kappa_{\mathrm{I}}^2} = \frac{3}{2}en_{\mathrm{e}}\overline{\nu}_{\mathcal{E}}\overline{\mathrm{T}}_{\mathrm{e}} \tag{15.4.21}$$

The left-hand side gives the ohmic power density absorbed by the bulk plasma, and the right-hand side gives the loss power density due to electron–neutral collisions. Since κ_{I} depends on the density n_{e} through (15.4.3), the energy balance relates n_{e} to T_{e} for a given current density.

Let us note the similarity of this energy balance to that in an inductive discharge (see Section 12.2). At low densities, $\kappa_{\mathrm{I}}^2 \ll 1$, the ohmic power is proportional to n_{e}, while at high densities, $\kappa_{\mathrm{I}}^2 \gg 1$, it varies inversely with n_{e}. This is the same behavior as shown in Figure 12.4 for the inductive power absorption. Therefore, similar to inductive discharges, there is a critical maintenance current density for atmospheric pressure rf discharges. Substituting (15.4.3) into (15.4.21) and let $n_{\mathrm{e}} \to 0$, we find

$$J_{\mathrm{crit}} = m\omega\epsilon_0\overline{\nu}_{\mathrm{m}}\left(\frac{6\overline{\mathrm{T}}_{\mathrm{e}}}{eM_{\mathrm{g}}}\right)^{1/2} \tag{15.4.22}$$

For high densities, $\kappa_{\mathrm{I}}^2 \gg 1$, we substitute $\kappa_{\mathrm{I}} = -\omega_{\mathrm{pe}}^2/\omega\overline{\nu}_{\mathrm{m}}$ and $\nu_{\mathcal{E}} = 2m\overline{\nu}_{\mathrm{m}}/M_{\mathrm{g}}$ into (15.4.21) to find

$$\frac{1}{2}\frac{J^2}{e^2 n_{\mathrm{e}}} = 3\frac{e}{M_{\mathrm{g}}}n_{\mathrm{e}}\overline{\mathrm{T}}_{\mathrm{e}} \tag{15.4.23}$$

Solving for n_{e}, we obtain

$$n_{\mathrm{e}} = \left(\frac{M_{\mathrm{g}}}{6e^3\overline{\mathrm{T}}_{\mathrm{e}}}\right)^{1/2} J \tag{15.4.24}$$

showing that at high densities, n_{e} is proportional to the rf current magnitude J and varies inversely with the square root of $\overline{\mathrm{T}}_{\mathrm{e}}$.

From (15.4.7), for high densities, the maximum sheath width is

$$s_{\mathrm{m}} = 2\,|\tilde{s}| = \frac{2e}{\omega n_{\mathrm{e}}}J \tag{15.4.25}$$

Using (15.4.24) for n_{e}, this gives

$$s_{\mathrm{m}} = \frac{2}{\omega}\left(\frac{6e\overline{\mathrm{T}}_{\mathrm{e}}}{M_{\mathrm{g}}}\right)^{1/2} \tag{15.4.26}$$

The sheath width varies inversely with the frequency and is proportional to the square root of $\overline{\mathrm{T}}_{\mathrm{e}}$.

15.4.2.2 Particle Balance

Due to the rf-oscillating electron temperature, the electron-activated processes with high activation energies $\mathcal{E}_a$ are strongly affected. Assuming Maxwellian rate coefficients of the form

$$K = K_0\,e^{-\mathcal{E}_a/T_e} \tag{15.4.27}$$

with the pre-exponential factor K_0 a weak function of T_e, then an rf-time averaged rate coefficient $\overline{K}$ is found by averaging over $T_e(t)$ using (15.4.20)

$$\overline{K} = \frac{1}{2\pi}\int_0^{2\pi} d\phi\, K_0(T_{e,max})\exp\left(-\frac{\mathcal{E}_a}{T_{e,max}\cos^2\phi}\right) \tag{15.4.28}$$

with $\phi = \omega t - \phi_0/2$ and K_0 evaluated at the maximum of $T_e(t)$. The integral in (15.4.28) yields

$$\overline{K} = K_0(T_{e,max})\,\mathrm{erfc}\left(\sqrt{\mathcal{E}_a/T_{e,max}}\right) \tag{15.4.29}$$

where erfc is the complementary error function. Non-Maxwellian effects can somewhat modify this result (see Lazzaroni et al., 2012a, appendix A). For $\mathcal{E}_a \gg T_{e,max}$, (15.4.29) has the simpler asymptotic form

$$\overline{K} = K_0(T_{e,max})\left(\frac{T_{e,max}}{\pi\mathcal{E}_a}\right)^{1/2}\exp\left(-\frac{\mathcal{E}_a}{T_{e,max}}\right) \tag{15.4.30}$$

Equation (15.4.30) indicates that $\overline{K}$ for highly activated processes should be evaluated at the maximum temperature $T_{e,max}$, rather than using (15.4.27) with an average temperature $\overline{T}_e$.

Let us examine the simplest particle balance in a homogeneous α-mode noble gas/trace gas discharge: (1) the dominant electron–ion pair production is Penning ionization of an electropositive trace gas T by noble gas metastables, and the dominant ion losses are to the electrodes. (2) Metastables are produced by direct electron excitation of noble gas atoms and are lost by Penning ionization. The metastable balance is then

$$\overline{K}_{ex} n_e n_g d = K_P n_m n_T l \tag{15.4.31}$$

where $\overline{K}_{ex}$ is the time-averaged rate coefficient given in (15.4.30), n_g is the ground state noble atom density, $d = l - s_m$ is the oscillating electron cloud width, K_P is the Penning rate coefficient, n_m is the metastable density, and $n_T \ll n_{He}$ is the trace gas density.

The ion balance is

$$K_P n_m n_T l = 2\Gamma_w \tag{15.4.32}$$

where Γ_w is the ion flux at the electrode wall. In the homogeneous model,

$$\Gamma_w = n_e u_s \tag{15.4.33}$$

where $u_s = u_B^2/(\nu_{mi}\lambda_D)$ is the ion loss velocity at the sheath edge, given by the high-pressure limit of (6.2.22), with $u_B = (e\overline{T}_e/M_i)^{1/2}$ the ion Bohm speed, ν_{mi} the ion–neutral collision frequency, and $\lambda_D = (\epsilon_0\overline{T}_e/en_e)^{1/2}$ the Debye length. Using u_B and λ_D to evaluate u_s and introducing the ion mobility $\mu_i = e/(M_i\nu_{mi})$, we obtain

$$\Gamma_w = \mu_i n_e^{3/2}\left(\frac{e\overline{T}_e}{\epsilon_0}\right)^{1/2} \tag{15.4.34}$$

Substituting (15.4.34) into the ion balance (15.4.32), and eliminating the Penning rate $K_P n_m n_T l$ using the metastable balance (15.4.31), we obtain

$$\overline{K}_{ex} n_g d = 2\mu_i \left(\frac{e n_e \overline{T}_e}{\epsilon_0} \right)^{1/2} \tag{15.4.35}$$

which gives the relation between n_e and $\overline{T}_e$ from the particle balances.

Substituting (15.4.30) for $\overline{K}_{ex}$ into (15.4.35), with $T_{e,max} \approx 2\overline{T}_e$ and $d = l - s_m$ yields

$$\exp\left(-\frac{\mathcal{E}_a}{2\overline{T}_e} \right) = \mathcal{F} \tag{15.4.36}$$

where

$$\mathcal{F} = \frac{\mu_i}{K_0(2\overline{T}_e) n_g (l - s_m)} \left(\frac{2\pi e \mathcal{E}_a n_e}{\epsilon_0} \right)^{1/2} \tag{15.4.37}$$

with $s_m(\overline{T}_e)$ given by (15.4.26) and $n_e(J, \overline{T}_e)$ given by (15.4.23). Taking the natural logarithm of both sides of (15.4.36) gives

$$\overline{T}_e = -\frac{\mathcal{E}_a}{2 \ln \mathcal{F}} \tag{15.4.38}$$

Let us note that $\ln \mathcal{F}$ is a weak (logarithmic) function of $\overline{T}_e$, so (15.4.38) can be solved iteratively for a specified rf current-density magnitude J by using a reasonable initial guess for $\overline{T}_e$ to evaluate $\mathcal{F}$ in (15.4.37).

The time-averaged total electron discharge power is

$$\overline{P}_e = \overline{p}_e d \cdot A \tag{15.4.39}$$

with A the discharge area. From (11.1.25), the dc voltage across each sheath is $\overline{V} = \frac{3}{8} V_{cap}$, with $V_{cap} = s_m J / \omega\epsilon_0$ from (15.4.8). Then, accounting for both sheaths, the total ion power is

$$\overline{P}_i = 2 \cdot \frac{3}{8} \frac{s_m J}{\omega \epsilon_0} \cdot e\Gamma_w \cdot A \tag{15.4.40}$$

Example 15.3 Let us determine the discharge equilibrium for an atmospheric pressure He/0.1%N_2 discharge. We consider the eight species: He, He*, He$^+$, He_2^*, He_2^+, N_2, N_2^+, and electrons. Selected rate constants for this chemistry are given in Table 15.1. The mobility of N_2^+ in He is $\mu_i = 0.0023$ m^2/V-s (Lazzaroni et al., 2012a). We take the following discharge input parameters:

- $p = 760$ Torr at 298.15 K ($n_g = 2.46 \times 10^{25}$ m^{-3}, $n_T = 2.46 \times 10^{22}$ m^{-3})
- $l = 1$ mm
- $f = 27.12$ MHz ($\omega = 1.704 \times 10^8$ s^{-1})
- $J = 400$ A/m^2

Using an initial guess of $\overline{T}_e = 1$ V and reaction 2 in Table 15.1, we find $\mathcal{F} = 8.8 \times 10^{-7}$ from (15.4.37), giving $\overline{T}_e = 0.93$ V from (15.4.38). Then, from reaction 1 in Table 15.1, $\nu_m = 1.2 \times 10^{12}$ s^{-1}. $J_{crit} = 117$ A/m^2 from (15.4.22), $n_e = 2.2 \times 10^{17}$ from (15.4.24), and $s_m = 0.134$ mm from (15.4.25). The capacitive voltage magnitude is $V_s = 36$ V and the total discharge voltage amplitude is $\tilde{V} = 62 - 54j$ V from (15.4.8), giving a discharge voltage magnitude $V_{rf} = |\tilde{V}| = 82$ V. The ion flux is $\Gamma_w = 3.0 \times 10^{19}$ m^{-2}-s^{-1} from (15.4.34), which gives the metastable helium density $n_m = 4.9 \times 10^{16}$ m^{-3}

Table 15.1 Selected Reaction Rate Constants for a He/N_2 Atmospheric Pressure Plasma

Number	Reaction	Rate Constant	Notes
1	$e + He \to He + e$	$4.937 \times 10^{-14} \times T_e^{0.2579}$	a,c
2	$e + He \to He^* + e$	$4.278 \times 10^{-13} \times T_e^{-2.151} \exp(-25.99/T_e)$	a,c
3	$e + He \to He^+ + 2e$	$3.558 \times 10^{-15} \times T_e^{0.4327} \exp(-25.4/T_e)$	a,c
4	$e + He^* \to He^+ + 2e$	$2.254 \times 10^{-13} \times T_e^{-0.1241} \exp(-5.725/T_e)$	a,c
5	$2He + He^* \to He_2^* + He$	2×10^{-46}	b,c
6	$2He + He^+ \to He_2^+ + He$	1.1×10^{-43}	b,c
7	$He + He_2^* \to 3He$	3.73×10^{-22}	a,c
8	$2He^* \to He_2^+ + e$	1.5×10^{-15}	a,c
9	$2He_2^* \to 2He + He_2^+ + e$	1.5×10^{-15}	a,c
10	$e + He_2^+ \to He + He^*$	$8.9 \times 10^{-15}(T_e/T_g)^{-1.5}$	a,c
11	$He^* + N_2 \to N_2^+ + He + e$	5×10^{-17}	a,c
12	$He_2^* + N_2 \to N_2^+ + 2He + e$	3×10^{-17}	a,c
13	$He_2^+ + N_2 \to N_2^+ + He_2^*$	1.4×10^{-15}	a,c
14	$e + N_2^+ \to 2N$	$4.8 \times 10^{-13}(T_e/T_g)^{-0.5}$	a,c
15	$e + N_2 \to N_2^+ + 2e$	$5.3 \times 10^{-15} \times T_e^{-0.936} \exp(-15.4/T_e)$	a,c
16	$N_2^+ + N_2 + He \to N_4^+ + He$	0.9×10^{-41}	b,d
17	$e + N_4^+ \to N_2 + N_2^*$	$2.6 \times 10^{-12}(T_e/T_g)^{-0.5}$	a,d

Notes. T_e in volts.
a) Rate constant in units of m^3/s.
b) Rate constant in units of m^6/s.
c) Lazzaroni et al. (2012a).
d) Cao and Johnsen (1991).

from (15.4.32). The electron and ion powers are 6700 W/m^2 and 130 W/m^2 from (15.4.39) and (15.4.40), respectively. Let us note that this solution assumes negligible densities for He^+, He_2^*, He_2^+, negligible electron–ion recombination (reaction 14), and negligible direct ionization of N_2 (reaction 15). These issues are examined in Problem 15.6.

15.4.3 Simulations and the $\alpha-\gamma$ Transition

Atmospheric pressure discharges have been simulated using various global (Park et al., 2008a, 2010; Liu et al., 2010; Lazzaroni et al., 2012b), fluid (Yuan and Raja, 2002; Stafford and Kushner, 2004; Shi and Kong, 2005; Wang et al., 2006; Balcon et al., 2008; Hong et al., 2008; Liu et al., 2009; Waskoenig, 2010; Waskoenig et al., 2010), particle-in-cell (Iza et al., 2007; Hong et al., 2008; Kawamura et al., 2014), and hybrid (Shi and Kong, 2006; Lazzaroni et al., 2012a,b; Ding et al., 2014; Kawamura et al., 2014) models. Reviews of simulation techniques with extensive references are given in Kim et al. (2005), Park et al. (2008a), and Lee et al. (2011b).

Global simulations allow the treatment of very complicated discharge chemistries, for example, helium/wet air, with $\mathcal{M} = 50$–100 different species and many hundreds of reactions. The absorbed electron power is specified as the input parameter, the electron temperature is uniform in space and does not vary with time during the rf cycle, and all particle densities are uniform in space. The effect of the oscillating T_e on the rate coefficients can be accounted for using (15.4.30). The surface losses are specified using various analytic models (Kim et al., 2006a), and the rf current, voltage, sheath width, and ion power are typically not determined. The equilibrium species densities and the electron temperature are found by numerically integrating a set of $\mathcal{M} + 1$ ordinary differential equations in time to a final steady state.

One-dimensional fluid models are solved numerically to determine time- and space-varying (x, t) discharge parameters. Typically, partial differential equations for particle and momentum balance are used for each species, as well as electron energy balance and Poisson's equation for the fields. The rate and transport coefficients can be based on Maxwellian distributions or, for electrons, on local Boltzmann equation analyses (see Chapter 19.2). The added complexity makes the numerics slower than global simulations, limiting the parameter space that can be explored.

One-dimensional particle-in-cell models obtain the fields, particle densities, and fluxes self-consistently from first principles, without making any assumptions about particle temperatures and velocity distributions. However, the slow numerics typically limit the simulations to only a few species. Fluid and PIC simulations are compared in Hong et al. (2008).

Various combinations of analytic, global, fluid, and particle-in-cell models are used for hybrid simulations. At their simplest, these assume spatial uniformity and combine the numerical solution of the species particle balances with analytical models of the discharge equilibrium. Simulation times can be fast, allowing exploration of complex chemistries (Tavant and Lieberman, 2016). More elaborate models use particle-in-cell/Monte Carlo methods to treat non-equilibrium electron transport along with fluid equations for the heavy particle dynamics (Sommerer and Kushner, 1992; Bogaerts et al., 1995).

15.4.3.1 α-Mode PIC Simulations

Figure 15.14 shows some α-mode results from a particle-in-cell simulation of a room temperature He/0.1%N_2 atmospheric pressure discharge with gap length $l = 1$ mm and an ion-induced secondary electron emission coefficient $\gamma_{se} = 0.25$ (Kawamura et al., 2014). The discharge is driven at 27.12 MHz by a low rf current density $J = 400$ A/m^2. Figure 15.14*a* and *b* gives the time-averaged charged particle and metastable densities and temperatures versus x, respectively. The two dotted vertical lines denote the positions of the maximal sheath edge locations at $x = s_m$ and $x = l - s_m$. The PIC results for the collision rates (not shown) reveal that the He* impact ionization of N_2 (i.e., Penning ionization) dominates over the electron impact ionization of N_2.

The electron temperature in Figure 15.14*b* is somewhat higher in the sheaths than in the bulk, leading to a large metastable generation within and near the plasma–sheath edge. Hence, the metastable density, and the resulting Penning ionization, occurs mainly within and near the sheath edge.

As seen in Figure 15.14*a*, $s_m \approx 0.3$ mm, compared to $s_m = 0.134$ mm from Example 15.3. To understand this, we note that the ion density falls by roughly 20% from the midplane to the sheath edge in the bulk, but decreases by almost a factor of three from the sheath edge to the electrode, indicating a Child law sheath variation rather than the homogeneous sheath of Example 15.3. Also, the PIC ion flux of $\Gamma_w = 1.9 \times 10^{19}$ m^{-2}-s^{-1} is somewhat lower than the flux of 3.0×10^{19} m^{-2}-s^{-1} in Example 15.3.

The PIC simulations were compared to a two-electron group (warm and hot) analytical/numerical hybrid model. A constant mobility Child law was used,

$$s_m = K_{CL} \frac{\mu_i J^2}{e\Gamma_w \omega^2 \epsilon_0} \tag{15.4.41}$$

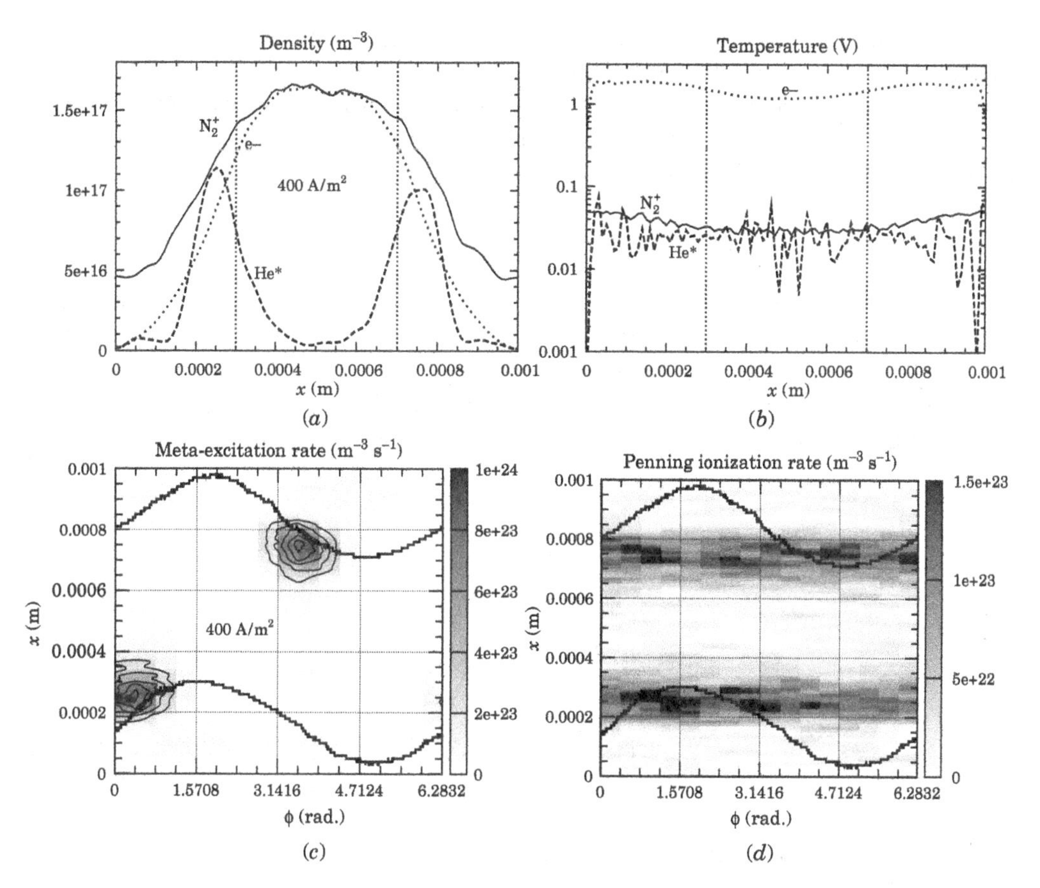

Figure 15.14 Atmospheric pressure He/0.1%N_2, α-mode PIC results for a 1-mm gap driven at 27.12 MHz with an rf current density of 400 A/m^2, showing (*a*) time-average densities versus position x, (*b*) time-averaged temperatures versus x, (*c*) the space–time variation of the helium metastable excitation rate, with space and time variation shown on the vertical and horizontal axes, respectively, and (*d*) the space–time variation of the Penning ionization rate; in (*a*) and (*b*), the maximum sheath width s_m is indicated by the vertical dotted lines; in (*c*) and (*d*), the dark solid curves show the sheath edge positions at the opposing walls. Source: Kawamura et al. (2014)/with permission of IOP Publishing.

The coefficient $K_{CL} = 3/4$ if all the ionization occurs in the bulk near the sheath edge and increases to 3/2 if all the ionization occurs uniformly within the sheath (Kawamura et al., 2014). This latter model gives $s_m \approx 0.25$ mm, better matching the PIC result.

Figure 15.14*c* shows a 2D contour plot of the rate for electron impact excitation of He to He* (metastable excitation). The vertical axis shows the spatial variation across the gap while the horizontal axis shows the time variation over an rf period. We see that the metastables are created when and where the electron density and temperature are both high. In particular, the excitation rate is a strong function of time during the rf period. However, as seen in Figure 15.14*c*, there is only one peak, near the maximum velocity of the expanding sheath edge. Because $T_e(t) \propto p_{ohm}(t) \propto E_p^2(t)$, with $E_p(t)$ the time-varying rf resistive field in the bulk plasma (15.4.20), would predict two peaks per rf cycle, separated by 180° in phase. However, in contrast to this homogeneous model result, there is also a dc ambipolar field near the plasma–sheath edge in the PIC simulations, $E_s \sim T_e/\lambda_{Ds}$ given by (6.2.20). The total field strength driving T_e is $|E_p| + |E_s|$ near the expanding sheath edge, and $|E_p| - |E_s|$ near the contracting sheath edge, giving a much larger T_e, and therefore metastable excitation rate, near the expanding sheath edge.

The metastables created in Figure 15.14*c* provide the ionization source shown in Figure 15.14*d*. While the metastable excitation is localized in time, the Penning ionization is nearly uniform in time because the metastable lifetime is much longer than the rf period.

15.4.3.2 γ-Mode PIC Simulations

Figure 15.15 shows the corresponding γ-mode results for a high rf current density $J = 2000$ A/m^2. The fivefold current density increase leads to a 20-fold increase in spatially averaged ion density. The electron temperature seen in Figure 15.15*b* drops in the bulk region and the warm bulk electrons become much cooler than the hot sheath electrons, indicating a γ-mode regime. The discharge is maintained by the hot sheath electrons, which, as shown in Figure 15.15*c*, mainly appear

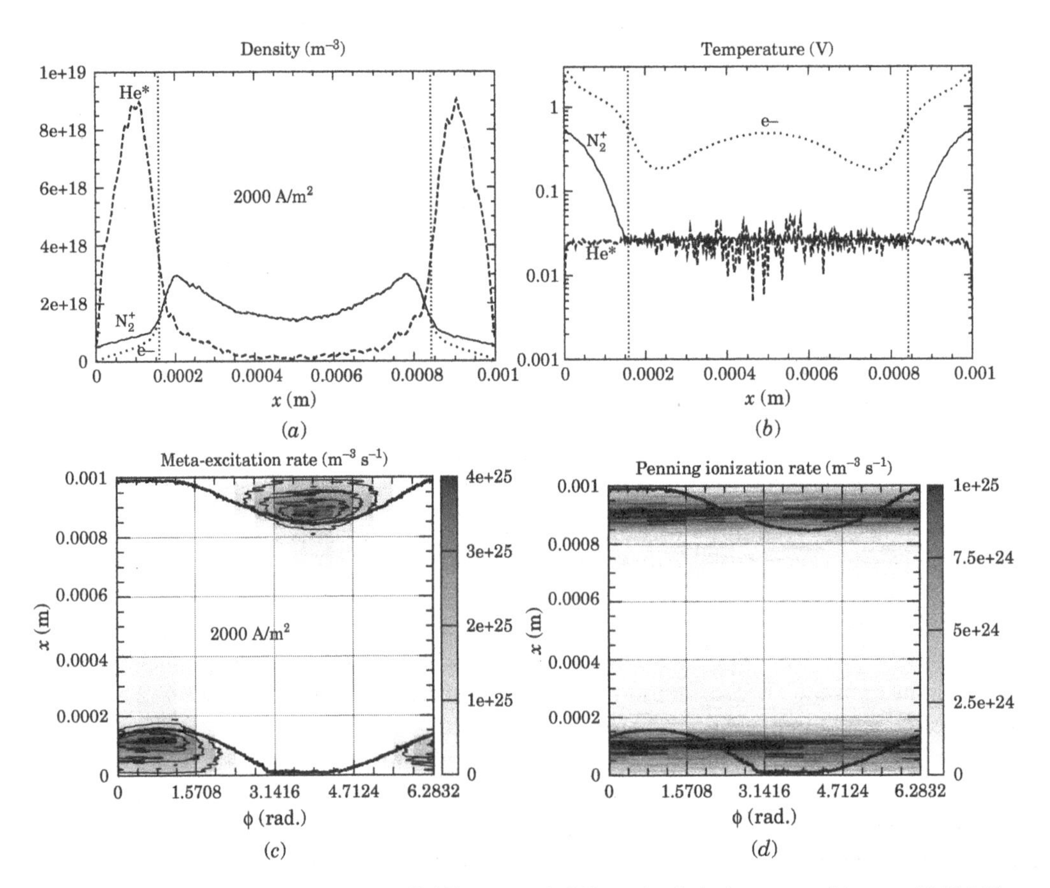

Figure 15.15 Atmospheric pressure He/0.1%N_2 γ-mode PIC results for a 1-mm gap driven at 27.12 MHz with an rf current density of 2000 A/m^2, showing (*a*) time-average densities versus position *x*, (*b*) time-averaged temperatures versus *x*, (*c*) the space–time variation of the helium metastable excitation rate, with space and time variation shown on the vertical and horizontal axes, respectively, and (*d*) the space–time variation of the Penning ionization rate; in (*a*) and (*b*), the maximum sheath width s_m is indicated by the vertical dotted lines; in (*c*) and (*d*), the dark solid curves show the sheath edge positions at the opposing walls. Source: Kawamura et al. (2014)/with permission of IOP Publishing.

at the time-maximum sheath width; the warm bulk electrons become too cold to create metastables. The avalanche multiplication of electrons in the sheaths concentrates the metastable densities in the sheath regions where the electrons are hot enough to excite the He atoms. Also, the higher electron and ion densities in the quasi-neutral core lead to increased electron–ion recombination, accounting for the dip in the densities seen in Figure 15.15*a* toward the center. In Figure 15.15*c*, the higher sheath field and lower bulk electron temperature at higher current density move most of the electron excitation to within the sheath regions.

As for low current densities, the Penning ionization rate seen in Figure 15.15*d* is independent of time. But at 2000 A/m^2, there is a very significant Penning ionization rate in the high-field sheath region near the time-phase of the maximum of the oscillating sheath width. These Penning-created electrons see a high sheath field, resulting in avalanche multiplication with a factor $\mathcal{M}_\mathrm{P}$, along with the multiplication $\mathcal{M}_\gamma$ induced by ion-bombardment secondary electrons. The combination of the two processes induces the collapse of the bulk electron temperature, resulting in the α–γ transition. Kawamura et al. (2014) give the transition condition as

$$\frac{2}{3}\mathcal{M}_\mathrm{P} + \frac{\gamma}{1+\gamma}\mathcal{M}_\gamma = 1 \tag{15.4.42}$$

Since the multiplication factors depend strongly (exponentially) on the peak electric field in the sheath,

$$E_\mathrm{peak} \approx \frac{2J}{\omega\epsilon_0} \tag{15.4.43}$$

and weakly on other discharge parameters, we expect the α–γ transition value of J to be proportional to the rf frequency. This will be seen in the experimental results presented below.

15.4.4 Experimental Results

There have been many atmospheric pressure discharge experiments to investigate various aspects of their operation and possible applications (Park et al., 2000, 2001; Moon et al., 2006; Shi et al., 2007; Liu et al., 2009; Waskoenig et al., 2010). Atmospheric pressure rf discharge configurations, physics, chemistry, and applications have also been extensively reviewed (Papadakis et al., 2011; Bruggeman and Brandenburg, 2013; Bruggeman et al., 2017; Tochikubo and Komuro, 2021).

Figure 15.16 shows the V–I characteristics for a 6-cm diameter helium atmospheric pressure discharge for various gap widths. The rf power was increased along the α-curves and was decreased along the γ-curves. In Figure 15.16*a* and *b*, at 1 and 2 mm gap size, the discharge spreads over the entire electrode area in an abnormal α-mode immediately after breakdown. In Figure 15.16*c* at 3 mm, the discharge starts in a normal α-mode and enters an abnormal α-mode as the power is increased. With further power increase, an α–γ transition jump occurs to a lower voltage γ-mode, with the discharge contracting from 3 to 0.42 cm radius (see Figure 15.6). At 4 mm gap, both the α and γ modes are normal (contracted). Dielectric barriers on the electrodes might be used to reduce or eliminate these discharge constrictions (Shi et al., 2007).

Figure 15.17 shows the α-mode V–I measurements (open symbols) of Liu et al. (2009) for a helium discharge with a 2-mm gap and a 2-cm electrode diameter, driven at 6.78, 13.56, and 27.12 MHz. Also shown (lines with solid symbols) are pure helium simulation results with secondary emission coefficient 0.25, from a two-electron group (warm and hot) analytical/numerical hybrid model (Ding et al., 2014). The last (highest current) measurement gives the α–γ transition, above which the discharge constricts as it enters the γ-mode. This transition current is seen to increase with increasing frequency, roughly as predicted by (15.4.43). The maintenance (lowest)

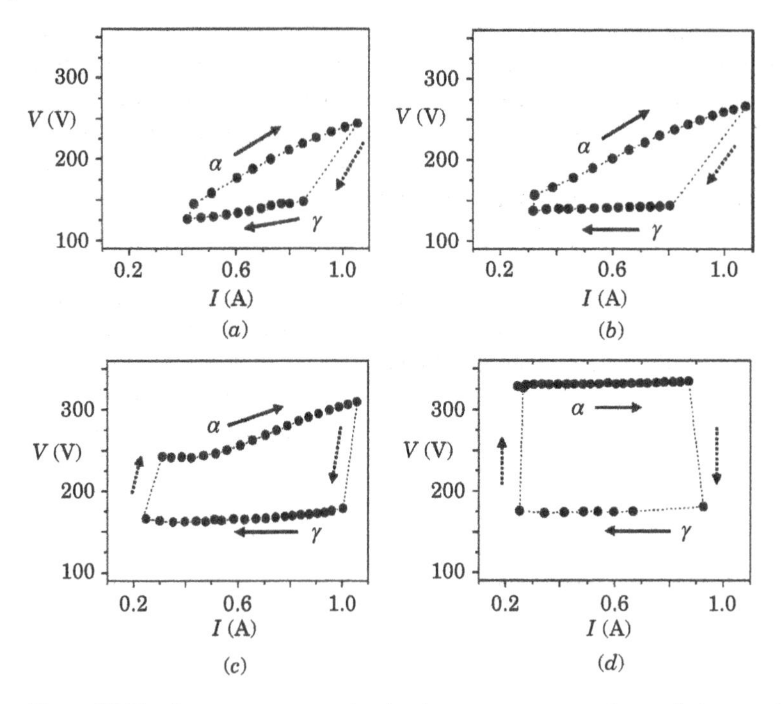

Figure 15.16 Root-mean-square (rms) voltage–current experimental characteristics for an atmospheric pressure, 6 cm diameter helium discharge driven at 13.56 MHz, with gaps of (*a*) 1 mm, (*b*) 2 mm, (*c*) 3 mm, and (*d*) 4 mm; at 4 mm, normal (constant voltage contracted discharge) α and γ regimes appear. Source: Moon et al. (2006)/with permission of AIP Publishing.

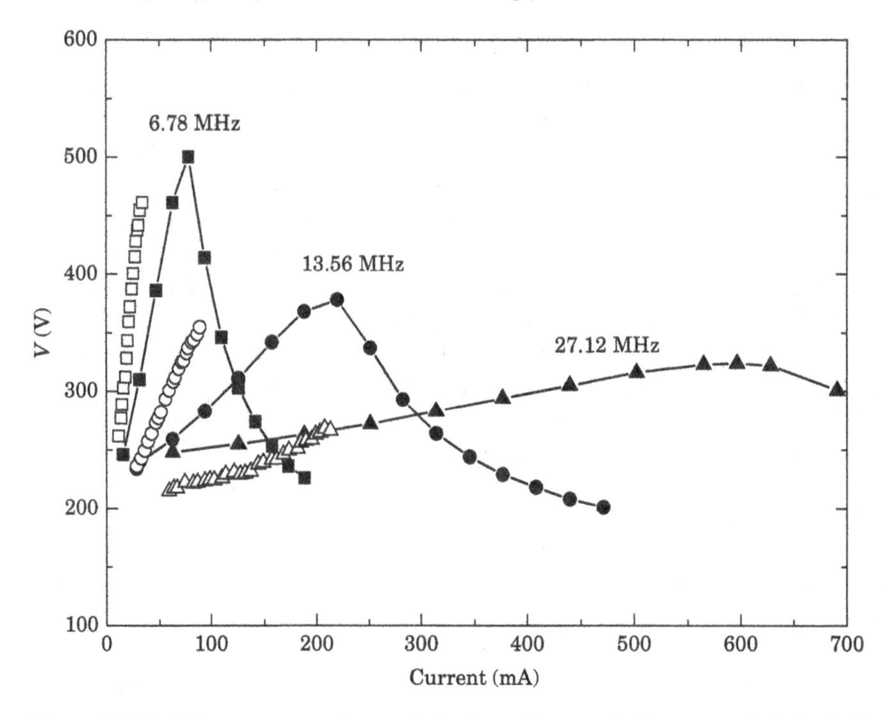

Figure 15.17 Voltage–current characteristics in an rf atmospheric pressure helium discharge with 2 mm gap; the lines with solid symbols give the hybrid simulation results, with the peak corresponding to the α–γ transition; the open symbols give the experimental α-mode results. Source: Ding et al. (2014)/with permission of IOP Publishing.

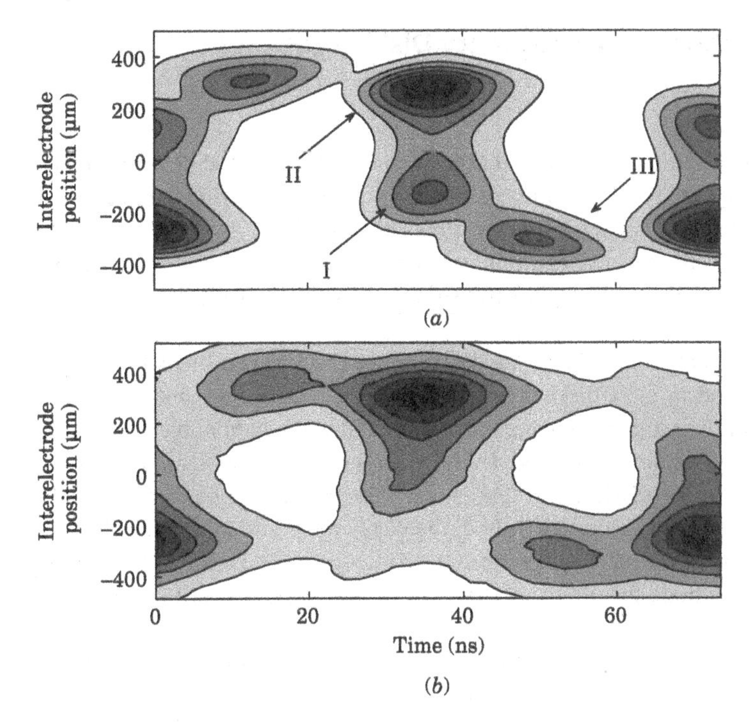

Figure 15.18 O ($^3P \rightarrow {}^3S°$) line emission pattern in a He/0.5%O_2 α-mode atmospheric pressure Penning discharge, within one 13.56 MHz rf cycle and within the 1 mm electrode gap, on a linear gray scale starting from zero; (*a*) fluid simulation and (*b*) phase-resolved optical emission spectroscopy (PROES) measurement. Source: Waskoenig et al. (2010)/with permission of IOP Publishing.

currents also increase strongly with increasing frequency, as indicated by (15.4.22). Similar comparisons between experimental V–I characteristics and hybrid model results were made for argon discharges, where gas heating could be important (Ding et al., 2014).

Figure 15.18 for a He/0.5%O_2 α-mode atmospheric pressure Penning discharge shows the space–time variations of the 844 nm O-atom emission line across the 1 mm gap and over one 13.56 MHz rf cycle, determined by (a) fluid simulations and (b) phase-resolved optical emission spectroscopy (PROES) measurements (Waskoenig et al., 2010). The discharge is in the α-mode but close to the α–γ transition. The emission patterns labeled "I," "II," and "III" refer to the time-phase of sheath expansion, sheath collapse, and maximum sheath width, respectively. Peaks "I" and "II" are associated with the time-phase of the maximum electric field in the bulk plasma; the field is strongest near the plasma–sheath edges, giving rise to the two peaks. Peak "III' arises from electron multiplication due to both secondary emission from the electrodes, and He* Penning ionization of O_2 within the time-phase of the high field region of the oscillating rf sheath. The high fields give rise to high electron energies, which, in regions of significant electron density, strongly excite the 844 nm emission line. As described previously for Figure 15.14c, the asymmetry between "I" and "II" is due to the dc (ambipolar) field, which adds to the bulk rf field during sheath expansion and subtracts from the bulk rf field during sheath collapse.

15.5 Atmospheric Pressure Low-Frequency Discharges

15.5.1 Discharge Regimes

Atmospheric pressure non-thermal discharges excited at frequencies of 50 Hz–100 kHz are widely used for materials processing applications, including ozone production, modifying surface properties of plastic films, and depositing barrier layers on films (Kogoma and Tanaka, 2021). Such discharges always include a dielectric layer or "barrier" on one or both electrodes to stabilize the discharge against transverse contraction followed by transition to a thermionically driven arc discharge (see Figure 14.2).

At these low frequencies, for excitation voltages modestly above the Paschen breakdown voltage, a pulsed, quasi-dc Townsend or glow discharge with reasonable transverse uniformity can sometimes be excited. The glow discharge regime is generally limited to the light noble gases, usually with small concentrations of Penning gases, e.g., He/N_2 with the Penning ionization reaction $He^* + N_2 \rightarrow e + N_2^+ + He$. Many of the discharge properties can then be understood from the analyses in Sections 14.1–14.3. In particular, although pulsed, the discharge is still maintained by ion bombardment-induced secondary electron emission from the cathode, which continually induces electron avalanches within the cathode sheath. In a narrow gap discharge, this leads to a relatively uniform transverse discharge.

At high overvoltages, when the excitation voltage significantly exceeds the Paschen voltage, then the gas breakdown is found to be independent of the cathode material, and much more rapid ($\sim$10 ns) than predicted by secondary emission from the cathode ($\sim$1 μs). The secondary emission breakdown mechanism of Section 14.3 does not apply, and the mechanism of "streamer breakdown" appears. A primary electron avalanche is first formed within the gap by a single random gas molecule ionization, perhaps due to a cosmic ray or stray photon ionization. The primary avalanche grows so large that it generates nearby secondary avalanches by photoionization of excited neutrals, near its head and tail. The charges so generated create a thin ionized conducting filament (streamer) connecting the cathode and anode. In a narrow gap geometry, the discharge region becomes filled with small radius filaments. Such discharges are called "dielectric barrier discharges" (DBDs) and were the first to find industrial application, namely, the production of ozone for purifying water.

15.5.2 Streamer Breakdown

The streamer theory was developed in the early 1940s when fast timescale experimental results indicated that the Townsend secondary emission mechanism could not account for the breakdown (see Meek and Craggs, 1953; Raether, 1964; Loeb, 2022 for original publications and older references). See Nijdam et al. (2020) for an extensive review of the physics and chemistry, including observations and measurements, simulations, and references. A good modern summary is given in Raizer (1991), Chapter 12. We give a simplified treatment, following this work.

The basic mechanism begins with the formation of a single primary electron avalanche, which we will take to be at the cathode. As shown schematically in Figure 15.19*a*, for a large applied electric field $\mathbf{E}_0$, the avalanche grows so rapidly that the positive ions cannot move, leaving behind a "fixed" positive ion space charge. As the avalanche grows, it expands radially due to radial electron diffusion and radial electron repulsion, giving rise to the cone-shaped charge distribution shown in the figure. All the electron charge is concentrated in a "cloud" at the head of the cone, with the positive charge concentrated in the volume behind the head. Note that the positive charge exerts an attractive force on the electrons, which can slow, and ultimately stop, their forward motion toward

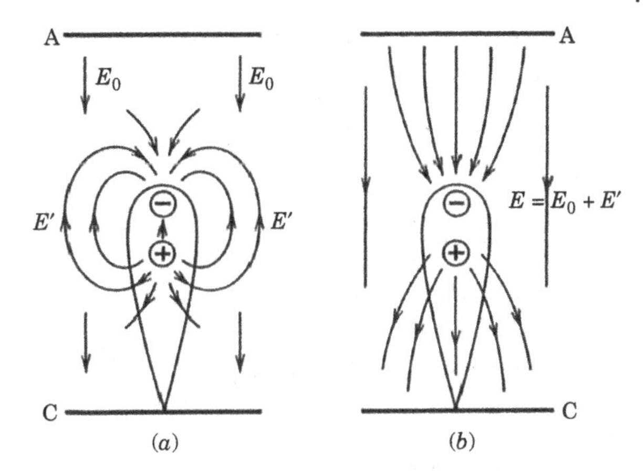

Figure 15.19 Electric fields in a cathode (C) – anode (A) gap containing a cathode-emitted electron avalanche; (*a*) external field $\mathbf{E}_0$ and space charge field of the avalanche $\mathbf{E}'$, shown separately; (*b*) the resulting field $\mathbf{E} = \mathbf{E}_0 + \mathbf{E}'$. Source: Raizer (1991)/with permission of Springer Nature.

the anode. The stopping condition, known as the *Meek criterion,* is that the electric field E'_+ *of the positive charge alone* becomes of order E_0, the applied field.

Let us consider an avalanche in an electropositive gas that almost reaches the anode, $x = l$, before being stopped. Similar to (14.3.1), the number N_e of electrons (and ions) in an avalanche of length x is given by

$$\frac{dN_e}{dx} = \alpha(E)N_e \tag{15.5.1}$$

with the solution

$$N_e = \exp[\alpha(E)x] \tag{15.5.2}$$

where α is the first Townsend coefficient, given approximately by (14.3.7), and E is the magnitude of the total electric field E. Note from (14.3.7) and Figure 14.3 that $\alpha(E)$ is a strongly (exponentially) increasing function of E.

Neglecting the electron electrostatic repulsion (see Raizer, 1991, section 12.2.6, for treatment of this effect), the avalanche head radius is determined by classical diffusion theory. The time-varying radial diffusion equation

$$\frac{\partial n_e}{\partial t} = D_e\left(\frac{\partial^2 n_e}{\partial r^2} + \frac{1}{r}\frac{\partial n_e}{\partial r}\right) \tag{15.5.3}$$

with an initial δ-function source at the avalanche axis $r = 0$ has the solution (Problem 15.8)

$$n_e = \frac{1}{4\pi D_e t}\exp(-r^2/4D_e t) \tag{15.5.4}$$

with $D_e = \mu_e \mathrm{T}_e$ the electron diffusion coefficient. Let $t = l/(\mu_e E_0)$, we obtain a mean radius at the avalanche head

$$R_+ = 2\left(\frac{\mathrm{T}_e l}{E_0}\right)^{1/2} \tag{15.5.5}$$

Making the approximation that all the avalanche positive charge is distributed within a sphere of radius R at the avalanche head, then the electric field of this positive charge at $r = R_+$ is

$$E_+ = \frac{eN_e}{4\pi\epsilon_0 R_+^2} \tag{15.5.6}$$

Substituting (15.5.2) for N_e and (15.5.5) for R_+ into (15.5.6), and setting $E_+ = E_0$, we obtain the Meek condition

$$\exp\left[\alpha(E_0)l\right] = \frac{16\pi\epsilon_0 \mathrm{T}_e l}{e} \tag{15.5.7}$$

Taking the natural logarithm,

$$\alpha(E_0)l = \ln\left(\frac{16\pi\epsilon_0 \mathrm{T}_e l}{e}\right) \tag{15.5.8}$$

which determines E_0 for a given gap l and temperature T_e. The argument of the logarithm is very large, so the dependence of the right-hand side on l and T_e is very weak.

Using nominal values $l = 1$ cm and $\mathrm{T}_e = 1.5$ V, we obtain $\alpha l = 17.6$, i.e., roughly 18 e-foldings of the avalanche, which gives $N_e \approx 4.2 \times 10^7$. For $l = 1$ cm, $\alpha = 176\ \mathrm{m}^{-1}$, giving $\alpha/n_g = 7.1 \times 10^{-23}\ \mathrm{m}^2$ at STP.

The avalanche head radius depends on the gas or gas mixture used. For pure helium, from Figure 14.3, $E/n_g \approx 15$ Td, corresponding to a breakdown field of approximately 3.7 kV/cm. At this field, (15.5.5) gives $R_+ \approx 0.4$ mm.

At the moment when the avalanche is stopped, as shown in Figure 15.19*a*, the electron and ion space charge form an electric dipole-like field structure $\mathbf{E}'$. This intensifies the total field $\mathbf{E} = \mathbf{E}_0 + \mathbf{E}'$ in the regions behind and in front of the avalanche, as shown in Figure 15.19*b*. Photons from the primary avalanche produce secondary avalanches in these intensified regions. The charges from these secondary avalanches create a thin ionized conducting filament ("streamer") connecting the cathode and anode.

More exact calculations give the Meek condition as $\alpha l \approx 18$–20 (Raizer, 1991). Streamer breakdown is a high *pl* phenomenon. This can be seen qualitatively using (14.3.7) for α, with the coefficients given in Table 14.1 extended beyond their true range of applicability into the low E/p regime. Let us recall from (14.3.5) that the Townsend breakdown condition $\alpha l = \ln(1 + 1/\gamma_{se})$ yields the Paschen breakdown curve (14.3.9). Therefore, similarly, the Meek breakdown condition $\alpha l = 19$ yields the streamer breakdown curve

$$V_b(\text{streamer}) = \frac{Bpl}{\ln Apl - \ln 19} \tag{15.5.9}$$

Equation (15.5.9) has an asymptote $V_b \to \infty$ when $pl \to 19/A$, with no solution below this value of *pl*. For helium with $A = 2.8$ Torr-cm, the asymptote is at $pl = 6.8$ Torr-cm. Above this value, the ratio of streamer-to-Paschen breakdown voltages is

$$\frac{V_b(\text{streamer})}{V_b(\text{Paschen})} = \frac{\ln \dfrac{Apl}{\ln(1 + 1/\gamma_{se})}}{\ln \dfrac{Apl}{19}} \tag{15.5.10}$$

which is a monotonically decreasing function of *pl*. With $\gamma_{se} = 0.25$, the streamer-to-Paschen voltage ratio is 7.4 at $pl = 10$ Torr-cm and 1.5 at $pl = 760$ Torr-cm.

As pointed out in the discussion of (14.3.10) in Section 14.3, the Paschen breakdown voltage for attaching gases such as air is high. The Paschen breakdown field is increased due to the electron attachment processes, which compete with the Townsend ionization processes at the lower values of E/p during the avalanche electron multiplication. This leads to a narrow difference between Paschen and streamer breakdown voltages. Figure 15.20 shows the streamer/Paschen breakdown ratio for helium and for air versus *pl* (Becker et al., 2005, figure 5.5.1); the streamer breakdown is for an assumed multiplication of 10^8. Note that for air, there is a very narrow region at atmospheric pressure where Paschen breakdown (and not streamer breakdown) might occur.

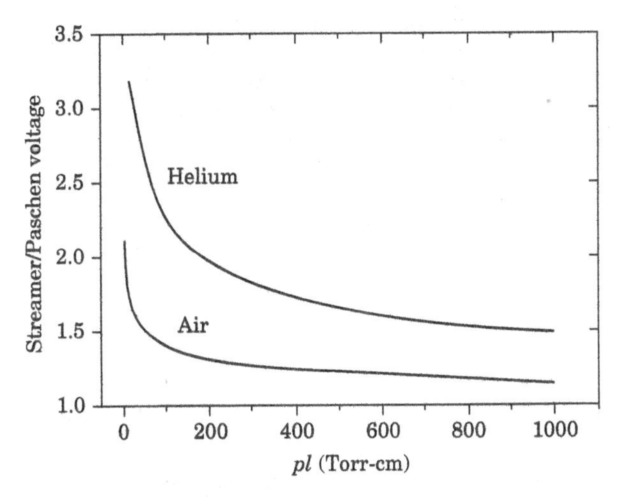

Figure 15.20 Streamer/Paschen breakdown voltage ratio versus pl; the streamer breakdown is calculated assuming a multiplication of 10^8. Source: Becker et al. (2005)/with permission of Taylor & Francis.

15.5.3 Glow Discharge Regime

Narrow gap, low-frequency driven, atmospheric pressure glow discharges have been observed experimentally, and modeled using fluid simulations, in the relatively narrow voltage region between Paschen and streamer breakdown (Yokoyama et al., 1990; Laroussi, 1996; Massines et al., 1998, 2003, 2005; Tochikubo et al., 1999; Golubovskii et al., 2003). A dielectric barrier on one or both electrodes is required to avoid a transition to an arc discharge. The usual gases used are helium and argon, often with a small percentage of Penning-ionizable gases, either deliberately introduced, or, for helium, by inadvertent contamination.

15.5.3.1 Experimental Results

A typical experimental system consists of an electrode–dielectric–gap–dielectric–electrode sandwich, usually symmetric about the midplane, driven by a low-frequency source in the range of approximately 1–100 kHz. At the lower frequencies, one short positive and one equal strength negative discharge current pulse are observed per cycle. At the higher frequencies, there can be symmetry breaking, with one large and one small pulse per cycle. The discharges are relatively transversely uniform, being stabilized against transverse contraction by dielectric layers, and, for helium, by high thermal conductivity, which inhibits gas heating instability. Below a kilohertz, discharge contraction effects become increasingly severe, limiting uniformity. At voltages near but below the Paschen breakdown voltage, weaker uniform discharges in the Townsend regime (see Figure 14.2) are found in nitrogen (Golubovskii et al., 2002; Massines et al., 2005; Naudé et al., 2005).

Figure 15.21 shows the measured discharge current and voltages for a 10 kHz, 1500 V glow discharge in helium (Massines et al., 1998). The $R = 2$ cm radius top and bottom electrodes were covered with alumina dielectric 0.6 mm thick, and the discharge gap was $l = 5$ mm. The measured series capacitance of the top and bottom dielectric layers was $C_d = 70$ pF, and the gap capacitance was $C \approx 2.2$ pF. The discharge ignites at a gap voltage $V_{ig} = 1550$ V, yielding two equal strength 90 mA current peaks per 10 kHz cycle. Each current pulse I has a rise time and initial decay time of about 1 μs, followed by a longer decay time tail. Note from the figure that the applied voltage is only

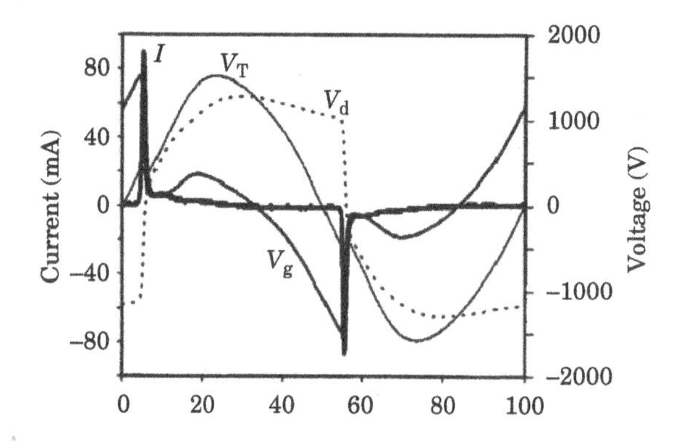

Figure 15.21 Time variation over one 10 kHz cycle of the measured values of applied voltage V_T, gas (gap) voltage V_g, memory (dielectric) voltage V_d, and discharge current I for a helium atmospheric pressure glow discharge. Source: Massines et al. (1998)/with permission of AIP Publishing.

$V_T = 400$ V when the discharge ignites; the additional voltage $V_d = 1150$ V is due to the charges stored on the top and bottom dielectric surfaces.

The ignition voltage corresponds to a breakdown field of 3.1×10^5 V/m, or $E/n_g \approx 13$ Td. As seen from Figure 14.3, $\alpha/n_g \approx 3 \times 10^{-24}$ m^2 for pure helium, yielding $\alpha \approx 74$ m^{-1}. The electron multiplication across the 5 mm gap is then $e^{\alpha l} \approx 1.4$, which is too small for Paschen breakdown. Therefore, it is believed that the helium is contaminated with a small fraction ($\sim 10^{-4}$–10^{-5}) of a low ionization energy gas, creating a Penning mixture that typically has an α at small E/n_g significantly higher that of pure helium alone (Massines et al., 1998; Golubovskii et al., 2003).

The total charge per discharge is measured to be $Q \approx 160$ nC. This corresponds to an average plasma density $n_{e0} = Q/(e\pi R^2 l) \approx 1.6 \times 10^{17}$ m^{-3}. Simulations (Massines et al., 1998; Golubovskii et al., 2003) show that n_e is not distributed equally along the gap. At the peak current density, n_e is high in the negative glow region near the cathode and is significantly lower in the positive column. The typical positive column density of $n_e \sim 1$–2×10^{16} m^{-3} extends over roughly half the gap length near the anode. The simulations also suggest that the discharge is in the subnormal glow regime. As shown in Figure 14.2, this regime has a negative differential resistance, giving instability and hysteresis for dc excitation. But the short pulse does not allow enough time for the discharge to transition into the normal glow regime.

15.5.3.2 Circuit Model

To understand these results, we examine the solution for the circuit shown in Figure 15.22. The discharge sandwich is assumed symmetric about the midplane and driven by an ideal voltage source, with $C_d = 70$ pF and $C = 2.2$ pF. The discharge is modeled by a time-varying gap resistance R, representing the resistance of the discharge during the discharge current pulse, in parallel with C. Considering a simple model, we take $R = \text{const} \ll (\omega C)^{-1}$ during the discharge pulse, and $R \gg (\omega C)^{-1}$ in the absence of the pulse. For comparison with the experimental results, we use a positive column resistance $R \approx l/(2\sigma A)$, with conductivity $\sigma \approx e^2 n_e/m\nu_m$. For an electron–neutral

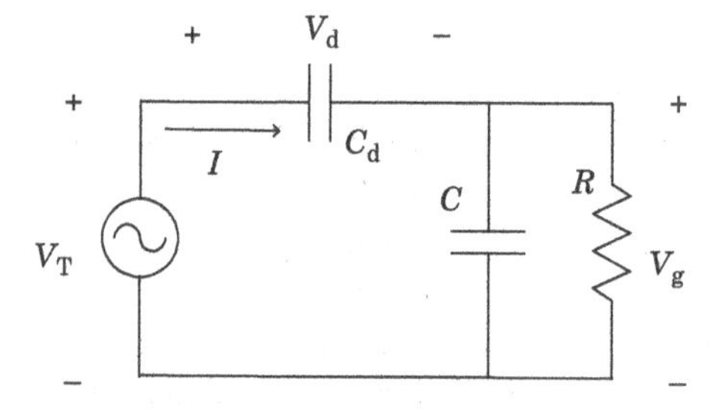

Figure 15.22 Simple circuit model for an atmospheric pressure glow discharge.

collision frequency $\nu_m = 2 \times 10^{12}$ s^{-1} and a typical positive column density $n_e \approx 2 \times 10^{16}$ m^{-3}, this gives $R \approx 14$ kΩ.

The source voltage is

$$V_T = V_{T0} \sin \phi \tag{15.5.11}$$

with $V_{T0} = 1500$ V and $\phi = \omega t$, shown as the dashed curve in Figure 15.23*a*. For this antisymmetric source voltage about $\phi = \pi$, the discharge currents and voltages for the phase interval $\pi < \phi < 2\pi$ must be the negative of those in the phase interval $0 < \phi < \pi$.

We decompose the gap voltage into two components as $V_g = V_C + V_R$, where V_C is the gap voltage in the absence of a discharge, which appears across C alone, and V_R is the voltage during the glow discharge pulse, which appears across R. The corresponding currents are I_C and I_R.

Let ϕ_{ig} be the ignition phase for the positive discharge current pulse. Then, for the preceding half-period, $\pi - \phi_{ig} < \phi < \phi_{ig}$, the gap voltage in the absence of the discharge can be written

$$V_C = V_{T0} \frac{C_d}{C_d + C} \sin \phi + V_{g0} \tag{15.5.12}$$

where V_{g0} is the gap voltage, and $-V_{g0}$ is the dielectric voltage, at phase $\phi = 0$. This is shown as the upper dotted curve in Figure 15.23*a*. Similarly, for the remaining phase interval $\phi_{ig} < \phi < \pi + \phi_{ig}$ prior to ignition of the negative pulse,

$$V_C = V_{T0} \frac{C_d}{C_d + C} \sin \phi - V_{g0} \tag{15.5.13}$$

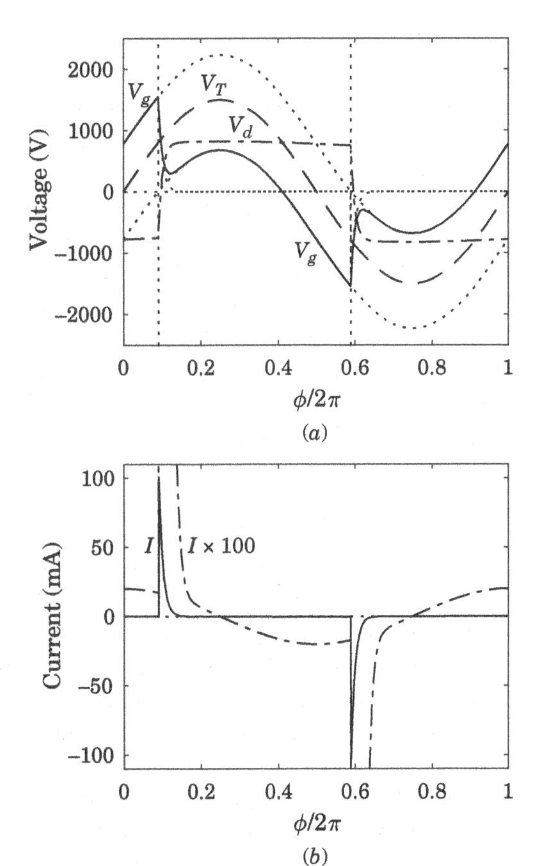

Figure 15.23 Circuit model results for the time variation over one 10 kHz cycle for a helium atmospheric pressure glow discharge; (*a*) source voltage V_T, gap voltage V_g, and dielectric voltage V_d; (*b*) gap current I.

which is shown as the lower dotted curve. The corresponding current is

$$I_C = C\frac{\mathrm{d}V_{g1}}{\mathrm{d}t} = \omega V_{T0}\frac{C_d C}{C_d + C}\cos\phi \tag{15.5.14}$$

When the plasma ignites at ϕ_{ig}, then the small resistance R modeling the discharge suddenly appears across the gap. C_d and C discharge through this resistor with a time constant $\tau = R(C_d + C) \approx 1$ μs, with $\omega\tau \ll 1$. The discharge voltage is then

$$V_R = V_{ig}\exp\left(-\frac{\phi - \phi_{ig}}{\omega\tau}\right), \qquad \phi > \phi_{ig} \tag{15.5.15}$$

with a corresponding discharge current $I_R = V_R/R$. Similarly, at $\pi + \phi_{ig}$, a negative voltage and current pulse appear. The voltages V_g and V_d must be continuous at ignition, as V_T is continuous and a discontinuity in V_d would imply an infinite current $I_d = C_d\,\mathrm{d}V_d/\mathrm{d}t$. This determines $V_{g0} = V_{ig}/2$.

For these parameters, the discharge ignites at $\phi_{ig}/2\pi = 0.089$. The gap voltage V_g is shown as the solid curve in Figure 15.23*a*, and the dielectric voltage $V_d = V_T - V_g$ is shown as the dot-dashed curve. The total current I in Figure 15.23*b* displays the two sharp current peaks seen in Figure 15.21, as well as a small reverse current after the discharge pulse (dot-dashed line), also seen experimentally. The maximum discharge current is $V_{ig}/R \approx 110$ mA, which compares reasonably well to the experimental result. The total charge just before the ignition is $Q = C_d V_{g0} + CV_{ig} \approx 110$ nC, corresponding to an initial density $n_e = 1.1 \times 10^{17}$ m^{-3}. This compares well to $Q = 160$ nC and $n_{e0} = 1.6 \times 10^{17}$ m^{-3} seen experimentally.

Further discussion of circuit models for helium and argon Penning discharges is given in Massines et al. (2005), and a more elaborate circuit model of the lower density Townsend regime in nitrogen is given in Naudé et al. (2005). Studies of the breakdown, glow discharge, and afterglow phases are found from 1D time-varying fluid simulations in Massines et al. (1998) and Golubovskii et al. (2003).

At frequencies significantly higher than 5–10 kHz, and particularly at the higher driving voltages, an asymmetric mode can appear, with one strong discharge current pulse, followed by one weaker pulse of the opposite sign, half a period later. This is because at higher frequencies, there is a higher density, residual quasi-neutral plasma in the gap region a half-period later, due to the strong discharge (see Problem 15.9). This essentially reduces the effective gap width for the following discharge. Because the breakdown voltage is proportional to the gap width, the following discharge breakdown is at a lower voltage. This produces a weaker discharge that decays more completely during the next half-period.

15.5.4 Filamentary Regime

At high overvoltages, streamer breakdown at a voltage V_{ig} occurs in low-frequency driven, narrow gap capacitive discharges with dielectric barriers on one or both electrodes. These are called *dielectric barrier discharges* (DBDs) or *silent discharges*. The DBD was invented by Siemens in 1857, when he first excited a discharge by placing the driving electrodes outside of a glass gas-filled tube. An early application was ozone production using air gas feed for water purification. This is still an important industrial application. DBD physics and chemistry modeling, especially with regard to ozone production, is described in Eliasson et al. (1987) and Eliasson and Kogelschatz (1991). Extensive databases exist for humid air discharge modeling, e.g., see Van Gaens and Bogaerts (2013, 2014) for argon/humid air. A review of the history, discharge physics, and applications is given by Kogelschatz (2003), see also Chapters 5 and 6 in Becker et al. (2005).

15.5.4.1 Discharge Properties

DBDs are typically operated with gaps of order $l = 1$ mm and frequencies $f = 50$ Hz–1 MHz, but operation at even higher frequencies is also used. The streamer breakdown mechanism leads to the formation of huge numbers of very short timescale ($\tau_f \sim 1$–10 ns), cylindrical microdischarge filaments per second, typically of order $r_f \approx 0.05$–0.1 mm in radius. As suggested by (15.5.5), the radius depends on the breakdown field, with electropositive gases typically having smaller fields, and therefore somewhat larger radii, than electronegative gases. The peak currents for a single microdischarge are of order $i_f \sim 0.1$ A, giving a charge $q_f \sim i_f \tau_f \sim 0.1$–1 nC. The energy deposited per microdischarge is typically $w_f \sim 1$ μJ. The filament densities and temperatures are high, $n_e \sim q_f/(e\pi r_f^2 l) \sim 10^{20}$–$10^{21}$ m^{-3}, $T_e \sim 1$–10 V, and the gas temperature is near room temperature.

The microdischarges usually occur randomly in the volume, but stationary spatial patterns are sometimes seen. Figure 15.24 shows an end-on photograph of a DBD discharge through a transparent electrode taken over 20 ms. The bright spots are the wider surface discharges formed at the microdischarge–dielectric interface. These spots are locally strongly charged, reducing the voltage drop across the microdischarge and extinguishing it before a transition to an arc discharge can occur.

DBDs have been used, or considered, for various materials processing applications, including thin film deposition and surface cleaning. However, as seen in the figure, the plasma is not locally uniform, leading to significant roughness after DBD surface processing. This can be tolerated for many industrial applications, e.g., SiO_2 deposition for oxygen permeation barriers on thin film plastic. But the inherent nonuniformity limits the use of DBDs for microelectronics, solar cell, and display panel processing.

15.5.4.2 DBD Circuit Model

The DBD circuit model is similar to the atmospheric pressure glow discharge circuit in Figure 15.22, but with the resistor R replaced by a constant voltage V_{ig}, during the period of microdischarge generation. Referring to Figure 15.22, the onset of DBD operation occurs at a source voltage magnitude

$$V_{T0min} = V_{ig}\left(1 + \frac{C}{C_d}\right) \tag{15.5.16}$$

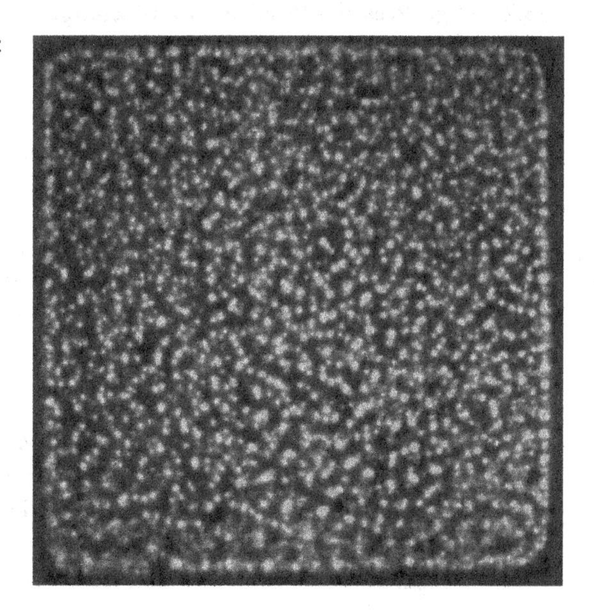

Figure 15.24 End-on photograph of dielectric barrier microdischarges in atmospheric pressure air; original size 6 cm × 6 cm; exposure time 20 ms. Source: Kogelschatz (2003)/with permission of Springer Nature.

which places a voltage $+V_{ig}$ across the gap. For $V_{T0} > V_{T0min}$, measurements show that microdischarges are generated during each portion of the increasing applied voltage lying between V_{T0min} and the maximum source voltage V_{T0}. During this time, the gap voltage magnitude remains at $+V_{ig}$. The generation vanishes during all, or almost all, of the subsequent decrease of the applied voltage magnitude. A half-cycle later, the discharge reignites with the opposite voltage polarities, i.e., with a gap voltage of $-V_{ig}$. Hence, there are two periods of generation per low-frequency cycle. We leave the circuit model solution for the student in Problem 15.10.

The power per unit area S dissipated in a DBD discharge is roughly proportional to the area density of microdischarges created per unit time, dN'_f/dt.

$$S \approx w_f \frac{dN'_f}{dt} \tag{15.5.17}$$

In the moderate range of power fluxes, $S \sim 1000\ \text{W/m}^2$, with $w_f = 1\ \mu\text{J}$, (15.5.17) implies $dN'_f/dt \sim 10^9$ microdischarges/m^2-s. S is given in terms of the circuit parameters shown in Figure 15.22 by (Manley, 1943; Kogelschatz, 2003)

$$S \approx \frac{4fC_d^2}{(C_d + C)A} V_{T0min}(V_{T0} - V_{T0min}), \qquad V_{T0} > V_{T0min} \tag{15.5.18}$$

with V_{T0min} given by (15.5.16) and A the discharge area (see Problem 15.10).

Problems

15.1 **Passive Bulk Discharge** Assuming that $S_{stoc} \ll S_{ohm,sh}$ and that $u_s \ll u_B$, the passive bulk model can be solved by simple substitutions, rather than requiring numerical solutions for some steps. Using $S_{stoc} = 0$ and (15.2.16) for u_s, determine the approximate equilibrium parameters for Example 15.1 in Section 15.2, and compare your results for V_{rf}, s_m, and n_0 with those given in the example. You should find 15–20% higher voltages and sheath widths, and 15–20% lower densities than those given in the example.

15.2 **High-Pressure Passive Bulk Discharge** Consider a 6 Torr argon discharge at 300 K with gap length 2.5 cm, driven at 40 A/m^2 by a 13.56 MHz rf power source. At this pressure, $S_{stoc} \ll S_{ohm,sh}$ and $u_s \ll u_B$, as in Problem 15.1. Using an initial guess $T_e = 2$ V for the bulk electron temperature,

(a) Determine μ_i, ν_{mi}, λ_{mi}, $\lambda_\mathcal{E}$, ν_m, $\mathcal{E}_c$, and u_B at this pressure and electron temperature.
(b) Find the discharge equilibrium parameters n_s and u_s, and show that $n_0 \approx 5.8 \times 10^{15}$ m^{-3}.
(c) Show that $S_{sh} \approx 30$ W/m^2, $s_m \approx 1.4$ mm, and $V_{rf} \approx 100$ V.
(d) Evaluate the Paschen voltage at this pressure and sheath width from (15.3.6), and show that V_{rf} somewhat exceeds the Paschen voltage. Therefore, at a current density somewhat less than 40 A/m^2, this passive bulk α-mode discharge makes a transition into the γ-mode (see Problem 15.3).

15.3 **Normal Current Density of a Gamma Mode Discharge** Consider the α–γ transition for the 6 Torr, 300 K argon discharge of Problem 15.2. As shown there, the sheath width $s_m \approx 1.4$ mm in the α-mode near the transition.

(a) Find the Paschen minimum voltage V_{min}, and show that the sheath width $s_{min} \approx 0.80$ mm at this minimum voltage. Since $s_m > s_{min}$, the α-mode discharge makes an abrupt transition from the high ps_m side of the Paschen curve into the γ-mode at the voltage minimum V_{min} and sheath width s_{min}.

(b) Proceeding as in Example 15.2 of Section 15.3 using an initial guess $T_e = 1.2$ V for the bulk electron temperature and $\mathcal{E}_c \approx 18$ V for the secondary electron energy loss per ionization, determine $\lambda_\mathcal{E}$, ν_m, and $\mathcal{M}_\gamma$ for this γ-mode discharge.

(c) The "normal" current density is the current density J at the Paschen minimum. Show that $J \approx 80\ \text{A/m}^2$, $n_s \approx 1.5 \times 10^{16}\ \text{m}^{-3}$, and $n_0 \approx 3.4 \times 10^{16}\ \text{m}^{-3}$.

(d) Find the secondary electron, ion, and bulk ohmic power fluxes.

(e) Note from the results of Problem 15.2 that the central plasma density abruptly increases by a factor of 5.7 when the discharge enters the γ-mode. At a fixed discharge current I_{rf} for an α-mode discharge of radius R_α filling its entire electrode area, the discharge also contracts abruptly upon entering the γ-mode. Find the contracted discharge radius.

15.4 Cathode Sheath Voltage and α–γ Transition Voltage The argon Paschen breakdown voltage and the corresponding α–γ transition voltage in Figure 15.11 are calculated assuming a uniform electric field in the sheath. For a better approximation, use expressions (14.3.15) and (14.3.16) for a linear field variation to repeat the calculation. Plot the resulting cathode sheath voltage and the corresponding α–γ transition voltage versus ps_m. Then, compare your results to those in Figure 15.11.

15.5 Atmospheric Pressure Discharge Maintenance Current and Voltage For an atmospheric pressure helium discharge driven at 13.56 MHz, with $\overline{\nu}_m = 1.3 \times 10^{13}\ \text{s}^{-1}$ and $\overline{T}_e =$ 1 V, find the maintenance current density J_{crit} and the discharge maintenance voltage $V_{D,crit}$.

15.6 Atmospheric Pressure Discharge Example Using the discharge parameters and equilibrium results of Example 15.3

(a) Show that the helium excimer/metastable density ratio is $n_{He_2^*}/n_{He^*} \sim 0.2$. Explain qualitatively the effect of the excimer Penning ionization (reaction 12 in Table 15.1) on the discharge equilibrium.

(b) Show that the He^+ and He_2^+ densities are negligible compared to the N_2^+ density.

(c) Show that electron–ion recombination loss (reaction 14 of Table 15.1) is small compared to ion losses to the walls.

(d) Show that direct ionization of N_2 (reaction 15 of Table 15.1) is small compared to Penning ionization by the helium metastable.

15.7 Low-Density Atmospheric Pressure Discharge Energy Balance Consider the 27.12 MHz atmospheric pressure He/0.1%N_2 discharge of Example 15.3, except that the rf current density is $J = 200\ \text{A/m}^2$.

(a) As shown in (15.4.22), there is a critical maintenance current density J_{crit} at which the electron density $n_e \to 0$. Estimate J_{crit} for an initial guess of $\overline{T}_e = 1$ V.

(b) For the general case of $n_e > 0$, show from (15.4.21) that

$$n_e = \left[\frac{M_g(J^2 - J_{crit}^2)}{6e^3\overline{T}_e}\right]^{1/2}$$

(c) Redo Example 15.3 at 200 A/m^2 using the above result for n_e.

15.8 Radial Electron Diffusion in a Discharge Avalanche Show that the solution of the avalanche head diffusion equation (15.5.3) is given by (15.5.4).

15.9 Density Decay in an Atmospheric Pressure Glow Discharge Consider the plasma density decay with time in the afterglow of the short discharge pulse of Figure 15.23. Assume that the electrons quickly cool and are at room temperature at all times within the afterglow.

(a) Show that the fundamental ambipolar diffusion decay time (5.2.11) is much longer than the 100 μs low-frequency period. Therefore, the plasma does not decay due to diffusive losses to the walls.

(b) Consider the density decay due to electron–N_2^+ recombination alone, with the rate constant K_{14} given in Table 15.1. Show that an initial density n_{e0} decays as

$$n_e(t) = \frac{n_{e0}}{1 + n_{e0}K_{14}t}$$

Note that for $n_{e0}K_{14}t \gg 1$, $n_e(t) \approx (K_{14}t)^{-1}$, independent of n_{e0}.

(c) For an initial density $n_{e0} = 1.1 \times 10^{17}$ m^{-3}, find the residual density $n_e(t)$ a half-period later ($t = 50$ μs), just before ignition of the next pulse.

15.10 Dielectric Barrier Discharge Circuit Model Consider the circuit model in Figure 15.22, but with the resistor R replaced by a constant voltage V_{ig}, during the part of the low-frequency cycle for microdischarge generation. During the part without generation, the circuit is just the series connection of the dielectric capacitor C_d and the gap capacitor C. Assume a source voltage

$$V_T = -V_{T0}\cos\phi$$

with $\phi = \omega t$. Note that the gap and dielectric voltage waveforms $V_g(t)$ and $V_d(t)$ must be continuous, as V_T is continuous and a discontinuity in V_d would imply an infinite capacitor current $I_d = C_d\,\mathrm{d}V_d/\mathrm{d}t$.

(a) Assuming gap and dielectric voltages during the part of the low-frequency cycle $0 < \phi < \phi_{ig}$ with no microdischarge generation to be

$$V_g = -V_{T0}\frac{C_d}{C_d + C}\cos\phi + V_{g0}$$

$$V_d = -V_{T0}\frac{C}{C_d + C}\cos\phi - V_{g0}$$

show that the time-phase ϕ_{ig} for the onset of microdischarge generation is given by

$$V_{ig} = -V_{T0}\frac{C_d}{C_d + C}\cos\phi_{ig} + V_{g0}$$

(b) Assuming that the gap voltage remains at V_{ig} during the microdischarge generation $\phi_{ig} < \phi < \pi$, show that $V_d = -V_{T0}\cos\phi - V_{ig}$ during this time.

(c) By symmetry, the voltage waveforms for $\pi < \phi < 2\pi$ are the negative of the voltage waveforms for $0 < \phi < \pi$. Using this condition, show that

$$V_{ig} = V_{T0}\frac{C_d}{C_d + C} - V_{g0}$$

(d) From the results in parts (a) and (c), show that

$$\cos\phi_{\mathrm{ig}} = 1 - \frac{2V_{\mathrm{ig}}}{V_{\mathrm{T0}}}\left(1 + \frac{C}{C_{\mathrm{d}}}\right)$$

(e) Sketch or plot $V_{\mathrm{T}}(\phi)$, $V_{\mathrm{g}}(\phi)$, and $V_{\mathrm{d}}(\phi)$ for $0 < \phi < 2\pi$, and verify that these voltages are continuous. Use the parameters $C_{\mathrm{d}} = C$ and $V_{\mathrm{T0}} = 6\,V_{\mathrm{ig}}$.

(f) Show that the discharge current is $I = \omega C_{\mathrm{T}} V_{\mathrm{T0}} \sin\phi$ during the period $0 < \phi < \phi_{\mathrm{ig}}$ of no microdischarge generation, and $I = \omega C_{\mathrm{d}} V_{\mathrm{T0}} \sin\phi$ during the period $\phi_{\mathrm{ig}} < \phi < \pi$ of microdischarge generation. Here $C_{\mathrm{T}} = C_{\mathrm{d}}C/(C_{\mathrm{d}} + C)$ is the capacitance of the series combination of C_{d} and C.

(g) The time-averaged discharge power per unit area is

$$S = \frac{1}{\pi A}\int_0^{\pi} \mathrm{d}\phi\, V_{\mathrm{T}}(\phi) I(\phi)$$

Using $V_{\mathrm{T}} = -V_{\mathrm{T0}}\cos\phi$ and I from part (f), show that S is given by (15.5.18), as first obtained by Manley (1943). Note from (15.5.16) that $V_{\mathrm{ig}} = V_{\mathrm{T0min}}\, C_{\mathrm{d}}/(C_{\mathrm{d}} + C)$, with V_{T0min} the minimum source voltage magnitude for DBD operation.

16

Etching

16.1 Etch Requirements and Processes

Plasma etching is a key process for removing material from surfaces. The process can be chemically selective, removing one type of material while leaving other materials unaffected, and can be anisotropic, removing material at the bottom of a trench while leaving the same material on the sidewalls unaffected. Plasma etching is the only commercially viable technology for anisotropic removal of material from surfaces. As such, it is an indispensable part of modern integrated circuit fabrication technology, as described in Chapter 1. For a more complete description of this area, the reader should consult other sources, e.g., Manos and Flamm (1989, chapters 1 and 2), and Donnelly and Kornblit (2013).

Although there are many other areas of application, nearly all modern developments in plasma etching have been driven by their potential for integrated circuit fabrication. In this chapter, we focus almost exclusively on this area, placing emphasis on the key concepts that determine etch rate, selectivity, and anisotropy in plasma etch processes. In this section, we introduce typical etch requirements and possible tradeoffs among them and describe the four types of plasma etch processes. In Section 16.2, some simple models of surface etching and discharge kinetics are described, a general chemical framework for plasma etching is introduced, and a brief summary of pattern transfer fidelity is given. In Section 16.3, the use of halogens to etch silicon is discussed. In particular, fluorine atom etching of silicon has been the most well-studied etch system, providing insight into other less well-characterized systems. In Section 16.4, some descriptions of silicon oxide and nitride etching, metal etching, and photoresist etching are given. Section 16.5 treats atomic layer etching (ALE), which can be used to remove monolayers of material in critical etch applications. Finally, charging and resulting device damage due to plasma etching is described in Section 16.6.

16.1.1 Plasma Etch Requirements

It is important to consider the entire set of processing requirements for a particular application. For pattern transfer by etching on a silicon wafer, these might include requirements on etch rate, anisotropy, selectivity, uniformity across the wafer, surface quality, and process reproducibility. Consider first the *etch rate* requirements for the typical set of films, shown in Figure 16.1a, consisting of 500 nm of resist, over 100 nm of polysilicon, and over 2 nm of "gate oxide" (e.g., oxynitride) on an epitaxial silicon wafer. For a single wafer process of commercial interest, the resist must be

Principles of Plasma Discharges and Materials Processing, Third Edition. Michael A. Lieberman and Allan J. Lichtenberg.

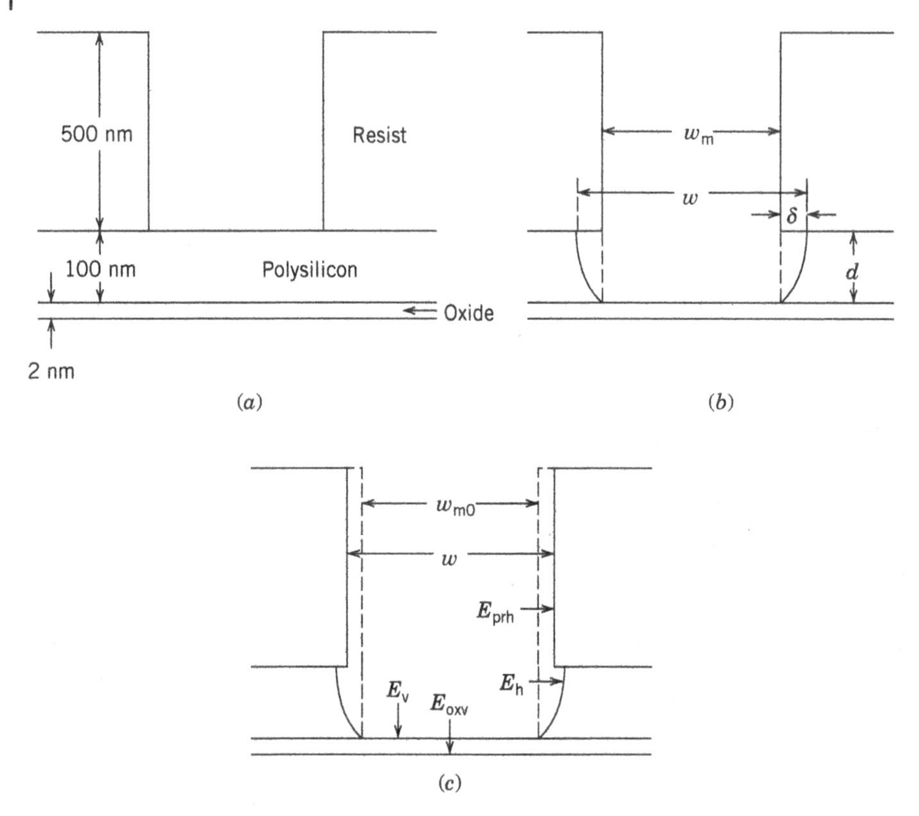

Figure 16.1 Calculation of plasma etch requirements: (*a*) a typical set of films; (*b*) anisotropy requirement for polysilicon etch; (*c*) uniformity requirement, including the effect of photoresist erosion.

stripped and the polysilicon must be etched in a few minutes. This leads to minimum etch rate requirements $E_{pr} = 250$ nm/min for the photoresist and $E_{poly} = 50$ nm/min for the polysilicon.

Next, consider the *selectivity* requirements for the polysilicon etch. For etch of the 100 nm polysilicon with the resist as a mask, a selectivity of

$$S = \frac{E_{poly}}{E_{pr}} \gg \frac{100\,\text{nm}}{500\,\text{nm}} = 0.2$$

is required in order to complete the polysilicon etch while not significantly eroding the resist. For this application, a selectivity of 2–3 might be acceptable. However, there is a second selectivity requirement. Due to a lack of *uniformity* across the wafer, it is necessary to overetch the polysilicon at some locations on the wafer in order to clear it from all unmasked regions. During overetch, the thin oxide is exposed to the etchant at some wafer locations. For an overetch of 20% (1.2 times the etch time required to clear a perfectly uniform polysilicon film using a perfectly uniform process), a polysilicon to gate oxide selectivity of

$$S = \frac{E_{poly}}{E_{ox}} \gg \frac{0.2 \times 100\,\text{nm}}{2\,\text{nm}} = 10$$

is required. Depending on the use, selectivities of 100–200 might be needed. Hence, selectivity to the underlying layer can be a significant issue in film removal.

Consider now, as shown in Figure 16.1*b*, the *anisotropy* required to etch a trench of width w into a film of thickness d, and let w_m be the minimum mask feature size that can be used (e.g., due to lithography limitations). The anisotropy for the film etch is defined as

$$a_h = \frac{E_v}{E_h} \tag{16.1.1}$$

where E_v and E_h are the vertical and horizontal etch rates, respectively. Assuming that the mask is not eroded, then after the etch, as shown in the figure, we find the relation

$$a_h = \frac{d}{\delta} \tag{16.1.2}$$

The maximum width of the trench is

$$w = w_m + 2\delta \tag{16.1.3}$$

Solving for δ and substituting this into (16.1.2), we obtain the anisotropy requirement

$$a_h \geq \frac{2d}{w - w_m} \tag{16.1.4}$$

As an example, for $w = 50$ nm, $d = 100$ nm, and $w_m = 25$ nm, we obtain $a_h \geq 8$. Even for $w_m \equiv 0$ (physically unreasonable), an anisotropy of $a_h \geq 4$ is required. Evidently, the smallest feature size that can be made has a width

$$w \approx \frac{2d}{a_h} \tag{16.1.5}$$

For etching of deep trenches ($d/w \gg 1$), the etch anisotropy requirements can be severe.

Consider now the impact of process *uniformity* on selectivity and anisotropy requirements. Referring to Figure 16.1c, we now also consider erosion of the mask sidewalls due to a horizontal etch rate E_{prh} and erosion of the underlying gate oxide sublayer due to a vertical etch rate E_{oxv}. Introducing the selectivities

$$s_{pr} = \frac{E_v}{E_{prh}} \tag{16.1.6a}$$

$$s_{ox} = \frac{E_v}{E_{oxv}} \tag{16.1.6b}$$

where E_v is the polysilicon vertical etch rate, we let t_{max} be the time required to completely etch the polysilicon layer at all unmasked locations on the wafer. Then,

$$t_{max} = f\frac{d}{E_v} \tag{16.1.7}$$

where f is an overetch nonuniformity factor due to film thickness variations,

$$d = d_0(1 \pm \alpha) \tag{16.1.8}$$

and etch rate variations across the wafer,

$$E_v = E_{v0}(1 \pm \beta) \tag{16.1.9}$$

Note here that across the wafer, thickness varies by a fraction α, and etch rate varies by a fraction β. Hence, t_{max} is determined by the maximum thickness and minimum etch rate

$$f = \frac{1 + \alpha}{1 - \beta} \approx 1 + \alpha + \beta \tag{16.1.10}$$

where we have assumed that $\alpha,\ \beta \ll 1$. Assuming also that $E_v \gg E_h$ in the polysilicon, the horizontal etch width δ can be estimated as the sum of two terms,

$$\delta \approx (E_h + E_{prh})t_{max} \tag{16.1.11}$$

where the horizontal etch rate in the resist mask leads to a horizontal etch of the polysilicon because the vertical etch in the polysilicon is fast compared to the horizontal erosion of the mask. Substituting (16.1.7) into (16.1.11), we find

$$\delta \approx fd\frac{E_h + E_{prh}}{E_v} \tag{16.1.12}$$

Substituting (16.1.12) in (16.1.3), we obtain

$$w \approx w_{m0} + 2fd\frac{E_h + E_{prh}}{E_v}$$

or

$$\frac{E_h + E_{prh}}{E_v} \approx \frac{w - w_{m0}}{2fd} \tag{16.1.13}$$

For example, let $w = 50$ nm, $w_{m0} = 25$ nm, $d = 100$ nm, and $\alpha = \beta = 0.1$, we find $(E_h + E_{prh})/E_v \approx 0.1$. Hence, as shown in Figure 16.2*a*, one can trade resist selectivity (E_{prh}) against etch anisotropy (E_h) within a triangular window in parameter space near the origin. For a deeper trench, the requirements on resist selectivity and etch anisotropy become even more severe.

Consider now the undesired etch of the gate oxide sublayer. In a worst-case analysis, the etch begins at a time

$$t_{min} = \frac{d}{E_v}\frac{1 - \alpha}{1 + \beta} \tag{16.1.14}$$

The maximum sublayer thickness etched is then

$$\begin{aligned}\delta_{ox} &= (t_{max} - t_{min})E_{oxv} \\ &= \tfrac{d}{E_v}2(\alpha + \beta)E_{oxv}\end{aligned} \tag{16.1.15}$$

This can be rewritten as

$$(\alpha + \beta)\frac{E_{oxv}}{E_v} = \frac{\delta_{ox}}{2d} \tag{16.1.16}$$

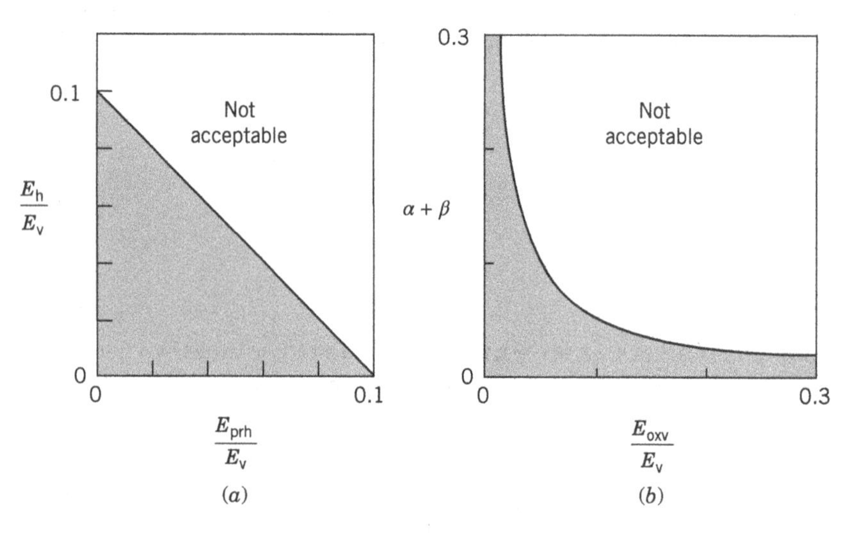

Figure 16.2 Acceptable trade-offs for plasma etching: (*a*) anisotropy versus photoresist selectivity and (*b*) uniformity versus oxide selectivity.

which shows that uniformity ($\alpha + \beta$) can be traded against sublayer selectivity (E_{oxv}/E_v), within the hyperbolic region near the origin, as shown in Figure 16.2*b*. As an example, for $d = 100$ nm and $\alpha = \beta = 0.1$, a selectivity to achieve $\delta_{ox} \leq 0.2$ nm of $s_{ox} = E_v/E_{oxv} \geq 200$ is required. For a more detailed estimation of the trade-offs among anisotropy, selectivity, and uniformity, the reviews by Flamm and Herb (1989), Donnelly and Kornblit (2013), and Arts et al. (2022) can be consulted.

16.1.2 Etch Processes

There are four basic low-pressure plasma processes commonly used to remove material from surfaces: sputtering, pure chemical etching, ion energy-driven etching, and ion inhibitor etching. Sputtering is the ejection of atoms from surfaces due to energetic ion bombardment. This process was described in Section 9.3 and is illustrated in Figure 16.3*a*. The discharge supplies energetic ions to the surface, with the ions typically having energies above a few hundred volts. Sputtering is a relatively unselective process since, from (9.3.14), the sputtering yield γ_{sput} at a given ion energy depends on the surface binding energy $\mathcal{E}_t$ and (weakly) on the masses of the targets and projectiles. Typically, γ_{sput} does not vary by more than a factor of 2–3 among different materials (see Table 9.2). Sputtering rates are generally low because the yield is typically on the order of one atom per incident ion, and ion fluxes incident on surfaces in discharges are often small compared to commercially significant rates for materials removal. Sputtering is, however, an anisotropic process, strongly sensitive to the angle of incidence of the ion (see Figure 9.5). The yield typically rises from its normal (0°) incidence value to some maximum value γ_{max} at θ_{max} and then falls to zero at grazing incidence (90°). Therefore, there is essentially no sidewall removal of material for ions normally incident on a substrate. However, because the sputtering yield peaks at $\theta_{max} \neq 0$, topographical patterns might not be faithfully transferred during sputter etching. Figure 16.4 shows ions at normal incidence on

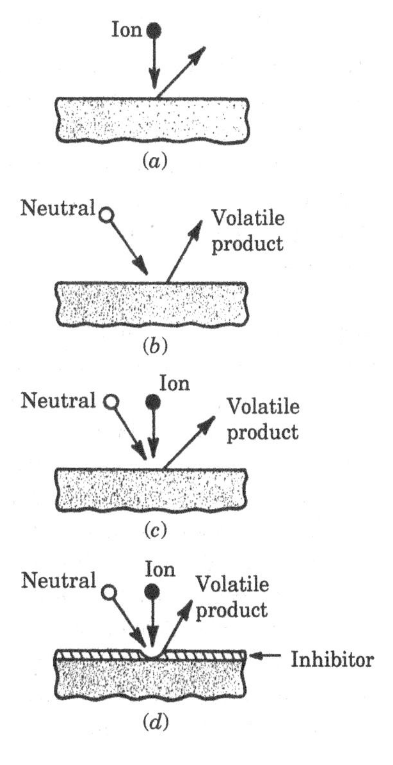

Figure 16.3 Four basic plasma etching processes: (*a*) sputtering; (*b*) pure chemical etching; (*c*) ion energy-driven etching; (*d*) ion-enhanced inhibitor etching. Source: Flamm and Herb (1989)/with permission of Elsevier.

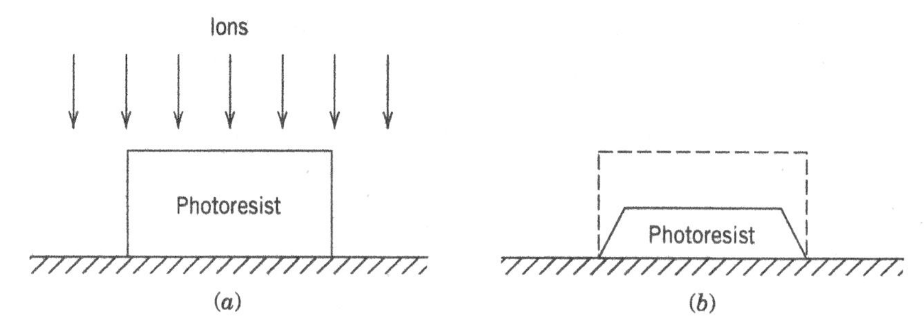

Figure 16.4 The development of facets due to sputtering of photoresist: (*a*) before sputtering and (*b*) after sputtering.

a step (*a*) before and (*b*) after sputtering. A facet has developed after sputtering due to the peaking of the yield at θ_{max}. Sputtering is the only one of the four etch processes that can remove involatile products from a surface. This is important for removing low fraction involatile components during film etching using other processes, e.g., the sputter removal of copper during etching of Al–2%Cu films. It is also important for sputter deposition processes, which are described in Section 17.4. For these applications, the mean free path of the sputtered atoms must be large enough to prevent redeposition on the substrate or target. Consequently, these processes are generally carried out at low pressure.

A second etch process is pure chemical etching, in which the discharge supplies gas-phase etchant atoms or molecules that chemically react with the surface to form gas-phase products. This process can be highly chemically selective. Some examples are

$$\mathrm{Si(s)} + 4\mathrm{F} \longrightarrow \mathrm{SiF_4(g)}$$

$$\text{photoresist} + \mathrm{O(g)} \longrightarrow \mathrm{CO_2(g)} + \mathrm{H_2O(g)}$$

Pure chemical etching, illustrated in Figure 16.3*b*, is almost invariably isotropic, since the gas-phase etchants arrive at the substrate with a near-uniform angular distribution. Therefore, unless the reaction is with a crystal having a rate depending on crystallographic orientation, one may expect a relatively isotropic etch rate. As described in Section 1.1, the etch products must be volatile. The etch rate can be quite large because the flux of etchants to the substrate can be high in processing discharges. However, etch rates are generally not limited by the rate of arrival of etchant atoms, but by one of a complex set of reactions at the surface leading to the formation of etch products. For example, for F-atom etching of silicon, there is considerable evidence that the rate-limiting step involves the reaction of an F^- ion generated at the surface with the fluorinated surface layer. We consider some of these issues in Section 16.3.

A third etch process, illustrated in Figure 16.3*c*, is *ion-enhanced energy-driven etching*, in which the discharge supplies both etchants (e.g., F atoms) and energetic ions to the surface. The combined effect of both etchant atoms and energetic ions in producing etch products can be much larger than that produced by either pure chemical etching or by sputtering alone, as is shown in Figure 1.4. This surprising result was first reported by Hosokawa et al. (1974). For etching of silicon with a high incident flux of F atoms, for example, a single 1 kV argon ion can cause the removal of as many as 25 silicon atoms (and 100 fluorine atoms) from the surface. Experiments suggest that the etching is chemical in nature, but with a reaction rate determined by the energetic ion bombardment. The etch rate generally increases with increasing ion energy above a threshold energy of a few volts. The etch product must be volatile, as for pure chemical etching. Because the energetic ions have a highly directional angular distribution when they strike the substrate, the etching can be highly anisotropic. However, ion energy-driven etching may have poor selectivity compared to pure chemical etching. The trade-off between anisotropy and selectivity is important in designing etch processes, as was shown earlier. The detailed mechanism for etch product formation and the rate of formation are not well understood. An empirical model is given in Section 16.2, and some proposed mechanisms are described in Section 16.3.

A fourth type of etch process, *ion-enhanced inhibitor etching*, illustrated in Figure 16.3*d*, involves the use of an inhibitor species (see Section 7.5). The discharge supplies etchants, energetic ions, and inhibitor precursor molecules that adsorb or deposit on the substrate to form a protective layer or polymer film. The etchant is chosen to produce a high chemical etch rate of the substrate in the absence of either ion bombardment or the inhibitor. The ion-bombarding flux prevents the inhibitor layer from forming or clears it as it forms, exposing the surface to the chemical etchant. Where the ion flux does not fall, the inhibitor protects the surface from the etchant. Inhibitor precursor molecules include CF_2, CF_3, CCl_2, and CCl_3 molecules, which can deposit on the substrate to form fluoro- or chloro-carbon polymer films. A classic example of an ion inhibitor plasma etch, described in more detail in Section 16.6, is the anisotropic etching of aluminum trenches or holes using CCl_4/Cl_2 or $CHCl_3/Cl_2$ discharges. Both Cl and Cl_2 rapidly etch aluminum, but the resulting etch is isotropic. The addition of carbon to the feedgas mix results in the formation of a protective chlorocarbon film on the surface. Ion bombardment clears the film from the trench bottom, allowing the etch process to proceed there. The same film on the sidewalls protects these from the etchant. With proper optimization, a highly anisotropic etch with vertical sidewalls can be formed. Ion inhibitor etching shares most other features of ion energy-driven etching. The process may not be as selective as pure chemical etching, and a volatile etch product must be formed. Contamination of the substrate and final removal of the protective inhibitor film are other issues that must be addressed for this etch process.

Except for sputtering, the etch chemistry must be chosen to yield a volatile product. Data such as that given in Table 7.5 can be used to determine etch product volatility. Table 16.1 gives a list of materials, along with possible etchant atom chemistries based on product volatility. In some cases, there is no satisfactory low-temperature chemistry available, e.g., copper has been etched in

Table 16.1 Etch Chemistries Based on Product Volatility

Material	Etchant Atoms
Si, Ge	F, Cl, Br
SiO_2	F, F+C
Si_3N_4, silicides	F
Al	Cl, Br
Cu	Cl ($T > 210°C$)
C, organics	O
W, Ta, Ti, Mo, Nb	F, Cl
Au	Cl
Cr	Cl, Cl+O
GaAs	Cl, Br
InP	Cl, C+H

chlorine only at elevated temperatures (see Section 16.4). In the following sections, we will explore a number of etch chemistries for various materials.

Although four etch processes have been distinguished, their use for a particular film etch often involves parallel or serial combinations of the processes, as has already been noted for Al–2%Cu etching. Consider, for example, the cutting of a vertical trench in a thick polysilicon layer that must stop with high selectivity at a silicon dioxide layer, as shown in Figure 16.1. This might be accomplished by a two-step process. The first step might be a fast, highly anisotropic ion energy-driven etch. Pure chemical etching in parallel at the sidewalls might determine the anisotropy of this process. After almost all of the polysilicon has been removed, the final step might be a slow, highly selective, but relatively isotropic etch to remove the remaining polysilicon with minimum etching of the underlying oxide. The small undercut produced by this step might be acceptable if the polysilicon that remains after the first step is thin enough.

16.2 Etching Kinetics

With the exception of the physical sputtering of elemental materials, the detailed mechanisms for plasma etch processes are not well understood. Simple empirical models that incorporate some of the key observations can provide insight into the use of various processes. In this section, kinetic models for surface etch processes are introduced in which known neutral and ion fluxes at the surface are used to determine the etch rate and anisotropy. These fluxes, in turn, must be found using a discharge model that accounts for the generation of both etchant atoms and bombarding ions. A general framework for the chemistry of etch processes is introduced; this will be elaborated in subsequent sections. Finally, pattern transfer fidelity during trench etches is briefly described.

16.2.1 Surface Kinetics

Consider first the example of an ion energy-driven process for O atom etching of a carbon substrate. We assume that the only reactions that occur are

$$\mathrm{O(g) + C(s)} \xrightarrow{K_\mathrm{a}} \mathrm{C{:}O} \tag{16.2.1}$$

$$\mathrm{C{:}O} \xrightarrow{K_\mathrm{d}} \mathrm{CO(g)} \tag{16.2.2}$$

$$\mathrm{ion + C{:}O} \xrightarrow{Y_\mathrm{i}K_\mathrm{i}} \mathrm{CO(g)} \tag{16.2.3}$$

Let θ be the fraction of surface sites (area density n_0') covered with C:O bonds. We assume Langmuir kinetics, as shown in Figure 16.5. All O atoms from the gas phase incident on the surface not covered with C:O are assumed to react immediately to form C:O. The rate-limiting etch step is assumed to be desorption of CO(g). The steady-state surface coverage is then found from

$$\frac{\mathrm{d}\theta}{\mathrm{d}t} = K_\mathrm{a} n_\mathrm{OS}(1-\theta) - K_\mathrm{d}\theta - Y_\mathrm{i}K_\mathrm{i}n_\mathrm{is}\theta = 0 \tag{16.2.4}$$

where n_OS and n_is are the neutral and ion densities at the surface and the plasma–sheath edge, respectively,

$$K_\mathrm{a} = \frac{1}{4}\frac{\overline{v}_\mathrm{O}}{n_0'} = \frac{1}{4}\left(\frac{8kT_\mathrm{O}}{\pi M_\mathrm{O}}\right)^{1/2}\frac{1}{n_0'}$$

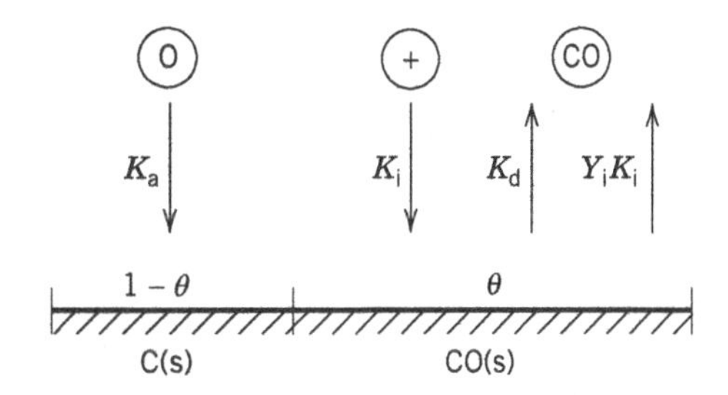

Figure 16.5 Surface etch model assuming Langmuir kinetics and rate-limiting desorption.

is the O-atom adsorption rate coefficient, K_d is the rate coefficient for thermal desorption of CO, Y_i is the yield of CO molecules desorbed per ion incident on a fully covered surface in the absence of other desorption mechanisms, and $K_i = u_B/n'_0 = (eT_e/M_i)^{1/2}/n'_0$ is the rate coefficient for ions incident on the surface.

For high ion energies, measurements (e.g., Steinbrüchel, 1989; Chang and Sawin, 1997; Wang and Wendt, 2001; and references therein) indicate that the yield is typically much greater than unity and scales as $Y_i \propto \sqrt{\mathcal{E}_i} - \sqrt{\mathcal{E}_{thr}}$, the same as for the sputtering yield (9.3.14). Solving (16.2.4) for θ, we obtain

$$\theta = \frac{K_a n_{OS}}{K_a n_{OS} + K_d + Y_i K_i n_{is}} \tag{16.2.5}$$

The flux of CO molecules leaving the surface is

$$\Gamma_{CO} = (K_d + Y_i K_i n_{is})\theta n'_0 \tag{16.2.6}$$

The vertical etch rate is

$$E_v = \frac{\Gamma_{CO}}{n_C} \quad (\text{m/s}) \tag{16.2.7}$$

where n_C is the carbon atom density of the substrate. Inserting (16.2.5) and (16.2.6) into (16.2.7), we obtain

$$E_v = \frac{n'_0}{n_C} \frac{1}{\dfrac{1}{K_d + Y_i K_i n_{is}} + \dfrac{1}{K_a n_{OS}}} \tag{16.2.8}$$

Assuming that the ions strike the substrate surface at normal incidence, then the ion flux incident on a vertical trench sidewall is zero. In this limit, we obtain a purely chemical horizontal etch rate:

$$E_h = \frac{n'_0}{n_C} \frac{1}{\dfrac{1}{K_d} + \dfrac{1}{K_a n_{OS}}} \tag{16.2.9}$$

The normalized etch rates $(n_C/n'_0)E_v/K_d$ and $(n_C/n'_0)E_h/K_d$ are plotted versus the normalized neutral atom density $K_a n_{OS}/K_d$ in Figure 16.6 in the regime $Y_i K_i n_{is} \gg K_d$, which is the usual regime for ion energy-driven etching. For $K_a n_{OS} \ll K_d$, the surface is starved for etchant atoms and both E_h and E_v are determined by the rate of arrival of O atoms to the surface, with $\theta \ll 1$. As $K_a n_{OS}$ is increased beyond K_d, the normalized horizontal (sidewall) etch rate saturates at 1 and $\theta \to 1$, while the vertical etch rate continues to increase linearly with n_{OS}, with $\theta \ll 1$. This is the neutral flux-limited regime of ion energy-driven etching. In turn, the normalized vertical etch rate saturates as $K_a n_{OS}$ is increased beyond $Y_i K_i n_{is}$. In this ion flux-limited regime, both vertical and horizontal surfaces are flooded with O atoms ($\theta \to 1$ for both surfaces), and the vertical etch rate is determined by the rate

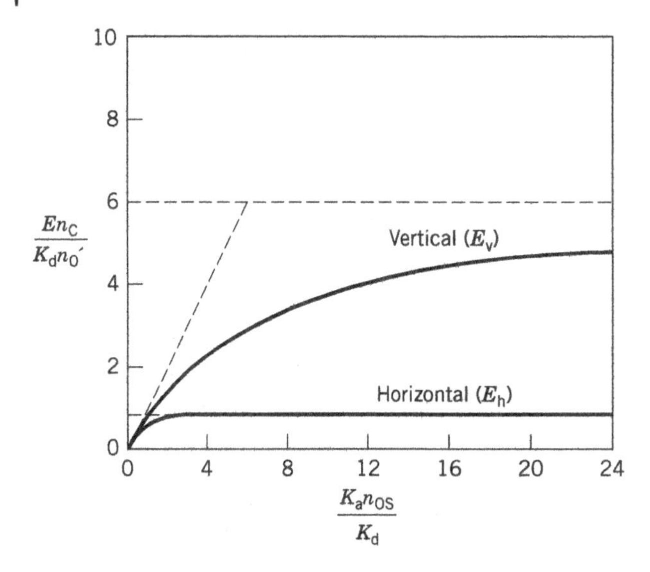

Figure 16.6 Normalized vertical (E_v) and horizontal (E_h) etch rates versus normalized gas-phase density n_{OS}, for $K_i Y_i n_{is}/K_d = 5$.

of arrival of energetic ions to the surface. The etch anisotropy in the regime $K_a n_{OS}$, $Y_i K_i n_{is} \gg K_d$ is

$$a_h = \frac{E_v}{E_h} = \frac{Y_i K_i n_{is}}{K_d} \frac{1}{1 + \dfrac{Y_i K_i n_{is}}{K_a n_{OS}}} \tag{16.2.10}$$

and has its maximum value $Y_i K_i n_{is}/K_d$ when $\theta \to 1$ for both horizontal and vertical surfaces, i.e., $K_a n_{OS} \gg Y_i K_i n_{is} \gg K_d$. In this ion flux-limited regime, high anisotropies can be achieved for high ion energies and fluxes (Y_i and n_{is} large) and low substrate temperatures (K_d small) provided n_{OS} are large enough. In the neutral flux-limited regime $K_d \ll K_a n_{OS} \ll Y_i K_i n_{is}$, the anisotropy is

$$a_h = \frac{K_a n_{OS}}{K_d} \tag{16.2.11}$$

independent of ion energy and flux.

In the usual ion-enhanced regime $Y_i K_i n_{is} \gg K_d$, the etch rate can be written in terms of the ion and neutral fluxes. Using $K_i = u_B/n_0'$ and $K_a = \frac{1}{4}\bar{v}_O/n_0'$ in (16.2.8), we obtain

$$\frac{1}{E_v} = n_C \left(\frac{1}{Y_i \Gamma_{is}} + \frac{1}{\Gamma_{OS}} \right) \tag{16.2.12}$$

Equation (16.2.12) shows that the ion and neutral fluxes and the yield (a function of ion energy) determine the ion-assisted etch rate.

Additional chemistry and physics can be incorporated into such etch models, including sputtering of carbon,

$$\Gamma_C = \gamma_{sput} K_i n_{is} n_0'$$

associative and normal desorption of O atoms,

$$C{:}O \longrightarrow C + O(g)$$

$$2\,C{:}O \longrightarrow 2C + O_2(g)$$

ion energy-driven desorption of O atoms

$$\text{ions} + \text{C:O} \longrightarrow \text{C} + \text{O(g)}$$

formation and desorption of CO_2 as an etch product; and the effect of nonzero ion angular bombardment of sidewall surfaces.

In some etch models, gas-phase etchants are first physisorbed. The physisorbed etchants subsequently react with the surface to form the etch product. The rate-limiting step is still chemisorption of the physisorbed etchants followed by desorption of the etch product. As an example, suppose that the initial step for F-atom etching of some material is physisorption of an F atom on the surface. For Langmuir kinetics, the adsorption–desorption steady state for physisorbed F atoms is described by

$$\frac{d\theta_p}{dt} = K_{ap} n_{FS}(1 - \theta_p) - K_{dp}\theta_p = 0 \tag{16.2.13}$$

which yields the F-atom coverage

$$\theta_p = \frac{K_{ap} n_{FS}}{K_{ap} n_{FS} + K_{dp}} \tag{16.2.14}$$

For physisorption, the activation energy for desorption is low; hence, $K_{dp} \gg K_{ap} n_{FS}$ for the usual discharge conditions so that

$$\theta_p \approx \frac{K_{ap} n_{FS}}{K_{dp}} \ll 1 \tag{16.2.15}$$

Because $\theta_p \propto n_{FS}$, the subsequent chemisorption and rate-limiting desorption steps for the etch product then yield the same dependence on n_{FS} as found in (16.2.8) and (16.2.9).

In some etch models, physisorbed (or gas phase) etchant atoms react *directly* with the surface to form the etch product, which is immediately desorbed. A possible example is the (non-ion assisted) F-atom etching of a fluorinated silicon layer (SiF_x, $x \approx 3$) (Winters and Coburn, 1992). Then, as in (9.4.29)–(9.4.31), the etch rate $E \propto K_r\theta_p$, with K_r the reaction rate coefficient. Since for physisorption the coverage $\theta_p \propto n_{FS}$ up to very high densities, there is no saturation of the etch rate with increasing n_{FS} in this model.

Although such ad hoc etch models can provide insight, they may not be faithful to the actual chemical physics for the etch process. Some of these issues will be addressed in Section 16.3, using the example of F atom etching of silicon, and a more complete model will be described.

16.2.2 Discharge Kinetics and Loading Effect

A general framework for electropositive and electronegative discharge modeling is given in Chapter 10 and applied to various discharges in succeeding chapters. Given the feed gas, gas pressure, power absorbed, and discharge geometry, the self-consistent ion-bombarding fluxes and energies and the neutral etchant densities and fluxes can be estimated using these methods. The complicated nature of the entire problem is illustrated at the end of Section 10.2 for the simplest discharge model of a uniform, electropositive plasma slab in the low-pressure regime. For electrode separation l and area A, we found the ion density (10.2.31) at the plasma–sheath edge and the neutral etchant density (10.2.40) at the surface, as required to determine the etch rates E_v and E_h using, for example, (16.2.8) and (16.2.9). The ion-bombarding energy depends on the type of discharge, as discussed in Chapters 11 and 12 (see also (10.2.4) and following discussion), and the yield scales with energy as in (9.3.14).

Consider, for example, the scaling behavior of the vertical etch rate E_v from (16.2.8) for a high-density source with $\mathcal{E}_i \approx 5T_e$. If P_{abs} is increased then both n_{is} and n_{OS} increase linearly with P_{abs}, and $\mathcal{E}_i$ is unchanged. In the usual ion-driven etch regime $Y_i K_i n_{is} \gg K_d$, we see that E_v increases linearly with P_{abs}. Since both n_{OS} and n_{is} increase, the etch regime (ion flux or neutral flux limited) is not altered. Now consider increasing n_g. From (10.2.30), T_e and therefore $\mathcal{E}_i$ fall slightly; n_{is} remains unchanged, and n_{OS} increases as $n_{OS} \propto n_g^{1-\mathcal{E}_{diss}/\mathcal{E}_{iz}} \propto n_g^{0.5-0.7}$. From Figure 16.6, we see that as n_{OS} increases, the system can enter the high etchant density regime where the etch rate is ion flux limited. Similar scaling laws can be determined for low-pressure capacitive rf discharges (see Problem 11.9) and for high-pressure discharges in which the ion transport is diffusion limited (see Problem 10.10). Systems containing substrate holders that are independently rf biased can be treated similarly.

Consider now the effect on O-atom density of additional O-atom loss at the electrodes due to recombination (with probability γ_{rec}) and due to etch reactions (with probability γ_r) on wafers having a total area A_w, in addition to pumping losses. As in Section 10.2, we assume a symmetric system with $n_{OS} \approx n_O$, $n_e \approx n_i$ (small negative ion density), and we also take γ_{rec}, $\gamma_r \ll 1$. Then, the O-atom rate equation (10.2.38) becomes

$$Al\frac{dn_O}{dt} = 2AlK_{diss}n_g n_i - S_p n_{OS} - (2A - A_w)\gamma_{rec}\frac{1}{4}\bar{v}_O n_{OS} - A_w\gamma_r\frac{1}{4}\bar{v}_O n_{OS} = 0 \tag{16.2.16}$$

which can be solved to obtain

$$n_{OS} = \frac{2Aln_g n_i K_{diss}}{S_p + A\gamma_{rec}\bar{v}_O/2 + A_w(\gamma_r - \gamma_{rec})\bar{v}_O/4} \tag{16.2.17}$$

This can also be written in the form

$$\frac{1}{n_{OS}} = \frac{1}{n_{OS}^{(0)}} + \frac{A_w(\gamma_r - \gamma_{rec})\bar{v}_O}{8Aln_g n_i K_{diss}} \tag{16.2.18}$$

where $n_{OS}^{(0)}$ is the etchant density in the absence of wafers ($A_w \equiv 0$). In the neutral flux-limited regime where the etch rate is proportional to n_{OS} and for the usual case that the reaction probability is large compared to the recombination probability, we see that (16.2.18) leads to a reduction in the etch rate. This is called the *loading effect.* We see from (10.2.37) and (16.2.18) that to minimize the loading effect, $n_g n_i l$ should be large. Depending on the etchant (O, F, Cl, etc.) and the wall and substrate materials, there can be considerable recombination on walls. For F atoms and most wall–substrate systems, it is generally the case that $\gamma_r \gg \gamma_{rec}$ and considerable loading effects are seen. For other etchants (e.g., Cl, Br, O), γ_{rec} can be of the order of or can exceed γ_r. For this case, very weak or even negative loading effects are seen.

The etchant atoms may not be created by a single dissociation reaction having an Arrhenius form, as assumed in (10.2.33). An example is the creation of F atoms in a CF_4 discharge (see Section 16.3).

16.2.3 Chemical Framework

Feedgas mixes for plasma etching are usually complex because of the conflicting requirements on etch rate, selectivity to mask and underlayer, and anisotropy. This is especially true for ion inhibitor processes where a balance must be struck among etchant, inhibitor, and ion fluxes to the substrate. Furthermore, the plasma itself dissociates the feedgas into other, usually more reactive, species. The feedgas and its dissociated products may include chemical constituents such as (Flamm, 1989, Chapter 2):

- Saturates: CF_4, CCl_4, CF_3Cl, COF_2, SF_6, etc.
- Unsaturates: CF, CF_2, CF_3, CCl_3, etc.
- Etchants: F, Cl, Br, O (for resist), F_2, Cl_2, Br_2, etc.
- Oxidants: O, O_2, etc.
- Reductants: H, H_2, etc.
- Nonreactive gases: N_2, Ar, He, etc.

These species react with each other in the gas phase and on the surface in reactions such as the following:

$$\text{e} + \text{saturate} \longrightarrow \text{unsaturate} + \text{etchant} + \text{e},$$
$$\text{etchant} + \text{substrate} \longrightarrow \text{volatile products},$$
$$\text{unsaturate} + \text{substrate} \longrightarrow \text{films}.$$

For some substrates (e.g., SiO_2), unsaturates can themselves be etchants:

$$\text{unsaturate} + \text{substrate} \longrightarrow \text{volatile products}.$$

At low pressures, three body reactions such as

$$\text{etchant} + \text{unsaturate}\,(+\text{M}) \longrightarrow \text{saturate}\,(+\text{M})$$

are not important in the gas phase, but may be important at surfaces or at high gas pressures. If oxidants or reductants are added to the feedgas, commonly O_2 or H_2, respectively, or gases that contain these atoms, then additional reactions can occur:

$$\text{oxidant} + \text{unsaturate} \longrightarrow \text{etchant} + \text{volatile product},$$
$$\text{reductant} + \text{etchant} \longrightarrow \text{volatile products}.$$

The ratio of etchant to unsaturate flux at the substrate is an important process parameter. As will be seen in the next section, a high ratio can lead to isotropic etching, while a low ratio can lead to film deposition. There can be an intermediate ratio in which inhibitor film can be deposited on sidewalls but cleared from trench bottoms by ion bombardment; this is the regime of anisotropic ion-enhanced inhibitor etching. Etchants (Cl_2, Br_2) can be added to the feedgas to increase the etchant/unsaturate ratio, and oxidants (O_2) can also be added to increase the ratio by burning unsaturates to produce etchants, e.g.,

$$CF_3 + O \rightarrow COF_2 + F$$

Conversely, feedgases with low F/C ratios compared to CF_4 (c-C_4F_8, C_3F_8, C_2F_4) and H_2 can be added to reduce the etchant/unsaturate ratio, pushing the system toward increased sidewall protection if that is desired. Inert gas additives are sometimes used to control discharge electrical and substrate thermal properties, dilute etchants, and alter gas-phase chemistry through mechanisms such as Penning ionization and excitation (see Section 8.4). Other additives are sometimes used in etch processes to break through protective oxide layers (e.g., Al_2O_3 for aluminum etching) and to scavenge contaminants (e.g., H_2O) (Flamm, 1989, chapter 2).

16.2.4 Pattern Transfer and Aspect-Ratio-Dependent Etching

Pattern transfer fidelity during etching of features such as trenches and vias (holes) can be a serious issue, in part due to required anisotropies, selectivities, and uniformities. Figure 16.7 illustrates some of these for trench etches. As described earlier, an ideal etch (*a*) has vertical sidewalls and a flat

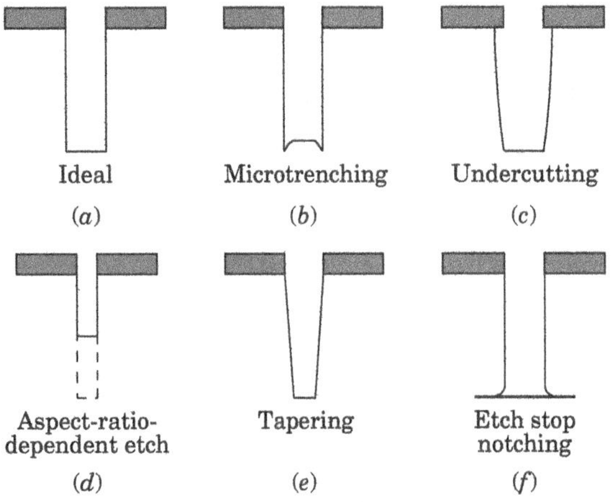

Ideal (a) Microtrenching (b) Undercutting (c) Aspect-ratio-dependent etch (d) Tapering (e) Etch stop notching (f)

Figure 16.7 Illustrations of pattern fidelity during trench etches; (*a*) ideal etch, (*b*) microtrenching, (*c*) mask undercutting, (*d*) aspect-ratio-dependent etch (ARDE), (*e*) tapering, and (*f*) etch stop notching.

bottom. However, glancing sidewall reflections of energetic ions can lead to microtrenches (*b*) in the bottom corners. A small isotropic etching component can lead to mask undercutting (*c*). Vertical etch rates in narrow trenches (*d*) can be less than in wider trenches (*a*), an aspect-ratio-dependent etch (ARDE) effect. Uneven polymer deposition on sidewalls can lead to trench tapering (*e*). Notching (*f*) is sometimes seen at the corners of etch stops, due to surface charging effects. Energetic ion reflections from the mask facets seen in Figure 16.4*b* can lead to "bowed" sidewalls. Some of these effects are considered briefly in the remainder of this chapter. A more thorough examination is given in Donnelly and Kornblit (2013).

16.2.4.1 Aspect-Ratio-Dependent Etching (ARDE)

As one quantitative example, a simple estimate of ARDE was given by Coburn and Winters (1989) for the etch rate at the bottom of a small circular hole of radius R and length L. They assumed a molecular flow regime for etchants ($R \ll \lambda_{\text{mfp}}$), no sidewall etching, and diffuse etchant sidewall reflectivity. Let κ_C be the "transmission probability" that a randomly directed molecule at one end of an open hole will exit the other end, and let s be the etch reaction probability. For an incoming etchant flux Γ_t at the top surface, the returning flux at the top due to sidewall reflection without reaching the bottom is $(1 - \kappa_C)\Gamma_t$, the returning flux at the top due to reflection from the bottom surface is $\kappa_C(1 - s)\Gamma_b$, and the bottom etching flux is $s\Gamma_b$. By flux conservation,

$$\Gamma_t - (1 - \kappa_C)\Gamma_t - \kappa_C(1 - s)\Gamma_b = s\Gamma_b \tag{16.2.19}$$

Assuming top and bottom etch rates $E_t \propto s\Gamma_t$ and $E_b \propto s\Gamma_b$, then solving (16.2.19) for Γ_b/Γ_t yields

$$\frac{E_b}{E_t} = \frac{\kappa_C}{\kappa_C + s(1 - \kappa_C)} \tag{16.2.20}$$

The transmission probability κ_C for molecular flow through an open tube of length L and arbitrary cross section is known as the *Clausing factor* (Clausing, 1932). A simple approximation (Hoffman et al., 1998, p. 38) is

$$\kappa_C \approx \frac{1}{1 + \frac{3}{16}\frac{PL}{A}} \tag{16.2.21}$$

with P and A the tube perimeter and cross-sectional area. For a cylindrical tube, $P/A = 2/R$. A more accurate graph for this case is given, for example, in Solmaz et al. (2020). For a long trench of width w, $P/A = 2/w$.

Note that κ_C is very small for a large aspect ratio cylindrical hole, $L/R \gg 1$. Equation (16.2.20) implies that for any reasonable etch probability, $s \gg \kappa_C$, the bottom-to-top etch rate ratio is much less than unity: $E_b/E_t = (3R)/(8sL)$. In Section 16.5 on atomic layer etching (ALE), we will examine one possible solution to the problem. As we will see in Chapter 17, the same issue arises for conformality in deposition processes, with, again, a possible solution based on atomic layer deposition (ALD).

16.3 Halogen Atom Etching of Silicon

One of the most important applications of plasma etching is the selective, anisotropic removal of patterned silicon or polysilicon films. Halogen atom etchants (F, Cl, Br) are almost always used for this purpose. In fact, F-atom etching of silicon is experimentally the most well-characterized surface etch process and is often used as a paradigm for describing plasma etch processes, as we do here. In this section, we first give a summary of pure chemical and ion-enhanced surface etch processes for F-atom etching. We then describe the discharge chemistry, concentrating on the well-studied CF_4 feedstock system. Finally, we describe silicon etching using other halogen atoms. For more detailed descriptions of silicon etching, the reader should consult the reviews by Flamm (1989, 1990), Winters and Coburn (1992), Donnelly and Kornblit (2013), and Donnelly (2017).

16.3.1 Pure Chemical F-Atom Etching

F atoms are known to spontaneously attack silicon and silicon dioxide in the absence of ion bombardment. The etch rates at high pressures were measured by Flamm et al. (1981) to have roughly an Arrhenius form over a wide range of temperatures and to depend linearly on the gas-phase F-atom density near the surface up to densities as high as 5×10^{15} cm^{-3}. For undoped silicon and for thermally grown silicon dioxide, the etch rates were fit to the relations

$$E_{Si}\ (\text{Å/min}) = 2.86 \times 10^{-12} n_{FS} T^{1/2} e^{-1248/T} \tag{16.3.1}$$

$$E_{SiO_2}\ (\text{Å/min}) = 0.61 \times 10^{-12} n_{FS} T^{1/2} e^{-1892/T} \tag{16.3.2}$$

where n_{FS} (cm^{-3}) is the F-atom density near the surface and T (K) is the surface temperature. The silicon-to-silicon dioxide selectivity is then

$$s = 4.66\, e^{644/T}. \tag{16.3.3}$$

At room temperature (300 K), and for a typical F atom density of 3×10^{14} cm^{-3}, $E_{Si} \approx 230$ Å/min, $E_{SiO_2} \approx 5.9$ Å/min, and $s \approx 40$. There is also good selectivity over Si_3N_4 and reasonable selectivities over resists.

The activation energy of 1248 K in (16.3.1) is also seen in molecular beam experiments (Winters and Coburn, 1992). However, the etch rate in (16.3.1) corresponds to a room temperature F-atom reaction probability $\gamma_r \equiv 4\Gamma_{SiF_4}/\Gamma_F \approx 0.0017$, where $\Gamma_{SiF_4} = E_{Si}\, n_{Si}$, $\Gamma_F = \frac{1}{4} n_{FS} \bar{v}_F$, $n_{Si} = 5.0 \times 10^{28}$ m^{-3}, and $\bar{v}_F = (8kT/\pi M_F)^{1/2}$ m/s. (Here, E_{Si} is in units of m/s.) At lower pressures with $n_{FS} = 10^{18}$–10^{20} m^{-3}, Ninomiya et al. (1985) measured $\gamma_r \approx 0.05$–0.1, depending on the F-atom flux, about 25–50 times larger. These large values are in reasonable agreement with molecular dynamics simulations (Humbird and Graves, 2004), giving $\gamma_r \approx 0.1$. On the other hand, some molecular beam experiments suggest intermediate values $\gamma_r \approx 0.01$–0.08.

A major source of these discrepancies is the evidence, over a wide variety of studies, that the reaction probability γ_r decreases with F-atom flux (Donnelly, 2017). To quantify this, we introduce the probability $\gamma_{Si/F} = \Gamma_{Si}(tot)/\Gamma_F(tot)$ of etching a silicon atom by an incident fluorine atom. Here, $\Gamma_{Si}(tot)$ is the silicon atom flux in all of the outgoing etch products, and $\Gamma_F(tot)$ is the fluorine atom flux in all of the incident etch precursors. Note that $\gamma_{Si/F}$ is independent of both the distribution of silicon-containing etch products (SiF_4, SiF_2, Si_2F_6, etc.) and the source of the incident fluorine atoms (F, F_2, NF_3, CF_4/O_2, SF_6, etc.). With SiF_4 as the sole etch product and F-atoms as the sole fluorine source, $\gamma_{Si/F} = \frac{1}{4}\gamma_r$.

Plotting $\gamma_{Si/F}$ versus $\Gamma_F(tot)$ for a variety of data gives the result shown in Figure 16.8. There is an almost 30-fold decrease in $\gamma_{Si/F}$ from the low flux molecular beam experiments and molecular dynamics simulations, to the high flux, flowing afterglow experiments of Flamm et al. (1981). These results span a variety of incident F-atom sources. The one exception, shown in the figure (upper right-hand corner), is the very high $\gamma_{Si/F}$ for high flux SF_6 feedstock. This was investigated by Arora et al. (2019), who showed that adsorbed sulfur acts as a catalyst to strongly increase $\gamma_{Si/F}$. They ascribe the enhancement to an increase in the weakly bound F-atom sticking coefficient and/or to a reduction in its desorption rate.

Discrepancies in the various measurements of $\gamma_{Si/F}$ could also be affected by carbon and oxygen surface contamination, residual ion bombardment from the F-atom (plasma) source, the existence of super-thermal F-atoms, and the unknown distribution of etch products (Ninomiya et al., 1985; Herrick et al., 2003).

The very high values of $\gamma_{Si/F}$ at high F-atom fluxes in SF_6 have led to its use for deep reactive ion etching of trenches and holes in silicon by a two-step cyclical gas exchange process, called the *Bosch process*. For the first step, a somewhat anisotropic ion-assisted etch is done in SF_6, lasting several seconds. For the second step, also typically seconds, a fluorocarbon plasma such as C_4F_8 is used to deposit a protective film on the sidewalls and the bottom. In the next and succeeding cycles, the SF_6 etch breaks through the bottom film to continue the downward etch, while the sidewall film significantly inhibits sidewall etching. Deep reactive ion etching processes are extensively used

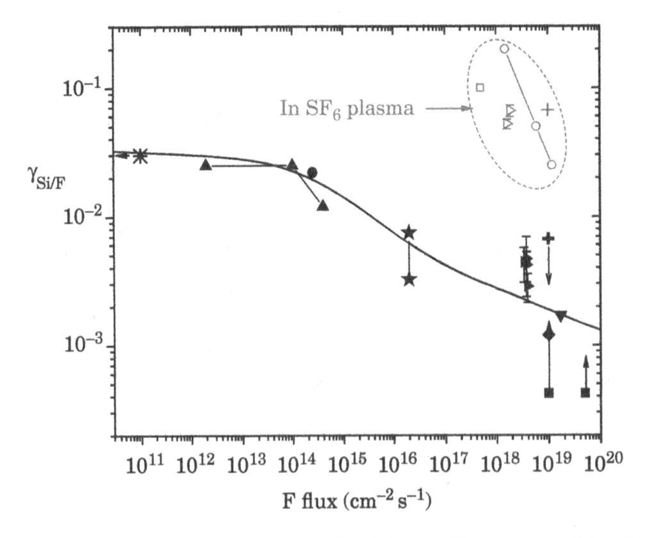

Figure 16.8 Probabilities $\gamma_{Si/F}$ of etching a silicon atom with a fluorine atom, versus the total fluorine atom flux $\Gamma_F(tot)$, from various published sources; Source: Donnelly (2017)/AIP Publishing/CC BY 4.0.

in microelectromechanical systems (MEMS) and wafer through-hole fabrications and can involve many hundreds of such cyclical steps.

The mechanism for pure chemical F-atom etching of silicon has been studied for over 30 years and is still not thoroughly understood. In the steady state, a fluorinated silicon SiF_x layer 2–5 monolayers thick forms at the surface. The fluorine-to-silicon ratio at the top of this layer is typically 3:1 (mostly SiF_3), and the ratio falls smoothly to zero at the SiF_x–Si interface. The layer thickness varies with etch conditions, typically the film is thin when the etch rate is high, and vice versa. The layer is stable at room temperature, i.e., if the incident flux of F atoms is terminated after the layer forms, then etching ceases. If the layer is then heated, it does not begin to decompose until temperatures of 300–400 °C have been reached. The decomposition products are SiF_2(g) and SiF_4(g), with the former being the most important. During etching, some etch product measurements indicate that roughly 65% of the etch product at room temperature is SiF_4(g), with Si_2F_6(g) and Si_3F_8(g) comprising the remaining product. There is still some controversy whether SiF_2(g) etch product is formed at room temperature; some experiments indicate a dependence of SiF_2(g) formation on the F-atom flux (Donnelly, 2017). As the temperature is raised, the SiF_4(g) percentage slowly increases to 80–90% of the total at 300 °C, with a corresponding reduction in Si_2F_6 and Si_3F_8. SiF_4 then begins to decrease and becomes a minor product above 600 °C. At high temperatures, the dominant etch product is SiF_2(g), which increases from 5% to 10% of the total at 300 °C up to 40–50% at 600 °C.

The formation of a steady-state SiF_x layer whose thickness varies inversely with the etch rate is suggested by a model in which diffusion of F atoms (or, as will be seen below, F^- ions) into the surface is balanced by an erosion of the surface due to the etching. Let n_F be the volume density of diffusing F atoms in the solid and E_{Si} (m/s) be the etch rate, then in the frame $x' = x - E_{Si}t$ moving with the etched surface at $x' = 0$, the flux vanishes:

$$\Gamma'_F = -D\frac{dn_F}{dx'} - n_F E_{Si} = 0 \tag{16.3.4}$$

For a crude model in which the diffusion coefficient D is a constant, independent of n_F, this can be solved to obtain

$$n_F = n_{F0} \exp\left(-\frac{E_{Si}x'}{D}\right) \tag{16.3.5}$$

yielding a layer having characteristic thickness D/E_{Si}. This type of model is suggestive of the more complicated kinetics, including both diffusion and reactions within the solid, that lead to the formation of the SiF_x layer.

At low F-atom fluxes, the measured linear dependence of the etch rate on n_{FS}, shown in Figure 16.8, is suggestive of weakly bound adsorbed F atoms as precursors to subsequent etch reactions. As pointed out in Section 16.2, the adsorption–desorption kinetics leads to a surface coverage for weakly adsorbed F atoms that are linear in the gas-phase F-atom density n_{FS} near the surface, as given by (16.2.15). The Arrhenius form of (16.3.1) at first sight suggests a single activated process for the etch reaction, but the belief is that this is probably fortuitous. The etch product distributions and the decomposition properties of the SiF_x layer differ greatly at low and at high temperatures, implying that the etch mechanisms differ also. In addition, weakly and strongly non-Arrhenius etch rates are seen for etching of silicon with F_2 and with XeF_2, respectively. The existence of an activation energy of 1248 K (≡ 0.108 V) for F-atom etching of silicon is not understood.

A significant feature of pure chemical etching using halogen atoms, known as the *doping effect,* is that the etch rate depends on the silicon doping levels, with n-type, or highly doped (n^+-type) silicon etching faster than intrinsic (i-type) or p-type. The dopants must be thermally activated in

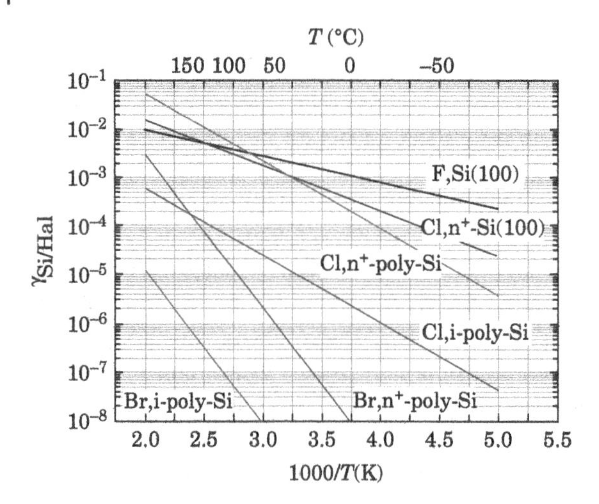

Figure 16.9 Probabilities $\gamma_{Si/Hal}$ of etching a silicon atom with a halogen atom, versus $1000/T(\mathrm{K})$, from various published sources. Source: Reproduced from Donnelly and Kornblit (2013)/with permission of AIP Publishing.

order to restore the crystalline structure. The effect is weak for F atoms, with a factor of two difference in etch rates, but is very strong for Cl atoms, where the etch rates can differ by several orders of magnitude. Figure 16.9 shows Arrhenius fits to the probability $\gamma_{Si/Hal}$ versus the Arrhenius temperature parameter $1000/T(\mathrm{K})$, for various halogen atom etch precursors, showing these effects.

The existence of a doping effect suggests that negative ion centers on the silicon surface play an important role in etch reactions. The electron affinity of an F atom in free space is $\mathcal{E}_{aff} \approx 3.45$ V, but near the surface, the affinity is increased by the energy[1] of the electrostatic image force (see (9.3.2)). For an F atom a distance $a_{eff} = 1$ Å from the SiF_x surface, (9.3.2) yields a large affinity $\mathcal{E}_{aff} \approx 3.45 + 3.60 \approx 7.05$ V. SiF_3 similarly has a large affinity. Hence, negative ion formation at the surface is favored. Winters and Haarer (1987) suggest that the rate-limiting etching step involves reaction of adsorbed F atoms at negative charge centers (areal density n'_-) on the SiF_x surface. The negative charge is supplied by electrons tunneling from the silicon substrate through the SiF_x layer. The variation of the charge density on the surface with doping level accounts for the etch rate variation in this model, with the room-temperature etch rate proportional to n'_-, to n_{FS} through (16.2.15), and to the surface density of SiF_3:

$$E_{Si} = K_r n'_- n'_{SiF_3} n_{FS} \tag{16.3.6}$$

where K_r is the rate constant. The linear dependence on n'_{SiF_3} arises because some fraction (roughly 1/2–1/3 for a fully fluorinated surface) of the adsorbed F atoms activated at the negative charge centers are presumed to attack the Si–SiF_3 bond holding an SiF_3 group to the surface, thus forming the SiF_4 etch product. Smaller concentrations of Si–SiF_2SiF_3 and Si–$SiF_2SiF_2SiF_3$ bonds are also attacked, yielding the observed Si_2F_6 and Si_3F_8 etch products in lesser concentrations. The remaining fraction (1/2–2/3) of the activated F atoms breaks Si–Si bonds within the layer, leading to the growth of the layer. The activation mechanism is likely to be the formation of negative ions,

$$\mathrm{e} + \mathrm{F{:}S} \rightarrow \mathrm{F^-{:}S}$$

Such a negative ion sees a strong image force directed into the surface, promoting lattice penetration. Once inside the lattice, any F atom produced by neutralization of F^- is likely to attack an

1 The energy for a surface having dielectric constant $\epsilon \gg 1$ is the same as that for a perfectly conducting surface (Ramo et al., 1994, chapter 1).

Si–Si bond. Winters and Coburn (1992) claim that the rate expression (16.3.6) is consistent with all experimental data on F-atom chemical etching of silicon and is also consistent with data on F_2 and XeF_2 etching. Additional information on the doping effect is given in Winters et al. (2007).

16.3.2 Ion Energy-Driven F Atom Etching

Etch rates for a given F-atom flux can be increased by a factor of 5–10 for sufficiently high fluxes (and energies) of bombarding ions. A single 1 kV Ar^+ ion can cause the removal of as many as 25 silicon atoms and 100 F atoms from the surface. The total surface concentration of fluorine in the SiF_x layer is reduced by up to a factor of two in the presence of ion bombardment. Furthermore, the etch product distribution changes, and, notably, a significant fraction of SiF_2(g) etch product is formed. Although the etch anisotropy can be as high as 5–10, this still implies a reasonably large pure chemical etch rate on trench sidewalls. Consequently, fluorine-based anisotropic silicon etches are not commonly used. Although inhibitor chemistries can be employed for sidewall protection and increased anisotropy, as we subsequently show, the contamination produced by the protective films is undesirable, and other halogen etch chemistries are generally used for strongly anisotropic etches.

Many mechanisms have been proposed to explain the enhanced etch rate due to ion energy-driven F atom etching of silicon, including the following:

1. *Formation of a damaged region that is more reactive to subsequently arriving fluorine.* However, experiments show that the energetic ion bombardment influences the fluorine that is present within the SiF_x layer at the moment of impact. Hence, this is probably not the mechanism for F-atom etching of silicon. However, it is known that lattice damage is an important mechanism for some systems, e.g., ion-enhanced XeF_2 etching of tungsten.
2. *Temperature increase due to etch reactions or ion bombardment.* The temperature rise is not large enough.
3. *Chemically enhanced physical sputtering.* For this proposed mechanism, the binding energies $\mathcal{E}_t$ of Si–SiF_2 and Si–SiF_4 bonds at some locations on the SiF_x surface are supposed to be much smaller than those of a pure silicon surface, yielding significant physical sputtering rates for SiF_x in the presence of ion bombardment. However, the binding energies would have to be of order 0.3–0.5 V, much smaller than the usual Si–Si or Si–F bond energies, in order that the species remain on the surface without thermally desorbing and yet be easily sputtered (e.g., see (9.3.14)). In addition, time-resolved etch rate measurements using modulated beams of ions show that ion-enhanced etching is much slower than predicted by a physical sputtering model. Hence, the evidence suggests that chemically enhanced physical sputtering is not the major contributor to ion-enhanced etching at the neutral–ion flux ratios typically found in etching discharges. This conclusion is supported by molecular dynamics simulations (Barone and Graves, 1995).
4. *Chemical reaction and desorption due to ion bombardment.* In this mechanism, sometimes called chemical sputtering, the ion bombardment causes a chemical reaction to occur that produces an easily desorbed etch product. When an energetic ion collides with and penetrates the SiF_x layer, producing a collision cascade as for physical sputtering (see Section 9.3), then a large number of Si–Si and Si–F bonds can be broken and reformed, leading to molecules such as SiF_4 and SiF_2 that are weakly bound to the surface. These molecules can thermally desorb during or after the collision cascade. This mechanism on the surface is similar to a reaction in the gas phase such as

$$\text{ion} + Si_2F_6 \longrightarrow SiF_4 + SiF_2 + \text{ion} \tag{16.3.7}$$

It is likely to be an important mechanism for ion energy-driven F-atom etching of silicon and is seen to be important in molecular dynamics simulations (Barone and Graves, 1995).

5. *Enhanced chemical etching.* In this mechanism, ion bombardment reduces the layer thickness, thus increasing the pure chemical etch rate given by (16.3.6), because n'_- increases. It is known experimentally that the pure chemical etch rate varies inversely with the layer thickness and that ion bombardment reduces this thickness. Hence, this mechanism can contribute to ion-enhanced etching. However, the etch rate enhancement is unlikely to exceed a factor of two for typical plasma etch conditions.

For the simple phenomenological models for ion-assisted etching described in Section 15.2, the increased etch rate is due to ion-enhanced desorption of etch products or ion-enhanced reaction of etchants with the surface. Gray et al. (1993) have developed a more complete model for Ar^+-enhanced F-atom etching of silicon (and silicon dioxide) that is consistent with high flux ion and atomic beam studies and with other data. The kinetics includes the following processes for silicon etching:

1. Physisorption and thermal desorption of F atoms

$$F(g) + S \underset{K_{d1}}{\overset{K_{a1}}{\rightleftharpoons}} F{:}S$$

2. Chemisorption of physisorbed F atoms at silicon dangling bond (Si*) sites

$$2\,F{:}S + Si^* \xrightarrow{K_{a2}} SiF_2{:}S$$

3. Ion-induced desorption of SiF_2

$$SiF_2{:}S \xrightarrow{Y_d K_i} SiF_2(g) + 2\,Si^*$$

4. Creation of SiF_4 by ion beam mixing followed by ion-induced desorption

$$2\,F{:}S + SiF_2{:}S \xrightarrow{Y_{mix} K_i} SiF_4(g) + 2\,Si^*$$

5. Physical sputtering of silicon

$$Si^* \xrightarrow{Y_{sput} K_i} Si(g) + Si^*$$

6. Chemical etching of silicon

$$2\,F{:}S + SiF_2{:}S \xrightarrow{K_r} SiF_4(g)$$

In this model, the ion bombardment increases the etch rate due to increases in the chemisorption site density n'_{Si^*} (mechanism 1), the rate of chemically enhanced physical sputtering of etch products (mechanism 3), and the rate of chemical reaction and thermal desorption of etch products (mechanism 4).

16.3.3 CF_4 Discharges

Because F_2 is difficult and dangerous to handle and itself etches silicon, generally leaving a rough and pitted surface, it is not often used as a feedstock. Common feedstock gases include CF_4, SF_6, and NF_3, along with low F/C feedstock additions such as C_2F_6. The most well-studied system is CF_4, which we use to illustrate the effects of gas- and surface-phase chemistry in discharges used for F-atom etching of silicon. The overall etch reaction in a CF_4 discharge is

$$4\,CF_4 + Si \longrightarrow 2\,C_2F_6 + SiF_4$$

Hence, the major effluent gases observed are CF_4, SiF_4, and C_2F_6.

CF_4 is a very stable tetrahedral molecule (symmetry group T_d with the carbon atom in the center) with an enthalpy of formation of −925 kJ/mol and a C–F bond distance of 1.3 Å. Its vibration frequencies ($\hbar\omega_{\text{vib}}/e$ in voltage units) are 0.054 V (doubly degenerate), 0.078 V (triply degenerate), 0.113 V (singly degenerate), and 0.159 V (triply degenerate), and its rotation constant is $B_{\text{rot}} = 2.4 \times 10^{-5}$ V (triply degenerate) for each of the three degrees of rotational freedom. The CF_3–F bond energy is 5.6 V. All excited states of CF_4 are repulsive; consequently, all electronic excitations of CF_4 are dissociative. In particular, the positive ion CF_4^+ is not stable.

The CF_3 radical is weakly bound and large, having pyramidal symmetry C_{3v}; the carbon atom is at the top of a flat pyramid with an equilateral triangular base of F atoms. The F–F bond distance is 4.1 Å and the C atom is 0.75 Å above the base. The CF_2–F bond energy is 3.8 V. CF_3 is electronegative with an electron affinity of 1.9 V, lower than the affinity 3.45 V of F atoms. The CF_3^- ion also has symmetry C_{3v}, with F–F bond distance and C atom height of 4.1 and 1.3 Å, respectively. The linear radical CF_2 has a CF–F bond energy of 5.8 V. The bond energy of the CF radical is 5.1 V. Some threshold energies and rate constants for important electron–CF_x reactions are given in Table 16.2. Reviews of electron collisions with CF_4 are given by Christophorou et al. (1996) and Christophorou and Olthoff (1999b). Toneli et al. (2019) give a complete CF_4-feedstock global model, including a detailed discussion of the chemistry. They identify 19 important species and give the important reactions among them, along with their rate coefficients.

Three-body neutral–neutral gas-phase recombination reactions among F atoms and CF_x radicals, as in (9.2.28), can be important because of the large size of the CF_x radicals and the large enthalpies of formation, in particular for C_2F_6 and C_2F_4 products. At the pressures of interest for etching, these reactions can be in the intermediate regime between low and high pressures. The effective two-body rate constant K_{AB} in the intermediate regime is, from (9.2.33),

$$K_{AB} = \frac{K_3' n_M}{1 + K_3' n_M / K_2} \tag{16.3.8}$$

where K_3' is defined by (9.2.36). Some values of K_2 and K_3' for important three-body association reactions are given in Table 16.3 (Plumb and Ryan, 1986). Important electron collision reactions in CF_4 discharges are reaction 1 in Table 16.2 for F^- and CF_3^- creation, reaction 4 for CF_3^+ creation, and reactions 2 and 3 for CF_3, CF_2, and F creation. The recombination reactions 1 and 4 in Table 16.3 can be strong at moderate gas pressures. In view of these and the 1:2.5 branching ratio between CF_3/CF_2 production from electron collisions with CF_4, the CF_2 radical density can much exceed the CF_3 radical density. The production of C_2F_6 and C_2F_5 by recombination reactions 4 and 5 is balanced by electron dissociation of these molecules into CF_3 and CF_2 products, by F-atom abstraction

Table 16.2 Selected Second-Order Reaction Rate Constants for Electron Impact Collisions in CF_4 Discharges

Number	Reaction	Rate Constant (cm³/s)	Source
1	$e + CF_4 \rightarrow CF_3 + F^-$ / $CF_3^- + F$	$4.6E{-}9\,T_e^{-3/2}\exp(-7/T_e)$	a)
2	$e + CF_4 \rightarrow CF_3 + F + e$	$2E{-}9\exp(-13/T_e)$	b)
3	$e + CF_4 \rightarrow CF_2 + 2F + e$	$5E{-}9\exp(-13/T_e)$	b)
4	$e + CF_4 \rightarrow CF_3^+ + F + 2e$	$1.5E{-}8\exp(-16/T_e)$	a)

Notes. T_e between 3 and 6 V. The notation E−9 means 10^{-9}.
a) Based on cross sections of Hayashi (1987).
b) Based on data of Plumb and Ryan (1986).

Table 16.3 Selected Values of Rate Constants K_2 and K_3' for Association Reactions in CF_4 Discharges

Number	Reaction	K_2 (cm^3/s)	K_3' (cm^6/s)
1	$F + CF_3 + M \rightarrow CF_4 + M$	2E–11	7.7E–27
2	$F + CF_2 + M \rightarrow CF_3 + M$	1.3E–11	3.0E–29
3	$CF + F + M \rightarrow CF_2 + M$	1E–11	3.2E–31
4	$CF_3 + CF_3 + M \rightarrow C_2F_6 + M$	8.3E–12	2.8E–23
5	$CF_2 + CF_3 + M \rightarrow C_2F_5 + M$	1E–12	2.3E–26

Notes. Here, M represents CF_4. The notation E–11 means 10^{-11}.
Source: Adapted from Plumb and Ryan (1986).

reactions, and by the flow of these multicarbon radicals to the walls. The latter can bring substantial carbon fluxes to the walls. As for O_2 discharges, negative ions are lost by positive–negative ion recombination, and, possibly, by electron impact detachment or associative detachment in the volume, and positive ions are lost by this same recombination in the volume and by flow or diffusion to the walls.

Etchant atoms can be lost to a surface by a two-step process of fast adsorption

$$F(g) + S \longrightarrow F{:}S$$

followed by recombination,

$$F(g) + F{:}S \longrightarrow F_2(g) + S$$

or by reaction with the surface, e.g.,

$$F{:}S + SiF_x \longrightarrow SiF_4(g)$$

Atoms that are not lost are desorbed back into the discharge. For fluorine atoms, as shown in Problem 9.11, the probability γ_{rec} that atoms adsorb and recombine on most surfaces is generally small at the substrate temperatures and atom densities characteristic of etching discharges. For example, F atoms incident on Al_2O_3, SiO_2, Pyrex, Teflon, stainless steel, Mo, Ni, and Al–0.1%Cu have recombination probabilities $\gamma_{rec} \sim 10^{-4}$–$10^{-3}$ at 300 K (Flamm, 1989). However, there are some exceptions: $\gamma_{rec} \gtrsim 0.01$, 0.05, and 0.2 for F atoms on Cu, brass, and Zn, respectively. For chlorine and oxygen atoms, recombination probabilities can be larger than those for fluorine atoms, up to ~0.1 for many surfaces. The reaction probability γ_r for F atoms is negligible on most surfaces, but can be significant for some surfaces, e.g., $\gamma_r \approx 0.0017$ and 1 for Si and BN, respectively. In many cases, except by design (i.e., substrates to be etched), etchant atoms incident on surfaces in processing discharges are recycled back into the discharge as atoms.

The behavior of free radical molecules incident on surfaces is more complicated due to their possible dissociation. On nonactive surfaces, free radicals can be adsorbed without dissociation. Their subsequent probabilities for recombination or reaction with the surface are generally small, as for etchant atoms. An example is the adsorption of CF_3 on SiO_2, which is nondissociative. The recombination and reaction probabilities are small, so most CF_3 radicals incident on SiO_2 desorb as CF_3 radicals.

However, on active surfaces such as pure silicon, CF_3 and CF_2 radicals generally dissociatively adsorb, producing a C atom and three (or two) F atoms that each bond to the silicon. Although these radicals deliver etchant atoms to the surface, they also deliver C atoms, which can form a protective film on the surface that inhibits the etch reaction. Similarly, CF_3 and CF_2 can dissociatively or nondissociatively adsorb on an SiF_x layer, leading to a buildup of carbon or polymer film. It is unlikely that the film will be removed from the surface in the absence of ion bombardment except as CF_4(g). Hence, the flux of CF_x radicals ($x < 4$) reduces the silicon etch rate. If Γ_{CF_x} is the *net* flux of CF_x adsorbed, Γ_F is the *net* flux of F atoms adsorbed, Γ_{SiF_4} is the flux of SiF_4 desorbed, and Γ_{CF_4} is the flux of CF_4 desorbed, then conservation of C atoms on the surface requires $\Gamma_{CF_4} = \Gamma_{CF_x}$, and conservation of F atoms on the surface requires

$$x\Gamma_{CF_x} + \Gamma_F = 4\Gamma_{CF_x} + 4\Gamma_{SiF_4} \tag{16.3.9}$$

Solving for the etch rate, we obtain

$$E_{Si} = \frac{\Gamma_{SiF_4}}{n_{Si}} = \frac{\Gamma_F - (4-x)\Gamma_{CF_x}}{4n_{Si}} \tag{16.3.10}$$

For $\Gamma_F < (4-x)\Gamma_{CF_x}$, there is net deposition of carbon and the etch rate is zero (Problem 16.3). For $x \approx 3$, we see that the condition for etching is $\Gamma_F > \Gamma_{CF_x}$.

Ion bombardment can shift the balance in (16.3.10) in one of two ways: (1) it can increase the ratio Γ_F/Γ_{CF_x} of net fluxes adsorbed. (2) It can lead to desorption of CF_y, $y < 4$, due to physical sputtering of CF_y polymer (and also of C) and due to ion energy-driven etching of CF_y polymer, in the same manner that ion energy-driven F atom etching of silicon leads to SiF_2 etch product release from the SiF_x surface. For desorption of CF_y, (16.3.10) is replaced by

$$E_{Si} = \frac{\Gamma_{SiF_4}}{n_{Si}} = \frac{\Gamma_F - (y-x)\Gamma_{CF_x}}{4n_{Si}} \tag{16.3.11}$$

For $y \leq x$, there is always etching.

Under fluorine-poor discharge conditions or energetic ion bombardment, spatially hollow profiles of CF and CF_2 radicals have been measured (Booth et al., 1999; Cunge and Booth, 1999). This suggests that the surfaces in CF_4 discharges are net sources of these radicals. The data and models (Zhang and Kushner, 2000) suggest that the mechanism involves the reaction of incoming CF_3^+ ions, multicarbon (C_2F_x etc.) radicals, and, possibly, CF_3 radicals with the fluorinated surface, producing the CF and CF_2 products.

Evidently, the ratio F/C of fluorine atoms to CF_x radicals in the discharge is an important process parameter in determining whether etching or film deposition occurs. Figure 16.10 gives an illustrative picture (not quantitative) of the boundary between etching and deposition as the F/C feedstock gas ratio and the bias voltage (ion-bombarding energy) are varied in typical fluorocarbon discharges. We describe methods for varying the F/C ratio below. The general trends indicated in this picture follow from (16.3.11). For F/C > 3, there is etching independent of bias voltage, and so both trench sidewalls and bottoms are etched. Although the horizontal (sidewall) rate is not ion assisted and can be small compared to the vertical rate, the sidewall is not protected by inhibitor film, and the etch anisotropy is not large. For 2 < F/C < 3, the sidewalls are protected by inhibitor film, but the ion bombardment exposes the trench bottoms to the etchants; this is the regime of highly anisotropic etching using fluorocarbon feedstocks. For F/C < 2, there is film deposition on both sidewall and bottom and etching ceases.

The loading effect is described in Section 16.2. For a fixed feedstock flow rate, the F-atom density in the discharge is depressed as the area of silicon being etched increases, due to the formation of SiF_4 etch product. Hence, the F/C ratio decreases and the equilibrium for the system shifts toward the left, as indicated in Figure 16.10. This can lead to polymer formation under heavy loading conditions.

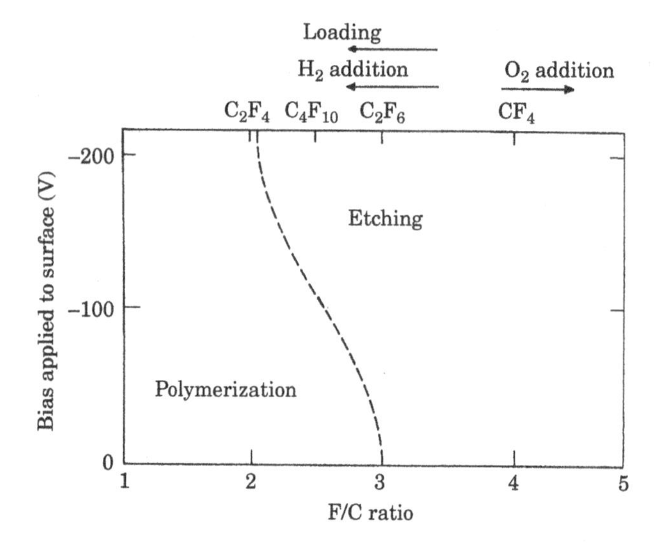

Figure 16.10 The influence of fluorine to carbon (F/C) ratio and electrode bias voltage on etching and polymerization processes in a fluorocarbon discharge. Source: Coburn and Winters (1979)/with permission of Springer Nature.

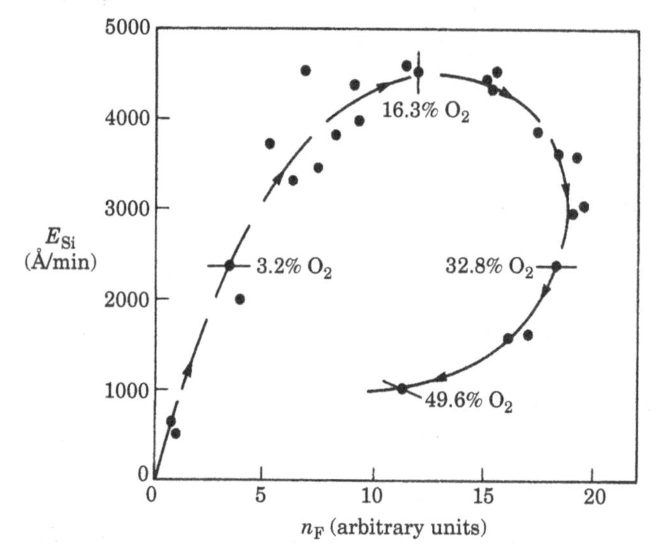

Figure 16.11 Locus of silicon etch rate E_{Si} and F-atom concentration n_F as the %O_2 is varied in a CF_4/O_2 feedstock mix. Source: Mogab et al. (1979)/with permission of American Institute of Physics.

16.3.4 O_2 and H_2 Feedstock Additions

Oxygen gas is often added to the feedstock. Figure 16.11 gives the variation of the silicon etch rate E_{Si} and the F atom concentration n_F versus %O_2 for a CF_4/O_2 feedstock mix in a capacitive rf discharge reactor (Mogab et al., 1979). O_2, CO_2, CO, and COF_2 are now seen in addition to CF_4, SiF_4, and C_2F_6 effluents. Three different regimes are observed:

1. Up to roughly 16% O_2, E_{Si} and n_F increase with %O_2.
2. Between 16 and 30% O_2, E_{Si} decreases with %O_2, although n_F continues to increase.
3. Above roughly 30% O_2, both E_{Si} and n_F decrease.

It is generally agreed that the first regime of increasing E_{Si} and n_F with $\%O_2$ is due to O atom (and, possibly, O_2 molecule) "burning" of CF_x unsaturates either in the gas phase or on the surface:

$$\begin{aligned} O + CF_3 &\longrightarrow COF_2 + F \\ O + CF_2 &\longrightarrow CO + 2F \\ &\longrightarrow COF + F \\ &\longrightarrow COF_2 \\ O + COF &\longrightarrow CO_2 + F \\ O + C &\longrightarrow CO \\ &\text{etc.} \end{aligned} \tag{16.3.12}$$

In addition to destroying CF_x radicals, many of these reactions produce F atoms, thus increasing n_F. Furthermore, the *net* flux Γ_O of O atoms (or O_2 molecules) adsorbed on the surface modifies the etch rate by removing adsorbed carbon from the surface:

$$E_{Si} = \frac{\Gamma_F + \Gamma_O - (y - x)\Gamma_{CF_x}}{4\,n_{Si}}, \qquad \Gamma_O < (y - x)\Gamma_{CF_x} \tag{16.3.13}$$

Hence, E_{Si} increases with $\%O_2$ because both Γ_F and Γ_O increase and because Γ_{CF_x} decreases. This shift in equilibrium to the right with O_2 addition is indicated in Figure 16.10.

The second regime of increasing n_F and decreasing E_{Si} is believed to be due to the competition of O atoms for chemisorption sites on the SiF_x lattice. Hence, in this regime, the surface layer becomes more "oxidelike," reducing the etch rate. In this regime, there is no C on the surface, and a crude model for the etch rate gives

$$E_{Si} = \frac{\Gamma_F}{4\,n_{Si}(1 + \eta_O\Gamma_O/\Gamma_F)} \tag{16.3.14}$$

where η_O gives the competitive efficiency for O atoms over F atoms to be adsorbed. The third regime in which both n_F and E_{Si} decrease is believed to be due to oxygen dilution effects, i.e., the flow of F atoms into the discharge is reduced by the $\%O_2$ in the feedstock.

The chemistry of CF_4/O_2 discharges is extremely complicated. In addition to the CF_4 reactions listed in Tables 16.2 and 16.3, and the neutral chemistry (16.3.12) in the gas phase and on the surface, electron dissociation of O_2, COF_2, and CO_2 is important, and some three-body gas-phase reactions, e.g.,

$$COF + F + M \longrightarrow COF_2 + M$$

might also be significant at high pressures. A fairly complete model of the gas-phase chemistry has been developed (Plumb and Ryan, 1986).

Hydrogen gas is sometimes added to the feedstock mix. The key additional reaction on the surface or in the gas phase is

$$H + F \longrightarrow HF$$

which reduces the F-atom concentration, thus shifting the equilibrium to the left in Figure 16.10, toward increasing polymer formation.

16.3.5 Cl-Atom Etching

16.3.5.1 Pure Chemical Etching

Chlorine atoms differ from fluorine atoms in two major respects for pure chemical silicon etching: (1) there are pronounced crystallographic effects, and (2) there is a large doping effect. Figure 16.12 shows the pure chemical etch rate E_{Si} at 400 K for doped silicon as a function of doping concentration n_D for various crystallographic conditions. The etch rates fit a generalized Arrhenius form (Ogryzlo et al., 1990):

$$E_{Si}(\text{Å/min}) = An_D^{\gamma} n_{Cl,S} T^{1/2} e^{-B/T} \tag{16.3.15}$$

where the parameters A, B, and γ are given in Table 16.4. The temperature variations are shown in Figure 16.9. The activation energy $\mathcal{E}_a = kB/e \approx 0.19$ V is roughly independent of doping level and crystallographic orientation. The very strong dependence on n_D indicates that Cl^- ions formed on the surface must play a critical role in Cl-atom etching, as was found for F^- ions. The dependence of the etch rate on crystallographic orientation can be ascribed to the different area densities of silicon atoms (or Si–Si bonds) at the surface. The 111 orientation has a higher density than the 100 orientation, which reduces Cl or Cl^- penetration into the lattice for 111, leading to a lower etch rate.

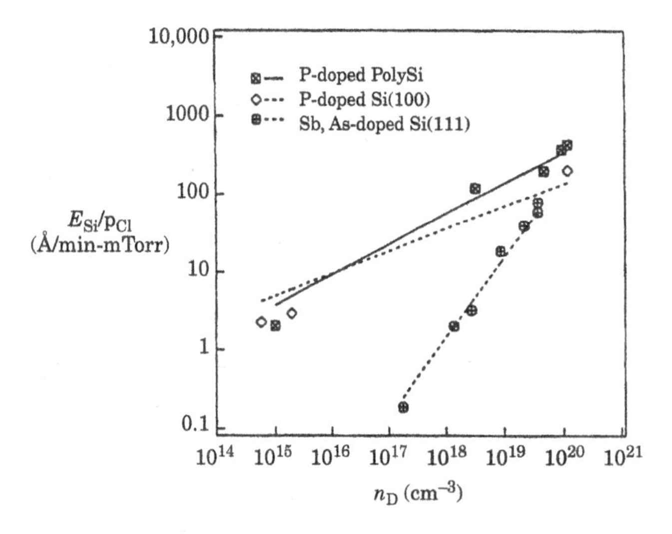

Figure 16.12 Etch rate E_{Si} versus doping level n_D and crystallographic orientation for Cl atom etching of n-type silicon at 400 K; p_{Cl} is the partial pressure of Cl atoms. Source: after Ogryzlo et al. (1990)/with permission of American Institute of Physics.

Table 16.4 Coefficients of the Modified Arrhenius Form for Cl Atom Etching of n-Type Silicon

Crystallographic Orientation	A (Å-cm$^{3+3\gamma}$/min-K$^{1/2}$)	B (K)	γ
Polysilicon	4×10^{-18}	2365	0.39
⟨100⟩	1.1×10^{-17}	2139	0.29
⟨111⟩	1.6×10^{-31}	2084	1.03

Source: Adapted from Flamm (1990).

Exposure of a pure silicon surface to Cl_2 leads to dissociative chemisorption which saturates at about one monolayer. Continued exposure can lead to a slow growing silicon chloride corrosion phase, but this regime is not of interest in typical etch applications. Etching rates are not significant for Cl_2 at room temperature. Hence, Cl_2 can and often does serve as a feedstock for Cl atom etching.

Exposure of a silicon surface to Cl atoms leads to the formation of an $SiCl_x$ layer several monolayers thick, thinner than that formed using F atoms. Using (16.3.15), pure chemical etch rates are found to be very small for Cl atoms on undoped or p-doped silicon, but can be substantial for heavily n-doped silicon. The etch products at room temperature are $SiCl_4$ and possibly Si_2Cl_6 and $SiCl_2$.

16.3.5.2 Ion-Assisted Etching

Many surface studies of ion-assisted etching have been with Cl_2 molecules, not Cl atoms. Etch products such as SiCl and $SiCl_2$ have been seen, in addition to $SiCl_4$ and Si_2Cl_6. There is general agreement that ion beam-induced mixing and recoil implantation of dissociated Cl_2 molecules on the surface lead to the formation of an $SiCl_x$ layer more than one monolayer thick, similar to that formed for pure Cl-atom etching. The ion-assisted etch yields (silicon atoms removed per incident ion) with Cl_2 are comparable to those seen using F_2, but are a factor of 5–10 lower than those seen for ion-assisted F atom etching. For example, as seen in Figure 16.13, yields of 3–5 have been observed for 1 kV Ar^+ ions with an adequate flux of Cl_2 molecules (Levinson et al., 1997). The threshold energy for Cl_2^+/Cl_2 silicon etching is about 20 V, in contrast to the approximately 40 V threshold for Ar^+/Cl_2.

For ion-assisted Cl-atom etching, molecular beam studies (Chang and Sawin, 1997) suggest that the etch yields for Cl atoms are two to three times higher than for Cl_2 molecules at the high neutral-to-ion flux ratios. Molecular dynamics simulations give a very small yield $Y = 0.06$ for 10 V Cl^+ ions, increasing to $Y = 0.14$ at 50 V (Barone and Graves, 1995). At 10 V, the saturated Cl surface coverage is about two monolayers thick, increasing to about three monolayers at 50 V. Hence, both Cl atoms and Cl_2 molecules can be important etchants for ion energy-driven etching.

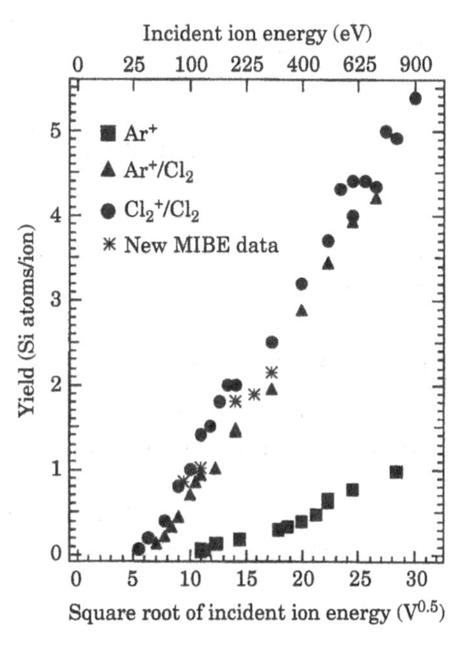

Figure 16.13 Cl_2-saturated, ion-assisted etching yields of silicon by Cl_2^+ and Ar^+; the pure sputtering yield by Ar^+ is also shown. Source: Levinson et al. (1997), with the permission of AIP Publishing.

It is also noteworthy that vacuum ultraviolet (VUV) radiation on the silicon surface can give a small etch rate in the presence of Cl and Cl_2, as described below.

As seen in Figure 16.9, Br atoms are even less reactive than Cl atoms. A very small room-temperature pure chemical etching is observed for heavily doped (n^+) silicon. At higher temperatures, etching is observed and a very large doping effect is seen. HBr is often added to Cl_2 or Cl_2/Ar gas feeds to reduce the formation of microtrenches at the bottom corners of etched trenches in ion-assisted silicon etch. This may be due to a slight roughening ($\lesssim 2$ nm) of the sidewalls due to H-atom isotropic etching. The roughening can convert specularly-reflected ions to a wide-angle scattering distribution.

16.3.5.3 Photon-Assisted Etching

Although Cl-atom pure chemical etching of p-type silicon (100) does not occur, Shin et al. (2012) observed a significant etch rate with negligible ion-bombarding energies in a high density ($n_e \sim 10^{12}$ cm^{-3}), 50 mTorr 1%Cl_2/Ar-pulsed discharge. They attributed this to VUV radiation on the surface in the presence of Cl and/or Cl_2. As seen in Figure 3.12, copious 105–107 nm VUV radiation can be emitted in argon-containing discharges. The photon-assisted etching was further examined by Zhu et al. (2014) in Cl_2, Br_2, HBr, and their mixtures, diluted with 50% argon by volume, in a 60 mTorr inductively coupled discharge. The ion-bombarding energy on the substrate could be controlled in the range of approximately 10–25 V. Below the ion-assisted etch threshold, the etch rate was independent of ion energy (except for a small H-atom etch rate in HBr-containing feedstocks). These sub-threshold etches were found to be induced mainly by VUV photons with wavelengths below about 130 nm. The measured silicon etch yields were $Y \gtrsim 1$. The etched surfaces were rough, which the authors attributed to surface contaminants remaining as micromasks during the etch.

Later measurements with an auxiliary controllable VUV source gave etching yields of 100s of silicon atoms per VUV photon (Du et al., 2022). Yields on the order of 100 had also previously been observed in 110–115 nm synchrotron radiation experiments with XeF_2-exposed silicon (Ney and Schwentner, 2006). These surprisingly large yields were hypothesized to be due to photocatalytic chain reactions. In chlorinated silicon, a possible proposed mechanism involved a desorbing $SiCl_x^-$ etch product, leaving behind a surface positive charge that catalyzed a large number of subsequent $SiCl_x$ etch product desorptions.

16.4 Other Etch Systems

In this section, we describe briefly some common etch systems. For a more thorough description, the reviews of Flamm (1989), Donnelly and Kornblit (2013), and Arts et al. (2022) should be consulted.

16.4.1 F and CF_x Etching of SiO_2

F atoms are known to etch SiO_2, although the pure chemical etch rate (16.3.2) is small and almost never significant in real etch systems. No more than a monolayer of fluorine is adsorbed on an SiO_2 surface. It is also known that CF_x radicals do not spontaneously etch SiO_2 and, furthermore, that these radicals do not dissociatively adsorb on SiO_2. Hence, there is essentially no pure chemical etching of SiO_2 in fluorocarbon plasmas, and all observed etching is ion energy driven.

Large ion-induced etch rates for SiO_2, $\gtrsim 2000$ Å/min, are seen for high ion-bombarding energies, $\gtrsim 500$ V, with both F atoms and CF_x radicals as the etchant species. The etching is anisotropic, and the etch rate correlates with the ion-bombarding energy and is independent of the substrate temperature. The loading effects are much smaller than those seen for F-atom etching of silicon. For F atoms, there is no selectivity for SiO_2 over silicon. Consequently, discharges rich in F atoms are generally not used to etch SiO_2 in the presence of silicon. High selectivity can be achieved for CF_x radical etchants that are produced using low F/C ratio fluorocarbon feedstocks, e.g., c-C_4F_8, or by adding hydrogen to saturated feedstocks, e.g., CF_4/H_2 mix. In both cases, the F-atom density is suppressed and a high density of CF_x unsaturates is created, as described in Section 16.3. Under these conditions, the etch products that are seen include SiF_4, SiF_2, $SiOF_2$, CO, CO_2, and COF_2. Figure 16.14 shows the variation of the gas phase densities n_F and n_{CF_2} and the SiO_2 etch rate E_{SiO_2} versus $\%H_2$ and $\%O_2$ added to a CF_4 parallel plate discharge (Flamm, 1989). For O_2 addition, n_F increases, n_{CF_2} is suppressed, and E_{SiO_2}, due to F atom etchants, increases with $\%O_2$ up to 30% O_2 addition. However, the etching is not selective over silicon. In contrast, with the addition of H_2, n_F is suppressed and n_{CF_2} increases, leading to a mild increase in E_{SiO_2} and a strong decrease in E_{Si} (not shown), with SiO_2/Si selectivity as high as 15:1. This is the regime of anisotropic selective etching of SiO_2 over silicon. The SiO_2 etch rate is observed to abruptly fall to zero above roughly 20% H_2 addition. As will be seen below, this is due to polymer film formation on the SiO_2 surface.

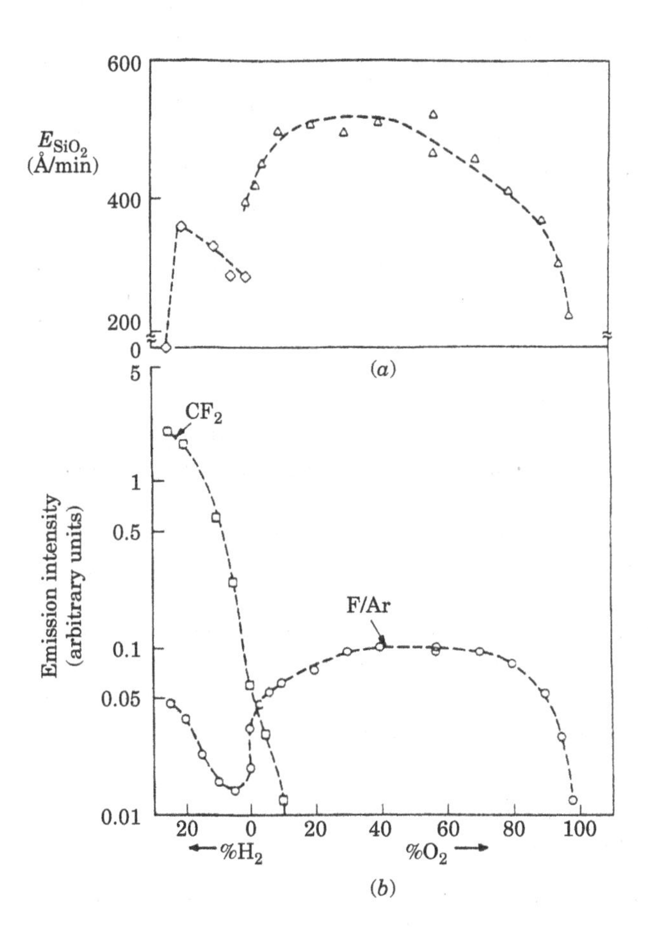

Figure 16.14 (a) SiO_2 etch rate and (b) plasma-induced emission for CF_2 and F/Ar actinometric emission ratio, versus $\%H_2$ and $\%O_2$ addition to a CF_4 parallel plate discharge. Source: Flamm (1989)/with permission of Springer Nature.

Under the action of ion bombardment and high-incident CF_x radical flux, an $SiC_xF_yO_z$ layer as thick as 10–20 Å forms on the SiO_2 surface. For lower radical fluxes, the layer is thinner. Ion beam mixing of adsorbed CF_x radicals is believed to play an important role in the formation of this layer. Under these conditions, the etch mechanisms are believed to be similar to those seen for F atom etching of silicon (Butterbaugh et al., 1991). The most important mechanism is probably the breaking and reforming of bonds within and on the surface of the $SiC_xF_yO_z$ layer due to the collision cascade produced when an energetic ion hits and penetrates the surface. This produces easily desorbable etch products that are weakly bound to the surface, such as SiF_4, SiF_2, CO, CO_2, COF_2, $SiOF_2$, and, possibly, O_2. A crucial point is that in addition to F atoms, adsorbed C atoms can here act as etchants, removing oxygen from the surface by reactions such as

$$\text{ion} + \text{C (s)} + \text{SiO}_2\text{ (s)} \longrightarrow \text{CO (g)} + \text{SiO (s)} + \text{ion}$$
$$\longrightarrow \text{CO}_2\text{ (g)} + \text{Si (s)} + \text{ion}$$

Hence, the presence of oxygen in the lattice impedes carbon buildup, allowing the surface to be etched. For the same conditions, carbon-containing polymer films as thick as 100–200 Å are observed to form on silicon as well as on nonactive surfaces. The film on silicon inhibits the etch reaction there, leading to the high observed selectivities for SiO_2/Si under unsaturate-rich conditions (see also Figure 16.10). Even for discharge conditions that do not lead to carbon-containing film deposition on silicon, (16.3.10) shows that the silicon etch rate can be small in CF_x-rich discharges, thus leading to high selectivities in the absence of polymer buildup on silicon surfaces. On the other hand, for very high unsaturate concentrations (> 20%H_2 in CF_4/H_2 mix), SiO_2 etching ceases due to the formation of thick polymer films on SiO_2 surfaces also, as seen in Figure 16.14. Similar polymer formation is seen for other feedstock gases used, such as C_2F_6, C_3F_6, C_4F_6, CHF_3, c-C_4F_8, c-C_5F_8, and their mixtures with H_2 and Ar. As shown in Figure 16.10, film deposition occurs on SiO_2 at low ion energies and etching at high ion energies, with the transition depending on the F/C ratio.

16.4.2 Si_3N_4 Etching

Silicon nitride is commonly used as a mask material for patterned oxidation of silicon, as a dielectric, and as a final passivation layer. There are two kinds of materials: Si_3N_4 produced by chemical vapor deposition (CVD) at high temperatures and that grown under plasma-enhanced conditions (PECVD) at temperatures less than 400 °C. The latter material does not necessarily have 3:4 Si/N stoichiometry and generally has a significant fraction of H atoms in the lattice. Etch rates for PECVD material are generally high compared to CVD Si_3N_4.

Pure chemical F-atom etching of Si_3N_4 can have selectivities of 5–10 over SiO_2, but is not selective over silicon. The etching is isotropic with an activation energy of about 0.17 V. Anisotropic ion energy-driven etching of Si_3N_4 is performed using low F/C ratio fluorocarbon feedstocks. There is little selectivity over SiO_2, but fairly high selectivities over silicon and resist can be attained. A brief summary is given in Donnelly and Kornblit (2013).

16.4.3 Fluorocarbon Plasma Etch Selectivities

The surface reaction mechanisms and polymer film formation in fluorocarbon plasma etching of Si, a-Si:H, SiO_2, and Si_3N_4 were studied experimentally by Schaepkens et al. (1999) and Standaert et al. (2004). For the same etching conditions, the fluorocarbon polymer film thickness was roughly found to increase with the material etched, from low thicknesses for SiO_2, to moderate thicknesses for Si_3N_4, to high thicknesses for Si. The differences in polymer thicknesses, etching behavior, and resulting etch selectivities, are attributed to the different abilities of the etched material to consume carbon during etching reactions. As shown in Figure 16.10, the film thicknesses and resulting etch selectivities are mainly controlled by the gas feed F/C ratio and the ion-bombarding energy on the etching surface.

Two etching regimes have been identified. For thin polymer formation, the ions penetrate through the polymer and drive the etching of the underlying material. For thick polymer formation, the ions stop within the polymer layer, releasing reactive fluorine atoms and other species that diffuse into the underlying material and drive the etch. When the polymer layer is too thick, then etching ceases and polymer deposition occurs, as described above. Sankaran and Kushner (2004) give a detailed surface reaction mechanism for the etching of Si and SiO_2 in fluorocarbon plasmas, which they use in their simulations of the feature-level etching of SiO_2 and porous SiO_2. Wang et al. (2017) give corresponding reaction mechanisms for fluorocarbon etching of Si_3N_4, which they use to examine quasi-ALE of Si_3N_4 and its selectivity to SiO_2.

16.4.4 Aluminum Etching

Aluminum is sometimes used as an interconnect material in less critical applications in integrated circuits because of its high electrical conductivity, excellent bondability and adherence to silicon and SiO_2, compatibility with CVD oxide and nitride, and ability to form both ohmic and Schottky contacts with silicon. Since AlF_3 is involatile, F atoms cannot be used to etch aluminum, and Cl_2 or Br_2 feedstocks are used instead. These vigorously and isotropically etch aluminum in the absence of ion bombardment. Molecular chlorine etches pure clean aluminum without a plasma; in fact, Cl_2 rather than Cl appears to be the primary etchant species for aluminum in etching discharges. The main etch products are Al_2Cl_6(g) at low temperatures ($\lesssim 200\ ^\circ$C) and $AlCl_3$(g) at higher temperatures.

An ion-enhanced etch with inhibitor chemistry is needed to anisotropically etch aluminum. For Cl_2 feedstock, additives such as CCl_4, $CHCl_3$, $SiCl_4$, and BCl_3 are used. Cl_2 and Cl do not etch Al_2O_3, even in the presence of ion bombardment. Processes used to break through the native oxide ($\sim$30 Å thick) and initiate the aluminum etch include physical sputtering due to the ion bombardment and additives such as CCl_4, $SiCl_4$, and BCl_3. Unsaturated radicals produced from these feedstock additives etch Al_2O_3 at slow rates.

Water vapor interferes with aluminum etching and must be excluded from the system or scavenged using water-seeking additives such as BCl_3 and $SiCl_4$. Copper is often added to aluminum to harden the material and increase its resistance to electromigration and hillock formation. A material such as Al/5%Cu is hard to etch because the copper chlorides are nearly involatile at room temperature; typical heat of vaporization is 15.4 kJ/mol and typical vapor pressure is 1 Torr at 572 °C. This material can be etched at temperatures considerably exceeding room temperature or in systems having high ion-bombarding fluxes, with consequent high physical sputtering rates for the copper. Aluminum chloride products can react with photoresist mask materials during etching. To reduce the deterioration of mask materials, the substrate temperature must be kept below 100–150 °C. Post etch corrosion due to $AlCl_3$ or Al_2Cl_6 deposits on the wafer can be a problem. These react with water vapor to generate HCl, which can corrode the aluminum and other structures on the wafer. Hence, a wet (HNO_3) or dry chemistry (fluorocarbon plasma) is used to remove any remaining aluminum chloride etch product from the surface. The dry process converts aluminum chlorides to unreactive aluminum fluorides.

16.4.5 Copper Etching

Copper is often used as an interconnect material, replacing aluminum in high performance integrated circuits, because its electrical conductivity is almost 60% higher and it has a significantly larger electromigration resistance. However, unlike aluminum, a commercially successful etching process has not been developed, due to the low vapor pressures of the copper halides. The copper fluorides are involatile and the copper chlorides have reasonable vapor pressures only at temperatures exceeding 200°C (Kulkarni and DeHoff, 2002; see also Table 7.5), limiting practical etching with organic photoresist masks.

Cl_2 etching has been the most well-studied process. Continuous exposure of copper to Cl_2 leads to a film of $CuCl_x, x \approx 1$, whose thickness grows linearly with time. This indicates that the film growth is limited by the sticking probability of Cl_2, or possibly by CuCl growth at the film–copper interface, rather than by diffusion of Cl_2 through the film (Winters, 1985). The dominant etch product below ~580°C is found to be the ring-shaped molecule Cu_3Cl_3, having an activation energy for desorption of 138 kJ/mol, roughly the sublimation enthalpy of Cu_3Cl_3(g) from CuCl(s). The evaporation rate of Cu_3Cl_3 from the CuCl surface is about that expected from vapor pressure considerations, in the absence of a steady Cl_2 flux incident on the surface. However, in the presence of a steady Cl_2 flux, the measured flux of Cu_3Cl_3 etch product is several orders of magnitude smaller than this evaporation rate. Various schemes have been investigated to increase this etch rate, such as the addition of infrared or ultraviolet radiation to the substrate during Cl_2 exposure. Various two-step processes have also been studied, in which Cl_2 is used to form a patterned CuCl film at low temperatures, and the CuCl film is subsequently removed by an additional gas phase or wet chemical process.

Copper etching in chlorine-containing plasmas has also been investigated. For example, Lee et al. (1997b) studied the copper etch mechanism in low-pressure CCl_4/N_2 plasmas. They found that chlorine atoms were the main etching species and that $CuCl_2$ was the main etch product formed on the surface. For a grounded substrate, the etch rate was found to rise abruptly with substrate temperature, with no etching below 190° C, and roughly 600 nm/min above 210° C.

Because a viable low-temperature copper etch process has not been developed, copper interconnects are usually patterned using a *damascene process*, in which a dielectric layer is first deposited on the substrate. The copper interconnect pattern is plasma-etched into the dielectric, and copper is then deposited over the entire surface using physical or chemical vapor deposition or electroplating. Finally, the excess copper on the surface is removed by chemical–mechanical polishing (CMP). Hence, in this process, dielectric etching replaces metal etching in forming the interconnects.

16.4.6 Resist Etching

Photoresist mask materials are primarily long-chain organic polymers consisting mostly of carbon and hydrogen. Oxygen plasmas are used to isotropically etch ("strip") resist mask materials from wafers and are also used for anisotropic pattern transfer into these materials in the so-called surface-imaged dry development schemes for photoresists. An active etchant for both applications is O atoms. Pure chemical etching of resists using O atoms is isotropic and highly selective over silicon and SiO_2. It is generally characterized by an activation energy of 0.2–0.6 V. In some cases, the etch rate does not have a simple Arrhenius form, with two activation energies depending on whether the substrate temperature is above or below the "glass transition temperature" T_g of the resist. For example, polymethyl methacrylate (PMMA) has T_g ~60–90 °C, with $\mathcal{E}_a \approx 0.2$ V for $T < T_g$ and $\mathcal{E}_a \approx 0.4$ V for $T > T_g$.

Pure chemical etch rates for many resists are low but can be enhanced by the addition of a few percent C_2F_6 or CF_4 to the feedstock mix. This may be due to F atom reactions with the resist to produce HF etch product, leaving unsaturated or radical sites on the polymer for subsequent O-atom attack. In some cases, the measured activation energy is lowered by a factor of as much as three with F-atom addition. Alternatively, it is known that small additions of fluorine atoms can increase the O-atom concentration in the discharge, thus increasing the etch rate. This may be due to a reduction in the O-atom recombination rate on the reactor walls because of F-atom chemisorption.

Ion-enhanced anisotropic etching in O_2 plasmas is used for dry development of surface-imaged photoresists, in which only a small fraction of the volume at the top of the photoresist layer is exposed to the light. There are a number of motivations for the use of surface-imaged resists. As lateral feature sizes continue to decrease, optical wavelengths to expose the patterns must also decrease. For decreased wavelengths, the depth of focus in the resist is also reduced. Hence, the pattern is not in accurate focus throughout the entire thickness of the resist. A second motivation is that optical reflections from the layer underlying the resist can be eliminated. These reflections can lead to photoresist exposure in regions not directly illuminated, with consequent pattern transfer that is not faithful to the original image. A final motivation is that resists become naturally opaque to light as the wavelength is reduced into the deep UV region below 2000–2200 Å; for wavelengths below these, surface imaging technology may be required.

A typical process flow is shown in Figure 16.15. First, the top 0.2 μm of a 1.5-μm-thick layer of photoresist is optically exposed to a pattern. Second, the resist is silylated by exposure to a

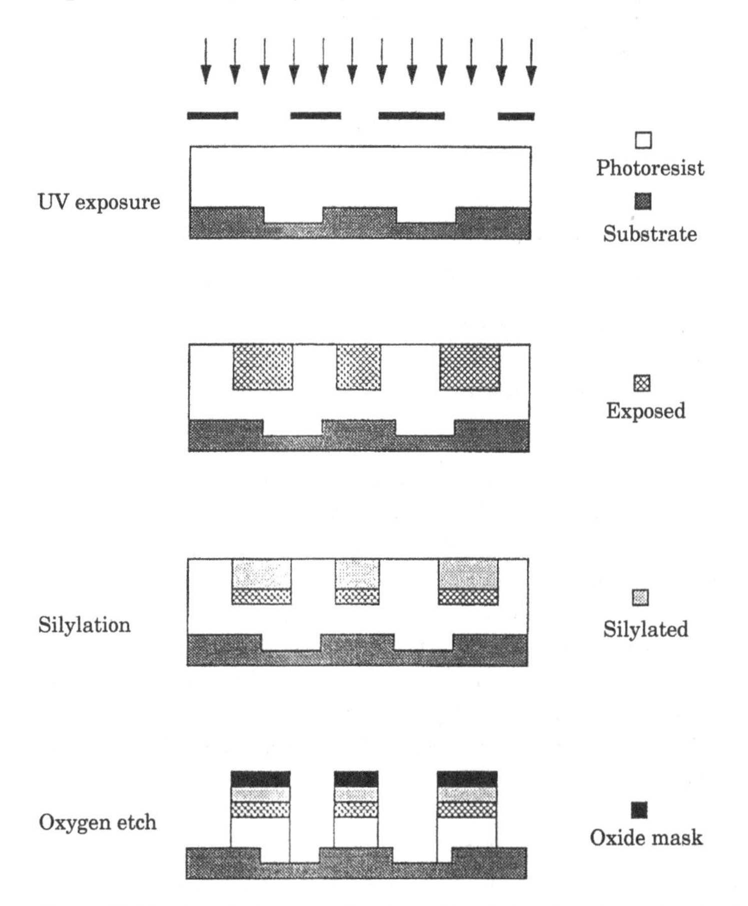

Figure 16.15 A typical process flow for a silylated surface imaged resist dry development scheme.

silicon-containing gas. The silicon is selectively absorbed into the exposed photoresist but is not absorbed into the unexposed photoresist. Finally, the photoresist is anisotropically etched in an O_2 plasma. The O atoms initially react with the exposed, silicon-containing surface layer to create an SiO_x mask that is impervious to subsequent ion-assisted O-atom etching. The unexposed, non-silicon-containing photoresist is anisotropically etched. The original surface image is therefore transferred into the entire thickness of the resist film. Clearly the unsilylated/silylated etch selectivity is a critical process parameter. Minimum selectivities of 10–20 are required.

The balance between ion bombardment and O-atom flux is delicate in this application. On the silylated areas, the O-atom flux must be large enough to oxidize the silylated layer to form the SiO_x mask, and the ion bombardment must be weak enough so that the mask is not physically sputtered away during the etch of the unsilylated areas. For the unsilylated areas, the ion energy and flux must be large enough to transfer the pattern with near vertical sidewalls into the resist. Any SiO_x sputtered onto the unsilylated areas may serve as a micro mask during the etch, leaving residues, often called "grass," on the unsilylated regions after the etch is completed. The requirement of anisotropic etching with low physical sputtering is severe and demands careful control of both ion energy and flux. Hence, high-density discharges, rather than rf diode discharges, which lack this control, are generally used. Fluorocarbon feedstock additions cannot be used to enhance the etch rate of unsilylated resist, or to prevent formation of residues or remove them during etch, because such additions lead to greatly enhanced etch rates for the SiO_x mask, and consequent reductions in unsilylated/silylated selectivity below the required minimum. In fact, trace fluorine contaminants have been found to have adverse effects on the selectivity.

16.4.7 Other Materials

At the inception of integrated circuit fabrication technology, the materials deposited and/or etched were mainly Ge, GeO_2, SiO_2, Si_3N_4, Al, Cu, and organic photoresists. This short list has expanded markedly as the technology has developed. Newer materials used or being considered include

semiconductors: GeSi, GaAs, InP, InAlAs, InGaAs, AlN, MoS_2, graphene
metals: Ti, Fe, Co, Ni, Ta, Ir, and Pt
oxides of Al, Ti, V, Co, Zn, Ga, Sr, Y, Zr, Ru, Hf, Ta, Pt, La, and Er
nitrides of Al, Ti, Zr, Nb, Ag, Hf, Ta, and W

and many ternary or other compounds. For example, developing magnetoresistive random access memory (MRAM) technology might use Ru, Ta, or TiN electrodes, CoFe, NiFe magnetic materials, and Al_2O_3 or NiO dielectrics. Ti and TiN can be etched in Cl_2 or BCl_3/Cl_2 feedstock plasmas. However, similar to copper, many of these materials do not form volatile etch products at reasonable processing temperatures. Then, as discussed in Section 16.1, ion sputtering is the main plasma-assisted etch mechanism. The etch rates are consequently low. Furthermore, sidewall deposition of the etch product in trenches and vias leads to aspect ratio dependent etch (ARDE) effects, with tapered rather than 90° sidewalls. The etch product can also stick to the chamber sidewalls, altering the wafer-to-wafer repeatability and/or generating particulates that lead to on-wafer damage, necessitating frequent chamber cleaning and/or disassembly.

When only small thicknesses of material need to be etched, some of these issues can be overcome by the technique of ALE. For example, Al_2O_3, HfO_2, and ZnO can be etched by thermal ALE. These processes are described in Section 16.5.

16.5 Atomic Layer Etching (ALE)

16.5.1 Introduction and History

Device fabrication with feature sizes of 20 nm or less requires high precision anisotropic and isotropic etching on the atomic scale. The etch must not alter the feature size or shape and must produce minimal induced damage. It must also be highly uniform and selective over the entire range of lengths from feature- to wafer-scale.

In the conventional anisotropic etch processes described in Sections 16.1–16.4, etchant atoms and energetic ions continually flow to the surface, with the vertical and horizontal etch rates depending on both fluxes, as seen in (16.2.8) and (16.2.9). However, at the $\lesssim$20 nm feature scale length, the etchant atom flux can be strongly depleted at the bottom of narrow trenches by sidewall shadowing and increased by sidewall reflections (see Figure 17.3 and accompanying discussion). Ion trajectories can be deflected by bottom and sidewall charging (see Figure 16.28 for an example). Typically, as shown in Figure 16.7, these competing effects lead to slower bottom etches in narrower trenches. These ARDE effects place severe limits on conventional etch processes. Similarly, at the wafer-scale length, ion fluxes tend to be center-high (see Figure 16.24), while etchant atom fluxes might be edge-high (see Section 9.5), leading to strong etch nonuniformities across the wafer. Uniformity must be strongly controlled for varying-size trenches etched at a single etch step across the entire wafer surface. Furthermore, etches must be highly selective, since overetches of 400% or more may be required. Typical precision etch requirements are a variability of only $\pm$0.5 nm, roughly a few atomic monolayers.

The key to overcoming these conventional anisotropic etch limitations is a cyclical two-step process called ALE. In an ideal ALE process, (1) the surface of the layer to be etched is first saturated with an etch precursor only one or a few monolayers thick, and then (2) the etch product is removed by a saturated ion bombardment. For high selectivity, the precursor should be chosen so that the (horizontal) etch rate due to the precursor alone (without ion bombardment) is near-zero. By saturating the surface with the etch precursor, and then saturating the surface to be etched with the ion bombardment, the limitations of ARDE and wafer-scale nonuniformities can be overcome. Furthermore, only on the order of a monolayer is etched per cycle, giving strong control over the precision of the etch. The cycling between etch precursor adsorption and ion bombardment etch steps is illustrated in Figure 16.16*a,b*, for a Cl_2 precursor on a silicon surface.

Unfortunately, energetic ion bombardment usually damages the underlying layer. The layer is amorphized and mixed with etch precursor by the energetic ions, which penetrate a short distance into the lattice. Typical damaged layer thicknesses are of order 2–6 monolayers for ion energies of 15–80 V. The cyclical ALE process then proceeds in the presence of this damaged layer, as illustrated in Figure 16.16*c,d*. It is these mixed layers that are cyclically etched.

Reviews of ion-assisted ALE are given in Agarwal and Kushner (2009), Lee et al. (2014), Oehrlein et al. (2015), Kanarik et al. (2015), and Arts et al. (2022). The earliest applications were to the etching of graphene and compound semiconductors. Matsuura et al. (1993) investigated $Si/Cl_2/Ar^+$ ALE in an ECR reactor, but the ion fluence was limited, so the etch did not proceed to saturation. Athavale and Economou (1996) used a remote helicon plasma reactor for $Si/Cl_2/Ar^+$ ALE, with its higher ion flux giving saturation of the etch and controllable monolayer removal. Agarwal and Kushner (2009) used a hybrid plasma simulation model, along with a two-dimensional Monte Carlo feature profile model, to simulate ALE processes for silicon etch in Ar/Cl_2 mixtures in an ICP reactor, and for SiO_2 etch in $Ar/c\text{-}C_4F_8$ mixtures in both ICP and MERIE reactors. They found good control of the etch on the atomic scale.

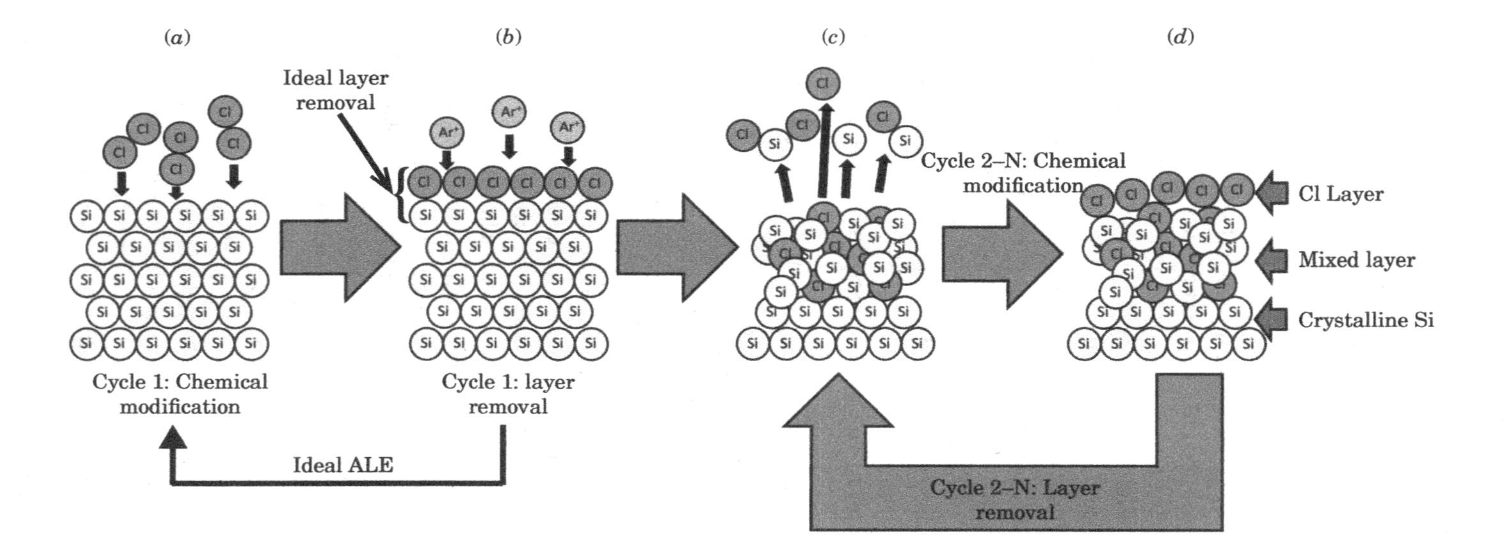

Figure 16.16 Ideal and non-ideal atomic layer etching of silicon by ion bombardment; in the ideal case (*a*), the surface is chlorinated with Cl_2; and (*b*) a single silicon monolayer is removed by a saturated Ar^+ bombardment etch step with ion energies $\mathcal{E}_i \lesssim 10-15$ V during each cycle, leaving a pristine silicon surface; however, for $\mathcal{E}_i > 15$ V, a damaged surface is left, and the etch is non-ideal; in subsequent etch cycles (c) with $\mathcal{E}_i > 10$ V, a chlorinated, damaged layer many monolayers thick is formed by ion bombardment during the previous cycle; a part of this layer is removed during each saturated etch step, leaving a chlorinated damaged surface. Source: Vella and Graves (2023)/with permission of AIP Publishing.

RF bias power pulsing, at typical on-and-off rates of 0.1–1000 kHz, is a precursor to ALE, as discussed briefly in Kanarik et al. (2015). During the on-time, the ion-bombarding energy is high, with ion-assisted etching occurring. During the off-time, the surface is resaturated with the etch precursor. However, in this scheme, resaturation also occurs during the on-time, so the on-time etch step is not self-limiting. Nevertheless, RF bias pulsing enables better control of etch uniformity, selectivity, and ion-induced damage, without the need for ALE reactor gas exchanges. Agarwal and Kushner (2009) simulated this process for SiO_2 etch in Ar/c-C_4F_8 mixtures in an ICP reactor. A non-sinusoidal (tailored voltage waveform) bias with a 10% voltage spike was used to give good ion energy discrimination during the etch phase. Although highly selective and having higher etch rates than a true ALE process, the etch did not saturate, due the continuous flow of etch precursor fluxes to the substrate during the etch step.

16.5.2 Experimental Results

Dorf et al. (2017) report experiments using an electron sheet beam to generate a $T_e \sim 0.3$ V low-temperature plasma above the wafer surface, giving a minimum ion energy < 3 V in the absence of rf bias. A 2 MHz bias source then gives control of the substrate ion-bombarding energy in the range of 2–50 V. A supplementary ICP source is used to provide the etch precursor flux needed for the ALE process. They investigated Si/Cl/Ar^+ ALE, using a Cl etch precursor generated in a Cl_2/Ar/N_2/H_2 ICP plasma. A 7:1 Ar/N_2 mixture at 5 mTorr was used for the ion bombardment step. Figure 16.17 shows (*a*) the measured etch depth per cycle versus the ion etch step time and (*b*) the measured etch depth per cycle versus the ion energy. In Figure 16.17*a*, for ion energies of $\mathcal{E}_i = 28.7$ and 46.2 V, with corresponding Cl exposure times per cycle of 6 s and 10 s, we see reproducible (different markers) etch step saturations. Figure 16.17*b*, at a Cl exposure time per cycle of 10 s and an etch time per cycle of 40 s, shows that the etch depth increases with energy, from 1.3 to 4.6 Å as $\mathcal{E}_i$ increases from 8 to 48 V. Note that the gas exchange time in these experiments was about 3 s.

At very low ion energies, $\mathcal{E}_i \sim 10$–15 V, the damaged layer is nonexistent or very thin. In this case, a near-ideal anisotropic ALE is possible. Dorf et al. (2017) measured the damaged layer thickness after long (600 s) exposures to 8 V ions, finding that it is similar to the native oxide layer thickness found on unprocessed silicon samples.

ALE can be highly selective if the right ion energy is chosen. Tan et al. (2015) measured the selectivity of Si/Cl_2/Ar^+ ALE over thermally grown SiO_2. The etch per cycle for the thermal oxide was found to be essentially zero for ion-bombarding energies below about 70 V, implying an essentially infinite Si/SiO_2 selectivity for the silicon ALE.

16.5.3 Molecular Dynamics Simulations

In molecular dynamics (MD) simulations of ion-assisted ALE, the classical time-dependent Newton laws are solved for a small assemblage of atoms modeling the crystal lattice to be etched, under successive impacts of either etch precursor neutrals or energetic ions. As an example for Si/Cl_2/Ar^+ ALE, an x–y–z crystal simulation size of $3.26 \times 3.26 \times 5.298$ $(nm)^3$ (about 2800 atoms) is used, with periodic boundary conditions along x and y. Empirical reactive bond potential energy functions are used to describe the time-varying Si–Si, Si–Cl, and Cl–Cl interactions. Interactions with Ar^+ are Auger-neutralized and also described by a simple pair potential. A review of the MD simulation technique and the various inter-atomic potentials that are used is given in Graves and Brault (2009). For a single MD simulation step at 300 K, a Cl_2 molecule or Ar^+ ion impacts the top ($z = 0$) surface,

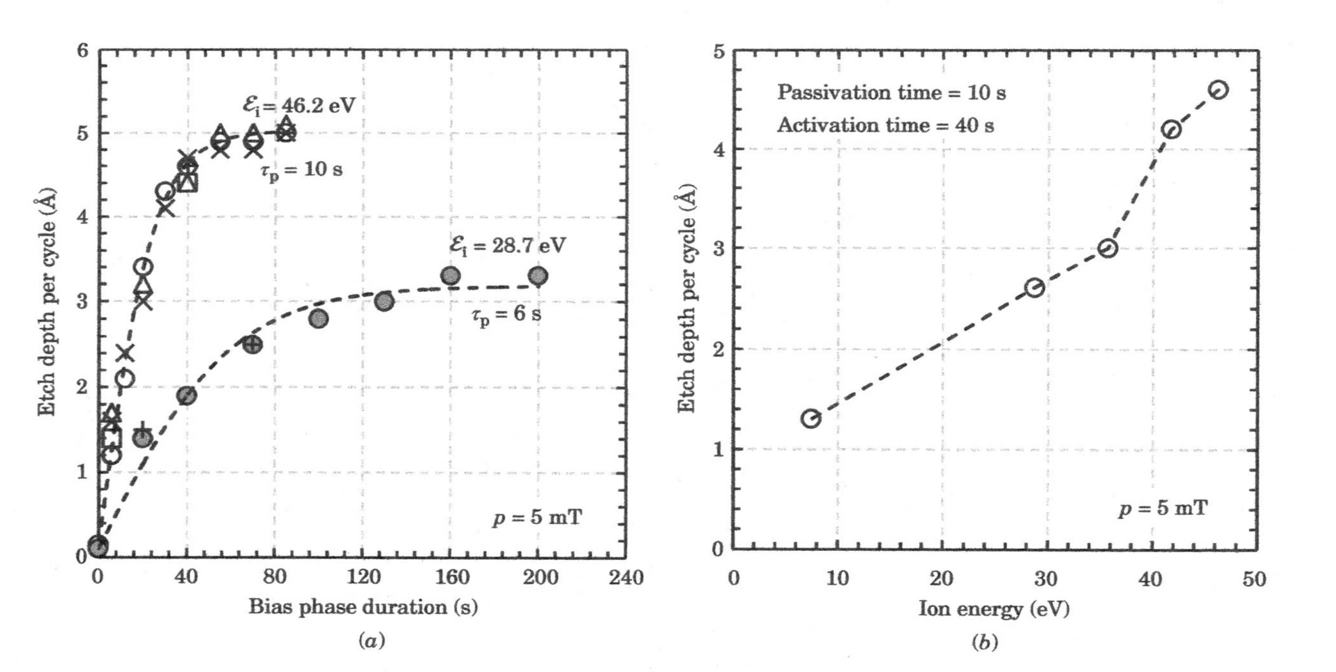

Figure 16.17 Measurements of Si/Cl/Ar^+ ALE, showing (*a*) the etch depth per cycle (Å) versus the etch step time t ("bias phase duration") for $\mathcal{E}_i = 28.7$ and 46.2 V ion energies, and (*b*) etch depth per cycle versus $\mathcal{E}_i$ for a chlorination time ("passivation time") of 10 s and an etch time per cycle ("activation time") of 40 s. Source: Dorf et al. (2017)/with permission of IOP Publishing.

and the resulting motions of all the atoms are followed for about 2 ps. The system is re-cooled and loosely bound products are removed. This procedure is repeated, first for thousands of Cl_2 impacts, and then for thousands of Ar^+ impacts, to gather the final simulation results for a single ALE etch cycle.

For continuous Si/Ar^+ sputtering of crystalline (100) silicon, Humbird et al. (2007) used MD simulations to determine the steady-state damaged layer silicon thickness Δ as a function of ion energy $\mathcal{E}_i$, finding approximately

$$\Delta \approx K_\Delta \left(\sqrt{\mathcal{E}_i} - \sqrt{\mathcal{E}_\Delta} \right), \qquad \mathcal{E}_i > \mathcal{E}_\Delta \tag{16.5.1}$$

with $K_\Delta \approx 0.36\ \text{nm/V}^{1/2}$ and $\mathcal{E}_\Delta \approx 9.6$ V.

MD simulations for $Si/Cl^+/Ar^+$ ALE were performed by Dorf et al. (2017), with ion energies as low as 25 V. Here, the etch precursor was Cl^+ ions, presumably generated in the main or a subsidiary chamber by a low-pressure, Cl_2-containing plasma. At 25 V, the chlorinated layer was found to be 1–2 nm thick, with another 0.5–1 nm of damaged silicon underneath.

For $Si/Cl_2/Ar^+$ ALE, Vella and Graves (2023) determined the ion energy dependence of the etch yield Y_{Si} for the (100) facet of crystalline silicon, and for the *initial* etch yield Y_{SiCl_x} at the top ($x = 0$) of a chlorinated damaged layer. The etch yield continually decreased as the etch penetrated deeper into the chlorinated layer, due to a decreasing n_{Cl}/n_{Si} ratio within the layer. At the beginning of the etch, they found an approximately constant atomic density, with a roughly parabolic chlorine atom density profile within the layer

$$n_{Cl}(z) \propto \left(1 - \frac{z}{\Delta}\right)^2, \qquad 0 < z < \Delta \tag{16.5.2}$$

The yields and the damaged layer thickness were found to approximately scale with the ion bombardment energy $\mathcal{E}_i$ with the form given by (9.3.14)

$$Y_{Si} \approx K_{Si} \left(\sqrt{\mathcal{E}_i} - \sqrt{\mathcal{E}_{Si}} \right), \qquad \mathcal{E}_i > \mathcal{E}_{Si} \tag{16.5.3}$$

$$Y_{SiCl_x} \approx K_{SiCl_x} \left(\sqrt{\mathcal{E}_i} - \sqrt{\mathcal{E}_{SiCl_x}} \right), \qquad \mathcal{E}_i > \mathcal{E}_{SiCl_x} \tag{16.5.4}$$

For the two etch yields, $K_{Si} \approx K_{SiCl_x} \approx 0.02$ with threshold energies $\mathcal{E}_{Si} \approx 49$ V and $\mathcal{E}_{SiCl_x} \approx 13$ V, estimated from extrapolations of data in Vella and Graves (2023).

We should note that the damaged layer is somewhat thicker in the presence of chlorination. Vella et al. (2022) find $\Delta \approx 3$ nm at $\mathcal{E}_i = 70$ V, in contrast to $\Delta \approx 2$ nm from (16.5.1) for unchlorinated silicon. Figure 16.18 shows a visualization of the damaged layer from MD simulations just after the

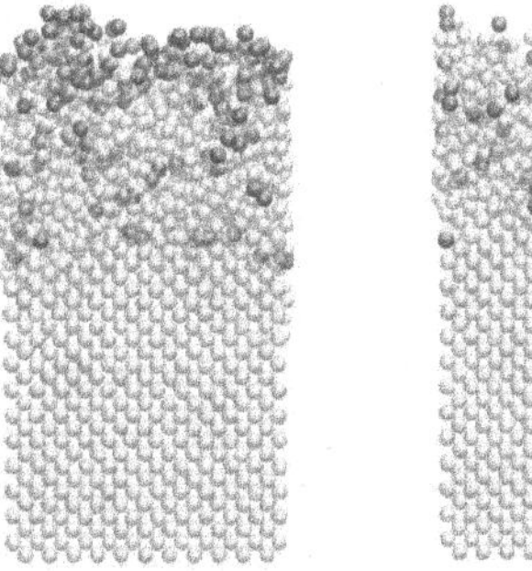

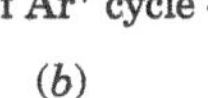

Figure 16.18 Visualization from a molecular dynamics simulation of $Si/Cl_2/Ar^+$ ALE, showing the damaged layer (a) just after the fourth chlorination step, and (b) just after the fourth Ar^+ etch step (*b*); $\mathcal{E}_i = 70$ V; the chlorine atoms are shown in dark grey, and the silicon atoms are shown semi-transparently. Source: Vella et al. (2022)/with permission of IOP Publishing.

fourth Cl_2 chlorination step (left panel) and just after the fourth Ar^+ etch step. A cyclic steady state has been established after the first few cycles. After chlorination, there is a decreasing Cl density versus position z into the layer. After the ion bombardment, some chlorine atoms remain in the layer.

16.5.4 Model of Atomic Layer Etching

We consider a simple model of the cyclical steady state for the well-studied system of $Si/Cl_2/Ar^+$ ALE etch. We first note that crystalline silicon has the cubic point group symmetry O_h with a lattice constant $a = 0.543$ nm. There are eight atoms per unit cell, giving an atomic density $n_0 = 8/a^3 = 5.0 \times 10^{22}$ cm^{-3}. The surface atom density on the (100) plane is $n_0' = 2/a^2 = 6.78 \times 10^{14}$ cm^{-2}, corresponding to a monolayer layer thickness of $a/4$. The unit cell contains four equally spaced monolayers.

Each etch cycle consists of a chlorination step followed by an ion-enhanced etch step. We assume a saturated chlorination step in the model. A similar approach could be used to understand the cyclical dynamics for an unsaturated chlorination.

16.5.4.1 Saturated Chlorination Step

Let us first consider the chlorination step for the cyclic steady-state ALE process shown in Figure 16.16c,d, due to a flux of Cl_2 molecules incident on the surface. We assume dissociative chemisorption on the surface, $Cl_2\,(g) + 2\,S \rightarrow 2\,Cl{:}S$. As seen from the Cl–Cl and Cl–Si bond strengths in Table 7.3, this reaction is strongly exothermic. (However, the chemisorption of a single Cl atom, $Cl_2\,(g) + S \rightarrow Cl{:}S + Cl\,(g)$, is also energetically possible.) The Cl desorption rate constant is $K_d \approx K_0 \exp(-\mathcal{E}_{desor}/T_g)$. For a typical $K_0 \sim 10^{13}$ s^{-1} and $\mathcal{E}_{desor} \sim 4$ V, the desorption rate for Cl (or products such as $SiCl_2$ and $SiCl_3$) at room temperature is negligible. Therefore, the chlorination is determined by the dissociative chemisorption alone. From (9.4.17) and (9.4.21),

$$\frac{dn'_{Cl}}{dt} \approx 2 \cdot \frac{1}{4} n_{Cl_2} \bar{v}_{Cl_2} s_0 (1-\theta)^2 \tag{16.5.5}$$

where n_{Cl_2} is the gas phase density near the surface, $\bar{v}_{Cl_2}$ is the Cl_2 molecule mean speed, $s_0 \sim 0.4$ is a typical zero-coverage sticking coefficient, and $\theta = n'_{Cl}/n'_0$ is the Cl atom surface coverage.

Introducing the normalized Cl_2 fluence $\varphi(t) = \Phi/n'_0$, with the Cl_2 fluence $\Phi(t) = \frac{1}{4}\int_0^t n_{Cl_2}(t')\bar{v}_{Cl_2}$, we obtain from (16.5.5) that

$$\frac{d\theta}{d\varphi} = 2s_0(1-\theta)^2 \tag{16.5.6}$$

Note that $\varphi = 1$ corresponds to a monolayer of Cl_2 molecules incident on the surface. The solution with an initial value $\theta = \theta_*$ at $t = 0$ is

$$\theta(\varphi) = \frac{\theta_* + 2s_0(1-\theta_*)\varphi}{1 + 2s_0(1-\theta_*)\varphi} \tag{16.5.7}$$

This indicates that θ smoothly increases from θ_* to unity (full saturation), with an initial φ-rate $2s_0(1-\theta_*)^2$ given by (16.5.6). For a typical Cl_2 pressure used of 25 mTorr, this chlorination timescale is very short, ~1 ms. However, we should realize that saturating the bottoms of narrow trenches or vias can take much longer. Due to the sidewall shadowing, only a small fraction of the precursor flux incident on the top of the trench is incident on the bottom. (See Problem 17.2 for a simple calculation.) Also, reactor gas exchange times are typically many seconds. Hence, in practical application to ALE of 3D features, the chlorination time per cycle is usually many seconds.

Chlorination can also be done using discharge-generated atomic chlorine on an unbiased substrate. In this case, there can be significant additional fluxes of Cl_2, along with Cl^+ and Cl_2^+ ions. With no rf bias, the maximum ion-bombarding energy is typically less than 20 V. For Cl_2^+ energies below this value, there is almost no ion-assisted etching (see Figure 16.13 and accompanying discussion). However, as described in Section 16.3, there can be a small ion-assisted etch rate for 10–20 V Cl^+ ions and due to VUV radiation on the surface. Additionally, as shown in Figure 16.9, at room temperature, there can be a significant spontaneous (isotropic) etch rate of silicon by chlorine atoms, with the rate depending strongly on the silicon doping. These second-order etch processes can prevent true saturation of the ALE chlorination step. The spontaneous Cl etching can be reduced by cooling the substrate. A significant advantage to using discharge-generated atomic chlorine (along with Cl_2^+ and Cl^+) is that the saturated surface coverage is typically 2–3 equivalent monolayers, which can give a significant increase in the etch per cycle, compared to Cl_2 chlorination. We leave the calculation of θ versus φ for Cl-atom chlorination as Problem 16.5 for the student.

16.5.4.2 Ion-Enhanced Etch Step

As shown in Figure 16.16, after chlorination of a pristine crystalline silicon layer, the first ion-assisted ALE etch step is usually special. The upper atomic silicon layer is removed, but, at the same time, a chlorinated silicon layer of thickness Δ is formed. During subsequent cycles, we assume the thickness Δ, atomic density n_0, and chlorine atom spatial profile within the chlorinated layer are relatively constant during the etch, but the chlorine content varies with time. In the frame moving with the etch, we use a parabolic Cl-atom volume density profile within the layer

$$n_{\mathrm{Cl}}(z,t) = n_{\mathrm{Cl0}}(t)\left(1 - \frac{z}{\Delta}\right)^2 \tag{16.5.8}$$

Here, $n_{\mathrm{Cl0}}(0)$ is the initial density at the beginning of the ion bombardment part of the etch cycle. Assuming a single "average" etch product $SiCl_x$, the number of Si-atoms per unit area, n'_{Si}, removed per second is

$$\frac{dn'_{\mathrm{Si}}(t)}{dt} = -\frac{1}{\alpha}\frac{dn'_{\mathrm{Cl0}}}{dt} = Y\Gamma_{\mathrm{i}} \tag{16.5.9}$$

where $\alpha = 3ax/(4\Delta)$, $n'_{\mathrm{Cl0}} = \frac{1}{4}a\,n_{\mathrm{Cl0}}$ (m^{-2}) is the area Cl atom density at the surface, Y is the chlorinated layer silicon etch yield, and Γ_{i} is the Ar^+ ion flux. The factor of 3 in α accounts for the average of the chlorine atom parabolic spatial profile.

We assume that Y varies linearly with the Cl surface coverage $\theta(t) = n'_{\mathrm{Cl0}}(t)/n'_0$

$$\begin{aligned} Y &= Y_1\theta + Y_0(1-\theta), \text{ for } Y > 0 \\ &= 0, \text{ otherwise} \end{aligned} \tag{16.5.10}$$

We let $Y_0 = K_{\mathrm{Si}}\left(\sqrt{\mathcal{E}_{\mathrm{i}}} - \sqrt{\mathcal{E}_{\mathrm{Si}}}\right)$ and $Y_1 = K_{\mathrm{SiCl}_x}\left(\sqrt{\mathcal{E}_{\mathrm{i}}} - \sqrt{\mathcal{E}_{\mathrm{SiCl}_x}}\right)$, as in (16.5.3) and (16.5.4), respectively. Substituting (16.5.10) into the second equality in (16.5.9), we obtain

$$\frac{d\theta}{dt} = -\frac{\alpha\Gamma_{\mathrm{i}}}{n'_0}\left[Y_1\theta + Y_0(1-\theta)\right] \tag{16.5.11}$$

Let us introduce the normalized Ar^+ fluence $\varphi_{\mathrm{i}}(t) = \int_0^t (\Gamma_{\mathrm{i}}/n'_0)\,dt'$ in place of the etch time t. In φ_{i}, the fluence $\Phi_{\mathrm{i}} = \int_0^t \Gamma_{\mathrm{i}}(t')\,dt'$ (total number of ions per unit area incident on the surface at time t) is normalized to the (100) silicon surface atom density n'_0. Introducing the difference yield $Y_{\mathrm{d}} = Y_1 - Y_0$, then (16.5.11) becomes

$$\frac{d\theta}{d\varphi_{\mathrm{i}}} = -\alpha(Y_{\mathrm{d}}\theta + Y_0) \tag{16.5.12}$$

The solution of (16.5.12) with initial condition $\theta = \theta_0$ at $\varphi_i = 0$ is

$$\theta = \left(\theta_0 + \frac{Y_0}{Y_d}\right) \exp\left(-\alpha Y_d \varphi_i\right) - \frac{Y_0}{Y_d} \tag{16.5.13}$$

We see that θ decays exponentially with a characteristic fluence scale size of $(\alpha Y_d)^{-1}$.

The number of silicon monolayers (ML) etched is $N_{Si} = \theta/x$. (Recall that a monolayer contains n'_0 atoms per unit area and has a thickness $a/4$.) From the first equality in (16.5.9), we then obtain

$$\frac{dN_{Si}}{dt} = -\frac{1}{\alpha}\frac{d\theta}{dt} \tag{16.5.14}$$

Integrating this with the initial condition that $N_{Si} = 0$ at $\theta = \theta_0$, we obtain the number of silicon monolayers etched as a function of the normalized ion fluence

$$N_{Si}(\varphi_i) = \frac{1}{\alpha}\left[\theta_0 - \theta(\varphi_i)\right] \tag{16.5.15}$$

There are two etch regimes.

16.5.4.3 Saturated Etch

For $\mathcal{E}_{Si} > \mathcal{E}_i > \mathcal{E}_{SiCl_x}$, Y_0 is negative. Then, from (16.5.13), $\theta \to \theta_* = -Y_0/Y_d$ as $\varphi_i \to \infty$. In this regime, the etch ceases after $(\theta_0 - \theta_*)/\alpha$ silicon monolayers are removed; there is true saturation of the etch.

16.5.4.4 Unsaturated Etch

For $\mathcal{E}_i > \mathcal{E}_{Si} > \mathcal{E}_{SiCl_x}$, Y_0 is positive. Then, $\theta \to \theta_* = 0$ when

$$\varphi_i \to \varphi_* = \frac{1}{\alpha Y_d} \ln\left(1 + \theta_0 \frac{Y_d}{Y_0}\right)$$

At the normalized fluence φ_*, all the chlorinated silicon is etched. The etch then continues into the unchlorinated silicon, with the sputtering yield Y_{Si}. This quasi-ALE process may still be quite useful for various etch applications.

To establish a cyclic steady state, we assume that the monolayer of adsorbed surface chlorine is redistributed into the chlorinated layer profile after a short ion fluence, $\varphi_i \sim 2$–5 ML, at the beginning of the etch step. Then, from conservation of Cl atoms within the layer

$$\frac{1}{3}\Delta\theta_0 n'_0 = \frac{1}{3}\Delta\theta_* n'_0 + \frac{a}{4} n'_0 \tag{16.5.16}$$

This gives the initial Cl surface coverage at the start of the etch

$$\theta_0 = \theta_* + \frac{3}{4}\frac{a}{\Delta} \tag{16.5.17}$$

One should note, however, that molecular dynamics studies indicate a considerable initial sputtering of Cl atoms due to ion bombardment; Vella et al. (2022) and Vella and Graves (2023) observed the order of 40% Cl atoms in the time-integrated etch product for 70–80 V ions, mainly emitted during the first five or so monolayers of ion fluence. This would modify the calculation of θ_0 in (16.5.17) and reduce the silicon etch per cycle.

Example 16.1 Figure 16.19 gives some model results, assuming no initial Cl sputtering, showing (*a*) the Cl surface coverage θ versus the normalized Cl_2 fluence φ during the saturated chlorination step, (*b*) the chlorine surface coverage θ (after the initial redistribution of the saturated surface chlorine into the chlorinated layer) versus the normalized ion fluence φ_i during the etch step, and

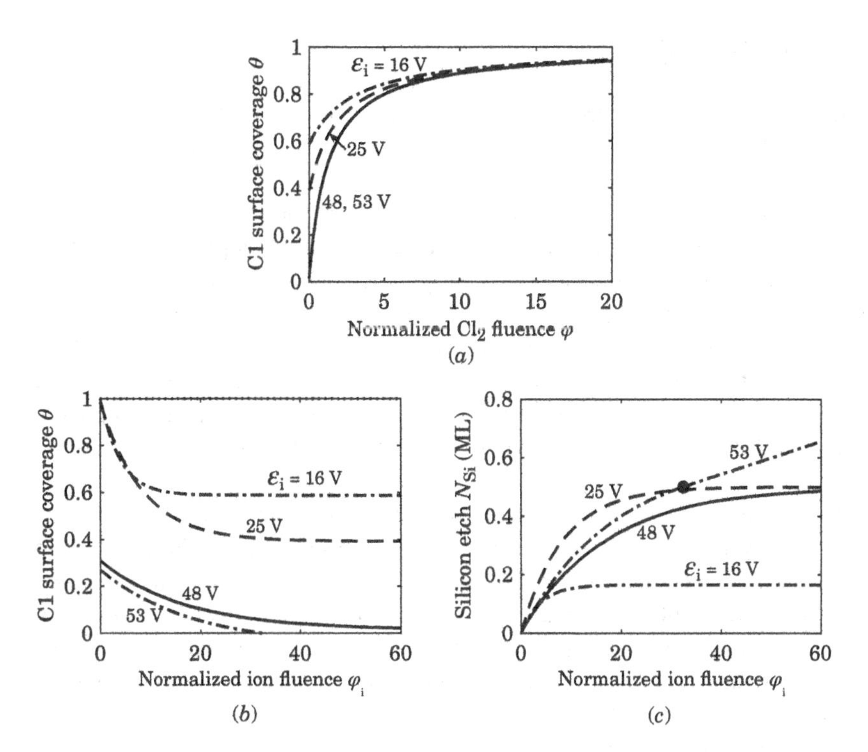

Figure 16.19 Model results for atomic layer etching of chlorinated silicon for various Ar^+ bombarding energies $\mathcal{E}_i$; for $\mathcal{E}_i = 16$, 25, and 48 V, a saturated chlorination step followed by a saturated ion-enhanced etch step is shown in the cyclical steady state; at 53 V, an unsaturated ion-enhanced etch step is shown; (*a*) gives the increase in Cl surface coverage θ versus the normalized Cl_2 fluence $\varphi = n_0'^{-1} \int_0^t \Gamma_{Cl_2}(t')\,dt'$ during the saturated chlorination step; (*b*) gives the decreasing Cl atom surface coverage θ (after an initial redistribution of the saturated surface chlorine into the chlorinated layer) versus the normalized ion fluence $\varphi_i = n_0'^{-1} \int_0^t \Gamma_i(t')\,dt'$ during the etch step; and (*c*) gives the silicon monolayers N_{Si} etched versus the normalized ion fluence φ_i during the etch step; the parameters used are $s_0 = 0.4$, $K_{Si} = K_{SiCl_x} = 0.02$, $\mathcal{E}_{Si} = 49$ V, $\mathcal{E}_{SiCl_x} = 13$ V, $x = 2$, $K_\Delta = 0.36$ nm/V$^{1/2}$, and $\mathcal{E}_\Delta = 9.6$ V; the solid dot in (*c*) gives the transition to silicon sputtering for $\mathcal{E}_i > \mathcal{E}_{Si}$.

(*c*) the silicon etched, N_{Si}, versus the normalized ion fluence φ_i during the etch step, for various ion energies $\mathcal{E}_i$. We have used $x = 2$ (an "average" etch product of $SiCl_2$). The energies $\mathcal{E}_i = 16$, 25, and 48 V are in the saturated chlorination and etch regimes. At all energies, the chlorination step in Figure 16.19a gives 95% surface coverage of Cl atoms after about $\varphi = 25$ monolayers of Cl_2 fluence. At 25 mTorr, this corresponds to about a 3 ms exposure to Cl_2. At 16 V, $\theta_0 = 1$ and $\theta_* = 0.59$, giving a Cl removal of 0.33 ML and a corresponding Si etch of only 0.165 ML per etch cycle. The chlorinated layer is only 0.3 nm thick. At 25 and 48 V, $\theta_0 = 0.98$, and 0.31, and $\theta_* = 0.39$, and 0.014, with $\Delta \approx 0.7$ and 1 nm, respectively. For these latter ion energies, an entire monolayer (6.78×10^{14} atoms/cm^2) of Cl is removed, with $N_{Si} \approx 0.5$ ML/cycle. The etch step time is about 5 seconds for a normalized ion fluence $\varphi_i = 40$ and a typical ion flux $\Gamma_i \sim 5 \times 10^{19}$ m^{-2}.

That the silicon etch per cycle saturates at 0.5 monolayers for ion energies between 25 and 48 V is a simple consequence of chlorine atom conservation: in the cyclical steady state with an average $SiCl_2$ etch product, a maximum amount of adsorbed chlorine of one monolayer per cycle can at most remove half a monolayer of silicon per cycle. The reader can explore this issue further in Problem 16.6.

The highest energy, $\mathcal{E}_i = 53$ V, is in the unsaturated etch regime, with $\theta_0 = 0.27$. After a normalized ion fluence of $\varphi_* = 32$ ML (solid dot in Figure 16.19c), the etch continues into the unchlorinated silicon layer.

The assumption of a single "average" etch product, e.g., $SiCl_2$, is made for simplicity in the model. Actually, time-varying distributions of etch products are seen during the etch step (Vella et al., 2022; Vella and Graves, 2023). Typically, Cl-atoms are seen early in time, followed by $SiCl_x$ products, with x decreasing to unity as the etch progresses. Because some ion flux can be diverted to feature sidewalls and some Cl atoms are evolved, small spontaneous and ion-enhanced sidewall etching can also occur. To minimize these undesired effects, the maximum ion fluence per etch step must not be chosen too large. Cryogenic temperatures can also be used to reduce the isotropic component of the sidewall etching, see Lill et al. (2023) for a review and discussion of low-temperature plasma etching.

The etch step is typically done at a very low argon gas pressure, of order 5 mTorr, to maintain the ion bombardment energy as monoenergetic as possible. In the absence of an rf bias voltage on the substrate, the minimum ion energy is of order $5.2\,T_e \sim 16$ V, for $T_e \sim 3$ V in a 5 mTorr argon discharge. This energy can be controllably increased by applying an rf bias. Of course, as discussed in Section 11.9, there is actually some ion-bombarding energy distribution on the substrate. The etch step must then be averaged over this distribution. To reduce this effect, bias voltage pulsing can be used, as described at the end of Section 11.6.

The maximum silicon etch of 0.5 ML per cycle can be increased if more than one chlorine monolayer can be adsorbed per cycle. As mentioned above, plasma-generated Cl atoms can be used to increase the adsorption to 2–3 monolayers. This can also be done at cryogenic substrate temperatures T_s (Lill et al., 2023), by physical adsorption of Cl_2 on top of a chemisorbed Cl-layer. T_s must be near but above the condensation or sublimation temperature at the operating gas pressure. If holes (vias) or trenches are present on the substrate, T_s must also lie above the corresponding capillary condensation temperatures (Atkins, 1986, section 7.6b).

16.5.4.5 Other Anisotropic ALE Processes

Another important process that has been studied is the anisotropic ALE of HfO_2 using BCl_3 precursor and energetic argon neutral atoms (Park et al., 2008b). HfO_2 is a high dielectric constant material with many applications to semiconductor device fabrication. Energetic argon neutrals, instead of ions, were used to reduce possible charge-related damage (see Section 16.6). These were generated from a 73 V Ar^+ ion beam by glancing incidence on a reflector. The ALE cycle consisted of a 20 s BCl_3 adsorption step and 10–30 s energetic-neutral induced, etch product desorption step, with intervening 3 s gas exchange steps. Adsorption saturation was observed above BCl_3 pressures of 220 mTorr. Saturation of the etch step was observed above an energetic-neutral dose of 1.5×10^{17} atoms/cm^2. The measured etch rate under saturation conditions was 1.2 Å/cycle.

Etching was not observed with Cl_2 precursor. Since the Hf–O bond energy is 791 kJ/mol, the authors attribute this to the much lower Cl–O bond energy of 272 kJ/mol, compared to the B–O bond energy of 806 kJ/mol (see Table 7.3). Therefore, volatile boron oxychloride formation is energetically favored, enabling the BCl_3 ALE process. Similar BCl_3 ALE processes were developed for TiO_2 (Park et al., 2009), ZrO_2 (Lim et al., 2009), and Al_2O_3 (Min et al., 2013).

In addition to energetic ions and neutrals, energetic photons can be used in one of the two anisotropic ALE steps. Photon-assisted continuous etches were already described in Section 16.3 for p-type silicon in chlorine discharges. As one ALE example, platinum-group metals such as ruthenium (Ru) are of considerable interest for integrated circuit fabrication. Coffey et al. (2021) describe a two-step ALE process for Ru in which the first step is 100–150 °C oxidation to form a RuO_2 film by simultaneous exposure to 110–160 nm VUV and 1 Torr O_2. The RuO_2 film thickness saturates after about ~5 min of exposure. In the second step, the oxidized film is isothermally removed by a 30 s exposure to formic acid (HCOOH) vapor. Approximately 1 Å thickness is removed per cycle.

16.5.5 Quasi-Atomic Layer Etching

Quasi-ALE is a process in which etch precursor adsorption or etch product removal (or both) are nonsaturating. In this case, the nonsaturating step *must be timed* to adsorb or remove a specific number of monolayers per cycle. Due to the nonsaturation, wafer- and feature-level nonuniformities are not completely compensated. As a result, aspect-ratio dependent etch (ARDE) effects can exist, i.e., when etching the bottoms of narrow trenches to a specific depth, wider trench bottoms will be overetched.

As an example, the preceding $Si/Cl_2/Ar^+$ ALE with $\mathcal{E}_i > 48$ V is quasi-ALE, because the etch continues into the silicon after the chlorinated layer is removed. However, this overetching occurs at the slow silicon sputtering rate, which is often acceptable, compared to the much larger overetching for a continuous etch process. Consequently, quasi-ALE can be highly useful in semiconductor device fabrication. Computational studies of 3D feature-level silicon structures were performed by Huard et al. (2017), and the various quasi-ALE nonidealities were characterized. The authors showed that significant benefits could be obtained from quasi-ALE, over that found for continuous etches.

A well-studied and characterized quasi-ALE process is $SiO_2/CF_x/Ar^+$. This is a two-step version of the continuous SiO_2 etch process described in Section 12.4. The first experimental demonstration was by Metzler et al. (2014) using C_4F_8 feedstock to generate the CF_x deposition precursor. A steady-state, 13.56 MHz inductively coupled discharge in 10 mTorr argon at fairly low power was used, with a substrate holder that could be biased by a 3.7 MHz auxiliary rf source. For the first (precursor deposition) step, timed injection releases a specific quantity of C_4F_8 into the discharge under the condition of no substrate rf bias. This nonsaturating step generates a controlled CF_x layer a few angstroms thick on the 25×25 mm^2, 10°C-cooled SiO_2 substrate. In the second (etch) step, a small applied rf bias voltage generates low-energy Ar^+ bombarding ions on the substrate, which remove the SiF_x and CO_y etch products. This step is saturated because the ion energy is chosen below the threshold energy for SiO_2 sputtering by Ar^+. Figure 16.20 shows the film thickness change versus time for the first eight etch cycles. The 1.5 s C_4F_8 first-step pulse introduces an approximately 4 Å increase in thickness due to the CF_x deposition. After a gas exchange (purge) time of 8.5 s, the second-step 10 V bias is applied to the substrate holder for 30 s. The ion-bombarding energy is about 20 V during this step. This suffices to cause reaction and removal of approximately 8 Å of fluorine- and carbon-containing etch products but does not etch the underlying SiO_2 surface. The saturation is clearly seen during this step. The cycle is then repeated, with the SiO_2 thickness decreasing by about 4 Å per cycle during the last 4–5 cycles in the cyclical steady state.

In further work, the use of CHF_3 feedstock was examined and shown to provide good selectivity for quasi-ALE of SiO_2 over silicon (Metzler et al., 2016). As for the continuous etching described in Section 16.4, the carbon-containing fluorocarbon film protects the silicon surface during the SiO_2 etch. Selectivity of quasi-ALE SiO_2 over Si_3N_4 was examined by Li et al. (2016). They found good selectivity of SiO_2 over Si_3N_4 for low ion energies, high CF_x film thicknesses, and short etch

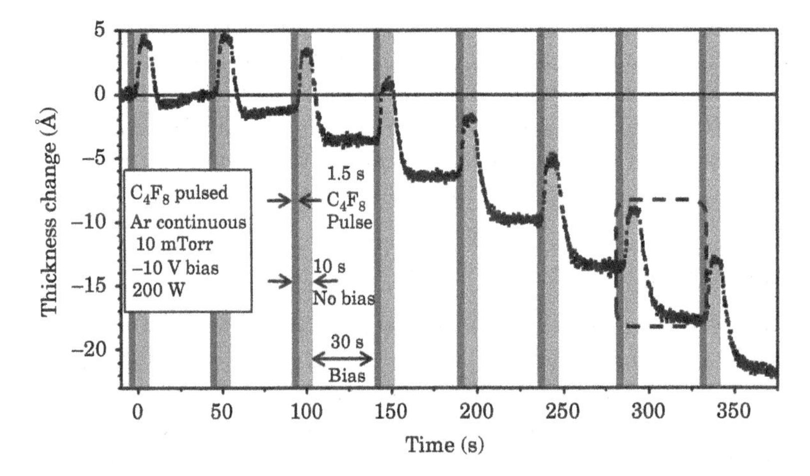

Figure 16.20 Measured film thickness change (Å) versus time for the $SiO_2/CF_x/Ar^+$ quasi-ALE etch process; eight etch cycles are shown. Source: Metzler et al. (2014)/with permission of AIP Publishing.

step times; in this case, as for silicon, the carbon-containing fluorocarbon film protects the Si_3N_4 surface. At higher ion energies, between the lower Ar^+ sputtering threshold of Si_3N_4 and the higher sputtering threshold for SiO_2, they found a significant selectivity of Si_3N_4 over SiO_2. Hence, the selectivity could be controlled by the quasi-ALE process parameters. Selective quasi-ALE of SiO_2 over Si_3N_4 for topographical structures was studied both using a hybrid simulation model, and by experiments in a dual-frequency (40 MHz/10 MHz) discharge, using the C_4F_8/Ar^+ two-step process (Wang et al., 2017). Significant aspect-ratio-dependent etch effects were observed, which the authors note may not be easy to overcome.

16.5.6 Thermal Atomic Layer Etching

As described in Sections 16.1–16.4, continuous thermal etching is an isotropic process in which a gas phase etch precursor is adsorbed onto the surface and reacts to form an easily desorbed etch product. Typically the adsorption surface coverage decreases with temperature, and the etch product desorption increases with temperature, giving a temperature "window" for this process to occur with high etch rates. Some examples given in Section 16.3 include F_2, Cl, F, and XeF_2 etching of polysilicon; the latter is an important process for MEMS device fabrication. Similarly, HF or HF/H_2O can be used to etch SiO_2. Although these etches are isotropic, there can be a significant post-etch surface roughness. There can also be strong ARDE effects, with bottoms of narrow trenches or vias etching much slower than exposed top surfaces, due to sidewall shadowing of the etch precursor.

To overcome these and other limitations of continuous isotropic etching, and to isotropically etch with monolayer precision, the technique of thermal ALE has been developed. This uses a set of thermally activated, self-limiting, cyclical reaction sequences. In broad outline, these are:

1. Ligand-exchange/fluorination
2. Conversion/ligand-exchange/fluorination
3. Oxidation/conversion/fluorination

16.5.6.1 Experimental Results

The dielectric Al_2O_3 has possible uses as a gate oxide, to passivate semiconductor surfaces, as an encapsulation barrier, and in metal corrosion protection. A well-studied, two-step ligand exchange/fluorination thermal ALE process is Al_2O_3 etching (Lee and George, 2015). The first step is etch precursor adsorption of the metal–ligand complex $Sn(acac)_2$, and its subsequent surface reaction. This ligand has a high affinity to aluminum and a low affinity to the carrier (tin) metal. The second step is the adsorption of HF to remove the monolayer of reacted material and re-fluorinate the surface. The two acetylacetonate (acac) ligands have the chemical structure $CH_3C(O)CHC(O)CH_3$. $Sn(acac)_2$ is a liquid at room temperature (25 °C) with a vapor pressure of 174 mTorr, and with a boiling point of 110 °C. A typical monolayer etch done at 200 °C used an $Sn(acac)_2$ fluence of 20 mTorr for 1 s, followed by an HF fluence of 80 mTorr for 1 s, with 30 s N_2 gas purges between steps 1 and 2. Both steps were found to be self-saturating.

The proposed $Al_2O_3/Sn(acac)_2/HF$ etch mechanism in the cyclical steady state is as follows. During the $Sn(acac)_2$ precursor adsorption step, the precursor reacts with the previously formed AlF_3 surface layer to form $Al(acac)_3$ and SnF(acac) etch products. Tin bonds strongly to fluorine because the bond enthalpy is high, 466.5 kJ/mol in the diatomic molecule. This surface reaction exposes the underlying Al_2O_3 surface. The $Al(acac)_3$ is volatile and quickly escapes. The SnF(acac) has some volatility, but at low temperatures, a significant fraction x of SnF(acac) can remain on the surface. The reaction is then

$$\begin{aligned} 2\,\mathrm{AlF_3\,(s)} + 6\,\mathrm{Sn(acac)_2\,(g)} &\rightarrow \mathrm{Al_2O_3\,(s)} + x\,\mathrm{SnF(acac)\,(s)} \\ &+ 2\,\mathrm{Al(acac)_3\,(g)} + (1-x)\,\mathrm{SnF(acac)\,(g)} \end{aligned} \tag{16.5.18}$$

where "(s)" and "(g)" denotes surface- and gas-phase species. From the measured mass changes, Lee and George (2015) estimate $x = 0.74$ at 150 °C, decreasing to $x = 0.15$ at 250 °C.

During the HF fluorination step, the HF removes the remaining SnF(acac) adsorbed on the surface and rapidly reacts with the underlying Al_2O_3 surface, producing a new AlF_3 monolayer and H_2O. The overall reaction is

$$x\,\mathrm{SnF(acac)\,(s)} + 6\,\mathrm{HF\,(g)} \rightarrow 2\,\mathrm{AlF_3}\,(s) + x\,\mathrm{SnF(acac)\,(g)} + 3\,\mathrm{H_2O\,(g)} \tag{16.5.19}$$

Note that AlF_3 (s) is a reaction intermediate for this thermal ALE.

Similar thermal atomic layer etches for HfO_2 and ZrO_2 have been measured (Lee et al., 2016). These etches at 200 °C, and their high selectivities to TiN, SiO_2, and Si_3N_4 are shown in Figure 16.21. The measured Al_2O_3, ZrO_2, and HfO_2 etch rates at this temperature were 0.23, 0.14, and 0.06 Å per cycle.

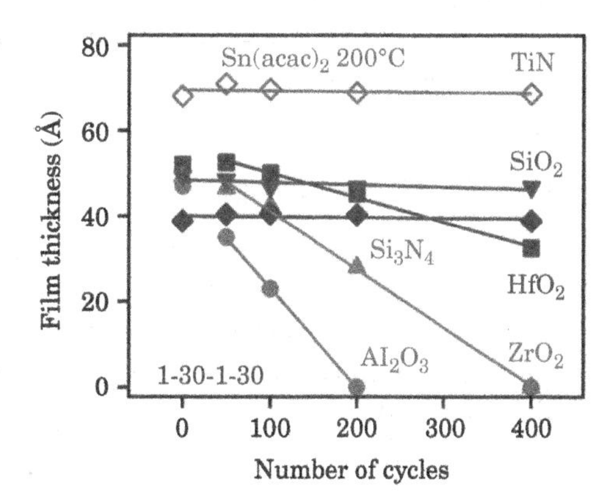

Figure 16.21 Film thickness etched (Å) versus number of thermal ALE etch cycles at 200 °C for various materials, using the two-step, $Sn(acac)_2$/HF cycle. Source: Lee et al. (2016)/with permission of American Chemical Society.

A similar, well-studied thermal ALE mechanism uses TMA (trimethylaluminum, $Al(CH_3)_3$ in place of $Sn(acac)_2$. The two reactions are:

$$2\,AlF_3\,(s) + 4\,Al(CH_3)_3\,(g) \rightarrow Al_2O_3\,(s) + 6\,AlF(CH_3)_2\,(g) \tag{16.5.20}$$

$$Al_2O_3\,(s) + 6\,HF\,(g) \rightarrow 2\,AlF_3\,(s) + 3\,H_2O\,(g) \tag{16.5.21}$$

HfO_2, ZrO_2, AlN, and AlF_3 are also etched using this process. DMAC (dimethylaluminum chloride, $AlCl(CH_3)_2$)) has also been used instead of TMA for these etches.

For thermal ALE of silicon dioxide, an SiO_2 monolayer is converted to Al_2O_3 through the strongly exothermic TMA reaction

$$3\,SiO_2\,(s) + 4\,Al(CH_3)_3\,(g) \rightarrow 2\,Al_2O_3\,(s) + 3\,Si(CH_3)_4\,(g) \tag{16.5.22}$$

Conversion (16.5.22) and ligand exchange (16.5.20) can be simultaneous, with the fluorination step (16.5.21) following. Hence, this is really a two-step process. Zinc oxide can also be etched using this method.

BCl_3 is used similarly to convert metal oxides, such as WO_3, into B_2O_3, which is subsequently removed by HF exposure. Other metal oxides that can be etched in this way include those of molybdenum, vanadium, tantalum, germanium, iron, tin, hafnium, niobium, and zirconium. Similarly, TiN can be oxidized with O_3, and the resulting TiO_2 monolayer can be removed with an HF fluorination.

In a more complex, three-step thermal ALE process, tungsten can be converted to WO_3 using O_2/O_3 gas mixtures. A second exposure to BCl_3 forms B_2O_3. The third exposure to HF removes the oxide layer. These and other aspects of thermal ALE are treated in more detail in the reviews of Ishikawa et al. (2017), Fang et al. (2018), and Fischer et al. (2021).

16.6 Substrate Charging

The flow of ions and electrons to patterned wafers during etching can charge features on the surface. This, in turn, can cause damage to underlying insulating films or can produce undesired distortions of ideal etch profiles. For example, consider the MOS transistor shown in Figure 16.22. The conducting gate electrode (usually polysilicon) is separated from the underlying (conducting) silicon substrate by a thin (2–20 nm) gate oxide. Charge collected on this gate generates an oxide electric field that can exceed the breakdown value, thus causing failure. Even if the breakdown field is not exceeded, the voltage produces a current flow through the oxide, which can generate defects, leading to oxide failure.

16.6.1 Gate Oxide Damage

For thin (2–20 nm) gate oxides, damage due to oxide breakdown is a concern for fields $E_{ox} \gtrsim 10$ MV/cm. In addition, thin oxides are not perfect insulators because electrons can tunnel through the oxide. The resulting flow of current weakens the oxide by causing charge trapping in the oxide and interface trap generation at the SiO_2/Si interface. There are various tunneling mechanisms, and we consider here only Fowler–Nordheim tunneling with

$$J_{FN} = KE_{ox}^2 \exp(-B/E_{ox}) \tag{16.6.1}$$

with $K \approx 20\ \mu\text{A/V}^2$ and $B \approx 250$ MV/cm. For thin oxides ($T_{ox} < 12$ nm), a hole-induced breakdown model has been found to reproduce oxide failure data quite well (Schuegraf and Hu, 1994). This model predicts the mean time t_{BD} for 50% of the devices to fail

$$t_{BD} = t_0 \exp(G/E_{ox}) \tag{16.6.2}$$

where $t_0 \approx 10^{-11}$ s and $G \approx 350$ MV/cm. This is often expressed as a mean flow of charge $Q_{BD} = J_{FN}A_{ox}t_{BD}$ through the oxide for 50% failure of the devices. Experimentally, it is found that a flow of as much as 1% of Q_{BD} through the device has deleterious effects, for example, producing a 5% reduction in transistor gain. Hence, it is desirable to limit the time-integrated flow of current through the oxide to below 1% of Q_{BD}.

16.6.2 Grounded Substrate

Let us consider the oxide voltage and current for the structure shown in Figure 16.22, with a gate oxide area A_{ox} and a field oxide area A_f; the total gate conductor area is $A_g = A_{ox} + A_f$. We assume that the silicon substrate is grounded. Let V_p and V_g be the plasma and gate conductor potentials with respect to ground. From (6.6.4), the plasma current flowing to the gate is

$$I_p = I_i - I_{e0} \exp[-(V_p - V_g)/\mathrm{T_e}] \tag{16.6.3}$$

where $I_i = J_iA_g \approx 0.61\, enu_BA_g$, $I_{e0} = J_{e0}A_g = \frac{1}{4}en\bar{v}_eA_g$, $u_B = (e\mathrm{T_e}/M)^{1/2}$, and $\bar{v}_e = (8e\mathrm{T_e}/\pi m)^{1/2}$. For a perfectly insulating gate oxide, $I_p = 0$ and from (16.6.3), $V_g = V_p - V_f$, where

$$V_f = \frac{\mathrm{T_e}}{2}\left(1 + \ln \frac{M}{2\pi m}\right) \tag{16.6.4}$$

is the floating potential. Note that V_f is independent of plasma density. For a gate oxide thickness T_{ox}, the electric field is V_g/T_{ox}. For example, if $V_g = 5$ V and $T_{ox} = 10$ nm, we find $E_{ox} = 5$ MV/cm, not sufficient for breakdown. It is usually the case that V_g is not large enough for breakdown.

From (16.6.1), the tunneling current can be written in terms of V_g as

$$I_{FN} = \frac{KA_{ox}V_g^2}{T_{ox}^2} \exp\left(-\frac{BT_{ox}}{V_g}\right) \tag{16.6.5}$$

Plotting I_p versus V_g and I_{FN} versus V_g on the same graph yields the equilibrium solution for the oxide current $I_p = I_{FN}$. This has been done by Shin and Hu (1996) and is shown in Figure 16.23 for typical plasma and gate parameters and various values of oxide thickness. The oxide I–V characteristic is similar to that of a diode and can approximated as a near-vertical line at the turn-on voltage, as seen in the figure. The plasma characteristic is also that of a diode, i.e., the ion saturation and transition regime of a Langmuir probe (see Figure 6.6). For thick oxides, the current flow

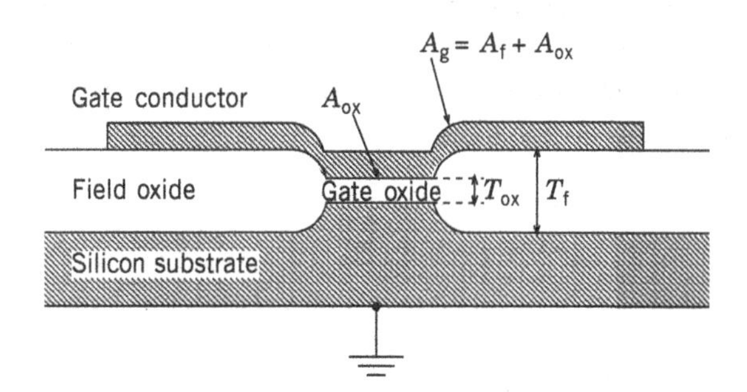

Figure 16.22 An antenna structure for an MOS transistor on a grounded silicon substrate.

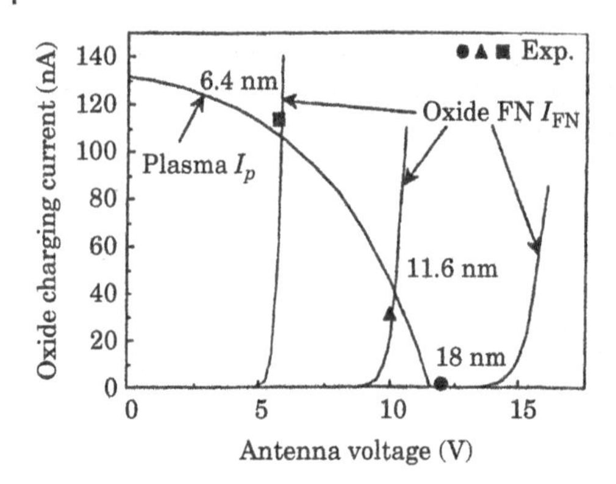

Figure 16.23 Plasma current I_p and oxide current I_{FN} versus antenna voltage V_g for various gate oxide thicknesses. Source: Shin and Hu (1996)/with permission of IOP Publishing.

is impeded; hence, the current is seen to decrease and the oxide voltage increase. For thin oxides which have a low impedance, the maximum current density that can flow through the oxide can be written as $(A_R + 1)J_i$, where $A_R = A_f/A_{ox}$ is called the *antenna ratio;* usually $A_R \gg 1$. Hence, the current is proportional to the antenna ratio in this regime. An important design consideration is to minimize the antenna ratio during processing.

16.6.3 Nonuniform Plasmas

The substrate is rarely grounded in plasma etch systems. For an isolated floating substrate, the oxide current vanishes, and $E_{ox} = 0$ from (16.6.1). Hence, there is apparently no damage. However, let us consider, following Cheung and Chang (1994), the more usual case of an entire wafer in a nonuniform plasma. The Boltzmann relation (2.4.16) requires that a radially decreasing density profile produces a radially decreasing plasma potential

$$V_p(r) = \mathrm{T_e} \ln \frac{n(r)}{n(0)} \tag{16.6.6}$$

We assume first, as shown in Figure 16.24*a*, that the conducting substrate is surrounded by insulator, such that no plasma currents can flow to its surface. There are two MOS transistors, at the wafer center and off-center, with thick gate oxides that are nearly insulating. The conducting gate potentials V_{g2} and V_{g1} at the wafer center and off-center are then

$$\begin{aligned} V_{g2} &= V_{p2} - V_f, \\ V_{g1} &= V_{p1} - V_f, \end{aligned} \tag{16.6.7}$$

where V_{p2} and V_{p1} are the center and off-center plasma potentials. We see that an open circuit voltage $V_{oc} = V_{p2} - V_{p1}$ appears across the series combination of the two gate oxides. This voltage capacitively divides across the oxides depending on their thicknesses and areas (Problem 16.7). The resulting voltages across the oxides may break down one or both of them.

Consider next thin gate oxides that are nearly conducting, as in Figure 16.24*b*. Since the gate potentials are equal, we have

$$V_g = V_{p2} - V_2 = V_{p1} - V_1 \tag{16.6.8}$$

where V_2 and V_1 are the plasma-to-gate electrode voltages at the center and off-center, respectively. A short circuit current flows from the center to the off-center transistor:

$$I_{sc} = I_{i2} - I_{e2}\, e^{-V_2/\mathrm{T_e}} = -(I_{i1} - I_{e1}\, e^{-V_1/\mathrm{T_e}}) \tag{16.6.9}$$

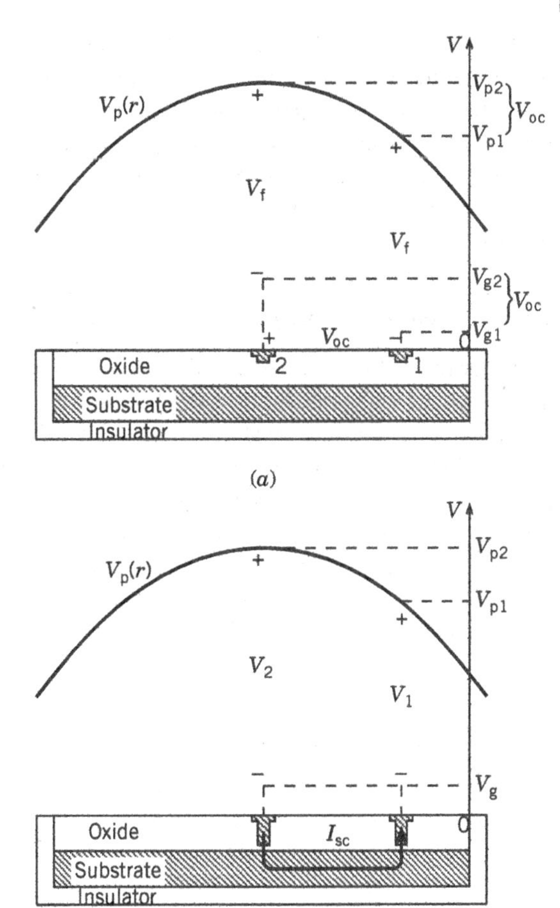

Figure 16.24 Gate oxide damage mechanisms in a nonuniform plasma; (*a*) thick oxide with insulated substrate, showing formation of an open circuit voltage V_{oc}; (*b*) thin oxide with insulated substrate, showing formation of a short circuit current I_{sc}. Source: Adapted from Cheung and Chang (1994).

Solving (16.6.8) and (16.6.9), we obtain

$$V_2 = \mathrm{T_e} \ln \frac{I_{e2} + I_{e1}\, e^{(V_{p2}-V_{p1})/\mathrm{T_e}}}{I_{i2} + I_{i1}} \tag{16.6.10}$$

$$V_1 = \mathrm{T_e} \ln \frac{I_{e2}\, e^{-(V_{p2}-V_{p1})/\mathrm{T_e}} + I_{e1}}{I_{i2} + I_{i1}} \tag{16.6.11}$$

$$I_{sc} = \frac{I_{i2}\left[e^{(V_{p2}-V_{p1})/\mathrm{T_e}} - 1\right]}{I_{e2}/I_{e1} + e^{(V_{p2}-V_{p1})/\mathrm{T_e}}} \tag{16.6.12}$$

Note that $I_{sc} \le I_{i2}$. For the limiting case of a large difference in plasma potentials, $V_{p2} - V_{p1} \gg \mathrm{T_e}$, we find $V_1 \approx \mathrm{T_e} \ln[I_{e1}/(I_{i2} + I_{i1})] \sim V_f$, $V_2 \approx V_{p2} - V_{p1} + V_1$, and $I_{sc} \approx I_{i2}$. Hence, the current flow is limited by the "back biased" plasma diode at the center to I_{i2}, and almost all the potential difference $V_{p2} - V_{p1}$ across the circuit is dropped across the plasma sheath at the center. This illustrates a general principle that can be used to analyze more complicated configurations, namely, that the sheaths at regions of more positive plasma potential are back biased and drop most of the potential difference, limiting the current there to the ion saturation value. The sheaths at regions of smaller plasma potential act as forward biased diodes, with a potential drop near, but somewhat smaller than, the floating potential V_f given in (16.6.4).

The substrate may not be completely surrounded by insulator. For a wafer whose substrate edge is exposed to the plasma, the substrate potential with respect to ground is $V_s \approx V_{p0} - V_f$, where V_{p0} is the plasma potential at the substrate edge. For thin (low impedance) oxides, $V_g \approx V_s$, and current flows from the plasma through both center and off-center gate oxides into the substrate and returns to the plasma at the substrate edge. If essentially the entire substrate area is exposed to the plasma, then $V_s \approx V_{p0} - V_f$. Plasma current flows into the wafer near its center and out near its edge (Problem 16.8).

16.6.4 Transient Damage During Etching

Consider the etching of a deposited polysilicon film to form gate electrodes. The gates (areas A_{ox}) are covered with a protective layer of photoresist; the uncovered areas are to be removed. As shown in Figure 16.25*a*, during most of the etch time, the film is continuous and plasma currents flow into the film near its center and out near its edge. Since these currents do not flow through the gate oxides, they are not damaged. Near the end of the etch time, the film generally clears first in the middle of the unpatterned areas. Then, we obtain the situation shown in Figure 16.25*b*, with isolated gates having large antenna ratios. Large currents can flow through the gate oxides, causing damage. Finally, after a sufficient overetch time, the polysilicon has been entirely cleared from the unprotected areas, greatly reducing the plasma current collected at the gates, and therefore, the damage.

16.6.5 Electron Shading Effect

The wafer topography itself can induce dc current flows through gate oxides on ungrounded substrates even in uniform plasmas (Hashimoto, 1994), causing damage. Figure 16.26 shows a moment

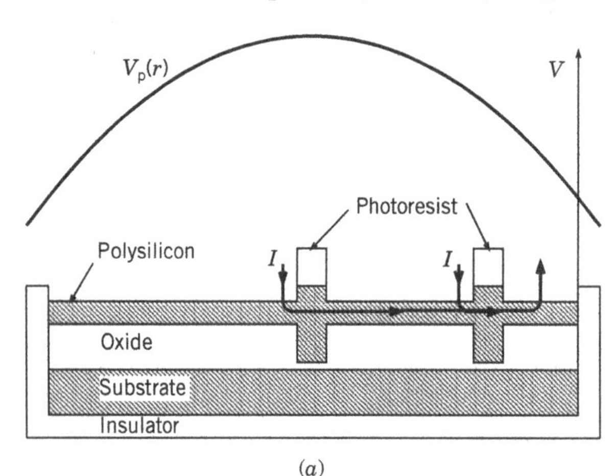

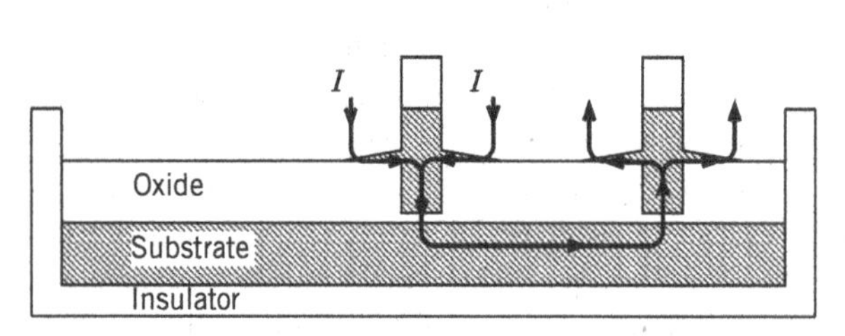

Figure 16.25 Transient damage of gate oxide during polysilicon etching in a nonuniform plasma; (*a*) during most of the etch time, the film is continuous and currents do not flow in the gate oxide; (*b*) near the end of the etch time, there are isolated gates with large antenna ratios, and large currents flow in the gate oxide.

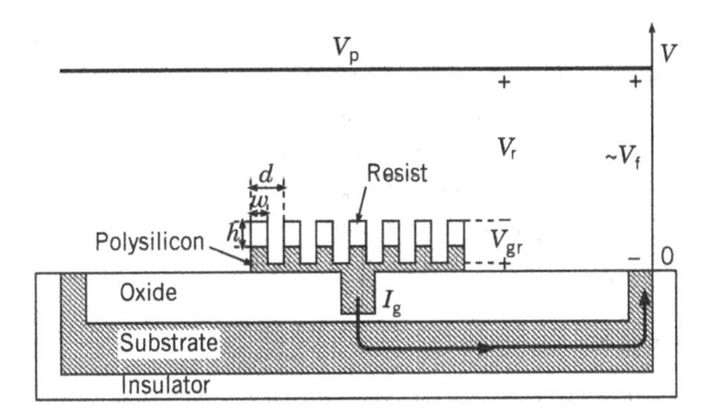

Figure 16.26 Gate oxide damage in a uniform plasma due to the electron shading effect. Adapted from Hashimoto, 1994.

near the endpoint for etching a pattern of conducting lines connected to a gate oxide. The lines are protected by a patterned photoresist layer. The plasma is uniform and the substrate edge is exposed to the plasma. It is found experimentally that a dc current can flow from the plasma into the pattern of lines, through the gate oxide, and return to the plasma at the exposed substrate surface. This effect can be understood as a result of the different fractions of incident ions and electrons that are absorbed by the insulating resist surface. The bombarding ions have strongly anisotropic velocities directed toward the wafer surface. Hence, the fraction of ions absorbed by the resist is roughly proportional to its top surface area. Electrons are also absorbed at the top surface. However, because they have an isotropic velocity distribution, electrons entering the trenches are additionally absorbed on the resist sidewalls.

Let us model the flow of ions and electrons to the resist surface using absorption fractions α_i and α_e for incident ions and electrons. We assume that $\alpha_e \geq \alpha_i$. The current flowing to the resist surface must vanish:

$$I_r = \alpha_i I_i - \alpha_e I_{e0}\, e^{-V_r/T_e} = 0 \tag{16.6.13}$$

which yields the plasma-to-resist potential

$$V_r = V_f + T_e \ln \frac{\alpha_e}{\alpha_i} \tag{16.6.14}$$

Since $\alpha_e/\alpha_i > 1$, we find that $V_r > V_f$. The current collected by the conducting gate is that not collected by the resist:

$$I_g = (1 - \alpha_i) I_i - (1 - \alpha_e) I_{e0}\, e^{-V_r/T_e} \tag{16.6.15}$$

Substituting (16.6.14) into (16.6.15) to eliminate V_r, we obtain

$$I_g = I_i(1 - \alpha_i/\alpha_e) \tag{16.6.16}$$

The factors α_i and α_e are difficult to determine theoretically. They depend not only on the geometrical factors of the resist (width w, height h) and the line spacing d (see Figure 16.26) but also on the gate-to-resist potential drop V_{gr}, and the ion and electron energy and angular distributions. A purely geometrical estimate for ions would give $\alpha_i \approx w/d$. For electrons, there is the same estimate from the resist lines and an additional contribution from the resist sidewalls with a solid angle factor θ/π, with $\tan\theta \approx 2h/(d - w)$. This yields $\alpha_e \approx \alpha_i + \theta(d - w)/\pi d$. However, V_{gr} will be positive, attracting electrons and repelling slowly moving ions, thus modifying these estimates. That this must be true is seen clearly for a thick (essentially insulating) gate oxide. Since no current can flow through the thick oxide, $I_g = 0$ in (16.6.16), which implies that $\alpha_e = \alpha_i$, i.e., the net electron shading effect vanishes. For this case, V_{gr} charges positive enough to attract sufficient electrons (and repel sufficient low energy ions), to just cancel any geometrical shading effects. For a high aspect ratio trench ($d - w \ll h$), V_{gr} can be many times T_e, and can even be a significant fraction of the ion-bombarding energy. On the other hand, for a thin (low impedance) gate oxide, the gate electrode voltage is equal to the substrate potential, $V_g = V_p - V_f$. Subtracting this from V_r given by (16.6.14), we find that $V_{gr} = T_e \ln(\alpha_e/\alpha_i)$. This is generally a small enough voltage to not much modify the geometrical factors. In this case, a current given by (16.6.16) flows through and can damage the gate oxide.

16.6.6 RF Biasing

We consider now an rf-biased substrate with an MOS transistor having a gate oxide area A_{ox} and thickness T_{ox}, and a field oxide area A_f and thickness T_f. The substrate is connected to the rf bias supply through a large blocking capacitor and is exposed to an assumed uniform plasma. We assume that the bias electrode area is small compared to the grounded area of the processing chamber. In this case, as described in Section 11.4, almost all of the applied rf bias voltage appears across the substrate-to-plasma interface, producing an rf current flow (A/m^2) from the substrate into the plasma. The current flows through the gate capacitance, resulting in an rf voltage V_{sg} across the gate oxide. To determine this voltage, we consider the example of a low-frequency bias (Section 11.8), for which the ion and electron conduction currents are large compared to the displacement current. The ion current J_i is almost constant during an rf cycle, while the electron current flows only during short periods when the time-varying voltage across the substrate sheath is near its minimum value. Since the electron pulse is short in duration, it suffices to assume that at the end of the pulse, the net negative charge on the gate is half of the total electron flow per rf period. By the same reasoning, just before the electron pulse, the net positive charge on the gate is also half of the magnitude of the total electron flow per rf period. Since the total electron and ion flows are equal, the total charge collected by the gate oscillates in time as a nearly sawtooth waveform with an amplitude

$$Q_{g0} = \frac{1}{2} J_i A_g \cdot \frac{2\pi}{\omega} \tag{16.6.17}$$

The current oscillation follows by differentiating the charge oscillation. This current is capacitive and does not directly damage the oxide. However, a voltage $V_{gs}(t) = Q_g(t)/C_g$ appears across the oxide. The amplitude of the voltage oscillation can be large enough to break down the gate oxide, and it can also produce an rf tunneling current through the oxide, which can cause damage.

To determine the voltage amplitude, we first write the gate capacitance as the sum of the field oxide and gate oxide capacitances:

$$C_g = C_f + C_{ox} = \frac{\epsilon A_f}{T_f} + \frac{\epsilon A_{ox}}{T_{ox}} = C_{ox}\left(1 + \frac{A_R}{T_R}\right) \tag{16.6.18}$$

Here, ϵ is the oxide dielectric constant and $T_R = T_f/T_{ox}$ is the field/gate oxide thickness ratio. The maximum voltage across the oxide is then found from (16.6.17) and (16.6.18) to be

$$V_{gs0} = \frac{Q_{g0}}{C_g} = \frac{\pi J_i T_{ox}}{\omega\epsilon} \frac{A_R + 1}{A_R/T_R + 1} \tag{16.6.19}$$

For $T_R \gg A_R \gg 1$, we find $V_{gs0} \propto A_R$; for $A_R \gg T_R \gg 1$, we find that $V_{gs0} = \pi J_i T_g/\omega\epsilon$, independent of A_R. The peak-to-peak voltage $2V_{gs0}$ is shown versus the antenna ratio in Figure 16.27 for a number of plasma densities with $T_e = 5$ V and $T_R = 50$ at 1 MHz, typical of gate oxide etching. We see the saturation effect at large-area ratios, and also a strong density dependence. From (16.6.19), the voltage is inversely proportional to the frequency. The higher is the density and the longer is the rf

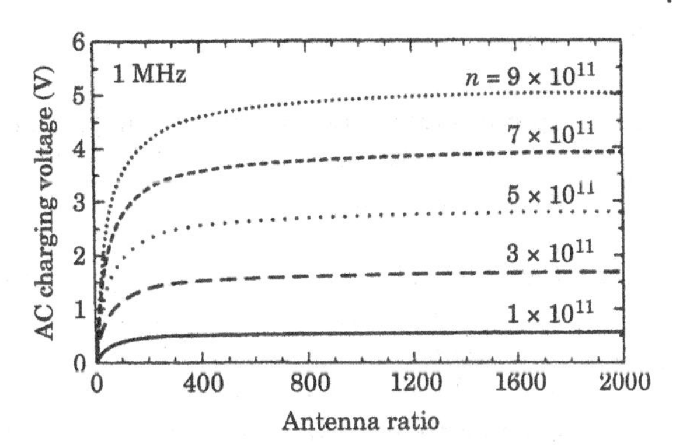

Figure 16.27 Peak-to-peak 1 MHz rf charging voltage versus antenna ratio A_R for various plasma densities, with electron temperature $T_e = 5$ V and a field/gate oxide thickness ratio $T_R = 50$. Source: Cheung and Chang (1994)/with permission of AIP Publishing.

period, the more ions are collected per period, and the larger is the peak charging voltage. Hence, it is undesirable to have too high a plasma density or too low a bias frequency.

16.6.7 Etch Profile Distortions

Distortions of ideal etch profiles such as undercut, tapered, or bowed sidewalls and microtrenches ("notches") at the bases of sidewalls are often observed after etching masked features, e.g., for etching of a pattern of polysilicon lines and spaces using chlorine plasmas. Deflection and subsequent scattering of incoming ions within trenches due to localized buildup of charge on insulating surfaces is believed to be a possible cause of these distortions (Kinoshita et al., 1996; Hwang and Giapis, 1997). Figure 16.28 illustrates the formation of a notch in polysilicon at the interface with the underlying SiO_2 insulating film. The notch is perpendicular to the unperturbed ion-bombarding direction and appears during the overetching step. It typically forms at the inner sidewall foot of the outermost trench when etching a series of trenches adjacent to an open area. The degree of notching depends on the plasma parameters such as the ion energy distribution, plasma density, and electron temperature, as well as the geometry and materials compositions. Most explanations ascribe the notching as being driven by a potential difference between the last polysilicon line, which attracts excess electrons at the side facing the open area, and the insulating trench bottom, which charges positively because the trench topography inhibits the collection of electron charge on the trench bottoms, compared to the collection on the open area. This leads to an electric field pointing from the trench bottom to the open area. This field can deflect incoming positive ions within the last trench into the trench corner nearest the open area, producing an enhanced etch rate there, leading to the formation of a notch. For a broad incoming ion energy distribution (see Figure 11.43), some low-energy ions can be deflected by of order 90°, as seen in the inset in Figure 16.28.

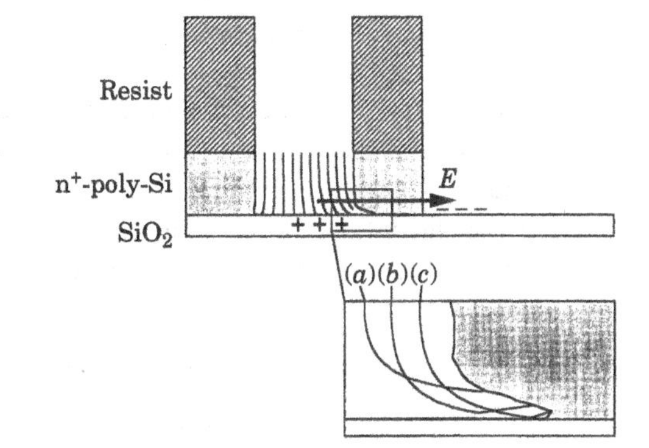

Figure 16.28 Location of the notch and the mechanisms proposed to contribute to the notching effect; (*a*) ion trajectory bending due to open area charging and direct bombardment of the polysilicon; (*b*) forward ion deflection due to SiO_2 charging under the etched area; (*c*) near grazing ion–SiO_2 surface collision, followed by forward scattering and bombardment of the notch apex Source: Hwang and Giapis (1997)/with permission of AIP Publishing.

Problems

16.1 Sputter Etching Estimate the maximum etch rate (Å/min) for physical sputtering of silicon using 600 V Ar^+ ions for an Ar^+ density n_{Ar^+} at the sheath edge of 10^{11} cm^{-3}. Use the data given in Table 9.2.

16.2 Free Radical Production in a CF_4 Discharge Consider a simplified mechanism for F-atom production in a CF_4 discharge, consisting of reactions 2 and 3 in Table 16.2 and reactions 1 and 2 in Table 16.3. Assume that the electron density is specified to be $n_e = 10^{10}$ cm^{-3} and that there is no other generation or loss of CF_4, CF_3, CF_2, and F than given by these reactions, i.e., no surface losses, etc. Assume that the rate constants for reactions 1 and 2 in Table 16.3 are the second order, i.e., for the high-pressure limit in which the reactions are independent of the concentration of the third molecule M (here CF_4).

(a) Write the differential equations for the densities of the four species, e.g., $dn_{CF_4}/dt = \cdots$.

(b) In the steady state, show that $n_F n_{CF_3}/n_{CF_4} = A(T)$ and that $n_{CF_2}/n_{CF_3} = B(T)$. Obtain A and B in terms of K_4, K_5, K_1, K_2, and n_e.

(c) If the initial concentration of CF_4 is n_0 and all other initial concentrations are zero at time $t = 0$, then find the equilibrium concentration ($t \to \infty$) of F atoms in terms of n_0, A, and B.

(d) For a CF_4 pressure of 10 Torr at 300 K, and silicon etching due to a flux of F atoms only, use (16.3.1) to estimate the initial etch rate (Å/min) when a piece of silicon is inserted into the equilibrium gas mixture.

16.3 Surface Model for Silicon Etch in a CF_4 Discharge Consider the following surface model for pure chemical silicon etch (no ion bombardment) in a CF_4 discharge. Let n_1 and n_2 be the gas-phase densities of CF_x radicals and F atoms near the surface, respectively, and let θ_1 and θ_2 be the fractions of the SiF_3 surface covered with CF_4 and SiF_4, respectively. Let K_{a1} and K_{a2} (cm^3/s) be the adsorption rate constants for CF_x radicals and F atoms, respectively, and let K_{d1} and K_{d2} be the desorption rate constants (s^{-1}) for $CF_4(g)$ and $SiF_4(g)$, respectively. Assume Langmuir kinetics with adsorption of CF_x and F on the SiF_3 surface only.

(a) In the steady state, give the two conservation equations for carbon and fluorine on the surface.

(b) Solve these to obtain the surfaces coverages θ_1 and θ_2.

(c) Find the silicon etch rate E_{Si} and plot the normalized etch rate per incident F atom, E_{Si}/n_2 (Å-cm^3/min) versus n_2/n_1 for $x = 3$, $K_{a1} = K_{a2} = 4 \times 10^{-14}$ cm^3/s, $n_0' = 7 \times 10^{14}$ cm^{-2}, $K_{d1} = K_{d2} = 10^{12}$ s^{-1}, and $n_{SiF_3} = 5 \times 10^{22}$ cm^{-3}. Assume that $n_1, n_2 \ll K_{d2}/K_{a1}$.

16.4 Comparison of Silicon and SiO_2 Loading Effects For the same reactor (volume Al, pumping speed S_p) and discharge conditions (gas density n_{CF_4}, plasma density n_i, and electron temperature T_e), SiO_2 etching in CF_4 discharges exhibits a smaller loading effect than silicon etching. Assume that the fractional dissociation ($n_F^{(0)}/n_{CF_4}$) is small in the absence of etching reactions. Assume that the overall reactions for silicon and SiO_2 etching are

$$4C + 16F + Si \longrightarrow SiF_4 + 2C_2F_6$$

$$C + 4F + SiO_2 \longrightarrow SiF_4 + CO_2$$

In both cases, assume that $\gamma_{rec} = 0$ on the walls and that $\gamma_r = 1$ on the substrates.

(a) Using (16.2.17) and (16.2.18), find $n_F^{(0)}/n_F$ in terms of $\bar{v}_F$, S_p, and the wafer area A_w.

(b) Find expressions for the etch rates (fluxes) Γ_{SiO_2} and Γ_{Si} as functions of n_F.

(c) For equal etch rates of SiO_2 and Si and for a silicon etch area $A_w(Si)$, find the SiO_2 etch area. Show that $A_w(SiO_2) \geq 4A_w(Si)$.

16.5 Atomic Layer Etching of Silicon Using Cl Atom Chlorination

(a) Repeat the analysis of (16.5.5)–(16.5.7) for the (non-dissociative) chemisorption of Cl atoms on the silicon surface, to obtain the chlorine surface coverage θ versus the normalized Cl atom fluence φ.

(b) Plot θ versus φ for a zero-coverage sticking coefficient of $s_0 = 0.4$, and compare to the dissociative chemisorption of Cl_2 shown in Figure 16.19*a*.

(c) For the same precursor fluxes of 10^{22} m^{-2}s^{-1}, compare the chlorination times of Cl and Cl_2 for 95% surface coverage.

16.6 Cryogenic Atomic Layer Etching of Silicon

Suppose that $\mathcal{M}_{Cl} > 1$ monolayers of Cl can be adsorbed on the surface at the end of each chlorination cycle, for example, by cooling the substrate to enable physisorption as well as chemisorption of Cl_2. For $\mathcal{E}_i = 48$ V:

(a) Find the required value of $\mathcal{M}_{Cl}$ that makes $\theta_0 = 1$ in (16.5.17).

(b) Calculate and plot $\theta(\varphi_i)$ from (16.5.13) and $N_{Si}(\varphi_i)$ from (16.5.15), comparing them to the $\mathcal{E}_i = 48$ V curves plotted in Figure 16.19*b* and Figure 16.19*c*, respectively.

16.7 Substrate Potential for a Thick Gate Oxide Show that the substrate potential with respect to ground for the system of Figure 16.24*a* with thick gate oxides is

$$V_s = V_{p1} - V_f + \frac{A_{ox2}/T_{ox2}}{A_{ox2}/T_{ox2} + A_{ox1}/T_{ox1}}(V_{p2} - V_{p1})$$

16.8 Potential for a Wafer Exposed to a Nonuniform Plasma For a parabolic plasma potential $V_p(r) = V_{p0} + \Delta V_p(1 - r^2/R^2)$, where R is the wafer radius, show that the (conducting) substrate potential with respect to ground is $V_s \approx V_{p0} - V_f - T_e \ln[T_e(1 - e^{-\Delta V_p/T_e})/\Delta V_p]$, and find the plasma current density $J_p(r)$ flowing into the wafer.

17

Deposition and Implantation

17.1 Introduction

Plasma-assisted deposition, implantation, and surface modification are important material processes for producing films on surfaces and modifying their properties. For example, as described in Chapter 1, the cycle of film and mask deposition, mask patterning, implantation or other modification, etching, and mask stripping is repeated many times during the manufacture of modern integrated circuit devices. Because device structures are sensitive to temperature, high-temperature deposition processes cannot be used in many cases. Fortunately, due to the nonequilibrium nature of low-pressure processing discharges, high-temperature films can be deposited at low temperatures. Furthermore, films can be deposited with improved properties, nonequilibrium chemical compositions, and crystal morphologies that are unattainable under equilibrium deposition conditions at any temperature. Unique films not found in nature can be deposited, e.g., diamond.

Consider two examples for integrated circuit fabrication. Most aluminum thin films (i.e., actually Al/Cu or Al/Si) used for interconnection are deposited on the wafer by physical sputtering from an aluminum or alloy target; this is essentially a room-temperature process. Although thermal evaporation sources can be used, it is more difficult to control film uniformity and composition with these sources. Another example is the final insulating "capping" layer on many devices, silicon nitride, which is deposited by plasma-enhanced chemical vapor deposition (PECVD) at temperatures near 300 °C. An equivalent non-plasma chemical vapor deposition (CVD) would require temperatures near 900 °C and therefore cannot be used because it would melt the aluminum, destroying the device. Furthermore, by varying the ion bombardment and other plasma parameters in PECVD of silicon nitride, the film composition, stress, and integrity can be controlled, greatly increasing its reliability as a capping layer. Let us note, however, that PECVD cannot replace CVD in some applications, e.g., most low-temperature PECVD films are amorphous and not crystalline. Crystalline films can more easily be achieved with CVD. Where high temperatures are allowed, CVD can be the method of choice for deposition of metals, dielectrics, and semiconducting films.

Ion implantation is another important process for semiconductor doping and has other uses, such as for surface hardening of materials. For silicon doping, ions such as boron, phosphorous, and arsenic are implanted. For surface hardening of metals, nitrogen or carbon is implanted. Conventional ion beam implanters are used for low-flux, high-energy implants. At high fluxes, particularly for low ion energies, and where mass-energy selection is not critical, plasma-immersion ion implantation (PIII) can be used to meet process requirements that are not attainable using conventional ion beam implanters. PIII processes have been developed for hardening medically implantable hip joints, for hardening tools and dies, and for doping semiconducting materials. Materials modifications through a combination of ion implantation and ion beam mixing of near-surface layers are also under development. As described in Chapter 16, ion beam mixing can also play a critical role in etch processes.

In this chapter, as in the previous, we focus on the area of integrated circuit processing to describe deposition, implantation, and other surface-modification processes. Recent general reviews by Gudmundsson et al. (2022) and Snyders et al. (2023) focus mainly on the various physical techniques that have been developed for plasma-enhanced thin film deposition. Seshan and Schepis (2018) describe thin film deposition technologies used in the semiconductor and photovoltaic industries. For thorough treatments of plasma-assisted deposition, implantation, and surface modification processes, the monographs of Konuma (1992), Smith (1995), Mahan (2000), Anders (2000), the collections of review articles edited by Vossen and Kern (1978, 1991), and references cited therein should be consulted.

The range of deposition processes in this chapter is broadly divided into three areas: PECVD, thermal and plasma-enhanced atomic layer deposition (thermal ALD and PEALD), and sputter deposition. PECVD is described in Section 17.2, using a well-known example of amorphous silicon (a-Si) deposition to introduce the discharge regime, gas-phase chemistry, and surface-reaction model. While specific to a-Si deposition, the discussion is relevant to PECVD for many other materials. PECVD of SiO_2 is also described to introduce a more complicated surface chemistry and to treat the issues of anisotropic deposition and conformality of deposition over topography, e.g., deposition in trenches. Almost all the discharges described in previous chapters are widely used for PECVD, with the exception of dc discharges, although the bulk of the deposition is done commercially with some form of multi- or single-wafer capacitive rf discharge reactor.

Atomic layer deposition (ALD), which is described in Section 17.3, is a key process for high precision thin film deposition on the monolayer length scale. It is highly important for depositions on three-dimensional feature sizes of 20 nm or less. Here, we focus first on the well-characterized example of thermal ALD of Al_2O_3. Next, we describe PEALD for the same material. Finally, we examine the key issue of ALD deposition conformality on three-dimensional nanostructures.

Sputter deposition, which is discussed in Section 17.4, includes both physical sputtering and reactive sputtering. In the former, atoms are sputtered from a target material and are transported to and deposited on a substrate. The mechanism of physical sputtering is described in Section 9.3, and some data are given in Table 9.2. Some issues related to sputtering uniformity are also considered in Section 14.5. In Section 17.4, we describe the influence of sputtered atom energy distributions on film properties. In reactive sputtering, a feedstock gas whose dissociation products chemically react with the target material is present in addition to the bombarding ions. Hence, the deposited film is a compound formed from the sputtered materials and the constituents of the reactive gases. In contrast to physical sputtering, where a model for the generation and transport of sputtered atoms from target to substrate is relatively straightforward, a reactive sputtering model involves surface reactions at both target and substrate in addition to sputtering at the target and deposition at the substrate. A simple model for this process is given to conclude Section 17.4. Sputtering discharges for depositing conducting films are generally dc-driven, usually dc planar magnetron sputtering discharges (see Section 14.5). For sputtering insulating films, capacitive rf discharges or rf-driven planar magnetron discharges are commonly used.

Ion implantation using PIII is described in Section 17.5. The basic principles for the process are given, a simple model for the dynamic high-voltage sheath formation is developed, and some applications to integrated circuit and other processing are described. PIII must generally be done in low-pressure ($p \lesssim 1$ mTorr) processing discharges in which the ion mean free path is comparable to or larger than the high-voltage sheath width, but there are some applications where higher pressures are desirable.

Other plasma-enhanced surface modification processes, not treated in this text, include low-temperature oxidation of silicon, plasma polymerization, and additional (non-PIII) nitriding and carbiding techniques. For example, good quality thin SiO_2 films have been grown on single-crystal silicon in oxygen discharges at substrate temperatures of 250–400 °C (Carl et al., 1991). The process is called *plasma anodization* because the substrate is generally biased positive with respect to the plasma, drawing a net dc current through the film as it grows. Oxidation kinetics can be explained by O^- transport-limited growth at the Si–SiO_2 interface. Contamination due to sputtering during film growth is an issue, so microwave and other high-density discharges having low sheath voltages are generally used.

17.2 Plasma-Enhanced Chemical Vapor Deposition

Chemical vapor deposition (CVD) consists of a thermally activated set of gas-phase and surface reactions that produce a solid product at a surface. In PECVD, the gas phase and often the surface reactions are controlled or strongly modified by the plasma properties. In place of thermal activation in CVD, the critical initial step in PECVD is electron impact dissociation of the feedstock gas. Since $T_e \sim$ 2–5 V in a low-pressure discharge easily suffices for feedstock dissociation and since T_e is much greater than the substrate (and heavy particle) temperature, the deposition can be carried out at temperatures much lower than for CVD. Because chemical reactions between neutral gas-phase precursor components are often required for PECVD, the discharge pressures used are in the range 0.1–10 Torr, considerably higher than those used for plasma-assisted etching. The neutral mean free paths are therefore small, of order 0.003–0.3 mm. The plasma densities are in the range 10^9–10^{11} cm^{-3}, and the fractional ionizations are low, of order 10^{-7}–10^{-4}. As for etching, the deposition is limited by either the feedstock gas flow rate and pressure or by the discharge power, depending on which is rate limiting. Surface activation energies for PECVD are often small, occasionally negative. Hence, deposition rates are usually not very sensitive to the substrate temperature T. However, film properties such as composition, stress, and morphology are generally strong functions of T. Consequently, T is usually optimized to achieve a desired set of film properties.

Deposited film uniformity is a critical issue for PECVD because of the high pressures, high flow rates, short mean free paths, high gas-phase reaction rates, and high surface sticking probabilities for some gas-phase deposition precursors (often, neutral radicals). This combination of factors makes it very difficult to achieve uniform precursor and ion fluxes across the substrate area. Hence, great care is required in the designing of the neutral transport system for flow of gases into and out of the reaction zone. Similarly, the variation of the power deposition per unit area in the discharge must be carefully controlled. For these reasons, rf-driven parallel-plate discharge geometries have been favored, although some depositions have been performed using high-density cylindrical discharges, such as ECRs, helicons, and rf inductive discharges (ICPs).

17.2.1 Amorphous Silicon

Amorphous silicon thin films are used in solar cells, for thin-film transistors for flat panel displays and for exposure drums for xerography. Whereas epitaxial (crystalline) silicon has a density of 2.33 g/cm^3, PECVD amorphous silicon grown using silane (SiH_4) discharges has a lower density, ~2.2 g/cm^3, due to the incorporation of 5–20% H atoms in the lattice. Hence, this material is usually denoted as a-Si:H. The hydrogen is required for this material to be semiconducting; the H-atoms terminate the dangling bonds in the amorphous material that would otherwise trap charge carriers. The material is inexpensive to make and easily deposited over large areas on a wide variety of substrates including glasses, metals, polymers, and ceramics. The feedstock gas in a capacitive rf discharge is typically SiH_4 at pressures of order 0.2–1 Torr, although $SiH_4/H_2/Ar$ mixes are sometimes used at somewhat higher pressures. Gas-phase additions such as B_2H_6 and PH_3 are used to grow p-type or n-type material, respectively. The rf power fluxes are typically 10–100 mW/cm^2, yielding deposition rates of 50–500 Å/min. The substrate temperatures are typically 25–400 °C, depending on the application. The activation energy for the deposition is low, 0.025–0.1 V, compared to 1.5 V for high-temperature CVD silicon deposition using SiH_4.

SiH_4 is a hazardous gas that reacts explosively with air or water vapor. The molecule is tetrahedral (symmetry group T_d, with the silicon atom in the center), having a heat of formation of 34.3 kJ/mol and a Si–H bond distance of 1.5 Å. The SiH_3–H bond energy is 3.9 V. The positive ion SiH_4^+ is unstable or weakly stable and has not been observed under typical discharge conditions; SiH_3^+ is normally observed. Both SiH_3 and SiH_2 radicals have a positive electron affinity; hence, silane discharges can be electronegative. The SiH_2–H, SiH–H, and Si–H bond energies are 3.0, 3.4, and 3.0 V, respectively. Some rate constants for significant (mostly two-body) gas-phase reactions are given in Table 17.1. A relatively complete $SiH_4/H_2/Ar$ gas-phase discharge model was introduced by Kushner (1988). This model includes over 35 electron impact reactions, 90 neutral–neutral reactions, 80 positive ion–neutral reactions, and a complete set of electron–ion and positive–negative ion recombination reactions. Updated cross sections and rate coefficients can be found in Perrin et al. (1996, 1998).

Table 17.1 Selected Reaction Rate Constants for SiH_4 Discharges

Number	Reaction	Rate Constant (cm^3/s)	Source
1	$e + SiH_4 \rightarrow SiH_3 + H + e$	$1.5E{-}8\exp(-10/T_e)$	a)
2	$e + SiH_4 \rightarrow SiH_2 + 2H + e$	$1.8E{-}9\exp(-10/T_e)$	a)
3	$e + SiH_4 \rightarrow SiH_3^- + H$	$1.5E{-}11\exp(-9/T_e)$	a)
4	$e + SiH_4 \rightarrow SiH_2^- + H_2$	$9E{-}12\exp(-9/T_e)$	a)
5	$e + SiH_4 \rightarrow SiH_3^+ + H + 2e$	$3.3E{-}9\exp(-12/T_e)$	a)
6	$e + SiH_4 \rightarrow SiH_2^+ + H_2 + 2e$	$4.7E{-}9\exp(-12/T_e)$	a)
7	$SiH_4 + H \rightarrow SiH_3 + H_2$	4E–13	b)
8	$SiH_4 + SiH_2 \rightarrow Si_2H_6^*$	1E–11	c)
9	$Si_2H_6^* \rightarrow Si_2H_4 + H_2$	5E6 /s	c)
10	$Si_2H_6^* + SiH_4 \rightarrow Si_2H_6 + SiH_4$	1E–10	c)
11	$SiH_4 + SiH_3 \rightarrow Si_2H_5 + H_2$	1.8E–15	c)
12	$SiH_3 + H \rightarrow SiH_2 + H_2$	1E–10	c)
13	$SiH_3 + SiH_3 \rightarrow SiH_2 + SiH_4$	7E–12	c)
14	$e + SiH_n^+ \rightarrow SiH_{n-1} + H$	$2.5E{-}7\,T_e^{-1/2}$	a)
15	$SiH_m^- + SiH_n^+ \rightarrow SiH_m + SiH_n$	5E–7	c)

Note. T_e in volts and $T \sim 500$–700 K for ions and neutrals. The notation E–8 means 10^{-8}.
a) Based on data in Kushner (1988).
b) Becerra and Walsh (1987).
c) Kushner (1988).

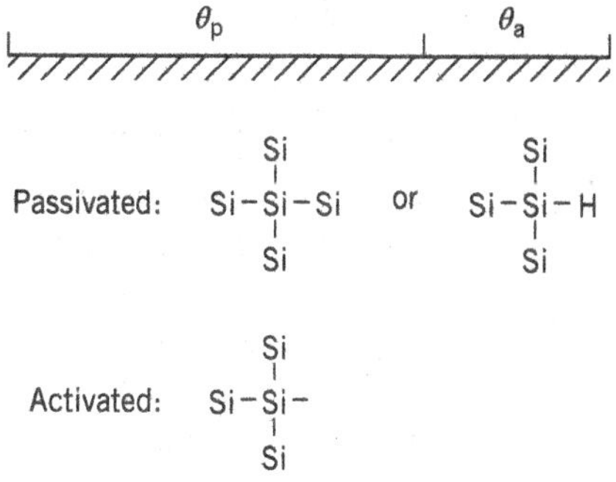

Figure 17.1 Surface coverage model for amorphous silicon deposition; θ_a and θ_p are the fractions of the surface that are active and passive, respectively.

θ_p θ_a

Passivated: Si–Si(Si)(Si)–Si or Si–Si(Si)(Si)–H

Activated: Si–Si(Si)(Si)–

There is considerable evidence (McCaughey and Kushner, 1989; Smith, 1995, Sec. 9.4.6) that SiH_3 and SiH_2 radicals are important precursors for film growth, that SiH_4 also participates in surface reactions, and that ion (SiH_3^+) bombardment plays a critical role in film growth. A simple model of the surface, shown in Figure 17.1, is that it consists of active sites, containing at least one dangling bond, and passive sites, containing either silicon or hydrogen atoms at all four bonds. The dangling bonds are created by ion bombardment, which also removes hydrogen from the surface. SiH_2 can insert itself into the lattice upon impact with the surface at either active or passive sites, leading to film growth in a manner similar to that of physical vapor deposition (i.e., as in (9.4.28)). Such films are generally of poor quality, having voids, undesired surface roughness, and other surface defects. Adsorbed SiH_3 radicals can diffuse along the surface but can insert into the lattice only at active sites, filling in the surface roughness and contributing to the growth of a smooth, high-quality film. SiH_4 adsorbed upon impact at active sites can lose an H atom, thus passivating the site. Based on these ideas, elaborated by McCaughey and Kushner, we let θ_a and θ_p be the fraction of the surface covered by active and passive sites, respectively, with $\theta_a + \theta_p = 1$. Then, the surface reactions can be represented as

$$SiH_3^+ + \theta_p \xrightarrow{K_i} \theta_a + Y_i H\,(g) \tag{17.2.1}$$

$$SiH_2 + \theta_a \xrightarrow{K_2} \theta_a \tag{17.2.2}$$

$$SiH_2 + \theta_p \xrightarrow{K_{2p}} \theta_p \tag{17.2.3}$$

$$SiH_3 + \theta_a \xrightarrow{K_3} \theta_p \tag{17.2.4}$$

$$SiH_4 + \theta_a \xrightarrow{K_4} \theta_p + SiH_3\,(g) \tag{17.2.5}$$

where Y_i is the yield of H atoms removed per incident ion, $K_i \approx u_B/n_0'$, $K_2 \approx \frac{1}{4}s_2\bar{v}_2/n_0'$, $K_3 \approx \frac{1}{4}\overline{M}s_3\bar{v}_3/n_0'$, $K_4 \approx \frac{1}{4}s_4\bar{v}_4/n_0'$, and $K_{2p} \approx \frac{1}{4}s_{2p}\bar{v}_2/n_0'$ are the rate constants, with s_2, s_3, and s_4 the sticking coefficients on the activated surface for SiH_2, SiH_3, and SiH_4, respectively, s_{2p} is the sticking coefficient for SiH_2 on the passivated surface, n_0' is the area density of sites, and $\overline{M}$ is the mean number of sites visited by a surface-diffusing SiH_3 radical before desorption. In the steady state, the rate of creation of active sites is

$$\frac{d\theta_a}{dt} = Y_i K_i n_{is}(1-\theta_a) - K_3 n_{3S}\theta_a - K_4 n_{4S}\theta_a = 0 \tag{17.2.6}$$

Solving for θ_a, we obtain

$$\theta_a = \frac{Y_i K_i n_{is}}{Y_i K_i n_{is} + K_3 n_{3S} + K_4 n_{4S}} \tag{17.2.7}$$

Note that SiH_2 adsorption and reaction do not affect θ_a in this model. The deposition rate follows from reactions (17.2.2), (17.2.3), and (17.2.4), which each deposit one silicon atom:

$$D_{Si} = (K_3 n_{3S} \theta_a + K_2 n_{2S}) \frac{n'_0}{n_{Si}} \tag{17.2.8}$$

For typical deposition processes, $Y_i \sim 5$–10, all ss are of order unity, $\overline{M} \sim 10$, and $K_4 n_{4S} \gg Y_i K_i n_{is} + K_3 n_{3S}$. Under these conditions, (17.2.7) yields

$$\theta_a \approx \frac{Y_i K_i n_{is}}{K_4 n_{4S}} \tag{17.2.9}$$

For typical discharge parameters, $n_{is}/n_{4S} \sim 10^{-4}$ and, therefore, $\theta_a \sim 10^{-2}$. The overall sticking coefficient for SiH_3 to react with the entire surface (active and passive) is then $s_3 \overline{M} \theta_a \sim 0.1$. Under typical conditions for film deposition, $n_{2S} \sim 10^{-2} n_{3S}$, such that the first term in (17.2.8), responsible for the "good" film deposition, is roughly 10 times larger than the second term, responsible for the "bad" film deposition. Clearly, from (17.2.8) and (17.2.9), good films are associated with high ion fluxes and energies, high SiH_3/SiH_2 ratios, and high SiH_3 surface diffusivities.

The preceding picture is oversimplified. For example, the reactions of H atoms at the surface,

$$\mathrm{H} + \theta_a \longrightarrow \theta_p$$

$$\mathrm{H} + \theta_p \longrightarrow \theta_a + \mathrm{H}_2\,(\mathrm{g})$$

can modify the overall surface dynamics. A more complete model of this type is presented by McCaughey and Kushner (1989). Other models have focused on the surface diffusion of SiH_3 and its reaction with the hydrogenated surface to remove dangling bonds, and on the role of sub-surface processes, such as H-atom diffusion, in bond formation (see Robertson, 2000 and references therein).

17.2.2 Silicon Dioxide and Conformality

SiO_2 can be grown by oxidation of bare silicon at 850–1100 °C using O_2 or H_2O gas. CVD oxide can also be deposited on substrates at 600–800 °C using SiH_4/O_2 or TEOS/O_2 feedstock gases and can be grown at still lower temperatures, 100–300 °C, using PECVD with the same feedstocks. TEOS (tetraethoxysilane), $Si(OC_2H_5)_4$, has the chemical structure shown in Figure 17.2, with C–O and

```
            H
            |
          H-C-H
            |
          H-C-H
            |
    H  H    O    H  H
    |  |    |    |  |
  H-C--C--O-Si-O--C--C-H        TEOS
    |  |    |    |  |
    H  H    O    H  H
            |
          H-C-H
            |
          H-C-H
            |
            H
```

Figure 17.2 Chemical structure of TEOS.

Si–O bond energies of 3.7 and 4.7 V, respectively. In contrast to silane, which is an explosive gas at room temperature, TEOS is a relatively inert liquid. Gases such as N_2 or Ar are often used as carriers of the vapor. Highly dilute TEOS/O_2 feedstock mixtures are usually used; a 1%TEOS/99%O_2 mixture is typical. Under these conditions, much of the gas-phase kinetics is dominated by O_2, and the discharge can be modeled as if it were a pure O_2 discharge. Highly oxygen-rich mixtures are required for good quality films because TEOS contains carbon and hydrogen, which the O_2 burns to form CO_2(g) and H_2O(g) effluents. If this is not done efficiently, then the films can have a substantial carbon and/or hydrogen content.

Oxide deposition using $SiH_4/Ar/N_2O$, $SiH_4/Ar/NO$, or $SiH_4/Ar/O_2$ gas mixtures can yield deposition rates of up to 2000 Å/min. The usual oxygen source is N_2O, as this produces copious oxygen atoms on dissociation and the best quality films. The deposition precursors are believed to be SiH_3, SiH_2, and O radicals created by electron impact dissociation of SiH_4 and the oxidant (N_2O, NO, or O_2). Film formation consists of surface reactions such as

$$SiH_3\,(g) + SiG_3(OH) \longrightarrow SiGH_2 + SiO_2 + H_2 \qquad (17.2.10)$$

where $G = \frac{1}{2}O$ is an oxygen atom that is shared with another surface silicon atom. Further oxygenation of the surface burns off most of the excess H atoms as H_2O (g). The final film typically has 2–9% H atoms. A relatively complete model of the gas phase and surface chemistry for SiO_2 film formation in $SiH_4/O_2/Ar$ discharges has been given by Meeks et al. (1998).

17.2.2.1 Conformality

The sticking probabilities of the precursors SiH_3 and SiH_2 in silane discharges tend to be high, e.g., $s \sim 0.35$. This tends to lead to nonconformal deposition on topographical features such as in trenches, i.e., the deposition rates at various points on the trench surface are different. To understand this, consider a simple model of deposition on the sidewalls and bottom of a trench of initial width w and depth d, due to a uniform isotropic source of precursors at the top of the trench, as shown in Figure 17.3*a*. Assume a unity sticking coefficient and ballistic transport of precursors within the trench, i.e., the mean free path for precursor collisions is much greater than w or d. Then, it can be shown (Problem 17.2) that the deposition flux Γ_{SiO_2} on the sidewall is

$$\Gamma_{SiO_2} \propto 1 - \cos\theta_s \qquad (17.2.11)$$

where θ_s is the angle subtended by the trench opening as seen at a position along the sidewall. Note that θ_s is 90° near the top of the trench and falls monotonically with depth along the sidewall. Hence, the maximum deposition rate is on the sidewall near the top of the trench. As deposition proceeds, as shown in Figure 17.3*b* and *c*, this can lead to the formation of a void or "keyhole" within the trench. This is undesirable for many applications. If the deposition is conformal, i.e., equal growth rates at all points within the trench, the keyhole is avoided and the trench completely fills with the insulating dielectric. To achieve conformal deposition, either the sticking probability should be small, leading to many precursor reflections within the feature, or precursors that stick with high probability should have high diffusion rates along the surface.

For SiO_2 deposition using silane-containing feedstocks, the sticking probabilities are high and the surface diffusion is not significant; consequently, the conformality of deposition is poor. Oxide deposition using highly dilute TEOS/O_2 feedstock at temperatures of 200–300 °C and pressures of 0.2–0.5 Torr leads to relatively low deposition rates, $\lesssim$ 500 Å/min, compared to silane-based deposition. However, the sticking coefficients for TEOS deposition precursors seem to be almost an order of magnitude smaller than for the silane precursors, e.g., $s \sim 0.045$ for TEOS, leading to good

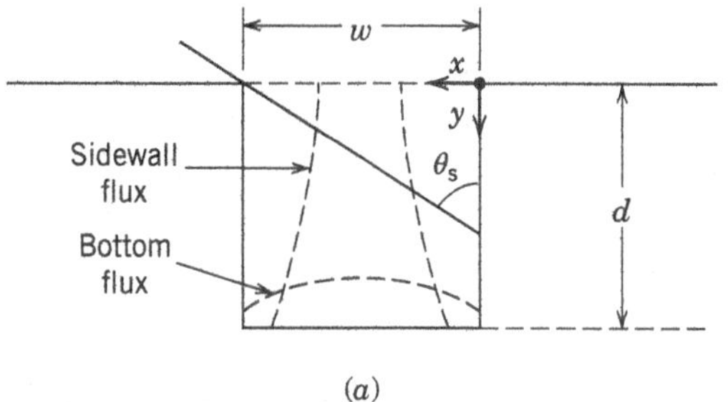

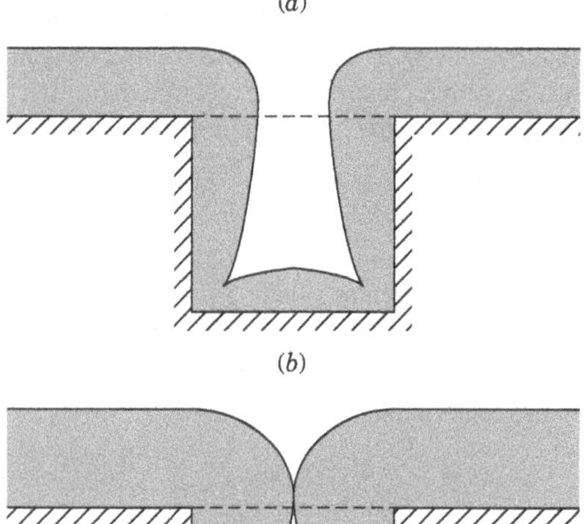

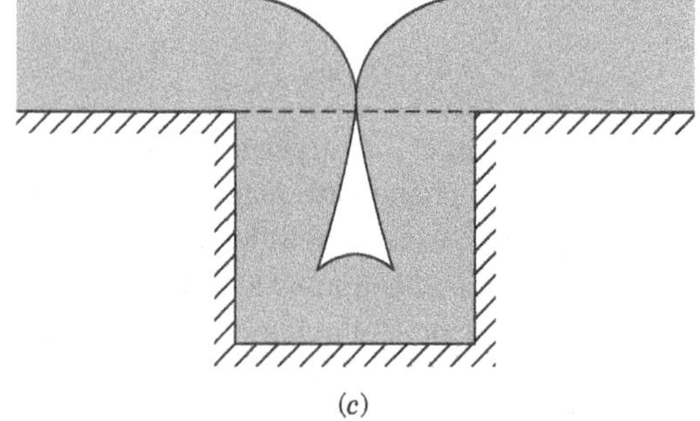

Figure 17.3 Nonconformal deposition within a trench, illustrating formation of a void as deposition proceeds: (*a*) before deposition, with the dashed lines giving the deposition flux incident on the sidewall and bottom; (*b*) midway during deposition; (*c*) just after the keyhole-shaped void has formed.

deposition conformality. TEOS precursors are believed to be species such as $Si(OC_2H_5)_n(OH)_{4-n}$ or $Si(OC_2H_5)_nO_{4-n}$, $n = 0$–3. These can be formed by electron impact dissociation, e.g., for $n = 1$–4

$$e + Si(OC_2H_5)_n(OH)_{4-n} \longrightarrow Si(OC_2H_5)_{n-1}(OH)_{4-n+1} + C_2H_4 + e \tag{17.2.12}$$

or by O-atom reactions with TEOS and its precursors, e.g.,

$$O + Si(OC_2H_5)_n(OH)_{4-n} \longrightarrow Si(OC_2H_5)_{n-1}(OH)_{4-n+1} + C_2H_4O \tag{17.2.13}$$

In highly dilute TEOS/O_2 mixtures, the latter reactions predominate. It is not known which precursors are present in the highest concentrations. The precursors adsorb on the growing film surface where reactions with adsorbed O atoms further fragment the precursor and further oxidize the carbon and hydrogen. This oxidation process on the surface may be the rate-limiting step in the deposition. It is also known that there can be significant directionality in the deposition process. Presumably, this is due to ion bombardment which enhances the vertical deposition rate. The measured TEOS deposition rates at moderate to high temperatures show a negative activation energy, $\mathcal{E}_a \sim -(0.1$–$0.2)$ V, i.e., the deposition rate increases as the substrate temperature is lowered. This can be interpreted in one of the two ways: either the desorption rate for TEOS precursors increases with increasing temperature, thus reducing the precursor coverage on the surface (Stout and Kushner, 1993), or there is increased surface recombination of O atoms as the temperature is increased, decreasing the gas-phase O-atom density (Cale et al., 1992).

The deposition chemistry and surface reactions for the former hypothesis can be described in a manner similar to that done previously for amorphous silicon deposition. The chemistry with the latter hypothesis can be described by the following three reactions:

1. O atoms oxidize TEOS precursors on the surface, leading to deposition. It is assumed that the TEOS fragments completely saturate the surface. Hence, the reaction rate is independent of the precursor surface coverage. The deposition rate by this reaction is given as

$$D_{SiO_2}^{(1)} \approx \frac{0.9 n_{OS}}{n_{SiO_2}} \text{cm/s} \tag{17.2.14}$$

   where the deposition rate constant 0.9 is determined by a fit to experimental data.
2. Oxygen ions also oxidize TEOS precursors, leading to a deposition rate

$$D_{SiO_2}^{(2)} \approx \frac{n_{O_2^+} u_B}{n_{SiO_2}} \text{cm/s} \tag{17.2.15}$$

3. Surface recombination of O atoms on wall (and deposition) surfaces to form nonreactive O_2 molecules reduces the available gas-phase O-atom concentration n_{OS} for reaction (17.2.13). The recombination probability $s_{rec}(T)$ on SiO_2 surfaces is activated but has a non-Arrhenius form (Greaves and Linnett, 1959), with $10^4 s_{rec} \approx 1.8, 2.7, 6.5$, and 50 at $T = 20, 127, 200$, and 394 °C, respectively. The flux of O atoms lost from the volume due to this process is

$$\Gamma_{rec} \approx 2 s_{rec}(T) \frac{1}{4} n_{OS} \bar{v}_O \tag{17.2.16}$$

   As T increases, Γ_{rec} increases, leading to a decrease in n_{OS} for a fixed generation rate, and, hence, a reduction in the deposition rate (17.2.14).

17.2.3 Silicon Nitride

Amorphous silicon nitride films were the first deposited on a large commercial scale using PECVD. They are used as a final encapsulating layer for integrated circuits because of their resistance to water vapor, salts, and other chemical contaminants. Other applications are as dielectric layers, masking layers, and passivation layers. The usual feedstock mix is SiH_4/NH_3 (Smith, 1995, Sec. 9.6.4). The film precursors are probably SiH_3, SiH_2, and NH radicals, created by electron impact dissociation. Other possible precursors, such as Si_2H_6, $Si(NH_2)_4$, and $Si(NH_2)_3$, have been observed in SiH_4/NH_3 discharges by mass spectroscopy. The deposition is normally carried out at 0.25–3 Torr at 250–500 °C, yielding deposition rates of 200–500 Å/min. The activation energy for the deposition rate is small and can even be negative, depending on discharge conditions. The stoichiometry of the films is SiN_xH_y (sometimes called p-Sinh), with $x \sim 1$–1.2 and $y \sim 0.2$–0.6. The hydrogen atoms are bonded in the lattice, and low hydrogen content is associated with high temperatures and high rf power fluxes. Below 300 °C, the hydrogen content is relatively constant. A key step in film formation is thermal- or ion-induced desorption of H or H_2 from the growing film. The film characteristics depend strongly on the hydrogen content, with high hydrogen content yielding undesired films. The mechanical stress can be controlled by varying electrical properties of the discharge, such as the rf driving frequency. This variation is associated with the ion-bombarding energy, which decreases (at fixed rf power) with increasing frequency, as described in Chapter 11.

The source of most of the hydrogen in the films has been identified as NH_3, not SiH_4. This has motivated the addition or use of other nitrogen sources such as N_2. Although SiH_4/N_2 can be used as a feedstock, the films are of poorer quality, the conformality is not as good, and the deposition rate is lower than with the use of NH_3 as the source of nitrogen. On the other hand, the films have much less hydrogen and are richer in nitrogen than those grown using NH_3. Some of these effects can be controlled using mixtures such as $SiH_4/NH_3/N_2$.

17.2.4 Large-Area PECVD

An important application of PECVD is the fabrication of thin film transistors (TFTs) or p-n junctions on large sheets of glass. TFT arrays are used as switches on active matrix liquid crystal displays, and pin junctions are used in thin film silicon solar cells. The rectangular glass substrate sizes can be as large as 2.94 m × 3.37 m. They are also quite thin, typically 0.7 mm for displays, and 0.5–4 mm for solar cells. The deposited materials include amorphous and microcrystalline silicon, sometimes doped with boron or phosphorus, silicon oxide, silicon nitride, and silicon oxynitride. Deposition temperatures are 200–350 °C, and reactor pressures are 0.3–3 Torr. Deposited film thickness is of order 0.5 μm, with required uniformities of ±5% or less.

Capacitively coupled reactors are usually used, having a 2–3-cm gap driven at 13.56 MHz. Standing wave effects (see Section 11.7) and gas flow effects (see Section 9.5) have to be carefully considered and properly compensated in such large reactors (Perrin et al., 2000; Choi and White, 2009). Large rf powers, ∼30 kW, and rf currents, ∼500 A, are required, presenting challenging rf power matching and power loss issues. For chamber wall cleaning, remote plasma sources can be used, with NF_3, SF_6, or F_2 feedstocks, producing highly reactive fluorine atoms that flow into the reactor chamber.

For a-Si:H and μc-Si deposition, feedstock mixes are typically SiH_4/H_2, sometimes doped with B_2H_6, PH_3, etc. Silicon oxide deposition usually uses SiH_4/O_2, and silicon nitride deposition uses $SiH_4/NH_3/N_2$. Noble gas diluents are sometimes added to improve deposited film properties.

As discussed in Sections 11.1 and 11.2, single-frequency capacitive reactors have only a limited number of parameters that can be varied to optimize the deposition process. These include the rf power, feedgas flow rates, chamber pressure, deposition temperature, and, perhaps, the discharge gap and rf frequency. The requirements of good film quality, adequate growth rates, and good uniformity over the entire 10 m^2 area often highly constrain the discharge parameters into a narrow operating window (Perrin et al., 2000).

17.3 Atomic Layer Deposition

17.3.1 Introduction

Integrated circuit fabrication with feature sizes of 20 nm or less often requires high-precision thin film deposition on the atomic scale. The deposited film must be highly conformal to the 3D features, with a controlled thickness, perhaps 10s of monolayers. It must be highly uniform and selective over the entire range of lengths from feature- to wafer-scale.

In the processes described in Sections 17.1 and 17.2, deposition precursors continually flow to the surface. For both high-temperature CVD and lower-temperature PECVD, at the $\lesssim$ 20 nm feature scale length, the precursor flux can be strongly depleted at the bottom of narrow trenches and small-diameter vias by sidewall shadowing and surface reactions, and increased by sidewall desorption. Typically, as shown in Figure 17.3, this leads to highly nonuniform coverage in narrow trenches or vias. These effects place severe limits on conventional continuous deposition processes. At the wafer-scale length, feedgas precursor fluxes might be edge-high due to gas flows (see Section 9.5), and plasma-generated precursors might be center-high (see Figure 16.24). These effects lead to deposition nonuniformities across the wafer.

One way to overcome these limitations is a cyclical two-step, thermally activated process called atomic layer deposition (ALD). In an ideal ALD process, (1) the surface is first saturated with a high fluence Φ_1 of a deposition precursor, typically yielding a precursor layer of order one monolayer thick. Here, $\Phi_1 = \int_0^{T_1} \Gamma_1(t)\,dt$ is the number of precursor molecules per unit area incident on the surface, with Γ_1 the precursor flux and T_1 the exposure time. The adsorbed precursor species must be chosen to passivate the surface, preventing multi-layer adsorption, i.e., the adsorption must be self-limiting. In the second step (2), saturated exposure of the adsorbed layer to a high fluence Φ_2 of a reactive gas converts it into the desired deposited material. This gas must react only with the adsorbed layer, and not with the underlying material, again giving a self-limiting reaction. Both fluences must be significantly greater than the monolayer surface density n'_0 and must be high enough to penetrate into and saturate the highest aspect ratio trenches and vias, allowing complete adsorption and reaction over the entire surface. With self-limited adsorption and reaction, the deposited film thickness is then insensitive to the fluences of precursor and reactant species. In typical ALD processes, less than or of order one monolayer of the desired material is deposited per ALD cycle, giving strong control over the precision of the deposition. Many ALD cycles are then used to build up the deposited film to the required thickness. By saturating the surface with the deposition precursor, and then saturating the surface with the reactant gas, the feature- and wafer-scale nonuniformities are largely compensated.

Some limitations of thermally activated ALD are the slow rates of deposition, 5–50 nm/min, and the high substrate temperatures that sometimes must be used with feedgas precursors and reactants. These can sometimes be overcome by plasma-assisted ALD (PEALD), in which a more reactive precursor or reactant species is produced by dissociation of the feedstock gas. However, this can also lead to reduced conformality and to damage of the deposited film.

As described in a review by Puurunen (2005), including the early history, ALD has been under research since the 1960s, with commercial applications since the 1980s. Development and use for integrated circuit manufacturing has been growing strongly over the last 25–30 years as feature sizes have continued to shrink. High dielectric constant gate oxides for three-dimensional finFET and tri-gate transistors, diffusion barriers for copper interconnects, and trench capacitor dielectrics are some of the critical fabrication steps using ALD. An edited monograph by Pinna and Knez (2012) describes applications to nanostructured materials deposition, including PEALD. A monograph by Kääriäinen et al. (2013) focuses on the basic principles and characteristics of ALD and also discusses nanotechnology applications. There are many reviews available for the interested student. In addition to Puurunen (2005), who also references earlier reviews, notable reviews are by George (2010), Johnson et al. (2014), and Kunene et al. (2022). Profijt et al. (2011) review PEALD, and ALD conformality issues are reviewed by Cremers et al. (2019).

In the following subsections, we first discuss the well-characterized example of thermally activated ALD of Al_2O_3. We then examine PEALD for this same oxide. Finally, we discuss the conformality for these ALD processes.

17.3.2 Thermal ALD of Al_2O_3

Early work used $AlCl_3$ for the feedgas precursor and H_2O for the oxidizing reactant. However, $AlCl_3$ is a solid at room temperature, giving rise to gasification and particulate issues, and its reaction product with H_2O contains HCl, which can etch substrate surfaces and the reactor chamber. Hence, the more commonly used precursor is trimethylaluminum (TMA, $Al(CH_3)_3$, although it exists as a dimer at room temperature). H_2O is the common oxidant, but O-atoms and O_3 have also been used. For TMA/H_2O chemistry in the cyclical steady state (George, 2010), the step 1 precursor subreaction at each OH-terminated surface site is

$$\text{Al–OH}\,(s) + \text{Al(CH}_3)_3\,(g) \rightarrow \text{AlO–Al(CH}_3)_2\,(s) + \text{CH}_4\,(g) \tag{17.3.1}$$

The step 2 subreaction at each CH_3-terminated site is

$$\text{Al–CH}_3\,(s) + \text{H}_2\text{O}\,(g) \rightarrow \text{Al–OH}\,(s) + \text{CH}_4\,(g) \tag{17.3.2}$$

As can be seen, the growing surface switches from hydroxyl-terminated to methyl-terminated each half cycle. The overall reaction to deposit one molecule of Al_2O_3 consists of two TMA subreactions (17.3.1) and three H_2O subreactions (17.3.2), all adjacent to each other on the surface. The overall reaction is strongly exothermic (see Problem 17.3)

$$2\,\text{Al(CH}_3)_3\,(g) + 3\,\text{H}_2\text{O}\,(g) \rightarrow \text{Al}_2\text{O}_3\,(s) + 6\,\text{CH}_4\,(g), \quad \Delta H_r \approx -1230\ \text{kJ/mol} \tag{17.3.3}$$

mainly due to the large Al–O bond energy (see Table 7.3).

The preceding description is somewhat oversimplified. For example, the number of CH_3 ligands bonded to AlO after TMA adsorption might not be two, as assumed in subreaction (17.3.1). The details of the deposition are also quite sensitive to the substrate temperature. Vandalon and Kessels (2017) used optical techniques to monitor the –OH and –CH_3 surface groups. They found that the TMA saturation is due to depletion of the undercoordinated surface O-atoms, and not to the large molecular size of TMA, i.e., steric hindrance. At $T_s = 150\,°C$, they also found persistent –CH_3 groups, as well as –OH, after the saturation exposure to H_2O. The –CH_3 groups were not incorporated into the deposited film. At 250 °C, there were almost no persistent –CH_3 groups. The data suggest a coverage-dependent H_2O reactivity at the lower temperatures, with an activation energy varying from 0.27 V at $\theta_{CH_3} = 1$ to 0.43 V at $\theta_{CH_3} = 0.66$. This implies a lower growth rate per cycle (GPC). Figure 17.4*a* gives some GPC data compiled from the published literature. The drop in GPC at low temperatures is mainly due to this effect. The drop at the higher temperatures may be due to increased OH desorption. Above 350 °C, decomposition of the TMA precursor may also play a role.

The deposited films are found to be virtually defect-free, with dielectric breakdown properties similar to thermally grown SiO_2. They are also highly conformal. The measured growth rates per cycle are 1.0–1.2 Å in the 200–300 °C temperature range. Note that the amorphous Al_2O_3 density is $n_{Al_2O_3} \approx 1.8 \times 10^{28}\ m^{-3}$, yielding a surface average of $n'_{Al_2O_3} = n^{2/3}_{Al_2O_3} \approx 7$ molecules per nm^2, and a monolayer thickness of $n^{-1/3}_{Al_2O_3} \approx 3.8$ Å. Measurements indicate that there are 7–9 O–H groups per nm^2 on alumina, but only around 4.5 Al atoms per nm^2 are typically deposited per cycle (Cremers et al., 2019). Below 200 °C, the mass density drops somewhat, due to increased O-atom and H-atom incorporation. At room temperature, there is up to 15 atomic percent of H-atoms in the deposited film.

Some materials that have been deposited by thermal ALD include (Cremers et al., 2019)

semiconductors: Ge and Si
metals: Ru, Pd, W, Ir, and Pt
oxides of Al, Ti, V, Fe, Zn, Y, Zr, Ru, Sn, and Hf
nitrides of B, Al, Si, Ti, Ta, and W

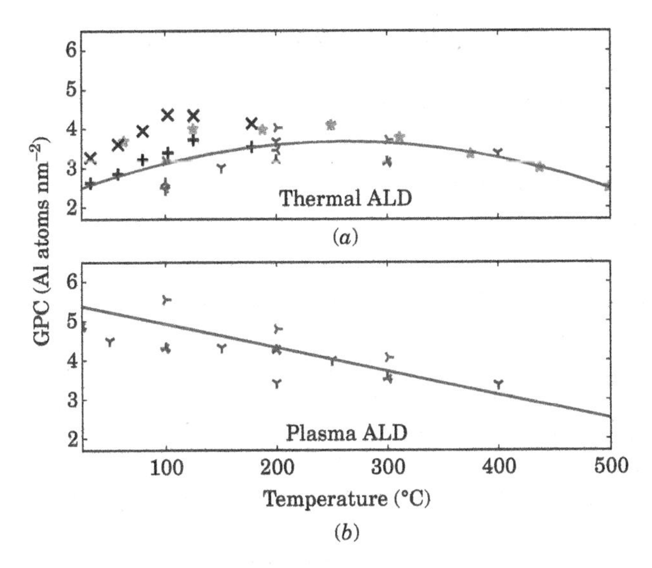

Figure 17.4 A compilation (different symbols) from the published literature of growth per cycle (GPC) in Al atoms per (nm)^2 versus substrate temperature for ALD of Al_2O_3 : (*a*) thermal ALD and (*b*) plasma-enhanced ALD. Source: Vandalon and Kessels (2017)/with permission of AIP Publishing.

A notable and important example is Si_3N_4 deposition using Si_2Cl_6 for the silicon precursor, coupled with NH_3 for the nitriding reactant (see Solmaz et al., 2020, for details and conformality simulations).

17.3.3 Plasma-Enhanced ALD of Al_2O_3

Figure 17.4*a* indicates that the thermal ALD growth per cycle decreases significantly with decreasing temperature below 200 °C. The lower thermal energy does not allow a full reaction of the H_2O oxidant with the $-CH_3$ bonds, giving a low –OH bond density on the surface. The reactivity of the oxidant can be greatly increased by using plasma-generated species such as O-atoms. These can be generated in a remote plasma source attached to a conventional ALD deposition reactor, or directly in an inductively or capacitively coupled plasma reactor. The resulting process is known as *plasma-enhanced atomic layer deposition* (PEALD).

For Al_2O_3 deposition, as for thermal ALD, the step 1 precursor subreaction is (17.3.1). The step 2 oxidation subreaction at each CH_3-terminated site is believed to be (Profijt et al., 2011)

$$\text{Al–CH}_3\,(\text{s}) + 4\,\text{O}\,(\text{g}) \rightarrow \text{Al–OH}\,(\text{s}) + \text{CO}_2\,(\text{g}) + \text{H}_2\text{O}\,(\text{g}) \tag{17.3.4}$$

although other surface reactions with the methylated surface may be important (Arts et al., 2022). Similar to thermal ALD, the growing surface switches from hydroxyl-terminated to methyl-terminated each half cycle. The overall reaction to deposit one molecule of Al_2O_3 consists of two TMA subreactions (17.3.1) and four O-oxidation subreactions (17.3.4), all adjacent to each other on the surface. The overall reaction is very strongly exothermic (see Problem 17.4)

$$2\,\text{Al(CH}_3)_3\,(\text{g}) + 16\,\text{O}\,(\text{g}) \rightarrow \text{Al}_2\text{O}_3\,(\text{s}) + 4\,\text{CO}_2\,(\text{g}) + 5\,\text{H}_2\text{O}\,(\text{g}) + 2\,\text{CH}_4\,(\text{g}) \tag{17.3.5}$$

Figure 17.4*b* gives a compilation of published growth rates for PEALD. As seen there, the growth per cycle (GPC) *increases* with decreasing temperature, giving high growth rates at room temperature. Although, as for thermal ALD, there is significant H-atom incorporation, the films are still of high quality. This process can be very useful for deposition on temperature-sensitive substrates (e.g., organic polymers).

Some deposited materials using PEALD include (Kessels et al., 2012; Cremers et al., 2019)

metals: Al, Ti, Co, Ni, Cu, Ru, Pd, Ag, Ta, Ir, and Pt
oxides of Al, Si, Ti, V, Co, Zn, Ga, Y, Zr, Ru, Hf, Ta, and Pt
nitrides of Al, Si, Ti, Zr, Nb, Mo, Hf, Ta, and W

and also some ternary compounds and semiconductors such as GaAs.

There are some significant challenges to the use of PEALD. One issue is its reduced feature-level conformality compared to thermal ALD. As we discuss in the next subsection, the oxidant species can recombine on saturated sidewall surfaces, reducing their fluxes to the deepest parts of the trenches or vias, and thus preventing saturation of the surface in a reasonable time. There is also the issue of plasma-induced damage. For remote plasma generation, there may be generation of other radical species besides the desired O, N, or H reactant. Additionally, for direct plasma reactor PEALD, the surface can be exposed to ion bombardment and VUV photons, which can cause damage or induce undesired surface reactions.

Ozone (O_3) is an oxidant that has also been used for some oxide and metal ALD processes, with a reactivity intermediate between H_2O and O-atoms. The TMA/O_3 ALD of Al_2O_3 was investigated by Kim et al. (2006b), who showed a dramatic increase in the GPC, from 0.08 to 0.2 nm/cycle, as the temperature was *decreased* from 300 °C to 30 °C. Some other ozone ALD processes are described in Cremers et al. (2019).

17.3.4 Conformality of ALD

17.3.4.1 Diffusion-Limited Regime

A simple analytic model for the reactant exposure fluence required to deposit a monolayer n'_0 in a small cylindrical hole of radius R and depth L was given by Gordon et al. (2003) in the diffusion-limited, molecular flow regime. A simplifying assumption is a highly reactive, step-function sticking coefficient s, which is unity for a surface previously saturated with the other reactant, and which abruptly becomes zero only when saturation coverage is reached. In this case, during the exposure to the incident reactant flux Γ_0 at the top of the hole, the reactive lower portion of the hole is separated from Γ_0 by a saturated hole of radius R and depth $z(t)$, having zero sticking coefficient. The reactant molecules diffusely reflect away from this saturated surface normal with a cosine law for the emitted flux (Lambert's cosine law), giving a diffusive flow.

The flux at depth z is given by the Clausing factor (16.2.21) for transmitted flux through this saturated hole

$$\Gamma(z) = \Gamma_0 \frac{1}{1 + \frac{3z}{8R}} \tag{17.3.6}$$

At time t within a time interval dt, an incremental depth dz at z is deposited on the sidewall surface $2\pi R\, dz$, with

$$dt = \frac{n'_0}{\Gamma(z)} \frac{2\pi R}{\pi R^2} dz \tag{17.3.7}$$

Integrating over the exposure time required to deposit a monolayer over the entire cylindrical sidewall, and changing variables from t to z, gives

$$T_1 = \int_0^{T_1} dt = \int_0^L dz \frac{n'_0}{\Gamma_0} \frac{2\pi R}{\pi R^2} \left(1 + \frac{3}{8}\frac{z}{R}\right) \tag{17.3.8}$$

Using (17.3.6) and evaluating the integral, we find

$$T_1 = \frac{n'_0}{\Gamma_0}\frac{2L}{R}\left(1 + \frac{3}{16}\frac{L}{R}\right) \tag{17.3.9}$$

After the sidewall deposition, a monolayer is deposited on the bottom in a time

$$T_2 = \frac{n'_0}{\Gamma_0}\left(1 + \frac{3}{8}\frac{L}{R}\right) \tag{17.3.10}$$

The total required time for this diffusion-limited deposition is $T_{\text{diff}} = T_1 + T_2$. Saturation of the top surface requires a time n'_0/Γ_0 or a fluence of $\Phi_0 = n'_0$. Saturation of the entire hole then requires a fluence

$$\Phi_{\text{diff}} = \Phi_0\left(1 + \frac{19}{8}\frac{L}{R} + \frac{3}{8}\frac{L^2}{R^2}\right) \tag{17.3.11}$$

We see that saturating a cylindrical hole of aspect ratio $L/R = 16$ requires about 100 times the fluence required to saturate the top surface. If this large fluence is not used, the deposition will be nonconformal.

17.3.4.2 Reaction-Limited Regime

The Gordon model was extended by Dendooven et al. (2009) to also account for reaction-limited kinetics using a Langmuir-type sticking coefficient (9.3.8), $s = s_0(1 - \theta)$, with s_0 a zero-coverage sticking coefficient and θ the reactant surface coverage. In this case, the reactant can either deposit or be diffusely re-emitted from the wall. The calculation was done numerically. The results are shown in Figure 17.5 for the TMA half-reaction for Al_2O_3 deposition for the Gordon step-function coverage model (top panel), and the Langmuir coverage model (following four panels) with $s_0 = 1$, 0.1, 0.01, and 0.001, respectively. A one second (nonsaturating) pulse of TMA at 0.3 Pa and 200 °C was used. As seen in this figure, the Gordon and Langmuir models agree fairly well for $s_0 = 1$, where the deposition is diffusion-limited, leading to a near step-function depth profile. However, as s_0 decreases, the deposition becomes reaction-limited, resulting in a relatively uniform depth profile.

Experimental measurements of the depth profile were performed on a macroscopic test structure consisting of two SiO_2-coated silicon wafers separated by a thin aluminum foil having a 5 mm × 20 mm slot cut into it. The aspect ratio was varied by changing the thickness of the foil. Figure 17.6*a* shows the experimental results for a TMA/purge/H_2O/purge cycle of 4/15/5/40 s. A comparison with the Langmuir model is given in Figure 17.6*b* for a 4 s TMA pulse at 200 °C with $s_0 = 0.1$. As seen, this choice of s_0 in the model gives a reasonable agreement with the measurements. However, sticking coefficients in the range of 0.001–0.9 (!) have been used in various simulation model studies (Cremers et al., 2019).

We can understand the reaction-limited regime as follows: for Langmuir kinetics with $s_0 \ll 1$, an initial surface coverage θ evolves in time according to

$$\frac{d\theta}{dt} = \frac{\Gamma_0}{n'_0}s_0(1 - \theta) \tag{17.3.12}$$

with n'_0 the deposited density per unit area and Γ_0 the incident reactive flux. For reaction-limited deposition $s_0 \ll 1$, the reactant is not depleted, and Γ_0 is the same at all surfaces within the hole. The solution of (17.3.12) is

$$\theta(t) = 1 - \exp\left(-\frac{s_0\Gamma_0}{n'_0}t\right) \tag{17.3.13}$$

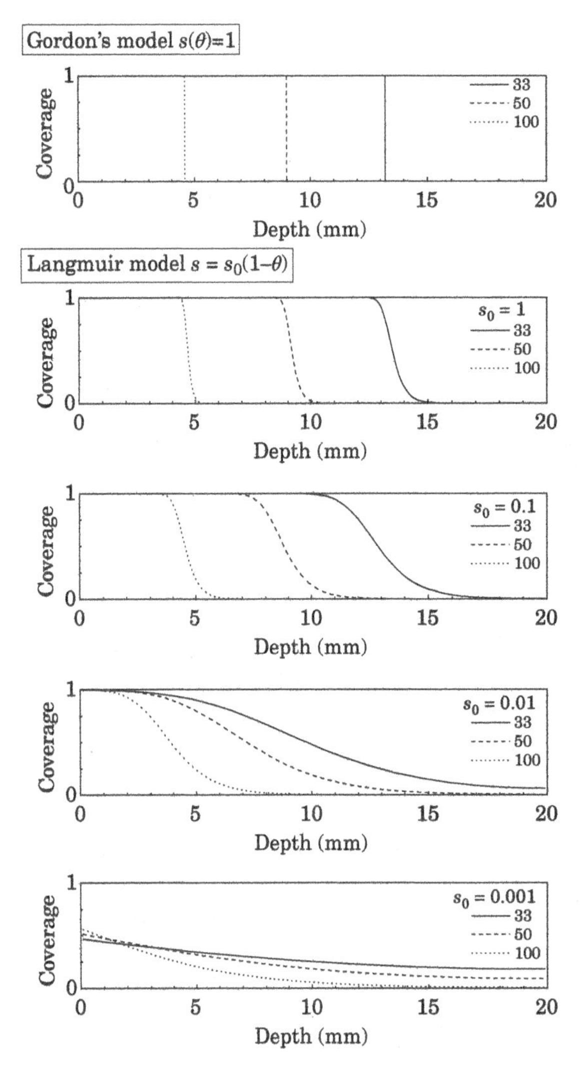

Figure 17.5 Top panel: Gordon's model simulation for Al_2O_3 deposition into rectangular test structures with aspect ratios $LP/(4A)$ of 33 (solid), 50 (dashed), and 100 (dotted); following four panels: Langmuir adsorption model simulations with $s_0 = 1, 0.1, 0.01$, and 0.001, respectively; L, P, and A are the depth, perimeter, and cross-sectional area of the hole. Source: Dendooven et al. (2009)/with permission of IOP Publishing.

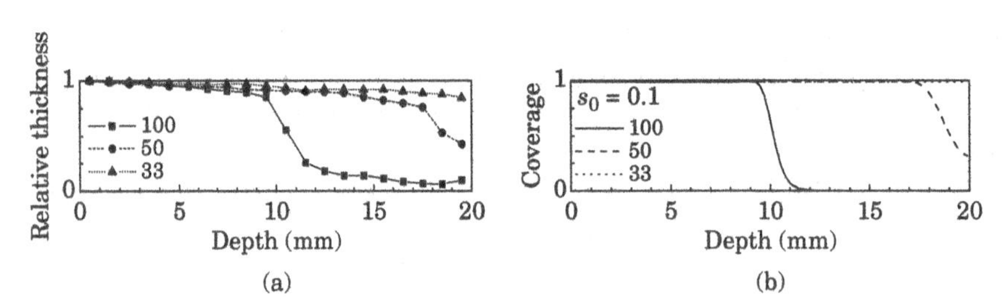

Figure 17.6 (*a*) Measured Al_2O_3 deposition depth profile measurements for 5 mm × 20 mm rectangular macroscopic hole structures with aspect ratios $LP/(4A)$ of 33 (solid), 50 (dashed), and 100 (dotted), and (*b*) corresponding Langmuir model simulations using a zero-coverage sticking coefficient $s_0 = 0.1$; L, P, and A are the depth, perimeter, and cross-sectional area of the hole, respectively. Source: Dendooven et al. (2009)/with permission of IOP Publishing.

For a final coverage θ_f, the reaction-limited exposure time is then

$$T_{\text{reac}} = \frac{n'_0}{s_0\Gamma_0} \ln \frac{1}{1-\theta_f} \tag{17.3.14}$$

Note that the exposure dose (fluence) $\Phi_{\text{reac}} = \Gamma_0 T_{\text{reac}}$ may be impractically large; for $s_0 = 0.001$ and 99% surface coverage, $\Phi_{\text{reac}} \approx 4600\, n'_0$.

A fit to a kinetic calculation (Fadeev and Rudenko, 2018) gives a good approximation for the transition between reaction- and diffusive-limited deposition

$$T_{\text{tot}} \approx \frac{n'_0}{\Gamma_0}\left(\frac{1}{s_0} \ln \frac{1}{1-\theta_f} + \frac{3}{8}\frac{L^2}{R^2}\right) \tag{17.3.15}$$

Note that this is effectively an ansatz of (17.3.14) and the dominant large aspect ratio term $3L^2/(8R^2)$ in (17.3.11).

17.3.4.3 Recombination-Limited Regime

A significant conformality issue that arises in PEALD, but not in thermal ALD, is radical recombination on the walls. O, H, and N radicals are typically used for one step of a PEALD process. Some surface recombination probabilities γ_{rec} are tabulated by Cremers et al. (2019). For example, for O(g) + O:S(s) → O_2(g), $\gamma_{\text{rec}} \approx 0.0021$ on Al_2O_3, and $\gamma_{\text{rec}} \approx 0.014$ on TiO_2. Metal PEALD processes using H radicals typically have much larger recombination probabilities; $\gamma_{\text{rec}} \approx 0.35$ on Ti, $\gamma_{\text{rec}} \approx 0.25$ on Cu, and $\gamma_{\text{rec}} \approx 0.25$ on Ni. For high recombination probabilities, we might expect that few radicals are able to penetrate into a deep hole, and the deposition becomes recombination-limited.

The effect of recombination on PEALD was examined by Knoops et al. (2010) using two-dimensional Monte Carlo simulations for a trench of depth L and width w, using fixed values of γ_{rec}, independent of the radical surface coverage θ. A zero-coverage sticking probability s_0 with Langmuir kinetics (9.3.8), $s = s_0(1-\theta)$, was used for the radical reaction with the surface. Figure 17.7*a* shows the normalized surface coverage versus the trench aspect ratio L/w without recombination losses ($\gamma_{\text{rec}} = 0$). This would be the normalized dose required for thermal ALD. The dose (fluence) required for >99% saturation of the trench surface has been normalized to the corresponding dose for a planar surface. We see the transition from diffusion-limited growth at large s_0 to reaction-limited growth at small s_0. In Figure 17.7*b*, the effects of a finite radical

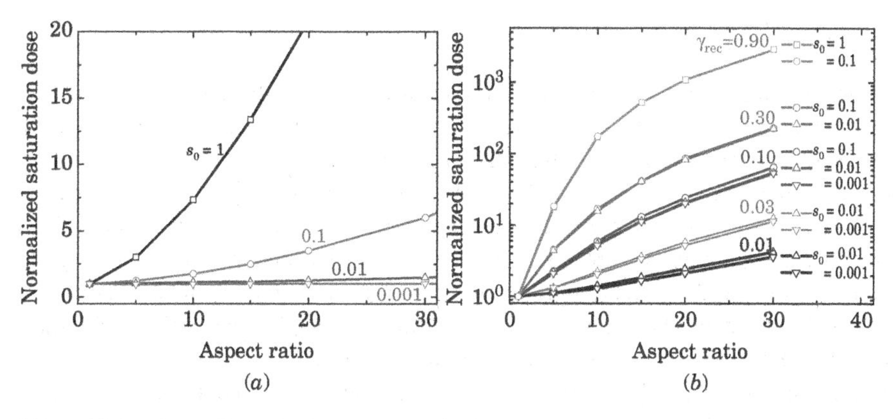

Figure 17.7 Monte Carlo simulation results for (*a*) the normalized saturation dose versus trench aspect ratio L/w for a recombination probability $\gamma_{\text{rec}} = 0$ and various zero-coverage reaction probabilities s_0, and (*b*) for values of γ_{rec} from 0.01 to 0.9 and various values of s_0; in (*b*), the trends with the same γ_{rec} value but different s_0 values overlap. Source: Knoops et al. (2010)/with permission of IOP Publishing.

surface recombination are shown. As can be seen, there is a strong dependence on $\gamma_{\rm rec}$ and a weak dependence on s_0. This suggests that the deposition is not limited by the reaction or diffusion rates, but by the recombination rate. Knoops et al. (2010) give a diagram showing, approximately, the division into diffusion-, reaction-, and recombination-dominated regimes.

To understand the recombination-dominated regime, we consider the radical depletion in a cylindrical hole of radius R and depth L for $s_0 \ll \gamma_{\rm rec}$. The axial flux Γ flowing from depth z to $z + {\rm d}z$ within the hole is depleted by the sidewall recombination losses

$$d\Gamma = -\gamma_{\rm rec}\Gamma_{\rm wall}\frac{2\pi R}{\pi R^2}dz \tag{17.3.16}$$

In the molecular flow regime, $\Gamma_{\rm wall}(z) \approx \Gamma(z) \approx \frac{1}{4}n(z)\overline{v}$, with n and $\overline{v}$ the radical density and mean speed. Using this and integrating (17.3.16) with $\Gamma(0) = \Gamma_0$ at the top surface $z = 0$ gives

$$\Gamma(z) \approx \Gamma_0 \exp\left(-\frac{2\gamma_{\rm rec}z}{R}\right) \tag{17.3.17}$$

Inserting $\Gamma(L)$ at the bottom of the hole in place of Γ_0 in (17.3.14), we obtain the recombination-limited exposure time

$$T_{\rm rec} \approx \frac{n_0'}{s_0\Gamma_0}\exp\left(\frac{2\gamma_{\rm rec}L}{R}\right)\ln\frac{1}{1-\theta_{\rm f}} \tag{17.3.18}$$

As we can see, $T_{\rm rec}$ increases exponentially with $2\gamma_{\rm rec}L/R$. This strongly limits the applicability of PECVD for $2\gamma_{\rm rec}L/R \gg 1$.

As one example, for PEALD of Ti using H atoms, $\gamma_{\rm rec} = 0.35$. Then, very large doses are required for saturation of large aspect ratios, making conformal deposition infeasible. This issue is not so severe with oxide or nitride depositions, which have smaller surface recombination probabilities. Conformal depositions with aspect ratios of 10–30 have been reported for metal oxide PEALD. For metals such as Ag PEALD, the aspect ratios for good conformality are typically limited to about 5. In contrast, thermal ALD conformality for aspect ratios of many thousands has been reported. These and other aspect of conformal ALD are reviewed by Cremers et al. (2019).

17.4 Sputter Deposition

17.4.1 Physical Sputtering

In physical sputter deposition, ions incident on a target physically sputter target atoms, which ballistically flow to and are deposited on a substrate. Argon ions at 500–1000 V are usually used. Because sputter yields are of order unity for almost all target materials, a very wide variety of pure metals, alloys, and insulators can be deposited. Physical sputtering, especially of elemental targets, is a well-understood process (see Section 9.3), enabling sputtering systems for various applications to be relatively easily designed. Reasonable deposition rates with excellent film uniformity, good surface smoothness, and adhesion can be achieved over large areas. Refractory materials can also be easily sputtered. Sputter deposition is highly nonconformal, although redeposition techniques by ion bombardment of the deposited film can improve the conformality.

At first sight, it might seem that when a multicomponent target is sputtered, the deposited film will have a different composition than the target due to the difference in sputtering yields of the components. However, when multicomponent targets are sputtered, because of the difference in sputtering yields, an altered layer forms at the target surface having a different composition than the target. In the steady state, in the absence of diffusion of components between the layer and the bulk target, the flux of atoms sputtered from the layer has the stoichiometry of the original target material. If the sticking coefficients of the components on the substrate are all the same, then the deposited film will have the composition of the bulk target material. Thus, alloy targets can be sputter deposited on substrates. However, targets such as ceramics or oxides having high vapor pressure constituents, e.g., O atoms, usually cannot be physically sputter deposited.

In the area of metal film deposition, sputtering is commonly used to deposit electrode and interconnection material. For example, various films have been deposited such as aluminum in integrated circuit devices; transition metal films such as iron, cobalt, and nickel for magnetic coatings; superconducting films such as niobium; reflective optical films such as aluminum, silver, and gold: corrosion-resistant films such as chromium; and films such as chromium for decorative purposes.

Assuming that all the sputtered material is deposited on the substrate, the deposition rate for physical sputtering is

$$D_{\mathrm{sput}} = \frac{\gamma_{\mathrm{sput}}\Gamma_{\mathrm{i}}}{n_{\mathrm{f}}}\frac{A_{\mathrm{t}}}{A_{\mathrm{s}}}\,\mathrm{cm/s} \tag{17.4.1}$$

where Γ_{i} is the incident ion flux (cm^{-2}-s^{-1}), n_{f} is the density of the deposited film (cm^{-3}), A_{t} (cm^2) is the target area sputtered, A_{s} (cm^2) is the substrate area on which the film is deposited, and γ_{sput} is the sputtering yield. An estimate for the sputtering yield in the linear cascade regime is given in (9.3.14), and some sputtering yields are tabulated in Table 9.2. For 1 kV argon ions with $A_{\mathrm{t}}/A_{\mathrm{s}} = 1$, $n_{\mathrm{f}} = 5 \times 10^{22}$ /cm$^{-3}\gamma_{\mathrm{sput}} = 1$, and an ion current density of 1 mA/cm^2, ($\Gamma_{\mathrm{i}} \approx 6.3 \times 10^{15}$ /cm^{-2}-s^{-1}), the deposition rate is 750 Å/min. Rf- or dc-driven planar magnetron discharges are usually used for sputtering; the operating pressure is generally 10^{-3}–10^{-2} Torr, which is low enough that the mean free path for sputtered atoms is larger than the separation between target and substrate.

Sputtered atoms are emitted with a cascade-type energy distribution (9.3.18). The maximum of this distribution occurs at $\mathcal{E} = \mathcal{E}_{\mathrm{t}}/2$, where $\mathcal{E}_{\mathrm{t}}$ is the surface binding energy of the target material. Since $\mathcal{E}_{\mathrm{t}} \sim 3$–6 V, the characteristic sputtered atom energies are 1.5–3 V. Atoms striking the substrate with these energies can produce some mixing and diffusion between incoming atoms and substrate materials, leading to enhanced bonding and adhesion.

The morphology of sputtered films is primarily influenced by the substrate temperature, which is usually independently controlled, and secondarily by the deposition pressure. The film morphology has been described by Thornton (1974, 1986) and is shown as a "structure zone diagram" in Figure 17.8. Let T_{m} be the melting temperature of a sputter deposited metal film, then at low pressures and at very low temperatures, $T/T_{\mathrm{m}} \lesssim 0.3$ (zone 1), the film consists of tapered columns with domed heads and significant voids between columns. These structures are formed by shadowing of atoms as they impinge on the growing film. The void fraction can be as high as 30%. For $0.3 \lesssim T/T_{\mathrm{m}} \lesssim 0.5$ (zone T), the films have a fibrous structure in which crystallites grow perpendicular to the substrate plane without significant voids ($\lesssim$ 5% by volume). The film surface is relatively smooth and the film is almost as dense as that of the bulk material. These properties are a result of ion-bombardment-induced surface mobility of deposited atoms on the substrate. This film morphology is desirable for many applications. For $0.5 \lesssim T/T_{\mathrm{m}} \lesssim 0.8$ (zone 2), thermally activated surface diffusion of deposited atoms leads to the appearance of columnar grains, which increase in diameter as T/T_{m} increases. For $0.8 \lesssim T/T_{\mathrm{m}} \lesssim 1$ (zone 3), volume diffusion of atoms within the film leads to a smooth, randomly oriented polycrystalline film. All of these zones are used in various sputtering applications.

Ion and fast neutral bombardment of the substrate can strongly influence film properties and are responsible for the desirable properties in zone T. Positive ions with energies of 20–30 V arise

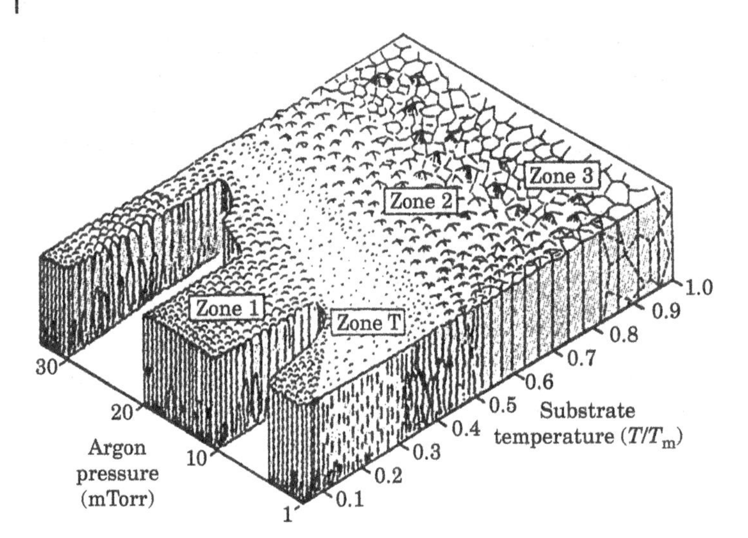

Figure 17.8 Structure zone diagram for the morphology of sputtered films. Source: Thornton (1986)/with permission of AIP Publishing.

from acceleration across the plasma–substrate sheath potential. The energy can be increased by applying a bias voltage (rf or, for conducting films, dc driven) to the substrate, and this is done in many commercial applications. Deposition pressure and ion bombardment also affect intrinsic film stress. High bombardment energies produce high compressive stresses due to recoil implantation. Generally, there is a transition from compressive to a generally more desirable tensile stress as the pressure is increased, with the transition pressure typically increasing with the atomic mass of the sputtered material (Konuma, 1992). This is believed to be due to a reduction in ion-bombarding energy as pressure is increased.

As seen from the above discussion, the number of physical parameters describing the film growth morphology is large and far exceeds the two-dimensional parametrization (T and p) used in the structure zone diagram of Figure 17.8. The two-dimensional parametrization can be modified to account for some of these additional parameters. In more physical terms, the pressure axis corresponds to the kinetic energy of bombarding ions and atoms per displaced atom in the growing film, and the temperature axis should additionally account for the potential energy of recombining ions and excited atoms incident on the surface. The vertical axis can also be used to indicate a film growth rate, or thickness per unit time. Such a modification has been given by Anders (2010). However, one must realize that all structure zone diagrams are crude representations of the complicated physics and chemistry that determine the film morphology.

17.4.2 Reactive Sputtering

For planar magnetron reactive sputtering, a feedstock gas whose dissociation products chemically react with the target is present in addition to the bombarding ions. The deposited film is a compound formed from the sputtered target materials and the reactive gas. A common application is the sputter deposition of films whose components have strongly different vapor pressures, and hence, sticking probabilities on the substrate. For example, physical sputtering of a SiO_2 target in argon can lead to deposition of a silicon-rich oxide film on the substrate. If O_2 gas is added to the system, then O atoms can be incorporated into the growing film to restore the 1 : 2 Si/O stoichiometry. A pure silicon target can also be used with O_2 gas to deposit SiO_2 films by reactive sputtering.

Planar magnetron discharge physics is described in Section 14.5. A general review on the physics and technology of magnetron sputtering, including reactive sputtering, is given in Gudmundsson (2020). Specific reviews on reactive sputtering can be found in Sproul et al. (2005) and Strijckmans et al. (2018).

Reactive sputtering is widely used to deposit dielectrics such as oxides and nitrides, as well as carbides and silicides. Due to their high hardness, good mechanical strength, and chemical inertness, thin transition metal nitride films are widely used as hard wear-resistant coatings, diffusion barriers, and optical and decorative coatings. Ceramics such as YBaCuO superconducting films can be sputter deposited from YBaCuO targets using O_2 as the reactive gas. Common reactive gases used for a wide variety of applications are O_2 and H_2O for O atoms, N_2 and NH_3 for N atoms, CH_4 and C_2H_2 for C atoms, and SiH_4 for Si atoms. Although ceramic or oxide targets can be used, they are not machinable and cannot handle high-power fluxes without cracking; hence, metal targets are most commonly used where high deposition rates and controllable film stoichiometry are desired.

In reactive sputtering, chemical reactions occur at both target and substrate, in addition to sputtering at the target and deposition at the substrate. There are two "modes" of operation for reactive sputtering of a metal target to deposit a compound film. For low ion flux and high gas flux, the target is covered (poisoned) by the compound. For high ion flux and low gas flux, the target remains metallic. Higher deposition rates are achieved in the "metallic mode" than in the "poisoned mode." For fixed ion flux, as the reactive gas flux is varied, there is a transition between the poisoned and metallic modes exhibiting hysteresis; i.e., the transition flux for increasing the flux to pass from the metallic to the poisoned mode is higher than the transition flux for decreasing the flux to pass from the poisoned to the metallic mode.

A simple model of reactive sputtering (Berg et al., 1989; Berg and Nyberg, 2005) makes it possible to understand the hysteresis and other properties of reactive sputter deposition. Let A_t and A_s be the target and substrate areas, θ_t and θ_s be the fractions of the target and substrate areas covered by the compound film, and γ_m and γ_c be the yields for sputtering the metal and the compound from the target. To simplify the calculation, we assume that the compound molecule is sputtered and is not split into its constituent atoms. In the steady state, the compound formation rate on the target must be equal to the sputtering rate of the compound from the target. Let Γ_i and Γ_r be the incident ion and reactive gas molecule fluxes, and let s_r be the sticking coefficient of a reactive molecule on the metal part of the target, then

$$n'_t \frac{d\theta_t}{dt} = i\Gamma_r s_r (1 - \theta_t) - \Gamma_i \gamma_c \theta_t = 0 \qquad (17.4.2)$$

where i is the number of atoms per molecule of reactive gas (e.g., $i = 2$ for O_2 gas). Sputtered compound molecules and metal atoms are assumed to be evenly deposited over the substrate surface. The coverage θ_s of compound on the substrate increases because reactive gas molecules are incident on the metallic part $(1 - \theta_s)$, and because a fraction $(1 - \theta_s)$ of the compound flux sputtered from the target is deposited on the metallic part of the substrate. Similarly, θ_s decreases because a fraction θ_s of the metal-atom flux sputtered from the target is deposited on the compound part of the substrate. Hence, accounting for the ratio of target and substrate areas, we obtain

$$\begin{aligned} n'_s \frac{d\theta_s}{dt} &= i\Gamma_r s_r (1 - \theta_s) + \Gamma_i \gamma_c \theta_t \left(\frac{A_t}{A_s}\right)(1 - \theta_s) \\ &\quad - \Gamma_i \gamma_m (1 - \theta_t) \left(\frac{A_t}{A_s}\right) \theta_s = 0 \end{aligned} \qquad (17.4.3)$$

The total number of reactive gas molecules per second that are consumed to form the compound deposited on the substrate is

$$\frac{dN_r}{dt} = \Gamma_r s_r[(1 - \theta_t)A_t + (1 - \theta_s)A_s] \tag{17.4.4}$$

and the target sputtering flux is

$$\Gamma_{sput} = \Gamma_i[\gamma_m(1 - \theta_t) + \gamma_c\theta_t] \tag{17.4.5}$$

Equations (17.4.2) and (17.4.3) can be simultaneously solved to determine the compound coverages θ_t and θ_s on the target and substrate as a function of the fluxes Γ_i and Γ_r, rate constants, and areas. Then, dN_r/dt and Γ_{sput} can be evaluated, exhibiting the hysteresis (Problem 17.6).

For reactive sputter deposition of TiN films at 10 mTorr using a titanium target and an Ar/N_2 gas mixture, the optical intensity of a titanium emission line (proportional to Γ_{sput}) is plotted for a fixed magnetron current (fixed Γ_i) in Figure 17.9 versus the input gas flow rate (equivalent to dN_r/dt + const where the constant is the number of reactive molecules per second removed by the pump). By controlling the input gas flow rate with a feedback system, the complete hysteresis curve could be traced out, as shown in Figure 17.9*a*. In the absence of feedback control, the yield exhibits hysteresis, jumping between the high and low deposition rate modes, as shown in Figure 17.9*b*.

The width (gas flow rate) of the hysteresis region widens significantly with increasing metal/compound sputtering yields γ_m/γ_c. For 500 V argon ions, $\gamma_m \approx 0.69$ for titanium, ≈ 0.42 for TiN, and ≈ 0.05 for TiO_2. Hence, the hysteresis region is particularly wide for oxides (Gudmundsson, 2020; Kubart et al., 2020). Stability can be better achieved by controlling the reactive gas partial pressure, rather than the flow rate (Sproul et al., 2005).

These types of models have been extended to account for additional effects at the surfaces, such as ion implantation into the target (see Strijckmans et al., 2018). They have also been applied to multicomponent reactive sputtering (Moradi et al., 1991).

17.5 Plasma-Immersion Ion Implantation

Ion implantation is a process in which an energetic ion beam is injected into the surface of a solid material with the result that the atomic composition and structure of the near-surface region of the target material is changed, and thereby also the properties of the material surface are changed. The process is routine in semiconductor device fabrication. Metallurgical implantation is an emerging technology; in this application, new surface alloys are created with improved resistance to wear, corrosion, and fatigue.

Conventional ion implantation is carried out in a vacuum environment in which an ion source is used to create an intense beam of ions of the species to be implanted. The ion beam is steered and accelerated through a potential of from tens to hundreds of kilovolts and transported to the target. Since the beam spot size is smaller than the wafer size, mechanical and electrostatic scanning are used to achieve dose uniformity. For some state-of-the-art semiconductor device structures, high angle tilting and rotation of the wafers are required to homogenize the dose uniformity on the sidewall area. This mechanical complexity significantly increases the physical size and cost of the implanter. The relatively low beam currents, limited by the source optics, lead to high costs for high-dose applications such as buried dielectric layers formed by implantation of oxygen (SIMOX), doping of thin-film transistors for active matrix flat panel displays, surface smoothing for optical coatings, ion beam mixing of thin films, and ion-assisted deposition. Lower-energy implantation

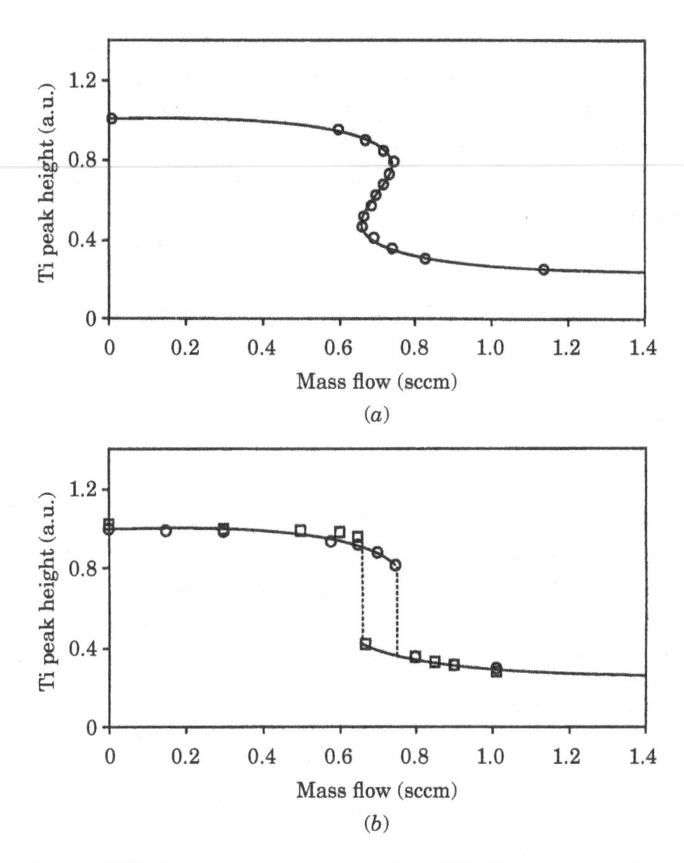

Figure 17.9 Reactive sputter deposition of TiN films, showing the optical emission signal for titanium versus the reactive gas flow rate (*a*) with and (*b*) without feedback control. Source: Berg et al. (1989)/with permission of AIP Publishing.

(energies less than 5 kV) can also be limited by ion beam optics if high doses are required. In plasma-immersion ion implantation (PIII), the intermediate stages of ion source, beam extraction, focusing, and scanning are omitted. The target is immersed in a plasma environment, and ions are extracted directly from the plasma and accelerated into the target by means of a series of negative high-voltage pulses applied to the target. Both metallurgical (Conrad et al., 1990) and semiconductor (Cheung, 1991) implantation processes have been demonstrated using PIII. Reviews of the area are given in Anders (2000) and Gupta (2011).

When a sudden negative voltage $-V_0$ is applied to the target, then, on the timescale of the inverse electron plasma frequency ω_{pe}^{-1}, electrons near the surface are driven away, leaving behind a uniform-density ion matrix sheath. The sheath thickness is a function of the applied voltage and the plasma density, as given in (6.3.4). Subsequently, on the timescale of the inverse ion plasma frequency, ions within the sheath are accelerated into the target. This, in turn, drives the sheath–plasma edge farther away, exposing new ions that are extracted. On a longer timescale, the system evolves toward a steady-state Child law sheath, with the sheath thickness given by (6.3.14). The Child law sheath is larger than the matrix sheath by a factor of order $(V_0/\mathrm{T_e})^{1/4}$, where $\mathrm{T_e}$ is the electron temperature. This steady state can be of interest in PIII for high-throughput implantations into conducting targets.

The matrix sheath and its time evolution determine the implantation current $J(t)$ and the energy distribution of implanted ions. The structures of the initial matrix sheath in one-dimensional planar, cylindrical, and spherical targets (Conrad, 1987) and two-dimensional wedge-shaped targets (Donnelly and Watterson, 1989) have been determined. In addition, analytical estimates of the sheath dynamics have been obtained (Lieberman, 1989c; Scheuer et al., 1990), and the self-consistent equations have been solved numerically to find the time evolution of the matrix sheath in planar geometry (Vahedi et al., 1991; Stewart and Lieberman, 1991; Emmert and Henry, 1992; Wood, 1993; Wang et al., 1993, and references therein). Plasma recovery after PIII was examined by Chung et al. (2012, 2013). Time-resolved measurements of plasma parameters in a PIII source have been made by Moreno et al. (2021). In this section, we first present some simple dynamical models for PIII sheath formation in the collisionless and collisional regimes and then briefly describe some experimentally investigated applications of PIII for integrated circuit fabrication and metallurgical surface modification.

17.5.1 Collisionless Sheath Model

Figure 17.10 shows the PIII geometry. The planar target is immersed in a uniform plasma of density n_0. At time $t = 0$, a voltage pulse of amplitude $-V_0$ and time width t_p is applied to the target, and the plasma electrons are driven away to form the matrix sheath, with the sheath edge at $x = s_0$. As time evolves (Figure 17.10*b*), ions are implanted, the sheath edge recedes, and a nonuniform, time-varying sheath forms near the target. The model assumptions are as follows:

1. The ion flow is collisionless. This is valid for sufficiently low gas pressures.
2. The electron motion is inertialess. This follows because the characteristic implantation timescale much exceeds ω_{pe}^{-1}.
3. The applied voltage V_0 is much greater than the electron temperature T_e; hence, the Debye length $\lambda_{De} \ll s_0$, and the sheath edge at s is abrupt.
4. During and after matrix sheath implantation, a quasistatic Child law sheath forms. The current demanded by this sheath is supplied by the uncovering of ions at the moving sheath edge and by the drift of ions toward the target at the Bohm (ion sound) speed $u_B = (eT_e/M)^{1/2}$. The assumption of Bohm flow will probably not be valid during the initial sheath expansion before

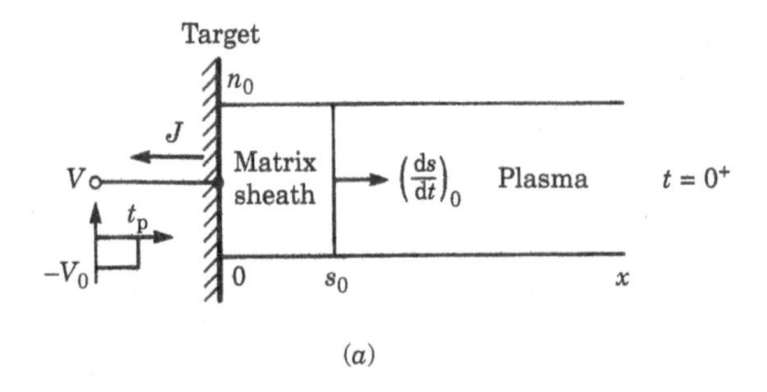

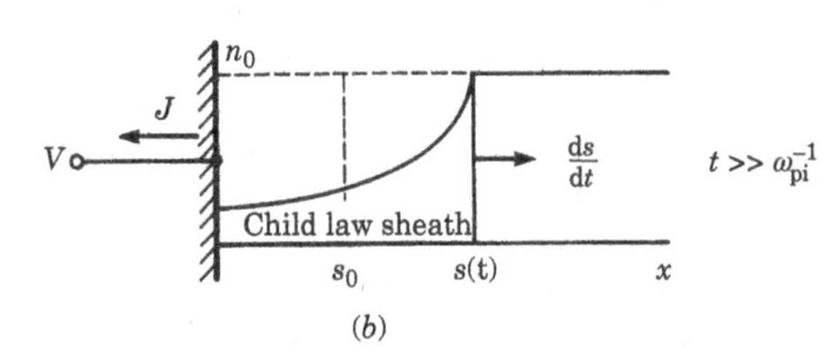

Figure 17.10 Planar PIII geometry (*a*) just after formation of the matrix sheath and (*b*) after evolution of the quasistatic Child law sheath.

the Bohm presheath has formed. The sheath dynamics using an alternative assumption, that the drift velocity of ions toward the target is zero, has been explored by Scheuer et al. (1990).

5. During the motion of an ion across the sheath, the electric field is frozen at its initial value, independent of time, except for the change in field due to the velocity of the moving sheath.

Assumptions 4 and 5 are approximations that permit an analytical solution to the sheath motion. These assumptions are justified post hoc by comparison with numerical results.

17.5.1.1 Sheath Motion

We assume that after a short transient, the ion matrix sheath evolves into a Child law sheath with time-varying current density and sheath thickness. The Child law current density J_c for a voltage V_0 across a sheath of thickness s is given by (6.3.12):

$$J_c = \frac{4}{9}\epsilon_0\left(\frac{2e}{M}\right)^{1/2}\frac{V_0^{3/2}}{s^2} \tag{17.5.1}$$

where ϵ_0 is the free space permittivity and e and M are the ion charge and mass. Equating J_c to the charge per unit time crossing the sheath boundary,

$$en_0\left(\frac{ds}{dt} + u_B\right) = J_c \tag{17.5.2}$$

we find the sheath velocity

$$\frac{ds}{dt} = \frac{2}{9}\frac{s_0^2 u_0}{s^2} - u_B \tag{17.5.3}$$

where

$$s_0 = \left(\frac{2\epsilon_0 V_0}{en_0}\right)^{1/2} \tag{17.5.4}$$

is the matrix sheath thickness and

$$u_0 = \left(\frac{2eV_0}{M}\right)^{1/2} \tag{17.5.5}$$

is the characteristic ion velocity. Integrating (17.5.3), we obtain

$$\tanh^{-1}\left(\frac{s}{s_c}\right) - \frac{s}{s_c} = \frac{u_B t}{s_c} + \tanh^{-1}\left(\frac{s_0}{s_c}\right) - \frac{s_0}{s_c} \tag{17.5.6}$$

where

$$s_c = s_0\left(\frac{2}{9}\frac{u_0}{u_B}\right)^{1/2} \tag{17.5.7}$$

is the steady-state Child law sheath thickness. Since $s_c \gg s_0$ and assuming $s_c \gg s$, we find by expanding (17.5.6), or by integrating (17.5.3) with $u_B \equiv 0$, that

$$\frac{s^3}{s_0^3} = \frac{2}{3}\omega_{pi} t + 1 \tag{17.5.8}$$

where $\omega_{pi} = (e^2 n_0/\epsilon_0 M)^{1/2} = u_0/s_0$ is the ion plasma frequency in the matrix sheath. The simplest implantation model assumes that (17.5.8) and (17.5.1) are valid for all times (Scheuer et al., 1990). Substituting (17.5.8) into (17.5.1) yields the implanting current density

$$J \equiv J_c = \frac{2}{9}\frac{en_0 u_0}{\left(1 + \frac{2}{3}\omega_{pi} t\right)^{2/3}}$$

However, the Child law is not valid early in time, where a matrix sheath exists. Substituting (17.5.7) into (17.5.8), we note that the timescale t_c for establishing the steady-state Child law sheath ($s = s_c$) is $t_c \approx (\sqrt{2}/9)\omega_{pi}^{-1}(2V_0/T_e)^{3/4}$. In the development that follows, we assume a rectangular voltage pulse, and we divide the implantation into two periods corresponding to matrix and Child law sheath implantations.

17.5.1.2 Matrix Sheath Implantation

Because the initial charge density in the matrix sheath is uniform, the initial electric field varies linearly with x: $E = (M/e)\omega_{pi}^2(x - s)$. Hence, the ion motion is

$$\frac{d^2x}{dt^2} = \omega_{pi}^2(x - s) \tag{17.5.9}$$

where x is the particle position. Approximating $s = s_0 + (ds/dt)_0 t$ in (17.5.9) and using (17.5.3) with $s = s_0$ and $u_B \ll u_0$, we obtain

$$\frac{d^2x}{dt^2} = \omega_{pi}^2(x - s_0) - \frac{2}{9}u_0\omega_{pi}^2 t \tag{17.5.10}$$

Integrating (17.5.10), we find

$$x - s_0 = (x_0 - s_0)\cosh\omega_{pi}t - \frac{2}{9}s_0\sinh\omega_{pi}t + \frac{2}{9}u_0 t \tag{17.5.11}$$

where we have let $x = x_0$ and $\dot{x} = 0$ at $t = 0$. (Choosing $\dot{x} \approx -u_B$, consistent with the sheath motion (17.5.3), yields a negligible correction to (17.5.11) because $u_B \ll u_0$.) Let $x = 0$ in (17.5.11), we can obtain the ion flight time t from

$$s_0 = (s_0 - x_0)\cosh\omega_{pi}t + \frac{2}{9}s_0\sinh\omega_{pi}t - \frac{2}{9}u_0 t \tag{17.5.12}$$

In a time interval between t and $t + dt$, ions from the interval between x_0 and $x_0 + dx_0$ are implanted. Differentiating x_0 in (17.5.12) with respect to time, we find

$$\frac{dx_0}{dt} = \frac{\omega_{pi}(s_0 - x_0)\sinh\omega_{pi}t + 29u_0(\cosh\omega_{pi}t - 1)}{\cosh\omega_{pi}t} \tag{17.5.13}$$

Using (17.5.12) to eliminate $s_0 - x_0$ in (17.5.13), we obtain the implantation current density $J = en_0 dx_0/dt$ as

$$\bar{J} = \frac{\sinh\bar{t}}{\cosh^2\bar{t}} + \frac{2}{9}\frac{1 + \bar{t}\sinh\bar{t} - \cosh\bar{t}}{\cosh^2\bar{t}} \tag{17.5.14}$$

where $\bar{J} = J/(en_0u_0)$ is the normalized current density and $\bar{t} = \omega_{pi}t$ is the normalized time. Equation (17.5.14) gives the implantation current density versus time for those ions in the initial matrix sheath $0 \le x_0 \le s_0$. Set $x_0 = s_0$ in (17.5.12), we obtain $\bar{t} \approx 2.7$. At this time, all matrix sheath ions are implanted; hence, we take (17.5.14) to reasonably approximate the current for $0 \le \bar{t} \le 2.7$. The left dashed curve in Figure 17.11 plots $\bar{J}$ versus $\bar{t}$. The maximum current density $\bar{J}_{max} \approx 0.55$ occurs at $\bar{t}_{max} \approx 0.95$, and $\bar{J}(2.7) \approx 0.19$.

17.5.1.3 Child Law Sheath Implantation

Consider now the implanted ions having initial positions at $x_0 > s_0$. The time t_s for the initial sheath edge at s_0 to reach x_0 is found from (17.5.8):

$$\omega_{pi}t_s = \frac{3}{2}\frac{x_0^3}{s_0^3} - \frac{3}{2} \tag{17.5.15}$$

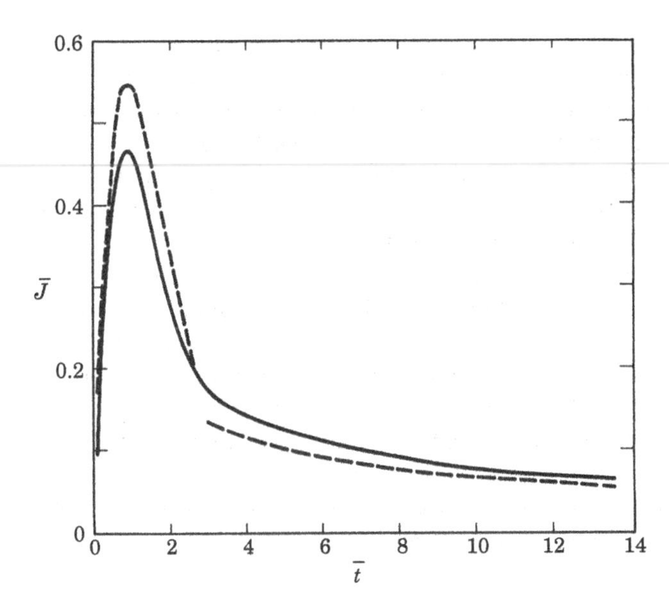

Figure 17.11 Normalized implantation current density $\bar{J} = J/(en_0u_0)$ versus normalized time $\bar{t} = \omega_{pi}t$. The dashed lines show the analytical solution for $\bar{t} < 2.7$ and $\bar{t} > 3.0$, and the solid line is the numerical solution of the fluid equations.

We expect this to be valid for $\bar{t} \gtrsim 2.7$. At time t_s, an ion at x_0 begins its flight across the sheath. The ion flight time is given by (Problem 6.2)

$$\omega_{pi}t' = \frac{3x_0}{s_0} \tag{17.5.16}$$

Hence, an ion at x_0 reaches the target at a time $t = t_s + t'$ given by

$$\bar{t} = \omega_{pi}t = \frac{3}{2}\frac{x_0^3}{s_0^3} - \frac{3}{2} + 3\frac{x_0}{s_0} \tag{17.5.17}$$

Differentiating (17.5.17), we obtain

$$\frac{dx_0}{dt} = \frac{u_0}{\frac{9}{2}(x_0^2/s_0^2) + 3} \tag{17.5.18}$$

The normalized implantation current density is thus

$$\bar{J} = \frac{1}{\frac{9}{2}(x_0^2/s_0^2) + 3} \tag{17.5.19}$$

Equations (17.5.17) and (17.5.19) give $\bar{J}(\bar{t})$ as a parametric function of x_0/s_0. If we set $x_0/s_0 = 1$, we find $\bar{t} = 3$ and $\bar{J}(3) = 2/15 \approx 0.133$. As $\bar{t} \to \infty$, $x_0 \to s_c \gg s_0$; hence $\bar{J}(\infty) \to (2/9)s_0^2/s_c^2$. Unnormalizing, we find $J(\infty) \to en_0u_B$, which correctly gives the steady-state Child law current density. The right dashed curve in Figure 17.11 shows the analytical results for $\bar{J}$ versus $\bar{t}$ for $\bar{t} \gtrsim 2.7$. We note that (17.5.14) and (17.5.19) do not smoothly join at $x_0 = s_0$, a consequence of the simplifying assumptions 4 and 5 that were used to solve for the sheath and ion motion.

The preceding analysis has been compared with numerical solutions of the nonlinear partial differential equations for the ion and electron motion (Lieberman, 1989c). The ion motion is collisionless, the electrons are in thermal equilibrium, and Poisson's equation relates the densities to the potential. Figure 17.11 shows a numerical solution for $V_0/T_e = 200$. We see that (17.5.14) for $\bar{t} \lesssim 2.7$ and (17.5.19) for $\bar{t} \gtrsim 2.7$ reasonably approximate the numerical results. The energy distribution of ions striking the target can be determined from the basic model. The analysis can also be applied to nonplanar geometries (Scheuer et al., 1990). The spatial structure and time evolution of the collisionless sheath have been measured by Cho et al. (1988) for planar targets at low voltages (~100 V) and by Shamim et al. (1991) for cylindrical and spherical targets at high voltages (~30 kV), obtaining good agreement with the collisionless model.

The effects of finite rise and fall times for the voltage pulse have been examined by Stewart and Lieberman (1991), based on a quasi-static Child law model. They obtained expressions for the time-varying sheath width and implantation current for a linear voltage rise and fall. The results agree well with particle-in-cell computer simulations for a finite risetime pulse (Anders, 2000, Figure 4.4). The preceding models neglect the flow of displacement current ($\epsilon_0 \partial E/\partial t$) during implantation. Displacement currents have been found to be important only for high-voltage implantations at low plasma densities, of the order of 10^8–10^9 cm^{-3} (Wood, 1993). The issues of sheath evacuation and replenishment during multiple pulses have been examined by Wood (1993). He found replenishment of the depleted ion region to occur on the timescale

$$\tau \sim \left(\frac{2\pi T_i}{T_e}\right)^{1/2} \frac{s}{u_B}$$

where s is the width of the sheath at the end of the voltage pulse. The effects of multiple ion masses and charge states have been examined by Qin et al. (1996). For singly charged ions, they found that an effective mass can be used in the models,

$$\sqrt{M_{\text{eff}}} = \sum_j \frac{n_j}{n} \sqrt{M_j}$$

where M_j and n_j are the mass and density of the jth type of ion, and n is the total ion density. Implantation inside pipes and holes gives rise to several additional concerns, including replenishment of ions between pulses, expulsion of sheath electrons, and possible overlap of the expanding sheaths (see Sheridan, 1996; Zeng et al., 1997 for details and models of these effects). Implantation of a dielectric film on a conduction substrate gives rise to a time-varying voltage across the dielectric and to surface charging. Both effects lower the sheath voltage drop, and therefore the ion energy, during an implantation pulse (Emmert, 1994; Linder and Cheung, 1996).

An important feature of PIII is secondary emission during ion implantation. The energetic ions striking the substrate release secondary electrons that accelerate across the sheath, gaining an energy eV_0 per electron. Since the secondary emission coefficients are large at the typical voltages used (e.g., $\gamma_{se} \sim 4.8$ for 20 kV N^+ ions on stainless steel), a large electron current and power must be supplied by the source. Furthermore, the high-energy electrons striking the vacuum chamber produce heat and X-rays, which lead to undesirable cooling and shielding requirements. Some techniques to suppress secondary emission effects have been investigated (Anders, 2000, Section 4.3).

17.5.2 Collisional Sheath Model

Ion collisions within the sheath at high gas pressures lead to reduced implantation energies and finite width energy and angular distributions for ions that greatly affect their implantation over topography, i.e., within trenches. The energy and angular distributions have been determined analytically and compared with particle in cell, Monte Carlo collision (PIC/MCC) simulations

(Vahedi et al., 1991, 1993b). The collisionless model assumptions 2, 3, and 5 are retained, but 1 is replaced by the assumption that the ion motion within the sheath is highly collisional, with charge transfer the dominant source of ion–neutral collisions. It is also assumed that the ion charge density n_s in the sheath is uniform in space but slowly varying in time, with $n_s(t) < n_0$, the bulk plasma density. A uniform distribution is seen experimentally for similar sheaths, such as the cathode sheaths in dc glow discharges (see Chapter 14) and is also seen in PIC/MCC simulations of collisional PIII.

To determine the energy distribution of the bombarding ions, the Maxwell equation

$$\frac{dE}{dx} = \frac{en_s}{\epsilon_0} \tag{17.5.20}$$

is integrated from the electrode surface at $x = 0$ to a position x within the sheath to obtain

$$E = \frac{en_s}{\epsilon_0}(x - s) \tag{17.5.21}$$

where the boundary condition $E = 0$ at $x = s$ has been used. Integrating again to determine the potential using $d\Phi/dx = -E$, we obtain

$$\Phi = -\frac{en_s}{2\epsilon_0}(s - x)^2 \tag{17.5.22}$$

where $\Phi = 0$ at $x = s$. Let $\Phi = -V_0$ at $x = 0$, we obtain the matrix sheath result

$$n_s = \frac{2\epsilon_0 V_0}{es^2} \tag{17.5.23}$$

The equation of motion of an ion starting from rest at $x = x_0$, after a charge transfer collision in the sheath, is

$$\frac{d^2x}{dt^2} = \frac{eE}{M} = \frac{2eV_0}{Ms^2}(x - s) \tag{17.5.24}$$

Assuming that s varies slowly in time, this can be integrated to obtain the ion velocity $u(x)$,

$$u^2 = \frac{u_0^2[(x^2 - x_0^2) - 2s(x - x_0)]}{s^2} \tag{17.5.25}$$

where u_0 is given by (17.5.5). The ion velocity at the target is then

$$u_t^2 = \frac{u_0^2(2sx_0 - x_0^2)}{s^2} \tag{17.5.26}$$

The distribution of ion flux $f(u_t)$ is determined by applying conservation of particles to obtain

$$f(u_t)\, du_t = \nu_{cx} n_s\, e^{-x_0/\lambda_i} dx_0 \tag{17.5.27}$$

where $\nu_{cx} = \bar{u}_t/\lambda_i$ is the ion–neutral charge-transfer collision frequency, $\bar{u}_t$ is the mean ion velocity near the target, and the exponential factor gives the probability that an ion created by charge transfer at x_0 will hit the target before a subsequent ion–neutral collision. Differentiating (17.5.26) to determine dx_0/du_t and substituting this into (17.5.27), we obtain, for $\lambda_i \ll s$,

$$f(u_t) \propto \frac{u_t}{(1 - u_t^2/u_0^2)^{1/2}} \exp\left\{\frac{s}{\lambda_i}\left[\left(1 - \frac{u_t^2}{u_0^2}\right)^{1/2} - 1\right]\right\}, \qquad u_t < u_0 \tag{17.5.28}$$

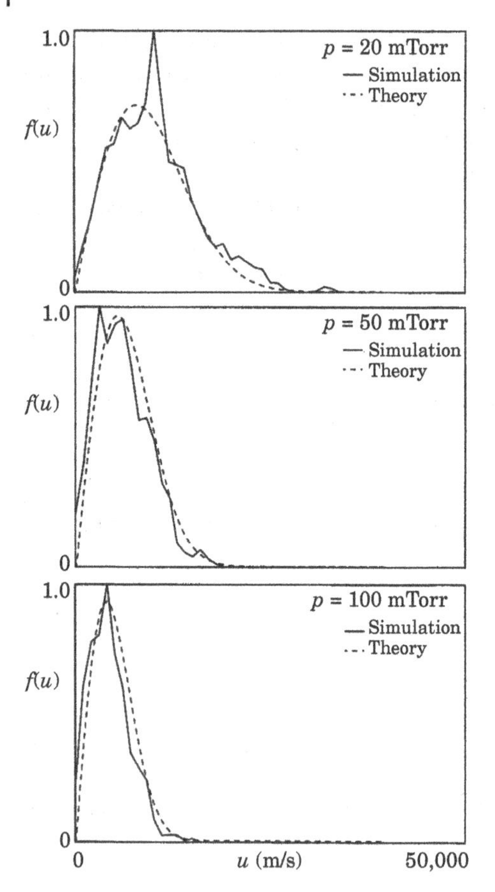

Figure 17.12 Ion velocity distribution at the target for a collisional sheath; the maximum velocity for collisionless acceleration to the target is roughly 5×10^4 m/s.

Figure 17.12 compares the analytical theory (17.5.28) and the computer simulation over a range of pressures (mean free path regimes). The mean ion velocity near the target is found from (17.5.28) to be

$$\bar{u}_t = \left(\frac{eV_0 \pi \lambda_i}{Ms} \right)^{1/2} \tag{17.5.29}$$

and the implantation current density is

$$J_t = en_s\bar{u}_t = \epsilon_0 \left(\frac{4\pi e \lambda_i}{M} \right)^{1/2} \frac{V_0^{3/2}}{s^{5/2}} \tag{17.5.30}$$

where the second equality follows by using (17.5.23) and (17.5.29). The scaling of the current density (17.5.30) is the same as found for the collisional rf sheath (11.2.54), showing the essential correspondence between the two sheaths. The energy distribution of fast neutrals generated by charge transfer processes has been modeled by Wang (1999).

Ion–neutral elastic scattering tends to isotropize the angular distribution of the impinging ions, leading to energetic ion bombardment of trench sidewalls for implantation over topography. Vahedi et al. (1993b) have modeled this process to obtain the ratio of trench sidewall to bottom fluxes,

$$\frac{\Gamma_h}{\Gamma_v} \approx \frac{0.34}{0.93 + \sigma_{cx}/\sigma_{sc}} \tag{17.5.31}$$

which is valid in the limits $\lambda_i \ll s$ and $\sigma_{sc} \ll \sigma_{cx}$, where σ_{sc} and σ_{cx} are the ion–neutral elastic scattering and charge-transfer cross sections, respectively.

17.5.3 Applications of PIII to Materials Processing

17.5.3.1 Semiconductor Processes

Figure 17.13 illustrates two PIII system configurations compatible with semiconductor thin-film processing requirements. To permit operation at pressures as low as 0.2 mTorr, ECR sources operating at 2.45 GHz provide the high ion density, $n_i \sim 10^{10}$–10^{11} cm^{-3}, to supply the required high implantation current. Hot filament sources, which are used for metallurgical implantation (see below), cannot be used because of contamination. The substrate is biased with a pulsed (2–30 kV, 1–3 μs), or, possibly, dc negative voltage to accelerate the ions toward the substrate surface.

With a diode configuration, shown in Figure 17.13*a*, gaseous sources such as Ar, N_2, BF_3, H_2O, and O_2 can be used to provide the ionization medium and the implanting ions. The diode configuration is most convenient for doping applications such as shallow junction formation and conformal doping of nonplanar device structures because many dopant gaseous sources are available. When metal-containing gases are used, e.g., WF_6, the diode configuration can operate as an ion-assisted chemical vapor deposition system. By adding another negatively biased target controlled by a separate power supply to form a triode configuration, as shown in Figure 17.13*b*, atoms from this target are sputtered into the plasma by the carrier gas plasma ions. Some of the emitted target atoms are ionized in the plasma and subsequently implanted into the substrate.

Secondary electron emission from the substrate, and from the target for triode systems, has significant negative consequences at the high ion fluxes and energies of PIII. The secondary electron current can be 5–10 times as large as the ion implantation current at high voltages in diode systems (30–100 kV) (Szapiro and Rocca, 1989). The secondary electrons are accelerated across the sheath and subsequently impinge on the chamber surfaces, which can be a serious X-ray hazard. Also, the PIII power source must supply the power and current, which leads to poor power efficiencies.

PIII has been applied, experimentally, to a number of semiconductor processes. For sub-100 nm p^+/n junction formation where boron implantation is used, preamorphization of the crystalline silicon together with large doses of boron minimizes the source/drain resistance. The final junction depth of these ultra-shallow junctions is dominated by diffusion of dopants during thermal activation, which greatly modifies the implanted depth profile. Because of the high-flux capability of

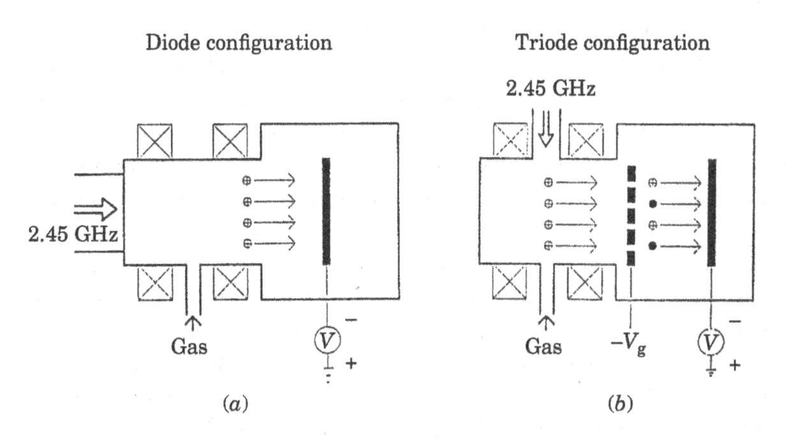

Figure 17.13 Schematic showing diode and triode configurations of PIII for semiconductor implantation.

PIII at low implantation energies, it is well suited for ultra shallow junction formation. The silicon is preamorphized with a 4 kV SiF_4 PIII implantation prior to a 2 kV BF_3 PIII implantation. After annealing at 1060 °C for 1 s, an extremely shallow junction depth of 80 nm is obtained with a sheet resistance of 447 Ω per square. Junctions with a total leakage current density at a reverse bias of −5 V lower than 30 nA/cm^2 have been fabricated (Pico et al., 1992), which compares to the state of the art using other technologies. B_2H_6 feedstock is also used in this application, and the boron ion energy distribution determined both from theory and experiments (Qin et al., 1992). Homogeneous boron doping of n-type crystalline silicon by PIII has been used to create high-efficiency solar cells (Lerat et al., 2016).

Another application has been to selective metal plating. Since copper has a very low electrical resistivity and good electromigration properties, it is an ideal conductor to replace aluminum for integrated circuit interconnects. However, plasma-assisted etching of copper has not been successful due to the lack of suitable volatile etch products. PIII has been used for selective and planarized plating of copper interconnects using palladium seeding, thus avoiding the need to etch the copper. A palladium sputtering target is immersed in the plasma and has an independently controlled negative bias to regulate the sputtering rate. The sputtered neutral palladium forms a continuous flux for deposition, while the Ar^+ and Pd^+ ions assist the penetration of deposited palladium into the substrate via ion beam mixing.

PIII has also been used to conformally dope silicon trenches. High packing densities of devices on silicon substrates are achievable by making use of vertical sidewalls for active transistor channels and as charge storage elements such as trench capacitors. Conventional implantation techniques have focused on multistep implants with collimated beams at controlled beam incidence angles. Taking advantage of the angular divergence of implanting ions in PIII at high gas pressures, conformal doping of high aspect ratio silicon trenches with BF_3 doping has been achieved. In these experiments, silicon trenches about 1 μm wide and 5 μm deep were implanted at −10 kV and a gas pressure of 5 mTorr, yielding a relatively uniform p^+/n junction depth on the top, bottom, and sidewalls of the trench.

Finally, various PIII surface modification processes have been developed for electronics applications. For example, He et al. (2013) used oxygen PIII to improve the surface work function of indium tin oxide films, which have important applications for the fabrication of organic light-emitting devices in the display industry.

17.5.3.2 Metallurgical Processes

PIII can also be used for metallurgical surface modification to improve wear, hardness, and corrosion resistance. In this context, the process has been called plasma source ion implantation (PSII). PSII can easily be used to implant nonplanar targets, e.g., tools and dies, with minimum shadowing and sputtering of the target (see Figure 1.5*b*). The latter can limit the retained dose of the implanted ion species. Ions have been implanted under batch processing conditions, with acceptable dose uniformities to the depths and concentrations required for surface modification, resulting in dramatic improvement in the life of manufacturing tools under actual industrial conditions (Conrad et al., 1990; Redsten et al., 1992). In a typical PSII process, the target is immersed in a nitrogen plasma of density $n_0 \sim 5 \times 10^9$ cm^{-3}. A series of 50 kV, 10 μs pulses at 100 kHz are applied to the target for minutes to hours. For these conditions, the initial matrix sheath thickness is 3 cm, and the Child law sheath thickness is 24 cm, but the pulse width is short enough that the Child law sheath does not have time to fully form. In the referenced work, the plasma is generated by a hot tungsten filament source, which is inserted into the chamber and biased at −(100–300) V. The filament emits electrons that are accelerated across the filament sheath into the plasma, where they subsequently ionize the background gas, which is typically at a pressure of 10^{-4} Torr. The dynamics of hot filament plasma sources is fairly well understood (Leung et al., 1976). Multipole magnets are required on the surface of the implantation chamber to confine the primary electrons (see Section 5.5 for a description of multipole magnetic confinement). Large-size ($\sim 1\ m^3$) PIII systems have been developed for metal and/or gas ion implantation into nonplanar targets as large as 40 cm in diameter (Song et al., 2018).

Problems

17.1 **Silane Discharge Model** Consider a simplified (uniform electron temperature) model for a high-pressure capacitive rf discharge in silane. Use the rate constants in Table 17.1. Assume that the silane density n_g is uniform and is much larger than all other densities, and ignore negative ions and all volume loss processes. The discharge parameters are $p = 200$ mTorr and $l = 3$ cm, and the gas is at room temperature.

(a) Assume that SiH_3 and SiH_2 are created by reactions 1 and 2 in Table 17.1, respectively, and that both species are lost to the electrode walls with unity sticking coefficient. Find the fluxes Γ_{SiH_3} and Γ_{SiH_2} at the electrodes, and find their ratio $\Gamma_{SiH_3}/\Gamma_{SiH_2}$, in terms of n_e, n_g, l, and the rate constants.

(b) By equating the volume rate of generation of positive ions (reactions 5 and 6) to the loss of ions to the discharge electrodes, determine T_e. Treat the SiH_3^+ and SiH_2^+ ions as identical, and assume an ion–neutral momentum transfer rate constant $K_{mi} \approx 10^{-9} cm^3/s$.

(c) Using your results in parts (a) and (b), determine values for Γ_{SiH_3}, Γ_{SiH_2}, Γ_{SiH_4}, and Γ_i at the electrode for an ion (and electron) density $n_0 = 3 \times 10^{10}\ cm^{-3}$ in the center of the discharge.

(d) Assuming $V_{rf} = 500$ V and collisional sheaths (see Section 11.2, and take $u_s = u_B$ in (11.2.53)), and assuming a reasonable value $\mathcal{E}_c = 100$ V for the collisional energy lost per electron–ion pair created, find the ion-bombarding energy $\mathcal{E}_{ic}$, given by (11.2.57), and the absorbed power per unit area S_{abs}.

17.2 **Deposition Rate Within a Trench** Consider deposition within a trench of width w and depth h, as shown in Figure 17.3a, due to an isotropic flux of precursors at the top of the trench having a sticking coefficient of unity. Assume ballistic transport within the trench, i.e., the mean free path for precursor collisions is much greater than w or h.

(a) Let dN be the number of precursor molecules incident on a differential width dy at a sidewall position y due to an isotropic flux emitted from a differential width dx at the top of the trench. Using precursor particle conservation, show that dN is proportional to $dx \cos\theta$ (emission by the source width) and to $dy \cos\theta'$ (reception by the sidewall width) and is inversely proportional to the distance r between the source and sidewall:

$$dN = A\frac{\cos\theta\ \cos\theta'\ dx\,dy}{r}$$

where $\cos\theta = x/r$, $\cos\theta' = y/r$, and A is a constant.

(b) Integrating the expression in part (a) from $x = 0$ to $x = w$, and assuming a uniform source at the top of the trench, show that $dN/dy \propto 1 - \cos\theta_s$, where $\cos\theta_s = y/(y^2 + w^2)^{1/2}$; note that θ_s is the angle subtended by the trench opening as seen at the position y along the sidewall.

(c) Using a similar procedure to that developed in parts (a) and (b), find an expression for the nonuniform deposition rate dN/dx versus x at the bottom of the trench.

17.3 Thermal Atomic Layer Deposition of Al_2O_3

(a) Summing two subreactions (17.3.1) that each produce the surface complex $AlO–Al(CH_3)_2$ (s), and three subreactions (17.3.2) on the resulting surface complex, show that one obtains the overall reaction (17.3.3) producing one molecule of deposited Al_2O_3.

(b) Using the data in Tables 7.1 and 7.2, show that the reaction enthalpy for the overall reaction (17.3.3) is approximately −1230 kJ/mol.

(c) Both H_2O and $Al(CH_3)_3$ are liquids at room temperature (25 °C), with saturated vapor pressures in a room temperature ALD reactor of 24 and 16 Torr, respectively. A typical exposure time to "fully saturate" a plane substrate surface has been measured to be $T =$ 10 ms. Calculate the fluences Φ_{H_2O} and $\Phi_{Al(CH_3)_3}$ (m^{-2}) and compare these to the surface density $n'_{Al_2O_3}$ of the deposited film.

17.4 Plasma-Enhanced Atomic Layer Deposition of Al_2O_3

(a) Summing two subreactions (17.3.1) that each produce the surface complex $AlO–Al(CH_3)_2$ (s), and four subreactions (17.3.4) on the resulting surface complex, show that one obtains the overall reaction (17.3.5) producing one molecule of deposited Al_2O_3.

(b) Using the data in Tables 7.1, 7.2, and 7.4, find the reaction enthalpy for the overall reaction (17.3.5).

17.5 Conformality for Plasma-Enhanced Atomic Layer Deposition

(a) For plasma-enhanced, atomic layer deposition (PEALD) with a zero-coverage sticking coefficient $s_0 = 0.01$ and a radical surface recombination coefficient $\gamma_{rec} = 0.1$, use (17.3.14) to find the fluence Φ_{reac} required (in monolayers) for 99% radical coverage of a plane surface of density n'_0 (m^2).

(b) Using (17.3.18), repeat part (a) for the bottom of a trench of depth-to-width ratio $L/w =$ 10, and $L/w = 20$. Compare your results to the more accurate Monte Carlo calculation in Figure 17.7*b*.

(c) Compare these results to the diffusion-limited fluence given by (17.3.11), and comment on which is more important.

17.6 Reactive Sputtering Deposition Consider the reactive sputtering model of Section 17.4 with equal target and substrate areas. Let $a = \gamma_m/\gamma_c$ and $Y = 2\Gamma_r s_r/\Gamma_i\gamma_c$ be the normalized sputtering coefficient and flux, respectively.

(a) Show using (17.4.2) and (17.4.3) that the surface coverages of the compound on the target and the substrate are given, respectively, by

$$\theta_t = \frac{Y}{1+Y}$$

$$\theta_s = \frac{Y^2+2Y}{Y^2+2Y+a}$$

(b) Show using (17.4.4) and (17.4.5) that the reactive gas flow and the sputtering flux are given, respectively, by

$$\frac{dN_r}{dt} \propto Y\left(\frac{1}{1+Y} + \frac{a}{Y^2+2Y+a}\right)$$

$$\Gamma_{sput} \propto \frac{a+Y}{1+Y}$$

(c) For the limiting case $a \gg 1$, graph dN_r/dt versus Y and Γ_{sput} versus Y. From these graphs, sketch Γ_{sput} versus dN_r/dt and show that the curve exhibits hysteresis similar to that shown in Figure 17.9. Is there hysteresis for the case $a \leq 1$? Prove your answer.

17.7 Collisionless PIII

(a) Derive the sheath motion (17.5.8) when a sudden negative voltage $-V_0$ is applied to the target by directly solving the differential equation (17.5.3) for the case $u_B \equiv 0$.

(b) Suppose a voltage $-V_0(t) = -\alpha t$ that varies linearly with time for $t > 0$ is applied to the target. Assuming a Child law sheath (17.5.1) and using (17.5.2) for the case $u_B \equiv 0$, determine the implantation current density $J_c(t)$ and sketch J_c versus t. Explain why the current density is singular at $t = 0+$ and suggest a value for the maximum current density at $t = 0+$.

17.8 Collisional PIII Consider the collisional sheath model for high-voltage implantation given in Section 17.5.

(a) Using (17.5.30) for J_t, along with the basic relation (17.5.2) (with $u_B \equiv 0$), show that the collisional uniform density (matrix) sheath expands as

$$s(t) = s_0(1 + \omega_0 t)^{2/7}$$

and find an expression for ω_0.

(b) Find an expression for $n_s(t)$.

18

Dusty Plasmas

18.1 Qualitative Description of Phenomena

Particulates or "dust" is an important constituent of planetary and astrophysical plasmas and has been extensively analyzed in that context (see Goertz, 1989). More recently these *dusty plasmas* have been found to be important in processing discharges. On the one hand, particulates can contaminate etching and deposition processes; on the other hand, the growth of particulates in discharges offers unique possibilities for powder synthesis and surface modification processes. These two aspects have led to a resurgence of interest in dusty plasmas. The two types of environments, that of large regions in space and of small laboratory discharges, have many common features and, of course, some features that are quite different. In this section, we focus our attention on the existence of particulates in processing-type discharges, keeping in mind the generality of some of the phenomena. The reader interested in a thorough treatment of dusty plasma physics and applications can consult textbooks such as Bouchoule (1999), Shukla and Mamun (2002), and Melzer (2019), as well as reviews by Merlino (2006, 2021), Ignatov (2005), Shukla and Eliasson (2009), Morfill and Ivlev (2009), and references therein.

Given the idealization of a single approximately spherical particle of known radius, orbital motion theory for a spherical probe immersed in a plasma can give a reasonable account of the equilibrium floating potential of the particle with respect to the plasma, and the equilibrium charge on the particle. The result is that the dust particle charges negatively with a potential of a few times the electron temperature, as required to repel the mobile electrons, such that the positive ion flux and electron flux to the particle are equal. We have already considered this situation with respect to cylindrical probes in Section 6.6. Given this potential, the charge is then determined by the capacitance of the particle with respect to the plasma.

Even if the sizes of the particles are uniform and known, this picture can change in a number of ways. (1) If the electrons are very energetic, or if the plasma is in an intense ultraviolet radiation environment, secondary electrons can be emitted from the particle surface, leading to a lower potential and charge. In an extreme situation, this can actually result in a reversal of the potential and a positively charged particle. (2) With a sufficiently large particle size and density, the particle charge density can be larger than the electron charge density in the plasma, leading to an *electronegative* equilibrium in which the "negative ions" are the negatively charged dust particles. This leads to more complicated equilibria and plasma stratification, as considered in Sections 10.3 and 10.4. However, as we shall see, the particles may concentrate near the plasma sheaths, leading to quite different electronegative structures than treated in those sections. (3) If the density of

Principles of Plasma Discharges and Materials Processing, Third Edition. Michael A. Lieberman and Allan J. Lichtenberg.

the dust grains is sufficiently high, their Debye spheres will overlap, modifying the equilibrium. Furthermore, this dusty plasma may have more potential than kinetic energy, such that it has a special character, known as a *strongly coupled plasma* (Ichimaru, 1982; Shukla and Eliasson, 2009; Morfill and Ivlev, 2009; Melzer, 2019). Such plasmas can exhibit liquid-like and crystal-like behavior, and these states are now under investigation.

The above properties are local. However, the global properties of the discharge lead to additional phenomena. The most obvious is the tendency for particulates to collect near the sheaths of a processing discharge. This results from a force balance in which the positive ion flow outward leads to an outward frictional force on the particulates, balancing the electric field force, which is directed inward on negatively charged particles. In Sections 10.3 and 10.4, we ignored the friction force between positive and negative ions, which usually is small, but could, in some instances, be significant (Deutsch and Räuchle, 1992). In dusty plasmas, the positive ion friction force tends to be dominant in the bulk plasma, pushing the particulates into the pre-sheath region where the electric field force is larger.

The mechanisms of particle formation are not as well understood as particulate charging and the plasma behavior with assumed particulate size and density. Particle formation in many systems can occur through successive steps (Boufendi and Bouchoule, 1994; Perrin and Hollenstein, 1999; Bouchoule, 1999; Melzer, 2019) of gas phase nucleation by growth of negative ion or neutral clusters, followed by a more rapid growth by coagulation of clusters, and then continued growth by surface deposition of neutral dissociation fragments with associated buildup of negative charge. Particles can also originate from fracture or sputtering of films deposited on the walls or the substrate. Although relatively well-posed calculations can be made, given the neutral chemistry and the seed particulates, many questions remain unanswered. In particular, the initial formation which apparently leads to a relatively uniform size of seed particles is not thoroughly understood, and the equilibrium size of the particulates is also not well understood.

In order to understand the behavior of dusty plasmas, in addition to the theory which we summarize below, there has been a significant effort to diagnose such discharges. Probe techniques are useful but have the same type of difficulties associated with other electronegative plasmas, in which the electrons are mobile compared to the negatively charged particles. A very useful technique has been laser scattering, since the laser wavelength can be adjusted to strongly scatter off of the particulates while the plasma is quite transparent to the laser light. Experiments of this nature were performed soon after the contamination effect of particulates was recognized, and before any detailed theory of discharge particulate formation. We will discuss these techniques briefly after our review of the theory.

Various methods for removing particles from processing discharges have been explored. The increasing knowledge of their formation and dynamics has led to new insights into methods of removal. A number of techniques for producing micro- and nano-particles in discharges have also been explored. We will consider some of these techniques, briefly, at the end of this chapter.

18.2 Particle Charging and Discharge Equilibrium

18.2.1 Equilibrium Potential and Charge

Consider a common situation of a dust particle acting like an isolated spherical probe within a plasma. The usual ordering is that the Debye length $\lambda_D \gg a$, the particulate radius. In the usual discharge with $T_e \gg T_i$, the Debye length for shielding around an isolated charged sphere with

$a \ll \lambda_D$ is $\lambda_D \approx \lambda_{Di} = (\epsilon_0 T_i/en_i)^{1/2}$ (see Problem 2.9). Following the methods used in Section 6.6 for ion collection by a cylindrical probe, the orbital motion limited (OML) theory for a spherical probe gives the well-known results for the electron and ion currents collected by the surface (Laframboise, 1966; Laframboise and Parker, 1973; see also Problem 6.8)

$$I_e = -I_{e0}\exp(\Phi_d/T_e) \tag{18.2.1}$$

$$I_i = I_{i0}(1 - \Phi_d/T_i) \tag{18.2.2}$$

where Φ_d is the potential of the probe (here, a particle) with respect to the plasma. For an assumed isotropic Maxwellian distribution of both electrons and ions entering the Debye sphere,

$$I_{e0} = \frac{1}{4} e n_e \bar{v}_e \cdot 4\pi a^2 \tag{18.2.3}$$

$$I_{i0} = \frac{1}{4} e n_i \bar{v}_i \cdot 4\pi a^2 \tag{18.2.4}$$

where $\bar{v}_e = (8eT_e/\pi m)^{1/2}$ and $\bar{v}_i = (8eT_i/\pi M)^{1/2}$. If the ion mean free path $\lambda_i > (1 - \Phi_d/T_i)a$, i.e., in a relatively collisionless plasma with a small collecting sphere, then the OML assumption is reasonably good (Annaratone et al., 1992), a condition which typically holds for dust particles in low-pressure discharges. However, in some cases, the assumption of isotropic Maxwellian ions is a rather poor approximation, due to presheath acceleration (see Sections 6.2 and 6.6). For more general distributions, I_{e0} and I_{i0} can be found in the literature (e.g., Laframboise, 1966; Whipple, 1981). In the other limiting case, the ions enter the Debye sphere radially (no transverse energy) at the Bohm velocity, which greatly enhances the collection area if $\lambda_D \gg a$ (see Section 6.6 and Allen et al., 1957).

In equilibrium, assuming no secondary or field emission of electrons from the surface, Φ_d can be found by setting the total current $I_e + I_i$ collected by the particle equal to zero. Doing this, using the approximations in (18.2.1)–(18.2.4), rearranging and taking the logarithm, we have

$$\Phi_d = -T_e\left[\ln\left(\frac{M}{m}\frac{T_e}{T_i}\frac{n_e^2}{n_i^2}\right)^{1/2} - \ln\left(1 - \frac{\Phi_d}{T_i}\right)\right] \tag{18.2.5}$$

Even for the approximate values of I_{e0} and I_{i0} in (18.2.3) and (18.2.4), the equation is transcendental, and solutions must be obtained numerically. An approximate analytic expression is given by Matsoukas and Russell (1995)

$$\Phi_d \approx -0.73\, T_e \ln\left(\frac{M}{m}\frac{T_i}{T_e}\frac{n_e^2}{n_i^2}\right)^{1/2} \tag{18.2.6}$$

typically $\Phi_d \sim -$ few T_e. Once the potential is known, the charge on the particle is straightforwardly obtained from

$$Q_d = C_d \Phi_d \tag{18.2.7}$$

where the particulate capacitance is (see Problem 2.9(d))

$$C_d = 4\pi\epsilon_0 a\left(1 + \frac{a}{\lambda_D}\right) \tag{18.2.8}$$

and for the usual ordering $a \ll \lambda_D$

$$C_d \approx 4\pi\epsilon_0 a \tag{18.2.9}$$

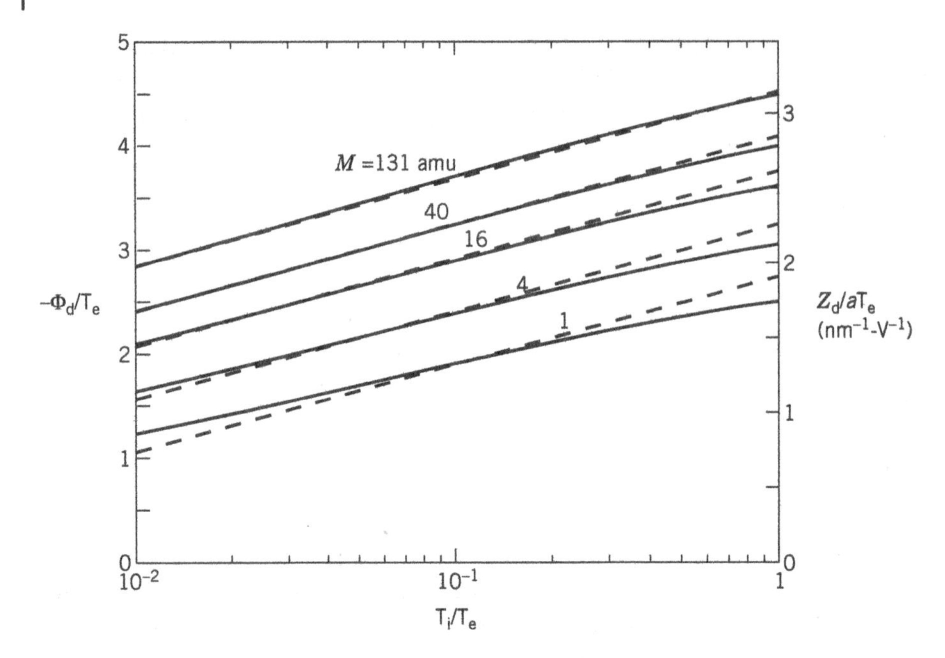

Figure 18.1 Normalized floating potential Φ_d/T_e versus ion-to-electron temperature ratio T_i/T_e, for different values of the ion mass; the right axis gives the corresponding value of the number of electrons on the dust particle, normalized to its radius a (in nm) times the electron temperature T_e (in volts); the solid lines correspond to numerical solutions; the dashed lines correspond to the approximate analytical solution. Source: Adapted from Matsoukas and Russell (1995).

Note again here that $\lambda_D \approx \lambda_{Di}$ for the usual ordering $T_i \ll T_e$. The two solutions (18.2.5) and (18.2.6) for Φ_d and solution (18.2.7) for $Z_d = -Q_d/e$ are shown in Figure 18.1 in normalized form for various temperature ratios and ion masses, assuming that $n_e = n_i$.

If a dust grain enters a plasma with some non-equilibrium initial conditions, the charge build-up can be obtained from

$$\frac{dQ}{dt} = I_e + I_i \tag{18.2.10}$$

Substituting Φ_d from (18.2.7) into I_e and I_i given by (18.2.1) and (18.2.2) and using these expressions in (18.2.10), we obtain

$$\frac{dQ}{dt} = -I_{e0} \exp\left(\frac{Q}{C_d T_e}\right) + I_{i0}\left(1 - \frac{Q}{C_d T_i}\right) \tag{18.2.11}$$

This equation can be solved numerically to obtain the dynamical build-up of charge. For an initially uncharged particle, the second term on the RHS in (18.2.11) is small compared to the first term. Dropping this term and expanding the exponential in the first term, we obtain the initial charge build-up dynamics

$$Q(t) = -I_{e0}\, t \tag{18.2.12}$$

Equating this to the equilibrium value $Q_d = C_d \Phi_d$ from (18.2.7), we obtain an estimate of the charging time $t = \tau$

$$\tau \sim \frac{C_d T_e}{I_{e0}} \tag{18.2.13}$$

where we have used $\Phi_d \sim -T_e$ from (18.2.5) or Figure 18.1. From (18.2.5)–(18.2.13), we see the essential scalings with a, n_e, and T_e. In particular, for $n_e \approx n_i$, $\Phi_d \propto T_e$ from (18.2.5), $Q_d \propto aT_e$ from (18.2.7), and $\tau \propto T_e^{1/2}/n_e a$ from (18.2.13). Consider for example an argon discharge with a 115 nm particle, $T_e = 2$ V and $T_i/T_e = 0.05$. Then, $Z_d = -Q_d/e \approx 477$ and $n_e\tau \approx 1.78 \times 10^4$ s-cm^{-3}. For a plasma with $n_e = 10^{10}$ cm^{-3}, we find that $\tau \approx 1.8$ μs, which is short compared to the growth time of the dust particles.

There are many physical effects that reduce the collection of negative charge on a dust grain. Since the charge collected is quantized in units of e, it is subject to shot noise fluctuations. Let $f_d(Z)$ be the distribution of charge number $Z = -Q/e$ on a collection of equal-size particles, then Matsoukas et al. (1996) show that f_d follows a Gaussian distribution for $Z \gg 1$ and $n_e \approx n_i$

$$f_d(Z) = \frac{1}{(2\pi\sigma_z^2)^{1/2}} \exp\left[-\frac{(Z - Z_d)^2}{2\sigma_z^2}\right] \tag{18.2.14}$$

where $Z_d = -Q_d/e$ is given by solving (18.2.5) for Φ_d and substituting into (18.2.7), and the standard deviation is (Problem 18.1)

$$\sigma_z \approx 0.5\, Z_d^{1/2} \tag{18.2.15}$$

Equation (18.2.15) gives a reasonable estimate for σ_z even for $Z_d \sim 1$. For this case of small Z_d, the charge on a grain can fluctuate to zero or positive, even though the average charge is negative.

Let us now consider the reduction in negative charge due to the effect of depleted plasma electrons. For example, in an rf discharge in silane, Boufendi et al. (1992) found a particle density of $n_d = 10^8$ cm^{-3} as measured by laser light scattering in a background ion density of $n_i = 5 \times 10^9$ cm^{-3}, measured by ion saturation current on a Langmuir probe. The typical particle radius, determined from electron microscopy, was 115 nm, which, as calculated above, gives $Z_d \approx 477$. For this Z_d and n_d, the plasma would be strongly depleted of electrons and most of the negative charge would reside on the particulates. Assuming a uniform density of dust grains and using quasineutrality together with (18.2.7) and (18.2.9), we have

$$n_e - n_d \frac{4\pi\epsilon_0 a}{e} \Phi_d = n_i \tag{18.2.16}$$

Solving for n_e and substituting the result into (18.2.5) yields

$$\Phi_d = -T_e \left[\ln\left(\frac{M}{m}\frac{T_e}{T_i}\right)^{1/2} - \ln\left(1 - \frac{\Phi_d}{T_i}\right) + \ln\left(1 + \frac{n_d}{n_i}\frac{4\pi\epsilon_0 a}{e}\Phi_d\right)\right] \tag{18.2.17}$$

Since $n_e > 0$, from (18.2.16), we see that $n_d(4\pi\epsilon_0 a/e)|\Phi_d| < n_i$ for a solution to exist. Approximate values in a strongly electronegative situation are given by the equality which, for the above example, gives $\Phi_d \approx -0.63$ V and $Z_d \approx 50$ (Problem 18.2).

One cannot, however, accurately calculate a new equilibrium using (18.2.1)–(18.2.5), with a self-consistent electron density from (18.2.16), since the basic assumption of an isolated particle collecting charge may be incorrect for large n_d. To see this, making the assumptions $n_i = 5 \times 10^9$ cm^{-3} and $T_i = 0.1$ V, we find a Debye length $\lambda_D = (\epsilon_0 T_i/en_i)^{1/2} = 3.3 \times 10^{-3}$ cm. However, the average separation of particles is given by $\Delta = (3/4\pi n_d)^{1/3} \approx 1.3 \times 10^{-3}$ cm. With $\Delta \lesssim \lambda_D$, the relation between charge and potential is modified, leading to a decreased charge on the particles from that which would be calculated using (18.2.9) and (18.2.17). Whipple et al. (1985) account for this under the assumption of a regular lattice of particles spaced by Δ. For example, with $\Delta/a = 100$ and $\lambda_D/\Delta = 10$, the equilibrium charge is approximately a tenth that given by (18.2.16). However, their analysis was performed for a space plasma with $T_i/T_e = 1$. For the laboratory plasma in the above example, with $T_i/T_e \sim 0.05$, the effect is much smaller.

Electrons can be emitted from dust grains due to field emission, electron, ion and metastable impact, ultraviolet (UV) photon absorption, and thermionic emission. Very small grains cannot collect negative charge because of field emission from the surface. The electric field at a smooth spherical surface is $E = Q_d/4\pi\epsilon_0 a^2$. Substituting for Q_d using (18.2.7) with C_d given from (18.2.9) and estimating $|\Phi_d| \approx 2\mathrm{T_e} = 4$ V from (18.2.5) or Figure 18.1, we find $E \approx 2\mathrm{T_e}/a$. Assuming a value for field emission to occur of $E \gtrsim 10^9$ V/m, then the particulate charge will not build up for particle radii $a \lesssim 2\mathrm{T_e}/E = 4$ nm. For non-spherical particles or particles with "bumpy" surfaces, field emission will occur for average radii greater than 4 nm. Let the onset of charging be at $a = 10$ nm, the number of negative charges at the onset is $Z_d \approx 4\pi\epsilon_0 a \cdot 2\mathrm{T_e}/e = 28$.

Secondary electron yields γ_{se} for ion or metastable impact depend on the particle material and the nature of the ion or metastable, and for electron impact depend on both the impact energy $\mathcal{E}_e$ and the particle material. For low-energy ions or metastables, $\gamma_{se} \approx 0.01$–0.1 (see Table 9.1). For electrons, $\gamma_{se} \approx 7.4\gamma_m(\mathcal{E}_e/\mathcal{E}_m)\exp[-2(\mathcal{E}_e/\mathcal{E}_m)^{1/2}]$, with the peak yield γ_m at energy $\mathcal{E}_m$ (Whetten, 1992). Typically, $\gamma_m \sim 1$–5 and $\mathcal{E}_m \sim 100$–500 V. The primary electrons can either be part of the thermal (e.g., Maxwellian) distribution or be energetic electrons generated by ions impinging on the capacitive discharge electrodes and accelerated by the high fields of the rf sheath. The yields from small particles can be significantly enhanced above the values for bulk materials, because scattered electrons escape more easily from a small particle than from a semi-infinite slab due to geometrical effects (Chow et al., 1994). Under some circumstances (e.g., high $\mathrm{T_e}$), the secondary emission current due to electron impact can be a significant fraction of the primary currents. A significant reduction in magnitude of the average negative charge on the particle is found by including these currents in the particle charge balance. The particle charge can even become positive under some conditions (Goertz, 1989).

Absorption of UV photons releases photoelectrons with yield γ_ν, i.e., $\Gamma_e = \gamma_\nu \Gamma_\nu$ with Γ_ν the photon flux. It is known that dust in space can charge positively due to UV exposure, and a laboratory plasma can be a strong source of UV due to electron impact excitation of neutrals. Many dusty plasma measurements have been conducted in SiH_4 discharges highly diluted with argon or helium. It is well known that up to 50% of the plasma power can be transferred to UV resonance radiation in these rare gas discharges. Since resonance radiation can be strongly trapped in the plasma, the UV flux internal to the plasma can be much larger than the emitted flux at the surface. Furthermore, the yield γ_ν can be higher for small particles than for bulk materials. A simulation model of charging with and without inclusion of UV and electron, ion and metastable impact emission of electrons shows a significant difference in particle charge distributions for the same plasma conditions. For example, for 1 nm particles without these processes and for the plasma conditions of Boufendi and Bouchoule (1994), the mean and standard deviation were approximately $-2e$ and e per particle, respectively, implying few positive particles. Including the additional processes, the mean was zero and the standard deviation was $1.5e$, implying equal numbers of positive and negative particles (Kortshagen and Bhandarkar, 1999).

18.2.2 Discharge Equilibrium

In the experimental situation described in our example, the electron density was found to fall by a factor of about 10 as the particles built up over time. The experimental results are shown as a function of time in Figure 18.2 for dust particle build-up in a 100 mTorr Ar/SiH_4 capacitive discharge (Boufendi et al., 1996) at a fixed 13.56 MHz driving voltage. In addition to the decrease in electron number density another important consequence, also shown in the figure, is the increase in the electron temperature. There can also be a modest increase in the positive ion density. Qualitatively

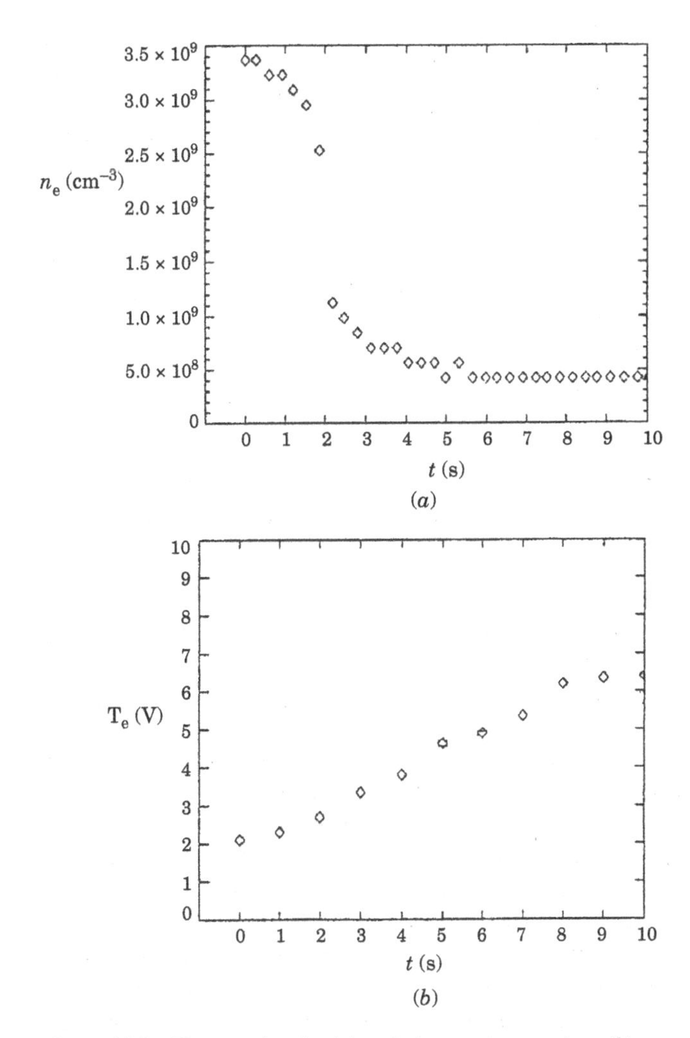

Figure 18.2 Electron density (a) and electron temperature (b) versus time in an 30 sccm argon + 1.2 sccm silane discharge; the plasma reactor is 13 cm diameter and 3 cm in height, driven at 13.56 MHz with a peak-to-peak voltage of 600 V, with a total pressure of 150 mTorr. Source: Boufendi et al. (1996)/with permission of AIP Publishing.

these changes can be understood in terms of quasineutrality and the plasma particle and energy balance. During the later stage of dusty discharge operation, with $n_e \ll n_i$, the quasineutrality condition (18.2.16) yields the scaling

$$n_i \propto n_d a \tag{18.2.18}$$

As described in Section 10.2, the particle balance is that the volume production due to ionization must equal the loss to the surface

$$K_{iz} n_e n_g \mathcal{V} = \Gamma_i A \tag{18.2.19}$$

where Γ_i is the particle loss flux, $\mathcal{V}$ is the volume, and A is the surface loss area. The loss area includes the discharge wall area A_w and the total surface area of the dust particles, $A = A_w + 4\pi a^2 n_d \mathcal{V}$. As the dust particles grow in size and number, they can become the main loss area. With $\Gamma_i \propto n_i$, we obtain from (18.2.19) that

$$K_{iz} \propto \frac{n_i}{n_e}(A_w + 4\pi a^2 n_d \mathcal{V}) \tag{18.2.20}$$

The electron power balance is that the heating power, taken here to be ohmic, must equal the electron power losses

$$\frac{1}{2}\frac{\tilde{J}_{rf}^2}{\sigma_{dc}}\mathcal{V} = \Gamma_i A e(\mathcal{E}_e + \mathcal{E}_c) \tag{18.2.21}$$

where $\sigma_{dc} = e^2 n_e / m\nu_m$ given by (4.2.22) is the dc plasma conductivity and $\mathcal{E}_e + \mathcal{E}_c$ is the electron energy lost per electron lost from the discharge. Hence, the LHS of (18.2.21) scales as $\tilde{J}_{rf}^2/\sigma_{dc} \propto \tilde{J}_{rf}^2/n_e$. For a voltage-driven capacitive discharge with $n_i \neq n_e$, set $n_s \sim n_i$ and $J_1 \sim \tilde{J}_{rf}$ in (11.2.23) and substituting this into (18.2.21), we have the scaling

$$\frac{\tilde{J}_{rf}^2}{\sigma_{dc}} \propto \tilde{V}_{rf}^{1/2}\frac{n_i}{n_e} \tag{18.2.22}$$

Hence from (18.2.21) and (18.2.22) with $\tilde{V}_{rf} = \text{const}$ and taking $\Gamma_i \propto n_i$, we obtain the scaling

$$n_e \propto \frac{1}{A_w + 4\pi a^2 n_d \mathcal{V}} \tag{18.2.23}$$

From (18.2.23), we see that the electron density falls as the particle surface area grows. Inserting (18.2.18) and (18.2.23) into (18.2.20), we obtain

$$K_{iz}(\mathrm{T_e}) \propto n_d a (A_w + 4\pi a^2 n_d \mathcal{V})^2 \tag{18.2.24}$$

Since $K_{iz} \propto e^{-\mathcal{E}_{iz}/\mathrm{T_e}}$ is an increasing function of $\mathrm{T_e}$, we see that $\mathrm{T_e}$ grows as n_d and a increase. The increase in $\mathrm{T_e}$ also leads to a resulting increase in the 7504 Å argon neutral emission during the dust formation, as seen experimentally. As will be seen in Section 18.4, during the final stage of particle growth $n_d \approx$ const and the particle radius grows slowly with time, as $a \propto t^{1/3}$. This leads to a modest increase in n_i given by (18.2.18), with a larger decrease in n_e. An additional effect as n_e decreases is an increase in the ohmic voltage drop across the bulk plasma. Since the total voltage across the discharge is fixed, this leads to a decrease in the sheath voltage drop, which modifies the scalings of K_{iz} and n_e. In the experiment, Boufendi et al. (1996) found a 40% increase in ion density in going through the transition for particulate formation. There was an accompanying increase in the bulk electric field $\tilde{E}_{rf} = \tilde{J}_{rf}/\sigma_{dc}$ from approximately 1–4 V/cm. Scalings similar to (18.2.23) and (18.2.24) can also be obtained for a capacitive discharge driven by a constant rf current (Problem 18.3).

18.3 Particulate Equilibrium

Given an equilibrium of a plane-parallel plasma discharge, with sheaths near both electrodes, the forces that act on the particulates are: gravity, neutral drag, ion drag, and electric fields. The ion drag is generally the principal outward force that balances the inward force of the electric field at the interface between the plasma and the sheaths. For horizontal electrodes, gravity adds to the ion force at the lower electrode and subtracts from the ion force at the upper electrode. Depending on the neutral flow, the neutral drag force may add, subtract, or be neutral with respect to the ion force.

Rather straightforwardly, the gravitational force on a particle is the product of the particle mass times the gravitational acceleration g

$$F_g = \frac{4}{3}\pi a^3 \rho_{md}\, g \tag{18.3.1}$$

where ρ_{md} is the mass density of the dust particle and $g = 9.8\ m/s^2$. The neutral drag force is determined from the first term in (2.3.14), which gives the time rate of momentum transfer per unit volume lost by neutrals having density n_g and flow velocity u_g impinging on particulates having density n_d

$$f_n = -M_g n_g \nu_{gd} u_g \tag{18.3.2}$$

The particulate flow velocity is assumed to be negligible and the neutral flow velocity is assumed to be small compared to the neutral thermal velocity. Approximating the interaction as hard sphere collisions, we write the collision frequency as $\nu_{gd} = n_d \sigma_{gd} \overline{v}_g$, where $\sigma_{gd} = \pi a^2$ and $\overline{v}_g = (8e\mathrm{T}_g/\pi M_g)^{1/2}$. Because the momentum lost by neutrals is gained by dust particles, we obtain the neutral drag force on a particle as

$$F_n = -f_n/n_d = M_g n_g \pi a^2 \overline{v}_g u_g \tag{18.3.3}$$

Similarly, the ion drag force is determined from the momentum transfer lost by ions impinging on particulates

$$f_i = -M n_i \nu_{id} u_i \tag{18.3.4}$$

where $\nu_{id} = n_d \langle \sigma_{id} v_i \rangle$ is the momentum transfer collision frequency for positive ions impinging on the negatively charged dust particles. The momentum transfer has two parts, one due to the transfer when an ion is collected and the other due to Coulomb scattering of ions by the particle. We write the rate constants $\langle \cdot \rangle$ for each process as the product of a cross section σ times an effective ion velocity

$$v_{ieff} = \left(u_i^2 + \frac{8e\mathrm{T}_i}{\pi M} \right)^{1/2} \tag{18.3.5}$$

which accounts for both the ion drift and thermal velocities. The cross section for collection can be written

$$\sigma = \pi b_c^2 \tag{18.3.6}$$

where b_c is the collection radius. This is found by equating the total ion current collected (18.2.2) to the product of the random thermal ion current flux times a collection area

$$I_{i0} \left(1 - \frac{\Phi_d}{\mathrm{T}_i} \right) = \frac{1}{4} e n_i \overline{v}_i \cdot 4\pi b_c^2 \tag{18.3.7}$$

Substituting (18.2.4) for I_{i0} into (18.3.7), we find

$$b_c = a \left(1 - \frac{\Phi_d}{\mathrm{T}_i} \right)^{1/2} \tag{18.3.8}$$

The momentum transfer cross section for Coulomb scattering is approximated by inserting (3.3.3) into (3.1.15), with the lower limit in (3.1.15) taken to be θ_{min} rather than zero, corresponding to a Coulomb potential that is cut off at a radius $r = \lambda_D$. The result is (Problem 3.5)

$$\sigma = \pi b_0^2 \ln \Lambda \tag{18.3.9}$$

where from (3.3.2)

$$b_0 = \frac{eQ_d}{2\pi\epsilon_0 M v_{ieff}^2} \tag{18.3.10}$$

and $\Lambda = 2/\theta_{\text{min}}$ is the Coulomb logarithm. For the Coulomb potential, (3.2.26) reduces to $b = b_0/\Theta$. Set $b = b_{\text{max}} = \lambda_\text{D}$ and $\Theta = \theta_{\text{min}}$ in this expression yields $\theta_{\text{min}} = b_0/\lambda_\text{D}$. Defining a minimum impact parameter b_{min} through the relation $\Lambda \equiv b_{\text{max}}/b_{\text{min}}$, then $\theta_{\text{min}} = b_0/\lambda_\text{D}$ corresponds to $b_{\text{min}} = b_0/2 = b_{90}$, where b_{90} is the impact parameter for a scattering angle of 90°. Accounting for both ion collection and Coulomb scattering, we then obtain the ion drag force

$$F_\text{i} = -f_\text{i}/n_\text{d} = Mn_\text{i}v_{\text{ieff}}u_\text{i}(\pi b_\text{c}^2 + \pi b_0^2 \ln \Lambda) \tag{18.3.11}$$

The direct collection term is often neglected in (18.3.11) but can be comparable to the Coulomb term for large dust particles.

For an isolated dust grain, the electric field E_{ext} acting on the dust grain is that produced by all the other charges in the system. Assuming a continuous model for the charge densities of positive ions, negative ions, electrons, and charged particulates, the electric field E_{ext} can be computed from the dusty plasma equilibrium alone. The electric field force acting on a particulate is

$$F_\text{E} = Q_\text{d}E_{\text{ext}} \tag{18.3.12}$$

We now compare the forces acting on the particulate, in an example. There are various models for determining b_{max} and b_{min} for the ion drag force, but since these quantities occur within a logarithm, varying their values modifies the numerical values only moderately. Northrop and Birmingham (1990) take $b_{\text{max}} = \lambda_{\text{De}}$, the electron Debye length, and $b_{\text{min}} = a$. These limits differ from the usual upper limit of $b_{\text{max}} = \lambda_{\text{Di}}$ and the usual lower limit of $b_{\text{min}} = b_{90}$. The upper limit of λ_{De} would be appropriate for ions which are accelerated in the presheath to the Bohm velocity, $\propto T_\text{e}^{1/2}$, and subsequently collisionally randomized. The choice of the lower limit in the impact parameter can make a difference in F_i, as seen in a practical example below. Winske and Jones (1994) discuss various methods of calculating F_i and compare them. They take an argon plasma with $n_\text{e} = n_\text{i} = 3 \times 10^9 \text{cm}^{-3}$, $\text{T}_\text{e} = 2\text{V}$, $\text{T}_\text{i} = 0.03$ V, $a = 0.3$ μm and $Z_\text{d} = 10^3$ (somewhat arbitrarily). With $n_\text{g} = 3 \times 10^{15}\text{cm}^{-3}$ (100 mTorr) they calculate an ion mobility of $\mu_\text{i} = 2 \times 10^3$ cm^2/V-s, and taking an electric field $E = 50$ V/cm in the pre-sheath region they obtain $u_\text{i} = \mu_\text{i}E \approx 10^5$ cm/s. Using these numbers, they obtain $\lambda_{\text{De}} = 190$ μm and $b_{90} = 3.5$ μm. Taking $\ln \Lambda = \ln \lambda_{\text{De}}/a$, we have $\ln \Lambda = 6.4$, or alternately taking $\ln \Lambda = \ln \lambda_{\text{De}}/b_{90}$, we have $\ln \Lambda = 4$ which is approximately 2/3 the first value. Other choices give values in the same range. Continuing the example, from (18.3.11), we have $F_\text{i} = 8 \times 10^{-13}$ N. This is to be compared with F_g ($\rho_{\text{md}} = 2.2$ g/cm^3) giving $F_\text{g} = 2.4 \times 10^{-15}$ N which is negligible. If one takes a (rather arbitrary) neutral flow velocity of $u_\text{g} = 10^3$ cm/s, then $F_\text{n} \approx 10^{-14}$ N, which is also small compared to F_i. The basic force balance, for this case, is between the ion drag force and the electric field force. Using $Q_\text{d} = -eZ_\text{d}$ in (18.3.12), we have $F_\text{E} = 8 \times 10^{-13}$ N, which approximately balances the ion drag force. In the plasma bulk, the ion drag force is generally large compared to the electric field force, while in the plasma sheath, the electric field force is large compared to the ion drag force. The result is typically a potential well for the dust grains in the interface region between the bulk plasma and the sheath.

For large dust grains, particularly in plasmas in which artificial particulates are introduced to obtain crystal structures, the gravity force (18.3.1) can become large due to the a^3 dependence. The particles are then not trapped near the upper electrode and tend to congregate deep in the high-field sheath region of the lower electrode. If there is a temperature gradient in the gas, there can also be a thermophoresis force. This arises because the momentum transfer of gas molecules is larger on the hot side of the particle than on the cold side. At high pressures, the force can be written as (Boeuf and Punset, 1999) $F_{\text{th}} = -(32/15)(a^2/v_\text{g})\kappa_\text{g}\nabla T_\text{g}$, where κ_g is the thermal conductivity of the gas. This force is generally smaller than the ion force acting to expel the dust grains.

For higher density plasmas, generated in inductive, electron cyclotron resonance (ECR), or helicon discharges, the particulate equilibria can be quite different from the lower density capacitive discharge conditions considered above. For example, Graves et al. (1994) studied an ECR discharge at a density of $n_i = 5 \times 10^{12}$ cm^{-3}. In this plasma, with a density three orders of magnitude larger than the plasma considered above, the ion drag dominates and the particulates are pushed deeply into the sheath or completely out of the plasma.

18.4 Formation and Growth of Dust Grains

The formation and growth of dust grains are not completely understood. General observations of precursors and of the change in number and size of the dust grains are made using infrared absorption, mass spectrometry, laser light scattering (LLS) and other techniques. The initial precursors to the formation of micron-sized dust grains in low-pressure discharges appear to be high mass, singly charged negative ion clusters, and in some regimes, neutral clusters. For example, Hollenstein et al. (1998) have observed negative ion clusters having masses up through 1300 amu in a 75 mTorr silane discharge. A mass of 1300 amu corresponds to a 1.2 nm-diameter particle with $j = 44$ silicon atoms.

The basic picture at low pressures or powers is that there is an initial stage of cluster formation in which feedstock monomers (possibly vibrationally excited) and/or feedstock dissociation fragments combine successively with singly negatively charged clusters. This stage has a rapid growth up to a certain critical size $j \approx 200$ ($\approx$2 nm diameter). The single negative charges are chemically bound and therefore are only weakly subject to field emission or other detachment processes. Since the negative clusters are confined, they may predominate over the buildup of neutral clusters, which are lost at diffusion rates, except at high power or pressure as described below. Cluster formation is succeeded by *coagulation* of the clusters when the number density of $j > 200$ clusters is sufficiently large. When the coagulated particulates exceed a diameter of around 10 nm, the build-up of negative charge on the particulates prevents further coagulation, and a slower accretion of mass by collisions with neutral fragments takes place, with accompanying negative charging.

At higher pressures or powers, the growth rate of clusters may be sufficiently fast that neutral loss times by diffusion may be slower than times for the initial build-up of the $j > 200$ clusters. In this case, the first stage may be primarily by neutral clusters, rather than singly negatively charged clusters. Alternatively, the clusters may fluctuate between primarily neutral and negatively charged (a small fraction of the time) during a neutral diffusion time, such that, on the average, there is an electrostatic potential confining them in the discharge (Fridman et al., 1996).

A simple model for negative ion cluster formation in silane begins with dissociative attachment

$$\mathrm{e} + \mathrm{SiH_4} \xrightarrow{K_{\mathrm{att}}} \mathrm{SiH_3^-} + \mathrm{H} \tag{18.4.1}$$

followed by a series of neutral silicon insertion reactions of the form

$$\mathrm{Si}_j\mathrm{H}_x^- + \mathrm{SiH}_y \xrightarrow{K_j} \mathrm{Si}_{j+1}\mathrm{H}_z^- + (\mathrm{H\ products}) \tag{18.4.2}$$

The neutrals can be SiH_4 (perhaps vibrationally excited) or its dissociation fragments. Various loss processes can compete with this growth; e.g., negative cluster recombination with positive ions, leading to the production of neutral clusters that may be lost from the system on a faster (diffusion) timescale. We consider these processes to illustrate the formation of high mass, negative ion clusters, assuming that the neutral molecule inserted is SiH_4 (Fridman et al., 1996; Gallagher, 2000).

Let n_j be the density of the jth negative cluster (containing j silicon atoms), then accounting for production and loss by the insertion reaction (18.4.2) and loss by recombination, we have the particle balance for the $(j+1)$st cluster

$$\frac{\mathrm{d}n_{j+1}}{\mathrm{d}t} = K_j n_g n_j - K_{j+1} n_g n_{j+1} - K_{rec} n_{j+1} n_i, \qquad j = 1, 2, \ldots \tag{18.4.3}$$

Here, n_g is the density primarily of the feedstock gas SiH_4. To find the steady-state solution to this reaction chain, we assume an SiH_3^- density n_1 determined by (18.4.1) and the competing recombination and insertion loss processes. Set $\mathrm{d}/\mathrm{d}t \equiv 0$ in (18.4.3) gives

$$\frac{n_{j+1}}{n_j} = \frac{K_j n_g}{K_{j+1} n_g + K_{rec} n_i} \tag{18.4.4}$$

As described in Section 8.4, $K_{rec} \approx 10^{-7}$ cm^3/s is roughly a constant independent of size. For small cluster sizes, estimates are that $K_j \approx K_0 \approx 3 \times 10^{-12}$ cm^3/s, independent of j. However, it is known that SiH_4 does not react significantly with growing film, which suggests that K_j decreases significantly above a certain j. This effect may be due to changes in electron affinity due to a weakening electric field at the surface as j increases (Gallagher, 2000) or to an increasing relaxation of vibrationally excited SiH_4 on the surface as j increases (Fridman et al., 1996). As one example, Gallagher gives the estimate

$$K_j \sim K_0 (1 - \mathrm{e}^{-100/j^{4/3}}) \tag{18.4.5}$$

Since $K_0 n_g \gg K_{rec} n_i$, we can pass from the discrete to the continuous limit

$$\frac{1}{n_j}\frac{\mathrm{d}n_j}{\mathrm{d}j} = \frac{n_{j+1} - n_j}{n_j}$$

to obtain, using (18.4.4)

$$\frac{1}{n_j}\frac{\mathrm{d}n_j}{\mathrm{d}j} = -\frac{K_{rec} n_i}{K_j n_g + K_{rec} n_i} \tag{18.4.6}$$

For $j < 100^{3/4} \approx 32$, $K_j = K_0$ from (18.4.5), and (18.4.6) can be integrated to obtain an exponentially decaying density

$$n_j = n_1 \mathrm{e}^{-j/j_0}$$

with $j_0 = 1 + K_0 n_g / K_{rec} n_i$. Depending on the discharge conditions, the decay can be fast or slow; such decays have been observed by Hollenstein et al. (1998). As j increases beyond $100^{3/4}$, K_j in (18.4.5) decreases below K_0 and the rate of decay of the density increases. Set $K_j n_g = K_{rec} n_i$ as the condition for a rapid drop in n_j and using (18.4.5), we obtain $j_{crit} = (100 K_0 n_g / K_{rec} n_i)^{3/4}$.

Evaluating these estimates for typical particle-producing discharge parameters $n_g = 3.3 \times 10^{16}$ cm^{-3} and $n_i = 2 \times 10^{10}$ cm^{-3}, we find $j_0 = 50$ and $j_{crit} = 590$. Hence, there is copious production of clusters with diameters exceeding 2 nm ($j = 200$). On the other hand, a lower-pressure discharge with $n_g = 7 \times 10^{15}$ cm^{-3} yields $j_{crit} \lesssim 170$. Lower-pressure discharges are less likely to generate significant particle densities. The reaction chain (18.4.3) can be solved for the time-varying cluster densities (Problem 18.4) to find the timescale for the j_{crit} density to approach its steady-state value: $\tau_{crit} \approx 1/K_{rec} n_i \approx 0.5$ ms in our example above.

More detailed numerical models (Choi and Kushner, 1993; Gallagher, 2000) allow for many more reactions in the negative-cluster particle balance and also examine the role of neutral clusters. The SiH_3 precursor density for neutral-cluster production is much larger than the SiH_3^- precursor density for negative clusters, but the dominant loss for small neutral clusters of diffusion to the walls is fast compared to the neutral-cluster generation rates. Hence, the jth neutral-cluster density decays rapidly with j, as seen both experimentally and in simulations. However, for sufficiently large j, the diffusion loss rate decreases and the neutral clusters can charge negatively by non-dissociative electron attachment

$$e + Si_jH_x \rightarrow Si_jH_x^-$$

This process (which is forbidden at small j; see Problem 3.16) along with positive-ion recombination with negative clusters tends to couple the neutral and negative cluster densities together for the larger js, as seen in the simulations.

Following precursor formation and nucleation to a large number density, the coagulation stage results in larger size particulates at densities much below that of the clusters. Coagulation is typically on a fast timescale compared to the subsequent particle growth. Figure 18.3 shows the typical time development of the particle radius a and number density n_d in a pure silane discharge (Courteille et al., 1996). Coagulation is seen as the initial sharp drop in n_d and increase in a at time $t \sim 2$ s, in which a increases to ~10 nm. Coagulation arises because the thermodynamic free energy of a distribution of small grains in a plasma is reduced when grains coalesce, due to a reduction in the total surface area and its associated free energy. During coagulation, the sum of the masses of the particles is conserved; hence, the average radius increases and the density decreases, with $n_d a^3 \sim$ const. The simplest kinetic description is a Brownian free molecular motion model involving the mutual collisions of assumed-neutral particulates due to their thermal motions. The dynamics can be understood qualitatively by considering the particles as equal-size, with

$$n_d(t)a^3(t) = \text{const} \tag{18.4.7}$$

The time evolution of n_d is due to the mutual collisions of the particles

$$\frac{dn_d}{dt} = -K_{dd}n_d^2 \tag{18.4.8}$$

where $K_{dd} = \sigma_{dd}\bar{v}_d$ is the rate constant. Using $\sigma_{dd} \approx \pi(2a)^2$ and $\bar{v}_d = (16kT_d/\pi m_d)^{1/2}$ (the reduced mass is $m_d/2$) with $m_d \propto a^3$, we obtain $K_{dd} \propto a^{1/2}$. Substituting n_d for a using (18.4.7), we obtain $K_{dd} \propto n_d^{-1/6}$. Using this in (18.4.8), we find

$$\frac{dn_d}{dt} = -Cn_d^{11/6} \tag{18.4.9}$$

where C is a constant. The solution to (18.4.9) is

$$n_d(t) = \left(n_{d0}^{-5/6} + \frac{5}{6}Ct\right)^{-6/5} \tag{18.4.10}$$

which is the Brownian free molecular motion result for neutral coagulation dynamics. Substituting this into (18.4.7), we obtain the time variation of the particle radii,

$$a(t) = a_0\left(1 + \frac{5}{6}n_{d0}^{5/6}Ct\right)^{2/5} \tag{18.4.11}$$

These time variations have been fit to the data in Figure 18.3 as the solid lines.

Various mechanisms are invoked to account for a reduced or zero negative charge on the coagulating clusters. Let us first note from (3.3.2) that the classical distance of closest approach of two singly negatively charged clusters is $b_0 = e/4\pi\epsilon_0\mathcal{E}_{clus}$, where $\mathcal{E}_{clus}$ is the relative energy in the center-of-mass system. For two clusters of radii a to make contact would require $2a \geq b_0$. At room temperature with $\mathcal{E}_{clus} \sim \frac{3}{2}T_g = 0.04$ V, this would imply $a \geq 20$ nm for coagulation to occur. However, 20 nm particles are already multiply charged, and experiments show that coagulation occurs

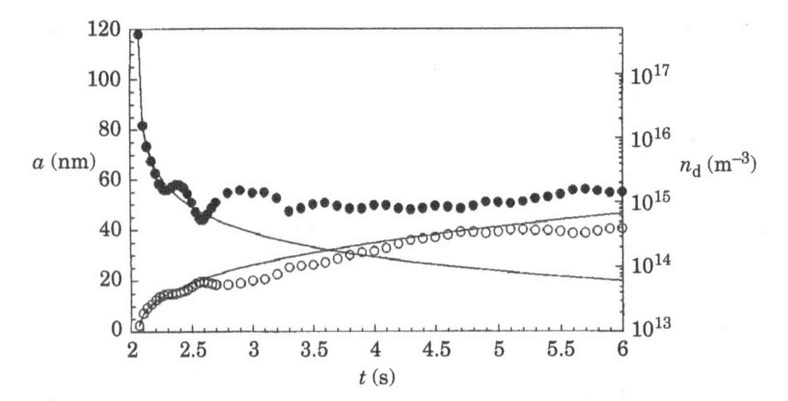

Figure 18.3 Time development of the particle radius a (open circles) and the number density n_d (solid circles) for early discharge times, obtained from Rayleigh scattering; solid lines show the best fit of the Brownian free molecular motion coagulation model. Source: Courteille et al. (1996)/with permission of AIP Publishing.

at smaller radii of a few nanometers. Hence, during coagulation either the clusters are hot (either directed or thermal energy), with $\mathcal{E}_{\text{clus}} \geq 0.2$ V, or they are mainly neutral. We have described in Section 18.2 various mechanisms for a reduced negative charge on small particulates. For example, charge fluctuations as in (18.2.15) can reduce the average Coulomb force. Kortshagen and Bhandarkar (1999) have studied a number of these mechanisms and concluded that the onset of coagulation is most likely due to a build-up of the dust particle density n_d to exceed the positive ion density n_i. As shown from (18.2.16), this would lead to an average reduced charge magnitude on the particles, facilitating coagulation (Problem 18.2b).

Coagulation stops when the particle size becomes sufficiently large that the equilibrium negative charge Q_d on the grains becomes greater than unity. This is consistent with the experimental data which also show that the particle density remains roughly constant thereafter, with a slow growth in the grain size. This subsequent growth of the charged particles is through standard deposition processes, at reasonably constant dust grain number. A simple calculation of this final growth stage is that the volume grows proportional to the rate of incoming neutral fragments, which is proportional to the area, giving

$$\frac{\mathrm{d}}{\mathrm{d}t}\left(n_{\text{sol}} \cdot \frac{4}{3}\pi a^3\right) = \Gamma_g^* s_g^* \cdot 4\pi a^2 \tag{18.4.12}$$

where n_{sol} is the solid density of the particle, $\Gamma_g^* = \frac{1}{4} n_g^* \overline{v}_g^*$ is the neutral fragment flux, and s_g^* is a neutral fragment sticking coefficient. The density n_g^* of the precursor fragments is determined by their production in the volume and their loss at the surfaces

$$K_{\text{diss}} n_e n_g \mathcal{V} = \Gamma_g^* (A_w + 4\pi a^2 n_d \mathcal{V}) \tag{18.4.13}$$

where $\mathcal{V}$ is the discharge volume and A_w is the surface area of the discharge. Solving (18.4.13) for Γ_g^* and substituting into (18.4.12), we obtain

$$\frac{da}{dt} = \frac{K_{\text{diss}} n_e n_g s_g^* \mathcal{V}}{n_{\text{sol}}} \frac{1}{A_w + 4\pi a^2 n_d \mathcal{V}} \tag{18.4.14}$$

Assuming a fixed n_e, for simplicity, (18.4.14) can be integrated to obtain

$$A_w \left(a - a_0\right) + n_d \mathcal{V} \frac{4\pi}{3}\left(a^3 - a_0^3\right) = \frac{K_{\text{diss}} n_e n_g s_g^* \mathcal{V}}{n_{\text{sol}}} t \tag{18.4.15}$$

For $A_w \gg 4\pi a^2 n_d \mathcal{V}$, (18.4.15) yields a linear growth of particle radius with time

$$a(t) = Ct + a_0 \tag{18.4.16}$$

where $C = K_{diss} n_e n_g s_g^* \mathcal{V} / n_{sol} A_w$. The growth rates are typically of order 1 nm/s. In the opposite limit of $A_w \ll 4\pi a^2 n_d \mathcal{V}$, we see a slowing of the growth due to a continual increase in the loss area, which depletes the neutral fragment density,

$$a(t) = \left[\frac{3K_{diss} n_e n_g s_g^*}{4\pi n_d n_{sol}} t + a_0^3 \right]^{1/3} \tag{18.4.17}$$

The predicted cube root dependence of the radius on time is in agreement with some measured results. Figure 18.4 shows the particle diameter versus the plasma on-time for a 75 mTorr silane discharge (Böhme et al., 1994), along with the cube root fit to the data.

For a high particle density, n_e and K_{diss} may not be fixed but may vary with a, as in (18.2.23). This leads to different time variations for $a(t)$ than given in (18.4.16) or (18.4.17). The growth eventually saturates. The saturation is not well understood but may, in part, be due to a change in the principal accretion species, and, in part, due to a change in the balance of forces on the growing particles that can produce losses of particles to the discharge walls. Particulates can also form in plasmas with more complicated chemistries than silane (see Hollenstein, 2000, for a discussion and references).

The particulates usually collect in a relatively narrow region at the interface between the plasma and the sheath as described in the previous section. For example, in a He/SiH_4 (5%) parallel plate capacitive discharge with 43 mm gap length, 600 mTorr pressure, and 30 sccm flow rate, Shiratani et al. (1994) obtained the particle size variation with space and time shown in Figure 18.5. The position is measured from the lower grounded electrode. Particularly, at the powered electrode plasma–sheath interface between 38 and 42 mm, we see an initially linear increase in the particle size, with an essentially saturated size after two seconds. The particles were also found to be relatively mono-dispersed (equal-sized), with a relatively constant, in time, particulate density of 10^7–10^8 cm^{-3}. The initial linear growth rate was calculated using the accretion of neutral radicals from the gas phase, as in (18.4.16).

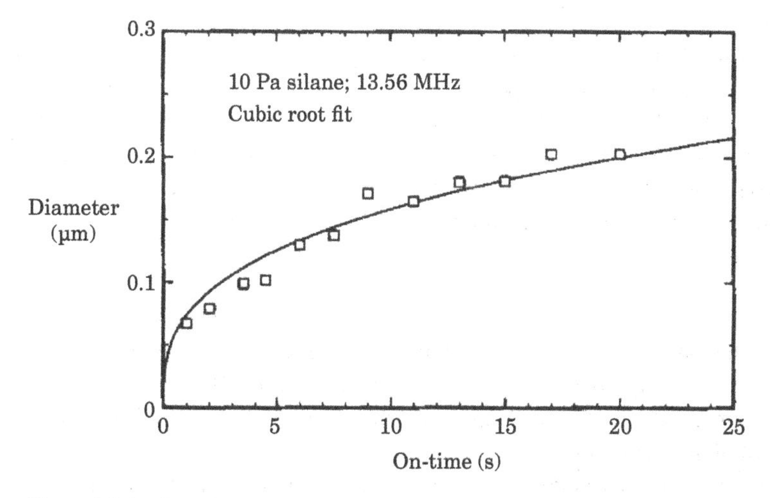

Figure 18.4 Particle diameter versus discharge on-time. Source: Adapted from Böhme et al. (1994).

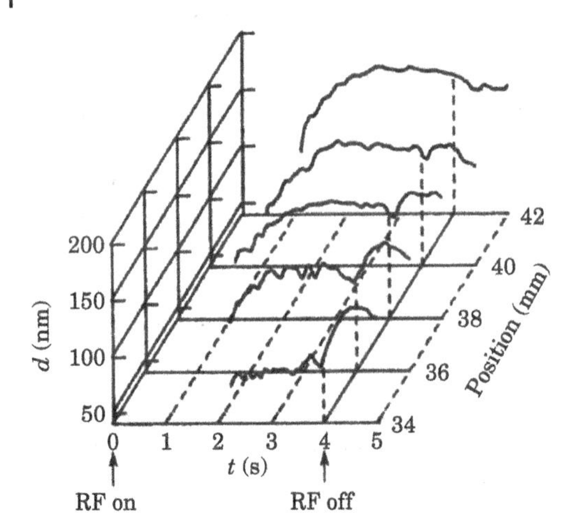

Figure 18.5 Time evolution of particulate size $d = 2a$ at 34–42 mm above the grounded electrode of a capacitive discharge after power turn-on; 43 mm electrode spacing, 40 W rf power, 4 s on-time in He/SiH_4 (5%) at 30 sccm and 600 mTorr. Source: Shiratani et al. (1994)/with permission of IEEE.

18.5 Physical Phenomena and Diagnostics

There are various physical phenomena connected with particulates that are interesting in their own right and may also be useful as diagnostics. We only consider these briefly and give a few key references for the reader who wishes to pursue a topic. For more details, the reviews by Merlino (2006, 2021), Ignatov (2005), Shukla and Eliasson (2009), and Morfill and Ivlev (2009) should be consulted.

18.5.1 Strongly Coupled Plasmas and Dust Crystals

Perhaps of most interest is the ability of the particulates to form crystalline structures at parameters easily accessible in laboratory plasmas. The fundamental coupling parameter is the ratio of the potential energy to the kinetic energy of the particulates. For a regular cubic lattice structure of particle separation Δ, the coupling parameter may be defined as

$$\mathcal{G} \equiv \frac{2Q_\mathrm{d}^2}{4\pi\epsilon_0 e\mathrm{T}_\mathrm{d}\Delta} \tag{18.5.1}$$

with T_d the particle temperature. Here $\mathcal{G}$ is for a one component plasma in which the neutralizing background is uniform and stationary. Thermodynamic arguments together with numerical computations have led to the determination of the transition to a crystalline structure at $\mathcal{G} = 171$ (see Ichimaru, 1982; Morfill and Thomas, 1996; Morfill and Ivlev, 2009). For a plasma with mobile background ions and electrons, Morfill and Thomas argue that the potential energy, and therefore the coupling parameter, should be reduced by a shielding factor, such that they take

$$\mathcal{G} \equiv \frac{2Q_\mathrm{d}^2}{4\pi\epsilon_0 e\mathrm{T}_\mathrm{d}\Delta}\, \mathrm{e}^{-\Delta/2\lambda_\mathrm{D}} \tag{18.5.2}$$

The large values of the coupling parameter required to obtain crystalline structures would only be realized in ordinary plasmas at very high densities or very low temperatures. However, the high value of $Z_\mathrm{d} = -Q_\mathrm{d}/e$ for the particulate grains allows dusty plasmas to enter this strongly coupled regime at lower densities and higher temperatures.

The study of the various phenomena associated with crystallization, such as transition parameters, grain boundaries, dislocations, annealing, and various wave phenomena, is clearly interesting in itself. The easy visualization of the crystal structure by laser scattering techniques has facilitated these studies and also led to ways of measuring basic particulate behavior. Observations of crystal structures have often been performed with artificial powders, where the size and density can be carefully controlled. See Zuzic et al. (1996) for an example of the observation of collective behavior and for other references. The study of crystal structures has also led to an appreciation of ion flow around the suspended dust grains, which tends to align the crystals into two-dimensional arrays, and can cause other interesting effects, such as plasma wakes behind the grains and dipole moments on the grains (e.g., see Melandso, 1997; Shukla and Eliasson, 2009).

18.5.2 Dust Acoustic Waves

An interesting phenomenon that occurs in dusty plasmas is that of dust acoustic waves, which can be used as a diagnostic. This has been studied at various levels of complexity, including dispersive effects and collisionless and collisional damping. Here, we present a simplified picture and related experiments following D'Angelo (1995) and Barkin et al. (1995). We assume a dusty plasma in which most of the negative charge resides on the dust. The relevant one-dimensional equations for the electrostatic waves, which are a generalization of the electron waves considered in Section 4.2, are

$$\frac{\partial n_d}{\partial t} + \frac{\partial}{\partial x}(n_d u_d) = 0 \tag{18.5.3}$$

$$n_d m_d \frac{\partial u_d}{\partial t} + n_d m_d u_d \frac{\partial u_d}{\partial x} + e\mathrm{T}_d \frac{\partial n_d}{\partial x} + Q_d n_d \frac{\partial \Phi}{\partial x} = 0 \tag{18.5.4}$$

$$e\mathrm{T}_i \frac{\partial n_i}{\partial x} + e n_i \frac{\partial \Phi}{\partial x} = 0 \tag{18.5.5}$$

$$n_i = Z_d n_d \tag{18.5.6}$$

where n_d and n_i are the densities of the dust and positive ion species, and T_d and T_i are their temperatures. Q_d is the negative charge on the dust grains of mass m_d, taken to be equal size, and u_d is their fluid velocity. Equations (18.5.3) and (18.5.4) are the continuity and momentum equations for the dust, (18.5.5) expresses the condition of Boltzmann equilibrium for ions, a good approximation compared to the slow dust motion, and (18.5.6) is the usual condition of charge neutrality. As in Section 4.2, we expand to first order around a zero-order steady-state solution, $n_d = n_{d0} + \tilde{n}_d$, $u_d = \tilde{u}_d$, and $\Phi = \tilde{\Phi}$, and eliminate the first-order quantities $\tilde{n}_d$, $\tilde{u}_d$, $\tilde{\Phi}$, assumed to vary as $e^{j(\omega t - kx)}$, to obtain the dispersion equation

$$\frac{\omega^2}{k^2} = \frac{e}{m_d}\left(\mathrm{T}_d + Z_d \mathrm{T}_i\right) \tag{18.5.7}$$

Measuring the wave phase velocity ω/k for known T_i, then Z_d/m_d can be determined. For a given density material, this relates particulate size to charge, which can be a useful diagnostic. Barkin et al. (1995) observed a dust acoustic wave for dust grains with $2a = 5$ μm, $m_d \approx 10^{-12}$ kg, and $Z_d \approx 4 \times 10^4$. They measured a wave velocity of $\omega/k \approx 9$ cm/s, which was close to the calculated value of approximately 8 cm/s.

18.5.3 Driven Particulate Motion

Another interesting experiment is to drive a sparse collection of dust grains ($Z_d n_d \ll n_i$) with a slow periodic electric field, which was experimentally done by Zuzic et al. (1996). In this situation, the particle oscillation motion is described by

$$m_d\ddot{x} - F_0(x) + m_d\dot{x}/\tau_{dg} = F_1 \cos \omega t \tag{18.5.8}$$

where $F_0(x)$ is the sum of the position-dependent forces acting on the particle, including electric field forces, ion and neutral drag forces, etc., $F_1 \cos \omega t$ is the applied drive, and $m_d\dot{x}/\tau_{dg}$ is the frictional force of the background neutral gas. Expanding $F_0(x)$ about the equilibrium $F_0(x_0) = 0$ for small friction, (18.5.8) has the resonant frequency

$$\omega_0^2 = -\frac{1}{m_d}\left(\frac{dF_0}{dx}\right)_{x_0} \tag{18.5.9}$$

Varying ω in (18.5.8) exhibits this resonance, which for known m_d determines the confining force gradient at equilibrium. Because the electric field varies rapidly compared to all the other forces at the particle equilibrium position just inside the sheath, most of the force gradient is due to the electric field gradient, which can be calculated from (18.5.9) by measuring ω_0. Furthermore, matching the theoretical frequency width of the resonance to the experiment determines the frictional drag time of the gas. This can be related to a standard theory of the friction on a sphere moving through a background gas, for which

$$\tau_{dg} = 2\sqrt{\pi}\rho_d a/3\rho_g c_s \tag{18.5.10}$$

where ρ_g is the neutral gas mass density and c_s is the gas sound speed. For known particulate grains, $\rho_d = 1.5$ g/cm^3 and $2a = 6.9$ μm, good agreement between the measured and calculated τ_{dg} was obtained. The measured frequency of $\omega_0 = 13.6$ s^{-1} (2.17 Hz) was reasonable, but the force gradient was not independently measured.

18.5.4 Laser Light Scattering

Laser light scattering (LLS) is an important *in situ* method to visualize particulate phenomena and to determine particle dynamics, sizes, and other parameters. Using linearly polarized LLS, the number density of particulates can be obtained from the calibrated intensity of the scattered light, or by attenuation of the transmitted light. The size of the particulates can be obtained from polarized LLS by comparing the scattering at 90° to the forward scattering, in the plane perpendicular to the direction of polarization of the electric field. Laser light scattering has been employed from the earliest measurements, e.g., Spears et al. (1986), Selwyn et al. (1989, 1991), Selwyn (1993, 1994), and has been a mainstay of observations, e.g., Boufendi et al. (1992) and Shiratani et al. (1996). If the particulates are predominantly equal-sized, as is often the case, then the number and size can be quite simply determined. We briefly describe the methods here.

There are two scattering regimes, depending on the parameter $2\pi a/\lambda$, with λ the optical wavelength. Typical wavelengths that have been used are 633 nm for HeNe lasers, 488 and 514.5 nm for Ar ion lasers, and a wide variety of wavelengths from UV lines at 351 and 356 nm through red lines at 647 and 676 nm for Kr ion lasers. For $2\pi a/\lambda \ll 1$, the scattering is in the Rayleigh regime in which the scatterers act as dipoles oriented along the direction of the electric field polarization, producing isotropic radiation in the plane perpendicular to the polarization direction. For $2\pi a/\lambda > 1$, the scattering is in the more complicated Mie regime (Mie, 1908), for which the radiation in the plane perpendicular to the E-field polarization is increasingly forward-scattered. The transition from Rayleigh to Mie scattering with increasing radius a is rather abrupt, due to an a^6/λ^4 dependence in the scattering formulae, leading to an easily identifiable particulate radius. An example shown in Figure 18.6 for an argon laser line of $\lambda = 488$ nm gives a breakaway from the isotropic

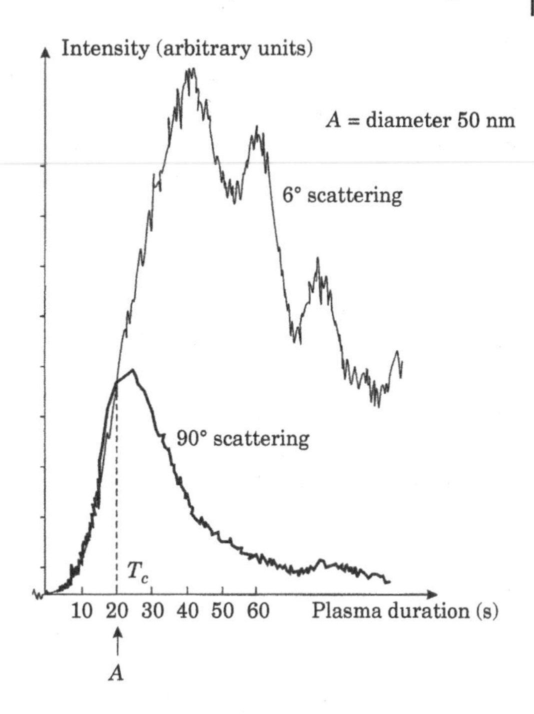

Figure 18.6 Scattered intensities versus plasma on-time at 6° and 90° from the incident direction, in the plane perpendicular to the direction of polarization of the electric field; the corresponding measured particle size is indicated at the Rayleigh–Mie transition. Source: Boufendi et al. (1999)/with permission of John Wiley & Sons.

(Rayleigh) scattering at $a = 50$ nm, i.e., $2\pi a/\lambda = 0.64$, which can be used as a standard factor. The number density can be obtained from an attenuation formula,

$$I(x) = I_0\, \mathrm{e}^{-C_{\mathrm{ext}} n'_{\mathrm{d}}} \tag{18.5.11}$$

where C_{ext}, known as the extinction parameter, is composed of a sum of absorption and scattering contributions,

$$C_{\mathrm{ext}} = C_{\mathrm{abs}} + C_{\mathrm{scat}} \tag{18.5.12}$$

and the line density $n'_{\mathrm{d}} = \int_0^{l_{\mathrm{b}}} n_{\mathrm{d}}(x)\,\mathrm{d}x$ is the particulate density integrated along the line of sight (beam length) of the laser beam. For equal-sized particles in the Rayleigh regime, C_{abs} and C_{scat} are given by (see Boufendi et al., 1999)

$$C_{\mathrm{abs}} = \frac{8\pi^2 a^3}{\lambda} \mathrm{Im}\left(\frac{N^2 - 1}{N^2 + 2}\right) \tag{18.5.13}$$

$$C_{\mathrm{scat}} = \frac{128\pi^5 a^6}{3\lambda^4} \left|\frac{N^2 - 1}{N^2 + 2}\right|^2 \tag{18.5.14}$$

where N is the complex index of refraction of the dust material. For $2\pi a/\lambda \ll 1$, the absorption term can dominate the extinction, but near the Rayleigh–Mie transition the scattering term generally dominates. If the dust grains are not equal-sized or if the index of refraction is not known, more measurements are needed and the computations become more complicated.

There can be a serious visualization issue for LLS of small particles. As seen in (18.5.13) and (18.5.14), the absorption and scattering parameters decrease strongly with the particle radius. For $a \gtrsim 50$ nm, visible light LLS can be quite useful. However, the absorbed and scattered light intensities are at least 1000 times smaller for $a \lesssim 5$ nm. For a review of LLS techniques and references to the original literature, the reader is referred to Boufendi et al. (1999) and Selwyn (1993, 1994).

18.5.4.1 LLS Visualizations

A classic example of LLS particulate visualization in a capacitive processing discharge is shown in Figure 18.7 (Selwyn et al., 1991). Three silicon wafers were placed on a single, 56 cm diameter rf-powered graphite electrode, with grounded reactor walls. A 200 mTorr argon plasma was driven at 13.56 MHz by 600 W power. As shown in Figure 18.7*a*, the laser beam was scanned by a rotating mirror through a wide angle near the plane of the powered-electrode plasma–sheath edge. The forward-scattered light was detected by a charge-coupled device (CCD) video camera. As seen in Figure 18.7*b*, a "ring" of particles is observed around the edge of each wafer, suspended near the plasma–sheath edge. A second "dome" of particles is suspended over the wafer centers. These structures may be formed around minima of the transversely-varying sheath widths across the powered electrode, which trap particles. Placing wafers in wells flush with the surface altered but did not eliminate these trapping phenomena, indicating the importance of surface material properties. The traps weakened and the feed gas flow carried particles out of the reactor when the power was slowly reduced from 600 to 50 W.

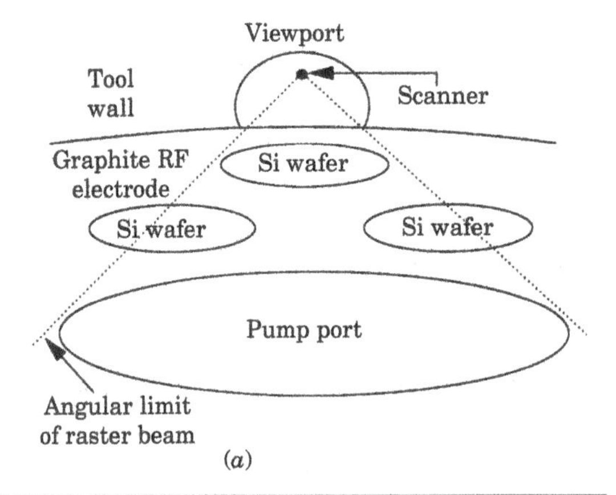

(*a*)

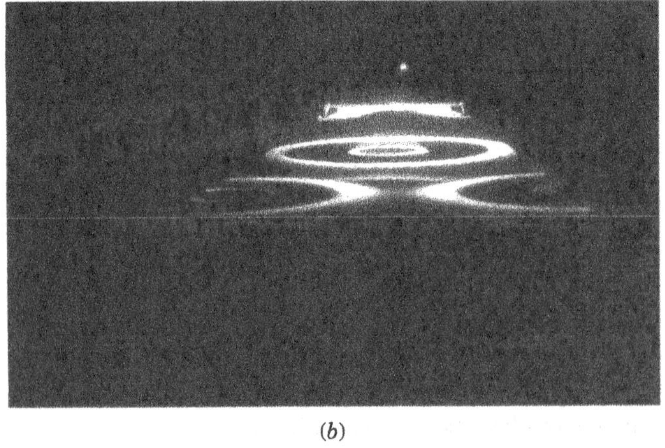

(*b*)

Figure 18.7 Visualization of particulates, showing (*a*) schematic of the electrode, wafers, and discharge features observed during forward direction LLS; the two foreground wafers are 5.7 cm diameter, the background wafer is 8.3 cm diameter; the dashed lines show the limits of the scanned laser beam; (*b*) photograph of scattered light from particulates in a 200 mTorr argon plasma; trapped particle structures including a dome and rings are seen over each of the wafers. Source: Selwyn et al. (1991)/AIP Publishing.

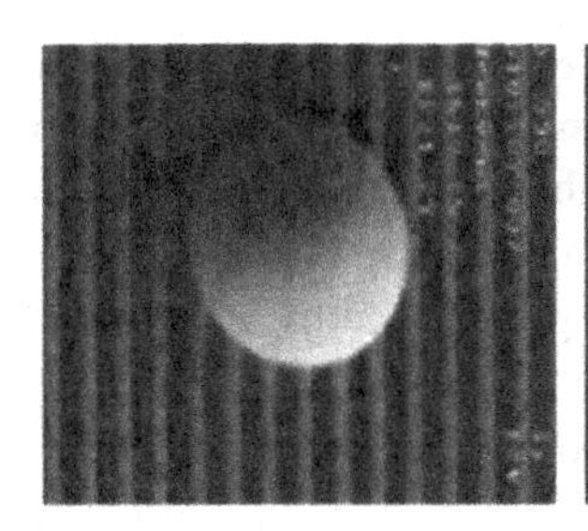
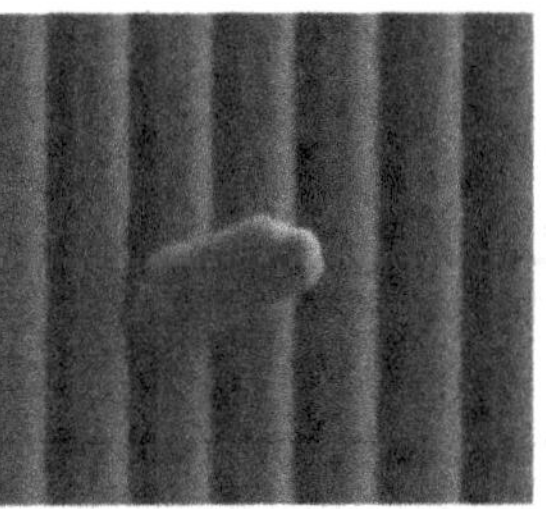

Figure 18.8 Examples of particulates on plasma-etched features. Source: Lee et al. (2014)/Reproduced with permission of IOP Publishing.

Particles falling onto 3D-etched features can lead to device failures, particularly when the particle size is of order the feature size. Some examples are shown in Figure 18.8. In plasma processing discharges, there are many possible sources of unwanted particles. These include formation by nucleation in the plasma volume as described in Section 18.4 for a silane plasma, aggregates of sputtered material from the wafer and the wafer holder, and flakes produced from residual films on the reactor wall surfaces. In addition to silane, other common processing feedgases that can nucleate particles include organo-silicon compounds such as $Si(OC_2H_5)_4$ (TEOS) and $O[Si(CH_3)_3]_2$ (HMDSO), fluorocarbons and hydrocarbons. Physical examination of the particulates seen in Figure 18.7*b* showed them to be silicon-carbon aggregates, presumably originating from sputtered wafer and electrode material. In cleaner discharges, particulates can also be produced by flaking of wall deposits, e.g., SiO_2 flakes from residual quartz films. Kasashima et al. (2015) observed numerous flakes generated by micro-arc discharges in mass production plasma processing equipment. They attributed these flakes to sudden changes induced by the micro-arc discharges in the plasma floating potential near deposited wall films, which produced impulsive forces on the film.

In addition to LLS, particulate size can be determined from electron microscopy of deposited particulates after the plasma has been turned off. The shapes of the particulates can be examined, particularly to see if they conform to the usual assumption of spherical grains, assumed in size calculations from *in-situ* LLS measurements. For small singly negatively-charged particles, the mass can also be determined from mass spectrometry of exiting particles, using a power-modulated discharge at a frequency such that the sheath collapses and negatively charged particles can escape to the analyzer orifice. By charging exiting neutrals, mass spectrometry can also be used to determine their mass. It is from measurements such as these that Hollenstein et al. (1998) have shown that below a particulate size of about 0.5 nm, the number density decreases exponentially with mass, as described in Section 18.4.

18.6 Removal or Production of Particulates

Particulates in processing plasmas are usually (but not always) unwanted. The early investigation of particulates (e.g., Selwyn et al., 1989, 1991) also involved studies of how to minimize their development or remove them before they can settle on and damage substrates that are being processed. Boufendi and Bouchoule (2002) give a review of this issue. There are two distinct situations to be considered: during the plasma on-time, and just after the plasma has been extinguished. During the on-time, as discussed in Section 18.3, electrostatic barriers at the plasma sheaths act to confine negatively charged particulates. However, the electrostatic forces might not be large enough to confine large-size particles against the forces of gravity or ion drag. Following Bouchoule (1999, p. 307), let us consider the balance of electrostatic and gravity forces in the sheath of a capacitive discharge.

We can estimate the sheath electric field as $E \sim V/s$, where V is the voltage across the sheath and s is the sheath thickness. Using the Child law (11.2.15) to substitute for s, we find $E \propto n_{\rm i}^{1/2}\mathrm{T}_{\rm e}^{1/4}V_{\rm rf}^{1/4}$. From (18.2.7) and (18.2.9), the mean charge on a particle of radius a is $|Q_{\rm d}| \propto a\Phi_{\rm d}$, with $\Phi_{\rm d} \approx 2\,\mathrm{T_e}$. From (18.3.12), the electric field force scales as $F_{\rm E} = |Q_{\rm d}E| \propto n_{\rm i}^{1/2}\mathrm{T}_{\rm e}^{5/4}V^{1/4}a$. In practical units, putting in the constants of proportionality, $F_{\rm E} \approx 3.6 \times 10^{-20} n_{\rm i}^{1/2}\mathrm{T}_{\rm e}^{5/4}V^{1/4}a$ N, with a in μm, $n_{\rm i}$ in m^{-3}, and $\mathrm{T_e}$ in V. The gravity force (18.3.1) is $F_{\rm g} \approx 1.6 \times 10^{-13}a^3$ N. Equating these two forces determines a maximum confined particle radius $a_{\max} \approx 4.8 \times 10^{-4} n_{\rm i}^{1/4}\mathrm{T}_{\rm e}^{5/8}V^{1/8}$. Using typical capacitive discharge parameters $n_{\rm i} \approx 10^{16}\ \mathrm{m}^{-3}$, $\mathrm{T_e} = 2$ V, and $V = 100$ V, Bouchoule gives the estimate $a_{\max} \approx 11$ μm. This value is well beyond the typical particle sizes in plasma reactors. Hence, we conclude that the gravity force is unable to push the particles through the electrostatic sheath barrier into the substrate.

When the plasma is extinguished, the charge on the particles relaxes toward a state of zero charge on the timescale τ given by (18.2.13). Generally, this time is short compared to the characteristic time required for the particle to transit the sheath. Hence, the particle trajectories can be found by assuming an essentially zero electric field force. However, there is some evidence that a small charge (positive or negative) can be left on the particles after the discharge is extinguished (Collins et al., 1996). This has been ascribed to the increase in the charge equilibration time τ in the late afterglow due to the decrease in plasma density. If τ exceeds the plasma decay time, then complete equilibration does not occur. Also, a conventional L-type capacitive matching network, as shown in Figure 11.49, can leave a considerable dc electric field remaining between the electrodes after the plasma is extinguished. This field can drive incompletely neutralized particles toward or away from the substrate holder, depending on the sign of their charge.

During the plasma on-time, particles can also be detrapped by the ion drag force acting at the sheath edge. In practical units, from (18.3.11), we estimate $F_{\rm i} \approx 3 \times 10^{-13}a^2(1 - 0.03\ln a)$ N, with a in μm. Balancing this force against the electric field force with $n_{\rm i} \approx 10^{16}\ \mathrm{m}^{-3}$, $\mathrm{T_e} = 2$ V, and $V = 100$ V, Bouchoule gives the estimate $a_{\max} \approx 20$ μm, which is also much larger than typical particle sizes in capacitive discharges. However, for high-density plasmas with $n_{\rm i} \sim 10^{17}$–$10^{18}\ \mathrm{m}^{-3}$, $a_{\max}$ can be much smaller and well within the range of particle sizes formed.

One method that has been explored to minimize particle contamination uses pulsed power modulation (see Section 10.6) at a frequency high enough that large particles cannot form during the on-time, but with the off-time long enough for the sheaths to collapse and negative ion clusters to escape to the walls. It was found experimentally (Bouchoule et al., 1991) that particulate formation could be significantly reduced or completely suppressed with a one second repetition time with an off-time of 8–10 ms. As discussed in Section 18.4 in connection with particle growth, a one second on-time is too short for large grains to form. The 8 ms off-time is sufficiently long for the sheath potential to collapse, such that the negatively charged precursor clusters can escape to the walls.

Other techniques include the use of high gas flows to sweep the particles out of the system, and the cutting of grooves in the substrate holder to guide the particles out of the system. Another technique that is being explored is to use laser beams to either break up particulates or supply a force that pushes them out of the active region. Effort has gone into destroying the particulates by use of high powered lasers (see Stoffels et al., 1994). The pulsed heating breaks up the particulates into sufficiently small pieces that they cannot hold charge, as described previously. The neutral small particulates can then be removed from the active region by weaker forces, e.g., gas flow. Another technique for particle removal, involving low power lasers, has been recently explored (see Annaratone, 1997), in which a laser produces a force which pushes the particles out of the active region. This force was estimated to be due to radiation pressure, but other explanations such as differential heating are also possible.

A growing area of interest is powder synthesis and surface modification processes using dusty plasmas. Powders of a given size can be produced as raw materials for industrial applications. Since the formation starts from a relatively uniform state, as described in Section 18.4, by removing the powder at a given time in its formation, a rather uniform grain size can be created, and the timing for precipitating the powder can then be chosen to fix the grain diameters. Nanostructured materials and coatings for wear, friction, and corrosion resistance can also be produced. Another important potential use that has been studied is the incorporation of powders into growing thin films, in which the small particles form a matrix for modifying the film properties. One example is the production of polymorphous silicon films, which contain nanometer-sized crystalline silicon islands within an amorphous a-Si:H base film. Experimental developments and modeling predictions for the use of this material to increase the efficiencies of pin junction silicon solar cells are described in Abolmasov et al. (2016).

The above discussion does not exhaust the possibilities for either removal, control, or production of particulates in plasma discharges. There is a fast- growing literature describing procedures and applications. For a more complete review of the possibilities and many references, the reader is referred to the various books and review articles mentioned in the introduction to this chapter.

Problems

18.1 Charge Fluctuations on Dust Grains Let f_d be the distribution of charge number $Z = -Q/e$ on a collection of equal-size particles. In the steady state, the rate R_z at which particles having charge numbers $Z + 1$ and $Z - 1$ are converted to particles having charge number Z must be equal to the rate at which particles having charge number Z are converted to particles having charge numbers $Z + 1$ and $Z - 1$

$$R_z = f_d(Z+1)\,I_i(Z+1) - f_d(Z-1)\,I_e(Z-1)$$
$$= f_d(Z)\,[I_i(Z) - I_e(Z)]$$

Here, I_e (which is negative) and I_i are given by substituting Φ from (18.2.5) into (18.2.1) and (18.2.2), respectively.

(a) Passing from the discrete to the continuous limit for $Z \gg 1$ by expanding $f_d(Z + 1)$ $I_i(Z + 1)$ and $f_d(Z - 1)I_e(Z - 1)$ to second order in a Taylor series around $Z = 0$, show that

$$\frac{d}{dZ}\left\{(I_i + I_e)f_d + \frac{1}{2}\frac{d}{dZ}\left[(I_i - I_e)f_d\right]\right\} = 0$$

(b) The equilibrium charge number Z_d is determined by the condition $I_i(Z_d) + I_e(Z_d) = 0$. Assume that the currents are slowly varying functions of Z near Z_d:$I^{-1}(dI/dZ) \ll f_d^{-1}(df_d/dZ)$, where $I = I_i,\ |I_e|$. Transforming from the variable Z to $Z_1 = Z - Z_d$, show that the result in (a) reduces to

$$\frac{d}{dZ_1}\left[(I_i' + I_e')Z_1 f_d + \frac{1}{2}(I_i - I_e)\frac{df_d}{dZ_1}\right] = 0$$

where the currents and their derivatives with respect to Z (denoted $'$) are evaluated at $Z = Z_d$.

(c) Show that the solution to the differential equation in (b) is

$$f_d = C\exp\left(-\frac{Z_1^2}{2\sigma_z^2}\right)$$

where

$$\sigma_z^2 = \frac{1}{2}\left(\frac{I_i - I_e}{I_i' + I_e'}\right)$$

(d) Using $\Phi_d \approx 2\mathrm{T_e}$, evaluate σ_z^2 and show that the standard deviation is $\sigma_z = Z_d^{1/2}/\sqrt{3}$.

18.2 Charging of Dust Grains

(a) Using (18.2.16) and (18.2.17) with $n_d = 10^8$ cm^{-3}, $n_i = 5 \times 10^9$ cm^{-3}, $\mathrm{T_e} = 2$ V, $\mathrm{T_i} = 0.1$ V, and $a = 115$ nm, show that $Z_d \approx 50$ and $\Phi_d \approx 0.63$ V.

(b) At the onset of coagulation for equal-size particles, an observer measures that $n_d = 10^{10}$ cm^{-3}, $n_i = 10^9$ cm^{-3}, $\mathrm{T_e} = 2.5$ V, $\mathrm{T_i} = 0.025$ V, and $a = 1$ nm. Find approximate values of Z_d, Φ_d, and n_e.

18.3 Current-Driven Dusty Capacitive Discharge Obtain the scalings for n_e and K_{iz} with n_d and a, analogous to (18.2.23) and (18.2.24), for an ohmically heated capacitive discharge driven by a constant rf current $\tilde{J}_{rf}$.

18.4 Transient Nucleation Model Consider the reaction chain (18.4.3) with $K_j = K_0$ for $j \lesssim 32$, as in (18.4.5). Assume a discharge with generation and loss of the SiH_3^- density n_1 according to

$$\frac{dn_1}{dt} = K_{att} n_e n_g - K_0 n_g n_1 - K_{rec} n_i n_1$$

(a) Introducing the Laplace transform,

$$\tilde{n}_{j+1}(s) = \int_0^\infty e^{-st}\, n_{j+1}(t)\, dt$$

and assuming that $n_j(t) = 0$ for $j \geq 1$, show that

$$\tilde{n}_1 = \frac{K_{att} n_e n_g}{s} \frac{1}{s + K_0 n_g + K_{rec} n_i}$$

$$\tilde{n}_{j+1} = \tilde{n}_j \frac{K_0 n_g}{s + K_0 n_g + K_{rec} n_i}$$

(b) From the results of (a), show that

$$\tilde{n}_{j+1} = \frac{K_{att} n_e n_g}{s} \frac{(K_0 n_g)^j}{(s + K_0 n_g + K_{rec} n_i)^{j+1}}$$

(c) Using the definite integral

$$\int_0^\infty e^{-st}\left(\frac{1}{k!} t^k e^{s_0 t}\right) dt = \frac{1}{(s - s_0)^{k+1}}$$

and the results of (b), show that the time-varying solutions for the negative ion clusters are

$$\frac{dn_{j+1}}{dt} = \frac{K_{att} n_e n_g}{j!} (K_0 n_g t)^j\, e^{-(K_0 n_g + K_{rec} n_i)t}$$

such that

$$n_{j+1}(t) = \int_0^t \frac{K_{att} n_e n_g}{j!} (K_0 n_g t')^j \, e^{-(K_0 n_g + K_{rec} n_i)t'} \, dt'$$

(d) Show that n_{j+1} has a maximum value at time $t = j/(K_0 n_g + K_{rec} n_i)$.

18.5 **Particle Growth in Capacitive Discharge** Assuming a constant dissociation rate coefficient K_{diss} and gas density n_g, and substituting (18.2.23) for n_e into (18.4.14), obtain the time variation of $a(t)$ in the limit of high particle densities $4\pi a^2 n_d \mathcal{V} \gg A_w$ in a voltage-driven capacitive discharge.

19

Kinetic Theory of Discharges

19.1 Basic Concepts

The Boltzmann equation (2.3.3)

$$\frac{\partial f_e}{\partial t} + \mathbf{v} \cdot \nabla f_e + \frac{\mathbf{F}}{m} \cdot \nabla_v f_e = \left. \frac{\partial f_e}{\partial t} \right|_c \tag{19.1.1}$$

determines the electron distribution function f_e. In the previous chapters, we have seen many examples of measured electron energy distribution functions (EEDFs) and self-consistent evaluations of f_e using particle-in-cell (PIC) kinetic simulations. And we have already discussed, at the end of Section 11.3, the use of Boltzmann term analysis to determine the electron heating, using PIC simulation, results for f_e. These results indicate that measured and PIC-simulated EEDFs in discharges are often strongly non-Maxwellian.

On the other hand, in the theory and modeling of plasma discharges, we have generally assumed the electrons to be in near-thermal equilibrium, with a Maxwellian distribution

$$f_e(v) = n_e \left(\frac{m}{2\pi e \mathrm{T}_e} \right)^{3/2} \exp\left(-\frac{mv^2}{2e\mathrm{T}_e} \right) \tag{19.1.2}$$

However, some rate constants and other discharge parameters depend sensitively on deviations from a Maxwellian distribution, e.g., for ionization

$$K_{iz} = 4\pi \int_{v_{iz}}^{\infty} \sigma_{iz}(v) f_e(v) v^3 \, dv$$

with only the high energy tail with $v \geq v_{iz} = (2e\mathcal{E}_{iz}/m)^{1/2}$ contributing significantly to K_{iz}. For example, a non-Maxwellian distribution with a reduced high-energy tail can yield a K_{iz} that is smaller by orders of magnitude than assuming a Maxwellian distribution. If a Maxwellian cannot be assumed, then f_e must be calculated by solving the Boltzmann equation. This solution is exceedingly difficult for discharges because the fields and interparticle collisions, which appear in the Boltzmann equation for each species, must be determined self-consistently, e.g., using Maxwell's equations for the fields. The resulting set of coupled nonlinear integro-differential equations in seven dimensions $(x, y, z, v_x, v_y, v_z, t)$ is intractable. Hence, various approximations are used, such as linearization around an assumed "zero-order" solution and spherical harmonic expansions in velocity space. Reviews of the various techniques and their applications to the analysis of various discharges are given in Kolobov and Godyak (1995), Kortshagen et al. (1996), and Aliev et al. (1997). A discussion of advanced procedures for solving the Boltzmann equation is given in Chapters 5 and 6 of Makabe and Petrović (2006).

Principles of Plasma Discharges and Materials Processing, Third Edition. Michael A. Lieberman and Allan J. Lichtenberg.

19.1.1 Two-Term Approximation

A common and very useful simplification is the two-term approximation, in which we expand the electron distribution function to first order in the deviation from isotropy, which we take to be cylindrically symmetric along the direction of anisotropy

$$f_e(\mathbf{r}, \mathbf{v}, t) \approx f_{e0}(\mathbf{r}, v, t) + \frac{\mathbf{v}}{v} \cdot \mathbf{f}_{e1}(\mathbf{r}, v, t) \tag{19.1.3}$$

Here, f_e is decomposed into the sum of an isotropic velocity part f_{e0}, depending on the speed $v = (v_x^2 + v_y^2 + v_z^2)^{1/2}$ only, and a small anisotropic velocity part, where the vector function $\mathbf{f}_{e1}$ defines the magnitude and direction of the anisotropic part of f_e, with $f_{e1} \ll f_{e0}$. For nonmagnetized plasmas, the direction is usually that of the field. Choosing the direction to be along z, we obtain

$$f_e(\mathbf{r}, v, \psi, t) \approx f_{e0}(\mathbf{r}, v, t) + \frac{v_z}{v} f_{e1}(\mathbf{r}, v, t) \tag{19.1.4}$$

where $v_z/v = \cos\psi$, with ψ the spherical polar angle in velocity space, and where $f_{e1} = |\mathbf{f}_{e1}|$ is not a function of ψ. The two-term approximation essentially corresponds to keeping the first two terms in a spherical harmonic expansion of the distribution function in velocity space.

The condition for this nearly isotropic f_e to hold is that the elastic scattering frequency $\nu_{el}(v)$ must be large compared to the characteristic frequencies for electron energy gain (e.g., from the field) and loss (e.g., due to inelastic collisions). This is often a good approximation for electrons in a weakly ionized plasma where electron–neutral collisions dominate and are mainly elastic over most of the energy range.

19.1.2 The Krook Collision Operator

Let us consider the collision term for a weakly anisotropic distribution with only elastic collisions between electrons and neutrals. The collision integral (B.4) in Appendix B can be written

$$\left.\frac{\partial f_e}{\partial t}\right|_c = \int d^3v_g \int_0^{2\pi} d\phi_1 \int_0^{\pi} (f_e' f_g' - f_e f_g)\, v I(v, \theta_1) \sin\theta_1 \, d\theta_1 \tag{19.1.5}$$

where we assume the neutrals are infinitely massive, so that $|\mathbf{v} - \mathbf{v}_g| = v = v'$, the electron speed, and $f_g' = f_g$. Because $v' = v$, we have $f_{e1}' = f_{e1}$ and $f_{e0}' = f_{e0}$. Substituting the expansion (19.1.4) into the factor in parentheses in (19.1.5), we find

$$(f_e' f_g' - f_e f_g) v = (f_{e0} f_g - f_{e0} f_g) v + f_{e1} f_g v_z' - f_{e1} f_g v_z = f_{e1} f_g (v_z' - v_z) \tag{19.1.6}$$

i.e., the first term on the right-hand side is zero. For infinitely massive neutrals, such that the electron energy is conserved in collisions, the scattering process yields (Holt and Haskell, 1965, Section 10.13)

$$v_z' = v_\perp \sin\theta_1 \cos\phi_1 + v_z \cos\theta_1$$

with $v_\perp = (v_x^2 + v_y^2)^{1/2}$ and θ_1 and ϕ_1 the scattering angles (see Figure 3.3). Substituting this and (19.1.6) into (19.1.5), we perform the ϕ_1 integration to obtain

$$\begin{aligned}\left.\frac{\partial f_e}{\partial t}\right|_c &= 2\pi \int f_g \, d^3v_g \int_0^{\pi} f_{e1} \cos\psi (\cos\theta_1 - 1)\, v I(v, \theta_1) \sin\theta_1 \, d\theta_1 \\ &\equiv -\nu_m f_{e1} \cos\psi \end{aligned} \tag{19.1.7}$$

with $\int f_g \, d^3v_g = n_g$ the neutral gas density and $\nu_m(v)$ the speed-dependent momentum transfer collision frequency

19.1.3 Two-Term Collisional Kinetic Equations

Writing the Boltzmann equation (19.1.1) for an unmagnetized plasma in one spatial dimension, we have

$$\frac{\partial f_{\mathrm{e}}}{\partial t} + v_z \frac{\partial f_{\mathrm{e}}}{\partial z} - \frac{e}{m} E_z \frac{\partial f_{\mathrm{e}}}{\partial v_z} = \left. \frac{\partial f_{\mathrm{e}}}{\partial t} \right|_{\mathrm{c}} \tag{19.1.8}$$

Using (19.1.4), we expand (19.1.8) in spherical harmonics to obtain in the lowest order

$$\begin{aligned} &\frac{\partial f_{\mathrm{e0}}}{\partial t} + \cos\psi \frac{\partial f_{\mathrm{e1}}}{\partial t} + v\cos\psi \frac{\partial f_{\mathrm{e0}}}{\partial z} + v\cos^2\psi \frac{\partial f_{\mathrm{e1}}}{\partial z} \\ &-\frac{e}{m} E_z \cos\psi \frac{\partial f_{\mathrm{e0}}}{\partial v} - \frac{e}{m} E_z \left[\frac{f_{\mathrm{e1}}}{v} + v\frac{\partial}{\partial v}\left(\frac{f_{\mathrm{e1}}}{v}\right)\cos^2\psi \right] = \left. \frac{\partial f_{\mathrm{e}}}{\partial t} \right|_{\mathrm{c}} \end{aligned} \tag{19.1.9}$$

where $\partial f_{\mathrm{e}}/\partial t|_{\mathrm{c}}$ is given by (19.1.7). Multiplying (19.1.9) by $\sin\psi$ and integrating over ψ from 0 to π, we obtain, after collecting terms

$$\frac{\partial f_{\mathrm{e0}}}{\partial t} + \frac{v}{3}\frac{\partial f_{\mathrm{e1}}}{\partial z} - \frac{e}{m} E_z \frac{1}{3v^2}\frac{\partial}{\partial v}(v^2 f_{\mathrm{e1}}) = 0 \tag{19.1.10}$$

Equation (19.1.10) gives the time rate of change of the isotropic part of the distribution, given the anisotropic part, and does not directly depend on the collisions. Multiplying (19.1.9) by $\sin\psi\cos\psi$ and integrating, as before, we obtain

$$\frac{\partial f_{\mathrm{e1}}}{\partial t} + v\frac{\partial f_{\mathrm{e0}}}{\partial z} - \frac{e}{m} E_z \frac{\partial f_{\mathrm{e0}}}{\partial v} = -\nu_{\mathrm{m}}(v) f_{\mathrm{e1}} \tag{19.1.11}$$

where, as in (19.1.7)

$$\nu_{\mathrm{m}}(v) = n_{\mathrm{g}} v\, 2\pi \int_0^{\pi} (1-\cos\theta_1) I(v,\theta_1) \sin\theta_1 \, d\theta_1 \tag{19.1.12}$$

is the momentum transfer collision frequency. Equation (19.1.11) gives the time rate of change of the anisotropic part of the distribution function, given the isotropic part. We see that ν_{m} is defined in the usual way

$$\nu_{\mathrm{m}} = n_{\mathrm{g}} \sigma_{\mathrm{m}}(v) v \tag{19.1.13}$$

where n_{g} is the neutral density and σ_{m} is the momentum transfer cross section.

The right-hand side of (19.1.10) is zero because the elastic collisions of the electrons are with infinitely massive neutrals. If the neutrals have a Maxwellian distribution and are not infinitely massive, then a collision term appears on the right-hand side of (19.1.10) (Holt and Haskell, 1965, Chapter 10; Smirnov, 1981, p. 66)

$$\frac{\partial f_{\mathrm{e0}}}{\partial t} + \frac{v}{3}\frac{\partial f_{\mathrm{e1}}}{\partial z} - \frac{e}{m} E_z \frac{1}{3v^2}\frac{\partial}{\partial v}(v^2 f_{\mathrm{e1}}) = C_{\mathrm{e0}} \tag{19.1.14}$$

For electron–neutral elastic collisions

$$C_{\mathrm{e0}} = C_{\mathrm{el}} = \frac{m}{M}\frac{1}{v^2}\frac{\partial}{\partial v}\left[v^3 \nu_{\mathrm{m}}(v) \left(f_{\mathrm{e0}} + \frac{e\mathrm{T}_{\mathrm{g}}}{mv}\frac{\partial f_{\mathrm{e0}}}{\partial v} \right) \right] \tag{19.1.15}$$

with T_{g} the neutral gas temperature. The first term in parentheses on the right-hand side of (19.1.15) accounts for elastic scattering energy losses, while the second term accounts for energy diffusion due to the nonzero gas temperature; this latter term is usually small.

If there are also energy losses due to inelastic collisions, then an additional term

$$C_{\mathrm{ex}} = -\nu_{\mathrm{ex}}(v) f_{\mathrm{e0}}(\mathbf{r}, v, t) + (v'/v)\nu_{\mathrm{ex}}(v') f_{\mathrm{e0}}(\mathbf{r}, v', t) \tag{19.1.16}$$

can be added to the right-hand side of (19.1.14), where $v'^2 = v^2 + 2e\mathcal{E}_{ex}/m$, $\nu_{ex}(v) = n_g\sigma_{ex}(v)v$ is the inelastic collision frequency, and $\mathcal{E}_{ex}$ is the electron energy lost in an inelastic collision. The first term in (19.1.16) accounts for the disappearance of electrons at speed v within the volume d^3v due to collisions which decrease v, and the second term represents the appearance of electrons within d^3v due to collisions which decrease v' to v. Electron vibrational (and rotational) collisional excitation of molecular gases can also lead to additional significant energy losses. Such losses have the general form of (19.1.16) for each vibrational (or rotational) level of interest. For most gases $\mathcal{E}_{vib} \ll \mathrm{T_e}$, and (19.1.16) can be expanded to the first order in $v' - v \approx e\mathcal{E}_{vib}/mv$ to obtain

$$C_{vib} \approx \frac{e\mathcal{E}_{vib}}{mv^2}\frac{\partial}{\partial v}\left(v\nu_{vib}(v)f_{e0}\right) \tag{19.1.17}$$

In general, (19.1.17) must be summed over vibrational (and rotational) levels to obtain the total energy loss. To account for ionization collisions, with the excess energy (exceeding $\mathcal{E}_{iz}$) equally shared among the incident and valence electrons, one can add the term

$$C_{iz} = -\nu_{iz}(v)f_{e0}(\mathbf{r}, v, t) + 4(v'/v)\nu_{iz}(v')f_{e0}(\mathbf{r}, v', t) \tag{19.1.18}$$

with $v'^2 = 2v^2 + 2e\mathcal{E}_{iz}/m$. The factor of four in (19.1.18) arises due to two electrons sharing the excess energy (Problem 19.3). Electron losses due to attachment can be introduced simply as

$$C_{att} = -\nu_{att}(v)f_{e0} \tag{19.1.19}$$

Coulomb collisions between charged particles can also be included. For electron–ion collisions, C_{el} in (19.1.15) can be modified by using $\nu_{ei} = n_i\sigma_{ei}(v)v$ and $\mathrm{T_i}$ in place of ν_m and $\mathrm{T_g}$, with

$$\sigma_{ei} = 4\pi\left(\frac{e^2}{4\pi\epsilon_0 mv^2}\right)^2 \ln\Lambda \tag{19.1.20}$$

with Λ as given in Section 3.3. For electron–electron collisions, the situation is more complicated due to the large energy transfers between electrons per collision. The collision term can be obtained from Fokker–Planck theory to be (Rosenbluth et al., 1957)

$$C_{ee} = \sigma_{ei}v^2\frac{\partial}{\partial v}\left[H(v)f_{e0} + \frac{v}{3}G(v)\frac{\partial f_{e0}}{\partial v}\right] \tag{19.1.21}$$

with

$$H = 4\pi\int_0^v f_{e0}(\mathbf{r}, v', t)v'^2\, dv' \tag{19.1.22}$$

and

$$G = 4\pi\left[\frac{1}{v^2}\int_0^v f_{e0}(\mathbf{r}, v', t)v'^4\, dv' + v\int_v^\infty f_{e0}(\mathbf{r}, v', t)v'\, dv'\right] \tag{19.1.23}$$

Note that C_{ee} depends quadratically on f_{e0}, such that inclusion of this term makes the kinetic equations nonlinear and therefore difficult to solve (see Shkarofsky et al., 1966, Chapter 7 for some approximate solutions).

Electron–neutral inelastic, ionization, and attachment collisions and electron–ion and electron–electron scattering also contribute to the momentum transfer collision frequency in (19.1.11), but these are generally small compared to electron–neutral elastic scattering in weakly ionized discharges. Including all collision terms, (19.1.11) and (19.1.14) are the fundamental kinetic equations for the electron distribution function in the limit that the anisotropy is small, $|f_{e1}| \ll |f_{e0}|$. The two-term equations can be modified to include the effect of a dc magnetic field force (Shkarofsky et al., 1966; Holt and Haskell, 1965), but we do not introduce this complication here. For a steady-state distribution with no spatial gradients or electric field, the left-hand side of (19.1.14) vanishes. If electron–neutral elastic scattering dominates, one can then set the right-hand side C_{el} in (19.1.15) equal to zero, to find that f_{e0} is a Maxwellian distribution at temperature T_g, i.e., the electrons and neutrals have equilibrated. However, this is rarely the situation in low-pressure gas discharges.

19.1.4 Diffusion and Mobility

Consider now a steady-state plasma with nonzero density gradient and dc electric field. Solving (19.1.11) for f_{e1} yields

$$f_{e1} = -\frac{1}{\nu_m}\left(v\frac{\partial f_{e0}}{\partial z} - \frac{e}{m}E_z\frac{\partial f_{e0}}{\partial v}\right) \tag{19.1.24}$$

Introducing the particle flux

$$\boldsymbol{\Gamma}_e = \int \mathbf{v}\left(f_{e0} + \frac{v_z}{v}f_{e1}\right) d^3v \tag{19.1.25}$$

and using spherical coordinates in velocity space, we see that the isotropic part of f_e does not contribute to the flux. From the anisotropic part, we obtain only a z component

$$\begin{aligned}\Gamma_{ez} &= 2\pi \int_0^\pi \sin\psi\, d\psi \cos^2\psi \int_0^\infty v f_{e1} v^2\, dv \\ &= \frac{4\pi}{3}\int_0^\infty v^3 f_{e1}\, dv\end{aligned} \tag{19.1.26}$$

Inserting f_{e1} from (19.1.24) into (19.1.26), we obtain

$$\Gamma_{ez} = -D_e\frac{dn_e}{dz} - \mu_e n_e E_z \tag{19.1.27}$$

where

$$D_e = \frac{4\pi}{3n_e}\int_0^\infty \frac{v^4}{\nu_m(v)} f_{e0}\, dv \tag{19.1.28}$$

is the diffusion coefficient, and

$$\mu_e = -\frac{4\pi e}{3mn_e}\int_0^\infty \frac{v^3}{\nu_m(v)}\frac{df_{e0}}{dv}\, dv \tag{19.1.29}$$

is the mobility. For a Maxwellian distribution, D_e and μ_e are related by the Einstein relation (5.1.9), as can be shown directly from (19.1.28) and (19.1.29) (Problem 19.4). These equations are important because they give the proper prescription for averaging over $\nu_m(v)$ to determine D_e and μ_e. However, the symmetric part of the distribution must be known.

19.1.5 Druyvesteyn Distribution

Consider the steady-state electron distribution function in a uniform plasma with a uniform steady electric field $E_z = E$ and with elastic collisions between electrons and neutral gas atoms. From (19.1.11), we have

$$f_{e1} = \frac{eE}{m\nu_m}\frac{df_{e0}}{dv} \tag{19.1.30}$$

For the ease of analysis, we take the gas temperature T_g to be negligible ($T_g \ll T_e$) and $m \ll M$ in (19.1.15), then (19.1.14) becomes

$$-\frac{eE}{3m}\frac{d(v^2 f_{e1})}{dv} = \frac{m}{M}\frac{d(v^3 \nu_m f_{e0})}{dv} \tag{19.1.31}$$

Integrating (19.1.31)

$$f_{e1} = -\frac{3m^2}{eEM} v \nu_m f_{e0} \tag{19.1.32}$$

and equating (19.1.30) and (19.1.32), we obtain

$$\frac{eE^2}{m\nu_m}\frac{df_{e0}}{dv} = -\frac{3m^2}{eM} v \nu_m f_{e0} \tag{19.1.33}$$

Integrating (19.1.33), we find

$$f_{e0} = A \exp\left[-\frac{3m^3}{e^2E^2M}\int_0^v v' \nu_m^2(v')\,dv'\right] \tag{19.1.34}$$

where A is a normalization constant determined by $\int f_{e0}\,d^3v = n_e$. For a constant collision frequency, $\nu_m(v) = \text{const}$, we obtain a Maxwellian distribution. For constant cross-section (hard sphere) collisions, $\sigma_m = \text{const}$ (constant mean free path), and using $\nu_m = n_g \sigma_m v$, we find that

$$f_{e0} = A e^{-Cv^4} \tag{19.1.35}$$

with C a constant, which is known as the *Druyvesteyn distribution* (Druyvesteyn and Penning, 1940). Many electron–neutral cross sections behave as hard sphere interactions at low energies (see Figure 3.9).

19.1.6 Electron Distribution in an RF Field

We consider the electron distribution in a uniform plasma with an rf electric field $E_z(t) = \text{Re}\,\tilde{E}\,e^{j\omega t}$, and with energy losses only due to elastic collisions of electrons with neutral gas atoms, in the frequency regime $\omega \gg \nu_\mathcal{E} \approx (2m/M)\nu_m$. Here, $\nu_\mathcal{E}$ is the electron energy loss frequency. This is a good approximation for noble gases, where the inelastic losses are small over most of the EEDF energy range. In this regime, the energy transferred by electrons to gas atoms over one rf period is small, and f_{e0} is independent of time. Introducing $f_{e1}(t) = \text{Re}\,\tilde{f}_{e1}\,e^{j\omega t}$ into (19.1.11), we obtain

$$\tilde{f}_{e1} = \frac{e\tilde{E}}{m(j\omega + \nu_m)}\frac{df_{e0}}{dv} \tag{19.1.36}$$

Substituting $E_z(t)$ and $f_{e1}(t)$ into (19.1.14) and time-averaging the resulting equation over an rf period (see Problem 4.5), we obtain, in analogy to (19.1.31), and with the same assumptions ($T_g \ll T_e$, $m/M \ll 1$)

$$-\frac{1}{2}\text{Re}\left[\frac{e\tilde{E}^*}{3m}\frac{d(v^2\tilde{f}_{e1})}{dv}\right] = \frac{m}{M}\frac{d(v^3\nu_m f_{e0})}{dv} \tag{19.1.37}$$

where $\tilde{E}^*$ is the complex conjugate of $\tilde{E}$. Integrating (19.1.37), we find

$$\text{Re}\,(\tilde{E}^*\tilde{f}_{e1}) = -\frac{6m^2}{eM} v \nu_m f_{e0} \tag{19.1.38}$$

Substituting (19.1.36) into (19.1.38), we obtain

$$\frac{e|\tilde{E}|^2\nu_m}{2m(\omega^2 + \nu_m^2)}\frac{df_{e0}}{dv} = -\frac{3m^2}{eM} v \nu_m f_{e0} \tag{19.1.39}$$

Equation (19.1.39) can be integrated to obtain

$$f_{e0} = A \exp\left[-\frac{6m^3}{e^2|\tilde{E}|^2 M}\int_0^v v'\left(\omega^2 + \nu_m^2(v')\right)\,dv'\right] \tag{19.1.40}$$

where A is the normalization constant, as in (19.1.34). We note that at high frequencies or low pressures, such that $\omega \gg \nu_m$, f_{e0} reduces to a Maxwellian distribution. Substituting (19.1.40) into (19.1.36) determines $\tilde{f}_{e1}$, the oscillating anisotropic part of f_e.

Comparing (19.1.33) and (19.1.39) for the case of ν_m = const, we see that these are the same equations, having the same solutions, provided we introduce an *effective* dc electric field in (19.1.39)

$$E_{eff} = \frac{|\tilde{E}|}{\sqrt{2}}\frac{\nu_m}{(\omega^2 + \nu_m^2)^{1/2}} \tag{19.1.41}$$

Note from the definitions (4.2.22) and (4.2.20) for the dc and rf plasma conductivities that

$$\sigma_{dc}E_{eff}^2 = \frac{1}{2}\mathrm{Re}\,\sigma_p|\tilde{E}|^2$$

i.e., the effective dc field gives the same ohmic power dissipation as the rf field. However, from (19.1.36), there is a phase delay between $\tilde{E}$ and $\tilde{f}_{e1}$, which induces a phase delay in the rf current density.

Unfortunately, for molecular gases with large, low-energy, vibrational excitation cross sections in the 0.1–3 V energy range, the inelastic energy loss frequency $\nu_\mathcal{E}$ can be comparable to the momentum transfer frequency ν_m. This is often the regime for many plasma processing discharges having complex gas feed chemistries. In this case, both f_{e0} and f_{e1} are time-modulated. The quadratic nonlinearity in (19.1.36) generates harmonics of the applied frequency ω, with $f_{e0} \propto E_z^2$ containing even harmonics, and $f_{e1} \propto E_z$ containing odd harmonics. Then the calculation of an effective field for $\omega \gtrsim (\nu_m, \nu_\mathcal{E})$ given in (19.1.41) is only a rough approximation. There are no simple analytic solutions in this regime. Numerical solutions were obtained by Wilhelm and Winkler (1979), and Winkler and Wilhelm (1980), see also Section 6.3.3 of Makabe and Petrović (2006).

19.1.7 Effective Electrical Conductivity

Let us consider a kinetic treatment of the rf conductivity σ_p for a given isotropic electron distribution function f_{e0}. We assume a plasma with uniform electron density n_e driven by an rf electric field $E(t) = \mathrm{Re}\,\tilde{E}\,e^{j\omega t}$. It is convenient here to normalize f_{e0} to unity rather than to n_e. With these assumptions, inserting (19.1.36) into (19.1.26), we obtain the rf current amplitude

$$\tilde{J} = -e\tilde{\Gamma}_e = -\frac{4\pi}{3}\frac{e^2 n_e \tilde{E}}{m}\int_0^\infty \frac{v^3\,dv}{j\omega + \nu_m(v)}\frac{df_{e0}}{dv} \tag{19.1.42}$$

with $\nu_m(v) = n_g\sigma_m(v)v$ the momentum transfer collision frequency. Hence, we find

$$\sigma_p = \frac{\tilde{J}}{\tilde{E}} = -\frac{4\pi}{3}\frac{e^2 n_e}{m}\int_0^\infty \frac{v^3\,dv}{j\omega + \nu_m(v)}\frac{df_{e0}}{dv} \tag{19.1.43}$$

In analogy to the fluid result (4.2.20), σ_p, is often expressed in the form

$$\sigma_p = \frac{\epsilon_0\omega_{pe}^2}{j\omega_{eff} + \nu_{eff}} = \frac{e^2 n_e}{m(j\omega_{eff} + \nu_{eff})} \tag{19.1.44}$$

where ω_{eff} and ν_{eff} are *effective radian and collision frequencies*, respectively. To determine these quantities, we equate (19.1.43) to (19.1.44) and take the real and imaginary parts. This yields two equations that can be simultaneously solved for ω_{eff} and ν_{eff}. There are two limiting cases when simple results can be obtained. For low frequencies, $\omega \ll \nu_{\text{eff}}$, we find

$$\frac{1}{\nu_{\text{eff}}} = -\frac{4\pi}{3}\int_0^\infty \frac{v^3\,\mathrm{d}v}{\nu_{\text{m}}(v)}\frac{\mathrm{d}f_{\text{e0}}}{\mathrm{d}v}$$

$$\frac{\omega_{\text{eff}}}{\omega} = -\frac{4\pi}{3}\nu_{\text{eff}}^2\int_0^\infty \frac{v^3\,\mathrm{d}v}{\nu_{\text{m}}^2(v)}\frac{\mathrm{d}f_{\text{e0}}}{\mathrm{d}v} \tag{19.1.45}$$

In the opposite limit of high frequencies, $\omega \gg \nu_{\text{eff}}$, we find

$$\nu_{\text{eff}} = -\frac{4\pi}{3}\int_0^\infty v^3\mathrm{d}v\,\nu_{\text{m}}(v)\frac{\mathrm{d}f_{\text{e0}}}{\mathrm{d}v}$$

$$\frac{\omega_{\text{eff}}}{\omega} = 1 \tag{19.1.46}$$

In the transition regime $\omega \sim \nu_{\text{eff}}$, the equations for ν_{eff} and ω_{eff} are coupled and must be solved simultaneously.

In both limiting cases, $\nu_{\text{eff}}/n_{\text{g}}$ and $\omega_{\text{eff}}/\omega$ are independent of the neutral gas density n_{g}; they depend only on the distribution function (e.g., T_{e} for a Maxwellian distribution) and the type of gas. However, in the transition regime, $\nu_{\text{eff}}/n_{\text{g}}$ and $\omega_{\text{eff}}/\omega$ are both explicit functions of n_{g}. Lister et al. (1996) have determined these for a Maxwellian distribution in argon, with the results given in Figure 19.1 for a 13.56 MHz driving frequency.

An alternate approach to determine ν_{eff} and σ_{p} is based on the use of Boltzmann term analysis, which is described at the end of Section 11.3. As shown there, the rate of change of electron momentum per unit volume due to electron–neutral elastic collisions is given by (11.3.5)

$$C_{\text{m}} = mn_{\text{e}}\int \mathrm{d}^3v\,v_x(\partial f_{\text{e}}/\partial t)_{\text{c}} = -mn_{\text{g}}n_{\text{e}}\int \mathrm{d}^3v\,\sigma_{\text{m}}vv_xf_{\text{e}}$$

with, here, f_{e} normalized to unity and with $\sigma_{\text{m}}(v)$ the electron–neutral momentum transfer cross section given in (3.1.15). Let us express C_{m} in terms of an effective collision frequency ν_{eff}

$$C_{\text{m}} = -m\nu_{\text{eff}}\Gamma_{\text{e}} \tag{19.1.47}$$

with $\Gamma_{\text{e}} = n_{\text{e}}u_{\text{e}}$ the macroscopic electron flux.

Let us consider some ways to determine ν_{eff}. The energy-dependent collision frequency is $\nu_{\text{m}}(v) = n_{\text{g}}\sigma_{\text{m}}(v)v$. For a zero-order isotropic electron distribution, an average momentum transfer frequency is then

$$\nu_{\text{m}} = \int \nu_{\text{m}}(v)f_{\text{e0}}(v)\,\mathrm{d}^3v \tag{19.1.48}$$

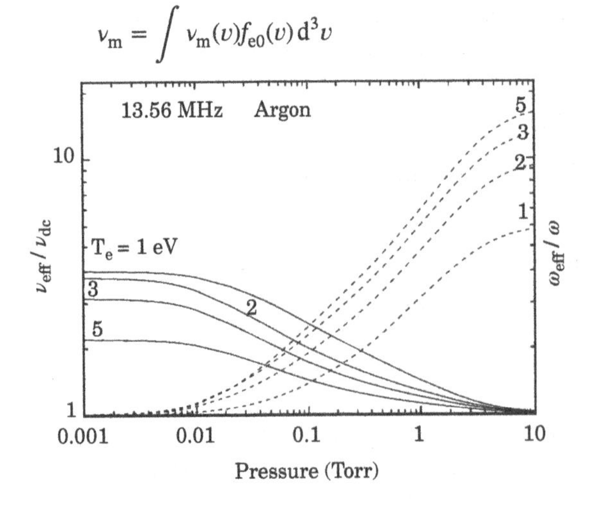

Figure 19.1 Variations of $\nu_{\text{eff}}/\nu_{\text{dc}}$ (solid lines) and $\omega_{\text{eff}}/\omega$ (dashed lines) as a function of pressure for different electron temperatures T_{e}; here, $\nu_{\text{dc}} = \nu_{\text{eff}}(\omega = 0)$. Source: Lister et al. (1996)/with permission of AIP Publishing.

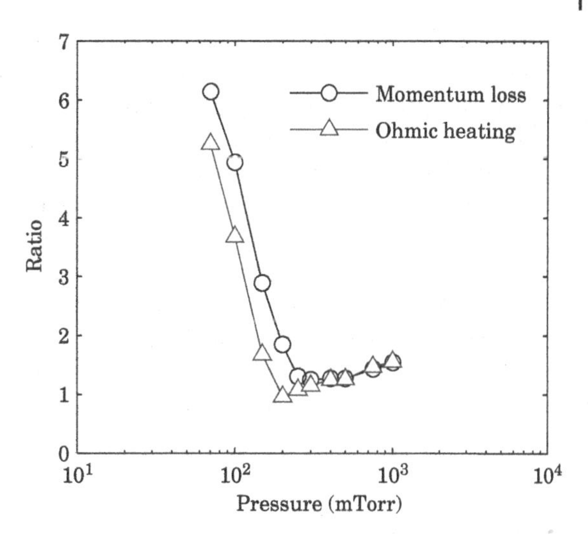

Figure 19.2 Ratios of $\nu_{\text{eff}}/\nu_{\text{m}}$ (open circles) and exact-to-approximate ohmic heating power (open triangles) versus argon gas pressure, determined by Boltzmann term analysis of PIC simulation results, for a 2-cm gap capacitive discharge driven by 25.6 A/m^2 at 13.56 MHz. Source: Lafleur et al. (2014a)/with permission of IOP Publishing.

As a rough estimate, we would expect $\nu_{\text{eff}} \approx \nu_{\text{m}}$. As we have seen in (19.1.44)–(19.1.46), the two-term approximation can be used for a better estimate to determine ν_{eff} for a general energy-dependent $\nu_{\text{m}}(v)$ and an isotropic, zero-order distribution $f_{\text{e}0}$. However, in general, $f_{\text{e}0}$ is not *a priori* known.

In PIC and other kinetic simulations, f_{e} can be directly evaluated, and Boltzmann term analysis can be used to determine C_{m}. Then, ν_{eff}, the plasma conductivity σ_{p}, and resulting ohmic heating power are also determined. This has been done by Lafleur et al. (2014a) for a 2-cm gap capacitive discharge driven by an rf current density of 25.6 A/m^2 at 13.56 MHz. (These are the same parameters as for Figure 11.15.) The results are shown in Figure 19.2. We see $\nu_{\text{eff}}/\nu_{\text{m}} \approx 1$ (open circles) at high pressures, increasing significantly as the pressure is reduced below 200 mTorr, with $\nu_{\text{eff}}/\nu_{\text{m}} \approx$ 6.3 at 70 mTorr. The normalized ohmic heating power displays the same behavior, but with a slightly smaller increase as the pressure is reduced. A somewhat similar behavior is seen in the two-term kinetic result for $\nu_{\text{eff}}/\nu_{\text{dc}}$ in Figure 19.1, with $\nu_{\text{eff}}/\nu_{\text{dc}} \approx 1$ at high pressures, decreasing to $\nu_{\text{eff}}/\nu_{\text{dc}} \approx 3$ at 70 mTorr. (But note here that the normalizing frequencies ν_{m} and ν_{dc} are not the same.)

There are some significant differences between the Boltzmann term analysis and the two-term kinetic results. We expect the Boltzmann term analysis to be more accurate because it accounts for the changing velocity variation of $f_{\text{e}}(\mathbf{v})$ as the pressure is varied. Typically, a fixed, isotropic $f_{\text{e}0}$, usually a Maxwellian, is assumed in the two-term analysis. The Boltzmann term analysis also accounts for time-varying and/or non-isotropic f_{e}s, which might be important at reduced pressures.

19.1.8 LXCat Database and Bolsig+ Solver

LXCat is an open-access website (https://www.lxcat.net/home) for electron and ion cross-section and transport coefficient data needed for modeling low temperature, weakly ionized plasmas. There are numerous cross-section data sets for various gases. The status of the LXCat project is summarized in a short review by Pitchford et al. (2017), and in an extensive review and tutorial by Carbone et al. (2021). These reviews list numerous databases contributed by different research groups around the world. The electron–neutral cross section data can be fed into an on-line solver Bolsig+ (Hagelaar and Pitchford, 2005) on the LXCat website, for the numerical solution of the Boltzmann equation in uniform electric fields. Alternately, the data and the Bolsig+ computer program can be downloaded for off-line use. Various other software tools can also be accessed.

The main purpose of Bolsig+ is to obtain electron transport and collision rate coefficients from cross-section data. Bolsig+ solves the two-term approximation of the Boltzmann equation for constant space and time dependence of a user-specified electric field. Electron–electron and electron–ion Coulomb collisions in the two-term approximation can also be incorporated into the calculation in the downloaded version. Spatially constant rf electric fields can also be treated in the high-frequency limit $\omega \gg \nu_\mathcal{E}$, where $\nu_\mathcal{E}$ is the electron energy relaxation frequency. For low-pressure noble gases, this limit may be a reasonable assumption for rf frequencies $\gtrsim 10$ MHz. But the reader should note that at high pressures and/or for molecular gases, this assumption may be reasonable only for discharges driven at and above microwave frequencies $\gtrsim 1$ GHz.

Maxwellian rate coefficients can also be computed from the cross-section data, and compared with the two-term Boltzmann equation solutions. This can give the student considerable insight into the effects of non-Maxwellian distributions on the modeling of the discharge chemistry and physics.

19.2 Local Kinetics

Although the solutions of the two-term approximation developed in Section 19.1 give some insight into the kinetic behavior, further approximations are necessary in order to account for the spatial variation. With both dc (ambipolar and/or heating) and rf (heating) fields spatially-varying, the electric field is generalized to

$$\mathbf{E}(\mathbf{r}, t) = \overline{\mathbf{E}}(\mathbf{r}) + \mathrm{Re}\,\tilde{\mathbf{E}}(\mathbf{r})\,\mathrm{e}^{j\omega t} \tag{19.2.1}$$

with $\overline{\mathbf{E}}_\mathrm{a} + \overline{\mathbf{E}}_\mathrm{h}$ the sum of ambipolar and dc heating fields. As in Section 19.1, we assume a dc discharge or an rf steady-state discharge in a frequency regime $\omega \gg (m/M)\nu_\mathrm{m}$, such that in both cases the isotropic part $f_{\mathrm{e}0}$ of the distribution is independent of time. For the anisotropic part, we write

$$\mathbf{f}_{\mathrm{e}1} = \bar{\mathbf{f}}_{\mathrm{e}1} + \mathrm{Re}\,\tilde{\mathbf{f}}_{\mathrm{e}1}\,\mathrm{e}^{j\omega t} \tag{19.2.2}$$

Inserting (19.2.1) and (19.2.2) along with the two-term expansion (19.1.3) into the Boltzmann equation (19.1.1), multiplying by $\sin\psi$, integrating over ψ from 0 to π, and performing the time averaging as in Section 19.1, we obtain (Problem 19.6)

$$\frac{v}{3}\nabla\cdot\bar{\mathbf{f}}_{\mathrm{e}1} - \frac{e}{3mv^2}\frac{\partial}{\partial v}\left[v^2\left(\overline{\mathbf{E}}\cdot\bar{\mathbf{f}}_{\mathrm{e}1} + \frac{1}{2}\mathrm{Re}\left(\tilde{\mathbf{E}}^*\cdot\tilde{\mathbf{f}}_{\mathrm{e}1}\right)\right)\right] = C_{\mathrm{e}0}(f_{\mathrm{e}0}) \tag{19.2.3}$$

which is a generalization of (19.1.14), but without time variation. Repeating the procedure, but multiplying by $\sin\psi\cos\psi$ before integrating, we find

$$\bar{\mathbf{f}}_{\mathrm{e}1} = -\frac{v}{\nu_\mathrm{m}}\nabla f_{\mathrm{e}0} + \frac{e\overline{\mathbf{E}}}{m\nu_\mathrm{m}}\frac{\partial f_{\mathrm{e}0}}{\partial v} \tag{19.2.4}$$

and

$$\tilde{\mathbf{f}}_{\mathrm{e}1} = \frac{e\tilde{\mathbf{E}}}{m(\nu_\mathrm{m} + j\omega)}\frac{\partial f_{\mathrm{e}0}}{\partial v} \tag{19.2.5}$$

Inserting (19.2.4) and (19.2.5) into (19.2.3), we obtain the kinetic equation for $f_{\mathrm{e}0}$

$$\begin{aligned} -\nabla\cdot\left(\frac{v^2}{3\nu_\mathrm{m}}\nabla f_{\mathrm{e}0}\right) &+ \frac{ve}{3m}\nabla\cdot\left(\frac{\overline{\mathbf{E}}}{\nu_\mathrm{m}}\frac{\partial f_{\mathrm{e}0}}{\partial v}\right) - \frac{e}{3mv^2}\frac{\partial}{\partial v}\left[-\frac{v^3}{\nu_\mathrm{m}}\overline{\mathbf{E}}\cdot\nabla f_{\mathrm{e}0}\right. \\ &\left. + \frac{ev^2}{m\nu_\mathrm{m}}\left(|\overline{\mathbf{E}}|^2 + \frac{|\tilde{\mathbf{E}}|^2}{2}\frac{\nu_\mathrm{m}^2}{\nu_\mathrm{m}^2+\omega^2}\right)\frac{\partial f_{\mathrm{e}0}}{\partial v}\right] = C_{\mathrm{e}0}(f_{\mathrm{e}0}) \end{aligned} \tag{19.2.6}$$

The first term describes the spatial diffusion of electrons, the second term describes the electron flux due to the dc electric field, the third term gives the diffusion cooling, and the fourth and fifth terms give the heating due to the dc and rf electric fields.

It is often convenient for discharge analysis to express (19.2.6) in terms of energy rather than velocity coordinates. Introducing

$$\mathcal{E} = mv^2/(2e) \tag{19.2.7}$$

such that $\mathrm{d}\mathcal{E} = (m/e)v\,\mathrm{d}v$, and introducing the electron energy probability function (EEPF) g_p as in (6.6.19), we obtain

$$\frac{2e}{3m}\frac{\mathcal{E}^{3/2}}{\nu_\mathrm{m}}\left[-\nabla^2 g_\mathrm{p} + \nabla\cdot\left(\overline{\mathbf{E}}\frac{\partial g_\mathrm{p}}{\partial\mathcal{E}}\right)\right] - \frac{2e}{3m}\frac{\partial}{\partial\mathcal{E}}$$
$$\times\left[-\frac{\mathcal{E}^{3/2}}{\nu_\mathrm{m}}(\overline{\mathbf{E}}\cdot\nabla)g_\mathrm{p} + \frac{\mathcal{E}^{3/2}}{\nu_\mathrm{m}}\left(|\overline{\mathbf{E}}|^2 + \frac{|\tilde{\mathbf{E}}|^2}{2}\frac{\nu_\mathrm{m}^2}{\nu_\mathrm{m}^2+\omega^2}\right)\frac{\partial g_\mathrm{p}}{\partial\mathcal{E}}\right] = 2\pi\left(\frac{2e}{m}\right)^{3/2}\mathcal{E}^{1/2}C_{\mathrm{e}0}(g_\mathrm{p}) \tag{19.2.8}$$

where $C_{\mathrm{e}0}(g_\mathrm{p})$ is the collision term transformed from $f_{\mathrm{e}0}$ to g_p and from v to $\mathcal{E}$ coordinates. Note that $g_\mathrm{p}(\mathcal{E}) = 2\pi(2e/m)^{3/2}f_{\mathrm{e}0}(v(\mathcal{E}))$ such that $\int g_\mathrm{p}\mathcal{E}^{1/2}\,\mathrm{d}\mathcal{E} = n_\mathrm{e}$.

In principle, (19.2.8) can be solved to determine the energy distribution function for all $\mathbf{r}$ and $\mathcal{E}$. However, solving this nonlinear integro-differential equation is difficult. In high-pressure discharges, however, where the electron motion is strongly collisional, the energy diffusion and collisional terms can be much stronger than the energy variations due to the spatial gradient terms. In this case, we can neglect all terms arising from the spatial inhomogeneity, including the ambipolar electric field[1] ($\mathbf{E}_\mathrm{a} = -\nabla\Phi_\mathrm{a} \equiv 0$) to obtain the *local approximation*

$$-\frac{\partial}{\partial\mathcal{E}}\left[\frac{2e}{3m}\frac{\mathcal{E}^{3/2}}{\nu_\mathrm{m}}\left(|\overline{\mathbf{E}}_\mathrm{h}|^2 + \frac{|\tilde{\mathbf{E}}|^2}{2}\frac{\nu_\mathrm{m}^2}{\nu_\mathrm{m}^2+\omega^2}\right)\frac{\partial g_\mathrm{p}}{\partial\mathcal{E}}\right] = 2\pi\left(\frac{2e}{m}\right)^{3/2}\mathcal{E}^{1/2}C_{\mathrm{e}0}(g_\mathrm{p}) \tag{19.2.9}$$

In this equation, g_p depends on $\mathbf{r}$ only through the dependences of $\overline{\mathbf{E}}_\mathrm{h}(\mathbf{r})$ and $\tilde{\mathbf{E}}(\mathbf{r})$ on $\mathbf{r}$. Hence, given the values of the field strengths at any point, (19.2.9) can be solved to determine the energy distribution at that point. Neglecting the ambipolar contribution to the dc field is equivalent to a model in which the space charge potential is zero everywhere inside the bulk plasma and falls sharply to a negative value at the walls (rectangular potential well).

We expect the local approximation to hold when the *energy relaxation length* $\lambda_\mathcal{E}$ (see Section 15.2 and also (19.4.38)) is small compared to the spatial inhomogeneity scale Λ of the discharge

$$\lambda_\mathcal{E} \ll \Lambda \tag{19.2.10}$$

where $\lambda_\mathcal{E}$ depends on both momentum transfer and energy loss collision processes

$$\lambda_\mathcal{E} \approx \left(\frac{\lambda_\mathrm{m}\lambda_\mathrm{inel}}{3}\right)^{1/2} \tag{19.2.11}$$

Here, λ_m is the total mean free path for momentum transfer and λ_inel is the mean free path accounting for all collisional energy loss processes, i.e.,

$$\lambda_\mathrm{inel}^{-1} = n_\mathrm{g}[(2m/M)\sigma_\mathrm{el} + \sigma_\mathrm{ex} + \sigma_\mathrm{iz} + \cdots] \tag{19.2.12}$$

The square root relation appears in (19.2.11) because the electron motion is diffusive between successive inelastic collisions. The requirement (19.2.10) for local behavior is most difficult to meet

1 Even for a spatially varying ambipolar field, then accounting for the ambipolar ion flux, the total ambipolar current density $\mathbf{J}_\mathrm{a}$ vanishes, and there is no $\mathbf{J}_\mathrm{a}\cdot\mathbf{E}$ contribution to (19.2.9).

for energies below $\mathcal{E}_{ex}$, where the only collisional energy losses are due to electron–neutral elastic scattering, and (19.2.10) becomes

$$(M/2m)^{1/2}\lambda_m \ll \Lambda \tag{19.2.13}$$

For argon with an average $\sigma_m \sim 10^{-19}$ m^2 from Figure 3.13, we obtain the condition $p\Lambda \gg 6$ Torr-cm. For a discharge scale length $\Lambda = 10$ cm, we find $p \gg 600$ mTorr. (Actually, the condition for local behavior near the Ramsauer minimum in argon is more severe.) Molecular gases typically have energy losses per momentum transfer collision a factor of 5–20 higher than atomic gases due to low-energy vibrational and rotational excitations, so we expect local behavior for $p \gg 60$ mTorr in these gases.

Example 19.1 Let us determine the EEPF in the positive column of a dc glow discharge in an atomic gas using local analysis. We neglect the ambipolar field (rectangular potential well approximation in the radial direction) and, as described in Section 14.2, we assume a long thin column with a constant axial heating field E. We examine energies only below the excitation threshold energy $\mathcal{E}_{ex}$, such that the collisional energy losses are only due to electron–neutral elastic scattering, $C_{e0} = C_{el}$. Assuming $T_g = 0$ and $m \ll M$ in the elastic collision energy loss term (19.1.15) as previously, transforming from v to $\mathcal{E}$ coordinates, and inserting this into (19.2.9), we obtain

$$-\frac{\partial}{\partial \mathcal{E}}\left(E^2\frac{\mathcal{E}^{3/2}}{\nu_m}\frac{\partial g_p}{\partial \mathcal{E}}\right) = \frac{3m^2}{Me}\frac{\partial}{\partial \mathcal{E}}\left(\mathcal{E}^{3/2}\nu_m g_p\right) \tag{19.2.14}$$

Integrating this twice with respect to $\mathcal{E}$ yields

$$g_p(\mathcal{E}) = g_{p0}\exp\left(-\frac{3m^2}{MeE^2}\int_0^{\mathcal{E}} \nu_m^2(\mathcal{E}')\,d\mathcal{E}'\right) \tag{19.2.15}$$

Consider hard sphere collisions, such that $\nu_m = n_g\sigma_m v = (2e\mathcal{E}/m)^{1/2}/\lambda_m$. Inserting this into (19.2.15) yields the Druyvesteyn EEPF

$$g_p = g_{p0}\exp\left(-\frac{3m}{M}\frac{\mathcal{E}^2}{E^2\lambda_m^2}\right) \tag{19.2.16}$$

equivalent to f_{e0} given in (19.1.35). Plotting g_p (log scale) versus $\mathcal{E}^2$ yields the dashed straight line shown in Figure 19.3.

We have thus far considered energies below the first excitation energy $\mathcal{E}_{ex}$ in an atomic gas, such that electron–neutral elastic scattering dominates the energy losses. In this case, $\nu_{\mathcal{E}} \ll \nu_m$, and thus

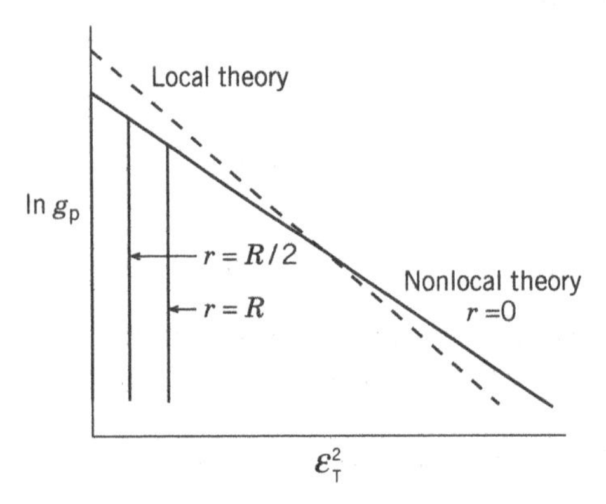

Figure 19.3 Schematic of electron energy probability function $g_p(\mathcal{E}, r)$ versus $\mathcal{E}_T^2 = (\mathcal{E} - \Phi(r))^2$ at a fixed heating field E and gas pressure p in the positive column of a glow discharge; (dashed line) local kinetics and (solid line) nonlocal kinetics.

the distribution function is almost isotropic. In the inelastic energy range above $\mathcal{E}_{ex}$ (typically for a relatively few electrons within the tail of the energy distribution), we still find that $\nu_{\mathcal{E}} \ll \nu_m$ for most gases, again yielding a nearly isotropic distribution. Hence, excluding high-energy electrons generated by such processes as secondary emission across high-voltage sheaths or injected electron beams, the two-term approximation (19.1.3) is valid over the entire range of energies. Provided that the energy relaxation length $\lambda_{\mathcal{E}}$ is much smaller than the discharge scale length Λ, local theory can be used to determine the distribution.

Example 19.2 Consider the EEPF in the positive column for electrons with energies exceeding the first inelastic energy threshold $\mathcal{E}_{ex}$. In this case, we can neglect elastic scattering energy losses and, hence, choose $C_{e0} = C_{ex}$ as given in (19.1.16). Furthermore, if f_{e0} decays rapidly with energy, then the second term in (19.1.16) is much smaller than the first term, and we can use $C_{ex} = -\nu_{ex} f_{e0}$. The kinetic equation (19.2.9) is then

$$\frac{d}{d\mathcal{E}}\left(\frac{\mathcal{E}^{3/2}}{\nu_m}\frac{dg_p}{d\mathcal{E}}\right) - \frac{3m\nu_{ex}}{2eE^2}\mathcal{E}^{1/2} g_p = 0 \tag{19.2.17}$$

Expanding the derivative and collecting terms, we obtain

$$\frac{d^2 g_p}{d\mathcal{E}^2} + \frac{\nu_m}{\mathcal{E}^{3/2}}\frac{d}{d\mathcal{E}}\left(\frac{\mathcal{E}^{3/2}}{\nu_m}\right)\frac{dg_p}{d\mathcal{E}} - \kappa^2 g_p = 0 \tag{19.2.18}$$

where

$$\kappa^2(\mathcal{E}) = \frac{3m\nu_m\nu_{ex}}{2eE^2\mathcal{E}} \tag{19.2.19}$$

We note that κ^{-1} is the characteristic energy for decay of g_p. In general, (19.2.18) must be integrated numerically. However, let us consider (Smirnov, 1981, p. 110) the weak field regime $\kappa\mathcal{E}_{ex} \gg 1$. Substituting $g_p = e^{S(\mathcal{E})}$ into (19.2.18), we obtain

$$\left(\frac{dS}{d\mathcal{E}}\right)^2 + \frac{d^2 S}{d\mathcal{E}^2} + \frac{\nu_m}{\mathcal{E}^{3/2}}\frac{d}{d\mathcal{E}}\left(\frac{\mathcal{E}^{3/2}}{\nu_m}\right)\frac{dS}{d\mathcal{E}} - \kappa^2 = 0 \tag{19.2.20}$$

For $\kappa\mathcal{E}_{ex} \gg 1$, the second and third terms in (19.2.20) are small; neglecting them, (19.2.20) can be integrated to yield

$$S(\mathcal{E}) = -\int_{\mathcal{E}_{ex}}^{\mathcal{E}} \left(\frac{3m\nu_m\nu_{ex}}{2eE^2\mathcal{E}}\right)^{1/2} d\mathcal{E} \tag{19.2.21}$$

Hence, we find

$$g_p(\mathcal{E}) = A\exp\left[-\int_{\mathcal{E}_{ex}}^{\mathcal{E}} \left(\frac{3m\nu_m\nu_{ex}}{2eE^2\mathcal{E}}\right)^{1/2} d\mathcal{E}\right], \qquad \mathcal{E} \gtrsim \mathcal{E}_{ex} \tag{19.2.22}$$

The constant A is found by joining the distribution functions (19.2.16) and (19.2.22) at the energy where $\nu_{ex} \approx (2m/M)\nu_m$, i.e., near the excitation threshold $\mathcal{E}_{ex}$.

19.3 Nonlocal Kinetics

For many discharges of interest in materials processing, the pressure is too low for the local approximation to hold. In these cases, a different approximation has been employed, called *nonlocal kinetics*. Originally developed to analyze low-pressure dc glow discharges, by Bernstein and Holstein (1954) and by Tsendin (1974), the technique has also been used to analyze capacitive and inductive rf discharges. Reviews of the methods and applications, including comparisons of the predictions with various experimental results, have been given by Kolobov and Godyak (1995) and Kortshagen et al. (1996). Here, following Kortshagen et al., we introduce the concept. Its use for determining the electron energy distribution in rf inductive and capacitive discharges is described in Sections 19.5 and 19.6, respectively.

The basic idea of the nonlocal approximation is that the total energy of electrons (sum of kinetic energy and potential energy in the ambipolar field) is the proper variable to describe the spatially inhomogeneous problem in a low-pressure discharge. In the absence of energy-loss collisions and without heating electric fields, the total electron energy is constant as the electrons bounce back and forth within the confining potential of the ambipolar field. Hence, the EEPF is a function of the total energy only. For weak energy losses and heating, the timescale for the bouncing motion of electrons can be much shorter than that of the energy losses or heating, such that the total energy is still approximately conserved over a bounce.

To simplify the derivation, we decompose the dc part of the electric field $\bar{\mathbf{E}}$ into the sum of mutually perpendicular ambipolar and heating components

$$\bar{\mathbf{E}} = \bar{\mathbf{E}}_a + \bar{\mathbf{E}}_h \tag{19.3.1}$$

with $\bar{\mathbf{E}}_a \cdot \bar{\mathbf{E}}_h = 0$ and $\bar{\mathbf{E}}_a = -\nabla\Phi(\mathbf{r})$. We also introduce the transformation to total electron energy coordinates

$$\begin{aligned} \mathcal{E}_T &= \mathcal{E} - \Phi(\mathbf{r}) \\ \mathbf{r}_T &= \mathbf{r} \end{aligned} \tag{19.3.2}$$

such that for any quantity $A\left(\mathcal{E}(\mathcal{E}_T, \mathbf{r}_T), \mathbf{r}(\mathbf{r}_T)\right) = A_T(\mathcal{E}_T, \mathbf{r}_T)$, we have the transformations

$$\nabla_T A_T = \nabla A - \bar{\mathbf{E}}_a \frac{\partial A}{\partial \mathcal{E}}, \qquad \frac{\partial A_T}{\partial \mathcal{E}_T} = \frac{\partial A}{\partial \mathcal{E}} \tag{19.3.3a}$$

or

$$\nabla_T = \nabla - \bar{\mathbf{E}}_a \frac{\partial}{\partial \mathcal{E}}, \qquad \frac{\partial}{\partial \mathcal{E}_T} = \frac{\partial}{\partial \mathcal{E}} \tag{19.3.3b}$$

We regroup the terms in the kinetic equation (19.2.8) to write it in the form

$$\begin{aligned} &\nabla \cdot \left[\frac{2e}{3m}\frac{\mathcal{E}^{3/2}}{\nu_m}\left(-\nabla g_p + \bar{\mathbf{E}}\frac{\partial g_p}{\partial \mathcal{E}}\right)\right] - \bar{\mathbf{E}} \cdot \frac{\partial}{\partial \mathcal{E}}\left[\frac{2e}{3m}\frac{\mathcal{E}^{3/2}}{\nu_m}\left(-\nabla g_p + \bar{\mathbf{E}}\frac{\partial g_p}{\partial \mathcal{E}}\right)\right] \\ &\quad - \frac{\partial}{\partial \mathcal{E}}\left[\frac{2e}{3m}\frac{\mathcal{E}^{3/2}}{\nu_m}\frac{|\tilde{\mathbf{E}}|^2}{2}\frac{\nu_m^2}{\nu_m^2+\omega^2}\frac{\partial g_p}{\partial \mathcal{E}}\right] = 2\pi\left(\frac{2e}{m}\right)^{3/2}\mathcal{E}^{1/2}C_{e0}(g_p) \end{aligned} \tag{19.3.4}$$

Substituting (19.3.1) and the transformations from (19.3.3) into (19.3.4), we obtain

$$\begin{aligned} &-\nabla_T \cdot \left(\frac{2e}{3m}\frac{\mathcal{E}^{3/2}}{\nu_m}\nabla_T g_T\right) - \frac{\partial}{\partial \mathcal{E}_T}\left[\frac{2e}{3m}\frac{\mathcal{E}^{3/2}}{\nu_m}\left(|\bar{\mathbf{E}}_h|^2 + \frac{|\tilde{\mathbf{E}}|^2}{2}\frac{\nu_m^2}{\nu_m^2+\omega^2}\right)\frac{\partial g_T}{\partial \mathcal{E}_T}\right] \\ &= 2\pi\left(\frac{2e}{m}\right)^{3/2}\mathcal{E}^{1/2}C_{e0}(g_T) \end{aligned} \tag{19.3.5}$$

where $\mathcal{E} = \mathcal{E}_T + \Phi(\mathbf{r}_T)$ and $g_T(\mathbf{r}_T, \mathcal{E}_T) \equiv g_p\left(\mathbf{r}_T, \mathcal{E}_T + \Phi(\mathbf{r}_T)\right)$ is the EEPF transformed to the total energy representation, and we have regrouped terms for compactness.

Since we expect g_T to be almost spatially independent, we introduce the expansion

$$g_T(\mathbf{r}_T, \mathcal{E}_T) = g_{T0}(\mathcal{E}_T) + g_{T1}(\mathbf{r}_T, \mathcal{E}_T) \tag{19.3.6}$$

where g_{T1} is a small correction due to the energy loss collisions and heating. Inserting this into (19.3.5) and neglecting higher order terms, e.g., all terms that do not contain a spatial derivative, we obtain

$$-\nabla_T \cdot \left(\frac{2e}{3m} \frac{\mathcal{E}^{3/2}}{\nu_m} \nabla_T g_{T1} \right) - \frac{\partial}{\partial \mathcal{E}_T} \left[\frac{2e}{3m} \frac{\mathcal{E}^{3/2}}{\nu_m} \left(|\overline{\mathbf{E}}_h|^2 + \frac{|\tilde{\mathbf{E}}|^2}{2} \frac{\nu_m^2}{\nu_m^2 + \omega^2} \right) \frac{\partial g_{T0}}{\partial \mathcal{E}_T} \right] = 2\pi \left(\frac{2e}{m} \right)^{3/2} \mathcal{E}^{1/2} C_{e0}(g_{T0}) \tag{19.3.7}$$

To determine the spatially independent part g_{T0}, we spatially average this kinetic equation over that part of the discharge which is accessible for electrons with a particular total energy. The spatial average of a space and energy-dependent quantity $A(\mathbf{r}_T, \mathcal{E}_T)$ is

$$\overline{A}(\mathcal{E}_T) = \frac{1}{\mathcal{V}_0} \int_{\mathcal{V}_{ac}} A(\mathbf{r}_T, \mathcal{E}_T)\, d\mathcal{V} \tag{19.3.8}$$

Here, $\mathcal{V}_0$ is the total discharge volume and $\mathcal{V}_{ac}(\mathcal{E}_T)$ is the accessible volume defined by $\mathcal{E}_T \geq -\Phi(\mathbf{r}_T)$ for all $\mathbf{r}_T$ in $\mathcal{V}_{ac}$. This definition of $\mathcal{V}_{ac}$ is illustrated in Figure 19.4 for a cylindrical discharge in the one-dimensional radial coordinate. Integrating (19.3.7) over $\mathcal{V}_{ac}$, the first term vanishes, i.e.,

$$\int_{\mathcal{V}_{ac}} \nabla_T \cdot \left(\frac{\mathcal{E}^{3/2}}{\nu_m} \nabla_T g_{T1} \right) d\mathcal{V} = \int_{S_{ac}} \frac{\mathcal{E}^{3/2}}{\nu_m} \nabla_T g_{T1} \cdot d\mathbf{A}_{ac} = 0 \tag{19.3.9}$$

because on the boundary S_{ac} of $\mathcal{V}_{ac}$, we have that $\mathcal{E} = \mathcal{E}_T + \Phi(\mathbf{r}_T) = 0$. We thus obtain an *averaged kinetic equation*

$$-\frac{d}{d\mathcal{E}_T} \left(\frac{2e}{3m} \overline{\mathcal{E}^{1/2} D_{\mathcal{E}}} \frac{dg_{T0}}{d\mathcal{E}_T} \right) = 2\pi \left(\frac{2e}{m} \right)^{3/2} \overline{\mathcal{E}^{1/2} C_{e0}} \tag{19.3.10}$$

where

$$\overline{\mathcal{E}^{1/2} D_{\mathcal{E}}} = \frac{1}{\mathcal{V}_0} \int_{\mathcal{V}_{ac}} \frac{2e}{3m} \frac{\mathcal{E}^{3/2}}{\nu_m} \left(|\overline{\mathbf{E}}_h|^2 + \frac{|\tilde{\mathbf{E}}|^2}{2} \frac{\nu_m^2}{\nu_m^2 + \omega^2} \right) d\mathcal{V} \tag{19.3.11}$$

is the spatial average of $\mathcal{E}^{1/2}$ times an *energy diffusion coefficient* $D_{\mathcal{E}}$. To understand this interpretation of $D_{\mathcal{E}}$, we introduce the square of the total heating field E_H

$$E_H^2 = |\overline{\mathbf{E}}_h|^2 + \frac{|\tilde{\mathbf{E}}|^2}{2} \frac{\nu_m^2}{\nu_m^2 + \omega^2}$$

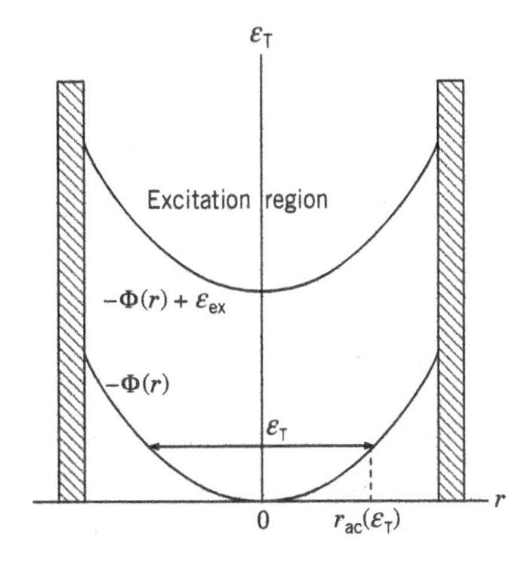

Figure 19.4 Schematic showing the definition of the accessible volume $\mathcal{V}_{ac} = \pi r_{ac}^2$ in an infinitely long cylindrical discharge.

such that $D_{\mathcal{E}} = 2e\mathcal{E}E_{\mathrm{H}}^2/3m\nu_{\mathrm{m}}$ from (19.3.11), which in turn can be written as

$$D_{\mathcal{E}} = \frac{(E_{\mathrm{H}}\lambda_{\mathrm{m}})^2\nu_{\mathrm{m}}}{3}$$

with $\lambda_{\mathrm{m}} = \upsilon/\nu_{\mathrm{m}}$. We see that $D_{\mathcal{E}}$ has the form of an energy diffusion coefficient, with $\Delta\mathcal{E} = E_{\mathrm{H}}\lambda_{\mathrm{m}}$ the random kick in energy and ν_{m}^{-1} the time between kicks.

Let us consider energies below the excitation energy $\mathcal{E}_{\mathrm{ex}}$. In this case, the collisional energy losses are assumed to be due to electron–neutral elastic scattering, $C_{\mathrm{e0}} = C_{\mathrm{el}}$, with $\mathrm{T_g} \ll \mathrm{T_e}$ and $m \ll M$, as previously. Then, analogous to (19.2.14) in Example 19.1 of Section 19.2, the right-hand side of (19.3.10) can be written as

$$2\pi\left(\frac{2e}{m}\right)^{3/2}\overline{\mathcal{E}^{1/2}C_{\mathrm{e0}}} = \frac{\mathrm{d}}{\mathrm{d}\mathcal{E}_{\mathrm{T}}}\left(\overline{\mathcal{E}^{1/2}F_{\mathcal{E}}}\,g_{\mathrm{T0}}\right) \tag{19.3.12}$$

where

$$\overline{\mathcal{E}^{1/2}F_{\mathcal{E}}} = \frac{1}{\mathcal{V}_0}\int_{\mathcal{V}_{\mathrm{ac}}}\frac{2m}{M}\mathcal{E}^{3/2}\nu_{\mathrm{m}}\,\mathrm{d}\mathcal{V} \tag{19.3.13}$$

with $F_{\mathcal{E}} = (2m/M)\nu_{\mathrm{m}}\mathcal{E}$ having the form of an energy friction coefficient. Hence, in the elastic scattering energy range, the nonlocal kinetic equation (19.3.10) can be written as

$$\frac{\mathrm{d}}{\mathrm{d}\mathcal{E}_{\mathrm{T}}}\left(\overline{\mathcal{E}^{1/2}D_{\mathcal{E}}}\frac{\mathrm{d}g_{\mathrm{T0}}}{\mathrm{d}\mathcal{E}_{\mathrm{T}}} + \overline{\mathcal{E}^{1/2}F_{\mathcal{E}}}\,g_{\mathrm{T0}}\right) = 0 \tag{19.3.14}$$

Example 19.3 Let us determine the EEPF in the positive column of a cylindrical dc glow discharge in the energy region below the excitation energy $\mathcal{E}_{\mathrm{ex}}$ from nonlocal analysis, using (19.3.14). As in Section 14.2, we assume a long thin column with a constant axial heating field E. We make the assumption that the ambipolar potential is parabolic, $\Phi(r) = -\Phi_0 r^2/R^2$, such that

$$\mathcal{E}(r) = \mathcal{E}_{\mathrm{T}} - \Phi_0 r^2/R^2 \tag{19.3.15}$$

with R the discharge radius. The accessible radius r_{ac} is determined from $\mathcal{E}(r) = 0$, which yields $r_{\mathrm{ac}} = (\mathcal{E}_{\mathrm{T}}/\Phi_0)^{1/2}R$. First evaluating $\overline{\mathcal{E}^{1/2}D_{\mathcal{E}}}$, we have from (19.3.11), assuming hard sphere collisions such that $\nu_{\mathrm{m}} = (2e\mathcal{E}/m)^{1/2}\lambda_{\mathrm{m}}^{-1}$, that

$$\overline{\mathcal{E}^{1/2}D_{\mathcal{E}}} = \frac{1}{\pi R^2}2\pi\int_0^{r_{\mathrm{ac}}} r\,\mathrm{d}r\,\frac{1}{3}\left(\frac{2e}{m}\right)^{1/2}E^2\lambda_{\mathrm{m}}\mathcal{E}(r) \tag{19.3.16}$$

Substituting $\mathcal{E}(r)$ from (19.3.15) into (19.3.16) and integrating, we obtain

$$\overline{\mathcal{E}^{1/2}D_{\mathcal{E}}} = \frac{1}{6}\left(\frac{2e}{m}\right)^{1/2}\frac{\mathcal{E}_{\mathrm{T}}^2}{\Phi_0}E^2\lambda_{\mathrm{m}} \tag{19.3.17}$$

Next evaluating $\overline{\mathcal{E}^{1/2}F_{\mathcal{E}}}$, we have from (19.3.13)

$$\overline{\mathcal{E}^{1/2}F_{\mathcal{E}}} = \frac{1}{\pi R^2}2\pi\int_0^{r_{\mathrm{ac}}} r\,\mathrm{d}r\,\frac{2m}{M}\left(\frac{2e}{m}\right)^{1/2}\frac{\mathcal{E}^2(r)}{\lambda_{\mathrm{m}}} \tag{19.3.18}$$

Substituting for $\mathcal{E}(r)$ and integrating, we obtain

$$\overline{\mathcal{E}^{1/2}F_{\mathcal{E}}} = \frac{1}{3}\frac{2m}{M}\left(\frac{2e}{m}\right)^{1/2}\frac{\mathcal{E}_{\mathrm{T}}^3}{\Phi_0}\frac{1}{\lambda_{\mathrm{m}}} \tag{19.3.19}$$

Inserting (19.3.17) and (19.3.19) into the nonlocal kinetic equation (19.3.14), we have

$$\frac{\mathrm{d}}{\mathrm{d}\mathcal{E}_{\mathrm{T}}}\left[\frac{1}{2}\frac{\mathcal{E}_{\mathrm{T}}^2}{\Phi_0}E^2\lambda_{\mathrm{m}}\frac{\mathrm{d}g_{\mathrm{T0}}}{\mathrm{d}\mathcal{E}_{\mathrm{T}}} + \frac{2m}{M}\frac{\mathcal{E}_{\mathrm{T}}^3}{\Phi_0}\frac{1}{\lambda_{\mathrm{m}}}g_{\mathrm{T0}}\right] = 0 \tag{19.3.20}$$

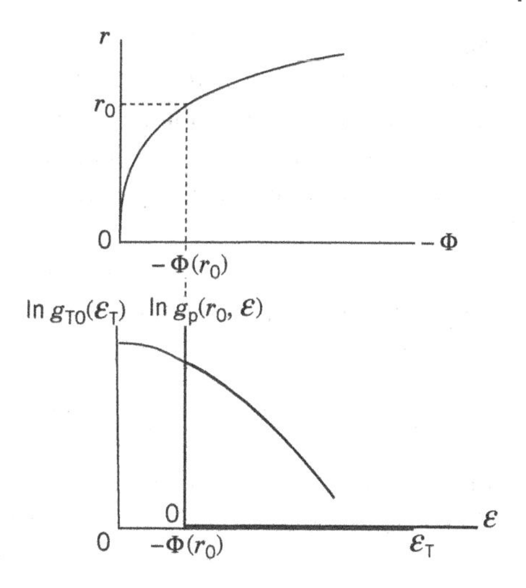

Figure 19.5 Illustrating the transformation of the EEPF g_{T0}, a function of the total energy $\mathcal{E}_T$, to the EEPF g_p, a function of the kinetic energy $\mathcal{E}$.

which integrated once with respect to $\mathcal{E}_T$ yields

$$\frac{dg_{T0}}{d\mathcal{E}_T} = -\frac{4m}{M}\frac{\mathcal{E}_T}{E^2\lambda_m^2}g_{T0} \tag{19.3.21}$$

Integrating again with respect to $\mathcal{E}_T$, we obtain

$$g_{T0} = g_{00}\exp\left(-\frac{2m}{M}\frac{\mathcal{E}_T^2}{E^2\lambda_m^2}\right) \tag{19.3.22}$$

Transforming back to $(r, \mathcal{E})$ coordinates,

$$g_p(\mathbf{r}, \mathcal{E}) = g_{T0}(\mathcal{E} - \Phi(\mathbf{r})) \tag{19.3.23}$$

we find the spatially varying EEPF

$$g_{p0}(r, \mathcal{E}) = g_{00}\exp\left[-\frac{2m}{M}\frac{1}{E^2\lambda_m^2}\left(\mathcal{E} + \Phi_0\frac{r^2}{R^2}\right)^2\right] \tag{19.3.24}$$

In Figure 19.3, we compare (19.3.24) (solid lines) with the result (19.2.16) (dashed line) determined by local kinetics. Both distribution functions have a Druyvesteyn energy distribution, but the coefficient in the nonlocal result is smaller than the corresponding coefficient in the local result. Hence, the nonlocal on-axis ($r = 0$) EEPF falls off with energy less steeply than the local EEPF given in (19.2.16). This relative enhancement of the tail is a characteristic feature of the effect of the space charge field on the EEPF.

Equation (19.3.23) is a *generalized Boltzmann relation*. As shown in Figure 19.5, at position r_0, the value of $-\Phi(r_0)$ is the minimum total energy needed by electrons to reach r_0. Hence, only electrons with $\mathcal{E}_T > -\Phi(r_0)$ form the distribution function $g_p(r_0, \mathcal{E})$. Therefore, g_p is just g_{T0} with the energies below $-\Phi(r_0)$ cut away. For the special case of a Maxwellian distribution $g_{T0} = A\,e^{-\mathcal{E}_T/T_e}$, (19.3.23) yields $g_p = A\,e^{-\mathcal{E}/T_e}\,e^{\Phi(r)/T_e}$. Integrating this g_p over energy yields the Boltzmann relation (2.4.16).

19.4 Quasilinear Diffusion and Stochastic Heating

Electron heating by time-varying fields is fundamental to rf and microwave discharges. In a *uniform* oscillating field $\tilde{\mathbf{E}}(t) = \mathrm{Re}\,\tilde{\mathbf{E}}_0\,e^{j\omega t}$, a single electron has a coherent velocity of motion that lags the phase of the electric field force $-e\tilde{\mathbf{E}}$ by 90°. Hence, the time-averaged power transferred from the field to the electron is zero. Electron collisions with other particles destroy the phase coherence of the motion, leading to a net transfer of power. For an ensemble of n electrons per unit volume, it is usual (see Section 4.2) to introduce the macroscopic current density $\tilde{\mathbf{J}} = -en\tilde{\mathbf{u}}$, with $\tilde{\mathbf{u}}$ the macroscopic electron velocity, and to relate the amplitudes of $\tilde{\mathbf{J}}$ and $\tilde{\mathbf{E}}$ through a local conductivity: $\tilde{\mathbf{J}}_0 = \sigma_\mathrm{p}\tilde{\mathbf{E}}_0$, where $\sigma_\mathrm{p} = e^2n/m(\nu_\mathrm{m} + j\omega)$ given by (4.2.20) is the plasma conductivity and ν_m is the electron collision frequency for momentum transfer. In this "fluid" approach, the average electron velocity $\tilde{\mathbf{u}}$ still oscillates coherently but lags the electric field by less than 90°, leading to an ohmic power transfer per unit volume

$$p_\mathrm{ohm} = \frac{1}{2}\mathrm{Re}\,\tilde{\mathbf{J}}_0 \cdot \tilde{\mathbf{E}}_0^* = \frac{1}{2}|\tilde{\mathbf{E}}_0|^2\,\mathrm{Re}\,(\sigma_\mathrm{p}) = \frac{1}{2}|\tilde{\mathbf{J}}_0|^2\,\mathrm{Re}\,(\sigma_\mathrm{p}^{-1})$$

Although the average velocity is coherent with the field, the fundamental mechanism that converts electric field energy to thermal energy is the breaking of the phase-coherent motion of individual electrons by collisions: the total force (electric field force plus that due to collisions) acting on an individual electron becomes spatially nonuniform and nonperiodic in time.

These observations suggest that a spatially *nonuniform* electric field by itself might lead to electron heating, even in the absence of interparticle collisions, provided that the electrons have thermal velocities sufficient to sample the field inhomogeneity. This phenomenon has been well-known in plasma physics since Landau (1946) demonstrated the collisionless damping of an electrostatic wave in a warm plasma and is variously referred to in the discharge literature as *collisionless, stochastic,* or *anomalous* heating. Such heating can be a basic feature of warm plasmas having space dispersion. The electron response ($\tilde{\mathbf{J}}$) at some point in the plasma is defined not only by the field ($\tilde{\mathbf{E}}$) at that point but also by an integrated effect over the neighboring space. Due to the spatial variation, the time-varying field seen by an individual "thermal" electron is nonperiodic. The electron can lose phase coherence with the field (which is strictly periodic), resulting in stochastic interaction with the field and collisionless heating.

In almost all discharges, the spatial variation of the time-varying field is strongly nonuniform, with a low field in the bulk of the plasma and one or more highly localized field regions (rf sheath, skin depth layer, etc.), usually near the plasma boundaries. An electron, being confined for many bounce times by the dc ambipolar and boundary sheath potentials in the discharge, interacts repeatedly with the high field regions, but interacts only weakly during its drift through the plasma bulk. This suggests the use of a dynamical "kick" model to investigate the energy transfer in which a ball bounces back and forth between a fixed and an oscillating wall. We have done this in Sections 11.1 and 11.2 to determine the stochastic heating of electrons in a capacitive discharge and will apply the result in Section 19.6 to determine the stochastic heating for a kinetic model of a capacitive discharge. We have also used the kick approximation to calculate electron cyclotron heating in Section 13.1. Here, we introduce the general kinetic approach to treat collisionless heating in discharges, known as *quasilinear theory*. Originally developed to describe the interaction of electrons with weak wave turbulence in hot plasmas (see, e.g., Nicholson, 1983, Chapter 10), the quasilinear theory has been applied to low-pressure rf and microwave discharges, which have sharply localized time-varying field regions excited by an external source. In this section, we use this theory to determine the collisionless heating in an inductive discharge.

In bounded discharges at very low pressures, electrons can bounce repeatedly back and forth as they interact with the heating fields, with collisional processes randomizing the localized interactions. Furthermore, the dynamics itself can cause phase randomization in the absence of collisions. A classification of the various regimes of heating and phase randomization, including the effects of collisions has been given by Kaganovich et al. (1996). A review of the kick model and its application to collisionless heating in rf and microwave discharges has been given by Lieberman and Godyak (1998).

19.4.1 Quasilinear Diffusion Coefficient

To incorporate stochastic heating into the kinetic theory of discharges, we must determine the energy diffusion coefficient using quasilinear theory. We follow the treatment given in Aliev et al. (1997). We assume a low-pressure discharge with a spatial scale length δ_{rf} for the rf heating fields, localized around a position z_0 within the discharge that is much smaller than the discharge scale length Λ or the electron mean free path λ_m. Under these conditions, it is possible to separate the fields and distribution function into two parts, having small and large spatial scales. The Boltzmann equation (19.1.1) for electrons, including the magnetic force, is

$$\frac{\partial f_e}{\partial t} + \mathbf{v} \cdot \nabla f_e - \frac{e}{m}(\mathbf{E} + \mathbf{v} \times \mathbf{B}) \cdot \nabla_v f_e = C_e(f_e) \tag{19.4.1}$$

where $C_e = \partial f_e / \partial t|_c$ is the electron collision term. We introduce the separation of space scales

$$f_e = \langle f_e \rangle + \tilde{f}_e, \qquad \mathbf{E} = \langle \mathbf{E} \rangle + \tilde{\mathbf{E}}, \qquad \mathbf{B} = \langle \mathbf{B} \rangle + \tilde{\mathbf{B}} \tag{19.4.2}$$

where $\langle\rangle$ denotes averaging over a scale length large compared to the characteristic size δ_{rf} of the heating layer, but small compared to either Λ or λ_m (whichever is smaller). We also make the weak field (quasilinear) assumption $\tilde{f}_e \ll \langle f_e \rangle$. Then, (19.4.1) can be separated into two parts

$$\frac{\partial \tilde{f}_e}{\partial t} + \mathbf{v} \cdot \nabla \tilde{f}_e - \frac{e}{m}(\langle \mathbf{E} \rangle + \mathbf{v} \times \langle \mathbf{B} \rangle) \cdot \nabla_v \tilde{f}_e = \frac{e}{m}(\tilde{\mathbf{E}} + \mathbf{v} \times \tilde{\mathbf{B}}) \cdot \nabla_v \langle f_e \rangle \tag{19.4.3}$$

$$\frac{\partial \langle f_e \rangle}{\partial t} + \mathbf{v} \cdot \nabla \langle f_e \rangle - \frac{e}{m}(\langle \mathbf{E} \rangle + \mathbf{v} \times \langle \mathbf{B} \rangle) \cdot \nabla_v \langle f_e \rangle = C_{QL} + C_e \tag{19.4.4}$$

where the quasilinear diffusion term is

$$C_{QL} = \frac{e}{m} \left\langle (\tilde{\mathbf{E}} + \mathbf{v} \times \tilde{\mathbf{B}}) \cdot \nabla_v \tilde{f}_e \right\rangle \tag{19.4.5}$$

In (19.4.3), the collision term C_e (of order v/λ_m) is omitted since it is small compared to the second term on the left-hand side (of order v/δ_{rf}). The quasilinear term C_{QL} describes the space-averaged interaction of the electrons with the small-scale rf fields.

To determine C_{QL}, we consider the simplest case of a homogeneous unbounded nonmagnetized plasma, with the rf fields concentrated near $z = z_0$. This corresponds to neglecting the ambipolar electric field, such that $\langle \mathbf{E} \rangle, \langle \mathbf{B} \rangle = 0$, and $\tilde{\mathbf{E}}, \tilde{\mathbf{B}}$ represent only the rf fields. We assume a sinusoidal steady state

$$\tilde{\mathbf{E}}(\mathbf{r}, t) = \mathrm{Re}\, \tilde{\mathbf{E}}(\mathbf{r})\, e^{j\omega t}, \qquad \tilde{\mathbf{B}}(\mathbf{r}, t) = \mathrm{Re}\, \tilde{\mathbf{B}}(\mathbf{r})\, e^{j\omega t}, \qquad \tilde{f}_e(\mathbf{r}, t) = \mathrm{Re} \tilde{f}_e(\mathbf{r})\, e^{j\omega t} \tag{19.4.6}$$

Then (19.4.3) reduces to

$$j\omega \tilde{f}_e + \mathbf{v} \cdot \nabla \tilde{f}_e = \frac{e}{m}(\tilde{\mathbf{E}} + \mathbf{v} \times \tilde{\mathbf{B}}) \cdot \nabla_v \langle f_e \rangle \tag{19.4.7}$$

Without loss of generality, we can assume that the spatial variation lies along the z-direction. Then to solve (19.4.7), we introduce the spatial Fourier transform along z for $\tilde{\mathbf{E}}$, $\tilde{\mathbf{B}}$, and $\tilde{f}_e$, e.g., for the electric field

$$\mathbf{E}_\mathbf{k} = \frac{1}{2\pi} \int dz\, \tilde{\mathbf{E}}(\mathbf{r})\, e^{j\mathbf{k}\cdot\mathbf{r}} \tag{19.4.8}$$

with $\mathbf{k} = \hat{z}k$ and the inverse transform

$$\tilde{\mathbf{E}}(\mathbf{r}) = \int \mathrm{d}k\, \mathbf{E}_{\mathbf{k}}\, \mathrm{e}^{-j\mathbf{k}\cdot\mathbf{r}} \tag{19.4.9}$$

Substituting (19.4.9) into (19.4.7), we obtain

$$\tilde{f}_{\mathrm{ek}} = -\frac{je}{m}\frac{\mathbf{E}_{\mathbf{k}} + \mathbf{v}\times\mathbf{B}_{\mathbf{k}}}{\omega - \mathbf{k}\cdot\mathbf{v}} \cdot \nabla_{\mathbf{v}}\langle f_{\mathrm{e}}\rangle \tag{19.4.10}$$

Inserting (19.4.10) into (19.4.5), we can evaluate the spatial average $\langle\rangle$ using the relation (for scalar functions $\tilde{A}_1$ and $\tilde{A}_2$)

$$\int \mathrm{d}z \tilde{A}_1(\mathbf{r})\tilde{A}_2^*(\mathbf{r}) = 2\pi \int \mathrm{d}k\, A_{1\mathbf{k}}A_{2\mathbf{k}}^* \tag{19.4.11}$$

The result for C_{QL} is

$$C_{\mathrm{QL}} = \delta(z - z_0)\frac{\partial}{\partial v_i}\left(D_{ij}\frac{\partial\langle f_{\mathrm{e}}\rangle}{\partial v_j}\right) \tag{19.4.12}$$

where subscripts i and j are indices that range over x, y, and z, and the delta function is a result of the space averaging and reflects the localization of the heating electric field to the region $z \approx z_0$. The quasilinear diffusion tensor is

$$D_{ij}(\mathbf{v}) = \frac{e^2\pi}{m^2}\int \mathrm{d}k\,(\mathbf{E}_{\mathbf{k}} + \mathbf{v}\times\mathbf{B}_{\mathbf{k}})_i(\mathbf{E}_{\mathbf{k}} + \mathbf{v}\times\mathbf{B}_{\mathbf{k}})_j^*\, \mathrm{Im}\left(\frac{1}{\omega - kv_z - j\nu}\right) \tag{19.4.13}$$

where the limit $\nu \to 0$ is to be taken. Using the standard relation

$$\lim_{\nu\to 0} \mathrm{Im}\left(\frac{1}{\omega - kv_z - j\nu}\right) = \pi\delta(\omega - kv_z) \tag{19.4.14}$$

corresponding to resonance between the rf field component having phase velocity ω/k and electrons moving with velocity v_z, D_{ij} can be written as

$$D_{ij}(\mathbf{v}) = \frac{e^2\pi^2}{m^2}\int \mathrm{d}k\,(\mathbf{E}_{\mathbf{k}} + \mathbf{v}\times\mathbf{B}_{\mathbf{k}})_i(\mathbf{E}_{\mathbf{k}} + \mathbf{v}\times\mathbf{B}_{\mathbf{k}})_j^*\, \delta(\omega - kv_z) \tag{19.4.15}$$

The rf electric and magnetic fields are related by Maxwell's equations (2.2.1):

$$\omega\mathbf{B}_{\mathbf{k}} = \mathbf{k}\times\mathbf{E}_{\mathbf{k}} \tag{19.4.16}$$

Using this, we transform the Lorentz force to obtain

$$\mathbf{E}_{\mathbf{k}} + \mathbf{v}\times\mathbf{B}_{\mathbf{k}} = (\mathbf{v}\cdot\mathbf{E}_{\mathbf{k}})\mathbf{k}/\omega - \mathbf{E}_{\mathbf{k}}(1 - \mathbf{k}\cdot\mathbf{v}/\omega) \tag{19.4.17}$$

Substituting this into (19.4.15), we note that the second term in (19.4.17) evaluates to zero and $\mathbf{k}/\omega$ evaluates to v_z^{-1} when the integration over the delta function is performed. Integrating

$$D_{zz}(\mathbf{v}) = \frac{e^2\pi^2}{m^2v_z^2}\int \mathrm{d}k\,|\mathbf{v}\cdot\mathbf{E}_{\mathbf{k}}|^2\, \delta(\omega - kv_z) \tag{19.4.18}$$

over the δ-function yields the final result

$$D_{zz} = \frac{e^2\pi^2}{m^2v_z^2|v_z|}|\mathbf{v}\cdot\mathbf{E}_{\mathbf{k}}(k = \omega/v_z)|^2 \tag{19.4.19}$$

All other components of the diffusion tensor vanish. This implies that the electrons receive a kick from the wave field directed along $\mathbf{k} = \hat{z}k$, independent of the direction of the electric field, even if the field is perpendicular to $\mathbf{k}$. As shown by Cohen and Rognlien (1996a,b), this is due to the presence of the rf magnetic force $-e\mathbf{v} \times \tilde{\mathbf{B}}$. For example, with $\tilde{\mathbf{E}} = \hat{y}\tilde{E}_y$ and $\tilde{\mathbf{B}} = -\hat{x}\tilde{B}_x$, the v_zs of electrons entering the heating zone are rotated into the y-direction by the magnetic force and kicked by the $\tilde{E}_y$-field. Upon leaving the zone, the v_y-kicks are rotated back into the z-direction. Although the rf magnetic field turns the direction of the kick, it does not change the energy transfer, because a magnetic field does no work on moving charges.

That the final kick is along the direction of the field variation can be seen from general dynamical considerations. The canonical momentum $P_y = mv_y - e\tilde{A}_y$ must be conserved during the electron motion into and out of the heating layer, because the Hamiltonian H is independent of y. Here, $\tilde{A}_y = j\tilde{E}_y/\omega$ is the vector potential for the rf fields. Hence, in the region outside the heating layer, where $\tilde{A}_y \approx 0$, we find that $v_y(t \to \infty) = v_y(t \to -\infty)$; i.e., there is no velocity kick along the y-direction.

19.4.2 Stochastic Heating

The time-averaged stochastic power flux $\overline{S}_{\text{stoc}}$ is found from the kinetic equation (19.4.4) for $\langle f_e \rangle$. Substituting (19.4.12) into this equation, assuming no spatial variation for $\langle f_e \rangle$, and dropping the electron collision term, we obtain

$$\left.\frac{\partial \langle f_e \rangle}{\partial t}\right|_{\text{QL}} = \delta(z - z_0)\frac{\partial}{\partial v_z}\left(D_{zz}\frac{\partial \langle f_e \rangle}{\partial v_z}\right) \tag{19.4.20}$$

Multiplying this by $\frac{1}{2}mv^2$ and integrating over z and over all velocities with $v_z \geq 0$, we obtain

$$\overline{S}_{\text{stoc}} = \frac{1}{2}m \int_{-\infty}^{\infty} dv_x \int_{-\infty}^{\infty} dv_y \int_0^{\infty} dv_z\,(v_x^2 + v_y^2 + v_z^2)\frac{\partial}{\partial v_z}\left(D_{zz}\frac{\partial \langle f_e \rangle}{\partial v_z}\right) \tag{19.4.21}$$

Integrating by parts with respect to v_z, the integrals over v_x^2 and v_y^2 vanish, yielding

$$\overline{S}_{\text{stoc}} = -m \int_{-\infty}^{\infty} dv_x \int_{-\infty}^{\infty} dv_y \int_0^{\infty} dv_z\, v_z D_{zz}\frac{\partial \langle f_e \rangle}{\partial v_z} \tag{19.4.22}$$

We will evaluate this for inductive heating in Section 19.5.

19.4.3 Relation to Velocity Kick Models

We now show that D_{zz} has the form of a velocity space diffusion coefficient $\propto \overline{(\Delta v_z)^2}/\tau$, integrated over the discharge length. This is most easily seen for a longitudinal field $\text{Re}\,E_z(z)\,e^{j\omega t + j\phi}$, where the rf magnetic force is absent. Assuming that the field is localized at one end of a discharge of length l and that $\delta_{\text{res}} \ll \lambda_{\text{m}} \ll l$, then the kick in velocity for an electron passing through the heating zone is found by integrating the acceleration $-eE_z/m$ over time

$$\Delta v_z = -\text{Re}\int_{-T_1}^{T_1} dt\,\frac{eE_z(z(t))}{m}\,e^{j\omega t + j\phi} \tag{19.4.23}$$

where the unperturbed motion is $z = v_z t$ and we choose T_1 such that $\delta_{\text{res}}/v_z \ll T_1 \ll \lambda_{\text{m}}/v_z$. Transforming the variable of integration from t to z and extending the integration limits to $\pm\infty$, we find

$$\Delta v_z = -\text{Re}\frac{e}{mv_z}\int_{-\infty}^{\infty} dz\,E_z(z)\,e^{j(\omega/v_z)z + j\phi} \tag{19.4.24}$$

The integral yields the Fourier transform $\mathbf{E}_k$ given by (19.4.8)

$$\Delta v_z = -\frac{2\pi e}{mv_z}\,\text{Re}\,E_k(k = \omega/v_z)\,e^{j\phi} \tag{19.4.25}$$

Squaring Δv_z and averaging over ϕ yields

$$\overline{(\Delta v_z)^2} = \frac{2\pi^2 e^2}{m^2 v_z^2}\left|E_k\left(k = \frac{\omega}{v_z}\right)\right|^2 \tag{19.4.26}$$

For a discharge length l, the time between collisions is $\tau = 2l/|v_z|$. Substituting for τ and $\overline{(\Delta v_z)^2}$ in (19.4.19), we find that D_{zz} can be expressed as

$$D_{zz} = \frac{\overline{(\Delta v_z)^2}}{\tau} \cdot l \tag{19.4.27}$$

19.4.4 Two-Term Kinetic Equations

If the electron elastic collision frequency ν_{el} is large compared to the characteristic frequencies for heating and collisional energy loss, then the two-term approximation (19.1.4) can be used. To determine the quasilinear diffusion coefficient for the spherically symmetric part f_{e0} of the distribution, we transform (19.4.12) to spherical velocity coordinates and average over solid angle. Putting $\langle f_{\mathrm{e}}\rangle = f_{\mathrm{e0}}$ in (19.4.12) and noting that $\partial f_{\mathrm{e0}}/\partial v_z = (v_z/v)(\mathrm{d}f_{\mathrm{e0}}/\mathrm{d}v) = \cos\psi(\mathrm{d}f_{\mathrm{e0}}/\mathrm{d}v)$, the (negative of the) velocity space flux is

$$\begin{aligned} -\Gamma_{\mathrm{v}} &= \hat{z} D_{zz}\frac{\partial f_{\mathrm{e0}}}{\partial v_z} \\ &= (\hat{v}\cos\psi - \hat{\psi}\sin\psi)\frac{e^2\pi^2}{m^2 v_z^2}\int \mathrm{d}k\,|\mathbf{v}\cdot\mathbf{E}_{\mathbf{k}}|^2\,\delta(\omega - k v_z)\cos\psi\frac{\mathrm{d}f_{\mathrm{e0}}}{\mathrm{d}v} \end{aligned} \tag{19.4.28}$$

where we have substituted for D_{zz} from (19.4.18). Taking the divergence of this flux in spherical velocity coordinates and averaging over solid angle, we obtain

$$C_{\mathrm{QL0}} = \frac{1}{v^2}\frac{\mathrm{d}}{\mathrm{d}v}\left(v^2 D_v \frac{\mathrm{d}f_{\mathrm{e0}}}{\mathrm{d}v}\right)\delta(z - z_0) \tag{19.4.29}$$

where

$$D_v = \frac{1}{v^2}\frac{e^2\pi^2}{m^2}\frac{1}{4\pi}\int \mathrm{d}k\int_0^{\pi}\sin\psi\,\mathrm{d}\psi\int_0^{2\pi}\mathrm{d}\phi\,|\mathbf{v}\cdot\mathbf{E}_{\mathbf{k}}|^2\,\delta(\omega - kv\cos\psi) \tag{19.4.30}$$

is the angle-averaged velocity diffusion coefficient.

19.4.5 Energy Relaxation Length

Let us consider the two-term kinetic equations (19.1.11) and (19.1.14) for a plasma half-space $z \geq 0$, with the heating zone centered at $z = 0$. In the steady state and with no macroscopic E-field, the equations are

$$v\frac{\partial f_{\mathrm{e0}}}{\partial z} = -\nu_{\mathrm{m}} f_{\mathrm{e1}} \tag{19.4.31}$$

$$\frac{v}{3}\frac{\partial f_{\mathrm{e1}}}{\partial z} = C_{\mathrm{e0}} + C_{\mathrm{QL0}} \tag{19.4.32}$$

Eliminating f_{e1} from these equations yields

$$-\frac{v^2}{3\nu_{\mathrm{m}}}\frac{\partial^2 f_{\mathrm{e0}}}{\partial z^2} = C_{\mathrm{e0}} + \delta(z)\frac{1}{v^2}\frac{\partial}{\partial v}\left(v^2 D_v \frac{\partial f_{\mathrm{e0}}}{\partial v}\right) \tag{19.4.33}$$

Then, for $z > 0$, f_{e0} evolves under the action of the electron–neutral collisions alone

$$-\frac{v^2}{3\nu_{\mathrm{m}}}\frac{\partial^2 f_{\mathrm{e0}}}{\partial z^2} = C_{\mathrm{e0}}, \qquad z > 0 \tag{19.4.34}$$

Integrating (19.4.33) over a small region Δz around $z = 0$, we can drop $C_{e0}\Delta z$ to obtain

$$-\frac{v^2}{3\nu_m}\frac{\partial f_{e0}}{\partial z}\bigg|_{z=0} = \frac{1}{2v^2}\frac{\partial}{\partial v}\left(v^2 D_v \frac{\partial f_{e0}}{\partial v}\right)_{z=0} \tag{19.4.35}$$

Equation (19.4.35) gives the boundary condition at $z = 0$ for the solution of (19.4.34) in the half-space $z > 0$.

Introducing an approximate form for the electron–neutral collision term

$$C_{e0} \approx -\nu_{inel} f_{e0} \tag{19.4.36}$$

then (19.4.34) can be integrated to obtain the spatial decay of the electron distribution

$$f_{e0} = F(v)\,e^{-z/\lambda_\mathcal{E}} \tag{19.4.37}$$

where

$$\lambda_\mathcal{E}(v) = \frac{v}{\sqrt{3\nu_m \nu_{inel}}} \tag{19.4.38}$$

is the *energy relaxation length.* Inserting (19.4.37) into (19.4.35) yields the equation for $F(v)$.

For a finite length discharge, with $l \ll \lambda_\mathcal{E}$, the spatial decay over the discharge length is weak and we can *bounce average* (19.4.33) to obtain a spatially independent kinetic equation. Averaging over the discharge length ($0 \le z \le l$), we obtain

$$C_{e0} + \frac{1}{v^2}\frac{d}{dv}\left(v^2 \frac{D_v}{l}\frac{df_{e0}}{dv}\right) = 0 \tag{19.4.39}$$

The bounce-averaged equation is commonly used to determine the energy distribution, e.g., as described in Section 19.6 for modeling capacitive discharges. However, at low pressures, an electron can interact coherently with the heating zone many times, leading to the phenomena of *bounce resonances,* neglected in the preceding analysis (see Aliev et al., 1997). The effects of the dc ambipolar field have also been neglected. The ambipolar field can trap low-energy electrons, confining them to the discharge center where they do not interact with the stochastic heating field. This can lead to two-temperature Maxwellian distributions (see Figures 11.9 and 19.10 for some examples from experiments and modeling).

In next section, we determine the stochastic heating and quasilinear diffusion for an inductive discharge heating field. Another application in the literature has been to surface wave discharges (Aliev et al., 1992). Quasilinear theory has been applied with some approximations to determine the heating in capacitive discharges (Aliev et al., 1997; Kaganovich, 2002). However, due to the strong fields in the sheaths of such discharges, the quasilinear ordering $f_{e1} \ll f_{e0}$ can break down. At very low pressures, electrons can be trapped in the rf wave field itself. If $\nu_m \lesssim \tilde{\nu}_b$, the bounce frequency of electrons in the rf wave field, then the quasilinear diffusion is reduced (Kaganovich et al., 2004).

19.5 Energy Diffusion in a Skin Depth Layer

19.5.1 Stochastic Heating

The theory of anomalous collisions in a thin transverse electric field layer, originally developed by Pippard (1949) to describe the high frequency skin resistance of metals at low temperatures, can be used to determine the collisionless heating within the skin depth layer in a low-pressure inductive discharge (Weibel, 1967; Turner, 1993; Godyak et al., 1993). The heating and energy diffusion can be found using the quasilinear analysis of the previous section. For our simplified treatment, we first assume that the electron distribution is Maxwellian and that the transverse electric field decays exponentially with distance z from the edge into the slab

$$\tilde{E}_y(z,t) = \tilde{E}_0 \, \mathrm{e}^{-|z|/\delta} \cos(\omega t + \phi) \tag{19.5.1}$$

with δ a constant skin depth. Substituting (19.5.1) into (19.4.8), we obtain the Fourier spectrum

$$E_k = \frac{E_0 \delta}{\pi(1 + k^2\delta^2)} \tag{19.5.2}$$

We use (19.4.18) to evaluate the diffusion tensor

$$\begin{aligned} D_{zz} &= \frac{e^2\pi^2 v_y^2}{m^2 v_z^2} \int \mathrm{d}k \, \frac{E_0^2\delta^2}{\pi^2(1+k^2\delta^2)^2} \, \delta(\omega - k v_z) \\ &= \frac{e^2 E_0^2 \delta^2}{m^2} \frac{v_y^2 |v_z|}{(v_z^2 + \omega^2\delta^2)^2} \end{aligned} \tag{19.5.3}$$

with $\delta(\omega - kv_z)$ the delta function. Substituting (19.5.3) into (19.4.22) and evaluating for a Maxwellian distribution (19.1.2), the v_x and v_y integrals can be done easily to obtain

$$\overline{S}_{\mathrm{stoc}} = n_{\mathrm{s}} \frac{e^2 E_0^2 \delta^2}{m} \int_0^\infty \mathrm{d}v_z \, v_z \frac{v_z^2}{(v_z^2 + \omega^2\delta^2)^2} f_{\mathrm{ez}} \tag{19.5.4}$$

with $f_{\mathrm{ez}} = (m/2\pi e\mathrm{T}_{\mathrm{e}})^{1/2} \, \mathrm{e}^{-m v_z^2/2e\mathrm{T}_{\mathrm{e}}}$. Substituting $\zeta = m v_z^2/2e\mathrm{T}_{\mathrm{e}}$ to evaluate the v_z integral, we obtain

$$\overline{S}_{\mathrm{stoc}} = \frac{m n_{\mathrm{s}}}{\overline{v}_{\mathrm{e}}} \left(\frac{e\tilde{E}_0 \delta}{m} \right)^2 \mathcal{I} \tag{19.5.5}$$

where

$$\mathcal{I}(\alpha) = \frac{1}{\pi} \int_0^\infty \mathrm{d}\zeta \, \mathrm{e}^{-\zeta} \frac{\zeta}{(\zeta + \alpha)^2} \tag{19.5.6}$$

with

$$\alpha = \frac{4\omega^2\delta^2}{\pi \overline{v}_{\mathrm{e}}^2} \tag{19.5.7}$$

proportional to the square of the ratio of sheath transit time to rf period. Let $\zeta' = \zeta + \alpha$, then $\mathcal{I}$ can be expressed as

$$\mathcal{I} = \frac{1}{\pi} \left[\mathrm{e}^{\alpha}(1 + \alpha)\mathrm{E}_1(\alpha) - 1 \right] \tag{19.5.8}$$

where the exponential integral

$$\mathrm{E}_1(\alpha) = \int_\alpha^\infty \mathrm{d}\zeta \, \frac{\mathrm{e}^{-\zeta}}{\zeta} \tag{19.5.9}$$

is tabulated in the literature and has the limits

$$\begin{aligned} \mathcal{I}(\alpha) &\approx -\tfrac{1}{\pi}(\ln\alpha + 1.58), & \alpha \ll 1 \\ &\approx \frac{1}{\pi\alpha^2}, & \alpha \gg 1 \end{aligned} \tag{19.5.10}$$

A graph of $\mathcal{I}$ versus α is given as the solid line in Figure 19.6. The nonlocal analyses (Weibel, 1967; Turner, 1993; Aliev et al., 1997) yield a non-exponential decay of the electric field and, consequently, corrections to the basic result (19.5.5), but these have been shown to be small (see Vahedi et al., 1995).

19.5.2 Effective Collision Frequency

Following Vahedi et al. (1995), we introduce an effective collision frequency ν_{stoc} by equating the stochastic heating (19.5.5) to an effective collisional heating power flux, defined in the following reasonable way

$$\begin{aligned}\overline{S}_{\text{stoc}} &= \frac{1}{2}\int_0^\infty dz\left(\tilde{E}_0 e^{-z/\delta}\right)^2 \frac{e^2 n_s}{m}\frac{\nu_{\text{stoc}}}{\nu_{\text{stoc}}^2+\omega^2} \\ &= \frac{1}{4}\frac{e^2 n_s \delta}{m}\frac{\nu_{\text{stoc}}}{\nu_{\text{stoc}}^2+\omega^2}\tilde{E}_0^2\end{aligned} \tag{19.5.11}$$

The resulting quadratic equation has two positive roots, one with $\nu_{\text{stoc}} > \omega$, and one with $\nu_{\text{stoc}} < \omega$. The choice is made on physical grounds. In the anomalous (nonlocal) regime $\alpha \ll 1$, the larger root is chosen, yielding

$$\nu_{\text{stoc}} \approx \frac{C_e \overline{v}_e}{4\delta} \tag{19.5.12}$$

with $C_e \approx 1/\mathcal{I}(\alpha)$. In the collisional (local) regime $\alpha \gg 1$, the smaller root is chosen

$$\nu_{\text{stoc}} \approx \frac{\overline{v}_e}{\delta\alpha} \tag{19.5.13}$$

A reasonable ansatz joining these two solutions is

$$\nu_{\text{stoc}} \approx \frac{\overline{v}_e}{4\delta}\left(\frac{1}{\mathcal{I}(\alpha)+\alpha/4}\right) \tag{19.5.14}$$

which is shown as the dashed line in Figure 19.6.

Electron–neutral collisions have been incorporated by Vahedi et al. (1995) using a fluid analysis, yielding an effective collision frequency that is approximately the sum of ohmic and stochastic collision frequencies, as follows from the form used in (19.5.11)

$$\nu_{\text{eff}} \approx \nu_m + \nu_{\text{stoc}} \tag{19.5.15}$$

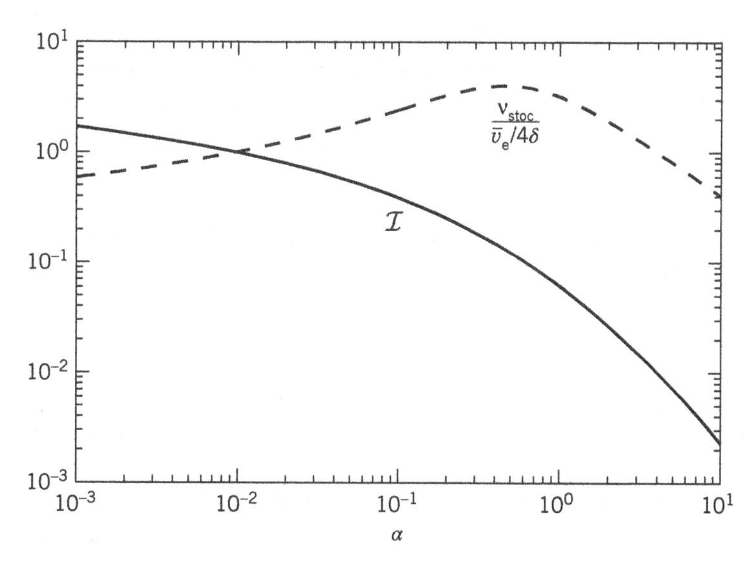

Figure 19.6 $\mathcal{I}$ and ν_{stoc} (normalized to $\overline{v}_e/4\delta$) versus α, where ν_{stoc} is defined in the following subsection.

Kinetic analyses (Tyshetskiy et al., 2002) yield additional collisional effects, including differences at low frequencies between plane and cylindrical geometry solutions, and possible "collisional cooling" at low frequencies and low collisionalities.

19.5.3 Energy Distribution

When calculating the stochastic heating (19.5.5), a Maxwellian distribution was assumed. At high densities, electron–electron collisions do tend to drive the distribution toward a Maxwellian, as is observed experimentally (see Section 19.6). Nevertheless, the actual form of the distribution is determined by a balance of all energy-dependent collisional processes, including quasilinear diffusion. To illustrate this, we apply the two-term kinetic equations (19.4.31) and (19.4.32) for an inductive heating zone at $z = 0$ in the plasma half-space $z \geq 0$. We first evaluate the diffusion coefficient (19.4.30). Substituting the inductive field (19.5.2) into (19.4.30), the ϕ integration and the k integration over the delta function can be easily done to obtain

$$D_v = \frac{e^2E_0^2\delta^2v^3}{4m^2}\int_0^\pi d\psi\,\sin\psi\frac{\sin^2\psi\cos^2\psi|\cos\psi|}{(v^2\cos^2\psi+\omega^2\delta^2)^2} \tag{19.5.16}$$

The substitution $\zeta = \cos^2\psi$ reduces this integral to elementary form, yielding the result

$$D_v = D_0\frac{\omega\delta}{v}\left[\left(1+\frac{2\omega^2\delta^2}{v^2}\right)\ln\left(1+\frac{v^2}{\omega^2\delta^2}\right)-2\right] \tag{19.5.17}$$

with $D_0 = e^2E_0^2\delta/4m^2\omega$. For low and high energies, D_v has the forms

$$\begin{aligned} D_v &\approx \frac{D_0}{6}\left(\frac{v}{\omega\delta}\right)^3, && v \ll \omega\delta \\ &\approx 2D_0\frac{\omega\delta}{v}\ln\frac{v}{\omega\delta}, && v \gg \omega\delta \end{aligned} \tag{19.5.18}$$

We assumed a constant mean free path λ_m with collision frequencies $\nu_m = v/\lambda_m$ and $\nu_{inel} = (2m/M)\nu_m$ in solving (19.4.33) to obtain (19.4.37) with $\lambda_\mathcal{E} = (M/6m)^{1/2}\lambda_m$. Substituting (19.4.37) into the boundary condition (19.4.35), we obtain

$$\left(\frac{2m}{3M}\right)^{1/2}vF = \frac{1}{2v^2}\frac{d}{dv}\left(v^2D_v\frac{dF}{dv}\right) \tag{19.5.19}$$

which determines the energy dependence of f_{e0}. In general, (19.5.19) must be solved numerically. An analytic solution can be found (Aliev et al., 1992) for the high energy regime $v \gg \omega\delta$, where we can approximate

$$D_v \approx D_{v0}\frac{\omega\delta}{v} \tag{19.5.20}$$

with $D_{v0} = 2D_0\ln(v/\omega\delta) \approx$ const. Substituting (19.5.20) into (19.5.19) and transforming from v to the energy variable $\mathcal{E} = mv^2/2e$, we obtain

$$\frac{1}{\mathcal{E}}\frac{d}{d\mathcal{E}}\left(\mathcal{E}\frac{dF}{d\mathcal{E}}\right) = \frac{1}{\mathcal{E}_0^2}F \tag{19.5.21}$$

with

$$\frac{1}{\mathcal{E}_0^2} = \frac{2e^2}{m^2\omega\delta D_{v0}}\left(\frac{2m}{3M}\right)^{1/2} \tag{19.5.22}$$

Equation (19.5.22) is Bessel's equation, and taking the solution that vanishes as $\mathcal{E} \to \infty$, we obtain

$$f_{e0} = F_0\,K_0(\mathcal{E}/\mathcal{E}_0)\,e^{-z/\lambda_\mathcal{E}} \tag{19.5.23}$$

where K_0 is the modified Bessel function of the second kind.

Consider now a finite system with $l \ll \lambda_\mathcal{E}$. Using a constant λ_m, the bounce-averaged kinetic equation (19.4.39) becomes

$$\frac{1}{\mathcal{E}}\frac{\mathrm{d}}{\mathrm{d}\mathcal{E}}\left(\mathcal{E}\frac{\mathrm{d}f_{\mathrm{e}0}}{\mathrm{d}\mathcal{E}}\right) = \frac{1}{\mathcal{E}_0^2}\frac{l}{2\lambda_\mathcal{E}}f_{\mathrm{e}0} \tag{19.5.24}$$

with the solution, as for (19.5.21)

$$f_{\mathrm{e}0} = F_0\,\mathrm{K}_0\left(\frac{\mathcal{E}}{\mathcal{E}_0}\left(\frac{l}{2\lambda_\mathcal{E}}\right)^{1/2}\right) \tag{19.5.25}$$

The bounce-averaged distribution (19.5.25) and the distribution (19.5.23) for the semi-infinite plasma have the same form for the energy dependence, but the bounce-averaged distribution is hotter due to the multiple interactions of the electrons with the skin layer field, and is uniform in z.

19.6 Kinetic Modeling of Discharges

19.6.1 Non-Maxwellian Global Models

As in the previous section, electron distributions are often non-Maxwellian at low plasma densities, with a Druyvesteyn-like distribution at high pressures and bi-Maxwellian distributions at low pressures. At higher densities, electron–electron collisions tend to drive the distribution toward a Maxwellian shape. Figure 11.9 shows measurements of a Druyvesteyn-like shape in a capacitive discharge for some pressure and driving voltage conditions.

The global model in Section 10.2 can be modified to treat non-Maxwellian distributions. For example, this has been done by Gudmundsson (2001) for an argon discharge with an energy distribution of the form

$$g_\mathrm{p}(\mathcal{E}) = g_x\frac{n_\mathrm{e}}{\mathrm{T}_\mathrm{eff}^{3/2}}\exp\left[-C_x(\mathcal{E}/\mathrm{T}_\mathrm{eff})^x\right] \tag{19.6.1}$$

for which the cases $x = 1$ and $x = 2$ correspond to Maxwellian and Druyvesteyn distributions, respectively. We summarize his calculation for the Druyvesteyn distribution here. Normalizing the distribution as $\int g_\mathrm{p}(\mathcal{E})\mathcal{E}^{1/2}\,\mathrm{d}\mathcal{E} = n_\mathrm{e}$ and with an average energy per electron $\overline{\mathcal{E}}_\mathrm{e} = \frac{3}{2}\mathrm{T}_\mathrm{eff}$, we obtain the coefficients $g_2 \approx 0.565$ and $C_2 \approx 0.243$. The various rate coefficients are evaluated as, e.g.,

$$K_\mathrm{iz}(\mathrm{T}_\mathrm{eff}) = \frac{1}{n_\mathrm{e}}\int_{\mathcal{E}_\mathrm{iz}}^\infty \sigma_\mathrm{iz}(\mathcal{E})(2e\mathcal{E}/m)^{1/2}\cdot\mathcal{E}^{1/2}g_\mathrm{p}(\mathcal{E})\,\mathrm{d}\mathcal{E} \tag{19.6.2}$$

with similar expressions for K_ex, K_el, etc. Using these results, the collisional energy loss $\mathcal{E}_\mathrm{c}(\mathrm{T}_\mathrm{eff})$ defined in (3.5.11) can be evaluated. The Bohm criterion for the ion velocity u_B entering the sheath can be written as (Amemiya, 1997)

$$\frac{e}{M}\frac{1}{u_\mathrm{B}^2} = \frac{1}{n_\mathrm{e}}\int_0^\infty \frac{g_\mathrm{p}}{2\mathcal{E}^{1/2}}\,\mathrm{d}\mathcal{E} \tag{19.6.3}$$

Evaluating this yields $u_{\mathrm{B}2} \approx 1.17\,(e\mathrm{T}_\mathrm{eff}/M)^{1/2}$. The average electron speed, as in (2.4.8), is found by integrating v over the distribution to be $\overline{v}_{\mathrm{e}2} = 1.03\,(8e\mathrm{T}_\mathrm{eff}/\pi M)^{1/2}$, and the average electron energy lost to the wall per electron lost to the wall, as in (2.4.11), is similarly evaluated to be $\mathcal{E}_\mathrm{e} \approx 1.80\,\mathrm{T}_\mathrm{eff}$. Approximating the electron wall flux as $\frac{1}{4}\overline{v}_\mathrm{e}n_\mathrm{e}(\mathrm{wall})$, and equating the ion and electron fluxes lost to an insulating wall, as in (6.2.14) and (6.2.15)

$$\frac{\overline{v}_\mathrm{e}}{4}\int_{V_\mathrm{s}}^\infty (\mathcal{E} - V_\mathrm{s})^{1/2}g_\mathrm{p}\,\mathrm{d}\mathcal{E} = u_\mathrm{B}n_\mathrm{e} \tag{19.6.4}$$

yields $V_{\mathrm{s}2} \approx 3.43\,\mathrm{T}_\mathrm{eff}$ for argon. The ion energy lost per ion lost to the wall is $\mathcal{E}_\mathrm{i} = Mu_\mathrm{B}^2/2e + V_\mathrm{s}$.

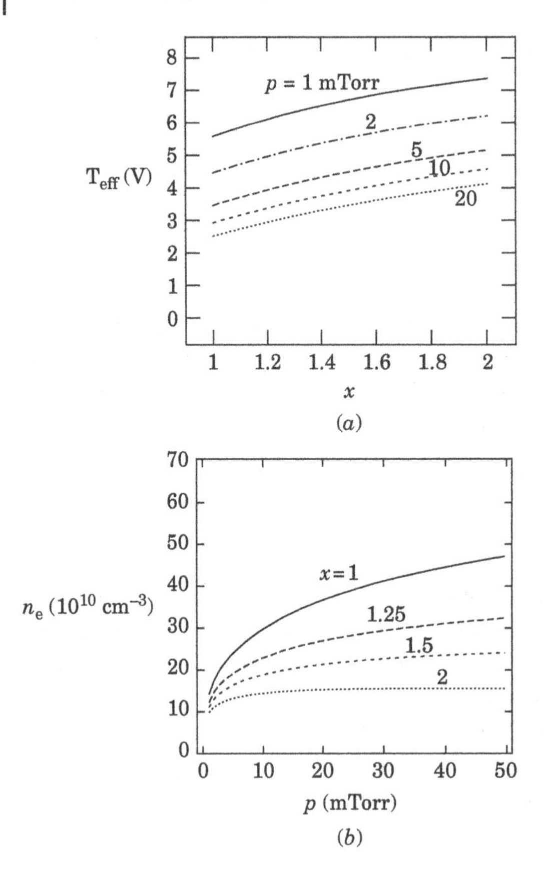

Figure 19.7 Comparison of global model results including Druyvesteyn ($x = 2$) and Maxwellian ($x = 1$) distributions; (*a*) T_{eff} versus energy exponent x at various pressures; (*b*) n_e versus p for various x's. Source: Gudmundsson (2001)/with permission of IOP Publishing.

With the preceding expressions, the particle and energy balance analysis of Section 10.2 yields the effective temperature T_{eff} and density n_e for a discharge with specified radius R, length l, pressure p, and absorbed power P_{abs}. Figure 19.7 shows the global model results for T_{eff} and n_e in an argon discharge for a typical geometry and power $R = 15.24$ cm, $l = 7.62$ cm, and $P_{abs} = 500$ W for various pressures p and assumed distributions of the form (19.6.1). As shown in Figure 19.7*a*, T_{eff} for a Druyvesteyn distribution ($x = 2$) is higher than for a Maxwellian distribution ($x = 1$) at a given p, since the Druyvesteyn energy tail is depleted. In spite of the higher T_{eff}, Figure 19.7*b* shows that n_e for the Druyvesteyn distribution (dotted line) lies considerably below n_e for the Maxwellian (solid line), especially at the higher pressures. This is mainly due to an increased energy loss factor $\mathcal{E}_c$ in (10.2.15) for the Druyvesteyn distribution, which has relatively more colder electrons that contribute to inelastic energy losses, compared to a Maxwellian distribution.

19.6.2 Inductive Discharges

Experiments in low-pressure inductive discharges in atomic gases give clear evidence that in the elastic energy range below $\mathcal{E}_{ex}$, the electron distribution function is a function solely of total electron energy (Kolobov et al., 1994; Kortshagen et al., 1995). Figure 19.8 shows data for the EEPF as a function of total energy $\mathcal{E}_T$ at different radial positions in a 100 W, 50 mTorr argon, planar inductive discharge (Kolobov et al., 1994). Note the absence of low-energy electrons at successively higher energies as the measurements are made further from the plasma center, indicating that these electrons are trapped by the internal ambipolar potential. Such data suggest that nonlocal theory can be

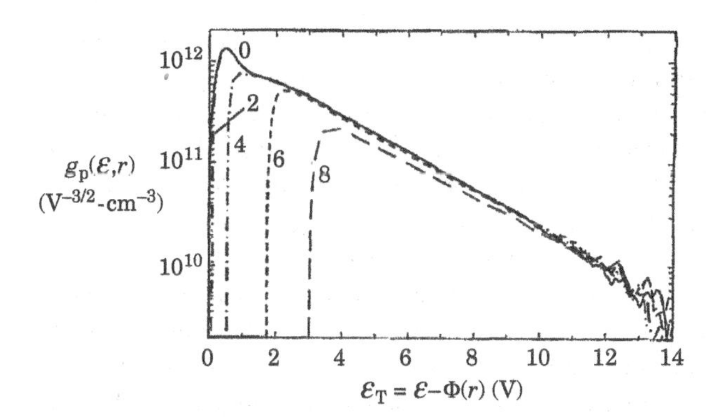

Figure 19.8 Experimental EEPF as a function of total energy $\mathcal{E}_T$ at different radial positions in a 100 W, 50 mTorr argon planar inductive discharge; curve labels correspond to radius (in cm) from the center at a fixed axial distance (4.4 cm) from the dielectric window. Source: Kolobov et al. (1994)/with permission of AIP Publishing.

used to model the distribution. This is especially attractive given the two-dimensional (r, z) nature of most inductive discharges, which would make a direct solution of the Boltzmann equation difficult. The nonlocal theory of inductive discharges has been developed by Kortshagen et al. (1995) and Kolobov and Hitchon (1995), see also Kolobov and Godyak (1995) for a review.

For a planar coil configuration (see Section 12.3) in a chamber of radius R and length l, the rf azimuthal heating field $\tilde{E}_\theta(r, z)$ typically peaks off axis at $r \sim R/2$ and decays axially away from the coil with a skin depth length ~1 cm. In contrast, the dc ambipolar potential $\Phi(r, z)$ forms a well with a maximum at $r = 0, z \sim l/2$, which traps electrons with total energy $\mathcal{E}_T$ less than the negative of the wall potential, $-\Phi_w$.

For slow electrons in the elastic energy range, the nonlocal kinetic equation (19.3.10) can be used to determine the EEPF g_{T0}. From (19.3.11), we have for ohmically heated electrons

$$\overline{\mathcal{E}^{1/2} D_\mathcal{E}} = \frac{1}{\mathcal{V}_0} \int_{\mathcal{V}_{ac}} \frac{2e}{3m} \frac{|\tilde{E}_\theta(r, z)|^2}{2} \frac{\mathcal{E}^{3/2} \nu_m}{\nu_m^2 + \omega^2} \, d\mathcal{V} \tag{19.6.5}$$

For the high densities typical of inductive discharges, electron–electron Coulomb collisions can be an important energy transfer mechanism. Hence, we write the energy transfer collision term as the sum of that due to Coulomb collisions and to electron–neutral elastic scattering: $C_{e0} \approx C_{ee} + C_{el}$. C_{ee} is given by the Fokker–Planck form (19.1.21), which can be expressed as the sum of a dynamical friction term F_{ee} proportional to H and a dynamical diffusion term D_{ee} proportional to G, where H and G are given by (19.1.22) and (19.1.23), respectively. Accounting also for the production I of low-energy electrons by inelastic energy losses of hotter electrons and by ionization (the second terms on the right-hand sides of (19.1.16) and (19.1.18), respectively), the kinetic equation (19.3.10) can be expressed in the form

$$-\frac{d}{d\mathcal{E}_T} \left[\overline{\mathcal{E}^{1/2}(D_\mathcal{E} + D_{ee})} \frac{dg_{T0}}{d\mathcal{E}_T} + \overline{\mathcal{E}^{1/2}(F_{ee} + F_\mathcal{E})} g_{T0} \right] = \overline{\mathcal{E}^{1/2} I} \tag{19.6.6}$$

Typically, because the slowest electrons are trapped in the center of the discharge by the ambipolar potential barrier, they cannot reach the region of high heating fields near the coil. Hence, the average energy diffusion coefficient $D_\mathcal{E}$ in (19.6.5) is small at low energies, such that the principal energy diffusion mechanism for the slowest electrons is from the D_{ee} term by Coulomb collisions. The more energetic electrons penetrate deeper into the edge rf field, and the average $D_\mathcal{E}$ is large for them. For a typical ionization fraction ($n_e/n_g \sim 10^{-4}$) in argon, the energy diffusion via Coulomb collisions is sufficiently strong that it results in a Maxwellian EEPF, as shown in Figure 19.8.

For two small groups of electrons, the nonlocal analysis is not very accurate: (1) free electrons with energies $\mathcal{E}_T > -\Phi_w$ and (2) trapped electrons in the inelastic energy range $\mathcal{E}_T > \mathcal{E}_{ex}$. While these electrons give small contributions to the plasma density and rf current, they determine important discharge parameters such as the ionization rate and the dc current density. One method of treating these electron groups is to solve the full spatially varying Boltzmann equation for them (Kolobov and Hitchon, 1995; Kortshagen and Heil, 1999). At high pressures, such a solution reveals a depletion of the EEPF tail in the region of highest dc potential where inelastic collisions occur. At low pressures, the EEPF tail is found to depend on the spatial coordinates for free electrons only. Also at low pressures, the electron heating is spatially separated from the ionization; the electron heating occurs near the coil, whereas the maximum ionization occurs near the maximum ambipolar potential where the rf field is absent.

In addition to solving (19.6.6), we also need to self-consistently solve the equations that determine the ambipolar potential $\Phi(r, z)$ and the heating field $\tilde{E}_\theta(r, z)$. This has been done using an iterative approach by Kortshagen et al. (1995), with the results given in Figure 19.9. The experimental results shown at various pressures are for a four turn, 13.56 MHz planar coil system in argon with $R =$ 7.5 cm and $l = 6$ cm. The theoretical results are those of the self-consistent nonlocal theory based on the use of (19.6.6). Similar good agreement between experiments and nonlocal theory is seen for the radially and axially varying EEPF. Comparisons with computer simulations also show good agreement (Kolobov et al., 1996).

19.6.3 Capacitive Discharges

In Section 11.3, we saw that the experimental electron energy distributions in capacitive discharges were generally non-Maxwellian (see Figure 11.9). Similar results have been found in PIC simulations. At low pressures (e.g., 10 mTorr), the distributions can be approximated by bi-Maxwellians, while at high pressures (e.g., 100 mTorr), the distribution falls more steeply above the excitation energy. These effects can be understood from a nonlocal kinetic analysis.

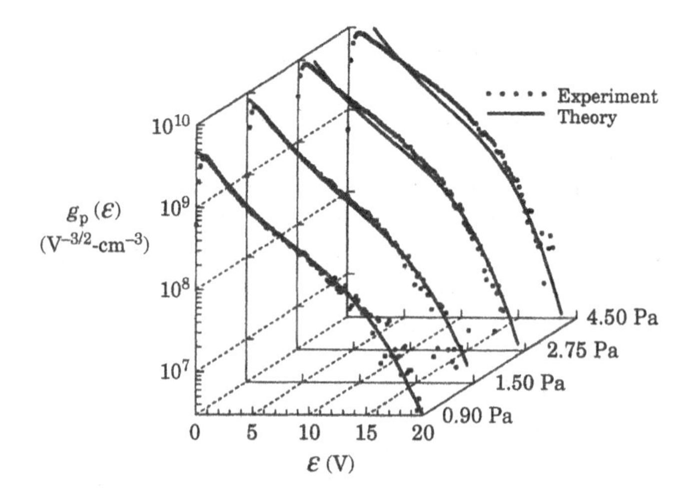

Figure 19.9 Comparison between measured and calculated EEPF in a planar inductive discharge; the measurements are performed in argon in the center of the discharge. Source: Adapted from Kortshagen et al. (1995).

Considering a plane-parallel discharge model of length l (varying in the x- direction), we assume an applied rf discharge voltage $V = V_{\rm rf} \cos \omega t$. The voltage across one of the sheaths is $V_{\rm s} = V_{\rm dc} + V_1 \cos \omega t$, where $V_1 = V_{\rm rf}/2$ and $V_{\rm dc}$ is the self-bias voltage. The width of this sheath is approximated by $s(t) = s_0(1 - \beta \cos \omega t)$, with $\beta = V_1/V_{\rm dc}$, where s_0 is related to the Child–Langmuir sheath width $s_{\rm m}$ by $s_{\rm m} \approx (1 + \beta)s_0$. We note that $\beta \approx 1$, so the important parameter is $\mathcal{E}_0 = V_{\rm dc} - V_1 \equiv V_{\rm dc}(1 - \beta)$. The sheath oscillation amplitude is generally much larger than that of the electron oscillations in the bulk of the plasma and is the main heating mechanism for those electrons that interact with the sheaths.

There are two physical processes that constrain less energetic electrons from being heated by the sheaths: the internal ambipolar potential $\Phi(x)$ that is a consequence of the plasma density profile, and the bulk oscillation of the plasma electrons due to the weak internal oscillating electric field. These effects can be captured in a model by assuming an internal square-well potential of height Φ_0, such that all electrons with the x-component of kinetic energy lower than Φ_0 do not interact with the large oscillating sheath fields.

In Section 19.3, we indicated the general method for nonlocal analysis. To perform an analytic calculation, we considered only ohmic heating balanced by electron–neutral elastic scattering energy losses. Here, starting from (19.3.5), we neglect elastic scattering losses, which are generally small, but include electron–neutral inelastic and ionization processes, $C_{\rm ex}$ and $C_{\rm iz}$, and electron–electron Coulomb collisions, $C_{\rm ee}$. The second terms on the right-hand sides of (19.1.16) for $C_{\rm ex}$ and (19.1.18) for $C_{\rm iz}$ can be grouped together to form a source term I that feeds the distribution at energy $\mathcal{E}_{\rm T}$. On the left-hand side of (19.3.5), we add the important stochastic heating term and the Fokker–Planck form for Coulomb collisions, which can be written as the divergence of a flux. The result, from (19.3.5), is

$$
\begin{aligned}
&-\frac{\partial}{\partial x}\left(\frac{2e}{3m}\frac{\mathcal{E}^{3/2}}{\nu_{\rm m}}\frac{\partial g_{\rm T}}{\partial x}\right) + \frac{2}{3}\frac{\partial}{\partial \mathcal{E}_{\rm T}}\left(\Gamma_{\rm ohm} + \Gamma_{\rm stoc} + \Gamma_{\rm ee}\right) \\
&= \mathcal{E}^{1/2}\left(-\nu_{\rm ex} g_{\rm T} - \nu_{\rm iz} g_{\rm T} + I\right)
\end{aligned}
\tag{19.6.7}
$$

where $\Gamma_j = -\mathcal{E}_{\rm T}^{1/2} D_{\mathcal{E}j}\, \partial g_{\rm T}/\partial \mathcal{E}_{\rm T}$ as in (19.3.10), with the ohmic diffusion coefficient in energy

$$
D_{\mathcal{E}\rm ohm} = \frac{e}{m}\frac{\mathcal{E}_{\rm T}}{\nu_{\rm m}}\frac{|\tilde{E}|^2}{2}\frac{\nu_{\rm m}^2}{\nu_{\rm m}^2 + \omega^2} \tag{19.6.8}
$$

and a stochastic diffusion coefficient $D_{\mathcal{E}\rm stoc}$ given approximately by

$$
D_{\mathcal{E}\rm stoc} \approx \left(\frac{m}{2e\mathcal{E}_{\rm T}}\right)^{1/2} \frac{\omega^2 s_0'^2 \beta^2 (\mathcal{E}_{\rm T}^2 - \Phi_0^2)}{d}\, \eta(\mathcal{E}_{\rm T} - \Phi_0) \tag{19.6.9}
$$

where $d = l - 2s_{\rm m}$ is the bulk plasma thickness and $\eta(\mathcal{E}_{\rm T} - \Phi_0)$ is a step function indicating that electrons are only heated if their energy is higher than Φ_0. We have already considered the stochastic heating for a Maxwellian distribution in Sections 11.1 and 11.2. It arises from the velocity kick obtained on reflection from the oscillating sheath

$$
\Delta v_x = 2\omega s_0' \sin \omega t \tag{19.6.10}
$$

which are assumed randomized over the phase ωt (Wang et al., 1998). The quantity s_0' is the amplitude of the large sheath oscillations with respect to the small amplitude bulk plasma oscillations due to the internal rf field: $s_0' \approx s_{\rm m}/2 - eE_{\rm rf}/m\omega^2$.

The form of the Coulomb flux is obtained by an expansion of (19.1.21), see for example, Cohen et al. (1980)

$$
\Gamma_{\rm ee} = -3\,\nu_{\rm ee} {\rm T_c}^{3/2}\left(g_{\rm T} + {\rm T_c}\frac{\partial g_{\rm T}}{\partial \mathcal{E}_{\rm T}}\right) \tag{19.6.11}
$$

Here, $\nu_{ee} = 5 \times 10^{-6}\,\mathrm{T_c}^{-3/2} n_{ave} \ln \Lambda$ is the electron self-collision frequency, with n_{ave} the average density of electrons in units of cm^{-3} and Λ given in Section 3.3. The form of the collision operator in (19.6.7) is a good approximation appropriate for energies lower than $\mathrm{T_c}^*$, the electron temperature at which electron–electron collisions dominate the kinetic equation and therefore generate a Maxwellian distribution.

Our next concern is the spatial averaging (19.3.8), also known as *bounce averaging*. This can be formally done with an arbitrary ambipolar potential, but leaves both an unknown potential function and complicated integrations. Furthermore, the formal averaging still leaves difficult questions concerning the detailed nature of the stochastic heating and the loss to the walls of energetic electrons. To solve these problems, a series of "reasonable" assumptions have been made (Wang et al., 1998), which we summarize here. Alternative assumptions can also be made to specify the complete problem, such as those used by Smirnov and Tsendin (1991) and Kaganovich and Tsendin (1992a,b), but make the space-averaged kinetic equation more difficult to solve. There is often a tradeoff between solvability and exactness of the defining equations. The formal approach as in (19.3.6) is to expand the distribution as $g_\mathrm{T} = g_\mathrm{T0}(\mathcal{E}_\mathrm{T}) + g_\mathrm{T1}(\mathcal{E}_\mathrm{T}, x)$ and then average over the bounce motion. A major simplification for the internal potential can be made by setting $\Phi(x) = \Phi_0$, a constant. To understand this, we note that the internal field in the bulk plasma is given by the Boltzmann relation $\Phi(x) = \mathrm{T_c} \ln n_e(x)/n_{e0}$, where $\mathrm{T_c}$ is the "temperature" of the low-energy part of the distribution. For low pressure, the potential is quite small and is relatively constant over most of the core plasma. The potential rises more steeply in the sheath region, that is, transiently occupied by electrons, due to the more rapid fall-off of the ion density. At the edge of the electron sheath, the potential rises very rapidly in the space charge region. The electrons escape in small bursts when the electron sheath has almost completely collapsed, as described in Section 11.1. Also, as described earlier, and seen in Figure 19.7, the electron oscillation is strongly nonlinear. This follows from the continuity of rf current (11.2.7), such that the decrease of electron density in the ion sheath region must be compensated by an increase in electron sheath velocity. Near the ion sheath–plasma interface, the electrons reflecting from the oscillating sheath do not gain or lose energy, as the oscillations are just compensated by the oscillating bulk plasma.

With the above approximations, averaging over the bulk distribution between the ion sheaths, we have the averaged nonlocal kinetic equation

$$\begin{aligned}
&-\frac{e|\tilde{E}|^2}{3m}\frac{\mathrm{d}}{\mathrm{d}\mathcal{E}_\mathrm{T}}\left(\frac{\mathcal{E}_\mathrm{T}^{3/2}}{\nu_\mathrm{m}}\frac{\nu_\mathrm{m}^2}{\nu_\mathrm{m}^2+\omega^2}\frac{\mathrm{d}g_\mathrm{T0}}{\mathrm{d}\mathcal{E}_\mathrm{T}}\right)\\
&\qquad -\frac{2}{3}\left(\frac{m}{2e}\right)^{1/2}\eta(\mathcal{E}_\mathrm{T}-\Phi_0)\frac{\omega^2 s_0'^2\beta^2}{d}\frac{\mathrm{d}}{\mathrm{d}\mathcal{E}_\mathrm{T}}(\mathcal{E}_\mathrm{T}^2-\Phi_0^2)\frac{\mathrm{d}g_\mathrm{T0}}{\mathrm{d}\mathcal{E}_\mathrm{T}}\\
&\qquad -2\nu_{ee}\mathrm{T_c}^{3/2}\frac{\mathrm{d}}{\mathrm{d}\mathcal{E}_\mathrm{T}}\left(g_\mathrm{T0}+\mathrm{T_c}\frac{\mathrm{d}g_\mathrm{T0}}{\mathrm{d}\mathcal{E}_\mathrm{T}}\right)\\
&\qquad = \mathcal{E}_\mathrm{T}^{1/2}\left(-\nu_\mathrm{esc}g_\mathrm{T0}-\nu_\mathrm{iz}g_\mathrm{T0}-\nu_\mathrm{ex}g_\mathrm{T0}+I\right)
\end{aligned} \tag{19.6.12}$$

In the right-hand side, we have introduced the escape frequency ν_esc (see Wang et al., 1998). This represents the particle loss to the walls, which occurs only for those electrons that have sufficient x-directed energy to overcome the potential rise between the electron sheath edge and the wall.

With given input quantities of pressure p, V_1, ω, and l, the unknown quantities, which must be determined along with g_T0, are $\mathrm{T_c}$, Φ_0, $\mathcal{E}_0 = V_\mathrm{dc} - V_1$, n_{e0}, and s_0. We need five physical conditions to determine the five unknown parameters in the equilibrium distribution:

1. The total electron escape rate is equal to the ionization rate.
2. The escape rate of ions is equal to that of electrons.
3. The flux leaving the plasma is equal to the flux crossing the sheath.
4. The total electron distribution cannot be heated by electron–electron collisions.
5. The space-and-time-averaged electron density in the sheath is equal to the space-averaged ion density in the sheath.

The differential equation (19.6.12) has been solved numerically (Wang et al., 1998, 1999). $g_{T0}(\mathcal{E}_T)$ has two constants to be determined, which can be taken as the value of g_{T0} and its logarithmic derivative at $\mathcal{E}_{max}$. $g_{T0}(\mathcal{E}_{max})$ is determined by the normalization of g_{T0}, since the kinetic equation is linear in g_{T0}. The derivative is determined by the requirement that there is no flux at $\mathcal{E}_T = 0$. The eigenvalue problem has two independent solutions, one which increases with $\mathcal{E}_T$, and one which decreases with $\mathcal{E}_T$. Since the integration is from $\mathcal{E}_T = \mathcal{E}_{max}$ to $\mathcal{E}_T = 0$, the error caused by an inaccurate choice of logarithmic derivative rapidly becomes unimportant if the value of $\mathcal{E}_{max}$ is large enough. This allows the "boundary" values at $\mathcal{E}_{max}$ to be chosen with considerable flexibility. The differential equation involves s_0, n_{e0}, V_{dc}, Φ_0, and T_c, which are determined recursively.

In Figure 19.10, we compare the nonlocal results to one-dimensional PIC simulations, and to the experimental results of Godyak and co-workers, shown in Figure 11.9 for $p = 10$ mTorr. The solid line shows the nonlocal distribution from (19.6.12), using the fully self-consistent theory, and the dashed line shows the theory result using the measured central density n_{e0}. The coarse and fine dotted lines show the experimental and PIC simulation results. The central density is higher for the fully self-consistent result than in the experiment, but is similar to the PIC result. The rather steep profiles of the theory, just below $\mathcal{E} = \Phi_0 = 3$ V, is caused by the abrupt turning-on of the stochastic heating due to the step function. The transition can be smoothed by turning on the stochastic heating more gradually, as is physically reasonable (Wang et al., 1999).

It is also possible to combine the nonlocal kinetic treatment of the electrons with the fluid theory for positive and negative ions, to treat electronegative discharges. As with electropositive discharges, nonlocal kinetics were required to give EEPFs in agreement with PIC simulations at low pressures (Wang et al., 1999).

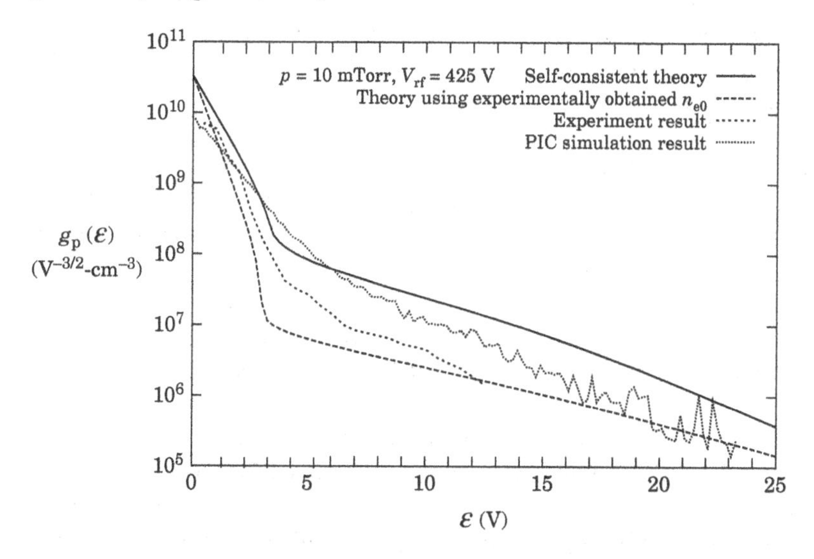

Figure 19.10 Comparison of EEPFs obtained experimentally and theoretically in an argon capacitive discharge at $p = 10$ mTorr and $V_{rf} = 425$ V. Source: Adapted from Wang et al. (1999).

Problems

19.1 Two-Term Expansion Procedure Carry out the ψ integrations of (19.1.9) to obtain (19.1.11) and (19.1.14).

19.2 Electron–Neutral Elastic Collision Term

(a) For no z-variation ($\partial/\partial z \equiv 0$) and $E_z = 0$, show by substituting (19.1.15) into (19.1.14) that the collision term C_{e0} does not yield a particle loss.

(b) Let $\nu_m(v) = \nu_{m0}$, a constant independent of v, show that C_{e0} yields an energy loss collision frequency $(2m/M)\nu_m$.

19.3 Excitation and Ionization Collision Terms

(a) Taking $f_{e1} = 0$ in (19.1.14) and $\nu_{ex} = \text{const}$ in (19.1.16), show that $\partial f_{e0}/\partial t = C_{ex}$ yields the results $\partial n_e/\partial t = 0$ and $\partial(n_e\mathcal{E}_e)/\partial t = -\nu_{ex}\mathcal{E}_{ex}n_e$.

(b) Taking $f_{e1} = 0$ in (19.1.14) and $\nu_{iz} = \text{const}$ in (19.1.16), show that $\partial f_{e0}/\partial t = C_{iz}$ yields the results $\partial n_e/\partial t = \nu_{iz}n_e$ and $\partial(n_e\mathcal{E}_e)/\partial t = -\nu_{iz}\mathcal{E}_{iz}n_e$.

19.4 Einstein Relation and Thermal Conductivity

(a) Using (19.1.28) and (19.1.29), show that the Einstein relation $D_e = \mu_e \mathrm{T}_e$ holds for a Maxwellian distribution, independent of the dependence of ν_m on v.

(b) For an electron distribution with a spatially-varying temperature $\mathrm{T}_e(z)$, the heat flow vector in (2.3.21) can be written as

$$\mathbf{q} = \int d^3v \, \frac{1}{2} m v^2 v_z f_e(v) = -\kappa_T \nabla \mathrm{T}_e$$

where κ_T is the thermal conductivity (here T_e is in units of volts). For a Maxwellian distribution with a constant collision frequency $\nu_m = \text{const}$, evaluate this integral using the two-term expansion (19.1.11) for f_{e1}, and determine κ_T. Comparing to the dc electrical conductivity $\sigma_{dc} = \mu_e n_e e$, with μ_e given by (19.1.29), show that $\kappa_T = 5\,\sigma_{dc}\mathrm{T}_e$. This is known as the Wiedemann–Franz law.

19.5 Effective Frequency and Collision Frequency For a Maxwellian distribution in a gas with a constant momentum transfer cross section, $\sigma_m(v) = \text{const}$, show from (19.1.45) and (19.1.46) that ν_{eff} in the high-frequency limit is $4/\pi$ times ν_{eff} in the low-frequency limit. Also show that ω_{eff} in the low-frequency limit is $(3\pi/8)\omega$.

19.6 Local Kinetics with RF Fields Derive (19.2.3)–(19.2.5) by inserting (19.2.1) and (19.2.2) along with the two-term expansion (19.1.3) into the Boltzmann equation (19.1.1).

19.7 Two-Term Quasilinear Diffusion Coefficient Obtain (19.4.30) by taking the divergence of (19.4.28) in spherical velocity coordinates and averaging over solid angle.

19.8 Stochastic Heating in an Inductive Discharge Since magnetic forces do no work on moving charges, they can be neglected in calculating the stochastic heating. Consider an electron with velocity $-v_z$ incident on the skin depth layer in an inductive discharge.

(a) Neglecting the rf magnetic force, show that the transverse velocity impulse

$$\Delta v_y = -\int_{-\infty}^{\infty} dt \frac{e\tilde{E}_y(z(t), t)}{m}$$

calculated by substituting (19.5.1) into the above expression (with $z(t) = -v_z t$ for $t < 0$ and $z(t) = v_z t$ for $t > 0$) is

$$\Delta v_y = \frac{2e\tilde{E}_0\delta}{m} \frac{v_z}{v_z^2 + \omega^2\delta^2} \cos\phi$$

(b) Averaging over a uniform distribution of initial electron phases ϕ, show that the energy change is

$$\Delta W = \frac{1}{2} m \langle (\Delta v_y)^2 \rangle_\phi = \frac{e^2\tilde{E}_0^2\delta^2}{m} \frac{v_z^2}{(v_z^2 + \omega^2\delta^2)^2}$$

which can be integrated over the particle flux to obtain the stochastic heating power

$$\overline{S}_{\text{stoc}} = \int_{-\infty}^{\infty} dv_x \int_{-\infty}^{\infty} dv_y \int_0^{\infty} dv_z \, v_z \Delta W(v_z) f_e$$

Compare this expression for $\overline{S}_{\text{stoc}}$ to (19.5.4) determined from quasilinear theory.

Appendix A

Collision Dynamics

The dynamics of a particle in a central force potential can be directly calculated. Using the center of mass coordinates, this corresponds to a collision between any two particles, in which their mutual interaction has a central force, e.g., the Coulomb and polarization potentials. The equation for the trajectory is straightforwardly calculated from the conservation of total energy and angular momentum. Referring to Figure A.1, these equations can be written

$$\frac{1}{2}m_{\rm R}\left(\dot{r}^2 + r^2\dot{\theta}^2\right) + U(r) = \frac{1}{2}m_{\rm R}v_0^2 \tag{A.1}$$

and

$$-m_{\rm R}r^2\dot{\theta} = m_{\rm R}bv_0 \tag{A.2}$$

where v_0 is the initial relative velocity, $m_{\rm R}$ is the reduced mass, and here θ is defined as shown in the figure. Substituting $\dot{\theta}$ from (A.2) into (A.1), we obtain an equation for $\dot{r}$ alone

$$\frac{1}{2}m_{\rm R}\dot{r}^2 = \frac{1}{2}m_{\rm R}v_0^2 - \left[U(r) + \frac{m_{\rm R}b^2v_0^2}{2r^2}\right] \tag{A.3}$$

where the term in brackets is an effective potential including the repulsive centrifugal potential $m_{\rm R}b^2v_0^2/2r^2$. For $U(r) \propto r^{-i}$, which includes the Coulomb and polarization potentials, this is readily solvable (see Goldstein, 1950). We form the trajectory equation by solving (A.3) for $\dot{r}$ and then dividing by $\dot{\theta}$ from (A.2), to obtain

$$\frac{dr}{d\theta} = \frac{\pm\left\{v_0^2 - 2m_{\rm R}\left[U(r) + m_{\rm R}b^2v_0^2 2r^2\right]\right\}^{1/2}}{bv_0/r^2} \tag{A.4}$$

where the minus sign is for the outward bound trajectory. The angle χ is then obtained by integrating (A.4),

$$\chi = \int_{r_{\rm min}}^{\infty} \frac{(bv_0/r^2)\,dr}{\left\{v_0{}^2 - \dfrac{2}{m_{\rm R}}\left[U(r) + \dfrac{m_{\rm R}b^2v_0^2}{2r^2}\right]\right\}^{1/2}} \tag{A.5}$$

and from Figure A.1

$$\Theta = \pi - 2\chi \tag{A.6}$$

We have used symmetry about the closest approach, to integrate (A.5) from the closest approach $r_{\rm min}$ to infinity, and then doubled the angle. The lower limit $r_{\rm min}$ is obtained from the condition that $\dot{r}$ changes sign, which from (A.3) is just

Principles of Plasma Discharges and Materials Processing, Third Edition. Michael A. Lieberman and Allan J. Lichtenberg.

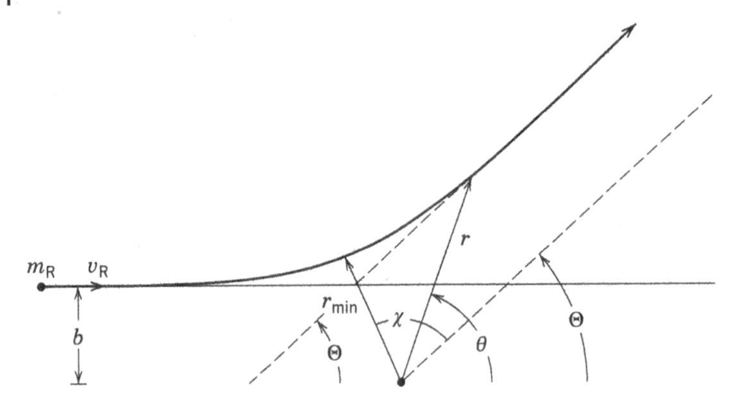

Figure A.1 Illustrating the exact classical calculation of the differential scattering cross section.

$$\frac{1}{2}m_R v_0^2 - \left[U(r_{min}) + \frac{m_R b^2 v_0^2}{2r_{min}^2}\right] = 0 \tag{A.7}$$

Although complete solutions of (A.5) are possible, they are not illuminating. However, for attractive potentials, it is often useful to distinguish between two cases: the case for which the centrifugal force serves as a barrier to deep penetration and the case for which this repulsive potential is overcome by the attractive potential. From (A.3), the transition occurs when $U(r) = -Cr^{-i}$, with $i = 2$. For $i < 2$, the effective potential

$$U_{eff}(r) = U(r) + \frac{m_R b^2 v_0^2}{2r^2} \tag{A.8}$$

is always repelling at the origin. These two cases are illustrated in Figure A.2, where $U_c(r)$ is the centrifugal potential. For $U(r) > 0$, the effective potential is, of course, always repulsive. An important characteristic of the potential in Figure A.2*b*, with $i > 2$, is that a resonance phenomenon can occur for an energy close to the value for which the force disappears (W_0 in the figure).

A.1 Coulomb Cross Section

For Coulomb collisions, there is a straightforward solution to the trajectory equation (A.5). Substituting into the Coulomb potential for an electron–ion collision

$$U(r) = -\frac{Ze^2}{4\pi\epsilon_0 r}$$

and defining a new variable $\rho = b/r$, we recast (A.5) with (A.6) into the form

$$\Theta = 2\int_0^{\rho_{max}} \frac{d\rho}{\left(1 - \frac{2Ze^2}{4\pi\epsilon_0 m_R v_0^2 b}\rho - \rho^2\right)^{1/2}} - \pi \tag{A.9}$$

where ρ_{max} is obtained from the solution of (A.7). Integrating (A.9), we have

$$\Theta = -2\cos^{-1}\left\{\frac{Ze^2/4\pi\epsilon_0 m_R v_0^2 b}{\left[1 + \left(Ze^2/4\pi\epsilon_0 m_R v_0^2 b\right)^2\right]^{1/2}}\right\} + \pi \tag{A.10}$$

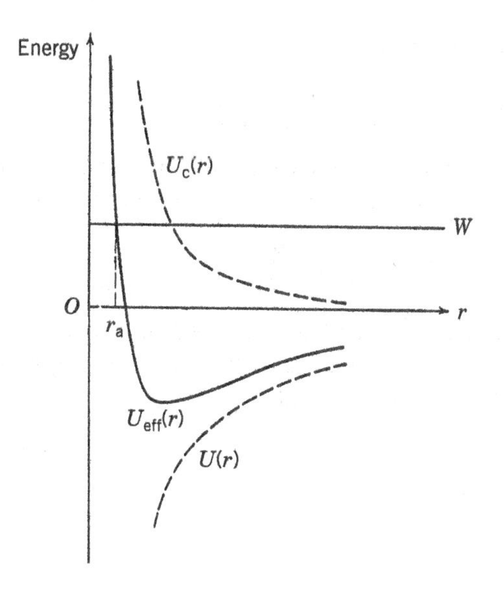

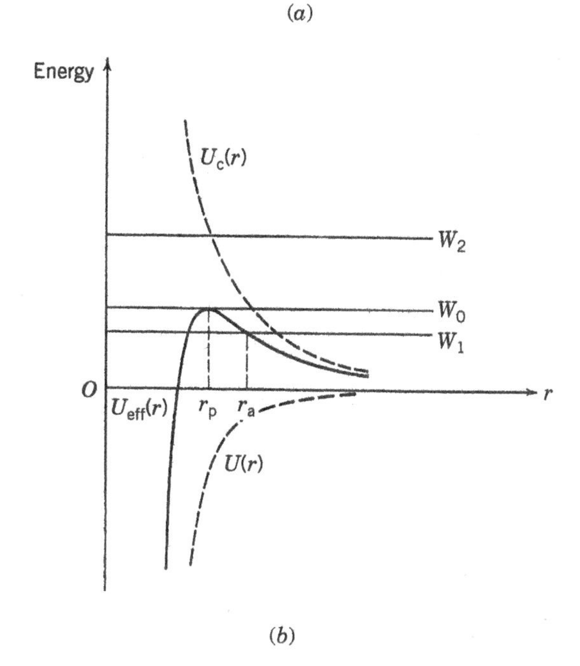

Figure A.2 The potential functions used for the calculation of elastic scattering in (*a*) an attractive inverse first power potential and (*b*) an attractive inverse third power potential.

which can be rewritten

$$\frac{Ze^2/4\pi\epsilon_0 m_R v_0^2 b}{[1+\left(Ze^2/4\pi\epsilon_0 m_R v_0^2 b\right)^2]^{1/2}} = \cos\left(\frac{\Theta}{2}-\frac{\pi}{2}\right) = \sin\frac{\Theta}{2} \tag{A.11}$$

Using a trigonometric identity, we have

$$\frac{Ze^2}{4\pi\epsilon_0 m_R v_0^2 b} = \tan\frac{\Theta}{2} \tag{A.12a}$$

or, solving for b,

$$b = \frac{Ze^2}{4\pi\epsilon_0 m_R v_0^2 \tan(\Theta/2)} \tag{A.12b}$$

From the definition of the differential cross section in (3.1.13), we have, after taking a derivative of (A.12b) and performing a few trigonometric manipulations,

$$\sigma(\Theta) = \frac{b}{\sin\Theta}\left|\frac{\mathrm{d}b}{\mathrm{d}\Theta}\right| = \frac{Z^2e^4}{(8\pi\epsilon_0)^2 m_R^2 v_0^4 \sin^4(\Theta/2)} \tag{A.13}$$

which is the well-known *Rutherford cross section* for Coulomb scattering. Because of the $\sin^4(\Theta/2)$ term in the denominator, the total scattering cross section is infinite, unless cut off by long-range shielding, as discussed in Section 3.3.

Appendix B

The Collision Integral

B.1 Boltzmann Collision Integral

We obtain the general form for the term $\partial f/\partial t|_c$, which occurs on the right-hand side (RHS) of the Boltzmann equation (2.3.3), known as the Boltzmann collision integral. Consider an elastic collision between incident and target particles having distributions f_1 and f_2 and velocities $\mathbf{v}_1$ and $\mathbf{v}_2$. The number of particles between $\mathbf{v}_1$ and $\mathbf{v}_1 + \mathrm{d}\mathbf{v}_1$ is

$$f_1(\mathbf{r}, \mathbf{v}_1, t)\, \mathrm{d}^3 v_1$$

and the number between $\mathbf{v}_2$ and $\mathbf{v}_2 + \mathrm{d}\mathbf{v}_2$ is

$$f_2(\mathbf{r}, \mathbf{v}_2, t)\, \mathrm{d}^3 v_2$$

The flux of incident particles in a coordinate system in which $\mathbf{v}_2$ is zero is

$$|\mathbf{v}_1 - \mathbf{v}_2| f_1\, \mathrm{d}^3 v_1$$

The differential cross section for scattering through angle θ_1 is $I\left(|\mathbf{v}_1 - \mathbf{v}_2|, \theta_1\right)$. The number of particles per unit time scattered out of the differential volume $\mathrm{d}^3 v_1\, \mathrm{d}^3 v_2$ and into the solid angle $\mathrm{d}\Omega$ is then

$$f_1 f_2\, |\mathbf{v}_1 - \mathbf{v}_2|\, \mathrm{d}^3 v_1\, \mathrm{d}^3 v_2\, I\, \mathrm{d}\Omega$$

Dividing by $\mathrm{d}^3 v_1$ and integrating over $\mathrm{d}^3 v_2$ and $\mathrm{d}\Omega$, we obtain all particles scattered out of the incident distribution f_1:

$$\left.\frac{\partial f_1}{\partial t}\right|_{\text{out}} = \int\!\!\int f_1 f_2\, |\mathbf{v}_1 - \mathbf{v}_2|\, \mathrm{d}^3 v_2\, I\, \mathrm{d}\Omega \tag{B.1}$$

The particles from the distributions f_1 and f_2 having velocities $\mathbf{v}_1$ and $\mathbf{v}_2$ are scattered to primed velocities $\mathbf{v}_1'$ and $\mathbf{v}_2'$ in distributions $f_1' \equiv f_1(\mathbf{r}, \mathbf{v}_1', t)$ and $f_2' \equiv f_2(\mathbf{r}, \mathbf{v}_2', t)$. The rate of scattering into f_1, from the reversibility of the equations of motion, is then

$$\left.\frac{\partial f_1}{\partial t}\right|_{\text{in}} = \int\!\!\int f_1' f_2'\, |\mathbf{v}_1' - \mathbf{v}_2'|\, \mathrm{d}^3 v_2'\, I'\, \mathrm{d}\Omega' \tag{B.2}$$

Finally, for elastic collisions, the relative velocity is conserved,

$$|\mathbf{v}_1' - \mathbf{v}_2'| = |\mathbf{v}_1 - \mathbf{v}_2| \tag{B.3}$$

Principles of Plasma Discharges and Materials Processing, Third Edition. Michael A. Lieberman and Allan J. Lichtenberg.

and the differential cross sections $I\,\mathrm{d}\Omega$ and $I'\,\mathrm{d}\Omega'$ in the primed and unprimed coordinates can be identified. Substituting (B.3) into (B.2) and subtracting (B.1) from (B.2), we have

$$\left.\frac{\partial f_1}{\partial t}\right|_{\mathrm{c}} = \int \mathrm{d}^3 v_2 \int_0^{2\pi} \mathrm{d}\phi_1 \int_0^{\pi} \left(f_1' f_2' - f_1 f_2\right) |\mathbf{v}_1 - \mathbf{v}_2| I \sin\theta_1\, \mathrm{d}\theta_1 \tag{B.4}$$

where we have written out $\mathrm{d}\Omega$ explicitly and noted that the integral over θ_1 is done before the integral over $\mathbf{v}_2$. The general form of (B.4) constitutes the RHS of (2.3.3).

It is not easy to evaluate (B.4) under the action of arbitrary forces on the LHS of (2.3.3). For small-angle Coulomb collisions, an expansion of (B.4) is possible to obtain the *Fokker–Planck collision integral* described in many texts on fully ionized plasmas (e.g., see Schmidt, 1979). For large-angle collisions with neutrals, a different expansion is usually used, which assumes a distribution close to equilibrium. We have already employed the resulting Krook collision operator in our formulation of the macroscopic equations in Section 2.3. We outline how this approximation is obtained below. First, however, we show that the general form (B.4) is satisfied by a Maxwellian distribution at equilibrium.

B.2 Maxwellian Distribution

At equilibrium, the distribution is stationary, $\partial f/\partial t|_{\mathrm{c}} = 0$, which is satisfied if

$$f_1' f_2' - f_1 f_2 = 0 \tag{B.5}$$

Taking the logarithm of (B.5), we have

$$\ln f_1' + \ln f_2' = \ln f_1 + \ln f_2 \tag{B.6}$$

Guessing a solution of the form

$$\ln f = -\xi^2 m v^2 + \ln C \tag{B.7}$$

for f_1, f_2, f_1', and f_2', and substituting into (B.6), we obtain

$$m_1 v_1'^2 + m_2 v_2'^2 = m_1 v_1^2 + m_2 v_2^2 \tag{B.8}$$

which expresses the conservation of energy in an elastic collision. Hence, (B.7) is a solution of (B.5). Taking the antilog of (B.7), we obtain the equilibrium distribution

$$f_1(\mathbf{v}) = f_2(\mathbf{v}) = C\,\mathrm{e}^{-\xi^2 m v^2} \tag{B.9}$$

which is the form assumed in (2.4.2). This gives, with the appropriate normalization, the Maxwellian distribution, with a common temperature for all the species, as in (2.4.7). We could also have included a function of the momentum in (B.7) and found a drifting Maxwellian at equilibrium. Equation (B.5) is clearly sufficient to satisfy $\partial f/\partial t\,|_{\mathrm{c}} = 0$. It is also necessary, which can be shown by use of the *Boltzmann H Theorem*, which states that the time derivative of the function

$$\mathrm{H} = \int f \ln f\, \mathrm{d}^3 v \tag{B.10}$$

which measures the randomness of the distribution, is zero if and only if (B.5) is satisfied. A more detailed account of the above material, including a derivation of the H theorem, can be found in Holt and Haskell (1965, chapter 5).

Appendix C

Diffusion Solutions for Variable Mobility Model

We consider the solution of the nonlinear low-pressure diffusion equation (5.3.6) in a plane-parallel system. The solution (5.3.7) obtained by Godyak and Maximov (see Godyak, 1986) is for a boundary condition of zero flux at the plasma center $x = 0$, corresponding to symmetric diffusion in an electropositive plasma over the region $-l/2 \le x \le l/2$. This solution has been generalized by Kouznetsov et al. (1996) to include an input flux, for a configuration in which (5.3.6) corresponds to an *electropositive* edge region of a discharge which has an additional negative ion species in an *electronegative* plasma core. Following Kouznetsov et al., we outline the solution here.

For the half-region $0 \le x \le l/2$, normalizing the position, density and potential variables as

$$\xi = \frac{2x}{l}, \qquad y = \frac{n}{n_0} \qquad \eta = -\frac{\Phi}{T_e} \tag{C.1}$$

then from (5.3.3), we have

$$\frac{d\eta}{d\xi} = -\frac{1}{y}\frac{dy}{d\xi} \tag{C.2}$$

with the integral of (C.2) yielding the Boltzmann relation

$$\eta = -\ln y \tag{C.3}$$

Substituting (C.2) into (5.3.6), we can eliminate n in favor of η to obtain

$$\frac{d^2\eta}{d\xi^2} - 2\left(\frac{d\eta}{d\xi}\right)^2 = 2\alpha\left(\frac{d\eta}{d\xi}\right)^{1/2} \tag{C.4}$$

where

$$\alpha = \frac{1}{2}\left(\frac{\pi l}{4\lambda_i}\right)^{1/2}\frac{\nu_{iz} l}{u_B} \tag{C.5}$$

with $u_B = (eT_e/M)^{1/2}$. With the substitution

$$\frac{d\eta}{d(\alpha^{2/3}\xi)} = \chi \tag{C.6}$$

(χ is a normalized electric field), (C.4) can be transformed to

$$\frac{d\chi}{2\chi^{1/2}(1+\chi^{3/2})} = d(\alpha^{2/3}\xi) \tag{C.7}$$

An integration yields

$$\frac{1}{6}\ln\frac{(\chi^{1/2}+1)^3}{\chi^{3/2}+1} + \frac{1}{\sqrt{3}}\tan^{-1}\frac{2\chi^{1/2}-1}{\sqrt{3}} = \alpha^{2/3}\xi + C_1 \tag{C.8}$$

Principles of Plasma Discharges and Materials Processing, Third Edition. Michael A. Lieberman and Allan J. Lichtenberg.

Eliminating $d(\alpha^{2/3}\xi)$ from (C.6) and (C.7) and integrating, we find

$$\frac{1}{3}\ln\left(1+\chi^{3/2}\right) = \eta + C_2 \tag{C.9}$$

Equations (C.8) and (C.9) provide a solution to (C.4) in a parametric form (χ serves as the parameter) with the two constants of integration C_1 and C_2 determined from boundary conditions.

For symmetric diffusion in a low-pressure electropositive discharge over the region $-l/2 \le x \le l/2$, we have a zero flux boundary condition at the plasma center $x = 0$, such that $dn/dx = 0$, $d\eta/d\xi = 0$ from (C.2), and $\chi = 0$ from (C.6). This yields $C_2 = 0$ from (C.9) and $C_1 = -\pi/6\sqrt{3}$ from (C.8). The density profile is found by substituting (C.3) into (C.9) to eliminate η, which yields

$$y = \left(1+\chi^{3/2}\right)^{-1/3} \tag{C.10}$$

or solving (C.10) for χ, we obtain

$$\chi = \left(y^{-3}-1\right)^{2/3} \tag{C.11}$$

Inserting (C.11) into (C.8) yields the density profile $y = n/n_0$ in (5.3.7), shown in Figure 5.2*b*. As described in (5.3.8), the profile is circular to a very good approximation.

When considering more general boundary conditions, it is convenient to have an expression for χ in terms of the ion drift velocity u_i. Using the normalizations (C.1) and substituting χ from (C.6) into (5.3.5), we obtain

$$\chi = \alpha^{-2/3}\frac{\pi l}{4\lambda_i}w^2 \tag{C.12}$$

where $w = u(x)/u_B$ is the normalized flow velocity. Substituting α from (C.5) into the preceding expression gives

$$\chi = a^{-2/3}w^2 \tag{C.13}$$

where

$$a = \frac{2\nu_{iz}\lambda_i}{\pi u_B} \tag{C.14}$$

is a normalized ionization rate. At the plasma sheath edge, the ions reach the Bohm speed, and their concentration is n_s. Using this boundary condition in (C.9) with χ obtained from (C.13) at $w = 1$ and η given by (C.3), we find

$$C_2 = \frac{1}{3}\ln\left(1+a^{-1}\right) + \ln y_s$$

where $y_s = n_s/n_0$. Substituting C_2 into (C.9) and using (C.3) and (C.13), we have

$$\frac{y_s}{y(\xi)} = \left[\frac{a + w^3(\xi)}{a+1}\right]^{1/3} \tag{C.15}$$

which gives the ratio of the density y at any position ξ to the density y_s at the sheath edge in terms of the ratio $w(\xi)$ of the ion drift velocity to the Bohm speed. In particular, for an electropositive region with an entering velocity u_{in} and a density n_0 at $x = 0$, the ratio becomes

$$h_l \equiv \frac{n_s}{n_0} = \left[\frac{a + w_0^3}{a+1}\right]^{1/3} \tag{C.16}$$

where $w_0 = u_{in}/u_B$. If u_{in} and the electron temperature T_e, and therefore ν_{iz} and u_B, are known, then (C.16) gives n_s/n_0. For the case of zero input flux, then $w_0 = 0$, and since normally $a \ll 1$, we have

$$h_l \approx a^{1/3} \tag{C.17}$$

where a, taken from (C.14), involves the temperature. We note that (C.17) gives $n/n(0)$ at $u = u_B$ in Figure 5.2*b*.

As we shall see in Chapter 10, an approximate value of the temperature can be found from an equilibrium calculation. An accurate, but complex, equation for the temperature can also be obtained from the present formalism. With χ obtained from (C.13) at $w = w_0$, we find from (C.8) that

$$C_1 = \frac{1}{6}\ln\frac{(a^{1/3}+w_0)^3}{a+w_0^3} + \frac{1}{\sqrt{3}}\tan^{-1}\frac{2w_0 - a^{1/3}}{\sqrt{3}a^{1/3}}.$$

Substituting C_1 into (C.8) and using (C.12), we obtain a general relation for $w(\xi)$, which, evaluated at the sheath edge where $w = 1$ and $\xi = 1$, gives

$$\frac{1}{6}\ln\frac{(a^{1/3}+1)^3}{a+1} + \frac{1}{\sqrt{3}}\tan^{-1}\frac{2-a^{1/3}}{\sqrt{3}a^{1/3}} - \frac{1}{6}\ln\frac{(a^{1/3}+w_0)^3}{a+w_0^3}$$
$$-\frac{1}{\sqrt{3}}\tan^{-1}\frac{2w_0 - a^{1/3}}{\sqrt{3}a^{1/3}} = \frac{\pi l}{4\lambda_i}a^{2/3} \tag{C.18}$$

For symmetric diffusion in an electropositive plasma with $w_0 = 0$, (C.18) can be solved for T_e. Because the density profile is found to be quite flat in the center (and consequently steeper at the edge), an approximation can be made which considerably simplifies the solution. Employing particle balance by equating the production of pairs by ionization to their loss to the wall, and approximating the density for production by ionization to be $n(x) \approx n_0$, we have

$$\nu_{iz} n_0 l/2 = u_B n_s \tag{C.19}$$

Using (C.19) in (C.17), we can solve for $h_l = n_s/n_0$ to obtain

$$h_l \approx \left(\frac{4\lambda_i}{\pi l}\right)^{1/2} \tag{C.20}$$

This expression for the ratio of edge-to-center density is close to that found by Godyak and Maximov (see Godyak, 1986) from the more exact relation (C.18).

References

Abada, H., P. Chabert, J. P. Booth, J. Robiche, and G. Cartry (2002), *J. Appl. Phys.* **92**, 4223.

Abolmasov, S., P. R. i. Cabarrocas, and P. Chatterjee (2016), *EPJ Photovoltaics* **7**, 70302.

Adrian, P. J., S. D. Baalrud, and T. Lafleur (2017), *Phys. Plasmas* **24**, 123505.

Agarwal, A., and M. J. Kushner (2009), *J. Vac. Sci. Technol.* **A27**, 37.

Ahn, T. H., K. Nakamura, and H. Sugai (1995), *Jpn. J. Appl. Phys.* **34**, L1405.

Alexandrov, A. F., L. S. Bogdankevich, and A. A. Rukhadze (1984), *Principles of Plasma Electrodynamics*, Springer, New York.

Aliev, Y. M., V. Y. Bychenkov, A. V. Maximov, and H. Schlüter (1992), *Plasma Sources Sci. Technol.* **1**, 126.

Aliev, Y. M., I. D. Kaganovich, and H. Schlüter (1997), *Phys. Plasmas* **4**, 2413.

Allen, J. E., R. L. F. Boyd, and P. Reynolds (1957), *Proc. Phys. Soc. London* **B70**, 297.

Allis, W. P., S. J. Buchsbaum, and A. Bers (1963), *Waves in Anisotropic Plasmas*, MIT Press, Cambridge, MA.

Amemiya, H. (1997), *J. Phys. Soc. Jpn.* **66**, 1335.

Anders, A., ed. (2000), *Handbook of Plasma Immersion Ion Implantation and Deposition*, Wiley, New York.

Anders, A. (2010), *Thin Solid Films* **518**, 4087.

Annaratone, B. M. (1997), *J. Phys. IV France* **7**, C4–155.

Annaratone, B. M., M. W. Allen, and J. E. Allen (1992), *J. Phys. D: Appl. Phys.* **25**, 417.

Arora, P., T. Nguyen, A. Chawla, S. K. Nam, and V. M. Donnelly (2019), *J. Vac. Sci. Technol.* **A37**, 061303.

Arslanbekov, R. R., A. A. Kudryavtsev, and I. A. Mouchan (1992), *Sov. Phys. Tech. Phys.* **37**, 395.

Arslanbekov, R. R., R. C. Tobin, and A. A. Kudryavtsev (1997), *J. Appl. Phys.* **81**, 554.

Arts, K., S. Hamaguchi, T. Ito, K. Karahashi, H. C. M. Knoops, A. J. M. Mackus, and W. M. M. Kessels (2022), *Plasma Sources Sci. Technol.* **31**, 103002.

Ashida, S., and M. A. Lieberman (1997), *Jpn. J. Appl. Phys.* **36**, 854.

Ashida, S., C. Lee, and M. A. Lieberman (1995), *J. Vac. Sci. Technol.* **A13**, 2498.

Ashida, S., M. R. Shim, and M. A. Lieberman (1996), *J. Vac. Sci. Technol.* **A14**, 391.

Asmussen Jr., J., T. A. Grotjohn, P. U. Mak, and M. A. Perrin (1997), *IEEE Trans. Plasma Sci.* **25**, 1196.

Athavale, S. D., and D. J. Economou (1996), *J. Vac. Sci. Technol.* **B14**, 3702.

Atkins, P. W. (1986), *Physical Chemistry*, 3rd ed., Freeman, New York.

Baalrud, S. D., C. C. Hegna, and J. D. Callen (2009), *Phys. Rev. Lett.* **103**, 205002.

Baalrud, S. D., T. Lafleur, W. Fox, and K. Germaschewski (2015), *Plasma Sources Sci. Technol.* **24**, 015034.

Baguer, N., and A. Bogaerts (2005), *J. Appl. Phys.* **98**, 033303.

Balcon, N., G. J. M. Hagelaar, and J. P. Boeuf (2008), *IEEE Trans. Plasma Sci.* **36** 2782.

Principles of Plasma Discharges and Materials Processing, Third Edition. Michael A. Lieberman and Allan J. Lichtenberg.

Banna, S., A. Agarwal, G. Cunge, M. Darnon, E. Pargon, and O. Joubert (2012), *J. Vac. Sci. Technol.* **A30**, 040801.
Bárdoš, L. (1996), *Surf. Coat. Technol.* **86–87**, 648.
Barkin, A., R. L. Merlino, and N. D'Angelo (1995), *Phys. Plasmas* **2**, 3563.
Barnett, C. F. (1989), in *A Physicist's Desk Reference*, H. L. Anderson, ed., American Institute of Physics, New York.
Barone, M. E., and D. B. Graves (1995), *J. Appl. Phys.* **78**, 6604.
Becerra, R., and R. Walsh (1987), *J. Phys. Chem.* **91**, 5765.
Becker, K. H., U. Kogelschatz, K. H. Schoenbach, and R. J. Barker (2005), *Non-Equillibrium Air Plasmas at Atmospheric Pressure*, IOP Publishing, Bristol.
Benoit-Cattin, P., and L. C. Bernard (1968), *J. Appl. Phys.* **39**, 5723.
Berezhnoj, S. V., C. B. Shin, U. Buddemeier, and I. Kaganovich (2000), *Appl. Phys. Lett.* **77**, 800.
Berg, S., and T. Nyberg (2005), *Thin Solid Films* **476**, 215.
Berg, S., H. O. Blom, M. Moradi, C. Nender, and T. Larsson (1989), *J. Vac. Sci. Technol.* **A7**, 1225.
Bernstein, I. B., and T. Holstein (1954), *Phys. Rev.* **94**, 1475.
Bird, R. B., W. E. Stewart, and E. N. Lightfoot (2002), *Transport Phenomena*, 2nd ed., Wiley, New York.
Bischoff, L., G. Hübner, I. Korolov, Z. Donkó, P. Hartmann, T. Gans, J. Held, V. Schulz-von der Gathen, Y. Liu, T. Mussenbrock and J. Schulze (2018), *Plasma Sources Sci. Technol.* **27**, 125009.
Blank, J. L. (1968), *Phys. Fluids* **11**, 1686.
Boeuf, J. P. (2014), *Front. Phys.* **2**, 74.
Boeuf, J. P., and C. Punset (1999), in *Dusty Plasmas*, A. Bouchoule, ed., Wiley, New York.
Bogaerts, A., R. Gijbels, and W. J. Goedheer (1995), *J. Appl. Phys.* **78** 2233.
Bohdansky, J., J. Roth, and H. L. Bay (1980), *J. Appl. Phys.* **51**, 2861.
Böhme, W., W. E. Köhler, M. Römheld, S. Veprek, and R. J. Seeböck (1994), *IEEE Trans. Plasma Sci.* **22**, 110.
Book, D. L. (1987), *NRL Plasma Formulary (Revised)*, Naval Research Laboratory, Washington, DC.
Booth, J. P., G. Cunge, P. Chabert, and N. Sadeghi (1999), *J. Appl. Phys.* **85**, 3097.
Booth, J. P., N. St. J. Braithwaite, A. Goodyear, and P. Barroy (2000), *Rev. Sci. Instrum.* **71**, 2722.
Born, M., and E. Wolf (1980), *Principles of Optics*, 6th ed., Pergamon, New York.
Boswell, R. W. (1970), *Phys. Lett. A* **33**, 457.
Boswell, R. W. (1984), *Plasma Phys. Control. Fusion* **26**, 1147.
Boswell, R. W., and F. F. Chen (1997), *IEEE Trans. Plasma Sci.* **25**, 1229.
Boswell, R. W., and D. Vender (1995), *Plasma Sources Sci. Technol.* **4**, 534.
Bouchoule, A. (1999), *Dusty Plasmas*, Wiley, New York.
Bouchoule, A., A. Plain, L. Boufendi, J. P. Blondeau, and C. Laure (1991), *J. Appl. Phys.* **70**, 1991.
Boufendi, L., and A. Bouchoule (1994), *Plasma Sources Sci. Technol.* **3**, 262.
Boufendi, L., and A. Bouchoule (2002), *Plasma Sources Sci. Technol.* **11**, A211.
Boufendi, L., A. Plain, J. P. Blondeau, A. Bouchoule, C. Laure, and M. Toogood (1992), *Appl. Phys. Lett.* **60**, 169.
Boufendi, L., A. Bouchoule, and T. Hbid (1996), *J. Vac. Sci. Technol.* **A14**, 572.
Boufendi, L., W. Stoffels, and E. Stoffels (1999), in *Dusty Plasmas*, A. Bouchoule, ed., Wiley, New York, p. 181.
Boyd, R. L. F., and J. B. Thompson (1959), *Proc. R. Soc.* **A252**, 102.
Boyle, P. C., A. R. Ellingboe, and M. M. Turner (2004a), *J. Phys. D: Appl. Phys.* **37**, 697.
Boyle, P. C., J. Robiche, and M. M. Turner (2004b), *J. Phys. D: Appl. Phys.* **37**, 1451.
Braithwaite, N. St. J., and J. E. Allen (1988), *J. Phys. D: Appl. Phys.* **21**, 1733.
Braithwaite, N. St. J., J. P. Booth, and G. Cunge (1996), *Plasma Sources Sci. Technol.* **5**, 677.

Braithwaite, N. St. J., T. E. Sheridan, and R. W. Boswell (2003), *J. Phys. D: Appl. Phys.* **36**, 2837.
Bransden, B. H., and C. J. Joachain (1983), *Physics of Atoms and Molecules*, Wiley, New York.
Bransden, B. H., and M. R. C. McDowell (1992), *Charge Exchange and the Theory of Ion-Atom Collisions*, Clarendon, Oxford, UK.
Brenning, N., J. T. Gudmundsson, D. Lundin, T. Minea, M. A. Raadu, and U. Helmersson (2016), *Plasma Sources Sci. Technol.* **25**, 125003.
Brenning, N., J. T. Gudmundsson, M. A. Raadu, T. J. Petty, T. Minea, and D. Lundin (2017), *Plasma Sources Sci. Technol.* **25**, 065024.
Brown, S. C. (1959), *Basic Data of Plasma Physics*, Technology Press and Wiley, New York.
Bruggeman, P., and R. Brandenburg (2013), *J. Phys. D: Appl. Phys.* **46**, 464001.
Bruggeman, P. J., F. Iza, and R. Brandenburg (2017), *Plasma Sources Sci. Technol.* **26**, 123002.
Brutschy, B., and H. Haberland (1979), *Phys. Rev. A* **19**, 2232.
Budden, K. G. (1966), *Radio Waves in the Ionosphere*, Cambridge University Press, Cambridge, UK.
Busch, C., and U. Kortshagen (1995), *Phys. Rev. E* **51**, 280.
Butterbaugh, J. W., D. C. Gray, and H. H. Sawin (1991), *J. Vac. Sci. Technol.* **B9**, 1461.
Cale, T. S., G. B. Raupp, and T. H. Gandy (1992), *J. Vac. Sci. Technol.* **A10**, 1128.
Cao, Y. S., and R. Johnsen (1991), *J. Chem. Phys.* **95**, 7356.
Carbone, E., W. Graef, G. Hagelaar, D. Boer, M. M. Hopkins, J. C. Stephens, B. T. Yee, S. Pancheshnyi, J. van Dijk, and L. Pitchford (2021), *Atoms* **9**, 16.
Carl, D. A., D. W. Hess, M. A. Lieberman, T. D. Nguyen, and R. Gronsky (1991), *J. Appl. Phys.* **70**, 3301.
Celik, Y., D. L. Crintes, D. Luggenholscher, and U. Czarnetzki (2009), *Plasma Phys. Control. Fusion* **51**, 124040.
Chabert, P. (2007), *J. Phys. D: Appl. Phys.* **40**, R63.
Chabert, P. (2016), *Plasma Sources Sci. Technol.* **25**, 025019.
Chabert, P., and M. M. Turner (2017), *J. Phys. D: Appl. Phys.* **50**, 23LT02.
Chabert, P., A. J. Lichtenberg, M. A. Lieberman, and A. M. Marakhtanov (2001), *Plasma Sources Sci. Technol.* **10**, 478.
Chabert, P., J. L. Raimbault, J. M. Rax, and A. Perret (2004a), *Phys. Plasmas* **11**, 4081.
Chabert, P., J. L. Raimbault, J. M. Rax, and M. A. Lieberman (2004b), *Phys. Plasmas* **11**, 1775.
Chang, J. P., and H. H. Sawin (1997), *J. Vac. Sci. Technol.* **A15**, 610.
Chantry, P. J. (1987), *J. Appl. Phys.* **62**, 1141.
Charles, C., and R. W. Boswell (1998), *Appl. Phys. Lett.* **84**, 350.
Charles, C., R. W. Boswell, and H. Kuwahara (1995), *Appl. Phys. Lett.* **67**, 40.
Chen, F. F. (1965), in *Plasma Diagnostic Techniques*, R. H. Huddlestone and S. L. Leonard, eds., Academic Press, New York.
Chen, F. F. (1984), *Introduction to Plasma Physics and Controlled Fusion*, 2nd ed., Plenum, New York.
Chen, F. F. (1991), *Plasma Phys. Control. Fusion* **33**, 339.
Chen, F. F. (1992), *J. Vac. Sci. Technol.* **A10**, 1389.
Chen, F. F., and R. W. Boswell (1997), *IEEE Trans. Plasma Sci.* **25**, 1245.
Cheung, N. W. (1991), *Nucl. Instrum. Methods* **55**, 811.
Cheung, K. P., and C. P. Chang (1994), *J. Appl. Phys.* **75**, 4415.
Cho, S. (1999), *Phys. Plasmas* **6**, 359.
Cho, M. H., N. Hershkowitz, and T. Intrator (1988), *J. Vac. Sci. Technol.* **A6**, 2978.
Choi, S. J., and M. J. Kushner (1993), *J. Appl. Phys.* **74**, 853.
Choi, S. J., and J. M. White (2009), *ECS Transactions* **25**, 701.
Chow, V. W., D. A. Mendis, and M. Rosenberg (1994), *IEEE Trans. Plasma Sci.* **22**, 179.
Christophorou, L. G., and J. K. Olthoff (1999a), *J. Phys. Chem. Ref. Data* **28**, 131.

Christophorou, L. G., and J. K. Olthoff (1999b), *J. Phys. Chem. Ref. Data* **28**, 967.
Christophorou, L. G., J. K. Olthoff, and M. V. V. S. Rao (1996), *J. Phys. Chem. Ref. Data* **25**, 1341.
Chung, K. J., J. M. Choe, G. H. Kim, and Y. S. Hwang (2012), *Thin Solid Films* **521**, 197.
Chung, K. J., B. Jung, G. H. Kim, and Y. S. Hwang (2013), *Thin Solid Films* **547**, 13.
Clausing, P. (1932), *Ann. Phys.* **404**, 961.
Cobine, J. D. (1958), *Gaseous Conductors*, Dover, New York.
Coburn, J. W., and M. Chen (1980), *J. Appl. Phys.* **51**, 3134.
Coburn, J. W., and E. Kay (1972), *J. Appl. Phys.* **43**, 4965.
Coburn, J. W., and H. F. Winters (1979), *J. Vac. Sci. Technol.* **16**, 391.
Coburn, J. W., and H. F. Winters (1989), *Appl. Phys. Lett.* **55**, 2730.
Coffey, B. M., H. C. Nallan, and J. C. Ekerdt (2021), *J. Vac. Sci. Technol.* **A39**, 012601.
Cohen, R. H., and T. D. Rognlien (1996a), *Plasma Sources Sci. Technol.* **5**, 442.
Cohen, R. H., and T. D. Rognlien (1996b), *Phys. Plasmas* **3**, 1839.
Cohen, R. H., I. B. Bernstein, J. J. Dorning, and G. Rowlands (1980), *Nucl. Fusion* **20**, 1421.
Collins, S. M., D. A. Brown, J. F. O'Hanlon, and R. N. Carlile (1996), *J. Vac. Sci. Technol.* **A14**, 634.
Colpo, P., T. Meziani, and F. Rossi (2005), *J. Vac. Sci. Technol.* **A23**, 270.
Conrad, J. R. (1987), *J. Appl. Phys.* **62**, 777.
Conrad, J. R., R. A. Dodd, S. Han, M. Madapura, J. Scheuer, K. Sridharam, and F. J. Worzala (1990), *J. Vac. Sci. Technol.* **A8**, 3146.
Cook, J. M., D. E. Ibbotson, P. D. Foo, and D. L. Flamm (1990), *J. Vac. Sci. Technol.* **A8**, 1820.
Corr, C. S., P. G. Steen, and W. G. Graham (2005), *Appl. Phys. Lett.* **86**, 141503.
Courteille, C., Ch. Hollenstein, J.-L. Dorier, P. Gay, W. Schwarzenbach, A. A. Howling, E. Bertran, G. Viera, R. Martins, and A. Macarico (1996), *J. Appl. Phys.* **80**, 2069.
Cremers, V., R. L. Puurunen, and J. Dendooven (2019), *Appl. Phys. Rev.* **6**, 021302.
Cunge, G., and J. P. Booth (1999), *J. Appl. Phys.* **85**, 3952.
Curreli, D., and F. F. Chen (2011), *Phys. Plasmas* **18**, 113501.
Curreli, D., and F. F. Chen (2014), *Plasma Sources Sci. Technol.* **23**, 064001.
Czarnetzki, U. (2013), *Phys. Rev. E* **88**, 063101.
Czarnetzki, U., T. Mussenbrock, and R. P. Brinkmann (2006), *Phys. Plasmas* **13**, 123503.
Czarnetzki, U., J. Schulze, E. Schüngel, and Z. Donkó (2011), *Plasma Sources Sci. Technol.* **20**, 024010.
D'Angelo, N. (1995), *J. Phys. D: Appl. Phys.* **28**, 1009.
Degeling, A. W., and R. W. Boswell (1997), *Phys. Plasmas* **4**, 2748.
Degeling, A. W., C. O. Jung, R. W. Boswell, and A. R. Ellingboe (1996), *Phys. Plasmas* **3**, 2788.
Degeling, A. W., T. E. Sheridan, and R. W. Boswell (1999), *Phys. Plasmas* **6**, 1641.
Dendooven, J., D. Deduytsche, J. Musschoot, R. L. Vanmeirhaeghe, and C. Detavernier (2009), *J. Electrochem. Soc.* **156**, P63.
Depla, D., S. Mahieu, and R. De Gryse (2009), *Thin Solid Films* **517**, 2825.
de Rosny, G., E. R. Mosburg Jr., J. R. Abelson, G. Devaud and R. C. Kerns (1983), *J. Appl. Phys.* **54**, 2272.
Derzsi, A., T. Lafleur, J. P. Booth, I. Korolov, and Z. Donkó (2016), *Plasma Sources Sci. Technol.* **25**, 015004.
Despiau-Pujo, E., and P. Chabert (2009), *Plasma Sources Sci. Technol.* **18**, 045028.
Deutsch, R., and E. Räuchle (1992), *Phys. Rev. A* **46**, 3442.
Ding, K., M. A. Lieberman, A. J. Lichtenberg, J. J. Shi, and J. Zhang (2014), *Plasma Sources Sci. Technol.* **23**, 065048.
Donkó, Z., J. Schulze, B. G. Heil, and U. Czarnetzki (2009), *J. Phys. D: Appl. Phys.* **42**, 025205.
Donkó, Z., J. Schulze, U. Czarnetzki, A. Derzsi, P. Hartmann, I. Korolov, and E. Schüngel (2012), *Plasma Phys. Control. Fusion* **54**, 124003.

Donnelly, V. M. (1989), in *Plasma Diagnostics*, Vol. **1**, O. Auciello and D. L. Flamm, eds., pp. 1–46, Academic Press, New York.
Donnelly, V. M. (2017), *J. Vac. Sci. Technol.* **A35**, 05C202.
Donnelly, V. M., and A. Kornblit (2013), *J. Vac. Sci. Technol.* **A31**, 050825.
Donnelly, I. J., and P. A. Watterson (1989), *J. Phys. D.* **22**, 90.
Dorf, L., J. C. Wang, S. Rauf, G. A. Monroy, Y. Zhang, A. Agarwal, J. Kenney, K. Ramaswamy, and K. Collins (2017), *J. Phys. D: Appl. Phys.* **50**, 274003.
Druyvesteyn, M. J., and F. M. Penning (1940), *Rev. Mod. Phys.* **12**, 87.
Du, L., D. J. Economou, and V. M. Donnelly (2022), *J. Vac. Sci. Technol.* **B40**, 022207.
Eckert, H. U. (1986), *Proceedings of the 2nd Annual International Conference on Plasma Chemistry and Technology*, H. Boening, ed., Technomic Publishing, Lancaster, PA.
Economou, D. (2014), *J. Phys. D: Appl. Phys.* **47**, 303001.
Edgell, W. F. (1961), in *Argon, Helium and the Rare Gases*, G. A. Cook, ed., Wiley, New York.
Eliasson, B., and U. Kogelschatz (1986), Basic data for modelling of electrical discharges in gases: oxygen, *Report KLR 86-11C*, Brown Boveri Konzernforschung, CH-5405 Baden.
Eliasson, B., and U. Kogelschatz (1991), *IEEE Trans. Plasma Sci.* **19**, 309.
Eliasson, B., M. Hirth, and U. Kogelschatz (1987), *J. Phys. D: Appl. Phys.* **20**, 1421.
Ellis, H. W., R. Y. Pai, E. W. McDaniel, E. A. Mason, and L. A. Viehland (1976), *At. Data Nucl. Data Tables* **17**, 177.
Ellis, H. W., E. W. McDaniel, D. L. Albritton, L. A. Viehland, S. L. Lin, and E. A. Mason (1978), *At. Data Nucl. Data Tables* **22**, 179.
Ellis, H. W., M. G. Thackston, E. W. McDaniel, and E. A. Mason (1984), *At. Data Nucl. Data Tables* **31**, 113.
Emmert, G. A. (1994), *J. Vac. Sci. Technol.* **B12**, 880.
Emmert, G. A., and M. A. Henry (1992), *J. Appl. Phys.* **71**, 113.
Fadeev, A. V., and K. V. Rudenko (2018), *Tech. Phys.* **63**, 1228.
Fang, C., Y. Cao, D. Wu, and A. Li (2018), *Prog. Nat. Sci. Mater. Int.* **28**, 667.
Fehsenfeld, F. C., A. L. Schmeltekopf, H. I. Schiff, and E. E. Ferguson (1967), *Planet. Space Sci.* **15**, 373.
Feldman, L. C., and J. W. Mayer (1986), *Fundamentals of Surface and Thin Film Analysis*, North-Holland, New York.
Fischer, A., A. Routzahn, S. M. George, and T. Lill (2021), *J. Vac. Sci. Technol.* **A39**, 030801.
Flamm, D. L. (1989), in *Plasma Etching: An Introduction*, D. M. Manos and D. L. Flamm, eds., Academic Press, New York.
Flamm, D. L. (1990), Mechanisms of silicon etching in fluorine- and chlorine-containing plasmas, *Report UCB/ERL M90/41*, College of Engineering, University of California, Berkeley, CA.
Flamm, D. L., and G. K. Herb (1989), in *Plasma Etching: An Introduction*, D. M. Manos and D. L. Flamm, eds., Academic Press, New York.
Flamm, D. L., V. M. Donnelly, and J. A. Mucha (1981), *J. Appl. Phys.* **52**, 3633.
Franklin, R. N. (1976), *Plasma Phenomena in Gas Discharges*, Clarendon, Oxford, UK.
Franklin, R. N. (2001), *J. Phys. D: Appl. Phys.* **34**, 1834.
Franklin, R. N. (2002), *J. Phys. D: Appl. Phys.* **35**, 2270.
Franklin, R. N. (2003), *J. Phys. D: Appl. Phys.* **36**, 2660.
Franklin, R. N. (2004), *J. Phys. D: Appl. Phys.* **37**, 1342.
Franklin, R. N., and J. R. Ockendon (1970), *J. Plasma Phys.* **4**, 371.
Franklin, R. N., and J. Snell (1994), *J. Phys. D: Appl. Phys.* **27**, 2102.
Franklin, R. N., and J. Snell (2000a), *J. Phys. D: Appl. Phys.* **33**, 2019.
Franklin, R. N., and J. Snell (2000b), *J. Plasma Phys.* **64**, 131.

Franklin, R. N., and J. Snell (2000c), *Phys. Plasmas* **7**, 3077.
Fridman, A. A., L. Boufendi, T. Hbid, B. V. Potapkin, and A. Bouchoule (1996), *J. Appl. Phys.* **79**, 1303.
Fruchtman, A. (2009), *Plasma Sources Sci. Technol.* **18**, 025033.
Fruchtman, A. (2017), *J. Phys. D: Appl. Phys.* **50**, 473002.
Fuller, E. N., P. D. Schettler, and J. C. Giddings (1966), *Ind. Eng. Chem.* **58**, 18; 83.
Fuller, N. C. M., M. V. Malyshev, V. M. Donnelly, and I. P. Herman (2000), *Plasma Sources Sci. Technol.* **9**, 116.
Gallagher, A. (2000), *Phys. Rev. E* **62**, 2690.
Gans, T., C. C. Lin, V. Schulz-von der Gathen, and H. F. Döble (2001a), *J. Phys. D: Appl. Phys.* **34**, L39.
Gans, T., V. Schulz-von der Gathen, and H. F. Döble (2001b), *Plasma Sources Sci. Technol.* **10**, 17.
Gans, T., V. Schulz-von der Gathen, and H. F. Döble (2004a), *Europhys. Lett.* **66**, 232.
Gans, T., V. Schulz-von der Gathen, and H. F. Döble (2004b), *Contrib. Plasma Phys.* **44**, 523.
Gans, T., D. O'Connell, J. Schulze, V. A. Kadetov, and U. Czarnetzki (2006a), *AIP Conf. Proc.* **876**, 260.
Gans, T., J. Schulze, D. O'Connell, U. Czarnetzki, R. Faulkner, A. R. Ellingboe, and M. M. Turner (2006b), *Appl. Phys. Lett.* **89**, 261502.
Gapanov, E. V., and M. E. Miller (1958), *Sov. Phys. JETP* **25**, 1273.
George, S. M. (2010), *Chem. Rev.* **110**, 111.
Gilland, J., R. Bruen, and N. Hershkowitz (1998), *Plasma Sources Sci. Technol.* **7**, 416.
Ginter, M. L., and R. Battino (1970), *J. Chem. Phys.* **52**, 4469.
Ginzburg, V. L. (1964), *The Propagation of Electromagnetic Waves in Plasma*, Pergamon, Oxford, UK.
Godyak, V. A. (1986), *Soviet Radio Frequency Discharge Research*, Delphic Associates, Falls Church, VA.
Godyak, V. A. (1990a), in *Plasma–Surface Interactions and Processing of Materials*, O. Auciello, A. Gras-Marti, J. A. Valles-Abarca, and D. L. Flamm, eds., pp. 95–134, Kluwer Academic, Boston, MA.
Godyak, V. A. (1990b), Private communication.
Godyak, V. A. (2003), *Plasma Phys. Control. Fusion* **45**, A399.
Godyak, V. A. (2011), *Plasma Sources Sci. Technol.* **20**, 025004.
Godyak, V. A. (2013), *J. Phys. D: Appl. Phys.* **46**, 283001.
Godyak, V. A. (2017), *Phys. Plasmas* **24**, 060702.
Godyak, V. A. (2020), *Phys. Plasmas* **27**, 013504.
Godyak, V. A. (2021), *J. Appl. Phys.* **129**, 041101.
Godyak, V. A., and B. M. Alexandrovich (2015), *J. Appl. Phys.* **118**, 233302.
Godyak, V. A., and B. M. Alexandrovich (2017), *Rev. Sci. Instrum.* **88**, 083512.
Godyak, V. A., and A. S. Khanneh (1986), *IEEE Trans. Plasma Sci.* **14**, 112.
Godyak, V. A., and V. I. Kolobov (1997), *Phys. Rev. Lett.* **79**, 4589.
Godyak, V. A., and R. B. Piejak (1990a), *Phys. Rev. Lett.* **65**, 996.
Godyak, V. A., and R. B. Piejak (1990b), *J. Vac. Sci. Technol.* **A8**, 3833.
Godyak, V. A., and R. B. Piejak (1997), *J. Appl. Phys.* **82**, 5944.
Godyak, V. A., and N. Sternberg (1990a), *IEEE Trans. Plasma Sci.* **18**, 159.
Godyak, V. A., and N. Sternberg (1990b), *Phys. Rev. A.* **42**, 2299.
Godyak, V. A., and N. Sternberg (2024), *J. Appl. Phys.* **135**, 013302.
Godyak, V. A., R. B. Piejak, and B. M. Alexandrovich (1991), *IEEE Trans. Plasma Sci.* **19**, 660.
Godyak, V. A., R. B. Piejak, and B. M. Alexandrovich (1992), *Plasma Sources Sci. Technol.* **1**, 36.
Godyak, V. A., R. B. Piejak, and B. M. Alexandrovich (1993), *J. Appl. Phys.* **73**, 3657.
Godyak, V. A., R. B. Piejak, and B. M. Alexandrovich (1994), *Plasma Sources Sci. Technol.* **3**, 169.
Godyak, V. A., R. B. Piejak, and B. M. Alexandrovich (1999), *J. Appl. Phys.* **85**, 703.
Godyak, V. A., R. B. Piejak, and B. M. Alexandrovich (2002), *Plasma Sources Sci. Technol.* **11**, 525.
Goertz, C. K. (1989), *Rev. Geophys.* **27**, 271.

Gogolides, E., and H. H. Sawin (1992), *J. Appl. Phys.* **72** 3988.
Gogolides, E., H. H. Sawin, and R. R. Brown (1992), *Chem. Eng. Sci.* **47** 3839.
Goldstein, H. (1950), *Classical Mechanics*, Addison-Wesley, Cambridge, MA.
Golubovskii, Y. B., V. A. Maiorov, J. Behnke, and J. F. Behnke (2002), *J. Phys. D: Appl. Phys.* **35**, 751.
Golubovskii, Y. B., V. A. Maiorov, J. Behnke, and J. F. Behnke (2003), *J. Phys. D: Appl. Phys.* **36**, 39.
Gordon, R. G., D. Hausmann, E. Kim, and J. Shepard (2003), *Chem. Vap. Deposition* **9**, 73.
Goto, H. H., H. D. Löwe, and T. Ohmi (1992), *J. Vac. Sci. Technol.* **A10**, 3048.
Gozadinos, G. (2001), *Collisionless Heating and Particle Dynamics in Radio-Frequency Capacitive Plasma Sheaths*, Thesis, Dublin City University, Dublin, Ireland.
Gozadinos, G., M. M. Turner, and D. Vender (2001a), *Phys. Rev. Lett.* **87**, 135004–1.
Gozadinos, G., M. M. Turner, D. Vender, and M. A. Lieberman (2001b), *Plasma Sources Sci. Technol.* **10**, 117.
Graves, D. B., and P. Brault (2009), *J. Phys. D: Appl. Phys.* **42**, 194011.
Graves, D. B., J. E. Daugherty, M. D. Kilgore, and R. K. Porteous (1994), *Plasma Sources Sci. Technol.* **3**, 433.
Gray, D. C., I. Tepermeister, and H. H. Sawin (1993), *J. Vac. Sci. Technol.* **B11**, 1243.
Greaves, J. C., and W. Linnett (1959), *Trans. Faraday Soc.* **55**, 1355.
Gudmundsson, J. T. (2001), *Plasma Sources Sci. Technol.* **10**, 76.
Gudmundsson, J. T. (2002a), *J. Phys. D: Appl. Phys.* **35**, 328.
Gudmundsson, J. T. (2002b), Notes on the electron excitation rate coefficients for argon and oxygen discharge, Report RH-21-2002, Science Institute, Univ. Iceland, Reykjavik, Iceland.
Gudmundsson, J. T. (2004), *J. Phys. D: Appl. Phys.* **37**, 2073.
Gudmundsson, J. T. (2020), *Plasma Sources Sci. Technol.* **29**, 113011.
Gudmundsson, J. T., and M. A. Lieberman (1998), *Plasma Sources Sci. Technol.* **7**, 83.
Gudmundsson, J. T., and M. A. Lieberman (2011), *Phys. Rev. Lett.* **107**, 045002.
Gudmundsson, J. T., and M. A. Lieberman (2015), *Plasma Sources Sci. Technol.* **24**, 035016.
Gudmundsson, J. T., and E. G. Thorsteinsson (2007), *Plasma Sources Sci. Technol.* **16**, 399.
Gudmundsson, J. T., A. M. Marakhtanov, K. K. Patel, V. P. Gopinath, and M. A. Lieberman (2000), *J. Phys. D: Appl. Phys.* **33**, 3010.
Gudmundsson, J. T., I. G. Kouznetsov, K. K. Patel, and M. A. Lieberman (2001), *J. Phys. D: Appl. Phys.* **34**, 1100.
Gudmundsson, J. T., N. Brenning, D. Lundin, and U. Helmersson (2012a), *J. Vac. Sci. Technol.* **A30**, 030801.
Gudmundsson, J. T., A. T. Hjartarson, and E. G. Thorsteinsson (2012b), *Vacuum* **86**, 808.
Gudmundsson, J. T., J. Krek, D. Q. Wen, E. Kawamura, and M. A. Lieberman (2021), *Plasma Sources Sci. Technol.* **30**, 125011.
Gudmundsson, J. T., A. Anders, and A. von Keudell (2022), *Plasma Sources Sci. Technol.* **31**, 083001.
Gupta, D. (2011), *Int. J. Adv. Technol.* **2**, 471.
Hagelaar, G. J. M., and L. C. Pitchford (2005), *Plasma Sources Sci. Technol.* **14**, 722.
Hagelaar, G. J. M., and N. Oudini (2011), *Plasma Phys. Control. Fusion* **53**, 124032.
Hagelaar, G. J. M., D. B. Mihailova, and J. van Dijk (2010), *J. Phys. D: Appl. Phys.* **43**, 465204.
Harrington, R. F. (1961), *Time-Harmonic Electromagnetic Fields*, McGraw-Hill, New York.
Hashimoto, K. (1994), *Jpn. J. Appl. Phys.* **33**, 6013.
Haugsjaa, P. O., and R. C. Amme (1970), *J. Chem. Phys.* **52**, 4874.
Hayashi, M. (1987), in *Swarm Studies and Inelastic Electron–Molecule Collisions*, L. C. Pitchford, B. V. McKoy, A. Chutjian, and S. Trajmar, eds., Springer, New York.
He, L., Z. Wu, Z. Li, J. Ju, Q. Ou, and R. Liang (2013), *J. Phys. D: Appl. Phys.* **46**, 175306.

He, L., F. He, J. Ouyang, and W. Dou (2020), *Phys. Plasmas* **27**, 123511.
Heald, M. A., and C. B. Wharton (1965), *Plasma Diagnostics with Microwaves*, Wiley, New York.
Hebner, G. A. (1996), *J. Appl. Phys.* **80**, 2624.
Heil, B. G., U. Czarnetzki, R. P. Brinkmann, and T. Mussenbrock (2008), *J. Phys. D: Appl. Phys.* **41**, 165202.
Helmersson, U., M. Lattemann, J. Bohlmark, A. P. Ehiasarian, and J. T. Gudmundsson (2006), *Thin Solid Films* **513**, 1.
Herrick, A., A. J. Perry, and R. W. Boswell (2003), *J. Vac. Sci. Technol.* **A21**, 955.
Hershkowitz, N. (1989), in *Plasma Diagnostics*, Vol. **1**, O. Auciello and D. L. Flamm, eds., Academic Press, New York.
Hershkowitz, N., K. N. Leung, and T. Romesser (1975), *Phys. Rev. Lett.* **35**, 277.
Herzberg, G. (1971), *The Spectra and Structures of Simple Free Radicals*, Dover, New York.
Hittorf, W. (1884), *Wiedemanns Ann. Phys.* **21**, 90.
Hoffman, D. M., B. Singh, and J. H. Thomas III, eds. (1998), *Handbook of Vacuum Science and Technology*, Academic Press, San Diego, CA.
Hollenstein, Ch. (2000), *Plasma Phys. Control. Fusion* **42**, R93.
Hollenstein, Ch., A. A. Howling, C. Courteille, D. Magni, S. M. Scholz, G. M. W. Kroesen, N. Simons, W. de Zeeuw, and W. Schwarzenbach (1998), *J. Phys. D: Appl. Phys.* **31**, 74.
Holt, H. E., and R. E. Haskell (1965), *Plasma Dynamics*, Macmillan, New York.
Hong, Y. J., M. Yoon, F. Iza, G. C. Kim, and J. K. Lee (2008), *J. Phys. D: Appl. Phys.* **41**, 245208.
Hopwood, J. A. (2000), in *Ionized Physical Vapor Deposition*, J. A. Hopwood, ed., pp. 181–207, Academic Press, San Diego, CA.
Hopwood, J., C. R. Guarnieri, S. J. Whitehair, and J. J. Cuomo (1993a), *J. Vac. Sci. Technol.* **A11**, 147.
Hopwood, J., C. R. Guarnieri, S. J. Whitehair, and J. J. Cuomo (1993b), *J. Vac. Sci. Technol.* **A11**, 152.
Horwitz, C. M. (1983), *Appl. Phys. Lett.* **43**, 977.
Hosokawa, N., R. Matsuzake, and T. Asamaki (1974), *Jpn. J. Appl. Phys.* **Suppl. 2, Pt. 1**, 435.
Howling, A. A., B. Legradic, M. Chesaux, and Ch. Hollenstein (2012), *Plasma Sources Sci. Technol.* **21**, 015005.
Hsu, C. C., M. A. Nierode, J. W. Coburn, and D. B. Graves (2006), *J. Phys. D: Appl. Phys.* **39**, 3272.
Huang, S., and J. T. Gudmundsson (2013), *Plasma Sources Sci. Technol.* **22**, 055020.
Huang, S., and J. T. Gudmundsson (2014), *Plasma Sources Sci. Technol.* **23**, 025015.
Huard, C. M., Y. Zhang, S. Sriraman, A. Patterson, K. Kanarik, and M. J. Kushner (2017), *J. Vac. Sci. Technol.* **A35**, 031306.
Huddlestone, R. H., and S. L. Leonard, eds. (1965), *Plasma Diagnostic Techniques*, Academic Press, New York.
Humbird, D., and D. B. Graves (2004), *J. Appl. Phys.* **96**, 791.
Humbird, D., D. B. Graves, A. A. E. Stevens, and W. M. M. Kessels (2007), *J. Vac. Sci. Technol.* **A25**, 1529.
Huo, C., D. Lundin, M. A. Raadu, A. Anders, J. T. Gudmundsson, and N. Brenning (2013), *Plasma Sources Sci. Technol.* **22**, 045005.
Hurlbatt, A., A. R. Gibson, S. Schröter, J. Bredin, A. P. S. Foote, P. Grondein, D. O'Connell, and T. Gans (2017), *Plasma Process. Polym.* **14**, 1600138.
Hussein, M. A., and G. A. Emmert (1990), *Phys. Fluids* **B2**, 218.
Hutchinson, M. H. R. (1980), *Appl. Phys.* **21**, 95.
Hutchinson, D. A. W., M. M. Turner, R. A. Doyle, and M. B. Hopkins (1995), *IEEE Trans. Plasma Sci.* **23**, 636.
Hwang, G. S., and K. P. Giapis (1997), *J. Vac. Sci. Technol.* **B15**, 70.
Hwang, G. S., and K. P. Giapis (1998), *Jpn. J. Appl. Phys.* **37**, 2291.

Hwang, H. J., Y. C. Kim, and C. W. Chung (2013), *Thin Solid Films* **547**, 9.

Ichimaru, S. (1982), *Rev. Mod. Phys.* **54**, 1017.

Ignatov, A. M. (2005), *Plasma Phys. Rep.* **31**, 46.

Ingold, J. H. (1997), *Phys. Rev. E* **56**, 5932.

Ishikawa, K., K. Karahashi, T. Ichiki, J. P. Chang, S. M. George, W. M. M. Kessels, H. J. Lee, S. Tinck, J. H. Um, and K. Kinoshita (2017), *Jpn. J. Appl. Phys.* **56**, 06HA02.

Iza, F., J. K. Lee, and M. Kong (2007), *Phys. Rev. Lett.* **99** 075004.

Iza, F., G. J. Kim, S. M. Lee, J. K. Lee, J. L. Walsh, Y. T. Zhang, and M. G. Kong (2008), *Plasma Process. Polym.* **5** 322.

Jackson, J. D. (1975), *Classical Electrodynamics*, 2nd ed., Wiley, New York.

Jaeger, F., A. J. Lichtenberg, and M. A. Lieberman (1972), *Plasma Phys.* **14**, 1073.

Jaffke, T., M. Meinke, R. Hashemi, L. G. Christophorou, and E. Illenberger (1992), *Chem. Phys. Lett.* **193**, 62.

Jo, Y. H., C. Cheon, H. Park, M. Y. Hur, and H. J. Lee (2022), *J. Korean Phys. Soc.* **80**, 787.

Johnson, E. V., T. Verbeke, J. C. Vanel, and J. P. Booth (2010), *J. Phys. D: Appl. Phys.* **43**, 412001.

Johnson, R. W., A. Hultqvist, and S. F. Bent (2014), *Mater. Today* **17**, 236.

Joyce, B. A., and C. T. Foxon (1984), in *Simple Processes at the Gas–Solid Interface*, C. H. Bamford, C. F. H. Tipper, and R. G. Compton, eds., Elsevier, Amsterdam.

Kääriäinen, T., D. Cameron, M. L. Kääriäinen, and A. Sherman (2013), *Atomic Layer Deposition: Principles, Characteristics, and Nanotechnology Applications*, 2nd ed., Wiley, New York.

Kagan, Yu. M., and V. I. Perel (1964), *Sov. Phy. Usp.* **6**, 767.

Kagan, Yu. M., R. I. Lyagushchenko, and S. I. Khvorostovskii (1975), *Sov. Phys. Tech. Phys.* **20**, 1164.

Kaganovich, I. D. (2001), *Phys. Plasmas* **8**, 2540.

Kaganovich, I. D. (2002), *Phys. Rev. Lett.* **89**, 265006.

Kaganovich, I. D., and L. D. Tsendin (1992a), *IEEE Trans. Plasma Sci.* **20**, 66.

Kaganovich, I. D., and L. D. Tsendin (1992b), *IEEE Trans. Plasma Sci.* **20**, 86.

Kaganovich, I. D., and L. D. Tsendin (1993), *Plasma Phys. Rep.* **19**, 645.

Kaganovich, I. D., V. I. Kolobov, and L. D. Tsendin (1996), *Appl. Phys. Lett.* **69**, 3818.

Kaganovich, I. D., B. N. Ramamurthi, and D. J. Economou (2000), *Appl. Phys. Lett.* **76**, 2844.

Kaganovich, I. D., O. V. Polomarov, and C. E. Theodosiou (2004), *Phys. Plasmas* **11**, 2399.

Kanarik, K. J., T. Lill, E. A. Hudson, S. Sriraman, S. Tan, J. Marks, V. Vahedi, and R. A. Gottscho (2015), *J. Vac. Sci. Technol.* **A33**, 020802.

Kannari, F., M. Obara, and T. Fujioka (1985), *J. Appl. Phys.* **57**, 4309.

Kasashima, Y., T. Motomura, N. Nabeoka, and F. Uesugi (2015), *Jpn. J. Appl. Phys.* **54**, 01AE02.

Katsch, H. M., A. Tewes, E. Quandt, A. Goehlich, T. Kawetzki, and H. F. Döbele (2000), *J. Appl. Phys.* **88**, 6232.

Kawamura, E., V. Vahedi, M. A. Lieberman, and C. K. Birdsall (1999), *Plasma Sources Sci. Technol.* **8**, R45.

Kawamura, E., M. A. Lieberman, and A. J. Lichtenberg (2006), *Phys. Plasmas* **13**, 053506.

Kawamura, E., M. A. Lieberman, A. J. Lichtenberg, and E. A. Hudson (2007), *J. Vac. Sci. Technol., A* **25**, 1456.

Kawamura, E., D. B. Graves, and M. A. Lieberman (2011), *Plasma Sources Sci. Technol.* **20**, 035009.

Kawamura, E., M. A. Lieberman, A. J. Lichtenberg, and D. B. Graves (2012), *Plasma Sources Sci. Technol.* **21**, 045014.

Kawamura, E., M. A. Lieberman, A. J. Lichtenberg, P. Chabert, and C. Lazzaroni (2014), *Plasma Sources Sci. Technol.* **23** 035014.

Kawamura, E., A. J. Lichtenberg, M. A. Lieberman, and A. M. Marakhtanov (2016), *Plasma Sources Sci. Technol.* **25**, 035007.

Kawamura, E., M. A. Lieberman, A. J. Lichtenberg, and P. Chabert (2020), *J. Vac. Sci. Technol.* **A38**, 023003.

Kawamura, E., M. A. Lieberman, A. J. Lichtenberg, and P. Chabert (2021), *Plasma Sources Sci. Technol.* **30**, 035001.

Kemaneci, E., E. Carbone, J. P. Booth, W. Graef, J. van Dijk, and G. Kroesen (2014), *Plasma Sources Sci. Technol.* **23**, 045002.

Kessels, E., H. Profijt, S. Potts, and R. van de Sanden (2012), in *Atomic Layer Deposition of Nanostructured Materials*, M. Pinna and M. Knez, eds., Wiley-VCH, Weinheim, Germany.

Kim, H. C., J. K. Lee, and J. W. Shon (2003), *Phys. Plasmas* **10**, 4545.

Kim, H. C., F. Iza, S. S. Yang, M. Radmilović-Radjenović, and J. K. Lee (2005), *J. Phys. D: Appl. Phys.* **38** R283.

Kim, S. K., S. W. Lee, C. S. Hwang, Y. S. Min, J. Y. Won, and J. Jeong (2006a), *J. Electrochem. Soc.* **153**, F69.

Kim, S., M. A. Lieberman, A. J. Lichtenberg, and J. T. Gudmundsson (2006b), *J. Vac. Sci. Technol.* **A24** 2025.

Kim, K. N., J. H. Lim, and G. Y. Yeom (2010), *Plasma Chem. Plasma Process.* **30**, 183.

Kimura, T., and K. Ohe (1999), *Plasma Sources Sci. Technol.* **8**, 553.

Kimura, T., A. J. Lichtenberg, and M. A. Lieberman (2001), *Plasma Sources Sci. Technol.* **10**, 430.

Kinoshita, T., M. Hane, and J. P. McVittie (1996), *J. Vac. Sci. Technol.* **B14**, 560.

Kitajima, T., Y. Takeo, Z. L. Petrović, and T. Makabe (2000), *Appl. Phys. Lett.* **77**, 489.

Klick, M. (1996), *J. Appl. Phys.* **79**, 3445.

Klick, M., W. Rehak, and M. Kammeyer (1997), *Jpn. J. Appl. Phys.* **36**, 4625.

Knoops, H. C. M., E. Langereis, M. C. M. van de Sanden, and W. M. M. Kessels (2010), *J. Electrochem. Soc.* **157**, G241.

Kogelschatz, U. (2003), *Plasma Chem. Plasma Process.* **23**, 1.

Kogoma, M., and K. Tanaka (2021), *Rev. Mod. Plasma Phys.* **5**, 3.

Kolobov, V. I., and D. Economou (1998), *Appl. Phys. Lett.* **72**, 656.

Kolobov, V. I., and V. A. Godyak (1995), *IEEE Trans. Plasma Sci.* **23**, 503.

Kolobov, V. I., and V. A. Godyak (2017), *Plasma Sources Sci. Technol.* **26**, 075013.

Kolobov, V. I., and W. N. G. Hitchon (1995), *Phys. Rev. E* **52**, 972.

Kolobov, V. I., and L. D. Tsendin (1995), *Plasma Sources Sci. Technol.* **4**, 551.

Kolobov, V. I., D. F. Beale, L. J. Mahoney, and A. E. Wendt (1994), *Appl. Phys. Lett.* **65**, 537.

Kolobov, V. I., G. J. Parker, and W. N. G. Hitchon (1996), *Phys. Rev. E* **53**, 1110.

Komachi, K. (1993), *J. Vac. Sci. Technol.* **A11**, 164.

Komori, A., T. Shoji, K. Miyamoto, J. Kawai, and Y. Kawai (1991), *Phys. Fluids* **B3**, 893.

Konuma, M. (1992), *Film Deposition by Plasma Techniques*, Springer, New York.

Kortshagen, U., and U. Bhandarkar (1999), *Phys. Rev. E* **60**, 887.

Kortshagen, U., and B. G. Heil (1999), *IEEE Trans. Plasma Sci.* **27**, 1297.

Kortshagen, U., I. Pukropski, and L. D. Tsendin (1995), *Phys. Rev. E* **51**, 6063.

Kortshagen, U., C. Busch, and L. D. Tsendin (1996), *Plasma Sources Sci. Technol.* **5**, 1.

Kossyi, I. A., A. Y. Kostinsky, A. A. Matveyev, and V. P. Silakov (1992), *Plasma Sources Sci. Technol.* **1**, 207.

Kota, G. P., J. W. Coburn, and D. B. Graves (1998), *J. Vac. Sci. Technol.* **A16**, 270.

Kota, G. P., J. W. Coburn, and D. B. Graves (1999), *J. Appl. Phys.* **85**, 74.

Kouznetsov, I. G., A. J. Lichtenberg, and M. A. Lieberman (1996), *Plasma Sources Sci. Technol.* **5**, 662.

Kouznetsov, I. G., A. J. Lichtenberg, and M. A. Lieberman (1999), *J. Appl. Phys.* **86**, 4142.
Krall, N. A., and A. W. Trivelpiece (1973), *Principles of Plasma Physics*, McGraw-Hill, New York.
Krishnakumar, E., and S. K. Srivastava (1992), *Int. J. Mass Spectrom. Ion Processes* **113**, 1.
Kubart, T., J. T. Gudmundsson, and D. Lundin (2020), in *High Power Impulse Magnetron Sputtering: Fundamentals, Technologies, Challenges, and Applications*, D. Lundin, T. Minea, and J. T. Gudmundsson, eds, Elsevier, Amsterdam, p. 223.
Kulkarni, N. S., and R. T. DeHoff (2002), *J. Electrochem. Soc.* **149**, G620.
Kunene, T. J., L. K. Tartibu, K. Ukoba, and T. C. Jen (2022), *Mater. Today: Proceed.* **62**, S95.
Kushner, M. J. (1988), *J. Appl. Phys.* **63**, 2532.
Kushner, M. J. (2003), *J. Appl. Phys.* **94**, 1436.
Kuypers, A. D., and H. J. Hopman (1990), *J. Appl. Phys.* **67**, 1229.
Lafleur, T., and A. Aanesland (2014), *Phys. Plasmas* **21**, 063510.
Lafleur, T., and J. P. Booth (2012), *J. Phys. D: Appl. Phys.* **45**, 395203.
Lafleur, T., and R. W. Boswell (2012a), *Phys. Plasmas* **19**, 053505.
Lafleur, T., and R. W. Boswell (2012b), *Phys. Plasmas* **19**, 023508.
Lafleur, T., and P. Chabert (2015a), *Plasma Sources Sci. Technol.* **24**, 025017.
Lafleur, T., and P. Chabert (2015b), *Plasma Sources Sci. Technol.* **24**, 044002.
Lafleur, T., R. W. Boswell, and J. P. Booth (2012a), *Appl. Phys. Lett.* **100**, 194101.
Lafleur, T., P. A. Delattre, E. V. Johnson, and J. P. Booth (2012b), *Appl. Phys. Lett.* **101**, 124104.
Lafleur, T., P. Chabert, and J. P. Booth (2014a), *Plasma Sources Sci. Technol.* **23**, 035010.
Lafleur, T., P. Chabert, M. M. Turner, and J. P. Booth (2014b), *Plasma Sources Sci. Technol.* **23**, 015016.
Laframboise, J. G. (1966), Theory of spherical and cylindrical Langmuir probes in a collisionless, Maxwellian plasma at rest, *UTIAS Report No. 100*, University of Toronto.
Laframboise, J. G., and L. W. Parker (1973), *Phys. Fluids* **16**, 629.
Landau, L. D. (1946), *J. Phys. (USSR)* **10**, 25.
Laroussi, M. (1996), *IEEE Trans. Plasma Sci.* **24**, 1188.
Lawton, S. A., and A. V. Phelps (1978), *J. Chem. Phys.* **69**, 1055.
Lazzaroni, C., M. A. Lieberman, A. J. Lichtenberg, and P. Chabert (2012a), *Plasma Sources Sci. Technol.* **21** 035013.
Lazzaroni, C., M. A. Lieberman, A. J. Lichtenberg, and P. Chabert (2012b), *J. Phys. D: Appl. Phys.* **45** 495204.
Lee, H. C. (2018), *Appl. Phys. Rev.* **5**, 011108.
Lee, Y., and S. M. George (2015), *ACS Nano* **9**, 2061.
Lee, C., and M. A. Lieberman (1995), *J. Vac. Sci. Technol.* **A13**, 368.
Lee, C., D. B. Graves, M. A. Lieberman, and D. W. Hess (1994), *J. Electrochem. Soc.* **141**, 1546.
Lee, S. K., S. S. Chun, C. Y. Hwang, and W. J. Lee (1997a), *Jpn. J. Appl. Phys., Part 1* **36**, 50.
Lee, Y. T., M. A. Lieberman, A. J. Lichtenberg, F. Bose, H. Baltes, and R. Patrick (1997b), *J. Vac. Sci. Technol. A* **15**, 113.
Lee, I., D. B. Graves, and M. A. Lieberman (2008), *Plasma Sources Sci. Technol.* **17**, 015018.
Lee, Y. S., H. S. Lee, and H. Y. Chang (2010a), *Thin Solid Films* **518**, 6682.
Lee, Y. S., H. S. Lee, S. H. Seo, and H. Y. Chang (2010b), *Appl. Phys. Lett.* **97**, 081503.
Lee, H. S., Y. S. Lee, S. H. Seo, and H. Y. Chang (2011a), *Thin Solid Films* **519**, 6955.
Lee, H. W., G. Y. Park, Y. S. Seo, Y. H. Im, S. B. Shim, and H. J. Lee (2011b), *J. Phys. D: Appl. Phys.* **44** 053001.
Lee, C. G. N., K. J. Kanarik, and R. A. Gottscho (2014), *J. Phys. D: Appl. Phys.* **47**, 273001.
Lee, Y., C. Huffman, and S. M. George (2016), *Chem. Mater.* **28**, 7657.

Lerat, J.-L., T. Desrues, J. Le Perchec, M. Coig, F. Milesi, F. Mazen, T. Michel, L. Roux, Y. Veschetti, and S. Dubois (2016), *Energy Procedia* **92**, 697.

Leung, K. N., T. K. Samec, and A. Lamm (1975), *Phys. Lett.* **51A**, 490.

Leung, K. N., G. R. Taylor, J. M. Barrick, S. L. Paul and R. E. Kribel (1976), *Phys. Lett.* **57A**, 145.

Levinson, J. A., E. S. G. Shaqfeh, M. Balooch, and A. V. Hamza (1997), *J. Vac. Sci. Technol.* **A15**, 1902.

Levitskii, S. M. (1951), *Zh. Tekh. Fiz.* **27**, 1000; *Sov. Phys. Tech. Phys.* **2**, 887.

Li, M., M. A. Vyvoda, and D. B. Graves (2000), in *Ionized Physical Vapor Deposition*, J. A. Hopwood, ed., Academic Press, San Diego, CA.

Li, C., D. Metzler, C. S. Lai, E. A. Hudson, and G. S. Oehrlein (2016), *J. Vac. Sci. Technol.* **A34**, 041307.

Li, W., Y. Liu, and G. Wang (2023), *Phys. Plasmas* **30**, 022103.

Liard, L., J. L. Raimbault, J. M. Rax, and P. Chabert (2007), *J. Phys. D: Appl. Phys.* **40**, 5192.

Lichtenberg, A. J., and M. A. Lieberman (1992), *Regular and Chaotic Motion*, Wiley, New York.

Lichtenberg, A. J., and M. A. Lieberman (2000), *J. Appl. Phys.* **87**, 7191.

Lichtenberg, A. J., V. Vahedi, M. A. Lieberman, and T. Rognlien (1994), *J. Appl. Phys.* **75**, 2339; **76**, 625.

Lichtenberg, A. J., I. G. Kouznetsov, Y. T. Lee, M. A. Lieberman, I. D. Kaganovich, and L. D. Tsendin (1997), *Plasma Sources. Sci. Technol.* **6**, 437.

Lichtenberg, A. J., M. A. Lieberman, I. G. Kouznetsov, and T. H. Chung (2000), *Plasma Sources. Sci. Technol.* **9**, 45.

Lieberman, M. A. (1988), *IEEE Trans. Plasma Sci.* **16**, 638.

Lieberman, M. A. (1989a), *IEEE Trans. Plasma Sci.* **17**, 338.

Lieberman, M. A. (1989b), *J. Appl. Phys.* **65**, 4168.

Lieberman, M. A. (1989c), *J. Appl. Phys.* **66**, 2926.

Lieberman, M. A. (2015), *Plasma Sources Sci. Technol.* **24**, 025009.

Lieberman, M. A., and S. Ashida (1996), *Plasma Sources Sci. Technol.* **5**, 145.

Lieberman, M. A., and V. A. Godyak (1998), *IEEE Trans. Plasma Sci.* **26**, 955.

Lieberman, M. A., and R. A. Gottscho (1994), in *Physics of Thin Films*, Vol. **18**, M. H. Francombe and J. L. Vossen, eds., Academic Press, New York.

Lieberman, M. A., and A. J. Lichtenberg (2010), *Plasma Sources Sci. Technol.* **19**, 065006.

Lieberman, M. A., and S. E. Savas (1990), *J. Vac. Sci. Technol.* **A8**, 1632.

Lieberman, M. A., A. J. Lichtenberg, and S. E. Savas (1991), *IEEE Trans. Plasma Sci.* **19**, 189.

Lieberman, M. A., J. P. Booth, P. Chabert, J. M. Rax, and M. M. Turner (2002), *Plasma Sources Sci. Technol.* **11**, 283.

Lieberman, M. A., J. Kim, J. P. Booth, P. Chabert, J. M. Rax, and M. M. Turner (2003), in *SEMI Technology Symposium Korea 2003 Proceedings*, Seoul, Korea.

Lieberman, M. A., A. J. Lichtenberg, E. Kawamura, T. Mussenbrock, and R. P. Brinkmann (2008), *Phys. Plasmas* **15**, 063505.

Lieberman, M. A., A. J. Lichtenberg, E. Kawamura, and P. Chabert (2016), *Phys. Plasmas* **23**, 013501.

Lill, T., I. L. Berry, M. Shen, J. Hoang, A. Fischer, T. Panagopoulos, J. P. Chang, and V. Vahedi (2023), *J. Vac. Sci. Technol.* **A41**, 023005.

Lim, W. S., J. B. Park, J. Y. Park, B. J. Park, and G. Y. Yeom (2009), *J. Nanosci. Nanotechnol.* **9**, 7379.

Linder, B. P., and N. W. Cheung (1996), *IEEE Trans. Plasma Sci.* **24**, 1383.

Lisovskii, V. A. (1998), *Tech. Phys.* **43**, 526.

Lisovskiy, V. A., J. P. Booth, K. Landry, D. Douai, V. Cassagne, and V. Yegorenkov (2006), *Phys. Plasmas* **13**, 103505.

Lister, G. G., Y.-M. Li, and V. A. Godyak (1996), *J. Appl. Phys.* **79**, 8993.

Liu, D. W., F. Iza, and M. G. Kong (2009), *Plasma Process. Polym.* **6** 446.

Liu, D. X., P. Bruggeman, F. Iza, M. Z. Rong, and M. G. Kong (2010), *Plasma Sources Sci. Technol.* **19** 025018.
Liu, Y. X., Y. R. Zhang, A. Bogaerts, and Y. N. Wang (2015), *J. Vac. Sci. Technol., A* **33**, 020801.
Lloyd, S., D. M. Shaw, M. Watanabe, and G. J. Collins (1999), *Jpn. J. Appl. Phys.* **38**, 4275.
Loeb, L. B. (2022), *Basic Processes of Gaseous Electronics*, UC Press, Berkeley, CA.
Lucken, R., V. Croes, T. Lafleur, J. L. Raimbault, A. Bourdon, and P. Chabert (2018), *Plasma Sources Sci. Technol.* **27**, 035004.
Lucken, R., A. Bourdon, M. A. Lieberman, and P. Chabert (2019), *Phys. Plasmas* **26**, 070702.
Lundin, D., T. Minea, and J. T. Gudmundsson, eds. (2020), *High Power Impulse Magnetron Sputtering: Fundamentals, Technologies, Challenges, and Applications*, Elsevier, Amsterdam.
Lymberopoulos, D. P., and D. J. Economou (1993), *J. Appl. Phys.* **73** 3668.
Lymberopoulos, D. P., and D. J. Economou (1995), *IEEE Trans. Plasma Sci.* **23** 573.
Mahan, J. E. (2000), *Physical Vapor Deposition of Thin Films*, Wiley, New York.
Makabe, T., and Z. L. Petrović (2006), *Plasma Electronics: Applications in Microelectronic Device Fabrication* Taylor and Francis, London.
Malyshev, M. V., and V. M. Donnelly (1997), *J. Vac. Sci. Technol.* **A15**, 550.
Malyshev, M. V., and V. M. Donnelly (1999), *Phys. Rev. E* **60**, 6016.
Malyshev, M. V., and V. M. Donnelly (2000a), *J. Appl. Phys.* **87**, 1642.
Malyshev, M. V., and V. M. Donnelly (2000b), *J. Appl. Phys.* **88**, 6207.
Malyshev, M. V., and V. M. Donnelly (2001), *J. Appl. Phys.* **90**, 1130.
Malyshev, M. V., N. C. M. Fuller, K. H. A. Bogart, V. M. Donnelly, and I. P. Herman (1999a), *Appl. Phys. Lett.* **74**, 1666.
Malyshev, M. V., V. M. Donnelly, J. I. Colonell, and S. Samukawa (1999b), *J. Appl. Phys.* **86**, 4813.
Manley, T. C. (1943), *Trans. Electrochem. Soc.* **84**, 83.
Manus, C. (1976), *Physica* **82C**, 165.
Manos, D. M., and H. F. Dylla (1989), in *Plasma Etching: An Introduction*, D. M. Manos and D. L. Flamm, eds., Academic Press, New York.
Manos, D. M., and D. L. Flamm, eds. (1989), *Plasma Etching: An Introduction*, Academic Press, New York.
Margot, J., F. Vidal, M. Chaker, T. W. Johnston, A. Aliouchouche, M. Tabbal, S. Delprat, O. Pauna, and D. Benhabib (2001), *Plasma Sources Sci. Technol.* **10**, 556.
Martines, E., R. Cavazzana, L. Cordaro, and M. Zuin (2021), *Appl. Sci.* **2021**, 11, 7444.
Massey, H. S. W., E. H. S. Burhop, and H. B. Gilbody (1969–74), *Electron and Ion Impact Phenomena*, 2nd ed., Clarendon, Oxford, UK.
Massines, F., A. Rabehi, P. Decomps, R. B. Gadri, P. Ségur, and C. Mayoux (1998), *J. Appl. Phys.* **83**, 2950.
Massines, F., P. Ségur, N. Gheradi, C. Khamphan, and A. Ricard (2003), *Surf. Coat. Technol.* **174–175**, 8.
Massines, F., N. Gherardi, N. Naudé, and P. Ségur (2005), *Plasma Phys. Control. Fusion* **47**, B577.
Matsoukas, T., and M. Russell (1995), *J. Appl. Phys.* **77**, 4285.
Matsoukas, T., M. Russell, and M. Smith (1996), *J. Vac. Sci. Technol.* **A14**, 624.
Matsuoka, M., and K. Ono (1988), *J. Vac. Sci. Technol.* **A6**, 25.
Matsuura, T., J. Murota, Y. Sawada, and T. Ohmi (1993), *Appl. Phys. Lett.* **63**, 2803.
Matthieussent, G., and J. Pelletier (1992), in *Microwave Excited Plasmas*, M. Moissan and J. Pelletier, eds., Elsevier, Amsterdam.
McCaughey, M. J., and M. J. Kushner (1989), *J. Appl. Phys.* **65**, 186.
McDaniel, E. W. (1964), *Collision Phenomena in Ionized Gases*, Wiley, New York.
McDaniel, E. W. (1989), *Atomic Collisions: Electron and Photon Projectiles*, Wiley, New York.

McDaniel, E. W., J. B. A. Mitchell, and M. E. Rudd (1993), *Atomic Collisions: Heavy Particle Projectiles*, Wiley, New York.
McNeil, J. R., G. J. Collins, K. B. Persson, and D. L. Franzen (1976), *Appl. Phys. Lett.* **28**, 207.
Meek, J. M., and J. D. Craggs (1953), *Electrical Breakdown in Gases*, Clarendon, Oxford.
Meeks, E., and J. W. Shon (1995), *IEEE Trans. Plasma Sci.* **23**, 539.
Meeks, E., R. S. Larson, P. Ho, C. Apblett, S. M. Han, E. Edelberg, and E. S. Aydil (1998), *J. Vac. Sci. Technol.* **A16**, 544.
Melandso, F. (1997), *Phys. Rev. E* **55**, 7495.
Melzer, A. (2019), *Physics of Dusty Plasmas*, Springer, Cham, Switzerland.
Merlino, R. L. (2006), *Plasma Phys. Appl.* **5**, 73.
Merlino, R. L. (2021), *Adv. Phys. X* **6**, 1.
Metze, A., D. W. Ernie, and H. J. Oskam (1986), *J. Appl. Phys.* **60**, 3081.
Metzler, D., R. L. Bruce, S. Engelmann, E. A. Joseph, and G. S. Oehrlein (2014), *J. Vac. Sci. Technol.* **A32**, 020603.
Metzler, D., C. Li, S. Engelmann, R. L. Bruce, E. A. Joseph, and G. S. Oehrlein (2016), *J. Vac. Sci. Technol.* **A34**, 01B101.
Meyyappan, M. (1996), *J. Vac. Sci. Technol., A* **14**, 2122.
Meziani, T., P. Colpo, and F. Rossi (2001), *Plasma Sources Sci. Technol.* **10**, 276.
Mie, G. (1908), *Ann. Phys.* **25**, 377.
Miller, P. A., and M. E. Riley (1997), *J. Appl. Phys.* **82**, 3689.
Milosavljević, V., and P. J. Cullen (2015), *EPL* **110**, 43001.
Milosavljević, V., N. MacGearailt, P. J. Cullen, S. Daniels, and M. M. Turner (2013), *J. Appl. Phys.* **113**, 163302.
Min, K. S., S. H. Kang, J. K. Kim, Y. I. Jhon, M. S. Jhon, and G. Y. Yeom (2013), *Microelectron. Eng.* **110**, 457.
Misium, G. R., A. J. Lichtenberg, and M. A. Lieberman (1989), *J. Vac. Sci. Technol.* **A7**, 1007.
Mitchell, J. B. A. (1990), *Phys. Rep.* **186**, 215.
Mogab, C. J., A. C. Adams, and D. L. Flamm (1979), *J. Appl. Phys.* **49**, 3796.
Moisan, M., and Z. Zakrzewski (1991), *J. Phys. D: Appl. Phys.* **24**, 1025.
Monahan, D. D., and M. M. Turner (2008), *Plasma Sources Sci. Technol.* **17**, 045003.
Moon, S. Y., W. Choe, and B. K. Kang (2004), *Appl. Phys. Lett.* **84** 188.
Moon, S. Y., J. K. Rhee, D. B. Kim, and W. Choe (2006), *Phys. Plasmas* **13**, 033502.
Moradi, M., C. Nender, S. Berg, H.-O. Blom, A. Belkind, and Z. Orban (1991), *J. Vac. Sci. Technol.* **A9**, 619.
Moreno, J., A. Khodaee, D. Okerstrom, M. P. Bradley, and L. Couëdel (2021), *Phys. Plasmas* **28**, 123523.
Morfill, G. E., and H. Thomas (1996), *J. Vac. Sci. Technol.* **A14**, 490.
Morfill, G. E., and A. V. Ivlev (2009), *Rev. Mod. Phys.* **81**, 1353.
Moroney, R. M., A. J. Lichtenberg, and M. A. Lieberman (1989), *J. Appl. Phys.* **66**, 1618.
Morris, M. A., M. Bowker, and D. A. King (1984), in *Simple Processes at the Gas–Solid Interface*, C. H. Bamford, C. F. H. Tipper, and R. G. Compton, eds., Elsevier, Amsterdam.
Mott-Smith, H. M., and I. Langmuir (1926), *Phys. Rev.* **28**, 727.
Musil, J., and F. Zacek (1970), *Plasma Phys.* **12**, 17.
Musil, J., and F. Zacek (1971), *Plasma Phys.* **13**, 471.
Mussenbrock, T., and R. P. Brinkmann (2006), *Appl. Phys. Lett.* **88**, 151503.
Mussenbrock, T., and R. P. Brinkmann (2007), *Plasma Sources Sci. Technol.* **16**, 377.
Mussenbrock, T., R. P. Brinkmann, M. A. Lieberman, A. J. Lichtenberg, and E. Kawamura (2008), *Phys. Rev. Lett.* **101**, 085004.

Naudé, N., J. P. Cambronne, N. Gherardi, and F. Massines (2005), *J. Phys. D: Appl. Phys.* **38**, 530.
Newman, S. M., A. J. Orr-Ewing, D. A. Newnham, and J. Ballard (2000), *J. Phys. Chem. A* **104**, 9467.
Ney, V., and N. Schwentner (2006), *J. Phys. Condens. Matter* **18**, S1603.
Niazi, K., A. J. Lichtenberg, M. A. Lieberman, and D. L. Flamm (1994), *Plasma Sources Sci. Technol.* **3**, 482.
Niazi, K., A. J. Lichtenberg, M. A. Lieberman (1995), *IEEE Trans. Plasma Sci.* **23**, 833.
Nicholson, D. R. (1983), *Introduction to Plasma Theory*, Wiley, New York.
Nijdam, S., J. Teunissen, and U. Ebert (2020), *Plasma Sources Sci. Technol.* **29**, 103001.
Ninomiya, K., K. Suzuki, S. Nishimatsu, and O. Okada (1985), *J. Appl. Phys.* **58**, 1177.
Northrop, T. G., and T. J. Birmingham (1990), *Planet. Space Sci.* **38**, 319.
Oehrlein, G. S., D. Metzler, and C. Li (2015), *ECS J. Solid State Sci. Technol.* **4**, N5041.
Ogryzlo, E. A., D. E. Ibbotson, D. L. Flamm, and J. A. Mucha (1990), *J. Appl. Phys.* **67**, 3115.
Overzet, L. J., and F. Y. Leong-Rousey (1995), *Plasma Sources Sci. Technol.* **4**, 432.
Panagopoulos, T., and D. J. Economou (1999), *J. Appl. Phys.* **85**, 3435.
Papadakis, A. P., S. Rossides, and A. C. Metaxas (2011), *Open Appl. Phys. J.* **4**, 45.
Park, J.-C., and B. Kang (1997a), *IEEE Trans. Plasma Sci.* **25**, 499.
Park, J.-C., and B. Kang (1997b), *IEEE Trans. Plasma Sci.* **25**, 1398.
Park, J., I. Henins, H. W. Hermann, G. S. Selwyn, J. Y. Heong, R. F. Hicks, D. Shim, and C. S. Chang (2000), *Appl. Phys. Lett.* **76** 288.
Park, J., I. Henins, H. W. Hermann, G. S. Selwyn, and R. F. Hicks (2001), *J. Appl. Phys.* **89** 20.
Park, G., H. Lee, G. Kim, and J. K. Lee (2008a), *Plasma Process. Polym.* **5** 569.
Park, S. D., W. S. Lim, B. J. Park, H. C. Lee, J. W. Bae, and G. Y. Yeom (2008b), *Electrochem. Solid-State Lett.* **11**, H71.
Park, J. B., W. S. Lim, S. D. Park, B. J. Park, and G. Y. Yeom (2009), *J. Korean Phys. Soc.* **54**, 976.
Park, G., Y. J. Hong, H. W. Lee, J. Y. Sim, and J. K. Lee (2010), *Plasma Process. Polym.* **7** 281.
Park, H., Y. Sakiyama, and H. J. Lee (2023), *Front. Phys.* **11**, 1137994.
Parker, J. V., J. C. Nickel, and R. W. Gould (1964), *Phys. Fluids* **7**, 1489.
Perret, A., P. Chabert, J. P. Booth, J. Jolly, J. Guillon, and Ph. Auvray (2003), *Appl. Phys. Lett.* **83**, 243.
Perrin, J., and Ch. Hollenstein (1999), in *Dusty Plasmas*, A. Bouchoule, ed., Wiley, New York, p. 77.
Perrin, J., O. Leroy, and M. C. Bordage (1996), *Contrib. Plasma Phys.* **36**, 3.
Perrin, J., M. Shiratani, P. Kae-Nune, H. Videlot, J. Jolly, and J. Guillon (1998), *J. Vac. Sci. Technol.* **A16**, 278.
Perrin, J., J. Schmitt, Ch. Hollenstein, A. Howling, and L., Sansonnens (2000), *Plasma Phys. Control. Fusion* **42**, B353.
Perry, A. J., D. Vender, and R. W. Boswell (1991), *J. Vac. Sci. Technol.* **B9**, 310.
Petrović, Z., and D. Marić (2004), private communication.
Peyerimhoff, S. D., and R. J. Buenker (1981), *Chem. Phys.* **57**, 279.
Phelps, A. V. (1985), Tabulations of cross sections and calculated transport and reaction coefficients for electron collisions with O_2, *JILA Information Center Report*, University of Colorado, Boulder, CO.
Phelps, A. V. (1999), *Plasma Sources Sci. Technol.* **8** B1.
Phelps, A. V., and Z. L. Petrović (1999), *Plasma Sources Sci. Technol.* **8** R21.
Pico, C. A., M. A. Lieberman, and N. W. Cheung (1992), *J. Electron. Mater.* **21**, 75.
Piejak, R. B., V. A. Godyak, and B. M. Alexandrovich (1992), *Plasma Sources Sci. Technol.* **1**, 179.
Piejak, R. B., V. A. Godyak, and B. M. Alexandrovich (1997), *J. Appl. Phys.* **81**, 3416.
Piejak, R. B., V. A. Godyak, R. Garner, B. M. Alexandrovich, and N. Sternberg (2004), *J. Appl. Phys.* **95**, 3785.
Piejak, R. B., J. Al-Kuzee, and N. St. J. Braithwaite (2005), *Plasma Sources Sci. Technol.* **14**, 734.

Pinna, N., and M. Knez (2012), *Atomic Layer Deposition of Nanostructured Materials*, Wiley-VCH, Weinheim, Germany.

Pippard, A. B. (1949), *Physica* **15**, 45.

Pitchford, L. C., L. L. Alves, K. Bartschat, S. F. Biagi, M.-C. Bordage, I. Bray, C. E. Brion, M. J. Brunger, L. Campbell, A. Chachereau, B. Chaudhury, L. G. Christophorou, E. Carbone, N. A. Dyatko, C. M. Franck, D. V. Fursa, R. K. Gangwar, V. Guerra, P. Haefliger, G. J. M. Hagelaar, A. Hoesl, Y. Itikawa, I. V. Kochetov, R. P. McEachran, W. L. Morgan, A. P. Napartovich, V. Puech, M. Rabie, L. Sharma, R. Srivastava, A. D. Stauffer, J. Tennyson, J. de Urquijo, J. van Dijk, L. A. Viehland, M. C. Zammit, O. Zatsarinny, and S. Pancheshnyi (2017), *Plasma Process. Polym.* **14**, 1600098.

Plumb, I. C., and K. R. Ryan (1986), *Plasma Chem. Plasma Process.* **6**, 205.

Popov, O. A. (1994), in *Physics of Thin Films*, Vol. **18**, M. H. Francombe and J. L. Vossen, eds., Academic Press, New York.

Popov, O. A., and V. A. Godyak (1985), *J. Appl. Phys.* **57**, 53.

Porteous, R. K., H. M. Wu, and D. B. Graves (1994), *Plasma Sources Sci. Technol.* **3**, 25.

Profijt, H. B., S. E. Potts, M. C. M. van de Sanden, and W. M. M. Kessels (2011), *J. Vac. Sci. Technol.* **A29**, 050801.

Proto, A., and J. T. Gudmundsson (2018), *Plasma Sources Sci. Technol.* **27**, 074002.

Proto, A., and J. T. Gudmundsson (2020), *J. Appl. Phys.* **128**, 113302.

Proto, A., and J. T. Gudmundsson (2021), *Plasma Sources Sci. Technol.* **30**, 065009.

Puurunen, R. L. (2005), *J. Appl. Phys.* **97**, 121301.

Qin, S., C. Chan, and N. E. McGruer (1992), *Plasma Sources Sci. Technol.* **1**, 1.

Qin, S., C. Chan, and Z. J. Jin (1996), *J. Appl. Phys.* **79**, 3432.

Raether, H. (1964), *Electron Avalanches and Breakdown in Gases*, Butterworths, Washington, DC.

Raimbault, J. L., and P. Chabert (2009), *Plasma Sources Sci. Technol.* **18**, 014017.

Raimbault, J. L., L. Liard, J. M. Rax, P. Chabert, A. Fruchtman, and G. Makrinich (2007), *Phys. Plasmas* **14**, 013503.

Raizer, Y. P. (1991), *Gas Discharge Physics*, Springer, New York.

Raizer, Y. P., M. N. Shneider, and N. A. Yatsenko (1995), *Radio Frequency Capacitive Discharges*, CRC Press, Boca Raton, FL.

Ramamurthi, B., and D. Economou (2002), *J. Vac. Sci. Technol.* **A20**, 467.

Ramo, S., J. R. Whinnery, and T. Van Duzer (1994), *Fields and Waves in Communication Electronics*, 3rd ed., Wiley, New York.

Rangwala, S. A., S. V. K. Kumar, E. Krishnakumar, and N. J. Mason (1999), *J. Phys. B: At. Mol. Opt. Phys.* **32**, 3795.

Rapp, D., and D. Briglia (1965), *J. Chem. Phys.* **43**, 1480.

Rapp, D., and W. E. Francis (1962), *J. Chem. Phys.* **37**, 2631.

Rauf, S. (2003), *IEEE Trans. Plasma Sci.* **31**, 471.

Rauf, S. (2005), *Plasma Sources Sci. Technol.* **14**, 329.

Rauf, S., K. Bera, and K. Collins (2008), *Plasma Sources Sci. Technol.* **17**, 035003.

Redsten, A. M., K. Sridharan, F. J. Worzala, and J. R. Conrad (1992), *J. Mater. Process. Technol.* **30**, 253.

Ricard, A. (1996), *Reactive Plasmas*, Société Française du Vide, Paris.

Riemann, K.-U. (1991), *J. Phys. D: Appl. Phys.* **24**, 493.

Riemann, K.-U. (1995), *IEEE Trans. Plasma Sci.* **23**, 709.

Riemann, K.-U. (1997), *Phys. Plasmas* **4**, 4158.

Robertson, J. (2000), *J. Appl. Phys.* **87**, 2608.

Robertson, S. (2013), *Plasma Phys. Control. Fusion* **55**, 093001.

Robiche, J., P. C. Boyle, M. M. Turner, and A. R. Ellingboe (2003), *J. Phys. D: Appl. Phys.* **36**, 1810.

Rosenbluth, M. N., W. M. MacDonald, and D. L. Judd (1957), *Phys. Rev.* **107**, 1.

Roth, J. R. (1994), *Industrial Plasma Engineering Vol. 1: Principles*, IOP Publishing, London.

Şahin, Ö., I. Tapan, E. N. Özmutlu, and R. Veenhof (2010), *J. Instrum. (JINST)* **5**, P05002.

Sankaran, A., and M. J. Kushner (2004), *J. Vac. Sci. Technol.* **A22**, 1242; 1260.

Sansonnens, L., A. A. Howling, and Ch. Hollenstein (2006), *Plasma Sources Sci. Technol.* **15**, 302.

Schaepkens, M., T. E. F. M. Standaert, N. R. Rueger, P. G. M. Sebel, G. S. Oehrlein, and J. M. Cook (1999), *J. Vac. Sci. Technol.* **A17**, 26.

Scheuer, J. T., M. Shamim, and J. R. Conrad (1990), *J. Appl. Phys.* **67**, 1241.

Schmidt, G. (1979), *Physics of High Temperature Plasmas*, 2nd ed., Academic Press, New York.

Schohl, S., H. A. J. Meijer, M. W. Ruf, and H. Hotop (1992), *Meas. Sci. Technol.* **3** 544.

Schuegraf, K. F., and C. M. Hu (1994), *Semicond. Sci. Technol.* **9**, 989.

Schulenberg, D. A., I. Korolov, Z. Donkó, A. Derzsi, and J. Schulze (2021), *Plasma Sources Sci. Technol.* **30**, 105003.

Schulze, J., T. Gans, D. O'Connell, U. Czarnetzki, A. R. Ellingboe, and M. M. Turner (2007), *J. Phys. D: Appl. Phys.* **40**, 7008.

Schulze, J., Z. Donkó, D. Luggenhölscher, and U. Czarnetzki (2009a), *Plasma Sources Sci. Technol.* **18**, 034011.

Schulze, J., E. Schüngel, and U. Czarnetzki (2009b), *J. Phys. D: Appl. Phys.* **42**, 092005.

Schulze, J., E. Schüngel, Z. Donkó, D. Luggenhölscher, and U. Czarnetzki (2010), *J. Phys. D: Appl. Phys.* **43**, 124016.

Schulze, J., A. Derzsi, K. Dittmann, T. Hemke, J. Meichsner, and Z. Donkó (2011a), *Phys. Rev. Lett.* **107**, 275001.

Schulze, J., E. Schüngel, U. Czarnetzki, M. Gebhardt, R. P. Brinkmann, and T. Mussenbrock (2011b), *Appl. Phys. Lett.* **98**, 031501.

Schulze, J., E. Schüngel, Z. Donkó, and U. Czarnetzki (2011c), *Plasma Sources Sci. Technol.* **20**, 015017.

Schulze, J., Z. Donkó, T. Lafleur, S. Wilczek, and R. P. Brinkmann (2018), *Plasma Sources Sci. Technol.* **27**, 055010.

Schwarz, S. E., and W. G. Oldham (1984), *Electrical Engineering: An Introduction*, Holt, Rinehart & Winston, New York.

Schweigert, I. V. (2004), *Phys. Rev. Lett.* **92**, 155011.

Selwyn, G. S. (1993), *Optical Diagnostic Techniques for Plasma Processing*, Monograph M11, American Vacuum Society Press, New York.

Selwyn, G. S. (1994), *Plasma Sources Sci. Technol.* **3**, 340.

Selwyn, G. S., J. Singh, and R. S. Bennett (1989), *J. Vac. Sci. Technol.* **A7**, 2758.

Selwyn, G. S., J. E. Heidenreich, and K. L. Haller (1991), *J. Vac. Sci. Technol.* **A9**, 2817.

Senn, G., J. D. Skalny, A. Stamatovic, N. J. Mason, P. Scheier, and T. D. Märk (1999), *Phys. Rev. Lett.* **82**, 5028.

Seshan, K., and D. Schepis, eds (2018), *Handbook of Thin Film Deposition*, 4th ed., Elsevier, Amsterdam.

Shamim, M., J. T. Scheuer, and J. R. Conrad (1991), *J. Appl. Phys.* **69**, 2904.

Sheridan, T. E. (1996), *Phys. Plasmas* **3**, 3507.

Sheridan, T. E. (1999), *J. Phys. D: Appl. Phys.* **32**, 1761.

Sheridan, T. E., M. J. Goeckner, and J. Goree (1991), *J. Vac. Sci. Technol.* **A9**, 688.

Shi, J. J., and M. G. Kong (2005), *J. Appl. Phys.* **97** 023306.

Shi, J. J., and M. G. Kong (2006), *Phys. Rev. Lett.* **96** 105009.

Shi, J. J., D. W. Liu, and M. G. Kong (2007), *IEEE Trans. Plasma Sci.* **35** 137.

Shin, H. C., and C. M. Hu (1996), *Semicond. Sci. Technol.* **11**, 463.

Shin, H., W. Zhu, V. M. Donnelly, and D. J. Economou (2012), *J. Vac. Sci. Technol.* **A30**, 021306.

Shiratani, M., T. Fukuzawa, and Y. Watanabe (1994), *IEEE Trans. Plasma Sci.* **22**, 103.

Shiratani, M., H. Kawasaki, T. Fukuzawa, and Y. Watanabe (1996), *J. Vac. Sci. Technol.* **A14**, 603.

Shkarofsky, I. P., T. W. Johnston, and M. P. Bachynski (1966), *The Particle Kinetics of Plasmas*, Addison-Wesley, Reading, MA.

Shukla, P. K., and B. Eliasson (2009), *Rev. Mod. Phys.* **81**, 25.

Shukla, P. K., and A. A. Mamun (2002), *Introduction to Dusty Plasma Physics*, IOP, Bristol.

Siegfried, D. E., and P. J. Wilber (1984), *AIAA J.* **22**, 1505.

Sigmund, P. (1981), in *Sputtering by Particle Bombardment I*, R. Behrisch, ed., Springer-Verlag, New York, Chapter 2.

Simon, A. (1959), *An Introduction to Thermonuclear Research*, Pergamon, New York.

Smirnov, B. M. (1977), *Introduction to Plasma Physics*, Mir, Moscow.

Smirnov, B. M. (1981), *Physics of Weakly Ionized Gases*, Mir, Moscow.

Smirnov, B. M. (1982), *Negative Ions*, McGraw-Hill, New York.

Smirnov, B. M. (1983), *Sov. Phys. Usp.* **26**, 31.

Smirnov, A. S., and L. D. Tsendin (1991), *IEEE Trans. Plasma Sci.* **19**, 130.

Smith, D. L. (1995), *Thin Film Deposition: Principles and Practice*, McGraw-Hill, New York.

Smith, H. B. (1998), *Phys. Plasmas* **5**, 3469.

Smolyakov, A. I., V. A. Godyak, and Y. O. Tyshetskiy (2001), *Phys. Plasmas* **8**, 3857.

Smolyakov, A. I., V. A. Godyak, and Y. O. Tyshetskiy (2003), *Phys. Plasmas* **10**, 2108.

Smullin, L. D., and P. Chorney (1958), *Proc. IRE* **46**, 360.

Snyders, R., D. Hegemann, D. Thiry, O. Zabeida, J. Klemberg-Sapieha, and L. Martinu (2023), *Plasma Sources Sci. Technol.* **32**, 074001.

Soberón, F., F. G. Marro, W. G. Graham, A. R. Ellingboe, and V. J. Law (2006), *Plasma Sources Sci. Technol.* **15**, 193.

Sobolewski, M. A. (2000), *Phys. Rev. E* **62**, 8540.

Solmaz, E., S. M. Ryu, J. Uh, and L. L. Raja (2020), *J. Vac. Sci. Technol.* **A38**, 052405.

Sommerer, T. J., and M. J. Kushner (1992), *J. Appl. Phys.* **71**, 1654.

Song, Y. P., D. Field, and D. F. Klemperer (1990), *J. Phys. D: Appl. Phys.* **23**, 673.

Song, F., F. Li, M. Zhu, L. Wang, B. Zhang, H. Gong, Y. Gan, and X. Jin (2018), *Plasma Sci. Technol.* **20**, 014013.

Spears, K. G., T. J. Robinson, and R. M. Roth (1986), *IEEE Trans. Plasma Sci.* **14**, 179.

Spitzer, L. (1956), *Physics of Fully Ionized Gases*, Interscience, New York.

Sproul, W. D., D. J. Christie, and D. C. Carter (2005), *Thin Solid Films* **491**, 1.

Stafford, D. S., and M. J. Kushner (2004), *J. Appl. Phys.* **96** 2451.

Standaert, T. E. F. M., C. Hedlund, E. A. Joseph, G. S. Oehrlein, and T. J. Dalton (2004), *J. Vac. Sci. Technol.* **A22**, 53.

Steinbrüchel, C. (1989), *Appl. Phys. Lett.* **55**, 1960.

Steinfeld, J. I. (1985), *Molecules and Radiation: An Introduction to Modern Molecular Spectroscopy*, 2nd ed., MIT Press, Cambridge, MA.

Stenzel, R. L. (1976), *Rev. Sci. Instrum.* **47**, 603.

Stenzel, R. L., and J. M. Urrutia (2021), *Rev. Sci. Instrum.* **92**, 111101.

Sternberg, N., and V. A. Godyak (2017), *Phys. Plasmas* **24**, 093504.

Sternberg, N., V. A. Godyak, and D. Hoffman (2006), *Phys. Plasmas* **13**, 063511.

Stevens, J. E., Y. C. Huang, R. L. Jarecki, and J. L. Cecchi (1992), *J. Vac. Sci. Technol.* **A10**, 1270.

Stewart, R. A., and M. A. Lieberman (1991), *J. Appl. Phys.* **70**, 3481.

Stix, T. H. (1992), *Waves in Plasmas*, American Institute of Physics, New York.

Stoffels, E., W. W. Stoffels, D. Vender, G. M. W. Kroesen, and F. J. de Hoog (1994), *IEEE Trans. Plasma Sci.* **22**, 116.

Stoffels, E., W. W. Stoffels, D. Vender, M. Kando, G. M. W. Kroesen, and F. J. de Hoog (1995), *Phys. Rev. E* **51**, 2425.

Stout, P. J., and M. J. Kushner (1993), *J. Vac. Sci. Technol.* **A11**, 2562.

Strijckmans, K., R. Schelfhout, and D. Depla (2018), *J. Appl. Phys.* **124**, 241101.

Su, C. H., and S. H. Lam (1963), *Phys. Fluids* **6**, 1479.

Su, T., and M. T. Bowers (1973), *Int. J. Mass Spectrom. Ion Phys.* **12**, 347.

Sudit, I. D., and F. F. Chen (1996), *Plasma Sources Sci. Technol.* **5**, 43.

Surendra, M., and M. Dalvie (1993), *Phys. Rev. E* **48**, 3914.

Surendra, M., and D. Vender (1994), *Appl. Phys. Lett.* **65**, 153.

Szapiro, B., and J. J. Rocca (1989), *J. Appl. Phys.* **65**, 3713.

Tan, S., W. Yang, K. J. Kanarik, T. Lill, V. Vahedi, J. Marks, and R. A. Gottscho (2015), *ECS J. Solid State Sci. Technol.* **4**, N5010.

Tang, X., and D. M. Manos (1999), *Plasma Sources Sci. Technol.* **8**, 594.

Tarey, R. D., B. B. Sahu, and A. Ganguli (2012), *Phys. Plasmas* **19**, 073520.

Tavant, A., and M. A. Lieberman (2016), *J. Phys. D: Appl. Phys.* **49** 465201.

Thompson, J. B. (1959), *Proc. Phys. Soc.* **73**, 818.

Thomson, J. J. (1912), *Philos. Mag.* **23**, 449.

Thomson, J. J. (1924), *Philos. Mag.* **47**, 337.

Thomson, J. J. (1927), *Philos. Mag.* **4**, 1128.

Thorne, A. P. (1988), *Spectrophysics*, Chapman & Hall, London.

Thornton, J. A. (1974), *J. Vac. Sci. Technol.* **11**, 666.

Thornton, J. A. (1986), *J. Vac. Sci. Technol.* **A4**, 3059.

Thornton, J. A., and A. S. Penfold (1978), in *Thin Film Processes*, J. L. Vossen and W. Kern, eds., Academic Press, New York.

Thorsteinsson, E. G., and J. T. Gudmundsson (2010a), *Plasma Sources Sci. Technol.* **19**, 015001.

Thorsteinsson, E. G., and J. T. Gudmundsson (2010b), *J. Phys. D: Appl. Phys.* **43**, 115201.

Thorsteinsson, E. G., and J. T. Gudmundsson (2010c), *J. Phys. D: Appl. Phys.* **43**, 115202.

Thorsteinsson, E. G., and J. T. Gudmundsson (2010d), *Plasma Sources Sci. Technol.* **19**, 055008.

Tochikubo, F., and A. Komuro (2021), *Jpn. J. Appl. Phys.* **60**, 040501.

Tochikubo, F., T. Chiba, and T. Watanabe (1999), *Jpn. J. Appl. Phys.* **38**, 5244.

Toneli, D. A., R. S. Pessoa, M. Roberto, and J. T. Gudmundsson (2019), *Plasma Sources Sci. Technol.* **28**, 025017.

Tonks, L., and I. Langmuir (1929), *Phys. Rev.* **34**, 876.

Trivelpiece, A. W., and R. W. Gould (1959), *J. Appl. Phys.* **30**, 1784.

Tsendin, L. D. (1974), *Sov. Phys. JETP* **39**, 805.

Tsendin, L. D. (1989), *Sov. Phys.–Tech. Phys.* **34**, 11.

Turner, M. M. (1993), *Phys. Rev. Lett.* **71**, 1844.

Turner, M. M. (1995), *Phys. Rev. Lett.* **75**, 1312.

Turner, M. M., and P. Chabert (2006a), *Appl. Phys. Lett.* **89**, 231502.

Turner, M. M., and P. Chabert (2006b), *Phys. Rev. Lett.* **96**, 205001.

Turner, M. M., and P. Chabert (2007), *Comput. Phys. Commun.* **177**, 88.

Turner, M. M., and P. Chabert (2014), *Appl. Phys. Lett.* **104**, 164102.

Turner, M. M., and M. B. Hopkins (1992), *Phys. Rev. Lett.* **69**, 3511.

Turner, M. M., and M. A. Lieberman (1999), *Plasma Sources Sci. Technol.* **8**, 313.

Tuszewski, M. (1996), *J. Appl. Phys.* **79**, 8967.

Tyshetskiy, Y. O., A. I. Smolyakov, and V. A. Godyak (2002), *Plasma Sources Sci. Technol.* **11**, 203.
Vahedi, V. (1993), *Modeling and Simulation of Rf Discharges Used for Plasma Processing*, Thesis, University of California, Berkeley, CA.
Vahedi, V., M. A. Lieberman, M. V. Alves, J. P. Verboncoeur, and C. K. Birdsall (1991), *J. Appl. Phys.* **69**, 2008.
Vahedi, V., C. K. Birdsall, M. A. Lieberman, G. DiPeso, and T. D. Rognlien (1993a), *Plasma Sources Sci. Technol.* **2**, 273.
Vahedi, V., R. A. Stewart, and M. A. Lieberman (1993b), *J. Vac. Sci. Technol.* **A11**, 1275.
Vahedi, V., M. A. Lieberman, G. DiPeso, T. D. Rognlien, and D. Hewett (1995), *J. Appl. Phys.* **78**, 1446.
Vandalon, V., and W. M. M. Kessels (2017), *J. Vac. Sci. Technol.* **A35**, 05C313.
Van Gaens, W., and A. Bogaerts (2013), *J. Phys. D: Appl. Phys.* **46**, 275201.
Van Gaens, W., and A. Bogaerts (2014), *J. Phys. D: Appl. Phys.* **47**, 079502.
Van Veldhuizen, E. M., and F. J. de Hoog (1984), *J. Phys. D* **17**, 953.
Vaughan, J. R. M. (1989), *IEEE Trans. Electron Dev.* **36**, 1963.
Vejby-Christensen, L., D. Kella, D. Mathur, H. B. Pedersen, H. T. Schmidt, and H. L. Andersen (1996), *Phys. Rev. A* **53**, 2371.
Vella, J. R., and D. B. Graves (2023), *J. Vac. Sci. Technol.* **A41**, 042601.
Vella, J. R., D. Humbird, and D. B. Graves (2022), *J. Vac. Sci. Technol.* **B40**, 023205.
Vender, D., and R. W. Boswell (1990), *IEEE Trans. Plasma Sci.* **18**, 725.
Vender, D., and R. W. Boswell (1992), *J. Vac. Sci. Technol., A* **10**, 1331.
Vender, D., W. W. Stoffels, E. Stoffels, G. M. W. Kroesen, and F. F. de Hoog (1995), *Phys. Rev. E* **51**, 2436.
Vidal, F., T. W. Johnston, J. Margot, M. Chaker, and O. Pauna (1999), *IEEE Trans. Plasma Sci.* **27**, 727.
Vinogradov, G. K., V. M. Menagarishvili, and S. Yoneyama (1998), *J. Vac. Sci. Technol.* **A16**, 3164.
Vossen, J. L., and W. Kern, eds. (1978), *Thin Film Processes*, Academic Press, New York.
Vossen, J. L., and W. Kern, eds. (1991), *Thin Film Processes II*, Academic Press, New York.
Waits, R. K. (1978), in *Thin Film Processes*, J. L. Vossen and W. Kern, eds., Academic Press, New York.
Walkup, R. E., K. L. Saenger, and G. S. Selwyn (1986), *J. Chem. Phys.* **84**, 2668.
Walsh, P. J. (1959), *Phys. Rev.* **116**, 511.
Wang, D. Z. (1999), *J. Appl. Phys.* **85**, 3949.
Wang, S. B., and A. E. Wendt (2001), *J. Vac. Sci. Technol.* **A19**, 2425.
Wang, D., T. Ma, and X. Deng (1993), *J. Appl. Phys.* **74**, 2986.
Wang, Z., A. J. Lichtenberg, and R. H. Cohen (1998), *IEEE Trans. Plasma Sci.* **26**, 59.
Wang, Z., A. J. Lichtenberg, and R. H. Cohen (1999), *Plasma Sources Sci. Technol.* **8**, 151.
Wang, Q., D. J. Economou, and V. M. Donnelly (2006), *J. Appl. Phys.* **100** 023301.
Wang, M., P. L. G. Ventzek, and A. Ranjan (2017), *J. Vac. Sci. Technol.* **A35**, 031301.
Warner, B. E., K. B. Persson, and G. J. Collins (1979), *J. Appl. Phys.* **50**, 5694.
Warren, R. (1955), *Phys. Rev.* **98**, 1658.
Waskoenig, J. (2010), Numerical simulations of the electron dynamics in single and dual radio-frequency driven atmospheric pressure plasmas and associated plasma chemistry in electronegative He-O_2 mixtures, PhD Thesis, Queen's University of Belfast.
Waskoenig, J., K. Niemi, N. Knake, L. M. Graham, S. Reuter, V. Schulz-von der Gathen, and T. Gans (2010), *Plasma Sources Sci. Technol.* **19** 045018.
Weibel, E. S. (1967), *Phys. Fluids* **10**, 741.
Wen, D. Q., J. Krek, J. T. Gudmundsson, E. Kawamura, M. A. Lieberman, and J. P. Verboncoeur (2021), *Plasma Sources Sci. Technol.* **30**, 105009.
Wen, D. Q., J. Krek, J. T. Gudmundsson, E. Kawamura, M. A. Lieberman, P. Zhang, and J. P. Verboncoeur (2023), *Plasma Sources Sci. Technol.* **32**, 064001.

Wendt, A. E. (1993), *2nd Workshop on High Density Plasmas and Applications*, AVS Topical Conference, August 5–6, 1993, San Francisco, CA.

Wendt, A. E., and W. N. G. Hitchon (1992), *J. Appl. Phys.* **71**, 4718.

Wendt, A. E., and M. A. Lieberman (1990), *J. Vac. Sci. Technol.* **A8**, 902.

Whetten, N. R. (1992), in *Handbook of Chemistry and Physics*, 73rd ed., D. R. Lide, ed., CRC Press, Boca Raton, FL.

Whipple, E. C. (1981), *Rep. Prog. Phys.* **44**, 1198.

Whipple, E. C., T. G. Northrop, and D. A. Mendis (1985), *J. Geophys. Res.* **90**, 7405.

Wild, C., and P. Koidl (1991), *J. Appl. Phys.* **69**, 2909.

Wilhelm, J., and R. Winkler (1979), *J. Phys.* **40**, C7–251.

Williamson, M. C., A. J. Lichtenberg, and M. A. Lieberman (1992), *J. Appl. Phys.* **72**, 3924.

Wilson, W. D., L. G. Haggmark, and J. P. Biersack (1977), *Phys. Rev. B* **15**, 2458.

Winkler, R., and J. Wilhelm (1980), *Comput. Phys. Commun.* **20**, 113.

Winske, D., and M. E. Jones (1994), *IEEE Trans. Plasma Sci.* **22**, 454.

Winters, H. F. (1985), *J. Vac. Sci. Technol., A* **3**, 786.

Winters, H. F., and J. W. Coburn (1992), *Surf. Sci. Rep.* **14**, 161.

Winters, H. F., and D. Haarer (1987), *Phys. Rev. B* **36**, 6613; (1988), **37**, 10379.

Winters, H. F., D. B. Graves, D. Humbird, and S. Tougaard (2007), *J. Vac. Sci. Technol.* **A25**, 96.

Wood, B. P. (1991), *Sheath Heating in Low Pressure Capacitive Radio Frequency Discharges*, Thesis, University of California, Berkeley, CA.

Wood, B. P. (1993), *J. Appl. Phys.* **73**, 4770.

Wood, B. P., M. A. Lieberman, and A. J. Lichtenberg (1995), *IEEE Trans. Plasma Sci.* **23**, 89.

Wu, A. C. F., M. A. Lieberman, and J. P. Verboncoeur (2007), *J. Appl. Phys.* **101**, 056105.

Yates, J. H., W. C. Ermler, N. W. Winter, P. A. Christiansen, Y. S. Lee, and K. S. Pitzer (1983), *J. Chem. Phys.* **79**, 6145.

Yatsenko, N. A. (1980), *Zh. Tekh. Fiz.* **50**, 2480; *Sov. Phys. Tech. Phys.* **25**, 1454.

Yatsenko, N. A. (1981), *Zh. Tekh. Fiz.* **51**, 1195; *Sov. Phys. Tech. Phys.* **26**, 678.

Yeom, G. Y., J. A. Thornton, and M. J. Kushner (1989a), *J. Appl. Phys.* **65**, 3816.

Yeom, G. Y., J. A. Thornton, and M. J. Kushner (1989b), *J. Appl. Phys.* **65**, 3825.

Yip, C. S., N. Hershkowitz, and G. Severn (2015), *Plasma Sources Sci. Technol.* **24**, 015018.

Yokoyama, T., M. Kogoma, T. Moriwaki, and S. Okazaki (1990), *J. Phys. D: Appl. Phys.* **23**, 1125.

You, S. J., T. T. Hai, M. Park, D. W. Kim, J. H. Kim, D. J. Seong, Y. H. Shin, S. H. Lee, G. Y. Park, J. K. Lee, and H. Y. Chang (2011), *Thin Solid Films* **519**, 6981.

Yuan, X., and L. L. Raja (2002), *Appl. Phys. Lett.* **81** 814.

Zalm, P. C. (1984), *J. Vac. Sci. Technol.* **B2**, 151.

Zangwill, A. (1988), *Physics at Surfaces*, Cambridge University Press, Cambridge, UK.

Zeng, X. C., A. G. Liu, T. K. Kwok, P. K. Chu, and B. Y. Tang (1997), *Phys. Plasmas* **4**, 4431.

Zhang, D., and M. J. Kushner (2000), *J. Vac. Sci. Technol.* **A18**, 2661.

Zhao, S. X., X. Xu, X. C. Li, and Y. N. Wang (2009), *J. Appl. Phys.* **105**, 083306.

Zhu, W., S. Sridhar, L. Liu, E. Hernandez, V. M. Donnelly, and D. J. Economou (2014), *J. Appl. Phys.* **115**, 203303.

Zuzic, M., H. M. Thomas, and G. E. Morfill (1996), *J. Vac. Sci. Technol.* **A14**, 496.

Index

a

Actinometry, 232–234
Adsorption, 257ff, 266–268
  chemisorption, 258–260
  dissociative, 258, 267
  physical, 267–268
  physisorption, 257–260
Affinity, 179, 200–202, 206–208, 578
Air discharge, 531, 550, 554, 555 (figure)
α–γ transition, 524ff
  atmospheric pressure, 543–547
  γ-mode model, 526–529
  low pressures, 533
  qualitative description, 524–526
  transition curve, 524, 527, 533
Anodization, 621
Argon
  collisional energy loss per electron-ion pair, 67 (figure)
  cross sections, 59 (figure), 62 (figure)
  discharge model, 281ff
  ion-neutral mean free path, 68
  passive bulk discharge model, 516–519
  probability of collision, 51 (figure)
  rate constants, 66 (table)
  Townsend coefficient α, 487 (figure)
Atmospheric pressure capacitive discharge, 534ff, 548ff
  α–γ transition, 543–547
  circuit model, 552, 555
  dielectric barrier (DBD), 555–556
  experiments, 545–547, 552
  filamentary regime, 555–556
  glow regime, 551–554
  homogeneous model, 536–541
  low-frequency driven, 548ff
  Penning ionization model, 540
  rf-driven, 534ff
  simulations, 543–545
  streamer breakdown, 548–550
Atom, 53ff
  degeneracy, 54
  electronic configurations, 55
  energy levels, 54–55
  fine structure, 55
  metastable, 56–58
  optical emission from, 55–56, 198–199, 229ff
  valence electrons, 55
Atomic layer deposition (ALD), 628ff
  conformality, 632–636
  ozone processes, 632
  plasma-enhanced, 631–632
  thermal, 630–631
Atomic layer etching (ALE), 595ff
  bias power pulsing, 597
  model of, 600–604
  molecular dynamics simulations, 597–600
  oxides, 604
  photon-assisted, 604
  quasi-ALE, 605–606
  thermal ALE, 606–608

b

Bohm
  velocity, 140
Bohr radius, 53
Bolsig+ Boltzmann equation solver, 689–690
Boltzmann constant, 29
Boltzmann equation, 25–26, 681
  Bolsig+ solver, 67, 229, 689–690
Boltzmann relation, 32, 292
  generalized, 697
Boltzmann term analysis, 310, 313, 362–365, 688–689

c

Capacitive discharge, 12–17, 329ff, 710ff
  asymmetric, 365ff, 381–383, 391ff
  Boltzmann term analysis, 310, 313, 362–365, 688–689
  dc bias voltage, 370–371
  dual frequency, 373ff, 400
  electrical asymmetry effect (EAE), 381–382
  electromagnetic effects, 383ff
  electronegative, 352–353
  excited neutrals in, 362, 363
  experiments, 305–306, 354ff, 388
  gas heating in, 362, 363
  homogeneous dual frequency model, 376–378
  homogeneous model, 330ff
  inhomogeneous model, 340ff
  ion bombarding energy, 394ff
  kinetic model, 710ff
  low frequency sheaths, 391ff
  magnetically enhanced (MERIE), 15, 401ff
  matching network, 406ff
  multi-frequency driven, 373ff
  nonlinear series resonance, 370ff
  scaling, 347–350, 366–367
  secondary electrons, 362, 363
  series resonance, 339
  simulations, 357ff, 388–391
  skin effects, 383ff
  spherical shell model, 367–368
  standing waves, 383ff
  surface waves, 383ff
  tailored voltage waveform, 382–383
  voltage-driven rf, 370ff, 373ff
CF_4 discharge, 580ff
  basic data, 580ff
  inhibitor film formation, 583
  rate constants, 581, 582 (table)
  silicon dioxide etching, 588–590
  silicon etching, 580ff
  surface kinetics, 582–585
Charge
  bound, 22
  free, 22
Charging, 608ff
  electron shading effect, 612–614
  gate oxide damage, 608–609
  in nonuniform plasmas, 610–612
  rf biasing effects, 614–615
  transient effects, 612
Chemical equilibrium, 184ff
  between phases, 187–190
  constant, 185–186, 224, 245–246
  heterogeneous, 187
  surface coverage, 190–191
  vapor pressure, 189 (table), 187–190
Chemical kinetics, 243ff
  gas-phase, 246ff
  surface, 253ff
Chemical potential, 182–184
Chemical reaction, 171–172, 184ff
  consecutiveff, 246
  elementary, 243ff
  equilibrium, 171–172
  opposing, 249
  with photon emission, 249ff
  rates, 221ff, 244–246
  stoichiometric, 184ff
  surface, 253ff
  three-body, 250ff
  three-body recombination, 252–253
Chemical vapor deposition (CVD), 590, 591
Chlorine discharges, 310ff
  etching in, 586–588, 591–592

instabilities in, 427
pulsed discharge, 321–322
Clausius-Clapeyron equation, 188
Collision
associative detachment, 215ff
associative ionization, 217–219, 224
atomic, 37ff
autodetachment, 200, 209
autodissociation, 199, 209
autoionization, 199
charge transfer, 60–62, 212ff
coulomb, 46–48, 718–720
deexcitation, 224
dissociation, 202ff
dissociative electron attachment, 206ff
dissociative ionization, 204ff
dissociative recombination, 205ff, 224
elastic, 37, 46ff, 211, 220
electron detachment, 209–210
electron-ion, 46–48
electron-neutral, 48–52
excitation, 59
heavy particle, 211ff, 221
inelastic, 37, 58–59
ionization, 58–59
metastable pooling, 217–219
molecular, 195ff
Penning ionization, 217–219
polar dissociation, 208–209
polarization scattering, 48–52, 220
positive-negative ion recombination, 213ff, 220
rearrangement of chemical bonds, 219ff
recombination, 213ff
small angle scattering, 44–48
three-body, 220–221, 225
transfer of excitation, 216ff
vibrational and rotational excitation, 210ff
Collision dynamics, 42ff, 717ff
adiabatic Massey criterion, 212
Arrhenius temperature dependence, 222
Boltzmann collision integral, 721ff
center-of-mass coordinates, 42–43
differential scattering, 39–41
energy transfer, 44
Franck-Condon principle, 202
Krook operator, 28, 682
small angle scattering, 44–46
Collision parameters
cross section, 37
frequency, 39
impact parameter, 39
mean free path, 38
probability of collision, 51
rate constant, 39
Collision terms, 683–685
Conductivity
dc plasma, 80
effective electrical, 687–688
plasma, 80
Confinement
magnetic, 75, 84–86, 123–133
Conservation
energy, 29, 76, 280, 295, 336ff, 343–346
equations, 25ff
momentum, 27–29
neutral radicals, 323ff
particles, 27, 284, 286ff, 295, 336ff
Continuity equation, 22
macroscopic, 25
Cross section, 37
argon, 59 (figure), 62 (figure)
Arrhenius, 65, 222
charge transfer, 60–62
differential, 39–41
excitation, 59
gas kinetic, 264 (table)
hard sphere, 38
ionization, 58–59
Langevin, 49–51, 220
LXCat database, 67, 226, 689–690
oxygen, 226 (figure)
Rutherford, 720
Thomson ionization, 58–59
total, 41
Current
conduction, 22, 89
displacement, 22–23, 89

Current (*contd.*)
  magnetization, 89
  polarization, 22
  total, 22, 79
Cyclotron frequency, 74–75

d

Dc discharge, 479ff
  anode sheath, 481
  breakdown, 486ff
  cathode fall thickness, 490 (table)
  cathode fall voltage, 490 (table)
  cathode sheath, 480, 485ff
  diffusion, 482ff
  Faraday dark space, 481, 491–492
  Paschen curve, 488 (figure), 486–489
  positive column, 479–480, 482ff
  Townsend coefficient α, 486–489
Debye length, 32–33, 35
Deposition, 619ff
  of amorphous silicon, 621–624
  chemical vapor (CVD), 619
  conformality, 625–626
  large area, 628
  reaction rates, 622 (table)
  of silicon dioxide, 624–627
  of silicon nitride, 627–628
Desorption, 257ff
  associative, 260
Detailed balancing, 222ff
Diagnostics
  hairpin resonator, 103–104
  magnetic (B-dot) probe, 104, 431–433
  microwave, 100ff
  optical, 672–674
  wall ion flux probe, 371
  wave, 100, 671
Dielectric constant, 78–80
  perpendicular, 89
  tensor, 90–92
Diffusion, 111ff, 685
  across a magnetic field, 123ff
  across multipoles, 129ff
  ambipolar, 112–113, 125
  Bohm, 128, 133
  boundary conditions, 113–114
  constant, 111, 685
  Einstein relation, 112
  in electronegative plasmas, 289ff
  Fick's law, 112
  Langmuir regime, 120–121
  low pressure, 119ff, 723ff
  of neutrals, 264–266
  nonambipolar, 126–129
  quasilinear, 697ff
  random walk, 112, 123, 136
  simulation in electronegative plasma, 306–313
  solutions, 113ff, 119ff
  steady state, 115–122
  variable mobility model, 119–120, 723ff
Discharge
  air, 531, 550, 554, 555 (figure)
  atmospheric pressure capacitive low-frequency, 548ff
  atmospheric pressure capacitive rf, 534ff
  dielectric barrier (DBD), 555–556
  electronegative model, 289ff
  electropositive model, 281ff
  high density, 17–18, 415ff
  high pressure, 8 (figure), 282
  high pressure capacitive, 513ff
  homogeneous sheath model, 332–335, 526–527, 536–541
  hot filament, 131, 650
  intermediate pressure, 282
  intermediate pressure capacitive, 514ff
  low pressure, 4, 282
  neutral radical density model, 288ff
  nonuniform density model, 286ff, 289ff, 297ff
  passive bulk model, 516–519
  Penning ionization model, 540
  series resonance, 108, 410
  typical parameters, 13 (table)
  uniform density model, 283ff, 297ff
Distribution function, 25
  Druyvesteyn, 685–686, 692, 697, 707ff
  electron, 9 (figure)

electron energy (EEDF), 31, 156–157, 691
electron energy probability (EEPF), 156–157
electron in rf field, 686–687
ion bombarding energy, 394ff
Maxwellian, 31, 722
Druyvesteyn distribution, 685–686, 707ff
Dual frequency capacitive discharge, 373ff, 400
Dusty plasmas, 655ff
crystals, 670
diagnostics, 670ff
discharge equilibrium, 660–662
driven particulate motion, 671–672
dust acoustic waves, 671
forces on particulates, 662ff
formation and growth, 665ff
particulate charging, 656–660
removal of particulates, 675ff
strongly coupled plasmas, 670

e

Edge-to-center density ratio, 122ff, 122 (figure), 514
in electronegative discharges, 299–301
Electrical asymmetry effect (EAE), 381–382
Electron cyclotron discharge, 18, 445ff
configurations, 445ff
coupling, 447–450
electron heating, 450ff
magnetic beach, 448
measurements, 462–463
plasma expansion, 458–462
simulations, 458
wave absorption, 453ff
Electron cyclotron resonance (ECR), 94
Electronegative discharge equilibrium, 289ff, 297ff
Boltzmann approximation, 293ff
global model, 297ff
nonuniform model, 301–304
pulsed discharges, 319ff
simulations, 306–313
uniform model, 297ff
Electron temperature, 30–32
calculation of, 284
Electropositive discharge equilibrium, 281ff
Emission
Auger electron, 256
secondary, 660
secondary electron, 254ff, 257 (table)
Energy, 172ff
average kinetic energy lost per particle lost, 31
collisional energy loss per electron-ion pair, 67–68
density, 26
Fermi, 254
Gibbs free, 175 (table), 182–184
ion bombarding, 7, 14, 282–283, 345–347, 349, 460–462
Energy diffusion coefficient, 696, 699ff
Energy relaxation length, 515ff, 535 (figure), 691–692, 702–703
for secondary electrons, 494, 527
Energy relaxation time, 535 (figure), 537
Enthalpy, 172ff
bond dissociation, 178 (table)
formation of gaseous atoms, 179 (table)
standard molar formation, 175 (table), 174–179
Entropy, 179ff
standard molar, 181
Equation of state
adiabatic, 29
isothermal, 29
perfect gas, 29, 172, 176, 183
Etching, 561ff
aluminum, 591
anisotropy, 3, 561–565
aspect-ratio dependent (ARDE), 574–575
Bosch process, 576
in CF_4 discharge, 580ff
chemical, 566, 575ff, 586–587
chemical framework, 572–573
chemistries, 567 (table)
chlorine atom, 586–588, 591–592
copper, 592
cryogenic, 604, 617

Etching (*contd.*)
discharge kinetics, 571–572
doping effect, 577–579
etchant atom flux dependence, 576
gas feedstocks, 572–573
inhibitor films, 567, 572–573
ion enhanced, 4, 566–567, 579ff, 587–588
isotropic, 3, 566
loading effect, 571–572, 583
for microfabrication, 3–4
O_2 and H_2 additions, 584–585
other etch systems, 588ff
other materials, 594
pattern transfer fidelity, 573–574
photon-assisted, 588
processes, 565ff
rates, 561–565, 582–583, 585, 586
requirements, 561ff
resist, 592–594
selectivity, 3, 561–565, 590–591
silicon by halogens, 4, 575ff
silicon dioxide, 588–590
silicon nitride, 590
silicon using chlorine, 586–588
sputter, 565–566
substrate charging damage, 608ff
surface kinetics, 568ff
trench, 1 (figure), 561ff
uniformity, 4, 561–565
Excimer, 199

f

Ferrites, 438–439
Flux
energy flux crossing a surface, 31
magnetic, 419–420, 459
particle flux crossing a surface, 31
Fragmentation, 261
Frequency
collision, 39
effective collision, 354, 419, 705–706
electron cyclotron, 74–75
electron gyration, 74–75
electron plasma, 78
ion cyclotron, 74–75
ion gyration, 74–75
ion plasma, 78
lower cutoff, 95
lower hybrid, 97
plasma, 78
upper cutoff, 95
upper hybrid, 95

g

Gas
chemical kinetics, 246ff
chemical potential, 182–184
depletion effects, 435, 473, 505
enthalpy of formation of gaseous atoms, 179 (table)
feedstocks for etching, 572–573
heating, 361–362, 435
perfect, 29, 172, 176, 183
showerhead flow, 270ff
Guiding center motion, 84ff, 90 (table)

h

Heat, 172ff
reversible, 173–174, 180–181
Helical resonator discharge, 429–430
Helicon discharge, 18, 464ff
absorption, 469ff
antenna coupling, 467–469
electron trapping effects, 473
modes, 464ff
neutral gas depletion, 473
Trivelpiece-Gould mode heating, 471–472
Helium
α–γ transition in, 529–531
He/N_2 rate constants, 541 (table)
Penning ionization model, 540
potential energy curves, 202 (figure)
High pressure capacitive discharge, 513ff
Hollow cathode discharges, 492ff
metal vapor in, 495
rf-driven, 497
sheath width effects, 496

Hydrogen
  atom, 53
  example of, 206
  potential energy curves, 200 (figure)
  probability of collision, 51 (figure)

i

Inductive discharge, 415ff, 430ff, 708–710
  anomalous skin depth, 418
  capacitive coupling, 423–425
  close-coupled, 437–438
  experiments, 310–311, 430ff, 440–441, 710
  ferrite-enhanced, 438–439
  high density, 415ff
  high pressure, 428
  hysteresis in, 425
  instabilities in, 425–427
  kinetic model, 708–710
  low density, 422–423
  low frequency, 436–437
  low pressure, 415ff
  matching network, 421–422
  planar coil, 430ff
  power absorption, 416ff, 422–423, 427
  simulations, 312, 434–435
  source configurations, 416
Intermediate pressure capacitive discharge, 514ff, 524ff
  energy relaxation length, 515–516
  experiments, 529–531, 533
  metastables and secondary electrons, 520–523
  passive bulk model, 516–519
  simulations, 519–523, 532–533
Ion bombarding energy, 394ff
Ionized physical vapor deposition, 507ff

k

Kinetic theory, 681ff
  capacitive discharge, 710ff
  inductive discharges, 708–710
  Krook collision operator, 682
  local kinetics, 690ff
  nonlocal kinetics, 693ff
  non-Maxwellian models, 707ff
  stochastic heating, 701, 703–704
  two-term approximation, 683ff
Kirchoff voltage law, 23

l

Langmuir isotherm, 190, 266–267
Lorentz force law, 24, 73
LXCat cross section database, 67, 226, 689–690

m

Macroscopic equations, 26–30
Macroscopic motion, 21
Macroscopic quantities, 26–30
Magnetically enhanced reactive ion etcher (MERIE), 15, 401ff
Magnetized plasma
  dielectric tensor, 90ff
  magnetic field expansion, 458–462
  resonances, 94ff
  waves, 92, 93, 94ff
Matching network
  capacitive discharge, 406ff
  electron cyclotron discharge, 447–448
  helicon discharge, 464
  inductive discharge, 421–422
Maxwellian distribution, 31, 722
  averaging over, 64ff
Maxwell's equations, 22–24
Measurements
  rf power, 408–409, 422
Mobility, 111–112, 685
  variable, 119–120
Molecule, 195ff
  electronic state, 195ff
  example of hydrogen, 206
  metastable, 209
  negative ion, 200–202
  optical emission, 198–199
  potential energy curves, 199, 200 (figure), 202 (figure)
  vibrational and rotational motion, 197ff

Momentum conservation, 27–29
Multipole magnetic confinement, 129ff, 434, 449–450

n

Navier-Stokes equation, 271
Neutral radical density
  pulsed discharge dynamics, 323ff
Nitrogen
  α–γ transition in, 532–533
Nonuniform plasmas
  damage due to, 610–612

o

Ohmic heating, 80–82, 337, 343–344, 416–418
  in planar magnetron, 503–504
Optical emission, 55–56, 198–199, 229ff
  phase-resolved (PROES), 234–237, 380 (figure), 547 (figure)
  trace rare gases spectroscopy, 234
  two-photon absorption laser-induced fluorescence (TALIF), 233
Oxygen
  actinometry, 233–234
  basic constants, 226 (table)
  collisional energy loss per electron-ion pair, 67 (figure)
  cross sections, 226 (figure)
  data set, 225–229
  discharge model, 304ff, 571–572
  electronegative discharge equilibrium, 304ff
  potential energy curves, 199 (figure)
  rate constants, 228 (table), 230 (table)
  surface recombination of, 627

p

Particle motion
  constant fields, 73ff
  cyclotron frequency, 74
  diamagnetic drifts, 124
  drifts, 75–76, 84ff, 124
  $E \times B$ drifts, 75–76, 124
  guiding center, 84ff, 90 (table)
  magnetic moment, 86–87
Paschen breakdown, 488 (figure), 486–489, 513–514, 524 (figure), 524–525, 532 (figure), 527–533, 534 (figure), 551 (figure)
Planar magnetron discharge, 18, 498ff
  HiPIMS, 504–505
  model, 501ff
  sputtering, 499, 503
  3D PIC simulation, 506
Planck's constant, 53
Plasma admittance, 332
  Plasma-enhanced chemical vapor deposition (PECVD), 621ff
    amorphous silicon, 621–624
    conformality, 625–626
    large area, 628
    silicon dioxide, 624–627
    silicon nitride, 627–628
Plasma equilibrium
  atmospheric pressure capacitive low-frequency, 548ff
  atmospheric pressure capacitive rf, 534ff
  dual frequency capacitive, 376–378
  electronegative, 289ff, 297ff
  electropositive, 281ff
  experiments and simulations, 304ff
  global models, 283ff, 297ff
  high pressure, 282
  high pressure capacitive, 513ff
  intermediate pressure, 282
  intermediate pressure capacitive, 516ff, 524ff
  low pressure, 282
  neutral radical density model, 288ff
  nonuniform density models, 286ff, 301–304
  uniform density model, 283ff, 297ff
Plasma heating, 280–281
  ohmic sheath, 350–351
  pressure, 310, 313, 351–352, 362–365
  secondary electron, 362, 363
  stochastic, 351–352
Plasma-immersion ion implantation (PIII), 5, 640ff
  applications, 649–651
  sheath models, 642–649
Plasma oscillations, 77–78

Poisson's equation, 23
Polarizability, 50 (table)
Potential
chemical, 182–184
definition of, 23
distributed, 458–462
floating, 142, 154, 282
plasma, 142, 153–156
Power measurements, 408–409, 422
Pressure, 27–29
vapor, 189 (table), 187–190
Probe diagnostics, 153ff
collisional effects, 159, 163–164
cylindrical, 158–160
double, 161–162
emissive, 162–163
hairpin resonator, 103–104
Langmuir, 153–155
planar, 155–156
probe circuit, 164–166
spherical, 655
time-varying fields, 166–167
wall ion flux, 371
Processing
batch, 13
of materials, 1ff
Pulsed discharges, 313ff
atmospheric pressure low frequency, 548ff
bias voltage pulsing, 383, 597
electropositive model, 314ff
negative ions, 319ff
neutral radicals, 323ff

q
Quantum number, 54–55
principal, 53
Quasi-neutrality, 34

r
Radiation
dipole, 55–56
optical, 198–199, 229ff
Radical, 4, 9 (figure)
Rate constant, 64ff, 221ff, 243ff, 246ff
argon, 65 (figure), 66 (table), 67
Arrhenius, 65, 70, 222
CF_4, 581, 582 (table)
He/N_2, 541 (table)
oxygen, 228 (table), 230 (table)
relation to equilibrium constant, 188–190, 222–225, 245–246
second order, 226
SiH_4, 622 (table)
third order, 229
Rogowski coil, 409

s
Saha equation, 225, 241
Secondary emission, 254ff, 257 (table), 660
Sheath, 10–12, 137ff
admittance, 332–335
Bohm criterion, 139ff, 147
capacitance, 334–335, 342–343
Child law, 145–147, 342, 348–349, 642–646
collisional, 143–144, 152–153, 348–349
collisionless, 138–139, 342
dc bias voltage of rf, 370–371
distributed, 458–462
dual frequency Child law, 378–380
dual frequency homogeneous, 374–376
in electronegative gases, 148–149
high voltage, 145ff, 283
homogeneous model, 332–335, 526–527, 536–541
matrix, 145, 644
multiple positive ion species, 149–152
nonlinear series resonance, 370ff
potential at floating wall, 142
presheath, 141–142
rf capacitive, 332–335, 340ff
rf collisional capacitive, 348–349
rf resistive, 391ff
series resonance, 108, 410
thickness, 142, 145–147, 152
voltage-driven rf, 370ff, 373ff
Showerhead gas flow, 270ff
SiH_4 discharge, 622–623
rate constants, 622 (table)
Simulation
hybrid, 458

Simulation (*contd.*)
particle-in-cell (PIC), 11–12, 24, 308–310, 312–313, 357ff, 506, 514, 519–523, 532–533, 542–545
Simulations
molecular dynamics (MD), 597–600
Skin depth, 386, 417–419, 703ff
Sound speed, 83, 84
Specific heat, 29
at constant pressure, 176
at constant volume, 180
Sputter deposition, 499, 503, 636ff
film morphology for, 638 (figure), 637–638
reactive, 638–640
uniformity of, 503
Sputtering, 261ff
dependence on angle, 261–262, 262 (figure), 565
energy distribution, 261, 637
reactive, 638–640
role in etching, 565–566, 579–580
yields, 263
Standing waves, 383ff
Sticking coefficient, 258
Stochastic heating, 701, 703–704
in capacitive discharge, 337–338, 344–345, 349, 354, 711
in inductive discharge, 418, 703ff
in rf magnetron, 403–404
Streamer breakdown, 514, 548–550, 551 (figure), 554
Meek condition, 549, 550
Surface kinetics, 263ff
Surface process, 68, 253ff
adsorption, 257ff, 568ff, 582–583
Auger emission, 256
desorption, 257ff, 568ff, 583
loss probability, 114, 264ff, 269ff
positive ion neutralization, 254ff
reactions, 268–269
recombination, 270, 572, 582, 627
Surface wave discharge, 473ff
capacitive, 383ff
cylindrical, 474–475
planar, 473–474
power balance, 475–476

t

Tailored voltage waveform, 382–383
TEOS discharge, 624–627
Thermal equilibrium, 6
distribution, 31
properties, 30ff
Thermodynamics, 171ff
first law, 172
properties of substances, 175 (table)
second law, 179–180
Townsend coefficient α, 487 (figure), 487 (table), 486–489, 549
Transformer, 420
ideal, 439

v

Velocity
Alfven, 98
average speed, 31
Bohm, 140
group, 457, 472, 477
phase, 98–100
sound, 83, 84
thermal, 31
Vlasov equation, 26

w

Wave
CMA diagram, 98–100
diagnostics, 100ff
dispersion, 83 (figure), 96, 97 (figure)
electromagnetic, 82–83, 383ff
electron cyclotron, 445ff
electrostatic, 83–84
extraordinary (x), 95, 457
helicon, 464ff

Landau damping, 84, 470–471
left hand polarized (LHP), 94, 448, 451
in magnetized plasma, 93ff
ordinary (o), 95
principal, 97 (table), 94–98
ray dynamics, 457–458
right hand polarized (RHP), 94, 448, 450–451
surface, 383ff, 473ff
trapping, 473
Trivelpiece-Gould (TG), 471–472
tunneling, 454–456
whistler, 457, 464
WKB approximation, 454
Wave-heated discharge, 445, 479
Work
function, 254, 257 (table)
pd$\mathcal{V}$, 173–174
reversible, 173–174, 180